W9-CHH-142

TANGENTS AND SECANTS

30. $\int \tan u\, du = \ln|\sec u| + C$

31. $\int \sec u\, du = \ln|\sec u + \tan u| + C$

32. $\int \tan^2 u\, du = \tan u - u + C$

33. $\int \sec^2 u\, du = \tan u + C$

34. $\int \sec u \tan u\, du = \sec u + C$

35. $\int \tan^3 u\, du = \frac{1}{2}\tan^2 u + \ln|\cos u| + C$

36. $\int \sec^3 u\, du = \frac{1}{2}\sec u \tan u + \frac{1}{2}\ln|\sec u + \tan u| + C$

37. $\int \tan^n u\, du = \frac{\tan^{n-1} u}{n-1} - \int \tan^{n-2} u\, du$

38. $\int \sec^n u\, du = \frac{\sec^{n-1} u \sin u}{n-1} + \frac{n-2}{n-1}\int \sec^{n-2} u\, du$

COTANGENTS AND COSECANTS

39. $\int \cot u\, du = \ln|\sin u| + C$

40. $\int \csc u\, du = \ln|\csc u - \cot u| + C$

41. $\int \cot^2 u\, du = -\cot u - u + C$

42. $\int \csc^2 u\, du = -\cot u + C$

43. $\int \csc u \cot u\, du = -\csc u + C$

44. $\int \cot^3 u\, du = -\frac{1}{2}\cot^2 u - \ln|\sin u| + C$

45. $\int \csc^3 u\, du = -\frac{1}{2}\csc u \cot u + \frac{1}{2}\ln|\csc u - \cot u| + C$

46. $\int \cot^n u\, du = -\frac{\cot^{n-1} u}{n-1} - \int \cot^{n-2} u\, du$

47. $\int \csc^n u\, du = -\frac{\csc^{n-1} u \cos u}{n-1} + \frac{n-2}{n-1}\int \csc^{n-2} u\, du$

HYPERBOLIC FUNCTIONS

48. $\int \sinh u\, du = \cosh u + C$

49. $\int \cosh u\, du = \sinh u + C$

50. $\int \tanh u\, du = \ln(\cosh u) + C$

51. $\int \coth u\, du = \ln|\sinh u| + C$

52. $\int \operatorname{sech} u\, du = \tan^{-1}(\sinh u) + C$

53. $\int \operatorname{csch} u\, du = \ln|\tanh \frac{1}{2}u| + C$

54. $\int \operatorname{sech}^2 u\, du = \tanh u + C$

55. $\int \operatorname{csch}^2 u\, du = -\coth u + C$

56. $\int \operatorname{sech} u \tanh u\, du = -\operatorname{sech} u + C$

57. $\int \operatorname{csch} u \coth u\, du = -\operatorname{csch} u + C$

58. $\int \sinh^2 u\, du = \frac{1}{4}\sinh 2u - \frac{1}{2}u + C$

59. $\int \cosh^2 u\, du = \frac{1}{4}\sinh 2u + \frac{1}{2}u + C$

60. $\int \tanh^2 u\, du = u - \tanh u + C$

61. $\int \coth^2 u\, du = u - \coth u + C$

62. $\int u \sinh u\, du = u \cosh u - \sinh u + C$

63. $\int u \cosh u\, du = u \sinh u - \cosh u + C$

(table continued at the back)

SALAS AND HILLE'S CALCULUS

ONE VARIABLE

SALAS AND HILLE'S CALCULUS

ONE VARIABLE

SEVENTH EDITION

REVISED BY

GARRET J. ETGEN

JOHN WILEY & SONS, INC.

New York Chichester Brisbane Toronto Singapore

In fond remembrance of
EINAR HILLE

ACQUISITIONS EDITOR	Ruth Baruth
DEVELOPMENTAL EDITOR	Nancy Perry
MARKETING MANAGER	Susan Elbe
PRODUCTION EDITOR	Charlotte Hyland
DESIGNER	Ann Marie Renzi
MANUFACTURING MANAGER	Susan Stetzer
ILLUSTRATION EDITOR	Sigmund Malinowski
ILLUSTRATION	Fine Line
CHAPTER TITLE DESIGN	Carol Grobe
COVER PHOTOGRAPH	Paul Silverman

This book was set in New Times Roman by Progressive Information Technologies and printed and bound by Von Hoffmann Press. The cover was printed by Phoenix Color Corp.

Recognizing the importance of preserving what has been written, it is a policy of John Wiley & Sons, Inc. to have books of enduring value published in the United States printed on acid-free paper, and we exert our best efforts to that end.

The paper in this book was manufactured by a mill whose forest management programs include sustained yield harvesting of its timberlands. Sustained yield harvesting principles ensure that the number of trees cut each year does not exceed the amount of new growth.

Library of Congress Cataloging in Publication Data:
Salas, Saturnino L.
Salas and Hille's calculus: one variable/revised by Garret J. Etgen.—7th ed.
p. cm.
Rev. ed. of: Calculus: one variable. 6th ed. 1990.
Includes index.
ISBN 0-471-58725-7
1. Calculus. I. Hille, Einar (deceased) II. Salas, Saturnino L. III. Etgen, Garret J., Calculus: one variable. IV. Title.
QA303.S17 1995
515—dc20 94-30556
CIP

Printed in the United States of America

10 9 8 7 6 5 4 3 2

Above all, this is a text on mathematics. The subject is calculus, and the emphasis is on the three basic concepts: limit, derivative, and integral.

This text is designed for a standard introductory single variable calculus sequence. Our fundamental goal in preparing the Seventh Edition has been to preserve and enhance the notable strengths that characterized previous editions, including:

- An emphasis on the mathematical exposition—an accurate, understandable treatment of the topics.
- A clear, concise approach. Basic ideas and important points are not obscured by excess verbiage.
- An appropriate level of rigor. Mathematical statements are careful and precise, and all important theorems are proved. This formality is presented in a way that is completely accessible to the beginning calculus student.
- A balance of theory and applications, illustrated by many examples and exercises.

At the same time, we recognize that with the rapid advances in computer technology and the current scrutiny of mathematics education at all levels, the teaching of calculus is undergoing a serious examination. Thus, an equally important and parallel goal of the Seventh Edition has been to incorporate modern technology and current trends without sacrificing the acknowledged strengths of the text.

FEATURES OF THE SEVENTH EDITION

Problem-Solving Skills and Real-World Applications

Over 1300 new problems have been added to the Seventh Edition.

- In order to develop students' problem-solving skills, we have significantly increased the number of problems at all levels. A large number of challenging and routine problems are now available in all exercise sets. Many additional medium-level problems are included to assist students in developing the understanding necessary to attack the challenging problems. In some problems, students are called upon to interpret and justify their answers to improve their analytical and communication skills.

- An even wider variety of real-world applications motivates students' study of mathematical topics.
- More illustrations have been added to exercise sets to provide students with visual support as they devise their problem-solving strategies.

Technology

Because the use of graphing calculators and/or computer algebra systems has increased in calculus courses, we have considerably expanded the application of technology in the text. We do not attempt to teach any particular technology and so use a generic approach. Technology problems are clearly designated with an icon (C) and may be skipped by instructors who prefer that their students not use calculators or computers.

- New technology-based examples appear within the chapter discussions of the material. These support the numerous exercises requiring the use of a graphics calculator or other graphing software located in the end-of-section problems sets.
- "Projects and Explorations Using Technology," a set of problems that requires a combination of approaches involving both analytical and technology skills, ends each chapter. As their title suggests, these problems are also suitable for use by students working in groups. A few of the problems introduce concepts to be developed later in the text, while others explore realistic applications of topics that have already been studied.

Increased Emphasis on Visualization

We recognize the importance of visualization in developing students' understanding of mathematical concepts. For that reason:

- All the artwork from the previous edition has been redrawn for increased clarity and understanding.
- Over 90 new figures have been added.
- Representations in three dimensions are now in full color for increased geometric understanding and include many new computer-generated figures of curves and surfaces in space.

Early Introduction of Differential Equations

A subsection on differential equations (separable equations) has been added to the Exponential Growth and Decay section of Chapter 7, allowing exercises on differential equations to be used throughout the rest of the book.

CONTENT AND ORGANIZATION CHANGES IN THE SEVENTH EDITION

In response to the evolutionary state of the current calculus curriculum, many changes have been made in organization and content to meet the needs of today's students and instructors.

Precalculus Review (Chapter 1)

- This material, which provides a brief yet comprehensive review of the precalculus topics basic to the study of calculus, has been rewritten and expanded.
- The real number system and the real line are now discussed; the section covering methods for solving inequalities in one variable has been reorganized; and the treatment of analytic geometry—the Cartesian coordinate system, straight lines, and the conic sections—has been expanded.
- The treatment of the function concept, the elementary functions, and graphing (including technology) has been reorganized and expanded.
- The treatment of one-to-one functions and inverses has been moved to Chapter 7, where it serves to connect logarithmic and exponential functions.
- The brief section on proofs now includes mathematical induction with examples and exercises.

Limits and Continuity (Chapter 2)

- To improve students' understanding of these critical topics, a few of the discussions have been expanded and some figures have been added.
- The section on limits now includes a technology approach, illustrated by examples and new exercises.
- The derivative in various forms is introduced in the exercises.

Differentiation and Applications of the Derivative (Chapters 3 and 4)

- The interpretations of the derivative as a rate of change are treated in one section rather than two, and the coverage includes examples from economics.
- Differentiation of inverse functions has been moved to Chapter 7.
- The derivative of rational powers is now approached through implicit differentiation.
- The treatment of applications of the derivative has been reorganized slightly to emphasize the two objectives of Chapter 4: optimization and curve sketching.

Integration and Applications of the Integral (Chapters 5 and 6)

- The interpretation of the definite integral as "area" has been expanded; a subsection on "signed area" has been added.
- The treatment of u-substitutions has been streamlined; for example, trigonometric integrals are now incorporated in the change of variables section rather than as a separate section.
- The introduction of Riemann sums serves to introduce and motivate the approach taken in the applications chapter.
- The treatments of the various applications of the definite integral have been expanded.

The Transcendental Functions (Chapter 7)

- Inverse functions and the calculus of inverse functions are treated here rather than in separate sections earlier in the text.
- Integration by parts has been moved to Chapter 8, Techniques of Integration.
- The treatment of applications of exponential and logarithmic functions now includes a subsection on differential equations (separable equations).
- The treatment of the inverse trig functions now includes the inverse secant function.

Techniques of Integration (Chapter 8)

- Integration by parts has been moved from Chapter 7 to Section 8.2, and the section on partial fractions now follows trig substitutions.
- Almost all the treatments have been expanded, with many new examples, increased coverage of reduction formulas, and a discussion of hyperbolic functions.

Conic Sections; Polar Coordinates and Parametric Equations (Chapter 9)

- This new chapter combines material from two chapters in the previous edition. The conic sections are treated in one section; translation of axes and rotation of axes have been moved to Chapter 1 or the Appendices.
- The treatment of arc length has been softened a little: the proof based on the least upper bound axiom has been replaced by an intuitive argument.

Sequences and Series (Chapters 10 and 11)

- The least upper bound axiom now serves as a prelude to sequences, and there is more emphasis on boundedness in the treatment of sequences.
- The treatment of indeterminate forms has been modified: The "other" indeterminate forms—differences, products, exponential forms—are treated in a separate subsection rather than integrating them with the $0/0$ and ∞/∞ forms.
- The treatment of power series has been expanded slightly; there are some new examples and figures, and the Lagrange form of the remainder is stated explicitly and used to derive bounds on the remainder.

FEATURES OF THE BOOK

Concise exposition The concepts of calculus are presented clearly and accurately without hand waving.

Theorems and proofs Highlighted theorems direct students to accurate mathematical statements. Most proofs are included to provide a high level of precision.

2.5 THE PINCHING THEOREM; TRIGONOMETRIC LIMITS ■ 111

■ 2.5 THE PINCHING THEOREM; TRIGONOMETRIC LIMITS

Figure 2.5.1 shows the graphs of three functions f, g, h. Suppose that, as suggested by the figure, for x close to c, f is trapped between g and h. (The values of these functions at c itself are irrelevant.) If, as x tends to c, both $g(x)$ and $h(x)$ tend to the same limit L, then $f(x)$ also tends to L. This idea is made precise in what we call *the pinching theorem.*

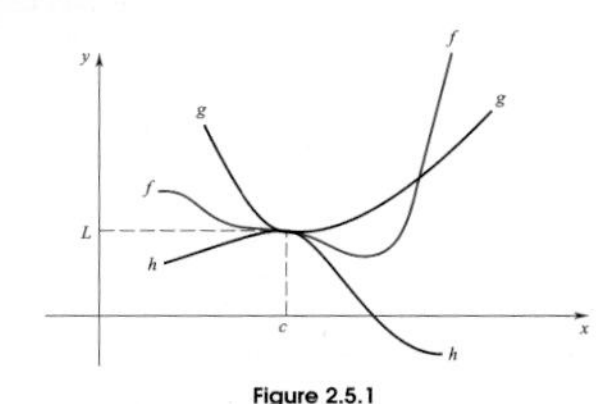

Figure 2.5.1

THEOREM 2.5.1 THE PINCHING THEOREM

Let $p > 0$. Suppose that, for all x such that $0 < |x - c| < p$,

$$h(x) \le f(x) \le g(x).$$

If

$$\lim_{x \to c} h(x) = L \quad \text{and} \quad \lim_{x \to c} g(x) = L,$$

then

$$\lim_{x \to c} f(x) = L.$$

PROOF Let $\epsilon > 0$. Let $p > 0$ be such that

$$\text{if} \quad 0 < |x - c| < p, \qquad \text{then} \quad h(x) \le f(x) \le g(x).$$

Choose $\delta_1 > 0$ such that

$$\text{if} \quad 0 < |x - c| < \delta_1, \qquad \text{then} \quad L - \epsilon < h(x) < L + \epsilon.$$

Choose $\delta_2 > 0$ such that

$$< |x - c| < \delta_2, \qquad \text{then} \quad L - \epsilon < g(x) < L + \epsilon.$$

$\delta_2\}$. For x satisfying $0 < |x - c| < \delta$, we have

$$L - \epsilon < h(x) \le f(x) \le g(x) < L + \epsilon,$$

$$|f(x) - L| < \epsilon. \quad \square$$

260 ■ CHAPTER 4 THE MEAN-VALUE THEOREM AND APPLICATIONS

Example 3 Figure 4.7.10 is a computer-generated graph of the function

$$f(x) = \frac{\cos x}{x}.$$

As $x \to 0^-$, $f(x) \to -\infty$; as $x \to 0^+$, $f(x) \to \infty$. The line $x = 0$ (the y-axis) is a vertical asymptote.

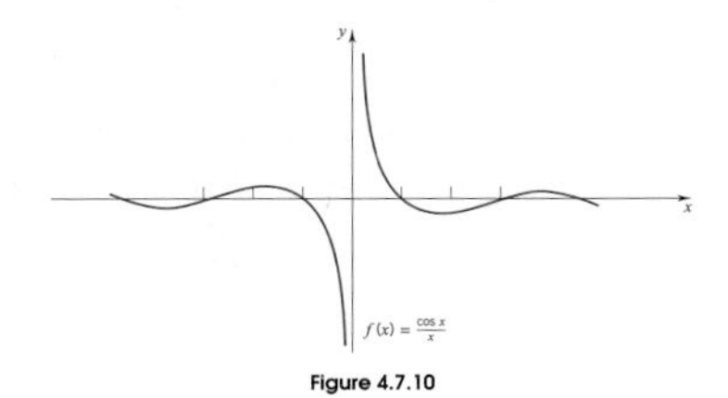

Figure 4.7.10

As $x \to \pm\infty$,

$$f(x) = \frac{\cos x}{x} \to 0.$$

This follows from the fact that

$$\left|\frac{\cos x}{x}\right| \le \frac{1}{|x|} \quad \text{for all} \quad x$$

and $1/|x| \to 0$ as $x \to \pm\infty$. Thus, the line $y = 0$ (the x-axis) is a horizontal asymptote. Note that f is an odd function $[f(-x) = -f(x)]$ so its graph is symmetric with respect to the origin. $\square$

Example 4 Find the vertical and horizontal asymptotes, if any, of the function

$$g(x) = \frac{x + 1 - \sqrt{x}}{x^2 - 2x + 1} = \frac{x + 1 - \sqrt{x}}{(x - 1)^2}.$$

SOLUTION The domain of g is $0 \le x < \infty$, $x \ne 1$. As $x \to 1$, $g(x) \to \infty$. Thus, the line $x = 1$ is a vertical asymptote. The behavior of g as $x \to \infty$ can be made more apparent by writing

$$g(x) = \frac{x + 1 - \sqrt{x}}{x^2 - 2x + 1} = \frac{x\left(1 + \frac{1}{x} - \frac{1}{\sqrt{x}}\right)}{x^2\left(1 - \frac{2}{x} + \frac{1}{x^2}\right)} = \frac{1 + \frac{1}{x} - \frac{1}{\sqrt{x}}}{x\left(1 - \frac{2}{x} + \frac{1}{x^2}\right)}.$$

Now, it is easy to see that $g(x) \to 0$ as $x \to \infty$. The line $y = 0$ (the x-axis) is a horizontal asymptote. $\square$

New examples To facilitate students' understanding, many examples have been revised and new examples have been added.

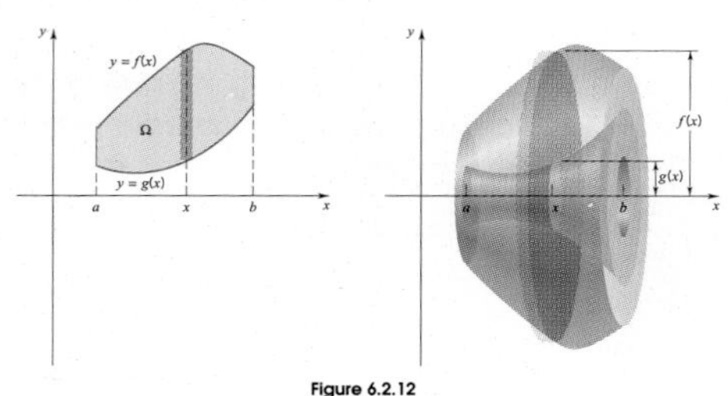

Figure 6.2.12

PROOF The cross section with coordinate x is a *washer* of outer radius $f(x)$, inner radius $g(x)$, and area

$$A(x) = \pi[f(x)]^2 - \pi[g(x)]^2 = \pi([f(x)]^2 - [g(x)]^2).$$

We can get the volume of the solid by integrating this function from a to b. ❑

Suppose now that the boundaries are functions of y rather than x (see Figure 6.2.13). By revolving Ω *about the y-axis,* we obtain a solid. It is clear from (6.2.4) that in this case

(6.2.6) $$V = \int_c^d \pi([F(y)]^2 - [G(y)]^2)\,dy.$$ (washer method about y-axis)

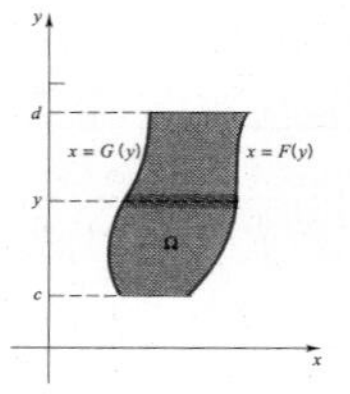

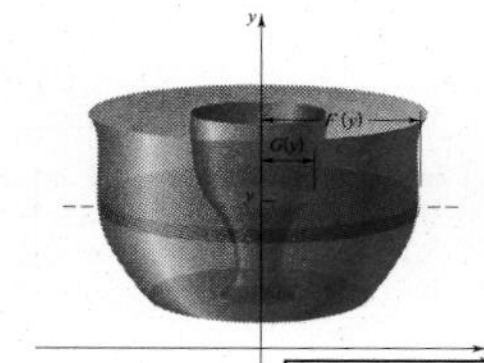

Figure 6.2.13

Improved visualization A completely new and expanded illustration program, including three-dimensional illustrations in full color, provides a better visual representation of concepts.

Example 2 A metal plate in the form of a trapezoid is affixed to a vertical dam as in Figure 6.6.5. The dimensions shown are given in meters; the weight density of water in the metric system is approximately 9800 newtons per cubic meter. Find the force on the plate.

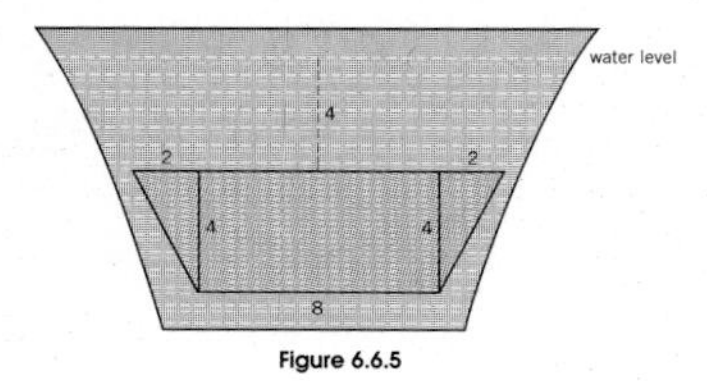

Figure 6.6.5

SOLUTION First we find the width of the plate x meters below the water level. By similar triangles (see Figure 6.6.6)

$$t = \tfrac{1}{2}(8 - x) \qquad \text{so that} \qquad w(x) = 8 + 2t = 16 - x.$$

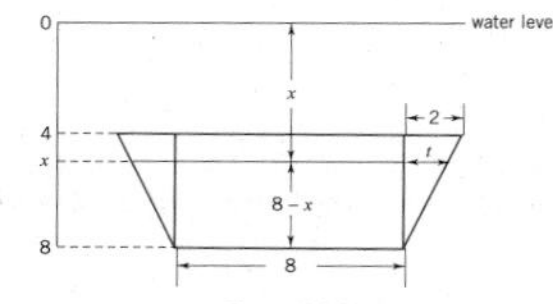

Figure 6.6.6

The force against the plate is

$$\int_4^8 9800x(16 - x)\,dx = 9800 \int_4^8 (16x - x^2)\,dx$$
$$= 9800\left[8x^2 - \tfrac{1}{3}x^3\right]_4^8 \cong 2{,}300{,}000 \text{ newtons.} \quad ❑$$

Real-world applications Students see how the concepts and methods of calculus connect with important problems in science and engineering.

New exercises The exercise sets have been revised and over 1300 new problems added, resulting in an improved balance between drill problems and more challenging exercises involving either theory or applications.

4.3 LOCAL EXTREME VALUES ■ 233

26. $f(x) = x + \cos 2x, \quad 0 < x < \pi.$
27. $f(x) = \sin^2 x - \sqrt{3}\sin x, \quad 0 < x < \pi.$
28. $f(x) = \sin^2 x, \quad 0 < x < 2\pi.$
29. $f(x) = \sin x \cos x - 3\sin x + 2x, \quad 0 < x < 2\pi$
30. $f(x) = 2\sin^3 x - 3\sin x, \quad 0 < x < \pi.$
31. Prove Theorem 4.3.4 by applying Theorem 4.2.3.
32. Prove the validity of the second-derivative test in the case that $f''(c) < 0$.
33. Find the critical numbers and the local extreme values of the polynomial

$$P(x) = x^4 - 8x^3 + 22x^2 - 24x + 4.$$

Then show that the equation $P(x) = 0$ has exactly two real roots, both positive.

34. A function f has derivative f' given by

$$f'(x) = x^3(x-1)^2(x+1)(x-2).$$

At what numbers x, if any, does f have a local maximum? A local minimum?

35. A polynomial function $p(x) = a_n x^n + a_{n-1}x^{n-1} + \cdots + a_1 x + a_0$ has critical numbers at $x = -1, 1, 2$, and 3, and corresponding values $p(-1) = 6$, $p(1) = 1$, $p(2) = 3$, and $p(3) = 1$. Sketch a possible graph for p if:
(a) n is odd, (b) n is even.
36. The quadratic function $f(x) = Ax^2 + Bx + C$ has a local minimum at $x = 2$ and passes through the points $(-1, 3)$ and $(3, -1)$. Find A, B, and C.
37. Determine a and b such that the function $f(x) = ax/(x^2 + b^2)$ has a local minimum at $x = -2$ and $f'(0) = 1$.
38. Let $f(x) = x^p(1-x)^q$, where $p \geq 2$ and $q \geq 2$ are integers.
(a) Show that the critical numbers of f are $x = 0$, $p/(p+q)$, and 1.
(b) Show that if p is even, then f has a local minimum at 0.
(c) Show that if q is even, then f has a local minimum at 1.
(d) Show that f has a local maximum at $p/(p+q)$ for all p and q.
39. Let

$$f(x) = \begin{cases} x^2 \sin(1/x), & x \neq 0 \\ 0, & x = 0. \end{cases}$$

In Exercise 67, Section 3.1, we saw that f is differentiable at 0 and that $f'(0) = 0$. Show that f has neither a local maximum nor a local minimum at 0.

40. Suppose that $C(x)$, $R(x)$, and $P(x)$ are the cost, revenue, s corresponding to the production and ertain item. Suppose, also, that C and functions. Then, since $P = R - C$, it fferentiable. Prove that if it is possible to maximize the profit by producing and selling x_0 items, then $C'(x_0) = R'(x_0)$. That is, the marginal cost equals the marginal revenue when the profit is maximized.
41. Let $y = f(x)$ be differentiable and suppose that the graph of f does not pass through the origin. Then the distance D from the origin to a point $P(x, f(x))$ on the graph is given by

$$D = \sqrt{x^2 + [f(x)]^2}.$$

Show that if D has a local extreme value at c, then the line through $(0, 0)$ and $(c, f(c))$ is perpendicular to the tangent line to the graph of f at c.

42. Prove that a polynomial of degree n has at most $n - 1$ local extreme values.
43. Let $f(x) = x^4 - 2x^2 - 3x + 2$.
(a) Show that f has exactly one critical number c in the interval $(1, 2)$.
(b) Use the bisection method (see Section 2.6) to approximate c to within $\frac{1}{16}$. Does f have a local maximum, a local minimum, or neither a maximum nor a minimum at c?
44. Let $f(x) = 2 + 20x + 4x^2 - x^4$.
(a) Show that f has exactly one critical number in the interval $(2, 3)$.
(b) Use the bisection method to approximate c to within $\frac{1}{16}$. Does f have a local maximum, a local minimum, or neither a maximum nor a minimum at c?
45. Let $f(x) = x^4 - 7x^2 + 2x - 3$.
(a) Show that f has exactly one critical number c in the interval $(2, 3)$.
(b) Use the Newton-Raphson method to approximate c; calculate x_3 and round your answer to four decimal places. Does f have a local maximum, a local minimum, or neither a maximum nor a minimum at c?
46. Let $f(x) = x\cos x$.
(a) Show that f has exactly one critical number in the interval $(0, \pi/2)$.
(b) Use the Newton-Raphson method to approximate c; calculate x_3 and round your answer to four decimal places. Does f have a local maximum, a local minimum, or neither a maximum nor a minimum at c?
47. Let $f(x) = \sin x + (x^2/2) - 2x$.
(a) Show that f has exactly one critical number in the interval $[2,3]$.
(b) Use the Newton-Raphson method to approximate c; calculate x_3 and round your answer to four decimal places. Does f have a local maximum, a local minimum, or neither a maximum nor a minimum at c?

In Exercises 48–51, use a graphing utility to graph the function f on the indicated interval. (a) Use the graph to estimate the critical numbers and the local extreme values; and (b) estimate the intervals on which f increases and the intervals on which f decreases. Round off your estimates to three decimal places.

Technology problems Problems marked by the icon encourage students to use technology as a tool to enhance understanding and problem-solving skills.

PROJECTS AND EXPLORATIONS USING TECHNOLOGY ■ 485

The inverse secant, $y = \sec^{-1}x$, is the inverse of $y = \sec x$, $x \in [0, \frac{1}{2}\pi) \cup (\frac{1}{2}\pi, \pi]$.

graph of $y = \sin^{-1} x$ (p. 462) graph of $y = \tan^{-1} x$ (p. 465)
graph of $y = \sec^{-1} x$ (p. 468)

$$\frac{d}{dx}(\sin^{-1}x) = \frac{1}{\sqrt{1-x^2}} \qquad \int \frac{dx}{\sqrt{a^2-x^2}} = \sin^{-1}\left(\frac{x}{a}\right) + C \quad (a > 0)$$

$$\frac{d}{dx}(\tan^{-1}x) = \frac{1}{1+x^2} \qquad \int \frac{dx}{a^2+x^2} = \frac{1}{a}\tan^{-1}\left(\frac{x}{a}\right) + C \quad (a \neq 0)$$

$$\frac{d}{dx}(\sec^{-1}x) = \frac{1}{|x|\sqrt{x^2-1}} \qquad \int \frac{dx}{x\sqrt{x^2-a^2}} = \frac{1}{a}\sec^{-1}\left(\frac{|x|}{a}\right) + C \quad (a > 0)$$

definition of the remaining inverse trigonometric functions (p. 471)

7.9 The Hyperbolic Sine and Cosine

$\sinh x = \frac{1}{2}(e^x - e^{-x})$, $\cosh x = \frac{1}{2}(e^x + e^{-x})$,

$$\frac{d}{dx}(\sinh x) = \cosh x, \qquad \frac{d}{dx}(\cosh x) = \sinh x.$$

graphs (pp. 475–476) basic identities (p. 477)

***7.10 The Other Hyperbolic Functions**

$$\tanh x = \frac{\sinh x}{\cosh x}, \qquad \coth x = \frac{\cosh x}{\sinh x},$$

$$\operatorname{sech} x = \frac{1}{\cosh x}, \qquad \operatorname{csch} x = \frac{1}{\sinh x}.$$

derivatives (p. 479) hyperbolic inverses (p. 481)
derivatives of hyperbolic inverses (p. 482)

■ PROJECTS AND EXPLORATIONS USING TECHNOLOGY

To do these exercises you will need a graphics calculator or a computer with graphing capability. The majority of these problems are open-ended so different approaches may be used to solve them. You should be aware that different approaches can result in slight variations in the answers. Round your numerical answers to at least four decimal places. The rounding method that your calculator or computer uses also may cause variations in answers.

7.1 The functions $f(x) = a \ln x$, where a is a constant, have a number of applications, one of which will be considered in a later exercise.
(a) Find the values of a for which the graph of f is tangent to the line $y = x$.
(b) For each real number a, how many solutions will there be of $f(x) = x$? What is the value of f' at each solution of $f(x) = x$?
(c) How many solutions are there to $f[f(x)] = x$? What is the value of f' at each of these solutions?
(d) Represent f as a logarithm function in another base.

7.2 Let $A(t)$ denote the area of the rectangle of width $2t$ that has its lower vertices on the x-axis and its upper vertices on the graph of

$$f(x) = e^{(1-x^4)/(2+x^2)}.$$

See the figure.

Chapter Highlights End-of-chapter lists stress important terms, ideas, and theorems.

Projects and Explorations Using Technology Special problem sets encourage deeper investigation of the material and can be used for cooperative learning activities.

SUPPLEMENTS

Student Aids

Answers to Odd-Numbered Exercises Answers to all the odd-numbered exercises are included at the back of the text.

Student Solutions Manual, Prepared by Garret J. Etgen, University of Houston This manual contains worked-out solutions to all the odd-numbered exercises and is available through your bookstore.

Instructor Aids

Instructor's Manual, by Garret J. Etgen and Sylvain Laroche This manual contains solutions to all the problems in the text.

Test Bank, by Sylvain Laroche A wide range of problems and their solutions are keyed to the text material and exercise sets.

Computerized Test Bank Available in both IBM and Macintosh formats, the Computerized Test Bank allows instructors to create, customize, and print a test containing any combination of questions from the test bank. Instructors can also edit the questions or add their own.

Technology Manuals

Discovering Calculus with Derive, by Jerry Johnson, University of Nevada-Reno, and Benny Evans, Oklahoma State University

- Derive instructions and tutorials
- Solved problems
- Practice problems
- Laboratory exercises

Discovering Calculus with Mathematica, by Cecilia A. Knoll, Florida Institute of Technology, Michael D. Shaw, Florida Institute of Technology, Jerry Johnson, and Benny Evans

- Mathematica introduction and commands
- Solved problems
- Exercises
- Laboratory projects

Discovering Calculus with Maple, by Kent Harris, Western Illinois University, and Robert J. Lopez, Rose-Hulman Institute of Technology

- Maple commands
- Example problems and step-by-step solutions
- Exercises

Discovering Calculus with Graphing Calculators, by Joan McCarter, Arizona State University

- Introductions to various calculators currently on the market (this manual is calculator nonspecific)
- Projects

- Additional exercises
- Critical thinking questions

ACKNOWLEDGMENTS

The revision of a text of this magnitude and stature requires a lot of encouragement and a lot of help. I was fortunate to have an ample supply of both from a variety of sources.

Each edition of this text was developed from those that preceded it. The present book owes much to the people who contributed to the first six editions. I am deeply indebted to them.

The reviewers of the seventh edition supplied detailed criticisms and valuable suggestions. I offer my sincere appreciation to the following individuals:

Linda Becerra
University of Houston

Jay Bourland
Colorado State University

Gary Crown
Wichita State University

Steve Davis
Davidson College

Anthony Dooley
University of New South Wales

William Fox
U.S. Military Academy—West Point

Barbara Gale
Prince Georges Community College

Pamela Gorkin
Bucknell University

Gary Itzkowitz
Rowan College of New Jersey

Harold Jacobs
East Stroudsburg University of Pennsylvania

Adam Lutoborski
Syracuse University

Douglas Mackenzie
University of New South Wales

Katherine Murphy
University of North Carolina-Chapel Hill

Jean E. Rubin
Purdue University

Stuart Smith
University of Toronto

I am especially grateful to Richard D. Byrd, University of Houston, who read the first draft of the manuscript letter by letter, made comments to match, read all the reviewers' comments, and then read the revised manuscript. This was a truly noble effort.

I am equally grateful to John Oman, University of Wisconsin—Oshkosh, who created the Projects and Explorations Using Technology problems.

I also want to thank Jun Ma, University of Houston, who provided detailed solutions for all the new exercises; there were more than 1300 of them.

I am deeply indebted to the editorial staff at John Wiley & Sons: Ruth Baruth, Mathematics Editor, who invited me to undertake this project in the first place and who provided encouragement when I needed encouragement and prodding when I needed prodding; Joan Carrafiello and Nancy Perry, Development Editors—Joan offered invaluable help in the beginning stages of this revision and Nancy continued it, along with exceptional organization and support; Charlotte Hyland, Production Editor, who was patient and understanding as she guided the project through the production stages; Sigmund Malinowski, who carefully directed the entirely new art program; and

Ann Marie Renzi, whose creativity produced the attractive interior design as well as the cover. This is an efficient, professional (and patient!) group.

Finally, I want to acknowledge the contributions of my wife, Charlotte. She supplied the sustenance and support that I had to have to complete this work.

Garret J. Etgen

CONTENTS

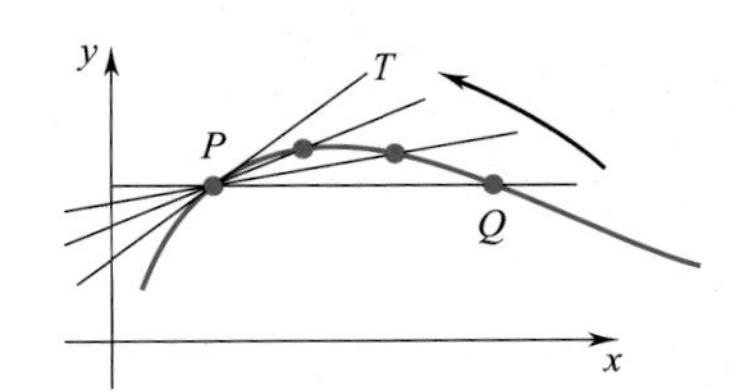

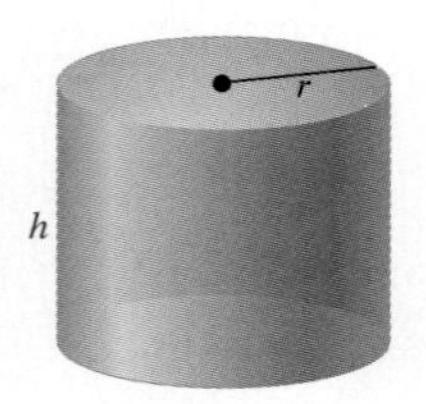

CHAPTER 3 DIFFERENTIATION 129

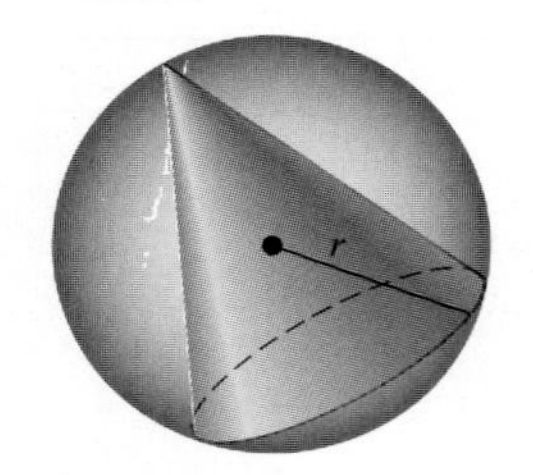

CHAPTER 4 THE MEAN-VALUE THEOREM AND APPLICATIONS 211

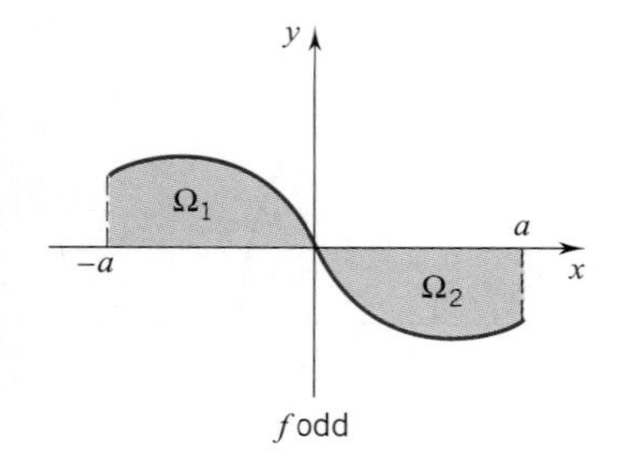

CHAPTER 5 INTEGRATION 277

CHAPTER 6 SOME APPLICATIONS OF THE INTEGRAL 347

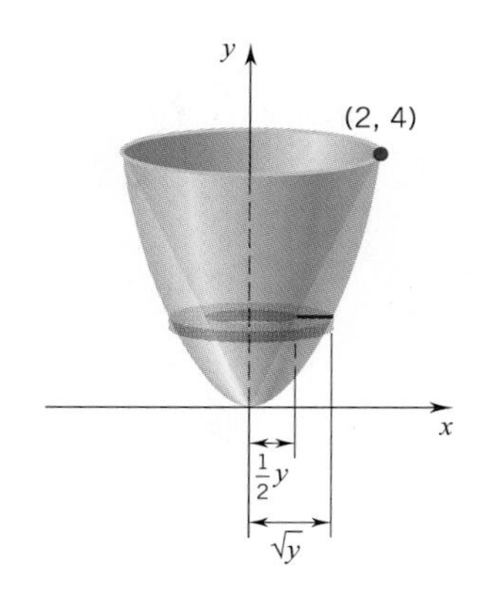

CHAPTER 7 THE TRANSCENDENTAL FUNCTIONS 395

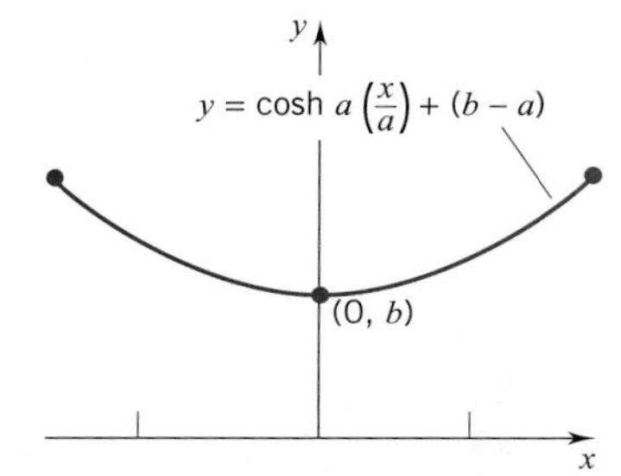

CHAPTER 8 TECHNIQUES OF INTEGRATION 489

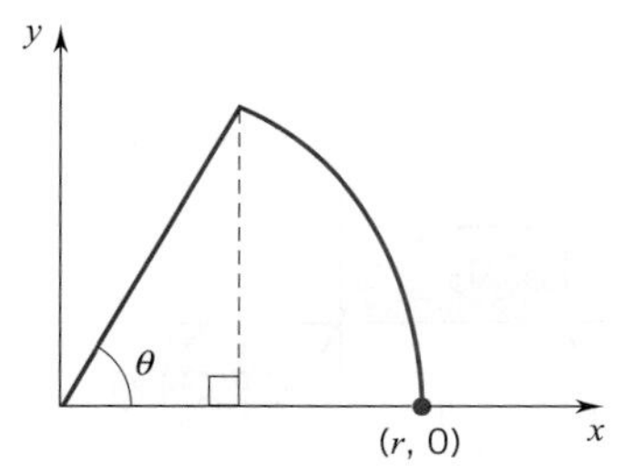

* Denotes optional section.

CHAPTER 9 CONIC SECTIONS; POLAR COORDINATES; PARAMETRIC EQUATIONS 549

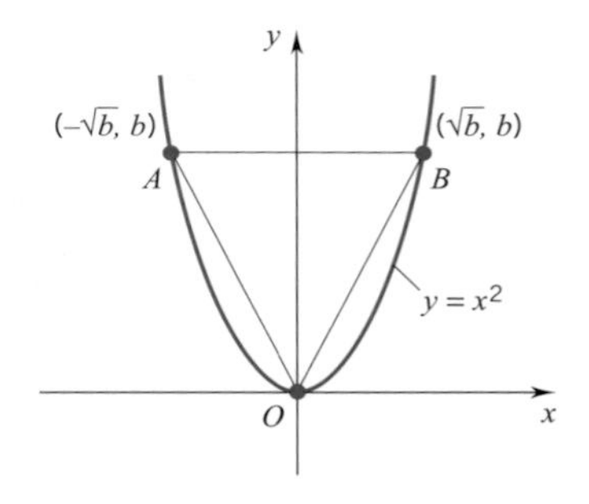

CHAPTER 10 SEQUENCES; INDETERMINATE FORMS; IMPROPER INTEGRALS 637

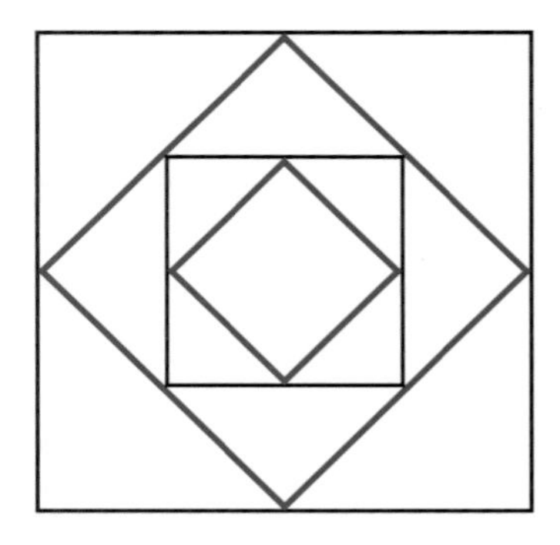

CHAPTER 11 INFINITE SERIES 693

SALAS AND HILLE'S CALCULUS

ONE VARIABLE

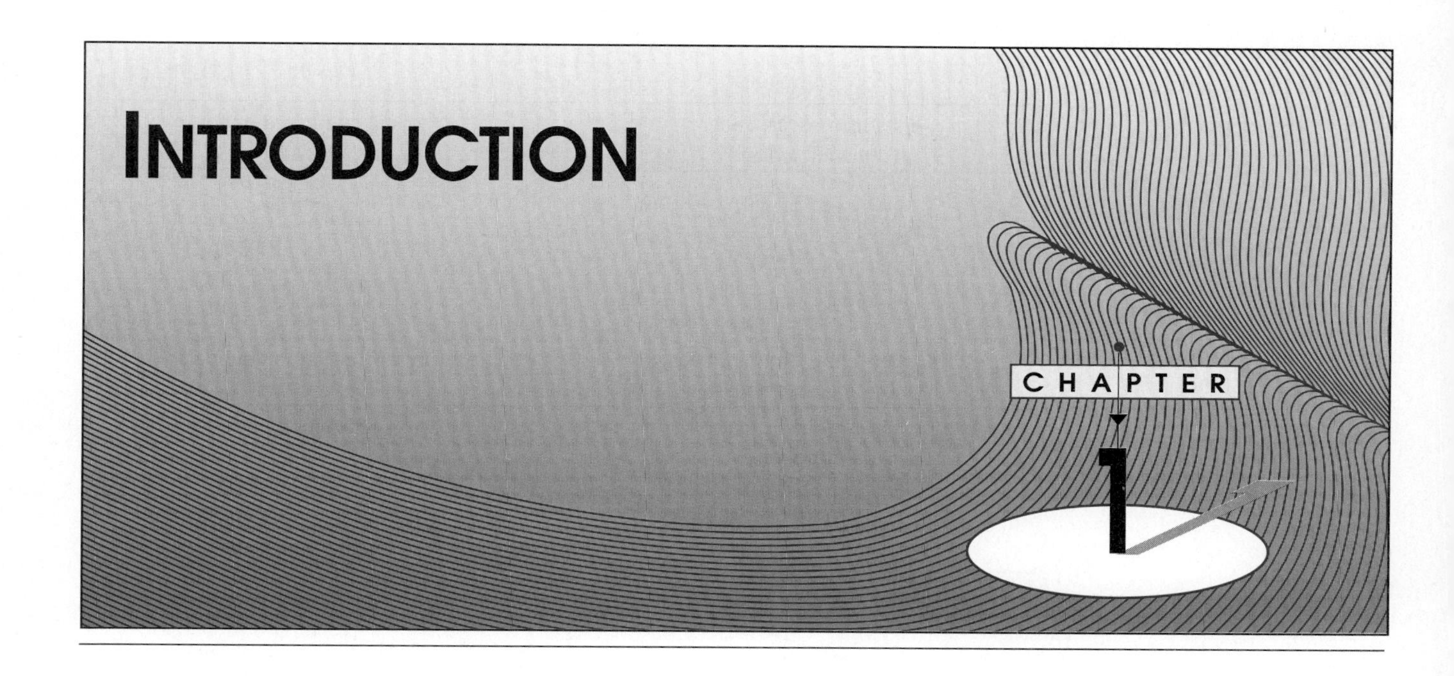

1.1 WHAT IS CALCULUS?

To a Roman in the days of the empire a "calculus" was a pebble used in counting and in gambling. Centuries later "calculare" came to mean "to compute," "to reckon," "to figure out." To the mathematician, physical scientist, and social scientist of today calculus is elementary mathematics (algebra, geometry, trigonometry) enhanced by *the limit process.*

Calculus takes ideas from elementary mathematics and extends them to a more general situation. Here are some examples. On the left-hand side you will find an idea from elementary mathematics; on the right, this same idea as extended by calculus.

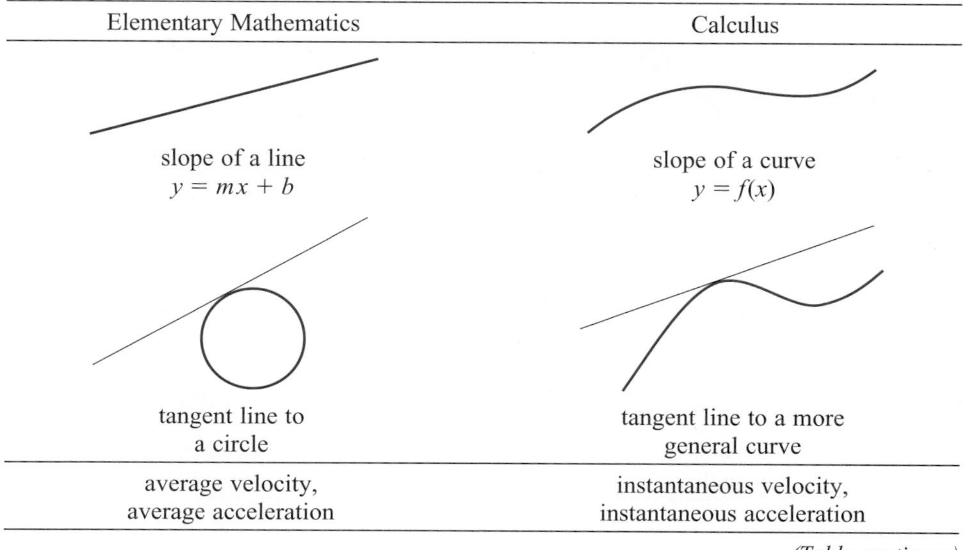

Elementary Mathematics	Calculus
slope of a line $y = mx + b$	slope of a curve $y = f(x)$
tangent line to a circle	tangent line to a more general curve
average velocity, average acceleration	instantaneous velocity, instantaneous acceleration

(Table continues)

distance moved under a constant velocity	distance moved under varying velocity
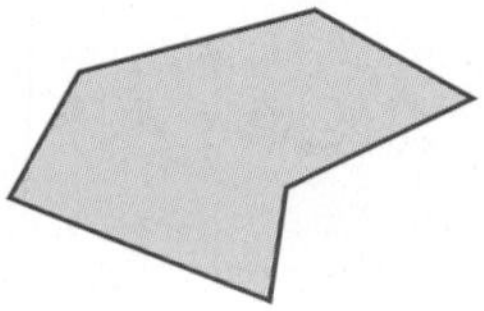 area of a region bounded by line segments	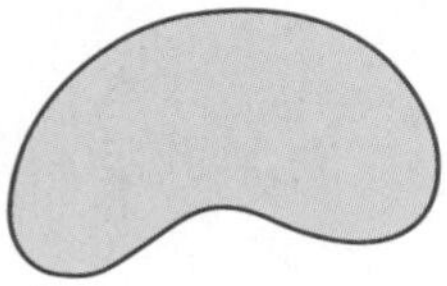 area of a region bounded by curves
sum of a finite collection of numbers $a_1 + a_2 + \cdots + a_n$	sum of an infinite series $a_1 + a_2 + \cdots + a_n + \cdots$
average of a finite collection of numbers	average value of a function on an interval
 length of a line segment	 length of a curve
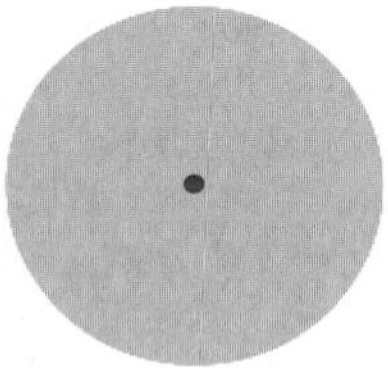 center of a circle	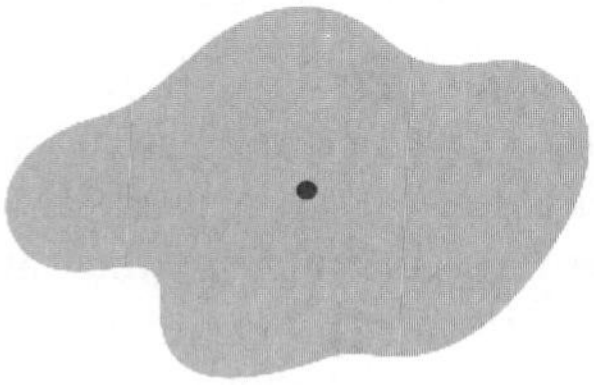 centroid of a region
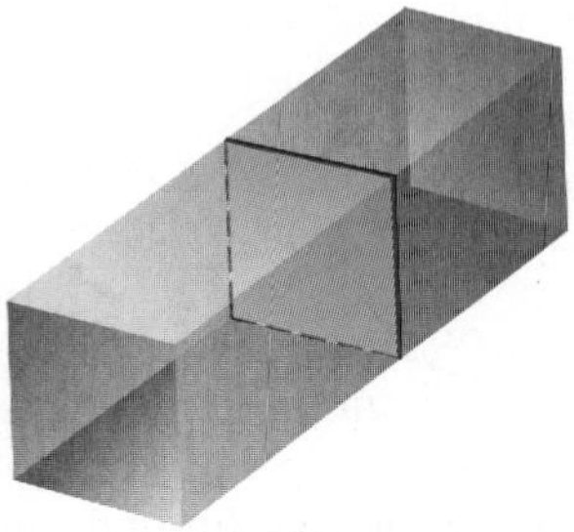 volume of a rectangular solid	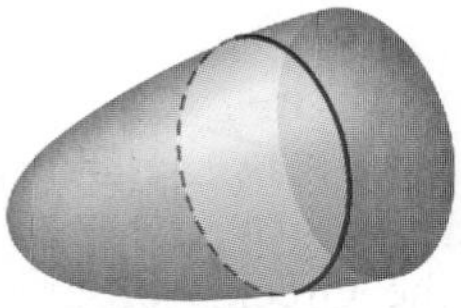 volume of a solid with a curved boundary
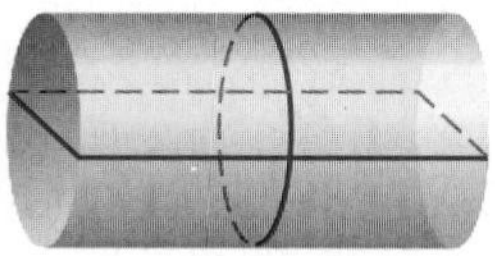 surface area of a cylinder	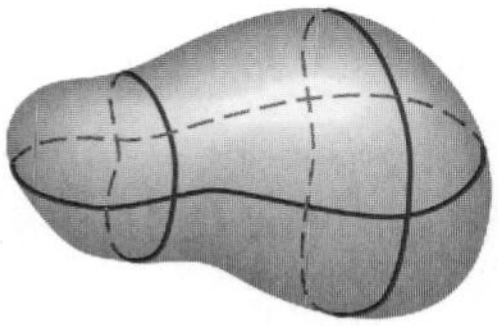 surface area of a more general solid

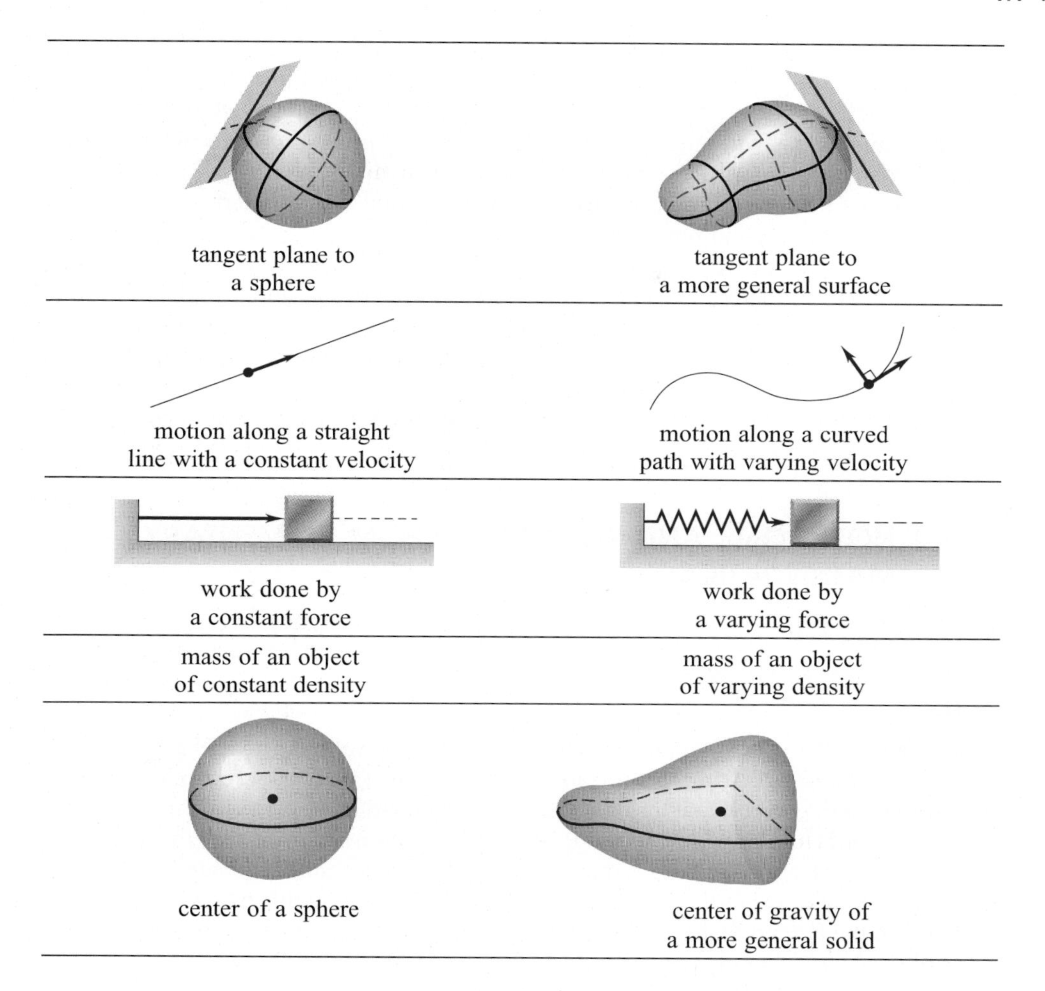

tangent plane to
a sphere
tangent plane to
a more general surface
motion along a straight
line with a constant velocity
motion along a curved
path with varying velocity
work done by
a constant force
work done by
a varying force
mass of an object
of constant density
mass of an object
of varying density
center of a sphere
center of gravity of
a more general solid

It is fitting to say something about the history of calculus. The origins can be traced back to ancient Greece. The ancient Greeks raised many questions (often paradoxical) about tangents, motion, area, the infinitely small, the infinitely large—questions that today are clarified and answered by calculus. Here and there the Greeks themselves provided answers (some very elegant), but mostly they provided only questions.

After the Greeks, progress was slow. Communication was limited, and each scholar was obliged to start almost from scratch. Over the centuries some ingenious solutions to calculus-type problems were devised, but no general techniques were put forth. Progress was impeded by the lack of a convenient notation. Algebra, founded in the ninth century by Arab scholars, was not fully systemized until the sixteenth century. Then, in the seventeenth century, Descartes established analytic geometry, and the stage was set.

The actual invention of calculus is credited to Sir Isaac Newton (1642–1727) and to Gottfried Wilhelm Leibniz (1646–1716), an Englishman and a German. Newton's invention is one of the few good turns that the great plague did mankind. The plague forced the closing of Cambridge University in 1665 and young Isaac Newton of Trinity College returned to his home in Lincolnshire for eighteen months of meditation, out of which grew his *method of fluxions,* his *theory of gravitation,* and his *theory of light*. The method of fluxions is what concerns us here. A treatise with this title was written by Newton in 1672, but it remained unpublished until 1736, nine years after his death. The new method (calculus to us) was first announced in 1687, but in vague general terms without symbolism, formulas, or applications. Newton himself seemed reluctant to publish anything tangible about his new method, and it is not surprising that the development on the Continent, in spite of a late start, soon overtook Newton and went beyond him.

Leibniz started his work in 1673, eight years after Newton. In 1675 he initiated the basic modern notation: dx and $\int$. His first publications appeared in 1684 and 1686. These made little stir in Germany, but the two brothers Bernoulli of Basel (Switzerland) took up the ideas and added profusely to them. From 1690 onward calculus grew rapidly and reached roughly its present state in about a hundred years. Certain theoretical subtleties were not fully resolved until the twentieth century.

■ 1.2 NOTATIONS AND FORMULAS FROM ELEMENTARY MATHEMATICS

The following outline is presented for review and easy reference.

Sets

A set is a well-defined collection of distinct objects. The objects in a set are called the *elements* or *members* of the set. Capital letters $A, B, C, \ldots$, are usually used to denote sets and lowercase letters $a, b, c, \ldots$, to denote the elements of a set.

For a collection S of objects to be a set it must be *well-defined* in the sense that given any object x whatsoever, it must be possible to determine whether or not x is an element of S. For example, the collection of capital letters on this page, the collection of states in the United States of America, and the collection of natural numbers are all examples of sets. On the other hand, suppose we wanted to discuss the collection of beautiful cities in England? Different people might have different collections. Thus, the collection of beautiful cities in England is not a set. Collections based on subjective judgments such as "all good football players," or "all intelligent adults" are not sets.

Notation

The object x is in the set A	$x \in A$
The object x is not in the set A	$x \notin A$
The set of all x for which property P holds (For example, $A = \{x: x \text{ is a vowel}\} = \{a, e, i, o, u\}$)	$\{x: P\}$
A is a subset of B (A is contained in B)	$A \subseteq B$
The union of A and B ($A \cup B = \{x: x \in A \text{ or } x \in B\}$)	$A \cup B$
The intersection of A and B ($A \cap B = \{x: x \in A \text{ and } x \in B\}$)	$A \cap B$
The empty set	$\varnothing$

These are the basic notions from set theory that you will need for this text.

Real Numbers

Our study of calculus is based on the real number system. The real number system consists of the set of real numbers together with the familiar arithmetic operations—addition, subtraction, multiplication, division—and certain other properties which are reviewed briefly here.

Classification

Natural numbers (*or positive integers*)	1, 2, 3, ...
Integers	0, 1, −1, 2, −2, ...
Rational numbers	$\{x: x = p/q$, where p, q are integers and $q \neq 0\}$†
Examples	$\frac{2}{5}, \frac{-19}{7}, \frac{4}{1}\ (=4)$.
Irrational numbers	real numbers that are not rational numbers.
Examples	$\sqrt{2}$, $\sqrt[3]{7}$, π, the solutions of the equation $3x^2 - 5 = 0$.

Decimal Representation

Every real number can be represented by a decimal. If $r = p/q$ is a rational number, then its decimal representation is found by dividing the denominator q into the numerator p. The resulting decimal expansion will either *terminate* (for example, $3/5 = 0.6$, $27/20 = 1.35$) or *repeat* (for example, $2/3 = 0.666666\ldots = 0.\overline{6}$, $15/11 = 1.363636\ldots = 1.\overline{36}$, $19/7 = 2.714285714285\ldots = 2.\overline{714285}$; the bar over the sequence of digits indicates that the sequence repeats indefinitely). The converse is also true: every repeating or terminating decimal represents a rational number.

The decimal expansion of an irrational number neither terminates nor repeats. For example,

$$\sqrt{2} = 1.414213562\ldots, \quad \pi = 3.141592653\ldots, \text{ and}$$
$$r = 0.10110111011110\ldots$$

are irrational numbers.

If we stop the decimal expansion of a given number at a certain decimal place, then the result is a rational number approximation for the number. For instance, $1.414 = 1414/1000$ and $3.14 = 314/100$ are common rational number approximations for $\sqrt{2}$ and π, respectively. More accurate approximations can be obtained by taking more decimal places in the expansions.

Geometric Representation

A fundamental concept in mathematics which connects the abstract notion of *real number* with the geometric notion of *point* is the representation of the real numbers as points on a straight line. This is done by choosing an arbitrary point O on a line to represent the number 0, and another point U (usually taken to the right of O) to represent the number 1. See Figure 1.2.1. The point O is called the *origin* and the distance between O and U determines a scale (a unit length). With O and U specified, each real number can be represented as a point on the line and, conversely,

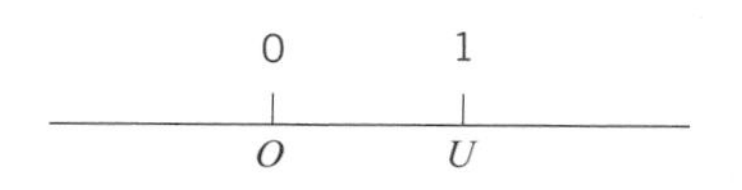

Figure 1.2.1

† Recall that division by 0 is undefined.

each point on the line represents a real number. A line representing the real numbers is called a *number line* or *real line*; the number associated with a point P on the line is called the *coordinate* of P. The *positive* numbers are identified with the points on the right side of O and the *negative* numbers with the points to the left of O. The point representing the number x $(x \neq 0)$ is x units from O if x is positive and $-x$ units from O if x is negative. In this context, we will frequently refer to real numbers as "points," and by this we will mean "points on a number line." Figure 1.2.2 shows some numbers plotted as points on a number line.

Figure 1.2.2

Order Properties

If a and b are real numbers, then *a is less than b*, denoted $a < b$, if $b - a$ is a positive number. This is equivalent to saying that *b is greater than a*, which is denoted $b > a$. Geometrically, $a < b$ if the point a lies to the left of the point b on a number line. The notation $a \leq b$ means that either $a < b$ or $a = b$ (equivalently, $b \geq a$).

The real numbers are *ordered* in the sense that if a and b are real numbers, then exactly one of the following holds:

$$a < b, \qquad a = b, \qquad a > b \qquad \text{(trichotomy)}.$$

The symbols $<, >, \leq, \geq$ are called *inequalities*. Inequalities satisfy the following properties:

(i) If $a < b$ and $b < c$, then $a < c$ (transitive property).

(ii) If $a < b$, then $a + c < b + c$ for all real numbers c.

(iii) If $a < b$ and $c < d$, then $a + c < b + d$.

(iv) If $a < b$ and $c > 0$, then $ac < bc$.

(v) If $a < b$ and $c < 0$, then $ac > bc$.

The corresponding properties hold for $>, \leq, \geq$. A crucial point is Property (v): If an inequality is multiplied by a negative quantity, then the "direction" of the inequality is reversed. Techniques for solving inequalities use these properties and are reviewed in Section 1.3.

Density

Between any two real numbers there are infinitely many rational numbers and infinitely many irrational numbers. In particular, *there is no smallest positive real number.*

Absolute Value

Two important properties of a real number a are its *sign* and its *size,* or *magnitude*. Geometrically, the sign of a tells us whether the point a is on the right or left of 0 on a number line. The magnitude of a is the distance between the point a and 0; 0 itself does not have a sign and its magnitude is 0. The magnitude of a is more commonly called *absolute value of a*, denoted $|a|$. The absolute value of a is also given by

$$|a| = \begin{cases} a & \text{if } a \geq 0 \\ -a & \text{if } a < 0. \end{cases}$$

Other characterizations $|a| = \max\{a, -a\}$; $|a| = \sqrt{a^2}$.

Geometric interpretations $|a|$ = distance between a and 0;

$|a - c|$ = distance between a and c.

Other properties

(i) $|a| = 0$ iff $a = 0$.†
(ii) $|-a| = |a|$.
(iii) $|ab| = |a|\,|b|$.
(iv) $|a + b| \leq |a| + |b|$ (the triangle inequality).††
(v) $\big||a| - |b|\big| \leq |a - b|$ (a variant of the triangle inequality).

Techniques for solving inequalities involving absolute values are also reviewed in Section 1.3.

Intervals

Suppose that $a < b$. The *open interval* (a, b) is the set of all numbers between a and b:

$$(a, b) = \{x\colon a < x < b\}.$$

The *closed interval* $[a, b]$ is the open interval (a, b) together with the endpoints a and b:

$$[a, b] = \{x\colon a \leq x \leq b\}.$$

There are seven other types of intervals:

$$(a, b] = \{x\colon a < x \leq b\}.$$

$$[a, b) = \{x\colon a \leq x < b\}.$$

$$(a, \infty) = \{x\colon a < x\}.$$

$$[a, \infty) = \{x\colon a \leq x\}.$$

$$(-\infty, b) = \{x\colon x < b\}.$$

$$(-\infty, b] = \{x\colon x \leq b\}.$$

$(-\infty, \infty)$ = set of real numbers.

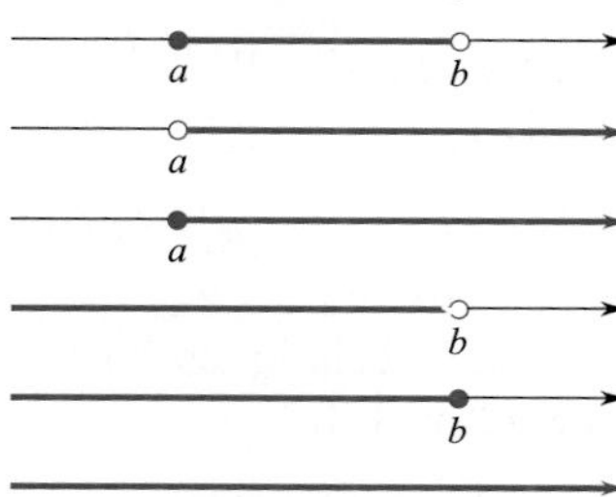

Interval notation is easy to remember: we use a square bracket to include an endpoint and a parenthesis when the endpoint is not included. On a number line, the inclusion or exclusion of an endpoint is signified with a solid "dot" or an open "dot," respectively. The symbols ∞ and $-\infty$, read "infinity" and "negative infinity" (or

---

† By "iff" we mean "if and only if." This expression is used so often in mathematics that it's convenient to have an abbreviation for it.

†† The absolute value of the sum of two numbers cannot exceed the sum of their absolute values. This is analogous to the fact that in a triangle the length of one side cannot exceed the sum of the lengths of the other two sides.

"minus infinity") are not real numbers. The ∞ symbol in the intervals just given indicates that the interval extends indefinitely in the positive direction; the $-\infty$ indicates that the interval extends indefinitely in the negative direction. Since ∞ and $-\infty$ are not real numbers, we do not have intervals of the form $[a, \infty]$ or $[-\infty, b)$, and so forth.

Boundedness

A set S of real numbers is said to be:

(i) *Bounded above* iff there exists a real number M such that

$$x \leq M \qquad \text{for all } x \in S;$$

M is called an *upper bound* for S.

(ii) *Bounded below* iff there exists a real number m such that

$$x \geq m \qquad \text{for all } x \in S;$$

m is called a *lower bound* for S.

(iii) *Bounded* iff it is bounded above and below.

For example, the intervals $(-\infty, 2]$ and $(-\infty, 2)$ are bounded above but not below. The set of positive integers is bounded below but not above. The intervals $[1, 5]$, $(-3, 7)$, and $(-5, \sqrt{2}]$ are bounded (these intervals are bounded above and below).

Algebra

Basic Factoring Formulas

$$(a + b)^2 = a^2 + 2ab + b^2$$
$$(a - b)^2 = a^2 - 2ab + b^2$$
$$(a + b)^3 = a^3 + 3a^2b + 3ab^2 + b^3$$
$$(a - b)^3 = a^3 - 3a^2b + 3ab^2 - b^3$$
$$a^2 - b^2 = (a - b)(a + b)$$
$$a^3 - b^3 = (a - b)(a^2 + ab + b^2)$$
$$a^3 + b^3 = (a + b)(a^2 - ab + b^2)$$

Polynomials and Rational Expressions

A *variable* (also called an *unknown*) is a symbol used to represent an arbitrary element in a given set. Lowercase letters near the end of the alphabet such as $t, u, x, y, \ldots,$ will normally be used to denote variables. In this text, the sets under consideration are usually sets of real numbers, and so variables are often referred to as *real variables.*

A polynomial in one variable is an expression of the form

$$P(x) = a_n x^n + a_{n-1} x^{n-1} + \cdots + a_2 x^2 + a_1 x + a_0$$

where $a_0, a_1, \ldots, a_n$ ($a_n \neq 0$) are real numbers called the *coefficients* of P, and n is a nonnegative integer called the *degree* of P. For example,

$$P(x) = 2x^4 - 5x^3 + \pi x - \sqrt{2}$$

is a polynomial of degree 4;

$$P(x) = 3x + 1 \qquad \text{and} \qquad P(x) = 6$$

are polynomials of degrees 1 and 0, respectively. In general, each nonzero real number is a polynomial of degree 0; 0 is also a polynomial, but no degree is assigned to it. The expression

$$F(x) = x^3 + 4x^{3/2} + \frac{1}{x^2}$$

is not a polynomial.

A number r is a *root* or *solution* of the polynomial equation

$$P(x) = a_n x^n + a_{n-1}x^{n-1} + \cdots + a_2x^2 + a_1x + a_0 = 0$$

iff

$$P(r) = a_n r^n + a_{n-1}r^{n-1} + \cdots + a_2r^2 + a_1r + a_0 = 0.$$

The number r is also called a *zero* of the polynomial P.

Recall the *factor theorem* which states that $x - r$ is a factor of a polynomial P iff r is a zero of P, that is

$$P(x) = (x - r)Q(x) \qquad \text{for some polynomial } Q$$

iff

$$P(r) = 0.$$

A polynomial of degree 1 has the form $P(x) = ax + b$, $a \neq 0$, and is called a *linear polynomial.* The number $r = -b/a$ is the (unique) zero of P.

A polynomial of degree 2 has the form $P(x) = ax^2 + bx + c$, $a \neq 0$, and is called a *quadratic polynomial.* The quadratic polynomial P has two zeros which are given by the *quadratic formula*:

$$r_1 = \frac{-b + \sqrt{b^2 - 4ac}}{2a}, \qquad r_2 = \frac{-b - \sqrt{b^2 - 4ac}}{2a},$$

and P can be written in terms of r_1 and r_2 as

$$P(x) = a(x - r_1)(x - r_2) \qquad \text{(verify this).}$$

This expression is called the *factored form* of P.

Quadratic polynomials can be used to illustrate a limitation of the set of real numbers, namely the existence of polynomials with real coefficients that do not have real zeros. In particular, if $b^2 - 4ac < 0$, then the zeroes of the quadratic polynomial P are not real numbers. This situation is remedied by introducing a new "number" i defined by $i^2 = -1$ (equivalently, i is a zero of the quadratic polynomial $P(x) = x^2 + 1$). The *complex numbers* consist of all numbers of the form $a + bi$, where a and b are real numbers. For example, $3 + 4i$, $-2 - 3i$, $0 + 6i = 6i$, and $\frac{1}{2} + 0i = \frac{1}{2}$ are complex numbers. The real numbers can be considered to be a subset of the complex numbers since any real number a can be written $a = a + 0i$.

The *Fundamental Theorem of Algebra* states that if $P(z)$ is a polynomial of degree n, $n \geq 1$, with complex coefficients, then P has a (complex) zero. Based on this fact, and using the factor theorem, it is easy to show that P has exactly n zeros. Thus, a polynomial of degree $n \geq 1$ with real coefficients has exactly n zeros, but some (or even all) of the zeros may be complex numbers. In Section 8.6, we will need to use the fact that a polynomial of degree $n \geq 1$ with real coefficients can be factored into a product of linear and quadratic factors corresponding to its real and complex zeros.

In this text we will be concerned almost exclusively with the set of real numbers, and so if the term "number" is used without qualification, it will be interpreted to mean "real number." You should be aware, however, that complex numbers can arise in the course of solving equations.

An expression of the form $P(x)/Q(x)$, where P and Q are polynomials, is called a *rational expression* (in one variable). For example,

$$\frac{1}{x-1}, \qquad \frac{2x^3+3x}{x^2-9}, \qquad \text{and} \qquad \frac{x}{x^2+1}$$

are rational expressions, while

$$\frac{x^{2/3}-2x}{\sqrt{x^3+2x-1}}$$

is not a rational expression. It is important to note that a rational expression $R(x) = P(x)/Q(x)$ is not defined at the points (numbers) x where $Q(x) = 0$. Also, $R(x) = 0$ at the points x where $P(x) = 0$.

Factorials

Let n be a positive integer. Then *n factorial*, denoted $n!$, is the product of the integers from 1 to n. Thus,

$$1! = 1, \quad 2! = 1 \cdot 2, \quad 3! = 1 \cdot 2 \cdot 3, \quad \ldots \quad n! = 1 \cdot 2 \cdot 3 \cdot \ \ldots \ \cdot (n-1) \cdot n.$$

Note that

$$1! = 1, \qquad 2! = 2, \qquad 3! = 6.$$

You can verify that $4! = 24$, $5! = 120$, and so on. Finally, it is convenient to define $0! = 1$.

Geometry

Elementary Figures

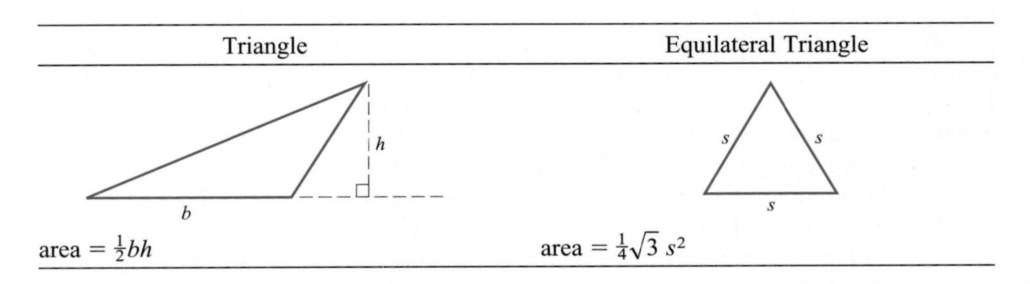

Triangle	Equilateral Triangle
area $= \frac{1}{2}bh$	area $= \frac{1}{4}\sqrt{3}\, s^2$

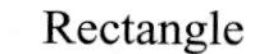

Rectangle

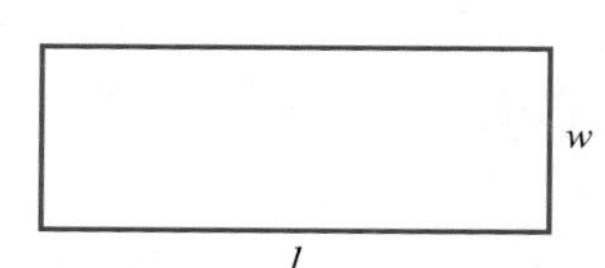

area = lw
perimeter = $2l + 2w$
diagonal = $\sqrt{l^2 + w^2}$

Rectangular Solid

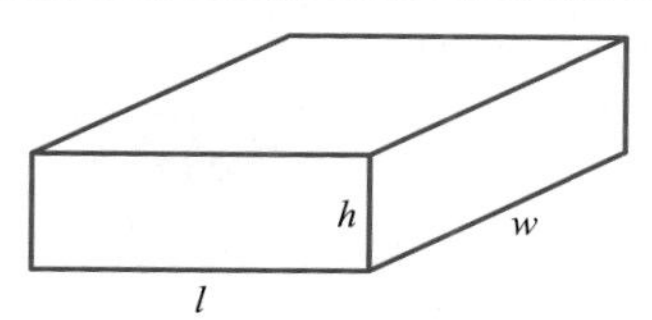

volume = lwh
surface area = $2lw + 2lh + 2wh$

Square

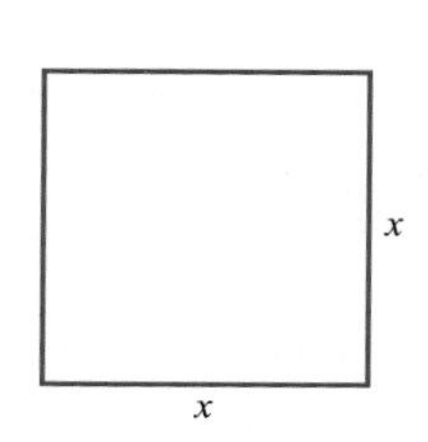

area = x^2
perimeter = $4x$
diagonal = $x\sqrt{2}$

Cube

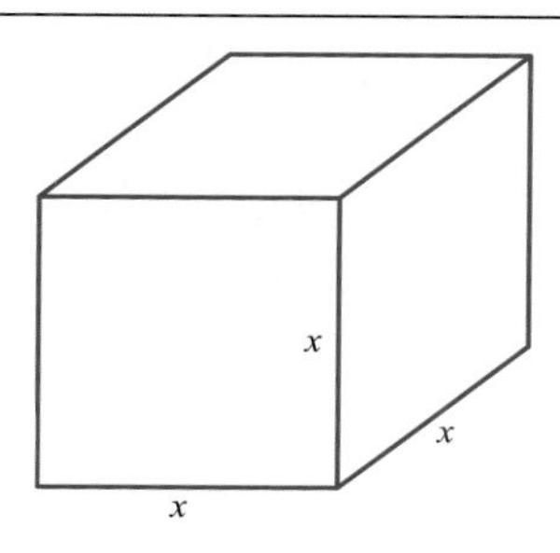

volume = x^3
surface area = $6x^2$

Circle

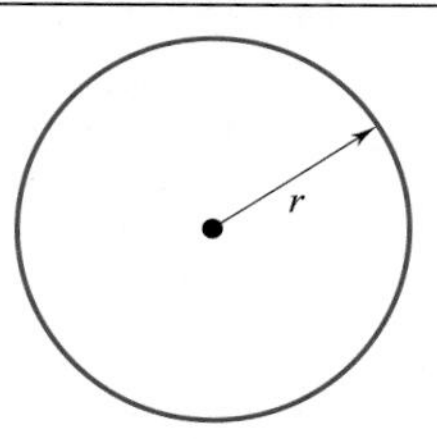

area = πr^2
circumference = $2\pi r$

Sphere

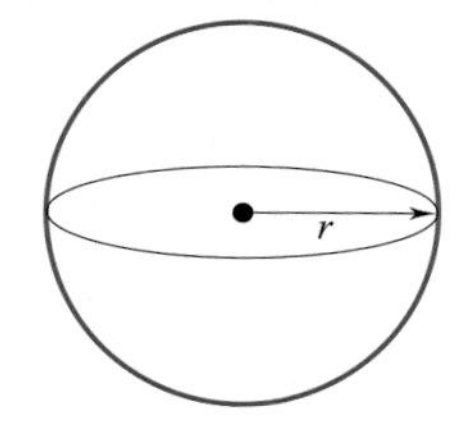

volume = $\frac{4}{3}\pi r^3$
surface area = $4\pi r^2$

Sector of a Circle: radius r, central angle θ measured in radians (see Section 1.6).

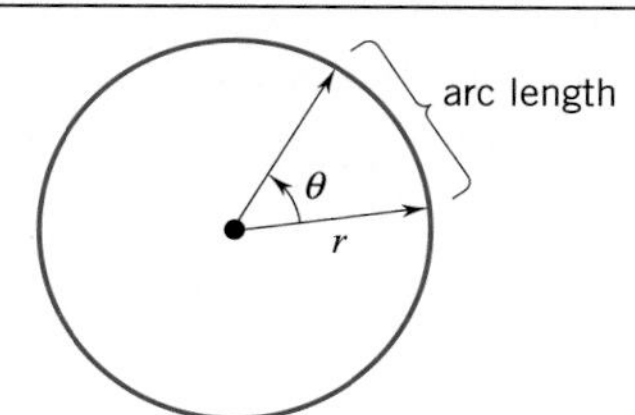

arc length = $r\theta$

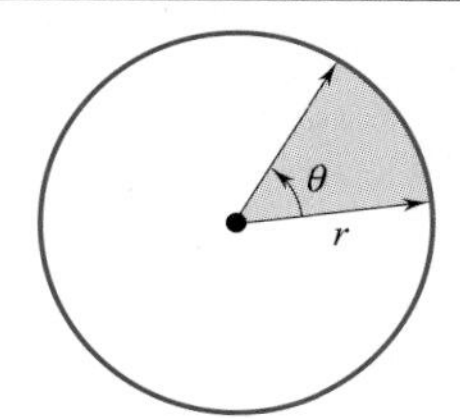

area = $\frac{1}{2}r^2\theta$

(Table continues)

Right Circular Cylinder	Right Circular Cone
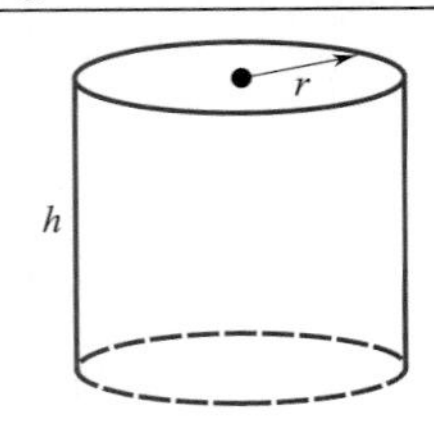	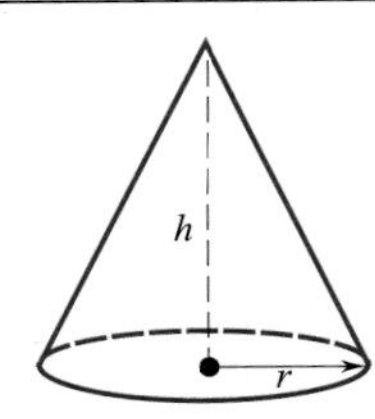
volume $= \pi r^2 h$ lateral area $= 2\pi rh$ total surface area $= 2\pi r^2 + 2\pi rh$	volume $= \frac{1}{3}\pi r^2 h$ slant height $= \sqrt{r^2 + h^2}$ lateral area $= \pi r\sqrt{r^2 + h^2}$ total surface area $= \pi r^2 + \pi r\sqrt{r^2 + h^2}$

EXERCISES 1.2

In Exercises 1–12, use the terms *integer*, *rational*, *irrational*, *complex* to classify the given number; use all the terms that apply.

1. $\frac{17}{7}$. **2.** -6.

3. $2.131313 \ldots = 2.\overline{13}$. **4.** $\sqrt{2} - 3$.

5. 0. **6.** $\pi - 2$.

7. $\sqrt[3]{8}$. **8.** 0.125.

9. $-\sqrt{9}$. **10.** $0.9999 \ldots = 0.\overline{9}$.

11. $0.21211211121111 \ldots$.

12. $13\frac{2}{7}$.

A method for writing the repeating decimal $3.\overline{135}$ in the form p/q, p, q integers, $q \neq 0$, is as follows:

Set $\qquad x = 3.135135 \ldots$.

Then $\qquad 1000x = 3135.135135 \ldots$.

Now subtract:

$$1000x - x = 3135.135135 \ldots - 3.135135 \ldots = 3132.$$

Thus

$$999x = 3132,$$

and

$$x = \frac{3132}{999} = \frac{116}{37}.$$

Use this method in Exercises 13–17 to express the given repeating decimal in the rational number form p/q.

13. $13.\overline{201}$.

14. $2.777 \ldots$. HINT: Multiply by 10.

15. $0.2323 \ldots$. **16.** $4.16\overline{3}$.

17. $5.252252 \ldots$.

18. Show that the repeating decimal $0.9999 \ldots$ represents the number 1. Note that $1 = 1.0$ also represents the number 1. Thus, it follows that a rational number may have more than one decimal representation.

In Exercises 19–24, replace the symbol $*$ by $<$, $>$, or $=$ to make the statement true.

19. $\frac{3}{4} * 0.75$. **20.** $0.33 * \frac{1}{3}$.

21. $\sqrt{2} * 1.414$. **22.** $4 * \sqrt{16}$.

23. $-\frac{2}{7} * -0.285714$. **24.** $\pi * \frac{22}{7}$.

In Exercises 25–31, write the given number without absolute value.

25. $|6|$. **26.** $|-4|$.

27. $|3 - 7|$. **28.** $|-5| - |8|$.

29. $|-5| + |-8|$. **30.** $|2 - \pi|$.

31. $|5 - \sqrt{5}|$.

In Exercises 32–41, indicate on a number line all the values of x that satisfy the given condition.

32. $x \geq 3$. **33.** $x \leq -\frac{3}{2}$.

34. $-2 \leq x \leq 3$. **35.** $x^2 < 16$.

36. $x^2 \geq 16$. **37.** $|x| \leq 0$.

38. $x^2 \geq 0$. **39.** $|x - 4| \leq 2$.

40. $|x + 1| > 3$. **41.** $|x + 3| \leq 0$.

In Exercises 42–50, sketch the given set on a number line.

42. $[3, \infty)$. **43.** $(-\infty, 2)$.

44. $(-4, 3]$. **45.** $[-2, 3] \cup [1, 5]$.

46. $(1, 4] \cup (4, 8)$. **47.** $[-2, 2] \cap [0, 4]$.

48. $[-3, \frac{3}{2}) \cap (\frac{3}{2}, \frac{5}{2}]$.

49. $(-\infty, -1) \cup (-2, \infty)$.

50. $(-\infty, 2) \cap [3, \infty)$.

In Exercises 51–57, determine whether the given set is bounded above, bounded below, or bounded. If a set is bounded above, give an upper bound; if it is bounded below, give a lower bound; if it is bounded, give an upper and a lower bound.

51. $\{0, 1, 2, 3, 4\}$.

52. $\{0, -1, -2, -3, \ldots\}$.

53. S is the set of even integers.

54. $S = \{x: x \leq 4\}$.

55. $S = \{x: x^2 > 3\}$.

56. $S = \left\{\dfrac{n-1}{n}: n = 1, 2, 3, \ldots\right\}$.

57. S is the set of rational numbers less than $\sqrt{2}$.

In Exercises 58–63, rewrite the given expression in factored form.

58. $9x^2 - 4$.

59. $x^2 - 10x + 25$.

60. $27x^3 - 8$.

61. $8x^6 + 64$.

62. $4x^4 + 4x^2 + 1$.

63. $4x^2 + 12x + 9$.

In Exercises 64–70, find the real zeros of the given quadratic polynomial.

64. $P(x) = x^2 - 9$.

65. $P(x) = x^2 - x - 2$.

66. $P(x) = 2x^2 - 5x - 3$.

67. $P(x) = x^2 - 6x + 9$.

68. $P(x) = x^2 + 8x + 16$.

69. $P(x) = x^2 - 2x + 2$.

70. $P(x) = x^2 - 2x + 5$.

In Exercises 71–75, determine all real numbers x for which the given rational expression $R(x)$ does not represent a real number. Also, determine all real numbers x for which $R(x) = 0$.

71. $R(x) = \dfrac{x-1}{x^2}$.

72. $R(x) = \dfrac{x^2 - 5x + 6}{x^2 - 4}$.

73. $R(x) = \dfrac{x^2 - 1}{x^2 + 1}$.

74. $R(x) = \dfrac{1}{x^2 - 2x + 8}$.

75. $R(x) = \dfrac{2x + 3}{x^2 + 2x + 5}$.

In Exercises 76–79, determine the value of the given expression.

76. $5!$.

77. $\dfrac{5!}{8!}$.

78. $\dfrac{8!}{3!5!}$.

79. $\dfrac{9!}{3!6!}$.

80. Prove that the sum of two rational numbers is a rational number.

81. Prove that the sum of a rational number and an irrational number is irrational.

82. Prove that the product of two rational numbers is a rational number.

83. What can you say about the product of a rational number and an irrational number?

84. Show by examples that the sum or product of two irrational numbers can either be a rational number or an irrational number.

85. Prove that $\sqrt{2}$ is an irrational number. HINT: Assume that $\sqrt{2} = p/q$ and that p and q have no common factors (other than 1 and -1). Square both sides of this equation and argue that both p and q must be divisible by 2. This contradicts the assumption.

86. Prove that $\sqrt{3}$ is irrational. HINT: Use the method of Exercise 85.

87. Let R be the set of all rectangles that have the same perimeter P. Show that the square (whose side necessarily has length $P/4$) is the element of R that has the largest area.

88. Show that if a circle and a square have the same perimeter, then the circle has a larger area than the square. Suppose a circle and an arbitrary rectangle have equal perimeters. Which has the larger area?

■ 1.3 INEQUALITIES

(All our work with inequalities is based on the order properties of real numbers as given on page 6.) In this section we work with the sort of inequalities that abound in calculus, inequalities that involve a variable.

To solve an equation in x is to find the set of numbers x for which the equation holds. To solve an inequality in x is to find the set of numbers x for which the inequality holds.

The way we solve an inequality is very similar to the way we solve an equation, but there is one important difference. We can maintain an inequality by adding the same number to both sides, or by subtracting the same number from both sides, or by multiplying or dividing both sides by the same *positive* number. But if we multiply or divide by a *negative* number, then the inequality is *reversed*:

$$x - 2 < 4 \quad \text{gives} \quad x < 6, \qquad x + 2 < 4 \quad \text{gives} \quad x < 2,$$

$$\tfrac{1}{2}x < 4 \quad \text{gives} \quad x < 8,$$

but

$$-\tfrac{1}{2}x < 4 \quad \text{gives} \quad x > -8.$$

↑ note

Example 1 Solve the inequality

$$-3(4 - x) < 12.$$

SOLUTION Multiplying both sides of the inequality by $-\frac{1}{3}$, we have

$$4 - x > -4 \qquad \text{(the inequality has been reversed).}$$

Subtracting 4, we get

$$-x > -8.$$

To isolate x, we multiply by -1. This gives

$$x < 8 \qquad \text{(the inequality has been reversed again).}$$

The solution set is the interval $(-\infty, 8)$. ❑

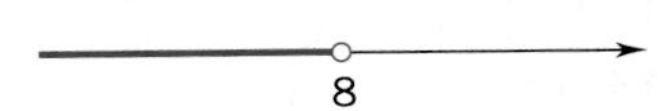

There are generally several ways to solve a given inequality. For example, the last inequality could have been solved this way:

$$-3(4 - x) < 12,$$

$$-12 + 3x < 12,$$

$$3x < 24 \qquad \text{(we added 12),}$$

$$x < 8 \qquad \text{(we divided by 3).}$$

The solution set, of course, is once again $(-\infty, 8)$.

The usual way to solve a quadratic inequality is to factor the quadratic.

Example 2 Solve the inequality

$$\tfrac{1}{5}(x^2 - 4x + 3) < 0.$$

SOLUTION First we eliminate the outside factor $\frac{1}{5}$ by multiplying through by 5: this gives

$$x^2 - 4x + 3 < 0.$$

Factoring the quadratic, we have

$$(x - 1)(x - 3) < 0.$$

The product $(x-1)(x-3)$ is zero at 1 and 3. We mark these points on a number line (Figure 1.3.1). The points 1 and 3 separate three intervals:

$$(-\infty, 1), \qquad (1, 3), \qquad (3, \infty).$$

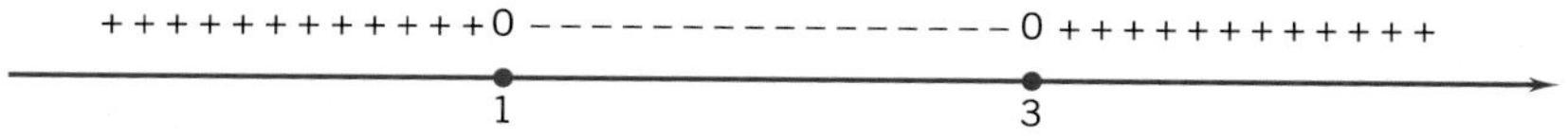

Figure 1.3.1

On each of these intervals the product $(x-1)(x-3)$ keeps a constant sign:

on $(-\infty, 1)$ [to the left of 1] sign of $(x-1)(x-3) = (-)(-) = +$;

on $(1, 3)$ [between 1 and 3] sign of $(x-1)(x-3) = (+)(-) = -$;

on $(3, \infty)$ [to the right of 3] sign of $(x-1)(x-3) = (+)(+) = +$.

The product $(x-1)(x-3)$ is negative only on $(1, 3)$. The solution set is the open interval $(1, 3)$. ❑

More generally, consider an expression of the form

$$(x-a_1)(x-a_2)\cdots(x-a_n) \qquad \text{with } a_1 < a_2 < \cdots < a_n.$$

Such an expression is zero at $a_1, a_2, \ldots, a_n$. It is positive on those intervals where an even number of terms are negative, and it is negative on those intervals where an odd number of terms are negative.

As an example, take the expression

$$(x+2)(x-1)(x-3).$$

This product is zero at $-2, 1, 3$. It is

negative on $(-\infty, -2)$, (3 negative terms)

positive on $(-2, 1)$, (2 negative terms)

negative on $(1, 3)$, (1 negative term)

positive on $(3, \infty)$. (0 negative terms)

See Figure 1.3.2.

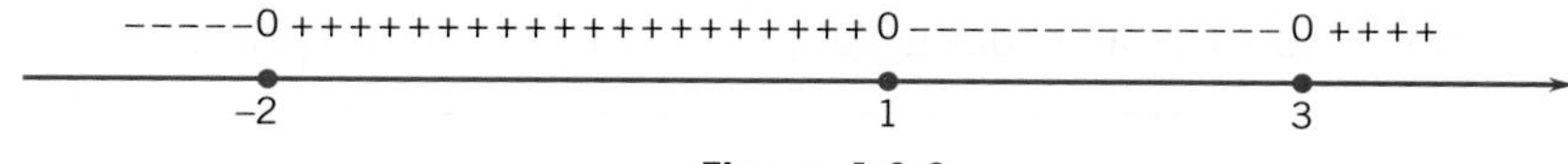

Figure 1.3.2

Example 3 Solve the inequality

$$(x+3)^5(x-1)(x-4)^2 < 0.$$

SOLUTION We view $(x+3)^5(x-1)(x-4)^2$ as the product of three factors: $(x+3)^5$, $(x-1)$, $(x-4)^2$. The product is zero at -3, 1, and 4. These points separate the intervals

$$(-\infty, -3), \qquad (-3, 1), \qquad (1, 4), \qquad (4, \infty).$$

On each of these intervals the product keeps a constant sign. It is

positive on $(-\infty, -3)$,	(2 negative factors)
negative on $(-3, 1)$,	(1 negative factor)
positive on $(1, 4)$,	(0 negative factors)
positive on $(4, \infty)$.	(0 negative factors)

See Figure 1.3.3.

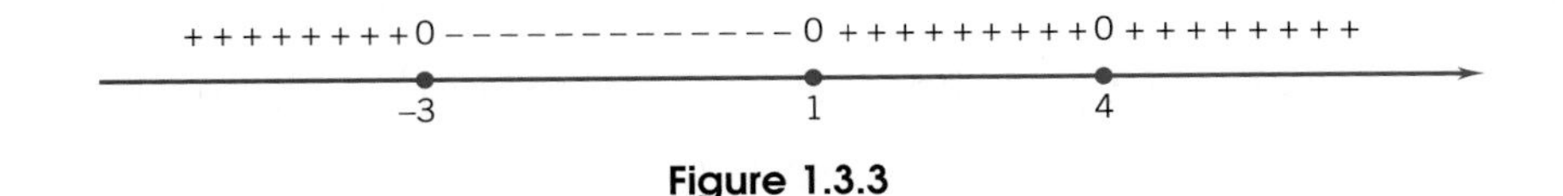

Figure 1.3.3

The solution set is the open interval $(-3, 1)$. ❑

Example 4 Solve the inequality

$$x^2 + 4x - 2 \leq 0.$$

SOLUTION To factor the quadratic we first complete the square by adding and subtracting 4:

$$x^2 + 4x + 4 - 4 - 2 \leq 0$$
$$(x^2 + 4x + 4) - 6 \leq 0$$
$$(x + 2)^2 - 6 \leq 0.$$

We can factor this quadratic as the difference of two squares:

$$(x + 2 + \sqrt{6})(x + 2 - \sqrt{6}) \leq 0.$$

The product is 0 at $-2 - \sqrt{6}$ and $-2 + \sqrt{6}$. It is negative in between. The solution set is the closed interval $[-2 - \sqrt{6}, -2 + \sqrt{6}]$. ❑

Remark The roots of the quadratic $x^2 + 4x - 2 = 0$ could have been obtained from the quadratic formula. ❑

In solving inequalities that involve quotients we use the fact that

(1.3.1)

$$\frac{a}{b} > 0 \quad \text{iff} \quad ab > 0 \qquad \text{and} \qquad \frac{a}{b} < 0 \quad \text{iff} \quad ab < 0.$$

Example 5 Solve the inequality

$$\frac{x + 2}{1 - x} < 1, \qquad (x \neq 1).$$

SOLUTION Note that we cannot multiply the given inequality by $1 - x$ since we don't know its sign. Subtracting 1 from both sides, we have

$$\frac{x+2}{1-x} - 1 < 0,$$

$$\frac{x+2-(1-x)}{1-x} < 0,$$

$$\frac{2x+1}{1-x} < 0,$$

$$(2x+1)(1-x) < 0, \qquad \text{[by (1.3.1)]}$$

$$(2x+1)(x-1) > 0 \qquad \text{(we multiplied by } -1\text{)},$$

$$(x+\tfrac{1}{2})(x-1) > 0 \qquad \text{(we divided by 2).}$$

The product $(x+\frac{1}{2})(x-1)$ is zero at $-\frac{1}{2}$ and 1. As you can check, it is positive on the open intervals $(-\infty, -\frac{1}{2})$ and $(1, \infty)$. Thus the solution set of our inequality is the set $(-\infty, -\frac{1}{2}) \cup (1, \infty)$; its graph is shown in the following figure. ❑

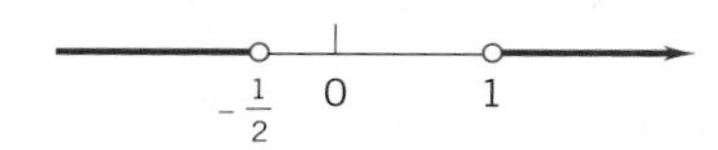

Example 6 Solve the inequality

$$\frac{(x+3)^5(x-1)}{(x-4)^2} < 0.$$

SOLUTION By (1.3.1) this inequality has the same solution as the inequality

$$(x+3)^5(x-1)(x-4)^2 < 0,$$

which we solved in Example 3. ❑

Inequalities and Absolute Value

Now we take up some inequalities that involve absolute values. With an eye toward Chapter 2 we introduce two Greek letters: δ (delta) and ϵ (epsilon).

Recall that for each real number a

(1.3.2)
$$|a| = \begin{cases} a, & \text{if } a \geq 0 \\ -a, & \text{if } a < 0, \end{cases} \qquad |a| = \max\{a, -a\}, \qquad |a| = \sqrt{a^2}.$$

We begin with the inequality

$$|x| < \delta$$

where δ is some positive number. To say that $|x| < \delta$ is to say that x lies within δ units of 0, or, equivalently, that x lies between $-\delta$ and δ. Thus

(1.3.3)
$$|x| < \delta \quad \text{iff} \quad -\delta < x < \delta.$$

To say that $|x - c| < \delta$ is to say that x lies within δ units of c, or, equivalently, that x lies between $c - \delta$ and $c + \delta$. Thus

(1.3.4)
$$|x - c| < \delta \quad \text{iff} \quad c - \delta < x < c + \delta.$$

Somewhat more delicate is the inequality

$$0 < |x - c| < \delta.$$

Here we have $|x - c| < \delta$ with the additional requirement that $x \neq c$. Consequently

(1.3.5)
$$0 < |x - c| < \delta \quad \text{iff} \quad c - \delta < x < c \quad \text{or} \quad c < x < c + \delta.$$

Thus, for example,

$|x| < \frac{1}{2}$ iff $-\frac{1}{2} < x < \frac{1}{2}$; *Solution:* $(-\frac{1}{2}, \frac{1}{2})$

$|x - 5| < 1$ iff $4 < x < 6$; *Solution:* $(4, 6)$

$0 < |x - 5| < 1$ iff $4 < x < 5$ or $5 < x < 6$. *Solution:* $(4, 5) \cup (5, 6)$. ❑

Example 7 Solve the inequality

$$|x + 2| < 3.$$

SOLUTION The inequality $|x + 2| < 3$ holds iff

$$|x - (-2)| < 3 \quad \text{iff} \quad -2 - 3 < x < -2 + 3 \quad \text{iff} \quad -5 < x < 1.$$

The solution set is the open interval $(-5, 1)$. ❑

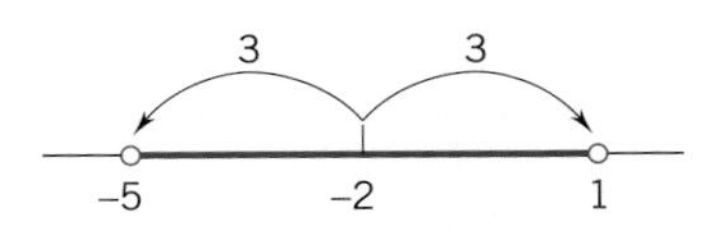

Example 8 Solve the inequality

$$|3x - 4| < 2.$$

SOLUTION Since

$$|3x - 4| = |3(x - \tfrac{4}{3})| = |3||x - \tfrac{4}{3}| = 3|x - \tfrac{4}{3}|,$$

the inequality can be rewritten

$$3|x - \tfrac{4}{3}| < 2.$$

This gives $|x - \tfrac{4}{3}| < \tfrac{2}{3}$. Therefore

$$\tfrac{4}{3} - \tfrac{2}{3} < x < \tfrac{4}{3} + \tfrac{2}{3},$$
$$\tfrac{2}{3} < x < 2.$$

The solution set is the open interval $(\tfrac{2}{3}, 2)$.

ALTERNATIVE SOLUTION The inequality

$$|3x - 4| < 2$$

is equivalent to

$$-2 < 3x - 4 < 2$$

by (1.3.3). Therefore

$$2 < 3x < 6 \qquad \text{(add 4 to both inequalities)}$$

and

$$\tfrac{2}{3} < x < 2 \qquad \text{(divide through by 3)}$$

as before. ❑

Let $\epsilon > 0$. If you think of $|a|$ as the distance between a and 0, then obviously

(1.3.6) $$|a| > \epsilon \quad \text{iff} \quad a > \epsilon \quad \text{or} \quad a < -\epsilon.$$

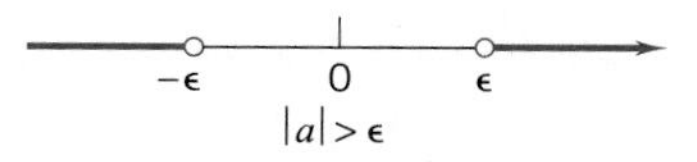

Example 9 Solve the inequality

$$|2x + 3| > 5.$$

SOLUTION As you saw, in general

$$|a| > \epsilon \quad \text{iff} \quad a > \epsilon \quad \text{or} \quad a < -\epsilon.$$

So here

$$2x + 3 > 5 \quad \text{or} \quad 2x + 3 < -5.$$

The first possibility gives $2x > 2$ and thus

$$x > 1.$$

The second possibility gives $2x < -8$ and thus

$$x < -4.$$

The total solution is therefore the union

$$(-\infty, -4) \cup (1, \infty). \quad \square$$

We come now to one of the fundamental inequalities of calculus: for all real numbers a and b,

(1.3.7)
$$|a + b| \leq |a| + |b|.$$

This is called the *triangle inequality* in analogy with the geometric maxim "in a triangle the length of each side is less than or equal to the sum of the lengths of the other two sides."

PROOF OF THE TRIANGLE INEQUALITY The key here is to think of $|x|$ as $\sqrt{x^2}$. Note first that

$$(a + b)^2 = a^2 + 2ab + b^2 \leq |a|^2 + 2|a||b| + |b|^2 = (|a| + |b|)^2.$$

Comparing the extremes of the inequality and taking square roots, we have

$$\sqrt{(a + b)^2} \leq |a| + |b|. \qquad \text{(Exercise 69)}$$

The result follows from observing that

$$\sqrt{(a + b)^2} = |a + b|. \quad \square$$

Here is a variant of the triangle inequality that also comes up in calculus: for all real numbers a and b,

(1.3.8)
$$\big||a| - |b|\big| \leq |a - b|.$$

The proof is left to you as an exercise.

EXERCISES 1.3

In Exercises 1–36, solve the given inequality and graph the solution set on a number line.

1. $2 + 3x < 5$.
2. $\frac{1}{2}(2x + 3) < 6$.
3. $16x + 64 \leq 16$.
4. $3x + 5 > \frac{1}{4}(x - 2)$.
5. $\frac{1}{2}(1 + x) < \frac{1}{3}(1 - x)$.
6. $3x - 2 \leq 1 + 6x$.
7. $x^2 - 1 < 0$.
8. $x^2 + 9x + 20 < 0$.
9. $x(x - 1)(x - 2) > 0$.
10. $x(2x - 1)(3x - 5) \leq 0$.
11. $x^3 - 2x^2 + x \geq 0$.
12. $x^2 - 4x + 4 \leq 0$.
13. $x^2 + 1 > 4x$.
14. $2x^2 + 9x + 6 \geq x + 2$.
15. $1 - 3x^2 < \frac{1}{2}(2 - x^2)$.
16. $6x^2 + 2x \leq (x - 1)^2$.
17. $\frac{1}{x} < x$.
18. $x + \frac{1}{x} \geq 0$.
19. $\frac{x}{x - 5} \geq 0$.
20. $\frac{x}{x + 5} < 0$.
21. $\frac{x}{x - 5} > \frac{1}{4}$.
22. $\frac{1}{3x - 5} < 2$.
23. $\frac{x^2 - 9}{x + 1} > 0$.
24. $\frac{x^2}{x^2 - 4} < 0$.
25. $x^3(x - 2)(x + 3)^2 < 0$.
26. $x^2(x - 3)(x + 4)^2 > 0$.
27. $x^2(x - 2)(x + 6) > 0$.
28. $7x(x - 4)^2 < 0$.

29. $\dfrac{2x}{x^2-4} > 0.$

30. $\dfrac{x^2-9}{3x} > 0.$

31. $\dfrac{1}{x-1} + \dfrac{4}{x-6} > 0.$

32. $\dfrac{3}{x-2} - \dfrac{5}{x-6} < 0.$

33. $\dfrac{2x-6}{x^2-6x+5} < 0.$

34. $\dfrac{2x+8}{x^2+8x+7} > 0.$

35. $\dfrac{x^2-4x+3}{x^2} > 0.$

36. $\dfrac{x^2-4x}{x+2} > 0.$

In Exercises 37–53, solve the given inequality and express the solution set in terms of intervals.

37. $|x| < 2.$

38. $|x| \geq 1.$

39. $|x| > 3.$

40. $|x-1| < 1.$

41. $|x-2| < \frac{1}{2}.$

42. $|x-\frac{1}{2}| < 2.$

43. $0 < |x| < 1.$

44. $0 < |x| < \frac{1}{2}.$

45. $0 < |x-2| < \frac{1}{2}.$

46. $0 < |x-\frac{1}{2}| < 2.$

47. $0 < |x-3| < 8.$

48. $|3x-5| < 3.$

49. $|2x+1| < \frac{1}{4}.$

50. $|5x-3| < \frac{1}{2}.$

51. $|2x+5| > 3.$

52. $|3x+1| > 5.$

53. $|5x-1| > 9.$

In Exercises 54–59, find an inequality of the form $|x-c| < \delta$, the solution of which is the given open interval.

54. $(-2, 2).$

55. $(-3, 3).$

56. $(0, 4).$

57. $(-3, 7).$

58. $(-4, 0).$

59. $(-7, 3).$

In Exercises 60–63, determine all values of $A > 0$ for which the statement is true.

60. If $|x-2| < 1$, then $|2x-4| < A$.

61. If $|x-2| < A$, then $|2x-4| < 3$.

62. If $|x+1| < A$, then $|3x+3| < 4$.

63. If $|x+1| < 2$, then $|3x+3| < A$.

64. Arrange the following in order: $1, x, \sqrt{x}, 1/x, 1/\sqrt{x}$, given that $x > 1$.

65. Redo Exercise 64 for the case $0 < x < 1$.

66. Compare

$$\sqrt{\frac{x}{x+1}} \quad \text{and} \quad \sqrt{\frac{x+1}{x+2}}$$

given that $x > 0$.

67. Let a and b have the same sign. If $a < b$, show that $(1/b) < (1/a)$.

68. Let a and b be nonnegative numbers. Show that if

$$a^2 \leq b^2 \quad \text{then} \quad a \leq b.$$

69. Let a and b be nonnegative numbers. Show that if

$$a \leq b \quad \text{then} \quad \sqrt{a} \leq \sqrt{b}.$$

70. Prove that for all real numbers a and b.

$$|a-b| \leq |a| + |b|.$$

71. Prove that for all real numbers a and b

$$\big||a|-|b|\big| \leq |a-b|.$$

HINT: Calculate $\big||a|-|b|\big|^2$ and use the fact that $ab \leq |a||b|$.

72. Show that $|a+b| = |a| + |b|$ iff $ab \geq 0$.

73. Show that if

$$0 \leq a \leq b \quad \text{then} \quad \frac{a}{1+a} \leq \frac{b}{1+b}.$$

74. Let a, b, c be nonnegative numbers. Show that

$$\text{if } a \leq b + c, \quad \text{then} \quad \frac{a}{1+a} \leq \frac{b}{1+b} + \frac{c}{1+c}.$$

75. Prove that if a and b are real numbers and $a < b$, then $a < (a+b)/2 < b$. The number $(a+b)/2$ is called the *arithmetic mean* of a and b. How are the three numbers, a, $(a+b)/2$, b, related on the number line?

76. Let a and b be nonnegative numbers with $a \leq b$. Prove that

$$a \leq \sqrt{ab} \leq \frac{a+b}{2} \leq b.$$

The number $\sqrt{ab}$ is called the *geometric mean* of a and b.

■ 1.4 COORDINATE PLANE; ANALYTIC GEOMETRY

Rectangular Coordinates

The relation between the real numbers and points on a straight line can be used to construct a coordinate system for the plane. In the plane, draw two number lines which are perpendicular to each other and intersect at their origins; let O be the point of intersection. It is customary to have one of the lines be horizontal with the positive numbers to the right of O and the other vertical with the positive numbers above O. The point O is called the *origin* and the number lines are called the *coordinate axes*. The horizontal axis is usually labeled the *x-axis* and the vertical axis is usually labeled the *y-axis.* The coordinate axes separate the plane into four regions called *quadrants.* The quadrants are numbered in the counterclockwise direction starting with the upper right. See Figure 1.4.1.

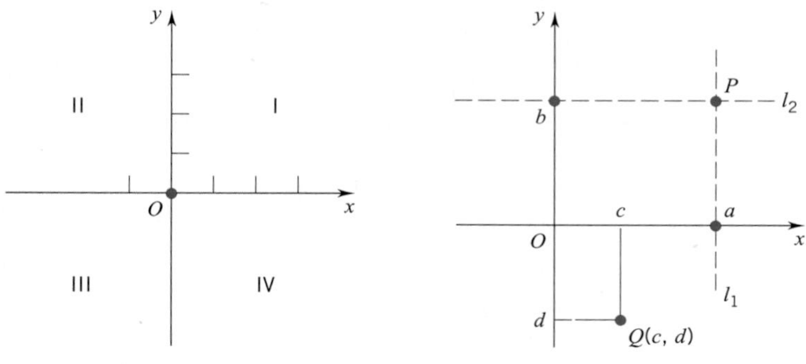

Figure 1.4.1

Figure 1.4.2

Choose any point P, $P \neq O$, in the plane. Let l_1 be the line through P parallel to the y-axis and l_2 the line through P parallel to the x-axis. Let a be the coordinate of the point of intersection of l_1 and the x-axis, and let b be the coordinate of the point of intersection of l_2 and the y-axis (Figure 1.4.2). Then P is identified with the *ordered pair* of real numbers (a, b). The ordered pair associated with the origin is $(0, 0)$. Conversely, starting with an ordered pair (c, d), we can reverse these steps and associate with it a unique point Q. If $P(a, b)$ is a point in the plane, then a and b are the *coordinates* of P with a the *x-coordinate* or *abscissa* and b the *y-coordinate* or *ordinate.* This coordinate system for the plane is called the *rectangular* or *Cartesian* coordinate system.

Distance and Midpoint Formulas

Let $P_0(x_0, y_0)$ and $P_1(x_1, y_1)$ be points in the plane. The distance $d(P_0, P_1)$ between P_0 and P_1 can be computed using the *Pythagorean Theorem*:

$$\text{Distance (Figure 1.4.3): } d(P_0, P_1) = \sqrt{(x_1 - x_0)^2 + (y_1 - y_0)^2}$$

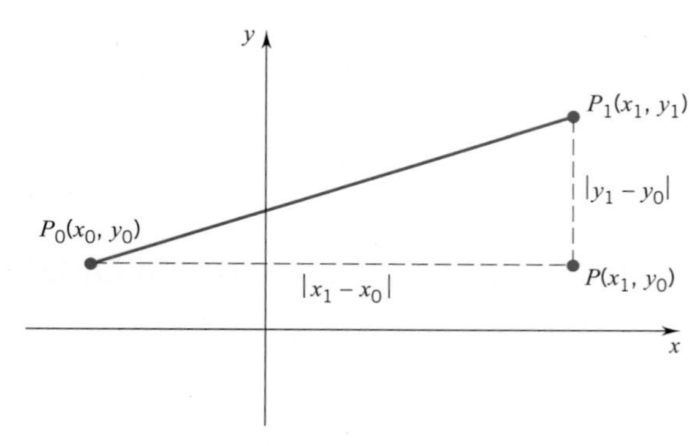

Figure 1.4.3

Let $M(x, y)$ be the midpoint of the line segment joining P_0 and P_1. Using the fact that the two triangles in Figure 1.4.4 are congruent, it can be shown that

$$x = \frac{x_0 + x_1}{2} \quad \text{and} \quad y = \frac{y_0 + y_1}{2}.$$

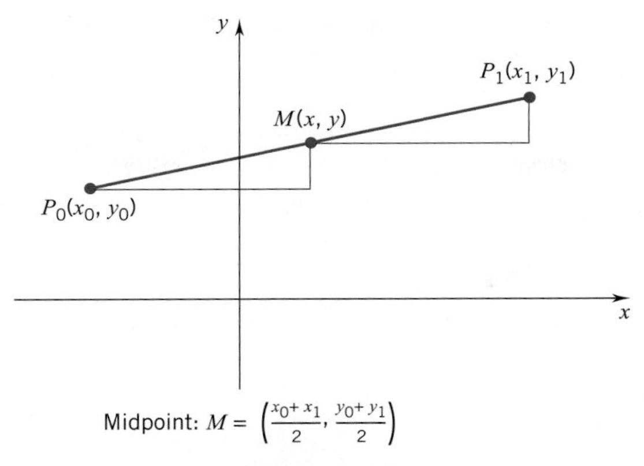

Figure 1.4.4

Lines

Let l be a straight line in the plane, and let $P_0(x_0, y_0)$ and $P_1(x_1, y_1)$ be any two distinct points on l.

(i) Slope The *slope* of l is given by

$$m = \frac{y_1 - y_0}{x_1 - x_0} \qquad \text{provided } x_1 \neq x_0.$$

Also,

$$m = \tan \theta,†$$

where θ, called the *inclination* of l, is the angle between the positively directed x-axis and l measured in the counterclockwise direction (Figure 1.4.5). If $x_1 = x_0$, then l is a vertical line; its slope is undefined and $\theta = \pi/2 = 90°$.

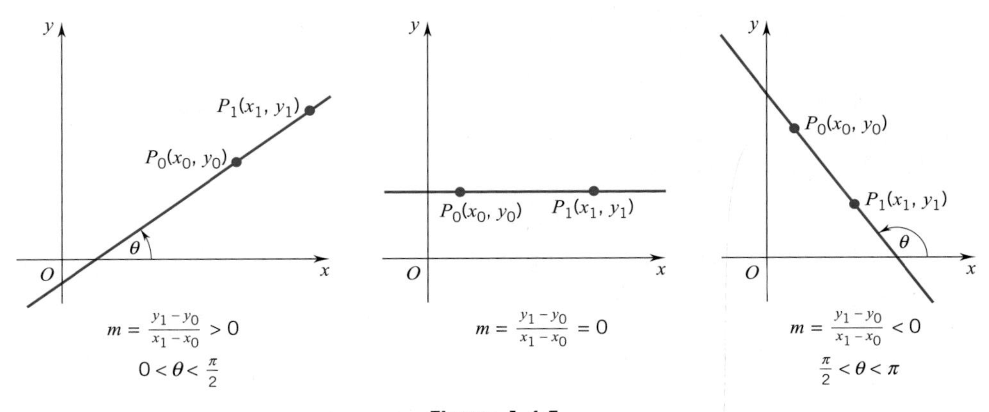

Figure 1.4.5

(ii) Intercepts If l is a nonvertical line, then it intersects the y-axis at a point $(0, b)$. The number b is called the *y-intercept* of l. Similarly, if l is not horizontal, then it will

† The trigonometric functions are reviewed in the Section 1.6.

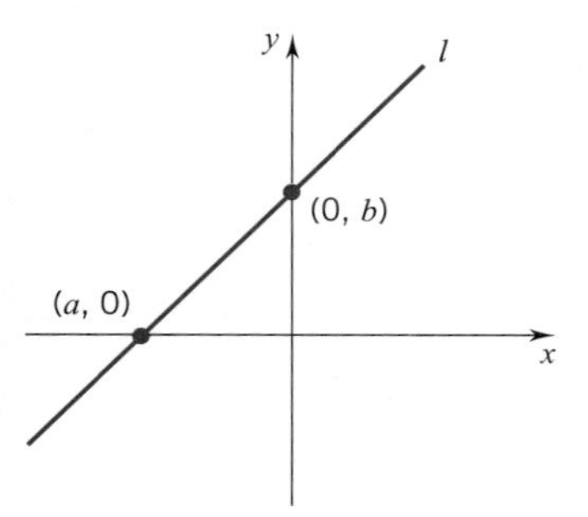

Figure 1.4.6

intersect the x-axis at a point $(a, 0)$. The number a is the *x-intercept* of l. These ideas are illustrated in Figure 1.4.6.

(iii) Equations An equation for l is an equation which involves two variables, say x and y, and has the property that a point $P(x_0, y_0)$ lies on l iff its coordinates, x_0 and y_0, satisfy the equation. The standard forms are:

vertical line	$x = a.$
horizontal line	$y = b.$
point-slope form	$y - y_0 = m(x - x_0).$
two-point form	$y - y_0 = \dfrac{y_1 - y_0}{x_1 - x_0}(x - x_0).$
slope-intercept form	$y = mx + b$ ($y = b$ when $x = 0$).
two intercept form	$\dfrac{x}{a} + \dfrac{y}{b} = 1.$
general form	$Ax + By + C = 0$ (A and B not both 0).

(iv) Parallel and Perpendicular Lines Given two nonvertical lines l_1 and l_2 with slopes m_1 and m_2, respectively, then l_1 and l_2 are:

$$\textit{parallel} \quad \text{iff} \quad m_1 = m_2,$$

$$\textit{perpendicular} \quad \text{iff} \quad m_1 m_2 = -1.$$

Example 1 Find the slope and y-intercept of each of the following lines:

$$l_1: 20x - 24y - 30 = 0, \qquad l_2: 2x - 3 = 0, \qquad l_3: 4y + 5 = 0.$$

SOLUTION The equation of l_1 can be written

$$y = \tfrac{5}{6}x - \tfrac{5}{4}.$$

This is in the form $y = mx + b$. The slope is $\frac{5}{6}$, and the y-intercept is $-\frac{5}{4}$.

The equation of l_2 can be written

$$x = \tfrac{3}{2}.$$

The line is vertical and the slope is not defined. Since the line does not cross the y-axis, it has no y-intercept.

The third equation can be written

$$y = -\tfrac{5}{4}.$$

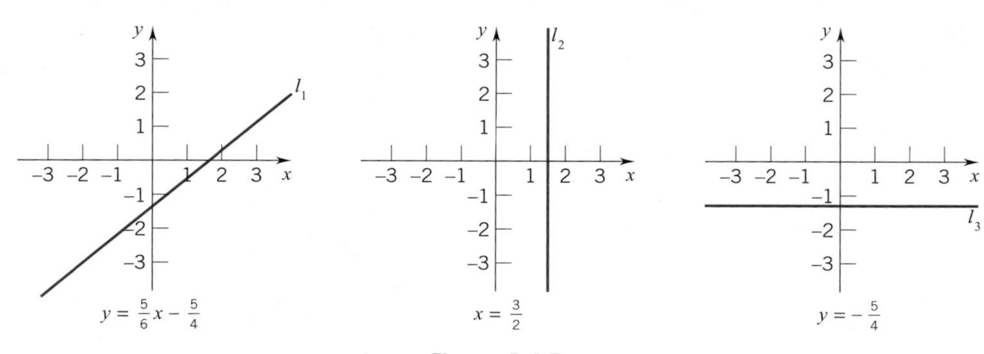

Figure 1.4.7

The line is horizontal. Its slope is 0 and the y-intercept is $-\frac{5}{4}$. The three lines are drawn in Figure 1.4.7. ❑

Example 2 Write an equation for the line l_2 that is parallel to

$$l_1 : 3x - 5y + 8 = 0$$

and passes through the point $P(-3, 2)$.

SOLUTION We can rewrite the equation of l_1 as

$$y = \tfrac{3}{5}x + \tfrac{8}{5}.$$

The slope of l_1 is $\frac{3}{5}$. The slope of l_2 must also be $\frac{3}{5}$. (*Remember*: For nonvertical parallel lines, $m_1 = m_2$.)

Since l_2 passes through $(-3, 2)$ with slope $\frac{3}{5}$, we can use the point-slope formula and write the equation as

$$y - 2 = \tfrac{3}{5}(x + 3). \quad ❑$$

Example 3 Write an equation for the line that is perpendicular to

$$l_1 : x - 4y + 8 = 0$$

and passes through the point $P(2, -4)$.

SOLUTION The equation for l_1 can be written

$$y = \tfrac{1}{4}x + 2.$$

The slope of l_1 is $\frac{1}{4}$. The slope of l_2 is therefore -4. (*Remember:* For nonvertical perpendicular lines, $m_1 m_2 = -1$.)

Since l_2 passes through $(2, -4)$ with slope -4, we can use the point-slope formula and write the equation as

$$y + 4 = -4(x - 2). \quad ❑$$

The Angle Between Two Lines

Figure 1.4.8 shows two intersecting lines l_1, l_2 with inclinations θ_1, θ_2. In the figure, $\theta_1 < \theta_2$. These two lines form angles, $\theta_2 - \theta_1 = \alpha$ and $180° - (\theta_2 - \theta_1) = 180° - \alpha$. If these two angles are equal, then $\alpha = 90°$ and the two lines are perpendicular. If α and $180° - \alpha$ are not equal, then the smaller of the two (the one less than 90°) is called the *angle between* l_1 *and* l_2.

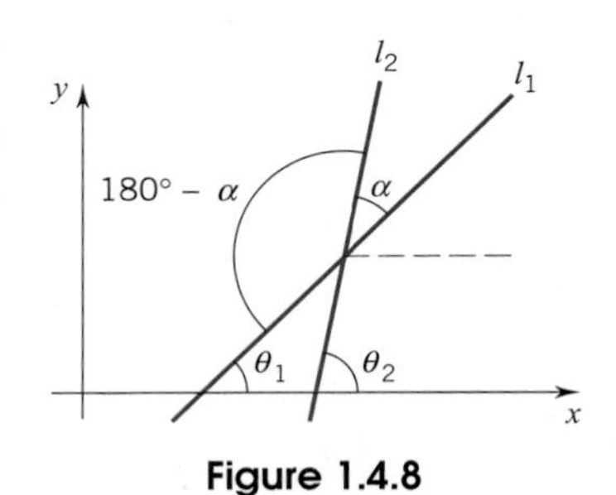

Figure 1.4.8

Example 4 Find the point where the lines

$$3x - 4y + 8 = 0 \qquad \text{and} \qquad 12x - 5y - 12 = 0$$

intersect, and then determine the angle between the lines.

SOLUTION To find the point of intersection, we solve the two equations simultaneously. As you can check, this gives $x = 8/3$, $y = 4$. The point of intersection is $P(8/3, 4)$.

To find the angle between the lines, we find the inclination of each line:

$$3x - 4y + 8 = 0 \qquad\qquad 12x - 5y - 12 = 0$$
$$4y = 3x + 8 \qquad\qquad 5y = 12x - 12$$
$$y = \tfrac{3}{4}x + 2. \qquad\qquad y = \tfrac{12}{5}x - \tfrac{12}{5}.$$
$$\tan\theta = \tfrac{3}{4} = 0.75 \qquad\qquad \tan\theta = \tfrac{12}{5} = 2.4$$
$$\theta \cong 37°. \qquad\qquad \theta \cong 67°.$$

The angle between the lines is approximately $67° - 37° = 30°$. See Figure 1.4.9. ❑

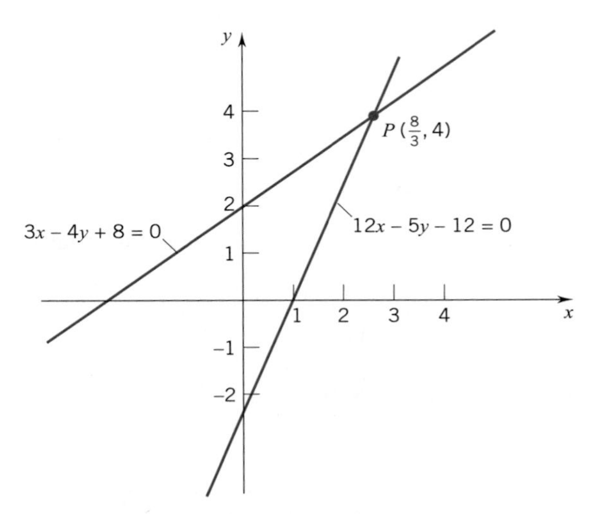

Figure 1.4.9

Distance Between a Point and a Line

Let l be a line and let P be a point which is not on l. It is easy to see that the point Q on l that is closest to P is the foot of the perpendicular from P to l. See Figure 1.4.10. The *distance between P and l*, denoted $d(P, l)$, is defined to be $d(P, Q)$.

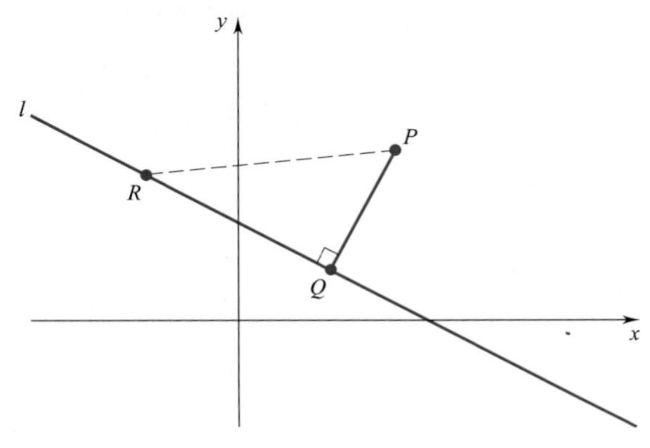

Figure 1.4.10

Example 5 Let l be the line with equation $Ax + By + C = 0$, with $A \neq 0$ and $B \neq 0$. Show that

$$d(O, l) = \frac{|C|}{\sqrt{A^2 + B^2}}.$$

SOLUTION Since l has slope $-A/B$, the line through the origin perpendicular to l has slope B/A and equation

$$y = \frac{B}{A}x \qquad \text{or} \qquad Bx - Ay = 0.$$

Solving the equations

$$\begin{aligned} Ax + By + C &= 0 \\ Bx - Ay \quad &= 0 \end{aligned}$$

simultaneously gives

$$x = \frac{-AC}{A^2 + B^2}, \qquad y = \frac{-BC}{A^2 + B^2}.$$

Thus the foot of the perpendicular from the origin to l is the point

$$Q\left(\frac{-AC}{A^2 + B^2}, \frac{-BC}{A^2 + B^2}\right)$$

and

$$\begin{aligned} d(O, l) = d(O, Q) &= \sqrt{\left(\frac{-AC}{A^2 + B^2}\right)^2 + \left(\frac{-BC}{A^2 + B^2}\right)^2} \\ &= \frac{1}{A^2 + B^2}\sqrt{A^2C^2 + B^2C^2} = \frac{|C|}{\sqrt{A^2 + B^2}}. \quad \square \end{aligned}$$

Translations

In Figure 1.4.11 we have drawn a rectangular coordinate system and marked a point $O'(x_0, y_0)$. Think of the Oxy system as a rigid frame and in your mind slide it along the plane, without letting it turn, so that the origin O falls on the point O' (Figure 1.4.12). Such a move, called a *translation*, produces a new coordinate system $O'XY$.

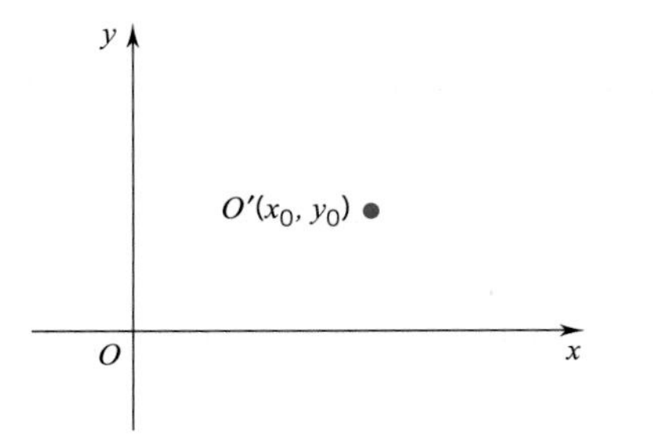

Figure 1.4.11

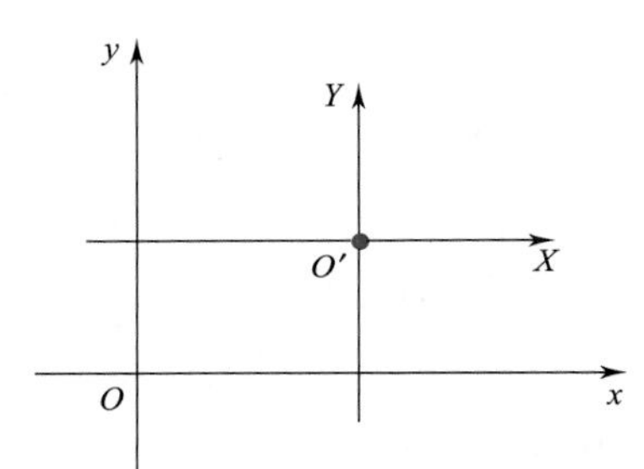

Figure 1.4.12

A point P now has two pairs of coordinates: a pair (x, y) with respect to the Oxy system and a pair (X, Y) with respect to the $O'XY$ system (Figure 1.4.13). To see the relation between these coordinates, note that, starting at O, we can reach P by first going to O' and then going on to P; thus

$$x = x_0 + X, \qquad y = y_0 + Y.$$

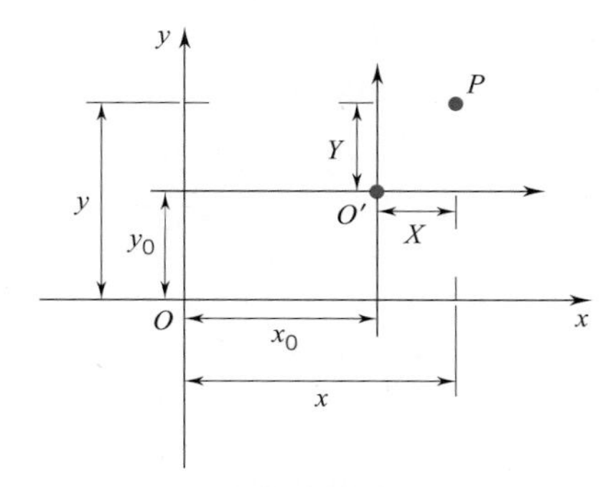

Figure 1.4.13

Translations are often used to simplify geometric arguments. In Example 5 we found the distance between the origin and a line l. Now we will derive the formula for the distance between a line and an arbitrary point by translating the coordinate system

DISTANCE BETWEEN A POINT AND A LINE

The distance between the line l: $Ax + By + C = 0$ and the point $P_1(x_1, y_1)$ is given by the formula

$$d(P_1, l) = \frac{|Ax_1 + By_1 + C|}{\sqrt{A^2 + B^2}}.$$

PROOF We translate the Oxy coordinate system to obtain a new system $O'XY$ with O' falling on the point P_1 (see Figure 1.4.14). The new coordinates are now related to the old coordinates as follows:

$$x = x_1 + X, \qquad y = y_1 + Y.$$

In the xy-system, l has equation

$$Ax + By + C = 0.$$

In the XY-system, l has equation

$$A(x_1 + X) + B(y_1 + Y) + C = 0.$$

We can write this last equation as

$$AX + BY + K = 0 \qquad \text{with} \quad K = Ax_1 + By_1 + C.$$

EXERCISES 1.4

In Exercises 1–6, find the distance between the given points.

1. $P_0(0, 5)$, $P_1(6, -3)$.
2. $P_0(2, 2)$, $P_1(5, 5)$.
3. $P_0(5, -2)$, $P_1(-3, 2)$.
4. $P_0(-5, 12)$, $O(0, 0)$.
5. $P_0(2, -2)$, $P_1(2, 4)$.
6. $P_0(2, 7)$, $P_1(-4, 7)$.

In Exercises 7–12, find the midpoint of the line segment $\overline{P_0P_1}$.

7. $P_0(2, 4)$, $P_1(6, 8)$.
8. $P_0(3, -1)$, $P_1(-1, 5)$.
9. $P_0(2, -3)$, $P_1(7, -3)$.
10. $P_0(5, -1)$, $P_1(5, 6)$.
11. $P_0(\sqrt{3}, 0)$, $P_1(0, \sqrt{3})$.
12. $P_0(a, 3)$, $P_1(3, a)$.

In Exercises 13–20, find the slope of the line through the given points.

13. $P_0(-2, 5)$, $P_1(4, 1)$.
14. $P_0(4, -3)$, $P_1(-2, -7)$.
15. $P_0(4, 2)$, $P_1(-3, 2)$.
16. $P_0(5, 0)$, $P_1(0, 5)$.
17. $P(a, b)$, $Q(b, a)$.
18. $P(4, -1)$, $Q(-3, -1)$.
19. $P(x_0, 0)$, $Q(0, y_0)$.
20. $O(0, 0)$, $P(x_0, y_0)$.

In Exercises 21–28, find the slope and y-intercept.

21. $y = 2x - 4$.
22. $6 - 5x = 0$.
23. $3y = x + 6$.
24. $y = 4x - 2$.
25. $4x = 1$.
26. $6y - 3x + 8 = 0$.
27. $7x - 3y + 4 = 0$.
28. $4y = 3$.

In Exercises 29–32, write an equation for the line with

29. slope 5 and y-intercept 2.
30. slope 5 and y-intercept -2.
31. slope -5 and y-intercept 2.
32. slope -5 and y-intercept -2.

In Exercises 33–34, write an equation for the horizontal line 3 units

33. above the x-axis.
34. below the x-axis.

In Exercises 35–36, write an equation for the vertical line 3 units

35. to the left of the y-axis.
36. to the right of the y-axis.

In Exercises 37–42, find an equation for the line that passes through the point $P(2, 7)$ and is

37. parallel to the x-axis.
38. parallel to the y-axis.
39. parallel to the line $3y - 2x + 6 = 0$.
40. perpendicular to the line $y - 2x + 5 = 0$.
41. perpendicular to the line $3y - 2x + 6 = 0$.
42. parallel to the line $y - 2x + 5 = 0$.

In Exercises 43–48, find the inclination of the line.

43. $x - y + 2 = 0$.
44. $6y + 5 = 0$.
45. $2x - 3 = 0$.
46. $x + y - 3 = 0$.
47. $3x + 4y - 12 = 0$.
48. $9x + 10y + 4 = 0$.

In Exercises 49–52, write an equation for the line with

49. inclination 30°, y-intercept 2.
50. inclination 60°, y-intercept 2.
51. inclination 120°, y-intercept 3.
52. inclination 135°, y-intercept 3.

In Exercises 53–56, determine the point(s) where the line intersects the circle.

53. $y = x$, $x^2 + y^2 = 1$.
54. $y = mx$, $x^2 + y^2 = 4$.
55. $4x + 3y = 24$, $x^2 + y^2 = 25$.
56. $y = mx + b$, $x^2 + y^2 = b^2$.

In Exercises 57–60, find the point where the lines intersect and determine the angle between the lines.

57. l_1: $4x - y - 3 = 0$, l_2: $3x - 4y + 1 = 0$.
58. l_1: $3x + y - 5 = 0$, l_2: $7x - 10y + 27 = 0$.
59. l_1: $4x - y + 2 = 0$, l_2: $19x + y = 0$.
60. l_1: $5x - 6y + 1 = 0$, l_2: $8x + 5y + 2 = 0$.

61. Find the distance between the line $5x + 12y + 2 = 0$ and (a) the origin. (b) the point $P(1, -3)$.

62. Find the distance between the line $2x - 3y + 1 = 0$ and (a) the origin. (b) the point $P(-2, 5)$.

63. Which of the points $(0, 1)$, $(1, 0)$, and $(-1, 1)$ is closest to l: $8x + 7y - 6 = 0$? Which is farthest from l?

64. Consider the triangle with vertices $A(2, 0)$, $B(4, 3)$, $C(5, -1)$. Which of these vertices is farthest from the opposite side?

65. Find the area of the triangle with vertices $(1, -2)$, $(-1, 3)$, $(2, 4)$.

66. Find the area of the triangle with vertices $(-1, 1)$, $(3, \sqrt{2})$, $(\sqrt{2}, -1)$.

In Exercises 67–74, identify the given equation as a conic section. If it is a circle, give the center and radius; if it is an ellipse, give the center; if it is a parabola, give the vertex; if it is a hyperbola, give the center.

67. $y^2 + 2y - x = 0$.
68. $x^2 + y^2 - 4x + 6y - 3 = 0$.
69. $2x^2 + 3y^2 - 8x + 6y + 5 = 0$.
70. $2x^2 + 2y^2 - 4x + 8y = -10$.
71. $y^2 - 4y - 4x^2 + 8x = 4$.
72. $4x^2 + 16x - 16y = 32$.
73. $4x^2 - y^2 - 24x - 4y + 16 = 0$.
74. $4x^2 + y^2 - 8x + 4y + 4 = 0$.

75. Determine the slope of the line that intersects the circle $x^2 + y^2 = 169$ only at the point (5, 12).

76. Find an equation for the line which is tangent to the circle $x^2 + y^2 - 2x + 6y - 15 = 0$ at the point (4, 1) on the circle. HINT: A line is tangent to a circle at a point P iff it is perpendicular to the radius at P.

77. The point $P(1, -1)$ is on the circle whose center is at $C(-1, 3)$. Find an equation for the tangent line to the circle at P.

The *perpendicular bisector* of the line segment $\overline{PQ}$ is the line which is perpendicular to $\overline{PQ}$ and passes through the midpoint of $\overline{PQ}$. In Exercises 78 and 79, find an equation for the perpendicular bisector of the line segment joining the given points.

78. $P(-1, 3)$, $Q(3, -4)$. **79.** $P(1, -4)$, $Q(4, 9)$.

In Exercises 80–83, the given points are the vertices of a triangle. State whether the triangle is *isoceles* (two sides equal), a right triangle, both of these, or neither of these.

80. $P_0(-4, 3)$, $P_1(-4, -1)$, $P_2(2, 1)$.

81. $P_0(-2, 5)$, $P_1(1, 3)$, $P_2(-1, 0)$.

82. $P_0(-2, -1)$, $P_1(0, 7)$, $P_2(3, 2)$.

83. $P_0(3, 4)$, $P_1(1, 1)$, $P_2(-2, 3)$.

84. An *equilateral triangle* is a triangle in which the three sides have equal length. If two of the vertices of an equilateral triangle are (0, 0) and (4, 3), find the third vertex. How many such triangles are there?

85. Show that the midpoint M of the hypotenuse of a right triangle is equidistant from the three vertices of the triangle. HINT: Introduce a coordinate system in which the sides of the triangle are on the coordinate axes; see the figure.

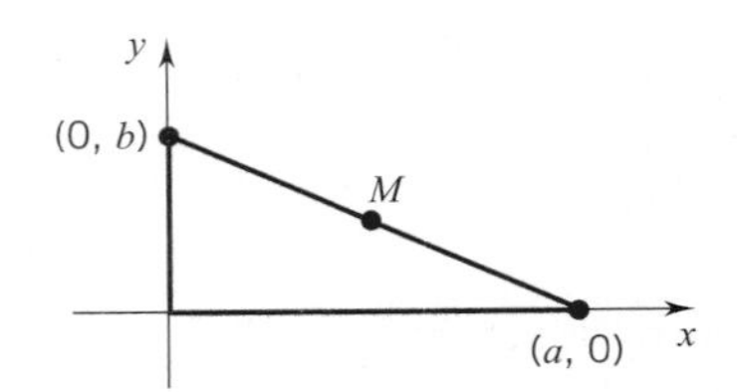

86. A *median* of a triangle is a line segment from a vertex to the midpoint of the opposite side. Find the lengths of the medians of the triangle with vertices $(-1, -2)$, (2, 1), $(4, -3)$.

87. The vertices of a triangle are (1, 0), (3, 4), and $(-1, 6)$. Find the point(s) where the medians of this triangle intersect.

88. Show that the medians of a triangle intersect in a single point (called the *centroid* of the triangle). HINT: Introduce a coordinate system such that one vertex is at the origin and one side is on the positive x-axis; see the figure.

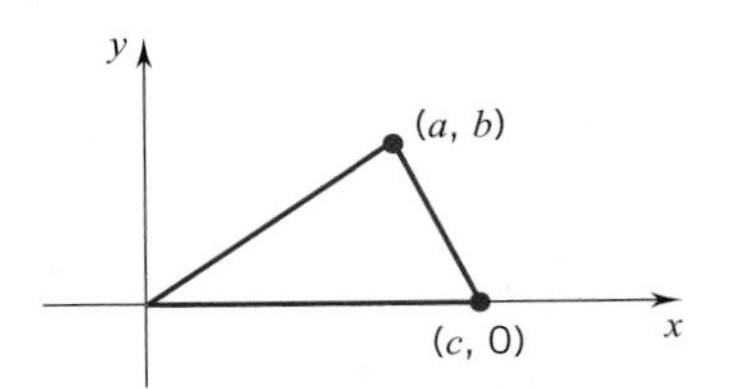

89. Prove that the diagonals of a parallelogram bisect each other. HINT: Introduce a coordinate system such that one vertex of the parallelogram is at the origin and one side is on the positive x-axis.

90. Let $P_1(x_1, y_1)$, $P_2(x_2, y_2)$, $P_3(x_3, y_3)$, $P_4(x_4, y_4)$, be the vertices of a quadrilateral. Show that the quadrilateral formed by joining the midpoints of adjacent sides is a parallelogram.

91. Temperature is normally measured either in degrees Fahrenheit (F) or in degrees Celsius (C). The relation between F and C is linear. The freezing point of water is 0°C or 32°F, and the boiling point of water is 100°C or 212°F. Find an equation giving the Fahrenheit temperature F in terms of the Celsius temperature C. Is there a temperature at which the Fahrenheit and Celsius temperatures are equal? If so, find it.

92. The relation between Fahrenheit temperature F and absolute (or Kelvin) temperature K is linear. If K = 273° when F = 32°, and K = 373° when F = 212°, express K in terms of F. Also, use Exercise 91 to determine the relation between Celsius temperature and absolute temperature. Is this relation linear?

■ 1.5 FUNCTIONS

The fundamental processes of calculus (called *differentiation* and *integration*) are processes applied to functions. To understand these processes and to be able to carry them out, you have to be thoroughly familiar with functions. Here we review some of the basic notions.

Function; Domain and Range

First we need a working definition for the term *function*. Let D and R be sets. A *function from D into R* is a rule that assigns to each element of D a unique element of R. The set D is called the *domain* of the function. Functions are usually denoted by letters such as f, g, F, G, and so on. If f is a function from D into R, then the element $y \in R$ that is assigned to the element $x \in D$ by f is denoted $f(x)$ (read "f of x"), and is called the *value of f at x*, or the *image of x under f*. The elements $x \in D$ and $y \in R$ are called *variables*, with x the *independent variable* and y the *dependent variable*. The set of values of f, that is $\{y \in R: y = f(x) \text{ for } x \in D\}$, is called the *range* of f. Throughout most of this text, D and R will be sets of real numbers, and functions from D into R are said to be *real-valued functions of a real variable*. In general, the rules for functions will be specified by equations, although there will be occasions when some other means, such as a set of equations, are used to define a function.

Here are some examples. We begin with the squaring function

$$f(x) = x^2, \qquad \text{for all real numbers } x.$$

The domain of f, denoted dom(f), is given explicitly as the set of all real numbers. Values of f can be found by substituting for x in the equation. For example,

$$f(4) = 4^2 = 16, \qquad f(-3) = (-3)^2 = 9, \qquad f(0) = 0^2 = 0.$$

As x runs through the real numbers, x^2 runs through all the nonnegative numbers. Therefore, the range of f is $[0, \infty)$. In abbreviated form we write

$$\text{dom}(f) = (-\infty, \infty) \qquad \text{and} \qquad \text{range}(f) = [0, \infty),$$

and say that f *maps* $(-\infty, \infty)$ *onto* $[0, \infty)$.

Now consider the function

$$g(x) = \sqrt{2x + 4}, \qquad x \in [0, 6].$$

The domain of g is given as the closed interval $[0, 6]$. At $x = 0$, g takes on the value 2:

$$g(0) = \sqrt{2 \cdot 0 + 4} = \sqrt{4} = 2,$$

and at $x = 6$, g has the value 4:

$$g(6) = \sqrt{2 \cdot 6 + 4} = \sqrt{16} = 4.$$

As x runs through the numbers from 0 to 6, $g(x)$ runs through the numbers from 2 to 4. Thus, the range of g is the closed interval $[2, 4]$. The function g maps $[0, 6]$ onto $[2, 4]$.

As suggested previously, the rule for a function might be specified by a set of equations rather than just one equation. For example, consider the function

$$h(x) = \begin{cases} x^2 & \text{if } x \geq 0 \\ 2x + 1 & \text{if } x < 0. \end{cases}$$

This function is said to be *piecewise defined*. The definition of h indicates that its domain is all real numbers. It can be verified that range(h) is all real numbers as well, although the vertification here is not as straightforward as in the first two examples. The function h maps $(-\infty, \infty)$ onto $(-\infty, \infty)$.

The *absolute value function* defined by

$$|x| = \begin{cases} x & \text{if } x \geq 0 \\ -x & \text{if } x < 0. \end{cases}$$

is a familiar example of a piecewise defined function. This function has domain $(-\infty, \infty)$ and range $[0, \infty)$.

Graphs of Functions

If f is a function with domain D, then the *graph* of f is the set of all points $P(x, f(x))$ in the plane, where $x \in D$. That is, the graph of f is the graph of the equation $y = f(x)$:

$$\text{graph of } f = \{(x, y)\colon x \in D \text{ and } y = f(x)\}.$$

An obvious method for sketching the graph of a function f is to plot some points $P(x, f(x))$, where $x \in \text{dom}(f)$. We plot enough points so that we can "see" what the graph is and then, typically, we connect the points with a "curve." Of course, if we can identify the curve in advance (for example, a straight line, a parabola, and so on), then it is much easier to draw its graph.

The graph of the squaring function

$$f(x) = x^2, \qquad x \in (-\infty, \infty)$$

is the parabola shown in Figure 1.5.1. The points that we plotted are indicated in the table and on the graph. The graph of the function

$$g(x) = \sqrt{2x + 4}, \qquad x \in [0, 6]$$

is the arc shown in Figure 1.5.2.

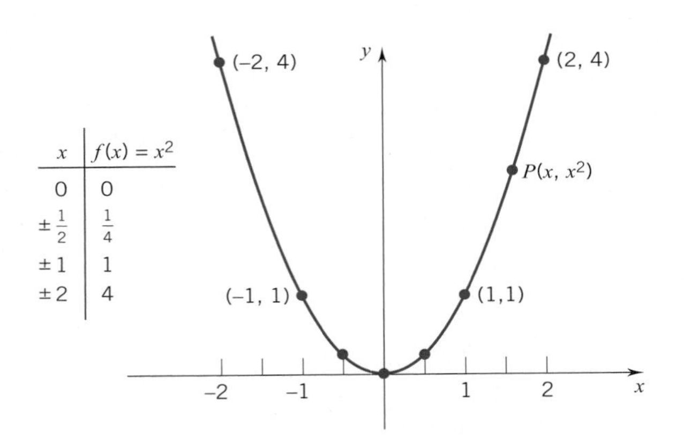

Figure 1.5.1

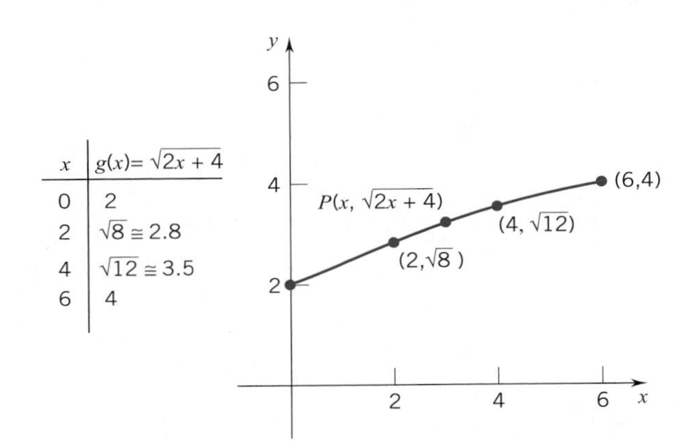

Figure 1.5.2

The graph of the piecewise defined function

$$h(x) = \begin{cases} x^2, & x \geq 0 \\ 2x + 1, & x < 0 \end{cases}$$

and the graph of the absolute value function are given in Figures 1.5.3 and 1.5.4.

Graphing calculators and computer software packages (that is, graphing utilities) can be valuable aids to sketching the graphs of functions, provided they are used properly. However, these technologies are not a substitute for learning the basic methods. We will not attempt to teach you how to use a graphing calculator or computer software in this text, but technology-based examples and exercises will occur throughout. You are encouraged to use these to enhance your understanding of the mathematical concepts and methods.

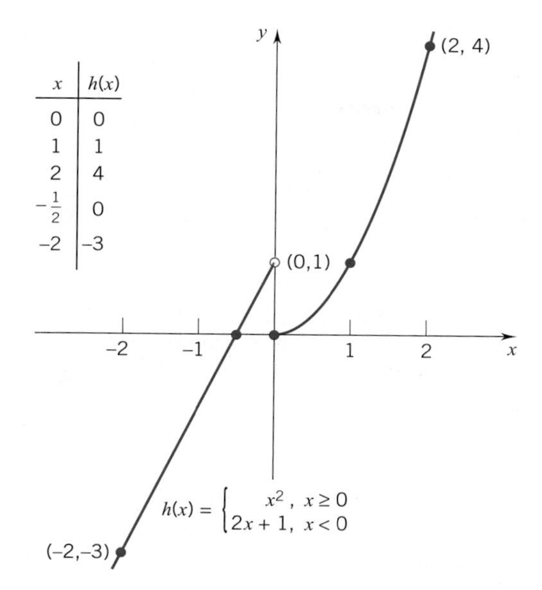

Figure 1.5.3

x	$\lvert x \rvert$
0	0
1	1
−1	1

(−1, 1) (1,1)

−1 1 x y

$$|x| = \begin{cases} x, & x \geq 0 \\ -x, & x < 0 \end{cases}$$

Figure 1.5.4

The next example illustrates why you must be careful when using a graphing utility to draw a graph.

Example 1 Let

$$f(x) = \tfrac{1}{2}x^3 - 2x + \tfrac{1}{2}.$$

The graph of f "looks like" the curve shown to the right.

A graphing utility was used to obtain the graphs of f shown in Figure 1.5.5(a)–(c).

$-5 \leq x \leq 5$, $x_{scl} = 1$
$-50 \leq y \leq 50$, $y_{scl} = 10$

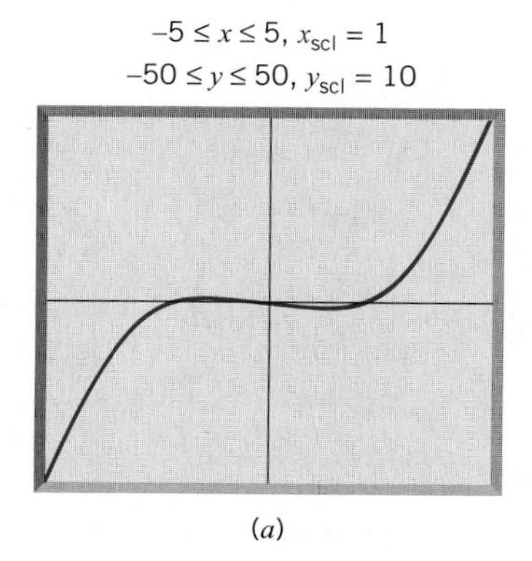

(*a*)

$-5 \leq x \leq 5$, $x_{scl} = 1$
$-0.5 \leq y \leq 0.5$, $y_{scl} = 0.1$

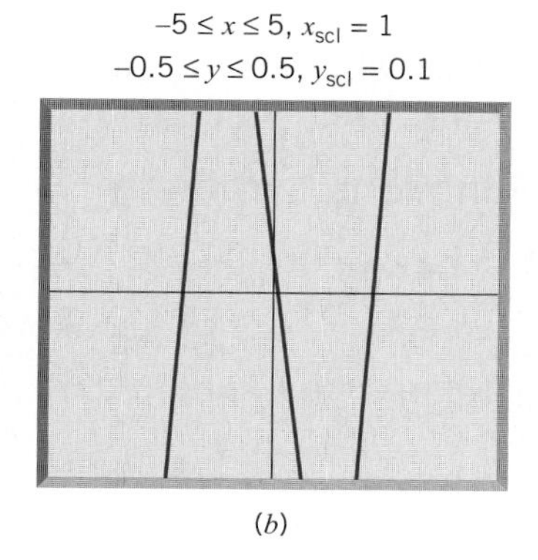

(*b*)

$-0.2 \leq x \leq 0.2$, $x_{scl} = 0.05$
$-1 \leq y \leq 1$, $y_{scl} = 0.1$

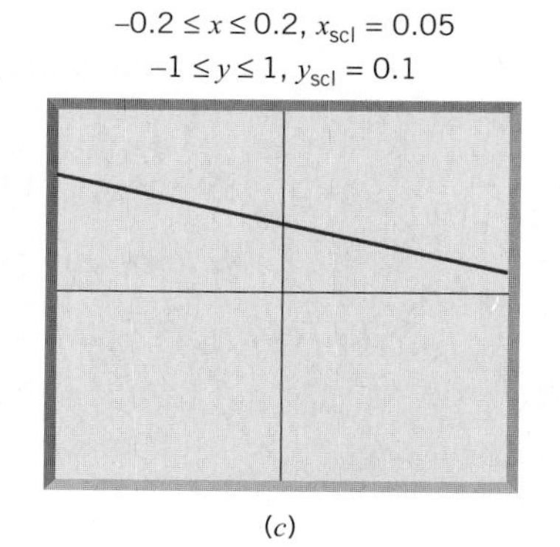

(*c*)

Figure 1.5.5

$-5 \le x \le 5$, $x_{scl} = 1$
$-10 \le y \le 10$, $Y_{scl} = 1$

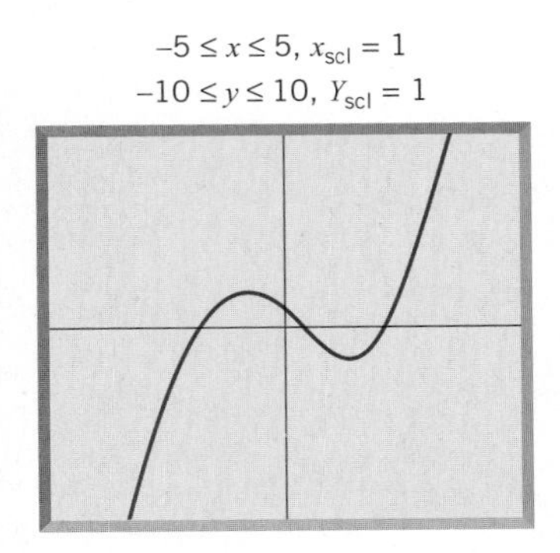

Figure 1.5.6

An accurate sketch of the graph of f is shown in Figure 1.5.6.

The difficulty with the graph in Figure 1.5.5(a) is that the y-scale, $[-50, 50]$, is too large; the "oscillation" near the origin is obscured by the large scale. On the other hand, it was reasonable to use this scale since the minimum and maximum values of f on the x-interval $[-5, 5]$ are approximately -50 and 50, respectively.

In Figure 1.5.5(b), the y-scale, $[-0.5, 0.5]$, is too small; the behavior of the graph for larger values of x and y has been eliminated. In Figure 1.5.5(c), the x-scale is too small; not enough of the graph is shown. The scaling in Figure 1.5.6 provides an accurate sketch of the graph.

This example illustrates that a basic knowledge of graphing, the general shapes of graphs of various functions, and so forth, is a prerequisite to the effective use of a graphing utility. The example also illustrates that you might have to "experiment" with several views of a graph before obtaining one with the desired accuracy. ❑

In general, the graph of a function is a "curve" in the plane; it is a geometric representation of the function. A graph provides a useful picture of the behavior of a function and, among other things, it can be used to determine the range. For example, it is clear from the graph of the function h in Figure 1.5.3 that range$(h) = (-\infty, \infty)$.

This geometric representation of a function raises a converse question, namely: Under what conditions is a given curve in the plane the graph of a function? By the so-called *vertical line test*, a curve in the plane is the graph of a function iff no vertical line intersects the curve in more than one point. Thus, we can conclude that the graphs of the circles, ellipses, and hyperbolas given at the end of Section 1.4 are not the graphs of functions. The curve shown in Figure 1.5.7 is the graph of a function, while the curve shown in Figure 1.5.8 is not.

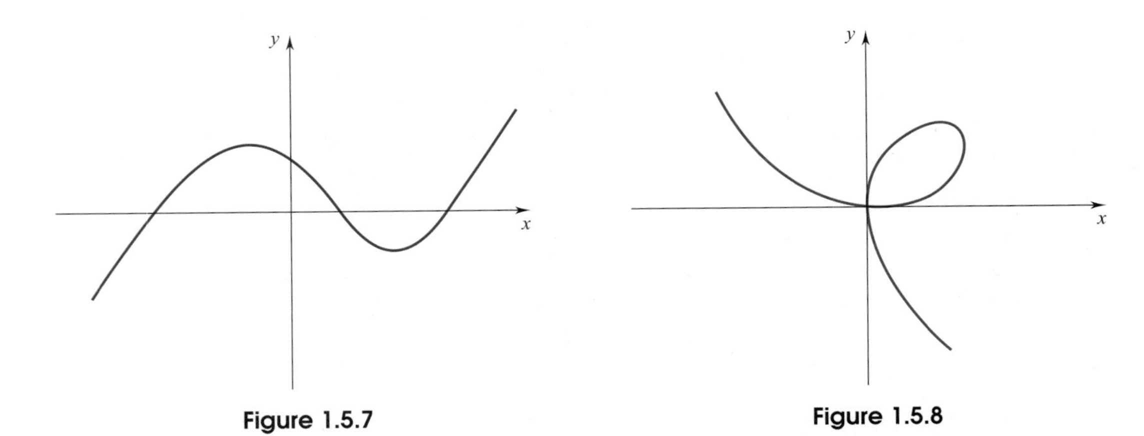

Figure 1.5.7 **Figure 1.5.8**

Even and Odd Functions; Symmetry

A function f is said to be *even* iff

$$f(-x) = f(x) \qquad \text{for all } x \in \text{dom}(f).$$

A function f is said to be *odd* iff

$$f(-x) = -f(x) \qquad \text{for all } x \in \text{dom}(f).$$

The graph of an even function is *symmetric about the y-axis* (Figure 1.5.9), and the graph of an odd function is *symmetric about the origin* (Figure 1.5.10).

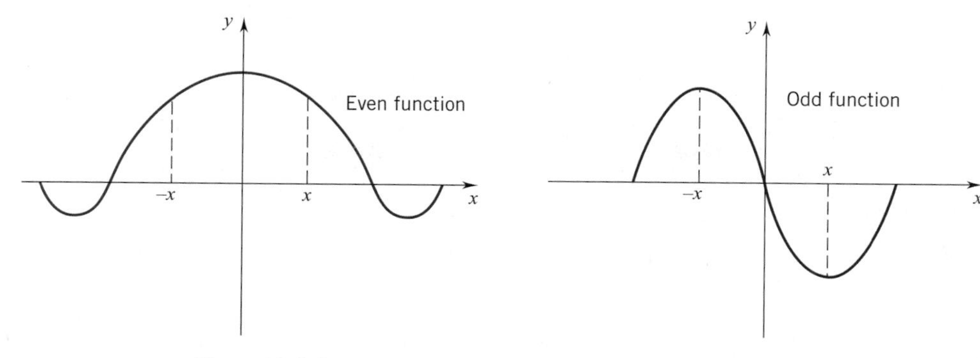

Figure 1.5.9

Figure 1.5.10

The squaring function $f(x) = x^2$ is even since

$$f(-x) = (-x)^2 = x^2 = f(x);$$

its graph is symmetric about the y-axis (see Figure 1.5.1). The function $f(x) = 4x - x^3$ is odd since

$$f(-x) = 4(-x) - (-x)^3 = -4x + x^3 = -(4x - x^3) = -f(x)$$

The graph of f is given in Figure 1.5.11. Note that it is symmetric with respect to the origin.

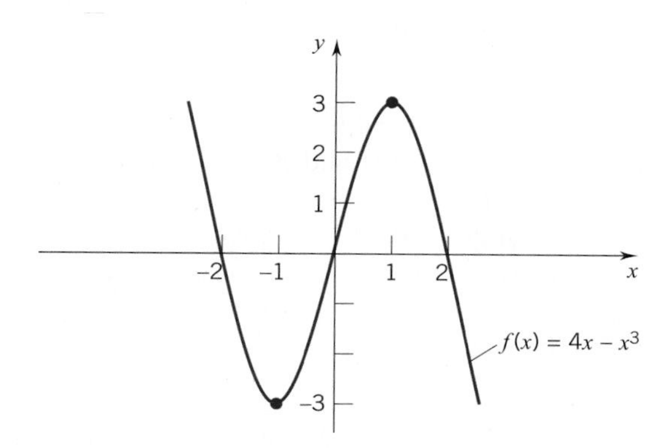

Figure 1.5.11

Convention on Domains

It will often be the case that the domain of a function will not be given explicitly. For example, you might be asked to consider

$$f(x) = x^3 + 1, \qquad \text{or} \qquad f(x) = \frac{1}{x - 2}, \qquad \text{or} \qquad f(x) = \sqrt{x} \qquad \text{(see Figure 1.5.12)}$$

without further explanation. The *convention* in such cases is to take as the domain the largest set of real numbers x for which $f(x)$ is itself a real number. The first function is defined for all real numbers x, and so we would take $\text{dom}(f) = (-\infty, \infty)$. The second function is defined for all real numbers except $x = 2$, and so $\text{dom}(f) = \{x: x \neq 2\} = (-\infty, 2) \cup (2, \infty)$. For the square-root function, $\text{dom}(f) = [0, \infty)$.

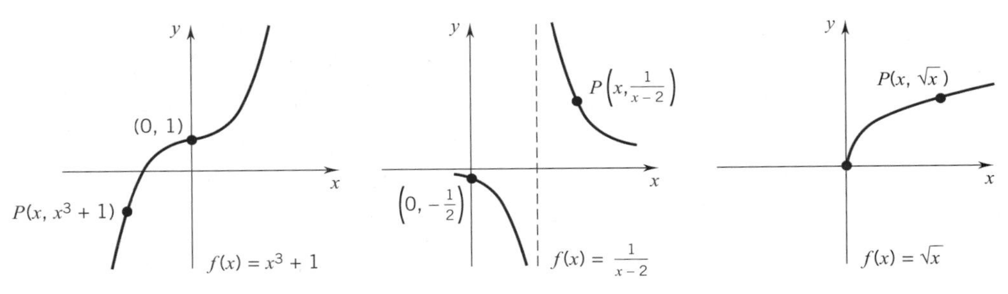

Figure 1.5.12

Example 2 Find the domain of each of the following functions:

$$\textbf{(a)}\ f(x) = \frac{x+1}{x^2+x-6}. \qquad \textbf{(b)}\ g(x) = \frac{\sqrt{4-x^2}}{x-1}.$$

SOLUTION **(a)** The function f is defined for all real numbers x such that $x^2 + x - 6 = (x+3)(x-2) \neq 0$. Thus, $\text{dom}(f) = \{x: x \neq -3, \ 2\} = (-\infty, \ -3) \cup (-3, \ 2) \cup (2, \infty)$.

(b) Since $\sqrt{4-x^2}$ is a real number iff $4 - x^2 \geq 0$, we must have $-2 \leq x \leq 2$. Also, g is not defined when $x - 1 = 0$, that is when $x = 1$. Thus,

$$\text{dom}(g) = \{x: -2 \leq x \leq 2 \quad \text{and} \quad x \neq 1\} = [-2, 1) \cup (1, 2]. \quad ❑$$

Example 3 Find the domain and range of the function:

$$f(x) = \frac{1}{\sqrt{2-x}} + 5.$$

SOLUTION First we look for the domain. Since $\sqrt{2-x}$ is real iff $2 - x \geq 0$, we need $x \leq 2$. But at $x = 2$, $\sqrt{2-x} = 0$ and its reciprocal is not defined. We must therefore restrict x to $x < 2$. The domain is $(-\infty, 2)$.

Now we look for the range. As x runs through $(-\infty, 2)$, $\sqrt{2-x}$ takes on all positive values and so does its reciprocal. The range of f is therefore $(5, \infty)$. ❑

Examples of Applications

Functions are used in a wide variety of disciplines as mathematical models which describe how variable quantities are related. In such cases, the application itself will usually dictate a domain for the function serving as the model.

Example 4 A candy manufacturer wants to make a tray by cutting four squares from the corners of a 12-by-20-inch rectangular sheet of cardboard and folding up the edges (see Figure 1.5.13). Express the volume V of the box as a function of the side x of the square and give the domain of the function.

SOLUTION From the figure, the dimensions of the tray are length $l = 20 - 2x$, width $w = 12 - 2x$, and height $h = x$. Thus,

$$V(x) = l \times w \times h = (20 - 2x)(12 - 2x)x = 240x - 64x^2 + 4x^3 \qquad \text{(cubic inches).}$$

While this cubic is defined for all values of x, the context of this problem dictates that $x > 0$, $12 - 2x > 0$, and $20 - 2x > 0$ (length, width, and height are all positive quantities). It follows from these inequalities that $0 < x < 6$. Therefore, the domain of this function is the interval $(0, 6)$. ❑

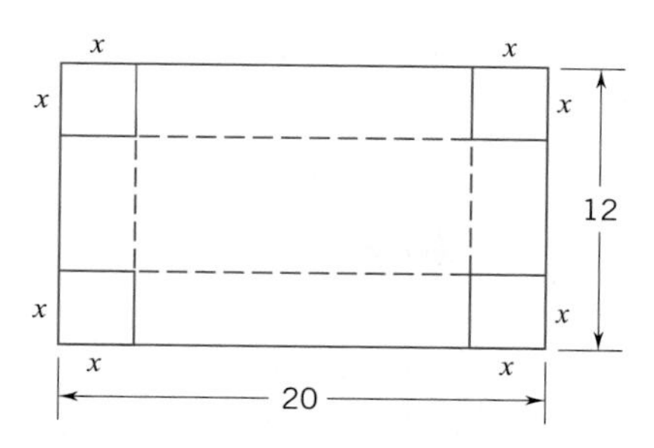

Figure 1.5.13

Example 5 A rectangle has an area of $k(>0)$ square units. Express the perimeter P of the rectangle as a function of the length x of one of its sides and give the domain of the function.

SOLUTION See Figure 1.5.14. Since the area $xy = k$, we have $y = k/x$. Now, the perimeter P of the rectangle is given by

$$P = 2x + 2y,$$

and so it follows that

$$P(x) = 2x + \frac{2k}{x}.$$

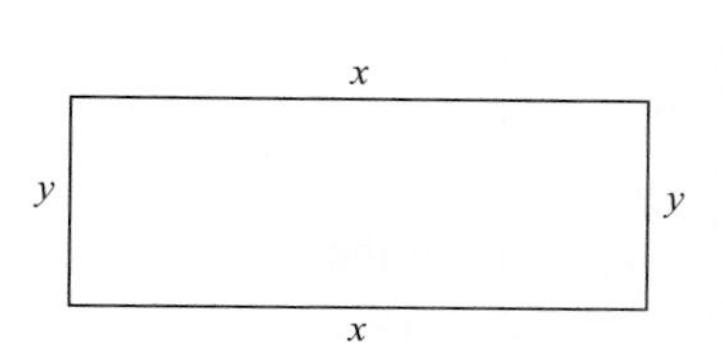

Figure 1.5.14

Since the length of a side of a rectangle cannot be negative, and since the area of the given rectangle is positive, we can conclude that $x > 0$. Also, there is no upper bound on the length of x. Thus, x satisfies $0 < x < \infty$ and $\text{dom}(P) = (0, \infty)$. ❑

EXERCISES 1.5

In Exercises 1–6, calculate (a) $f(0)$, (b) $f(1)$, (c) $f(-2)$, (d) $f(3/2)$.

1. $f(x) = 2x^2 - 3x + 2$.
2. $f(x) = \dfrac{2x - 1}{x^2 + 4}$.
3. $f(x) = \sqrt{x^2 + 2x}$.
4. $f(x) = |x + 3| - 5x$.
5. $f(x) = \dfrac{2x}{|x + 2| + x^2}$.
6. $f(x) = 1 - \dfrac{1}{(x + 1)^2}$.

In Exercises 7–10, calculate (a) $f(-x)$, (b) $f(1/x)$, (c) $f(a + b)$.

7. $f(x) = x^2 - 2x$.
8. $f(x) = \dfrac{x}{x^2 + 1}$.
9. $f(x) = \sqrt{1 + x^2}$.
10. $f(x) = \dfrac{x}{|x^2 - 1|}$.

In Exercises 11 and 12, calculate (a) $f(a + h)$ and (b) $[f(a + h) - f(a)]/h$, $h \neq 0$.

11. $f(x) = 2x^2 - 3x$.
12. $f(x) = \dfrac{1}{x - 2}$.

In Exercises 13–18, find the number(s) x, if any, where f takes on the value 1.

13. $f(x) = |2 - x|$.
14. $f(x) = \sqrt{1 + x}$.
15. $f(x) = x^2 + 4x + 5$.
16. $f(x) = 4 + 10x - x^2$.
17. $f(x) = \dfrac{2}{\sqrt{x^2 - 5}}$.
18. $f(x) = \dfrac{x}{|x|}$.

In Exercises 19–30, find the domain and range of the given function.

19. $f(x) = |x|$.
20. $g(x) = x^2 - 1$.
21. $f(x) = 2x - 3$.
22. $g(x) = \sqrt{x} + 5$.
23. $f(x) = \dfrac{1}{x^2}$.
24. $g(x) = \dfrac{4}{x}$.

25. $f(x) = \sqrt{1 - x}$.

26. $g(x) = \sqrt{x - 3}$.

27. $f(x) = \sqrt{7 - x} - 1$.

28. $g(x) = \sqrt{x - 1} - 1$.

29. $f(x) = \dfrac{1}{\sqrt{2 - x}}$.

30. $g(x) = \dfrac{1}{\sqrt{4 - x^2}}$.

In Exercises 31–44, find the domain and sketch the graph of the given function.

31. $f(x) = 1$.

32. $f(x) = -1$.

33. $f(x) = 2x$.

34. $f(x) = 2x + 1$.

35. $f(x) = \frac{1}{2}x$.

36. $f(x) = -\frac{1}{2}x$.

37. $f(x) = \frac{1}{2}x + 2$.

38. $f(x) = -\frac{1}{2}x - 3$.

39. $f(x) = \sqrt{4 - x^2}$.

40. $f(x) = \sqrt{9 - x^2}$.

41. $f(x) = x^2 - x - 6$.

42. $f(x) = |x - 1|$.

43. $f(x) = |x| + x$.

44. $f(x) = \dfrac{x^2 - 1}{x - 1}$.

In Exercises 45–48, sketch the graph and specify the domain and range of the given function.

45. $f(x) = \begin{cases} -1 & x < 0 \\ 1, & x > 0. \end{cases}$

46. $f(x) = \begin{cases} x^2, & x \leq 0 \\ 1 - x, & x > 0. \end{cases}$

47. $f(x) = \begin{cases} 1 + x, & 0 \leq x \leq 1 \\ x, & 1 < x < 2 \\ \frac{1}{2}x + 1, & 2 \leq x. \end{cases}$

48. $f(x) = \begin{cases} x^2, & x < 0 \\ -1, & 0 < x < 2 \\ x, & 2 < x. \end{cases}$

In Exercises 49–52, state whether the given curve is the graph of a function. If it is, find the domain and range.

49.

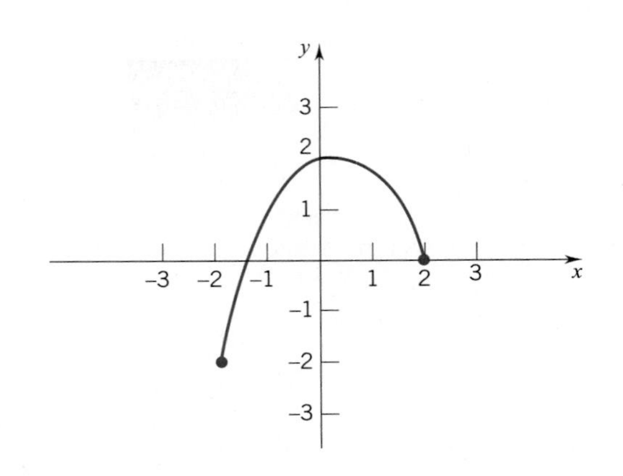

50.

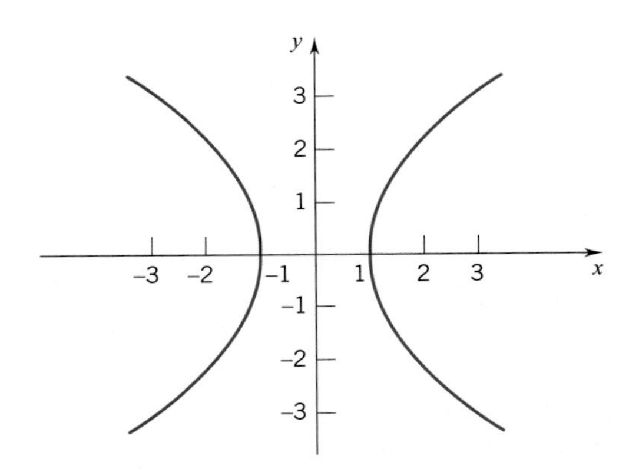

51.

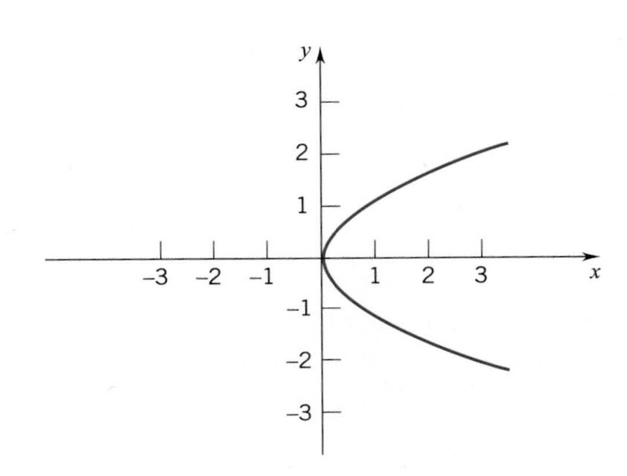

52.

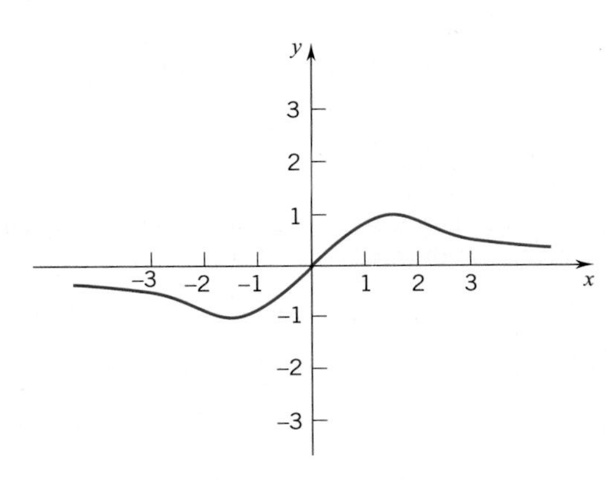

In Exercises 53–58, determine whether the function is odd, even, or neither.

53. $f(x) = x^3$.

54. $f(x) = x^2 + 1$.

55. $g(x) = x(x - 1)$.

56. $g(x) = x(x^2 + 1)$.

57. $f(x) = \dfrac{x^2}{1 - |x|}$.

58. $F(x) = x + \dfrac{1}{x}$.

59. The graph of $f(x) = \frac{1}{3}x^3 + \frac{1}{2}x^2 - 12x - 6$ looks like

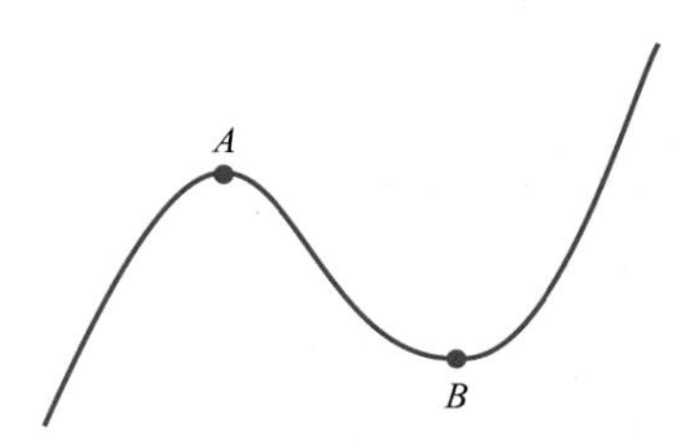

(a) Use a graphing utility to sketch an accurate graph of f.
(b) Find the zero(s) of f accurate to three decimal places.
(c) Find the coordinates of the points marked A and B, accurate to three decimal places.

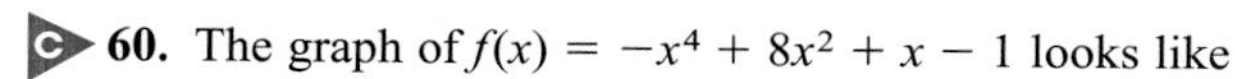

60. The graph of $f(x) = -x^4 + 8x^2 + x - 1$ looks like

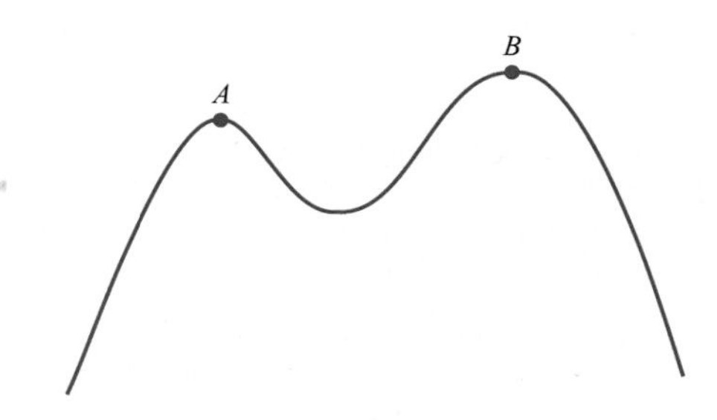

(a) Use a graphing utility to sketch an accurate graph of f.
(b) Find the zero(s) of f, if any. Use three decimal place accuracy.
(c) Find the coordinates of the points marked A and B, accurate to three decimal places.

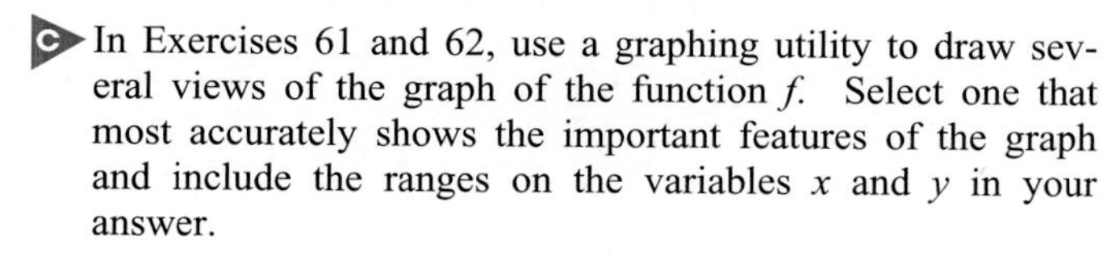

In Exercises 61 and 62, use a graphing utility to draw several views of the graph of the function f. Select one that most accurately shows the important features of the graph and include the ranges on the variables x and y in your answer.

61. $f(x) = |x^3 - 3x^2 - 24x + 4|$.

62. $f(x) = \sqrt{x^3 - 8}$.

In Exercises 63–80, determine a formula for the function that is described and give its domain.

63. Express the area of a circle as a function of its circumference.

64. Express the volume of a sphere as a function of its surface area.

65. Express the volume of a cube as a function of the area of one of its faces.

66. Express the volume of a cube as a function of its total surface area.

67. Express the surface area of a cube as a function of the length of a diagonal of one face.

68. Express the volume of a cube as a function of one of its diagonals.

69. Express the area of an equilateral triangle as a function of the length of a side.

70. A right triangle with hypotenuse c is rotated around one of its legs to form a cone (see the figure). If x is the length of the other leg, express the volume of the cone as a function of x.

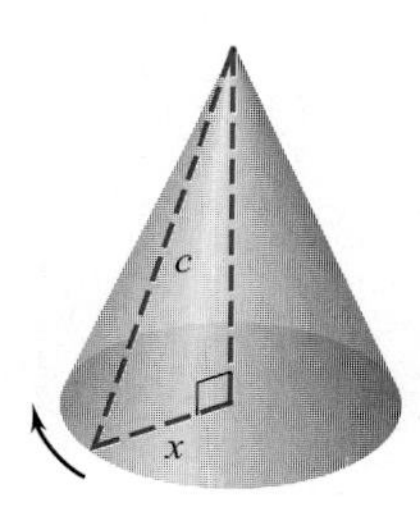

71. U.S. Post Office regulations require that the length plus the girth of a package for mailing cannot exceed 108 inches. A rectangular box with a square end whose edge length is x is designed to exactly meet the regulation (see the figure). Express the volume of the box as a function of x.

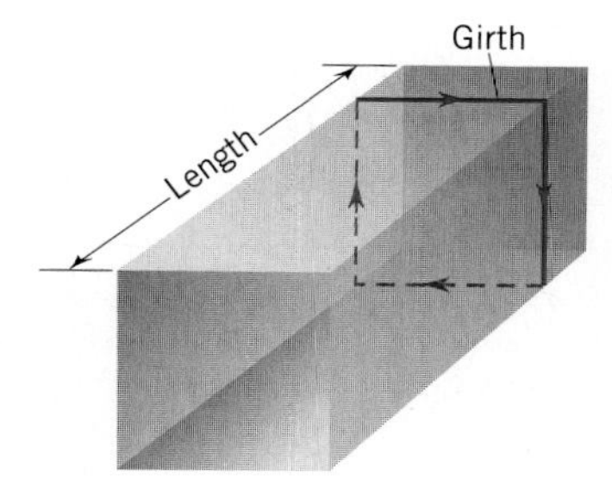

72. Suppose that a cylindrical mailing container exactly meets the U.S. Post Office regulations given in Exercise

71 (see the figure). Express the volume of the container as a function of the radius of the end.

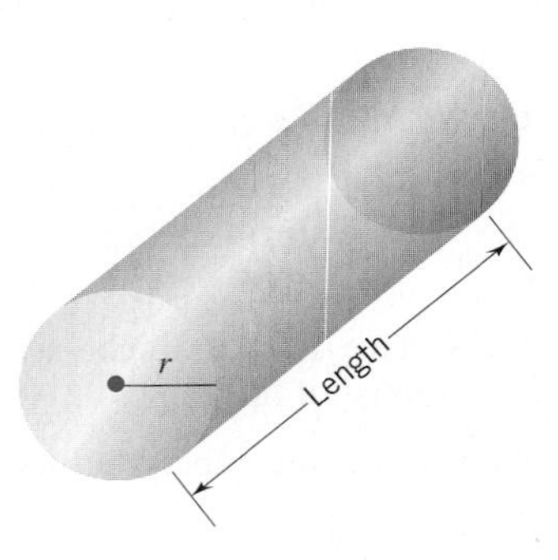

73. A Norman window is a window in the shape of a rectangle surmounted by a semicircle (see the figure). If the perimeter of the window is 15 feet, express the area as a function of its width x.

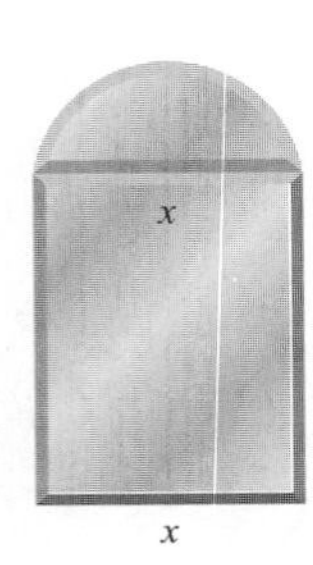

74. A window has the shape of a rectangle surmounted by an equilateral triangle. If the perimeter of the window is 15 feet, express the area as a function of one side of the equilateral triangle.

75. A rectangular beam of width x is cut from a circular log of diameter d (see the figure). Express the cross sectional area of the beam as a function of x.

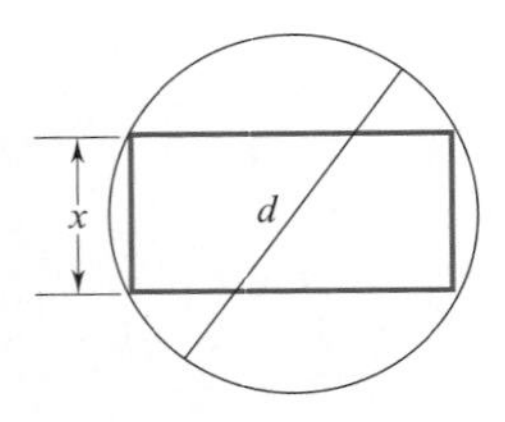

76. A rectangular sheet of metal with a perimeter of 30 inches is rolled into a cylinder of height h. Express the volume of the cylinder as a function of h.

77. Express the area of the rectangle shown in the figure as a function of the x-coordinate of the point P.

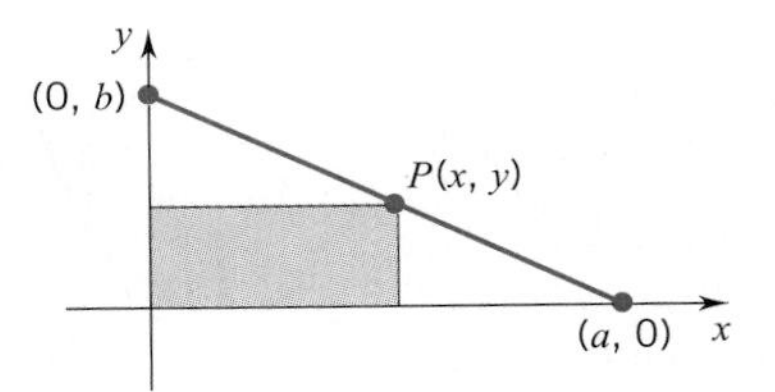

78. A right triangle is formed by the coordinate axes and a line through the point (2, 5) (see the figure). Express the area of the triangle as a function of the x-intercept.

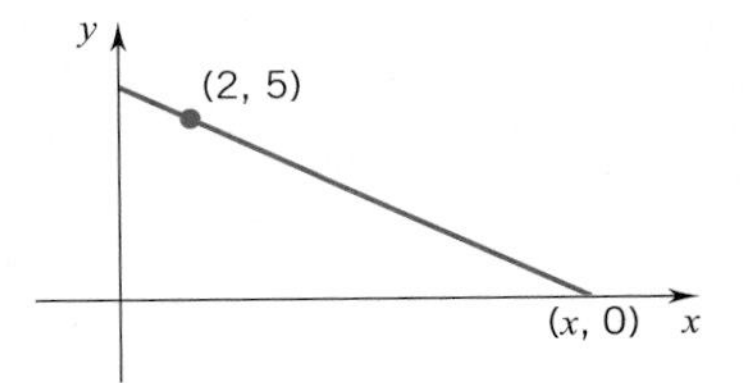

79. A string 28 inches long is to be cut into two pieces, one piece to form a square and the other to form a circle. Express the total area enclosed by the square and circle as a function of the perimeter of the square.

80. A tank in the shape of an inverted cone is being filled with water (see the figure). Express the volume of water in the tank as a function of its height.

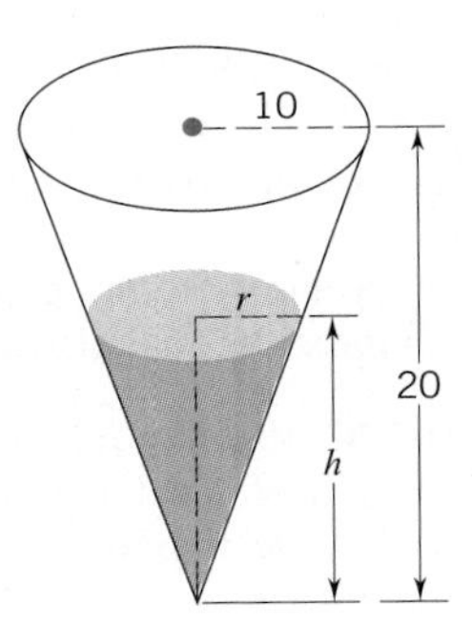

■ 1.6 THE ELEMENTARY FUNCTIONS

The basic types of functions that will be used throughout this text are polynomial functions, rational functions, trigonometric functions, and exponential and logarithmic functions. These functions are commonly called the *elementary functions*. We review polynomial, rational, and trigonometric functions here. Exponential and logarithmic functions are defined and studied in Chapter 7.

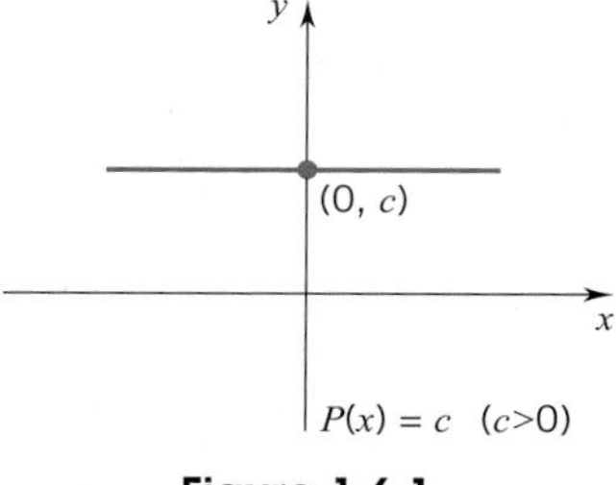

Figure 1.6.1

Polynomial Functions

Polynomials were introduced in Section 1.2. A *polynomial function* is a function defined by an equation of the form

$$P(x) = a_n x^n + a_{n-1}x^{n-1} + \cdots + a_1 x + a_0,$$

where the coefficients $a_n, a_{n-1}, \ldots, a_1, a_0$ are real numbers, $a_n \neq 0$, and n, the degree of P, is a nonnegative integer. The function $P(x) = 0$ is also a polynomial function, but no degree is assigned to it.

The domain of a polynomial function P is $(-\infty, \infty)$. It can be shown that the range of P is $(-\infty, \infty)$ if its degree is odd. If $n > 1$ is even, then the range of P is an interval of the form $[b, \infty)$ when $a_n > 0$, and an interval of the form $(-\infty, b]$ when $a_n < 0$. Finally, if $P(x) = c$ is constant, then the range of P is a point.

Polynomial functions of degree 0 are the nonzero constant functions $P(x) = c$ (constant), $c \neq 0$. Their graphs are horizontal lines $|c|$ units above or below the x-axis, depending on the sign of c (Figure 1.6.1). The x-axis is the graph of $P(x) = 0$.

Polynomial functions of degree 1 have the form $P(x) = ax + b$, $a \neq 0$, and are called *linear polynomials* since their graphs are straight lines with (nonzero) slope a and y-intercept b (Figure 1.6.2).

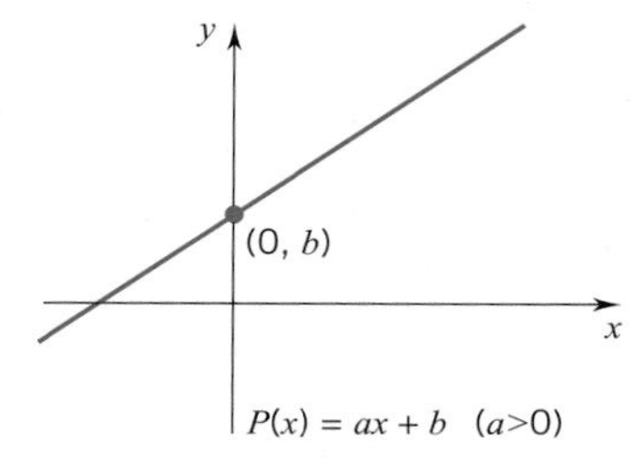

Figure 1.6.2

Figure 1.6.3

Polynomial functions of degree 2 have the form $P(x) = ax^2 + bx + c$, $a \neq 0$, and are often called *quadratic functions*. The graph of a quadratic function is a parabola which opens up if $a > 0$ and down if $a < 0$ (Figure 1.6.3). The location of the graph with respect to the x- and y-axes depends upon the coefficients a, b, and c. For example, if $a > 0$, then three possibilities for the location of the graph of $P(x) = ax^2 + bx + c$ with respect to the x-axis exist. These possibilities are indicated in Figure 1.6.4.

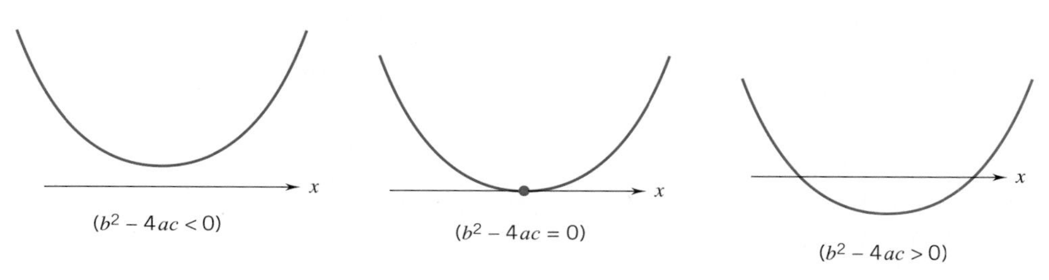

Figure 1.6.4

Polynomial functions of degree 3 have the form $P(x) = ax^3 + bx^2 + cx + d$, $a \neq 0$. These functions are also called *cubics*. In general, the graph of a cubic polynomial has one of the following two shapes, again determined by the sign of a (Figure 1.6.5). Also, we have not tried to locate these graphs with respect to the coordinate axes because there are so many possibilities. Our purpose is simply to indicate the two typical shapes.

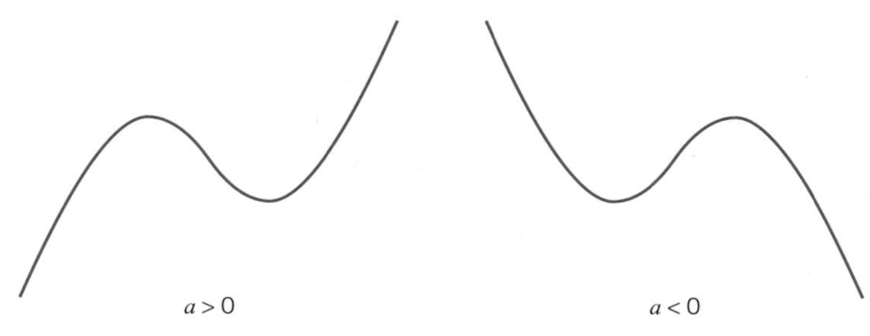

Figure 1.6.5

Polynomial functions become more complicated as their degrees increase. In Chapter 4 we show how to use techniques of calculus to analyze polynomial functions and to draw accurate graphs.

Rational Functions

Rational expressions were also defined in Section 1.2. A *rational function* is simply a function of the form

$$R(x) = \frac{P(x)}{Q(x)},$$

where P and Q are polynomials. Note that every polynomial P is a rational function because $P(x) = P(x)/1$ is the quotient of two polynomials. Since division by 0 is not allowed, a rational function $R = P/Q$ is not defined at those points x (if any) where $Q(x) = 0$; R is defined at all other points. Thus, $\text{dom}(R) = \{x\colon Q(x) \neq 0\}$.

Rational functions are more difficult to analyze and graph than polynomials. For example, it is important to analyze the behavior of a rational function $R = P/Q$ in the vicinity of a zero of Q, as well as the behavior of R for large values of x, both positive and negative. If P and Q have no common factors (P/Q is in "lowest terms"), then the zeros of Q correspond to *vertical asymptotes* of the graph of R; the existence of *horizontal asymptotes* depends on the behavior of R for large positive and large negative values of x (that is, as $x \to \pm\infty$). Vertical and horizontal asymptotes are defined and treated in detail in Chapter 4.

A sketch of the graph of

$$R(x) = \frac{1}{x^2 - 4x + 4} = \frac{1}{(x-2)^2}$$

is shown in Figure 1.6.6. Note that $x = 2$ is a vertical asymptote and the x-axis is a horizontal asymptote.

The graph of

$$R(x) = \frac{x^2}{x^2 - 1} = \frac{x^2}{(x-1)(x+1)}$$

is sketched in Figure 1.6.7. The lines $x = 1$ and $x = -1$ are vertical asymptotes of the graph and $y = 1$ is a horizontal asymptote.

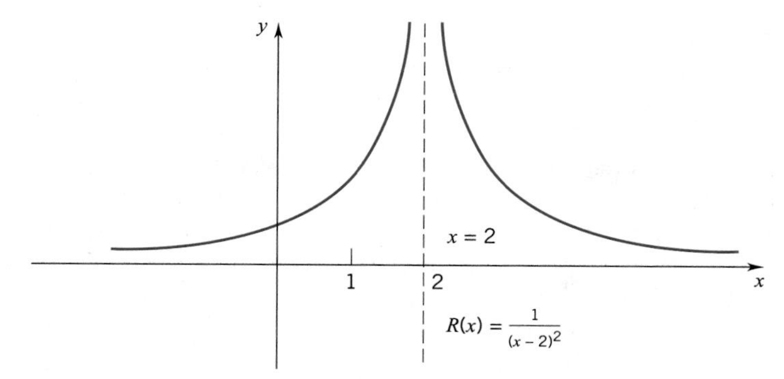

Figure 1.6.6

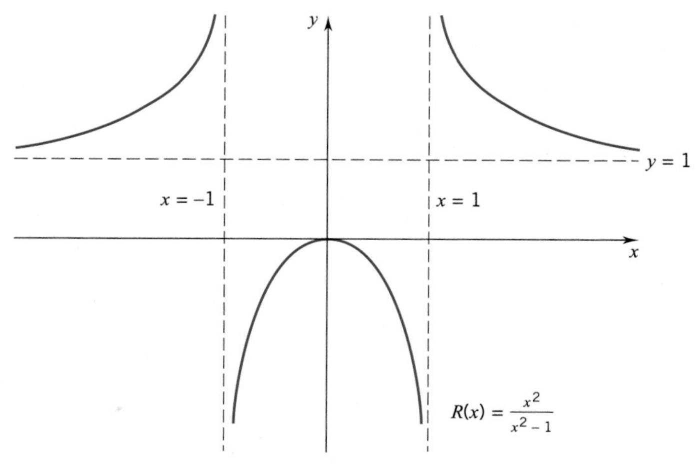

Figure 1.6.7

Trigonometric Functions

We assume that the reader is already familiar with trigonometry, and so we present here a brief review of the definitions, basic properties, and graphs of the trigonometric functions.

Angles An angle in the plane is generated by rotating a ray (half-line) about its endpoint. The starting position of the ray is called the *initial side* of the angle, and the final position of the ray is called the *terminal side.* The point of intersection of the initial and terminal sides is called the *vertex*. An angle is said to be *positive* if it is generated by a counterclockwise rotation and *negative* if the rotation is clockwise (see Figure 1.6.8).

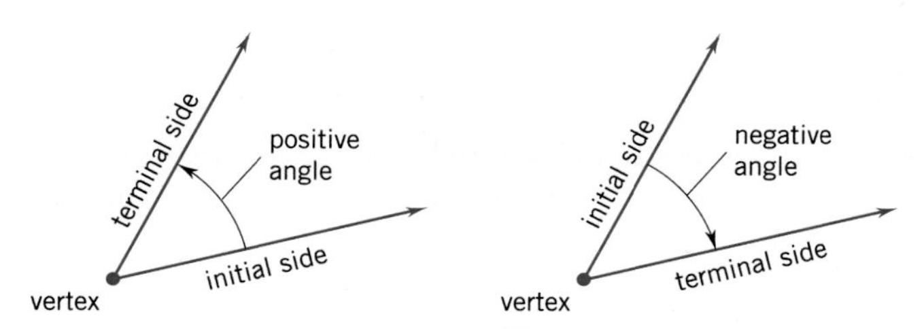

Figure 1.6.8

Degree measure, traditionally used to measure angles, has a serious drawback. It is artificial; there is no intrinsic connection between a degree and the geometry of a circle. Why choose 360° for one complete revolution? Why not 100°? or 400°?

There is another way of measuring angles that is more natural and lends itself better to the methods of calculus; measuring angles in *radians*. Let θ be an angle in the plane. Introduce a coordinate system such that the vertex of θ is at the origin and the initial side is the positive x-axis. In this context, θ is said to be in *standard position*. Let $P(1, 0)$ be the point of intersection of the initial side with the unit circle: $x^2 + y^2 = 1$. As the initial side is rotated onto the terminal side of θ, the point P traces an arc on the unit circle. Let s be the length of that arc. The *radian measure* of θ is s if θ is a positive angle and it is $-s$ if θ is a negative angle. Thus, the radian measure of θ is the signed length of the arc that θ (in standard position) subtends on the unit circle (see Figure 1.6.9). If there is no rotation, then the radian measure of θ is 0.

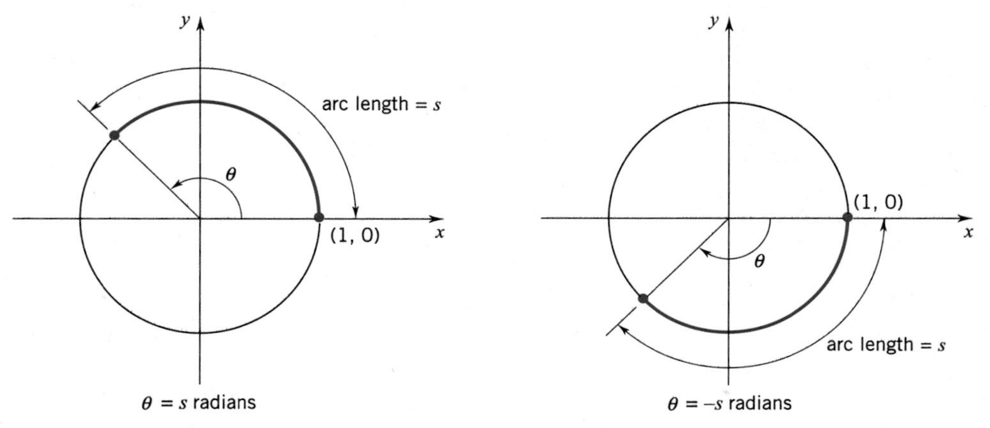

Figure 1.6.9

Since the circumference of the unit circle is 2π, the measure of the angle corresponding to one complete revolution in the counterclockwise direction is 2π radians; half a revolution (a straight angle) comprises π radians; and a quarter revolution (a right angle) comprises $\pi/2$ radians (see Figure 1.6.10).

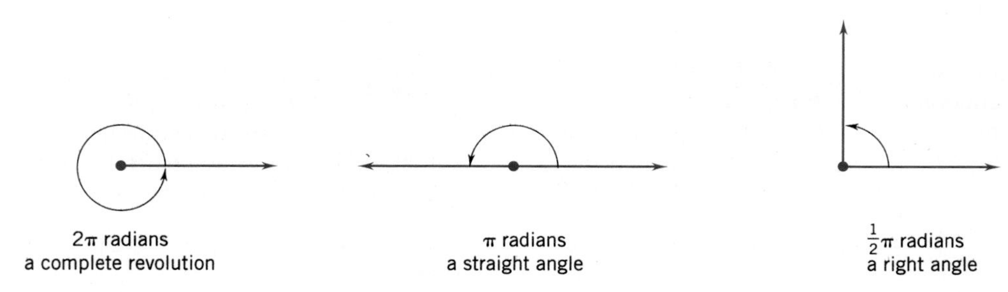

Figure 1.6.10

The conversion of degrees to radians and vice versa is made by noting that if A is the measure of an angle in degrees and x is the measure of the same angle in radians, then

$$\frac{A}{360} = \frac{x}{2\pi}.$$

In particular,

$$1 \text{ radian} = \frac{360}{2\pi} \text{ degrees} \cong 57.30°$$

and

$$1° = \frac{2\pi}{360} \text{ radian} \cong 0.0175 \text{ radian}$$

The following table gives some common angles measured in both degrees and radians.

degrees	0°	30°	45°	60°	90°	120°	135°	150°	180°	270°	360°
radians	0	$\pi/6$	$\pi/4$	$\pi/3$	$\pi/2$	$2\pi/3$	$3\pi/4$	$5\pi/6$	π	$3\pi/2$	2π

Definitions of the Trigonometric Functions As mentioned at the beginning of this section, we are concerned with real-valued functions of a real variable and so we use the unit circle and the radian measure of angles as opposed to right triangles and degree measure to define the trigonometric functions. Let θ be an angle in standard position and let $P(x, y)$ be the point of intersection of the terminal side of θ and the unit circle (see Figure 1.6.11).

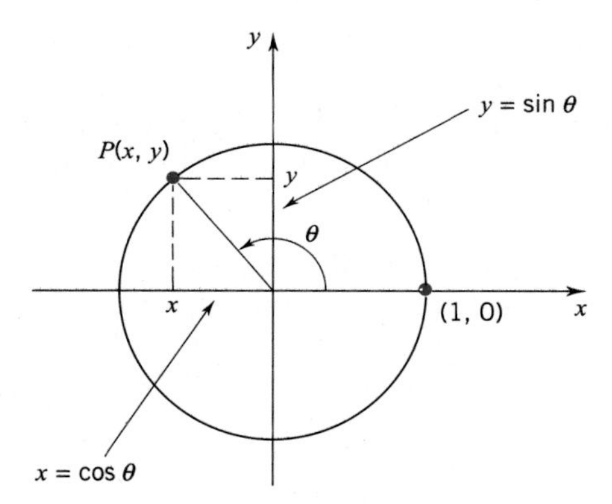

Figure 1.6.11

The six trigonometric functions are defined as follows:

sine: $\sin\theta = y$

cosine: $\cos\theta = x$

tangent: $\tan\theta = \dfrac{y}{x} = \dfrac{\sin\theta}{\cos\theta}, \quad x \neq 0$

cosecant: $\csc\theta = \dfrac{1}{y} = \dfrac{1}{\sin\theta}, \quad y \neq 0$

secant: $\sec\theta = \dfrac{1}{x} = \dfrac{1}{\cos\theta}, \quad x \neq 0$

cotangent: $\cot\theta = \dfrac{x}{y} = \dfrac{\cos\theta}{\sin\theta}, \quad y \neq 0.$

The sine and cosine functions are defined for all real numbers. That is, both sine and cosine have domain $(-\infty, \infty)$. The domains of the tangent and secant functions are all real numbers except for those values of θ for which $x = 0$, namely $\theta = (2n + 1)\pi/2$, for any integer n. The domains of the cosecant and cotangent functions are all real numbers except for those values of θ for which $y = 0$, namely $\theta = n\pi$, for any integer n.

Values We concentrate on the sine, cosine, and tangent functions since cosecant, secant, and cotangent are simply the reciprocals of these functions. The values of the trigonometric functions for the angles 0 (=0°), $\pi/6$ (=30°), $\pi/4$ (=45°), $\pi/3$ (=60°), $\pi/2$ (=90°), and selected multiples up to 2π, are given in the following table. You should already have these values memorized.

	0	$\pi/6$	$\pi/4$	$\pi/3$	$\pi/2$	$2\pi/3$	$3\pi/4$	$5\pi/6$	π	$3\pi/2$	2π
$\sin\theta$	0	$\frac{1}{2}$	$\frac{\sqrt{2}}{2}$	$\frac{\sqrt{3}}{2}$	1	$\frac{\sqrt{3}}{2}$	$\frac{\sqrt{2}}{2}$	$\frac{1}{2}$	0	-1	0
$\cos\theta$	1	$\frac{\sqrt{3}}{2}$	$\frac{\sqrt{2}}{2}$	$\frac{1}{2}$	0	$-\frac{1}{2}$	$-\frac{\sqrt{2}}{2}$	$-\frac{\sqrt{3}}{2}$	-1	0	1
$\tan\theta$	0	$\frac{\sqrt{3}}{3}$	1	$\sqrt{3}$	Undefined	$-\sqrt{3}$	-1	$-\frac{\sqrt{3}}{3}$	0	Undefined	0

In general, the (approximate) values of the trigonometric functions for any angle θ can be obtained with a hand calculator or from a table of values.

Periodicity A function f is *periodic* if there is a number p, $p \neq 0$, such that $f(x + p) = f(x)$ whenever x and $x + p$ are in the domain of f. The smallest positive number with this property is called the *period* of f. It is easy to verify from the definitions of the sine and cosine functions that they are periodic with period 2π.

sine

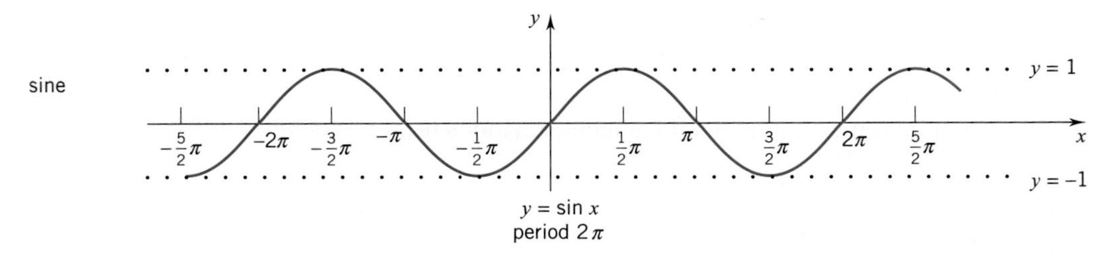

$y = \sin x$
period 2π

cosine

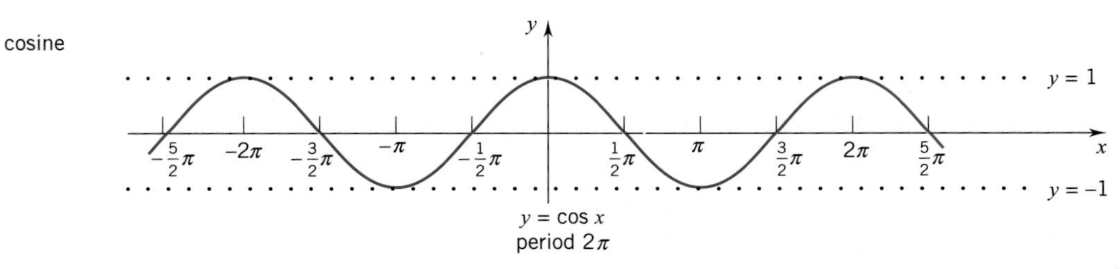

$y = \cos x$
period 2π

Figure 1.6.12

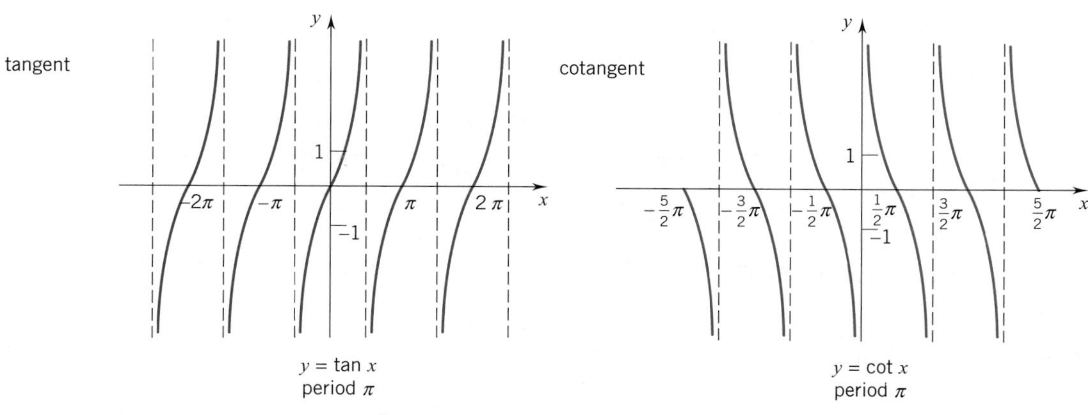

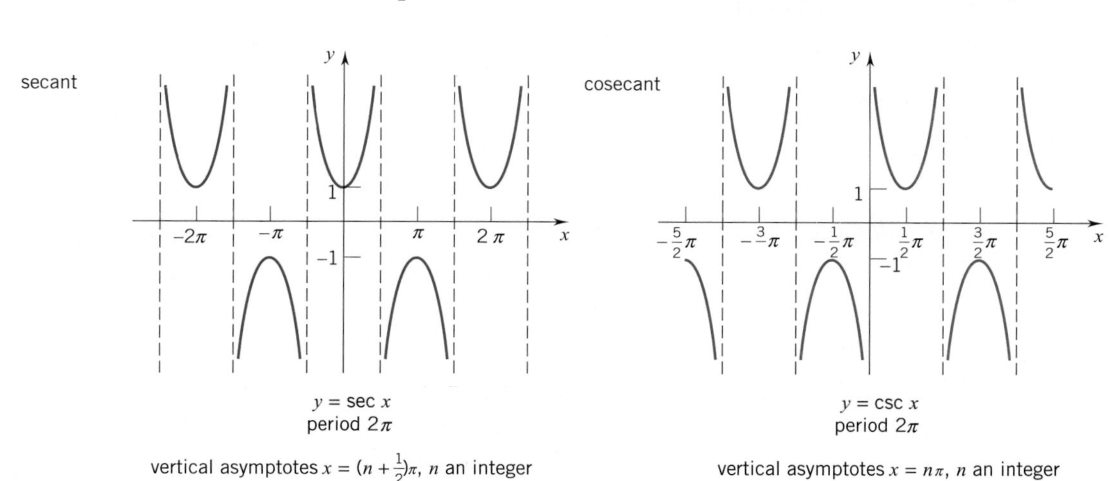

Figure 1.6.13

That is,

$$\sin(\theta + 2\pi) = \sin\theta \qquad \text{and} \qquad \cos(\theta + 2\pi) = \cos\theta \quad \text{for all } \theta.$$

It follows that the secant and cosecant functions are also periodic with period 2π. It can be verified that the tangent and cotangent functions have period π.

Graphs The graphs of the six trigonometric functions are shown in Figures 1.6.12 and 1.6.13. Since we normally use x and y to represent the independent and dependent variables when drawing a graph, we follow that convention here, replacing θ by x and letting $y = \sin x$, $y = \cos x$, and so forth.

Important Identities We assume that you are familiar with the common trigonometric identities. The most important of these are listed here for reference.

(i) *unit circle*

$$\sin^2\theta + \cos^2\theta = 1, \qquad \tan^2\theta + 1 = \sec^2\theta, \qquad 1 + \cot^2\theta = \csc^2\theta.$$

(ii) *addition formulas*

$$\sin(\alpha + \beta) = \sin\alpha\cos\beta + \cos\alpha\sin\beta,$$
$$\sin(\alpha - \beta) = \sin\alpha\cos\beta - \cos\alpha\sin\beta,$$
$$\cos(\alpha + \beta) = \cos\alpha\cos\beta - \sin\alpha\sin\beta,$$
$$\cos(\alpha - \beta) = \cos\alpha\cos\beta + \sin\alpha\sin\beta.$$

(iii) *even/odd functions*

$$\sin(-\theta) = -\sin\theta$$
$$\cos(-\theta) = \cos\theta.$$

Thus, the sine function is an odd function; its graph is symmetric with respect to the origin. The cosine function is an even function; its graph is symmetric with respect to the y-axis. See Figure 1.6.12.

(iv) *double-angle formulas*

$$\sin 2\theta = 2\sin\theta\cos\theta,$$
$$\cos 2\theta = \cos^2\theta - \sin^2\theta = 2\cos^2\theta - 1 = 1 - 2\sin^2\theta.$$

(v) *half-angle formulas*

$$\sin^2\theta = \frac{1}{2}(1 - \cos 2\theta), \qquad \cos^2\theta = \frac{1}{2}(1 + \cos 2\theta).$$

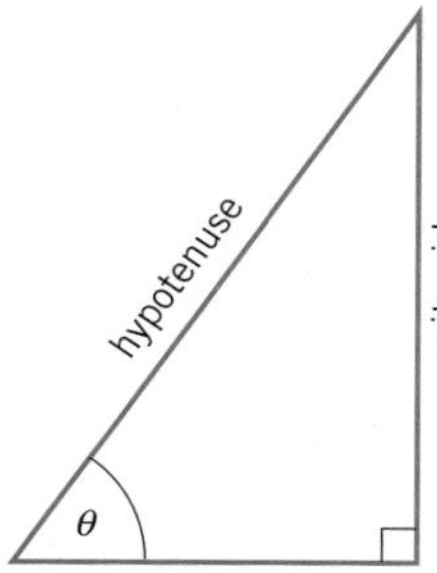

Figure 1.6.14

In Terms of a Right Triangle For angles θ between 0 and $\pi/2$, the trigonometric functions can also be defined as ratios of the sides of a right triangle (see Figure 1.6.14).

$$\sin\theta = \frac{\text{opposite side}}{\text{hypotenuse}}, \qquad \csc\theta = \frac{\text{hypotenuse}}{\text{opposite side}},$$
$$\cos\theta = \frac{\text{adjacent side}}{\text{hypotenuse}}, \qquad \sec\theta = \frac{\text{hypotenuse}}{\text{adjacent side}},$$
$$\tan\theta = \frac{\text{opposite side}}{\text{adjacent side}}, \qquad \cot\theta = \frac{\text{adjacent side}}{\text{opposite side}}.$$

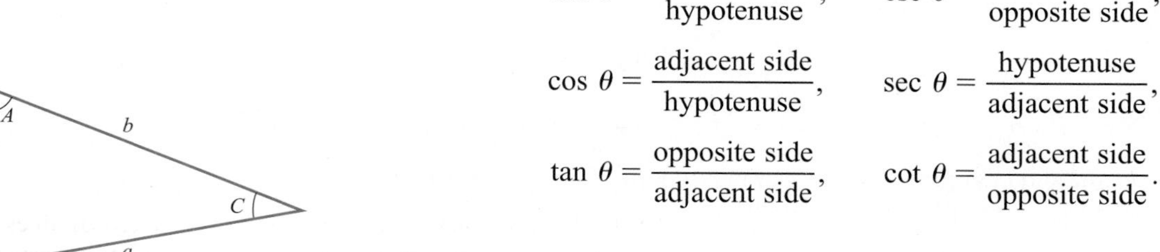

Figure 1.6.15

Arbitrary Triangles Let a, b, and c be the sides of an arbitrary triangle and let A, B and C be the corresponding angles. (See Figure 1.6.15.)

area: $\frac{1}{2}ab\sin C = \frac{1}{2}ac\sin B = \frac{1}{2}bc\sin A.$

law of sines: $\frac{\sin A}{a} = \frac{\sin B}{b} = \frac{\sin C}{c}.$

law of cosines: $a^2 = b^2 + c^2 - 2bc\cos A.$
$b^2 = a^2 + c^2 - 2ac\cos B.$
$c^2 = a^2 + b^2 - 2ab\cos C.$

EXERCISES 1.6

In Exercises 1–10, state whether the given function is a polynomial function, a rational function (but not a polynomial), or neither a polynomial function nor a rational function. If the function is a polynomial, state its degree.

1. $f(x) = 3.$

2. $f(x) = 1 + \frac{1}{2}x.$

3. $g(x) = \frac{1}{x}.$

4. $h(x) = \frac{x^2 - 4}{\sqrt{2}}.$

5. $F(x) = \frac{x^3 - 3x^{3/2} + 2x}{x^2 - 1}.$

6. $f(x) = 5x^4 - \pi x^2 + \frac{1}{2}.$

7. $f(x) = \sqrt{x}(\sqrt{x} + 1).$

8. $g(x) = \frac{x^2 - 2x - 8}{x + 2}.$

9. $f(x) = \frac{\sqrt{x^2 + 1}}{x^2 - 1}.$

10. $h(x) = \frac{\sin^2 x + \cos^2 x}{x^3 + 8}.$

In Exercises 11–16, determine the domain of the given function and sketch its graph.

11. $f(x) = 3x - \frac{1}{2}.$

12. $f(x) = \frac{1}{x + 1}.$

13. $g(x) = x^2 - x - 6.$

14. $F(x) = x^3 - x.$

15. $f(x) = \frac{1}{x^2 - 4}.$

16. $g(x) = x + \frac{1}{x}.$

In Exercises 17–22, convert the given degree measure into radian measure.

17. $225°.$

18. $-210°.$

19. $-300°.$

20. $450°.$

21. $15°.$

22. $3°.$

In Exercises 23–28, convert the given radian measure into degree measure.

23. $-\frac{3\pi}{2}.$

24. $\frac{5\pi}{4}.$

25. $\frac{5\pi}{3}.$

26. $-\frac{11\pi}{6}.$

27. $2.$

28. $-\sqrt{3}.$

In Exercises 29–36, determine the exact value(s) of x in the interval $[0, 2\pi)$ which satisfy the given equation.

29. $\sin x = \frac{1}{2}.$

30. $\cos x = -\frac{1}{2}.$

31. $\tan \frac{1}{2}x = 1.$

32. $\sqrt{\sin^2 x} = 1.$

33. $\cos x = \frac{\sqrt{2}}{2}.$

34. $\sin 2x = -\frac{\sqrt{3}}{2}.$

35. $\cos 2x = 0.$

36. $\tan x = -\sqrt{3}.$

In Exercises 37–46, find the approximate values of the given expression. Use four decimal place accuracy.

37. $\sin 51°.$

38. $\cos 17°.$

39. $\sin(2.352).$

40. $\cos(-13.461).$

41. $\tan 72.4°.$

42. $\cot(-13.5°).$

43. $\tan(11.249).$

44. $\cot(7.311).$

45. $\sec(4.360).$

46. $\csc(-9.725).$

In Exercises 47–52, find the solutions of the given equation that are in the interval $[0, 2\pi)$. Express your answers in radians and use four decimal place accuracy.

47. $\sin x = 0.5231.$

48. $\cos x = -0.8243.$

49. $\tan x = 6.7192.$

50. $\cot x = -3.0649.$

51. $\sec x = -4.4073.$

52. $\csc x = 10.260.$

In Exercises 53–58, determine the domain and range of the given function.

53. $f(x) = |\sin x|.$

54. $g(x) = \sin^2 x + \cos^2 x.$

55. $f(x) = 2\cos 3x.$

56. $F(x) = 1 + \sin x.$

57. $f(x) = 1 + \tan^2 x.$

58. $h(x) = \sqrt{\cos^2 x}.$

In Exercises 59–64, sketch the graph of the given function.

59. $f(x) = 3\sin 2x.$

60. $f(x) = 1 + \sin x.$

61. $g(x) = 1 - \cos x.$

62. $F(x) = \tan \frac{1}{2}x.$

63. $f(x) = \sqrt{\sin^2 x}.$

64. $g(x) = -2\cos x.$

In Exercises 65–70, determine whether the given function is odd, even, or neither.

65. $f(x) = \sin 3x.$

66. $g(x) = \tan x.$

67. $f(x) = 1 + \cos 2x.$

68. $g(x) = \sec x.$

69. $f(x) = x^3 + \sin x.$

70. $h(x) = \frac{\cos x}{x^2 + 1}.$

71. Suppose that l_1 and l_2 are two nonvertical lines. If $m_1 m_2 = -1$, then l_1 and l_2 intersect at right angles. Show that if l_1 and l_2 do not intersect at right angles, then the angle α between l_1 and l_2 is given by the formula:

$$\tan \alpha = \left| \frac{m_2 - m_1}{1 + m_1 m_2} \right|.$$

HINT: Use the identity

$$\tan(\theta - \phi) = \frac{\tan \theta - \tan \phi}{1 + \tan \theta \tan \phi}.$$

72. For angles θ between 0 and $\pi/2$, verify that the definition of the trigonometric functions in terms of the unit circle and the definitions in terms of a right triangle are equivalent. HINT: Introduce a coordinate system with one of the angles of a right triangle in standard position. See the figure.

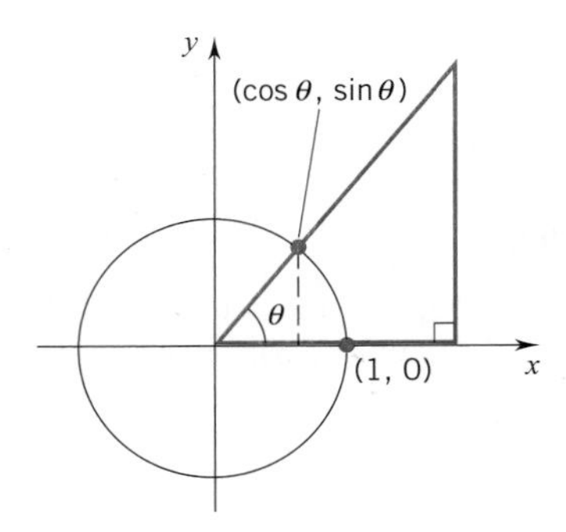

In Exercises 73 and 74, consider a general triangle with angles A, B, C and corresponding sides a, b, c. For example, see Figure 1.6.15.

73. Prove the law of sines:

$$\frac{\sin A}{a} = \frac{\sin B}{b} = \frac{\sin C}{c}.$$

HINT: Drop a perpendicular from one vertex to the opposite side and use the two right triangles that are formed.

74. Prove the law of cosines:

$$a^2 = b^2 + c^2 - 2bc \cos A.$$

HINT: Drop a perpendicular from angle B to side b, and use trigonometry and the Pythagorean theorem.

75. Prove that the area A of an isosceles triangle is given by $A = \frac{1}{2}a^2 \sin \theta$, where a is the length of the equal sides and θ is the angle included between them.

76. (a) Use a graphing utility to graph the polynomial functions

$$f(x) = x^4 + 2x^3 - 5x^2 - 3x + 1 \quad \text{and}$$
$$g(x) = -x^4 + x^3 + 4x^2 - 3x + 2.$$

(b) Based on your graphs in part (a), make a conjecture as to the general shape of the graph of a polynomial function of degree 4 (a *quartic* polynomial).

(c) Now graph the quartic polynomials

$$f(x) = x^4 - 4x^2 + 4x + 2 \quad \text{and} \quad g(x) = -x^4$$

and compare these graphs with your conjectures in part (b). The methods of calculus will help us distinguish between the various possibilities for the graph of a quartic polynomial. Can you determine one property that the graphs of all quartic polynomials of the form

$$P(x) = x^4 + ax^3 + bx^2 + cx + d$$

share? What about quartics of the form $Q(x) = -x^4 + ax^3 + bx^2 + cx + d$?

77. (a) Use a graphing utility to graph the polynomial functions

$$f(x) = x^5 - 7x^3 + 6x + 2 \quad \text{and}$$
$$g(x) = -x^5 + 5x^3 - 3x - 3.$$

(b) Based on your graphs in part (a), make a conjecture as to the general shape of the graph of a polynomial function of degree 5 (a *quintic* polynomial).

(c) Now graph

$$P(x) = x^5 + ax^4 + bx^3 + cx^2 + dx + e$$

for several choices of a, b, c, d, and e. (For example, try $a = b = c = d = e = 0$.) How do these graphs compare with your graph of f from part (a)?

78. (a) Use a graphing utility to graph $f_A(x) = A \cos x$ for several values of A; use both positive and negative values. Compare your graphs with the graph of $f(x) = \cos x$.

(b) Now graph $f_B(x) = \cos Bx$ for several values of B. Since the cosine function is even, it is sufficient to use only positive values for B. Use some values between 0 and 1 and some values greater than 1. Again, compare your graphs with the graph of $f(x) = \cos x$.

(c) Describe the effects that the coefficients A and B have on the graph of the cosine function.

79. Let $f_n(x) = x^n$, $n = 1, 2, 3, \ldots$.

(a) Using a graphing utility, draw the graphs of f_n for $n = 2, 4, 6$, in one coordinate system, and in another coordinate system draw the graphs of f_n for $n = 1, 3, 5$.

(b) Based on your results in part (a), make a sketch of the graph of f_n when n is even, and when n is odd.

(c) For a given positive integer k, compare the graphs of f_k and f_{k+1} on $[0, 1]$ and on $[1, \infty)$.

80. (a) Use a graphing utility to draw the graphs of

$$f(x) = \frac{1}{x} \sin x \quad \text{and}$$

$$g(x) = x \sin\left(\frac{1}{x}\right) \quad \text{for } -\frac{\pi}{2} \le x \le \frac{\pi}{2}, \quad x \ne 0.$$

(b) Use the "zoom in" feature to examine the behavior of the graphs of f and g at the origin. What can you say about the values of $f(x)$ and $g(x)$ when x is close to 0?

■ 1.7 COMBINATIONS OF FUNCTIONS

As we will see in our study of calculus, functions often arise as combinations of other functions. In this section we review the basic methods for combining two or more functions.

Algebraic Combinations of Functions

Let f and g be real-valued functions of a real variable. Then the sum, difference, product, and quotient of f and g are defined in a natural way in terms of the sum, difference, product, and quotient of real numbers. These algebraic combinations of f and g are defined as follows:

$$\begin{aligned} &\textit{sum:} && [f+g](x) = f(x) + g(x). \\ &\textit{difference:} && [f-g](x) = f(x) - g(x). \\ &\textit{product:} && [f\cdot g](x) = f(x)\cdot g(x). \\ &\textit{quotient:} && \left[\frac{f}{g}\right](x) = \frac{f(x)}{g(x)}. \end{aligned}$$

According to these definitions, $(f+g)(x)$, $(f-g)(x)$, and $(f\cdot g)(x)$ exist only when both $f(x)$ and $g(x)$ exist. Thus, $f+g$, $f-g$, and $f\cdot g$ each have domain $\text{dom}(f) \cap \text{dom}(g)$. The definition of f/g requires the additional condition that $g(x) \neq 0$.

Example 1 Let

$$f(x) = x^3 + x - 2 \qquad \text{and} \qquad g(x) = \tfrac{1}{2}x - \sqrt{x}.$$

Determine (a) $f+g$, (b) $f\cdot g$, and (c) f/g, and give the domain in each case.

SOLUTION Since f is a polynomial, $\text{dom}(f) = (-\infty, \infty)$. You can verify that $\text{dom}(g) = [0, \infty)$.

(a) $(f+g)(x) = f(x) + g(x) = (x^3 + x - 2) + (\tfrac{1}{2}x - \sqrt{x}) = x^3 + \tfrac{3}{2}x - 2 - \sqrt{x}$

and

$$\text{dom}(f+g) = (-\infty, \infty) \cap [0, \infty) = [0, \infty).$$

(b)
$$\begin{aligned} (f\cdot g)(x) &= f(x)\cdot g(x) = (x^3 + x - 2)(\tfrac{1}{2}x - \sqrt{x}) \\ &= \tfrac{1}{2}x^4 + \tfrac{1}{2}x^2 - x - x^3\sqrt{x} - x\sqrt{x} + 2\sqrt{x} \end{aligned}$$

and $\text{dom}(f\cdot g) = [0, \infty)$. Note that this product can also be written:

$$(f\cdot g)(x) = \tfrac{1}{2}x^4 + \tfrac{1}{2}x^2 - x - x^{7/2} - x^{3/2} + 2x^{1/2}.$$

(c)
$$\left(\frac{f}{g}\right)(x) = \frac{f(x)}{g(x)} = \frac{x^3 + x - 2}{\tfrac{1}{2}x - \sqrt{x}}.$$

To find the domain of f/g we start with $\text{dom}(f) \cap \text{dom}(g) = [0, \infty)$ and impose the additional restriction that $g(x) \neq 0$. Now $g(x) = 0$ implies

$$\tfrac{1}{2}x = \sqrt{x} \qquad \text{or} \qquad \tfrac{1}{4}x^2 = x.$$

It follows from the second equation that $g(x) = 0$ for $x = 0$ and $x = 4$. Thus

$$\text{dom}(f/g) = \{x\colon x \in [0, \infty) \text{ and } x \neq 0, 4\} = (0, 4) \cup (4, \infty). \quad \square$$

With α and β real numbers, we can also form *scalar multiples* of functions

$$[\alpha f](x) = \alpha \cdot f(x)$$

and *linear combinations* of functions

$$[\alpha f + \beta g](x) = \alpha \cdot f(x) + \beta \cdot g(x).$$

It should be clear that $\text{dom}(\alpha f) = \text{dom}(f)$ and $\text{dom}(\alpha f + \beta g) = \text{dom}(f) \cap \text{dom}(g)$.

Example 2 Let

$$f(x) = x^2 - 3x + \frac{2}{x} \quad \text{and} \quad g(x) = x - \frac{1}{x} + \sqrt{x+1}.$$

Then

$$\text{dom}(f) = \{x\colon x \neq 0\} \quad \text{and} \quad \text{dom}(g) = \{x\colon x \geq -1 \text{ and } x \neq 0\}.$$

Now

$$(3f)(x) = 3\left(x^2 - 3x + \frac{2}{x}\right) = 3x^2 - 9x + \frac{6}{x}$$

with $\text{dom}(3f) = \{x\colon x \neq 0\}$ and

$$\begin{aligned}(4f - 3g)(x) &= 4\left(x^2 - 3x + \frac{2}{x}\right) - 3\left(x - \frac{1}{x} + \sqrt{x+1}\right)\\ &= 4x^2 - 15x + \frac{11}{x} - 3\sqrt{x+1}\end{aligned}$$

with $\text{dom}(4f - 3g) = \text{dom}(f) \cap \text{dom}(g) = [-1, 0) \cup (0, \infty)$ (verify this). ❑

Composition of Functions

We have seen how to combine functions algebraically. There is another way to combine functions called *composition.* To describe it, we begin with two functions f and g, and a number x in the domain of g. By applying g to x we get the number $g(x)$. If $g(x)$ is in the domain of f, then we can apply f to $g(x)$ and thereby obtain the number $f(g(x))$.

What is $f(g(x))$? It is the result of first applying g to x and then applying f to $g(x)$. The idea is illustrated in Figure 1.7.1. This new function—it takes x in the domain of g such that $g(x)$ is in the domain of f, and assigns to it the value $f(g(x))$—is called the *composition* of f and g and is denoted by the symbol $f \circ g$ (see Figure 1.7.2).

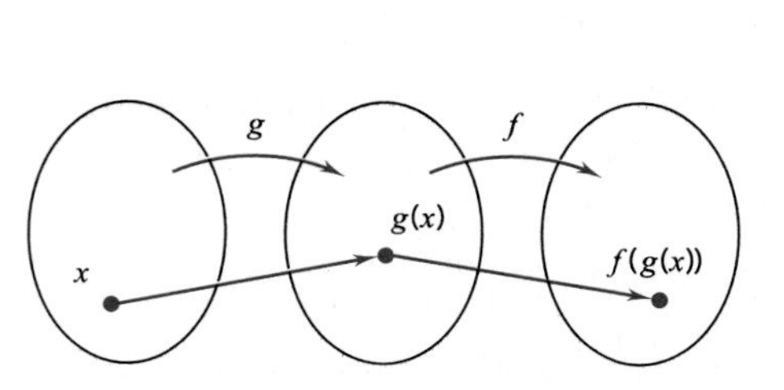

Figure 1.7.1

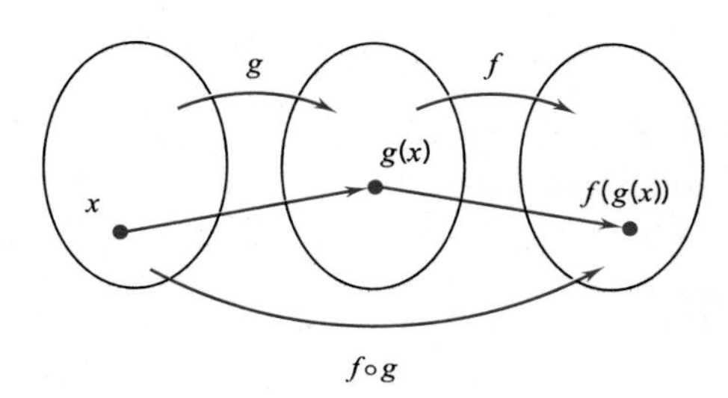

Figure 1.7.2

DEFINITION 1.7.1 COMPOSITION

Let f and g be functions and let $D = \{x \in \text{dom}(g)\colon g(x) \in \text{dom}(f)\}$. The *composition* $f \circ g$ (read "f circle g") is the function defined on D by

$$(f \circ g)(x) = f(g(x)).$$

Example 3 Suppose that

$$g(x) = x^2 \qquad \text{(the squaring function)}$$

and

$$f(x) = x + 3. \qquad \text{(the function that adds 3)}$$

Then

$$(f \circ g)(x) = f(g(x)) = g(x) + 3 = x^2 + 3.$$

In other words, $f \circ g$ is the function that *first* squares and *then* adds 3. Since f and g each have domain the set of all real numbers, the domain of $f \circ g$ is all real numbers.

On the other hand,

$$(g \circ f)(x) = g(f(x)) = [f(x)]^2 = (x + 3)^2$$

Thus, $g \circ f$ is the function that *first* adds 3 and *then* squares; $\text{dom}(g \circ f)$ is all real numbers. Notice that $g \circ f$ is *not* the same as $f \circ g$. ❑

Example 4 Let

$$g(x) = \sqrt{3 - x}, \qquad \text{dom}(g) = (-\infty, 3]$$

and

$$f(x) = x^2 - 1, \qquad \text{dom}(f) = (-\infty, \infty).$$

Then

$$(f \circ g)(x) = f(g(x)) = \left(\sqrt{3 - x}\right)^2 - 1 = (3 - x) - 1 = 2 - x$$

and $\text{dom}(f \circ g) = \{x \in \text{dom}(g)\colon g(x) \in \text{dom}(f)\} = \text{dom}(g)$ since $\text{dom}(f)$ is all real numbers. Thus, $\text{dom}(f \circ g) = (-\infty, 3]$.

Also,

$$(g \circ f)(x) = g(f(x)) = \sqrt{3 - (x^2 - 1)} = \sqrt{4 - x^2}$$

and $\text{dom}(g \circ f) = \{x \in \text{dom}(f)\colon f(x) \in (-\infty, 3]\} = \{x\colon x^2 - 1 \le 3\} = [-2, 2]$. ❑

It is possible to form the composition of more than two functions. For example, the triple composition $f \circ g \circ h$ consists of first h, then g, and then f:

$$(f \circ g \circ h)(x) = f[g(h(x))].$$

We can go on in this manner with as many functions as we like.

Example 5 If

$$f(x) = \frac{1}{x}, \qquad g(x) = x^2 + 1, \qquad h(x) = \cos x,$$

then

$$(f \circ g \circ h)(x) = f[g(h(x))] = \frac{1}{g(h(x))} = \frac{1}{[h(x)]^2 + 1} = \frac{1}{\cos^2 x + 1}. \quad \square$$

Example 6 Find functions f and g such that $f \circ g = F$ if

$$F(x) = (x + 1)^5.$$

A SOLUTION The function consists of first adding 1 and then taking the fifth power. We can therefore set

$$g(x) = x + 1 \qquad \text{(adding 1)}$$

and

$$f(x) = x^5. \qquad \text{(taking the fifth power)}$$

As you can see,

$$(f \circ g)(x) = f(g(x)) = [g(x)]^5 = (x + 1)^5. \quad \square$$

Example 7 Find three functions f, g, h such that $f \circ g \circ h = F$ if

$$F(x) = \frac{1}{|x| + 3}.$$

A SOLUTION F takes the absolute value, adds 3, and then inverts. Let h take the absolute value:

$$\text{set} \quad h(x) = |x|.$$

Let g add 3:

$$\text{set} \quad g(x) = x + 3.$$

Let f do the inverting:

$$\text{set} \quad f(x) = \frac{1}{x}.$$

With this choice of f, g, h, we have

$$(f \circ g \circ h)(x) = f(g(h(x))) = \frac{1}{g(h(x))} = \frac{1}{h(x) + 3} = \frac{1}{|x| + 3}. \quad \square$$

EXERCISES 1.7

In Exercises 1–8, let $f(x) = 2x^2 - 3x + 1$ and $g(x) = x^2 + 1/x$. Determine the indicated values.

1. $(f + g)(2)$.

2. $(f - g)(-1)$.

3. $(fg)(-2)$.

4. $\left(\frac{f}{g}\right)(1)$.

5. $(2f - 3g)\left(\frac{1}{2}\right)$.

6. $\left(\frac{f + 2g}{f}\right)(-1)$.

7. $(f \circ g)(1)$.

8. $(g \circ f)(1)$.

In Exercises 9–12, determine $f + g$, $f - g$, fg, and f/g, and give the domain of each.

9. $f(x) = 2x - 3, \quad g(x) = 2 - x.$

10. $f(x) = x^2 - 1, \quad g(x) = x + \dfrac{1}{x}.$

11. $f(x) = \sqrt{x - 1}, \quad g(x) = x - \sqrt{x + 1}.$

12. $f(x) = \sin^2 x, \quad g(x) = \cos 2x.$

13. Given that $f(x) = x + 1/\sqrt{x}$ and $g(x) = \sqrt{x} - 2/\sqrt{x}$, find (a) $6f + 3g$, (b) $f - g$, (c) f/g.

14. Given that

$$f(x) = \begin{cases} 1 - x, & x \leq 1 \\ 2x - 1, & x > 1, \end{cases} \quad g(x) = \begin{cases} 0, & x < 2 \\ -1, & x \geq 2, \end{cases}$$

find $f + g$, $f - g$, and fg. HINT: Break up the domains of the two functions in the same manner.

In Exercises 15–22, sketch the graph with f and g as shown in the figure.

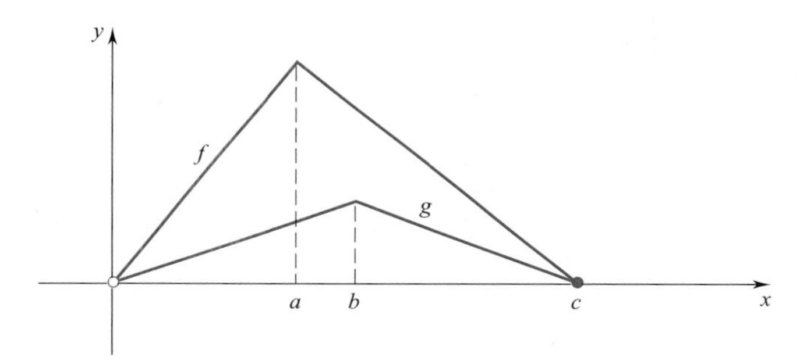

15. $2g$.

16. $\frac{1}{2}f$.

17. $-f$.

18. $0 \cdot g$.

19. $-2g$.

20. $f + g$.

21. $f - g$.

22. $f + 2g$.

In Exercises 23–32, form the composition $f \circ g$ and give the domain.

23. $f(x) = 2x + 5, \quad g(x) = x^2.$

24. $f(x) = x^2, \quad g(x) = 2x + 5.$

25. $f(x) = \sqrt{x}, \quad g(x) = x^2 + 5.$

26. $f(x) = x^2 + x, \quad g(x) = \sqrt{x}.$

27. $f(x) = \dfrac{1}{x}, \quad g(x) = \dfrac{x - 2}{x}.$

28. $f(x) = \dfrac{1}{x - 1}, \quad g(x) = x^2.$

29. $f(x) = \dfrac{1}{\sqrt{x} - 3}, \quad g(x) = (x^2 - 1)^2.$

30. $f(x) = \dfrac{1}{x} - \dfrac{1}{x + 1}, \quad g(x) = \dfrac{x}{x^2 - 1}.$

31. $f(x) = \sqrt{1 - x^2}, \quad g(x) = \cos 2x.$

32. $f(x) = \sqrt{1 - x}, \quad g(x) = 2 \cos x$ for $x \in [0, 2\pi]$.

In Exercises 33–36, form the composition $f \circ g \circ h$ and give the domain.

33. $f(x) = 4x, \quad g(x) = x - 1, \quad h(x) = x^2.$

34. $f(x) = x - 1, \quad g(x) = 4x, \quad h(x) = x^2.$

35. $f(x) = \dfrac{1}{x}, \quad g(x) = \dfrac{1}{2x + 1}, \quad h(x) = x^2.$

36. $f(x) = \dfrac{x + 1}{x}, \quad g(x) = \dfrac{1}{2x + 1}, \quad h(x) = x^2.$

In Exercises 37–40, find f such that $f \circ g = F$ given that

37. $g(x) = \dfrac{1 + x^2}{1 + x^4}, \quad F(x) = \dfrac{1 + x^4}{1 + x^2}.$

38. $g(x) = x^2, \quad F(x) = ax^2 + b.$

39. $g(x) = 3x, \quad F(x) = 2 \sin 3x.$

40. $g(x) = -x^2, \quad F(x) = \sqrt{a^2 + x^2}.$

In Exercises 41–44, find g such that $f \circ g = F$ given that

41. $f(x) = x^3, \quad F(x) = \left(1 - \dfrac{1}{x^4}\right)^2.$

42. $f(x) = x + \dfrac{1}{x}, \quad F(x) = a^2x^2 + \dfrac{1}{a^2x^2}.$

43. $f(x) = x^2 + 1, \quad F(x) = (2x^3 - 1)^2 + 1.$

44. $f(x) = \sin x, \quad F(x) = \sin \dfrac{1}{x}.$

In Exercises 45–49, find $f \circ g$ and $g \circ f$.

45. $f(x) = \sqrt{x}, \quad g(x) = x^2.$

46. $f(x) = 3x + 1, \quad g(x) = x^2.$

47. $f(x) = 1 - x^2, \quad g(x) = \sin x.$

48. $f(x) = 2x, \quad g(x) = \frac{1}{2}.$

49. $f(x) = x^3 + 1, \quad g(x) = \sqrt[3]{x - 1}.$

50. Suppose that f and g are odd functions. What can you conclude about fg? Justify your answer.

51. Suppose that f and g are even functions. What can you conclude about fg? Justify your answer.

52. Suppose that f is an even function and g is an odd function. What can you conclude about fg? Justify your answer.

53. Suppose that

$$f(x) = \begin{cases} x, & 0 \leq x \leq 1 \\ 1, & x > 1. \end{cases}$$

How is f defined for $x < 0$ if (a) f is even? (b) f is odd?

54. Suppose that $f(x) = x^2 - x$ for $x \geq 0$. How is f defined for $x < 0$ if (a) f is even? (b) f is odd?

55. Given that the function f is defined for all real numbers, show that the function $g(x) = f(x) + f(-x)$ is an even function.

56. Show that every function that is defined for all real numbers can be written as the sum of an even function and an odd function.

57. For $x \neq 0, 1$, define

$$f_1(x) = x, \qquad f_2(x) = \frac{1}{x}, \qquad f_3(x) = 1 - x,$$

$$f_4(x) = \frac{1}{1-x}, \qquad f_5(x) = \frac{x-1}{x}, \qquad f_6(x) = \frac{x}{x-1}.$$

This family of functions is *closed* under composition; that is, the composition of any two of these functions is again one of these functions. Tabulate the results of composing these functions one with the other by filling in the table shown in the figure. To indicate that $f_i \circ f_j = f_k$, write "f_k" in the ith row, jth column. We have already made two entries in the table. Check out these two entries and then fill in the rest of the table.

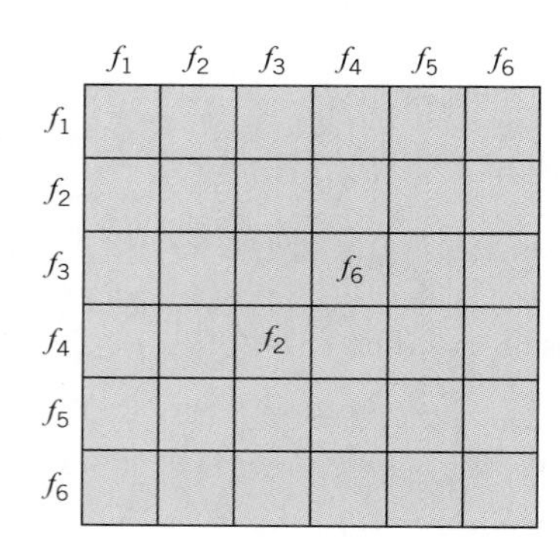

	f_1	f_2	f_3	f_4	f_5	f_6
f_1						
f_2						
f_3				f_6		
f_4			f_2			
f_5						
f_6						

58. Find the minimal family of functions that contains the given functions and is closed under composition.
(a) $f(x) = 1 - x$. (b) $f(x) = 1 + x$. (c) $f(x) = -x^3$.

Two functions f and g are *inverses* of each other if

$$(f \circ g)(x) = x \qquad \text{for all } x \in \text{dom}(g)$$

and

$$(g \circ f)(x) = x \qquad \text{for all } x \in \text{dom}(f).$$

For example, $f(x) = 2x + 3$ and $g(x) = \frac{1}{2}(x - 3)$ are inverses of each other since

$$(f \circ g)(x) = f(g(x)) = 2[\tfrac{1}{2}(x-3)] + 3 = x$$

$$(g \circ f)(x) = g(f(x)) = \tfrac{1}{2}([2x+3] - 3) = x.$$

In Exercises 59–62, verify that the given functions are inverses of each other.

59. $f(x) = 3x - 5, \quad g(x) = \frac{1}{3}(x + 5)$.

60. $f(x) = x^3, \quad g(x) = \sqrt[3]{x}$.

61. $f(x) = \dfrac{1}{x+1}, \quad g(x) = \dfrac{1-x}{x}$.

62. $f(x) = \sqrt[5]{x+1}, \quad g(x) = x^5 - 1$.

In Exercises 63 and 64, use the family of functions $F(x) = (x - a)^2 + b$.

63. (a) Choose a value for b and, using a graphing utility, graph F for several different values of a. Be sure to choose both positive and negative values. Compare your graphs with the graph of $f(x) = x^2$, and describe the effect that varying a has on the graph of F.
(b) Now fix a value of a and graph F for several values of b; again, use both positive and negative values. Compare your graphs with the graph of $f(x) = x^2$, and describe the effect on that varying b has on the graph of F.

64. For any values of a and b, the graph of F is a parabola which opens up. Find values for a and b such that the parabola will have x-intercepts at $-\frac{3}{2}$ and 2. Verify your result algebraically.

In Exercises 65 and 66, let $f(x) = x^3 - 3x + 1$.

65. (a) Using a graphing utility, graph $f(x - a)$ for $a = -2, -1, 0, 1, 2$. Describe the effect that varying a has on the graph of f.
(b) Graph $f(bx)$ for $b = -2, -1, 0, 1, 2, 3$. Describe the effect that varying b has on the graph of f.
(c) Graph $f(x) + c$ for $c = -2, -1, 0, 1, 2$. Describe the effect that varying c has on the graph of f.

66. Use your graphing utility to find values for a, b, and c such that the graph of $f(b[x - a]) + c$ has x-intercepts at $-2, \frac{1}{2}$, and 3. Verify your result algebraically.

In Exercises 67 and 68, let $f(x) = \sin x$

67. (a) Using a graphing utility, graph Af for $A = -3, -2, -1, 2, 3$. Compare your graphs with the graph of f.
(b) Now graph $f(Bx)$ for $B = -3, -2, -\frac{1}{2}, \frac{1}{3}, \frac{1}{2}, 2, 3$. Compare your graphs with the graph of f.

68. (a) Using a graphing utility, graph $f(x - C)$ for $C = -\frac{\pi}{2}, -\frac{\pi}{4}, \frac{\pi}{3}, \frac{\pi}{2}, \pi, 2\pi$. Compare your graphs with the graph of f.
(b) Now graph $Af(Bx - C)$ for several values of A, B, and C. Describe the effects of A, B, and C.

■ 1.8 A NOTE ON MATHEMATICAL PROOF; MATHEMATICAL INDUCTION

The notion of proof goes back to Euclid's *Elements,* and the rules of proof have changed little since they were formulated by Aristotle. We work in a deductive system where truth is argued on the basis of assumptions, definitions, and previously proved results. We cannot claim that such and such is true without clearly stating the basis on which we make that claim.

A theorem is an implication; it consists of a hypothesis and a conclusion:

if (hypothesis) . . . , then (conclusion) . . . .

Here is an example:

if a and b are positive numbers, then ab is positive.

A common mistake is to ignore the hypothesis and persist with the conclusion: to insist, for example, that $ab > 0$ just because a and b are numbers.

Another common mistake is to confuse a theorem

if A, then B

with its converse

if B, then A.

The fact that a theorem is true does not mean that its converse is true: While it is true that

if a and b are positive numbers, then ab is positive,

it is *not* true that

if ab is positive, then a and b are positive numbers;

[$(-2)(-3)$ is positive, but -2 and -3 are not positive].

A third, more subtle mistake is to presume that the hypothesis of a theorem represents the only condition under which the conclusion is true. There may well be other conditions under which the conclusion is true. Thus, for example, not only is it true that

if a and b are positive numbers, then ab is positive

but it is also true that

if a and b are negative numbers, then ab is positive.

In the event that a theorem

if A, then B

and its converse

if B, then A

are both true, then we can write

A if and only if B or more briefly A iff B.

We know, for example, that

$$\text{if} \quad x \geq 0, \qquad \text{then} \quad |x| = x;$$

we also know that

$$\text{if} \quad |x| = x, \qquad \text{then} \quad x \geq 0.$$

We can summarize by writing

$$x \geq 0 \qquad \text{iff} \qquad |x| = x.$$

A final point. One way of proving

$$\text{if } A \text{, then } B$$

is to assume that

$$\text{(1)} \qquad A \text{ holds and } B \text{ does not hold}$$

and then arrive at a contradiction. The contradiction is taken to indicate that (1) is false and therefore

$$\text{if } A \text{ holds, then } B \text{ must hold.}$$

Some of the theorems of calculus are proved by this method.

Calculus provides procedures for solving a wide range of problems in the physical and social sciences. The fact that these procedures give us answers that seem to make sense is comforting, but it is only because we can prove our theorems that we can be confident in the results. Accordingly, your study of calculus should include the study of some proofs.

Mathematical Induction

Mathematical induction is a method of proof which can be used to show that certain propositions are true for all positive integers n. The method is based upon the following axiom:

1.8.1 AXIOM OF INDUCTION

Let S be a set of positive integers. If

(A) $1 \in S$, and

(B) $k \in S$ implies that $k + 1 \in S$,

then all the positive integers are in S.

You can think of the axiom of induction as a kind of "domino theory." If the first domino falls (Figure 1.8.1), and if each domino that falls causes the next one to fall, then, according to the axiom of induction, each domino will fall.

While we cannot prove that this axiom is valid (axioms are by their very nature assumptions and therefore not subject to proof), we can argue that it is *plausible*.

Let's assume that we have a set S that satisfies conditions (A) and (B). Now let's choose a positive integer m and "argue" that $m \in S$.

From (A) we know that $1 \in S$. Since $1 \in S$, we know from (B) that $1 + 1 \in S$, and thus that $(1 + 1) + 1 \in S$, and so on. Since m can be obtained from 1 by adding 1 successively $(m - 1)$ times, it *seems clear* that $m \in S$. ❑

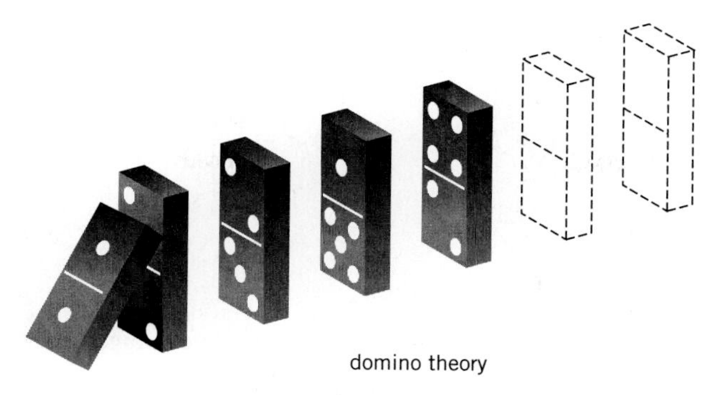

Figure 1.8.1

To prove that a given proposition is true for *all* positive integers n, we let S be the set of positive integers for which the proposition is true. We prove first that $1 \in S$, that is, that the proposition is true for $n = 1$. Next we assume that the proposition is true for some positive integer k, and prove that it is also true for $k + 1$, that is, $k \in S$ implies that $k + 1 \in S$. Then, by the axiom of induction, it follows that S contains the set of positive integers and so the proposition is true for all positive integers.

Example 1 Show that

$$1 + 2 + 3 + \cdots + n = \frac{n(n+1)}{2}$$

for all positive integers n.

SOLUTION Let S be the set of positive integers n for which

$$1 + 2 + 3 + \cdots + n = \frac{n(n+1)}{2}.$$

Then $1 \in S$ since

$$1 = \frac{1(1+1)}{2}.$$

Next, assume that the positive integer $k \in S$, that is, assume

$$1 + 2 + 3 + \cdots + k = \frac{k(k+1)}{2}.$$

Now consider the sum of the first $k + 1$ integers:

$$\begin{aligned}
1 + 2 + 3 + \cdots + k + (k+1) &= [1 + 2 + 3 + \cdots k] + (k+1) \\
&= \frac{k(k+1)}{2} + (k+1) \qquad \text{(by the induction hypothesis)} \\
&= \frac{k(k+1) + 2(k+1)}{2} \\
&= \frac{(k+1)(k+2)}{2},
\end{aligned}$$

and so $k + 1 \in S$. Thus, by the axiom of induction, we can conclude that all positive integers are in S, that is

$$1 + 2 + 3 + \cdots + n = \frac{n(n+1)}{2}$$

for all positive integers n. ❑

Example 2 Show that, if $x \geq -1$, then

$$(1 + x)^n \geq 1 + nx \qquad \text{for all positive integers } n.$$

SOLUTION Take $x \geq -1$ and let S be the set of positive integers n for which

$$(1 + x)^n \geq 1 + nx.$$

Since

$$(1 + x)^1 \geq 1 + 1 \cdot x,$$

you can see that $1 \in S$.

Assume now that $k \in S$. By the definition of S,

$$(1 + x)^k \geq 1 + kx.$$

Since

$$(1 + x)^{k+1} = (1 + x)^k(1 + x) \geq (1 + kx)(1 + x) \qquad \text{(explain)}$$

and

$$(1 + kx)(1 + x) = 1 + (k + 1)x + kx^2 \geq 1 + (k + 1)x,$$

it follows that

$$(1 + x)^{k+1} \geq 1 + (k + 1)x$$

and thus that $k + 1 \in S$.

We have shown that

$$1 \in S \qquad \text{and that} \qquad k \in S \quad \text{implies} \quad k + 1 \in S.$$

By the axiom of induction, all the positive integers are in S. ❑

EXERCISES 1.8

In Exercises 1–10, show that the statement holds for all positive integers n.

1. $2n \leq 2^n$.
2. $1 + 2n \leq 3^n$.
3. $n(n + 1)$ is divisible by 2.
   HINT: $(k + 1)(k + 2) = k(k + 1) + 2(k + 1)$.
4. $1 + 3 + 5 + \cdots + (2n - 1) = n^2$.
5. $1^2 + 2^2 + 3^2 + \cdots + n^2 = \frac{1}{6}n(n + 1)(2n + 1)$.
6. $1^3 + 2^3 + 3^3 + \cdots + n^3 = (1 + 2 + 3 + \cdots + n)^2$.
   HINT: Use Example 1.
7. $1^3 + 2^3 + \cdots + (n - 1)^3 < \frac{1}{4}n^4 < 1^3 + 2^3 + \cdots + n^3$.
8. $1^2 + 2^2 + \cdots + (n - 1)^2 < \frac{1}{3}n^3 < 1^2 + 2^2 + \cdots + n^2$.
9. $\dfrac{1}{\sqrt{1}} + \dfrac{1}{\sqrt{2}} + \dfrac{1}{\sqrt{3}} + \cdots + \dfrac{1}{\sqrt{n}} > \sqrt{n}$.
10. $\dfrac{1}{1 \cdot 2} + \dfrac{1}{2 \cdot 3} + \dfrac{1}{3 \cdot 4} + \cdots + \dfrac{1}{n(n + 1)} = \dfrac{n}{n + 1}$.
11. For what integers n is $3^{2n+1} + 2^{n+2}$ divisible by 7? Prove that your answer is correct.
12. For what integers n is $9^n - 8n - 1$ divisible by 64? Prove that your answer is correct.

13. Find a simplifying expression for the product

$$\left(1-\frac{1}{2}\right)\left(1-\frac{1}{3}\right)\cdots\left(1-\frac{1}{n}\right)$$

and verify its validity for all integers $n \geq 2$.

14. Find a simplifying expression for the product

$$\left(1-\frac{1}{2^2}\right)\left(1-\frac{1}{3^2}\right)\cdots\left(1-\frac{1}{n^2}\right)$$

and verify its validity for all integers $n \geq 2$.

15. Prove that an N-sided convex polygon has $\frac{1}{2}N(N-3)$ diagonals, $N > 3$.

16. Prove that the sum of the angles in an N-sided convex polygon is $(N-2)180°$, $N > 2$.

17. Prove that all sets with n elements have 2^n subsets. Count the empty set $\varnothing$ and the whole set as subsets.

■ CHAPTER HIGHLIGHTS

1.1 What Is Calculus?

Calculus as elementary mathematics (algebra, geometry, trigonometry enhanced by the limit process).

Sir Isaac Newton (p. 3)   Gottfried Wilhelm Leibniz (p. 3)

1.2 Notations and Formulas from Elementary Mathematics

sets
- notation (p. 4)

real numbers
- real line (p. 6)   order properties (p. 6)   absolute value (p. 7)
- intervals (p. 7)   boundedness (p. 8)

algebra
- factoring formulas (p. 8)
- polynomials and rational expressions (p. 8)

geometry
- elementary figures (pp. 10–12)

1.3 Inequalities

To solve an inequality is to find the set of numbers for which the inequality holds.

We can maintain an inequality by adding the same quantity to both sides, or by subtracting the same quantity from both sides, or by multiplying or dividing both sides by the same *positive* quantity. But if we multiply or divide both sides by a *negative* quantity, then the inequality is *reversed.*

Inequalities and Absolute Value

$$|x-c| < \delta \quad \text{iff} \quad c-\delta < x < c+\delta$$

$$0 < |x-c| < \delta \quad \text{iff} \quad c-\delta < x < c \quad \text{or} \quad c < x < c+\delta$$

$$|x| > \epsilon \quad \text{iff} \quad x > \epsilon \quad \text{or} \quad x < -\epsilon$$

$$|a+b| \leq |a| + |b|, \ \big||a| - |b|\big| \leq |a-b|$$

1.4 Coordinate Plane; Analytic Geometry

rectangular coordinates (p. 22)
distance and midpoint formulas (p. 22)
lines
- slope (p. 23)   intercepts (p. 23)   equations (p. 24)
- parallel and perpendicular lines (p. 24)   angle between two lines (p. 25)

distance between a point and a line (p. 28)

■ PROJECTS AND EXPLORATIONS USING TECHNOLOGY

To do these exercises you will need a graphics calculator or a computer with graphing capability. The majority of these problems are open-ended so different approaches may be used to solve them. You should be aware that different approaches can result in slight variations in the answers. Round your numerical answers to at least four decimal places. The rounding method that your calculator or computer uses also may cause variations in answers.

1.1 When economic analysts study such systems as the stock market or the seasonal sales of some product, they often try to smooth out short-term fluctuations in order to determine long-term trends. One method for doing this is to study a *moving average* that averages the current value with the value from the previous period. For example, if $f(x)$ denotes the sales of a certain product for the current period, then the function g, defined by

$$g(x) = \frac{f(x-1) + f(x)}{2}, \tag{1}$$

is such a moving average.

(a) Draw the graphs of the functions f and g for each of the following functions:

$$f(x) = x^4 - 3x^2 + 0.1x + 1 \qquad \text{and} \qquad f(x) = \frac{x^3 - 6}{x^4 + 3}$$

(b) For each function f in part (a), find and compare the domains and ranges of f and g.

(c) For each function f in part (a), find the points of intersection of the graphs of f and g. What characteristic do the points of intersection have?

A function f is *one-to-one* if $f(x_1) \neq f(x_2)$ whenever x_1, x_2 are in the domain of f and $x_1 \neq x_2$. Geometrically, a function f is one-to-one if it has the property that no horizontal line intersects its graph in more than one point.

(d) Based on the examples just given, would you conjecture that g will be one-to-one if f is one-to-one?

1.2 Sometimes it might be preferable to weight a moving average more heavily in favor of recent values. For example, the function

$$h(x) = \frac{f(x-1) + 2f(x)}{3} \tag{2}$$

gives twice as much weight to the current value versus the value for the preceding period.

(a) Let g and h be the moving averages defined by (1) and (2). Draw the graphs of f, g, and h for the two functions given in Exercise 1.

(b) For each of the functions in Exercise 1, compare the intersections of h and f to those of g and f.

(c) Answer questions (b) and (d) in Exercise 1 for h and the two functions given in Exercise 1.

(d) Suppose that the sales of a new product are modeled by the function

$$f(x) = 2x + 0.1x \sin(\pi x),$$

where $f(x)$ is measured in thousands of units and x is in months since the introduction of the product. Draw the graphs of f, g, and h.

(e) For a given problem, it is not always obvious which moving average is better. Explain why (1) might be better if the time period is "one day," while (2) might be better if the time period is "one year."

1.3 Another method for analyzing functions is to study the *change function*

$$c(x) = f(x) - f(x-1).$$

(a) Graph the functions $f(x)$ and $c(x)$ for each of the following functions:

$$f(x) = \sqrt{1+x} \qquad \text{and} \qquad f(x) = |x^2 - 7x + 2|.$$

(b) What is the meaning in terms of f if, for all x, $c(x) > 0$? $c(x) < 0$? $c(x) = 0$?

(c) Explain why the change function c may be one-to-one even when f is not one-to-one.

(d) Give an example of a function f such that f is even and its change function c is odd.

1.4 Suppose that some oil has been spilled in water and has formed a circular oil slick. Suppose, in addition, that one minute after the spill, the radius of the slick is 2 meters and 3 minutes after the spill, the radius is 6 meters.

(a) Express the radius r of the spill as a function of time t if it is observed that the radius is increasing at a constant rate. Was the radius 0 at time $t = 0$?

(b) Using composition of functions, express the area A and circumference C of the spill as functions of time.

(c) Apply the change function defined in Exercise 3 to each of the functions r, A, and C. What do they tell us about the spill?

(d) Suppose that the problem is modified to consider a gas that has been released in the atmosphere and has a spherical shape. Use the given data to answer parts (a)–(c) for the radius r, the volume V, and the surface area S of the sphere.

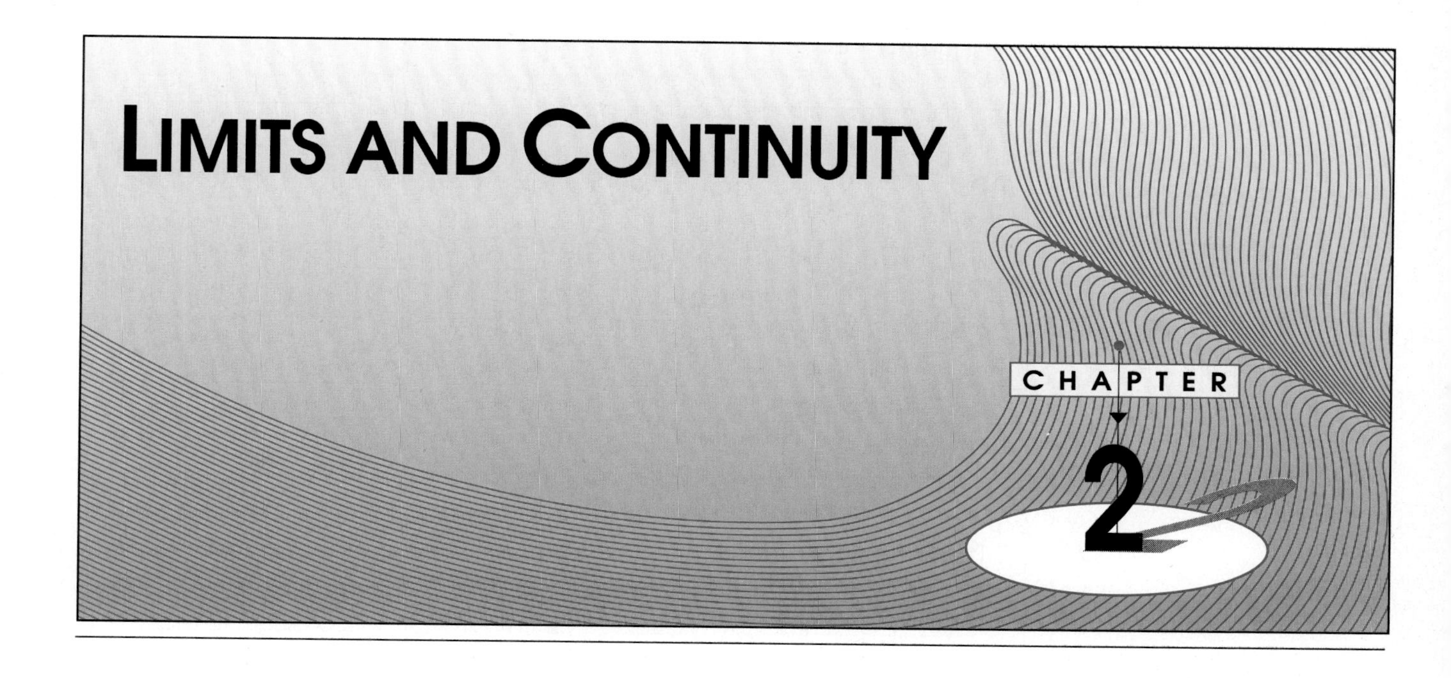

LIMITS AND CONTINUITY

2.1 THE IDEA OF LIMIT

Introduction

We could begin by saying that limits are important in calculus, but that would be a major understatement. *Without limits calculus simply does not exist. Every single notion of calculus is a limit in one sense or another.*

What is instantaneous velocity? It is the limit of average velocities.
What is the slope of a curve? It is the limit of slopes of secant lines. See Figure 2.1.1.
What is the length of a curve? It is the limit of the lengths of polygonal paths.
What is the sum of an infinite series? It is the limit of finite sums.
What is the area of a region bounded by a curve? It is the limit of areas of regions bounded by line segments. See Figure 2.1.2

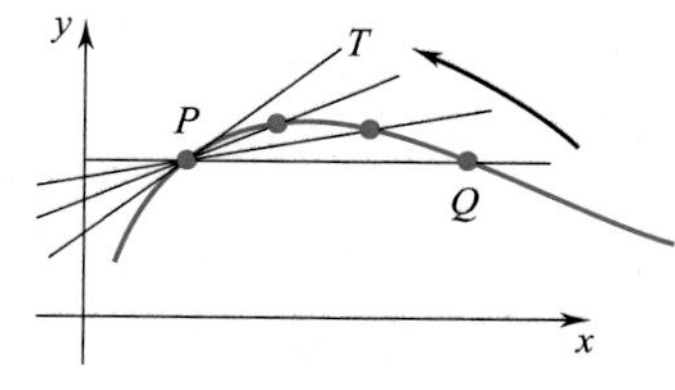

Figure 2.1.1

The Idea of Limit

We begin with a number L and a function f defined *near the number c but not necessarily defined at c itself.* A rough translation of

$$\lim_{x \to c} f(x) = L \qquad \textit{(the limit of } f(x) \textit{ as } x \textit{ tends to } c \textit{ is } L)$$

might read

as x approaches c, f(x) approaches L

or, equivalently,

when x is close to c but different from c, f(x) is close to L.

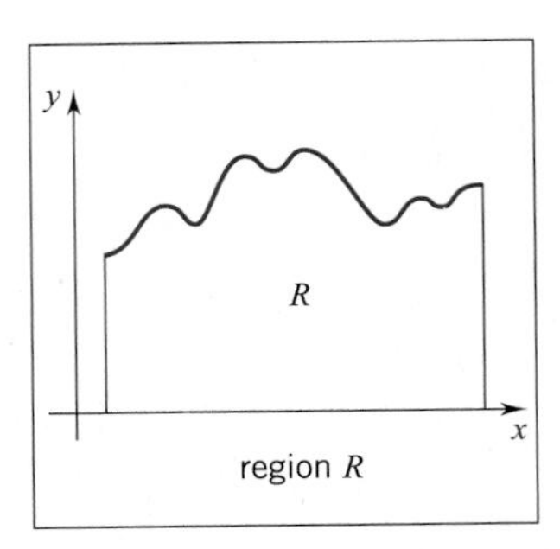

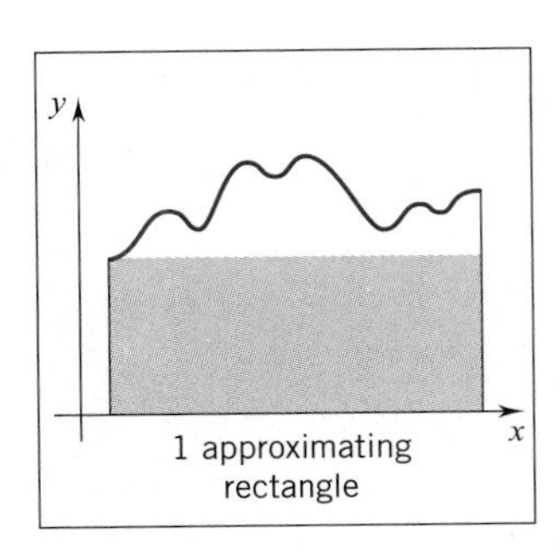

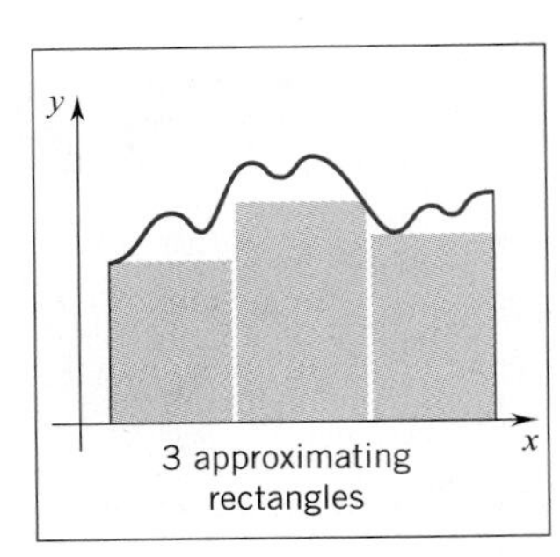

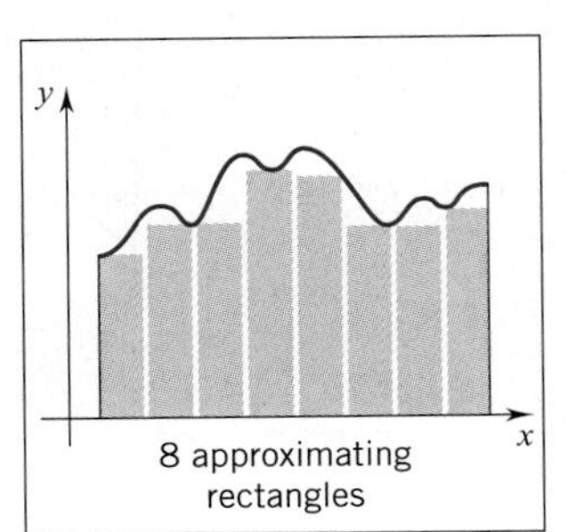

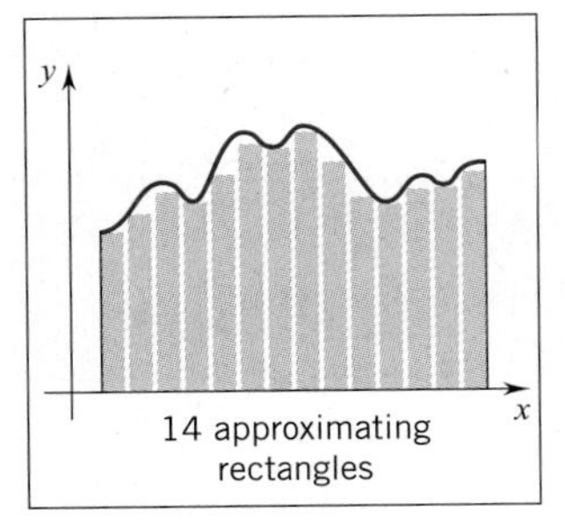

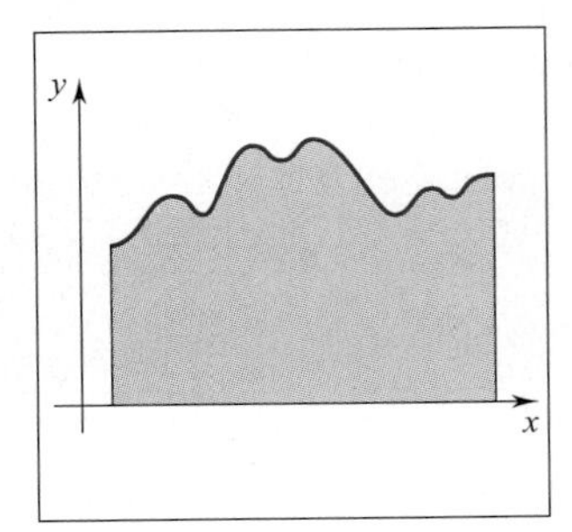

Figure 2.1.2

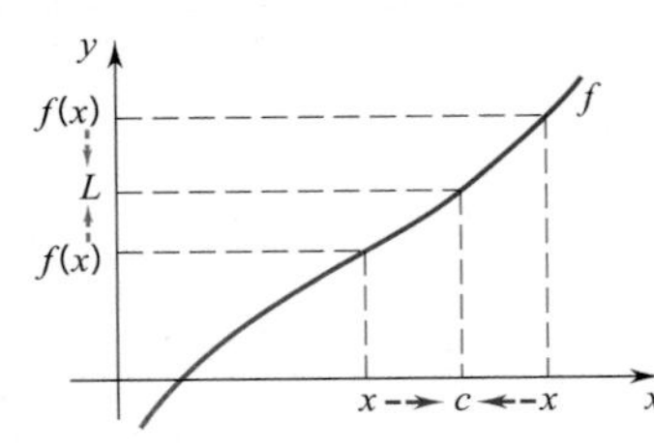

Figure 2.1.3

We illustrate the idea in Figure 2.1.3. The curve represents the graph of a function f. The number c appears on the x-axis, the limit L on the y-axis. As x approaches c along the x-axis, $f(x)$ approaches L along the y-axis.

In taking the limit as x approaches c, it does not matter whether f is defined at c and if so, how it is defined there. The only thing that matters is the values that f takes on when x is *near* c. For example, in Figure 2.1.4 the graph of f is a broken curve defined peculiarly at c, and yet

$$\lim_{x \to c} f(x) = L$$

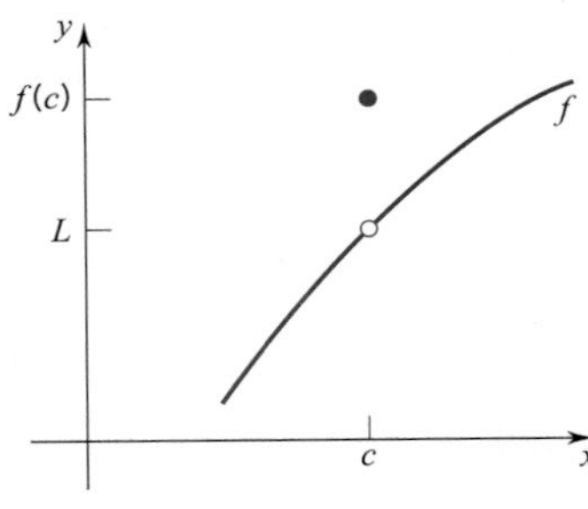
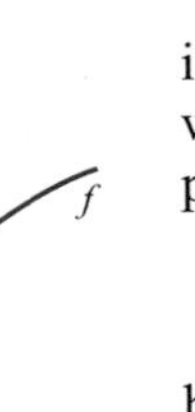

Figure 2.1.4

because, as suggested in Figure 2.1.5,

as x approaches c, $f(x)$ approaches L.

Numbers x near c fall into two natural categories: those that lie to the left of c and those that lie to the right of c. We write

$$\lim_{x \to c^-} f(x) = L \quad \text{(the left-hand limit of } f(x) \text{ as } x \text{ tends to } c \text{ is } L)$$

to indicate that

as x approaches c from the left, $f(x)$ approaches L.

Similarly,

$$\lim_{x \to c^+} f(x) = L \quad \text{(the right-hand limit of } f(x) \text{ as } x \text{ tends to } c \text{ is } L)$$

indicates that

as x approaches c from the right, $f(x)$ approaches L.†

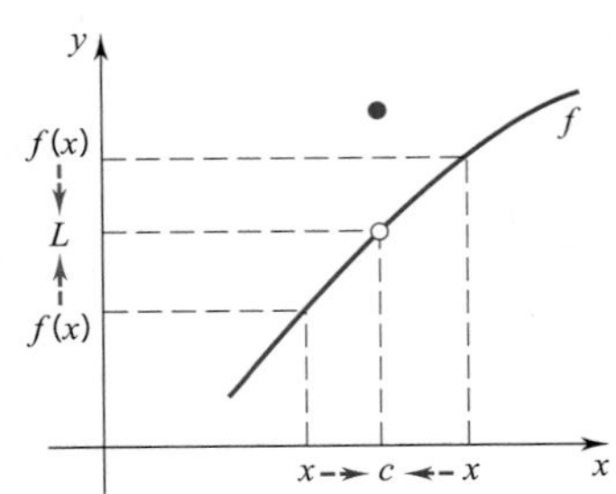

Figure 2.1.5

† The left-hand limit is sometimes written $\lim_{x \uparrow c} f(x)$ and the right-hand limit, $\lim_{x \downarrow c} f(x)$.

For an example, see the graph of the function f in Figure 2.1.6. As x approaches 5 from the left, $f(x)$ approaches 2. Thus,

$$\lim_{x \to 5^-} f(x) = 2.$$

As x approaches 5 from the right, $f(x)$ approaches 4, so

$$\lim_{x \to 5^+} f(x) = 4.$$

Note that $\lim_{x \to 5} f(x)$ does not exist since it is not true that there is a single number L with the property that $f(x)$ is close to L *whenever* x is close to 5 ($f(x)$ is close to 2 when x is close to 5 and is less than 5; $f(x)$ is close to 4 when x is close to 5 and is greater than 5).

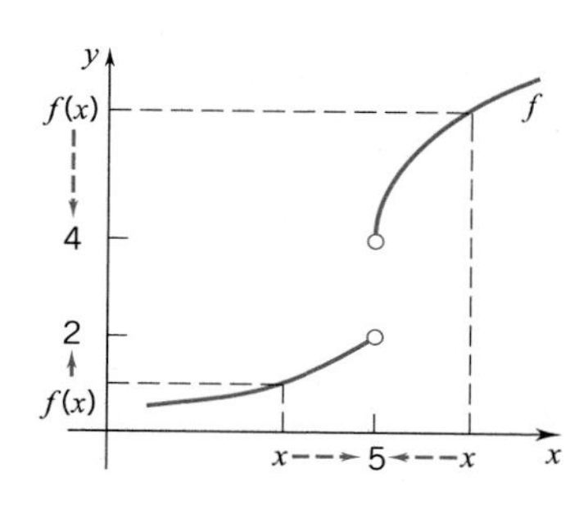

Figure 2.1.6

The left- and right-hand limits are called *one-sided limits.* The relationship between the *limit of f as x approaches c* and the two *one-sided limits of f as x approaches c* should be intuitively clear, namely

$$\lim_{x \to c} f(x) = L$$

if and only if both

$$\lim_{x \to c^-} f(x) = L \qquad \textit{and} \qquad \lim_{x \to c^+} f(x) = L.$$

Example 1 For the function f graphed in Figure 2.1.7

$$\lim_{x \to (-2)^-} f(x) = 5 \qquad \text{and} \qquad \lim_{x \to (-2)^+} f(x) = 5$$

and therefore

$$\lim_{x \to -2} f(x) = 5.$$

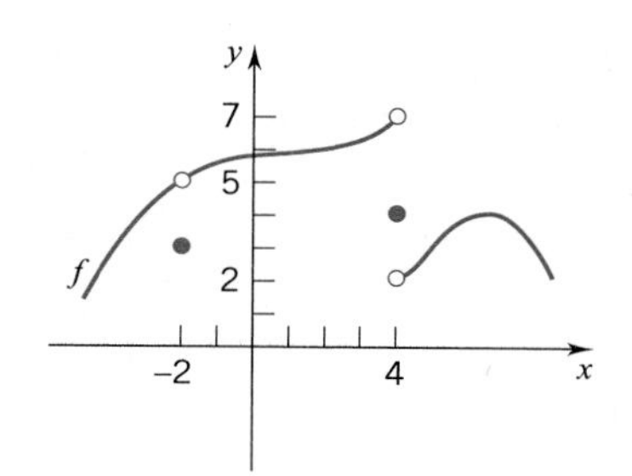

Figure 2.1.7

For the limit, it does not matter that $f(-2) = 3$.

Examining the graph of f near $x = 4$, we find that

$$\lim_{x \to 4^-} f(x) = 7 \qquad \text{whereas} \qquad \lim_{x \to 4^+} f(x) = 2.$$

Since these one-sided limits are different,

$$\lim_{x \to 4} f(x) \quad \text{does not exist.} \quad \square$$

Example 2 Consider the function f graphed in Figure 2.1.8. As x approaches 3 from either side, $f(x)$ becomes arbitrarily large.† Becoming arbitrarily large, $f(x)$ does not approach any fixed number L. Therefore

$$\lim_{x \to 3} f(x) \quad \text{does not exist.}$$

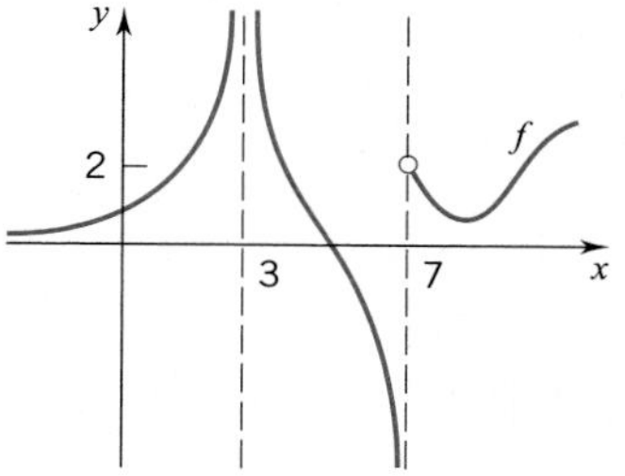

Now let's focus on what happens near 7. As x approaches 7 from the left, $f(x)$ becomes arbitrarily large negative (i.e., less than any preassigned number).†† Under these circumstances $f(x)$ can't possibly approach a fixed number. Thus

$$\lim_{x \to 7^-} f(x) \quad \text{does not exist.}$$

Figure 2.1.8

† This particular behavior is symbolized by writing $f(x) \to \infty$ as $x \to 3$; the line $x = 3$ is a *vertical asymptote* of the function. Vertical and horizontal asymptotes are treated in Section 4.7.

†† This behavior is symbolized by writing $f(x) \to -\infty$ as $x \to 7^-$; the line $x = 7$ is a vertical asymptote of f.

Since the left-hand limit at $x = 7$ does not exist,

$$\lim_{x \to 7} f(x) \quad \text{does not exist.}$$

(Note that the right-hand limit of f at $x = 7$ does exist; $\lim_{x \to 7^+} f(x) = 2$.) ❑

Example 3 Let f be the function defined by $f(x) = \dfrac{|x|}{x}$ for $x \neq 0$. This function can also be written equivalently as

$$f(x) = \begin{cases} 1, & \text{if } x > 0 \\ -1, & \text{if } x < 0. \end{cases} \qquad \text{(Figure 2.1.9)}$$

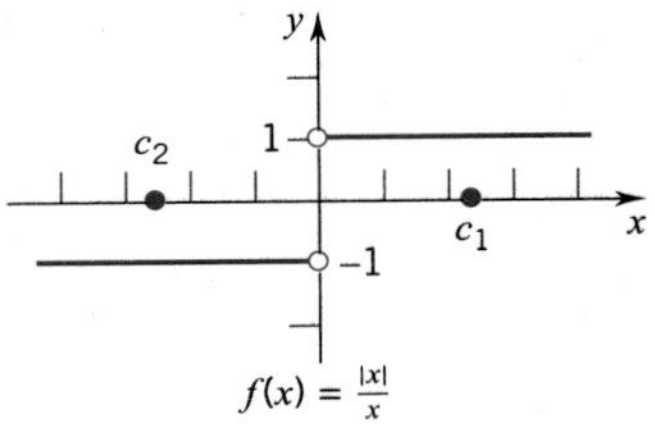

Figure 2.1.9

We consider $\lim_{x \to c} f(x)$ for several values of c.

For $c = 0$,

$$\lim_{x \to 0^-} f(x) = \lim_{x \to 0^-} (-1) = -1$$

and

$$\lim_{x \to 0^+} f(x) = \lim_{x \to 0^+} (1) = 1.$$

Thus,

$$\lim_{x \to 0} \frac{|x|}{x} \quad \text{does not exist.}$$

If $c = c_1$ is any positive number, then $f(x)$ "is close to 1" (in fact, $f(x) = 1$) when x is close to c_1. Therefore, when $c_1 > 0$

$$\lim_{x \to c_1} \frac{|x|}{x} = 1.$$

A similar analysis shows that when $c = c_2$ is a negative number, then

$$\lim_{x \to c_2} \frac{|x|}{x} = -1. \quad ❑$$

Our next example is somewhat exotic. It is an adaptation of an example known as the Dirichlet function (P. G. Lejeune Dirichlet, 1805–1859).

Example 4 Let f be the function defined as follows:

$$f(x) = \begin{cases} 7, & x \text{ rational} \\ 4, & x \text{ irrational.} \end{cases} \qquad \text{(Figure 2.1.10)}$$

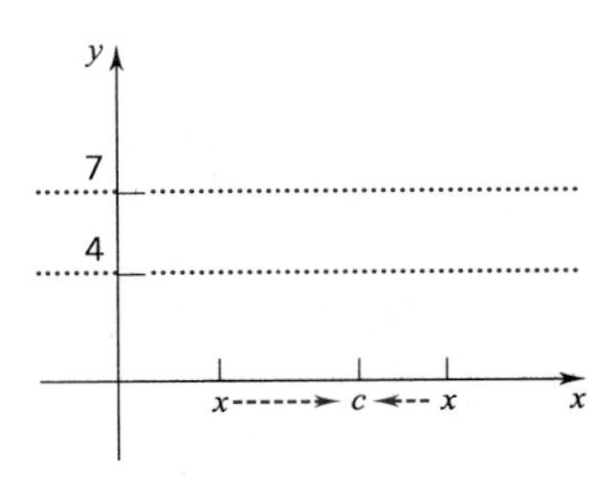

Figure 2.1.10

Now let c be any real number. As x approaches c, x passes through both rational and irrational numbers. As this happens, $f(x)$ jumps wildly back and forth between 7 and 4, and thus cannot stay close to any fixed number L. Therefore, $\lim_{x \to c} f(x)$ does not exist. ❑

We go on with examples. Where a graph has not been provided, you may find it useful to sketch one.

Example 5

$$\lim_{x \to 3} (2x + 5) = 11.$$

As x approaches 3, $2x$ approaches 6, and $2x + 5$ approaches 11. ❑

Example 6

$$\lim_{x \to 2} (2x^2 - 3x + 5) = 7.$$

When x is close to 2, $2x^2$ is close to 8, $3x$ is close to 6, and $2x^2 - 3x + 5$ is close to $8 - 6 + 5 = 7$. ❑

Example 7

$$\lim_{x \to 3} \frac{1}{x - 1} = \frac{1}{2}.$$

As x approaches 3, $x - 1$ approaches 2, and $1/(x - 1)$ approaches $\frac{1}{2}$. ❑

Example 8

$$\lim_{x \to 2} \frac{1}{x - 2} \quad \text{does not exist.}$$

Here the function $f(x) = 1/(x - 2)$ is not defined at $x = 2$. However, if x is close to 2 and is greater than 2, then $1/(x - 2)$ is a large positive number ($f(x) \to \infty$ as $x \to 2^+$); if x is close to 2 and is less than 2, then $1/(x - 2)$ is a large negative number ($f(x) \to -\infty$ as $x \to 2^-$). See the table of values and Figure 2.1.11.

x	1.5	1.9	1.99	1.999	2	2.001	2.01	2.1	2.5
$f(x)$	−2	−10	−100	−1000		1000	100	10	2

❑

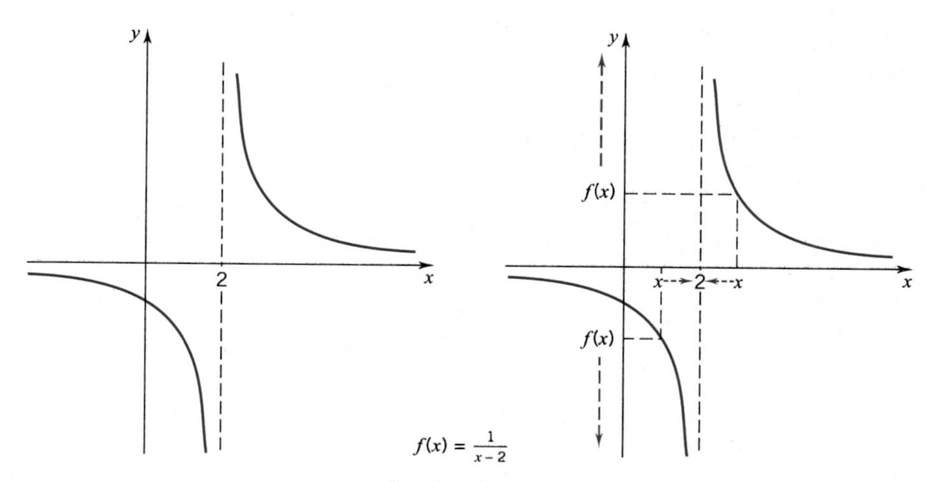

Figure 2.1.11

Before going on to the next examples, remember that, in taking the limit as x approaches a given number c, it does not matter whether the function is defined at the number c and if so, how it is defined there. The only thing that matters is how the function is defined *near* the number c.

Example 9

$$\lim_{x\to 4} \frac{x^2 - 2x - 8}{x - 4} = 6.$$

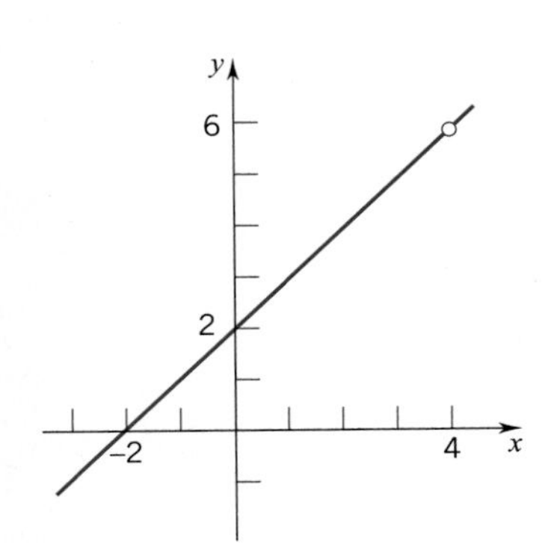

Figure 2.1.12

At $x = 4$, the function is undefined: both numerator and denominator are 0 and 0/0 makes no sense. But that doesn't matter. For all $x \neq 4$, and therefore *for all x near* 4,

$$\frac{x^2 - 2x - 8}{x - 4} = \frac{(x + 2)(x - 4)}{x - 4} = x + 2.$$

Therefore,

$$\lim_{x\to 4} \frac{x^2 - 2x - 8}{x - 4} = \lim_{x\to 4} (x + 2) = 6.$$

The graph of $f(x) = (x^2 - 2x - 8)/(x - 4)$ is shown in Figure 2.1.12. ❑

Example 10

$$\lim_{x\to -3} \frac{2x + 6}{x + 3} = 2.$$

At $x = -3$, the function is undefined: both numerator and denominator are 0. But, as we said before, that doesn't matter. For all $x \neq -3$, and therefore for all x near -3,

$$\frac{2x + 6}{x + 3} = \frac{2(x + 3)}{x + 3} = 2 \qquad \text{so that} \qquad \lim_{x\to -3} \frac{2x + 6}{x + 3} = \lim_{x\to -3} 2 = 2.$$

Sketch the graph of $f(x) = \dfrac{2x + 6}{x + 3}$. ❑

Example 11

$$\lim_{x\to 2} \frac{x - 2}{x^2 - 4x + 4} \quad \text{does not exist.}$$

This result does not follow from the fact that the function is not defined at 2, but rather from the fact that for all $x \neq 2$, and therefore for all x near 2,

$$\frac{x - 2}{x^2 - 4x + 4} = \frac{x - 2}{(x - 2)^2} = \frac{1}{x - 2},$$

and, as you saw before,

$$\lim_{x\to 2} \frac{1}{x - 2} \quad \text{does not exist.}$$ ❑

Example 12

$$\text{If } f(x) = \begin{cases} 3x - 1, & x \neq 2 \\ 45, & x = 2, \end{cases} \qquad \text{then} \quad \lim_{x\to 2} f(x) = 5.$$

It doesn't matter to us that $f(2) = 45$. For $x \neq 2$, and thus for all x near 2,

$$f(x) = 3x - 1 \qquad \text{so that} \qquad \lim_{x\to 2} f(x) = \lim_{x\to 2} (3x - 1) = 5.$$ ❑

Example 13 Let

$$f(x) = \begin{cases} 1 - x^2, & x < 1 \\ 2x, & x > 1. \end{cases}$$

Then

$$\lim_{x \to 1} f(x) \quad \text{does not exist}$$

and

$$\lim_{x \to 1.05} f(x) = 2.10.$$

The first limit does not exist since the one-sided limits are different:

$$\lim_{x \to 1^-} f(x) = \lim_{x \to 1^-} (1 - x^2) = 0 \qquad \text{and} \qquad \lim_{x \to 1^+} f(x) = \lim_{x \to 1^+} (2x) = 2.$$

The second limit does exist because for values of x sufficiently near 1.05, the values of the function are computed using the rule $2x$ whether x is to the left or the right of 1.05, and

$$\lim_{x \to 1.05} 2x = 2.10. \quad \square$$

The preceding examples illustrate two intuitive methods for evaluating the limit of a function f at a number c: (1) using the graph of f (see Examples 1–4), and (2) "analyzing" the behavior of f for values of x close to c (Examples 5–13). We conclude this section with a third method which might be termed *numerical* or *experimental*. This method can be used when the intuitive approaches are difficult (or perhaps impossible) to apply. Typical numerical problems will require a calculator or a computer; calculators/computers with graphing capabilities can be used to support numerical results.

Example 14 Let $f(x) = (\sin x)/x$. If we try to evaluate f at 0, we get the meaningless ratio 0/0. However, f is defined for all $x \neq 0$, and so we can consider

$$\lim_{x \to 0} \frac{\sin x}{x}.$$

We select a sequence of numbers which approaches 0 from the right, and a sequence of numbers which approaches 0 from the left. Using a calculator, we evaluate f at these numbers. The results are tabulated in Table 2.1.1.

■ Table 2.1.1

(Left side)		(Right side)	
x (radians)	$\frac{\sin x}{x}$	x (radians)	$\frac{\sin x}{x}$
−1	0.84147	1	0.84147
−0.5	0.95885	0.5	0.95885
−0.1	0.99833	0.1	0.99833
−0.01	0.99998	0.01	0.99998
−0.001	0.99999	0.001	0.99999

Based on these calculations, it appears that

$$\lim_{x \to 0^+} \frac{\sin x}{x} = 1 \qquad \text{and} \qquad \lim_{x \to 0^-} \frac{\sin x}{x} = 1.$$

Thus, we are led to conclude

$$\lim_{x \to 0} \frac{\sin x}{x} = 1.$$

The graphs shown in Figure 2.1.13 support this conclusion. First, we graphed the function with the x-interval $-10 \leq x \leq 10$ (part a). Then we zoomed in to obtain the graphs in 2.1.13(b) and 2.1.13(c). The fact that this limit is 1 is established rigorously in Section 2.5. ❑

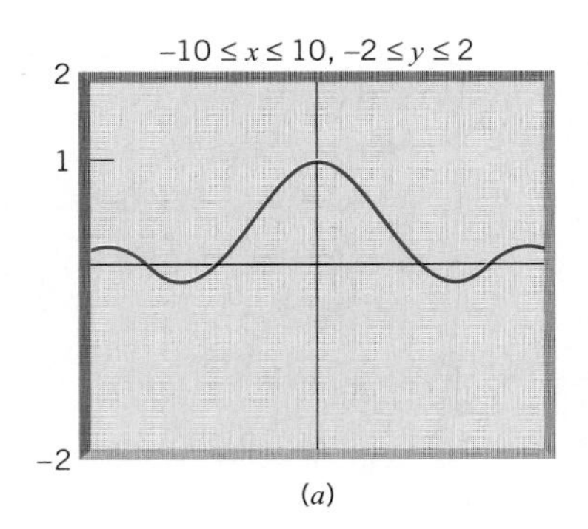

(a)

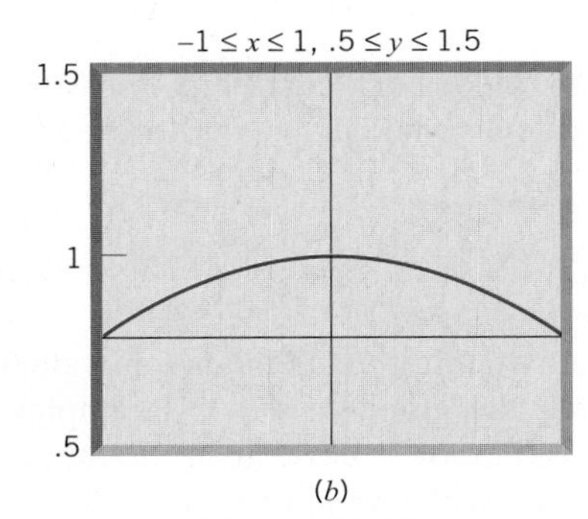

(b)

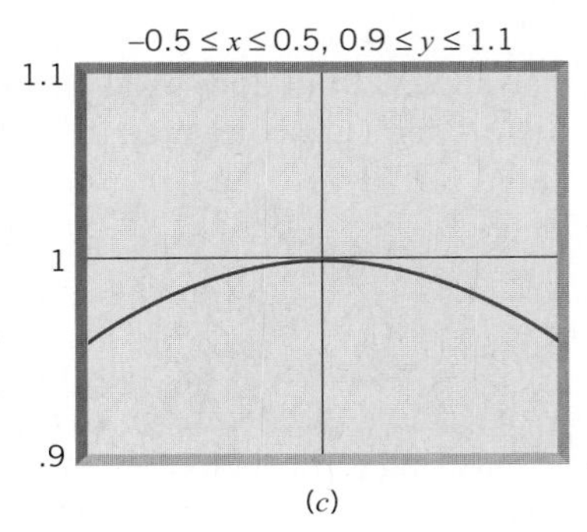

(c)

Figure 2.1.13

Example 15 In the function $f(x) = 2^x$, the independent variable x appears as an exponent. This is an example of an *exponential function.* At this stage, an expression like 2^x is defined only when x is a rational number; *logarithms* are needed to define irrational powers.† However, we can use a limit to try to estimate a number like $2^{\sqrt{2}}$. We take a sequence of rational numbers $x_1, x_2, x_3, \ldots$ which approaches $\sqrt{2}$ and evaluate $2^{x_1}, 2^{x_2}, 2^{x_3}, \ldots$ Then $2^{\sqrt{2}}$ is defined to be the "limit," if it exists. Recall that

$$\sqrt{2} = 1.414213562 \ldots .$$

Using the values $x = 1.4, 1.41, 1.414, \ldots$, we get the results shown in Table 2.1.2.

■ **Table 2.1.2**

x	1	1.4	1.41	1.414	1.4142	1.41421	1.414213
2^x	2	2.639016	2.657372	2.664750	2.665119	2.665138	2.665143

It appears, therefore, that $2^{\sqrt{2}} = \lim_{x \to (\sqrt{2})^-} 2^x \cong 2.6651$ (to four decimal places). The value of $2^{\sqrt{2}}$ given by a hand calculator is 2.665144143. The graph of $f(x) = 2^x$,

† Logarithmic and exponential functions are defined and studied in Chapter 7.

$0 \le x \le 2$, shown in Figure 2.14(a), was obtained on a graphics calculator. The estimates found in Table 2.1.2 can be confirmed by repeatedly zooming in on $\sqrt{2}$. See Figure 2.14(b). ❑

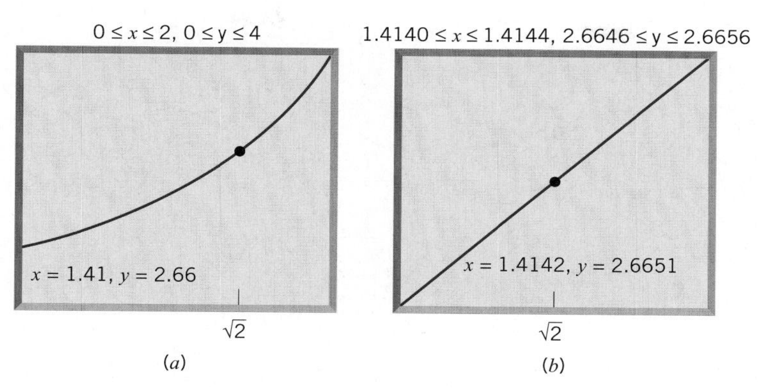

Figure 2.1.14

If you have found all this very imprecise, you are absolutely right. It is imprecise, but it need not remain so. One of the great triumphs of calculus has been its capacity to formulate limit statements with precision, but for this you have to wait until Section 2.2.

EXERCISES 2.1

In Exercises 1–12 you are given a number c and the graph of a function f. Use the graph of f to find

(a) $\lim_{x \to c^-} f(x)$. (b) $\lim_{x \to c^+} f(x)$. (c) $\lim_{x \to c} f(x)$. (d) $f(c)$

1. $c = 2$.

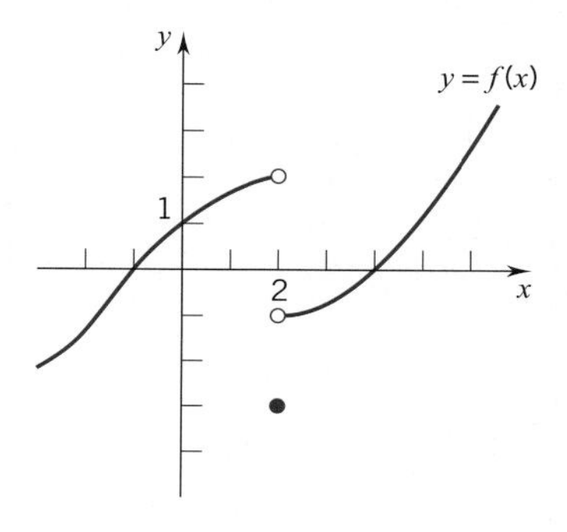

2. $c = 3$.

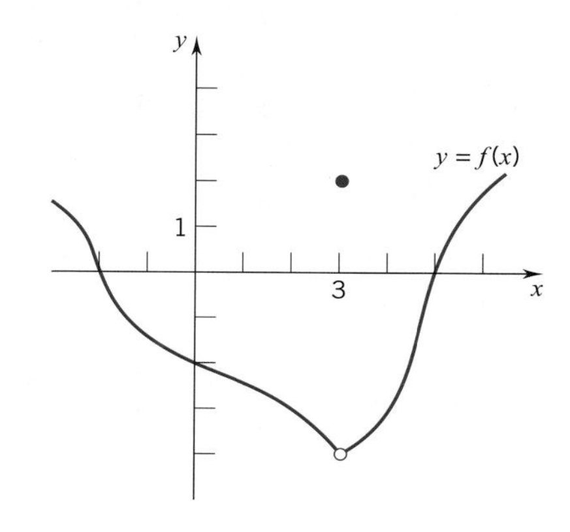

3. $c = 3$.

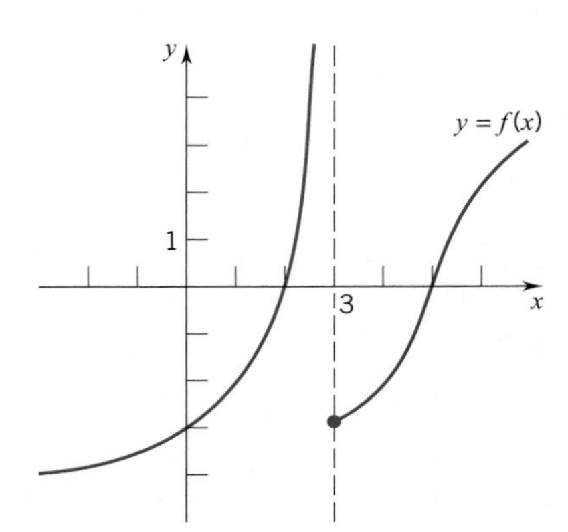

4. $c = 4$.

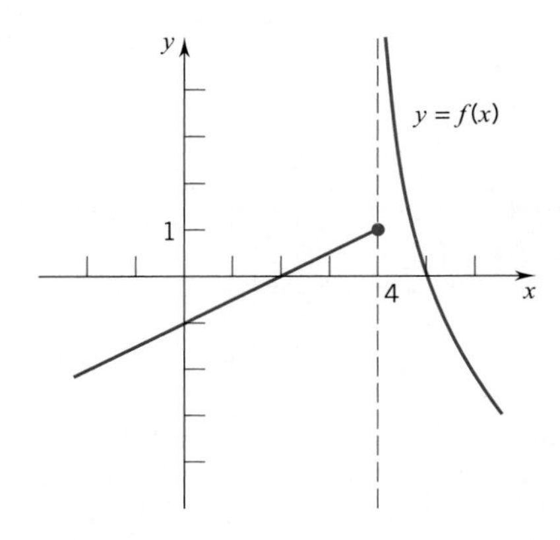

5. $c = -2$.

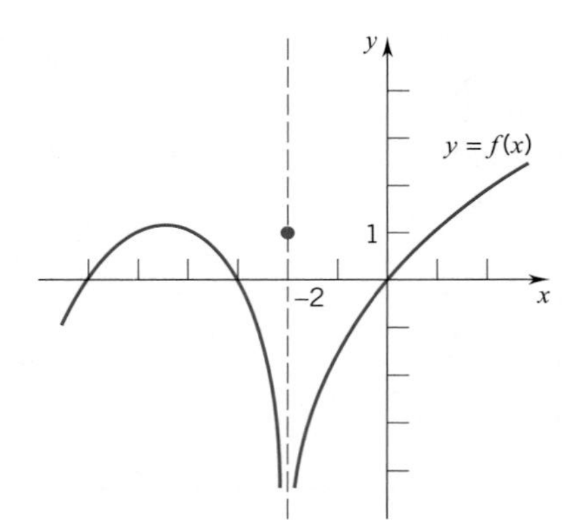

6. $c = 1$.

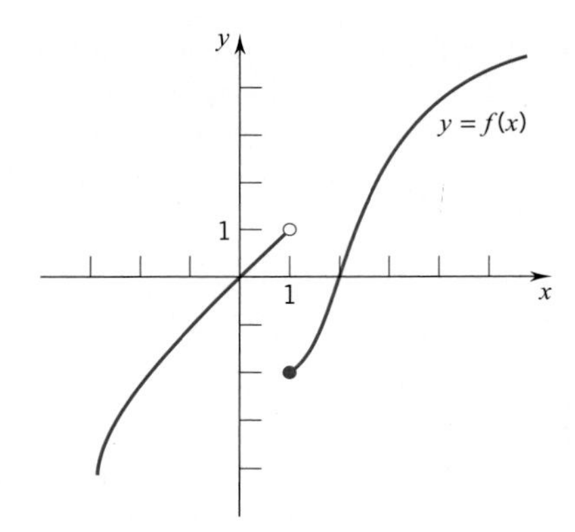

7. $c = 1$.

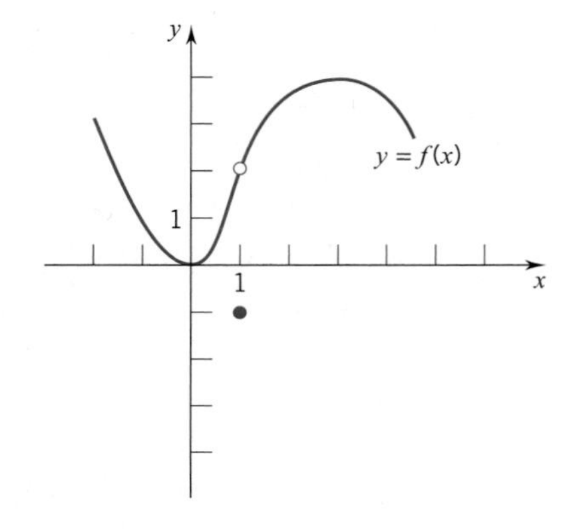

8. $c = -1$.

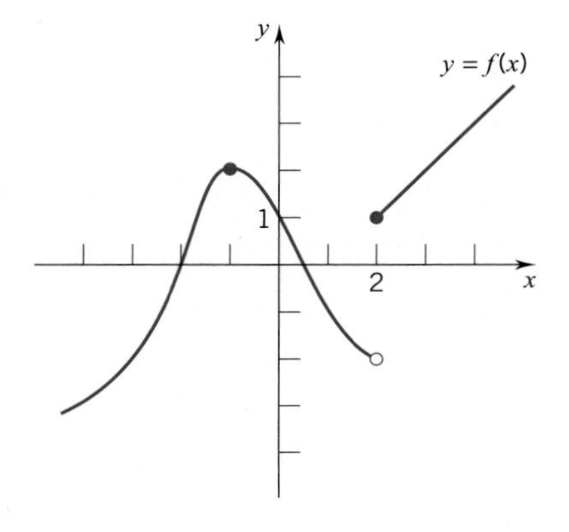

9. $c = 3$.

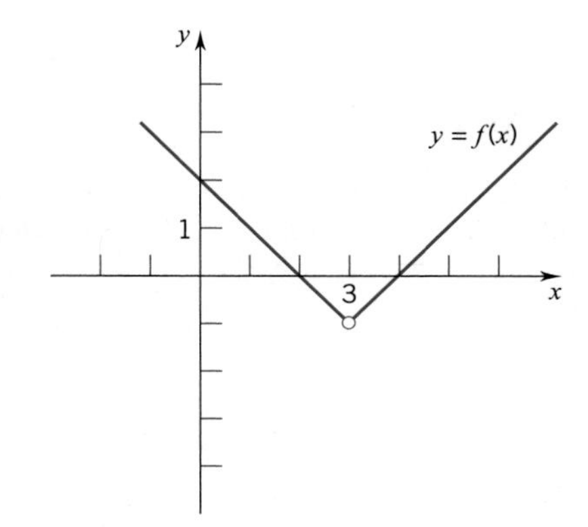

10. $c = 3$.

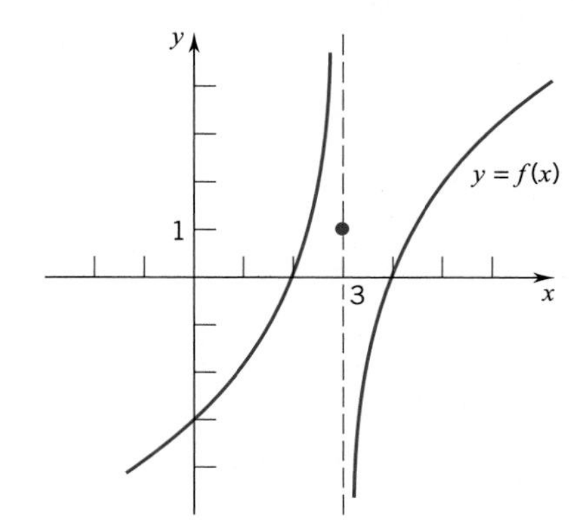

11. $c = 2$.

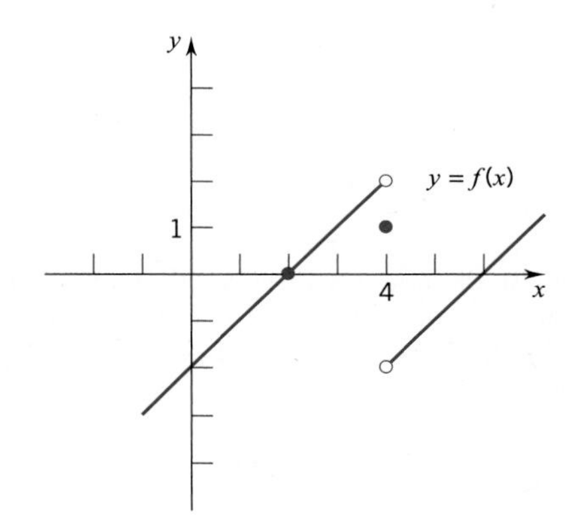

12. $c = -2$.

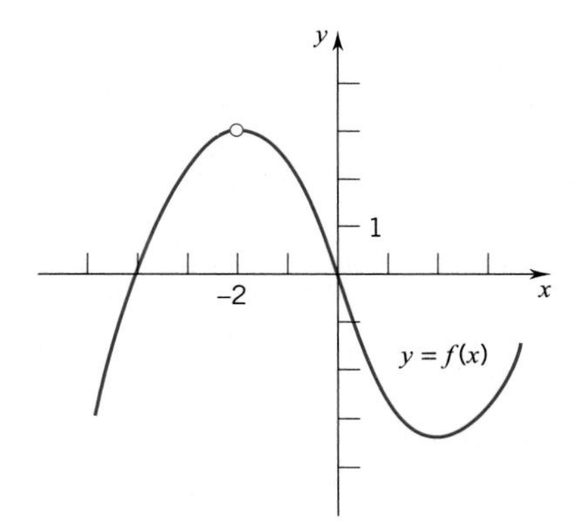

In Exercises 13 and 14, state the values of c for which $\lim_{x \to c} f(x)$ does not exist.

13.

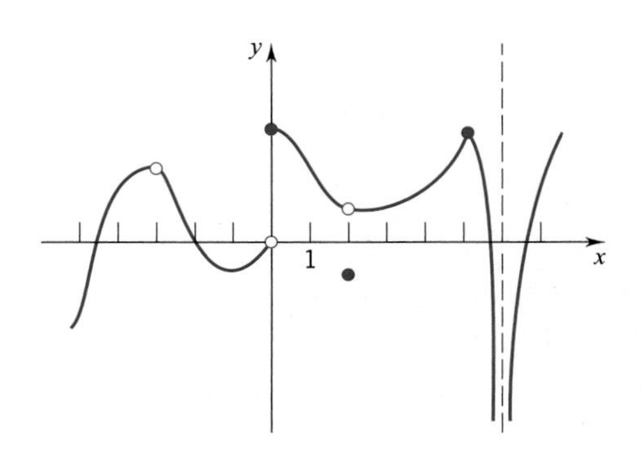

14.

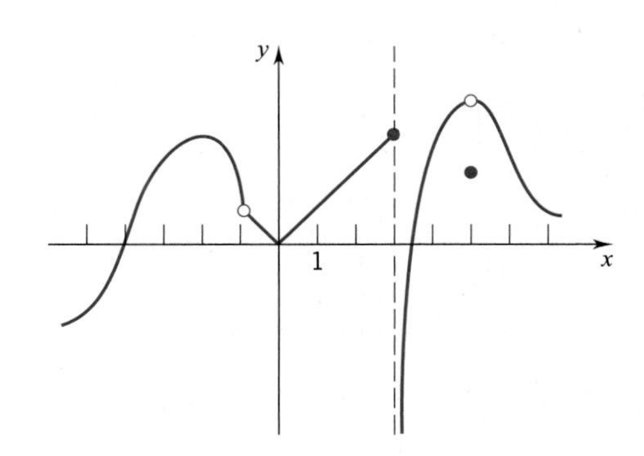

In Exercises 15–50, decide on intuitive grounds whether or not the indicated limit exists, and evaluate the limit if it does exist.

15. $\lim_{x\to 0}(2x-1)$.

16. $\lim_{x\to 1}(2-5x)$.

17. $\lim_{x\to -2} x^2$.

18. $\lim_{x\to 4}\sqrt{x}$.

19. $\lim_{x\to -3}(|x|-2)$.

20. $\lim_{x\to 0}\dfrac{1}{|x|}$.

21. $\lim_{x\to 1}\dfrac{3}{x+1}$.

22. $\lim_{x\to -1}\dfrac{4}{x+1}$.

23. $\lim_{x\to -1}\dfrac{2}{x+1}$.

24. $\lim_{x\to 2}\dfrac{1}{3x-6}$.

25. $\lim_{x\to 3}\dfrac{2x-6}{x-3}$.

26. $\lim_{x\to 3}\dfrac{x^2-6x+9}{x-3}$.

27. $\lim_{x\to 3}\dfrac{x-3}{x^2-6x+9}$.

28. $\lim_{x\to 2}\dfrac{x^2-3x+2}{x-2}$.

29. $\lim_{x\to 2}\dfrac{x-2}{x^2-3x+2}$.

30. $\lim_{x\to 1}\dfrac{x-2}{x^2-3x+2}$.

31. $\lim_{x\to 0}\left(x+\dfrac{1}{x}\right)$.

32. $\lim_{x\to 1}\left(x+\dfrac{1}{x}\right)$.

33. $\lim_{x\to 0}\dfrac{2x-5x^2}{x}$.

34. $\lim_{x\to 3}\dfrac{x-3}{6-2x}$.

35. $\lim_{x\to 1}\dfrac{x^2-1}{x-1}$.

36. $\lim_{x\to 1}\dfrac{x^3-1}{x-1}$.

37. $\lim_{x\to 1}\dfrac{x^3-1}{x+1}$.

38. $\lim_{x\to 1}\dfrac{x^2+1}{x^2-1}$.

39. $\lim_{x\to 0} f(x);\quad f(x)=\begin{cases}1, & x\neq 0\\ 3, & x=0.\end{cases}$

40. $\lim_{x\to 1} f(x);\quad f(x)=\begin{cases}3x, & x<1\\ 3, & x>1.\end{cases}$

41. $\lim_{x\to 4} f(x);\quad f(x)=\begin{cases}x^2, & x\neq 4\\ 0, & x=4.\end{cases}$

42. $\lim_{x\to 0} f(x);\quad f(x)=\begin{cases}-x^2, & x<0\\ x^2, & x>0.\end{cases}$

43. $\lim_{x\to 0} f(x);\quad f(x)=\begin{cases}x^2, & x<0\\ 1+x, & x>0.\end{cases}$

44. $\lim_{x\to 1} f(x);\quad f(x)=\begin{cases}2x, & x<1\\ x^2+1, & x>1.\end{cases}$

45. $\lim_{x\to 0} f(x);\quad f(x)=\begin{cases}2, & x \text{ rational}\\ -2, & x \text{ irrational}.\end{cases}$

46. $\lim_{x\to 1} f(x);\quad f(x)=\begin{cases}2x, & x \text{ rational}\\ 2, & x \text{ irrational}.\end{cases}$

47. $\lim_{x\to 2} f(x);\quad f(x)=\begin{cases}3x, & x<1\\ x+2, & x\geq 1.\end{cases}$

48. $\lim_{x\to 0} f(x);\quad f(x)=\begin{cases}2x, & x\leq 1\\ x+1, & x>1.\end{cases}$

49. $\lim_{x\to 1}\dfrac{\sqrt{x^2+1}-\sqrt{2}}{x-1}$.

50. $\lim_{x\to 5}\dfrac{\sqrt{x^2+5}-\sqrt{30}}{x-5}$.

Consider function f defined on an interval I and a point $a\in I$. Let $h\neq 0$ be a number such that $a+h\in I$. The line determined by the points $(a,f(a))$ and $(a+h,f(a+h))$ is called a *secant line* (see the figure).

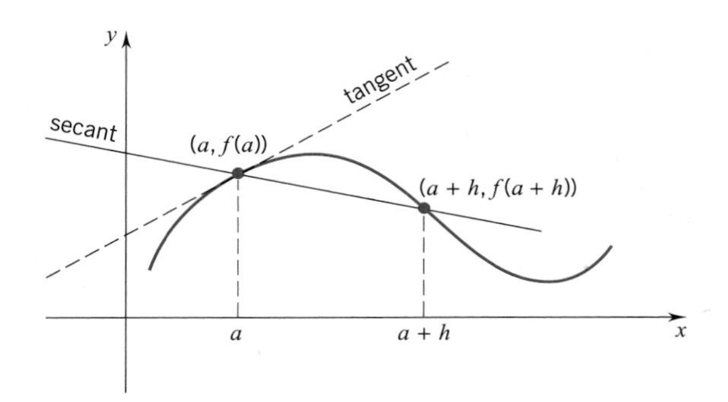

The slope of the secant line is given by

$$m_{\text{sec}}=\frac{f(a+h)-f(a)}{h}.$$

If we let h approach 0, then the secant line will approach the dashed line as a limiting position. The *tangent line* to the graph of f at the point $(a,f(a))$ is defined to be the line through $(a,f(a))$ with slope

$$\lim_{h\to 0}\frac{f(a+h)-f(a)}{h}\qquad\text{(provided this limit exists).}$$

In Exercises 51–57, calculate $(f(a+h)-f(a))/h$, determine whether $\lim_{h\to 0}(f(a+h)-f(a))/h$ exists, and give an equation for the tangent line to the graph of f at $(a,f(a))$ when the limit does exist.

51. $f(x)=x^2,\quad a=2$.

52. $f(x)=x^3+1,\quad a=1$.

53. $f(x)=1-2x+x^2,\quad a=-1$.

54. $f(x)=1/x,\quad a=2$.

55. $f(x)=\sqrt{x},\quad a=1$.
HINT: Rationalize the numerator.

56. $f(x)=|x|,\quad a=0$.
HINT: Consider the cases $h>0$ and $h<0$ separately.

57. $f(x)=\sqrt[3]{x},\quad a=0$.

Exercises 58 and 59 are concerned with the function $f(x)=\sin(1/x)$ and its limit as x approaches 0.

58. (a) Show that $f(1/(n\pi))=0$ for every nonzero integer n.
(b) Show that $f(1/(2n\pi+\pi/2))=1$ for every integer n.

(c) Show that $f(1/(2n\pi + 3\pi/2)) = -1$ for every integer n.

59. (a) Plot the points $(x, f(x))$ on the graph of f with $x = \frac{1}{\pi}$, $\frac{1}{2\pi}, \frac{1}{3\pi}, \frac{1}{4\pi}, \frac{1}{(\pi/2)}, \frac{1}{(5\pi/2)}, \frac{1}{(9\pi/2)}, \frac{1}{(3\pi/2)}, \frac{1}{(7\pi/2)}, \frac{1}{(11\pi/2)}$.

(b) Does $\lim_{x\to 0^+} f(x)$ exist? What would you conjecture about $\lim_{x\to 0^-} f(x)$?

(c) Sketch the graph of f for $-\pi \leq x \leq \pi$.

In Exercises 60–65, estimate $\lim_{x\to a} f(x)$ by creating a table of values. Then use a graphing utility to zoom in on the graph of f near $x = a$ to justify or improve your guess.

60. Estimate

$$\lim_{x\to 0} \frac{\cos x - 1}{x} \qquad \text{(radian measure)}$$

after evaluating the quotient at $x = \pm 1, \pm 0.1, \pm 0.01, \pm 0.001$.

61. Estimate

$$\lim_{x\to 0} \frac{\tan 2x}{x} \qquad \text{(radian measure)}$$

after evaluating the quotient at $x = \pm 1, \pm 0.1, \pm 0.01, \pm 0.001$.

62. Estimate

$$\lim_{x\to 0} \frac{x - \sin x}{x^3} \qquad \text{(radian measure)}$$

after evaluating the quotient at $x = \pm 1, \pm 0.1, \pm 0.01, \pm 0.001, \pm 0.0001$.

63. Estimate

$$\lim_{x\to 1} \frac{x^{3/2} - 1}{x - 1}$$

after evaluating the quotient at $x = 0.9, 0.99, 0.999, 0.9999$ and $x = 1.1, 1.01, 1.001, 1.0001$.

64. Estimate

$$\lim_{x\to 0} \frac{2^x - 1}{x}$$

after evaluating the quotient at $x = \pm 1, \pm 0.1, \pm 0.01, \pm 0.001, \pm 0.0001$.

65. Estimate

$$\lim_{x\to 0^+} (1 + x)^{1/x}$$

by evaluating $(1 + x)^{1/x}$ at $x = 1, 0.1, 0.001, 0.0001, 0.00001, 0.000001, 0.0000001$.

66. Estimate the value of 3^π by taking a sequence $x_1, x_2, x_3, \ldots,$ of rational numbers which approach π and evaluating $3^{x_1}, 3^{x_2}, 3^{x_3}, \ldots.$ Recall $\pi \cong 3.14159265$.

■ 2.2 DEFINITION OF LIMIT

Our work with limits in Section 2.1 was informal. It is time to be more precise.

Let f be a function and let c be a real number. We do not require that f be defined at c, but we do require that f be defined at least on a set of the form $(c - p, c) \cup (c, c + p)$ with $p > 0$. (This guarantees that we can form $f(x)$ for all $x \neq c$ that are "sufficiently close" to c.)

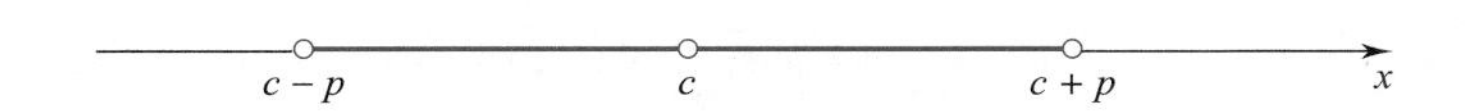

To say that

$$\lim_{x\to c} f(x) = L$$

means we can make $f(x)$ be as close to L as we want simply by choosing x close enough to c, $x \neq c$. That is:

Informal Statement	**ϵ, δ Statement**
$\lvert f(x) - L \rvert$ can be made arbitrarily small	you pick $\epsilon > 0$; $\lvert f(x) - L \rvert$ can be made less than ϵ
simply by requiring that	simply by requiring that
$\lvert x - c \rvert$ be sufficiently small but different from zero.	$\lvert x - c \rvert$ satisfy an inequality of the form $0 < \lvert x - c \rvert < \delta$ for δ sufficiently small.

Putting the various pieces together in compact form we have the following precise definition.

DEFINITION 2.2.1 THE LIMIT OF A FUNCTION

Let f be a function defined at least on some set of the form

$$(c - p, c) \cup (c, c + p), \text{ with } p > 0.$$

Then

$$\lim_{x \to c} f(x) = L \quad \text{iff}$$

for each $\epsilon > 0$, there exists $\delta > 0$ such that

$$\text{if} \quad 0 < |x - c| < \delta, \qquad \text{then} \quad |f(x) - L| < \epsilon.$$

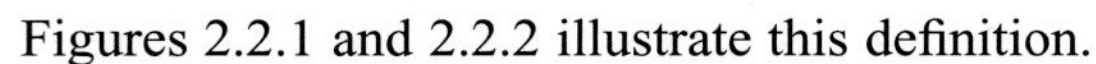

Figures 2.2.1 and 2.2.2 illustrate this definition.

In general, the choice of δ depends upon the previous choice of ϵ. We do not require that there exists a number δ which "works" for *all* ϵ, but rather, that for *each* ϵ there exists a δ which "works" for it.

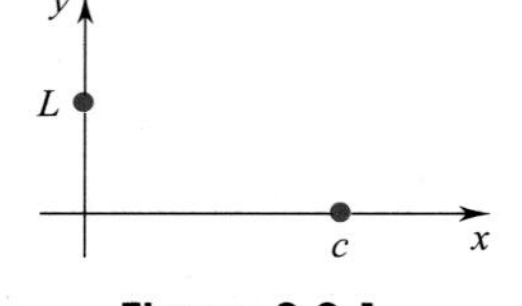

Figure 2.2.1

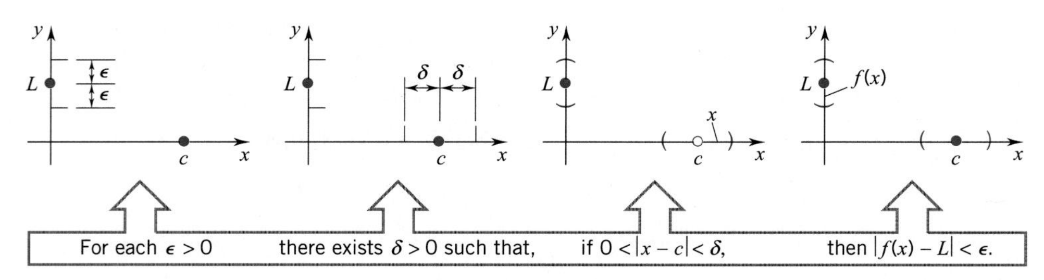

Figure 2.2.2

In Figure 2.2.3, we give two choices of ϵ and for each we display a suitable δ. For a δ to be suitable, all the points within δ of c (with the possible exception of c itself) must be taken by the function f to within ϵ of L. In part (b) of the figure we began with a smaller ϵ and had to use a smaller δ.

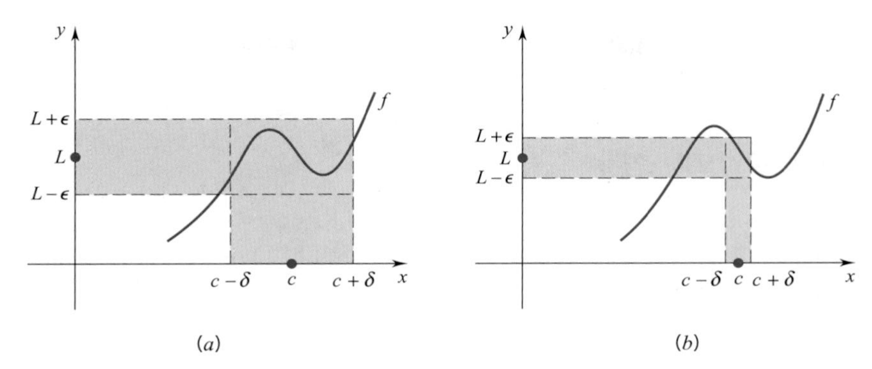

Figure 2.2.3

The δ of Figure 2.2.4 is too large for the given ϵ. In particular the points marked x_1 and x_2 in the figure are not taken by the function f to within ϵ of L.

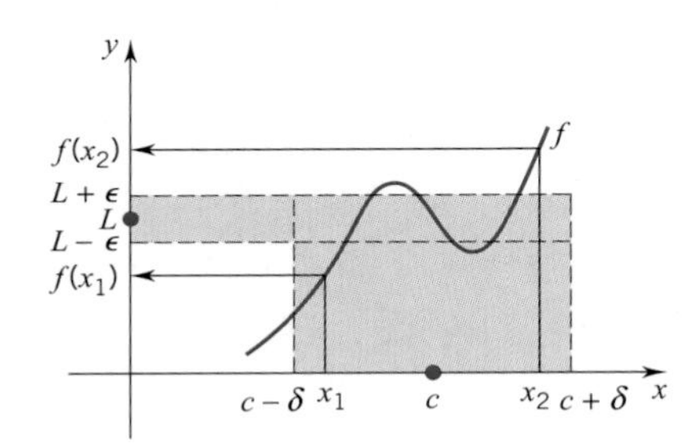

Figure 2.2.4

Next we apply the ϵ, δ definition of limit to a variety of functions. If you have never run across ϵ, δ arguments before, you may find them confusing at first. It usually takes a little while for the ϵ, δ idea to take hold.

Example 1 Show that

$$\lim_{x \to 2} (2x - 1) = 3. \qquad \text{(Figure 2.2.5)}$$

Finding a δ. Let $\epsilon > 0$. We seek a number $\delta > 0$ such that,

$$\text{if} \quad 0 < |x - 2| < \delta, \qquad \text{then} \quad |(2x - 1) - 3| < \epsilon.$$

What we have to do first is establish a connection between

$$|(2x - 1) - 3| \qquad \text{and} \qquad |x - 2|.$$

The connection is simple:

$$|(2x - 1) - 3| = |2x - 4|$$

so that

$$(*) \qquad |(2x - 1) - 3| = 2|x - 2|.$$

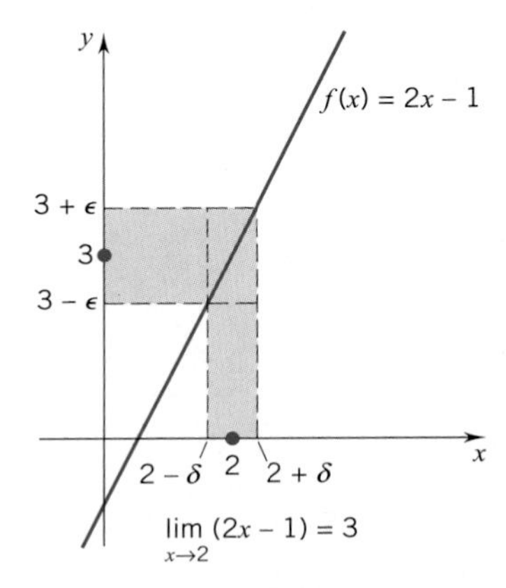

Figure 2.2.5

To make $|(2x-1)-3|$ less than ϵ, we need to make $2|x-2|<\epsilon$, which can be accomplished by making $|x-2|<\epsilon/2$. This suggests that we choose $\delta=\frac{1}{2}\epsilon$.

Showing that the δ "works." If $0<|x-2|<\frac{1}{2}\epsilon$, then $2|x-2|<\epsilon$ and, by (∗), $|(2x-1)-3|<\epsilon$. ❑

Remark In Example 1 we chose $\delta=\frac{1}{2}\epsilon$, but we could have chosen *any* positive number δ such that $\delta\le\frac{1}{2}\epsilon$. In general, if a certain δ^* "works" for a given ϵ, then any $\delta<\delta^*$ will also work. ❑

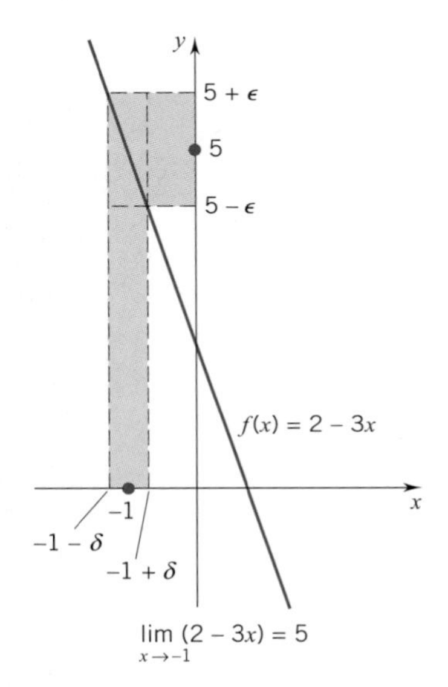

Figure 2.2.6

Example 2 Show that

$$\lim_{x\to-1}(2-3x)=5. \qquad \text{(Figure 2.2.6)}$$

Finding a δ. Let $\epsilon>0$. We seek a number $\delta>0$ such that

$$\text{if}\quad 0<|x-(-1)|<\delta,\qquad\text{then}\quad |(2-3x)-5|<\epsilon.$$

To find a connection between

$$|x-(-1)|\qquad\text{and}\qquad|(2-3x)-5|,$$

we simplify both expressions:

$$|x-(-1)|=|x+1|$$

and

$$|(2-3x)-5|=|-3x-3|=|-3||x+1|=3|x+1|.$$

We can conclude that

$$(**)\qquad |(2-3x)-5|=3|x-(-1)|.$$

We can make the expression on the left less than ϵ by making $|x-(-1)|$ less than $\epsilon/3$. This suggests that we set $\delta=\frac{1}{3}\epsilon$.

Showing that the δ "works." If $0<|x-(-1)|<\frac{1}{3}\epsilon$, then $3|x-(-1)|<\epsilon$ and, by (∗∗), $|(2-3x)-5|<\epsilon$. ❑

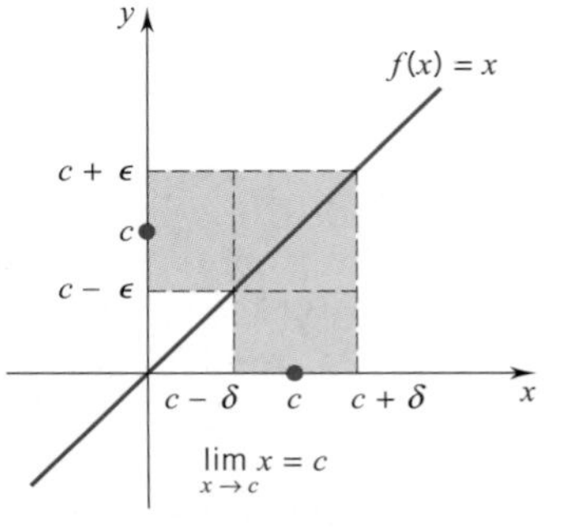

Figure 2.2.7

Example 3 Let c be any number. Then

$$\textbf{(2.2.2)}\qquad \boxed{\lim_{x\to c}x=c.} \qquad \text{(Figure 2.2.7)}$$

PROOF Let $\epsilon>0$. We must find $\delta>0$ such that

$$\text{if}\quad 0<|x-c|<\delta,\qquad\text{then}\quad |x-c|<\epsilon.$$

Obviously we can choose $\delta=\epsilon$. ❑

Example 4 Let c be any number. Then

(2.2.3)

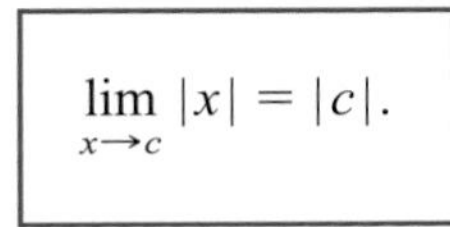

$$\lim_{x \to c} |x| = |c|.$$

(Figure 2.2.8)

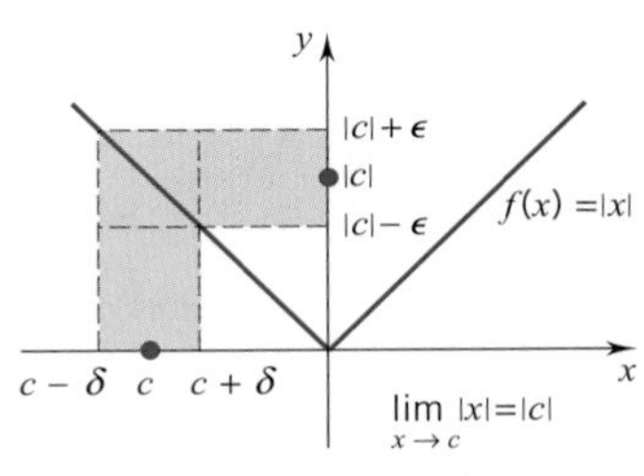

Figure 2.2.8

PROOF Let $\epsilon > 0$. We seek $\delta > 0$ such that

$$\text{if} \quad 0 < |x - c| < \delta, \qquad \text{then} \quad \big||x| - |c|\big| < \epsilon.$$

Since

$$\big||x| - |c|\big| \leq |x - c|,$$

(Recall: $\big||a| - |b|\big| \leq |a - b|$)

we can choose $\delta = \epsilon$; for

$$\text{if} \quad 0 < |x - c| < \epsilon, \qquad \text{then} \quad \big||x| - |c|\big| < \epsilon. \quad \square$$

Example 5

(2.2.4)

$$\boxed{\lim_{x \to c} a = a.}$$

(Figure 2.2.9)

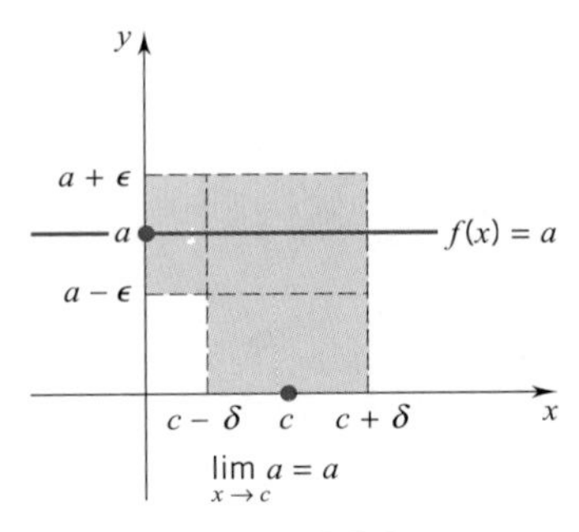

Figure 2.2.9

PROOF Here we are dealing with the constant function

$$f(x) = a.$$

Let $\epsilon > 0$. We must find $\delta > 0$ such that

$$\text{if} \quad 0 < |x - c| < \delta, \qquad \text{then} \quad |a - a| < \epsilon.$$

Since $|a - a| = 0$, we always have

$$|a - a| < \epsilon$$

no matter how δ is chosen; in short, any positive number will do for δ. $\square$

Usually ϵ, δ arguments are carried out in two stages. First we do a little algebraic scratchwork, labeled "finding a δ" in some of the preceding examples. This scratchwork involves working backward from $|f(x) - L| < \epsilon$ to find a $\delta > 0$ sufficiently small so that we can begin with the inequality $0 < |x - c| < \delta$ to arrive at $|f(x) - L| < \epsilon$. This first stage is just preliminary, but it shows us how to proceed in the second stage. The second stage consists of showing that the δ "works" by verifying that for our choice of δ the implication

$$\text{if} \quad 0 < |x - c| < \delta, \qquad \text{then} \quad |f(x) - L| < \epsilon$$

is true. The next two examples are more complicated and therefore may give you a better feeling for this idea of working backward to find a δ.

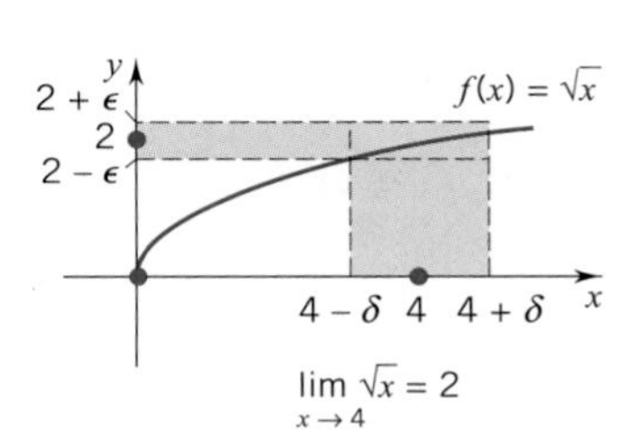

Figure 2.2.10

Example 6

$$\lim_{x \to 4} \sqrt{x} = 2. \qquad \text{(Figure 2.2.10)}$$

Finding a δ. Let $\epsilon > 0$. We seek $\delta > 0$ such that

$$\text{if} \quad 0 < |x - 4| < \delta, \qquad \text{then} \quad |\sqrt{x} - 2| < \epsilon.$$

First we want a relation between $|x - 4|$ and $|\sqrt{x} - 2|$. To be able to form $\sqrt{x}$ at all we need $x \geq 0$. To ensure this we must have $\delta \leq 4$. (Why?)

Remembering that we must have $\delta \leq 4$, let's go on. If $x \geq 0$, then we can form $\sqrt{x}$ and write

$$x - 4 = (\sqrt{x})^2 - 2^2 = (\sqrt{x} + 2)(\sqrt{x} - 2).$$

Taking absolute values, we have

$$|x - 4| = |\sqrt{x} + 2||\sqrt{x} - 2|.$$

Since $|\sqrt{x} + 2| > 1$, we have

$$|\sqrt{x} - 2| < |x - 4|.$$

This last inequality suggests that we simply set $\delta = \epsilon$. But remember the previous requirement $\delta \leq 4$. We can meet all requirements by setting δ = minimum of 4 and ϵ.

Showing that the δ "works." Let $\epsilon > 0$. Choose $\delta = \min\{4, \epsilon\}$ and assume that

$$0 < |x - 4| < \delta.$$

Since $\delta \leq 4$, we have $x \geq 0$ and so $\sqrt{x}$ is defined. Now, we can write

$$|x - 4| = |\sqrt{x} + 2||\sqrt{x} - 2|.$$

Since $|\sqrt{x} + 2| > 1$, we can conclude that

$$|\sqrt{x} - 2| < |x - 4|.$$

Since $|x - 4| < \delta$ and $\delta \leq \epsilon$, it does follow that

$$|\sqrt{x} - 2| < \epsilon. \quad \square$$

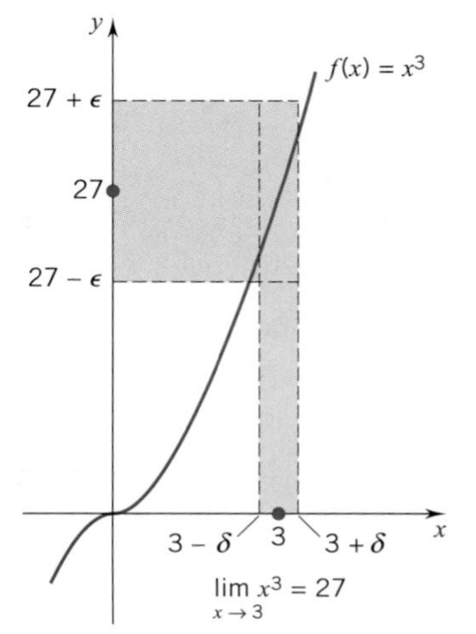

Figure 2.2.11

Example 7

$$\lim_{x \to 3} x^3 = 27. \qquad \text{(Figure 2.2.11)}$$

Finding a δ. Let $\epsilon > 0$. We seek $\delta > 0$ such that

$$\text{if} \quad 0 < |x - 3| < \delta, \qquad \text{then} \quad |x^3 - 27| < \epsilon.$$

The needed connection between $|x - 3|$ and $|x^3 - 27|$ is found by factoring:

$$x^3 - 27 = (x - 3)(x^2 + 3x + 9)$$

and thus

$$(*) \qquad |x^3 - 27| = |x - 3||x^2 + 3x + 9|.$$

At this juncture we need to get a handle on the size of $|x^2 + 3x + 9|$ for x close to 3. For convenience let's take x within one unit of 3.

If $|x-3|<1$, then $2<x<4$ and

$$\begin{aligned}|x^2+3x+9| &\leq |x^2|+|3x|+|9| \\ &= |x|^2+3|x|+9 \\ &< 16+12+9=37.\end{aligned}$$

It follows by (*) that

(**) if $|x-3|<1$, then $|x^3-27|<37|x-3|$.

If, in addition, $|x-3|<\epsilon/37$, then it will follow that

$$|x^3-27|<37(\epsilon/37)=\epsilon.$$

This means that we can take $\delta=$ minimum of 1 and $\epsilon/37$.

Showing that the δ "works." Let $\epsilon>0$. Choose $\delta=\min\{1,\epsilon/37\}$ and assume that

$$0<|x-3|<\delta.$$

Then

$$|x-3|<1 \qquad \text{and} \qquad |x-3|<\epsilon/37.$$

By (**)

$$|x^3-27|<37|x-3|,$$

and, since $|x-3|<\epsilon/37$, we have

$$|x^3-27|<37(\epsilon/37)=\epsilon. \quad \square$$

There are many different ways of formulating the same limit statement. Sometimes one formulation is more convenient, sometimes another. In any case, it is useful to recognize that the following are equivalent:

(2.2.5)

(i) $\lim_{x\to c} f(x)=L$	(ii) $\lim_{h\to 0} f(c+h)=L$
(iii) $\lim_{x\to c} (f(x)-L)=0$	(iv) $\lim_{x\to c} \lvert f(x)-L\rvert=0$.

The equivalence of these four statements is obvious. For example, the equivalence of (i) and (ii) can be seen in Figure 2.2.12: simply think of h as being the signed distance from c to x. Then $x=c+h$, and x approaches c iff h approaches 0. Also, it is a good exercise in ϵ, δ technique to prove that (i) is equivalent to (ii).

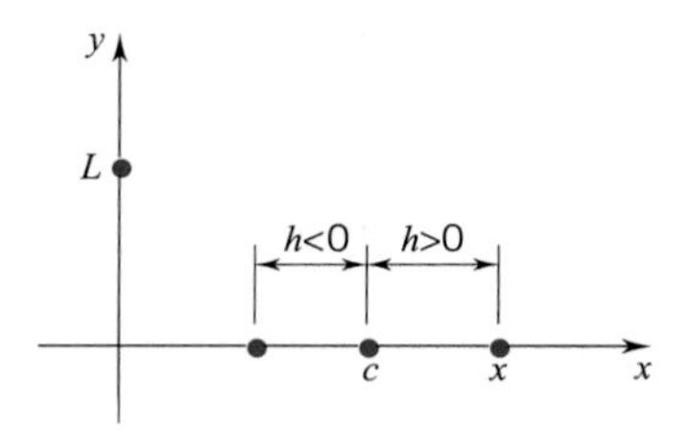

Figure 2.2.12

Example 8 For $f(x)=x^3$, we have

$$\lim_{x\to 2} x^3=8, \qquad \lim_{h\to 0}(2+h)^3=8,$$

$$\lim_{x\to 2}(x^3-8)=0, \qquad \lim_{x\to 2}|x^3-8|=0. \quad \square$$

We come now to the ϵ, δ definitions of one-sided limits. These are just the usual ϵ, δ statement, except that for a left-hand limit, the δ has to "work" only for x to the left of c, and for a right-hand limit, the δ has to "work" only for x to the right of c.

DEFINITION 2.2.6 LEFT-HAND LIMIT

Let f be a function defined at least on an interval of the form $(c - p, c)$, with $p > 0$. Then

$$\lim_{x \to c^-} f(x) = L \quad \text{iff}$$

for each $\epsilon > 0$ there exists $\delta > 0$ such that

$$\text{if} \quad c - \delta < x < c, \qquad \text{then} \quad |f(x) - L| < \epsilon.$$

DEFINITION 2.2.7 RIGHT-HAND LIMIT

Let f be a function defined at least on an interval of the form $(c, c + p)$, with $p > 0$. Then

$$\lim_{x \to c^+} f(x) = L \quad \text{iff}$$

for each $\epsilon > 0$ there exists $\delta > 0$ such that

$$\text{if} \quad c < x < c + \delta, \qquad \text{then} \quad |f(x) - L| < \epsilon.$$

As indicated in Section 2.1, one-sided limits give us a simple way of determining whether or not a (two-sided) limit exists:

(2.2.8)
$$\lim_{x \to c} f(x) = L \quad \text{iff} \quad \lim_{x \to c^-} f(x) = L \quad \text{and} \quad \lim_{x \to c^+} f(x) = L.$$

The proof is left as an exercise. The proof is easy because any δ that "works" for the limit will work for both one-sided limits, and any δ that "works" for both one-sided limits will work for the limit.

You should also understand that one-sided limits arise naturally; that is, independent of (two-sided) limits. For example, suppose the domain of a function f is the interval (a, b). Then the only limit that we can consider at $x = a$ is the right-hand limit, $\lim_{x \to a^+} f(x)$. Similarly, at $x = b$, the only limit we can consider is the left-hand limit, $\lim_{x \to b^-} f(x)$. For the function $f(x) = \sqrt{x - 4}$, the only limit at $x = 4$ that makes sense is the right-hand limit ($\sqrt{x - 4}$ is not defined for $x < 4$), and

$$\lim_{x \to 4^+} \sqrt{x - 4} = 0.$$

Example 9 If f is the function defined by

$$f(x) = \begin{cases} 2x + 1, & x \leq 0 \\ x^2 - x, & x > 0, \end{cases} \qquad \text{(Figure 2.2.13)}$$

then $\lim_{x\to 0} f(x)$ does not exist.

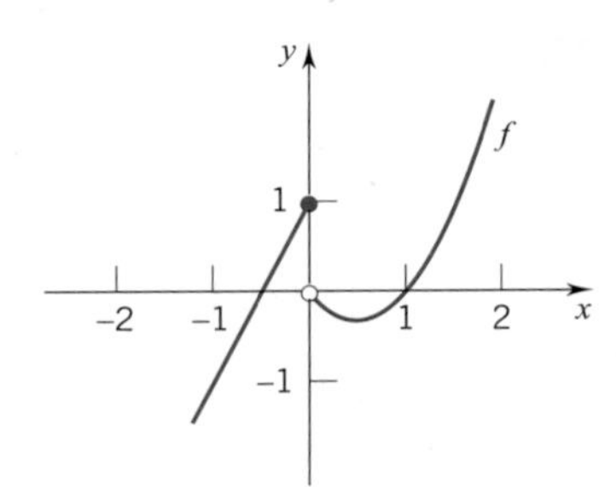

Figure 2.2.13

PROOF The left- and right-hand limits at 0 are

$$\lim_{x\to 0^-} f(x) = \lim_{x\to 0^-} (2x + 1) = 1 \quad \text{and} \quad \lim_{x\to 0^+} f(x) = \lim_{x\to 0^+} (x^2 - x) = 0.$$

Since these one-sided limits are different, $\lim_{x\to 0} f(x)$ does not exist. ❑

Example 10 If g is the function defined by

$$g(x) = \begin{cases} 1 + x^3, & x < 1 \\ 3, & x = 1 \\ 4 - 2x, & x > 1, \end{cases} \qquad \text{(Figure 2.2.14)}$$

then $\lim_{x\to 1} g(x) = 2$.

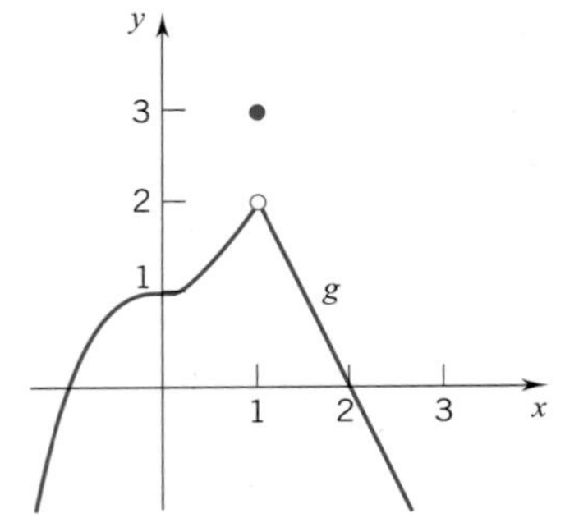

Figure 2.2.14

PROOF The left- and right-hand limits at 1 are

$$\lim_{x\to 1^-} g(x) = \lim_{x\to 1^-} (1 + x^2) = 2 \quad \text{and} \quad \lim_{x\to 1^+} g(x) = \lim_{x\to 1^+} (4 - 2x) = 2.$$

Thus, $\lim_{x\to 1} g(x) = 2$. NOTE: it does not matter that $g(1) \neq 2$. ❑

EXERCISES 2.2

In Exercises 1–22, decide in the manner of Section 2.1 whether or not the indicated limit exists. Evaluate the limits that do exist.

1. $\lim_{x\to 1} \dfrac{x}{x + 1}$.

2. $\lim_{x\to 0} \dfrac{x^2(1 + x)}{2x}$.

3. $\lim_{x\to 0} \dfrac{x(1 + x)}{2x^2}$.

4. $\lim_{x\to 4} \dfrac{x}{\sqrt{x} + 1}$.

5. $\lim_{x\to 4} \dfrac{x}{\sqrt{x + 1}}$.

6. $\lim_{x\to -1} \dfrac{1 - x}{x + 1}$.

7. $\lim_{x\to 1} \dfrac{x^4 - 1}{x - 1}$.

8. $\lim_{x\to 2} \dfrac{x}{|x|}$.

9. $\lim_{x\to 0} \dfrac{x}{|x|}$.

10. $\lim_{x\to 1} \dfrac{x^2 - 1}{x^2 - 2x + 1}$.

11. $\lim_{x\to -2} \dfrac{|x|}{x}$.

12. $\lim_{x\to 9} \dfrac{x - 3}{\sqrt{x} - 3}$.

13. $\lim_{x\to 3^+} \dfrac{x + 3}{x^2 - 7x + 12}$.

14. $\lim_{x\to 0^-} \dfrac{x}{|x|}$.

15. $\lim_{x\to 1^+} \dfrac{\sqrt{x - 1}}{x}$.

16. $\lim_{x\to 3^-} \sqrt{9 - x^2}$.

17. $\lim_{x\to 2^+} f(x)$ if $f(x) = \begin{cases} 2x - 1, & x \leq 2 \\ x^2 - x, & x > 2. \end{cases}$

18. $\lim_{x\to -1^-} f(x)$ if $f(x) = \begin{cases} 1, & x \leq -1 \\ x + 2, & x > -1. \end{cases}$

19. $\lim_{x\to 2} f(x)$ if $f(x) = \begin{cases} 3, & x \text{ an integer} \\ 1, & \text{otherwise.} \end{cases}$

20. $\lim_{x\to 3} f(x)$ if $f(x) = \begin{cases} x^2, & x < 3 \\ 7, & x = 3 \\ 2x + 3, & x > 3. \end{cases}$

21. $\lim_{x\to \sqrt{2}} f(x)$ if $f(x) = \begin{cases} 3, & x \text{ an integer} \\ 1, & \text{otherwise.} \end{cases}$

22. $\lim_{x\to 2} f(x)$ if $f(x) = \begin{cases} x^2, & x \leq 1 \\ 5x, & x > 1. \end{cases}$

23. Which of the δ's displayed in the figure "works" for the given ϵ?

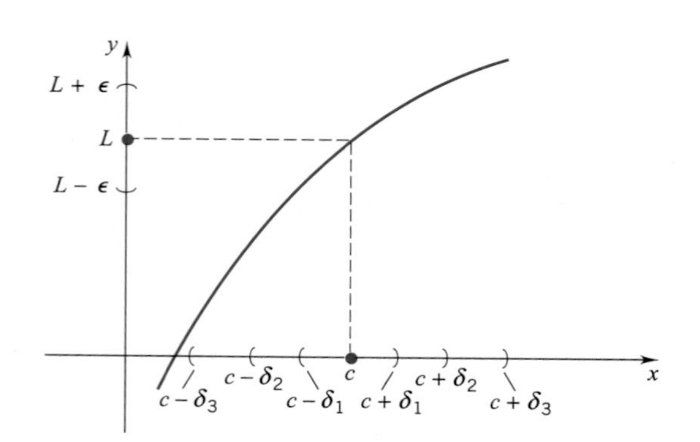

24. For which of the ϵ's given in the figure does the specified δ work?

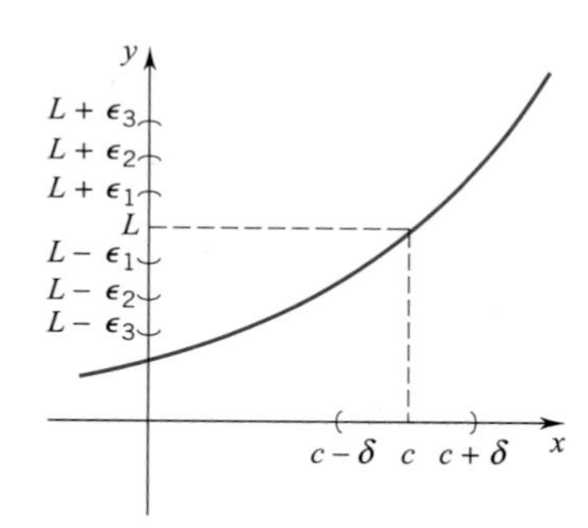

For the limits in Exercises 25–28, find the largest δ that "works" for the given ϵ.

25. $\lim_{x\to 1} 2x = 2;\ \epsilon = 0.1.$

26. $\lim_{x\to 4} 5x = 20;\ \epsilon = 0.5.$

27. $\lim_{x\to 2} \frac{1}{2}x = 1;\ \epsilon = 0.01.$

28. $\lim_{x\to 2} \frac{1}{5}x = \frac{2}{5};\ \epsilon = 0.1.$

In Exercises 29–34, give an ϵ, δ proof for the following limits.

29. $\lim_{x\to 4} (2x - 5) = 3.$

30. $\lim_{x\to 2} (3x - 1) = 5.$

31. $\lim_{x\to 3} (6x - 7) = 11.$

32. $\lim_{x\to 0} (2 - 5x) = 2.$

33. $\lim_{x\to 2} |1 - 3x| = 5.$

34. $\lim_{x\to 2} |x - 2| = 0.$

35. Let f be some function of which you know only that

$$\text{if}\quad 0 < |x - 3| < 1, \qquad \text{then}\quad |f(x) - 5| < 0.1.$$

Which of the following statements are necessarily true?

(a) If $|x - 3| < 1$, then $|f(x) - 5| < 0.1$.
(b) If $|x - 2.5| < 0.3$, then $|f(x) - 5| < 0.1$.
(c) $\lim_{x\to 3} f(x) = 5$.
(d) If $0 < |x - 3| < 2$, then $|f(x) - 5| < 0.1$.
(e) If $0 < |x - 3| < 0.5$, then $|f(x) - 5| < 0.1$.
(f) If $0 < |x - 3| < \frac{1}{4}$, then $|f(x) - 5| < \frac{1}{4}(0.1)$.
(g) If $0 < |x - 3| < 1$, then $|f(x) - 5| < 0.2$.
(h) If $0 < |x - 3| < 1$, then $|f(x) - 4.95| < 0.05$.
(i) If $\lim_{x\to 3} f(x) = L$, then $4.9 \leq L \leq 5.1$.

36. Suppose that $|A - B| < \epsilon$ for each $\epsilon > 0$. Prove that $A = B$. HINT: Consider what happens if $A \neq B$ and $\epsilon = \frac{1}{2}|A - B|$.

In Exercises 37–38, give the four limit statements displayed in 2.2.5, taking

37. $f(x) = \dfrac{1}{x - 1}, \quad c = 3.$

38. $f(x) = \dfrac{x}{x^2 + 2}, \quad c = 1.$

In Exercises 39–44, assume that x is a fixed number and find

$$\lim_{h\to 0} \frac{f(x + h) - f(x)}{h}$$

for the given function f.

39. $f(x) = 5x + 2.$

40. $f(x) = 2x^2 - 3x.$

41. $f(x) = 4x + 5x^2.$

42. $f(x) = x^3 + 2x + 4.$

43. $f(x) = \dfrac{1}{x + 1}.$

44. $f(x) = \sqrt{x + 2}.$ HINT: rationalize the numerator.

45. Prove that

(2.2.9)
$$\lim_{x\to c} f(x) = 0 \quad \text{iff} \quad \lim_{x\to c} |f(x)| = 0.$$

46. (a) Prove that

$$\text{if}\quad \lim_{x\to c} f(x) = L \quad \text{then} \quad \lim_{x\to c} |f(x)| = |L|.$$

(b) Show that the converse is false. Give an example where

$$\lim_{x\to c} |f(x)| = |L| \qquad \text{and} \qquad \lim_{x\to c} f(x) = M \neq L.$$

and then give an example where

$$\lim_{x\to c} |f(x)| \ \text{exists} \quad \text{but} \quad \lim_{x\to c} f(x) \ \text{does not exist.}$$

In Exercises 47–50, $\lim_{x\to a} f(x) = L$ and ϵ are given. Use a graphing utility to graph f, then zoom in at $x = a$ to find a δ

such that $|f(x) - a| < \epsilon$ whenever $|x - a| < \delta$.
HINT: graph f, $L + \epsilon$ and $L - \epsilon$.

47. $\lim_{x \to 2} (x^2 + x - 3) = 3$; $\epsilon = 0.01$.

48. $\lim_{x \to 1} \dfrac{x - 1}{\sqrt{x} - 1} = 2$; $\epsilon = 0.1$.

49. $\lim_{x \to 0} \dfrac{\sin 3x}{x} = 3$; $\epsilon = 0.01$.

50. $\lim_{x \to 1} \tan\left(\dfrac{\pi}{4} x\right) = 1$; $\epsilon = 0.1$.

51. Give an ϵ, δ proof that statement (i) of 2.2.5 is equivalent to (ii).

52. Give an ϵ, δ proof of (2.2.8).

53. (a) Show that $\lim_{x \to c} \sqrt{x} = \sqrt{c}$ for $c > 0$.

HINT: If x and c are positive, then

$$0 \le |\sqrt{x} - \sqrt{c}| = \frac{|x - c|}{\sqrt{x} + \sqrt{c}} < \frac{1}{\sqrt{c}} |x - c|.$$

(b) Show that $\lim_{x \to 0^+} \sqrt{x} = 0$.

In Exercises 54–57, give an ϵ, δ proof for the following limits.

54. $\lim_{x \to 2} x^2 = 4$.

55. $\lim_{x \to 1} x^3 = 1$.

56. $\lim_{x \to 3} \sqrt{x + 1} = 2$.

57. $\lim_{x \to 3^-} \sqrt{3 - x} = 0$.

58. Prove that if

$$g(x) = \begin{cases} x, & x \text{ rational} \\ 0, & x \text{ irrational,} \end{cases}$$

then $\lim_{x \to 0} g(x) = 0$.

59. The function f defined by

$$f(x) = \begin{cases} 1, & x \text{ rational} \\ 0, & x \text{ irrational} \end{cases}$$

is called the *Dirichlet function.* Prove that for no number c does $\lim_{x \to c} f(x)$ exist.

In Exercises 60–62, prove the given limit statement.

60. $\lim_{x \to c^-} f(x) = L$ iff $\lim_{h \to 0} f(c - |h|) = L$.

61. $\lim_{x \to c^+} f(x) = L$ iff $\lim_{h \to 0} f(c + |h|) = L$.

62. $\lim_{x \to c} f(x) = L$ iff $\lim_{x \to c^-} f(x) = \lim_{x \to c^+} f(x) = L$.

63. Suppose that $\lim_{x \to c} f(x) = L$.

(a) Prove that if $L > 0$, then $f(x) > 0$ for all $x \neq c$ in an interval of the form $(c - \gamma, c + \gamma)$, $\gamma > 0$.
HINT: Use an ϵ, δ argument, setting $\epsilon = L$.

(b) Prove that if $L < 0$, then $f(x) < 0$ for all $x \neq c$ in an interval of the form $(c - \gamma, c + \gamma)$, $\gamma > 0$.

64. Prove or give a counterexample: If $f(c) > 0$ and $\lim_{x \to c} f(x)$ exists, then $f(x) > 0$ for all x in an interval of the form $(c - \gamma, c + \gamma)$, $\gamma > 0$.

65. Suppose that $f(x) \le g(x)$ for all $x \in (c - p, c + p)$, $p > 0$, except possibly at c itself.

(a) Prove that $\lim_{x \to c} f(x) \le \lim_{x \to c} g(x)$, provided each of these limits exist.

(b) If $f(x) < g(x)$ on $(c - p, c + p)$, except possibly at c itself, does it follow that $\lim_{x \to c} f(x) < \lim_{x \to c} g(x)$?

66. Prove that if $\lim_{x \to c} f(x) = L$, then there is a number $\delta > 0$ and a number B such that $|f(x)| < B$ if $0 < |x - c| < \delta$.
HINT: It suffices to prove that $L - 1 < f(x) < L + 1$ for $0 < |x - c| < \delta$.

In Exercises 2.1, the slope of the tangent line to the graph of a function f at the point $(a, f(a))$ was defined to be

$$\lim_{h \to 0} \frac{f(a + h) - f(a)}{h} \qquad \text{provided this limit exists.}$$

The equivalence of (i) and (ii) in (2.2.5) implies that the slope of the tangent line is also given by

$$\lim_{x \to a} \frac{f(x) - f(a)}{x - a} \qquad \text{provided this limit exists.}$$

In Exercises 67–70, determine if $\lim_{x \to a} (f(x) - f(a))/(x - a)$ exists, and give an equation for the tangent line to the graph of f at $(a, f(a))$ when the limit does exist.

67. $f(x) = 2x^2 - 3x$, $a = 2$.

68. $f(x) = x^3 + 1$, $a = -1$.

69. $f(x) = \sqrt{x}$, $a = 4$.

70. $f(x) = 1/(x + 1)$, $a = 1$.

C In Exercises 71–74, estimate $\lim_{x \to a} (f(x) - f(a))/(x - a)$ numerically by evaluating the quotient at $a \pm 0.1$, $a \pm 0.01$, $a \pm 0.001$. Use a graphing utility to support or improve your numerical result.

71. $f(x) = \cos \pi x$, $a = \frac{1}{3}$.

72. $f(x) = \sqrt{x}$, $a = 1$.

73. $f(x) = 3^x$, $a = 0$.

74. $f(x) = 2^x$, $a = 1$.

■ 2.3 SOME LIMIT THEOREMS

As you saw in the last section, it can be rather tedious to apply the ϵ, δ limit test to individual functions. By proving some general theorems about limits we can avoid some of this repetitive work. The theorems themselves, of course (at least the first ones), will have to be proved by ϵ, δ methods.

We begin by showing that, if a limit exists, it is unique.

THEOREM 2.3.1 THE UNIQUENESS OF A LIMIT

$$\text{If} \quad \lim_{x\to c} f(x) = L \text{ and } \lim_{x\to c} f(x) = M, \qquad \text{then} \quad L = M.$$

PROOF We show that $L = M$ by proving that the assumption that $L \neq M$ leads to the false conclusion

$$|L - M| < |L - M|.$$

Let's assume that $L \neq M$. Then $|L - M|/2 > 0$. Since

$$\lim_{x\to c} f(x) = L,$$

we know that there exists $\delta_1 > 0$ such that

$$\text{(1)} \qquad \text{if} \quad 0 < |x - c| < \delta_1, \qquad \text{then} \quad |f(x) - L| < \frac{|L - M|}{2}.$$

(Here we are using $|L - M|/2$ as ϵ.)

Also, since

$$\lim_{x\to c} f(x) = M,$$

we know that there exists $\delta_2 > 0$ such that

$$\text{(2)} \qquad \text{if} \quad 0 < |x - c| < \delta_2, \qquad \text{then} \quad |f(x) - M| < \frac{|L - M|}{2}.$$

(Again, using $|L - M|/2$ as ϵ.)

Now let x_1 be a number that satisfies the inequality

$$0 < |x_1 - c| < \text{minimum of } \delta_1 \text{ and } \delta_2.$$

Then by (1) and (2)

$$|f(x_1) - L| < \frac{|L - M|}{2} \qquad \text{and} \qquad |f(x_1) - M| < \frac{|L - M|}{2}.$$

It follows that

$$\begin{aligned}|L - M| &= |[L - f(x_1)] + [f(x_1) - M]| \\ &\leq |L - f(x_1)| + |f(x_1) - M| \\ &= |f(x_1) - L| + |f(x_1) - M| < \frac{|L - M|}{2} + \frac{|L - M|}{2} = |L - M|. \quad \square\end{aligned}$$

(by the triangle inequality ↑)

($|a| = |-a|$ ↑)

THEOREM 2.3.2

If $\lim_{x \to c} f(x) = L$ and $\lim_{x \to c} g(x) = M$, then

$$\textbf{(i)} \lim_{x \to c} [f(x) + g(x)] = L + M,$$

$$\textbf{(ii)} \lim_{x \to c} [\alpha f(x)] = \alpha L \qquad \text{for each real } \alpha,$$

$$\textbf{(iii)} \lim_{x \to c} [f(x)g(x)] = LM.$$

PROOF Let $\epsilon > 0$. To prove (i) we must show that there exists $\delta > 0$ such that

$$\text{if} \quad 0 < |x - c| < \delta, \qquad \text{then} \quad |[f(x) + g(x)] - [L + M]| < \epsilon.$$

Note that

$$(*) \qquad |[f(x) + g(x)] - [L + M]| = |[f(x) - L] + [g(x) - M]| \le |f(x) - L| + |g(x) - M|.$$

We make $|[f(x) + g(x)] - [L + M]|$ less than ϵ by making $|f(x) - L|$ and $|g(x) - M|$ each less than $\frac{1}{2}\epsilon$. Since $\epsilon > 0$, we know that $\frac{1}{2}\epsilon > 0$. Since

$$\lim_{x \to c} f(x) = L \qquad \text{and} \qquad \lim_{x \to c} g(x) = M,$$

we know that there exist positive numbers δ_1 and δ_2 such that

$$\text{if} \quad 0 < |x - c| < \delta_1, \qquad \text{then} \quad |f(x) - L| < \tfrac{1}{2}\epsilon$$

and

$$\text{if} \quad 0 < |x - c| < \delta_2, \qquad \text{then} \quad |g(x) - M| < \tfrac{1}{2}\epsilon.$$

Now we set δ = minimum of δ_1 and δ_2 and note that, if $0 < |x - c| < \delta$, then

$$|f(x) - L| < \tfrac{1}{2}\epsilon \qquad \text{and} \qquad |g(x) - M| < \tfrac{1}{2}\epsilon,$$

and thus by $(*)$

$$|[f(x) + g(x)] - [L + M]| < \epsilon.$$

In summary, by setting $\delta = \min\{\delta_1, \delta_2\}$, we found that,

$$\text{if} \quad 0 < |x - c| < \delta, \qquad \text{then} \quad |[f(x) + g(x)] - [L + M]| < \epsilon.$$

Thus (i) is proved. For proofs of (ii) and (iii), see the supplement at the end of this section. ❑

If you are wondering about $\lim_{x \to c} [f(x) - g(x)]$, note that $f(x) - g(x) = f(x) + (-1)g(x)$, and so the result

(2.3.3)

$$\lim_{x \to c} [f(x) - g(x)] = L - M$$

follows from (i) and (ii).

The results of Theorem 2.3.2 are easily extended (by mathematical induction) to any finite collection of functions; namely, if

$$\lim_{x \to c} f_1(x) = L_1, \qquad \lim_{x \to c} f_2(x) = L_2, \quad . \; . \; . \; , \lim_{x \to c} f_n(x) = L_n,$$

then

(2.3.4)
$$\lim_{x \to c} [\alpha_1 f_1(x) + \alpha_2 f_2(x) + \cdots + \alpha_n f_n(x)] = \alpha_1 L_1 + \alpha_2 L_2 + \cdots + \alpha_n L_n$$

and

(2.3.5)
$$\lim_{x \to c} [f_1(x) f_2(x) \cdots f_n(x)] = L_1 L_2 \cdots L_n.$$

Let $P(x) = a_n x^n + \cdots + a_1 x + a_0$ be a polynomial and let c be any real number. Then it follows that

(2.3.6)
$$\lim_{x \to c} P(x) = P(c).$$

PROOF We already know that

$$\lim_{x \to c} x = c. \qquad \text{(see (2.2.2) on page 84)}$$

Applying (2.3.5) to $f(x) = x$ multiplied k times by itself, we have

$$\lim_{x \to c} x^k = c^k \qquad \text{for each positive integer } k.$$

We also know from (2.2.4) on page 85 that $\lim_{x \to c} a_0 = a_0$. It follows now from (2.3.4) that

$$\lim_{x \to c} (a_n x^n + \cdots + a_1 x + a_0) = a_n c^n + \cdots + a_1 c + a_0;$$

that is,

$$\lim_{x \to c} P(x) = P(c). \quad \square$$

In words, this result says that the limit of a polynomial $P(x)$ as x approaches c not only exists, but is actually the value of P at c. Functions that have this property are said to be *continuous at c*. Thus, polynomial functions are continuous at $x = c$ for each number c. Continuity will be treated in detail in the next section.

Examples

$$\lim_{x\to 1}(5x^2 - 12x + 2) = 5(1)^2 - 12(1) + 2 = -5,$$

$$\lim_{x\to 0}(14x^5 - 7x^2 + 2x + 8) = 14(0)^5 - 7(0)^2 + 2(0) + 8 = 8,$$

$$\lim_{x\to -1}(2x^3 + x^2 - 2x - 3) = 2(-1)^3 + (-1)^2 - 2(-1) - 3 = -2. \quad \square$$

We come now to reciprocals and quotients.

THEOREM 2.3.7

$$\text{If } \lim_{x\to c} g(x) = M \quad \text{with} \quad M \neq 0, \qquad \text{then} \quad \lim_{x\to c}\frac{1}{g(x)} = \frac{1}{M}.$$

PROOF See the supplement at the end of this section. ❑

Examples

$$\lim_{x\to 4}\frac{1}{x^2} = \frac{1}{16}, \qquad \lim_{x\to 2}\frac{1}{x^3 - 1} = \frac{1}{7}, \qquad \lim_{x\to -3}\frac{1}{|x|} = \frac{1}{|-3|} = \frac{1}{3}. \quad \square$$

Once you know that reciprocals present no trouble, quotients become easy to handle.

THEOREM 2.3.8

$$\text{If } \lim_{x\to c} f(x) = L \text{ and } \lim_{x\to c} g(x) = M \text{ with } M \neq 0, \quad \text{then} \quad \lim_{x\to c}\frac{f(x)}{g(x)} = \frac{L}{M}.$$

PROOF The key here is to observe that the quotient can be written as a product:

$$\frac{f(x)}{g(x)} = f(x)\,\frac{1}{g(x)}.$$

With

$$\lim_{x\to c} f(x) = L \qquad \text{and} \qquad \lim_{x\to c}\frac{1}{g(x)} = \frac{1}{M},$$

the product rule (part (iii) of Theorem 2.3.2) gives

$$\lim_{x\to c}\frac{f(x)}{g(x)} = L\,\frac{1}{M} = \frac{L}{M}. \quad \square$$

As an immediate consequence of this theorem on quotients you can see that, if $R = P/Q$ is a rational function (quotient of polynomials) and $Q(c) \neq 0$, then

(2.3.9)
$$\lim_{x \to c} R(x) = \lim_{x \to c} \frac{P(x)}{Q(x)} = \frac{P(c)}{Q(c)} = R(c).$$

This result says that a rational function is *continuous* at the numbers c where $Q(c) \neq 0$.

Examples

$$\lim_{x \to 2} \frac{3x - 5}{x^2 + 1} = \frac{6 - 5}{4 + 1} = \frac{1}{5}, \qquad \lim_{x \to 3} \frac{x^3 - 3x^2}{1 - x^2} = \frac{27 - 27}{1 - 9} = 0. \quad \square$$

There is no point looking for a limit that does not exist. The next theorem gives a condition under which a quotient does not have a limit.

THEOREM 2.3.10

If $\lim_{x \to c} f(x) = L$ with $L \neq 0$ and $\lim_{x \to c} g(x) = 0$, then $\lim_{x \to c} \frac{f(x)}{g(x)}$ does not exist.

PROOF Suppose, on the contrary, that there exists a real number K such that

$$\lim_{x \to c} \frac{f(x)}{g(x)} = K.$$

Then

$$L = \lim_{x \to c} f(x) = \lim_{x \to c} \left[g(x) \cdot \frac{f(x)}{g(x)} \right] = \lim_{x \to c} g(x) \cdot \lim_{x \to c} \frac{f(x)}{g(x)} = 0 \cdot K = 0,$$

which contradicts our assumption that $L \neq 0$. $\square$

Examples From Theorem 2.3.10 you can see that

$$\lim_{x \to 1} \frac{x^2}{x - 1}, \qquad \lim_{x \to 2} \frac{3x - 7}{x^2 - 4}, \qquad \text{and} \qquad \lim_{x \to 0} \frac{5}{x}$$

all fail to exist. $\square$

Suppose $\lim_{x \to c} f(x) = L$ and $\lim_{x \to c} g(x) = M$. Combining Theorems 2.3.8 and 2.3.10, we have

$$\lim_{x \to c} \frac{f(x)}{g(x)} = \begin{cases} L/M & \text{if } M \neq 0 \\ \text{does not exist} & \text{if } M = 0 \text{ and } L \neq 0. \end{cases}$$

This raises the question: What happens if $L = M = 0$? This is the most difficult and the most interesting case. A limit of a quotient in which both the numerator and the denominator approach 0 is called an *indeterminate form of type* 0/0; as you will see in the examples below, the limit may or may not exist. At this stage, we will use algebraic methods to analyze limits of this type. We will learn other, more powerful, methods later.

Example 1 Evaluate the limits that exist:

(a) $\lim_{x\to 3}\dfrac{x^2-x-6}{x-3}$, **(b)** $\lim_{x\to 4}\dfrac{(x^2-3x-4)^2}{x-4}$, **(c)** $\lim_{x\to -1}\dfrac{x+1}{(2x^2+7x+5)^2}$.

SOLUTION In each case both numerator and denominator tend to zero, and so we have to be careful.

(a) First we factor the numerator:

$$\frac{x^2-x-6}{x-3}=\frac{(x+2)(x-3)}{x-3}.$$

For $x\neq 3$,

$$\frac{x^2-x-6}{x-3}=x+2.$$

Thus

$$\lim_{x\to 3}\frac{x^2-x-6}{x-3}=\lim_{x\to 3}(x+2)=5.$$

(b) Note that

$$\frac{(x^2-3x-4)^2}{x-4}=\frac{[(x+1)(x-4)]^2}{x-4}=\frac{(x+1)^2(x-4)^2}{x-4}$$

so that, for $x\neq 4$,

$$\frac{(x^2-3x-4)^2}{x-4}=(x+1)^2(x-4).$$

It follows then that

$$\lim_{x\to 4}\frac{(x^2-3x-4)^2}{x-4}=\lim_{x\to 4}(x+1)^2(x-4)=0.$$

(c) Since

$$\frac{x+1}{(2x^2+7x+5)^2}=\frac{x+1}{[(2x+5)(x+1)]^2}=\frac{x+1}{(2x+5)^2(x+1)^2},$$

you can see that, for $x\neq -1$,

$$\frac{x+1}{(2x^2+7x+5)^2}=\frac{1}{(2x+5)^2(x+1)}.$$

By Theorem 2.3.10

$$\lim_{x\to -1}\frac{1}{(2x+5)^2(x+1)}\quad\text{does not exist,}$$

and therefore

$$\lim_{x\to -1}\frac{x+1}{(2x^2+7x+5)^2}\quad\text{does not exist either.}\quad \square$$

Example 2 Evaluate the following limits:

(a) $\lim_{x \to 2} \dfrac{1/x - 1/2}{x - 2}$ **(b)** $\lim_{x \to 9} \dfrac{x - 9}{\sqrt{x} - 3}$.

SOLUTION

(a) For $x \neq 2$,

$$\frac{1/x - 1/2}{x - 2} = \frac{\dfrac{2 - x}{2x}}{x - 2} = \frac{-(x - 2)}{2x(x - 2)} = \frac{-1}{2x}.$$

Thus,

$$\lim_{x \to 2} \frac{1/x - 1/2}{x - 2} = \lim_{x \to 2} \left[\frac{-1}{2x}\right] = -\frac{1}{4}.$$

(b) First, we "rationalize" the denominator:

$$\frac{x - 9}{\sqrt{x} - 3} = \frac{x - 9}{\sqrt{x} - 3} \cdot \frac{\sqrt{x} + 3}{\sqrt{x} + 3} = \frac{(x - 9)(\sqrt{x} + 3)}{x - 9} = \sqrt{x} + 3 \qquad (x \neq 9).$$

Alternatively, we could have factored the numerator:

$$\frac{x - 9}{\sqrt{x} - 3} = \frac{(\sqrt{x} - 3)(\sqrt{x} + 3)}{\sqrt{x} - 3} = \sqrt{x} + 3 \qquad (x \neq 9)$$

and arrived at the same result. Either way,

$$\lim_{x \to 9} \frac{x - 9}{\sqrt{x} - 3} = \lim_{x \to 9} [\sqrt{x} + 3] = 6. \quad \square$$

Remark In this section we have phrased everything in terms of two-sided limits. Although we won't stop here to prove it, *all these results carry over to one-sided limits.* ❑

EXERCISES 2.3

1 Given that

$$\lim_{x \to c} f(x) = 2, \qquad \lim_{x \to c} g(x) = -1, \qquad \lim_{x \to c} h(x) = 0,$$

evaluate the limits that exist. If the limit does not exist, explain why.

(a) $\lim_{x \to c} [f(x) - g(x)]$. (b) $\lim_{x \to c} [f(x)]^2$.

(c) $\lim_{x \to c} \dfrac{f(x)}{g(x)}$. (d) $\lim_{x \to c} \dfrac{h(x)}{f(x)}$.

(e) $\lim_{x \to c} \dfrac{f(x)}{h(x)}$. (f) $\lim_{x \to c} \dfrac{1}{f(x) - g(x)}$.

2. Given that

$$\lim_{x \to c} f(x) = 3, \qquad \lim_{x \to c} g(x) = 0, \qquad \lim_{x \to c} h(x) = -2,$$

evaluate the limits that exist. If the limit does not exist, explain why.

(a) $\lim_{x \to c} [3f(x) + 2h(x)]$. (b) $\lim_{x \to c} [h(x)]^3$.

(c) $\lim_{x \to c} \dfrac{h(x)}{x - c}$. (d) $\lim_{x \to c} \dfrac{g(x)}{h(x)}$.

(e) $\lim_{x \to c} \dfrac{4}{f(x) - h(x)}$. (f) $\lim_{x \to c} [3 + g(x)]^2$.

3. When asked to evaluate

$$\lim_{x \to 4} \left(\frac{1}{x} - \frac{1}{4}\right)\left(\frac{1}{x - 4}\right),$$

I reply that the limit is zero since $\lim_{x \to 4} [1/x - \frac{1}{4}] = 0$ and cite Theorem 2.3.2 (limit of a product) as justification. Verify that the limit is actually $-\frac{1}{16}$ and identify my error.

4. When asked to evaluate

$$\lim_{x\to 3} \frac{x^2 + x - 12}{x - 3},$$

I say that the limit does not exist since $\lim_{x\to 3} (x - 3) = 0$ and cite Theorem 2.3.10 (limit of a quotient) as justification. Verify that the limit is actually 7 and identify my error.

In Exercises 5–38, evaluate the limits that exist.

5. $\lim_{x\to 2} 3$.

6. $\lim_{x\to 3} (5 - 4x)^2$.

7. $\lim_{x\to -4} (x^2 + 3x - 7)$.

8. $\lim_{x\to -2} 3|x - 1|$.

9. $\lim_{x\to \sqrt{3}} |x^2 - 8|$.

10. $\lim_{x\to -1} \dfrac{x^2 + 1}{3x^5 + 4}$.

11. $\lim_{x\to 0} \left(x - \dfrac{4}{x}\right)$.

12. $\lim_{x\to 5} \dfrac{2 - x^2}{4x}$.

13. $\lim_{x\to 0} \dfrac{x^2 + 1}{x - 1}$.

14. $\lim_{x\to 0} \dfrac{x^2}{x^2 + 1}$.

15. $\lim_{x\to 2} \dfrac{x}{x^2 - 4}$.

16. $\lim_{h\to 0} h\left(1 - \dfrac{1}{h}\right)$.

17. $\lim_{h\to 0} h\left(1 + \dfrac{1}{h}\right)$.

18. $\lim_{x\to 2} \dfrac{x - 2}{x^2 - 4}$.

19. $\lim_{x\to 2} \dfrac{x^2 - 4}{x - 2}$.

20. $\lim_{x\to -2} \dfrac{(x^2 - x - 6)^2}{x + 2}$.

21. $\lim_{x\to 4} \dfrac{\sqrt{x} - 2}{x - 4}$.

22. $\lim_{x\to 1} \dfrac{x - 1}{\sqrt{x} - 1}$.

23. $\lim_{x\to 1} \dfrac{x^2 - x - 6}{(x + 2)^2}$.

24. $\lim_{x\to -2} \dfrac{x^2 - x - 6}{(x + 2)^2}$.

25. $\lim_{h\to 0} \dfrac{1 - 1/h^2}{1 - 1/h}$.

26. $\lim_{h\to 0} \dfrac{1 - 1/h^2}{1 + 1/h^2}$.

27. $\lim_{h\to 0} \dfrac{1 - 1/h}{1 + 1/h}$.

28. $\lim_{h\to 0} \dfrac{1 + 1/h}{1 + 1/h^2}$.

29. $\lim_{t\to -1} \dfrac{t^2 + 6t + 5}{t^2 + 3t + 2}$.

30. $\lim_{x\to 2^+} \dfrac{\sqrt{x^2 - 4}}{x - 2}$.

31. $\lim_{t\to 0} \dfrac{t + a/t}{t + b/t}$.

32. $\lim_{x\to 1} \dfrac{x^2 - 1}{x^3 - 1}$.

33. $\lim_{x\to 1} \dfrac{x^5 - 1}{x^4 - 1}$.

34. $\lim_{h\to 0} h^2\left(1 + \dfrac{1}{h}\right)$.

35. $\lim_{h\to 0} h\left(1 + \dfrac{1}{h^2}\right)$.

36. $\lim_{x\to -4} \left(\dfrac{3x}{x + 4} + \dfrac{8}{x + 4}\right)$.

37. $\lim_{x\to -4} \left(\dfrac{2x}{x + 4} + \dfrac{8}{x + 4}\right)$.

38. $\lim_{x\to -4} \left(\dfrac{2x}{x + 4} - \dfrac{8}{x + 4}\right)$.

39. Evaluate the limits that exist.

(a) $\lim_{x\to 4} \left(\dfrac{1}{x} - \dfrac{1}{4}\right)$.

(b) $\lim_{x\to 4} \left[\left(\dfrac{1}{x} - \dfrac{1}{4}\right)\left(\dfrac{1}{x - 4}\right)\right]$.

(c) $\lim_{x\to 4} \left[\left(\dfrac{1}{x} - \dfrac{1}{4}\right)(x - 2)\right]$.

(d) $\lim_{x\to 4} \left[\left(\dfrac{1}{x} - \dfrac{1}{4}\right)\left(\dfrac{1}{x - 4}\right)^2\right]$.

40. Evaluate the limits that exist.

(a) $\lim_{x\to 3} \dfrac{x^2 + x + 12}{x - 3}$.

(b) $\lim_{x\to 3} \dfrac{x^2 + x - 12}{x - 3}$.

(c) $\lim_{x\to 3} \dfrac{(x^2 + x - 12)^2}{x - 3}$.

(d) $\lim_{x\to 3} \dfrac{x^2 + x - 12}{(x - 3)^2}$.

41. Given that $f(x) = x^2 - 4x$, evaluate the limits that exist.

(a) $\lim_{x\to 4} \dfrac{f(x) - f(4)}{x - 4}$.

(b) $\lim_{x\to 1} \dfrac{f(x) - f(1)}{x - 1}$.

(c) $\lim_{x\to 3} \dfrac{f(x) - f(1)}{x - 3}$.

(d) $\lim_{x\to 3} \dfrac{f(x) - f(2)}{x - 3}$.

42. Given that $f(x) = x^3$, evaluate the limits that exist.

(a) $\lim_{x\to 3} \dfrac{f(x) - f(3)}{x - 3}$.

(b) $\lim_{x\to 3} \dfrac{f(x) - f(2)}{x - 3}$.

(c) $\lim_{x\to 3} \dfrac{f(x) - f(3)}{x - 2}$.

(d) $\lim_{x\to 1} \dfrac{f(x) - f(1)}{x - 1}$.

43. Show by example that $\lim_{x\to c} [f(x) + g(x)]$ can exist even if $\lim_{x\to c} f(x)$ and $\lim_{x\to c} g(x)$ do not exist.

44. Show by example that $\lim_{x\to c} [f(x)g(x)]$ can exist even if $\lim_{x\to c} f(x)$ and $\lim_{x\to c} g(x)$ do not exist.

Exercises 45–51: True or false? Justify your answers.

45. If $\lim_{x\to c} [f(x) + g(x)]$ exists but $\lim_{x\to c} f(x)$ does not exist, then $\lim_{x\to c} g(x)$ does not exist.

46. If $\lim_{x\to c} [f(x) + g(x)]$ and $\lim_{x\to c} f(x)$ exist, then it can happen that $\lim_{x\to c} g(x)$ does not exist.

47. If $\lim_{x\to c} \sqrt{f(x)}$ exists, then $\lim_{x\to c} f(x)$ exists.

48. If $\lim_{x\to c} f(x)$ exists, then $\lim_{x\to c} \sqrt{f(x)}$ exists.

49. If $\lim_{x\to c} f(x)$ exists, then $\lim_{x\to c} \dfrac{1}{f(x)}$ exists.

50. If $f(x) \le g(x)$ for all $x \ne c$, then $\lim_{x \to c} f(x) \le \lim_{x \to c} g(x)$.

51. If $f(x) < g(x)$ for all $x \ne c$, then $\lim_{x \to c} f(x) < \lim_{x \to c} g(x)$.

52. (a) Verify that
$\max\{f(x), g(x)\} = \frac{1}{2}\{[f(x) + g(x)] + |f(x) - g(x)|\}$.
(b) Find a similar expression for $\min\{f(x), g(x)\}$.

53. Let $h(x) = \min\{f(x), g(x)\}$ and $H(x) = \max\{f(x), g(x)\}$. Show that

$$\text{if} \quad \lim_{x \to c} f(x) = L \quad \text{and} \quad \lim_{x \to c} g(x) = L,$$

$$\text{then} \quad \lim_{x \to c} h(x) = L \quad \text{and} \quad \lim_{x \to c} H(x) = L.$$

HINT: Use Exercise 52.

54. (*The stability of limit*) Given a function f defined on an interval $I = (c - p, c + p)$, $p > 0$. Suppose the function g is also defined on I and $g(x) = f(x)$ except, possibly, at a finite set of points $x_1, x_2, \ldots, x_n$ in I.
(a) Show that if $\lim_{x \to c} f(x) = L$, then $\lim_{x \to c} g(x) = L$.
(b) Show that if $\lim_{x \to c} f(x)$ does not exist, then $\lim_{x \to c} g(x)$ does not exist.

55. (a) Suppose that $\lim_{x \to c} f(x) = 0$ and $\lim_{x \to c} [f(x)g(x)] = 1$. Prove that $\lim_{x \to c} g(x)$ does not exist.
(b) Suppose that $\lim_{x \to c} f(x) = L \ne 0$ and $\lim_{x \to c} [f(x)g(x)] = 1$. Does $\lim_{x \to c} g(x)$ exist, and if so, what is it?

56. Suppose f is a function with the following property: $\lim_{x \to c} [f(x) + g(x)]$ does not exist whenever $\lim_{x \to c} g(x)$ does not exist. Prove that $\lim_{x \to c} f(x)$ *does* exist.

57. Fix a number x and evaluate

$$\lim_{h \to 0} \frac{f(x + h) - f(x)}{h}$$

for
(a) $f(x) = x$. (b) $f(x) = x^2$.
(c) $f(x) = x^3$. (d) $f(x) = x^4$.
(e) Guess the limit for $f(x) = x^n$, where n is any positive integer.

58. Fix a number $x \ne 0$ and repeat Exercise 57 for
(a) $f(x) = \dfrac{1}{x}$. (b) $f(x) = \dfrac{1}{x^2}$. (c) $f(x) = \dfrac{1}{x^3}$.
(d) Guess the limit for $f(x) = 1/x^n$, where n is any positive integer.
(e) What will this limit be for $f(x) = x^n$, where n is any nonzero integer? Does your "formula" also hold for $n = 0$?

*SUPPLEMENT TO SECTION 2.3

PROOF OF THEOREM 2.3.2 (ii)

We consider two cases: $\alpha \ne 0$ and $\alpha = 0$. If $\alpha \ne 0$, then $\epsilon/|\alpha| > 0$ and, since

$$\lim_{x \to c} f(x) = L$$

we know that there exists $\delta > 0$ such that,

$$\text{if} \quad 0 < |x - c| < \delta, \qquad \text{then} \quad |f(x) - L| < \frac{\epsilon}{|\alpha|}.$$

From the last inequality we obtain

$$|\alpha||f(x) - L| < \epsilon \quad \text{and thus} \quad |\alpha f(x) - \alpha L| < \epsilon.$$

The case $\alpha = 0$ was treated before in (2.2.4). ❑

PROOF OF THEOREM 2.3.2 (iii)

We begin with a little algebra:

$$\begin{aligned} |f(x)g(x) - LM| &= |[f(x)g(x) - f(x)M] + [f(x)M - LM]| \\ &\le |f(x)g(x) - f(x)M| + |f(x)M - LM| \\ &= |f(x)||g(x) - M| + |M||f(x) - L| \\ &\le |f(x)||g(x) - M| + (1 + |M|)|f(x) - L|. \end{aligned}$$

Now let $\epsilon < 0$. Since $\lim_{x\to c} f(x) = L$ and $\lim_{x\to c} g(x) = M$, we know:

(1) That there exists $\delta_1 > 0$ such that, if $0 < |x - c| < \delta_1$, then

$$|f(x) - L| < 1 \qquad \text{and thus} \qquad |f(x) < 1 + |L|.$$

(2) That there exists $\delta_2 > 0$ such that

$$\text{if } 0 < |x - c| < \delta_2, \qquad \text{then} \qquad |g(x) - M| < \left(\frac{\frac{1}{2}\epsilon}{1 + |L|}\right).$$

(3) That there exists $\delta_3 > 0$ such that

$$\text{if } \; 0 < |x - c| < \delta_3, \qquad \text{then} \qquad |f(x) - L| < \left(\frac{\frac{1}{2}\epsilon}{1 + |M|}\right).$$

We now set $\delta = \min\{\delta_1, \delta_2, \delta_3\}$ and observe that, if $0 < |x - c| < \delta$, then

$$\begin{aligned} |f(x)g(x) - LM| &< |f(x)|\,|g(x) - M| + (1 + |M|)|f(x) - L| \\ &< \underbrace{(1 + |L|)}_{\text{by (1)}}\underbrace{\left(\frac{\frac{1}{2}\epsilon}{1 + |L|}\right)}_{\text{by (2)}} + (1 + |M|)\underbrace{\left(\frac{\frac{1}{2}\epsilon}{1 + |M|}\right)}_{\text{by (3)}} = \epsilon. \qquad \square \end{aligned}$$

PROOF OF THEOREM 2.3.7

For $g(x) \neq 0$.

$$\left|\frac{1}{g(x)} - \frac{1}{M}\right| = \frac{|g(x) - M|}{|g(x)||M|}.$$

Choose $\delta_1 > 0$ such that

$$\text{if } \; 0 < |x - c| < \delta_1, \qquad \text{then } \; |g(x) - M| < \frac{|M|}{2}.$$

For such x,

$$|g(x)| > \frac{|M|}{2} \quad \text{so that} \quad \frac{1}{|g(x)|} < \frac{2}{|M|}$$

and thus

$$\left|\frac{1}{g(x)} - \frac{1}{M}\right| = \frac{|g(x) - M|}{|g(x)||M|} \leq \frac{2}{|M|^2}|g(x) - M|.$$

Now let $\epsilon > 0$ and choose $\delta_2 > 0$ such that

$$\text{if } \; 0 < |x - c| < \delta_2, \qquad \text{then } \; |g(x) - M| < \frac{|M|^2}{2}\epsilon.$$

Setting $\delta = \min\{\delta_1, \delta_2\}$, we find that

$$\text{if } \; 0 < |x - c| < \delta, \qquad \text{then } \; \left|\frac{1}{g(x)} - \frac{1}{M}\right| < \epsilon. \qquad \square$$

■ 2.4 CONTINUITY

In ordinary language, to say that a certain process is "continuous" is to say that it goes on without interruption and without abrupt changes. In mathematics the word "continuous" has much the same meaning.

The idea of continuity is so important to calculus and its applications that we discuss it with some care. First we treat the idea of *continuity at a point* (or number) c, and then we discuss *continuity on an interval.*

Continuity at a Point

DEFINITION 2.4.1

Let f be a function defined at least on an open interval $(c - p, c + p)$ with $p > 0$. Then f is *continuous at* c iff

$$\lim_{x \to c} f(x) = f(c).$$

Remark Recall that in the definition of "limit of f at c" (Definition 2.2.1) we did not require that f be defined at c itself. In contrast, the definition of "continuity at c" requires f to be defined at c. Thus, according to this definition, a function f is continuous at a point c iff:

(i) f is defined at c,

(ii) $\lim_{x \to c} f(x)$ exists,

(iii) $\lim_{x \to c} f(x) = f(c)$.

The function f is said to be *discontinuous at* c if it is not continuous there. ❑

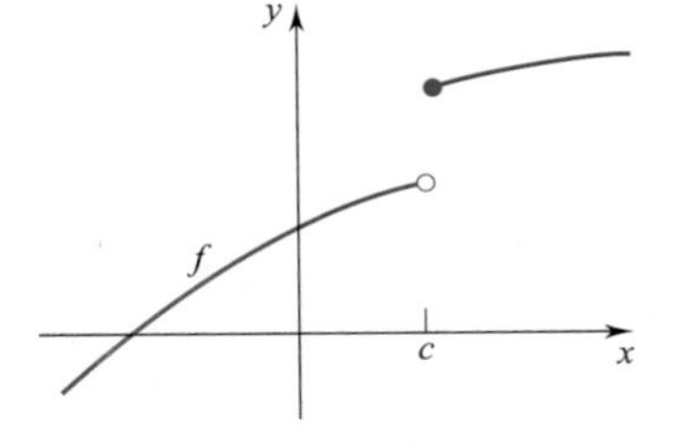

Figure 2.4.1

If the domain of f contains an interval $(c - p, c + p)$ (so that f is defined at c), then f can fail to be continuous at c only for one of two reasons: either $f(x)$ does not have a limit as x tends to c, or it does have a limit, but the limit is not $f(c)$. In the latter case, f is said to have a *removable discontinuity at* c; the discontinuity can be removed by redefining f at c. If the limit is L, define f at c to be L.

The function depicted in Figure 2.4.1 is discontinuous at c because it does not have a limit at c. Note that $\lim_{x \to c^-} f(x)$ and $\lim_{x \to c^+} f(x)$ each exist, but they are not equal. A discontinuity of this particular type is called a *jump discontinuity.*

The function depicted in Figure 2.4.2 does have a limit at c. It is discontinuous at c only because its limit at c is not its value at c. The discontinuity is removable; it can be removed by lowering the dot into place (by redefining f at c).

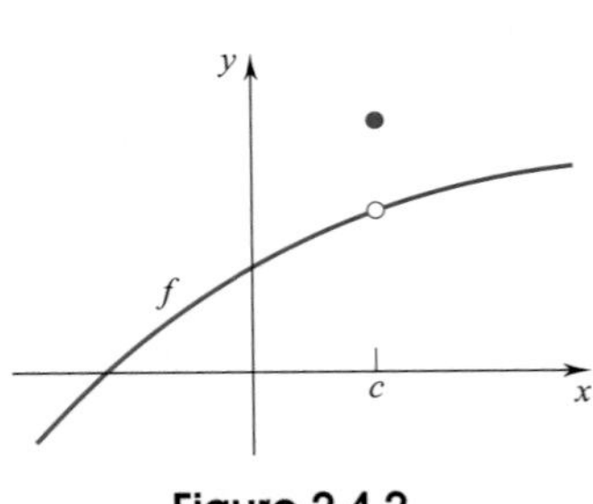

Figure 2.4.2

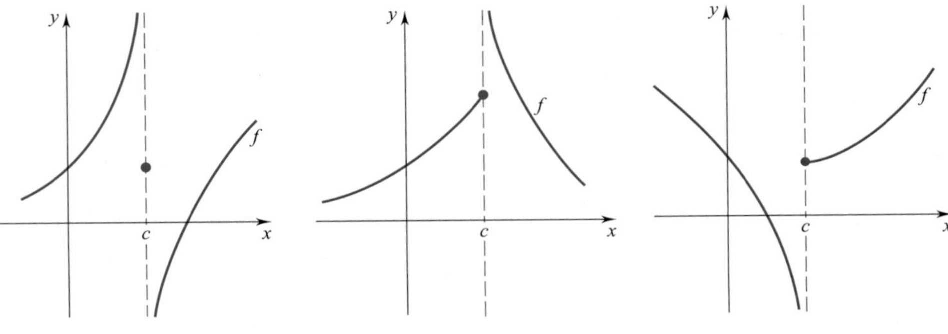

Figure 2.4.3

Additional examples of functions that fail to be continuous at a point c are shown in Figure 2.4.3. In each of these examples, at least one of the one-sided limits

$$\lim_{x \to c^-} f(x) \qquad \text{or} \qquad \lim_{x \to c^+} f(x)$$

fails to exist.

In Figure 2.4.4, we have tried to suggest the Dirichlet function

$$f(x) = \begin{cases} 1, & x \text{ rational} \\ 0, & x \text{ irrational.} \end{cases}$$

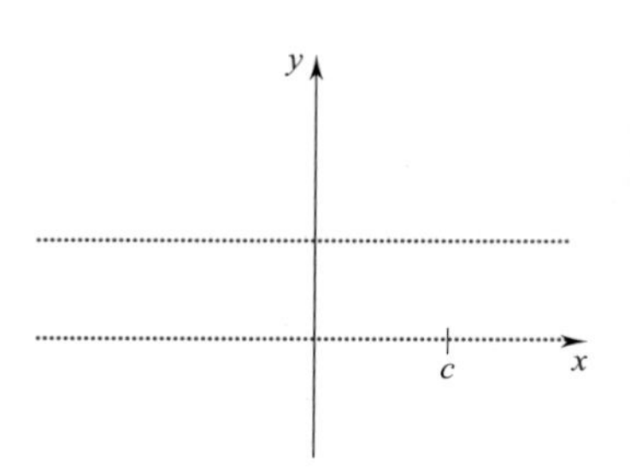

Figure 2.4.4

At no point c does f have a limit. It is therefore everywhere discontinuous.

Most of the functions that you have encountered so far are continuous at each point of their domains. In particular, this is true for polynomials P,

$$\lim_{x \to c} P(x) = P(c), \qquad \text{(see 2.3.6)}$$

for rational functions (quotients of polynomials) $R = P/Q$,

$$\lim_{x \to c} R(x) = \lim_{x \to c} \frac{P(x)}{Q(x)} = \frac{P(c)}{Q(c)} = R(c) \qquad \text{provided} \quad Q(c) \neq 0, \qquad \text{(see 2.3.9)}$$

and for the absolute-value function,

$$\lim_{x \to c} |x| = |c|. \qquad \text{(see 2.2.3)}$$

In Exercise 53, Section 2.2, you were asked to show that

$$\lim_{x \to c} \sqrt{x} = \sqrt{c} \qquad \text{for each } c > 0.$$

This makes the square-root function continuous at each positive number. We discuss later what happens for $c = 0$.

With f and g continuous at c, we have

$$\lim_{x \to c} f(x) = f(c), \qquad \lim_{x \to c} g(x) = g(c)$$

and thus, by the limit theorems,

$$\lim_{x \to c} [f(x) + g(x)] = f(c) + g(c), \qquad \lim_{x \to c} [\alpha f(x)] = \alpha f(c) \quad \text{for each real } \alpha$$

$$\lim_{x \to c} [f(x)g(x)] = f(c)g(c), \qquad \text{and, if } g(c) \neq 0, \ \lim_{x \to c} f(x)/g(x) = f(c)/g(c).$$

In summary, we have the following theorem:

THEOREM 2.4.2

If f and g are continuous at c, then

(i) $f + g$ is continuous at c **(ii)** αf is continuous at c for each real α,
(iii) fg is continuous at c, **(iv)** f/g is continuous at c provided $g(c) \neq 0$.

Parts (i) and (iii) can be extended to any finite number of functions. Also, since $f - g = f + (-1)g$, it follows from (i) and (ii) that $f - g$ is continuous at c.

Example 1 The function

$$F(x) = 3|x| + \frac{x^3 - x}{x^2 - 5x + 6} + 4$$

is continuous at all real numbers other than 2 and 3. You can see this by noting that

$$F = 3f + g/h + k$$

where

$$f(x) = |x|,\ g(x) = x^3 - x,\ h(x) = x^2 - 5x + 6, \text{ and } k(x) = 4.$$

Since f, g, h, and k are everywhere continuous, F is continuous except at 2 and 3, the numbers where h takes on the value 0 (that is, the numbers where F is not defined). ❑

Since continuity at a point c is defined in terms of a limit, there is an ϵ, δ version of the definition. A direct translation of

$$\lim_{x \to c} f(x) = f(c)$$

into ϵ, δ terms reads like this: for each $\epsilon > 0$ there exists $\delta > 0$ such that

$$\text{if} \quad 0 < |x - c| < \delta \qquad \text{then} \quad |f(x) - f(c)| < \epsilon.$$

In the case of continuity at c, the restriction $0 < |x - c|$ is unnecessary. We can allow $|x - c| = 0$ because then $x = c$, $f(x) = f(c)$, and thus $|f(x) - f(c)| = 0$. Being 0, $|f(x) - f(c)|$ is certainly less than ϵ.

Thus, an ϵ, δ characterization of continuity at c reads as follows:

(2.4.3)

$$f \text{ is continuous at } c \quad \text{iff} \quad \begin{cases} \text{for each } \epsilon > 0 \text{ there exists } \delta > 0 \text{ such that} \\ \text{if} \quad |x - c| < \delta \quad \text{then} \quad |f(x) - f(c)| < \epsilon. \end{cases}$$

In simple intuitive language

$$f \text{ is continuous at } c \quad \text{iff} \quad \text{for } x \text{ close to } c,\ f(x) \text{ is close to } f(c).$$

We are now ready to take up the continuity of composite functions. Remember the defining formula: $(f \circ g)(x) = f(g(x))$.

THEOREM 2.4.4

If g is continuous at c and f is continuous at $g(c)$, then the composition $f \circ g$ is continuous at c.

The idea here is simple: with g continuous at c, we know that

$$\text{for } x \text{ close to } c, \quad g(x) \text{ is close to } g(c);$$

from the continuity of f at $g(c)$, we know that

$$\text{with } g(x) \text{ close to } g(c), \quad f(g(x)) \text{ is close to } f(g(c)).$$

In summary,

$$\text{with } x \text{ close to } c, \quad f(g(x)) \text{ is close to } f(g(c)).$$ ❑

The argument we just gave is too vague to be a proof. Here in contrast is a proof. We begin with $\epsilon > 0$. We must show that there exists some number $\delta > 0$ such that

$$\text{if} \quad |x - c| < \delta, \qquad \text{then} \quad |f(g(x)) - f(g(c))| < \epsilon.$$

In the first place, we observe that, since f is continuous at $g(c)$, there does exist a number $\delta_1 > 0$ such that

$$(1) \qquad \text{if} \quad |t - g(c)| < \delta_1, \qquad \text{then} \quad |f(t) - f(g(c))| < \epsilon.$$

With $\delta_1 > 0$, we know from the continuity of g at c that there exists a number $\delta > 0$ such that

$$(2) \qquad \text{if} \quad |x - c| < \delta, \qquad \text{then} \quad |g(x) - g(c)| < \delta_1.$$

Combining (2) and (1) we have what we want: by (2)

$$\text{if} \quad |x - c| < \delta, \qquad \text{then} \quad |g(x) - g(c)| < \delta_1$$

so that by (1)

$$|f(g(x)) - f(g(c))| < \epsilon. \quad \square$$

This proof is illustrated in Figure 2.4.5. The numbers within δ of c are taken by g to within δ_1 of $g(c)$, and then by f to within ϵ of $f(g(c))$.

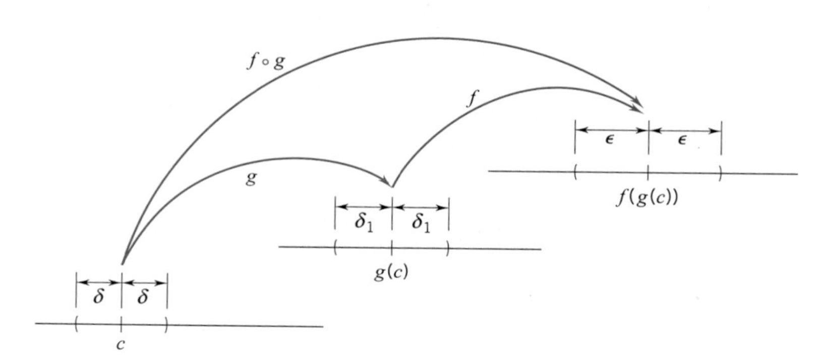

Figure 2.4.5

It is time to look at some examples.

Example 2 The function

$$F(x) = \sqrt{\frac{x^2 + 1}{(x - 8)^3}}$$

is continuous at all numbers greater than 8. To see this, note that $F = f \circ g$, where

$$f(x) = \sqrt{x} \qquad \text{and} \qquad g(x) = \frac{x^2 + 1}{(x - 8)^3}.$$

Now, take any $c > 8$. Since g is a rational function and g is defined at c, g is continuous at c. Also, since $g(c)$ is positive and f is continuous at each positive number, f is continuous at $g(c)$. By Theorem 2.4.4, F is continuous at c. $\square$

The continuity of composites holds for any finite number of functions. The only requirement is that each function be continuous *where it is applied.*

Example 3 The function

$$F(x) = \frac{1}{5 - \sqrt{x^2 + 16}}$$

is continuous everywhere except at $x = \pm 3$, where it is not defined. To see this, note that $F = f \circ g \circ k \circ h$, where

$$f(x) = \frac{1}{x}, \qquad g(x) = 5 - x, \qquad k(x) = \sqrt{x}, \qquad h(x) = x^2 + 16$$

and observe that each of these functions is being evaluated only where it is continuous. In particular, g and h are continuous everywhere, f is being evaluated only at nonzero numbers, and k is being evaluated only at positive numbers. ❑

Just as we considered one-sided limits, we can consider one-sided continuity.

DEFINITION 2.4.5 ONE-SIDED CONTINUITY

A function f is called

$$\textit{continuous from the left at } c \quad \text{iff} \quad \lim_{x \to c^-} f(x) = f(c).$$

It is called

$$\textit{continuous from the right at } c \quad \text{iff} \quad \lim_{x \to c^+} f(x) = f(c).$$

In Figure 2.4.6 we have an example of continuity from the right at 0; in Figure 2.4.7 we have an example of continuity from the left at 1.

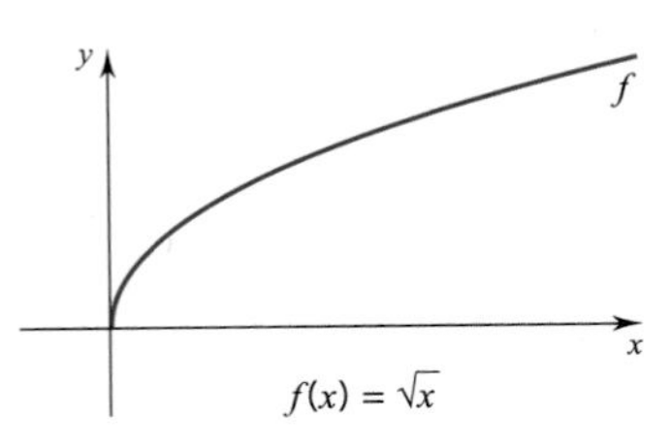

Figure 2.4.6

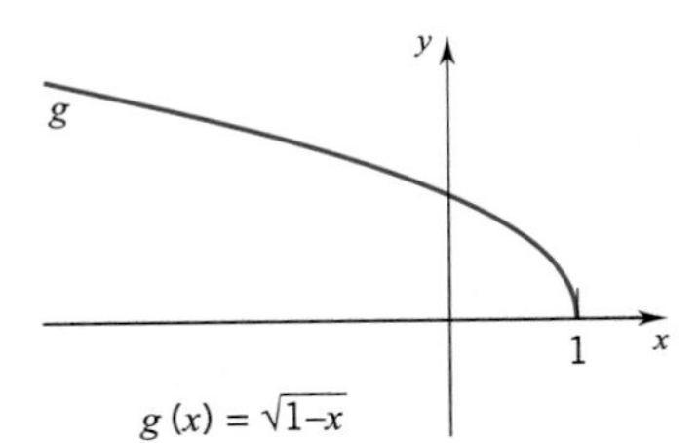

Figure 2.4.7

It is clear from (2.2.8) that a function f is continuous at c iff it is continuous from both sides at c. Thus

(2.4.6) f is continuous at c iff $f(c)$, $\lim_{x \to c^-} f(x)$, $\lim_{x \to c^+} f(x)$ all exist and are equal.

We apply this result to piecewise-defined functions in the following examples.

Example 4 Determine the discontinuities, if any, of the following function:

$$f(x) = \begin{cases} 2x + 1, & x \leq 0 \\ 1, & 0 < x \leq 1 \\ x^2 + 1, & x > 1. \end{cases} \qquad \text{(Figure 2.4.8)}$$

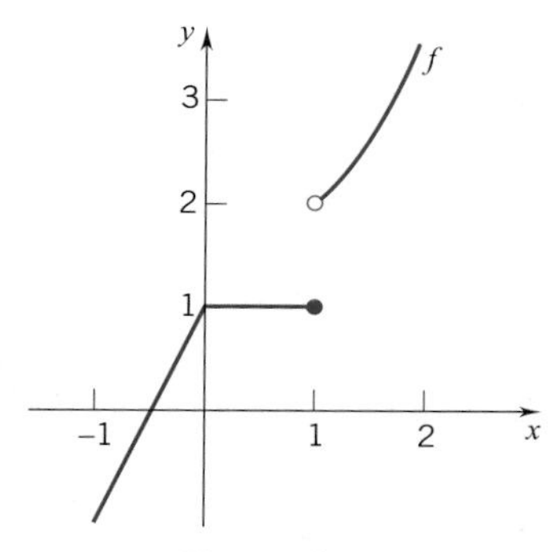

Figure 2.4.8

SOLUTION Clearly, f is continuous at each point in the open intervals $(-\infty, 0)$, $(0, 1)$, and $(1, \infty)$ since f is a polynomial in each of these intervals. Thus, we only have to check the behavior of f at $x = 0$ and $x = 1$. The figure suggests that f is continuous at 0 and discontinuous at 1. Indeed, at $x = 0$, $f(0) = 1$,

$$\lim_{x \to 0^-} f(x) = \lim_{x \to 0^-} (2x + 1) = 1 \quad \text{and} \quad \lim_{x \to 0^+} f(x) = \lim_{x \to 0^+} (1) = 1,$$

so f is continuous at 0. At $x = 1$,

$$\lim_{x \to 1^-} f(x) = \lim_{x \to 1^-} (1) = 1 \quad \text{and} \quad \lim_{x \to 1^+} f(x) = \lim_{x \to 1^+} (x^2 + 1) = 2.$$

Thus, f is discontinuous at 1; in fact, f has a jump discontinuity at this point. ❑

Example 5 Determine the discontinuities of the following function:

$$f(x) = \begin{cases} x^3, & x \leq -1 \\ x^2 - 2, & -1 < x < 0 \\ 3 - x, & 0 \leq x < 2 \\ \dfrac{4x - 1}{x - 1}, & 2 \leq x < 4 \\ \dfrac{15}{7 - x}, & 4 < x < 7 \\ 5x + 2, & 7 \leq x. \end{cases}$$

SOLUTION It should be clear that f is continuous at each point in the open intervals $(-\infty, -1)$, $(-1, 0)$, $(0, 2)$, $(2, 4)$, $(4, 7)$, and $(7, \infty)$. All we have to check is the behavior of f at $x = -1, 0, 2, 4$, and 7. To do so, we apply (2.4.6).

At $x = -1$, f is continuous since $f(-1) = (-1)^3 = -1$,

$$\lim_{x \to -1^-} f(x) = \lim_{x \to -1^-} x^3 = -1, \quad \text{and} \quad \lim_{x \to -1^+} f(x) = \lim_{x \to -1^+} (x^2 - 2) = -1.$$

Our findings at the other four points are displayed in the chart below. Try to verify each entry.

c	$f(c)$	$\lim_{x \to c^-} f(x)$	$\lim_{x \to c^+} f(x)$	Conclusion
0	3	-2	3	Discontinuous
2	7	1	7	Discontinuous
4	Not defined	5	5	Discontinuous
7	37	Does not exist	37	Discontinuous

Note that the discontinuity at 4 is removable; if we define $f(4) = 5$, then f is continuous at 4. The discontinuities at $x = 0$ and $x = 2$ are jump discontinuities. ❑

Continuity on Intervals

Let (a, b) be an open interval. A function f is *continuous on* (a, b) if it is continuous at each point $c \in (a, b)$. For a function defined on a closed interval $[a, b]$, the most continuity that we can expect is:

1. Continuity on the open interval (a, b),
2. Continuity from the right at a,
3. Continuity from the left at b.

A function f that fulfills these requirements is said to be *continuous on* $[a, b]$. These definitions can be extended in an obvious way to functions defined on half-open intervals and on infinite intervals.

Functions that are continuous on a closed interval $[a, b]$ have some important special properties not shared by functions in general. We discuss two such properties in Section 2.6.

The concept of *continuity on an interval* also has a geometric interpretation, namely a function f defined on an interval I is continuous on that interval if its graph has no "holes" or "jumps." That is, f is continuous on I if its graph is a "connected" curve. In practical terms, this means that you can draw the graph of f without lifting your pencil from the paper. Look back over the examples in this section to verify this.

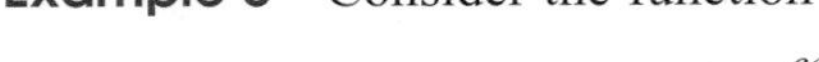

$f(x) = \sqrt{1 - x^2}$

Figure 2.4.9

Example 6 Consider the function

$$f(x) = \sqrt{1 - x^2}.$$

The graph of f is the semicircle shown in Figure 2.4.9. The function is continuous on $[-1, 1]$ because it is continuous on the open interval $(-1, 1)$, continuous from the right at -1, and continuous from the left at 1. ❑

Example 7 Let g be the function defined by

$$g(x) = \sqrt{x^2 - 4}.$$

Note that $\text{dom}(g) = \{x : x^2 - 4 \geq 0\}$. You can verify that this is the same as the set $(-\infty, -2] \cup [2, \infty)$. The function g is continuous on each of these intervals since it is continuous on each of the open intervals $(-\infty, -2)$ and $(2, \infty)$, continuous from the left at -2, and continuous from the right at 2. ❑

EXERCISES 2.4

1. The graph of a function f is given in the figure.
   (a) At which points is f discontinuous?
   (b) For each point of discontinuity found in (a), determine whether f is continuous from the right, from the left, or neither.
   (c) Which, if any, of the points of discontinuity found in (a) is removable? Which, if any, is a jump discontinuity?

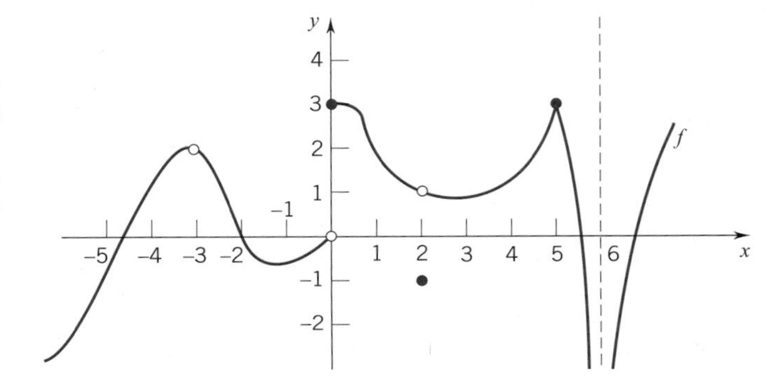

2. The graph of a function g is given in the figure. Determine the intervals on which g is continuous.

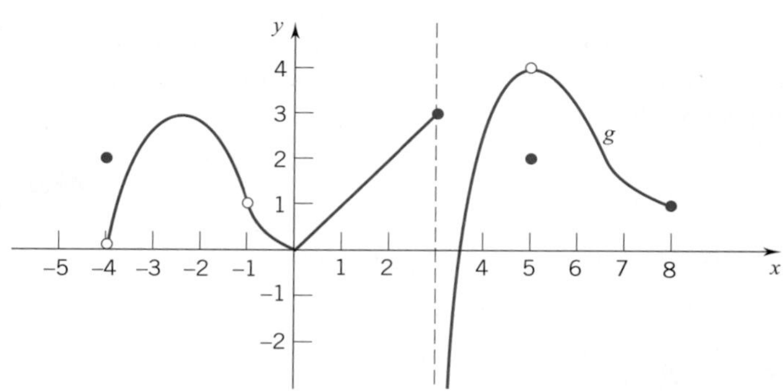

In Exercises 3–16, determine whether or not the function is continuous at the indicated point. If not, determine whether the discontinuity is a removable discontinuity, a jump discontinuity, or neither.

3. $f(x) = x^3 - 5x + 1; \quad x = 2.$

4. $g(x) = \sqrt{(x-1)^2 + 5}; \quad x = 1.$

5. $f(x) = \sqrt{x^2 + 9}; \quad x = 3.$

6. $f(x) = |4 - x^2|; \quad x = 2.$

7. $f(x) = \begin{cases} x^2 + 4, & x < 2 \\ x^3, & x \geq 2; \end{cases} \quad x = 2.$

8. $h(x) = \begin{cases} x^2 + 5, & x < 2 \\ x^3, & x \geq 2; \end{cases} \quad x = 2.$

9. $g(x) = \begin{cases} x^2 + 4, & x < 2 \\ 5, & x = 2 \\ x^3, & x > 2; \end{cases} \quad x = 2.$

10. $g(x) = \begin{cases} x^2 + 5, & x < 2 \\ 10, & x = 2 \\ 1 + x^3, & x > 2; \end{cases} \quad x = 2.$

11. $f(x) = \begin{cases} -1, & x < 0 \\ 0, & x = 0 \\ 1, & x > 0; \end{cases} \quad x = 0.$

12. $f(x) = \begin{cases} 1 - x, & x < 1 \\ 1, & x = 1 \\ x^2 - 1, & x > 1; \end{cases} \quad x = 1.$

13. $h(x) = \begin{cases} \dfrac{x^2 - 1}{x + 1}, & x \neq -1 \\ -2, & x = -1; \end{cases} \quad x = -1.$

14. $g(x) = \begin{cases} \dfrac{1}{x + 1}, & x \neq -1 \\ 0, & x = -1; \end{cases} \quad x = -1.$

15. $f(x) = \begin{cases} -x^2, & x < 0 \\ 1 - \sqrt{x}, & x \geq 0; \end{cases} \quad x = 0.$

16. $g(x) = \dfrac{x(x+1)(x+2)}{\sqrt{(x-1)(x-2)}}; \quad x = -2.$

In Exercises 17–30, sketch the graph and classify the discontinuities (if any) as being a removable discontinuity, a jump discontinuity, or neither.

17. $f(x) = \dfrac{x^2 - 4}{x - 2}.$

18. $f(x) = \dfrac{x - 3}{x^2 - 9}.$

19. $f(x) = |x - 1|.$

20. $h(x) = |x^2 - 1|.$

21. $g(x) = \begin{cases} x - 1, & x < 1 \\ 0, & x = 1 \\ x^2, & x > 1. \end{cases}$

22. $g(x) = \begin{cases} 2x - 1, & x < 1 \\ 0, & x = 1 \\ x^2, & x > 1. \end{cases}$

23. $f(x) = \max\,\{x, x^2\}$

24. $f(x) = \min\,\{x, x^2\}.$

25. $f(x) = \begin{cases} -1, & x < -1 \\ x^3, & -1 \leq x \leq 1 \\ 1, & 1 < x. \end{cases}$

26. $g(x) = \begin{cases} 1, & x \leq -2 \\ \frac{1}{2}x, & -2 < x < 4 \\ \sqrt{x}, & 4 \leq x. \end{cases}$

27. $h(x) = \begin{cases} 1, & x \leq 0 \\ x^2, & 0 < x < 1 \\ 1, & 1 \leq x < 2 \\ x, & 2 \leq x. \end{cases}$

28. $g(x) = \begin{cases} -x^2, & x < -1 \\ 3, & x = -1 \\ 2 - x, & -1 < x \leq 1 \\ x^2, & 1 < x. \end{cases}$

29. $f(x) = \begin{cases} 2x + 9, & x < -2 \\ x^2 + 1, & -2 < x \leq 1 \\ 3x - 1, & 1 < x < 3 \\ x + 6, & 3 < x. \end{cases}$

30. $g(x) = \begin{cases} x + 7, & x < -3 \\ |x - 2|, & -3 < x < -1 \\ x^2 - 2x, & -1 < x < 3 \\ 2x - 3, & 3 \leq x. \end{cases}$

Each of the functions f in Exercises 31–34 is defined everywhere except at $x = 1$. Where possible, define f at 1 so that it becomes continuous at 1.

31. $f(x) = \dfrac{x^2 - 1}{x - 1}$.

32. $f(x) = \dfrac{1}{x - 1}$.

33. $f(x) = \dfrac{x - 1}{|x - 1|}$.

34. $f(x) = \dfrac{(x - 1)^2}{|x - 1|}$.

35. Let $f(x) = \begin{cases} x^2, & x < 1 \\ Ax - 3, & x \geq 1. \end{cases}$ Find A given that f is continuous at 1.

36. Let $f(x) = \begin{cases} A^2x^2, & x \leq 2 \\ (1 - A)x, & x > 2. \end{cases}$ For what values of A is f continuous at 2?

37. Give necessary and sufficient conditions on A and B for the function

$$f(x) = \begin{cases} Ax - B, & x \leq 1 \\ 3x, & 1 < x < 2 \\ Bx^2 - A, & 2 \leq x \end{cases}$$

to be continuous at $x = 1$ but discontinuous at $x = 2$.

38. Give necessary and sufficient conditions on A and B for the function in Exercise 37 to be continuous at $x = 2$ but discontinuous at $x = 1$.

In Exercises 39–42, define the function at 5 so that it becomes continuous at 5.

39. $f(x) = \dfrac{\sqrt{x + 4} - 3}{x - 5}$.

40. $f(x) = \dfrac{\sqrt{x + 4} - 3}{\sqrt{x - 5}}$.

41. $f(x) = \dfrac{\sqrt{2x - 1} - 3}{x - 5}$.

42. $f(x) = \dfrac{\sqrt{x^2 - 7x + 16} - \sqrt{6}}{(x - 5)\sqrt{x + 1}}$.

In Exercises 43–45, at what points (if any) is the given function continuous?

43. $f(x) = \begin{cases} 1, & x \text{ rational} \\ 0, & x \text{ irrational.} \end{cases}$

44. $g(x) = \begin{cases} x, & x \text{ rational} \\ 0, & x \text{ irrational.} \end{cases}$

45. $g(x) = \begin{cases} 2x, & x \text{ an integer} \\ x^2, & \text{otherwise.} \end{cases}$

46. Three important functions in science and engineering are:

1. The *Heaviside function* $H_c(x) = \begin{cases} 0, & x < c \\ 1, & x \geq c, \end{cases}$

   where c is a given number.

2. The *unit pulse function*

   $P_{\epsilon,c}(x) = \frac{1}{\epsilon}[H_c(x) - H_{c+\epsilon}(x)]$

3. The *Dirac delta function* $\delta_c(x) = \lim_{\epsilon \to 0} P_{\epsilon,c}(x)$.

(a) Graph the functions H_c, $P_{\epsilon,c}$, and δ_c.
(b) Determine where each of the functions is continuous.
(c) Find $\lim_{x \to c^-} H_c(x)$ and $\lim_{x \to c^+} H_c(x)$. What can you say about $\lim_{x \to c} H(x)$?

47. (*Important*) Prove that

$$f \text{ is continuous at } c \quad \text{iff} \quad \lim_{h \to 0} f(c + h) = f(c).$$

48. (*Important*) Let f and g be continuous at c. Prove that if:
(a) $f(c) > 0$, then there exists $\delta > 0$ such that $f(x) > 0$ for all $x \in (c - \delta, c + \delta)$.
(b) $f(c) < 0$, then there exists $\delta > 0$ such that $f(x) < 0$ for all $x \in (c - \delta, c + \delta)$.
(c) $f(c) < g(c)$, then there exists $\delta > 0$ such that $f(x) < g(x)$ for all $x \in (c - \delta, c + \delta)$.

49. Suppose that f has a discontinuity at c which is not removable. Let g be a function such that $g(x) = f(x)$ except, possibly, at a finite set of points $x_1, x_2, \ldots, x_n$. Show that g also has a nonremovable discontinuity at c.

50. (a) Prove that if f is continuous everywhere, then $|f|$ is continuous everywhere.
(b) Give an example to show that the continuity of $|f|$ does not imply the continuity of f.
(c) Give an example of a function f such that f is continuous nowhere, but $|f|$ is continuous everywhere.

51. Suppose the function f has the property that there exists a number B such that

$$|f(x) - f(c)| \leq B|x - c|$$

for all x in the interval $(c - p, c + p)$, $p > 0$. Prove that f is continuous at c.

52. Suppose the function f has the property that

$$|f(x) - f(t)| \leq |x - t|$$

for each pair of points x, t in the interval (a, b). Prove that f is continuous on (a, b).

53. Prove that if

$$\lim_{h \to 0} \frac{f(c + h) - f(c)}{h} = L$$

exists, then f is continuous at c.

54. Suppose that the function f is continuous on $(-\infty, \infty)$. Show that f can be written

$$f = f_e + f_o,$$

where f_e is an even function which is continuous on $(-\infty, \infty)$ and f_o is an odd function which is continuous on $(-\infty, \infty)$. HINT: See Exercise 56, Section 1.7.

■ 2.5 THE PINCHING THEOREM; TRIGONOMETRIC LIMITS

Figure 2.5.1 shows the graphs of three functions f, g, h. Suppose that, as suggested by the figure, for x close to c, f is trapped between g and h. (The values of these functions at c itself are irrelevant.) If, as x tends to c, both $g(x)$ and $h(x)$ tend to the same limit L, then $f(x)$ also tends to L. This idea is made precise in what we call *the pinching theorem.*

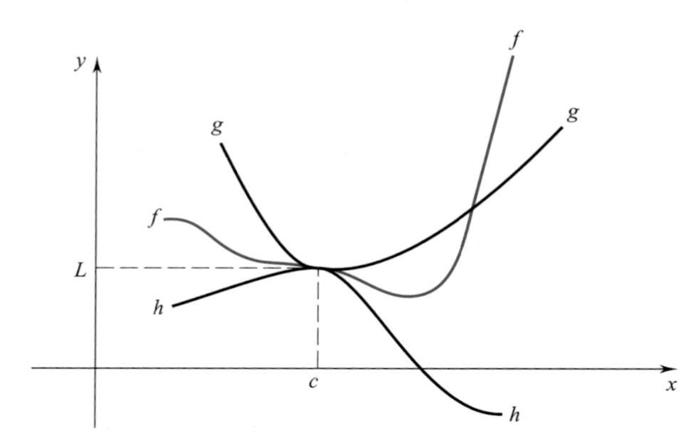

Figure 2.5.1

THEOREM 2.5.1 THE PINCHING THEOREM

Let $p > 0$. Suppose that, for all x such that $0 < |x - c| < p$,

$$h(x) \leq f(x) \leq g(x).$$

If

$$\lim_{x \to c} h(x) = L \quad \text{and} \quad \lim_{x \to c} g(x) = L,$$

then

$$\lim_{x \to c} f(x) = L.$$

PROOF Let $\epsilon > 0$. Let $p > 0$ be such that

$$\text{if} \quad 0 < |x - c| < p, \quad \text{then} \quad h(x) \leq f(x) \leq g(x).$$

Choose $\delta_1 > 0$ such that

$$\text{if} \quad 0 < |x - c| < \delta_1, \quad \text{then} \quad L - \epsilon < h(x) < L + \epsilon.$$

Choose $\delta_2 > 0$ such that

$$\text{if} \quad 0 < |x - c| < \delta_2, \quad \text{then} \quad L - \epsilon < g(x) < L + \epsilon.$$

Let $\delta = \min\{p, \delta_1, \delta_2\}$. For x satisfying $0 < |x - c| < \delta$, we have

$$L - \epsilon < h(x) \leq f(x) \leq g(x) < L + \epsilon,$$

and thus

$$|f(x) - L| < \epsilon. \quad \square$$

Remark With obvious modifications, the pinching theorem also holds for one-sided limits. We do not spell out the details here because we will be working with two-sided limits throughout this section. ❑

We come now to some trigonometric limits. All calculations are based on radian measure.

As our first application of the pinching theorem we prove that

(2.5.2)
$$\lim_{x \to 0} \sin x = 0.$$

PROOF To follow the argument, see Figure 2.5.2.†

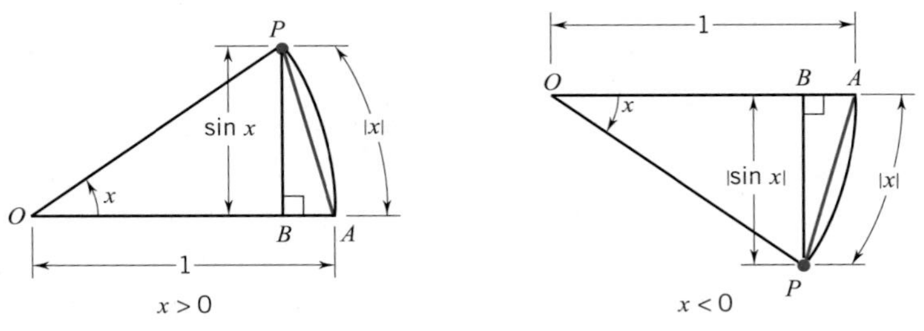

Figure 2.5.2

For small $x \neq 0$

$$0 < |\sin x| = \text{length of } \overline{BP} < \text{length of } \overline{AP} < \text{length of } \widehat{AP} = |x|.$$

Thus, for such x

$$0 < |\sin x| < |x|.$$

Since

$$\lim_{x \to 0} 0 = 0 \quad \text{and} \quad \lim_{x \to 0} |x| = 0.$$

we know from the pinching theorem that

$$\lim_{x \to 0} |\sin x| = 0 \quad \text{and therefore} \quad \lim_{x \to 0} \sin x = 0. \quad ❑$$

From this it follows readily that

(2.5.3)
$$\lim_{x \to 0} \cos x = 1.$$

PROOF In general, $\cos^2 x + \sin^2 x = 1$. For x close to 0, the cosine is positive and we have

$$\cos x = \sqrt{1 - \sin^2 x}.$$

As x tends to 0, $\sin x$ tends to 0, $\sin^2 x$ tends to 0, and therefore $\cos x$ tends to 1. ❑

† Recall that in a circle of radius 1, a central angle of x radians subtends an arc of length $|x|$.

Next we show that the sine and cosine are everywhere continuous; which is to say, for all real numbers c,

(2.5.4)
$$\lim_{x \to c} \sin x = \sin c \qquad \text{and} \qquad \lim_{x \to c} \cos x = \cos c.$$

PROOF Take any real number c. By (2.2.5) we can write

$$\lim_{x \to c} \sin x \quad \text{as} \quad \lim_{h \to 0} \sin (c + h).$$

This form of the limit suggests that we use the addition formula

$$\sin (c + h) = \sin c \cos h + \cos c \sin h.$$

Since $\sin c$ and $\cos c$ are constants, we have

$$\begin{aligned} \lim_{h \to 0} \sin (c + h) &= (\sin c)(\lim_{h \to 0} \cos h) + (\cos c)(\lim_{h \to 0} \sin h) \\ &= (\sin c)(1) + (\cos c)(0) = \sin c. \end{aligned}$$

The proof that $\lim_{x \to c} \cos x = \cos c$ is left to you. ❑

The remaining trigonometric functions

$$\tan x = \frac{\sin x}{\cos x}, \qquad \cot x = \frac{\cos x}{\sin x}, \qquad \sec x = \frac{1}{\cos x}, \qquad \csc x = \frac{1}{\sin x}$$

are all continuous where defined. Justification? They are all quotients of continuous functions.

We turn now to two limits, the importance of which will become clear in Chapter 3:

(2.5.5)
$$\lim_{x \to 0} \frac{\sin x}{x} = 1 \qquad \text{and} \qquad \lim_{x \to 0} \frac{1 - \cos x}{x} = 0.$$

Remark These results were suggested in Section 2.1 using a numerical approach. See Example 14 and Exercise 60 in Section 2.1. ❑

PROOF We show that

$$\lim_{x \to 0} \frac{\sin x}{x} = 1$$

by using some simple geometry and the pinching theorem. For any x satisfying $0 < x < \pi/2$ (see Figure 2.5.3), length of $\overline{PB} = \sin x$, length of $\overline{OB} = \cos x$, and length $\overline{OA} = 1$. Also, since triangle OAQ is a right triangle, we have $\tan x = \overline{QA}/1$ or

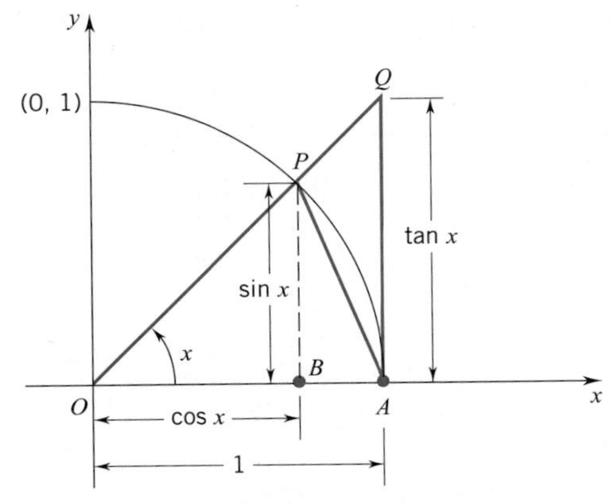

Figure 2.5.3

$\overline{QA} = \tan x$. Now

$$\text{Area of triangle } OAP = \tfrac{1}{2}(1)\sin x = \tfrac{1}{2}\sin x$$

$$\text{Area of sector } OAP = \tfrac{1}{2}(1)^2 x = \tfrac{1}{2}x$$

$$\text{Area of triangle } OAQ = \tfrac{1}{2}(1)\tan x = \tfrac{1}{2}\frac{\sin x}{\cos x}.$$

Since triangle $OAP \subseteq$ sector $OAP \subseteq$ triangle OAQ (and these are all proper containments), we have

$$\tfrac{1}{2}\sin x < \tfrac{1}{2}x < \tfrac{1}{2}\frac{\sin x}{\cos x}$$

$$1 < \frac{x}{\sin x} < \frac{1}{\cos x}$$

$$\cos x < \frac{\sin x}{x} < 1.$$

(taking reciprocals, see Exercise 67, Section 1.3)

This inequality was derived for $x > 0$, but since

$$\cos(-x) = \cos x \qquad \text{and} \qquad \frac{\sin(-x)}{-x} = \frac{-\sin x}{-x} = \frac{\sin x}{x},$$

this inequality also holds for $x < 0$.

We can now apply the pinching theorem. Since

$$\lim_{x\to 0} \cos x = 1 \qquad \text{and} \qquad \lim_{x\to 0} 1 = 1,$$

we can conclude that

$$\lim_{x\to 0} \frac{\sin x}{x} = 1.$$

We can now show that

$$\lim_{x\to 0} \frac{1 - \cos x}{x} = 0.$$

For small $x \neq 0$, $\cos x$ is close to 1 and so $\cos x \neq -1$. Therefore, we can write

$$\begin{aligned}\frac{1-\cos x}{x} &= \left(\frac{1-\cos x}{x}\right)\left(\frac{1+\cos x}{1+\cos x}\right)^{\dagger}\\ &= \frac{1-\cos^2 x}{x(1+\cos x)}\\ &= \frac{\sin^2 x}{x(1+\cos x)}\\ &= \left(\frac{\sin x}{x}\right)\left(\frac{\sin x}{1+\cos x}\right).\end{aligned}$$

Now, since

$$\lim_{x\to 0}\frac{\sin x}{x} = 1 \qquad \text{and} \qquad \lim_{x\to 0}\frac{\sin x}{1+\cos x} = 0,$$

it follows that

$$\lim_{x\to 0}\frac{1-\cos x}{x} = 0. \quad \square$$

We are now in a position to evaluate a variety of trigonometric limits.

Example 1 Find

$$\lim_{x\to 0}\frac{\sin 4x}{x}.$$

SOLUTION You know that

$$\lim_{x\to 0}\frac{\sin x}{x} = 1.$$

From this it follows that

$$\lim_{x\to 0}\frac{\sin 4x}{4x} = 1. \qquad \text{(see Exercise 34)}$$

Now,

$$\lim_{x\to 0}\frac{\sin 4x}{x} = \lim_{x\to 0} 4\left(\frac{\sin 4x}{4x}\right) = 4\lim_{x\to 0}\frac{\sin 4x}{4x} = 4. \quad \square$$

Example 2 Find

$$\lim_{x\to 0} x\cot 3x.$$

SOLUTION We begin by writing

$$x\cot 3x = x\,\frac{\cos 3x}{\sin 3x}.$$

† This "trick" is a fairly common procedure with trigonometric expressions. It is much like using "conjugates" to revise algebraic expressions:

$$\frac{3}{4+\sqrt{2}} = \frac{3}{4+\sqrt{2}}\cdot\frac{4-\sqrt{2}}{4-\sqrt{2}} = \frac{3(4-\sqrt{2})}{14}.$$

Since

$$\lim_{x\to 0} \frac{\sin x}{x} = 1 \qquad \text{and equivalently} \qquad \lim_{x\to 0} \frac{x}{\sin x} = 1,$$

we would like to "pair off" $\sin 3x$ with $3x$. Thus, we write

$$x \frac{\cos 3x}{\sin 3x} = \frac{1}{3}\left(\frac{3x}{\sin 3x}\right)(\cos 3x).$$

Now we take the limit:

$$\lim_{x\to 0} x \cot 3x = \frac{1}{3}\left(\lim_{x\to 0} \frac{3x}{\sin 3x}\right)\left(\lim_{x\to 0} \cos 3x\right) = \frac{1}{3}(1)(1) = \frac{1}{3}. \quad \square$$

Example 3 Find

$$\lim_{x\to \pi/4} \frac{\sin(x - \frac{1}{4}\pi)}{(x - \frac{1}{4}\pi)^2}.$$

SOLUTION

$$\frac{\sin(x - \frac{1}{4}\pi)}{(x - \frac{1}{4}\pi)^2} = \left[\frac{\sin(x - \frac{1}{4}\pi)}{(x - \frac{1}{4}\pi)}\right] \cdot \frac{1}{x - \frac{1}{4}\pi}.$$

We know that

$$\lim_{x\to \pi/4} \frac{\sin(x - \frac{1}{4}\pi)}{x - \frac{1}{4}\pi} = 1.$$

Since $\lim_{x\to \pi/4} (x - \frac{1}{4}\pi) = 0$, you can see by Theorem 2.3.10 that

$$\lim_{x\to \pi/4} \frac{\sin(x - \frac{1}{4}\pi)}{(x - \frac{1}{4}\pi)^2} \quad \text{does not exist.} \quad \square$$

Example 4 Find

$$\lim_{x\to 0} \frac{x^2}{\sec x - 1}.$$

SOLUTION The evaluation of this limit requires a little imagination. Since both the numerator and denominator approach zero as x approaches zero, it is not clear what happens to the fraction; the limit is an indeterminate form of type 0/0. However, we can rewrite the fraction in a more amenable form by using the trick just introduced:

$$\begin{aligned}
\frac{x^2}{\sec x - 1} &= \frac{x^2}{\sec x - 1}\left(\frac{\sec x + 1}{\sec x + 1}\right) \\
&= \frac{x^2(\sec x + 1)}{\sec^2 x - 1} = \frac{x^2(\sec x + 1)}{\tan^2 x} \\
&= \frac{x^2 \cos^2 x\,(\sec x + 1)}{\sin^2 x} \\
&= \left(\frac{x}{\sin x}\right)^2 (\cos^2 x)(\sec x + 1).
\end{aligned}$$

Since each of these expressions has a limit as x tends to 0, the fraction we began with has a limit:

$$\lim_{x\to 0} \frac{x^2}{\sec x - 1} = \lim_{x\to 0} \left(\frac{x}{\sin x}\right)^2 \cdot \lim_{x\to 0} \cos^2 x \cdot \lim_{x\to 0} (\sec x + 1)$$
$$= (1)(1)(2) = 2. \quad \square$$

EXERCISES 2.5

In Exercises 1–32, evaluate the limits that exist.

1. $\lim_{x\to 0} \frac{\sin 3x}{x}$.

2. $\lim_{x\to 0} \frac{2x}{\sin x}$.

3. $\lim_{x\to 0} \frac{3x}{\sin 5x}$.

4. $\lim_{x\to 0} \frac{\sin 3x}{2x}$.

5. $\lim_{x\to 0} \frac{\sin 4x}{\sin 2x}$.

6. $\lim_{x\to 0} \frac{\sin 3x}{5x}$.

7. $\lim_{x\to 0} \frac{\sin x^2}{x}$.

8. $\lim_{x\to 0} \frac{\sin x^2}{x^2}$.

9. $\lim_{x\to 0} \frac{\sin x}{x^2}$.

10. $\lim_{x\to 0} \frac{\sin^2 x^2}{x^2}$.

11. $\lim_{x\to 0} \frac{\sin^2 3x}{5x^2}$.

12. $\lim_{x\to 0} \frac{\tan^2 3x}{4x^2}$.

13. $\lim_{x\to 0} \frac{2x}{\tan 3x}$.

14. $\lim_{x\to 0} \frac{4x}{\cot 3x}$.

15. $\lim_{x\to 0} x \csc x$.

16. $\lim_{x\to 0} \frac{\cos x - 1}{2x}$.

17. $\lim_{x\to 0} \frac{x^2}{1 - \cos 2x}$.

18. $\lim_{x\to 0} \frac{x^2 - 2x}{\sin 3x}$.

19. $\lim_{x\to 0} \frac{1 - \sec^2 2x}{x^2}$.

20. $\lim_{x\to 0} \frac{1}{2x \csc x}$.

21. $\lim_{x\to 0} \frac{2x^2 + x}{\sin x}$.

22. $\lim_{x\to 0} \frac{1 - \cos 4x}{9x^2}$.

23. $\lim_{x\to 0} \frac{\tan 3x}{2x^2 + 5x}$.

24. $\lim_{x\to 0} x^2(1 + \cot^2 3x)$.

25. $\lim_{x\to 0} \frac{\sec x - 1}{x \sec x}$.

26. $\lim_{x\to \pi/4} \frac{1 - \cos x}{x}$.

27. $\lim_{x\to \pi/4} \frac{\sin x}{x}$.

28. $\lim_{x\to 0} \frac{\sin^2 x}{x(1 - \cos x)}$.

29. $\lim_{x\to \pi/2} \frac{\cos x}{x - \frac{1}{2}\pi}$.

30. $\lim_{x\to \pi} \frac{\sin x}{x - \pi}$.

31. $\lim_{x\to \pi/4} \frac{\sin (x + \frac{1}{4}\pi) - 1}{x - \frac{1}{4}\pi}$.

32. $\lim_{x\to \pi/6} \frac{\sin (x + \frac{1}{3}\pi) - 1}{x - \frac{1}{6}\pi}$.

33. Show that $\lim_{x\to c} \cos x = \cos c$ for all real numbers c.

34. Show that

$$\text{if} \quad \lim_{x\to 0} f(x) = L \quad \text{then} \quad \lim_{x\to 0} f(ax) = L \text{ for all } a \neq 0.$$

HINT: Let $\epsilon > 0$. If $\delta_1 > 0$ "works" for the first limit, then $\delta = \delta_1/|a|$ "works" for the second limit.

In Exercises 35–38, determine whether $\lim_{h\to 0} (f(a + h) - f(a))/h$ exists, and write an equation for the tangent line to the graph of f at $(a, f(a))$ when the limit does exist.

35. $f(x) = \sin x$, $a = \pi/4$. HINT: Use the addition formula for the sine function.

36. $f(x) = \cos x$, $a = \pi/3$.

37. $f(x) = \cos 2x$, $a = \pi/6$.

38. $f(x) = \sin 3x$, $a = \pi/2$.

39. Prove that $\lim_{x\to 0} x \sin (1/x) = 0$. HINT: Use the pinching theorem.

40. Prove that $\lim_{x\to \pi} (x - \pi) \cos^2[1/(x - \pi)] = 0$.

41. Prove that $\lim_{x\to 1} |x - 1| \sin x = 0$.

42. Let f be the Dirichlet function

$$f(x) = \begin{cases} 1 & x \text{ rational} \\ 0 & x \text{ irrational.} \end{cases}$$

Prove that $\lim_{x\to 0} xf(x) = 0$.

43. Prove that if there is a number B such that $|f(x)| \leq B$ for all $x \neq 0$, then $\lim_{x\to 0} xf(x) = 0$. NOTE: Exercises 39 and 42 are special cases of this general result.

44. Prove that if there is a number B such that $|f(x)/x| \leq B$ for all $x \neq 0$, then $\lim_{x\to 0} f(x) = 0$.

45. Prove that if there is a number B such that $|f(x) - L|/|x - c| \leq B$ for all $x \neq c$, then $\lim_{x\to c} f(x) = L$.

■ 2.6 TWO BASIC THEOREMS

A function that is continuous on an interval does not "skip" any values, and thus its graph is an "unbroken curve." There are no "holes" in it and no "jumps." This is the idea behind the *intermediate-value theorem.*

THEOREM 2.6.1 THE INTERMEDIATE-VALUE THEOREM

If f is continuous on $[a, b]$ and C is a number between $f(a)$ and $f(b)$, then there is at least one number c in the interval $[a, b]$ for which $f(c) = C$.

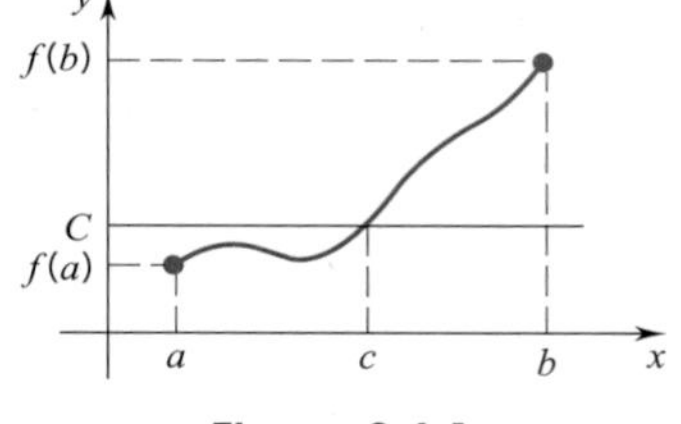

Figure 2.6.1

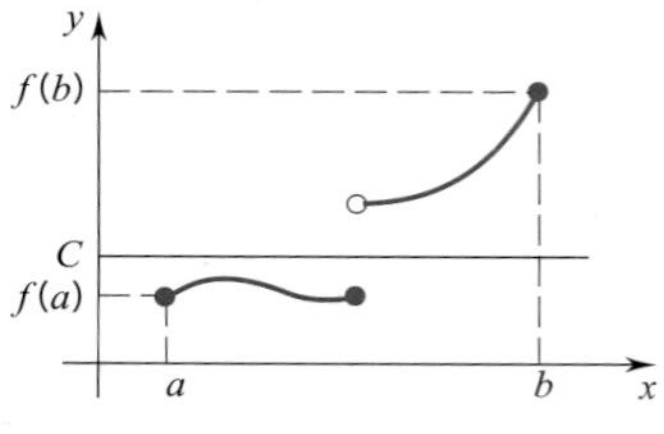

Figure 2.6.2

We illustrate the theorem in Figure 2.6.1. What can happen in the discontinuous case is illustrated in Figure 2.6.2. There the number C has been "skipped." You can find a proof of the intermediate-value theorem in Appendix B. We will assume the result and use it.

We apply the intermediate-value theorem to the important problem of approximating the zeros of a function. In particular, suppose that f is continuous on $[a, b]$ and that either

$$f(a) < 0 < f(b) \quad \text{or} \quad f(b) < 0 < f(a). \qquad \text{(Figure 2.6.3)}$$

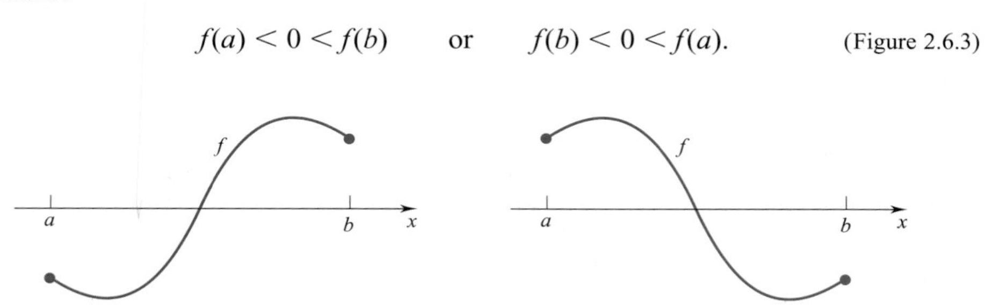

Figure 2.6.3

Then, by the intermediate-value theorem, we know that the equation $f(x) = 0$ has at least one root in $[a, b]$. For simplicity let's assume that there is only one such root and call it c. How can we estimate the location of c? The intermediate-value theorem itself gives us no clue. The simplest method of finding the approximate location of c is called the *bisection method.*

The Bisection Method

The bisection method is an iterative process. A basic step is iterated (carried out repeatedly) until the goal is reached, in this case an approximation for c that is as accurate as we wish.

It is standard practice to label the elements of successive iterations by using as subscripts $n = 1, 2, 3$, and so forth. We begin by setting $u_1 = a$ and $v_1 = b$. We now bisect $[u_1, v_1]$. If c is the midpoint of $[u_1, v_1]$, our search is over. If not, then c lies in one of the halves of $[u_1, v_1]$. We call that half $[u_2, v_2]$. If c is the midpoint of $[u_2, v_2]$, then our search is over. If not, then c lies in one of the halves of $[u_2, v_2]$. We call that half $[u_3, v_3]$ and continue. The first three iterations for a particular function are depicted in Figure 2.6.4.

After n bisections we are examining the midpoint m_n of the interval $[u_n, v_n]$. We can thus be certain that

$$|c - m_n| \le \frac{1}{2}(v_n - u_n) = \frac{1}{2}\left(\frac{v_{n-1} - u_{n-1}}{2}\right) = \cdots = \frac{b - a}{2^n}.$$

Thus m_n approximates c within $(b - a)/2^n$. If we want m_n to approximate c within a

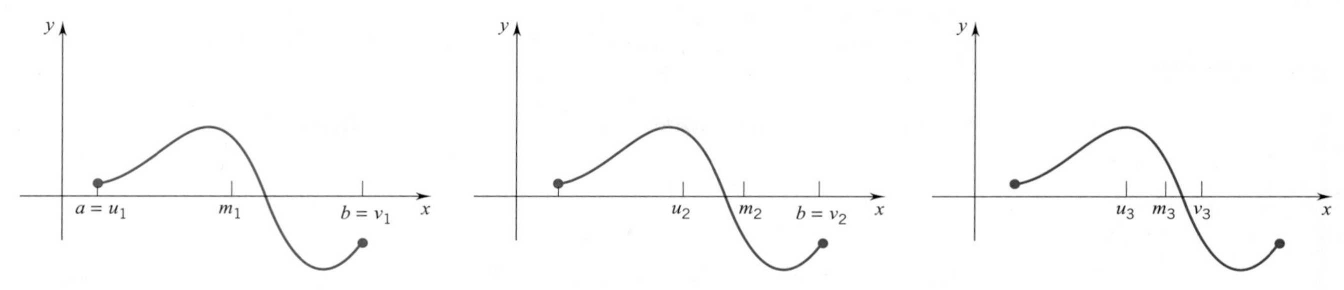

Figure 2.6.4

given ϵ, then we must carry out the iteration to the point where

$$\frac{b-a}{2^n} < \epsilon.$$

Example 1 The function $f(x) = x^2 - 2$ is zero at $x = \sqrt{2}$. Since $f(1) < 0$ and $f(2) > 0$, we can estimate $\sqrt{2}$ by applying the bisection method to f on the interval $[1, 2]$. Let's say we want a numerical estimate accurate to within 0.001. To achieve this degree of accuracy we must carry out the iteration to the point where

$$\frac{2-1}{2^n} < 0.001,$$

or, equivalently, to the point where $2^n > 1000$. As you can check, this condition is met by taking $n \geq 10$. Thus 10 bisections will suffice.

$u_1 = 1,\ v_1 = 2,\ m_1 = 1.5$:
$\quad f(u_1) < 0, \quad f(v_1) > 0, \quad f(m_1) = (1.5)^2 - 2 = 0.25 > 0,$
$u_2 = 1,\ v_2 = 1.5,\ m_2 = 1.25$:
$\quad f(u_2) < 0, \quad f(v_2) > 0, \quad f(m_2) = (1.25)^2 - 2 = -0.4375 < 0,$
$u_3 = 1.25,\ v_3 = 1.5,\ m_3 = 1.375$:
$\quad f(u_3) < 0, \quad f(v_3) > 0, \quad f(m_3) = (1.375)^2 - 2 = -0.109375 < 0,$
$u_4 = 1.375,\ v_4 = 1.5,\ m_4 = 1.4375$:
and so on.

The complete computations are summarized in Table 2.6.1. To increase readability, all the entries in the table have been rounded off to 5 decimal places.

■ **Table 2.6.1**

n	u_n	v_n	m_n	$f(m_n)$
1	1.00000	2.00000	1.50000	0.25000
2	1.00000	1.50000	1.25000	−0.43750
3	1.25000	1.50000	1.37500	−0.10938
4	1.37500	1.50000	1.43750	0.06641
5	1.37500	1.43750	1.40625	−0.02246
6	1.40625	1.43750	1.42187	0.02171
7	1.40625	1.42187	1.41406	−0.00043
8	1.41406	1.42187	1.41797	0.01064
9	1.41406	1.41797	1.41601	0.00508
10	1.41406	1.41601	1.41503	0.00231

We can conclude that $m_{10} = 1.41503$ approximates $\sqrt{2}$ within 0.001. ❑

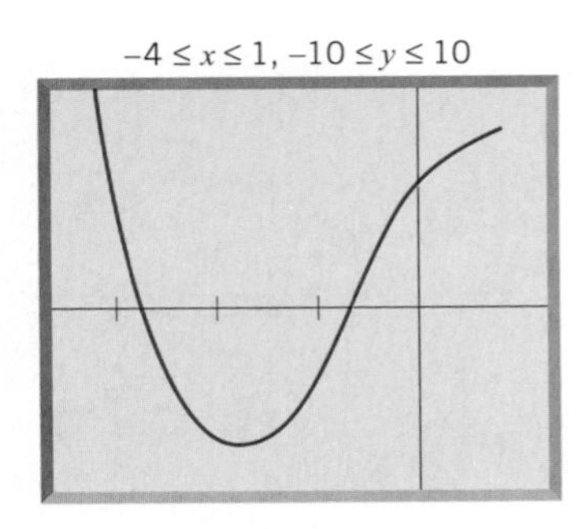

Figure 2.6.5

Example 2 Let $f(x) = \frac{1}{2}x^4 - 3x^2 + 4x + 4$. Use the intermediate-value theorem to show that f has at least two zeros in the interval $[-3, 0]$. Sketch the graph of f and find approximations for the two zeros.

SOLUTION The values of f at the integers from -3 to 0 are as follows:

x	-3	-2	-1	0
$f(x)$	5.5	-8	-2.5	4

Thus f has a zero r_1 in the interval $[-3, -2]$ and a zero r_2 in the interval $[-1, 0]$.

The graph of f shown in Figure 2.6.5 was obtained with a graphics calculator. Using the zoom feature, we found that approximate values for these zeros are: $r_1 \cong -2.80$ and $r_2 \cong -0.68$. ❑

Boundedness Properties

Suppose that f is a function which is defined on an interval I. Then f is said to be *bounded on I* if there exist numbers k and K such that

$$k \le f(x) \le K \quad \text{for all} \quad x \in I.$$

Equivalently, f is bounded on I if there exists a number B such that $|f(x)| \le B$ for all $x \in I$. The function f is *unbounded on I* if it is not bounded.

The function f has a maximum value on I if there is a number $d \in I$ such that $f(d) \ge f(x)$ for all $x \in I$. The value $f(d) = M$ is called the *maximum value of f on I.* Similarly, f has a minimum value on I if there is a number $c \in I$ such that $f(c) \le f(x)$ for all $x \in I$; the value $f(c) = m$ is called the *minimum value of f on I.* The maximum and minimum values of f, if they exist, are called the *extreme values of f on I.*

The trigonometric functions sine and cosine are bounded on $(-\infty, \infty)$ since

$$-1 \le \sin x \le 1 \qquad \text{and} \qquad -1 \le \cos x \le 1$$

for all x. Also, 1 is the maximum value of both sine and cosine, and -1 is the minimum value of both functions: $\sin(\pi/2) = 1$, $\cos(0) = 1$, $\sin(3\pi/2) = -1$, and $\cos(\pi) = -1$. The sine and cosine functions also illustrate the fact that the maximum and/or minimum values of a function may be attained at more than one point on the interval I; $\sin x$ and $\cos x$ attain their extreme values at infinitely many points on $(-\infty, \infty)$. See Figure 2.6.6.

The tangent function, $\tan x$, is unbounded on $(-\pi/2, \pi/2)$ and is bounded on $(-\pi/4, \pi/4)$. See Figure 2.6.7.

If f has a maximum value M and a minimum value m on I, then clearly f is bounded on I since

$$m \le f(x) \le M$$

for all $x \in I$. On the other hand, a function can be bounded on an interval I without having a maximum value or a minimum value on the interval. For example, $\tan x$ has neither a maximum nor a minimum value on $(-\pi/4, \pi/4)$ (see Figure 2.6.7).

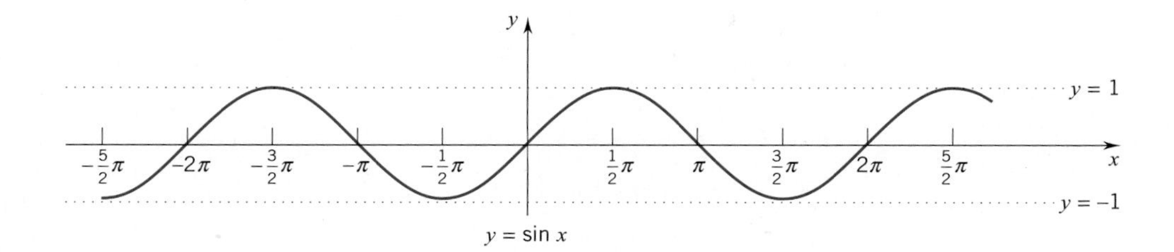

$y = \sin x$

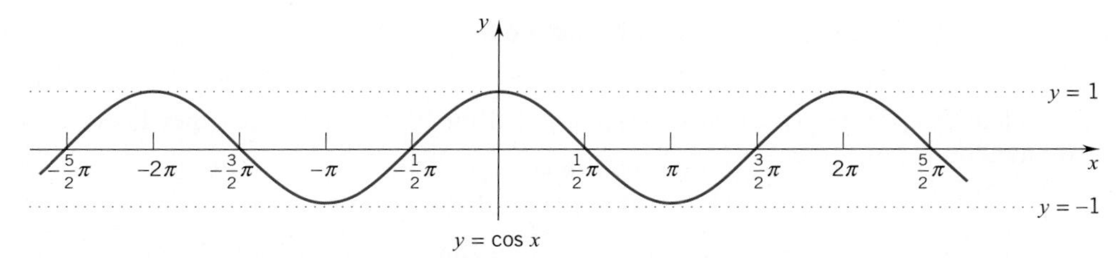

$y = \cos x$

Figure 2.6.6

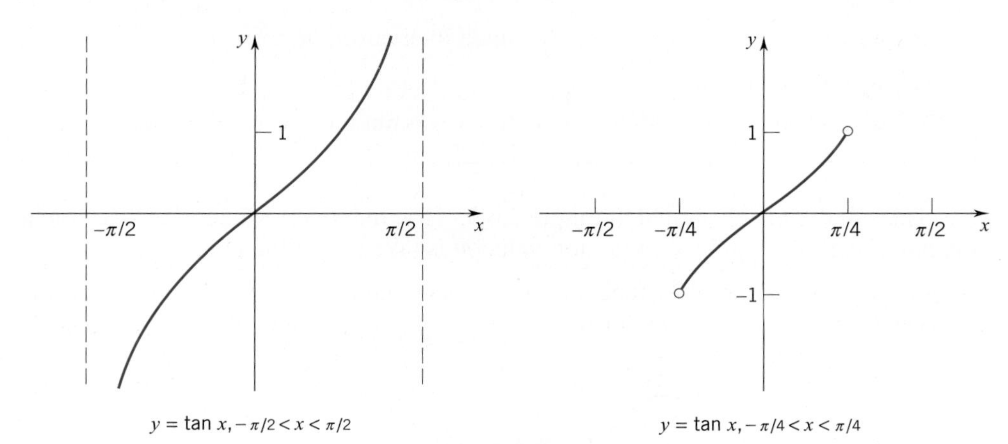

$y = \tan x, -\pi/2 < x < \pi/2$

$y = \tan x, -\pi/4 < x < \pi/4$

Figure 2.6.7

Example 3 Let

$$g(x) = \begin{cases} 1/x^2, & x > 0 \\ 0, & x = 0. \end{cases} \qquad \text{(Figure 2.6.8)}$$

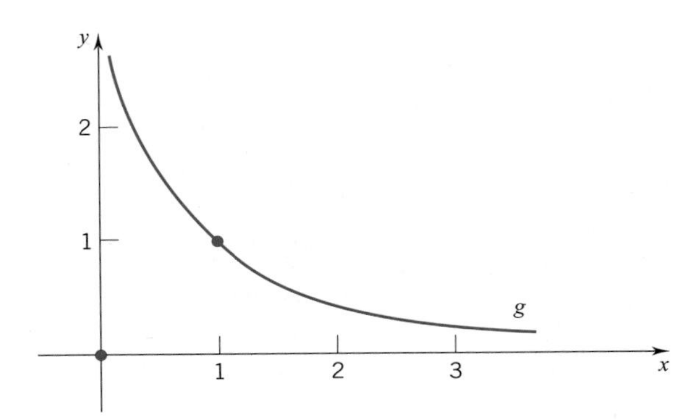

Figure 2.6.8

It is clear from the graph that g is unbounded on $[0, \infty)$. On the other hand, g is bounded on $[1, \infty)$, since

$$0 < g(x) \leq 1$$

for each x on this interval. Note that g has a maximum value on $[1, \infty)$, $g(1) = 1$, but no minimum value. Why? ❑

> **THEOREM 2.6.2 BOUNDEDNESS: EXTREME-VALUE THEOREM**
>
> Let f be continuous on the closed, bounded interval $[a, b]$. Then
>
> **(1)** f is bounded on $[a, b]$,
>
> **(2)** f attains both its maximum value M and its minimum value m on $[a, b]$.

The situation is illustrated in Figure 2.6.9. The maximum value M is taken on at the point marked d, and the minimum value m is taken on at the point marked c.

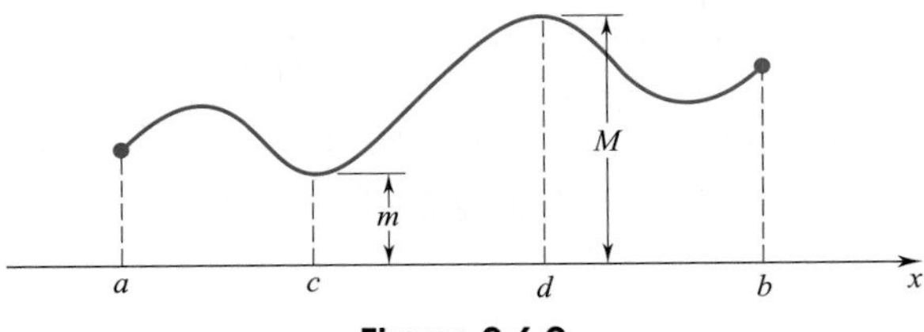

Figure 2.6.9

It is interesting to note that the full hypothesis of this theorem is needed. That is, if either the continuity requirement on the function f, or the requirement that the interval be closed and bounded, is dropped, then the conclusion of the theorem does not follow. As an example, we can take the function

$$f(x) = \begin{cases} 3, & x = 1 \\ x, & 1 < x < 5 \\ 3, & x = 5. \end{cases}$$

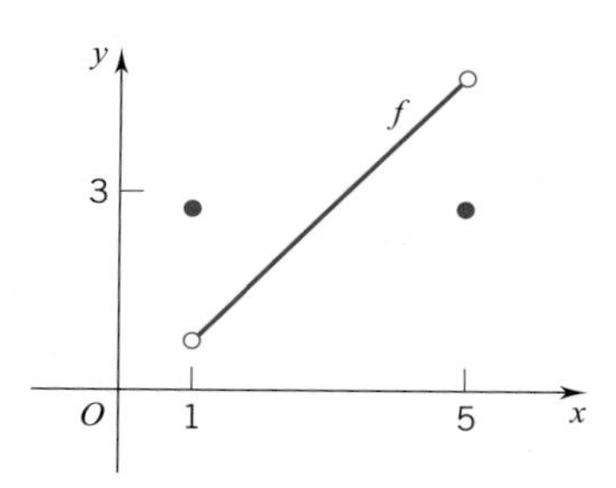

Figure 2.6.10

The graph of f is pictured in Figure 2.6.10. The function is defined on $[1, 5]$, but it takes on neither a maximum nor a minimum value.

If, instead of dropping the continuity requirement, we drop the requirement that the interval be of the form $[a, b]$, then again the result fails. For example, as previously illustrated, the tangent function, $\tan x$, is continuous and unbounded on $(-\pi/2, \pi/2)$; on the other hand, $\tan x$ is continuous and bounded on $(-\pi/4, \pi/4)$, but has neither a maximum value nor a minimum value on this interval (see Figure 2.6.7). The function g in Example 3 has a maximum value on $[1, \infty)$, but no minimum value.

A proof of the maximum-minimum theorem appears in Appendix B. Techniques for determining the maximum and minimum values of functions are described in Chapter 4. These techniques require an understanding of "differentiation," the subject to which we devote Chapter 3.

EXERCISES 2.6

In Exercises 1–16, sketch the graph of a function f that is defined on $[0, 1]$ and meets the given conditions (if possible).

1. f is continuous on $[0, 1]$, minimum value 0, maximum value $\frac{1}{2}$.

2. f is continuous on $[0, 1)$, minimum value 0, no maximum value.

3. f is continuous on $(0, 1)$, takes on the values 0 and 1, but does not take on the value $\frac{1}{2}$.

4. f is continuous on $[0, 1]$, takes on the values -1 and 1, but does not take on the value 0.

5. f is continuous on $[0, 1]$, maximum value 1, minimum value 1.

6. f is continuous on $[0, 1]$ and nonconstant, takes on no integer values.

7. f is continuous on $[0, 1]$, takes on no rational values.

8. f is not continuous on $[0, 1]$, takes on both a maximum value and a minimum value and every value in between.

9. f is continuous on $[0, 1]$, takes on only two distinct values.

10. f is continuous on $(0, 1)$, takes on only three distinct values.

11. f is not continuous on $(0, 1)$ and the range of f is an open, bounded interval.

12. f is not continuous on $(0, 1)$ and the range of f is a closed, bounded interval.

13. f is continuous on $(0, 1)$ and the range of f is an unbounded interval.

14. f is continuous on $[0, 1]$ and the range of f is an unbounded interval.

15. f is not continuous on $[0, 1]$ and the range of f is $[0, \infty)$.

16. f is continuous on $[0, 1)$ and the range of f is a closed, bounded interval.

In Exercises 17–19 use the intermediate-value theorem to show that there is a root of the given equation in the indicated interval.

17. $2x^3 - 4x^2 + 5x - 4 = 0, \quad [1, 2].$

18. $2\cos x - x^2 = 0, \quad [0, \pi/2].$

19. $\dfrac{1}{x-1} + \dfrac{1}{x-4} = 0, \quad (1, 4).$

20. Given that $\lim f(x) = 0$ and that g is bounded on $(c - p, c + p)$ with $p > 0$, prove that

$$\lim_{x \to c} f(x)g(x) = 0.$$

21. *(Fixed-point property)* Show that if f is continuous and $0 \le f(x) \le 1$ for all $x \in [0, 1]$, then there exists at least one point c in $[0, 1]$ for which $f(c) = c$.
HINT: Apply the intermediate-value theorem to the function $g(x) = x - f(x)$.

22. Let n be a positive integer.
(a) Prove that if $0 \leq a < b$, then $a^n < b^n$.
HINT: Use mathematical induction.
(b) Prove that every nonnegative real number x has a unique nonnegative nth root $x^{1/n}$. HINT: The existence of $x^{1/n}$ can be seen by applying the intermediate-value theorem to the function $f(t) = t^n$ for $t \geq 0$. The uniqueness follows from part (a).

23. Suppose that f is bounded on the interval $(-p, p)$ with $p > 0$. Define g on $(-p, p)$ by

$$g(x) = \begin{cases} xf(x), & x \neq 0 \\ 0, & x = 0. \end{cases}$$

Prove that g is continuous at 0.

24. Given that f and g are continuous functions on $[a, b]$, and that $f(a) < g(a)$ and $g(b) < f(b)$, show that there exists at least one number $c \in (a, b)$ such that $f(c) = g(c)$. HINT: Consider $f(x) - g(x)$.

25. The intermediate-value theorem can be used to prove that each polynomial equation of odd degree

$$x^n + a_{n-1}x^{n-1} + \cdots + a_1x + a_0 \qquad \text{with } n \text{ odd}$$

has at least one real root. Prove that the cubic equation

$$x^3 + ax^2 + bx + c = 0$$

has at least one real root.

26. Assume that at any given instant in time, the temperature on the earth's surface varies continuously with position. Prove that there is at least one pair of points diametrically opposite each other on the equator where the temperature is the same. HINT: Form a function that relates the temperature at diametrically opposite points on the equator.

27. Let $\mathscr{C}$ denote the set of all circles whose radii are less than or equal to 10 inches. Prove that there is at least one member of $\mathscr{C}$ whose area is exactly 250 square inches.

28. Fix a positive number P. Let $\mathscr{R}$ denote the set of all rectangles having perimeter P. Prove that there is a member of $\mathscr{R}$ that has maximum area. What are the dimensions of the rectangle that has maximum area? HINT: Express the area of an arbitrary element of R as a function of the length of one of its sides.

29. Given a circle C of radius R. Let $\mathscr{F}$ denote the set of all rectangles that can be inscribed in C. Prove that there is a member of $\mathscr{F}$ that has maximum area.

(C) In Exercises 30–33, use the bisection method to approximate the root of $f(x) = 0$ on the given interval to within 0.001.

30. $f(x) = 5 - x^2$; $[-3, -2]$.

31. $f(x) = x^2 - 3$; $[1, 2]$.

32. $f(x) = x^3 + x - 10$; $[1, 4]$.

33. $f(x) = \cos x - x$; $[0, 1]$.

(C) In Exercises 34–37, use the intermediate-value theorem to locate the zeros of the given function. Then use a graphing utility to approximate the zeros to within 0.001.

34. $f(x) = 2x^3 + 5x - 7$.

35. $f(x) = x^3 - 5x + 3$.

36. $f(x) = x^5 - 3x + 1$.

37. $f(x) = x^3 - 2\sin x + \frac{1}{2}$.

(C) In Exercises 38–41, use a graphing utility to graph f on the given interval. Is f bounded? If so, estimate its maximum and minimum values.

38. $f(x) = \dfrac{x^3 - 8x + 6}{4x + 1}$; $[0, 5]$.

39. $f(x) = \dfrac{2x}{1 + x^2}$; $[-2, 2]$.

40. $f(x) = \dfrac{\sin 2x}{x^2}$; $[-\pi/2, \pi/2]$.

41. $f(x) = \dfrac{1 - \cos x}{x^2}$; $[-2, 2]$.

■ CHAPTER HIGHLIGHTS

2.1 The Idea of Limit

Intuitive Interpretation of $\lim_{x \to c} f(x) = L$:

as x approaches c, $f(x)$ approaches L

or, equivalently,

for x close to c but different from c, $f(x)$ is close to L.

Restricting our attention to values of x to one side of c, we have *one-sided limits*:

$$\lim_{x \to c^-} f(x) \qquad \text{and} \qquad \lim_{x \to c^+} f(x).$$

2.2 Definition of Limit

limit of a function (p. 82) left-hand limit (p. 88) right-hand limit (p. 88)
four formulations of the limit statement (p. 87)

2.3 Some Limit Theorems

uniqueness of limit (p. 92) stability of limit (p. 101)

The "arithmetic of limits": if $\lim_{x\to c} f(x) = L$ and $\lim_{x\to c} g(x) = M$, then

$$\lim_{x\to c} [f(x) + g(x)] = L + M, \qquad \lim_{x\to c} [\alpha f(x)] = \alpha L \quad \text{for each real } \alpha,$$

$$\lim_{x\to c} [f(x)g(x)] = LM, \qquad \lim_{x\to c} \frac{f(x)}{g(x)} = \frac{L}{M} \quad \text{provided } M \neq 0.$$

If $M = 0$ and $L \neq 0$, then $\lim_{x\to c} f(x)/g(x)$ does not exist.

The same results hold for one-sided limits.

2.4 Continuity

continuity at c (p. 102) removable discontinuity, jump discontinuity (p. 102)
one-sided continuity (p. 106) continuity on intervals (p. 108)

The sum, difference, product, quotient, and composition of continuous functions are continuous where defined. Polynomials and the absolute value function are everywhere continuous. Rational functions are continuous on their domains. The square-root function is continuous at all positive numbers and continuous from the right at 0.

2.5 The Pinching Theorem; Trigonometric Limits

pinching theorem (p. 111) $\quad \lim_{x\to 0} \frac{\sin x}{x} = 1, \quad \lim_{x\to 0} \frac{1 - \cos x}{x} = 0.$

Each trigonometric function is continuous on its domain.

2.6 Two Basic Theorems

intermediate-value theorem (p. 118) bisection method (p. 118)
bounded functions (p. 120) maximum value (p. 120)
minimum value (p. 120) extreme values (p. 120)
boundedness; extreme-value theorem (p. 122)

■ PROJECTS AND EXPLORATIONS USING TECHNOLOGY

To do these exercises you will need a graphics calculator or a computer with graphing capability. The majority of these problems are open-ended so different approaches may be used to solve them. You should be aware that different approaches can result in slight variations in the answers. Round your numerical answers to at least four decimal places. The rounding method that your calculator or computer uses also may cause variations in answers.

2.1 Let

$$F(x) = \frac{\sin(\tan x) - \tan(\sin x)}{x^7},$$

and consider the limit

$$\lim_{x\to 0} F(x).$$

(a) Explain why F is defined for values of x near 0. Then explain why

$$\lim_{x \to 0} [\sin(\tan x) - \tan(\sin x)] = 0.$$

(b) Use your calculator to evaluate F for values of x near 0 (be sure to use both positive and negative values) and approximate the limit.

(c) Graph F and approximate the limit. Zoom in as close as you can and note the "hole" in the graph at $x = 0$.

(d) If you zoom in too close, the picture becomes distorted. Explain why your calculator or computer has trouble with this graph.

2.2 The *greatest integer function,* denoted by $[\![x]\!]$, has the set of all real numbers for its domain and, for each real number x, $[\![x]\!]$ is the greatest integer that is less than or equal to x. For example,

$$[\![1.75]\!] = 1, \quad [\![4]\!] = 4, \quad [\![-2.3]\!] = -3, \quad [\![-\tfrac{1}{2}]\!] = -1, \quad \text{and} \quad [\![-6]\!] = -6.$$

(a) Find the points of discontinuity of $f(x) = [\![x]\!]$. Classify the discontinuities of f as being removable, jump, or neither. Suppose that f is discontinuous at the number a. Is f continuous from the right at a? From the left?

(b) Let

$$g(x) = [\![3 \sin(x^2)]\!], \qquad x \in [-\pi, \pi].$$

Find the points of discontinuity of g. Then use the theorems on the composition of continuous functions to verify your answers.

(c) Let

$$h(x) = 3 \sin([\![x^2]\!]), \qquad x \in [-\pi, \pi].$$

Find the points of discontinuity of h. Again, use the theorems on the composition of continuous functions to verify your answers.

2.3 A number x is a *fixed point* of a function f if $f(x) = x$.

(a) Let $f(x) = \cos x$. Use the continuity of f and the intermediate value theorem to explain why f must have a fixed point on the interval $[0, 1]$. HINT: Consider $F(x) = f(x) - x$.

(b) Show that f has exactly one fixed point on $[0, 1]$ and use the bisection method to approximate it, accurate to two decimal places. HINT: Verify first that f has a fixed point on the interval $[\frac{1}{2}, 1]$.

(c) Starting with $x_1 = f(1)$, generate the values $x_2 = f(x_1)$, $x_3 = f(x_2)$, . . . , $x_n = f(x_{n-1})$, ... Note that these values appear to approach the fixed point found in part (b).

(d) Repeat parts (a)–(c) with the function

$$g(x) = \frac{3}{x^2 + 1}, \qquad x \in [1, 2].$$

This time, observe that the values $x_1 = g(2)$, $x_2 = g(x_1)$, $x_3 = g(x_2)$, . . . , do not approach the fixed point.

(e) Draw the graph of each of these functions near their respective fixed points. The graph of $f(x) = \cos x$ intersects the line $y = x$ at a "flatter" angle than $g(x) = 3/(x^2 + 1)$ does. This is one of the properties of functions that will be studied in the following chapters.

2.4 The following modification of the bisection method for finding the zeros of a function is sometimes used. The line connecting the endpoints of the graph of the function on the interval is calculated and its x-intercept is used instead of the midpoint. For example, if

$$f(x) = x^3 + x - 9,$$

then f has a zero in the interval $[1, 2]$. The line joining the endpoints $(1, -7)$ and $(2, 1)$ is $y = 8x - 15$ and its x-intercept is $15/8$. See the figure.

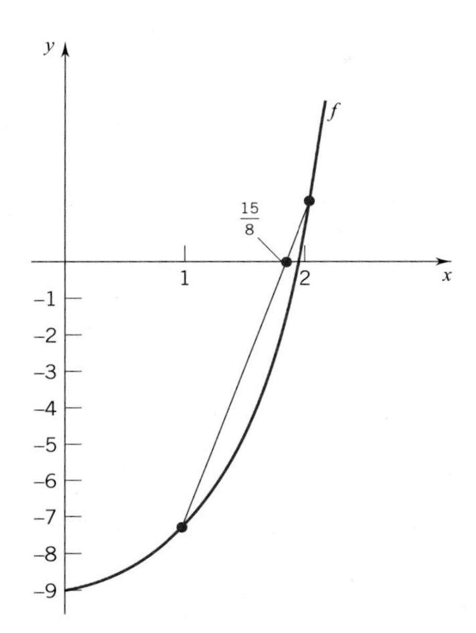

(a) Continuing with this example, calculate the next three x-intercepts. Notice that these values approach the zero of f very quickly, and that only one endpoint of the interval changes. Can you explain this from the ''shape'' of the graph of f near the zero? A difference between this method and the bisection method is that the lengths of the intervals for the bisection method must approach zero while the lengths of the intervals approach a positive constant with this method.

(b) While this method requires more work per step, it has the advantage of providing better approximations in fewer steps. How many bisection steps would be required to get an answer as accurate as that found in part (a)?

(c) A hybrid method uses the following modification of the method introduced in this exercise. If this method changes the same endpoint twice in succession, then one step of the midpoint method is done. Try the hybrid method on the function f just given. The hybrid method takes advantage of the good feature of the bisection method (the lengths of the intervals go to zero) and the modified bisection method (fewer steps are required).

(d) Suppose that a mistake is made and you start the hybrid method on the interval $[2, 3]$ instead of $[1, 2]$. Show that the procedure can still be used.

(e) Find an example of a function that has zeros in an interval but where the calculated line is horizontal so that the method terminates. Explain under what conditions (if any) the bisection method might terminate.

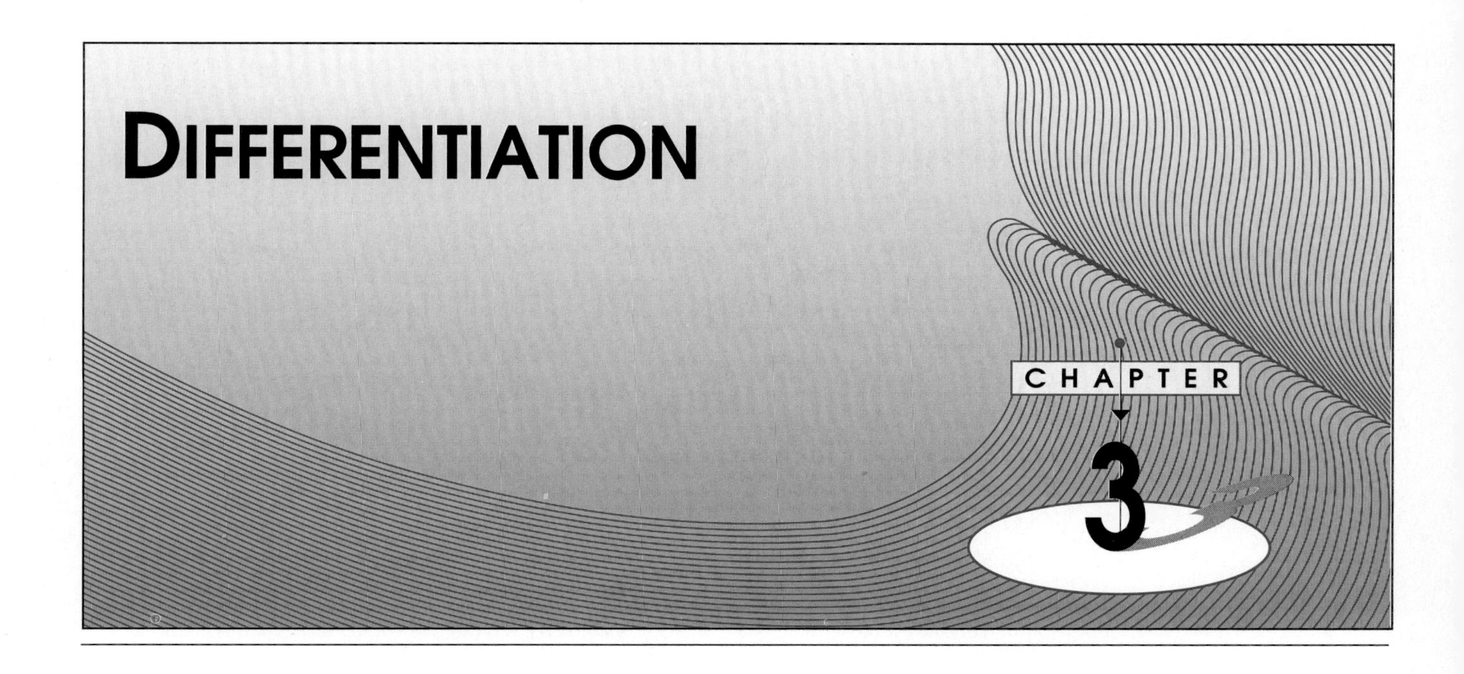

3.1 THE DERIVATIVE

Introduction: The Tangent to a Curve

We begin with a function f, and on its graph we choose a point $(x, f(x))$. See Figure 3.1.1. What line, if any, should be called tangent to the graph at that point?

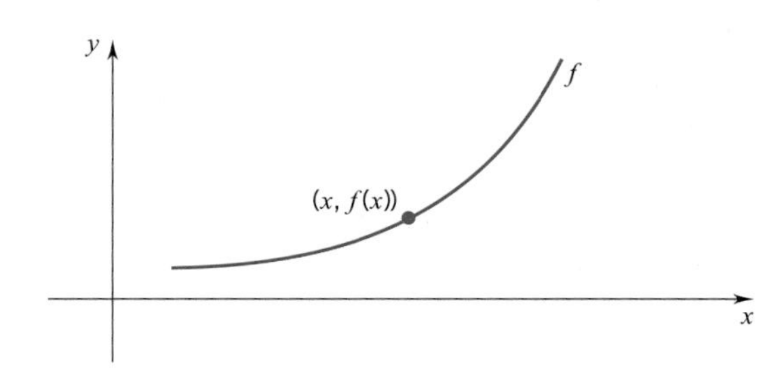

Figure 3.1.1

To answer this question, we choose a small number $h \neq 0$ and on the graph mark the point $(x + h, f(x + h))$. Now we draw the secant line that passes through these two points. The situation is pictured in Figure 3.1.2, first with $h > 0$ and then with $h < 0$.

As h tends to zero from the right (see the figure), the secant line tends to the limiting position indicated by the dashed line, and it tends to the same limiting position as h tends to zero from the left. The line at this limiting position is what we call "the tangent to the graph at the point $(x, f(x))$."

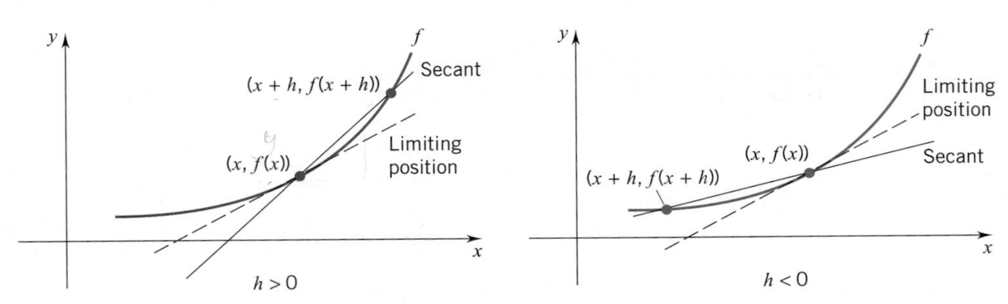

Figure 3.1.2

Since the approximating secant lines have slopes

$$(*) \qquad \frac{f(x+h)-f(x)}{h} \qquad \text{(verify this)}$$

you can expect the tangent line, the limiting position of these secants, to have slope

$$(**) \qquad \lim_{h\to 0}\frac{f(x+h)-f(x)}{h}.$$

While $(*)$ measures the steepness of the line that passes through the points $(x, f(x))$ and $(x+h, f(x+h))$, $(**)$ measures the steepness of the graph at $(x, f(x))$ and is called the "slope of the graph" at that point.

Derivatives and Differentiation

In the introduction we spoke informally about tangent lines and gave a geometric interpretation to limits of the form

$$\lim_{h\to 0}\frac{f(x+h)-f(x)}{h}.$$

Here we begin the systematic study of such limits, what mathematicians call the theory of *differentiation*.

DEFINITION 3.1.1

A function f is said to be *differentiable at* x iff

$$\lim_{h\to 0}\frac{f(x+h)-f(x)}{h} \quad \text{exists.}$$

If this limit exists, it is called the *derivative of* f *at* x, and is denoted by $f'(x)$.† The function f is a *differentiable function* if it is differentiable at each x in its domain.

Geometrically,

$$f'(x) = \lim_{h\to 0}\frac{f(x+h)-f(x)}{h}$$

† This prime notation goes back to the French mathematician Joseph-Louis Lagrange (1736–1813). Other notations are introduced later.

is the *slope of the graph at the point* $(x, f(x))$. The line that passes through the point $(x, f(x))$ with slope $f'(x)$ is called the *tangent line to the graph of f at* $(x, f(x))$. This is the line that best approximates the graph of f near the point $(x, f(x))$.

It is time to consider some examples.

Example 1 We begin with the quadratic function

$$f(x) = x^2. \qquad \text{(Figure 3.1.3)}$$

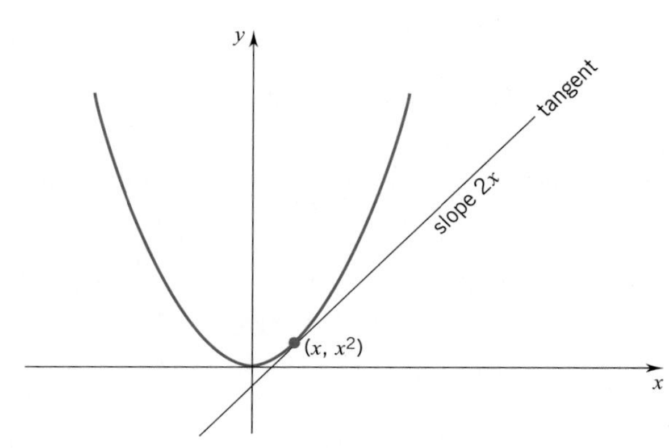

Figure 3.1.3

The domain of f is all real numbers. Fix any number x. To find $f'(x)$ we form the *difference quotient*

$$\frac{f(x+h) - f(x)}{h} = \frac{(x+h)^2 - x^2}{h}, \qquad h \neq 0.$$

Since

$$\frac{(x+h)^2 - x^2}{h} = \frac{x^2 + 2xh + h^2 - x^2}{h} = \frac{2xh + h^2}{h} = 2x + h,$$

we have

$$\frac{f(x+h) - f(x)}{h} = 2x + h,$$

and therefore

$$f'(x) = \lim_{h \to 0} \frac{f(x+h) - f(x)}{h} = \lim_{h \to 0} (2x + h) = 2x. \quad \square$$

Example 2 The domain of a linear function

$$f(x) = mx + b$$

is all real numbers. For any number x,

$$f'(x) = m.$$

In other words, the slope is constantly m. To verify this, note that for $h \neq 0$

$$\frac{f(x+h) - f(x)}{h} = \frac{[m(x+h) + b] - [mx + b]}{h} = \frac{mh}{h} = m.$$

It follows that

$$f'(x) = \lim_{h \to 0} \frac{f(x+h) - f(x)}{h} = \lim_{h \to 0} m = m.$$

This result can also be verified geometrically: $f'(x)$ is the slope of the tangent line to the graph of the linear function $f(x) = mx + b$ at the point $(x, f(x))$; but the tangent line to the graph of a linear function is simply the line itself (you should verify this). Thus, it follows that $f'(x) = m$. ❑

It is important to note that in order to apply the definition of the derivative

$$f'(x) = \lim_{h \to 0} \frac{f(x+h) - f(x)}{h},$$

x and $x + h$ must be in the domain of f and, since the limit involved is a two-sided limit, f must be defined on some open interval containing x. Also, in calculating the limit, x is fixed and the "variable" is h.

In our next example we will be dealing with the square-root function. Although this function is defined for all $x \geq 0$, you can expect a derivative only for $x > 0$.

Example 3 The square-root function

$$f(x) = \sqrt{x}, \qquad x \geq 0 \qquad \text{(Figure 3.1.4)}$$

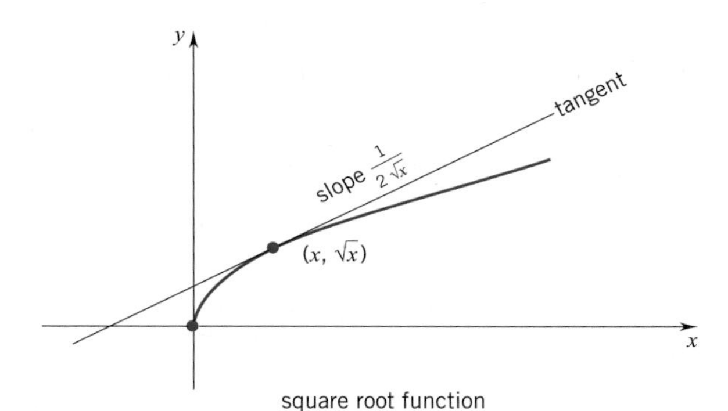

Figure 3.1.4

has derivative

$$f'(x) = \frac{1}{2\sqrt{x}}, \qquad \text{for } x > 0.$$

To verify this, fix an $x > 0$ and form the difference quotient

$$\frac{f(x+h) - f(x)}{h} = \frac{\sqrt{x+h} - \sqrt{x}}{h}, \qquad h \neq 0.$$

To simplify this expression, we multiply both the numerator and the denominator by $\sqrt{x+h} + \sqrt{x}$. This gives

$$\frac{f(x+h) - f(x)}{h} = \left(\frac{\sqrt{x+h} - \sqrt{x}}{h}\right)\left(\frac{\sqrt{x+h} + \sqrt{x}}{\sqrt{x+h} + \sqrt{x}}\right)$$

$$= \frac{(x+h) - x}{h(\sqrt{x+h} + \sqrt{x})} = \frac{1}{\sqrt{x+h} + \sqrt{x}}.$$

Thus,

$$f'(x) = \lim_{h\to 0} \frac{f(x+h) - f(x)}{h} = \frac{1}{2\sqrt{x}} \qquad \text{for } x > 0. \quad \square$$

To *differentiate* a function is to find its derivative. As illustrated by the preceding examples, when we differentiate a function f we get a new function f' called the *derivative* of f. The domain of f' is necessarily a subset of the domain of f. In particular,

$$\text{dom}(f') = \left\{ x \in \text{dom}(f)\colon \lim_{h\to 0} \frac{f(x+h) - f(x)}{h} \text{ exists} \right\}.$$

Example 4 Here we differentiate the function

$$f(x) = \frac{1}{x}, \qquad x \neq 0.$$

(See Figure 3.1.5.) We form the difference quotient

$$\frac{f(x+h) - f(x)}{h} = \frac{\dfrac{1}{x+h} - \dfrac{1}{x}}{h}$$

and simplify:

$$\frac{\dfrac{1}{x+h} - \dfrac{1}{x}}{h} = \frac{\dfrac{x}{x(x+h)} - \dfrac{x+h}{x(x+h)}}{h} = \frac{\dfrac{-h}{x(x+h)}}{h} = \frac{-1}{x(x+h)}.$$

Thus

$$f'(x) = \lim_{h\to 0} \frac{f(x+h) - f(x)}{h} = \lim_{h\to 0} \frac{-1}{x(x+h)} = -\frac{1}{x^2}. \quad \square$$

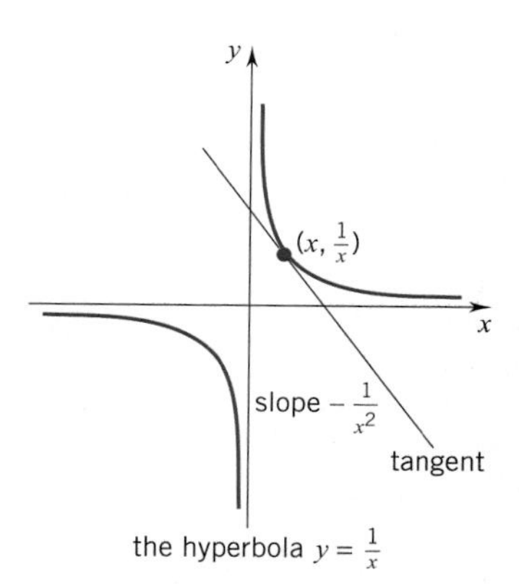

Figure 3.1.5

Evaluating Derivatives

Example 5 Find $f'(-2)$ given that

$$f(x) = 2 + 3x - x^2.$$

SOLUTION We let $x = -2$ in the definition:

$$\begin{aligned} f'(-2) &= \lim_{h\to 0} \frac{f(-2+h) - f(-2)}{h} \\ &= \lim_{h\to 0} \frac{[2 + 3(-2+h) - (-2+h)^2] - [2 + 3(-2) - (-2)^2]}{h} \\ &= \lim_{h\to 0} \frac{[2 - 6 + 3h - (4 - 4h + h^2)] - (-8)]}{h} \\ &= \lim_{h\to 0} \frac{7h - h^2}{h} = \lim_{h\to 0} (7 - h) = 7. \end{aligned}$$

Alternatively, we can calculate $f'(x)$ for an arbitrary x:

$$\begin{aligned} f'(x) &= \lim_{h\to 0} \frac{f(x+h)-f(x)}{h} \\ &= \lim_{h\to 0} \frac{[2+3(x+h)-(x+h)^2]-[2+3x-x^2]}{h} \\ &= \lim_{h\to 0} \frac{[2+3x+3h-(x^2+2xh+h^2)]-[2+3x-x^2]}{h} \\ &= \lim_{h\to 0} \frac{[3h-2xh-h^2]}{h} = \lim_{h\to 0}(3-2x-h) = 3-2x. \end{aligned}$$

Then, substituting -2 for x, we have

$$f'(-2) = 3 - 2(-2) = 7. \quad \square$$

You might be wondering why we showed you the alternate solution in Example 5. Suppose that, in addition to calculating $f'(-2)$, we also wanted to find $f'(0), f'(\tfrac{1}{2})$, and $f'(4)$? Obviously, it would be tedious to form each of the difference quotients and calculate the limits individually. On the other hand, since

$$f'(x) = 3 - 2x,$$

we have $f'(0) = 3 - 2(0) = 3$, $f'(\tfrac{1}{2}) = 3 - 2(\tfrac{1}{2}) = 2$, and $f'(4) = 3 - 2(4) = -5$. The advantage of having a ''general'' expression for f' should now be clear.

Example 6 Find $f'(-3)$ and $f'(1)$ given that

$$f(x) = \begin{cases} x^2, & x \le 1 \\ 2x - 1, & x > 1. \end{cases} \qquad \text{(see Figure 3.1.6)}$$

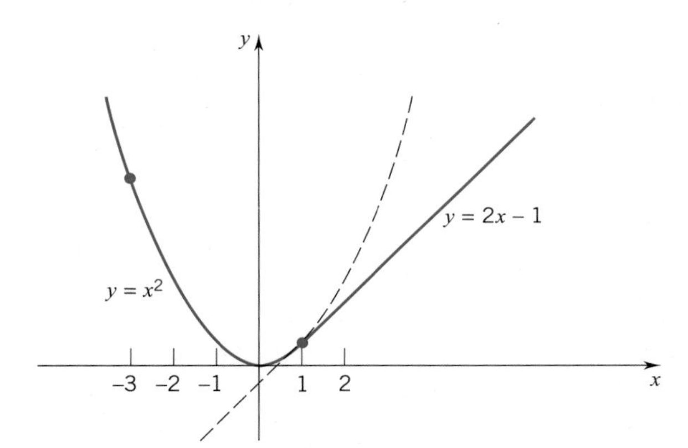

Figure 3.1.6

SOLUTION By definition

$$f'(-3) = \lim_{h\to 0} \frac{f(-3+h)-f(-3)}{h}.$$

However, for all x sufficiently close to -3, $f(x) = x^2$, and from Example 1, we know that $f'(x) = 2x$. Thus, $f'(-3) = 2(-3) = -6$.

Now let's find

$$f'(1) = \lim_{h \to 0} \frac{f(1+h) - f(1)}{h}.$$

Since f is not defined by the same formula on both sides of 1, we will evaluate this limit by taking one-sided limits. Note that $f(1) = 1^2 = 1$.

To the left of 1, $f(x) = x^2$. Thus

$$\lim_{h \to 0^-} \frac{f(1+h) - f(1)}{h} = \lim_{h \to 0^-} \frac{(1+h)^2 - 1}{h}$$

$$= \lim_{h \to 0^-} \frac{(1 + 2h + h^2) - 1}{h} = \lim_{h \to 0^-} (2 + h) = 2.$$

To the right of 1, $f(x) = 2x - 1$. Thus

$$\lim_{h \to 0^+} \frac{f(1+h) - f(1)}{h} = \lim_{h \to 0^+} \frac{[2(1+h) - 1] - 1}{h} = \lim_{h \to 0^+} 2 = 2.$$

The limit of the difference quotient exists and is 2:

$$f'(1) = \lim_{h \to 0} \frac{f(1+h) - f(1)}{h} = 2. \quad \square$$

Tangent Lines and Normal Lines

We begin with a function f and choose a point (x_0, y_0) on the graph. If f is differentiable at x_0, then, as you know, the tangent line through this point has slope $f'(x_0)$. To get an equation for this tangent line, we use the point-slope formula

$$y - y_0 = m(x - x_0).$$

In this case $m = f'(x_0)$ and the equation becomes

(3.1.2)

$$y - y_0 = f'(x_0)(x - x_0).$$

The line through (x_0, y_0) that is perpendicular to the tangent is called the *normal line*. Since the slope of the tangent is $f'(x_0)$, the slope of the normal is $-1/f'(x_0)$, provided of course that $f'(x_0) \neq 0$. (*Remember*: For perpendicular lines, neither of which is vertical, $m_1 m_2 = -1$.) As an equation for this normal line we have

(3.1.3)

$$y - y_0 = -\frac{1}{f'(x_0)}(x - x_0), \quad \text{provided } f'(x_0) \neq 0.$$

Some tangents and normals are pictured in Figure 3.1.7.

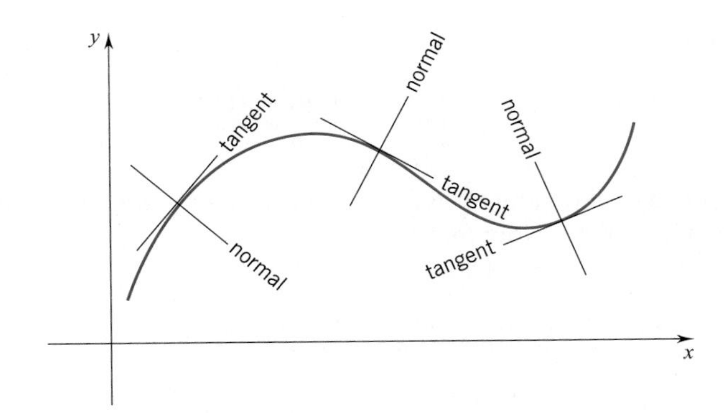

Figure 3.1.7

Example 7 Find equations for the tangent and normal to the graph of

$$f(x) = x^2$$

at the point $(-3, f(-3)) = (-3, 9)$.

SOLUTION From Example 1, we know that

$$f'(x) = 2x.$$

At the point $(-3, 9)$ the slope is $f'(-3) = -6$. As an equation for the tangent we have

$$y - 9 = -6(x - (-3)), \quad \text{which simplifies to} \quad y - 9 = -6(x + 3).$$

An equation for the normal can be written

$$y - 9 = \tfrac{1}{6}(x + 3). \quad \square$$

If $f'(x_0) = 0$, Equation (3.1.3) does not apply. If $f'(x_0) = 0$, the tangent line at (x_0, y_0) is horizontal (equation $y = y_0$) and the normal line is vertical (equation $x = x_0$). In Figure 3.1.8 you can see several instances of a horizontal tangent. In the first two cases the point of tangency is the origin, the tangent line is the x-axis ($y = 0$), and the normal line is the y-axis ($x = 0$).

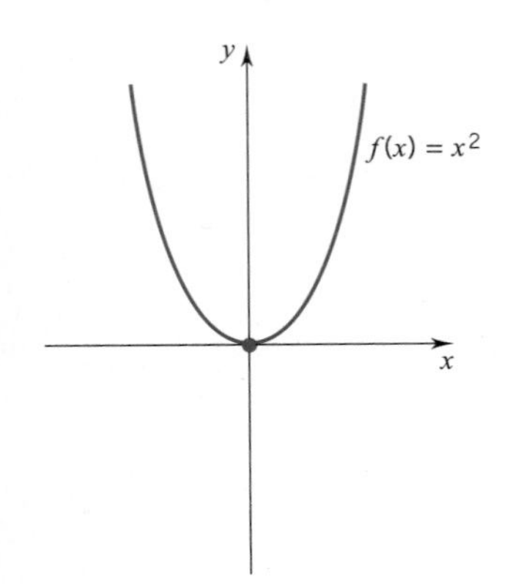

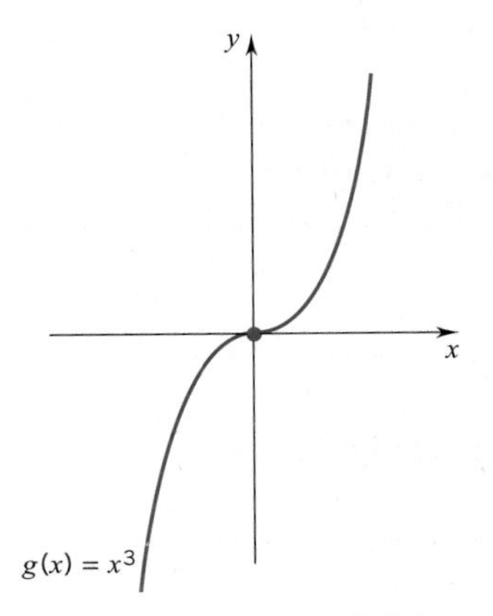

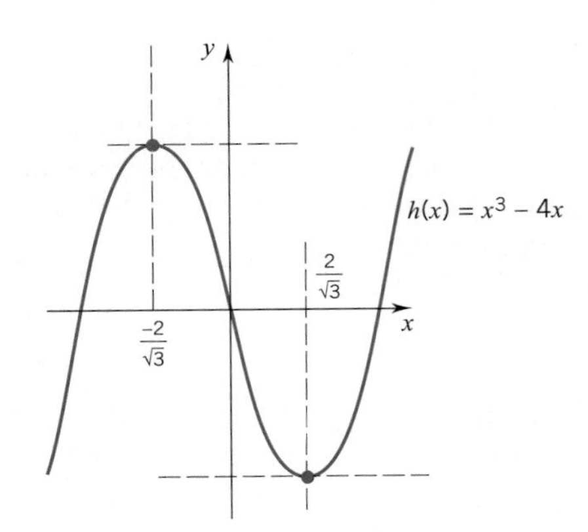

Figure 3.1.8

A Note on Vertical Tangents

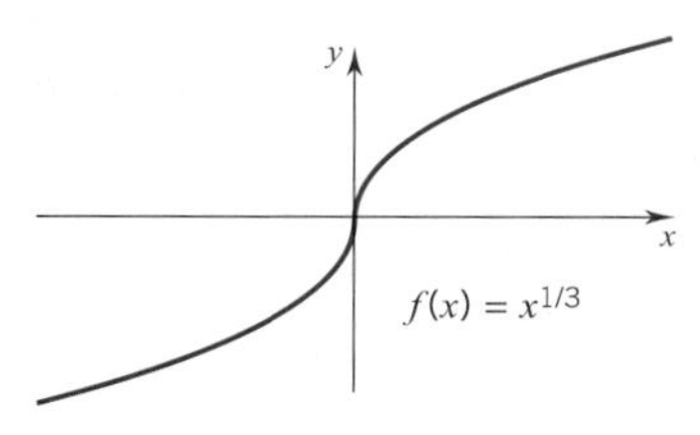

Figure 3.1.9

It is also possible for the graph of a function to have a vertical tangent. Figure 3.1.9 shows the graph of the cube-root function $f(x) = x^{1/3}$. The difference quotient at $x = 0$,

$$\frac{f(0+h) - f(0)}{h} = \frac{h^{1/3} - 0}{h} = \frac{1}{h^{2/3}},$$

increases without bound as h tends to zero. The graph has a vertical tangent at the origin (the line $x = 0$) and it has a horizontal normal (the line $y = 0$).

Vertical tangents are mentioned here only in passing. They are discussed in detail in Section 4.7.

Differentiability and Continuity

Differentiability, like continuity, is a so-called "point concept": a function f is differentiable at the point x iff

$$\lim_{h \to 0} \frac{f(x+h) - f(x)}{h} \quad \text{exists.}$$

Also like continuity, we say that f is differentiable on the open interval (a, b) if it is differentiable at each $x \in (a, b)$. Similarly, f is differentiable on $(-\infty, b)$, (a, ∞), or $(-\infty, \infty)$ if it is differentiable at each x in the indicated interval.

For example, $f(x) = x^2$ and $f(x) = mx + b$ are each differentiable on $(-\infty, \infty)$, while $f(x) = \sqrt{x}$ is differentiable on $(0, \infty)$. The function $f(x) = 1/x$ is differentiable on $(-\infty, 0)$ and on $(0, \infty)$.

We can extend the definition of differentiability to include either of the endpoints of an interval [a, b] by introducing one-sided derivatives. This is developed in the Exercises. Finally, as stated in Definition 3.1.1, a function f is said to be a *differentiable function* if it is differentiable at each point in its domain.

A function can be continuous at some number x without being differentiable there. For example, the absolute-value function $f(x) = |x|$ is continuous at 0 (it is everywhere continuous), but it is not differentiable at 0:

$$\frac{f(0+h) - f(0)}{h} = \frac{|0+h| - |0|}{h} = \frac{|h|}{h} = \begin{cases} -1, & h < 0 \\ 1, & h > 0 \end{cases}$$

so that

$$\lim_{h \to 0^-} \frac{f(0+h) - f(0)}{h} = -1, \qquad \lim_{h \to 0^+} \frac{f(0+h) - f(0)}{h} = 1$$

and thus

$$\lim_{h \to 0} \frac{f(0+h) - f(0)}{h} \quad \text{does not exist.} \quad ❑$$

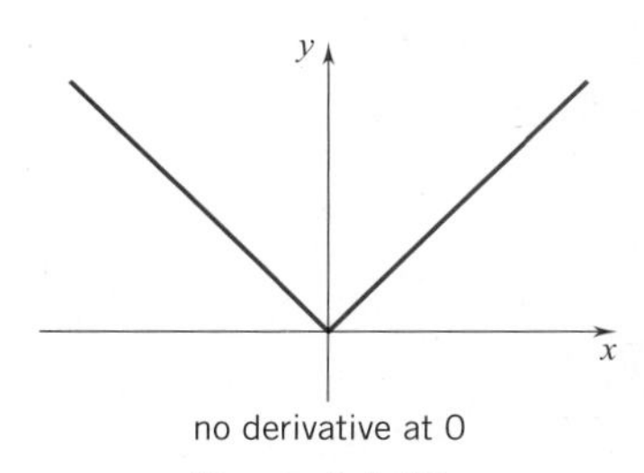

no derivative at 0

Figure 3.1.10

In Figure 3.1.10 we have drawn the graph of the absolute-value function. The lack of differentiability at 0 is evident geometrically. At the point $(0, 0) = (0, f(0))$ the graph changes direction abruptly and there is no single tangent line. You should also note that the absolute-value function *is* differentiable on $(-\infty, 0)$, $f'(x) = -1$ on this interval; and it *is* differentiable on $(0, \infty)$ with $f'(x) = 1$.

You can see a similar change of direction in the graph of

$$f(x) = \begin{cases} x^2, & x \le 1 \\ \tfrac{1}{2}x + \tfrac{1}{2}, & x > 1 \end{cases} \qquad \text{(Figure 3.1.11)}$$

no derivative at 1

Figure 3.1.11

at the point (1, 1). Once again, f is everywhere continuous (verify this), but it is not differentiable at 1:

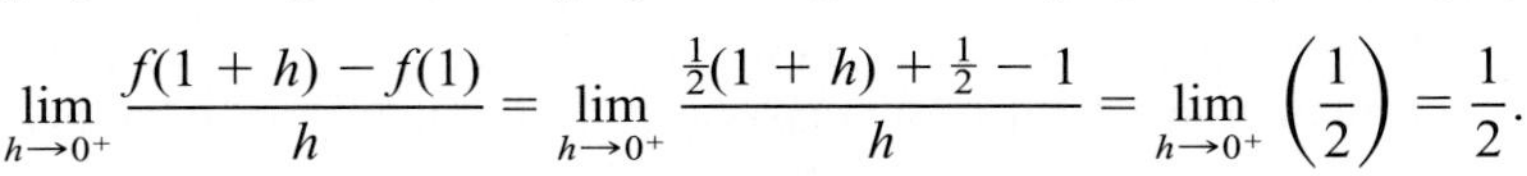

$$\lim_{h\to 0^-} \frac{f(1+h)-f(1)}{h} = \lim_{h\to 0^-} \frac{(1+h)^2-1}{h} = \lim_{h\to 0^-} \frac{h^2+2h}{h} = \lim_{h\to 0^-} (h+2) = 2,$$

$$\lim_{h\to 0^+} \frac{f(1+h)-f(1)}{h} = \lim_{h\to 0^+} \frac{\frac{1}{2}(1+h)+\frac{1}{2}-1}{h} = \lim_{h\to 0^+} \left(\frac{1}{2}\right) = \frac{1}{2}.$$

Since these one-sided limits are different, the two-sided limit

$$\lim_{h\to 0} \frac{f(1+h)-f(1)}{h} \quad \text{does not exist.} \quad \square$$

Contrast these two illustrations with Example 6.

A graphing utility was used to obtain the graph of

$$f(x) = |x^3 - 6x^2 + 8x| + 3$$

shown in Figure 3.1.12. It appears that f is differentiable except, possibly, at $x = 0$, $x = 2$, and $x = 4$ where abrupt changes in direction seem to occur. By zooming in near the point $(2, f(2))$ you can confirm that the left-hand and right-hand limits of the difference quotient both exist at $x = 2$, but they are not equal. See Figure 3.1.13. A similar situation occurs at $x = 0$ and $x = 4$. Thus, f fails to be differentiable at $x = 0$, 2, and 4.

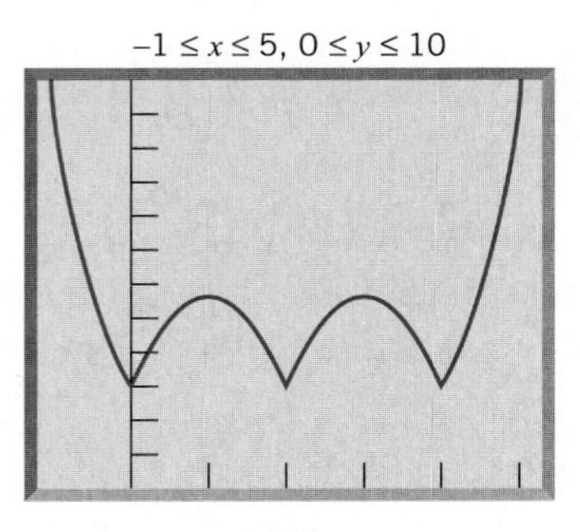

Figure 3.1.12

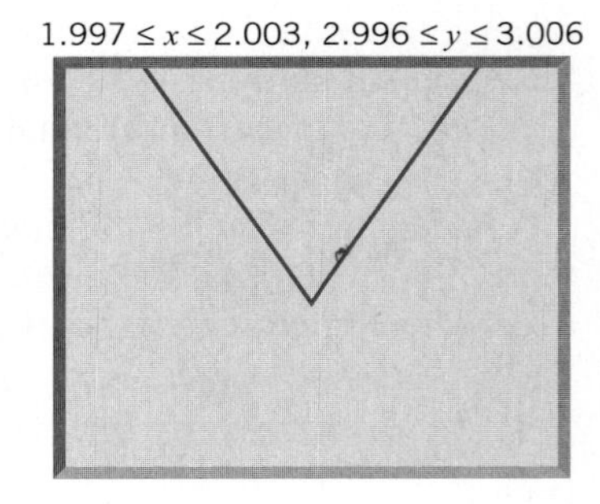

Figure 3.1.13

Although not every continuous function is differentiable, every differentiable function is continuous.

THEOREM 3.1.4

If f is differentiable at x, then f is continuous at x.

PROOF For $h \ne 0$ and $x + h$ in the domain of f,

$$f(x+h) - f(x) = \frac{f(x+h)-f(x)}{h} \cdot h.$$

With f differentiable at x,

$$\lim_{h\to 0} \frac{f(x+h)-f(x)}{h} = f'(x).$$

Since $\lim_{h\to 0} h = 0$, we have

$$\lim_{h\to 0} [f(x+h)-f(x)] = \left[\lim_{h\to 0} \frac{f(x+h)-f(x)}{h}\right]\cdot\left[\lim_{h\to 0} h\right] = f'(x)\cdot 0 = 0.$$

It follows that

$$\lim_{h\to 0} f(x+h) = f(x) \qquad \text{(explain)}$$

and thus (by Exercise 47, Section 2.4) f is continuous at x. See, also, Exercise 53, Section 2.4. ❑

EXERCISES 3.1

In Exercises 1–12, differentiate the given function by forming a difference quotient

$$\frac{f(x+h)-f(x)}{h}$$

and taking the limit as h tends to 0.

1. $f(x) = 4$.
2. $f(x) = c$.
3. $f(x) = 2 - 3x$.
4. $f(x) = 4x + 1$.
5. $f(x) = 5x - x^2$.
6. $f(x) = 2x^3 + 1$.
7. $f(x) = x^4$.
8. $f(x) = 1/(x+3)$.
9. $f(x) = \sqrt{x-1}$.
10. $f(x) = x^3 - 4x$.
11. $f(x) = 1/x^2$.
12. $f(x) = 1/\sqrt{x}$.

In Exercises 13–18, find $f'(2)$ by forming the difference quotient

$$\frac{f(2+h)-f(2)}{h}$$

and taking the limit as $h \to 0$.

13. $f(x) = (3x-7)^2$.
14. $f(x) = 7x - x^2$.
15. $f(x) = 9/(x+4)$.
16. $f(x) = 5 - x^4$.
17. $f(x) = x + \sqrt{2x}$.
18. $f(x) = \sqrt{6-x}$.

In Exercises 19–24, find equations for the tangent and normal to the graph of f at the point $(a, f(a))$.

19. $f(x) = x^2$; $a = 2$.
20. $f(x) = \sqrt{x}$; $a = 4$.
21. $f(x) = 5x - x^2$; $a = 4$.
22. $f(x) = 5 - x^3$; $a = 2$.
23. $f(x) = 1/(x+2)$; $a = -3$.
24. $f(x) = 1/x^2$; $a = -2$.

25. The graph of a function f is shown in the figure below.

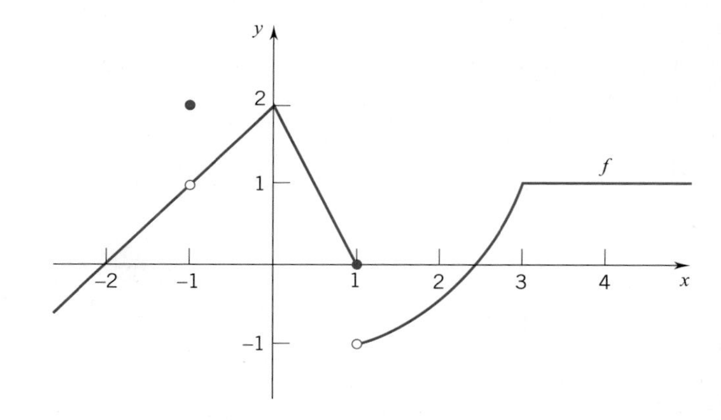

(a) For which numbers c does f fail to be continuous? For each discontinuity, state whether it is a removable discontinuity, a jump discontinuity, or neither.
(b) At which numbers c is f continuous but not differentiable?

26. Repeat Exercise 25 for the function g whose graph is shown in the figure below.

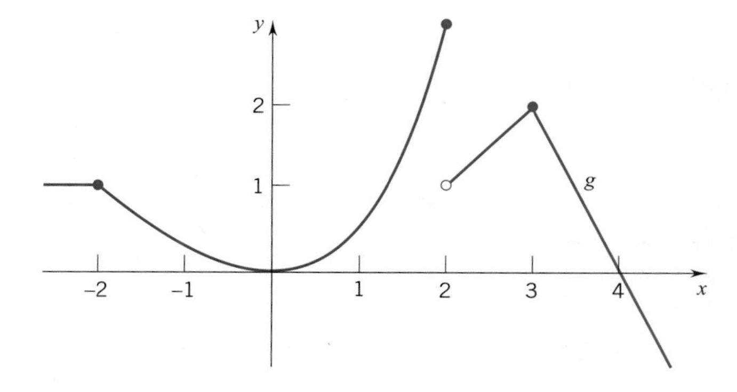

In Exercises 27–32, draw the graph of each of the following functions and indicate where it is not differentiable.

27. $f(x) = |x + 1|$.

28. $f(x) = |2x - 5|$.

29. $f(x) = \sqrt{|x|}$.

30. $f(x) = |x^2 - 4|$.

31. $f(x) = \begin{cases} x^2, & x \le 1 \\ 2 - x, & x > 1. \end{cases}$

32. $f(x) = \begin{cases} x^2 - 1, & x \le 2 \\ 3, & x > 2. \end{cases}$

In Exercises 33–36, find $f'(c)$ if it exists.

33. $f(x) = \begin{cases} 4x, & x < 1 \\ 2x^2 + 2, & x \ge 1; \end{cases}$ $c = 1$.

34. $f(x) = \begin{cases} 3x^2, & x \le 1 \\ 2x^3 + 1, & x > 1; \end{cases}$ $c = 1$.

35. $f(x) = \begin{cases} x + 1, & x \le -1 \\ (x + 1)^2, & x > -1; \end{cases}$ $c = -1$.

36. $f(x) = \begin{cases} -\frac{1}{2}x^2, & x < 3 \\ -3x, & x \ge 3; \end{cases}$ $c = 3$.

Sketch the graph of the derivative of the function with the given graph.

37.

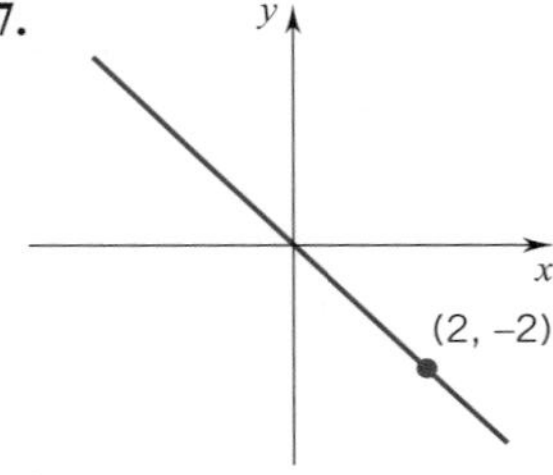

38.

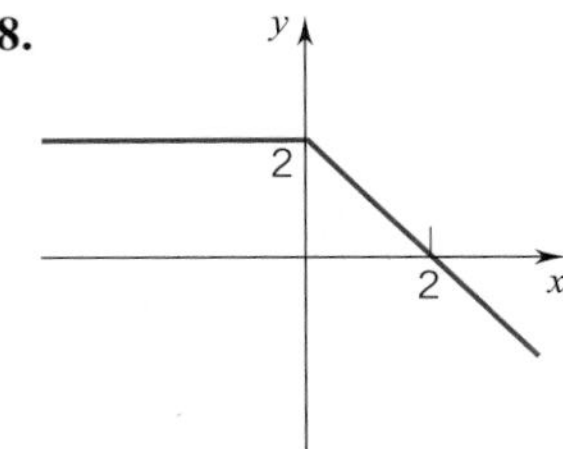

39.

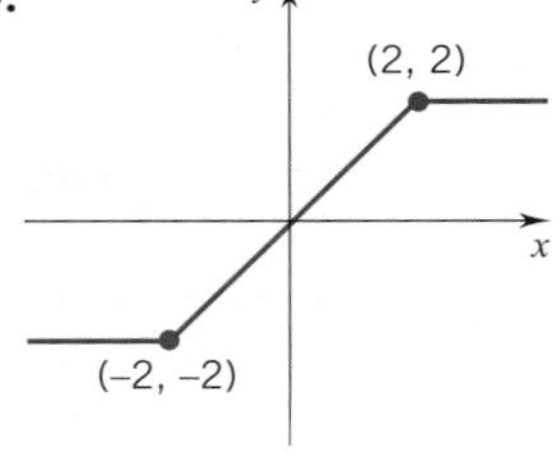

40.

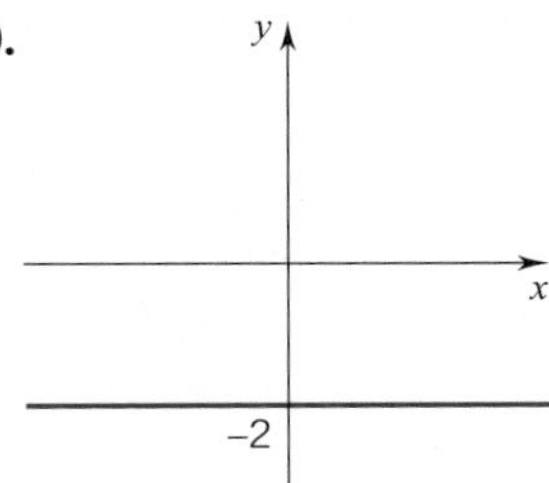

41.

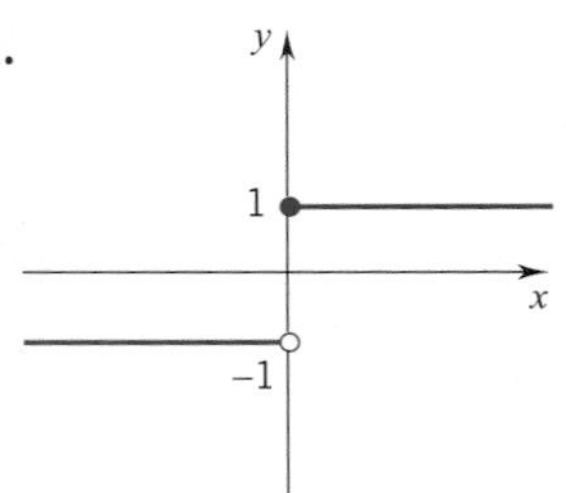

42.

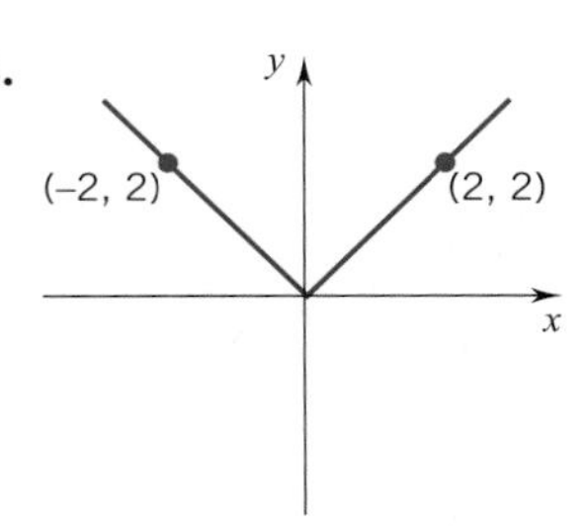

In Exercises 43–48, each limit represents the derivative of a function f at a value c. Determine f and c in each case.

43. $\lim_{h \to 0} \dfrac{(1 + h)^2 - 1}{h}$.

44. $\lim_{h \to 0} \dfrac{(-2 + h)^3 + 8}{h}$.

45. $\lim_{h \to 0} \dfrac{\sqrt{4 + h} - 2}{h}$.

46. $\lim_{h \to 0} \dfrac{(8 + h)^{1/3} - 2}{h}$.

47. $\lim_{h \to 0} \dfrac{\cos(\pi + h) + 1}{h}$.

48. $\lim_{h \to 0} \dfrac{\sin(\frac{\pi}{6} + h) - \frac{1}{2}}{h}$.

49. Show that

$$f(x) = \begin{cases} x^2, & x \le 1 \\ 2x, & x > 1 \end{cases}$$

is not differentiable at $x = 1$. HINT: Use Theorem 3.1.4.

50. Find A and B given that the function

$$f(x) = \begin{cases} x^3, & x \le 1 \\ Ax + B, & x > 1 \end{cases}$$

is differentiable at $x = 1$.

51. Let

$$f(x) = \begin{cases} (x + 1)^2, & x \le 0 \\ (x - 1)^2, & x > 0. \end{cases}$$

(a) Compute $f'(x)$ for $x \ne 0$.
(b) Show that f is not differentiable at $x = 0$.

52. Given that f is differentiable at c, let g be the function defined by

$$g(x) = \begin{cases} f(x), & x \le c \\ f'(c)(x - c) + f(c), & x > c. \end{cases}$$

(a) Show that g is differentiable at c. What is $g'(c)$?
(b) Suppose that the graph of f is as shown in the figure below. Sketch the graph of g. Note that this exercise is a generalization of Exercise 50.

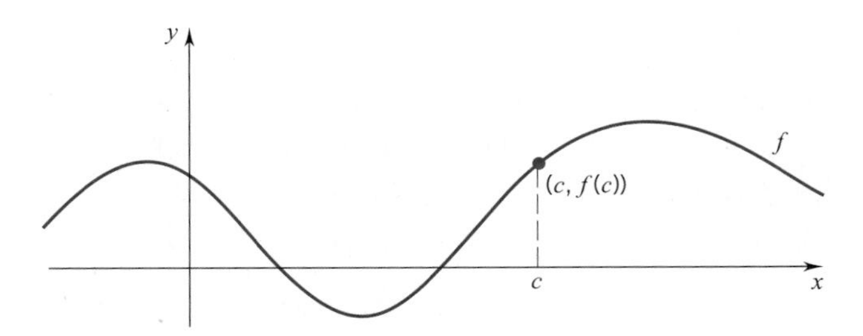

In Exercises 53–58, give an example of a function f that is defined for all real numbers and satisfies the given conditions.

53. $f'(x) = 0$ for all real x.

54. $f'(x) = 0$ for all $x \ne 0$; $f'(0)$ does not exist.

55. $f'(x)$ exists for all $x \ne -1$; $f'(-1)$ does not exist.

56. $f'(x)$ exists for all $x \neq \pm 1$; neither $f'(1)$ nor $f'(-1)$ exists.

57. $f'(1) = 2$ and $f(1) = 7$.

58. $f'(x) = 1$ for $x < 0$ and $f'(x) = -1$ for $x > 0$.

59. Suppose that f is an odd function $(f(-x) = -f(x))$. Prove that if f is differentiable, then f' is an even function.

60. Suppose that f is an even function. Prove that if f is differentiable, then f' is an odd function.

(*Important*) Given a function f and a value x in the domain of f. The *left-hand derivative of f at x* and the *right-hand derivative of f at x*, denoted $f'_-(x)$ and $f'_+(x)$, respectively, are defined by

$$f'_-(x) = \lim_{h\to 0^-} \frac{f(x+h) - f(x)}{h}$$

and

$$f'_+(x) = \lim_{h\to 0^+} \frac{f(x+h) - f(x)}{h}$$

provided these limits exist.

61. Let $f(x) = \begin{cases} x^2 - x & x \le 2 \\ 2x - 2 & x > 2. \end{cases}$

(a) Show that f is continuous at 2.
(b) Find $f'_-(2)$ and $f'_+(2)$.
(c) Is f differentiable at 2?

62. Let $f(x) = x\sqrt{x}$, $x \ge 0$.

(a) Calculate $f'(x)$ for any $x > 0$.
(b) Calculate $f'_+(0)$.

63. Let $f(x) = \sqrt{1-x}$ for $0 \le x \le 1$.

(a) Calculate $f'(x)$ for any $x \in (0, 1)$.
(b) Find $f'_+(0)$, if it exists.
(c) Find $f'_-(1)$, if it exists.

64. Let $f(x) = \begin{cases} 1 - x^2 & x \le 0 \\ x^2 & x > 0. \end{cases}$

(a) Find $f'_-(0)$, if it exists.
(b) Find $f'_+(0)$, if it exists.
(c) Is f differentiable at 0? HINT: Sketch the graph of f.

65. Given a function f and a number c in the domain of f, prove that if $f'_-(c)$ and $f'_+(c)$ both exist and are equal, then f is differentiable at c.

66. Let $f(x) = \begin{cases} x & x \text{ rational} \\ 0 & x \text{ irrational} \end{cases}$ and

$g(x) = \begin{cases} x^2 & x \text{ rational} \\ 0 & x \text{ irrational.} \end{cases}$

Both functions are continuous at 0 and are discontinuous at any $x \neq 0$. (See Exercise 44, Section 2.4.)

(a) Can either function be differentiable at a value x where $x \neq 0$? Explain.
(b) Show that f is not differentiable at 0.
(c) Show that g is differentiable at 0 and give $g'(0)$.

67. Let $f(x) = \begin{cases} x \sin(1/x), & x \neq 0 \\ 0, & x = 0 \end{cases}$ and

$g(x) = \begin{cases} x^2 \sin(1/x), & x \neq 0 \\ 0, & x = 0. \end{cases}$

The graphs of f and g are indicated in the figures below.

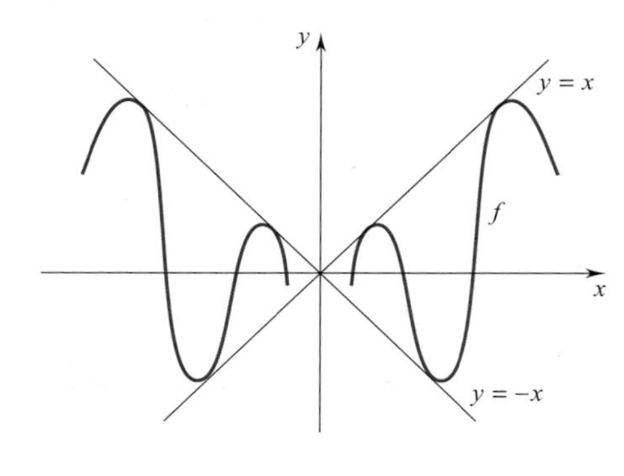

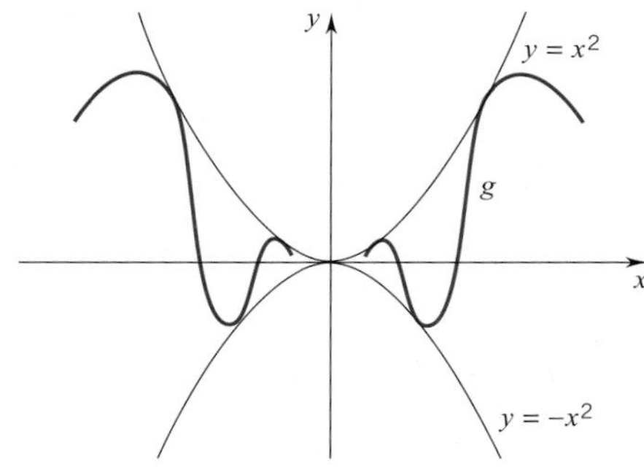

(a) Show that f and g are both continuous at 0.
(b) Show that f is not differentiable at 0.
(c) Show that g is differentiable at 0 and give $g'(0)$.

(*Important*) According to Definition 3.1.1, the derivative of f at c is given by

$$f'(c) = \lim_{h\to 0} \frac{f(c+h) - f(c)}{h}$$

provided this limit exists. Setting $x = c + h$, we can write

(3.1.5) $$f'(c) = \lim_{x\to c} \frac{f(x) - f(c)}{x - c}$$

(see (2.2.5), Section 2.2). This alternative definition of the derivative has advantages in certain situations. Convince yourself of the equivalence of both definitions by calculating $f'(c)$ by both methods in Exercises 68–72.

68. $f(x) = x^3 + 1$; $c = 2$.

69. $f(x) = x^2 - 3x$; $c = 1$.

70. $f(x) = \sqrt{1+x}$; $c = 3$.

71. $f(x) = x^{1/3}$; $c = -1$.

72. $f(x) = \dfrac{1}{x+2}$; $c = 0$.

73. Let $f(x) = x^{5/2}$ and consider the difference quotient

$$D(h) = \frac{f(2+h) - f(2)}{h}.$$

(a) Use a graphing utility to graph D for $-1 \le h \le 1$.
(b) Use the zoom feature to obtain successive magnifications of the graph of D near $h = 0$. Estimate $f'(2)$ accurate to three decimal places.
(c) Calculate D at $h = \pm 0.001$ and compare your result with your estimate in (b).

74. Repeat Exercise 73 for $f(x) = x^{2/3}$.

75. Let $f(x) = 4x - x^3$.
(a) Find an equation for the tangent line T to the graph of f at $x_0 = \frac{3}{2}$; write your equation in the form $T(x) = f'(\frac{3}{2})(x - \frac{3}{2}) + f(\frac{3}{2})$.
(b) Use a graphing utility to graph f and T together.
(c) Notice that the tangent line is a good approximation to the graph of f for values of x close to $x = \frac{3}{2}$. Determine an interval on which $|f(x) - T(x)| < 0.01$.

76. Let $f(x) = \sqrt{x-1}$ and $x_0 = 2$. Repeat Exercise 75.

■ 3.2 SOME DIFFERENTIATION FORMULAS

Calculating the derivative of

$$f(x) = (x^3 + 2x - 3)(4x^2 + 1) \qquad \text{or} \qquad f(x) = \frac{6x^2 - 1}{x^4 + 5x + 1}$$

by forming the proper difference quotient

$$\frac{f(x+h) - f(x)}{h}$$

and then taking the limit as h tends to 0 is somewhat laborious. Here we derive some general formulas that enable us to calculate such derivatives quite quickly and easily.

We begin by pointing out that constant functions have derivative identically 0:

(3.2.1) $$\text{if } f(x) = \alpha, \quad \alpha \text{ any constant,} \quad \text{then} \quad f'(x) = 0 \quad \text{for all real } x$$

and the identity function $f(x) = x$ has constant derivative 1:

(3.2.2) $$\text{if} \quad f(x) = x, \qquad \text{then} \quad f'(x) = 1 \qquad \text{for all real } x.$$

PROOF For $f(x) = \alpha$,

$$f'(x) = \lim_{h \to 0} \frac{f(x+h) - f(x)}{h} = \lim_{h \to 0} \frac{\alpha - \alpha}{h} = \lim_{h \to 0} 0 = 0.$$

For $f(x) = x$,

$$f'(x) = \lim_{h \to 0} \frac{f(x+h) - f(x)}{h} = \lim_{h \to 0} \frac{(x+h) - x}{h} = \lim_{h \to 0} \frac{h}{h} = \lim_{h \to 0} 1 = 1. \quad \square$$

Remark These results can also be verified geometrically. The graph of a constant function $f(x) = \alpha$ is a horizontal line and the slope of a horizontal line is 0. The graph of the identity function $f(x) = x$ is the graph of the line $y = x$, which has slope 1. ☐

THEOREM 3.2.3 DERIVATIVES OF SUMS AND SCALAR MULTIPLES

Let α be a real number. If f and g are differentiable at x, then $f + g$ and αf are differentiable at x. Moreover,

$$(f + g)'(x) = f'(x) + g'(x) \quad \text{and} \quad (\alpha f)'(x) = \alpha f'(x).$$

PROOF To verify the first formula note that

$$\frac{(f + g)(x + h) - (f + g)(x)}{h} = \frac{[f(x + h) + g(x + h)] - [f(x) + g(x)]}{h}$$
$$= \frac{f(x + h) - f(x)}{h} + \frac{g(x + h) - g(x)}{h}.$$

By definition

$$\lim_{h\to 0} \frac{f(x + h) - f(x)}{h} = f'(x), \qquad \lim_{h\to 0} \frac{g(x + h) - g(x)}{h} = g'(x).$$

Thus

$$\lim_{h\to 0} \frac{(f + g)(x + h) - (f + g)(x)}{h} = f'(x) + g'(x),$$

which means that

$$(f + g)'(x) = f'(x) + g'(x).$$

To verify the second formula we must show that

$$\lim_{h\to 0} \frac{(\alpha f)(x + h) - (\alpha f)(x)}{h} = \alpha f'(x).$$

This follows directly from the fact that

$$\frac{(\alpha f)(x + h) - (\alpha f)(x)}{h} = \frac{\alpha f(x + h) - \alpha f(x)}{h} = \alpha\left[\frac{f(x + h) - f(x)}{h}\right]. \quad ❑$$

Remark In this section and the next few sections we will be presenting rules and formulas for calculating derivatives. It will be to your advantage to memorize them. To do so, it sometimes helps to verbalize the rule or formula. For example, Theorem 3.2.3 says that

"the derivative of the sum of two functions is the sum of the derivatives," and

"the derivative of a constant times a function is that constant times the derivative of the function." ❑

Since $f - g = f + (-1)g$, it follows that if f and g are differentiable at x, then $f - g$ is differentiable at x, and

$$(f - g)'(x) = f'(x) - g'(x).$$

"The derivative of the difference of two functions is the difference of the derivatives."

The results of Theorem 3.2.3 are easily extended, by mathematical induction, to any finite collection of functions. That is, if $f_1, f_2, \ldots, f_n$ are each differentiable at x, and $\alpha_1, \alpha_2, \ldots, \alpha_n$ are real numbers, then $\alpha_1 f_1 + \alpha_2 f_2 + \cdots + \alpha_n f_n$ is differentiable at x and

(3.2.4) $$(\alpha_1 f_1 + \alpha_2 f_2 + \cdots + \alpha_n f_n)'(x) = \alpha_1 f_1'(x) + \alpha_2 f_2'(x) + \cdots + \alpha_n f_n'(x).$$

THEOREM 3.2.5 THE PRODUCT RULE

If f and g are differentiable at x, then so is their product, and

$$(fg)'(x) = f(x)g'(x) + g(x)f'(x).$$

"The derivative of the product of two functions is the first function times the derivative of the second plus the second function times the derivative of the first."

PROOF We form the difference quotient

$$\frac{(fg)(x+h) - (fg)(x)}{h} = \frac{f(x+h)g(x+h) - f(x)g(x)}{h}$$

and rewrite it as

$$f(x+h)\left[\frac{g(x+h) - g(x)}{h}\right] + g(x)\left[\frac{f(x+h) - f(x)}{h}\right].$$

(Here we have added and subtracted $f(x+h)g(x)$ in the numerator and then regrouped the terms so as to display difference quotients for f and g.) Since f is differentiable at x, we know that f is continuous at x (Theorem 3.1.4) and thus

$$\lim_{h \to 0} f(x+h) = f(x).$$

Since

$$\lim_{h \to 0} \frac{g(x+h) - g(x)}{h} = g'(x) \qquad \text{and} \qquad \lim_{h \to 0} \frac{f(x+h) - f(x)}{h} = f'(x),$$

we obtain

$$\lim_{h \to 0} \frac{(fg)(x+h) - (fg)(x)}{h} = f(x)g'(x) + g(x)f'(x). \quad \square$$

Using the product rule, it is not hard to prove that

(3.2.6)

for each positive integer n

$$p(x) = x^n \quad \text{has derivative} \quad p'(x) = nx^{n-1}.$$

In particular,

$$p(x) = x \quad \text{has derivative} \quad p'(x) = 1 \cdot x^0 = 1,$$
$$p(x) = x^2 \quad \text{has derivative} \quad p'(x) = 2x,$$
$$p(x) = x^3 \quad \text{has derivative} \quad p'(x) = 3x^2,$$
$$p(x) = x^4 \quad \text{has derivative} \quad p'(x) = 4x^3,$$

and so on.

PROOF OF (3.2.6) We proceed by mathematical induction on n. If $n = 1$, then we have the identity function

$$p(x) = x,$$

which we know satisfies

$$p'(x) = 1 = 1 \cdot x^0.$$

This means that the formula holds for $n = 1$.

We suppose now that the result holds for $n = k$, that is, we assume that if $p(x) = x^k$, then $p'(x) = kx^{k-1}$, and go on to show that it holds for $n = k + 1$. We let

$$p(x) = x^{k+1}$$

and note that

$$p(x) = x \cdot x^k.$$

Applying the product rule (3.2.5) and our induction hypothesis, we obtain

$$p'(x) = x \cdot kx^{k-1} + x^k \cdot 1 = (k + 1)x^k.$$

This shows that the formula holds for $n = k + 1$.

By the axiom of induction the formula holds for all positive integers n. ❑

The formula for differentiating polynomials follows from (3.2.4) and (3.2.6):

(3.2.7)

If $\quad P(x) = a_n x^n + a_{n-1}x^{n-1} + \cdots + a_2 x^2 + a_1 x + a_0,$

then $\quad P'(x) = na_n x^{n-1} + (n - 1)a_{n-1}x^{n-2} + \cdots + 2a_2 x + a_1.$

For example,

$$P(x) = 12x^3 - 6x - 2 \quad \text{has derivative} \quad P'(x) = 36x^2 - 6$$

and

$$Q(x) = \tfrac{1}{4}x^4 - 2x^2 + x + 5 \quad \text{has derivative} \quad Q'(x) = x^3 - 4x + 1.$$

Example 1 Differentiate

$$F(x) = (x^3 + 2x - 3)(4x^2 + 1)$$

and find $F'(-1)$.

SOLUTION We have a product $F(x) = f(x)g(x)$ with

$$f(x) = x^3 - 2x + 3 \qquad \text{and} \qquad g(x) = 4x^2 + 1.$$

The product rule gives

$$\begin{aligned} F'(x) &= f(x)g'(x) + g(x)f'(x) \\ &= (x^3 - 2x + 3)(8x) + (4x^2 + 1)(3x^2 - 2) \\ &= 8x^4 - 16x^2 + 24x + 12x^4 - 5x^2 - 2 \\ &= 20x^4 - 21x^2 + 24x - 2. \end{aligned}$$

Setting $x = -1$, we have

$$F'(-1) = 20(-1)^4 - 21(-1)^2 + 24(-1) - 2 = 20 - 21 - 24 - 2 = -27. \quad \square$$

Example 2 Differentiate

$$F(x) = (ax + b)(cx + d), \qquad \text{(where } a, b, c, d \text{ are constants).}$$

SOLUTION We have a product $F(x) = f(x)g(x)$ with

$$f(x) = ax + b \qquad \text{and} \qquad g(x) = cx + d.$$

Again we use the product rule

$$F'(x) = f(x)g'(x) + g(x)f'(x).$$

In this case

$$F'(x) = (ax + b)c + (cx + d)a = 2acx + bc + ad.$$

We can also do the problem without using the product rule by first carrying out the multiplication:

$$F(x) = acx^2 + bcx + adx + bd$$

and then differentiating:

$$F'(x) = 2acx + bc + ad.$$

The result is the same. $\square$

We come now to reciprocals.

THEOREM 3.2.8 THE RECIPROCAL RULE

If g is differentiable at x and $g(x) \neq 0$, then $1/g$ is differentiable at x and

$$\left(\frac{1}{g}\right)'(x) = -\frac{g'(x)}{[g(x)]^2}.$$

PROOF Since g is differentiable at x, g is continuous at x (Theorem 3.1.4). Since $g(x) \neq 0$, we know that $1/g$ is continuous at x, and thus that

$$\lim_{h \to 0} \frac{1}{g(x+h)} = \frac{1}{g(x)}.$$

For h different from 0 and sufficiently small, $g(x + h) \neq 0$ (the continuity of g at x and the fact that $g(x) \neq 0$ guarantee this—see Exercise 48, Section 2.4). Now the difference quotient is:

$$\frac{1}{h}\left[\frac{1}{g(x+h)} - \frac{1}{g(x)}\right] = \frac{1}{h}\left[\frac{g(x) - g(x+h)}{g(x+h)g(x)}\right]$$
$$= -\left[\frac{g(x+h) - g(x)}{h}\right]\frac{1}{g(x+h)}\frac{1}{g(x)}.$$

As h tends to zero, the right-hand side (and thus the left) tends to

$$-\frac{g'(x)}{[g(x)]^2}. \quad \square$$

From this last result it is easy to see that the formula for the derivative of a positive integer power, x^n, also applies to negative integer powers; namely,

(3.2.9)

> for each negative integer n
>
> $p(x) = x^n$ has derivative $p'(x) = nx^{n-1}$.

This formula holds for all x except, of course, at $x = 0$ where no negative power is even defined. In particular, for $x \neq 0$,

$$p(x) = x^{-1} \quad \text{has derivative} \quad p'(x) = (-1)x^{-2} = -x^{-2},$$
$$p(x) = x^{-2} \quad \text{has derivative} \quad p'(x) = -2x^{-3},$$
$$p(x) = x^{-3} \quad \text{has derivative} \quad p'(x) = -3x^{-4},$$

and so on.

PROOF OF (3.2.9) Note that

$$p(x) = \frac{1}{g(x)} \quad \text{where} \quad g(x) = x^{-n} \text{ and } -n \text{ is a positive integer.}$$

The rule for reciprocals gives

$$p'(x) = -\frac{g'(x)}{[g(x)]^2} = -\frac{(-nx^{-n-1})}{x^{-2n}} = (nx^{-n-1})x^{2n} = nx^{n-1}. \quad \square$$

Remark An obvious question now is: What about the derivative of $p(x) = x^0$? Since $x^0 = 1$ for all $x \neq 0$ (0^0 is another indeterminate form; it will be treated in Chapter 10), it follows that $p'(x) = 0$ for all $x \neq 0$. But,

$$0 = 0 \cdot x^{-1} \quad \text{for all} \quad x \neq 0,$$

and so the formula also holds for $n = 0$. Thus,

> for *any* integer n
>
> $p(x) = x^n$ has derivative $p'(x) = nx^{n-1}$.

$\square$

Example 3 Differentiate

$$f(x) = \frac{5}{x^2} - \frac{6}{x}$$

and find $f'(\frac{1}{2})$.

SOLUTION To apply (3.2.9) we write

$$f(x) = 5x^{-2} - 6x^{-1}.$$

Differentiation gives

$$f'(x) = -10x^{-3} + 6x^{-2}.$$

Back in fractional notation

$$f'(x) = -\frac{10}{x^3} + \frac{6}{x^2}.$$

Setting $x = \frac{1}{2}$, we have

$$f'(\tfrac{1}{2}) = -\frac{10}{(\frac{1}{2})^3} + \frac{6}{(\frac{1}{2})^2} = -80 + 24 = -56. \quad \square$$

Example 4 Differentiate

$$f(x) = \frac{1}{ax^2 + bx + c}, \qquad \text{(where } a, b, c \text{ are constants).}$$

SOLUTION Here we have a reciprocal $f(x) = 1/g(x)$ with

$$g(x) = ax^2 + bx + c.$$

The reciprocal rule (3.2.8) gives

$$f'(x) = -\frac{g'(x)}{[g(x)]^2} = -\frac{2ax + b}{[ax^2 + bx + c]^2}. \quad \square$$

Finally we come to quotients in general.

THEOREM 3.2.10 THE QUOTIENT RULE

If f and g are differentiable at x and $g(x) \neq 0$, then the quotient f/g is differentiable at x and

$$\left(\frac{f}{g}\right)'(x) = \frac{g(x)f'(x) - f(x)g'(x)}{[g(x)]^2}$$

"The derivative of a quotient is the denominator times the derivative of the numerator minus the numerator times the derivative of the denominator all divided by the square of the denominator."

The proof of the quotient rule is left to you as an exercise. Also, you should note that the reciprocal rule is a special case of the quotient rule (take $f(x) = 1$). Therefore it is not necessary for you to memorize the reciprocal rule.

From the quotient rule you can see that all rational functions (quotients of polynomials) are differentiable wherever they are defined.

Example 5 Differentiate

$$F(x) = \frac{ax + b}{cx + d} \qquad \text{(where } a, b, c, d \text{ are constants).}$$

SOLUTION We are dealing with a quotient $F(x) = f(x)/g(x)$. The quotient rule,

$$F'(x) = \frac{g(x)f'(x) - f(x)g'(x)}{[g(x)]^2},$$

gives

$$F'(x) = \frac{(cx + d) \cdot a - (ax + b) \cdot c}{(cx + d)^2} = \frac{ad - bc}{(cx + d)^2}. \quad \square$$

Example 6 Differentiate

$$F(x) = \frac{6x^2 - 1}{x^4 + 5x + 1}.$$

SOLUTION Again we are dealing with a quotient $F(x) = f(x)/g(x)$. The quotient rule gives

$$F'(x) = \frac{(x^4 + 5x + 1)(12x) - (6x^2 - 1)(4x^3 + 5)}{(x^4 + 5x + 1)^2}$$

$$= \frac{-12x^5 + 4x^3 + 30x^2 + 12x + 5}{(x^4 + 5x + 1)^2}. \quad \square$$

Example 7 Find $f'(0)$, $f'(1)$, and $f'(2)$ for the function

$$f(x) = \frac{5x}{1 + x}.$$

SOLUTION First we find a general expression for $f'(x)$, and then we evaluate it at 0, 1, and 2. Using the quotient rule, we get

$$f'(x) = \frac{(1 + x)5 - 5x(1)}{(1 + x)^2} = \frac{5}{(1 + x)^2}.$$

This gives

$$f'(0) = \frac{5}{(1 + 0)^2} = 5, \qquad f'(1) = \frac{5}{(1 + 1)^2} = \frac{5}{4}, \qquad f'(2) = \frac{5}{(1 + 2)^2} = \frac{5}{9}. \quad \square$$

Example 8 Find $f'(-1)$ for

$$f(x) = \frac{x^2}{ax^2 + b} \qquad \text{(where } a \text{ and } b \text{ are constants).}$$

SOLUTION We first find $f'(x)$ in general by the quotient rule:

$$f'(x) = \frac{(ax^2 + b)2x - x^2(2ax)}{(ax^2 + b)^2} = \frac{2bx}{(ax^2 + b)^2}.$$

Now we evaluate f' at -1:

$$f'(-1) = -\frac{2b}{(a+b)^2}. \quad \square$$

Remark Some expressions are easier to differentiate if we rewrite them in more convenient form. For example, we can differentiate

$$f(x) = \frac{x^5 - 2x}{x^2} = \frac{x^4 - 2}{x}, \qquad x \neq 0,$$

by the quotient rule, or we can write

$$f(x) = (x^4 - 2)x^{-1}$$

and use the product rule; even better, we can notice that

$$f(x) = x^3 - 2x^{-1}$$

and proceed from there:

$$f'(x) = 3x^2 + 2x^{-2}. \quad \square$$

EXERCISES 3.2

In Exercises 1–20, find the derivative of the given function.

1. $F(x) = 1 - x.$

2. $F(x) = 2(1 + x).$

3. $F(x) = 11x^5 - 6x^3 + 8.$

4. $F(x) = \dfrac{3}{x^2}.$

5. $F(x) = ax^2 + bx + c$; a, b, c constant.

6. $F(x) = \dfrac{x^4}{4} - \dfrac{x^3}{3} + \dfrac{x^2}{2} - \dfrac{x}{1}.$

7. $F(x) = -\dfrac{1}{x^2}.$

8. $F(x) = \dfrac{(x^2 + 2)}{x^3}.$

9. $G(x) = (x^2 - 1)(x - 3).$

10. $F(x) = x - \dfrac{1}{x}.$

11. $G(x) = \dfrac{x^3}{1 - x}.$

12. $F(x) = \dfrac{ax - b}{cx - d}$; a, b, c, d constant.

13. $G(x) = \dfrac{x^2 - 1}{2x + 3}.$

14. $G(x) = \dfrac{7x^4 + 11}{x + 1}.$

15. $G(x) = (x - 1)(x - 2).$

16. $G(x) = \dfrac{2x^2 + 1}{x + 2}.$

17. $G(x) = \dfrac{6 - 1/x}{x - 2}.$

18. $G(x) = \dfrac{1 + x^4}{x^2}.$

19. $G(x) = (9x^8 - 8x^9)\left(x + \dfrac{1}{x}\right).$

20. $G(x) = \left(1 + \dfrac{1}{x}\right)\left(1 + \dfrac{1}{x^2}\right).$

In Exercises 21–26, find $f'(0)$ and $f'(1)$.

21. $f(x) = \dfrac{1}{x - 2}.$

22. $f(x) = x^2(x + 1).$

23. $f(x) = \dfrac{1 - x^2}{1 + x^2}.$

24. $f(x) = \dfrac{2x^2 + x + 1}{x^2 + 2x + 1}.$

25. $f(x) = \dfrac{ax + b}{cx + d}$; a, b, c, d constant.

26. $f(x) = \dfrac{ax^2 + bx + c}{cx^2 + bx + a}$; a, b, c constant.

In Exercises 27–30, find $f'(0)$ given that $h(0) = 3$ and $h'(0) = 2$.

27. $f(x) = xh(x).$

28. $f(x) = 3x^2h(x) - 5x.$

29. $f(x) = h(x) - \dfrac{1}{h(x)}.$

30. $f(x) = h(x) + \dfrac{x}{h(x)}.$

In Exercises 31–36, find an equation for the tangent to the graph of f at the point $(a, f(a))$.

31. $f(x) = \dfrac{x}{x + 2}$; $a = -4.$

32. $f(x) = (x^3 - 2x + 1)(4x - 5)$; $a = 2.$

33. $f(x) = (x^2 - 3)(5x - x^3)$; $a = 1.$

34. $f(x) = x^2 - \dfrac{10}{x}$; $a = -2.$

35. $f(x) = 6/x^2$; $a = 3.$

36. $f(x) = \dfrac{3x + 5}{7x - 3}$; $a = 1.$

In Exercises 37–42, find the points where the tangent to the curve is horizontal.

37. $f(x) = (x - 2)(x^2 - x - 11)$.

38. $f(x) = x^2 - \dfrac{16}{x}$.

39. $f(x) = \dfrac{5x}{x^2 + 1}$.

40. $f(x) = (x + 2)(x^2 - 2x - 8)$.

41. $f(x) = x + \dfrac{4}{x^2}$.

42. $f(x) = \dfrac{x^2 - 2x + 4}{x^2 + 4}$.

In Exercises 43–46, find the points where the tangent to the curve

43. $f(x) = -x^2 - 6$ is parallel to the line $y = 4x - 1$.

44. $f(x) = x^3 - 3x$ is perpendicular to the line $5y - 3x - 8 = 0$.

45. $f(x) = x^3 - x^2$ is perpendicular to the line $5y + x + 2 = 0$.

46. $f(x) = 4x - x^2$ is parallel to the line $2y = 3x - 5$.

47. Find the area of the triangle formed by the x-axis and the lines tangent and normal to the curve $f(x) = 6x - x^2$ at the point $(5, 5)$.

48. Find the area of the triangle formed by the x-axis and the lines tangent and normal to the curve $f(x) = 9 - x^2$ at the point $(2, 5)$.

49. Determine the coefficients A, B, C so that the curve $f(x) = Ax^2 + Bx + C$ will pass through the point $(1, 3)$ and be tangent to the line $4x + y = 8$ at the point $(2, 0)$.

50. Determine A, B, C, D so that the curve $f(x) = Ax^3 + Bx^2 + Cx + D$ will be tangent to the line $y = 3x - 3$ at the point $(1, 0)$ and tangent to the line $y = 18x - 27$ at the point $(2, 9)$.

51. Find the value(s) of x where the tangent line to the graph of the quadratic function $f(x) = ax^2 + bx + c$ has a horizontal tangent line. NOTE: This gives a way to find the vertex of the parabola $y = ax^2 + bx + c$.

52. Find conditions on a, b, c, and d that will guarantee that the graph of the cubic polynomial $p(x) = ax^3 + bx^2 + cx + d$ has:
(a) Exactly two horizontal tangents.
(b) Exactly one horizontal tangent.
(c) No horizontal tangents.

53. Find the value(s) of c, if any, where the tangent line to the graph of $f(x) = x^3 - x$ at $(c, f(c))$ is parallel to the secant line through $(-1, f(-1))$ and $(2, f(2))$.

54. Find the value(s) of c, if any, where the tangent line to the graph of $f(x) = x/(x + 1)$ at $(c, f(c))$ is parallel to the secant line through $(1, f(1))$ and $(3, f(3))$.

55. Let $f(x) = 1/x$, $x > 0$. Show that the triangle that is formed by *any* tangent line to the graph of f and the coordinate axes has an area of 2 square units.

56. Let $f(x) = x^3$.
(a) Determine an equation for the tangent line to the graph of f at $x = c$, where $c \neq 0$.
(b) Determine whether the tangent line found in (a) intersects the graph of f at a point other than (c, c^3). If it does, determine the x-coordinate of the point of intersection.

57. Given two functions f and g, show that if f and $f + g$ are differentiable, then g is differentiable. Give an example to show that the differentiability of $f + g$ does not imply that f and g are each differentiable.

58. Given two functions f and g, if f and $f \cdot g$ are differentiable, does it follow that g is differentiable? If not, find a condition that will imply that g is differentiable when both f and $f \cdot g$ are differentiable.

59. Prove the validity of the quotient rule.
HINT: $f/g = f \cdot (1/g)$.

60. Verify that, if f, g, h are differentiable, then

$$(fgh)'(x) = f'(x)g(x)h(x) + f(x)g'(x)h(x) + f(x)g(x)h'(x).$$

HINT: Apply the product rule to $[f(x)g(x)]h(x)$.

61. Use the result in Exercise 60 to find the derivative of $F(x) = (x^2 + 1)(1 + (1/x))(2x^3 - x + 1)$.

62. Use the result in Exercise 60 to find the derivative of $G(x) = \sqrt{x}[1/(1 + 2x)](x^2 + x - 1)$.

63. Use the product rule (3.2.5) to show that, if f is differentiable, then

$$g(x) = [f(x)]^2 \quad \text{has derivative} \quad g'(x) = 2f(x)f'(x).$$

64. Show that, if f is differentiable, then

$$g(x) = [f(x)]^n \text{ has derivative } g'(x) = n[f(x)]^{n-1}f'(x)$$

for each nonzero integer n. HINT: Mimic the inductive proof of (3.2.6).

65. Use the result in Exercise 64 to find the derivative of $g(x) = (x^3 - 2x^2 + x + 2)^3$.

66. Use the result in Exercise 64 to find the derivative of $g(x) = [x^2/(1 + 2x)]^{10}$.

67. Let $f(x) = x^2 - 4x + 2$ on $[-1, 3]$, and let $g(x) = x^2 - 4x + 2$ on $(-\infty, \infty)$.
(a) Find $f'_+(-1)$ and $f'_-(3)$ (see the definition in Exercises 3.1).
(b) Find $g'(x)$ for any x, and evaluate g' at -1 and at 3. How do these results compare with the results in part (a)?

68. Suppose that g is a differentiable function on an open interval I. Let $[a, b]$ be a closed interval contained in I and let f be the function defined on $[a, b]$ by

$$f(x) = g(x) \qquad \text{for all} \quad x \in [a, b].$$

Prove that $f'_+(a) = g'(a)$ and $f'_-(b) = g'(b)$. This result provides an alternative method for calculating one-sided derivatives. When it can be applied, it is usually easier to use than the definition in Exercises 3.1.

69. Find A and B given that the derivative of

$$f(x) = \begin{cases} Ax^3 + Bx + 2, & x \leq 2 \\ Bx^2 - A, & x > 2 \end{cases}$$

is continuous for all real x. HINT: f itself must be continuous.

70. Find A and B given that the derivative of

$$f(x) = \begin{cases} Ax^2 + B, & x < -1 \\ Bx^5 + Ax + 4, & x \geq -1 \end{cases}$$

is continuous for all real x.

71. Let $f(x) = \sin x$.

(a) Approximate $f'(x)$ at $x = 0$, $x = \pi/6$, $x = \pi/4$, $x = \pi/3$, and $x = \pi/2$ using the difference quotient

$$\frac{f(x+h) - f(x)}{h}$$

with $h = \pm 0.001$.

(b) Compare the approximate values of $f'(x)$ found in (a) with the values of $\cos x$ at each of these points.

(c) Use the results in (b) to guess a formula for the derivative of the sine function.

72. Repeat Exercise 71 with $f(x) = \cos x$, using $\sin x$ in part (b) and the cosine function in part (c).

73. Let $f(x) = 2^x$.

(a) Approximate $f'(x)$ at $x = 0$, $x = 1$, $x = 2$, and $x = 3$ using the difference quotient

$$\frac{f(x+h) - f(x)}{h}$$

with $h = \pm 0.001$.

(b) Calculate the ratio $f'(x)/f(x)$ for each of the values of x in (a).

(c) Use the results in (b) to guess a formula for the derivative of $f(x) = 2^x$.

74. Let $f(x) = x^4 + x^3 - 5x^2 + 2$.

(a) Use a graphing utility to graph f on the interval $[-4, 4]$ and estimate the x-coordinates of the points where the tangent line to the graph of f is horizontal.

(b) Use a graphing utility to graph $|f|$. Are there any points where f is not differentiable? If so, estimate where f fails to be differentiable.

■ 3.3 THE *d/dx* NOTATION; DERIVATIVES OF HIGHER ORDER

The *d/dx* Notation

So far we have indicated the derivative by a prime, but there are other notations that are widely used, particularly in science and engineering. The most popular of these is the double-*d* notation of Leibniz. In the Leibniz notation the derivative of a function y is indicated by writing

$$\frac{dy}{dx}, \quad \frac{dy}{dt}, \quad \text{or} \quad \frac{dy}{dz}, \quad \text{and so forth,}$$

depending upon whether the letter x, t, or z, and so on, is being used for the elements of the domain of y. For instance, if y is initially defined by

$$y(x) = x^3,$$

then the Leibniz notation gives

$$\frac{dy(x)}{dx} = 3x^2.$$

Usually writers drop the (x) and simply write

$$y = x^3 \quad \text{and} \quad \frac{dy}{dx} = 3x^2.$$

The symbols

$$\frac{d}{dx}, \quad \frac{d}{dt}, \quad \frac{d}{dz}, \quad \text{and so forth,}$$

are also used as prefixes before expressions to be differentiated. For example,

$$\frac{d}{dx}(x^3 - 4x) = 3x^2 - 4, \qquad \frac{d}{dt}(t^2 + 3t + 1) = 2t + 3, \qquad \frac{d}{dz}(z^5 - 1) = 5z^4.$$

In the Leibniz notation the differentiation formulas are:

$$\frac{d}{dx}[f(x) + g(x)] = \frac{d}{dx}[f(x)] + \frac{d}{dx}[g(x)], \qquad \frac{d}{dx}[\alpha f(x)] = \alpha\frac{d}{dx}[f(x)],$$

$$\frac{d}{dx}[f(x)g(x)] = f(x)\frac{d}{dx}[g(x)] + g(x)\frac{d}{dx}[f(x)],$$

$$\frac{d}{dx}\left[\frac{1}{g(x)}\right] = -\frac{1}{[g(x)]^2}\frac{d}{dx}[g(x)],$$

$$\frac{d}{dx}\left[\frac{f(x)}{g(x)}\right] = \frac{g(x)\dfrac{d}{dx}[f(x)] - f(x)\dfrac{d}{dx}[g(x)]}{[g(x)]^2}.$$

Often functions f and g are replaced by u and v and the x is left out altogether. Then the formulas look like this:

$$\frac{d}{dx}(u + v) = \frac{du}{dx} + \frac{dv}{dx}, \qquad \frac{d}{dx}(\alpha u) = \alpha\frac{du}{dx},$$

$$\frac{d}{dx}(uv) = u\frac{dv}{dx} + v\frac{du}{dx},$$

$$\frac{d}{dx}\left(\frac{1}{v}\right) = -\frac{1}{v^2}\frac{dv}{dx}, \qquad \frac{d}{dx}\left(\frac{u}{v}\right) = \frac{v\dfrac{du}{dx} - u\dfrac{dv}{dx}}{v^2}.$$

The only way to develop a feeling for this notation is to use it. Below we work out some examples.

Example 1 Find

$$\frac{dy}{dx} \quad \text{for} \quad y = \frac{3x - 1}{5x + 2}.$$

SOLUTION We use the quotient rule:

$$\frac{dy}{dx} = \frac{(5x + 2)\dfrac{d}{dx}(3x - 1) - (3x - 1)\dfrac{d}{dx}(5x + 2)}{(5x + 2)^2}$$

$$= \frac{(5x + 2)(3) - (3x - 1)(5)}{(5x + 2)^2} = \frac{11}{(5x + 2)^2}. \quad \square$$

Example 2 Find

$$\frac{dy}{dx} \quad \text{for} \quad y = (x^3 + 1)(3x^5 + 2x - 1).$$

SOLUTION Here we use the product rule:

$$\begin{aligned}\frac{dy}{dx} &= (x^3+1)\frac{d}{dx}(3x^5+2x-1) + (3x^5+2x-1)\frac{d}{dx}(x^3+1)\\ &= (x^3+1)(15x^4+2) + (3x^5+2x-1)(3x^2)\\ &= (15x^7+15x^4+2x^3+2) + (9x^7+6x^3-3x^2)\\ &= 24x^7+15x^4+8x^3-3x^2+2. \quad \square\end{aligned}$$

Example 3 Find

$$\frac{d}{dt}\left(t^3 - \frac{t}{t^2-1}\right).$$

SOLUTION

$$\begin{aligned}\frac{d}{dt}\left(t^3 - \frac{t}{t^2-1}\right) &= \frac{d}{dt}(t^3) - \frac{d}{dt}\left(\frac{t}{t^2-1}\right)\\ &= 3t^2 - \left[\frac{(t^2-1)(1)-t(2t)}{(t^2-1)^2}\right] = 3t^2 + \frac{t^2+1}{(t^2-1)^2}. \quad \square\end{aligned}$$

Example 4 Find

$$\frac{du}{dx} \quad \text{for} \quad u = x(x+1)(x+2).$$

SOLUTION You can think of u as

$$[x(x+1)](x+2) \quad \text{or as} \quad x[(x+1)(x+2)].$$

From the first point of view,

$$\begin{aligned}\frac{du}{dx} &= [x(x+1)](1) + (x+2)\frac{d}{dx}[x(x+1)]\\ &= x(x+1) + (x+2)[x(1)+(x+1)(1)]\\ &= x(x+1) + (x+2)(2x+1). \qquad (*)\end{aligned}$$

From the second point of view,

$$\begin{aligned}\frac{du}{dx} &= x\frac{d}{dx}[(x+1)(x+2)] + (x+1)(x+2)(1)\\ &= x[(x+1)(1)+(x+2)(1)] + (x+1)(x+2)\\ &= x(2x+3) + (x+1)(x+2). \qquad (**)\end{aligned}$$

Both $(*)$ and $(**)$ can be multiplied out to give

$$\frac{du}{dx} = 3x^2 + 6x + 2.$$

Alternatively, this same result can be obtained by first carrying out the multiplication and then differentiating:

$$u = x(x+1)(x+2) = x(x^2+3x+2) = x^3+3x^2+2x$$

so that

$$\frac{du}{dx} = 3x^2 + 6x + 2. \quad \square$$

Example 5 Evaluate dy/dx at $x = 0$ and $x = 1$ given that

$$y = \frac{x^2}{x^2 - 4}.$$

SOLUTION

$$\frac{dy}{dx} = \frac{(x^2 - 4)2x - x^2(2x)}{(x^2 - 4)^2} = -\frac{8x}{(x^2 - 4)^2}.$$

At $x = 0$,

$$\frac{dy}{dx} = -\frac{8 \cdot 0}{(0^2 - 4)^2} = 0; \qquad \text{at } x = 1, \quad \frac{dy}{dx} = -\frac{8 \cdot 1}{(1^2 - 4)^2} = -\frac{8}{9}. \quad \square$$

Remark The notations

$$\left.\frac{dy}{dx}\right|_{x=a} \quad \text{and} \quad \frac{dy}{dx}(a)$$

are sometimes used to emphasize the fact that we are evaluating the derivative dy/dx at $x = a$. Thus, in Example 5, we have

$$\left.\frac{dy}{dx}\right|_{x=0} = 0 \quad \text{or} \quad \frac{dy}{dx}(0) = 0,$$

and

$$\left.\frac{dy}{dx}\right|_{x=1} = -\tfrac{8}{9} \quad \text{or} \quad \frac{dy}{dx}(1) = -\tfrac{8}{9}. \quad \square$$

Derivatives of Higher Order

As we noted in Section 3.1, when we differentiate a function f we get a new function f', the derivative of f, whose domain is a subset of the domain of f. Now suppose that f' can be differentiated. If we calculate $(f')'$, we get the *second derivative of f*, which is denoted f''. So long as we have differentiability, we can continue in this manner, forming the *third derivative of f*, denoted f''', and so on. However, the prime notation is not used beyond the third derivative. For the *fourth derivative of f*, we write $f^{(4)}$ and more generally, the nth derivative of f is denoted $f^{(n)}$. The functions f', f'', f''', $f^{(4)}, \ldots, f^{(n)}$, are called the derivatives of f of *orders* $1, 2, 3, 4, \ldots, n$, respectively. For example, if $f(x) = x^5$, then

$$f'(x) = 5x^4, \qquad f''(x) = 20x^3, \qquad f'''(x) = 60x^2, \qquad f^{(4)}(x) = 120x, \qquad f^{(5)}(x) = 120.$$

In this case, all higher derivatives are identically zero. As a variant of this notation you can write $y = x^5$ and then

$$y' = 5x^4, \qquad y'' = 20x^3, \qquad y''' = 60x^2, \qquad \text{and so on.}$$

Since each polynomial P has a derivative P' that is in turn a polynomial, and each rational function Q has a derivative Q' that is in turn a rational function, polynomials and rational functions have derivatives of all orders. In the case of a polynomial of degree n, derivatives of order greater than n are all identically zero. (Explain.)

In the Leibniz notation the derivatives of higher order are written

$$\frac{d^2y}{dx^2} = \frac{d}{dx}\left(\frac{dy}{dx}\right), \frac{d^3y}{dx^3} = \frac{d}{dx}\left(\frac{d^2y}{dx^2}\right), \ldots, \frac{d^ny}{dx^n} = \frac{d}{dx}\left(\frac{d^{n-1}y}{dx^{n-1}}\right), \ldots$$

or

$$\frac{d^2}{dx^2}[f(x)] = \frac{d}{dx}\left[\frac{d}{dx}[f(x)]\right], \qquad \frac{d^3}{dx^3}[f(x)] = \frac{d}{dx}\left[\frac{d^2}{dx^2}[f(x)]\right],$$

$$\ldots, \qquad \frac{d^n}{dx^n}[f(x)] = \frac{d}{dx}\left[\frac{d^{n-1}}{dx^{n-1}}[f(x)]\right], \ldots.$$

Below we work out some examples.

Example 6 If

$$f(x) = x^4 - 3x^{-1} + 5,$$

then

$$f'(x) = 4x^3 + 3x^{-2} \qquad \text{and} \qquad f''(x) = 12x^2 - 6x^{-3}. \quad \square$$

Example 7

$$\frac{d}{dx}(x^5 - 4x^3 + 7x) = 5x^4 - 12x^2 + 7$$

so that

$$\frac{d^2}{dx^2}(x^5 - 4x^3 + 7x) = \frac{d}{dx}(5x^4 - 12x^2 + 7) = 20x^3 - 24x$$

and

$$\frac{d^3}{dx^3}(x^5 - 4x^3 + 7x) = \frac{d}{dx}(20x^3 - 24x) = 60x^2 - 24. \quad \square$$

Example 8 Finally we consider $y = x^{-1}$. In the Leibniz notation

$$\frac{dy}{dx} = -x^{-2}, \qquad \frac{d^2y}{dx^2} = 2x^{-3}, \qquad \frac{d^3y}{dx^3} = -6x^{-4}, \qquad \frac{d^4y}{dx^4} = 24x^{-5}, \ldots$$

Based on these calculations, we are led to the general result

$$\frac{d^ny}{dx^n} = (-1)^n n!\, x^{-n-1} \qquad \text{[Recall that } n! = n(n-1)(n-2)\cdots 3\cdot 2\cdot 1.]$$

In the prime notation

$$y' = -x^{-2}, \qquad y'' = 2x^{-3}, \qquad y''' = -6x^{-4}, \qquad y^{(4)} = 24x^{-5}$$

and

$$y^{(n)} = (-1)^n n!\, x^{-n-1}. \quad \square$$

EXERCISES 3.3

In Exercises 1–10, find dy/dx.

1. $y = 3x^4 - x^2 + 1.$

2. $y = x^2 + 2x^{-4}.$

3. $y = x - \dfrac{1}{x}.$

4. $y = \dfrac{2x}{1 - x}.$

5. $y = \dfrac{x}{1 + x^2}.$

6. $y = x(x - 2)(x + 1).$

7. $y = \dfrac{x^2}{1 - x}.$

8. $y = \left(\dfrac{x}{1+x}\right)\left(\dfrac{2-x}{3}\right)$.

9. $y = \dfrac{x^3+1}{x^3-1}$.

10. $y = \dfrac{x^2}{(1+x)}$.

In Exercises 11–26, find the indicated derivative.

11. $\dfrac{d}{dx}(2x-5)$.

12. $\dfrac{d}{dx}(5x+2)$.

13. $\dfrac{d}{dx}[(3x^2 - x^{-1})(2x+5)]$.

14. $\dfrac{d}{dx}[(2x^2 + 3x^{-1})(2x - 3x^{-2})]$.

15. $\dfrac{d}{dt}\left(\dfrac{t^2+1}{t^2-1}\right)$.

16. $\dfrac{d}{dt}\left(\dfrac{2t^3+1}{t^4}\right)$.

17. $\dfrac{d}{dt}\left(\dfrac{t^4}{2t^3-1}\right)$.

18. $\dfrac{d}{dt}\left[\dfrac{t}{(1+t)^2}\right]$.

19. $\dfrac{d}{du}\left(\dfrac{2u}{1-2u}\right)$.

20. $\dfrac{d}{du}\left(\dfrac{u^2}{u^3+1}\right)$.

21. $\dfrac{d}{du}\left(\dfrac{u}{u-1} - \dfrac{u}{u+1}\right)$.

22. $\dfrac{d}{du}[u^2(1-u^2)(1-u^3)]$.

23. $\dfrac{d}{dx}\left(\dfrac{x^2}{1-x^2} - \dfrac{1-x^2}{x^2}\right)$.

24. $\dfrac{d}{dx}\left(\dfrac{3x^4+2x+1}{x^4+x-1}\right)$.

25. $\dfrac{d}{dx}\left(\dfrac{x^3+x^2+x+1}{x^3-x^2+x-1}\right)$.

26. $\dfrac{d}{dx}\left(\dfrac{x^3+x^2+x-1}{x^3-x^2+x+1}\right)$.

In Exercises 27–30, evaluate dy/dx at $x = 2$.

27. $y = (x+1)(x+2)(x+3)$.

28. $y = (x+1)(x^2+2)(x^3+3)$.

29. $y = \dfrac{(x-1)(x-2)}{(x+2)}$.

30. $y = \dfrac{(x^2+1)(x^2-2)}{x^2+2}$.

In Exercises 31–36, find the second derivative.

31. $f(x) = 7x^3 - 6x^5$.

32. $f(x) = 2x^5 - 6x^4 + 2x - 1$.

33. $f(x) = \dfrac{x^2-3}{x}$.

34. $f(x) = x^2 - \dfrac{1}{x^2}$.

35. $f(x) = (x^2-2)(x^{-2}+2)$.

36. $f(x) = (2x-3)\left(\dfrac{2x+3}{x}\right)$.

In Exercises 37–42, find d^3y/dx^3.

37. $y = \frac{1}{3}x^3 + \frac{1}{2}x^2 + x + 1$.

38. $y = (1+5x)^2$.

39. $y = (2x-5)^2$.

40. $y = \frac{1}{6}x^3 - \frac{1}{4}x^2 + x - 3$.

41. $y = x^3 - \dfrac{1}{x^3}$.

42. $y = \dfrac{x^4+2}{x}$.

In Exercises 43–48, find the indicated derivatives.

43. $\dfrac{d}{dx}\left[x\dfrac{d}{dx}(x - x^2)\right]$.

44. $\dfrac{d^2}{dx^2}\left[(x^2-3x)\dfrac{d}{dx}(x + x^{-1})\right]$.

45. $\dfrac{d^4}{dx^4}[3x - x^4]$.

46. $\dfrac{d^5}{dx^5}[ax^4 + bx^3 + cx^2 + dx + e]$, $\quad a, \ldots, e$ constant.

47. $\dfrac{d^2}{dx^2}\left[(1+2x)\dfrac{d^2}{dx^2}(5 - x^3)\right]$.

48. $\dfrac{d^3}{dx^3}\left[\dfrac{1}{x}\dfrac{d^2}{dx^2}(x^4 - 5x^2)\right]$.

49. Find a quadratic polynomial p such that $p(1) = 3$, $p'(1) = -2$, and $p''(1) = 4$.

50. Find a cubic polynomial p such that $p(-1) = 0$, $p'(-1) = 3$, $p''(-1) = -2$, and $p'''(-1) = 6$.

51. Let $f(x) = x^n$, where n is a positive integer.
(a) Find $f^{(k)}(x)$ for $k = n$.
(b) Find $f^{(k)}(x)$ when $k > n$.
(c) Find $f^{(k)}(x)$ when $k < n$.

52. Given the polynomial function
$p(x) = a_n x^n + a_{n-1}x^{n-1} + \cdots + a_1 x + a_0$:
(a) Find $(d^n/dx^n)[p(x)]$
(b) What is $(d^k/dx^k)[p(x)]$ when $k > n$?

53. Let $f(x) = \begin{cases} x^2, & x \geq 0 \\ 0, & x < 0. \end{cases}$
(a) Show that f is differentiable at 0 and give $f'(0)$.
(b) Determine $f'(x)$ for all x.
(c) Show that $f''(0)$ does not exist.
(d) Sketch the graph of f.

54. Let $g(x) = \begin{cases} x^3, & x \geq 0 \\ 0, & x < 0. \end{cases}$
(a) Show $g'(0)$ and $g''(0)$ both exist and give their values.
(b) Determine $g'(x)$ and $g''(x)$ for all x.
(c) Show that $g'''(0)$ does not exist.
(d) Sketch the graph of g.

55. Show that in general

$$(fg)''(x) \neq f(x)g''(x) + f''(x)g(x).$$

56. Verify the identity

$$f(x)g''(x) - f''(x)g(x) = \frac{d}{dx}[f(x)g'(x) - f'(x)g(x)].$$

In Exercises 57–60, determine the values of x for which (a) $f''(x) = 0$, (b) $f''(x) > 0$, (c) $f''(x) < 0$.

57. $f(x) = x^3$.

58. $f(x) = x^4$.

59. $f(x) = x^4 + 2x^3 - 12x^2 + 1$.

60. $f(x) = x^4 + 3x^3 - 6x^2 - x$.

61. Prove by mathematical induction that

$$\text{if} \quad y = x^{-1}, \qquad \text{then} \quad \frac{d^n y}{dx^n} = (-1)^n n!\, x^{-n-1}.$$

62. Calculate y', y'', y''' for $y = 1/x^2$. Use these results to guess a formula for $y^{(n)}$ for any positive integer n, and then prove your conjecture using mathematical induction.

63. If f and g are differentiable functions, then

$$(f \cdot g)' = f'g + fg'$$

by the product rule.
(a) Show that if f and g are twice differentiable, then

$$(f \cdot g)'' = f''g + 2f'g' + fg''.$$

(b) Let f and g have derivatives of order 3. Show that

$$(f \cdot g)''' = f'''g + 3f''g' + 3f'g'' + fg'''.$$

64. Let f and g have derivatives of order n. The results in Exercise 63 suggest that the expansion of $(f \cdot g)^n$ parallels the expansion of $(a + b)^n$ by the binomial theorem. That is,

$$(*) \quad (f \cdot g)^{(n)} = f^{(n)}g + nf^{(n-1)}g' + \cdots + \binom{n}{k} f^{(n-k)}g^{(k)} + \cdots + nf'g^{(n-1)} + fg^{(n)}$$

where $\binom{n}{k} = n!/(k!(n-k)!)$ is the coefficient of $a^{n-k}b^k$ in the expansion of $(a + b)^n$, called the kth binomial coefficient. Formula $(*)$ is known as *Leibniz's rule.* Use mathematical induction to prove this formula.

65. Let u, v, w be differentiable functions of x. Express the derivative of the product uvw in terms of the functions u, v, w, and their derivatives.

66. Let $f(x) = \frac{1}{2}x^3 - 3x^2 + 3x + 3$.
(a) Calculate $f'(x)$.
(b) Use a graphing utility to graph f and f' together.
(c) Find the x-coordinates of the points where the tangent line to the graph of f is horizontal by finding the zeros of f'. Approximate the zeros of f' with three decimal place accuracy.

67. Let $f(x) = x^3 + x^2 - 4x + 1$.
(a) Calculate $f'(x)$.
(b) Use a graphing utility to graph f and f' together.
(c) What can you say about the graph of f when $f'(x) < 0$? What can you say about the graph of f when $f'(x) > 0$?

68. Let $f(x) = x^3 - x$.
(a) Find an equation for the tangent line to the graph of f at the point $(1, 0)$.
(b) Use a graphing utility to graph f and the tangent line together. Notice that the graph and the tangent line intersect at another point (a, b).
(c) Find the point (a, b).

69. Let $f(x) = \frac{1}{3}x^3 - 3x^2 + 4x + 1$.
(a) Find an equation for the tangent line to the graph of f at the point $(0, 1)$.
(b) Use a graphing utility to graph f and the tangent line together. Show that the graph and the tangent line intersect at another point (a, b).
(c) Find the point (a, b).

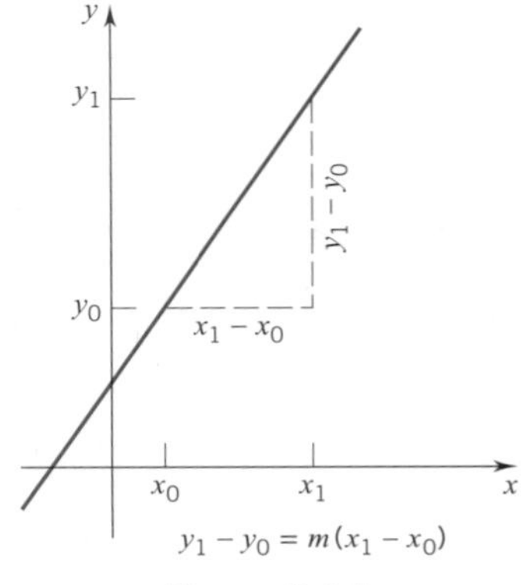

Figure 3.4.1

■ 3.4 THE DERIVATIVE AS A RATE OF CHANGE

In the case of a linear function $y = mx + b$ the graph is a straight line and the slope m measures the steepness of the line by giving the rate of climb of the line, *the rate of change of y with respect to x.*

As x changes from x_0 to x_1, y changes m times as much:

$$y_1 - y_0 = m(x_1 - x_0). \qquad \text{(Figure 3.4.1)}$$

Thus the slope $m = (y_1 - y_0)/(x_1 - x_0)$ gives the change in y per unit change in x.

In the more general case of a differentiable function $y = f(x)$, the difference quotient

$$\frac{f(x+h)-f(x)}{(x+h)-h} = \frac{f(x+h)-f(x)}{h}, \quad h \neq 0,$$

gives the *average rate of change of y (or f) with respect to x*. Geometrically, this ratio is the slope of the secant line through $(x, f(x))$ and $(x+h, f(x+h))$. The limit as h approaches zero is the derivative $dy/dx = f'(x)$, which can be interpreted as the *instantaneous rate of change of y with respect to x*, or as the slope of the tangent line to the graph of f at the point $(x, f(x))$. Since the graph is a curve, the rate of change of y with respect to x can vary from point to point. At $x = x_1$ (see Figure 3.4.2) the rate of change of y with respect to x is $f'(x_1)$; the steepness of the graph is that of a line of slope $f'(x_1)$. At $x = x_2$ the rate of change of y with respect to x is $f'(x_2)$; the steepness of the graph is that of a line of slope $f'(x_2)$. At $x = x_3$ the rate of change of y with respect to x is $f'(x_3)$; the steepness of the graph is that of a line of slope $f'(x_3)$.

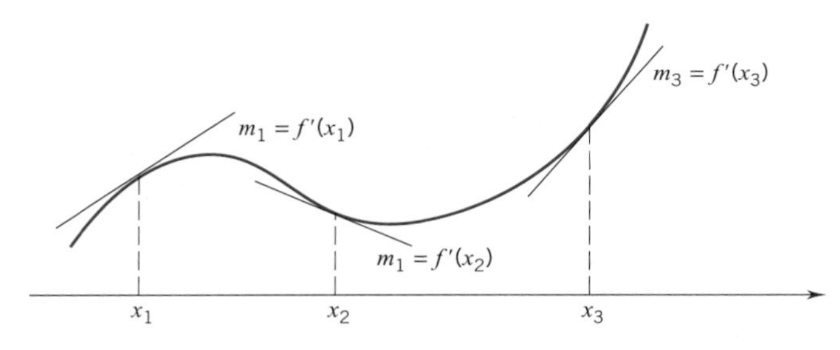

Figure 3.4.2

From your understanding of slope it should be apparent to you when examining Figure 3.4.2 that $f'(x_1) > 0$, $f'(x_2) < 0$, and $f'(x_3) > 0$, and that, in general, a function increases on any interval where the derivative remains positive and decreases on any interval where the derivative remains negative. We will take up this matter carefully in Chapter 4. Right now we look at rates of change in the context of some simple geometric figures.

Example 1 A square of side s has area $A = s^2$. Now suppose that s increases. As s increases from 1 to 2, A increases by 3 square units:

$$2^2 - 1^2 = 4 - 1 = 3. \qquad \text{(See Figure 3.4.3.)}$$

As s increases from 2 to 3, A increases by 5 square units:

$$3^2 - 2^2 = 9 - 4 = 5.$$

As s increases from 3 to 4, A increases by 7 square units:

$$4^2 - 3^2 = 16 - 9 = 7,$$

and so on. Continued unit changes in s cause ever-increasing changes in A.

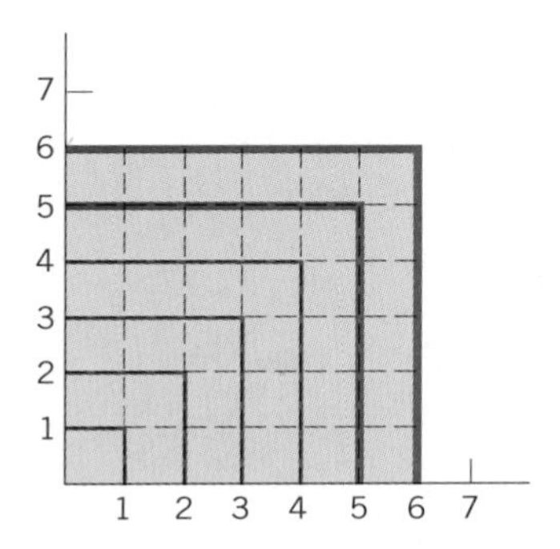

Figure 3.4.3

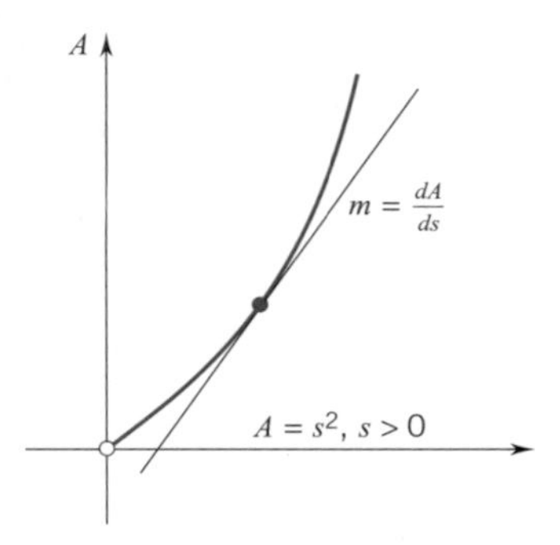

Figure 3.4.4

In Figure 3.4.4 we have plotted A against s. The instantaneous rate of change of A with respect to s is the derivative

$$\frac{dA}{ds} = 2s.$$

This appears in Figure 3.4.4 as the slope of the tangent line. At $s = 1$, $dA/ds = 2$; thus, at $s = 1$, A is increasing at the rate of a linear function of slope 2. At $s = 2$, $dA/ds = 4$; thus, at $s = 2$, A is increasing at the rate of a linear function of slope 4, and so on. At $s = k$, $dA/ds = 2k$ and A is increasing at the rate of a linear function of slope $2k$. ❑

Example 2 Suppose that we have a right circular cylinder of changing dimensions (Figure 3.4.5). When the base radius is r and the height is h, the cylinder has volume

$$V = \pi r^2 h.$$

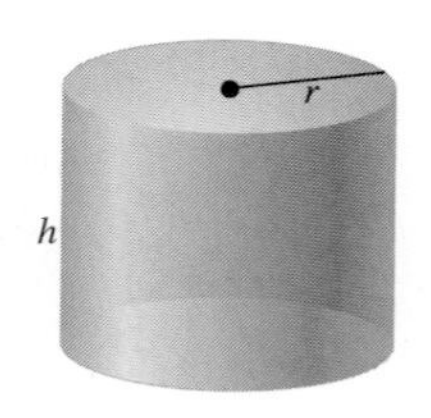

Figure 3.4.5

If r remains constant while h changes, then V can be viewed as a function of h. The rate of change of V with respect to h is the derivative

$$\frac{dV}{dh} = \pi r^2.$$

If h remains constant while r changes, then V can be viewed as a function of r. The rate of change of V with respect to r is the derivative

$$\frac{dV}{dr} = 2\pi r h.$$

Suppose now that r is changing but V is being kept constant. How is h changing with respect to r? To answer this we express h in terms of r and V:

$$h = \frac{V}{\pi r^2} = \frac{V}{\pi} r^{-2}.$$

Since V is being held constant, h is now a function of r. The rate of change of h with respect to r is the derivative

$$\frac{dh}{dr} = -\frac{2V}{\pi} r^{-3} = -\frac{2(\pi r^2 h)}{\pi} r^{-3} = -\frac{2h}{r}. \quad ❑$$

The derivative as a rate of change is one of the fundamental ideas of calculus. We will come back to it again and again.

Velocity and Acceleration

Suppose that an object is moving along a straight line and that, for each time t during a certain time interval, the object has position (coordinate) $x(t)$. Then, at time $t + h$, the position of the object is $x(t + h)$, and $x(t + h) - x(t)$ is the distance that the object traveled during the time period t to $t + h$. The ratio

$$\frac{x(t+h) - x(t)}{(t+h) - t} = \frac{x(t+h) - x(t)}{h}$$

gives the *average velocity* of the object during this time period. If

$$\lim_{h \to 0} \frac{x(t+h) - x(t)}{h} = x'(t)$$

exists, then $x'(t)$ gives the (*instantaneous*) *rate of change of position with respect to time*. This rate of change of position is called the *velocity* of the object; in symbols

(3.4.1)
$$v(t) = x'(t).$$

If the velocity function is itself differentiable, then its rate of change with respect to time is called the *acceleration*; in symbols,

(3.4.2)
$$a(t) = v'(t) = x''(t).$$

In the Leibniz notation,

(3.4.3)
$$v = \frac{dx}{dt} \quad \text{and} \quad a = \frac{dv}{dt} = \frac{d^2x}{dt^2}.$$

The *speed* is by definition the absolute value of the velocity:

(3.4.4)
$$\text{speed at time } t = |v(t)|.$$

1. Positive velocity indicates motion in the positive direction (x is increasing). Negative velocity indicates motion in the negative direction (x is decreasing).
2. Positive acceleration indicates increasing velocity (increasing speed in the positive direction or decreasing speed in the negative direction). Negative acceleration indicates decreasing velocity (decreasing speed in the positive direction or increasing speed in the negative direction).
3. It follows from (2) that, if the velocity and acceleration have the same sign, the object is speeding up, but if the velocity and acceleration have opposite signs, the object is slowing down.

Example 3 An object moves along the x-axis, its position at each time t given by the function

$$x(t) = t^3 - 12t^2 + 36t - 27.$$

Let's study the motion from time $t = 0$ to time $t = 9$.

The object starts out 27 units to the left of the origin:

$$x(0) = 0^3 - 12(0)^2 + 36(0) - 27 = -27$$

and ends up 54 units to the right of the origin:

$$x(9) = 9^3 - 12(9)^2 + 36(9) - 27 = 54.$$

We can find the velocity function by differentiating the position function:

$$v(t) = x'(t) = 3t^2 - 24t + 36 = 3(t-2)(t-6).$$

It is clear that

$$v(t) \text{ is } \begin{cases} \text{positive,} & \text{for } 0 \le t < 2 \\ 0, & \text{at } t = 2 \\ \text{negative,} & \text{for } 2 < t < 6 \\ 0, & \text{at } t = 6 \\ \text{positive,} & \text{for } 6 < t \le 9. \end{cases}$$

We can interpret all this as follows: The object begins by moving to the right [$v(t)$ is positive for $0 \le t < 2$]; it comes to a stop at time $t = 2$ [$v(2) = 0$]; it then moves left [$v(t)$ is negative for $2 < t < 6$]; it stops at time $t = 6$ [$v(6) = 0$]; it then moves right and keeps going right [$v(t) > 0$ for $6 < t \le 9$].

We can find the acceleration by differentiating the velocity:

$$a(t) = v'(t) = 6t - 24 = 6(t-4).$$

Obviously

$$a(t) \text{ is } \begin{cases} \text{negative,} & \text{for } 0 \le t < 4 \\ 0, & \text{at } t = 4 \\ \text{positive,} & \text{for } 4 < t \le 9. \end{cases}$$

At the beginning the velocity decreases, reaching a minimum at time $t = 4$. Then the velocity starts to increase and continues to increase.

Figure 3.4.6 shows a diagram for the sign of the velocity and a comparable diagram for the sign of the acceleration. Combining the two diagrams we have a brief description of the motion in convenient form.

The direction of the motion at each time $t \in [0, 9]$ is represented schematically in Figure 3.4.7.

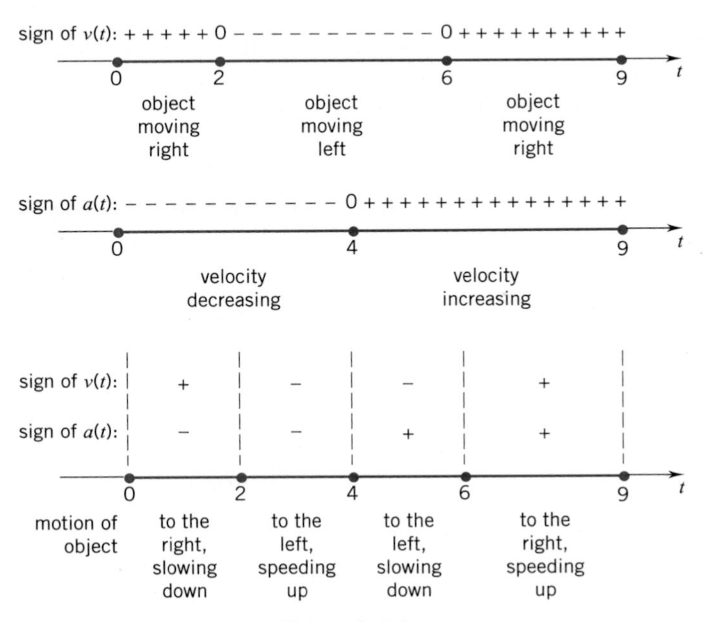

Figure 3.4.6

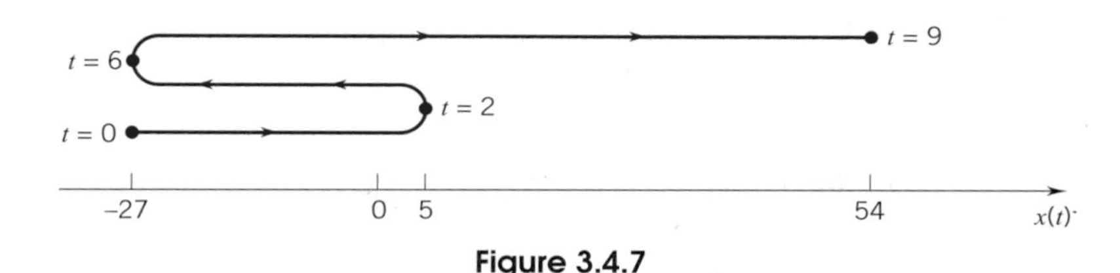

Figure 3.4.7

A better way to represent the motion is to graph x as a function of t, as we did in Figure 3.4.8. The velocity $v(t) = x'(t)$ then appears as the slope of the tangent to the curve. Note, for example, that at $t = 2$ and $t = 6$, the tangent is horizontal and the slope is 0. This reflects the fact that $v(2) = 0$ and $v(6) = 0$. ❑

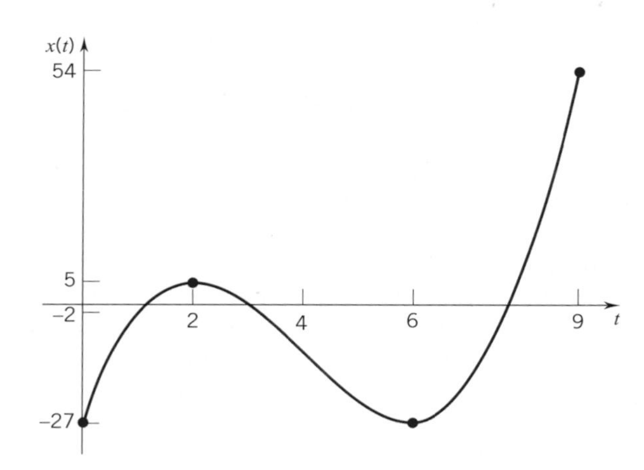

Figure 3.4.8

Before going on, a few words about units. The units of velocity and acceleration depend upon the units used to measure distance and the units used to measure time. The units of velocity are units of distance per unit time:

feet per second, meters per second, miles per hour, and so forth.

The units of acceleration are units of distance per unit time per unit time:

feet per second per second, meters per second per second,
miles per hour per hour, and so forth.

Free Fall (Near the Surface of the Earth)

Imagine an object (for example, a rock or an apple) falling to the ground (Figure 3.4.9). We will assume that the object is in *free fall*: namely, that the gravitational pull on the object is constant throughout the fall and that there is no air resistance.†

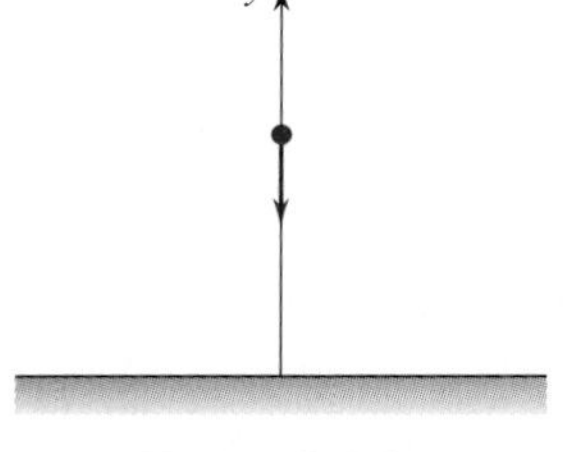

Figure 3.4.9

† In practice, neither of these conditions is ever fully met. Gravitational attraction near the surface of the earth does vary somewhat with altitude, and there is always some air resistance. Nevertheless, in the setting within which we will be working the results that we obtain are correct within close approximation.

Galileo's formula for free fall gives the height of the object at each time t of the fall:

(3.4.5) $$y(t) = -\tfrac{1}{2}gt^2 + v_0 t + y_0.\dagger$$

Let's examine this formula. First, the formula assumes that the positive y direction is up. Next, since $y(0) = y_0$, the constant y_0 represents the height of the object at time $t = 0$. This is called the *initial position*. Differentiation gives

$$y'(t) = -gt + v_0.$$

Since $y'(0) = v_0$, the constant v_0 gives the velocity of the object at time $t = 0$. This is called *the initial velocity*. A second differentiation gives

$$y''(t) = -g.$$

This indicates that the object falls with constant negative acceleration $-g$. (Why negative?)

There is a minor technicality that we should mention here. Although not explicitly stated, there is the implication that the height function $y(t)$ given by (3.4.5) is defined only for $t \geq 0$. Therefore, to be precise, the initial velocity $v(0) = y'(0)$ and the initial acceleration $a(0) = y''(0)$ should be defined as right-handed derivatives (see Exercises 3.1 for the definition of one-sided derivatives). However, it is easy to verify that the right-hand derivative of y at $t = 0$ is v_0 and the right-hand derivative of y' at $t = 0$ is $-g$, and so the values given here as ordinary (two-sided) derivatives are correct.

The constant g is a *gravitational constant*. If time is measured in seconds and distance in feet, then g is approximately 32 feet per second per second; if time is measured in seconds and distance in meters, then g is approximately 9.8 meters per second per second.†† In making numerical calculations, we will take g as 32 feet per second per second or as 9.8 meters per second per second. Equation (3.4.5) then reads

$$y(t) = -16t^2 + v_0 t + y_0 \qquad \text{(distance in feet)}$$

or

$$y(t) = -4.9t^2 + v_0 t + y_0 \qquad \text{(distance in meters).}$$

Example 4 A stone is dropped from a height of 98 meters. In how many seconds does it hit the ground? What is the speed at the instant of impact?

SOLUTION Here $y_0 = 98$ and $v_0 = 0$. Consequently, by (3.4.5),

$$y(t) = -4.9t^2 + 98.$$

To find t at the moment of impact, we set $y(t) = 0$. This gives

$$-4.9t^2 + 98 = 0, \qquad t^2 = 20, \qquad t = \pm\sqrt{20} = \pm 2\sqrt{5}.$$

† Galileo Galilei (1564–1642), a great Italian astronomer and mathematician, is popularly known today for his early experiments with falling objects. His astronomical observations led him to support the Copernican view of the solar system. For this he was brought before the Inquisition.

†† The value of this constant varies with latitude and elevation. It is approximately 32 feet per second per second at the equator at elevation zero. In Greenland it is about 32.23.

We can disregard the negative value because the stone was not dropped before time $t = 0$. We conclude that it takes $2\sqrt{5} \cong 4.47$ seconds for the stone to hit the ground.

The velocity at impact is the velocity at time $t = 2\sqrt{5}$. Since

$$v(t) = y'(t) = -9.8t,$$

we have

$$v(2\sqrt{5}) = -(19.6)\sqrt{5}.$$

(One might argue that we should use the left-hand derivative here; you can verify that the left-hand derivative of y at $t = 2\sqrt{5}$ is $-(19.6)\sqrt{5}$. If we were to use one-sided derivatives, what do you think the right-hand derivative of y at $t = 2\sqrt{5}$ would be?) The speed at impact is $|v(2\sqrt{5})| = (19.6)\sqrt{5} \cong 43.8$ meters per second. ❑

Example 5 An explosion causes debris to rise vertically with an initial velocity of 72 feet per second.

(a) In how many seconds does it attain maximum height?

(b) What is this maximum height?

(c) What is the speed of the debris as it reaches a height of 32 feet (i) going up? (ii) coming back down?

SOLUTION The basic equation in this case is

$$y(t) = -16t^2 + v_0 t + y_0.$$

Here $y_0 = 0$ (it starts at ground level) and $v_0 = 72$ (the initial velocity is 72 feet per second). The equation of motion is therefore

$$y(t) = -16t^2 + 72t.$$

Differentiation gives

$$v(t) = y'(t) = -32t + 72.$$

The maximum height is attained when the velocity is 0. This occurs at time $t = \frac{72}{32} = \frac{9}{4}$. Since $y(\frac{9}{4}) = 81$, the maximum height attained is 81 feet.

To answer part (c) we must first find those times t for which $y(t) = 32$. Since

$$y(t) = -16t^2 + 72t,$$

the condition $y(t) = 32$ gives

$$16t^2 - 72t + 32 = 0.$$

This quadratic has two solutions, $t = \frac{1}{2}$ and $t = 4$. Since $v(\frac{1}{2}) = 56$ and $v(4) = -56$, the velocity going up is 56 feet per second and the velocity coming down is -56 feet per second. In each case the speed is 56 feet per second. ❑

Economics

In business and economics one is often interested in how changes in such variables as production, supply, or price will affect other variables such as cost, revenue, or profit. If f is a function that describes the relationship between a pair of these variables, then the term *marginal* is used to specify the derivative of f.

For example, suppose $C = C(x)$ represents the cost of producing x units of a certain commodity. Although in reality x is a nonnegative integer, in theory and practice it is convenient to assume that C is defined for all x in some interval, and that C is differentiable. The derivative $C'(x)$ is called the *marginal cost*.

Originally, economists defined the marginal cost at a production level x to be $C(x + 1) - C(x)$, which is the cost of producing one additional unit of the commodity. Since

$$C(x + 1) - C(x) = \frac{C(x + 1) - C(x)}{1} \cong \lim_{h \to 0} \frac{C(x + h) - C(x)}{h} = C'(x),$$

it follows that the marginal cost $C'(x)$ at the production level x is approximately the cost of producing the $(x + 1)$-st unit.

Similarly, if $R = R(x)$ is the revenue received for selling x units of the commodity, then $R'(x)$ is called the *marginal revenue*. The marginal revenue at a sales level x is approximately the revenue obtained by selling one additional unit.

If $C = C(x)$ and $R = R(x)$ are the cost and revenue functions associated with producing and selling x units of the commodity, then

$$P(x) = R(x) - C(x)$$

is called the *profit function*. The values of x (if any) at which $C(x) = R(x)$, that is "cost" = "revenue," are called *break-even points*. The derivative, P', is the *marginal profit*. In Chapter 4 we show that the maximum profit occurs when marginal cost equals marginal revenue.

Example 6 A manufacturer of computer components determines that the total cost C of producing x components per week is

$$C(x) = 2000 + 50x - \frac{x^2}{20} \text{ (dollars).}$$

What is the marginal cost at the production level of 20 units? What is the exact cost of producing the 21st component?

SOLUTION The marginal cost at a production level x is given by the derivative

$$C'(x) = 50 - \frac{x}{10}.$$

Thus, the marginal cost at the production level of 20 components is

$$C'(20) = 50 - \frac{20}{10} = 48 \text{ (dollars).}$$

The exact cost of producing the 21st component is

$$C(21) - C(20) = 2000 + 50(21) - \frac{(21)^2}{20} - \left[2000 + 50(20) - \frac{(20)^2}{20}\right]$$

$$= 3027.95 - 2980 = 47.95 \text{ (dollars).} \quad \square$$

Example 7 A manufacturer of digital watches determines that the cost and revenue functions involved in producing and selling x watches are:

$$C(x) = 1200 + 13x$$

and

$$R(x) = 75x - \frac{x^2}{2},$$

respectively. Find the profit function and determine the break-even points. Find the marginal profit and determine the production/sales level at which the marginal profit is zero.

SOLUTION The profit function P is given by

$$P(x) = R(x) - C(x) = 75x - \frac{x^2}{2} - (1200 + 13x) = 62x - \frac{x^2}{2} - 1200.$$

To find the break-even points, set $P(x) = 0$:

$$62x - \frac{x^2}{2} - 1200 = 0$$

$$x^2 - 124x + 2400 = 0$$

$$(x - 24)(x - 100) = 0.$$

Thus, the break-even points are $x = 24$ and $x = 100$.

The marginal profit is given by the derivative:

$$P'(x) = 62 - x$$

and $P'(x) = 0$ when $x = 62$.

Figure 3.4.10 gives a graphical representation of the cost and revenue functions, the break-even points, and indicates the regions of profit and loss. ❑

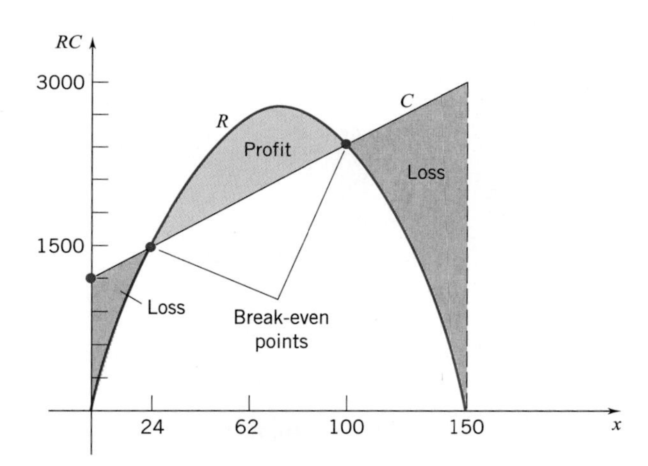

Figure 3.4.10

EXERCISES 3.4

1. Find the rate of change of the area of a circle with respect to the radius r. What is the rate when $r = 2$?
2. Find the rate of change of the volume of a cube with respect to the length s of a side. What is the rate when $s = 4$?
3. Find the rate of change of the area of a square with respect to the length z of a diagonal. What is the rate when $z = 4$?
4. Find the rate of change of $y = 1/x$ with respect to x at $x = -1$.
5. Find the rate of change of $y = [x(x + 1)]^{-1}$ with respect to x at $x = 2$.
6. Find the values of x for which the rate of change of $y = x^3 - 12x^2 + 45x - 1$ with respect to x is zero.
7. Find the rate of change of the volume of a sphere with respect to the radius r.

8. Find the rate of change of the surface area of a sphere with respect to the radius r. What is this rate of change when $r = r_0$? How must r_0 be chosen so that the rate of change is 1?

9. Find x_0 given that the rate of change of $y = 2x^2 + x - 1$ with respect to x at $x = x_0$ is 4.

10. Find the rate of change of the area A of a circle with respect to
(a) the diameter d. (b) the circumference C.

11. Find the rate of change of the volume V of a cube with respect to
(a) the length w of a diagonal on one of the faces.
(b) the length z of one of the diagonals of the cube.

12. The dimensions of a rectangle are changing in such a way that the area of the rectangle remains constant. Find the rate of change of the height h with respect to the base b.

13. The area of a sector in a circle is given by the formula $A = \frac{1}{2}r^2\theta$ where r is the radius and θ is the central angle measured in radians.
(a) Find the rate of change of A with respect to θ if r remains constant.
(b) Find the rate of change of A with respect to r if θ remains constant.
(c) Find the rate of change of θ with respect to r if A remains constant.

14. The total surface area of a right circular cylinder is given by the formula $A = 2\pi r(r + h)$ where r is the radius and h is the height.
(a) Find the rate of change of A with respect to h if r remains constant.
(b) Find the rate of change of A with respect to r if h remains constant.
(c) Find the rate of change of h with respect to r if A remains constant.

15. For what value of x is the rate of change of

$$y = ax^2 + bx + c \quad \text{with respect to } x$$

the same as the rate of change of

$$z = bx^2 + ax + c \quad \text{with respect to } x?$$

Assume that a, b, c are constant with $a \neq b$.

16. Find the rate of change of the product $f(x)g(x)h(x)$ with respect to x at $x = 1$ given that

$$f(1) = 0, \quad g(1) = 2, \quad h(1) = -2,$$
$$f'(1) = 1, \quad g'(1) = -1, \quad h'(1) = 0.$$

In Exercises 17–24, an object moves along a coordinate line, its position at each time $t \geq 0$ given by $x(t)$. Find the position, velocity, acceleration, and speed at time t_0.

17. $x(t) = 4 + 3t - t^2; \quad t_0 = 5.$

18. $x(t) = 5t - t^3; \quad t_0 = 3.$

19. $x(t) = t^3 - 6t; \quad t_0 = 2.$

20. $x(t) = 4t^2 - 3t + 6; \quad t_0 = 1.$

21. $x(t) = \dfrac{18}{t + 2}; \quad t_0 = 1.$

22. $x(t) = \dfrac{2t}{t + 3}; \quad t_0 = 3.$

23. $x(t) = (t^2 + 5t)(t^2 + t - 2); \quad t_0 = 1.$

24. $x(t) = (t^2 - 3t)(t^2 + 3t); \quad t_0 = 2.$

In Exercises 25–30, an object moves along a coordinate line, its position at each time $t \geq 0$ given by $x(t)$. Determine when, if ever, the object changes direction.

25. $x(t) = t^3 - 3t^2 + 3t.$

26. $x(t) = t + \dfrac{3}{t + 1}.$

27. $x(t) = t + \dfrac{5}{t + 2}.$

28. $x(t) = t^4 - 4t^3 + 4t^2.$

29. $x(t) = \dfrac{t}{t^2 + 8}.$

30. $x(t) = 3t^5 + 10t^3 + 15t.$

Objects A, B, C move along the x-axis. Their positions from time $t = 0$ to time $t = t_3$ have been graphed in the figure below.

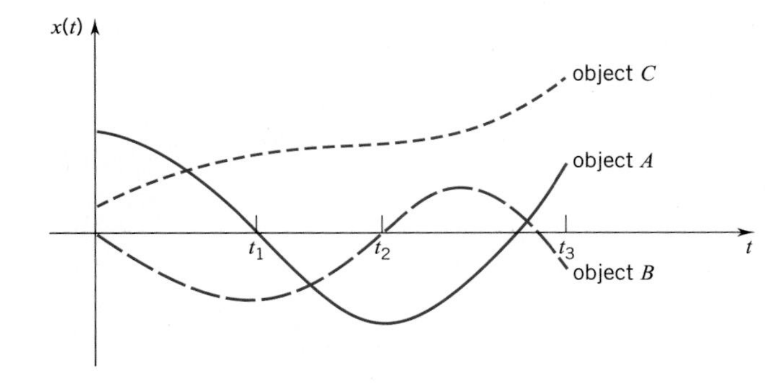

In Exercises 31–40, find:

31. Which object begins furthest to the right?

32. Which object finishes furthest to the right?

33. Which object has the greatest speed at time t_1?

34. Which object maintains the same direction during the time interval $[t_1, t_3]$?

35. Which object begins by moving left?

36. Which object finishes moving left?

37. Which object changes direction at time t_2?

38. Which object speeds up during the time interval $[0, t_1]$?

39. Which object slows down during the time interval $[t_1, t_2]$?

40. Which object changes direction during the time interval $[t_2, t_3]$?

An object moves along the x-axis, its position at each time $t \geq 0$ given by $x(t)$. In Exercises 41–48, determine the time interval(s), if any, during which the object satisfies the given condition.

41. $x(t) = t^4 - 12t^3 + 28t^2$; moving right.

42. $x(t) = t^3 - 12t^2 + 21t$; moving left.

43. $x(t) = 5t^4 - t^5$; speeding up.

44. $x(t) = 6t^2 - t^4$; slowing down.

45. $x(t) = t^3 - 6t^2 - 15t$; moving left and slowing down.

46. $x(t) = t^3 - 6t^2 - 15t$; moving right and slowing down.

47. $x(t) = t^4 - 8t^3 + 16t^2$; moving right and speeding up.

48. $x(t) = t^4 - 8t^3 + 16t^2$; moving left and speeding up.

In Exercises 49–60, neglect air resistance. For the numerical calculations, take g as 32 feet per second per second or as 9.8 meters per second per second.

49. An object is dropped and hits the ground 6 seconds later. From what height was it dropped?

50. Supplies are dropped from a helicopter and seconds later hit the ground at 98 meters per second. How high was the helicopter?

51. An object is projected vertically upward from ground level with velocity v_0. What is the height in meters attained by the object?

52. An object projected vertically upward from ground level returns to earth in 8 seconds. What was the initial velocity in feet per second?

53. An object projected vertically upward passes every height less than the maximum twice, once on the way up, once on the way down. Show that the speed is the same in each direction. Assume that height is measured in feet.

54. An object is projected vertically upward from the ground. Show that it takes the object the same amount of time to reach its maximum height as it takes for it to drop from that height back to the ground. Assume that height is measured in meters.

55. A rubber ball is thrown straight down from a height of 224 feet at a speed of 80 feet per second. If the ball always rebounds with one-fourth of its impact speed, what will be the speed of the ball the third time it hits the ground?

56. A ball is thrown straight up from ground level. How high will the ball go if it reaches a height of 64 feet in 2 seconds?

57. A stone is thrown upward from ground level. The initial speed is 32 feet per second. (a) In how many seconds will the stone hit the ground? (b) How high will it go? (c) With what minimum speed should the stone be thrown so as to reach a height of 36 feet?

58. To estimate the height of a bridge a man drops a stone into the water below. How high is the bridge (a) if the stone hits the water 3 seconds later? (b) if the man hears the splash 3 seconds later? (Use 1080 feet per second as the speed of sound.)

59. A falling stone is at a certain instant 100 feet above the ground. Two seconds later it is only 16 feet above the ground. (a) From what height was it dropped? (b) If it was thrown down with an initial speed of 5 feet per second, from what height was it thrown? (c) If it was thrown upward with an initial speed of 10 feet per second, from what height was it thrown?

60. A rubber ball is thrown straight down from a height of 4 feet. If the ball rebounds with one-half of its impact speed and returns exactly to its original height before falling again, how fast was it thrown originally?

In Exercises 61–64, a cost function for a certain commodity is given. Find the marginal cost function at a production level of 100 units and compare that with the actual cost of producing the 101st unit.

61. $C(x) = 200 + 0.02x + 0.0001x^2, \quad x > 0.$

62. $C(x) = 1000 + 2x + 0.02x^2 + 0.0001x^3, \quad x \geq 0.$

63. $C(x) = 200 + 0.01x + \dfrac{100}{x}, \quad x > 0.$

64. $C(x) = 2000 + 2\sqrt{x}, \quad x \geq 0.$

65. A manufacturer of electric motors estimates that the cost of producing x motors per day is given by

$$C(x) = 1000 + 25x - \frac{x^2}{10}, \quad 0 \leq x \leq 200.$$

Find the marginal cost of producing 10 motors and compare it with the exact cost of producing the 11th motor.

66. The cost and revenue functions for the production of x units of a certain commodity are

$$C(x) = 4x + 1400 \quad \text{and}$$

$$R(x) = 20x - \frac{x^2}{50} \quad \text{for} \quad 0 \leq x \leq 1000.$$

(a) Find the profit function and determine the break-even points.
(b) Find the marginal profit and determine the production level at which the marginal profit is zero.
(c) Sketch the cost and revenue functions in the same coordinate system and indicate the regions of profit and loss. What production level do you think will produce the maximum profit?

Economists often study costs, revenues, and profits in terms of averages of these quantities. The average cost $\overline{C}$ (cost per unit), average revenue $\overline{R}$ (revenue per unit), and average profit $\overline{P}$ (profit per unit) are given by

$$\overline{C}(x) = \frac{C(x)}{x}, \qquad \overline{R}(x) = \frac{R(x)}{x}, \qquad \text{and} \qquad \overline{P}(x) = \frac{P(x)}{x},$$

where C, R, and P are the cost, revenue, and profit functions. The marginal average cost, marginal average revenue, and marginal average profit functions are the corresponding derivatives of $\overline{C}$, $\overline{R}$, and $\overline{P}$.

67. Calculate the marginal average cost functions for the given cost functions.
(a) $C(x) = 200 + 0.02x + 0.0001x^2$.
(b) $C(x) = 200 + 0.01x + \dfrac{100}{x}$.

68. Calculate the marginal average cost, marginal average revenue, and marginal average profit functions corresponding to the cost and revenue functions given in Exercise 66.

69. An object moves along the x-axis. Its position at time t is given by

$$x(t) = t^3 - 7t^2 + 10t + 5, \quad 0 \le t \le 5.$$

(a) Determine the velocity function v. Use a graphing utility to graph v.
(b) Use the graph to estimate when the object is moving to the right and when it is moving to the left.
(c) Use the graphing utility to graph the speed $|v|$ of the object. Estimate when the object stops. Estimate the maximum speed for $1 \le t \le 4$.
(d) Determine the acceleration a and use the graphing utility to graph it. Estimate when the object is speeding up and when it is slowing down.

70. The cost and revenue functions for the production of a certain commodity are

$$C(x) = 4 + 0.75x \quad \text{and}$$

$$R(x) = \frac{10x}{1 + 0.25x^2} \quad \text{for } x \ge 0,$$

where x is measured in hundreds of units.
(a) Using a graphing utility, graph the cost and revenue functions together. Estimate the break-even points accurate to two decimal places.
(b) Graph the profit function and estimate the production level that will yield the maximum profit. Use two decimal place accuracy. How many units should be produced to maximize the profit?

■ 3.5 THE CHAIN RULE

In this section we take up the differentiation of composite functions. Until we get to Theorem 3.5.7, our approach is completely intuitive—no real definitions, no proofs, just informal discussion. Our purpose is to give you some experience with the standard computational procedures and some insight into why these procedures work. Theorem 3.5.7 puts it all on a sound footing.

Suppose that y is a differentiable function of u:

$$y = f(u)$$

and u in turn is a differentiable function of x:

$$u = g(x).$$

Then y is a composite function of x:

$$y = f(u) = f(g(x)) = (f \circ g)(x).$$

Does y have a derivative with respect to x? Yes it does, and it's given by a formula that is easy to remember:

(3.5.1)
$$\boxed{\frac{dy}{dx} = \frac{dy}{du}\frac{du}{dx}.}$$

This formula, known as the *chain rule*, says that "the rate of change of y with respect to x is the rate of change of y with respect to u times the rate of change of u with respect to x." (Plausible as all this sounds, remember that we have proved nothing. All we have done is asserted that the composition of differentiable functions is differentiable and given you a formula, a formula that needs justification and is justified at the end of this section.)

Before using the chain rule in elaborate computations, let's confirm its validity in some simple instances.

If $y = 2u$ and $u = 3x$, then $y = 6x$. Clearly

$$\frac{dy}{dx} = 6 = 2 \cdot 3 = \frac{dy}{du}\frac{du}{dx},$$

and so, in this case certainly, the chain rule is confirmed:

$$\frac{dy}{dx} = \frac{dy}{du}\frac{du}{dx}.$$

If $y = u^3$ and $u = x^2$, then $y = (x^2)^3 = x^6$. This time

$$\frac{dy}{dx} = 6x^5, \qquad \frac{dy}{du} = 3u^2 = 3x^4, \qquad \frac{du}{dx} = 2x$$

and once again

$$\frac{dy}{dx} = 6x^5 = 3x^4 \cdot 2x = \frac{dy}{du}\frac{du}{dx}.$$

Example 1 Find dy/dx by the chain rule given that

$$y = \frac{u-1}{u+1} \qquad \text{and} \qquad u = x^2.$$

SOLUTION

$$\frac{dy}{du} = \frac{(u+1)1 - (u-1)1}{(u+1)^2} = \frac{2}{(u+1)^2} \qquad \text{and} \qquad \frac{du}{dx} = 2x$$

so that

$$\frac{dy}{dx} = \frac{dy}{du}\frac{du}{dx} = \left[\frac{2}{(u+1)^2}\right]2x = \frac{4x}{(x^2+1)^2}. \quad ❑$$

Remark We would have obtained the same result without the chain rule by first writing y as a function of x and then differentiating:

$$\text{with} \quad y = \frac{u-1}{u+1} \qquad \text{and} \qquad u = x^2 \quad \text{we have} \quad y = \frac{x^2-1}{x^2+1}$$

and therefore

$$\frac{dy}{dx} = \frac{(x^2+1)2x - (x^2-1)2x}{(x^2+1)^2} = \frac{4x}{(x^2+1)^2}. \quad ❑$$

If f is a differentiable function of u and u is a differentiable function of x, then, according to (3.5.1),

(3.5.2)

$$\frac{d}{dx}[f(u)] = \frac{d}{du}[f(u)]\frac{du}{dx}.$$

(All we have done here is written y as $f(u)$.) This is the formulation of the chain rule that we will use most frequently in later work.

Suppose now that we were asked to calculate

$$\frac{d}{dx}[(x^2-1)^{100}].$$

We could expand $(x^2-1)^{100}$ into polynomial form by using the binomial theorem or, if we were masochistic, by repeated multiplication, but we would have a terrible mess on our hands; for example, $(x^2-1)^{100}$ has 101 terms. With (3.5.2) we can derive a formula that will render such calculations almost trivial.

Assuming (3.5.2), we can prove that, if u is a differentiable function of x and n is an integer, then

(3.5.3)

$$\boxed{\frac{d}{dx}(u^n) = nu^{n-1}\frac{du}{dx}.}$$

PROOF Set $f(u) = u^n$. Then

$$\frac{d}{dx}(u^n) = \underset{\uparrow\,(3.5.2)}{=}\; \frac{d}{du}(u^n)\frac{du}{dx} = nu^{n-1}\frac{du}{dx}. \quad \square$$

To calculate

$$\frac{d}{dx}[(x^2-1)^{100}]$$

we set $u = x^2 - 1$. Then by our formula

$$\frac{d}{dx}[(x^2-1)^{100}] = 100(x^2-1)^{99}\frac{d}{dx}(x^2-1) = 100(x^2-1)^{99}2x = 200x(x^2-1)^{99}.$$

Remark While it is clear that the chain rule is the only practical way to calculate the derivative of $f(x) = (x^2-1)^{100}$, you do have a choice in differentiating a similar but simpler function such as $g(x) = (x^2-1)^5$. Calculating the derivative of g by the chain rule, we get

$$\frac{d}{dx}[(x^2-1)^5] = 5(x^2-1)^4\frac{d}{dx}(x^2-1) = 5(x^2-1)^4 2x = 10x(x^2-1)^4.$$

On the other hand, if we were to first expand the expression $(x^2-1)^5$, we would get

$$g(x) = x^{10} - 5x^8 + 10x^6 - 10x^4 + 5x^2 - 1$$

and then

$$g'(x) = 10x^9 - 40x^7 + 60x^5 - 40x^3 + 10x.$$

As a final answer, this is correct, but it is somewhat unwieldy. To simplify, notice that $10x$ is a factor of g':

$$g'(x) = 10x(x^8 - 4x^6 + 6x^4 - 4x^2 + 1),$$

but note that the expression in parentheses is the binomial expansion for $(x^2 - 1)^4$ so that

$$g'(x) = 10x(x^2 - 1)^4$$

as we saw earlier. The point is, the chain rule gave us the neat, compact result immediately, without the need for expansions and factoring. ❑

Here are more examples of a similar sort.

Example 2

$$\frac{d}{dx}\left[\left(x + \frac{1}{x}\right)^{-3}\right] = -3\left(x + \frac{1}{x}\right)^{-4}\frac{d}{dx}\left(x + \frac{1}{x}\right) = -3\left(x + \frac{1}{x}\right)^{-4}\left(1 - \frac{1}{x^2}\right).$$

❑

Example 3

$$\frac{d}{dx}[1 - (2 + 3x)^2]^3 = 3[1 - (2 + 3x)^2]^2\frac{d}{dx}[1 - (2 + 3x)^2].$$

Since

$$\frac{d}{dx}[1 - (2 + 3x)^2] = -2(2 + 3x)\frac{d}{dx}(2 + 3x) = -2(2 + 3x)3 = -6(2 + 3x),$$

we have

$$\begin{aligned}\frac{d}{dx}[1 - (2 + 3x)^2]^3 &= 3[1 - (2 + 3x)^2]^2[-6(2 + 3x)]\\ &= -18(2 + 3x)[1 - (2 + 3x)^2]^2.\end{aligned}$$ ❑

Example 4 Calculate the derivative of

$$f(x) = 4x(x^2 + 3)^3.$$

SOLUTION Here we need to use the product rule and the chain rule:

$$\begin{aligned}\frac{d}{dx}[4x(x^2 + 3)^3] &= 4x\frac{d}{dx}[(x^2 + 3)^3] + (x^2 + 3)^3\frac{d}{dx}(4x)\\ &= 4x[3(x^2 + 3)^2\,2x] + 4(x^2 + 3)^3\\ &= 24x^2(x^2 + 3)^2 + 4(x^2 + 3)^3\\ &= (x^2 + 3)^2(28x^2 + 12) = 4(x^2 + 3)^2(7x^2 + 3).\end{aligned}$$ ❑

The formula

$$\frac{dy}{dx} = \frac{dy}{du}\frac{du}{dx}$$

can easily be extended to more variables. For example, if x itself depends on s, then we have

(3.5.4)
$$\boxed{\frac{dy}{ds} = \frac{dy}{du}\frac{du}{dx}\frac{dx}{ds}.}$$

If, in addition, s depends on t, then

(3.5.5)
$$\frac{dy}{dt} = \frac{dy}{du}\frac{du}{dx}\frac{dx}{ds}\frac{ds}{dt},$$

and so on. Each new dependence adds a new link to the chain.

Example 5 Find dy/ds given that

$$y = 3u + 1, \qquad u = x^{-2}, \qquad x = 1 - s.$$

SOLUTION

$$\frac{dy}{du} = 3, \qquad \frac{du}{dx} = -2x^{-3}, \qquad \frac{dx}{ds} = -1.$$

Therefore

$$\frac{dy}{ds} = \frac{dy}{du}\frac{du}{dx}\frac{dx}{ds} = (3)(-2x^{-3})(-1) = 6x^{-3} = 6(1-s)^{-3}. \quad \square$$

Example 6 Find dy/dt at $t = 9$ given that

$$y = \frac{u+2}{u-1}, \qquad u = (3s-7)^2, \qquad s = \sqrt{t}.$$

SOLUTION As you can check,

$$\frac{dy}{du} = -\frac{3}{(u-1)^2}, \qquad \frac{du}{ds} = 6(3s-7), \qquad \frac{ds}{dt} = \frac{1}{2\sqrt{t}}.\dagger$$

At $t = 9$, we have $s = 3$ and $u = 4$ so that

$$\frac{dy}{du} = -\frac{3}{(4-1)^2} = -\frac{1}{3}, \qquad \frac{du}{ds} = 6(9-7) = 12, \qquad \frac{ds}{dt} = \frac{1}{2\sqrt{9}} = \frac{1}{6}.$$

Thus, at $t = 9$,

$$\frac{dy}{dt} = \frac{dy}{du}\frac{du}{ds}\frac{ds}{dt} = \left(-\frac{1}{3}\right)(12)\left(\frac{1}{6}\right) = -\frac{2}{3}. \quad \square$$

So far we have worked entirely in Leibniz's notation. What does the chain rule look like in prime notation?

Let's go back to the beginning. Once again let y be a differentiable function of u:

$$y = f(u),$$

and let u be a differentiable function of x:

$$u = g(x).$$

---

† It was shown in Section 3.1 that

$$\frac{d}{dx}(\sqrt{x}) = \frac{1}{2\sqrt{x}}.$$

Then

$$y = f(u) = f(g(x)) = (f \circ g)(x)$$

and, according to the chain rule (as yet unproved),

$$\frac{dy}{dx} = \frac{dy}{du}\frac{du}{dx}.$$

Since

$$\frac{dy}{dx} = \frac{d}{dx}[(f \circ g)(x)] = (f \circ g)'(x), \qquad \frac{dy}{du} = f'(u) = f'(g(x)), \qquad \frac{du}{dx} = g'(x),$$

the chain rule can be written

(3.5.6)

$$\boxed{(f \circ g)'(x) = f'(g(x))g'(x).}$$

The chain rule in prime notation says that

"the derivative of a composition $f \circ g$ at x is the derivative of f at $g(x)$ times the derivative of g at x."

In Leibniz's notation the chain rule *appears* seductively simple, to some even obvious. "After all, to prove it, all you have to do is cancel the du's":

$$\frac{dy}{dx} = \frac{dy}{\cancel{du}}\frac{\cancel{du}}{dx}.$$

Of course, this is just nonsense. What would one cancel from

$$(f \circ g)'(x) = f'(g(x))g'(x)?$$

Although Leibniz's notation is useful for routine calculations, mathematicians generally turn to the prime notation when precision is required.

It is time for us to be precise. How do we know that the composition of differentiable functions is differentiable? What assumptions do we need? Precisely under what circumstances is it true that

$$(f \circ g)'(x) = f'(g(x))g'(x)?$$

The following theorem provides the definitive answer.

THEOREM 3.5.7 THE CHAIN-RULE THEOREM

If g is differentiable at x and f is differentiable at $g(x)$, then the composition $f \circ g$ is differentiable at x and

$$(f \circ g)'(x) = f'(g(x))g'(x).$$

A proof of this theorem appears in the supplement to this section. The argument is not as easy as "canceling" the du's.

We conclude this section with an application that involves the chain rule.

Example 7 Gravel is being poured by a conveyor onto a conical pile at the constant rate of 60π cubic feet per minute. Frictional forces within the pile are such that the height is always two-thirds of the radius. How fast is the radius of the pile changing at the instant it is 5 feet?

SOLUTION The formula for the volume V of a right circular cone of radius r and height h is

$$V = \tfrac{1}{3}\pi r^2 h.$$

However, in this case we are told that $h = \tfrac{2}{3}r$, and so we have

$$(*) \qquad V = \tfrac{2}{9}\pi r^3.$$

Since gravel is being poured onto the pile, the volume, and hence the radius, are functions of time t. We are given that $dV/dt = 60\pi$ and we want to find dr/dt at the instant when $r = 5$. Differentiating $(*)$ with respect to t using the chain rule, we get

$$\frac{dV}{dt} = \frac{dV}{dr} \cdot \frac{dr}{dt} = (\tfrac{2}{3}\pi r^2)\frac{dr}{dt}.$$

Solving for dr/dt and using the fact that $dV/dt = 60\pi$, we find that

$$\frac{dr}{dt} = \frac{180\pi}{2\pi r^2} = \frac{90}{r^2}.$$

Now, when $r = 5$,

$$\frac{dr}{dt} = \frac{90}{(5)^2} = \frac{90}{25} = 3.6.$$

Thus, the radius is increasing at the rate of 3.6 feet per minute at the instant the radius is 5 feet. ❑

EXERCISES 3.5

In Exercises 1–6, differentiate the given function: (a) by expanding before differentiation, (b) by using the chain rule. Then reconcile the results.

1. $f(x) = (x^2 + 1)^2$.

2. $f(x) = (x^3 - 1)^2$.

3. $f(x) = (2x + 1)^3$.

4. $f(x) = (x^2 + 1)^3$.

5. $f(x) = (x + x^{-1})^2$.

6. $f(x) = (3x^2 - 2x)^2$.

In Exercises 7–26, differentiate the given function.

7. $f(x) = (1 - 2x)^{-1}$.

8. $f(x) = (1 + 2x)^5$.

9. $f(x) = (x^5 - x^{10})^{20}$.

10. $f(x) = \left(x^2 + \dfrac{1}{x^2}\right)^3$.

11. $f(x) = \left(x - \dfrac{1}{x}\right)^4$.

12. $f(x) = \left(x + \dfrac{1}{x}\right)^3$.

13. $f(x) = (x - x^3 - x^5)^4$.

14. $f(t) = \left(\dfrac{1}{1 + t}\right)^4$.

15. $f(t) = (t^2 - 1)^{100}$.

16. $f(t) = (t - t^2)^3$.

17. $f(t) = (t^{-1} + t^{-2})^4$.

18. $f(x) = \left(\dfrac{4x + 3}{5x - 2}\right)^3$.

19. $f(x) = \left(\dfrac{3x}{x^2 + 1}\right)^4$.

20. $f(x) = [(2x + 1)^2 + (x + 1)^2]^3$.

21. $f(x) = (x^4 + x^2 + x)^2$.

22. $f(x) = (x^2 + 2x + 1)^3$.

23. $f(x) = \left(\dfrac{x^3}{3} + \dfrac{x^2}{2} + \dfrac{x}{1}\right)^{-1}$.

24. $f(x) = [(6x + x^5)^{-1} + x]^2$.

25. $f(x) = [(x + x^{-1})^2 - (x^2 + x^{-2})^{-1}]^3$.

26. $f(x) = [(x^{-1} + 2x^{-2})^3 + 3x^{-3}]^4$.

In Exercises 27–30, find dy/dx at $x = 0$.

27. $y = \dfrac{1}{1 + u^2}$, $\quad u = 2x + 1$.

28. $y = u + \dfrac{1}{u}$, $\quad u = (3x + 1)^4$.

29. $y = \dfrac{2u}{1 - 4u}$, $\quad u = (5x^2 + 1)^4$.

30. $y = u^3 - u + 1, \quad u = \dfrac{1-x}{1+x}.$

In Exercises 31 and 32, find dy/dt.

31. $y = \dfrac{1-7u}{1+u^2}, \quad u = 1 + x^2, \quad x = 2t - 5.$

32. $y = 1 + u^2, \quad u = \dfrac{1-7x}{1+x^2}, \quad x = 5t + 2.$

In Exercises 33 and 34, find dy/dx at $x = 2$.

33. $y = (s+3)^2, \quad s = \sqrt{t-3}, \quad t = x^2.$

34. $y = \dfrac{1+s}{1-s}, \quad s = t - \dfrac{1}{t}, \quad t = \sqrt{x}.$

Given that

$$f(0) = 1, \quad f'(0) = 2, \quad f(1) = 0, \quad f'(1) = 1, \quad f(2) = 1, \quad f'(2) = 1,$$

$$g(0) = 2, \quad g'(0) = 1, \quad g(1) = 1, \quad g'(1) = 0, \quad g(2) = 2, \quad g'(2) = 1,$$

$$h(0) = 1, \quad h'(0) = 2, \quad h(1) = 2, \quad h'(1) = 1, \quad h(2) = 0, \quad h'(2) = 2,$$

evaluate the following.

35. $(f \circ g)'(0)$.

36. $(f \circ g)'(1)$.

37. $(f \circ g)'(2)$.

38. $(g \circ f)'(0)$.

39. $(g \circ f)'(1)$.

40. $(g \circ f)'(2)$.

41. $(f \circ h)'(0)$.

42. $(f \circ h \circ g)'(1)$.

43. $(g \circ f \circ h)'(2)$.

44. $(g \circ h \circ f)'(0)$.

In Exercises 45–48, find $f''(x)$.

45. $f(x) = (x^3 + x)^4$

46. $f(x) = (x^2 - 5x + 2)^{10}$.

47. $f(x) = \left(\dfrac{x}{1-x}\right)^3.$

48. $f(x) = \sqrt{x^2+1} \quad \left(\text{Recall that } \dfrac{d}{dx}[\sqrt{x}] = \dfrac{1}{2\sqrt{x}}\right).$

In Exercises 49–52, find the indicated derivative.

49. $\dfrac{d}{dx}[f(x^2+1)].$

50. $\dfrac{d}{dx}\left[f\left(\dfrac{x-1}{x+1}\right)\right].$

51. $\dfrac{d}{dx}[[f(x)]^2 + 1].$

52. $\dfrac{d}{dx}\left[\dfrac{f(x)-1}{f(x)+1}\right].$

In Exercises 53–56, determine the values of x for which
(a) $f'(x) = 0$. (b) $f'(x) > 0$. (c) $f'(x) < 0$.

53. $f(x) = (1 + x^2)^{-2}$.

54. $f(x) = (1 - x^2)^2$.

55. $f(x) = x(1 + x^2)^{-1}$.

56. $f(x) = x(1 - x^2)^3$.

An object moves along a coordinate line, its position at each time $t \geq 0$ given by $x(t)$. In Exercises 57–60, determine when the object changes direction.

57. $x(t) = (t+1)^2(t-9)^3$.

58. $x(t) = t(t-8)^3$.

59. $x(t) = (t^3 - 12t)^4$.

60. $x(t) = (t^2 - 8t + 15)^3$.

61. A function L has the property that $L'(x) = 1/x$ for $x \neq 0$. Determine the derivative of $L(x^2 + 1)$.

62. Let f and g be differentiable functions such that $f'(x) = g(x)$ and $g'(x) = f(x)$, and let

$$H(x) = [f(x)]^2 - [g(x)]^2.$$

Find $H'(x)$.

63. Let f and g be differentiable functions such that $f'(x) = g(x)$ and $g'(x) = -f(x)$, and let

$$T(x) = [f(x)]^2 + [g(x)]^2.$$

Find $T'(x)$.

64. Let f be a differentiable function. Use the chain rule to show that:
(a) if f is even, then f' is odd.
(b) if f is odd, then f' is even.
(Also see Exercises 59 and 60, Section 3.1.)

65. The number a is a *double zero* (or a zero of *multiplicity* 2) of the polynomial function $p(x)$ if

$$p(x) = (x-a)^2 q(x) \qquad \text{and} \qquad q(a) \neq 0.$$

Prove that a is a double zero of $p(x)$ iff a is a zero of both $p(x)$ and $p'(x)$ and $p''(a) \neq 0$. HINT: Recall the factor theorem; see Section 1.2.

66. The number a is a *triple zero* (or a zero of *multiplicity* 3) of the polynomial function $p(x)$ if

$$p(x) = (x-a)^3 q(x) \qquad \text{and} \qquad q(a) \neq 0.$$

Prove that a is a triple zero of $p(x)$ iff a is a zero of $p(x)$, $p'(x)$, and $p''(x)$, and $p'''(a) \neq 0$.

67. The number a is a *zero of multiplicity k* of the polynomial function $p(x)$ if

$$p(x) = (x-a)^k q(x) \qquad \text{and} \qquad q(a) \neq 0.$$

Use the results in Exercises 65 and 66 to state a theorem about a zero of multiplicity k.

68. Two functions f and g are said to be *inverse functions* if

$$(f \circ g)(x) = x \qquad \text{and} \qquad (g \circ f)(x) = x.$$

For example, $f(x) = 3x + 2$ and $g(x) = \frac{1}{3}x - \frac{2}{3}$ are inverse functions. Let f and g be differentiable. Show that if f and g are inverse functions, then

$$f'(y) = \frac{1}{g'(x)},$$

where $y = g(x)$, provided $g'(x) \neq 0$.

69. An object is moving along the curve $y = x^3 - 3x + 5$ so that its x-coordinate at time t is $x = 2t^2 - t + 2$, $t \geq 0$. At what rate is the y-coordinate changing when $t = 2$?

70. An equilateral triangle of side length x and altitude h has area A given by

$$A = \frac{\sqrt{3}}{4}x^2 \qquad \text{where} \qquad x = \frac{2\sqrt{3}}{3}h.$$

Find the rate of change of A with respect to h and determine the rate of change of A when $h = 2\sqrt{3}$.

71. Air is being pumped into a spherical balloon in such a way that its radius is increasing at the constant rate of 2 centimeters per second. What is the rate of change of the balloon volume at the instant the radius is 10 centimeters? (The volume V of a sphere of radius r is $V = \frac{4}{3}\pi r^3$.)

72. Air is being pumped into a spherical balloon at the constant rate of 200 cubic centimeters per second. How fast is the surface area of the balloon changing at the instant the radius is 5 centimeters? (The surface area S of a sphere of radius r is $S = 4\pi r^2$.)

73. If an object of mass m has speed v, then its *kinetic energy*, KE, is given by

$$\text{KE} = \tfrac{1}{2}mv^2.$$

Suppose that v is a function of time. What is the rate of change of KE with respect to t?

74. Newton's Law of Gravitational Attraction states that if two bodies are at a distance r apart, then the force F exerted by one body on the other is given by

$$F(r) = \frac{k}{r^2},$$

where k is a positive constant. Suppose that, as a function of time, the distance between the two bodies is given by

$$r(t) = 49t - 4.9t^2, \qquad 0 \le t \le 10.$$

(a) Find the rate of change of F with respect to t.
(b) Show that $(F \circ r)'(3) = -(F \circ r)'(7)$.

*SUPPLEMENT TO SECTION 3.5

To prove Theorem 3.5.7 it is convenient to use the formulation of derivative given in (3.1.5), Exercises 3.1.

THEOREM 3.5.8

The function f is differentiable at x iff

$$\lim_{t\to x} \frac{f(t) - f(x)}{t - x} \quad \text{exists.}$$

If this limit exists, it is $f'(x)$.

Proof

For each t in the domain of f, $t \neq x$, define

$$G(t) = \frac{f(t) - f(x)}{t - x}.$$

The theorem follows from observing that f is differentiable at x iff

$$\lim_{h\to 0} G(x + h) \quad \text{exists,}$$

and noting that

$$\lim_{h\to 0} G(x + h) = L \quad \text{iff} \quad \lim_{t\to x} G(t) = L. \quad \square$$

Proof of Theorem 3.5.7

By Theorem 3.5.8 it is enough to show that

$$\lim_{t\to x} \frac{f(g(t)) - f(g(x))}{t - x} = f'(g(x))g'(x).$$

We begin by defining an auxiliary function F on the domain of f by setting

$$F(y) = \begin{cases} \dfrac{f(y) - f(g(x))}{y - g(x)}, & y \neq g(x) \\ f'(g(x)), & y = g(x). \end{cases}$$

F is continuous at $g(x)$ since

$$\lim_{y \to g(x)} F(y) = \lim_{y \to g(x)} \frac{f(y) - f(g(x))}{y - g(x)},$$

and the right-hand side is (by Theorem 3.5.8) $f'(g(x))$, which is the value of F at $g(x)$. For $t \neq x$,

$$\text{(1)} \qquad \frac{f(g(t)) - f(g(x))}{t - x} = F(g(t)) \left[\frac{g(t) - g(x)}{t - x} \right].$$

To see this we note that, if $g(t) = g(x)$, then both sides are 0. If $g(t) \neq g(x)$, then

$$F(g(t)) = \frac{f(g(t)) - f(g(x))}{g(t) - g(x)},$$

so that again we have equality.

Since g, being differentiable at x, is continuous at x and since F is continuous at $g(x)$, we know that the composition $F \circ g$ is continuous at x. Thus

$$\lim_{t \to x} F(g(t)) = F(g(x)) = f'(g(x)).$$

↑ by our definition of F

This, together with Equation (1), gives

$$\lim_{t \to x} \frac{f(g(t)) - f(g(x))}{t - x} = f'(g(x))g'(x). \quad ❑$$

■ 3.6 DIFFERENTIATING THE TRIGONOMETRIC FUNCTIONS

An outline review of trigonometry—definitions, identities, and graphs—appears in Chapter 1. As discussed there, the calculus of the trigonometric functions is simplified by the use of radian measure. Therefore, we use radian measure throughout our work and refer to degree measure only in passing.

The derivative of the sine function is the cosine function:

(3.6.1)

$$\boxed{\frac{d}{dx}(\sin x) = \cos x.}$$

PROOF Fix any number x. For $h \neq 0$

$$\begin{aligned} \frac{\sin(x + h) - \sin x}{h} &= \frac{[\sin x \cos h + \cos x \sin h] - [\sin x]}{h} \\ &= \sin x \frac{\cos h - 1}{h} + \cos x \frac{\sin h}{h}. \end{aligned}$$

Now, as shown in Section 2.5, (2.5.5),

$$\lim_{h \to 0} \frac{\cos h - 1}{h} = 0 \qquad \text{and} \qquad \lim_{h \to 0} \frac{\sin h}{h} = 1.$$

Since $\sin x$ and $\cos x$ remain constant as h approaches zero, it follows that

$$\begin{aligned}\lim_{h\to 0}\frac{\sin(x+h)-\sin x}{h} &= \lim_{h\to 0}\left(\sin x\,\frac{\cos h-1}{h}+\cos x\,\frac{\sin h}{h}\right)\\ &= \sin x\left(\lim_{h\to 0}\frac{\cos h-1}{h}\right)+\cos x\left(\lim_{h\to 0}\frac{\sin h}{h}\right).\end{aligned}$$

Thus

$$\lim_{h\to 0}\frac{\sin(x+h)-\sin x}{h} = (\sin x)(0)+(\cos x)(1) = \cos x. \quad \square$$

The derivative of the cosine function is the negative of the sine function:

(3.6.2)
$$\boxed{\frac{d}{dx}(\cos x) = -\sin x.}$$

PROOF Fix any number x. For $h \neq 0$

$$\cos(x+h) = \cos x\cos h - \sin x\sin h.$$

Therefore

$$\begin{aligned}\lim_{h\to 0}\frac{\cos(x+h)-\cos x}{h} &= \lim_{h\to 0}\frac{[\cos x\cos h-\sin x\sin h]-[\cos x]}{h}\\ &= \cos x\left(\lim_{h\to 0}\frac{\cos h-1}{h}\right)-\sin x\left(\lim_{h\to 0}\frac{\sin h}{h}\right)\\ &= -\sin x. \quad \square\end{aligned}$$

Example 1 To differentiate

$$f(x) = \cos x\sin x$$

we use the product rule:

$$\begin{aligned}f'(x) &= \cos x\frac{d}{dx}(\sin x)+\sin x\frac{d}{dx}(\cos x)\\ &= \cos x(\cos x)+\sin x(-\sin x) = \cos^2 x-\sin^2 x. \quad \square\end{aligned}$$

We come now to the tangent function. Since $\tan x = \sin x/\cos x$, we have

$$\frac{d}{dx}(\tan x) = \frac{\cos x\dfrac{d}{dx}(\sin x)-\sin x\dfrac{d}{dx}(\cos x)}{\cos^2 x} = \frac{\cos^2 x+\sin^2 x}{\cos^2 x} = \frac{1}{\cos^2 x} = \sec^2 x.$$

The derivative of the tangent function is the secant squared:

(3.6.3)
$$\boxed{\frac{d}{dx}(\tan x) = \sec^2 x.}$$

The derivatives of the other trigonometric functions are as follows:

(3.6.4)

$$\frac{d}{dx}(\cot x) = -\csc^2 x,$$

$$\frac{d}{dx}(\sec x) = \sec x \tan x,$$

$$\frac{d}{dx}(\csc x) = -\csc x \cot x.$$

The verification of these formulas is left as an exercise.

It is time for some sample problems.

Example 2 Find $f'(\pi/4)$ for

$$f(x) = x \cot x.$$

SOLUTION We first find $f'(x)$ by the product rule:

$$f'(x) = x\frac{d}{dx}(\cot x) + \cot x\frac{d}{dx}(x) = -x\csc^2 x + \cot x.$$

Now we evaluate f' at $\pi/4$:

$$f'(\pi/4) = -\frac{\pi}{4}(\sqrt{2})^2 + 1 = 1 - \frac{\pi}{2}. \quad \square$$

Example 3 Find

$$\frac{d}{dx}\left[\frac{1 - \sec x}{\tan x}\right].$$

SOLUTION By the quotient rule,

$$\frac{d}{dx}\left[\frac{1 - \sec x}{\tan x}\right] = \frac{\tan x\frac{d}{dx}(1 - \sec x) - (1 - \sec x)\frac{d}{dx}(\tan x)}{\tan^2 x}$$

$$= \frac{\tan x(-\sec x \tan x) - (1 - \sec x)(\sec^2 x)}{\tan^2 x}$$

$$= \frac{\sec x(\sec^2 x - \tan^2 x) - \sec^2 x}{\tan^2 x}$$

$$= \frac{\sec x - \sec^2 x}{\tan^2 x} = \frac{\sec x(1 - \sec x)}{\tan^2 x}. \quad \square$$

(recall $\sec^2 x - \tan^2 x = 1$) ↑

Example 4 Find an equation for the tangent to the curve

$$y = \cos x$$

at the point where $x = \pi/3$.

SOLUTION Since $\cos \pi/3 = \frac{1}{2}$, the point of tangency is $(\pi/3, \frac{1}{2})$. To find the slope of the tangent, we evaluate the derivative

$$\frac{dy}{dx} = -\sin x$$

at $x = \pi/3$. This gives $m = -\sqrt{3}/2$. An equation for the tangent can be written

$$y - \frac{1}{2} = -\frac{\sqrt{3}}{2}\left(x - \frac{\pi}{3}\right). \quad \square$$

Example 5 An object moves along the x-axis, its position at each time t given by the function

$$x(t) = t + 2 \sin t.$$

Determine those times t from 0 to 2π when the object is moving to the left.

SOLUTION The object is moving to the left only when its velocity $v(t) < 0$. Since

$$v(t) = x'(t) = 1 + 2 \cos t,$$

the object is moving to the left only when $\cos t < -\frac{1}{2}$. As you can check, the only t in $[0, 2\pi]$ for which $\cos t < -\frac{1}{2}$ are those t that lie between $2\pi/3$ and $4\pi/3$. Thus from time $t = 0$ to $t = 2\pi$, the object is moving left only during the time interval $(2\pi/3, 4\pi/3)$. $\square$

The Chain Rule and the Trigonometric Functions

If f is a differentiable function of u and u is a differentiable function of x, then, as you saw in Section 3.5,

$$\frac{d}{dx}[f(u)] = \frac{d}{du}[f(u)]\frac{du}{dx}.$$

Written in this form, the derivatives of the six trigonometric functions appear as follows:

(3.6.5)

$$\frac{d}{dx}(\sin u) = \cos u\frac{du}{dx}, \qquad \frac{d}{dx}(\cos u) = -\sin u\frac{du}{dx},$$

$$\frac{d}{dx}(\tan u) = \sec^2 u\frac{du}{dx}, \qquad \frac{d}{dx}(\cot u) = -\csc^2 u\frac{du}{dx},$$

$$\frac{d}{dx}(\sec u) = \sec u \tan u\frac{du}{dx}, \qquad \frac{d}{dx}(\csc u) = -\csc u \cot u\frac{du}{dx}.$$

Example 6

$$\frac{d}{dx}(\cos 2x) = -\sin 2x\frac{d}{dx}(2x) = -2 \sin 2x. \quad \square$$

Example 7

$$\frac{d}{dx}[\sec(x^2+1)] = \sec(x^2+1)\tan(x^2+1)\frac{d}{dx}(x^2+1)$$
$$= 2x\sec(x^2+1)\tan(x^2+1). \quad \square$$

Example 8

$$\frac{d}{dx}(\sin^3 \pi x) = \frac{d}{dx}(\sin \pi x)^3$$
$$= 3(\sin \pi x)^2 \frac{d}{dx}(\sin \pi x)$$
$$= 3(\sin \pi x)^2 \cos \pi x \frac{d}{dx}(\pi x)$$
$$= 3(\sin \pi x)^2 \cos \pi x\,(\pi) = 3\pi \sin^2 \pi x \cos \pi x. \quad \square$$

Our treatment of the trigonometric functions has been based entirely on radian measure. When degrees are used, the derivatives of the trigonometric functions contain the extra factor $\frac{1}{180}\pi \cong 0.0175$.

Example 9 Find

$$\frac{d}{dx}(\sin x^\circ).$$

SOLUTION Since $x^\circ = \frac{1}{180}\pi x$ radians,

$$\frac{d}{dx}(\sin x^\circ) = \frac{d}{dx}\left(\sin \frac{1}{180}\pi x\right) = \frac{1}{180}\pi \cos \frac{1}{180}\pi x = \frac{1}{180}\pi \cos x^\circ. \quad \square$$

The extra factor $\frac{1}{180}\pi$ is a disadvantage, particularly in problems where it occurs repeatedly. This tends to discourage the use of degree measure in theoretical work.

EXERCISES 3.6

In Exercises 1–12, differentiate the given function.

1. $y = 3\cos x - 4\sec x$.
2. $y = x^2 \sec x$.
3. $y = x^3 \csc x$.
4. $y = \sin^2 x$.
5. $y = \cos^2 t$.
6. $y = 3t^2 \tan t$.
7. $y = \sin^4 \sqrt{u}$.
8. $y = u \csc u^2$.
9. $y = \tan x^2$.
10. $y = \cos \sqrt{x}$.
11. $y = [x + \cot \pi x]^4$.
12. $y = [x^2 - \sec 2x]^3$.

In Exercises 13–24, find the second derivative.

13. $y = \sin x$.
14. $y = \cos x$.
15. $y = \dfrac{\cos x}{1 + \sin x}$.
16. $y = \tan^3 2\pi x$.
17. $y = \cos^3 2u$.
18. $y = \sin^5 3t$.
19. $y = \tan 2t$.
20. $y = \cot 4u$.
21. $y = x^2 \sin 3x$.
22. $y = \dfrac{\sin x}{1 - \cos x}$.
23. $y = \sin^2 x + \cos^2 x$.
24. $y = \sec^2 x - \tan^2 x$.

In Exercises 25–30, find the indicated derivative.

25. $\dfrac{d^4}{dx^4}(\sin x)$.
26. $\dfrac{d^4}{dx^4}(\cos x)$.
27. $\dfrac{d}{dt}\left[t^2 \dfrac{d^2}{dt^2}(t \cos 3t)\right]$.
28. $\dfrac{d}{dt}\left[t \dfrac{d}{dt}(\cos t^2)\right]$.
29. $\dfrac{d}{dx}[f(\sin 3x)]$.
30. $\dfrac{d}{dx}[\sin(f(3x))]$.

In Exercises 31–36, find an equation for the tangent to the curve at $x = a$.

31. $y = \sin x$; $a = 0$.
32. $y = \tan x$; $a = \pi/6$.
33. $y = \cot x$; $a = \pi/6$.
34. $y = \cos x$; $a = 0$.

35. $y = \sec x; \quad a = \pi/4.$ **36.** $y = \csc x; \quad a = \pi/3.$

In Exercises 37–46, determine the numbers x between 0 and 2π where the tangent to the curve is horizontal.

37. $y = \cos x.$ **38.** $y = \sin x.$

39. $y = \sin x + \sqrt{3}\cos x.$ **40.** $y = \cos x - \sqrt{3}\sin x.$

41. $y = \sin^2 x.$ **42.** $y = \cos^2 x.$

43. $y = \tan x - 2x.$ **44.** $y = 3\cot x + 4x.$

45. $y = 2\sec x + \tan x.$ **46.** $y = \cot x - 2\csc x.$

An object moves along the x-axis, its position at each time t given by $x(t)$. In Exercises 47–52, determine those times from $t = 0$ to $t = 2\pi$ when the object is moving to the right with increasing speed.

47. $x(t) = \sin 3t.$ **48.** $x(t) = \cos 2t.$

49. $x(t) = \sin t - \cos t.$ **50.** $x(t) = \sin t + \cos t.$

51. $x(t) = t + 2\cos t.$ **52.** $x(t) = t - \sqrt{2}\sin t.$

In Exercises 53–56, find dy/dt (a) by using (3.5.4) and (b) by writing y as a function of t and then differentiating.

53. $y = u^2 - 1, \quad u = \sec x, \quad x = \pi t.$

54. $y = [\frac{1}{2}(1 + u)]^3, \quad u = \cos x, \quad x = 2t.$

55. $y = [\frac{1}{2}(1 - u)]^4, \quad u = \cos x, \quad x = 2t.$

56. $y = 1 - u^2, \quad u = \csc x, \quad x = 3t.$

57. It can be shown by induction that the nth derivative of the sine function is given by the formula

$$\frac{d^n}{dx^n}(\sin x) = \begin{cases} (-1)^{(n-1)/2}\cos x, & n \text{ odd} \\ (-1)^{n/2}\sin x, & n \text{ even}. \end{cases}$$

Persuade yourself that this formula is correct and obtain a similar formula for the nth derivative of the cosine function.

58. Verify the following differentiation formulas.

(a) $\dfrac{d}{dx}(\cot x) = -\csc^2 x.$

(b) $\dfrac{d}{dx}(\sec x) = \sec x \tan x.$

(c) $\dfrac{d}{dx}(\csc x) = -\csc x \cot x.$

59. Let $f(x) = \begin{cases} x\sin\left(\dfrac{1}{x}\right), & x \neq 0 \\ 0, & x = 0 \end{cases}$ and

$$g(x) = \begin{cases} x^2\sin\left(\dfrac{1}{x}\right), & x \neq 0 \\ 0, & x = 0. \end{cases}$$

In Exercise 67, Section 3.1, you were asked to show that f is continuous at 0, but not differentiable there, and that g is differentiable at 0. Both f and g are differentiable for any $x \neq 0$.

(a) Find $f'(x)$ and $g'(x)$ for $x \neq 0$.
(b) Show that $g'(x)$ is not continuous at 0.

60. Let $f(x) = \begin{cases} \cos x, & x \geq 0 \\ ax + b, & x < 0. \end{cases}$

(a) Determine a and b so that f is differentiable at 0.
(b) Using the values found in part (a), sketch the graph of f.

Let $g(x) = \begin{cases} \sin x, & 0 \leq x \leq 2\pi/3 \\ ax + b, & 2\pi/3 < x \leq 2\pi. \end{cases}$

(c) Determine a and b so that g is differentiable at $2\pi/3$.
(d) Using the values found in part (c), sketch the graph of g.

61. Let $y = A\sin\omega t + B\cos\omega t$, where A, B, and ω are constants. Show that y satisfies the equation

$$\frac{d^2y}{dx^2} + \omega^2 y = 0.$$

62. A simple pendulum consists of a mass m swinging at the end of a (massless) rod or wire of length L (see the figure). The angular displacement θ of the pendulum at time t is given by

$$\theta(t) = a\cos(\omega t + \phi),$$

where a, ω, and ϕ are constants.

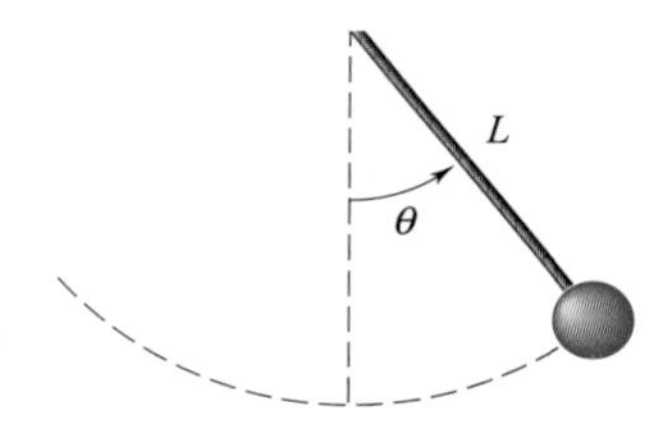

(a) Show that θ satisfies the equation $d^2\theta/dt^2 + \omega^2\theta = 0$. (Compare this result with the result in Exercise 61.)
(b) Show that θ can be written in the form

$$\theta(t) = A\sin\omega t + B\cos\omega t,$$

where A, B, and ω are constants.

63. An isosceles triangle has equal sides of length c, and the angle between them is x radians. Express the area A of the triangle as a function of x and find the rate of change of A with respect to x.

64. A triangle has sides of length a and b, and the angle between them is x radians. If a and b are constants, find the rate of change of the third side c with respect to x. HINT: Use the law of cosines.

65. A 13-foot ladder leans against a vertical wall. If the bottom of the ladder is being pulled away from the wall at the rate of 2 feet per second, how fast is the angle between the ladder and the ground changing at the instant the bottom of the ladder is 5 feet from the wall?

66. A lighthouse is on an island 1 mile from a straight shoreline. Suppose that the light is turning at the rate of 2

revolutions per minute in the counterclockwise direction.
(a) How fast is the light beam moving along the shore at the instant it makes an angle of 45° with the shoreline?
(b) How fast is the light beam moving at the instant the angle is 90°?

67. A man starts at a point A and walks 40 feet north. He then turns and walks due east at 4 feet per second. If a searchlight placed at A follows him, at what rate is the light turning at the end of 6 seconds?

C 68. Let θ be measured in degrees.
(a) Estimate

$$\lim_{\theta \to 0} \frac{\sin\theta}{\theta}$$

by letting $\theta = 5$, 1, 0.1, 0.01, and 0.001 degrees.
(b) Find the value of $\pi/180$ and compare the result with the limit found in part (a). This gives another demonstration of the reason for using radian measure rather than degree measure.

C 69. Let $f(x) = \cos x^2$. Determine

$$f'(0) = \lim_{h \to 0} \frac{f(h) - f(0)}{h}$$

numerically by letting $h = \pm 0.1$, ± 0.01, and ± 0.001. Compare your answer to what you obtain by the chain rule.

C 70. Let $f(x) = \cos^2 x$. Determine

$$f'(\tfrac{1}{4}\pi) = \lim_{h \to 0} \frac{f(\frac{1}{4}\pi + h) - f(\frac{1}{4}\pi)}{h}$$

numerically by letting $h = \pm 0.1$, ± 0.01, and ± 0.001. Compare your answer to what you obtain by the chain rule.

C 71. Let $f(x) = x^2 - \sin 3x$.
(a) Use a graphing utility to graph f for $-\pi \le x \le \pi$.
(b) Estimate the zeros of f with two decimal place accuracy.
(c) Estimate with two decimal place accuracy the x-coordinates of the points where the tangent line to the graph of f is horizontal.

C 72. Let $f(x) = \sin x$, $-\pi \le x \le \pi$.
(a) Using a graphing utility to graph f and $p_1(x) = x$ together. Notice that $p_1 = x$ is a good approximation for $\sin x$ near 0.
(b) Compare the graphs of f and $p_3(x) = x - \frac{1}{6}x^3$.
(c) Compare the graphs of f and $p_5(x) = x - \frac{1}{6}x^3 + \frac{1}{120}x^5$.

C 73. Let $f(x) = \cos x$, $-\pi \le x \le \pi$. Repeat Exercise 72 for f and $p_0(x) = 1$, $p_2(x) = 1 - \frac{1}{2}x^2$, and $p_4(x) = 1 - \frac{1}{2}x^2 + \frac{1}{24}x^4$. The use of polynomials to approximate the trigonometric functions (and other functions) is one of the main topics in Chapter 11.

■ 3.7 IMPLICIT DIFFERENTIATION; RATIONAL POWERS

The functions that we have been studying up to this point have been defined by expressing the dependent variable *explicitly* in terms of the independent variable. For instance

$$y = (x^2 - 1)^3, \qquad y = x^2 \sec x, \qquad y = \begin{cases} x^2 + 1, & x \le 0 \\ 1 - x, & x > 0, \end{cases}$$

or, in general, $y = f(x)$. In this section we consider functional relationships between two variables in which the dependent variable is not given explicitly in terms of the independent variable.

For example, suppose you know that y is a function of x that satisfies the equation

$$3x^3y - 4y - 2x + 1 = 0.$$

We say that this equation defines y *implicitly* as a function of x. We want to develop a method for finding the derivative of y with respect to x when the functional relationship between x and y is given implicitly. Continuing with this example, one way to find dy/dx is to solve first for y:

$$(3x^3 - 4)y - 2x + 1 = 0,$$

$$(3x^3 - 4)y = 2x - 1,$$

$$y = \frac{2x - 1}{3x^3 - 4},$$

and then differentiate:

$$\frac{dy}{dx} = \frac{(3x^3-4)2-(2x-1)(9x^2)}{(3x^3-4)^2} = \frac{6x^3-8-18x^3+9x^2}{(3x^3-4)^2}$$

$$(1) \qquad = -\frac{12x^3-9x^2+8}{(3x^3-4)^2}.$$

It is also possible to find dy/dx without first solving the equation for y. The technique is called *implicit differentiation.*

Returning to the original equation

$$3x^3y - 4y - 2x + 1 = 0,$$

we differentiate both sides of the equation with respect to x (remembering that y is a function of x). We have

$$\frac{d}{dx}(3x^3y - 4y - 2x + 1) = \frac{d}{dx}(0)$$

$$\underbrace{\left(3x^3\frac{dy}{dx} + 9x^2y\right)}_{\text{by the product rule}} - 4\frac{dy}{dx} - 2 = 0$$

$$(3x^3-4)\frac{dy}{dx} = 2 - 9x^2y,$$

$$(2) \qquad \frac{dy}{dx} = \frac{2-9x^2y}{3x^3-4}.$$

The answer looks different from what we obtained before because this time y appears on the right-hand side. To satisfy yourself that the two answers are really the same all you have to do is substitute

$$y = \frac{2x-1}{3x^3-4}$$

in (2). This gives

$$\frac{dy}{dx} = \frac{2-9x^2\left(\dfrac{2x-1}{3x^3-4}\right)}{3x^3-4} = \frac{6x^3-8-18x^3+9x^2}{(3x^3-4)^2} = -\frac{12x^3-9x^2+8}{(3x^3-4)^2}$$

which is the answer (1) we obtained before.

Implicit differentiation is particularly useful where it is inconvenient (or impossible) to first solve the given equation for y.

Example 1 Use implicit differentiation to find dy/dx.

(a) $2xy - y^3 + 1 = x + 2y$. **(b)** $\cos(x-y) = (2x+1)^2y$.

SOLUTION **(a)** Differentiating implicitly with respect to x, we have

$$\underbrace{2x\frac{dy}{dx} + 2y}_{\text{(by the product rule)}} - \underbrace{3y^2\frac{dy}{dx}}_{\text{(by the chain rule)}} = 1 + 2\frac{dy}{dx}$$

$$[2x - 3y^2 - 2]\frac{dy}{dx} = 1 - 2y$$

and

$$\frac{dy}{dx} = \frac{1 - 2y}{2x - 3y^2 - 2}.$$

(b) Differentiate implicitly with respect to x:

$$\underbrace{-\sin(x - y)\left[1 - \frac{dy}{dx}\right]}_{\text{(by the chain rule)}} = (2x + 1)^2 \frac{dy}{dx} + 2(2x + 1)(2)y$$

$$[\sin(x - y) - (2x + 1)^2]\frac{dy}{dx} = 4(2x + 1)y + \sin(x - y).$$

Thus,

$$\frac{dy}{dx} = \frac{4(2x + 1)y + \sin(x - y)}{\sin(x - y) - (2x + 1)^2}. \quad \square$$

Example 2 Find the slope of the curve

$$x^3 + y^3 = 1 + 3xy^2 \quad \text{at} \quad (2, -1).$$

SOLUTION Differentiating this equation implicitly with respect to x gives

$$3x^2 + 3y^2\frac{dy}{dx} = 3x\left(2y\frac{dy}{dx}\right) + 3y^2$$

or

$$x^2 + y^2\frac{dy}{dx} = 2xy\frac{dy}{dx} + y^2.$$

Solving for dy/dx, we have

$$\frac{dy}{dx} = \frac{y^2 - x^2}{y^2 - 2xy}.$$

At $x = 2$ and $y = -1$, we find that

$$\frac{dy}{dx} = \frac{(-1)^2 - 2^2}{(-1)^2 - 2(2)(-1)} = -\frac{3}{5}.$$

The slope is $-\frac{3}{5}$. □

We can also find higher derivatives by implicit differentiation.

Example 3 Find d^2y/dx^2 given that

$$y^3 - x^2 = 4.$$

SOLUTION Differentiation with respect to x gives

$$3y^2\frac{dy}{dx} - 2x = 0$$

and

$$(*) \qquad \frac{dy}{dx} = \frac{2x}{3y^2}.$$

Differentiating again with respect to x, using the quotient rule, we get

$$\frac{d^2y}{dx^2} = \frac{d(dy/dx)}{dx} = \frac{3y^2(2) - 2x(6y(dy/dx))}{(3y^2)^2} = \frac{6y^2 - 12xy(dy/dx)}{9y^4}.$$

Now, substituting (*) into this equation gives

$$\frac{d^2y}{dx^2} = \frac{6y^2 - 12xy(2x/3y^2)}{9y^4}$$

which, as you can check, simplifies to

$$\frac{d^2y}{dx^2} = \frac{6y^3 - 8x^2}{9y^5}. \quad \square$$

A Theoretical Aside

In the preceding discussion and examples, we assumed that the given equations determined y as a differentiable function of x. However, this is not always true; proper interpretations have to be made.

For example, if we differentiate $x^2 + y^2 = -9$ implicitly, we find that

$$2x + 2y\frac{dy}{dx} = 0 \qquad \text{and therefore} \qquad \frac{dy}{dx} = -\frac{x}{y}.$$

However, the result is meaningless. It is meaningless because there is no real-valued function y of x that satisfies the equation $x^2 + y^2 = -9$. Implicit differentiation can be applied meaningfully to an equation in x and y only if there is a differentiable function y of x that satisfies the equation.

Now consider the equation $x^2 + y^2 = 9$. The graph is a circle of radius 3 centered at the origin (Figure 3.7.1). Clearly this graph does not define y as a function of x (remember the vertical line test). If we solve this equation for y, we get

$$y = \pm\sqrt{9 - x^2} \qquad -3 \leq x \leq 3.$$

Although this equation does not define y as a function of x,

$$y = \sqrt{9 - x^2} \quad \text{(Figure 3.7.2)} \qquad \text{and} \qquad y = -\sqrt{9 - x^2} \quad \text{(Figure 3.7.3)}$$

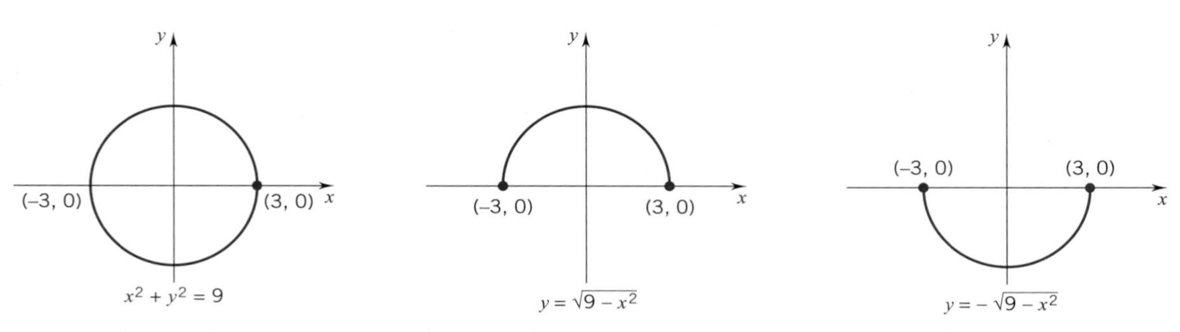

Figure 3.7.1

Figure 3.7.2

Figure 3.7.3

are functions of x on $[-3, 3]$ that satisfy the original equation. In fact, there are an infinite number of functions $y = f(x)$ that satisfy the given equation:

$$y = -\sqrt{9 - x^2} \quad \text{for} \quad 0 \le x \le 3 \qquad \text{and} \qquad y = \begin{cases} -\sqrt{9 - x^2}, & -3 \le x < 0 \\ 0\sqrt{9 - x^2}, & \le x \le 3 \end{cases}$$

are other examples. However, the functions $y = \sqrt{9 - x^2}$ and $y = -\sqrt{9 - x^2}$ are the only two that are continuous on $[-3, 3]$ and differentiable on $(-3, 3)$.

The point of this discussion is that an equation involving x and y does not always define y implicitly as a function of x, and, on the other hand, such an equation may give rise to more than one function. Also, as noted before, implicit differentiation can be applied to an equation in x and y only if there is at least one differentiable function y of x that satisfies the equation.

Rational Powers

We have shown that the differentiation formula

$$\frac{d(x^n)}{dx} = nx^{n-1}$$

holds for all integers n. Here we show that this formula also holds for rational exponents p/q.

(3.7.1)

$$\boxed{\frac{d(x^{p/q})}{dx} = \frac{p}{q} x^{(p/q)-1}.}$$

PROOF Let $y = x^{p/q}$, where p and q are nonzero integers. Then $y^q = x^p$. Differentiating this equation implicitly with respect to x, we have

$$qy^{q-1}\frac{dy}{dx} = px^{p-1}$$

and

$$\frac{dy}{dx} = \frac{p}{q}\frac{x^{p-1}}{y^{q-1}} = \frac{p}{q}x^{p-1}y^{1-q}.$$

Now, replacing y by $x^{p/q}$, we have

$$\frac{dy}{dx} = \frac{p}{q}x^{p-1}(x^{p/q})^{(1-q)} = \frac{p}{q}x^{p-1}x^{(p/q)-p}$$

$$= \frac{p}{q}x^{(p/q)-1}. \quad \square$$

Here are some simple examples:

$$\frac{d}{dx}(x^{2/3}) = \frac{2}{3}x^{-1/3}, \qquad \frac{d}{dx}(x^{5/2}) = \frac{5}{2}x^{3/2}, \qquad \frac{d}{dx}(x^{-7/9}) = -\frac{7}{9}x^{-16/9}.$$

If u is a differentiable function of x, then, by the chain rule,

(3.7.2)
$$\frac{d}{dx}(u^{p/q}) = \frac{p}{q}\, u^{(p/q)-1}\frac{du}{dx}.$$

The result holds on every open x-interval where $u^{(p/q)-1}$ is defined.

Example 4

(a) $\dfrac{d}{dx}[(1 + x^2)^{1/5}] = \frac{1}{5}(1 + x^2)^{-4/5}(2x) = \frac{2}{5}x(1 + x^2)^{-4/5}$.

(b) $\dfrac{d}{dx}[(1 - x^2)^{1/5}] = \frac{1}{5}(1 - x^2)^{-4/5}(-2x) = -\frac{2}{5}x(1 - x^2)^{-4/5}$.

(c) $\dfrac{d}{dx}[(1 - x^2)^{1/4}] = \frac{1}{4}(1 - x^2)^{-3/4}(-2x) = -\frac{1}{2}x(1 - x^2)^{-3/4}$.

You should verify that the first statement holds for all real x, the second for all $x \neq \pm 1$, the third only for $x \in (-1, 1)$. ❑

Example 5

$$\begin{aligned}\frac{d}{dx}\left[\left(\frac{x}{1+x^2}\right)^{1/2}\right] &= \frac{1}{2}\left(\frac{x}{1+x^2}\right)^{-1/2}\frac{d}{dx}\left(\frac{x}{1+x^2}\right)\\ &= \frac{1}{2}\left(\frac{x}{1+x^2}\right)^{-1/2}\frac{(1+x^2)(1) - x(2x)}{(1+x^2)^2}\\ &= \frac{1}{2}\left(\frac{1+x^2}{x}\right)^{1/2}\frac{1-x^2}{(1+x^2)^2}\\ &= \frac{1-x^2}{2x^{1/2}(1+x^2)^{3/2}}.\end{aligned}$$

The result holds for all $x > 0$. ❑

EXERCISES 3.7

In Exercises 1–10, use implicit differentiation to obtain dy/dx in terms of x and y.

1. $x^2 + y^2 = 4$.

2. $x^3 + y^3 - 3xy = 0$.

3. $4x^2 + 9y^2 = 36$.

4. $\sqrt{x} + \sqrt{y} = 4$.

5. $x^4 + 4x^3y + y^4 = 1$.

6. $x^2 - x^2y + xy^2 + y^2 = 1$.

7. $(x - y)^2 - y = 0$.

8. $(y + 3x)^2 - 4x = 0$.

9. $\sin(x + y) = xy$.

10. $\tan xy = xy$.

In Exercises 11–16, express d^2y/dx^2 in terms of x and y.

11. $y^2 + 2xy = 16$.

12. $x^2 - 2xy + 4y^2 = 3$.

13. $y^2 + xy - x^2 = 9$.

14. $x^2 - 3xy = 18$.

15. $4\tan y = x^3$.

16. $\sin^2 x + \cos^2 y = 1$.

In Exercises 17–20, find dy/dx and d^2y/dx^2 at the given point.

17. $x^2 - 4y^2 = 9$; $(5, 2)$.

18. $x^2 + 4xy + y^3 + 5 = 0$; $(2, -1)$.

19. $\cos(x + 2y) = 0$; $(\pi/6, \pi/6)$.

20. $x = \sin^2 y$; $(\frac{1}{2}, \pi/4)$.

In Exercises 21–26, find equations for the tangent and normal at the indicated point.

21. $2x + 3y = 5$; $(-2, 3)$.

22. $9x^2 + 4y^2 = 72$; $(2, 3)$.

23. $x^2 + xy + 2y^2 = 28$; $(-2, -3)$.

24. $x^3 - axy + 3ay^2 = 3a^3$; (a, a).

25. $x = \cos y$; $\left(\frac{1}{2}, \frac{\pi}{3}\right)$

26. $\tan xy = x$; $\left(1, \frac{\pi}{4}\right)$.

Find dy/dx in Exercises 27–34.

27. $y = (x^3 + 1)^{1/2}$.

28. $y = (x + 1)^{1/3}$.

29. $y = x\sqrt{x^2 + 1}$.

30. $y = x^2\sqrt{x^2 + 1}$.

31. $y = \sqrt[4]{2x^2 + 1}$.

32. $y = (x + 1)^{1/3}(x + 2)^{2/3}$.

33. $y = \sqrt{2 - x^2}\sqrt{3 - x^2}$.

34. $y = \sqrt{(x^4 - x + 1)^3}$.

Compute the indicated derivative in Exercises 35–40.

35. $\frac{d}{dx}\left(\sqrt{x} + \frac{1}{\sqrt{x}}\right)$.

36. $\frac{d}{dx}\left(\sqrt{\frac{3x + 1}{2x + 5}}\right)$.

37. $\frac{d}{dx}\left(\frac{x}{\sqrt{x^2 + 1}}\right)$.

38. $\frac{d}{dx}\left(\frac{\sqrt{x^2 + 1}}{x}\right)$

39. $\frac{d}{dx}\left(\sqrt[3]{x} + \frac{1}{\sqrt[3]{x}}\right)$.

40. $\frac{d}{dx}\left(\sqrt{\frac{ax + b}{cx + d}}\right)$, a, b, c, d constant.

41. (*Important*) Show the general form of the graph.
(a) $f(x) = x^{1/n}$, n a positive even integer.
(b) $f(x) = x^{1/n}$, n a positive odd integer.
(c) $f(x) = x^{2/n}$, n an odd integer greater than 1.

Find the second derivative in Exercises 42–46.

42. $y = \sqrt{a^2 + x^2}$.

43. $y = \sqrt[3]{a + bx}$.

44. $y = x\sqrt{a^2 - x^2}$.

45. $y = \sqrt{x}\tan\sqrt{x}$.

46. $y = \sqrt{x}\sin\sqrt{x}$.

47. Show that all normals to the circle $x^2 + y^2 = r^2$ pass through the center of the circle.

48. Determine the x-intercept of the tangent to the parabola $y^2 = x$ at the point where $x = a$.

The angle between two curves is the angle between their tangents at the point of intersection. If the slopes are m_1 and m_2, then the angle of intersection α can be obtained from the formula

$$\tan\alpha = \left|\frac{m_2 - m_1}{1 + m_1 m_2}\right|. \quad \text{(Exercise 71, Section 1.6)}$$

49. At what angles do the parabolas $y^2 = 2px + p^2$ and $y^2 = p^2 - 2px$ intersect?

50. At what angles does the line $y = 2x$ intersect the curve $x^2 - xy + 2y^2 = 28$?

51. The curves $y = x^2$ and $x = y^3$ intersect at the points $(1, 1)$ and $(0, 0)$. Find the angle between the curves at each of these points.

52. Find the angles at which the circles $(x - 1)^2 + y^2 = 10$ and $x^2 + (y - 2)^2 = 5$ intersect.

Two curves are said to be *orthogonal* if at each point of intersection, the angle between them is a right angle. In Exercises 53 and 54, show that the given curves are orthogonal.

53. The hyperbola $x^2 - y^2 = 5$ and the ellipse $4x^2 + 9y^2 = 72$.

54. The ellipse $3x^2 + 2y^2 = 5$ and $y^3 = x^2$.
HINT: The curves intersect at $(1, 1)$ and $(-1, 1)$.

Two families of curves are said to be *orthogonal trajectories* (of each other) if each member of one family is orthogonal to each member of the other family. In Exercises 55 and 56, show that the given families of curves are orthogonal trajectories.

55. The family of circles $x^2 + y^2 = r^2$ and the family of lines $y = mx$.

56. The family of parabolas $x = ay^2$ and the family of ellipses $x^2 + \frac{1}{2}y^2 = b$.

57. Find equations for the tangents to the ellipse $4x^2 + y^2 = 72$ that are perpendicular to the line $x + 2y + 3 = 0$.

58. Find equations for the normals to the hyperbola $4x^2 - y^2 = 36$ that are parallel to the line $2x + 5y - 4 = 0$.

59. The curve defined by the equation $(x^3 + y^2)^2 = x^2 - y^2$ is called a *lemniscate*. The graph is shown in the figure. Find the four points on the curve at which the tangent line is horizontal.

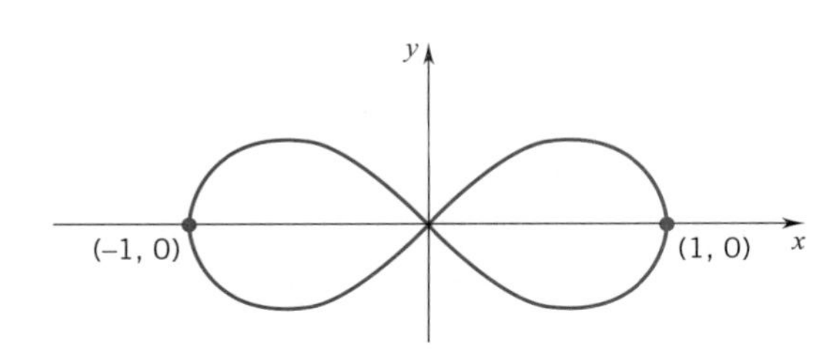

60. The curve defined by the equation $x^{2/3} + y^{2/3} = a^{2/3}$ is called an astroid. The graph is shown in the figure.

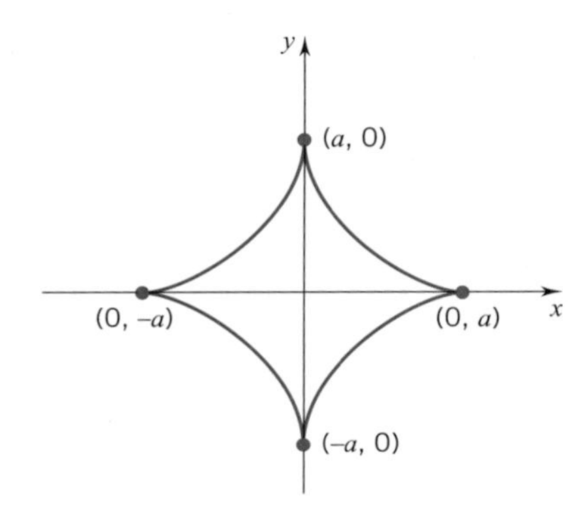

(a) Find the slope of the tangent line to the graph at an arbitrary point (x_1, y_1), $x_1 \neq 0$.
(b) At what points on the curve is the slope of the tangent line 0, 1, or -1?

61. Prove that the sum of the x- and y-intercepts of any tangent line to the graph of

$$x^{1/2} + y^{1/2} = c^{1/2}$$

is constant and equal to c.

62. A circle of radius 1 with center on the y-axis is inscribed in the parabola $y = 2x^2$. See the figure. Find the points of intersection. HINT: The circle has equation $x^2 + (y - a)^2 = 1$, and the circle and parabola have the same tangent lines at the points of intersection.

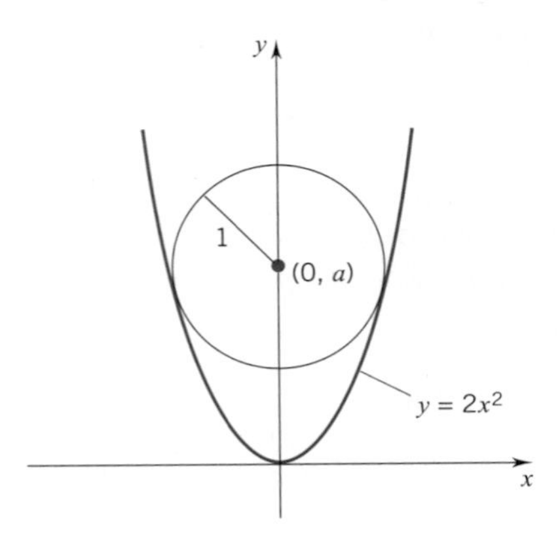

63. The function y defined by

$$y = \begin{cases} \sqrt{1 - x^2} & -1 \le x < -\frac{1}{2} \\ -\sqrt{1 - x^2} & -\frac{1}{2} \le x \le \frac{1}{2} \\ \sqrt{1 - x^2} & \frac{1}{2} < x \le 1. \end{cases}$$

satisfies the equation $x^2 + y^2 = 1$

(a) Sketch the graph of this function.
(b) Use implicit differentiation to find the slope of the tangent line to the graph of y at the points $(-\sqrt{\frac{3}{2}}, \frac{1}{2})$, $(\sqrt{\frac{3}{2}}, \frac{1}{2})$ and $(0, -1)$.
(c) Find $y'_+(-\frac{1}{2})$, $y'_-(\frac{1}{2})$.

64. Determine

$$\lim_{h \to 0} \frac{(1 + h)^{98.6} - 1}{h}$$

numerically by letting $h = \pm 0.1, \pm 0.01, \pm 0.001$. Confirm your answer by other means.

65. Let $f(x) = x^{3/4}$. Determine

$$f'(16) = \lim_{h \to 0} \frac{f(16 + h) - f(16)}{h}$$

numerically by letting $h = \pm 0.1$, ± 0.01, ± 0.001. Compare your answer to what you obtain by the methods of this section.

66. Determine

$$\lim_{x \to 32} \frac{x^{2/5} - 4}{x^{4/5} - 16}$$

numerically. Confirm your answer by other means.

67. A graphing utility in parametric mode can be used to graph some equations in x and y. Sketch the graph of the equation $x^2 + y^2 = 4$ by first setting $x = t$, $y = \sqrt{4 - t^2}$ and then setting $x = t$, $y = -\sqrt{4 - t^2}$.

68. Use a graphing utility to sketch the graph of $(2 - x)y^2 = x^3$. This curve is called a *cissoid.*

69. (a) Use a graphing utility to sketch the graph of $x^4 = x^2 - y^2$. This curve is called a *figure eight curve.*
(b) Find the x-coordinates of the points on the graph where the tangent line is horizontal.

■ 3.8 RATES OF CHANGE PER UNIT TIME

Earlier you saw that velocity is the rate of change of position with respect to time t, and acceleration the rate of change of velocity with respect to time t. In this section we work with other quantities that vary with time. The fundamental point is this: *If Q is a quantity that varies with time, then the derivative dQ/dt gives the rate of change of that quantity with respect to time.*

Example 1 A spherical balloon is expanding. If the radius is increasing at the rate of 2 inches per minute, at what rate is the volume increasing when the radius is 5 inches?

SOLUTION

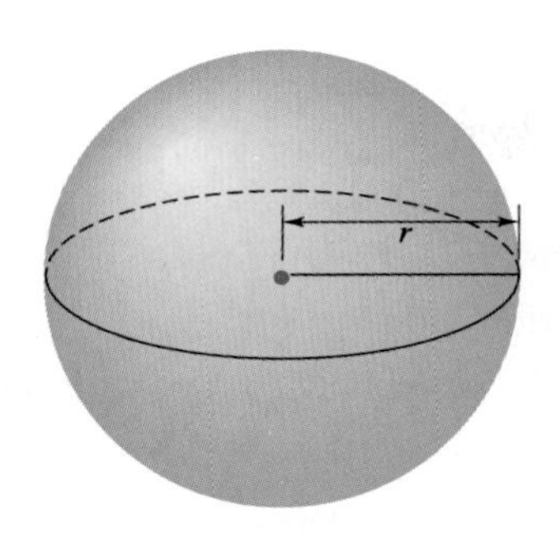

Find $\dfrac{dV}{dt}$ when $r = 5$ inches

given that $\dfrac{dr}{dt} = 2$ in./min.

$V = \frac{4}{3}\pi r^3$. (volume of a sphere of radius r)

Since r and V are functions of t, we differentiate the equation $V = \frac{4}{3}\pi r^3$ implicitly with respect to t. We have

$$\frac{dV}{dt} = 4\pi r^2 \frac{dr}{dt}.$$

Substituting $r = 5$ and $dr/dt = 2$, we find that

$$\frac{dV}{dt} = 4\pi(5)^2(2) = 200\pi.$$

When the radius is 5 inches, the volume is increasing at the rate of 200π cubic inches per minute. ❑

Example 2 An object is moving in the clockwise direction around the unit circle $x^2 + y^2 = 1$. As it passes through the point $(\frac{1}{2}, \frac{1}{2}\sqrt{3})$, its y-coordinate is decreasing at the rate of 3 units per second. At what rate is the x-coordinate changing at this point?

SOLUTION

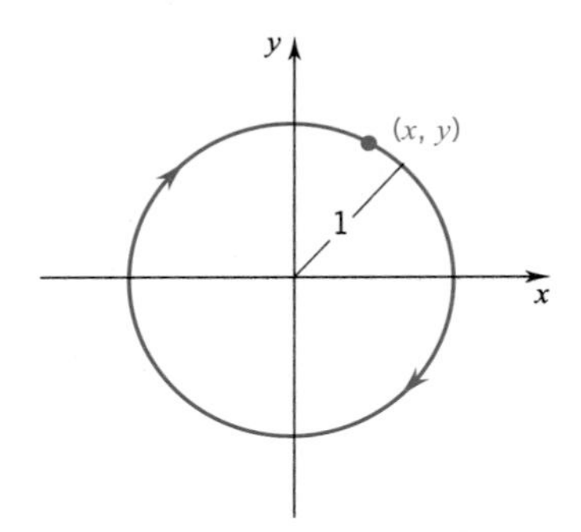

Find $\dfrac{dx}{dt}$ when $x = \frac{1}{2}$ and $y = \frac{1}{2}\sqrt{3}$

given that $\dfrac{dy}{dt} = -3$ units/sec.

$x^2 + y^2 = 1$. (equation of circle)

Differentiating $x^2 + y^2 = 1$ with respect to t, we have

$$2x\frac{dx}{dt} + 2y\frac{dy}{dt} = 0 \qquad \text{and thus} \qquad x\frac{dx}{dt} + y\frac{dy}{dt} = 0.$$

Substitution of $x = \frac{1}{2}$, $y = \frac{1}{2}\sqrt{3}$, and $dy/dt = -3$ gives

$$\frac{1}{2}\frac{dx}{dt} + \frac{1}{2}\sqrt{3}(-3) = 0 \qquad \text{so that} \qquad \frac{dx}{dt} = 3\sqrt{3}.$$

As the particle passes through the point $(\frac{1}{2}, \frac{1}{2}\sqrt{3})$, the x-coordinate is increasing at the rate of $3\sqrt{3}$ units per second. ❑

As you must have noticed, the first two sample problems were solved by the same general method, a method that we recommend to you for solving problems of this type:

Step 1. Draw a diagram, where relevant, and indicate the quantities that vary.

Step 2. Specify in mathematical form the rate of change you are looking for, and record all given information.

Step 3. Find an equation involving the variable whose rate of change is to be found.

Step 4. Differentiate with respect to time t the equation found in step 3.

Step 5. State the final answer in coherent form, specifying the units that you are using.

Example 3 A water trough with vertical cross section in the shape of an equilateral triangle is being filled at a rate of 4 cubic feet per minute. Given that the trough is 12 feet long, how fast is the level of the water rising at the instant that the water reaches a depth of $1\frac{1}{2}$ feet?

SOLUTION Let x be the depth of the water in feet and V the volume of water in cubic feet.

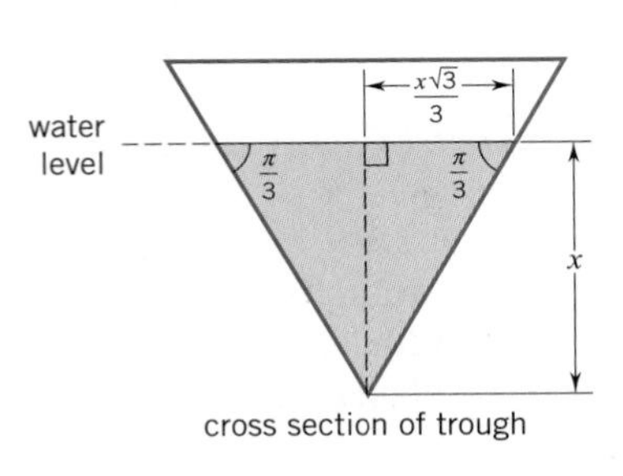

cross section of trough

Find $\dfrac{dx}{dt}$ when $x = \frac{3}{2}$

given that $\dfrac{dV}{dt} = 4$ ft³/min.

$$\text{Area of cross section} = \frac{1}{2}\left(\frac{2x}{\sqrt{3}}\right)x = \frac{\sqrt{3}}{3}x^2.$$

$$\text{Volume of water} = 12\left(\frac{\sqrt{3}}{3}x^2\right) = 4\sqrt{3}x^2.$$

Differentiation of $V = 4\sqrt{3}x^2$ with respect to t gives

$$\frac{dV}{dt} = 8\sqrt{3}x\frac{dx}{dt}.$$

Substituting $x = \frac{3}{2}$ and $dV/dt = 4$, we have

$$4 = 8\sqrt{3}\left(\frac{3}{2}\right)\frac{dx}{dt} \qquad \text{and thus} \qquad \frac{dx}{dt} = \frac{1}{9}\sqrt{3}.$$

At the instant that the water reaches a depth of $1\frac{1}{2}$ feet, the water level is rising at the rate of $\frac{1}{9}\sqrt{3}$ feet per second (about 0.19 feet per second). ❑

Example 4 Two ships, one heading west and the other east, approach each other on parallel courses 8 nautical miles apart.† Given that each ship is cruising at 20 nautical miles per hour (knots), at what rate is the distance between them diminishing when they are 10 nautical miles apart?

† The international nautical mile measures 6080 feet. Nautical speed is measured in knots—nautical miles per hour.

SOLUTION Let y be the distance between the ships measured in nautical miles. Since the ships are moving in opposite directions at the rate of 20 knots each, the distance x (see figure) is decreasing at the rate of 40 knots. Thus, we want to

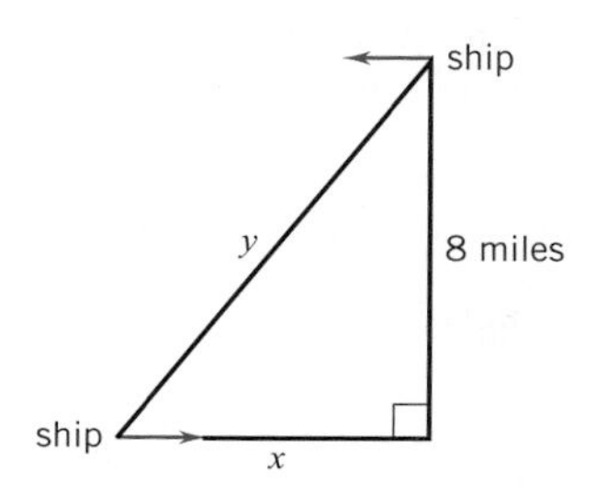

find $\dfrac{dy}{dt}$ when $y = 10$

given that $\dfrac{dx}{dt} = -40$ knots.

(Note that dx/dt is taken as negative since x is decreasing.)

$$x^2 + 8^2 = y^2. \qquad \text{(Pythagorean theorem)}$$

Differentiating $x^2 + 8^2 = y^2$ with respect to t, we find that

$$2x\frac{dx}{dt} + 0 = 2y\frac{dy}{dt} \qquad \text{and consequently} \qquad x\frac{dx}{dt} = y\frac{dy}{dt}.$$

When $y = 10$, $x = 6$. (Why?) Substituting $x = 6$, $y = 10$, and $dx/dt = -40$, we have

$$6(-40) = 10\frac{dy}{dt} \qquad \text{so that} \qquad \frac{dy}{dt} = -24.$$

(Note that dy/dt is negative since y is decreasing.) When the two ships are 10 miles apart, the distance between them is diminishing at the rate of 24 knots. ❑

Example 5 A conical paper cup 8 inches across the top and 6 inches deep is full of water. The cup springs a leak at the bottom and loses water at the rate of 2 cubic inches per minute. How fast is the water level dropping at the instant when the water is exactly 3 inches deep?

SOLUTION We begin with a diagram that represents the situation after the cup has been leaking for a while (Figure 3.8.1). We label the radius and height of the remaining "cone of water" r and h. We can relate r and h by similar triangles (Figure 3.8.2). Let h be the depth of the water measured in inches.

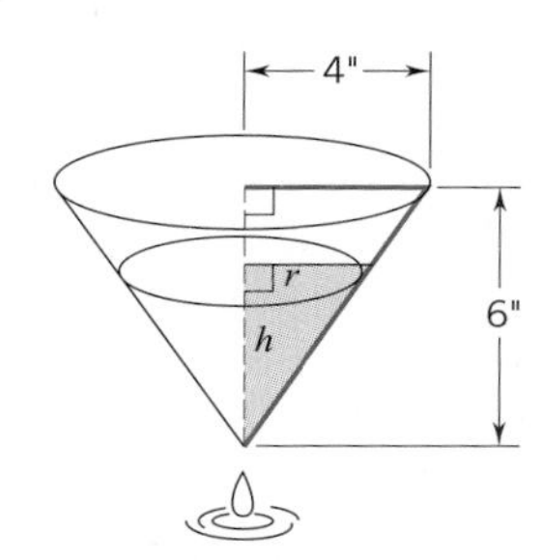

Figure 3.8.1

$$\text{Find } \frac{dh}{dt} \text{ when } h = 3 \quad \text{given that } \frac{dV}{dt} = -2 \text{ in.}^3/\text{min.}$$

$$V = \frac{1}{3}\pi r^2 h \quad \text{(volume of a cone)} \qquad \text{and} \qquad \frac{r}{h} = \frac{4}{6} \quad \text{(similar triangles).}$$

Using the second equation to eliminate r from the first equation, we have

$$V = \frac{1}{3}\pi\left(\frac{2h}{3}\right)^2 h = \frac{4}{27}\pi h^3.$$

Differentiation with respect to t gives

$$\frac{dV}{dt} = \frac{4}{9}\pi h^2 \frac{dh}{dt}.$$

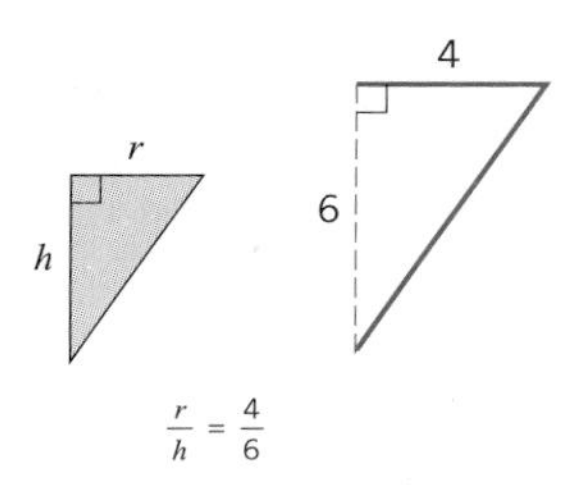

Figure 3.8.2

Substituting $h = 3$ and $dV/dt = -2$, we have

$$-2 = \frac{4}{9}\pi(3)^2\frac{dh}{dt} \qquad \text{and thus} \qquad \frac{dh}{dt} = -\frac{1}{2\pi}.$$

At the instant when the water is exactly 3 inches deep, the water level is dropping at the rate of $1/2\pi$ inches per minute (about 0.16 inches per minute). ❑

Example 6 A balloon leaves the ground 500 feet away from an observer and rises vertically at the rate of 140 feet per minute. At what rate is the angle of inclination of the observer's line of sight increasing at the instant when the balloon is exactly 500 feet above the ground?

SOLUTION Let x be the altitude of the balloon and θ the angle of inclination of the observer's line of sight.

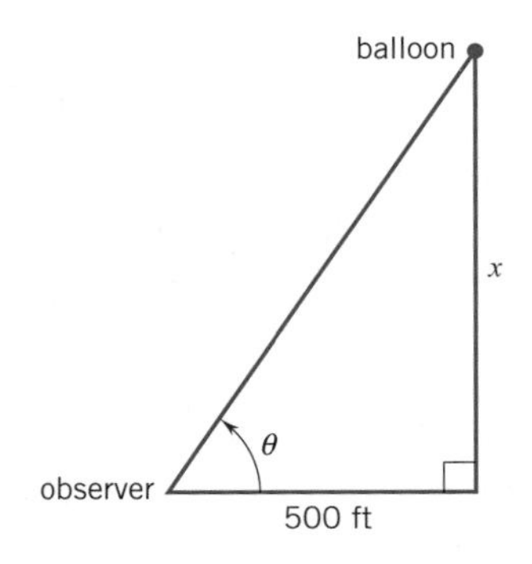

$$\text{Find } \frac{d\theta}{dt} \text{ when } x = 500 \text{ feet,}$$

$$\text{given that } \frac{dx}{dt} = 140 \text{ ft/min.}$$

$$\tan\theta = \frac{x}{500}.$$

Differentiation with respect to t gives

$$\sec^2\theta\,\frac{d\theta}{dt} = \frac{1}{500}\frac{dx}{dt}.$$

When $x = 500$, the triangle is isosceles, and therefore $\sec\theta = \sqrt{2}$. Substituting $\sec\theta = \sqrt{2}$ and $dx/dt = 140$, we have

$$(\sqrt{2})^2\frac{d\theta}{dt} = \frac{1}{500}(140) \qquad \text{and therefore} \qquad \frac{d\theta}{dt} = 0.14.$$

At the instant when the balloon is exactly 500 feet above the ground, the inclination of the observer's line of sight is increasing at the rate of 0.14 radians per minute (about 8 degrees per minute). ❑

Example 7 A ladder 13 feet long is leaning against the side of a building. If the foot of the ladder is pulled away from the building at the rate of 0.1 foot per second, how fast is the angle formed by the ladder and the ground changing at the instant when the top of the ladder is 12 feet above the ground?

SOLUTION

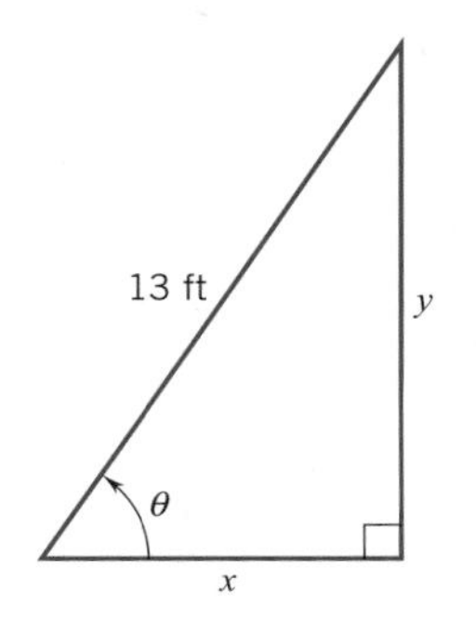

$$\text{Find } \frac{d\theta}{dt} \text{ when } y = 12 \text{ feet}$$

$$\text{given that } \frac{dx}{dt} = 0.1 \text{ ft/sec.}$$

$$\cos\theta = \frac{x}{13}.$$

Differentiation with respect to t gives

$$-\sin\theta\,\frac{d\theta}{dt} = \frac{1}{13}\frac{dx}{dt}.$$

When $y = 12$, $\sin\theta = \frac{12}{13}$. Substituting $\sin\theta = \frac{12}{13}$ and $dx/dt = 0.1$, we have

$$-\left(\frac{12}{13}\right)\frac{d\theta}{dt} = \frac{1}{13}(0.1) \qquad \text{and thus} \qquad \frac{d\theta}{dt} = -\frac{1}{120}.$$

At the instant that the top of the ladder is 12 feet above the ground, the angle formed by the ladder and the ground is decreasing at the rate of $\frac{1}{120}$ radians per second (about half a degree per second). ❑

EXERCISES 3.8

1. A point moves along the straight line $x + 2y = 2$. Find (a) the rate of change of the y-coordinate, given that the x-coordinate is increasing at a rate of 4 units per second; (b) the rate of change of the x-coordinate, given that the y-coordinate is decreasing at a rate of 2 units per second.

2. A particle is moving in circular orbit $x^2 + y^2 = 25$. As it passes through the point $(3, 4)$, its y-coordinate is decreasing at the rate of 2 units per second. How is the x-coordinate changing?

3. A particle is moving along the parabola $y^2 = 4(x + 2)$. As it passes through the point $(7, 6)$ its y-coordinate is increasing at the rate of 3 units per second. How fast is the x-coordinate changing at this instant?

4. A particle is moving along the parabola $4y = (x + 2)^2$ in such a way that its x-coordinate is increasing at the constant rate of 2 units per second. How fast is the particle's distance to the point $(-2, 0)$ changing at the instant it is at the point $(2, 4)$?

5. A particle is moving around the ellipse $4x^2 + 16y^2 = 64$. At any time t its x- and y- coordinates are given by $x = 4\cos t$, $y = 2\sin t$. At what rate is the particle's distance to the origin changing at an arbitrary time t? At what rate is the distance to the origin changing when $t = \pi/4$?

6. A particle is moving along the curve $y = x\sqrt{x}$, $x \geq 0$. At time t its coordinates are $(x(t), y(t))$. Find the points on the curve, if any, at which both coordinates are changing at the same rate.

7. A heap of rubbish in the shape of a cube is being compacted. Given that the volume decreases at a rate of 2 cubic meters per minute, find the rate of change of an edge of the cube when the volume is exactly 27 cubic meters. What is the rate of change of the surface area of the cube at that instant?

8. At a certain instant the dimensions of a rectangle are a and b. These dimensions are changing at the rates m, n, respectively. Find the rate at which the area is changing.

9. The volume of a spherical balloon is increasing at a constant rate of 8 cubic feet per minute. How fast is the radius increasing when the radius is exactly 10 feet? How fast is the surface area increasing at that instant?

10. A spherical balloon has an initial volume of V_0 cubic meters of helium. Suppose that the gas is being released from the balloon at the rate of $1/(1 + t^2)$ cubic meters per second, $t \geq 0$.

(a) What is the rate of change of the radius of the balloon at any time t? At what rate is the radius changing when $t = 3$?

(b) What is the rate of change of the surface area of the balloon at any time t? At what rate is the surface area changing when $t = 3$?

11. An isosceles triangle has equal sides of length 10 centimeters. Let θ denote the angle included between the two equal sides.

(a) Express the area A of the triangle as a function of θ (radians).

(b) Suppose that θ is increasing at the rate of $10°$ per minute. How fast is A changing at the instant $\theta = 60°$?

(c) At what value of θ will the triangle have maximum area?

12. At a certain instant the side of an equilateral triangle is α centimeters long and increasing at the rate of k centimeters per minute. How fast is the area increasing?

13. The perimeter of a rectangle is fixed at 24 centimeters. If the length l of the rectangle is increasing at the rate of 1 centimeter per second, for what value of l does the area of the rectangle start to decrease?

14. The dimensions of a rectangle are changing in such a way that the perimeter is always 24 inches. Show that, at the instant when the area is 32 square inches, the area is either increasing or decreasing 4 times as fast as the length is increasing.

15. A rectangle is inscribed in a circle of radius 5 inches. If the length of the rectangle is decreasing at the rate of 2 inches per second, how fast is the area changing at the instant when the length is 6 inches? HINT: A diagonal of the rectangle is a diameter of the circle.

16. A boat is held by a bow line that is wound about a bollard 6 feet higher than the bow of the boat. If the boat is drifting away at the rate of 8 feet per minute, how fast is the line unwinding when the bow is exactly 30 feet from the bollard?

17. Two boats are racing with constant speed toward a finish marker, boat A sailing from the south at 13 mph and boat B approaching from the east. When equidistant from the marker the boats are 16 miles apart and the distance between them is decreasing at the rate of 17 mph. Which boat will win the race?

18. A ladder 13 feet long is leaning against a wall. If the foot of the ladder is pulled away from the wall at the rate of 0.5 foot per second, how fast will the top of the ladder be dropping at the instant when the base is 5 feet from the wall?

19. A tank contains 1000 cubic feet of natural gas at a pressure of 5 pounds per square inch. Find the rate of change of the volume if the pressure decreases at a rate of 0.05 pound per square inch per hour. (Assume Boyle's law: *pressure* $\times$ *volume* $=$ *constant*.)

20. The adiabatic law for the expansion of air is $PV^{1.4} = C$. At a given instant the volume is 10 cubic feet and the pressure 50 pounds per square inch. At what rate is the pressure changing if the volume is decreasing at a rate of 1 cubic foot per second?

21. A man standing 3 feet from the base of a lamppost casts a shadow 4 feet long. If the man is 6 feet tall and walks away from the lamppost at a speed of 400 feet per minute, at what rate will his shadow lengthen? How fast is the tip of his shadow moving?

22. A light is attached to the wall of a building 64 feet above the ground. A ball is dropped from the same height, but 20 feet away from the side of the building. The height y of the ball at time t is given by $y(t) = 64 - 16t^2$. How fast is the shadow of the ball moving along the ground after 1 second? After 3 seconds? After 4 seconds? (Explain.)

23. An object that weighs 150 pounds on the surface of the earth will weigh $150(1 + \frac{1}{4000}r)^{-2}$ pounds when it is r miles above the surface. If the object's altitude is increasing at the rate of 10 miles per second, how fast is its weight decreasing at the instant it is 400 miles above the surface?

24. In the special theory of relativity the mass of a particle moving at velocity v is

$$\frac{m}{\sqrt{1 - v^2/c^2}}$$

where m is the mass at rest and c is the speed of light. At what rate is the mass changing when the particle's velocity is $\frac{1}{2}c$ and the rate of change of the velocity is $0.01\,c$ per second?

25. Water is dripping through the bottom of a conical cup 4 inches across and 6 inches deep. Given that the cup loses half a cubic inch of water per minute, how fast is the water level dropping when the water is 3 inches deep?

26. Water flows from a faucet into a hemispherical basin 14 inches in diameter at a rate of 2 cubic inches per second. How fast does the water rise (a) when the water is exactly halfway to the top? (b) just as it runs over? (The volume of a spherical segment is given by $\pi rh^2 - \frac{1}{3}\pi h^3$ where r is the radius of the sphere and h is the depth of the segment.)

27. The base of an isosceles triangle is 6 feet. If the altitude is 4 feet and increasing at the rate of 2 inches per minute, at what rate is the vertex angle changing?

28. As a boy winds up the cord, his kite is moving horizontally at a height of 60 feet with a speed of 10 feet per minute. How fast is the inclination of the cord changing when its length is 100 feet?

29. A revolving searchlight $\frac{1}{2}$ mile from shore makes 1 revolution per minute. How fast is the light traveling along the straight beach at the instant it passes over a shorepoint 1 mile away from the shorepoint nearest to the searchlight?

30. Two cars, car A traveling east at 30 mph and car B traveling north at 22.5 mph, are heading toward an intersection I. At what rate is angle IAB changing at the instant when cars A and B are 300 feet and 400 feet, respectively, from the intersection?

31. A rope 32 feet long is attached to a weight and passed over a pulley 16 feet above the ground. The other end of the rope is pulled away along the ground at the rate of 3 feet per second. At what rate is the angle between the rope and the ground changing at the instant when the weight is exactly 4 feet off the ground?

32. A slingshot is made by fastening the two ends of a 10-inch rubber strip 6 inches apart. If the midpoint of the strip is drawn back at the rate of 1 inch per second, at what rate is the angle between the segments of the strip changing 8 seconds later?

33. A balloon is released 500 feet away from an observer. If the balloon rises vertically at the rate of 100 feet per minute and at the same time the wind is carrying it horizontally away from the observer at the rate of 75 feet per minute, at what rate is the angle of inclination of the observer's line of sight changing 6 minutes after the balloon has been released?

34. A searchlight is trained on a plane that flies directly above the light at an altitude of 2 miles at a speed of 400 miles

per hour. How fast must the light be turning 2 seconds after the plane passes directly overhead?

(*Important*) Angular Velocity; Uniform Circular Motion
As a particle moves along a circle of radius r, it effects a change in the central angle marked θ in the figure. The angle θ is measured in radians. The time rate of change of θ, $\omega = d\theta/dt$, is called the *angular velocity* of the particle. Circular motion with constant positive angular velocity is called *uniform circular motion.*

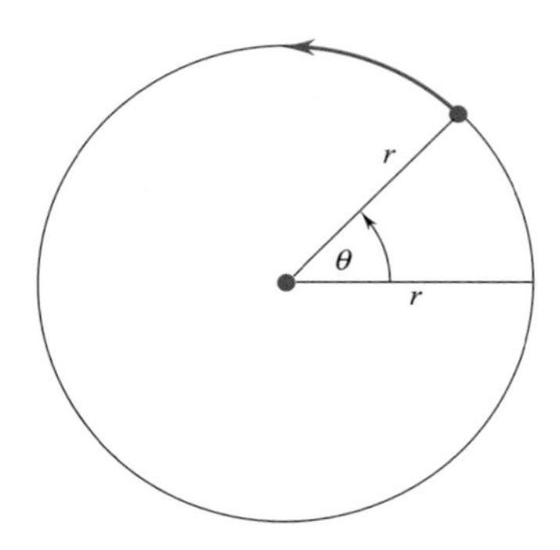

35. A particle in uniform circular motion traces out a circular arc. The time rate of change of the length of that arc is called the *speed* of the particle. What is the speed of a particle that moves along a circle of radius r with constant positive angular velocity ω?

36. The kinetic energy KE of a particle of mass m is given by

$$\text{KE} = \tfrac{1}{2}mv^2$$

where v is the speed. What is the kinetic energy of the particle in Exercise 35?

37. A point P moves uniformly along the circle $x^2 + y^2 = r^2$ with angular velocity ω. Find the xy-coordinates of P at time t given that the motion starts at time $t = 0$ with $\theta = \theta_0$.

38. With P as in Exercise 37 find the velocity and acceleration of the projection of P onto†
(a) the x-axis. (b) the y-axis.

39. The figure shows a sector in a circle of radius r. The sector is the union of triangle T and segment S. Suppose that the radius vector rotates counterclockwise with a constant angular velocity of ω radians per second. Show that the area of the sector changes at a constant rate, but the area of T and the area of S do not change at a constant rate.

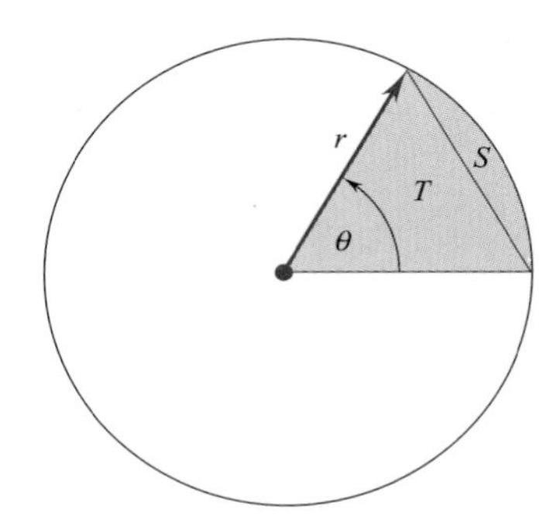

40. Take S and T as in Exercise 39. In general the area of S and the area of T change at different rates. There is, however, one value of θ between 0 and π at which both areas have the same instantaneous rate of change. Find this value of θ.

■ 3.9 DIFFERENTIALS; NEWTON-RAPHSON APPROXIMATIONS

Differentials

Let f be a differentiable function. In Figure 3.9.1, you can see the graph of f and below it the graph of the tangent line at the point $(x, f(x))$. As the figure suggests, for small h, $h \neq 0$, $f(x + h) - f(x)$, the change in f from x to $x + h$, can be approximated by the product $f'(x)h$:

(3.9.1)
$$f(x + h) - f(x) \cong f'(x)h.$$

† The *projection* of $P(x, y)$ onto the x-axis is the point $Q(x, 0)$; the projection of $P(x, y)$ onto the y-axis is the point $R(0, y)$.

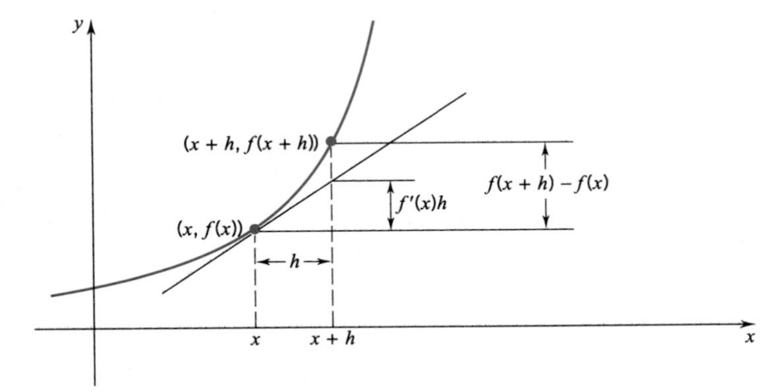

Figure 3.9.1

How good is this approximation? It is good in the sense that, for small h, the difference between the two quantities,

$$[f(x+h)-f(x)]-f'(x)h,$$

is small compared to h. How small? Small enough that the ratio

$$\frac{[f(x+h)-f(x)]-f'(x)h}{h}$$

tends to 0 as h tends to 0:

$$\lim_{h\to 0}\frac{[f(x+h)-f(x)]-f'(x)h}{h}=\lim_{h\to 0}\frac{f(x+h)-f(x)}{h}-\lim_{h\to 0}\frac{f'(x)h}{h}$$
$$=f'(x)-f'(x)=0.$$

The quantities $f(x+h)-f(x)$ and $f'(x)h$ have names:

DEFINITION 3.9.2

Let $h \neq 0$. The difference $f(x+h)-f(x)$ is called the *increment of f from x to $x+h$*, and is denoted by Δf:

$$\Delta f = f(x+h)-f(x).\dagger$$

The product $f'(x)h$ is called the *differential of f at x with increment h*, and is denoted by df:

$$df = f'(x)h.$$

Display (3.9.1) says that, for small h, Δf, and df are approximately equal:

$$\boxed{\Delta f \cong df.}$$

† The symbol Δ is the capital of the Greek letter δ, which corresponds to the English letter d; Δf is read "delta f."

How approximately equal are they? Enough so that the ratio

$$\frac{\Delta f - df}{h}$$

tends to 0 as h tends to 0.

Let's see what all this amounts to in a very simple case. The area of a square of side x is given by the function

$$f(x) = x^2, \qquad x > 0.$$

If the length of each side is increased from x to $x + h$, the area increases from $f(x)$ to $f(x + h)$. The change in area is the increment Δf:

$$\begin{aligned}\Delta f &= f(x + h) - f(x)\\ &= (x + h)^2 - x^2\\ &= (x^2 + 2xh + h^2) - x^2\\ &= 2xh + h^2.\end{aligned}$$

As an estimate for this change we can use the differential

$$df = f'(x)h = 2xh. \qquad \text{(Figure 3.9.2)}$$

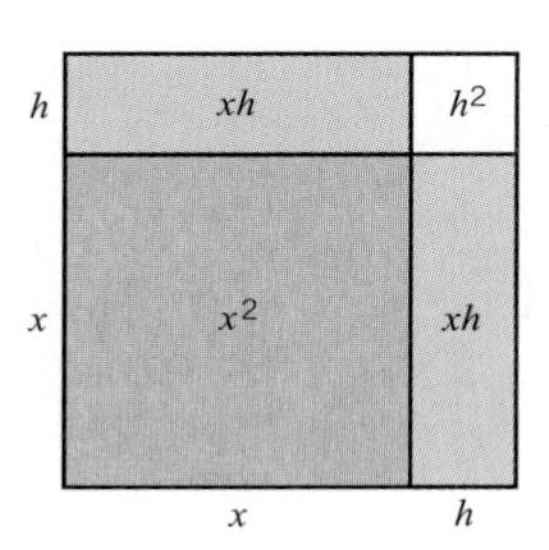

Figure 3.9.2

The error of this estimate, the difference between the actual change Δf and the estimated change df, is the difference

$$\Delta f - df = h^2.$$

As promised, the error is small compared to h in the sense that

$$\frac{\Delta f - df}{h} = \frac{h^2}{h} = h$$

tends to 0 as h tends to 0.

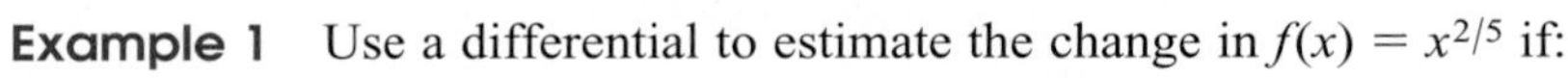

Example 1 Use a differential to estimate the change in $f(x) = x^{2/5}$ if:

(a) x is increased from 32 to 34. **(b)** x is decreased from 1 to $\frac{9}{10}$.

SOLUTION Since $f'(x) = \frac{2}{5}x^{-3/5} = 2/(5x^{3/5})$, we have

$$df = f'(x)h = \frac{2}{5x^{3/5}}h.$$

For part (a) we set $x = 32$ and $h = 2$. The differential then becomes

$$df = \frac{2}{5(32)^{3/5}}(2) = \frac{4}{40} = 0.10.$$

A change in x from 32 to 34 increases the value of f by approximately 0.10. Checking this result with a hand calculator, we have

$$\Delta f = f(34) - f(32) = (34)^{2/5} - (32)^{2/5} \cong 4.0982 - 4 = 0.0982.$$

For part (b) we set $x = 1$ and $h = -\frac{1}{10}$. In this case, the differential is

$$df = \frac{2}{5(1)^{3/5}}\left(-\frac{1}{10}\right) = -\frac{2}{50} = -0.04.$$

A change in x from 1 to $\frac{9}{10}$ decreases the value of f by approximately 0.04. Checking this result with a hand calculator, we have

$$\Delta f = f(0.9) - f(1) = (0.9)^{2/5} - (1)^{2/5} \cong 0.9587 - 1 = -0.0413. \quad \square$$

Example 2 Use a differential to estimate $\sqrt{104}$.

SOLUTION We know $\sqrt{100}$. What we need is an estimate for the increase of

$$f(x) = \sqrt{x}$$

when x increases from 100 to 104. Here

$$f'(x) = \frac{1}{2\sqrt{x}} \qquad \text{and thus} \qquad df = f'(x)h = \frac{h}{2\sqrt{x}}.$$

With $x = 100$ and $h = 4$, df becomes

$$\frac{4}{2\sqrt{100}} = \frac{1}{5} = 0.2.$$

A change in x from 100 to 104 increases the value of the square root by approximately 0.2. It follows that

$$\sqrt{104} \cong \sqrt{100} + 0.2 = 10 + 0.2 = 10.2.$$

As you can check, $(10.2)^2 = 104.04$, so that we are not far off. $\square$

Example 3 Use a differential to approximate $\cos 40°$.

SOLUTION Let $f(x) = \cos x$. We know that $\cos 45° = \cos \pi/4 = \sqrt{2}/2$. Converting 40° to radians, we have

$$40° = 45° - 5° = \frac{\pi}{4} - \left(\frac{\pi}{180}\right)5 = \frac{\pi}{4} - \frac{\pi}{36} \text{ radians.}$$

We will use a differential to estimate the change in $\cos x$ when x decreases from $\pi/4$ to $(\pi/4) - (\pi/36)$. Now,

$$f'(x) = -\sin x \qquad \text{and} \qquad df = f'(x)h = -h \sin x.$$

With $x = \pi/4$ and $h = -\pi/36$, df is given by

$$df = -\left(-\frac{\pi}{36}\right)\sin\left(\frac{\pi}{4}\right) = \frac{\pi}{36}\frac{\sqrt{2}}{2} = \frac{\pi\sqrt{2}}{72} \cong 0.0617.$$

A decrease in x from $\pi/4$ to $(\pi/4) - (\pi/36)$ increases the value of the cosine function by approximately 0.0617. Therefore

$$\cos 40° \cong \cos 45° + 0.0617 \cong 0.7071 + 0.0617 = 0.7688.$$

Checking this result with a hand calculator, we find that $\cos 40° \cong 0.7660$. $\square$

Newton-Raphson Method

Figure 3.9.3 shows the graph of a function f. Since the graph of f crosses the x-axis at $x = c$, the number c is a solution (root) of the equation $f(x) = 0$. In the setup of Figure 3.9.3, we can approximate c as follows: Start at a point x_1 (see the figure). The tangent line at $(x_1, f(x_1))$ intersects the x-axis at a point x_2, which is closer to c than

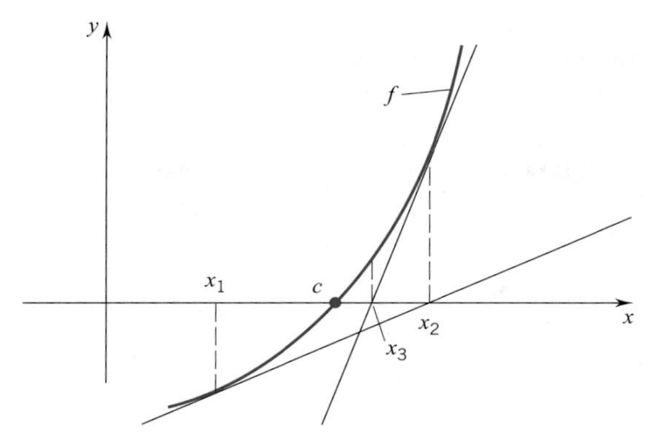

Figure 3.9.3

x_1. The tangent line at $(x_2, f(x_2))$ intersects the x-axis at a point x_3, which is closer to c than x_2. Continuing in this manner, we will obtain better and better approximations $x_4, x_5, \ldots, x_n$ to the root c.

There is an algebraic connection between x_n and x_{n+1} that we now develop. The tangent line at $(x_n, f(x_n))$ has equation

$$y - f(x_n) = f'(x_n)(x - x_n).$$

The x-intercept of this line, x_{n+1}, can be found by setting $y = 0$:

$$0 - f(x_n) = f'(x_n)(x_{n+1} - x_n).$$

Solving this equation for x_{n+1}, we have

(3.9.3)
$$x_{n+1} = x_n - \frac{f(x_n)}{f'(x_n)}.$$

This method for locating a root of an equation $f(x) = 0$ is called the *Newton-Raphson method.* The method does not work in all cases. First, there are some conditions that must be placed on the function f. Clearly, f must be differentiable on some interval that contains the root c. Also, note that if $f'(x_n) = 0$ for some n, then the tangent line at $(x_n, f(x_n))$ is horizontal and the next approximation x_{n+1} cannot be calculated. See Figure 3.9.4. Thus, we assume that $f'(x) \neq 0$ in some interval containing c.

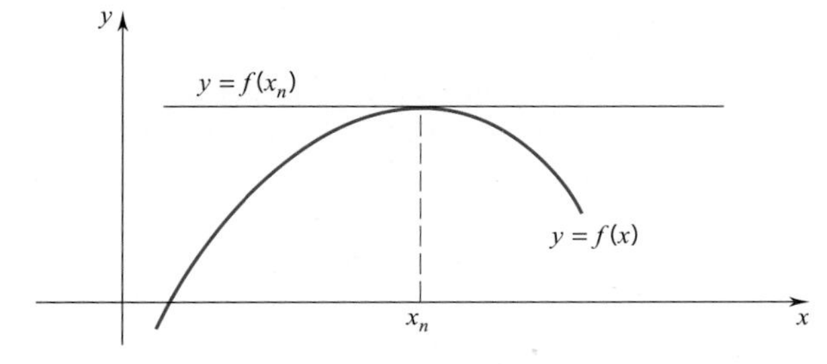

Figure 3.9.4

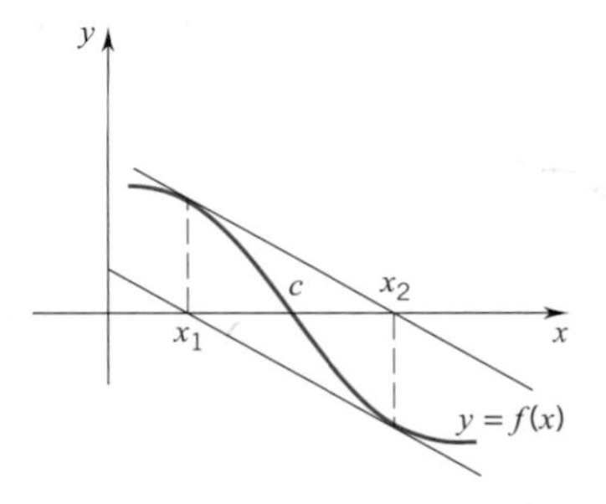

Figure 3.9.5

The method can also fail if proper care is not taken in choosing the first approximation x_1. For example, it can happen that the first approximation x_1 produces a second approximation x_2 which, in turn, gives the same x_1 as the third approximation, and so on—the approximations simply alternate between x_1 and x_2. See Figure 3.9.5. Another type of difficulty can arise if $f'(x_1)$ is close to zero. In this case, the second approximation x_2 could be worse than x_1, the third approximation x_3 could be worse than x_2, and so forth. See Figure 3.9.6.

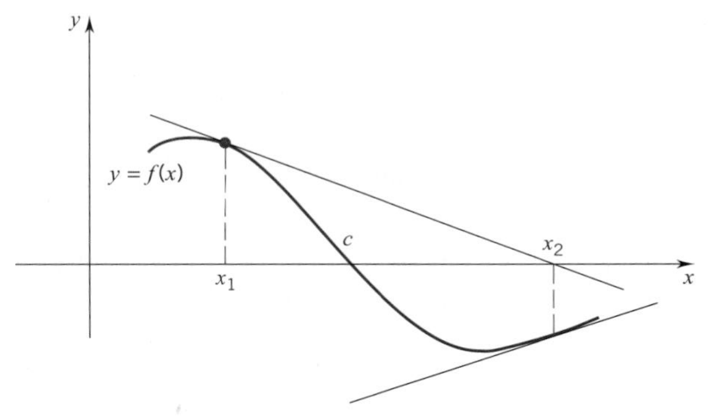

Figure 3.9.6

There is a condition that guarantees that the Newton-Raphson method will work. Suppose that f is twice differentiable and that $f(x)f''(x) > 0$ on the open interval I joining c and x_1. If $f(x) > 0$ on I, then the graph of f is concave up† on I and we have the situation pictured in Figure 3.9.7. On the other hand, if $f(x) < 0$ on I, then the graph of f is concave down on I and we have the situation pictured in Figure 3.9.8. In either case, the sequence of approximations $x_1, x_2, x_3, \ldots$ will "converge" to the root c.

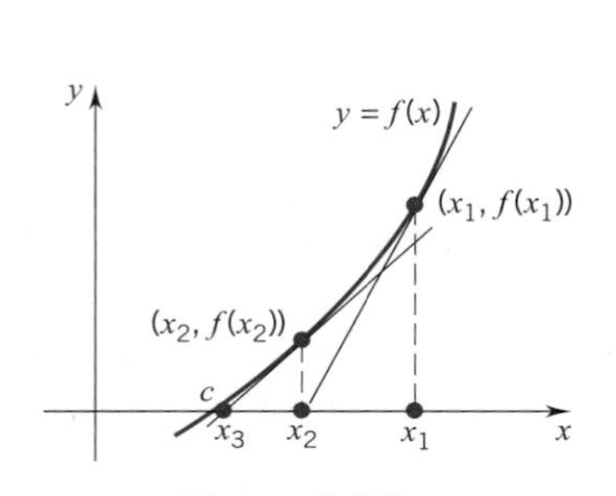

Figure 3.9.7

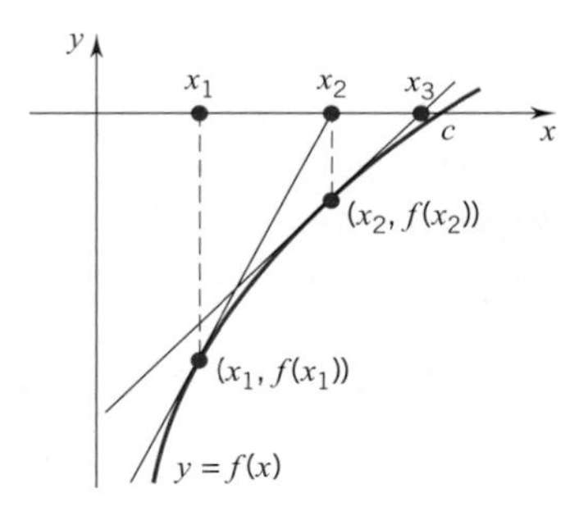

Figure 3.9.8

Example 4 The number $\sqrt{3}$ is a root of the equation $x^2 - 3 = 0$. We will estimate $\sqrt{3}$ by applying the Newton-Raphson method to the function $f(x) = x^2 - 3$ starting at $x_1 = 2$. (As you can check, $f(x)f''(x) > 0$ on $(\sqrt{3}, 2)$ and therefore we can be sure that the method applies.) Since $f'(x) = 2x$, the Newton-Raphson formula gives

$$x_{n+1} = x_n - \left(\frac{x_n^2 - 3}{2x_n}\right) = \frac{x_n^2 + 3}{2x_n}.$$

† The concavity of a graph and the role of the second derivative are treated in Chapter 4.

Successive calculations with this formula (using a calculator) are given in the following table:

n	x_n	$x_{n+1} = \dfrac{x_n^2 + 3}{2x_n}$
1	2	1.75000
2	1.75000	1.73214
3	1.73214	1.73205

Since $(1.73205)^2 = 2.999997$, the method has generated a very accurate estimate of $\sqrt{3}$ in only three steps. ❑

EXERCISES 3.9

1. Use a differential to estimate the change in the volume of a cube caused by an increase h in the length of each side. Interpret geometrically the error of your estimate $\Delta V - dV$.

2. Use a differential to estimate the area of a ring of inner radius r and width h. What is the exact area?

In Exercises 3–10, use differentials to estimate the value of the indicated expression. Then compare your estimate with the result given by a calculator.

3. $\sqrt[3]{1010}$. **4.** $\sqrt{125}$.

5. $\sqrt[4]{15}$. **6.** $1/\sqrt{24}$.

7. $\sqrt[5]{30}$. **8.** $(26)^{2/3}$.

9. $(33)^{3/5}$. **10.** $(33)^{-1/5}$.

In Exercises 11–14, use differentials to estimate the indicated expression. (Remember to convert to radian measure.) Compare your estimate with the result given by a calculator.

11. $\sin 46°$. **12.** $\cos 62°$.

13. $\tan 28°$. **14.** $\sin 43°$.

15. Estimate $f(2.8)$ given that $f(3) = 2$ and $f'(x) = (x^3 + 5)^{1/5}$.

16. Estimate $f(5.4)$ given that $f(5) = 1$ and $f'(x) = \sqrt[3]{x^2 + 2}$.

17. Find the approximate volume of a thin cylindrical sheet with open ends given that the inner radius is r, the height is h, and the thickness is t.

18. The diameter of a steel ball is measured to be 16 centimeters, with a maximum error of 0.3 centimeters. Estimate by differentials the maximum error in (a) the surface area when calculated by the formula $S = 4\pi r^2$; (b) the volume when calculated by the formula $V = \frac{4}{3}\pi r^3$.

19. A box is to be constructed in the form of a cube to hold 1000 cubic feet. Use a differential to estimate how accurately the inner edge must be made so that the volume will be correct to within 3 cubic feet.

20. Use differentials to estimate the values of x for which
(a) $\sqrt{x+1} - \sqrt{x} < 0.01$.
(b) $\sqrt[4]{x+1} - \sqrt[4]{x} < 0.002$.

21. A hemispherical dome with radius 50 feet will be given a coat of paint 0.01 inch thick. The contractor for the job wants to estimate the number of gallons of paint that will be needed. Use differentials to obtain an estimate (there are 231 cubic inches in a gallon). HINT: Approximate the change in the volume of the hemisphere corresponding to an increase of 0.01 inch in the radius.

22. Assume that the earth is a sphere of radius 4000 miles. The volume of ice at the north and south poles is estimated to be 8 million cubic miles. Suppose that this ice melts and the water produced distributes uniformly over the surface of the earth. Estimate the depth of the added water at any point on the earth.

23. For oscillations of small amplitude, the relationship between the period P of one complete oscillation and the length L of a simple pendulum (see Exercise 62, Section 3.6) is given by the equation

$$P = 2\pi\sqrt{\frac{L}{g}},$$

where g is the (constant) acceleration of gravity. Show that a small change dL in the length of a pendulum produces a change dP in the period that satisfies

$$\frac{dP}{P} = \frac{1}{2}\frac{dL}{L}.$$

24. Suppose that the pendulum in a pendulum clock has length 90 centimeters. Use the result in Exercise 23 to determine how the length of the pendulum should be adjusted if the clock is losing 15 seconds per hour?

25. A pendulum has length 3.26 feet and goes through one complete oscillation in 2 seconds. Use Exercise 23 to find the approximate change in P if the pendulum is lengthened by 0.01 foot.

26. As a metal cube is heated, the length of each edge increases $\frac{1}{10}\%$ per degree increase in temperature. Show by differentials that the surface area increases about $\frac{2}{10}\%$ per degree and the volume increases about $\frac{3}{10}\%$ per degree.

27. We are trying to determine the area of a circle by measuring the diameter. How accurately must we measure the diameter if our estimate is to be correct within 1%?

28. Estimate by differentials how precisely x must be determined if (a) x^n is to be accurate within 1%; (b) $x^{1/n}$ is to be accurate within 1%.

In Exercises 29–36, use the Newton-Raphson method to estimate a root of the equation $f(x) = 0$ starting at the indicated value of x: (a) Express x_{n+1} in terms of x_n. (b) Give x_4 rounded off to five decimal places and evaluate f at that approximation.

29. $f(x) = x^2 - 24$; $x_1 = 5$.

30. $f(x) = x^2 - 17$; $x_1 = 5$.

31. $f(x) = x^3 - 25$; $x_1 = 3$.

32. $f(x) = x^5 - 30$; $x_1 = 2$.

33. $f(x) = \cos x - x$; $x_1 = 1$.

34. $f(x) = \sin x - x^2$; $x_1 = 1$.

35. $f(x) = 2 \sin x - x$; $x_1 = 2$.

36. $f(x) = x^3 - 4x + 1$; $x_1 = 2$.

37. The function $f(x) = x^{1/3}$ is 0 at $x = 0$. Show that the Newton-Raphson method applied to f starting at *any* value $x_1 \neq 0$ fails to generate values that approach the solution $x = 0$. Describe the sequence $x_1, x_2, x_3, \ldots$, that the method generates.

38. Let $f(x) = 2x^3 - 3x^2 - 1$.
(a) Prove that the equation $f(x) = 0$ has a root in the interval $(1, 2)$. HINT: Use the intermediate-value theorem.
(b) Show that the Newton-Raphson method with $x_1 = 1$ fails to generate values that approach the root in $(1, 2)$.
(c) Estimate the root by starting at $x_1 = 2$. Determine x_4 rounded off to five decimal places, and evaluate $f(x_4)$.

39. The function $f(x) = x^4 - 2x^2 - \frac{17}{16}$ has two zeros, one at a, where $0 < a < 2$, and the other at $-a$ since f is an even function.
(a) Show that the Newton-Raphson methods fails if $x_1 = \frac{1}{2}$ is used as the initial estimate for finding the root a. Describe the sequence $x_1, x_2, x_3, \ldots$ that the method generates.
(b) Estimate the root a by starting at $x_1 = 2$. Determine x_4 rounded off to five decimal places, and evaluate $f(x_4)$.

40. Let $f(x) = x^2 - a$, where $a > 0$. The roots of the equation $f(x) = 0$ are $\pm\sqrt{a}$.
(a) Show that if $x_1 > 0$ is any initial estimate for $\sqrt{a}$, then the Newton-Raphson method gives the iteration formula

$$x_{n+1} = \frac{1}{2}\left(x_n + \frac{a}{x_n}\right), \qquad n \geq 1.$$

(b) Let $a = 5$. Beginning with $x_1 = 2$, use the formula in part (a) to calculate x_4 rounded off to five decimal places and evaluate $f(x_4)$.

41. Let $f(x) = x^k - a$, where k is a positive integer and $a > 0$. The number $a^{1/k}$ is a root of the equation $f(x) = 0$.
(a) Show that if $x_1 > 0$ is any initial estimate for $a^{1/k}$, then the Newton-Raphson method gives the iteration formula

$$x_{n+1} = \frac{1}{k}\left[(k-1)x_n + \frac{a}{x_n^{k-1}}\right].$$

Note that for $k = 2$, this formula reduces to the formula given in part (a) of Exercise 40.
(b) Use the formula in part (a) to approximate $\sqrt[3]{23}$. Begin with $x_1 = 3$ and calculate x_4 rounded off to five decimal places. Evaluate $f(x_4)$.

Let f be a function that is defined on an interval I. A number $c \in I$ is a *fixed point* of f if $f(c) = c$ (see Exercise 21, Section 2.6). A given function may or may not have a fixed point. If f is differentiable, then the Newton-Raphson method applied to the function $F(x) = f(x) - x$ can be used to approximate the fixed points of f.

42. Let $f(x) = 2x^3 - 4x - 3$.
(a) Show that f has a fixed point on $[1, 2]$. HINT: Let $F(x) = f(x) - x$ and use the intermediate-value theorem.
(b) Use the Newton-Raphson method to approximate the fixed point of f. Begin with $x_1 = 2$ and calculate x_4 rounded off to four decimal places; evaluate $f(x_4)$.

43. Let $f(x) = \frac{1}{2}\cos x$.
(a) Show that f has a fixed point on $[0, \pi/2]$.
(b) Use the Newton-Raphson method to approximate the fixed point of f. Begin with $x_1 = 0$ and calculate x_4 rounded off to four decimal places; evaluate $f(x_4)$.

44. Let $f(x) = \frac{2}{3}\sin x + 1$.
(a) Show that f has a fixed point on $[0, 2]$.
(b) Use the Newton-Raphson method to approximate the fixed point of f. Begin with $x_1 = 2$ and calculate x_4 rounded off to four decimal places; evaluate $f(x_4)$.

Let g be a function defined at least on some open interval containing the number 0. We say that g is of *smaller order than* h, or that $g(h)$ is *little-o*(h), and write $g(h) = o(h)$, iff $g(h)$ is small enough compared with h that

$$\lim_{h\to 0} \frac{g(h)}{h} = 0, \qquad \text{or equivalently,} \qquad \lim_{h\to 0} \frac{g(h)}{|h|} = 0.$$

45. Which of the following statements is true?

(a) $h^3 = o(h)$. (b) $\dfrac{h^2}{h-1} = o(h)$. (c) $h^{1/3} = o(h)$.

46. Show that, if $g(h) = o(h)$, then $\lim_{h\to 0} g(h) = 0$.

47. Show that, if $g_1(h) = o(h)$ and $g_2(h) = o(h)$, then

$$g_1(h) + g_2(h) = o(h) \qquad \text{and} \qquad g_1(h)g_2(h) = o(h).$$

48. The figure to the right shows the graph of a differentiable function f and a line with slope m that passes through the point $(x, f(x))$. The vertical separation at $x + h$ between the line with slope m and the graph of f has been labeled $g(h)$.
(a) Calculate $g(h)$.
(b) Show that, of all lines that pass through $(x, f(x))$, the tangent line is the line that best approximates the graph of f near the point $(x, f(x))$ by showing that

$$g(h) = o(h) \qquad \text{iff} \qquad m = f'(x).$$

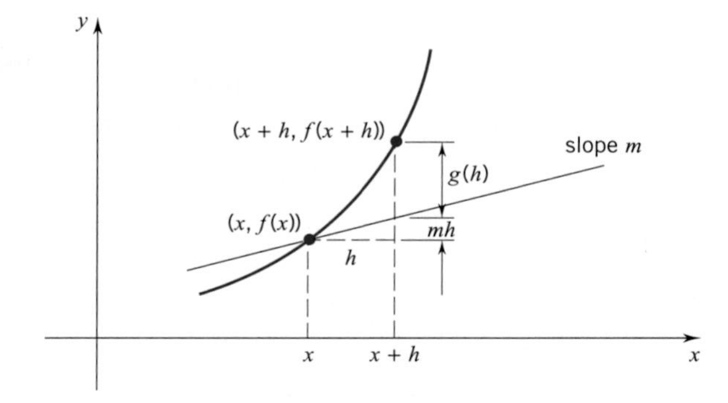

■ CHAPTER HIGHLIGHTS

3.1 The Derivative

derivative of f at x: $\lim_{h\to 0} \dfrac{f(x+h) - f(x)}{h}$ (p. 130)

tangent line (p. 135) normal line (p. 135) vertical tangent (p. 137)

If f is differentiable at x, then f is continuous at x (p. 138). The converse is false.

left-hand derivative of f at x: $\lim_{h\to 0^-} \dfrac{f(x+h) - f(x)}{h}$.

right-hand derivative of f at x: $\lim_{h\to 0^+} \dfrac{f(x+h) - f(x)}{h}$.

alternative definition of the derivative at c: $\lim_{h\to 0} \dfrac{f(x) - f(c)}{x - c}$.

3.2 Some Differentiation Formulas

If $f(x) = \alpha$, α a constant, then $f'(x) = 0$. If $f(x) = x$, then $f'(x) = 1$.

$$(f+g)'(x) = f'(x) + g'(x), \qquad (\alpha f)'(x) = \alpha f'(x),$$

$$(fg)'(x) = f(x)g'(x) + g(x)f'(x), \qquad \left(\frac{f}{g}\right)'(x) = \frac{g(x)f'(x) - f(x)g'(x)}{[g(x)]^2} \qquad (g(x) \neq 0).$$

For any integer n, $p(x) = x^n$ has derivative $p'(x) = nx^{n-1}$.

If $p(x) = a_n x^n + a_{n-1}x^{n-1} + \cdots + a_2x^2 + a_1x + a_0$ is a polynomial, then

$$p'(x) = na_nx^{n-1} + (n-1)a_{n-1}x^{n-2} + \cdots + 2a_2x + a_1.$$

3.3 The *d/dx* Notation; Derivatives of Higher Order

Another notation for the derivative is the *double-d* notation of Leibniz: if $y = f(x)$, then

$$\frac{dy}{dx} = f'(x).$$

Higher derivatives are calculated by repeated differentiation. For example, the second derivative of $y = f(x)$ is the derivative of the derivative:

$$f''(x) = [f'(x)]' \qquad \text{or} \qquad \frac{d^2y}{dx^2} = \frac{d}{dx}\left(\frac{dy}{dx}\right).$$

3.4 The Derivative as a Rate of Change

dy/dx gives the rate of change of y with respect to x

Velocity and Acceleration

velocity, acceleration, speed (p. 161)
free fall (p. 163)

If velocity and acceleration have the same sign, the object is *speeding up.* If the velocity and acceleration have opposite signs, the object is *slowing down.*

Free Fall (Near the Surface of the Earth)

Economics

marginal cost, marginal revenue, marginal profit, break-even points.

3.5 The Chain Rule

Various forms of the chain rule:

$$\frac{dy}{dx} = \frac{dy}{du}\frac{du}{dx}, \qquad \frac{d}{dx}[f(u)] = \frac{d}{du}[f(u)]\frac{du}{dx}, \qquad (f \circ g)'(x) = f'(g(x))g'(x).$$

If u is a differentiable function of x and n is an integer, then

$$\frac{d}{dx}(u^n) = nu^{n-1}\frac{du}{dx}.$$

3.6 Differentiating the Trigonometric Functions

$$\frac{d}{dx}(\sin u) = \cos u\frac{du}{dx}, \qquad \frac{d}{dx}(\cos u) = -\sin u\frac{du}{dx},$$

$$\frac{d}{dx}(\tan u) = \sec^2 u\frac{du}{dx}, \qquad \frac{d}{dx}(\cot u) = -\csc^2 u\frac{du}{dx},$$

$$\frac{d}{dx}(\sec u) = \sec u \tan u\frac{du}{dx}, \qquad \frac{d}{dx}(\csc u) = -\csc u \cot u\frac{du}{dx}.$$

3.7 Implicit Differentiation; Rational Powers

finding dy/dx from an equation in x and y without first solving the equation for y

If u is a differentiable function of x and r is a rational number, then

$$\frac{d}{dx}(u^r) = ru^{r-1}\frac{du}{dx}.$$

3.8 Rates of Change per Unit Time

If a quantity Q varies with time, then the derivative dQ/dt gives the rate of change of that quantity with respect to time.

A 5-step procedure for solving *related rate problems* is outlined on p. 194.

angular velocity; uniform circular motion.

3.9 Differentials; Newton-Raphson Approximations

increment: $\Delta f = f(x + h) - f(x)$ differential: $df = f'(x)h$

$\Delta f \cong df$ in the sense that $\dfrac{\Delta f - df}{h}$ tends to 0 as $h \to 0$

Newton-Raphson method (p. 202).

■ PROJECTS AND EXPLORATIONS USING TECHNOLOGY

To do these exercises you will need a graphics calculator or a computer with graphing capability. The majority of these problems are open-ended so different approaches may be used to solve them. You should be aware that different approaches can result in slight variations in the answers. Round your numerical answers to at least four decimal places. The rounding method that your calculator or computer uses also may cause variations in answers.

3.1 When designing the curved shape of a car fender, the following type of problem must be solved: Find two cubic polynomials p and q, with p defined on $[1, 3]$, q defined on $[3, 9]$, and such that:

$$p(1) = 3, \quad p(3) = 7, \quad q(3) = 7, \quad q(9) = -2, \quad p'(3) = q'(3),$$
$$p''(3) = q''(3), \quad p''(1) = 0 \quad \text{and} \quad q''(9) = 0.$$

The conditions $p(1) = 3$, $p(3) = 7$, $q(3) = 7$, and $q(9) = -2$ state that the curved fender passes through the points $(1, 3)$, $(3, 7)$, and $(9, -2)$. The condition $p'(3) = q'(3)$ ensures that curve is "smooth" at the point where the graphs of p and q meet. The condition $p''(3) = q''(3)$ is also a smoothness condition and, further, it ensures that certain forces on the fender will not be troublesome. The last conditions $p''(1) = 0$ and $q''(9) = 0$ help in the manufacture of the fender; they imply that the fender will be "straight" at the ends. A curve that satisfies these conditons is called a *cubic spline.*

(a) Let $p(x) = ax^3 + bx^2 + cx + d$ and $q(x) = \alpha x^3 + \beta x^2 + \gamma x + \delta$. Write the system of linear equations generated by the specified conditions and find the solutions of the system.
(b) Draw the graphs of p and q on their respective intervals and verify visually that the resulting curve on $[1, 9]$ is continuous and smooth.
(c) Let

$$F(x) = \begin{cases} p(x), & 1 \le x \le 3 \\ q(x), & 3 < x \le 9. \end{cases}$$

Show that F, F', and F'' are continuous on $[1, 9]$.

3.2 Let $g(a)$ be the largest (rightmost) zero of

$$f_a(x) = 20x^3 - 40x^2 + 25x - 5 + a.$$

For example, the graph of $f_0(x)$ is shown in the figure and $g(0) = 1$.

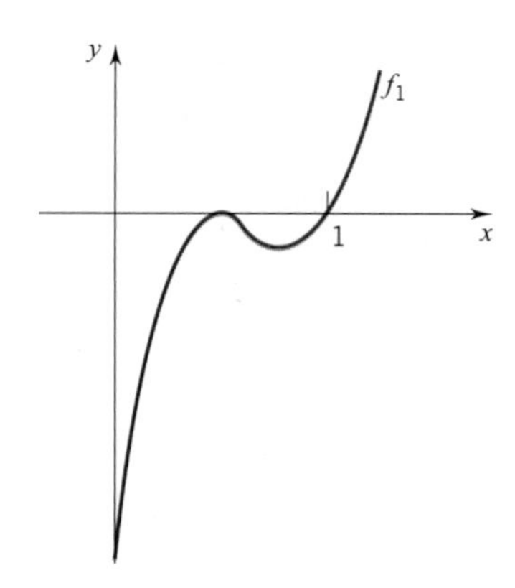

(a) Draw the graphs of f for $a = 0.1, 0.2, 0.3, \ldots, 0.9, 1$. Use Newton-Raphson iteration to approximate the values of $g(a)$ for these values of a.
(b) The function g has a discontinuity on $[0, 1]$. Based on your results in part (a), find the point of discontinuity.
(c) Using difference quotients, approximate the graph of g' on $[0, 1]$.
(d) By calculating additional values of g near the point of discontinuity, determine whether the left- and right-hand limits of g' exist at the point of discontinuity.
(e) Use difference quotients on g' to approximate g'' on $[0, 1]$.
(f) Make a table of values of g' and $f' \circ g$. Can you determine a relation between these two functions?

3.3 Suppose that we want to draw the graph of the equation

$$(x^2 + y^2 + 3x)^2 = 2(x^2 + y^2).$$

(a) Use implicit differentiation to find dy/dx.
(b) Assign some specific values to y, and for each such value graph the function

$$f(x) = (x^2 + y^2 + 3x)^2 - 2(x^2 + y^2).$$

Then record the approximate values of the zeros of f.
(c) Assign some specific values to x, and for each such value graph the function

$$g(y) = (x^2 + y^2 + 3x)^2 - 2(x^2 + y^2).$$

Then record the approximate values of the zeros of g.
(d) Use the information from parts (b) and (c) to find values a, b, c, and d such that the graph of the equation lies inside the rectangle R given by

$$R\colon a \leq x \leq b, \quad c \leq y \leq d.$$

(e) In how many points can a vertical line intersect the graph of the equation? In how many points can a horizontal line intersect the graph?
(f) Use Newton-Raphson iteration and the values from parts (b) and (c) to calculate and plot some points on the graph.
(g) Evaluate dy/dx at each of the points calculated in part (f) and sketch a short segment of the tangent line at each of these points. Use this information to sketch the graph of the equation (if necessary find additional points and tangent lines).
(h) Use a graphing utility to graph the equation and compare with your sketch from part (g).

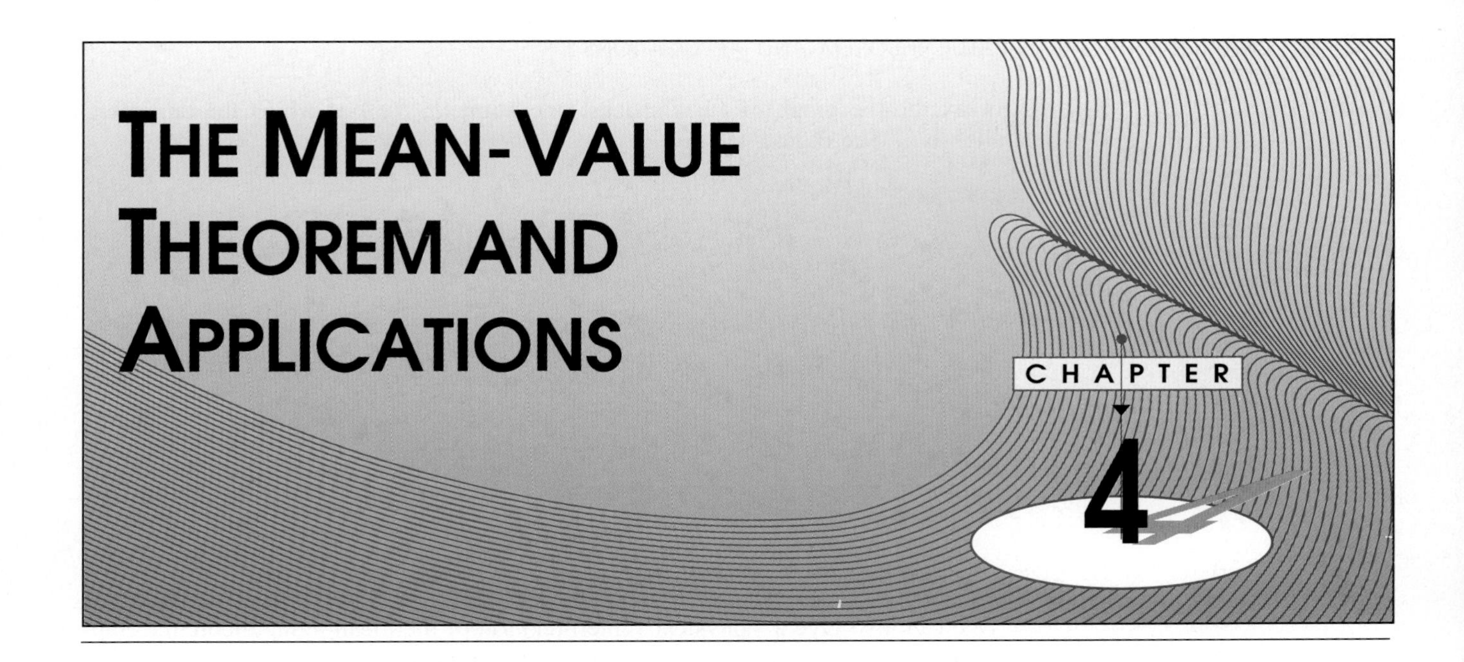

4.1 THE MEAN-VALUE THEOREM

In this section we prove a result known as *the mean-value theorem.* First stated by the French mathematician Joseph-Louis Lagrange† (1736–1813), this theorem has come to permeate the theoretical structure of all calculus.

THEOREM 4.1.1 THE MEAN-VALUE THEOREM

If f is differentiable on the open interval (a, b) and continuous on the closed interval $[a, b]$ then there is at least one number c in (a, b) for which

$$f'(c) = \frac{f(b) - f(a)}{b - a}$$

or, equivalently,

$$f(b) - f(a) = f'(c)(b - a).$$

The number

$$\frac{f(b) - f(a)}{b - a}$$

is the slope of the line l that passes through the points $(a, f(a))$ and $(b, f(b))$. To say that there is at least one number c for which

$$f'(c) = \frac{f(b) - f(a)}{b - a}$$

† Lagrange, whom you encountered earlier in connection with the prime notation for differentiation, was born in Turin, Italy. He spent twenty years as mathematician in residence at the court of Frederick the Great. Later he taught at the renowned École Polytechnique in France.

is to say the the graph of f has at least one point $(c, f(c))$ at which the tangent is parallel to l. See Figure 4.1.1.

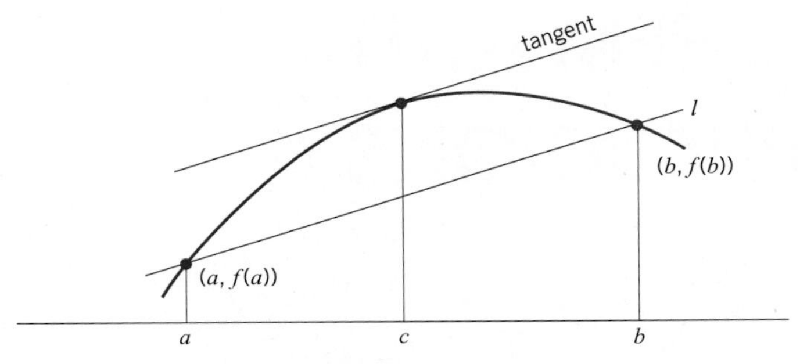

Figure 4.1.1

We can also give a "physical" interpretation of the mean-value theorem. Suppose that an object is moving in one direction along a straight line and that, during a certain time interval $t = a$ to $t = b$, its position at time t is denoted by $x(t)$ (see Section 3.4). Assume that the position function $x(t)$ is continuous on $[a, b]$ and differentiable on (a, b). Now, $x(b) - x(a)$ is the distance that the object travels during the time interval $[a, b]$ and

$$\frac{x(b) - x(a)}{b - a}$$

is the average velocity of the object over this interval. By the mean-value theorem, there is at least one time $t = c$ for which

$$x'(c) = \frac{x(b) - x(a)}{b - a}.$$

But, $x'(c) = v(c)$ is the instantaneous velocity of the object at $t = c$. Thus, the mean-value theorem tells us that there is at least one time $t = c$ at which the velocity at that instant equals the average velocity over the whole interval. Think about this when you are driving your car. For example, if you drive 240 miles in 4 hours, then your average velocity is 60 miles per hour and, by the mean-value theorem, your speedometer must have registered 60 miles per hour at least once during your trip.

We will prove the mean-value theorem in steps. First we will show that if a function f has a nonzero derivative at some point x_0, then, close to x_0, $f(x)$ is greater than $f(x_0)$ on one side of x_0 and less than $f(x_0)$ on the other side of x_0.

THEOREM 4.1.2

Let f be differentiable at x_0. If $f'(x_0) > 0$, then

$$f(x_0 - h) < f(x_0) < f(x_0 + h)$$

for all positive h sufficiently small. If $f'(x_0) < 0$, then

$$f(x_0 + h) < f(x_0) < f(x_0 - h)$$

for all positive h sufficiently small.

PROOF We take the case $f'(x_0) > 0$ and leave the other case to you. By the definition of the derivative,

$$\lim_{k \to 0} \frac{f(x_0 + k) - f(x_0)}{k} = f'(x_0).$$

With $f'(x_0) > 0$ we can use $f'(x_0)$ itself as ϵ and conclude that there exists $\delta > 0$ such that

$$\text{if} \quad 0 < |k| < \delta, \qquad \text{then} \quad \left| \frac{f(x_0 + k) - f(x_0)}{k} - f'(x_0) \right| < f'(x_0).$$

For such k we have

$$-f'(x_0) < \frac{f(x_0 + k) - f(x_0)}{k} - f'(x_0) < f'(x_0)$$

and it follows that

$$\frac{f(x_0 + k) - f(x_0)}{k} > 0. \qquad \text{(why?)}$$

Now let $0 < h < \delta$. Replacing k with h, we have

$$\frac{f(x_0 + h) - f(x_0)}{h} > 0 \qquad \text{and so} \qquad f(x_0) < f(x_0 + h).$$

Replacing k with $-h$, we have

$$\frac{f(x_0 - h) - f(x_0)}{-h} > 0 \qquad \text{and so} \qquad f(x_0 - h) < f(x_0). \quad \square$$

Next we prove a special case of the mean-value theorem, known as Rolle's theorem (after the French mathematician Michel Rolle (1652–1719) who first announced the result in 1691). In Rolle's theorem we make the additional assumption that $f(a)$ and $f(b)$ are both 0. (See Figure 4.1.2.) In this case the line joining $(a, f(a))$ to $(b, f(b))$ is horizontal. (It is the x-axis.) The conclusion is that there is a point $(c, f(c))$ at which the tangent is horizontal.

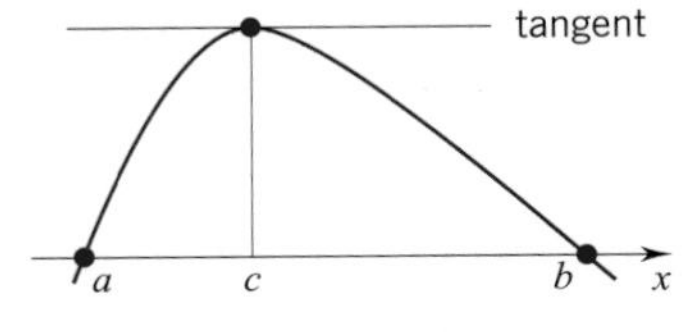

Figure 4.1.2

> **THEOREM 4.1.3 ROLLE'S THEOREM**
>
> Let f be differentiable on the open interval (a, b) and continuous on the closed interval $[a, b]$. If $f(a)$ and $f(b)$ are both 0, then there is at least one number c in (a, b) for which
>
> $$f'(c) = 0.$$

PROOF If f is constantly 0 on $[a, b]$, the result is obvious. If f is not constantly 0 on $[a, b]$, then either f takes on some positive values or some negative values. We assume the former and leave the other case to you.

Since f is continuous on $[a, b]$, f must take on a maximum value at some point c of $[a, b]$. (Theorem 2.6.2.) This maximum value, $f(c)$, must be positive. Since $f(a)$ and $f(b)$ are both 0, c cannot be a and it cannot be b. This means that c must lie in the open interval (a, b) and therefore $f'(c)$ exists. Now $f'(c)$ cannot be greater than 0 and it cannot be less than 0 because each of these conditions would imply that f takes on values greater than $f(c)$. (This follows from Theorem 4.1.2.) We conclude therefore that $f'(c) = 0$. ❑

Remark Rolle's theorem is sometimes stated as:

> Let g be differentiable on the open interval (a, b) and continuous on the closed interval $[a, b]$. If $g(a) = g(b)$, then there is at least one number c in (a, b) for which
> $$g'(c) = 0.$$

In Exercise 38, you are asked to show that this version of Rolle's theorem is an immediate consequence of the given version. ❑

Rolle's theorem has some useful applications independent of the mean-value theorem.

Example 1 Show that the polynomial function $p(x) = 2x^3 + 5x - 1$ has exactly one real zero.

SOLUTION From Exercise 25, Section 2.6, we know that p has at least one real zero (by the Fundamental Theorem of Algebra [see Section 1.2], p has at most three real zeros). Suppose that p has more than one real zero. In particular, suppose that $p(a) = p(b) = 0$, where a and b are real numbers and $a \neq b$. Without loss of generality, assume that $a < b$. Since a polynomial is differentiable everywhere, p is differentiable on (a, b) and continuous on $[a, b]$. Thus, by Rolle's theorem, there is a number c in (a, b) such that $p'(c) = 0$. But

$$p'(x) = 6x^2 + 5 \geq 5 \qquad \text{for all} \quad x,$$

and so $p'(x) \neq 0$ for all x. Thus, we have a contradiction and we can conclude that p has exactly one real zero. ❑

We are now ready to give a proof of the mean-value theorem.

PROOF OF THE MEAN-VALUE THEOREM. We will create a function g that satisfies the conditions of Rolle's theorem and is so related to f that the conclusion $g'(c) = 0$ leads to the conclusion

$$f'(c) = \frac{f(b) - f(a)}{b - a}.$$

It is not hard to see that

$$g(x) = f(x) - \left[\frac{f(b) - f(a)}{b - a}(x - a) + f(a)\right]$$

is exactly such a function. Geometrically $g(x)$ is represented in Figure 4.1.3. The line that passes through $(a, f(a))$ and $(b, f(b))$ has the equation

$$y = \frac{f(b) - f(a)}{b - a}(x - a) + f(a).$$

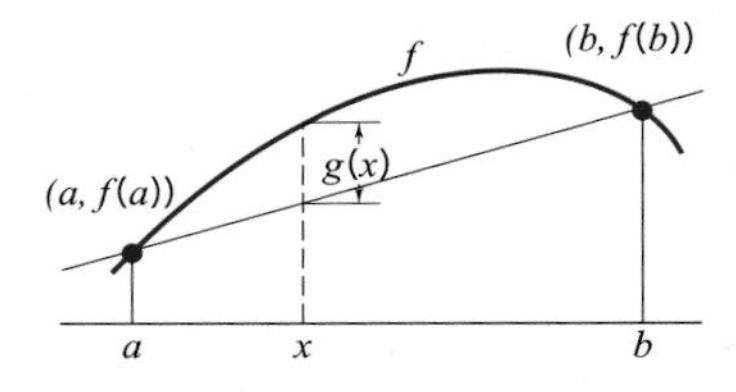

Figure 4.1.3

[This is not hard to verify. The slope is right, and, when $x = a$, $y = f(a)$.] The difference

$$g(x) = f(x) - \left[\frac{f(b) - f(a)}{b - a}(x - a) + f(a)\right]$$

is simply the vertical separation between the graph of f and the line in question.

If f is differentiable on (a, b) and continuous on $[a, b]$, then so is g. As you can check, $g(a)$ and $g(b)$ are both 0. Therefore, by Rolle's theorem, there is at least one number c in (a, b) for which $g'(c) = 0$. Since in general

$$g'(x) = f'(x) - \frac{f(b) - f(a)}{b - a},$$

in particular

$$g'(c) = f'(c) - \frac{f(b) - f(a)}{b - a}.$$

With $g'(c) = 0$, we have

$$f'(c) = \frac{f(b) - f(a)}{b - a}. \quad \square$$

The hypothesis of the mean-value theorem requires f to be differentiable on the open interval (a, b) and continuous on the closed interval $[a, b]$. As you would expect, if either of these conditions fails to hold, then the conclusion of the theorem may not hold. This is illustrated in the Exercises.

EXERCISES 4.1

In Exercises 1–4, show that f satisfies the conditions of Rolle's theorem on the indicated interval and find all numbers c on the interval such that $f'(c) = 0$.

1. $f(x) = x^3 - x$; $[0, 1]$.

2. $f(x) = x^4 - 2x^2 - 8$; $[-2, 2]$.

3. $f(x) = \sin 2x$; $[0, 2\pi]$.

4. $f(x) = x^{2/3} - 2x^{1/3}$; $[0, 8]$.

In Exercises 5–10, verify that f satisfies the conditions of the mean-value theorem on the indicated interval and find all numbers c that satisfy the conclusion of the theorem.

5. $f(x) = x^2$; $[1, 2]$.

6. $f(x) = 3\sqrt{x} - 4x$; $[1, 4]$.

7. $f(x) = x^3$; $[1, 3]$.

8. $f(x) = x^{2/3}$; $[1, 8]$.

9. $f(x) = \sqrt{1 - x^2}$; $[0, 1]$.

10. $f(x) = x^3 - 3x$; $[-1, 1]$.

11. Determine whether the function $f(x) = \sqrt{1 - x^2}/(3 + x^2)$ satisfies the conditions of Rolle's theorem on the interval $[-1, 1]$. If so, find the values of c such that $f(c) = 0$.

12. The function $f(x) = x^{2/3} - 1$ takes on the value 0 at $x = -1$ and at $x = 1$.
(a) Show that $f'(x) \neq 0$ for all $x \in (-1, 1)$.
(b) Explain why this function does not contradict Rolle's theorem.

13. Does there exist a differentiable function f that satisfies $f(0) = 2$, $f(2) = 5$, and $f'(x) \leq 1$ on $(0, 2)$? If not, why not?

14. Does there exist a differentiable function f that has the value 1 only when $x = 0$, 2, and 3, and $f'(x) = 0$ only when $x = -1$, 3/4, and 3/2? If not, why not?

15. Sketch the graph of

$$f(x) = \begin{cases} 2x + 2, & x \le -1 \\ x^3 - x, & x > -1 \end{cases}$$

and compute the derivative. Determine whether f satisfies the conditions of the mean-value theorem on the interval $[-3, 2]$ and, if so, find the values of c guaranteed by the theorem.

16. Sketch the graph of

$$f(x) = \begin{cases} 2 + x^3, & x \le 1 \\ 3x, & x > 1 \end{cases}$$

and compute the derivative. Determine whether f satisfies the conditions of the mean-value theorem on the interval $[-1, 2]$ and, if so, find the values of c guaranteed by the theorem.

17. Consider the quadratic function $f(x) = Ax^2 + Bx + C$. Show that, for any interval $[a, b]$, the value of c that satisfies the conclusion of the mean-value theorem is $(a + b)/2$, the midpoint of the interval.

18. Set $f(x) = x^{-1}$, $a = -1$, $b = 1$. Verify that there is no number c for which

$$f'(c) = \frac{f(b) - f(a)}{b - a}.$$

Explain how this does not violate the mean-value theorem.

19. Exercise 18 with $f(x) = |x|$.

20. Graph the function $f(x) = |2x - 1| - 3$ and compute the derivative. Verify that $f(-1) = 0 = f(2)$ and yet $f'(x)$ is never 0. Explain how this does not violate Rolle's theorem.

21. Show that the equation $6x^4 - 7x + 1 = 0$ does not have more than two distinct real roots. (Use Rolle's theorem.)

22. Show that the equation $6x^5 + 13x + 1 = 0$ has exactly one real root. (Use Rolle's theorem and the intermediate-value theorem.)

23. Show that the equation $x^3 + 9x^2 + 33x - 8 = 0$ has exactly one real root.

24. (a) Let f be differentiable on (a, b). Prove that if $f'(x) \neq 0$ for all $x \in (a, b)$, then f has at most one zero in (a, b).
(b) Let f be twice differentiable on (a, b). Prove that if $f''(x) \neq 0$ for all $x \in (a, b)$, then f has at most two zeros in (a, b).

25. Let $P(x) = a_n x^n + \cdots + a_1 x + a_0$ be a nonconstant polynomial. Show that between any two consecutive roots of the equation $P'(x) = 0$ there is at most one root of the equation $P(x) = 0$.

26. Let f be twice differentiable. Show that, if the equation $f(x) = 0$ has n distinct real roots, then the equation $f'(x) = 0$ has at least $n - 1$ distinct real roots and the equation $f''(x) = 0$ has at least $n - 2$ distinct real roots.

27. Recall that a number a is a fixed point of a function f if $f(a) = a$ (see Exercises 3.9). Prove that if f is differentiable on an interval I and $f'(x) < 1$ for all $x \in I$, then f has at most one fixed point in I. HINT: Assume that f has two fixed points in I and let $g(x) = f(x) - x$.

28. Show that the equation $x^3 + ax + b = 0$ has exactly one real root if $a \ge 0$ and at most one real root between $-\frac{1}{3}\sqrt{3}|a|$ and $\frac{1}{3}\sqrt{3}|a|$ if $a < 0$.

29. Given that $|f'(x)| \le 1$ for all real numbers x, show that $|f(x_1) - f(x_2)| \le |x_1 - x_2|$ for all real numbers x_1 and x_2.

30. Let f be differentiable on an open interval I. Prove that, if $f'(x) = 0$ for all x in I, then f is constant on I.

31. Let f be differentiable on (a, b) with $f(a) = f(b) = 0$ and $f'(c) = 0$ for some c in (a, b). Show by example that f need not be continuous on $[a, b]$.

32. Prove that for all real x and y
(a) $|\cos x - \cos y| \le |x - y|$.
(b) $|\sin x - \sin y| \le |x - y|$.

33. Let f be differentiable on (a, b) and continuous on $[a, b]$.
(a) Prove that if there is a constant M such that $f'(x) \le M$ for all $x \in (a, b)$, then

$$f(b) \le f(a) + M(b - a).$$

(b) Prove that if there is a constant m such that $f'(x) \ge m$ for all $x \in (a, b)$, then

$$f(b) \ge f(a) + m(b - a).$$

(c) Parts (a) and (b) imply that if there exists a constant L such that $|f'(x)| \le L$ on (a, b), then

$$f(a) - L(b - a) \le f(b) \le f(a) + L(b - a).$$

Verify this result.

34. Let f and g be differentiable functions that satisfy the condition $f(x)g'(x) - g(x)f'(x) \neq 0$ for all x in some interval I. Suppose that $f(a) = f(b) = 0$, where $a, b \in I$, $a < b$, and $f(x) \neq 0$ for all $x \in (a, b)$. Prove that if $g(a) \neq 0$ and $g(b) \neq 0$, then $g(x) = 0$ for exactly one $x \in (a, b)$. HINT: Suppose $g(x) \neq 0$ for all $x \in (a, b)$ and consider $h = f/g$. Then reverse the roles of f and g.

35. Let $f(x) = \cos x$ and $g(x) = \sin x$ on $I = (-\infty, \infty)$. Prove that the zeros of f and g separate each other on I; that is, prove that between two consecutive zeros of $\cos x$ there is exactly one zero of $\sin x$ and conversely. HINT: Use Exercise 34.

36. (*Important*) Use the mean-value theorem to show that, if f is continuous at x and $x + h$ and differentiable in between, then

$$f(x + h) - f(x) = f'(x + \theta h)h$$

for some number θ between 0 and 1. (In some texts this is how the mean-value theorem is stated.)

37. Let $h > 0$. Suppose that f is continuous on $[x_0 - h, x_0 + h]$ and differentiable on $(x_0 - h, x_0) \cup (x_0, x_0 + h)$. Show that if

$$\lim_{x \to x_0^-} f'(x) = \lim_{x \to x_0^+} f'(x) = L,$$

then f is differentiable at x_0 and $f'(x_0) = L$. HINT: Use Exercise 36.

38. Let f be continuous on $[a, b]$ and differentiable on (a, b). Without assuming the mean-value theorem, prove that if $f(a) = f(b)$, then there is at least one number $c \in (a, b)$ for which $f'(c) = 0$. HINT: Suppose $f(a) = f(b) = k$ and let $g(x) = f(x) - k$.

39. An object moves along a coordinate line, its position at time t given by $f(t)$ where f is a differentiable function. (a) What does Rolle's theorem say about the motion? (b) What does the mean-value theorem say about the motion?

40. A certain tollway is 120 miles long and the speed limit is 65 miles per hour. If a driver's entry ticket at one end of the tollway is stamped 12 noon and she exits at the other end at 1:45 P.M., should she be given a speeding ticket? Explain.

41. The results of an investigation of a car accident showed that the driver applied his brakes and skidded 280 feet in 6 seconds. If the speed limit on the street where the accident occurred was 30 miles per hour, was the driver exceeding the speed limit at the instant he applied his brakes? Explain. HINT: 30 miles per hour = 44 feet per second.

42. Let f be a differentiable function. Apply the mean-value theorem to f with $a = x$ and $b = x + \Delta x$ to show that

$$\Delta f = f(x + \Delta x) - f(x) = f'(c)\Delta x,$$

where $x < c < x + \Delta x$. Compare this result with the result in Section 3.9. If f' is continuous and Δx is small, then $f'(c) \cong f'(x)$ and we have

$$\Delta f \cong f'(x)\Delta x$$

which is the result in Section 3.9.

43. Use the mean-value theorem to estimate $\sqrt{65}$. HINT: Let $f(x) = \sqrt{x}$, $a = 64$, $b = 65$, and use Exercise 42.

44. Let $f(x) = x^3 - 3x^2 + 6x - 12$.
(a) Show that the equation $f(x) = 0$ has exactly one solution.
(b) Find an interval that contains the solution.
(c) Use the Newton-Raphson method to approximate the solution. Calculate x_3 and round off to four decimal places.

45. Repeat Exercise 44 for the function $f(x) = 1 + 4x - 2 \cos x$.

In Exercises 46–47, show that the given function satisfies the hypotheses of Rolle's theorem on the indicated interval. Use a graphing utility to graph f' and estimate the number(s) c where $f'(c) = 0$. Round off your estimates to three decimal places.

46. $f(x) = 2x^3 + 3x^2 - 3x - 2$; $[-2, 1]$.

47. $f(x) = 1 - x^3 - \cos(\pi x/2)$; $[0, 1]$.

Suppose that the function f satisfies the hypotheses of the mean-value theorem on an interval $[a, b]$. We can find the numbers c that satisfy the conclusion of the mean-value theorem by finding the zeros of the function g given by

$$g(x) = f'(x) - \frac{f(b) - f(a)}{b - a}.$$

In Exercises 48–50 use a graphing utility to graph the function g corresponding to the given function f on the indicated interval. Estimate the zeros of g accurate to three decimal places. For each zero c in the interval, graph the tangent line to the graph of f at $(c, f(c))$, and graph the line through $(a, f(a))$ and $(b, f(b))$. Verify that they are parallel.

48. $f(x) = x^2 + 1/x$; $[2, 4]$.

49. $f(x) = x^4 - 7x^2 + 2$; $[1, 3]$.

50. $f(x) = x \cos x + 4 \sin x$; $[-\pi/2, \pi/2]$.

■ 4.2 INCREASING AND DECREASING FUNCTIONS

To place our discussion on a solid footing we begin with a definition.

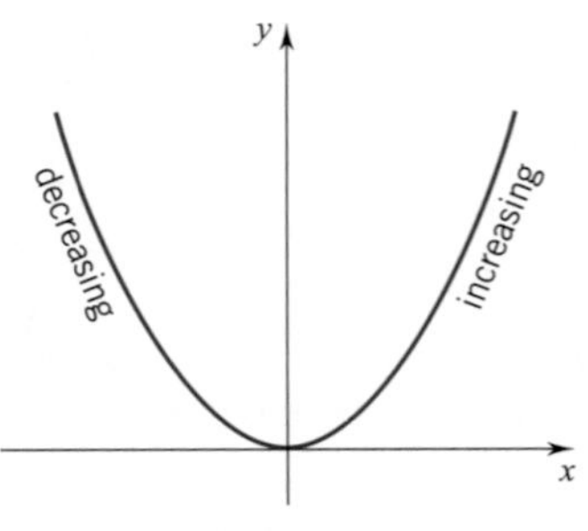

Figure 4.2.1

DEFINITION 4.2.1

A function f is said to:

(i) *increase* on the interval I iff for every two numbers x_1, x_2 in I,

$$x_1 < x_2 \quad \text{implies} \quad f(x_1) < f(x_2);$$

(ii) *decrease* on the interval I iff for every two numbers x_1, x_2 in I,

$$x_1 < x_2 \quad \text{implies} \quad f(x_1) > f(x_2).$$

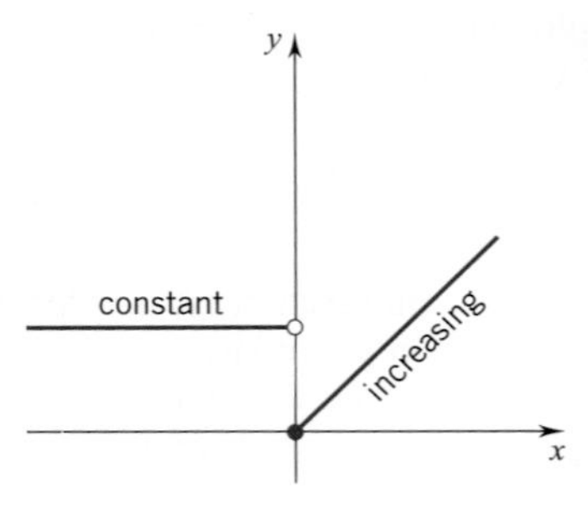

Figure 4.2.2

Preliminary Examples

(a) The quadratic function

$$f(x) = x^2 \qquad \text{(Figure 4.2.1)}$$

decreases on $(-\infty, 0]$ and increases on $[0, \infty)$.

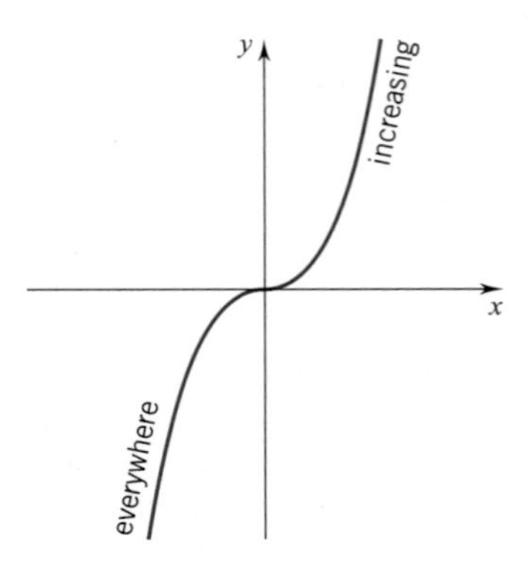

Figure 4.2.3

(b) The function

$$f(x) = \begin{cases} 1, & x < 0 \\ x, & x \geq 0 \end{cases} \qquad \text{(Figure 4.2.2)}$$

is constant on $(-\infty, 0)$, it neither increases nor decreases; f increases on $[0, \infty)$.

(c) The cubic function

$$f(x) = x^3 \qquad \text{(Figure 4.2.3)}$$

is everywhere increasing.

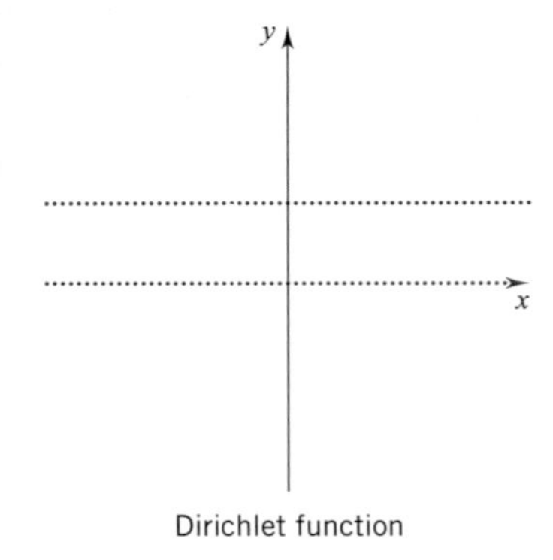

Dirichlet function

Figure 4.2.4

(d) In the case of the Dirichlet function

$$g(x) = \begin{cases} 1, & x \text{ rational} \\ 0, & x \text{ irrational} \end{cases} \qquad \text{(Figure 4.2.4)}$$

there is no interval on which the function increases and no interval on which the function decreases. On every interval the function jumps back and forth between 0 and 1 an infinite number of times. ❑

If f is a differentiable function, then we can determine the intervals on which f increases and the intervals on which f decreases by examining the sign of the first derivative.

THEOREM 4.2.2

Let f be differentiable on an open interval I.

(i) If $f'(x) > 0$ for all x in I, then f increases on I.

(ii) If $f'(x) < 0$ for all x in I, then f decreases on I.

(iii) If $f'(x) = 0$ for all x in I, then f is constant on I.

PROOF Choose any two points x_1 and x_2 in I with $x_1 < x_2$. Since f is differentiable on I, it is continuous on I. Therefore we know that f is differentiable on (x_1, x_2) and continuous on $[x_1, x_2]$. By the mean-value theorem there is a number c in (x_1, x_2) for which

$$f'(c) = \frac{f(x_2) - f(x_1)}{x_2 - x_1}.$$

In (i), $f'(c) > 0$ and we have

$$\frac{f(x_2) - f(x_1)}{x_2 - x_1} > 0.$$

Thus $f(x_1) < f(x_2)$. In (ii), $f'(c) < 0$ and we have

$$\frac{f(x_2) - f(x_1)}{x_2 - x_1} < 0.$$

Thus $f(x_1) > f(x_2)$. In (iii), $f'(c) = 0$ and we have

$$\frac{f(x_2) - f(x_1)}{x_2 - x_1} = 0.$$

Thus $f(x_1) = f(x_2)$. ❑

Remark In Section 3.2 we showed that if $f(x) = \alpha$, constant, on an open interval I, then $f'(x) = 0$ for all $x \in I$. Part (iii) of Theorem 4.2.2 gives the converse: if $f'(x) = 0$ for all x in an open interval I, then $f(x) = \alpha$ for some constant α. Combining these two statements, we have:

Suppose that f is differentiable on an open interval I. Then

$f(x) = \alpha$ (α a constant) for all $x \in I$ iff

$f'(x) = 0$ for all $x \in I$.

❑

Theorem 4.2.2 is useful, but it doesn't tell the complete story. Look, for example, at the function $f(x) = x^2$. The derivative $f'(x) = 2x$ is negative for x in $(-\infty, 0)$, zero at $x = 0$, and positive for x in $(0, \infty)$. Theorem 4.2.2 assures us that

f decreases on $(-\infty, 0)$ and increases on $(0, \infty)$, but actually

$$f \text{ decreases on } (-\infty, 0] \text{ and increases on } [0, \infty).$$

To get these stronger results we need a theorem that works for closed intervals too.

To extend Theorem 4.2.2 so that it works for an arbitrary interval I, the only additional condition we need is continuity at the endpoint(s).

THEOREM 4.2.3

Let f be continuous on an arbitrary interval I and differentiable on the interior of I.

(i) If $f'(x) > 0$ for all x in the interior of I, then f increases on all of I.
(ii) If $f'(x) < 0$ for all x in the interior of I, then f decreases on all of I.
(iii) If $f'(x) = 0$ for all x in the interior of I, then f is constant on all of I.

A proof of this result is not hard to construct. As you can verify, a word-for-word repetition of our proof of Theorem 4.2.2 also works here.

It is time for some examples.

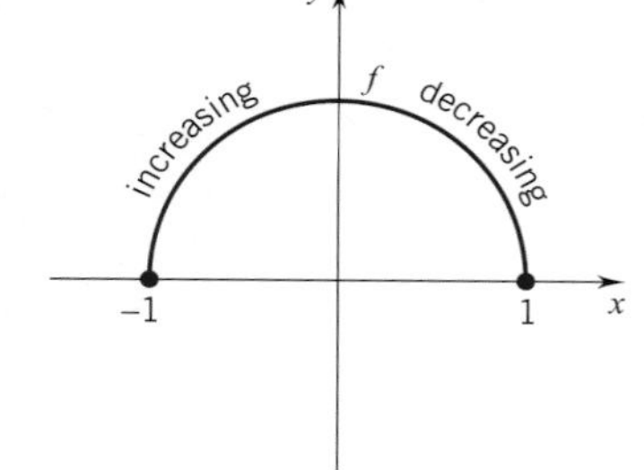

Figure 4.2.5

Example 1 The function

$$f(x) = \sqrt{1 - x^2}$$

has derivative

$$f'(x) = -\frac{x}{\sqrt{1 - x^2}}.$$

Since $f'(x) > 0$ for all x in $(-1, 0)$ and f is continuous on $[-1, 0]$, f increases on $[-1, 0]$. Since $f'(x) < 0$ for all x in $(0, 1)$ and f is continuous on $[0, 1]$, f decreases on $[0, 1]$. The graph of f is a semicircle. (Figure 4.2.5) ❑

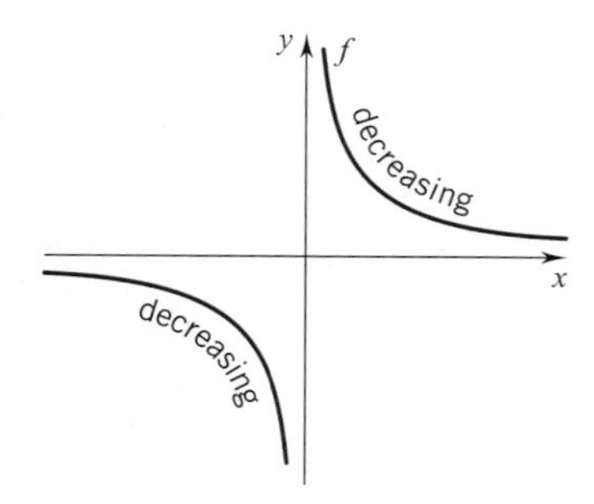

Figure 4.2.6

Example 2 The function

$$f(x) = \frac{1}{x}$$

is defined for all $x \neq 0$. The derivative

$$f'(x) = -\frac{1}{x^2}$$

is negative for all $x \neq 0$. Thus the function f decreases both on $(-\infty, 0)$ and on $(0, \infty)$. (Figure 4.2.6) ❑

Example 3 The function

$$g(x) = \tfrac{4}{5}x^5 - 3x^4 - 4x^3 + 22x^2 - 24x + 6$$

is a polynomial. It is therefore everywhere continuous and everywhere differentiable. Differentiation gives

$$\begin{aligned} g'(x) &= 4x^4 - 12x^3 - 12x^2 + 44x - 24 \\ &= 4(x^4 - 3x^3 - 3x^2 + 11x - 6) \\ &= 4(x+2)(x-1)^2(x-3). \end{aligned}$$

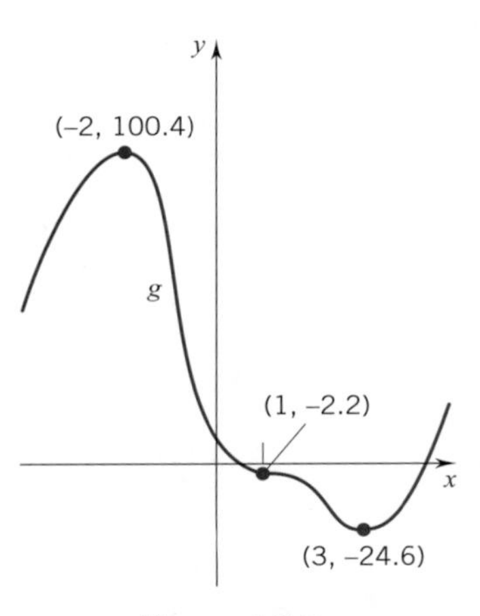

Figure 4.2.7

The derivative g' takes on the value 0 at -2, at 1, and at 3. These numbers determine four intervals on which g' keeps a constant sign (see Section 1.3):

$$(-\infty, -2), \qquad (-2, 1), \qquad (1, 3), \qquad (3, \infty).$$

The sign of g' on these intervals and the consequences for g are as follows:

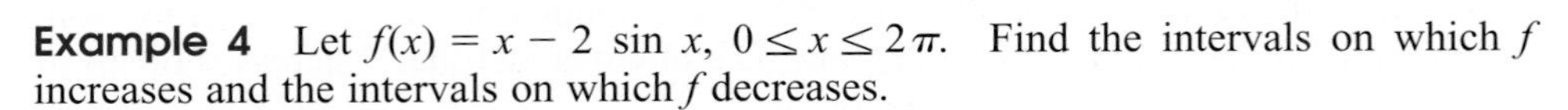

Since g is everywhere continuous, g increases on $(-\infty, -2]$, decreases on $[-2, 3]$, and increases on $[3, \infty)$. (Figure 4.2.7) ❑

Example 4 Let $f(x) = x - 2 \sin x$, $0 \leq x \leq 2\pi$. Find the intervals on which f increases and the intervals on which f decreases.

SOLUTION The derivative of f is

$$f'(x) = 1 - 2 \cos x.$$

Setting $f'(x) = 0$, we have

$$1 - 2 \cos x = 0$$

$$\cos x = \frac{1}{2}.$$

The solutions are $x = \pi/3$ and $x = 5\pi/3$. Thus, f' has constant sign on the intervals $(0, \pi/3)$, $(\pi/3, 5\pi/3)$, and $(5\pi/3, 2\pi)$. The sign of f' and the behavior of f are recorded below.

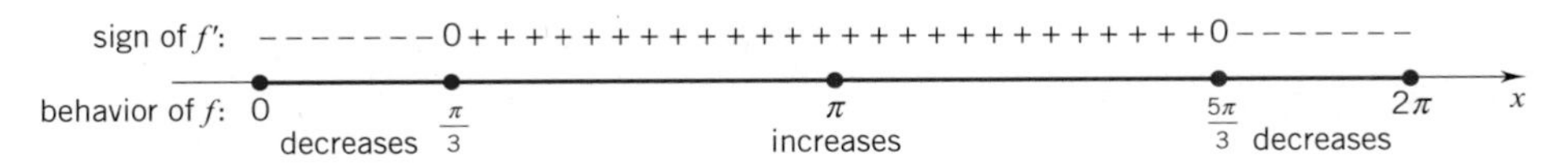

Since f is continuous on $[0, 2\pi]$, f decreases on $[0, \pi/3]$, increases on $[\pi/3, 5\pi/3]$, and decreases on $[5\pi/3, 2\pi]$. (Figure 4.2.8) ❑

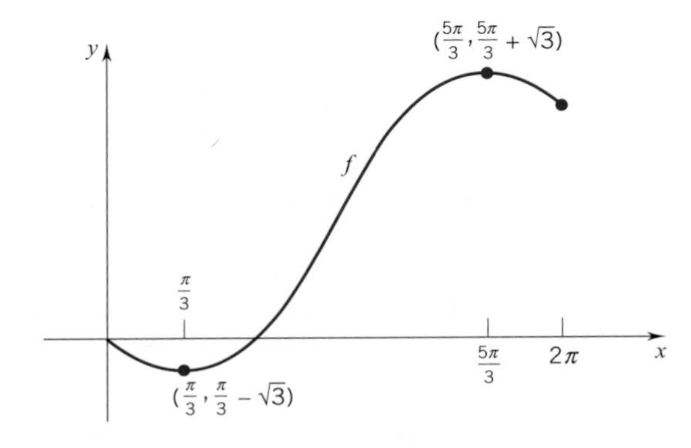

Figure 4.2.8

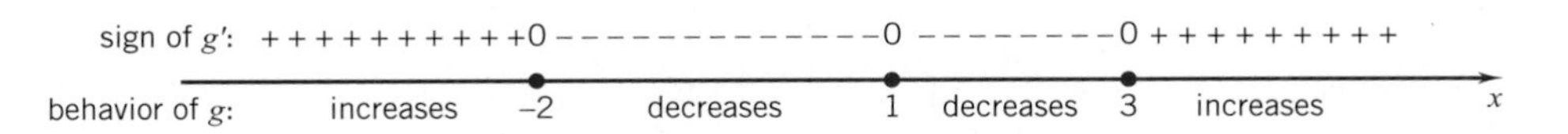

While Theorems 4.2.2 and 4.2.3 have wide applicability, there are also some limitations. If, for example, f is discontinuous at some point in its domain, then Theorems 4.2.2 and 4.2.3 do not tell the whole story.

Example 5 The function

$$f(x) = \begin{cases} x^3, & x < 1 \\ \frac{1}{2}x + 2, & x \geq 1 \end{cases}$$

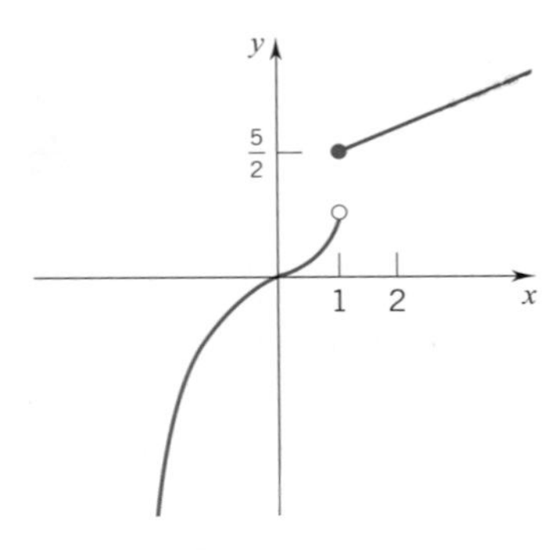

Figure 4.2.9

is graphed in Figure 4.2.9. Obviously there is a discontinuity at $x = 1$. Differentiation gives

$$f'(x) = \begin{cases} 3x^2, & x < 1 \\ \text{does not exist}, & x = 1 \\ \frac{1}{2}, & x > 1. \end{cases}$$

Since $f'(x) > 0$ on $(-\infty, 0) \cup (0, 1)$, we know from Theorem 4.2.2 that f increases on $(-\infty, 1)$. Since $f'(x) > 0$ on $(1, \infty)$ and f is continuous on $[1, \infty)$, we know from Theorem 4.2.3 that f increases on $[1, \infty)$. The obvious fact that f increases on all of $(-\infty, \infty)$ is not derivable from the theorems.

Now consider the function

$$g(x) = \begin{cases} \frac{1}{2}x + 2, & x < 1 \\ x^3, & x \geq 1, \end{cases}$$

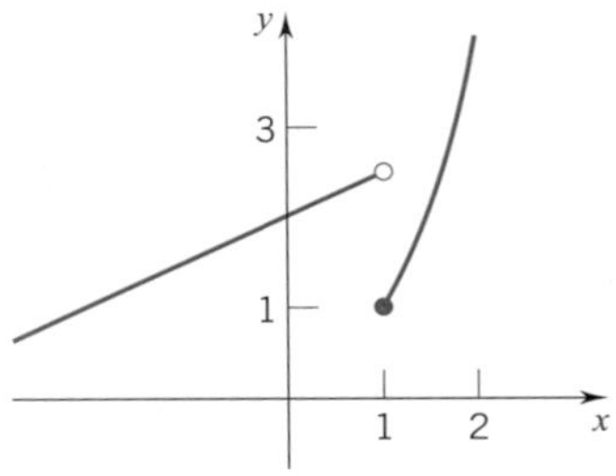

Figure 4.2.10

which is graphed in Figure 4.2.10. Again, there is a discontinuity at $x = 1$. The derivative of g is

$$g'(x) = \begin{cases} \frac{1}{2}, & x < 1 \\ \text{does not exist}, & x = 1 \\ 3x^2, & x > 1, \end{cases}$$

so g increases on $(-\infty, 1)$ and on $[1, \infty)$ but, clearly, g does not increase on $(-\infty, \infty)$. ❑

Equality of Derivatives

If two differentiable functions differ by a constant,

$$f(x) = g(x) + C,$$

then their derivatives are equal:

$$f'(x) = g'(x).$$

The converse is also true. In fact, we have the following theorem.

THEOREM 4.2.4

(i) Let I be an open interval. If $f'(x) = g'(x)$ for all x in I, then f and g differ by a constant on I.

(ii) Let I be an arbitrary interval. If $f'(x) = g'(x)$ for all x in the interior of I and f and g are continuous on I, then f and g differ by a constant on I.

PROOF Set $H = f - g$. For the first assertion apply (iii) of Theorem 4.2.2 to H. For the second assertion apply (iii) of Theorem 4.2.3 to H. We leave the details as an exercise. ❑

We illustrate the theorem in Figure 4.2.11. At points with the same x-coordinate the slopes are equal, and thus the curves have the same steepness. The separation between the curves remains constant; the curves are "parallel."

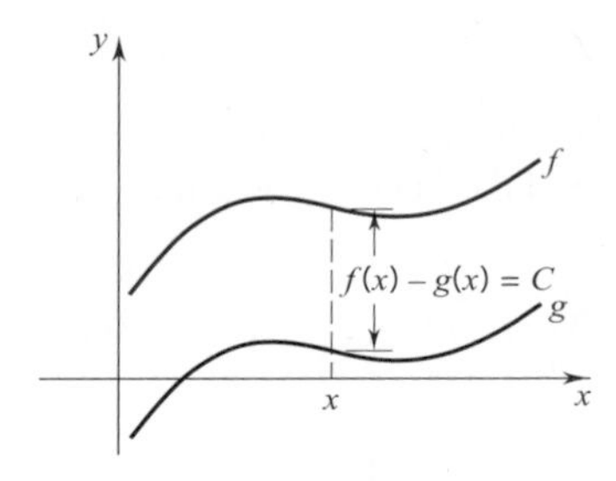

Figure 4.2.11

Example 6 Find f given that

$$f'(x) = 6x^2 - 7x - 5 \quad \text{for all real } x \quad \text{and} \quad f(2) = 1.$$

SOLUTION It is not hard to find a function that has $6x^2 - 7x - 5$ as its derivative:

$$\frac{d}{dx}\left(2x^3 - \frac{7}{2}x^2 - 5x\right) = 6x^2 - 7x - 5.$$

By Theorem 4.2.4 we know that $f(x)$ differs from $2x^3 - \frac{7}{2}x^2 - 5x$ only by some constant C. Thus we can write

$$f(x) = 2x^3 - \tfrac{7}{2}x^2 - 5x + C.$$

To evaluate C we use the fact that $f(2) = 1$. Since $f(2) = 1$ and also

$$f(2) = 2(2)^3 - \tfrac{7}{2}(2)^2 - 5(2) + C = 16 - 14 - 10 + C = -8 + C,$$

we have $-8 + C = 1$, and therefore $C = 9$. Thus

$$f(x) = 2x^3 - \tfrac{7}{2}x^2 - 5x + 9. \quad ❑$$

EXERCISES 4.2

In Exercises 1–27, find the intervals on which f increases and the intervals on which f decreases.

1. $f(x) = x^3 - 3x + 2$.

2. $f(x) = x^3 - 3x^2 + 6$.

3. $f(x) = x + \dfrac{1}{x}$.

4. $f(x) = (x - 3)^3$.

5. $f(x) = x^3(1 + x)$.

6. $f(x) = x(x + 1)(x + 2)$.

7. $f(x) = (x + 1)^4$.

8. $f(x) = 2x - \dfrac{1}{x^2}$.

9. $f(x) = \dfrac{1}{|x - 2|}$.

10. $f(x) = \dfrac{x}{1 + x^2}$.

11. $f(x) = \dfrac{x^2 + 1}{x^2 - 1}$.

12. $f(x) = \dfrac{x^2}{x^2 + 1}$.

13. $f(x) = |x^2 - 5|$.

14. $f(x) = x^2(1 + x)^2$.

15. $f(x) = \dfrac{x - 1}{x + 1}$.

16. $f(x) = x^2 + \dfrac{16}{x^2}$.

17. $f(x) = \left(\dfrac{1 - \sqrt{x}}{1 + \sqrt{x}}\right)^7$.

18. $f(x) = \sqrt{\dfrac{2 + x}{1 + x}}$.

19. $f(x) = \sqrt{\dfrac{1 + x^2}{2 + x^2}}$.

20. $f(x) = |x + 1||x - 2|$.

21. $f(x) = \sqrt{\dfrac{3 - x}{x}}$.

22. $f(x) = x + \sin x, \quad 0 \le x \le 2\pi$.

23. $f(x) = x - \cos x, \quad 0 \le x \le 2\pi$.

24. $f(x) = \cos^2 x, \quad 0 \le x \le \pi$.

25. $f(x) = \cos 2x + 2\cos x, \quad 0 \le x \le \pi$.

26. $f(x) = \sin^2 x - \sqrt{3}\sin x, \quad 0 \le x \le \pi$.

27. $f(x) = \sqrt{3}\,x - \cos 2x, \quad 0 \le x \le \pi$.

In Exercises 28–36, find f given the following information.

28. $f'(x) = x^2 - 1$ for all real x, $f(0) = 1$.

29. $f'(x) = x^2 - 1$ for all real x, $f(1) = 2$.

30. $f'(x) = 2x - 5$ for all real x, $f(2) = 4$.

31. $f'(x) = 5x^4 + 4x^3 + 3x^2 + 2x + 1$ for all real x, $f(0) = 5$.

32. $f'(x) = 4x^{-3}$ for $x > 0$, $f(1) = 0$.

33. $f'(x) = x^{1/3} - x^{1/2}$ for $x > 0$, $f(0) = 1$.

34. $f'(x) = x^{-5} - 5x^{-1/5}$ for $x > 0$, $f(1) = 0$.

35. $f'(x) = 2 + \sin x$ for all real x, $f(0) = 3$.

36. $f'(x) = 4x + \cos x$ for all real x, $f(0) = 1$.

In Exercises 37–40, find the intervals on which f increases and the intervals on which f decreases.

37. $f(x) = \begin{cases} x + 7, & x < -3 \\ |x + 1|, & -3 \leq x < 1 \\ 5 - 2x, & 1 \leq x. \end{cases}$

38. $f(x) = \begin{cases} (x - 1)^2, & x < 1 \\ 5 - x, & 1 \leq x < 3 \\ 7 - 2x, & 3 \leq x. \end{cases}$

39. $f(x) = \begin{cases} 4 - x^2, & x < 1 \\ 7 - 2x, & 1 \leq x < 3 \\ 3x - 10, & 3 \leq x. \end{cases}$

40. $f(x) = \begin{cases} x + 2, & x < 0 \\ (x - 1)^2, & 0 < x < 3 \\ 8 - x, & 3 < x < 7 \\ 2x - 9, & 7 < x \\ 6, & x = 0, 3, 7. \end{cases}$

In Exercises 41–44, f is a differentiable function and the graph of its derivative f' is given. If $f(0) = 1$, give a rough sketch of the graph of f.

41.

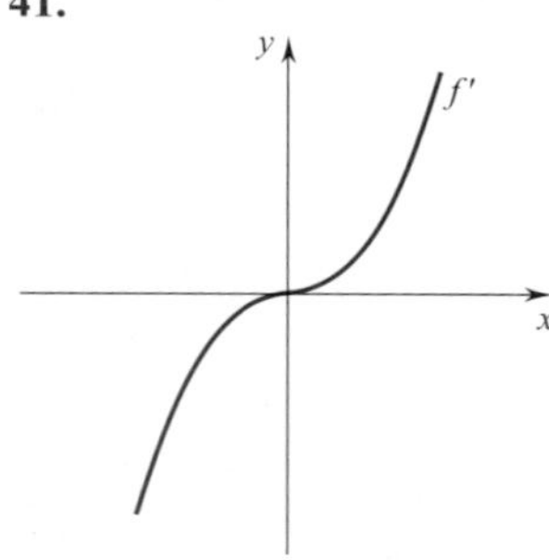

42.

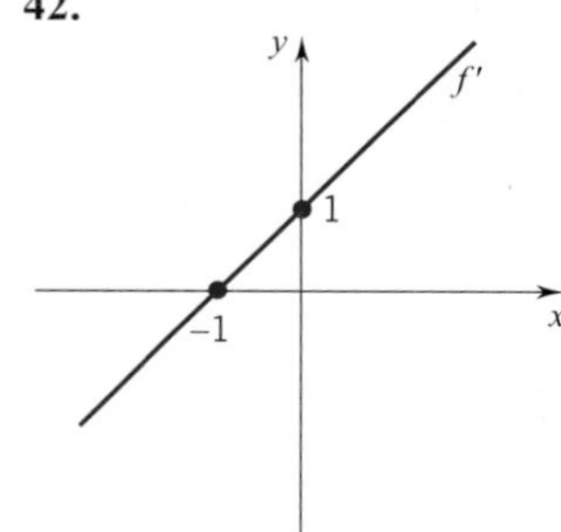

43.

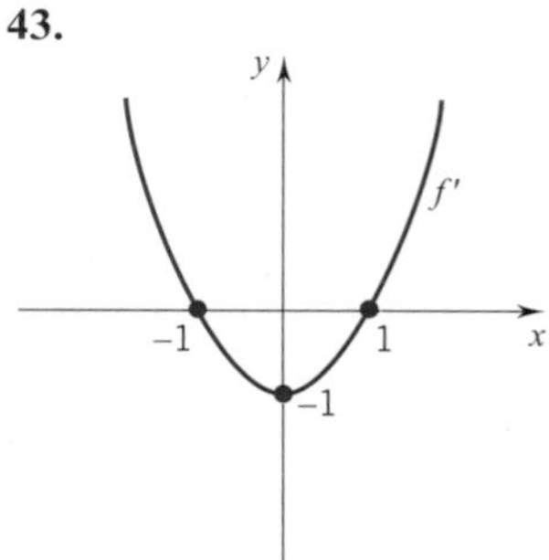

44.

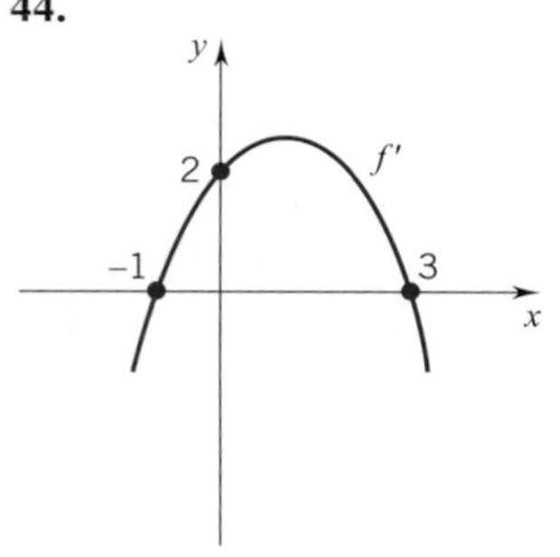

In Exercises 45 and 46, the graph of a continuous function f is given. Sketch the graph of f'.

45.

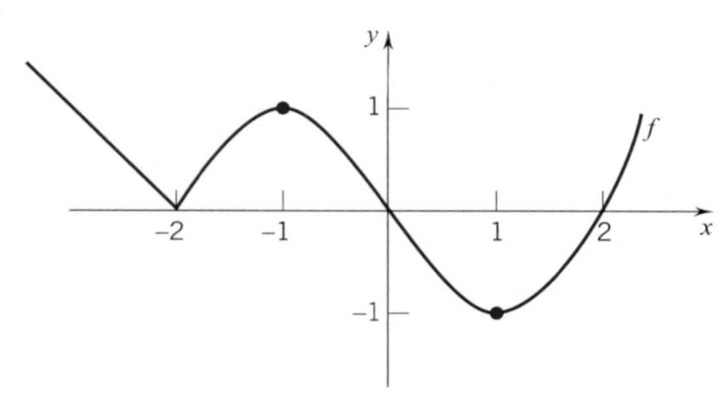

46.

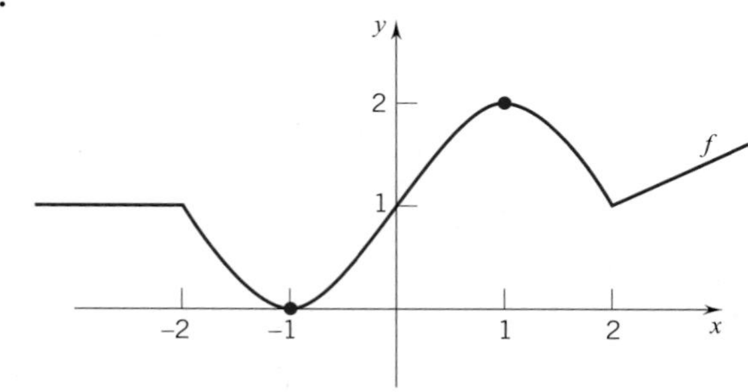

In Exercises 47–50, sketch the graph of a differentiable function f that satisfies the given conditions, if possible. If it is not possible, explain why.

47. $f(x) > 0$ for all x, $f(0) = 1$, and $f'(x) < 0$ for all x.

48. $f(1) = -1$, $f'(x) < 0$ for all $x \neq 1$, and $f'(1) = 0$.

49. $f(-1) = 4$, $f(2) = 2$, and $f'(x) > 0$ for all x.

50. f has x-intercepts only at $x = 1$ and $x = 2$, $f(3) = 4$, and $f(5) = -1$.

In Exercises 51–54, an object is moving along a straight line with its position at time t given by $x(t)$. Determine the time interval(s) when the object is moving to the right, determine the time interval(s) when it is moving to the left, and find the time(s) when it changes direction. Also, find the time interval(s) when the velocity of the object is increasing and when it is decreasing. Draw a figure that illustrates the motion of the object. See Example 3, Section 3.4.

51. $x(t) = t^3 - 6t^2 + 9t + 2$.

52. $x(t) = (2t - 1)(t - 1)^2$.

53. $x(t) = 2 \sin 3t$, $t \in [0, \pi]$.

54. $x(t) = 3 \cos (2t + \frac{\pi}{4})$, $t \in [0, 2\pi]$.

55. Suppose that the function f increases on (a, b) and on (b, c). Suppose further that $\lim_{x \to b^-} f(x) = M$, $\lim_{x \to b^+} f(x) = N$ and $f(b) = L$. For which values of L, if any, can you conclude that f increases on (a, c) if (a) $M < N$? (b) $M > N$? (c) $M = N$?

56. Suppose that the function f increases on (a, b) and decreases on (b, c). Under what conditions can you conclude that f increases on $(a, b]$ and decreases on $[b, c)$?

57. Give an example of a function f that satisfies all of the following conditions: (i) f is defined for all real x, (ii) f' is positive wherever it exists, (iii) there is no interval on which f increases.

58. Prove Theorem 4.2.4.

59. Let $f(x) = \sec^2 x$ and $g(x) = \tan^2 x$ on the interval $I = (-\pi/2, \pi/2)$.
(a) Show that $f'(x) = g'(x)$ for all $x \in I$.
(b) The result in part (a) implies, by Theorem 4.2.4, that $f - g = C$, constant, on $(-\pi/2, \pi/2)$. Find the value of C.

60. Let L be a differentiable function on $(0, \infty)$ such that $L'(x) = 1/x$ and $L(1) = 0$. This function is studied in detail in Chapter 7. Prove that for any two positive numbers a and b, $L(ab) = L(a) + L(b)$. HINT: Begin by showing that, for any $x > 0$, $L(ax)$ and $L(x)$ have the same derivative.

61. Let f and g be differentiable functions such that $f'(x) = -g(x)$ and $g'(x) = f(x)$ for all x.
(a) Prove that $f^2(x) + g^2(x) = C$, constant.
(b) Suppose there is a number a such that $f(a) = 1$ and $g(a) = 0$. Use this information to find the value of C. Can you name a pair of functions f, g that have these properties?

62. Prove that if f is differentiable on an interval I and $f'(x) \geq 0$ (≤ 0) on I with $f'(x) = 0$ for at most finitely many values of $x \in I$, then f increases (decreases) on I.

63. Let $f(x) = x - \sin x$.
(a) Prove that f is increasing on $(-\infty, \infty)$.
(b) Use the result in part (a) to show that $\sin x < x$ on $(0, \infty)$ and $\sin x > x$ on $(-\infty, 0)$.

64. Let f and g be differentiable functions on the interval $(-c, c)$ such that $f(0) = g(0)$.
(a) Prove that if $f'(x) > g'(x)$ for all $x \in (0, c)$, then $f(x) > g(x)$ for all $x \in (0, c)$. HINT: Set $F = f - g$ and use Theorem 4.2.2.
(b) Prove that if $f'(x) > g'(x)$ for all $x \in (-c, 0)$, then $f(x) < g(x)$ for all $x \in (-c, 0)$.

65. Prove that $\tan x > x$ for all $x \in (0, \pi/2)$. HINT: Use Exercise 64(a).

66. Prove that $1 - x^2/2 < \cos x$ for all $x \in (0, \infty)$. HINT: Use Exercises 63 and 64.

67. Let $n > 1$ be an integer. Prove that $(1 + x)^n > 1 + nx$ for all $x \in (0, \infty)$.

68. Prove that $x - x^3/6 < \sin x$ for all $x \in (0, \infty)$.

C **69.** It follows from Exercises 63 and 68 that

$$x - \frac{1}{6}x^3 < \sin x < x \qquad \text{for } x > 0.$$

Use this result to estimate $\sin 4°$.

C In Exercises 70–73, use a graphing utility to graph the function f and its derivative f' on the indicated interval. Estimate the zeros of f' correct to three decimal places. Then estimate the subintervals on which f increases and the subintervals on which f decreases.

70. $f(x) = 2x^3 - x^2 - 13x - 6$; $[-3, 4]$.

71. $f(x) = 3x^4 - 10x^3 - 4x^2 + 10x + 9$; $[-2, 5]$.

72. $f(x) = x^4 + 3x^3 - 2x^2 + 4x + 4$; $[-5, 3]$.

73. $f(x) = x \cos x - 3 \sin 2x$; $[0, 6]$.

Energy of a Falling Body (Near the Surface of the Earth)

If we lift an object, we counteract the force of gravity. In so doing, we increase what physicists call the *gravitational potential energy* of the object. The gravitational potential energy, GPE, of an object is defined by

$$\text{GPE} = \text{weight} \times \text{height}.$$

Since the weight of an object of mass m is mg (we take this from physics), where g is the gravitational constant, we can write

$$\text{GPE} = mgy,$$

where y is the height.

If we lift an object and release it, the object drops. As it drops, it loses height and therefore loses gravitational potential energy, but it picks up speed. The speed with which the object falls gives the object another form of energy called *kinetic energy*, the energy of motion. The kinetic energy KE of an object in straight-line motion is given by

$$\text{KE} = \frac{1}{2}mv^2$$

where v is the velocity of the object.

74. Prove the Law of Conservation of Energy: GPE + KE = C, constant. HINT: Differentiate the expression GPE + KE and use the fact that $dv/dt = -g$.

75. An object initially at rest falls freely from height y_0. Show that the speed of the object at height y is given by

$$|v| = \sqrt{2g(y_0 - y)}.$$

76. According to the results in Section 3.4, the position of an object that falls from rest from a height y_0 is given by

$$y(t) = -\tfrac{1}{2}gt^2 + y_0.$$

Calculate the speed of the object starting from this equation and show that the result is equivalent to the result found in Exercise 75.

77. A bobsled run descends from a height of 150 meters from its starting point to the bottom of the run. Suppose that a bobsled starts from rest at the top and slides down the run without friction. Find the speed of the bobsled at the bottom of the run (use $g = 9.8$ meters per second per second). (NOTE: To solve this problem by calculating the forces and acceleration would require a detailed knowledge of the path of the run down the hill. The Law of Conservation of Energy provides a simple method for finding the speed at the bottom.)

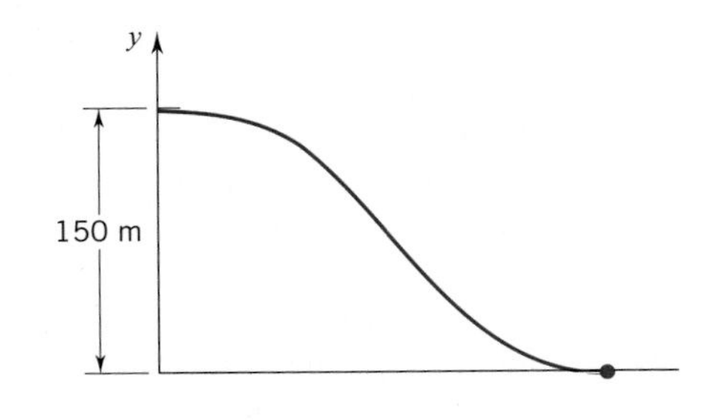

■ 4.3 LOCAL EXTREME VALUES

In many problems of economics, engineering, and physics it is important to determine how large or how small a certain quantity can be. If the problem admits a mathematical formulation, it is often reducible to the problem of finding the maximum or minimum value of some function.

In this section we consider maximum and minimum values for functions defined on an *open interval* or on a *union of open intervals*. We will take up functions defined on closed or half-closed intervals in the next section. We begin with a definition.

DEFINITION 4.3.1 LOCAL EXTREME VALUES

A function f is said to have a *local maximum at c* iff

$$f(c) \geq f(x) \qquad \text{for all } x \text{ sufficiently close to } c.\dagger$$

A function f is said to have a *local minimum at c* iff

$$f(c) \leq f(x) \qquad \text{for all } x \text{ sufficiently close to } c.$$

The local maxima and local minima of f are called the *local extreme values of f*.

We illustrate these notions in Figure 4.3.1. A careful look at the figure suggests that local maxima and minima occur only at points where the tangent is horizontal $[f'(c) = 0]$ or where there is no tangent line $[f'(c)$ does not exist$]$. This is indeed the case.

† What do we mean by saying that "such and such is true *for all x sufficiently close to c*"? We mean that it is true for all x in some open interval $(c - \delta, c + \delta)$ centered at c.

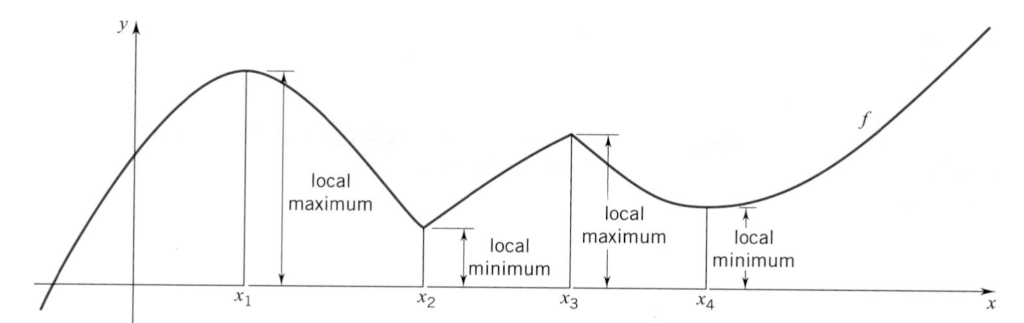

Figure 4.3.1

THEOREM 4.3.2

If f has a local maximum or minimum at c, then either

$$f'(c) = 0 \quad \text{or} \quad f'(c) \text{ does not exist.}$$

PROOF We are given that f has a local extreme value at c. Suppose that $f'(c)$ exists. If $f'(c) > 0$ or $f'(c) < 0$, then, by Theorem 4.1.2, there must be numbers x_1 and x_2 that are arbitrarily close to c which satisfy

$$f(x_1) < f(c) < f(x_2).$$

This makes it impossible for a local maximum or a local minimum to occur at c. Therefore, if $f'(c)$ exists, then it must have the value 0. The only other possibility is that $f'(c)$ does not exist. ❑

Based on this result, we have the following important definition:

DEFINITION 4.3.3

Given a function f. The numbers c in the domain of f for which either

$$f'(c) = 0 \quad \text{or} \quad f'(c) \text{ does not exist}$$

are called *critical numbers of f*.†

In view of Theorem 4.3.2, when searching for the local maxima and minima of a function f, the only numbers that we need to consider are the critical numbers of f.

We illustrate the technique for finding local maxima and minima by some examples. In each example the first step will be to find the critical numbers. We begin with very simple cases.

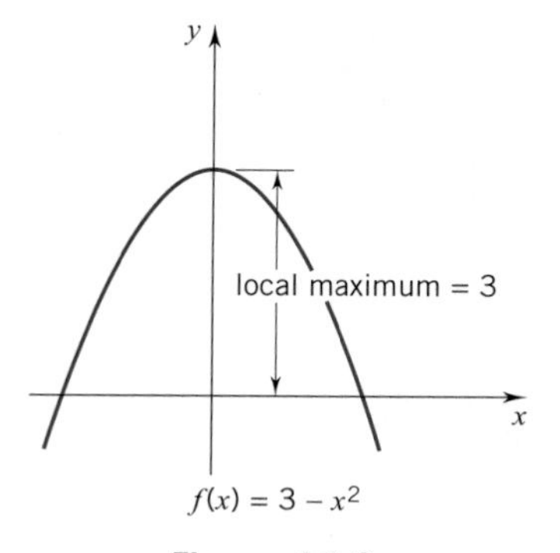

Figure 4.3.2

Example 1 For

$$f(x) = 3 - x^2 \qquad \text{(Figure 4.3.2)}$$

† The terms *critical values* and *critical points* are also used.

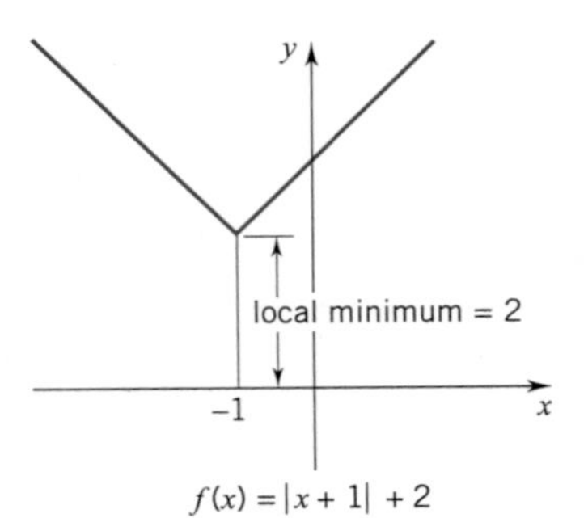

$f(x) = |x + 1| + 2$
Figure 4.3.3

the derivative

$$f'(x) = -2x$$

exists everywhere. Since $f'(x) = 0$ only at $x = 0$, 0 is the only critical number. The number $f(0) = 3$ is obviously a local maximum since the graph of f is a parabola that opens down. ❑

Example 2 In the case of

$$f(x) = |x + 1| + 2 = \begin{cases} -x + 1, & x < -1 \\ x + 3, & x \geq -1, \end{cases} \qquad \text{(Figure 4.3.3)}$$

differentiation gives

$$f'(x) = \begin{cases} -1, & x < -1 \\ 1, & x > -1. \end{cases}$$

This derivative is never 0. It fails to exist only at -1. The number -1 is the only critical number. The value $f(-1) = 2$ is a local minimum. ❑

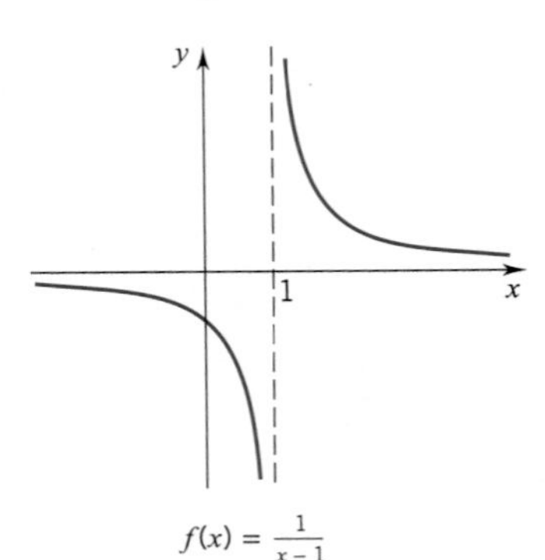

$f(x) = \frac{1}{x-1}$
Figure 4.3.4

Example 3 Figure 4.3.4 shows the graph of the function

$$f(x) = \frac{1}{x - 1}.$$

The domain is $(-\infty, 1) \cup (1, \infty)$. The derivative

$$f'(x) = -\frac{1}{(x - 1)^2}$$

exists throughout the domain of f and is never 0. Thus there are no critical numbers. In particular, 1 is not a critical number of f because 1 is not in the domain of f. Since f has no critical numbers, there are no local extreme values. ❑

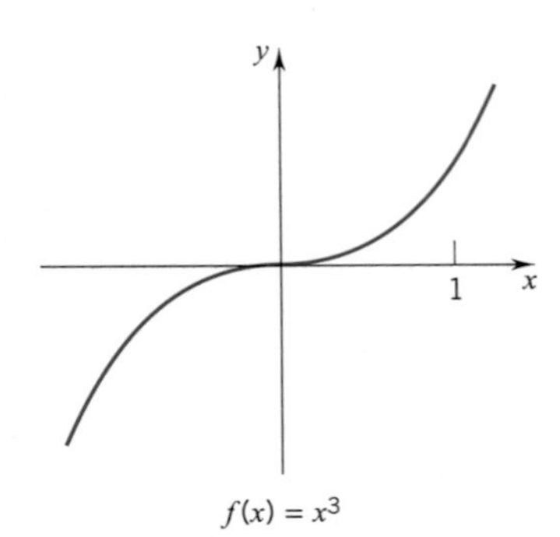

$f(x) = x^3$
Figure 4.3.5

CAUTION The fact that c is a critical number of f does not guarantee that $f(c)$ is a local extreme value. This is illustrated in the next two examples.

Example 4 In the case of the function

$$f(x) = x^3, \qquad \text{(Figure 4.3.5)}$$

the derivative $f'(x) = 3x^2$ is 0 at 0, but $f(0) = 0$ is not a local extreme value. This function is everywhere increasing. ❑

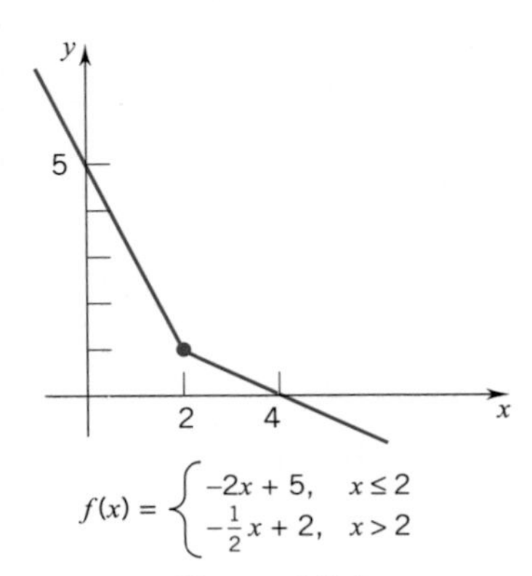

$f(x) = \begin{cases} -2x + 5, & x \leq 2 \\ -\frac{1}{2}x + 2, & x > 2 \end{cases}$
Figure 4.3.6

Example 5 The function

$$f(x) = \begin{cases} -2x + 5 & x < 2 \\ -\frac{1}{2}x + 2 & x \geq 2 \end{cases} \qquad \text{(Figure 4.3.6)}$$

is everywhere decreasing. Although 2 is a critical number ($f'(2)$ does not exist), $f(2) = 1$ is not a local extreme value. ❑

There are two widely used tests for determining the behavior of a function at a critical number. The first test (given in Theorem 4.3.4) requires that we examine the sign of the first derivative on both sides of the critical number. The second test (given in Theorem 4.3.5) requires that we examine the sign of the second derivative at the critical number itself.

THEOREM 4.3.4 THE FIRST DERIVATIVE TEST

Suppose that c is a critical number of f and that f is continuous at c. If there is a positive number δ such that:

(i) $f'(x) > 0$ for all x in $(c - \delta, c)$ and $f'(x) < 0$ for all x in $(c, c + \delta)$, then $f(c)$ is a local maximum. (Figures 4.3.7 and 4.3.8.)

(ii) $f'(x) < 0$ for all x in $(c - \delta, c)$ and $f'(x) > 0$ for all x in $(c, c + \delta)$, then $f(c)$ is a local minimum. (Figures 4.3.9 and 4.3.10.)

(iii) $f'(x)$ keeps constant sign on $(c - \delta, c) \cup (c, c + \delta)$, then $f(c)$ is not a local extreme value.

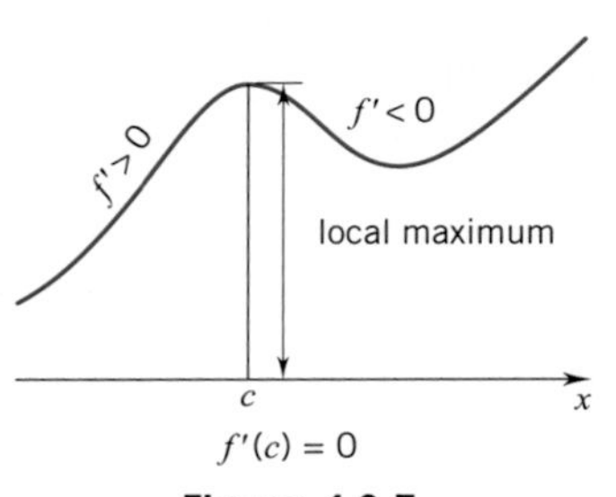

$f'(c) = 0$

Figure 4.3.7

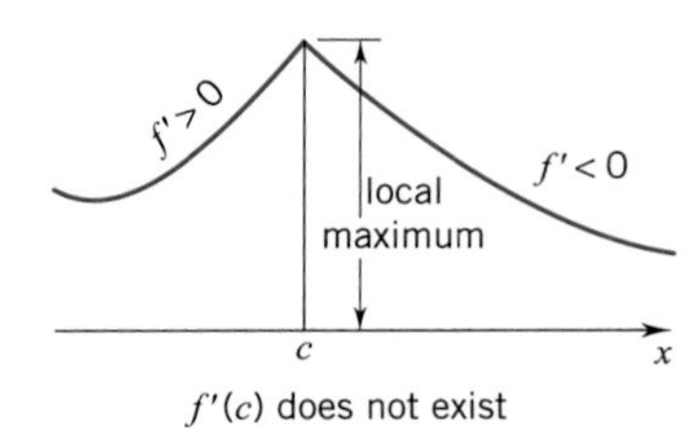

$f'(c)$ does not exist

Figure 4.3.8

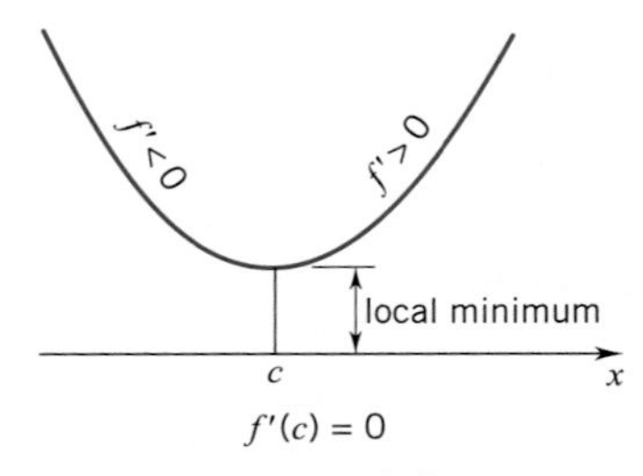

$f'(c) = 0$

Figure 4.3.9

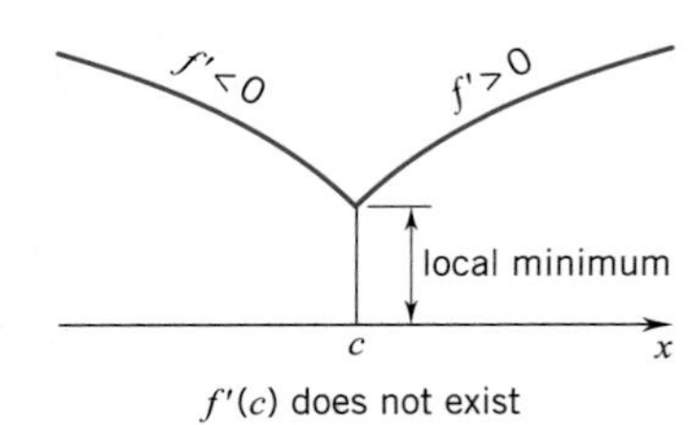

$f'(c)$ does not exist

Figure 4.3.10

PROOF The result is a direct consequence of Theorem 4.2.3. The details of the proof are left to you as Exercise 31. ❑

Example 6 The function

$$f(x) = x^4 - 2x^3, \qquad \text{all real } x$$

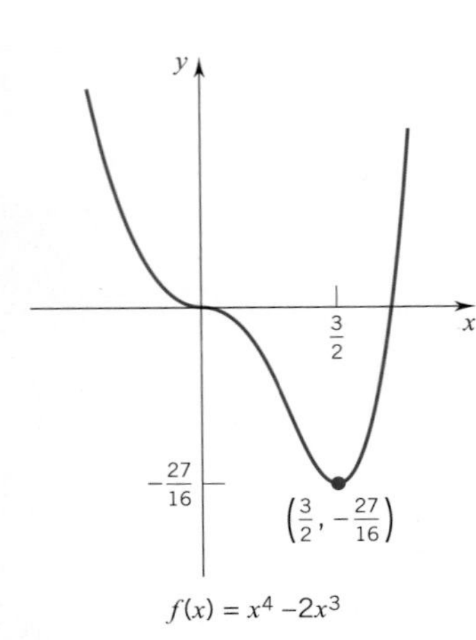

Figure 4.3.11

has derivative

$$f'(x) = 4x^3 - 6x^2 = 2x^2(2x - 3), \qquad \text{all real } x.$$

The only critical numbers are 0 and $\frac{3}{2}$. The sign of f' is recorded below.

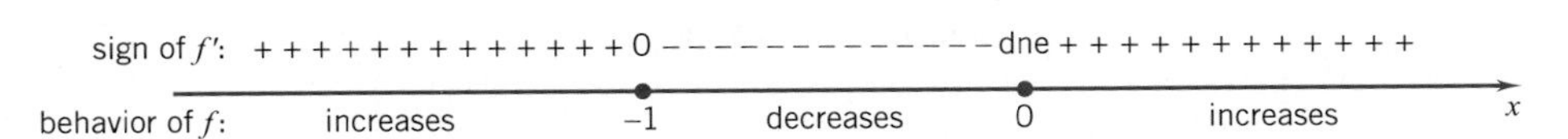

Since f' keeps the same sign on both sides of 0, $f(0) = 0$ is not a local extreme value. However, $f(\frac{3}{2}) = -\frac{27}{16}$ is a local minimum. The graph of f appears in Figure 4.3.11. ❑

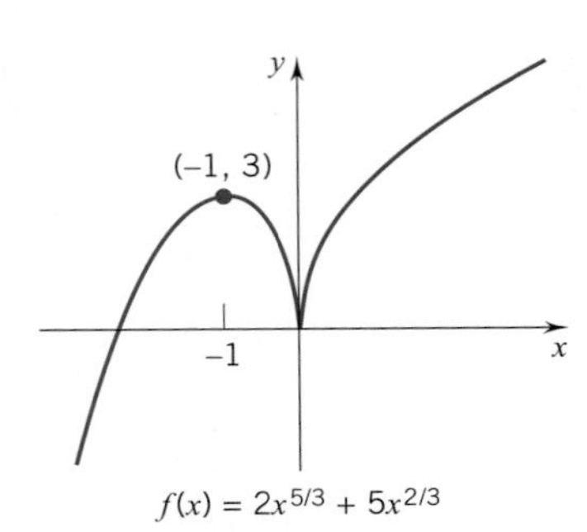

Figure 4.3.12

Example 7 The function

$$f(x) = 2x^{5/3} + 5x^{2/3},$$

is defined for all real x. The derivative of f is given by

$$f'(x) = \frac{10}{3}x^{2/3} + \frac{10}{3}x^{-1/3} = \frac{10}{3}x^{-1/3}(x + 1), \qquad x \neq 0.$$

Since $f'(-1) = 0$ and $f'(0)$ does not exist, the critical numbers are -1 and 0. The sign of f' is recorded below. (To save space in the diagram we write "dne" for "does not exist.")

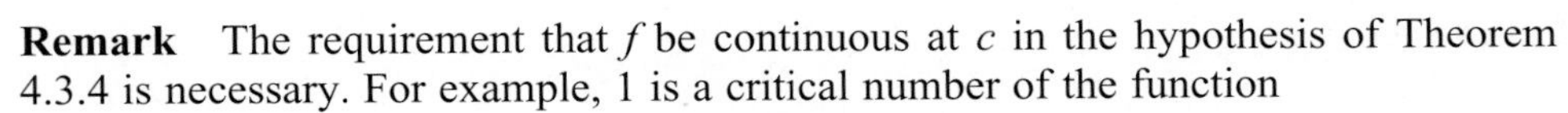

In this case $f(-1) = 3$ is a local maximum and $f(0) = 0$ is a local minimum. The graph appears in Figure 4.3.12. ❑

Remark The requirement that f be continuous at c in the hypothesis of Theorem 4.3.4 is necessary. For example, 1 is a critical number of the function

$$f(x) = \begin{cases} 1 + 2x, & x \leq 1 \\ 5 - x, & x > 1 \end{cases} \qquad \text{(Figure 4.3.13)}$$

since $f'(1)$ does not exist. Note that $f'(x) = 2 > 0$ to the left of 1 and $f'(x) = -1 < 0$ to the right of 1, but $f(1) = 3$ is not a local maximum of f. ❑

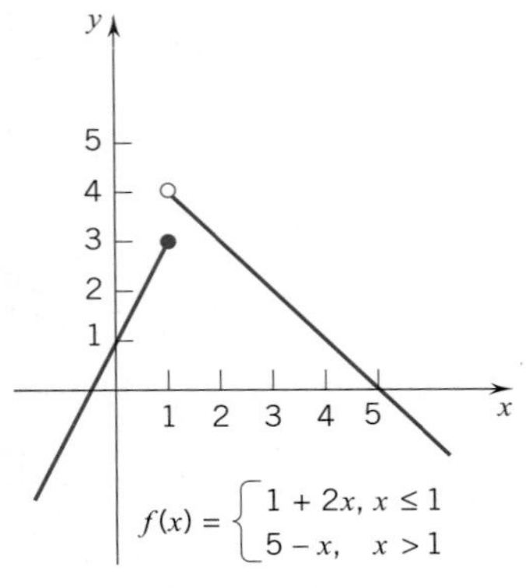

Figure 4.3.13

In some situations it might be difficult to determine the sign of the first derivative on one side or the other of a critical number c. For example, $x = \frac{2}{3}$ is a critical number of the polynomial function

$$p(x) = \tfrac{3}{5}x^5 + x^4 - \tfrac{1}{3}x^3 - \tfrac{1}{2}x^2 - \tfrac{2}{3}x + 1$$

since

$$p'(x) = 3x^4 + 4x^3 - x^2 - x - \tfrac{2}{3} \qquad \text{and} \qquad p'(\tfrac{2}{3}) = 0.$$

Examining the sign of p' on either side of $\frac{2}{3}$ will be difficult, especially since we do not know what the other critical numbers of p are.

If c is a critical number for a function f and if f is twice differentiable at c, then the following test is sometimes easier to apply.

THEOREM 4.3.5 THE SECOND-DERIVATIVE TEST

Suppose that $f'(c) = 0$ and that $f''(c)$ exists.

If $f''(c) > 0$, then $f(c)$ is a local minimum value.
If $f''(c) < 0$, then $f(c)$ is a local maximum value.

PROOF We handle the case $f''(c) > 0$. The other is left as an exercise. Since f'' is the derivative of f', we see from Theorem 4.1.2 that there exists $\delta > 0$ such that, if

$$c - \delta < x_1 < c < x_2 < c + \delta,$$

then

$$f'(x_1) < f'(c) < f'(x_2).$$

Since $f'(c) = 0$, we have

$$f'(x) < 0 \quad \text{for } x \text{ in } (c - \delta, c) \qquad \text{and} \qquad f'(x) > 0 \quad \text{for } x \text{ in } (c, c + \delta).$$

By the first-derivative test $f(c)$ is a local minimum. ❑

Continuing with the illustration begun earlier,

$$p''(x) = 12x^3 + 12x^2 - 2x - 1$$

and

$$p''(\tfrac{2}{3}) = 12(\tfrac{2}{3})^3 + 12(\tfrac{2}{3})^2 - 2(\tfrac{2}{3}) - 1 = \tfrac{59}{9} > 0.$$

Thus p has a local minimum at $x = \frac{2}{3}$.

Example 8 For

$$f(x) = 2x^3 - 3x^2 - 12x + 5$$

we have

$$f'(x) = 6x^2 - 6x - 12 = 6(x^2 - x - 2) = 6(x - 2)(x + 1)$$

and

$$f''(x) = 12x - 6.$$

The critical numbers are 2 and -1; the first derivative is 0 at each of these points. Since $f''(2) = 18 > 0$ and $f''(-1) = -18 < 0$, we can conclude from the second-derivative test that $f(2) = -15$ is a local minimum and $f(-1) = 12$ is a local maximum. ❑

We can apply the first-derivative test even at points where the function is not differentiable, provided that it is continuous there. On the other hand, the second-derivative test can only be applied at points where the function is twice differentiable and then only if the second derivative is different from zero. This is a drawback.

Take, for example, the function $f(x) = x^{4/3}$. We have $f'(x) = \frac{4}{3}x^{1/3}$ so that

$$f'(0) = 0, \qquad f'(x) < 0 \quad \text{for } x < 0, \qquad f'(x) > 0 \quad \text{for } x > 0.$$

By the first-derivative test, $f(0) = 0$ is a local minimum. We cannot get this information from the second-derivative test because $f''(x) = \frac{4}{9}x^{-2/3}$ does not exist at $x = 0$.

To show what happens if the second derivative equals zero at a critical number c, consider the functions

$$f(x) = x^3, \qquad g(x) = x^4 \qquad \text{and} \qquad h(x) = -x^4. \qquad \text{(Figure 4.3.14)}$$

In the case of the function $f(x) = x^3$, we already know that $x = 0$ is a critical point and that f has neither a local maximum nor a local minimum value at 0 (see Example 4). Note that $f''(x) = 6x$ and $f''(0) = 0$. In the case of $g(x) = x^4$, we have $g'(x) = 4x^3$ so that

$$g'(0) = 0, \qquad g'(x) < 0 \quad \text{for } x < 0, \qquad g'(x) > 0 \quad \text{for } x > 0.$$

By the first-derivative test, $g(0) = 0$ is a local minimum. But here again the second-derivative test is of no avail: $g''(x) = 12x^2$ is 0 at $x = 0$.

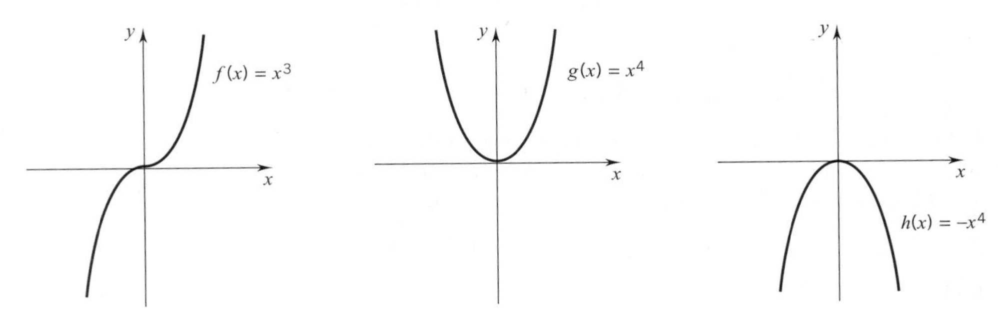

Figure 4.3.14

Finally, since the function h is the negative of the function g, it is clear that $h'(0) = h''(0) = 0$ and h has a local maximum at $x = 0$. ❑

EXERCISES 4.3

In Exercises 1–30, find the critical numbers of f and the local extreme values.

1. $f(x) = x^3 + 3x - 2.$

2. $f(x) = 2x^4 - 4x^2 + 6.$

3. $f(x) = x + \dfrac{1}{x}.$

4. $f(x) = x^2 - \dfrac{3}{x^2}.$

5. $f(x) = x^2(1 - x).$

6. $f(x) = (1 - x)^2(1 + x).$

7. $f(x) = \dfrac{1 + x}{1 - x}.$

8. $f(x) = \dfrac{2 - 3x}{2 + x}.$

9. $f(x) = \dfrac{2}{x(x + 1)}.$

10. $f(x) = |x^2 - 16|.$

11. $f(x) = x^3(1 - x)^2.$

12. $f(x) = \left(\dfrac{x - 2}{x + 2}\right)^3.$

13. $f(x) = (1 - 2x)(x - 1)^3.$

14. $f(x) = (1 - x)(1 + x)^3.$

15. $f(x) = \dfrac{x^2}{1 + x}.$

16. $f(x) = \dfrac{|x|}{1 + |x|}.$

17. $f(x) = |x - 1||x + 2|.$

18. $f(x) = x\sqrt[3]{1 - x}.$

19. $f(x) = x^2\sqrt[3]{2 + x}.$

20. $f(x) = \dfrac{1}{x + 1} - \dfrac{1}{x - 2}.$

21. $f(x) = |x - 3| + |2x + 1|.$

22. $f(x) = x^{7/3} - 7x^{1/3}.$

23. $f(x) = x^{2/3} + 2x^{-1/3}.$

24. $f(x) = \dfrac{x^3}{x + 1}.$

25. $f(x) = \sin x + \cos x, \quad 0 < x < 2\pi.$

26. $f(x) = x + \cos 2x, \quad 0 < x < \pi.$

27. $f(x) = \sin^2 x - \sqrt{3} \sin x, \quad 0 < x < \pi.$

28. $f(x) = \sin^2 x, \quad 0 < x < 2\pi.$

29. $f(x) = \sin x \cos x - 3 \sin x + 2x, \quad 0 < x < 2\pi$

30. $f(x) = 2 \sin^3 x - 3 \sin x, \quad 0 < x < \pi.$

31. Prove Theorem 4.3.4 by applying Theorem 4.2.3.

32. Prove the validity of the second-derivative test in the case that $f''(c) < 0$.

33. Find the critical numbers and the local extreme values of the polynomial

$$P(x) = x^4 - 8x^3 + 22x^2 - 24x + 4.$$

Then show that the equation $P(x) = 0$ has exactly two real roots, both positive.

34. A function f has derivative f' given by

$$f'(x) = x^3(x - 1)^2(x + 1)(x - 2).$$

At what numbers x, if any, does f have a local maximum? A local minimum?

35. A polynomial function $p(x) = a_n x^n + a_{n-1} x^{n-1} + \cdots + a_1 x + a_0$ has critical numbers at $x = -1, 1, 2,$ and 3, and corresponding values $p(-1) = 6$, $p(1) = 1$, $p(2) = 3$, and $p(3) = 1$. Sketch a possible graph for p if: (a) n is odd, (b) n is even.

36. The quadratic function $f(x) = Ax^2 + Bx + C$ has a local minimum at $x = 2$ and passes through the points $(-1, 3)$ and $(3, -1)$. Find A, B, and C.

37. Determine a and b such that the function $f(x) = ax/(x^2 + b^2)$ has a local minimum at $x = -2$ and $f'(0) = 1$.

38. Let $f(x) = x^p(1 - x)^q$, where $p \geq 2$ and $q \geq 2$ are integers.
(a) Show that the critical numbers of f are $x = 0$, $p/(p + q)$, and 1.
(b) Show that if p is even, then f has a local minimum at 0.
(c) Show that if q is even, then f has a local minimum at 1.
(d) Show that f has a local maximum at $p/(p + q)$ for all p and q.

39. Let

$$f(x) = \begin{cases} x^2 \sin(1/x), & x \neq 0 \\ 0, & x = 0. \end{cases}$$

In Exercise 67, Section 3.1, we saw that f is differentiable at 0 and that $f'(0) = 0$. Show that f has neither a local maximum nor a local minimum at 0.

40. Suppose that $C(x)$, $R(x)$, and $P(x)$ are the cost, revenue, and profit functions corresponding to the production and sale of x units of a certain item. Suppose, also, that C and R are differentiable functions. Then, since $P = R - C$, it follows that P is differentiable. Prove that if it is possible to maximize the profit by producing and selling x_0 items, then $C'(x_0) = R'(x_0)$. That is, the marginal cost equals the marginal revenue when the profit is maximized.

41. Let $y = f(x)$ be differentiable and suppose that the graph of f does not pass through the origin. Then the distance D from the origin to a point $P(x, f(x))$ on the graph is given by

$$D = \sqrt{x^2 + [f(x)]^2}.$$

Show that if D has a local extreme value at c, then the line through $(0, 0)$ and $(c, f(c))$ is perpendicular to the tangent line to the graph of f at c.

42. Prove that a polynomial of degree n has at most $n - 1$ local extreme values.

C 43. Let $f(x) = x^4 - 2x^2 - 3x + 2$.
(a) Show that f has exactly one critical number c in the interval $(1, 2)$.
(b) Use the bisection method (see Section 2.6) to approximate c to within $\frac{1}{16}$. Does f have a local maximum, a local minimum, or neither a maximum nor a minimum at c?

C 44. Let $f(x) = 2 + 20x + 4x^2 - x^4$.
(a) Show that f has exactly one critical number in the interval $(2, 3)$.
(b) Use the bisection method to approximate c to within $\frac{1}{16}$. Does f have a local maximum, a local minimum, or neither a maximum nor a minimum at c?

C 45. Let $f(x) = x^4 - 7x^2 + 2x - 3$.
(a) Show that f has exactly one critical number c in the interval $(1, 2)$.
(b) Use the Newton-Raphson method to approximate c; calculate x_3 and round your answer to four decimal places. Does f have a local maximum, a local minimum, or neither a maximum nor a minimum at c?

C 46. Let $f(x) = x \cos x$.
(a) Show that f has exactly one critical number in the interval $(0, \pi/2)$.
(b) Use the Newton-Raphson method to approximate c; calculate x_3 and round your answer to four decimal places. Does f have a local maximum, a local minimum, or neither a maximum nor a minimum at c?

C 47. Let $f(x) = \sin x + (x^2/2) - 2x$.
(a) Show that f has exactly one critical number in the interval $[2,3]$.
(b) Use the Newton-Raphson method to approximate c; calculate x_3 and round your answer to four decimal places. Does f have a local maximum, a local minimum, or neither a maximum nor a minimum at c?

C In Exercises 48–51, use a graphing utility to graph the function f on the indicated interval. (a) Use the graph to estimate the critical numbers and the local extreme values; and (b) estimate the intervals on which f increases and the intervals on which f decreases. Round off your estimates to three decimal places.

48. $f(x) = 3x^3 - 7x^2 - 14x + 24$; $[-3, 4]$.

49. $f(x) = |3x^3 + x^2 - 10x + 2| + 3x$; $[-4, 4]$.

50. $f(x) = x^4 - 6x^2 + 9x + 3$; $[-5, 5]$.

51. $f(x) = \dfrac{8 \sin 2x}{1 + \frac{1}{2}x^2}$; $[-3, 3]$.

In Exercises 52 and 53, the derivative f' of a function f is given. Use a graphing utility to graph f' on the indicated interval. Estimate the critical numbers of f and determine whether f has a local maximum, a local minimum, or neither a maximum nor a minimum at each. Round off your estimates to three decimal places.

52. $f'(x) = 2x^3 + x^2 - 4x + 3$; $[-4, 4]$.

53. $f'(x) = \sin^2 x + 2 \sin 2x$; $[-2, 2]$.

■ 4.4 ENDPOINT AND ABSOLUTE EXTREME VALUES

Endpoint Maxima and Minima

For functions defined on an open interval or on the union of open intervals the critical numbers are those where the derivative is 0 or the derivative does not exist. For functions defined on a closed or half-closed interval

$$[a, b], \quad [a, b), \quad (a, b], \quad [a, \infty), \quad \text{or} \quad (-\infty, b]$$

or on a union of such intervals, the *endpoints* of the domain (a and b in the case of $[a, b]$, a in the case of $[a, b)$ or $[a, \infty)$, and so forth) are also called *critical numbers*.

Endpoints can give rise to what are called *endpoint maxima* and *endpoint minima*. See, for example, Figures 4.4.1, 4.4.2, 4.4.3, and 4.4.4.

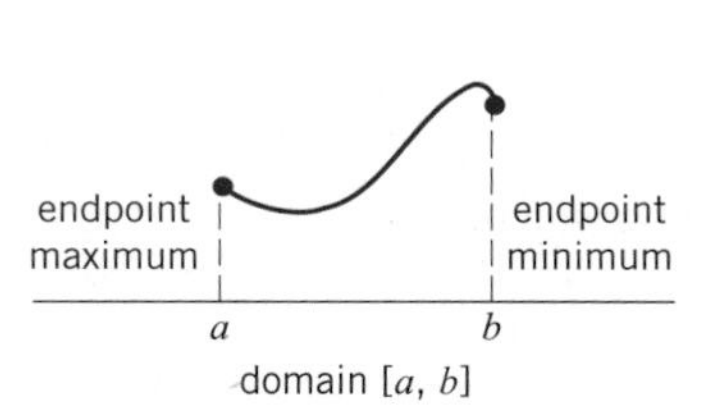

Figure 4.4.1

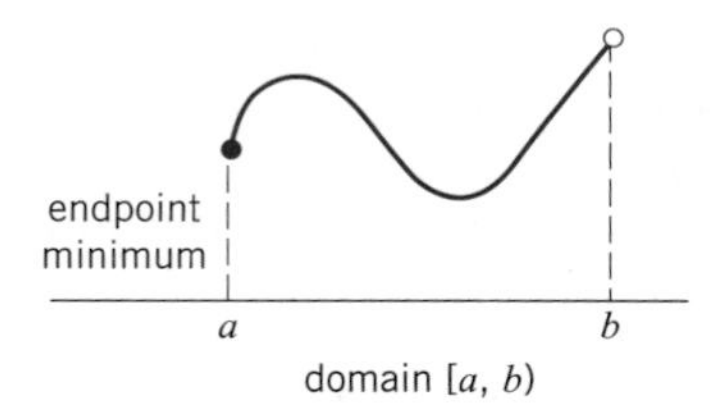

Figure 4.4.2

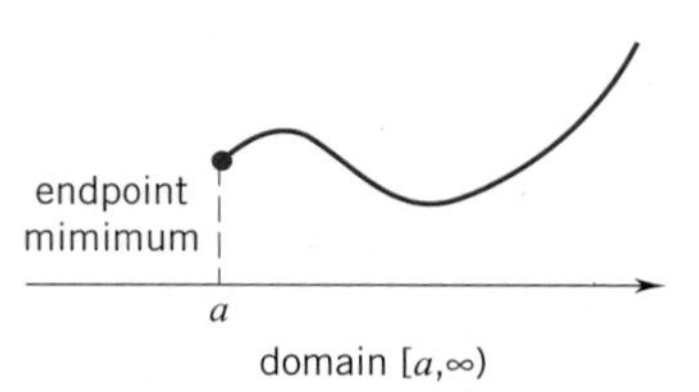

Figure 4.4.3

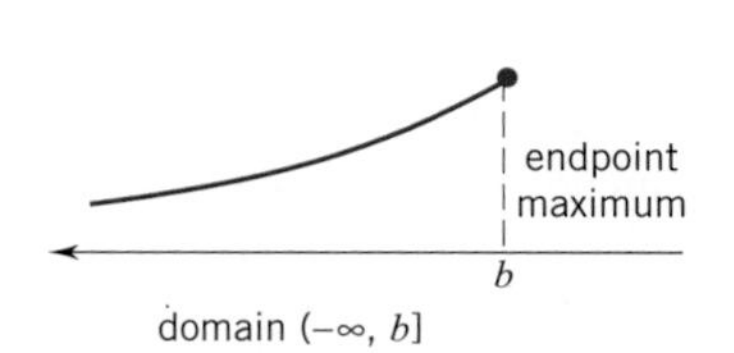

Figure 4.4.4

DEFINITION 4.4.1 ENDPOINT EXTREME VALUES

If c is an endpoint of the domain of f, then f is said to have an *endpoint maximum* at c iff

$$f(c) \geq f(x) \qquad \text{for all } x \text{ in the domain of } f \text{ sufficiently close to } c.$$

It is said to have an *endpoint minimum* at c iff

$$f(c) \leq f(x) \qquad \text{for all } x \text{ in the domain of } f \text{ sufficiently close to } c.$$

Endpoints are usually tested either by calculating a one-sided derivative at the endpoint, or by examining the sign of the derivative at nearby points. Suppose, for example, that a is the left endpoint and f is continuous from the right at a. If $f'_+(a) < 0$, or if $f'(x) < 0$ for all x sufficiently close to a, then f decreases on an interval of the form $[a, a + \delta)$ and therefore $f(a)$ must be an endpoint maximum. (Figure 4.4.1.) On the other hand, if $f'_+(a) > 0$, or if $f'(x) > 0$ for all x sufficiently close to a, then f increases on an interval of the form $[a, a + \delta)$ and so $f(a)$ must be an endpoint minimum. (Figure 4.4.2.) Similar reasoning can be applied to right endpoints.

Absolute Maxima and Minima

Whether or not a function f has a local or endpoint extreme value at some point depends entirely on the behavior of f for x close to that point. Absolute extreme values, which we define below, depend on the behavior of the function on its entire domain.

We begin with a number d in the domain of f. Here d can be an interior point or an endpoint.

DEFINITION 4.4.2 ABSOLUTE EXTREME VALUES

A function f is said to have an *absolute maximum at d* iff

$$f(d) \geq f(x) \qquad \text{for all } x \text{ in the domain of } f;$$

f is said to have an *absolute minimum at d* iff

$$f(d) \leq f(x) \qquad \text{for all } x \text{ in the domain of } f.$$

These values are called the *absolute extreme values of f.*

A function can be continuous on an interval (or even differentiable there) without taking on an absolute maximum or an absolute minimum. (The functions depicted in Figures 4.4.2 and 4.4.3 have no absolute maximum. The function depicted in Figure 4.4.4 has no absolute minimum.) All we can say in general is that, if f takes on an absolute extreme value, then it does so at a critical point.

There are some special cases, however, where the continuity of f does guarantee the existence of absolute extreme values. Recall Theorem 2.6.2 which states:

> If f is continuous on a closed, bounded interval [a, b], then
> **(i)** f is bounded on $[a, b]$, and
> **(ii)** f attains its (absolute) maximum value M and its (absolute) minimum value m on $[a, b]$.

In this case the absolute extreme values of f on $[a, b]$ can be determined as follows:

1. Find the critical numbers $c_1, c_2, \ldots, c_n$ of f in the open interval (a, b).
2. Calculate $f(a), f(c_1), f(c_2), \ldots, f(b)$.
3. The largest of the numbers found in step 2 is the absolute maximum of f and the smallest is the absolute minimum of f.

Example 1 Find the critical numbers and classify all extreme values of the function f defined on $[-\frac{1}{2}, \frac{3}{2}]$ by

$$f(x) = x^2 - 2|x| + 2 = \begin{cases} x^2 + 2x + 2, & -\frac{1}{2} \le x < 0 \\ x^2 - 2x + 2, & 0 \le x \le \frac{3}{2}. \end{cases}$$

SOLUTION Since f is continuous on a closed, bounded interval, we know that it has an absolute maximum and an absolute minimum. This function is differentiable on the open interval $(-\frac{1}{2}, \frac{3}{2})$, except at $x = 0$:

$$f'(x) = \begin{cases} 2x + 2, & -\frac{1}{2} < x < 0 \\ 2x - 2, & 0 < x < \frac{3}{2}. \end{cases} \qquad \text{(verify this)}$$

This makes 0 a critical number. Since $f'(x) = 0$ at $x = 1$, 1 is a critical number. The endpoints $-\frac{1}{2}$ and $\frac{3}{2}$ are also critical numbers. (Note that $2x + 2 = 0$ when $x = -1$, but we have ignored this value since it is not in the domain of f.)

The sign of f' and behavior of f are as follows:

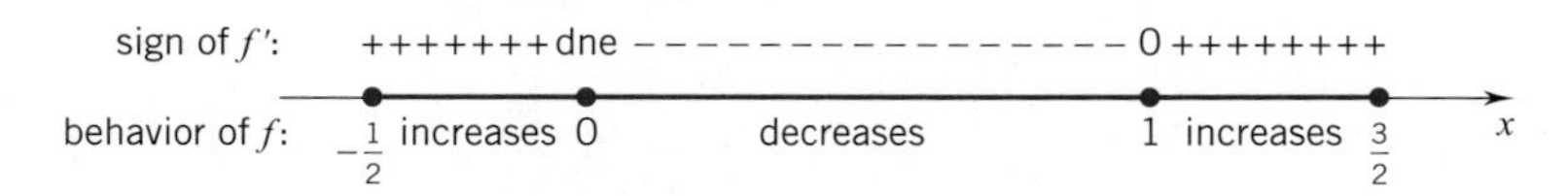

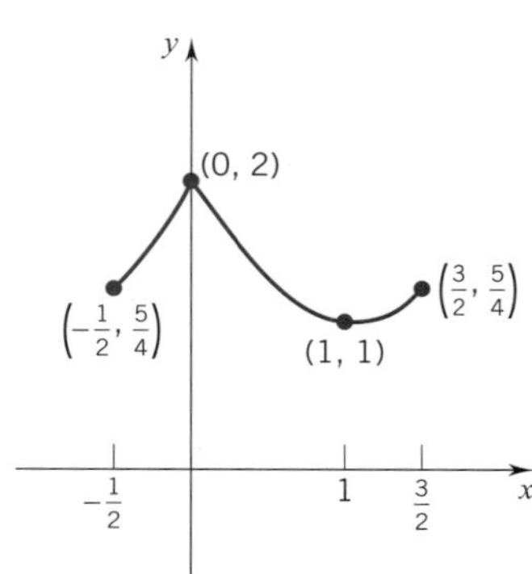

Figure 4.4.5

Therefore

$$f(-\tfrac{1}{2}) = \tfrac{1}{4} - 1 + 2 = \tfrac{5}{4} \quad \text{is an endpoint minimum;}$$
$$f(0) = 2 \quad \text{is a local maximum;}$$
$$f(1) = 1 - 2 + 2 = 1 \quad \text{is a local minimum;}$$
$$f(\tfrac{3}{2}) = \tfrac{9}{4} - 3 + 2 = \tfrac{5}{4} \quad \text{is an endpoint maximum.}$$

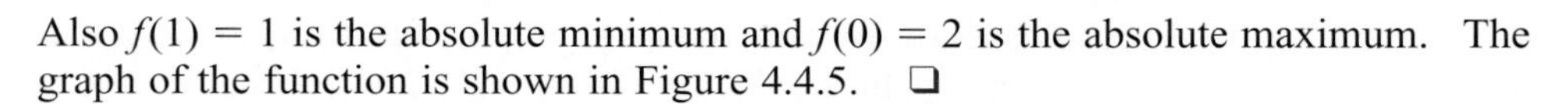

Also $f(1) = 1$ is the absolute minimum and $f(0) = 2$ is the absolute maximum. The graph of the function is shown in Figure 4.4.5. ❑

Behavior of f as $x \to \pm\infty$†

We now state four definitions. Once you grasp the first one, the others become transparent.

To say that

$$\text{as} \quad x \to \infty, \quad f(x) \to \infty$$

is to say that, *as x increases without bound, f(x) becomes arbitrarily large*. In formal terms, the definition is: Given any positive number M, there exists a positive number K such that

$$\text{if} \quad x \geq K, \qquad \text{then} \quad f(x) \geq M.$$

Thus, for example, as $x \to \infty$,

$$x^2 \to \infty, \qquad \sqrt{1 + x^2} \to \infty, \qquad \tan\left(\frac{\pi}{2} - \frac{2}{x^2}\right) \to \infty. \quad \square$$

To say that

$$\text{as} \quad x \to \infty, \quad f(x) \to -\infty$$

is to say that, *as x increases without bound, f(x) becomes arbitrarily large negative*. Again in formal terms, the definition is: Given any negative number M, there exists a positive number K such that

$$\text{if} \quad x \geq K, \qquad \text{then} \quad f(x) \leq M.$$

Thus, for example, as $x \to \infty$,

$$-x^4 \to -\infty, \qquad 1 - \sqrt{x} \to -\infty, \qquad \tan\left(\frac{1}{x^2} - \frac{\pi}{2}\right) \to -\infty. \quad \square$$

To say that

$$\text{as} \quad x \to -\infty, \quad f(x) \to \infty$$

is to say that, *as x decreases without bound, f(x) becomes arbitrarily large:* given any positive number M, there exists a negative number K such that,

$$\text{if} \quad x \leq K, \qquad \text{then} \quad f(x) \geq M.$$

Thus, for example, as $x \to -\infty$,

$$x^2 \to \infty, \qquad \sqrt{1 - x} \to \infty, \qquad \tan\left(\frac{\pi}{2} - \frac{1}{x^2}\right) \to \infty. \quad \square$$

Finally, to say that

$$\text{as} \quad x \to -\infty, \quad f(x) \to -\infty$$

is to say that, *as x decreases without bound, f(x) becomes arbitrarily large negative*: given any negative number M, there exists a negative number K such that,

$$\text{if} \quad x \leq K, \qquad \text{then} \quad f(x) \leq M.$$

† We will use the notation $x \to \pm\infty$ to mean $x \to \infty$ or $x \to -\infty$.

Thus, for example, as $x \to -\infty$,

$$x^3 \to -\infty, \qquad -\sqrt{1-x} \to -\infty, \qquad \tan\left(\frac{1}{x^2} - \frac{\pi}{2}\right) \to -\infty. \quad \square$$

Remark As you can readily see, $f(x) \to -\infty$ iff $-f(x) \to \infty$. $\square$

Suppose now that P is a nonconstant polynomial:

$$P(x) = a_n x^n + a_{n-1}x^{n-1} + \cdots + a_1 x + a_0 \qquad (a_n \neq 0,\ n \geq 1).$$

When $|x|$ is large—that is, when x is either a large positive number or a large negative number—the leading term $a_n x^n$ dominates. Thus, what happens to $P(x)$ as $x \to \pm\infty$ depends entirely on what happens to $a_n x^n$. (For confirmation, do Exercise 41.)

Example 2

(a) As $x \to \infty$, $3x^4 \to \infty$, and therefore $3x^4 - 100x^3 + 2x - 5 \to \infty$.

(b) As $x \to -\infty$, $5x^3 \to -\infty$, and therefore $5x^3 + 12x^2 + 80 \to -\infty$. $\square$

Finally, we point out that, if $f(x) \to \infty$, then f cannot have an absolute maximum value and, if $f(x) \to -\infty$, then f cannot have an absolute minimum value.

The following theorem is sometimes helpful in finding absolute extreme values of a continuous function defined on an arbitrary interval I.

> **THEOREM 4.4.3**
>
> Let f be continuous on an interval I and suppose that f has exactly one local extreme value in the interior of I, say at $x = c$.
>
> If $f(c)$ is a local maximum, then $f(c)$ is the absolute maximum of f on I.
> If $f(c)$ is a local minimum, then $f(c)$ is the absolute minimum of f on I.

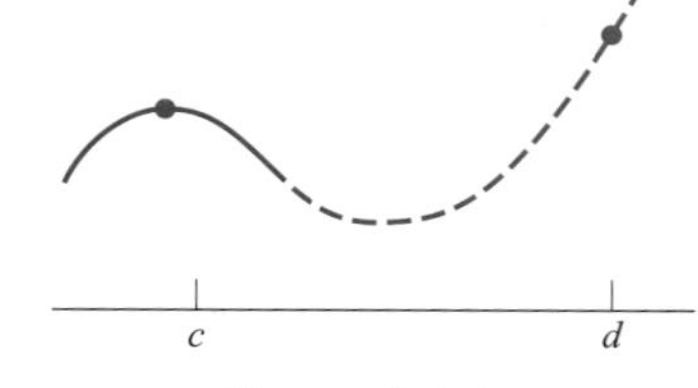

Figure 4.4.6

An intuitive proof of this theorem goes as follows: Suppose that f has a local maximum at $x = c$. If $f(c)$ is not the absolute maximum of f, then there is a point d such that $f(d) \geq f(c)$. But this would imply that there exists a local minimum of f between $x = c$ and $x = d$. See Figure 4.4.6.

Example 3 Find the critical numbers and classify all the extreme values of

$$f(x) = 6\sqrt{x} - x\sqrt{x}.$$

SOLUTION The domain is $[0, \infty)$. Therefore the endpoint 0 is a critical number.
This function will be easier to differentiate if we rewrite it using fractional exponents:

$$f(x) = 6x^{1/2} - x^{3/2}.$$

Now, on $(0, \infty)$,

$$f'(x) = 3x^{-1/2} - \tfrac{3}{2}x^{1/2} = \frac{3(2-x)}{2\sqrt{x}}. \qquad \text{(verify this)}$$

Since $f'(x) = 0$ at $x = 2$, we see that 2 is an interior critical number.

The sign of f' and the behavior of f are as follows:

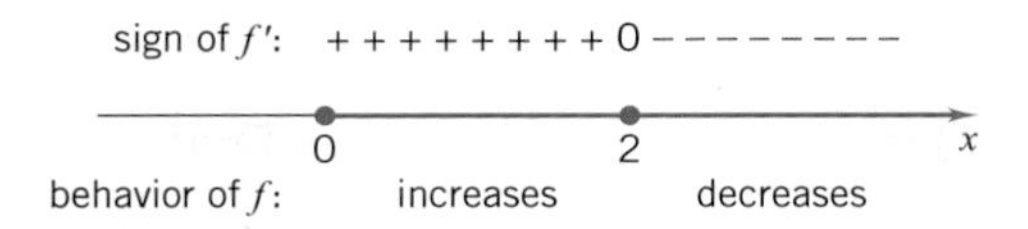

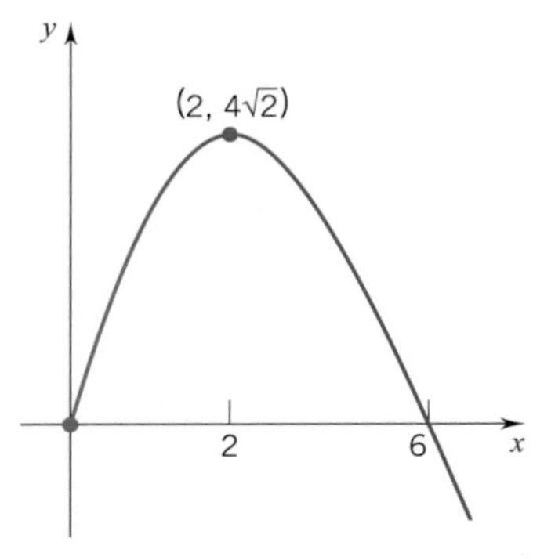

Figure 4.4.7

Therefore,

$$f(0) = 0 \text{ is an endpoint minimum;}$$
$$f(2) = 6\sqrt{2} - 2\sqrt{2} = 4\sqrt{2} \text{ is a local maximum.}$$

Since $f(x) = \sqrt{x}(6 - x)$, it is easy to see that $f(x) \to -\infty$ as $x \to \infty$. Thus the function has no absolute minimum value. Also, since f has exactly one local extreme value on $(0, \infty)$, $f(2) = 4\sqrt{2}$ is the absolute maximum value of f by Theorem 4.4.3. The graph of f appears in Figure 4.4.7. ❑

A Summary on Finding All the Extreme Values (Local, Endpoint, and Absolute) of a Continuous Function *f*

1. Find the critical numbers. These are the endpoints of the domain (if any) and the interior numbers c at which $f'(c) = 0$ or $f'(c)$ does not exist.
2. Test each endpoint of the domain by examining the sign of the first derivative nearby.
3. Test each interior critical number c by examining the sign of the first derivative on both sides of c (first-derivative test) or by checking the sign of the second derivative at c itself (second-derivative test).
4. If f is defined on (a, ∞) for some number a or on $(-\infty, b)$ for some number b, determine the behavior of f as $x \to \infty$ or as $x \to -\infty$.
5. Determine whether any of the endpoint extremes and local extremes are absolute extremes; use Theorems 2.6.2 and Theorem 4.4.3 when applicable.

Example 4 Find the critical numbers and classify all the extreme values of

$$f(x) = \tfrac{1}{4}(x^3 - \tfrac{3}{2}x^2 - 6x + 2), \qquad x \in [-2, \infty).$$

SOLUTION The left endpoint -2 is a critical number. To find the interior critical numbers, we differentiate:

$$f'(x) = \tfrac{1}{4}(3x^2 - 3x - 6) = \tfrac{3}{4}(x + 1)(x - 2).$$

Since $f'(x) = 0$ at $x = -1$ and $x = 2$, the numbers -1 and 2 are interior critical numbers.

The sign of f' and the behavior of f are:

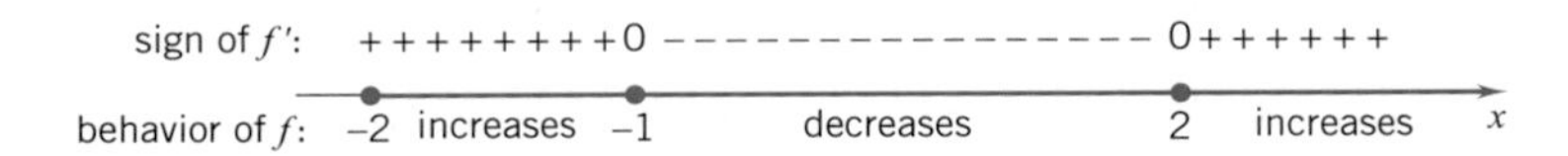

We can see from the sign of f' that

$$f(-2) = \tfrac{1}{4}(-8 - 6 + 12 + 2) = 0 \quad \text{is an endpoint minimum;}$$
$$f(-1) = \tfrac{1}{4}(-1 - \tfrac{3}{2} + 6 + 2) = \tfrac{11}{8} \quad \text{is a local maximum;}$$
$$f(2) = \tfrac{1}{4}(8 - 6 - 12 + 2) = -2 \quad \text{is a local minimum.}$$

The function takes on no absolute maximum value since $f(x) \to \infty$ as $x \to \infty$; $f(2) = -2$ is the absolute minimum value of f. The graph of f is shown in Figure 4.4.8. ❑

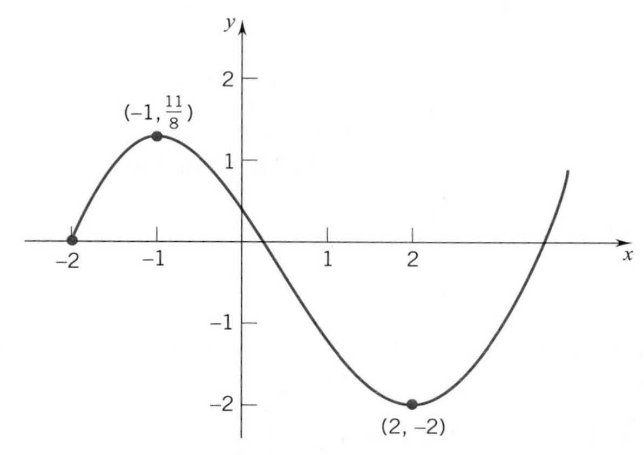

Figure 4.4.8

Example 5 Find the critical numbers and classify all the extreme values of

$$f(x) = \sin x - \sin^2 x, \qquad x \in [0, 2\pi].$$

SOLUTION The endpoints 0 and 2π are critical numbers. On $(0, 2\pi)$

$$\begin{aligned} f'(x) &= \cos x - 2 \sin x \cos x \\ &= \cos x(1 - 2 \sin x). \end{aligned}$$

Setting $f'(x) = 0$, we have

$$\cos x(1 - 2 \sin x) = 0.$$

This equation is satisfied when $\cos x = 0$, which gives $x = \pi/2$ and $x = 3\pi/2$; and when $\sin x = \frac{1}{2}$, which gives $x = \pi/6$ and $x = 5\pi/6$. Thus, the numbers $\pi/6$, $\pi/2$, $5\pi/6$, and $3\pi/2$ are interior critical numbers.

The sign of f' and the behavior of f are as follows:

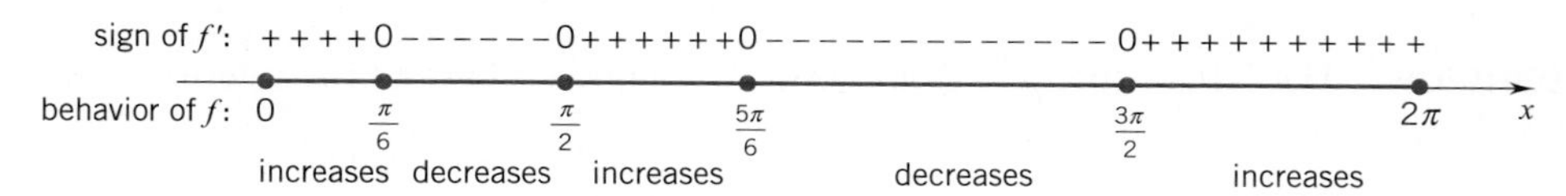

Therefore,

$$f(0) = 0 \quad \text{is an endpoint minimum;}$$
$$f(\pi/6) = \tfrac{1}{4} \quad \text{is a local maximum;}$$
$$f(\pi/2) = 0 \quad \text{is a local minimum;}$$
$$f(5\pi/6) = \tfrac{1}{4} \quad \text{is local maximum;}$$
$$f(3\pi/2) = -2 \quad \text{is a local minimum;}$$
$$f(2\pi) = 0 \quad \text{is an endpoint maximum}$$

Also, $f(\pi/6) = f(5\pi/6) = \frac{1}{4}$ is the absolute maximum of f and $f(3\pi/2) = -2$ is the absolute minimum. The graph of the function is shown in Figure 4.4.9. ❑

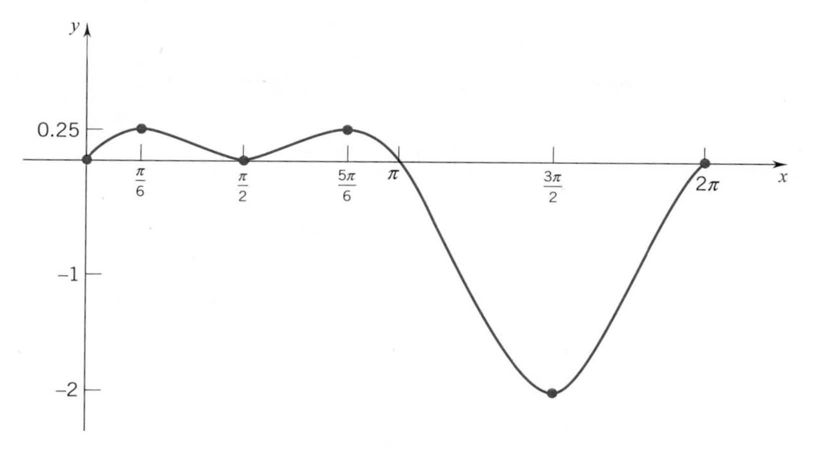

Figure 4.4.9

In the next section we will illustrate some practical applications of the theory that we have developed up to this point.

EXERCISES 4.4

In Exercises 1–30, find the critical numbers and classify the extreme values.

1. $f(x) = \sqrt{x+2}$.

2. $f(x) = (x-1)(x-2)$.

3. $f(x) = x^2 - 4x + 1, \quad x \in [0, 3]$.

4. $f(x) = 2x^2 + 5x - 1, \quad x \in [-2, 0]$.

5. $f(x) = x^2 + \dfrac{1}{x}$.

6. $f(x) = x + \dfrac{1}{x^2}$.

7. $f(x) = x^2 + \dfrac{1}{x}, \quad x \in [\frac{1}{10}, 2]$.

8. $f(x) = x + \dfrac{1}{x^2}, \quad x \in [1, \sqrt{2}]$.

9. $f(x) = (x-1)(x-2), \quad x \in [0, 2]$.

10. $f(x) = (x-1)^2(x-2)^2, \quad x \in [0, 4]$.

11. $f(x) = \dfrac{x}{4+x^2}, \quad x \in [-3, 1]$.

12. $f(x) = \dfrac{x^2}{1+x^2}, \quad x \in [-1, 2]$.

13. $f(x) = (x - \sqrt{x})^2$.

14. $f(x) = x\sqrt{4 - x^2}$.

15. $f(x) = x\sqrt{3 - x}$.

16. $f(x) = \sqrt{x} - \dfrac{1}{\sqrt{x}}$.

17. $f(x) = 1 - \sqrt[3]{x-1}$.

18. $f(x) = (4x-1)^{1/3}(2x-1)^{2/3}$.

19. $f(x) = \sin^2 x - \sqrt{3}\cos x, \quad 0 \le x \le \pi$.

20. $f(x) = \cot x + x, \quad 0 < x \le \dfrac{2\pi}{3}$.

21. $f(x) = 2\cos^3 x + 3\cos x, \quad 0 \le x \le \pi$.

22. $f(x) = \sin 2x - x, \quad 0 \le x \le \pi$.

23. $f(x) = \tan x - x, \quad -\dfrac{\pi}{3} \le x < \dfrac{\pi}{2}$.

24. $f(x) = \sin^4 x - \sin^2 x, \quad 0 \le x \le \dfrac{2\pi}{3}$.

25. $f(x) = \begin{cases} -2x, & 0 \le x < 1 \\ x - 3, & 1 \le x \le 4 \\ 5 - x, & 4 < x \le 7. \end{cases}$

26. $f(x) = \begin{cases} x + 9, & -8 \le x < -3 \\ x^2 + x, & -3 \le x \le 2 \\ 5x - 4, & 2 < x < 5. \end{cases}$

27. $f(x) = \begin{cases} x^2 + 1, & -2 \le x < -1 \\ 5 + 2x - x^2, & -1 \le x \le 3 \\ x - 1, & 3 < x < 6. \end{cases}$

28. $f(x) = \begin{cases} 2 - 2x - x^2, & -2 \le x \le 0 \\ |x - 2|, & 0 < x < 3 \\ \frac{1}{3}(x-2)^3, & 3 \le x \le 4. \end{cases}$

29. $f(x) = \begin{cases} |x + 1|, & -3 \le x < 0 \\ x^2 - 4x + 2, & 0 \le x < 3 \\ 2x - 7, & 3 \le x < 4. \end{cases}$

30. $f(x) = \begin{cases} -x^2, & 0 \le x < 1 \\ -2x, & 1 < x < 2 \\ -\frac{1}{2}x^2, & 2 \le x \le 3. \end{cases}$

In Exercises 31–34, sketch the graph of a differentiable function satisfying the given conditions, if possible. If it is not possible, state why.

31. Domain is $(-\infty, \infty)$, local maximum at -1, local minimum at 1, $f(3) = 6$ is the absolute maximum, no absolute minimum.

32. Domain is $[0, \infty)$, $f(0) = 1$ is the absolute minimum, local maximum at 4, local minimum at 7, no absolute maximum.

33. Domain is $[-3, 3]$, $f(-3) = f(3) = 0$, $f(x) \neq 0$ for all $x \in (-3, 3)$, $f'(-1) = f'(1) = 0$, $f'(x) > 0$ if $|x| > 1$, $f'(x) < 0$ if $|x| < 1$.

34. Domain is $(-1, 3]$, local minimum at 2, $f(1) = 5$ is the absolute maximum, no absolute minimum.

35. Show that the cubic polynomial $p(x) = x^3 + ax^2 + bx + c$ has no extreme values iff $a^2 \leq 3b$.

36. Let r be a rational number, $r > 1$, and let f be the function defined by

$$f(x) = (1 + x)^r - (1 + rx) \qquad \text{for} \quad x \geq -1.$$

Show that 0 is a critical number of f and show that $f(0) = 0$ is the absolute minimum of f.

37. Suppose that c is a critical number for f and $f'(x) > 0$ for $x \neq c$. Prove that, if $f(c)$ is a local maximum, then f is not continuous at c.

38. What can be said about the function f continuous on $[a, b]$, if for some c in (a, b), $f(c)$ is both a local maximum and a local minimum?

39. Suppose that f is continuous on its domain $[a, b]$ and $f(a) = f(b)$. Prove that f has at least one critical point in (a, b).

40. Suppose that $c_1 < c_2$ and that both $f(c_1)$ and $f(c_2)$ are local maxima. Prove that, if f is continuous on $[c_1, c_2]$, then, for some c in (c_1, c_2), $f(c)$ is a local minimum.

41. Let P be a nonconstant polynomial with positive leading coefficient:

$$P(x) = a_n x^n + a_{n-1}x^{n-1} + \cdots + a_1 x + a_0 \qquad (n \geq 1,\ a_n > 0).$$

Clearly, as $x \to \infty$, $a_n x^n \to \infty$. Show that, as $x \to \infty$, $P(x) \to \infty$ by showing that, given any positive number M, there exists a positive number K such that, if $x \geq K$, then $f(x) \geq M$.

42. An object moving along a straight line with position function $x(t)$ given by

$$x(t) = A \sin(\omega t + \phi_0),$$

where A, ω, and ϕ_0 are constants, is said to be in *simple harmonic motion.* For example, consider the motion of a mass m suspended by a spring and oscillating about its equilibrium position.

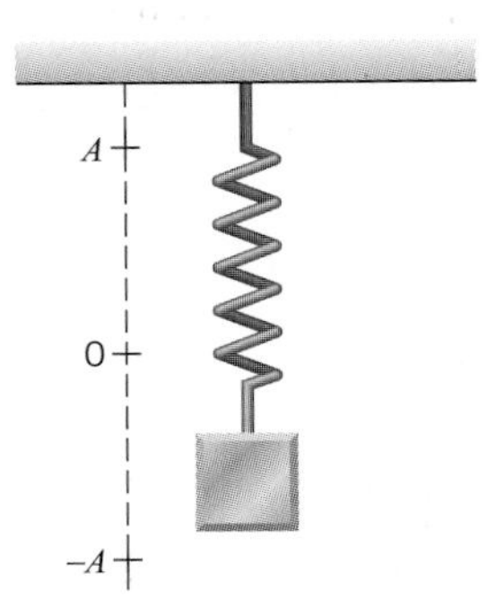

(a) Show that the position function $x(t)$ satisfies

$$x''(t) + \omega^2 x(t) = 0.$$

(b) Find the absolute maximum and the absolute minimum values of $x(t)$.

43. Of all rectangles with a given diagonal of length c, prove that the square has the largest area.

44. The sum of two numbers is 16. Find the numbers if the sum of their cubes is an absolute minimum.

45. A piece of wire of length L will be cut into two pieces, one piece to form a square and the other piece to form an equilateral triangle. How should the wire be cut so as to:
(a) Maximize the sum of the areas of the square and triangle?
(b) Minimize the sum of the areas of the square and triangle?

In Exercises 46–49, use a graphing utility to graph the function f on the indicated interval. Estimate the critical numbers of f and classify all the extreme values. Round off your estimates to three decimal places.

46. $f(x) = x^4 - 7x^2 + 10x + 3$; $[-3, 3]$.

47. $f(x) = x^3 - 4x + 2x \sin x$; $[-2.5, 3]$.

48. $f(x) = 5x^{2/3} + 3x^{5/3} + 1$; $[-3, 1]$.

49. $f(x) = x \cos 2x - \cos^2 x$; $[-\pi, \pi]$.

■ 4.5 SOME MAX-MIN PROBLEMS

The techniques of the last two sections can be brought to bear on a wide variety of max-min problems. The *key* idea in solving such problems is to express the quantity to be maximized or minimized as a function of one variable. If the function is differentiable, we can then differentiate and analyze the results.

Example 1 An architect wants to design a window in the shape of a rectangle capped by a semicircle. If the perimeter of the window is constrained to be 24 feet, what dimensions should the architect choose for the window in order to admit the greatest amount of light?

SOLUTION The maximum amount of light will be admitted when the area of the window is a maximum. As indicated in Figure 4.5.1, we let x be the radius of the semicircular part and y the height of the rectangular part. Then the area A of the window is given by

$$A = \tfrac{1}{2}\pi x^2 + 2xy.$$

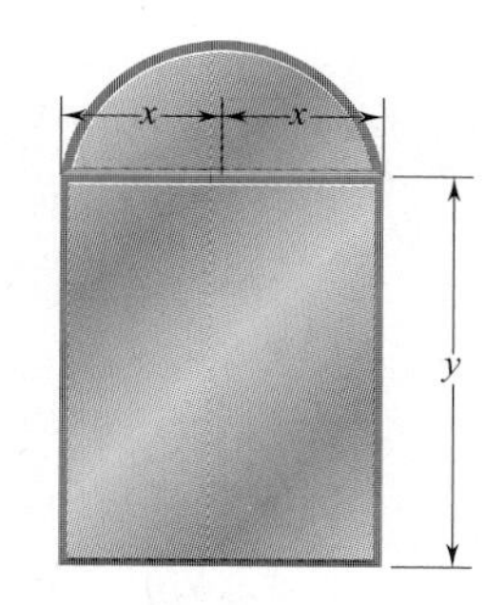

Figure 4.5.1

In order to apply the methods of the preceding sections, we need to express A as a function of one variable only.

Since the perimeter is 24 feet, we have

$$2x + 2y + \pi x = 24$$

and so

$$y = \frac{24 - (2+\pi)x}{2} = 12 - \frac{(2+\pi)}{2}x.$$

Now the area can be expressed in terms of x alone:

$$\begin{aligned} A(x) &= \tfrac{1}{2}\pi x^2 + 2x\left[12 - \frac{(2+\pi)}{2}x\right] \\ &= \tfrac{1}{2}\pi x^2 + 24x - (2+\pi)x^2 = 24x - \left(2 + \tfrac{1}{2}\pi\right)x^2 \end{aligned}$$

Also, since x and y represent lengths and A is an area, these variables cannot be negative. Therefore, it follows that $0 \le x \le 24/(2+\pi)$.

Our problem can now be formulated as: Find the absolute maximum of the function

$$A(x) = 24x - (2 + \tfrac{1}{2}\pi)\,x^2, \qquad x \in \left[0, \frac{24}{2+\pi}\right].$$

The derivative

$$A'(x) = 24 - 2\,(2 + \tfrac{1}{2}\pi)\,x = 24 - (4+\pi)x$$

is 0 only at $x = 24/(4+\pi)$. Since $A'(x) > 0$ for $0 < x < 24/(4+\pi)$ and $A'(x) < 0$ for $24/(4+\pi) < x < 24/(2+\pi)$, it follows that $A[24/(4+\pi)]$ is a local maximum. (NOTE: We could just as easily have used the second-derivative test to find that A has a local maximum at this value.) Also, since A has only one local extreme value on the interval, we can conclude, by Theorem 4.4.3, that $A[24/(4+\pi)]$ is the absolute maximum.

The dimensions of the window that will admit the greatest amount of light are:

radius of semicircular part $\quad x = \dfrac{24}{4+\pi} \cong 3.36$ feet,

height of rectangular part $\quad y = 12 - \left(\dfrac{2+\pi}{2}\right)\left(\dfrac{24}{4+\pi}\right) = \dfrac{24}{4+\pi},$

the height of the rectangle equals the radius of the semicircle. ❑

This example illustrates a basic strategy for solving max-min problems.

Strategy

1. Draw a representative figure and assign labels to the relevant quantities involved in the problem.
2. Identify the quantity that is to be maximized or minimized and find a formula for it.
3. Express the quantity to be maximized or minimized as a function of one variable only; use the conditions given in the problem to eliminate variables.
4. Determine the domain of the function found in (3).
5. Apply the techniques of the preceding sections to find the extreme value(s).

Steps 3 and 4 are the key steps in this procedure.

Example 2 A soft-drink manufacturer wants to fabricate cylindrical cans for its product. The can is to have a volume of 12 fluid ounces, which is approximately 22 cubic inches. Find the dimensions of the can that will require the least amount of material. See Figure 4.5.2.

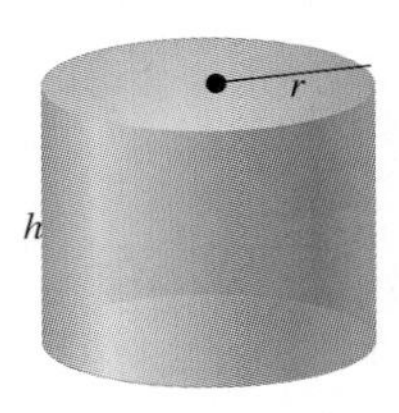

Figure 4.5.2

Let r be the radius of the can and h its height. The total surface area (top, bottom, lateral area) of a right circular cylinder of radius r and height h is

$$S = 2\pi r^2 + 2\pi rh,$$

and this is the quantity that we want to minimize.

Now, since the volume $V = \pi r^2 h$ has to be 22 cubic inches, we have

$$\pi r^2 h = 22$$

and

$$h = \frac{22}{\pi r^2}.$$

Also, since r and h must both be positive, it follows that $0 < r < \infty$.

Thus, we want to minimize the function

$$\begin{aligned} S(r) &= 2\pi r^2 + 2\pi r\left(\frac{22}{\pi r^2}\right) \\ &= 2\pi r^2 + \frac{44}{r}, \qquad r \in (0, \infty). \end{aligned}$$

(Key steps completed.)

The derivative of S is

$$\frac{dS}{dr} = 4\pi r - \frac{44}{r^2}.$$

To find the critical numbers of S, set $dS/dr = 0$:

$$4\pi r - \frac{44}{r^2} = 0$$

$$r^3 = \frac{11}{\pi}$$

and

$$r = \sqrt[3]{\frac{11}{\pi}}$$

is the only critical number. Since

$$\frac{d^2S}{dr^2} = 4\pi + \frac{88}{r^3} > 0$$

at $r = \sqrt[3]{11/\pi}$, S has a local minimum at this point. By Theorem 4.4.3, $S(\sqrt[3]{11/\pi})$ is the absolute minimum value of S. Therefore, the dimensions of the can that will require the least amount of material are:

$$\text{radius} \qquad r = \sqrt[3]{\frac{11}{\pi}} \cong 1.5 \text{ inches},$$

$$\text{height} \qquad h = \frac{22}{\pi(11/\pi)^{2/3}} = 2\sqrt[3]{\frac{11}{\pi}} \cong 3 \text{ inches};$$

the height of the cylindrical can should be twice the radius. ❑

Example 3 An isosceles triangle has base 6 and height 12. Find the maximum possible area of a rectangle that can be placed inside the triangle with one side on the base of the triangle.

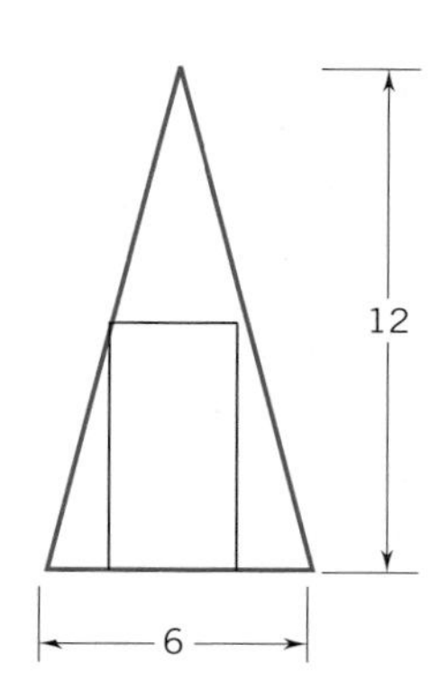

Figure 4.5.3

SOLUTION Figure 4.5.3 shows the isosceles triangle with a typical rectangle placed inside. In Figure 4.5.4, we have introduced a rectangular coordinate system. With x and y as indicated in Figure 4.5.4, we want to maximize the area A, which is given by

$$A = 2xy.$$

Now, since the point (x, y) lies on the line through $(0, 12)$ and $(3, 0)$, it can be shown that $y = 12 - 4x$. Also, since x and y are both nonnegative, we have $0 \leq x \leq 3$. Thus, our problem is:

Maximize the function

$$A(x) = 2x(12 - 4x) = 24x - 8x^2, \qquad x \in [0, 3].$$

(Key steps completed.)

The derivative

$$A'(x) = 24 - 16x$$

is 0 only at $x = 24/16 = 3/2$. Since $A'(x) > 0$ for $0 < x < 3/2$ and $A'(x) < 0$ for $3/2 < x < 3$, it follows that A has a local maximum and hence an absolute maximum at $x = 3/2$. If $x = 3/2$, then $y = 6$ and the maximum possible area is $A = 2(3/2)6 = 18$ square units. ❑

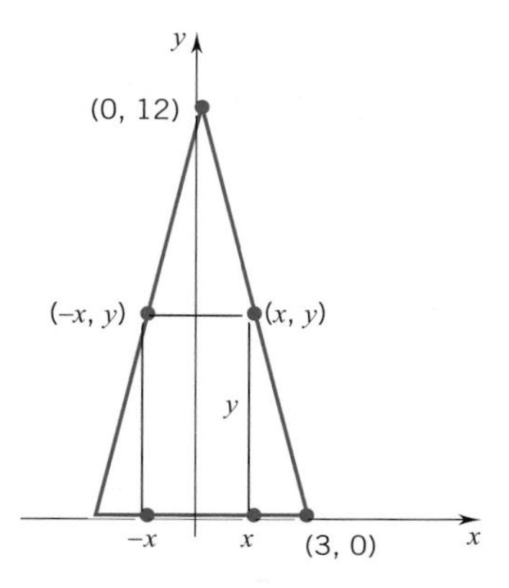

Figure 4.5.4

Example 4 A state highway department plans to construct a new road between town A and town B. Town A lies on an abandoned road that runs east-west. Town B is 3 miles north of the point on that road that is 5 miles east of A. The engineering division proposes that the road be constructed by restoring a section of the old road from A up to a point P and joining it to a new road that connects P and B. If the cost of restoring the old road is \$200,000 per mile and the cost of the new road is \$400,000 per mile, how much of the old road should be restored in order to minimize the department's costs?

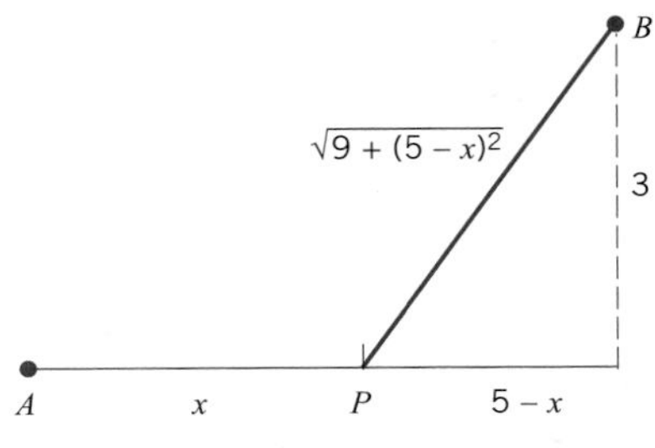

Figure 4.5.5

SOLUTION Figure 4.5.5 shows the geometry of the problem. Notice that we have chosen a straight line joining P and B rather than some curved path (recall that the shortest distance between two points is a straight line). We let x be the amount of old road that will be restored. Then

$$\sqrt{9 + (5 - x)^2} = \sqrt{34 - 10x + x^2}$$

is the length of the new road. The total cost of constructing the two sections of the road is

$$C(x) = 200{,}000x + 400{,}000[34 - 10x + x^2]^{1/2}, \qquad 0 \leq x \leq 5.$$

We want to find the value of x that minimizes this function.

(*Key steps completed.*)

Differentiation gives

$$\begin{aligned} C'(x) &= 200{,}000 + 400{,}000\left(\tfrac{1}{2}\right)[34 - 10x + x^2]^{-1/2}(2x - 10) \\ &= 200{,}000 + \frac{400{,}000(x - 5)}{[34 - 10x + x^2]^{1/2}}, \qquad 0 < x < 5. \end{aligned}$$

Setting $C'(x) = 0$, we find that

$$1 + \frac{2(x - 5)}{[34 - 10x + x^2]^{1/2}} = 0$$

$$2(x - 5) = -[34 - 10x + x^2]^{1/2}$$

$$4(x^2 - 10x + 25) = 34 - 10x + x^2$$

$$3x^2 - 30x + 66 = 0.$$

$$x^2 - 10x + 22 = 0.$$

Now, by the quadratic formula, we have

$$x = \frac{10 \pm \sqrt{100 - 4(22)}}{2} = 5 \pm \sqrt{3}.$$

The value $x = 5 + \sqrt{3}$ is not in the domain of our function; the value we want is $x = 5 - \sqrt{3}$. You can verify that the sign of C' is:

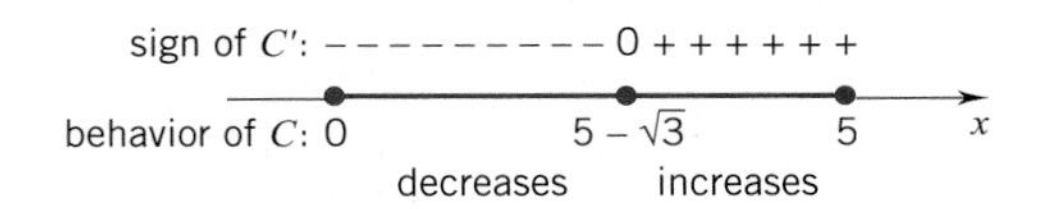

Thus $x = 5 - \sqrt{3} \cong 3.27$ gives the absolute minimum value for C. The highway department will minimize its costs by restoring 3.27 miles of the old road.

You can check that the minimum cost is $\$200{,}000(5 + 3\sqrt{3}) \cong \$2{,}039{,}230$. Using all of the old road, the cost would be exactly \$2,200,000, while using none it would cost $\$400{,}000\sqrt{34} \cong \$2{,}332{,}400$. The most expensive option costs almost \$300,000 more than the cheapest option. ❑

Example 5 A right circular cylinder is inscribed in a cone of height H and radius R. Determine the dimensions of the cylinder with the largest possible volume. What is the maximum volume?

SOLUTION Let r be the radius of the cylinder and h its height. We want to find the maximum value of

$$V = \pi r^2 h.$$

Using similar triangles (see Figure 4.5.6), we see that

$$\frac{H}{R} = \frac{H-h}{r},$$

and so

$$h = H - \frac{Hr}{R}.$$

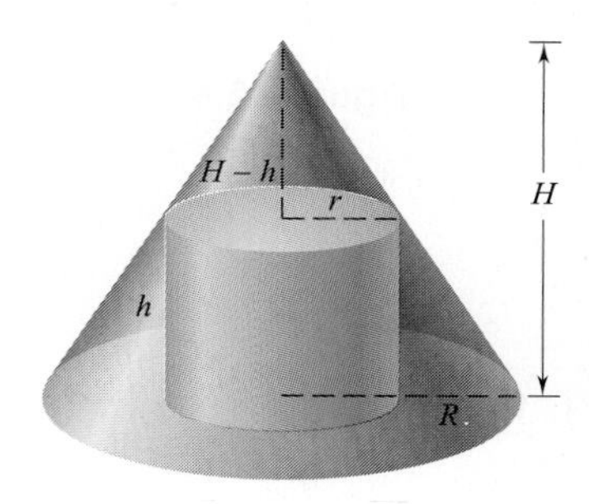

Figure 4.5.6

Since r and h are both nonnegative, it follows that $0 \le r \le R$. Thus, we want to maximize the function

$$V(r) = \pi r^2 \left[H - \frac{Hr}{R}\right] = \pi r^2 H - \frac{\pi r^3 H}{R}, \qquad r \in [0, R].$$

(*Key steps completed.*)

The endpoints, 0 and R, are critical numbers of V, and it is easy to see that $V(0) = V(R) = 0$ is the absolute minimum of V. The derivative of V on $(0, R)$ is

$$\frac{dV}{dr} = 2\pi r H - \frac{3\pi r^2 H}{R}.$$

Setting $dV/dr = 0$, we have

$$\pi r H \left(2 - \frac{3r}{R}\right) = 0$$

so

$$r = 0 \qquad \text{and} \qquad r = \frac{2R}{3}.$$

The only critical number in $(0, R)$ is $r = 2R/3$. By examining the sign of the first derivative, or by calculating the second derivative and evaluating at $r = 2R/3$, you can verify that V has a local maximum and hence an absolute maximum at this value. The dimensions of the cylinder with the largest possible volume are $r = 2R/3$ and $h = H/3$; the maximum volume is $V = 4\pi R^2 H/27$. ❑

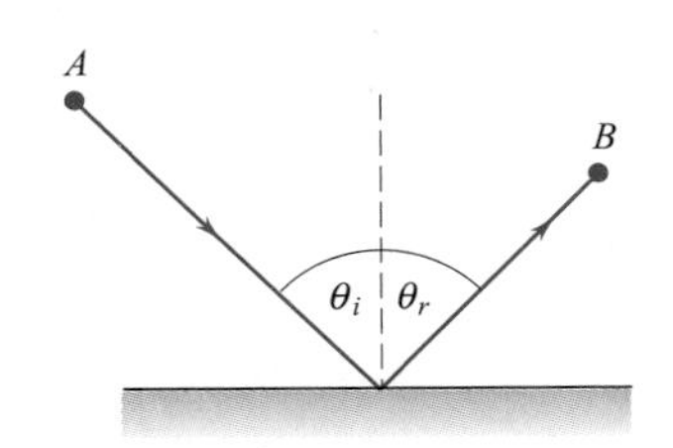

Figure 4.5.7

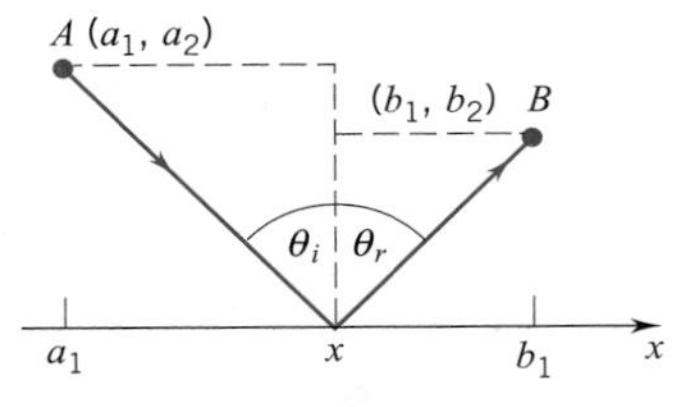

Figure 4.5.8

Example 6 (*The angle of incidence equals the angle of reflection.*) Figure 4.5.7 depicts light from a point A reflected to a point B by a mirror. Two angles have been marked: the *angle of incidence*, θ_i, and the *angle of reflection*, θ_r. Experiment shows that $\theta_i = \theta_r$. Derive this result by postulating that the light that travels from A to the mirror and then to B follows the shortest possible path.†

SOLUTION We can write the length of the path as a function of x: in the setup of Figure 4.5.8,

$$L(x) = \sqrt{(x - a_1)^2 + a_2^2} + \sqrt{(x - b_1)^2 + b_2^2}, \qquad x \in [a_1, b_1].$$

Differentiation gives

$$L'(x) = \frac{x - a_1}{\sqrt{(x - a_1)^2 + a_2^2}} + \frac{x - b_1}{\sqrt{(x - b_1)^2 + b_2^2}}.$$

We can therefore see that

$$\begin{aligned} L'(x) = 0 \quad & \text{iff} \quad \frac{x - a_1}{\sqrt{(x - a_1)^2 + a_2^2}} = \frac{b_1 - x}{\sqrt{(x - b_1)^2 + b_2^2}} \\ & \text{iff} \quad \sin\theta_i = \sin\theta_r \qquad \text{(see the figure)} \\ & \text{iff} \quad \theta_i = \theta_r. \end{aligned}$$

That $L(x)$ is minimal when $\theta_i = \theta_r$ can be seen from noting that $L''(x)$ is always positive:

$$L''(x) = \frac{a_2^2}{[(x - a_1)^2 + a_2^2]^{3/2}} + \frac{b_2^2}{[(x - b_1)^2 + b_2^2]^{3/2}} > 0. \quad \square$$

We must admit that there is a simpler way to do Example 6 that requires no calculus. Can you find it?

Now we will work out a problem in which the function to be maximized is defined, not on an interval or union of intervals, but on a discrete set of points, in this case a finite collection of integers.

Example 7 A manufacturing plant has a capacity of 25 articles per week. Experience has shown that n articles per week can be sold at a price of p dollars each where $p = 110 - 2n$ and the cost of producing n articles is $600 + 10n + n^2$ dollars. How many articles should be made each week to give the largest profit?

SOLUTION The profit (P dollars) on the sale of n articles is

$$P = \text{revenue} - \text{cost} = np - (600 + 10n + n^2).$$

With $p = 110 - 2n$, this simplifies to

$$P = 100n - 600 - 3n^2.$$

In this problem n must be an integer, and thus it makes no sense to differentiate P with respect to n. The formula shows that P is negative if n is less than 8 or greater than

† This is a special case of Fermat's *Principle of Least Time*, which says that, of all (neighboring) paths, light chooses the one that demands the least time. If light passes from one medium to another, the geometrically shortest path is not necessarily the path of least time.

25. By direct calculation we construct Table 4.5.1. The table shows that the largest profit comes from setting production at 17 articles per week.

■ Table 4.5.1

n	P	n	P	n	P
8	8	14	212	20	200
9	57	15	225	21	177
10	100	16	232	22	148
11	137	17	233	23	113
12	168	18	228	24	72
13	193	19	217	25	25

We can avoid such massive computation by considering the function

$$f(x) = 100x - 600 - 3x^2, \qquad 8 \le x \le 25.$$

This function is differentiable with respect to x, and for integral values of x it agrees with P. Differentiation of f gives

$$f'(x) = 100 - 6x.$$

Obviously, $f'(x) = 0$ at $x = \frac{100}{6} = 16\frac{2}{3}$. Since $f'(x) > 0$ on $(8, 16\frac{2}{3})$, f increases on $[8, 16\frac{2}{3}]$. Since $f'(x) < 0$ on $(16\frac{2}{3}, 25)$, f decreases on $[16\frac{2}{3}, 25]$. The largest value of f corresponding to an integer value of x will therefore occur at $x = 16$ or $x = 17$. Direct calculation of $f(16)$ and $f(17)$ shows that the choice of $x = 17$ is correct. ❑

EXERCISES 4.5

1. Find the greatest possible value of xy given that x and y are both positive and $x + y = 40$.
2. Find the dimensions of the rectangle of perimeter 24 that has the largest area.
3. A rectangular garden 200 square feet in area is to be fenced off against rabbits. Find the dimensions that will require the least amount of fencing if one side of the garden is already protected by a barn.
4. Find the largest possible area for a rectangle with base on the x-axis and upper vertices on the curve $y = 4 - x^2$.
5. Find the largest possible area for a rectangle inscribed in a circle of radius 4.
6. The cross section of a beam is in the form of a rectangle of length l and width w. (See the figure.) Assuming that the strength of the beam varies directly with w^2l, what are the dimensions of the strongest beam that can be sawed from a round log of diameter 3 feet?

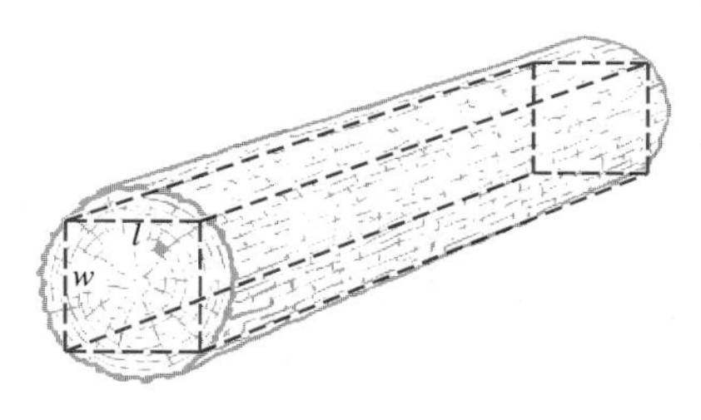

7. A rectangular playground is to be fenced off and divided in two by another fence parallel to one side of the playground. Six hundred feet of fencing is used. Find the dimensions of the playground that will enclose the greatest total area.
8. A rectangular warehouse will have 5000 square feet of floor space and will be separated into two rectangular rooms by an interior wall. The cost of the exterior walls is $150 per linear foot and the cost of the interior wall is

\$100 per linear foot. Find the dimensions that will minimize the cost of building the warehouse.

9. Rework Example 1, this time assuming that the semicircular portion of the window admits only one-third as much light per square foot as does the rectangular portion of the window.

10. One side of a rectangular field is bounded by a straight river. The other three sides are bounded by straight fences. The total length of the fence is 800 feet. Determine the dimensions of the field given that its area is a maximum.

11. Find the coordinates of P that maximize the area of the rectangle shown in the figure below.

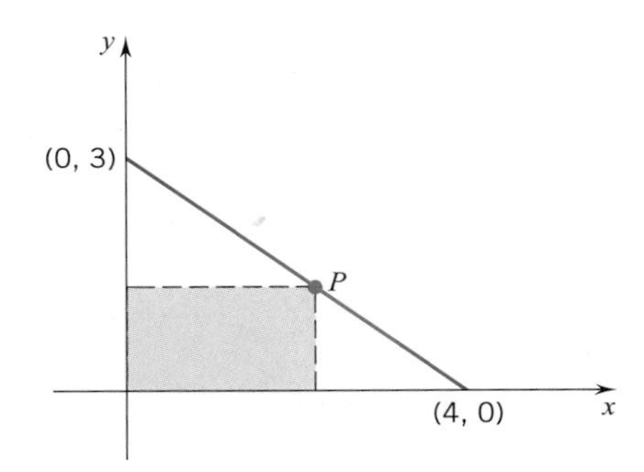

12. The base of a triangle is on the x-axis, one side lies along the line $y = 3x$, and the third side passes through the point $(1, 1)$. What is the slope of the third side if the area of the triangle is to be a minimum?

13. A triangle is formed by the coordinate axes and a line through the point $(2, 5)$ as in the figure below. Determine the slope of this line if the area of the triangle is to be a minimum.

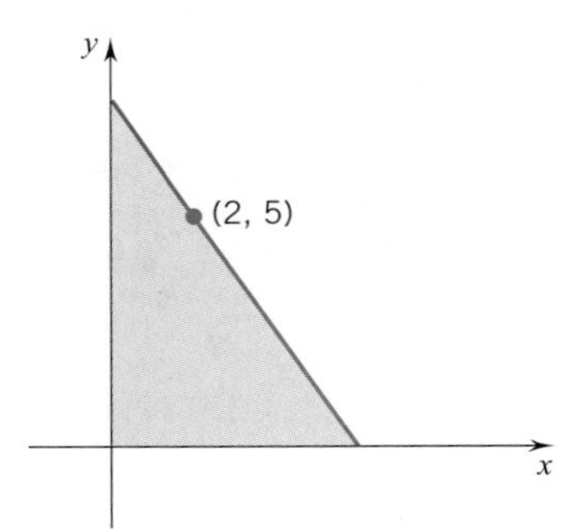

14. In the setting of Exercise 13 determine the slope of the line if the area is to be a maximum.

15. What are the dimensions of the base of the rectangular box of greatest volume that can be constructed from 100 square inches of cardboard if the base is to be twice as long as it is wide? Assume that the box has a top.

16. Exercise 15 under the assumption that the box has no top.

17. Find the dimensions of the isosceles triangle of largest area with perimeter 12.

18. Find the point(s) on the parabola $y = \frac{1}{8}x^2$ closest to the point $(0, 6)$.

19. Find the point(s) on the parabola $x = y^2$ closest to the point $(0, 3)$.

20. Find A and B given that the function $y = Ax^{-1/2} + Bx^{1/2}$ has a minimum value of 6 at $x = 9$.

21. A pentagon with a perimeter of 30 inches is to be constructed by adjoining an equilateral triangle to a rectangle. Find the dimensions of the rectangle and triangle that will maximize the area of the pentagon.

22. A 10-foot section of gutter is made from a 12-inch-wide strip of sheet metal by folding up 4-inch strips on each side so that they make the same angle with the bottom of the gutter. Determine the depth of the gutter that has the greatest carrying capacity. *Caution*: there are two ways to sketch the trapezoidal cross section, as shown in the figure below.

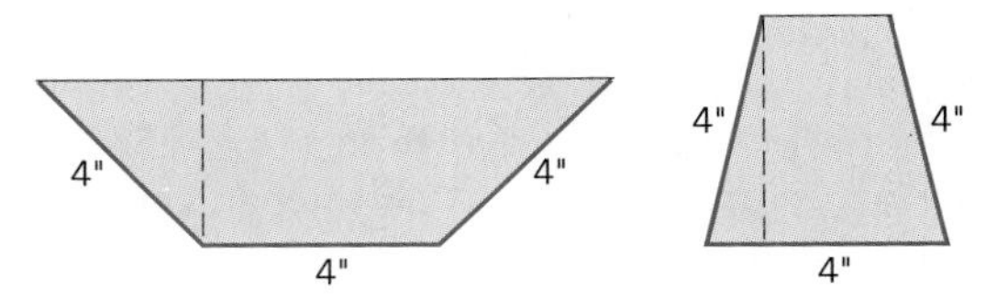

23. From a rectangular piece of cardboard of dimensions 8×15, four congruent squares are to be cut out, one at each corner. (See the figure below.) The remaining crosslike piece is then to be folded into an open box. What size squares should be cut out if the volume of the resulting box is to be a maximum?

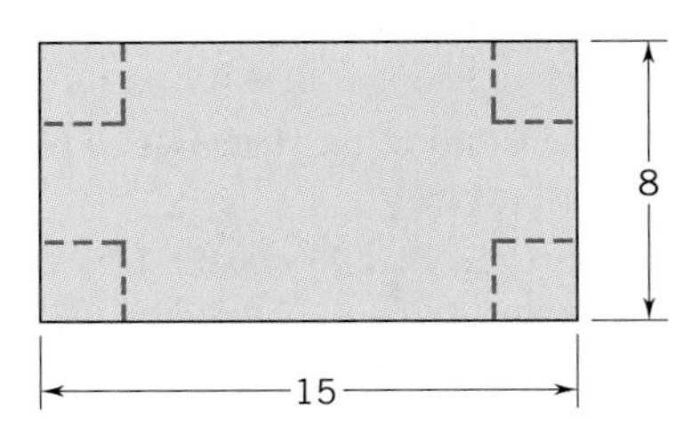

24. A page is to contain 81 square centimeters of print. The margins at the top and bottom are to be 3 centimeters each and, at the sides, 2 centimeters each. Find the most economical dimensions given that the cost of a page varies directly with the perimeter of the page.

25. Let ABC be a triangle with vertices $A = (-3, 0)$, $B = (0, 6)$, $C = (3, 0)$. Let P be a point on the line segment that joins B to the origin. Find the position of P that minimizes the sum of the distances between P and the vertices.

26. Solve Exercise 25 with $A = (-6, 0)$, $B = (0, 3)$, $C = (6, 0)$.

27. An 8-foot-high fence is located 1 foot from a building. Determine the length of the shortest ladder that can be leaned against the building and touch the top of the fence.

28. Two hallways, one 8 feet wide and the other 6 feet wide, meet at right angles. Determine the length of the longest ladder that can be carried horizontally from one hallway into the other.

29. A rectangular banner has a red border and a white center. The width of the border at top and bottom is 8 inches and along the sides it is 6 inches. The total area is 27 square feet. What should be the dimensions of the banner if the area of the white center is to be a maximum?

30. Find the absolute maximum value of $y = x(r^2 + x^2)^{-3/2}$.

31. A string 28 inches long is to be cut into two pieces, one piece to form a square and the other to form a circle. How should the string be cut so as to (a) maximize the sum of the two areas? (b) minimize the sum of the two areas?

32. What is the maximum volume for a rectangular box (square base, no top) made from 12 square feet of cardboard?

33. The figure below shows a cylinder inscribed in a right circular cone of height 8 and base radius 5. Find the dimensions of the cylinder if its volume is to be a maximum.

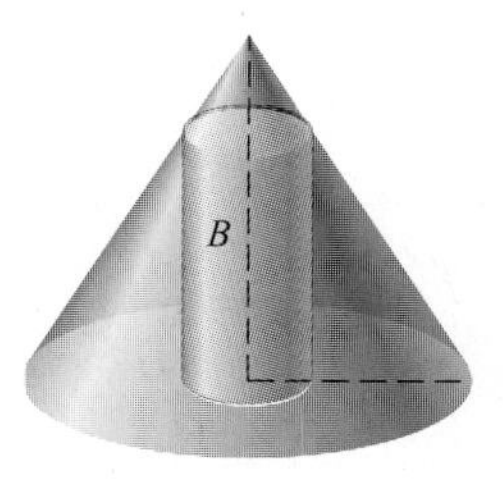

34. A variant of Exercise 33. This time find the dimensions of the cylinder if the area of its curved surface is to be a maximum.

35. A rectangular box with square base and top is to be made to contain 1250 cubic feet. The material for the base costs 35 cents per square foot, for the top 15 cents per square foot, and for the sides 20 cents per square foot. Find the dimensions that will minimize the cost of the box.

36. What is the largest possible area for a parallelogram inscribed in a triangle ABC in the manner of the figure below?

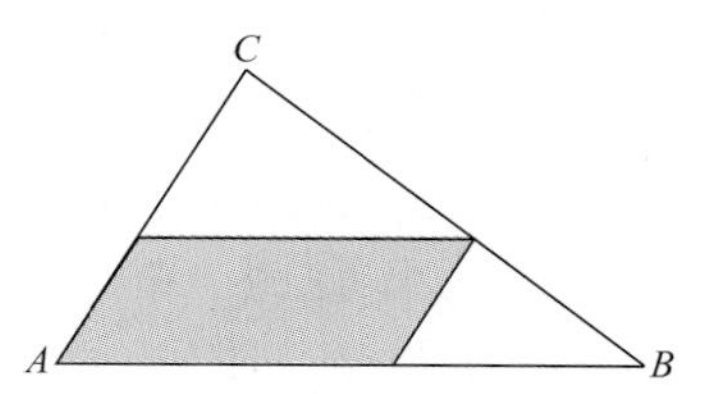

37. Find the dimensions of the isosceles triangle of least area that circumscribes a circle of radius r.

38. What is the maximal possible area for a triangle inscribed in a circle of radius r?

39. The figure below shows a right circular cylinder inscribed in a sphere of radius R. Find the dimensions of the cylinder if its volume is to be a maximum.

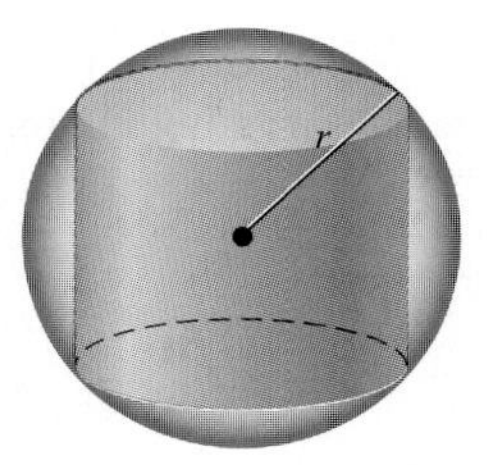

40. A variant of Exercise 39. This time find the dimensions of the right circular cylinder if the area of its curved surface is to be a maximum.

41. A right circular cone is inscribed in a sphere of radius R as in the figure below. Find the dimensions of the cone if its volume is to be a maximum.

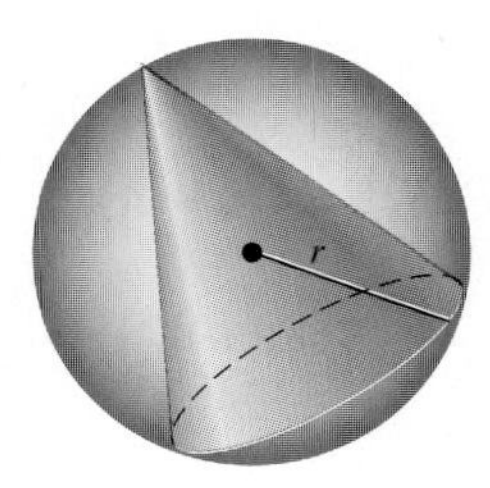

42. What is the largest possible volume for a right circular cone of slant height a?

43. A power line is needed to connect a power station on the shore of a river to an island 4 kilometers downstream and 1 kilometer offshore. Find the minimum cost for such a line given that it costs \$50,000 per kilometer to lay wire under water and \$30,000 per kilometer to lay wire under ground.

44. A tapestry 7 feet high hangs on a wall. The lower edge is 9 feet above an observer's eye. How far from the wall should the observer stand to obtain the most favorable view? Namely, what distance from the wall maximizes the visual angle of the observer? HINT: Use the formula for $\tan(A - B)$.

45. A body of weight W is dragged along a horizontal plane by means of a force P whose line of action makes an angle θ with the plane. The magnitude of the force is given by the equation

$$P = \frac{mW}{m \sin \theta + \cos \theta}$$

where m denotes the coefficient of friction. For what value of θ is the pull a minimum?

46. If a projectile is fired from O so as to strike an inclined plane that makes a constant angle α with the horizontal, its range is given by the formula

$$R = \frac{2v^2 \cos \theta \sin (\theta - \alpha)}{g \cos^2 \alpha},$$

where v and g are constants and θ is the angle of elevation. Calculate θ for maximum range.

47. The lower edge of a movie theater screen 30 feet high is 6 feet above an observer's eye. How far from the screen should the observer sit to obtain the most favorable view? Namely, what distance from the screen maximizes the visual angle of the observer?

48. A local bus company offers charter trips to Blue Mountain Museum at a fare of \$37 per person if 16 to 35 passengers sign up for the trip. The company does not charter trips for fewer than 16 passengers. The bus has 48 seats. If more than 35 passengers sign up, then the fare for every passenger is reduced by 50 cents for each passenger in excess of 35 that signs up. Determine the number of passengers that generates the greatest revenue for the bus company.

49. The Hotwheels Rent-A-Car Company derives an average net profit of \$12 per customer if it services 50 customers or less. If it services over 50 customers, then the average net profit is decreased by 6 cents for each customer over 50. What number of customers produces the greatest total net profit for the company?

■ 4.6 CONCAVITY AND POINTS OF INFLECTION

We begin with a picture of the graph of a function f, Figure 4.6.1. To the left of c_1 and between c_2 and c_3, the graph "curves up" (we call it *concave up*); between c_1 and c_2, and to the right of c_3, the graph "curves down" (we call it *concave down*). These terms deserve a precise definition.

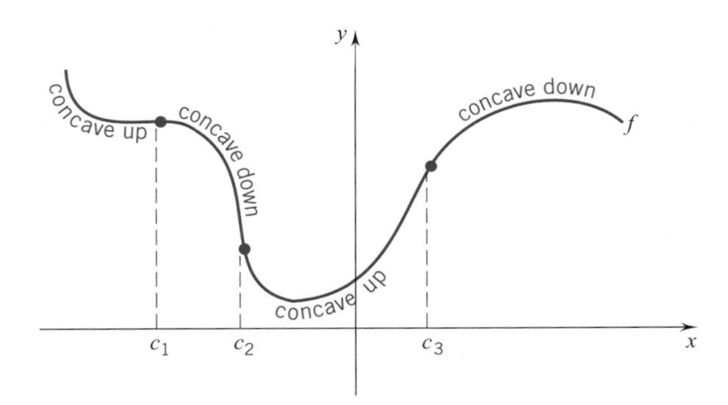

Figure 4.6.1

> **DEFINITION 4.6.1 CONCAVITY**
>
> Let the function f be differentiable on an open interval I. The graph of f is said to be *concave up* on I iff f' increases on I; it is said to be *concave down* on I iff f' decreases on I.

In other words, the graph is concave up on an open interval where the slope increases and concave down on an open interval where the slope decreases.

There is also a geometric interpretation of concavity: The graph of f is concave up on an interval I if the tangent lines to the graph are *below* the curve; the graph of f is concave down on I if the tangent lines to the graph are *above* the curve. You can verify this interpretation using Figure 4.6.1.

Points that join arcs of opposite concavity are called *points of inflection.* The graph in Figure 4.6.1 has three of them: $(c_1, f(c_1))$, $(c_2, f(c_2))$, $(c_3, f(c_3))$. Here is a formal definition.

DEFINITION 4.6.2 POINT OF INFLECTION

Let the function f be continuous at $x = c$. The point $(c, f(c))$ is called a *point of inflection* iff there exists $\delta > 0$ such that the graph of f is concave in one sense on $(c - \delta, c)$ and concave in the opposite sense on $(c, c + \delta)$.

Example 1 The graph of the quadratic function $f(x) = x^2 - 4x + 3$ is concave up everywhere since the derivative $f'(x) = 2x - 4$ is everywhere increasing. (See Figure 4.6.2.) ❑

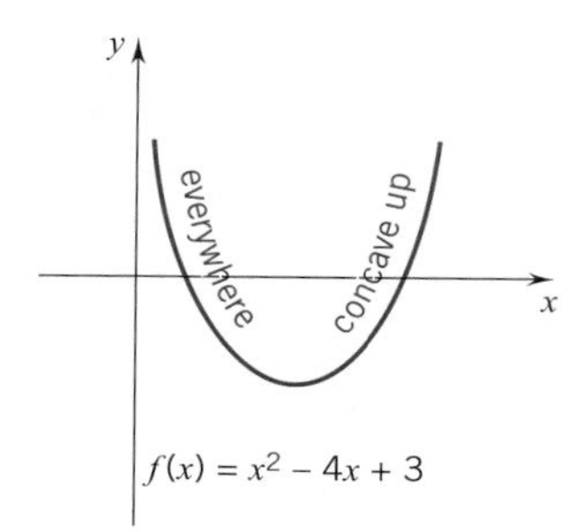

Figure 4.6.2

Example 2 Consider the cubic function $f(x) = x^3$. You can verify that

$$f'(x) = 3x^2 \qquad \text{decreases on } (-\infty, 0] \text{ and increases on } [0, \infty).$$

Thus, the graph of f is concave down on $(-\infty, 0)$ and concave up on $(0, \infty)$. The origin, $(0, f(0)) = (0, 0)$, is a point of inflection. (See Figure 4.6.3.) ❑

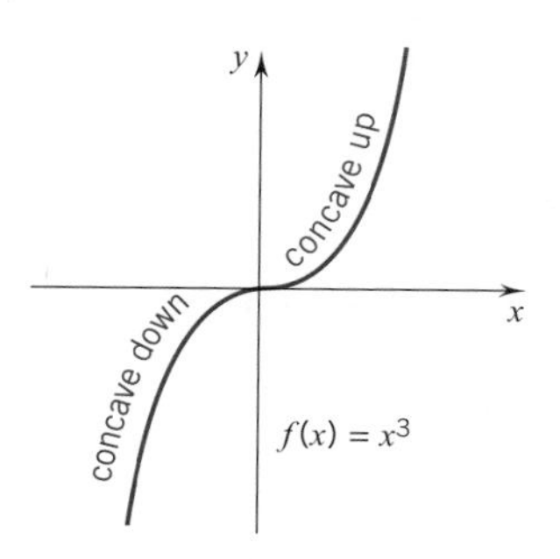

Figure 4.6.3

If f is twice differentiable, then, remembering that $f'' = (f')'$, we can determine the concavity of the graph by looking at the sign of the second derivative.

THEOREM 4.6.3

Let f be twice differentiable on an open interval I.

(i) If $f''(x) > 0$ for all x in I, then f' increases on I and the graph of f is concave up.

(ii) If $f''(x) < 0$ for all x in I, then f' decreases on I and the graph of f is concave down.

PROOF Apply the proof of Theorem 4.2.2 to the function f'. ❑

The following result gives us a way of identifying possible points of inflection.

THEOREM 4.6.4

If the point $(c, f(c))$ is a point of inflection, then either

$$f''(c) = 0 \quad \text{or} \quad f''(c) \quad \text{does not exist.}$$

PROOF Suppose that $(c, f(c))$ is a point of inflection. Let's assume that the graph of f is concave up to the left of c and concave down to the right of c. The other case can be handled in a similar manner.

In this situation f' increases on an interval $(c - \delta, c)$ and decreases on an interval $(c, c + \delta)$.

Suppose now that $f''(c)$ exists. Then f' is continuous at c. It follows that f' increases on the half-open interval $(c - \delta, c]$ and decreases on the half-open interval $[c, c + \delta)$.† This says that f' has a local maximum at c. Since, by assumption, $f''(c)$ exists, $f''(c) = 0$. (Theorem 4.3.2 applied to f'.)

We have shown that if $f''(c)$ exists, then $f''(c) = 0$. The other possibility, of course, is that $f''(c)$ does not exist (see Example 4 below). ❑

Example 3 For the function

$$f(x) = x^3 - 6x^2 + 9x + 1 \qquad \text{(Figure 4.6.4)}$$

we have

$$f'(x) = 3x^2 - 12x + 9 = 3(x^2 - 4x + 3)$$

and

$$f''(x) = 6x - 12 = 6(x - 2).$$

Since $f''(x) = 0$ only at $x = 2$, f'' has constant sign on $(-\infty, 2)$ and on $(2, \infty)$. The sign of f'' on these intervals and the consequences for the graph of f are as follows:

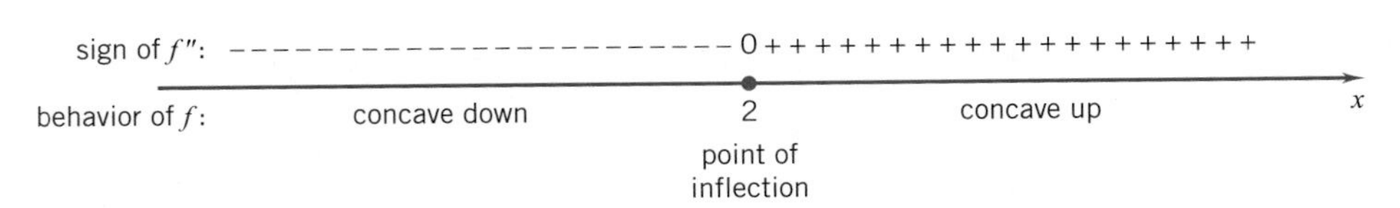

Figure 4.6.4

The point $(2, f(2)) = (2, 3)$ is a point of inflection. ❑

Example 4 For

$$f(x) = 3x^{5/3} - 5x \qquad \text{(Figure 4.6.5)}$$

we have

$$f'(x) = 5x^{2/3} - 5 \qquad \text{and} \qquad f''(x) = \tfrac{10}{3}x^{-1/3}.$$

The second derivative does not exist at $x = 0$. Since

$$f''(x) \text{ is } \begin{cases} \text{negative,} & \text{for } x < 0 \\ \text{positive,} & \text{for } x > 0, \end{cases}$$

† See Exercise 56, Section 4.2

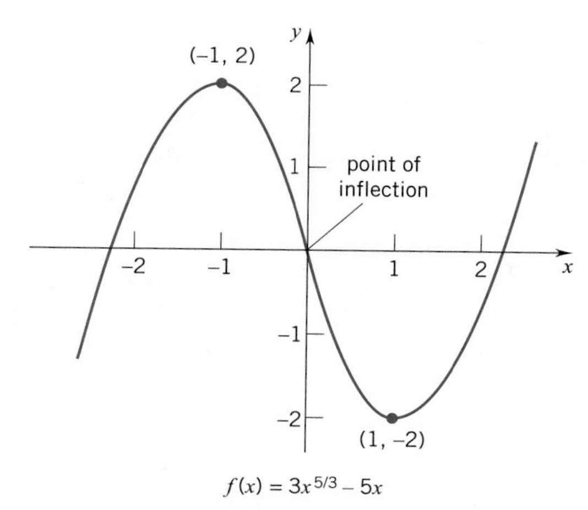

$f(x) = 3x^{5/3} - 5x$

Figure 4.6.5

the graph is concave down on $(-\infty, 0)$ and concave up on $(0, \infty)$. Since f is continuous at 0, the point $(0, f(0)) = (0, 0)$ is a point of inflection. ❑

CAUTION The fact that

$$f''(c) = 0 \quad \text{or} \quad f''(c) \text{ does not exist}$$

does not guarantee that $(c, f(c))$ is a point of inflection. As you can verify, the function $f(x) = x^4$ satisfies $f''(0) = 0$, but the graph is always concave up and there are no points of inflection. If f is discontinuous at c, then $f''(c)$ does not exist, but $(c, f(c))$ cannot be a point of inflection. A point of inflection occurs at c iff f is continuous at c and the point $(c, f(c))$ joins arcs of opposite concavity. ❑

Example 5 Determine the concavity and find the points of inflection (if any) of the graph of

$$f(x) = x + \cos x, \qquad x \in [0, 2\pi].$$

SOLUTION For $x \in (0, 2\pi)$, we have

$$f'(x) = 1 - \sin x$$

and

$$f''(x) = -\cos x.$$

Since $f''(x) = 0$ at $x = \pi/2$ and $3\pi/2$, f'' has constant sign on $(0, \pi/2)$, $(\pi/2, 3\pi/2)$, and $(3\pi/2, 2\pi)$. The sign of f'' on these intervals and the consequences for f are as follows:

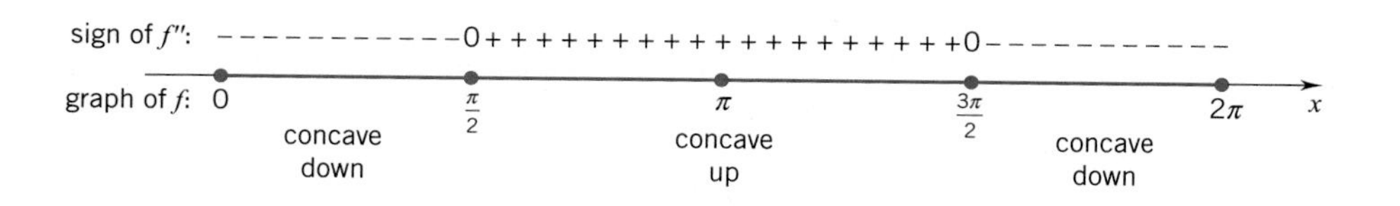

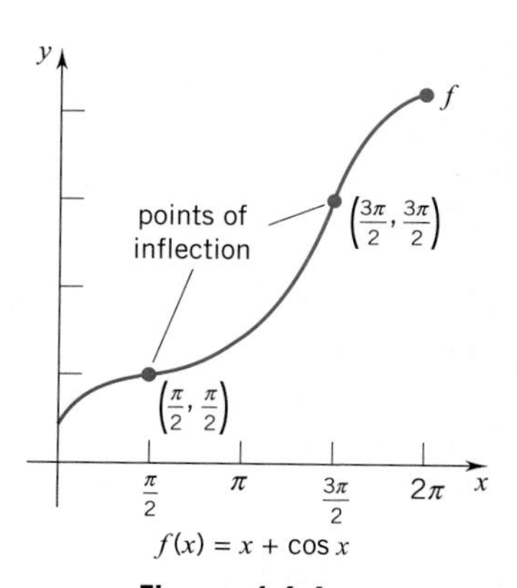

$f(x) = x + \cos x$

Figure 4.6.6

The points $(\pi/2, f(\pi/2)) = (\pi/2, \pi/2)$ and $(3\pi/2, f(3\pi/2)) = (3\pi/2, 3\pi/2)$ are points of inflection. The graph of f is shown in Figure 4.6.6. ❑

EXERCISES 4.6

1. The graph of a function f is given in the figure below. (a) Determine the intervals on which f increases and the intervals on which f decreases; (b) determine the intervals on which the graph of f is concave up, the intervals on which the graph is concave down, and give the x-coordinates of the points of inflection.

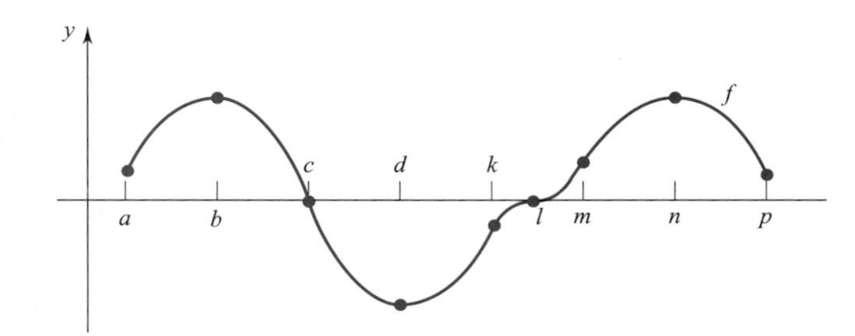

2. Repeat Exercise 1 for the function g whose graph is given in the figure below.

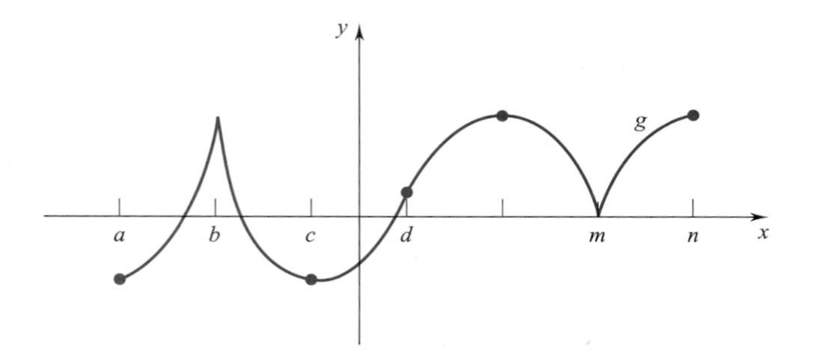

In Exercises 3–20, describe the concavity of the graph of f and find the points of inflection (if any).

3. $f(x) = \dfrac{1}{x}$.

4. $f(x) = x + \dfrac{1}{x}$.

5. $f(x) = x^3 - 3x + 2$.

6. $f(x) = 2x^2 - 5x + 2$.

7. $f(x) = \dfrac{1}{4}x^4 - \dfrac{1}{2}x^2$.

8. $f(x) = x^3(1 - x)$.

9. $f(x) = \dfrac{x}{x^2 - 1}$.

10. $f(x) = \dfrac{x + 2}{x - 2}$.

11. $f(x) = (1 - x)^2(1 + x)^2$.

12. $f(x) = \dfrac{6x}{x^2 + 1}$.

13. $f(x) = \dfrac{1 - \sqrt{x}}{1 + \sqrt{x}}$.

14. $f(x) = (x - 3)^{1/5}$.

15. $f(x) = (x + 2)^{5/3}$.

16. $f(x) = x\sqrt{4 - x^2}$.

17. $f(x) = \sin^2 x, \quad x \in [0, \pi]$.

18. $f(x) = 2\cos^2 x - x^2, \quad x \in [0, \pi]$.

19. $f(x) = x^2 + \sin 2x, \quad x \in [0, \pi]$.

20. $f(x) = \sin^4 x, \quad x \in [0, \pi]$.

In Exercises 21–28, find: (a) the intervals on which f increases or decreases; (b) the local maxima and minima; (c) the intervals on which the graph of f is concave up and the intervals on which it is concave down; and (d) the points of inflection. Use this information to sketch the graph of f.

21. $f(x) = x^3 - 9x$.

22. $f(x) = 3x^4 + 4x^3 + 1$.

23. $f(x) = \dfrac{2x}{x^2 + 1}$.

24. $f(x) = x^{1/3}(x - 6)^{2/3}$.

25. $f(x) = x + \sin x, \quad x \in [-\pi, \pi]$.

26. $f(x) = \sin x + \cos x, \quad x \in [0, 2\pi]$.

27. $f(x) = \begin{cases} x^3, & x < 1 \\ 3x - 2, & x \geq 1. \end{cases}$

28. $f(x) = \begin{cases} 2x + 4, & x \leq -1 \\ 3 - x^2, & x > -1. \end{cases}$

In Exercises 29–32, sketch the graph of a continuous function f that satisfies the given conditions.

29. $f''(x) > 0$ if $x < 0$ or if $x > 2$, $f''(x) < 0$ if $0 < x < 2$; $f'(0) = f'(4) = 0$, $f'(x) < 0$ if $x < 4$ and $x \neq 0$, $f'(x) > 0$ if $x > 4$; $f(0) = 0$, $f(2) = -2$, $f(4) = -5$.

30. $f''(x) > 0$ if $|x| > 2$, $f''(x) < 0$ if $|x| < 2$; $f'(0) = 0$, $f'(x) > 0$ if $x < 0$, $f'(x) < 0$ if $x > 0$; $f(0) = 1$, $f(-2) = f(2) = \frac{1}{2}$, $f(x) > 0$ for all x and f is an even function.

31. $f''(x) < 0$ if $x < 0$, $f''(x) > 0$ if $x > 0$; $f'(-1) = f'(1) = 0$, $f'(0)$ does not exist, $f'(x) > 0$ if $|x| > 1$, $f'(x) < 0$ if $|x| < 1$ $(x \neq 0)$; $f(-1) = 2$, $f(1) = -2$, f is an odd function.

32. $f''(x) < 0$ if $x < -3$ or if $0 < x < 4$; $f''(x) > 0$ if $-3 < x < 0$ or if $x > 4$; $f'(-1) = f'(2) = 0$, $f'(x) < 0$ if $x < -1$ or if $x > 2$, $f'(x) > 0$ if $-1 < x < 2$; $f(-1) = -2$, $f(0) = 0$, $f(2) = 2$, $f(x) < 0$ if $x < 0$, $f(x) > 0$ if $x > 0$.

33. Find d given that $(d, f(d))$ is a point of inflection of the graph of

$$f(x) = (x - a)(x - b)(x - c).$$

34. Find c given that the graph of $f(x) = cx^2 + x^{-2}$ has a point of inflection at $(1, f(1))$.

35. Find a and b given that the graph of $f(x) = ax^3 + bx^2$ passes through $(-1, 1)$ and has a point of inflection at $x = \frac{1}{3}$.

36. Determine A and B so that the curve

$$y = Ax^{1/2} + Bx^{-1/2}$$

will have a point of inflection at $(1, 4)$.

37. Determine A and B so that the curve

$$y = A\cos 2x + B\sin 3x$$

will have a point of inflection at $(\pi/6, 5)$

38. Find necessary and sufficient conditions on A and B for $f(x) = Ax^2 + Bx + C$:
(a) to decrease between A and B with graph concave up.
(b) to increase between A and B with graph concave down.

39. Find a function f such that $f'(x) = 3x^2 - 6x + 3$ and $(1, -2)$ is a point of inflection of the graph of f.

40. Prove that the graph of a polynomial of degree n has at most $n-2$ points of inflection.

41. Given the cubic polynomial $p(x) = x^3 + ax^2 + bx + c$.
(a) Prove that the graph of p has exactly one point of inflection.
(b) Show that p has local extreme values iff $a^2 \geq 3b$.

42. Prove that if the cubic polynomial $p(x) = x^3 + ax^2 + bx + c$ has a local maximum and a local minimum, then the point of inflection is the midpoint of the line segment connecting the local extrema on the graph.

C In Exercises 43–46, use a graphing utility to graph the function f on the indicated interval. (a) Estimate the intervals where the graph of f is concave up and the intervals where it is concave down. (b) Estimate the x-coordinate of each point of inflection. Round off your estimates to three decimal places.

43. $f(x) = x^4 - 5x^2 + 3$; $[-4, 4]$.

44. $f(x) = x\sin x$; $[-2\pi, 2\pi]$.

45. $f(x) = 1 + x^2 - 2x\cos x$; $[-\pi, \pi]$.

46. $f(x) = x^{2/3}(x^2 - 4)$; $[-5, 5]$.

C In Exercises 47–48, the second derivative of a function f is given. Use a graphing utility to graph f'' on the indicated interval. (a) Estimate the intervals where the graph of f is concave up and the intervals where it is concave down. (b) Estimate the x-coordinate of each point of inflection. Round off your estimates to three decimal places.

47. $f''(x) = x^3 + \frac{1}{2}x^2 - \frac{16}{3}x + 2$; $[-3, 2.5]$.

48. $f''(x) = 2x\sin x - 2$; $[-\pi, \pi]$.

■ 4.7 VERTICAL AND HORIZONTAL ASYMPTOTES; VERTICAL TANGENTS AND CUSPS

Vertical and Horizontal Asymptotes

In Figure 4.7.1 you can see the graph of the function

$$f(x) = \frac{1}{|x - c|} \quad \text{for } x \text{ close to } c.$$

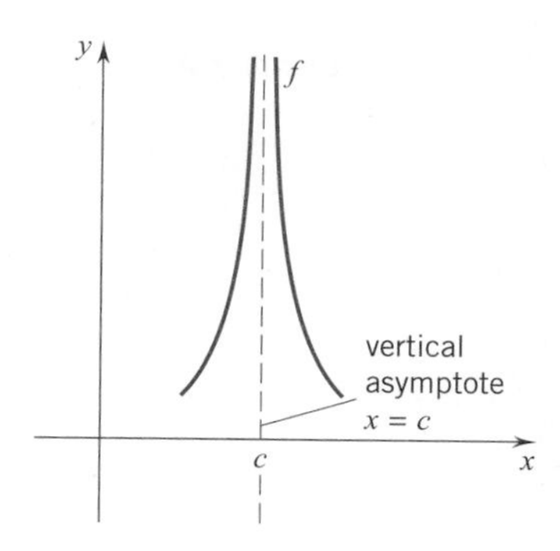

Figure 4.7.1

As $x \to c$, $f(x) \to \infty$, that is, given any positive number M there exists a positive number δ such that

$$\text{if} \quad 0 < |x - c| < \delta, \qquad \text{then} \quad f(x) \geq M.$$

The line $x = c$ is called a *vertical asymptote.* Figure 4.7.2 shows the graph of

$$g(x) = -\frac{1}{|x - c|} \quad \text{for } x \text{ close to } c.$$

In this case, we see that $g(x) \to -\infty$ as $x \to c$. Again the line $x = c$ is called a *vertical asymptote.*

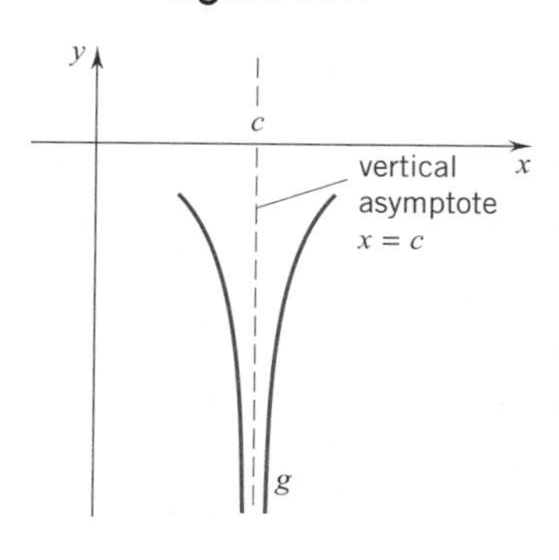

Figure 4.7.2

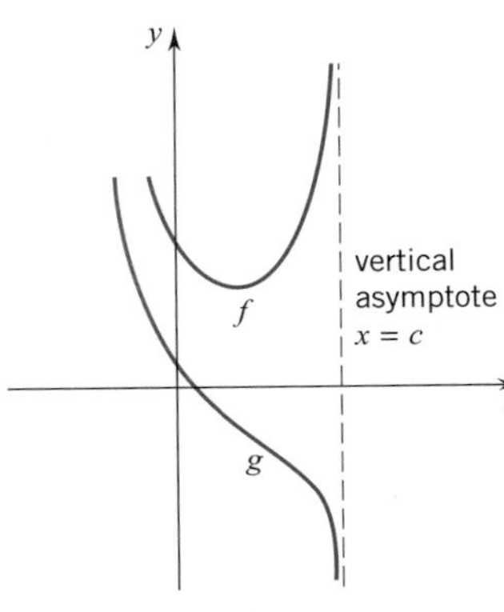

Figure 4.7.3

Vertical asymptotes can also arise from one-sided behavior. With f and g as in Figure 4.7.3, we write

$$\text{as} \quad x \to c^-, \quad f(x) \to \infty \quad \text{and} \quad g(x) \to -\infty.$$

With f and g as in Figure 4.7.4, we write

$$\text{as} \quad x \to c^+, \quad f(x) \to \infty \quad \text{and} \quad g(x) \to -\infty.$$

In each case the vertical line $x = c$ is a vertical asymptote for both functions. To summarize, the line $x = c$ is a vertical asymptote for a function f if any one of the following conditions holds: $f(x) \to \infty$ or $-\infty$ as $x \to c$; $f(x) \to \infty$ or $-\infty$ as $x \to c^-$; or $f(x) \to \infty$ or $-\infty$ as $x \to c^+$. Typically, to locate the vertical asymptotes for a function f, find the values $x = c$ at which f is discontinuous and examine the behavior of f as x approaches c.

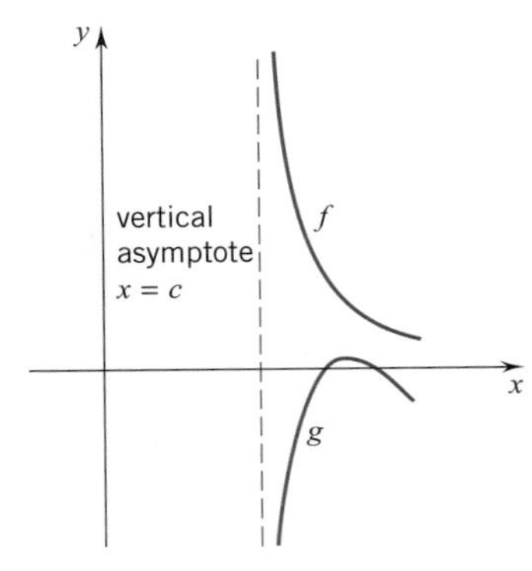

Figure 4.7.4

Example 1 The function

$$f(x) = \frac{3x + 6}{x^2 - 2x - 8} = \frac{3(x + 2)}{(x + 2)(x - 4)}$$

is defined and continuous everywhere except $x = 4$ and $x = -2$. As $x \to 4^+$, $f(x) \to \infty$; as $x \to 4^-$, $f(x) \to -\infty$. Thus, the line $x = 4$ is a vertical asymptote. Since $\lim_{x \to -2} f(x) = -\frac{1}{2}$ exists, f does not have a vertical asymptote at $x = -2$. See Figure 4.7.5. ❑

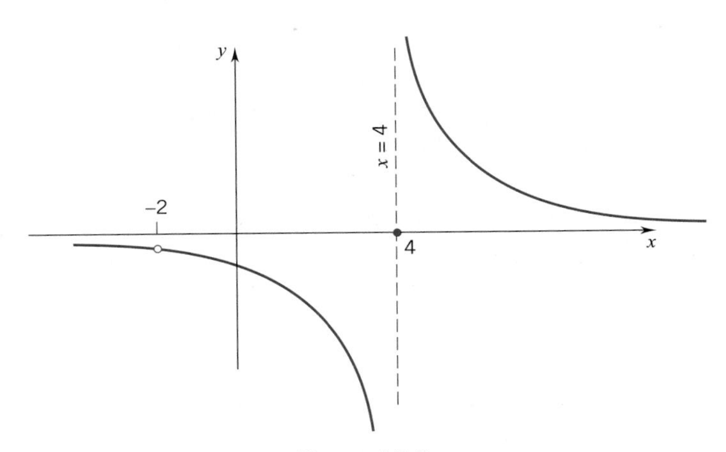

Figure 4.7.5

Recall that $f(x) = \tan x \to \infty$ as $x \to (\pi/2)^-$ and $\tan x \to -\infty$ as $x \to (\pi/2)^+$. Thus the line $x = \pi/2$ is a vertical asymptote for the tangent function. Indeed, the lines $x = (2n + 1)\pi/2$, $n = 0, \pm 1, \pm 2, \ldots$, are all vertical asymptotes for the tangent function (Figure 4.7.6).

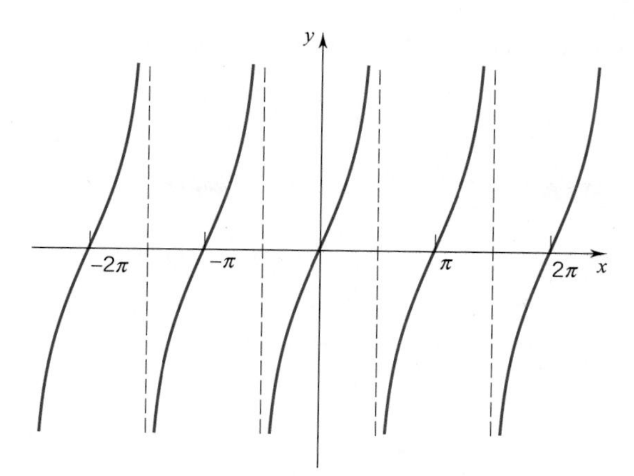

Figure 4.7.6

It is also possible for a function to have a *horizontal asymptote.* If there is a number L such that $f(x) \to L$ as $x \to \infty$, or if $f(x) \to L$ as $x \to -\infty$, then the horizontal line $y = L$ is called a *horizontal asymptote.* See Figures 4.7.7. and 4.7.8. To be precise, $f(x) \to L$ as $x \to \infty$ means that given a positive number ϵ there is a positive number K such that if $x \geq K$, then $|f(x) - L| < \epsilon$. Similarly, $f(x) \to L$ as $x \to -\infty$ means that given a positive number ϵ there is a negative number M such that if $x \leq M$, then $|f(x) - L| < \epsilon$.

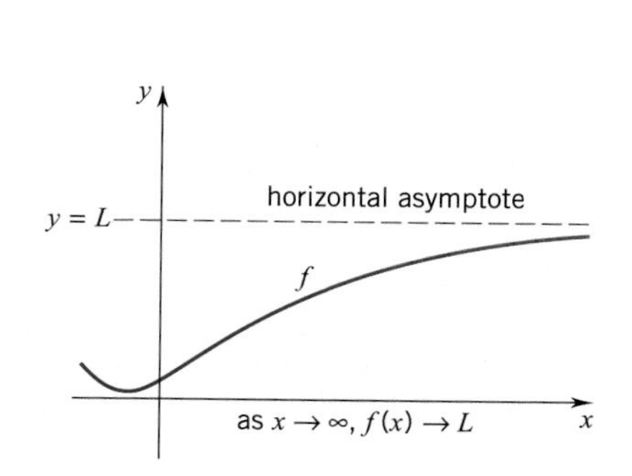

Figure 4.7.7

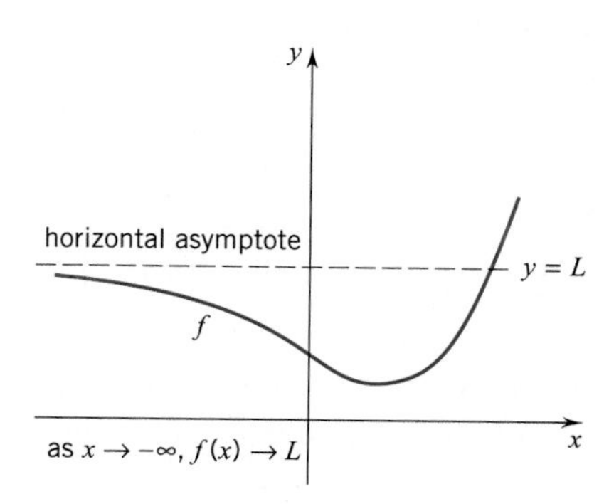

Figure 4.7.8

Example 2 Figure 4.7.9 shows the graph of the function

$$f(x) = \frac{x}{x-2}.$$

As $x \to 2^-, f(x) \to -\infty$; as $x \to 2^+, f(x) \to \infty$. The line $x = 2$ is a vertical asymptote. As $x \to \pm\infty$,†

$$f(x) = \frac{x}{x-2} = \frac{x}{x\left(1 - \dfrac{2}{x}\right)} = \frac{1}{1 - \dfrac{2}{x}} \to 1.$$

The line $y = 1$ is a horizontal asymptote. ❑

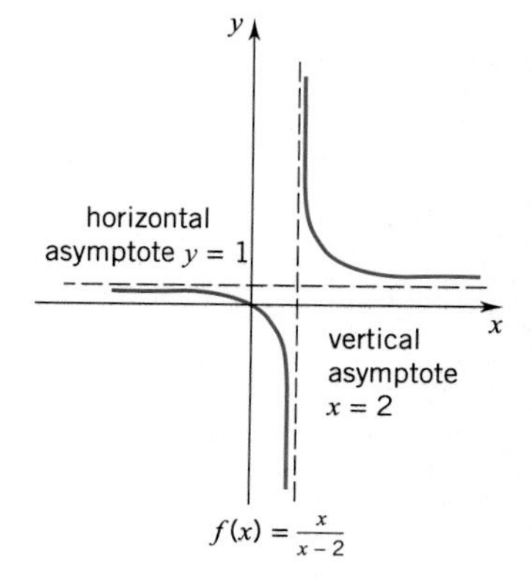

Figure 4.7.9

† Recall that $x \to \pm\infty$ means $x \to \infty$ or $x \to -\infty$.

Example 3 Figure 4.7.10 is a computer-generated graph of the function

$$f(x) = \frac{\cos x}{x}.$$

As $x \to 0^-, f(x) \to -\infty$; as $x \to 0^+, f(x) \to \infty$. The line $x = 0$ (the y-axis) is a vertical asymptote.

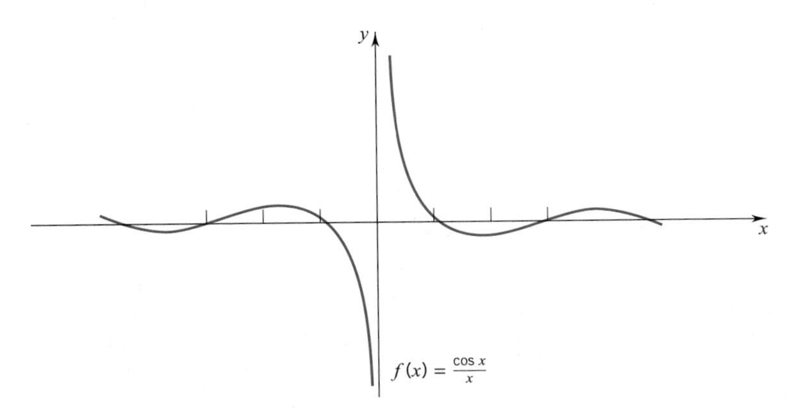

Figure 4.7.10

As $x \to \pm\infty$,

$$f(x) = \frac{\cos x}{x} \to 0.$$

This follows from the fact that

$$\left|\frac{\cos x}{x}\right| \leq \frac{1}{|x|} \qquad \text{for all} \quad x$$

and $1/|x| \to 0$ as $x \to \pm\infty$. Thus, the line $y = 0$ (the x-axis) is a horizontal asymptote. Note that f is an odd function $[f(-x) = -f(x)]$ so its graph is symmetric with respect to the origin. ❑

Example 4 Find the vertical and horizontal asymptotes, if any, of the function

$$g(x) = \frac{x + 1 - \sqrt{x}}{x^2 - 2x + 1} = \frac{x + 1 - \sqrt{x}}{(x-1)^2}.$$

SOLUTION The domain of g is $0 \leq x < \infty$, $x \neq 1$. As $x \to 1$, $g(x) \to \infty$. Thus, the line $x = 1$ is a vertical asymptote. The behavior of g as $x \to \infty$ can be made more apparent by writing

$$g(x) = \frac{x + 1 - \sqrt{x}}{x^2 - 2x + 1} = \frac{x\left(1 + \dfrac{1}{x} - \dfrac{1}{\sqrt{x}}\right)}{x^2\left(1 - \dfrac{2}{x} + \dfrac{1}{x^2}\right)} = \frac{1 + \dfrac{1}{x} - \dfrac{1}{\sqrt{x}}}{x\left(1 - \dfrac{2}{x} + \dfrac{1}{x^2}\right)}.$$

Now, it is easy to see that $g(x) \to 0$ as $x \to \infty$. The line $y = 0$ (the x-axis) is a horizontal asymptote. ❑

The technique suggested in Examples 2 and 4—factoring out the highest power of x from the numerator and the denominator—can be used to prove the following general result concerning the behavior of a rational function as $x \to \pm\infty$. Let

$$R(x) = \frac{p(x)}{q(x)} = \frac{a_n x^n + \cdots + a_1 x + a_0}{b_k x^k + \cdots + b_1 x + b_0}, \quad a_n \neq 0,\ b_k \neq 0,$$

be a rational function. Then,

(4.7.1)
$$\text{as} \quad x \to \pm\infty, \begin{cases} R(x) \to 0, & \text{if } n < k \\ R(x) \to \dfrac{a_n}{b_n}, & \text{if } n = k \\ R(x) \to \pm\infty, & \text{if } n > k. \end{cases}$$

Example 5 The function

$$f(x) = \frac{5 - 3x^2}{1 - x^2}$$

is continuous everywhere except $x = \pm 1$.

The line $x = 1$ is a vertical asymptote:

$$\text{as} \quad x \to 1, \quad 5 - 3x^2 \to 2 \qquad \text{and} \qquad 1 - x^2 \to 0.$$

In particular:

$$\text{as} \quad x \to 1^-, \quad 1 - x^2 \text{ is positive so } f(x) \to \infty;$$
$$\text{as} \quad x \to 1^+, \quad 1 - x^2 \text{ is negative so } f(x) \to -\infty.$$

The line $x = -1$ is a vertical asymptote:

$$\text{as} \quad x \to -1, \quad 5 - 3x^2 \to 2 \qquad \text{and} \qquad 1 - x^2 \to 0.$$

In particular:

$$\text{as} \quad x \to -1^+, \quad 1 - x^2 \text{ is positive so } f(x) \to \infty;$$
$$\text{as} \quad x \to -1^-, \quad 1 - x^2 \text{ is negative so } f(x) \to -\infty.$$

Since f is a rational function, we can use (4.7.1) to investigate the behavior as $x \to \pm\infty$. We have $n = k = 2$ and so $f(x) \to -3/-1 = 3$ as $x \to \pm\infty$. Thus, the line $y = 3$ is a horizontal asymptote. The graph of f is shown in Figure 4.7.11. Note that f is an even function $[f(-x) = f(x)]$; its graph is symmetric with respect to the y-axis. ❑

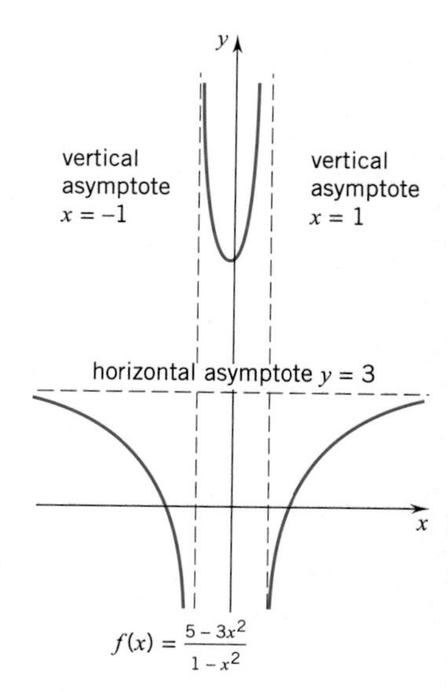

Figure 4.7.11

Vertical Tangents; Vertical Cusps

(For the remainder of this section assume that f is continuous at $x = c$ and differentiable for $x \neq c$.)

We say that the graph of f has a *vertical tangent* at the point $(c, f(c))$ iff

$$\text{as} \quad x \to c, \quad f'(x) \to \infty \quad \text{or} \quad f'(x) \to -\infty.$$

Examples

(a) The function $f(x) = x^{1/3}$ has a vertical tangent at the point (0,0) since

$$f'(x) = \frac{1}{3}x^{-2/3} \to \infty \quad \text{as} \quad x \to 0 \qquad \text{(Figure 4.7.12a).}$$

(b) The function $g(x) = (2 - x)^{1/5}$ has a vertical tangent at the point (2, 0) since

$$g'(x) = -\frac{1}{5}(2 - x)^{-4/5} \to -\infty \quad \text{as} \quad x \to 2 \qquad \text{(Figure 4.7.12b).}$$

❑

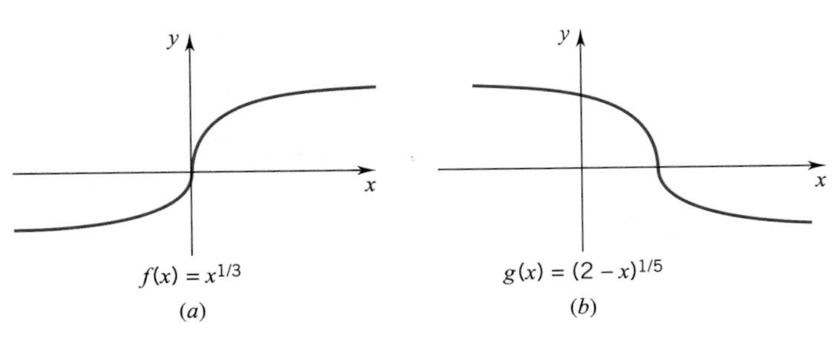

Figure 4.7.12

On occasion you will see a graph become almost vertical and then virtually double back on itself. Such a pattern signals the presence of what is known as a "vertical cusp." We say that the graph of f has a *vertical cusp* at $(c, f(c))$ iff

$$\text{as} \quad x \to c^-, \quad f'(x) \to -\infty \qquad \text{and} \qquad \text{as} \quad x \to c^+, \quad f'(x) \to \infty,$$

or

$$\text{as} \quad x \to c^-, \quad f'(x) \to \infty \qquad \text{and} \qquad \text{as} \quad x \to c^+, \quad f'(x) \to -\infty.$$

Examples

(a) The function $f(x) = x^{2/3}$ has a vertical cusp at (0, 0): $f'(x) = \frac{2}{3}x^{-1/3}$ and

$$\text{as} \quad x \to 0^-, \quad f'(x) \to -\infty, \qquad \text{as} \quad x \to 0^+, \quad f'(x) \to \infty \quad \text{(Figure 4.7.13}a\text{).}$$

(b) The function $g(x) = 2 - (x - 1)^{2/5}$ has a vertical cusp at (1, 2): $g'(x) = -\frac{2}{5}(x - 1)^{-3/5}$ and

$$\text{as} \quad x \to 1^-, \quad g'(x) \to \infty, \qquad \text{as} \quad x \to 1^+, \quad g'(x) \to -\infty. \quad \text{(Figure 4.7.13}b\text{).}$$

❑

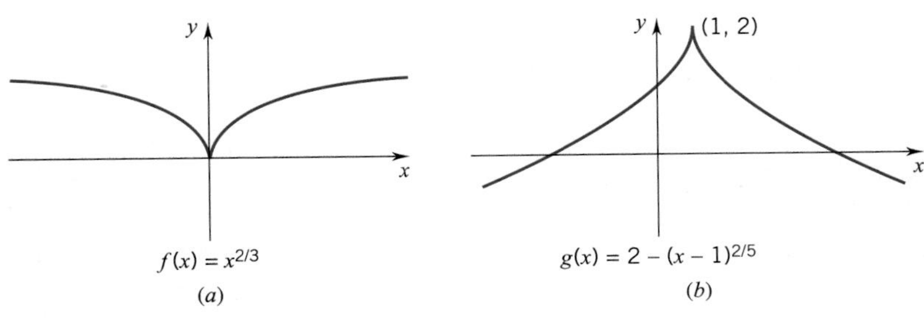

Figure 4.7.13

The fact that $f'(c)$ does not exist does *not* mean that the graph of f has either a vertical tangent or a vertical cusp at $(c, f(c))$. Unless the conditions spelled out earlier are met, the graph of f can simply be making a "corner" at $(c, f(c))$. For example, the function

$$f(x) = |x^3 - 1|$$

has derivative

$$f'(x) = \begin{cases} -3x^2, & x < 1 \\ 3x^2, & x > 1. \end{cases}$$

At $x = 1$, $f'(x)$ does not exist. As $x \to 1^-$, $f'(x) \to -3$, and as $x \to 1^+$, $f'(x) \to 3$. There is no vertical tangent here and no vertical cusp. For the graph, see Figure 4.7.14.

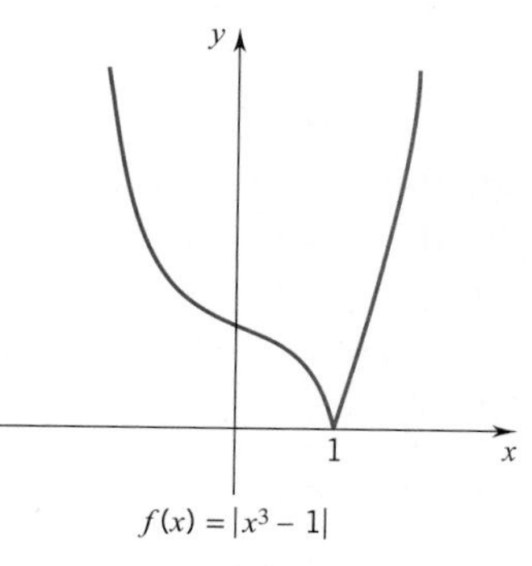

Figure 4.7.14

EXERCISES 4.7

1. The graph of a function f is given in the figure below.

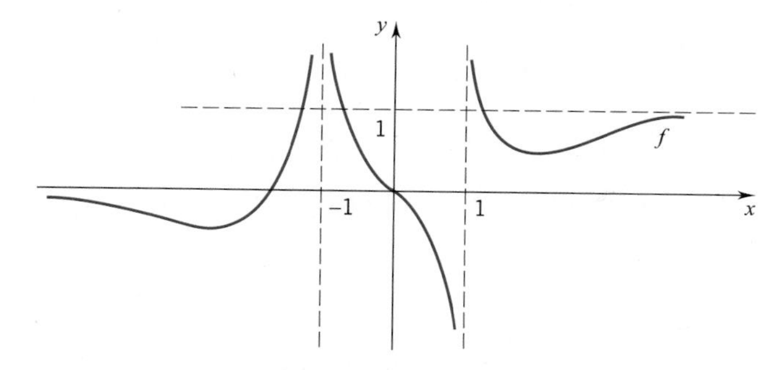

(a) As $x \to -1$, $f(x) \to$?
(b) As $x \to 1^-$, $f(x) \to$?
(c) As $x \to 1^+$, $f(x) \to$?
(d) As $x \to \infty$, $f(x) \to$?
(e) As $x \to -\infty$, $f(x) \to$?
(f) Give the equations of the vertical asymptotes, if any.
(g) Give the equations of the horizontal asymptotes, if any.

2. The graph of a function g is given in the figure below.

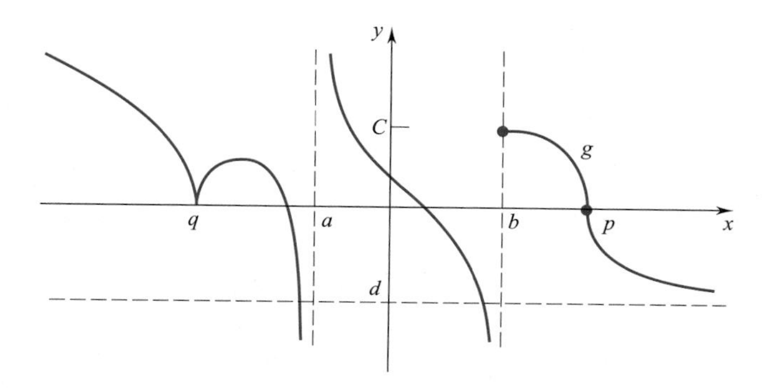

(a) As $x \to \infty$, $g(x) \to$?
(b) As $x \to b^+$, $g(x) \to$?
(c) Give the equations of the vertical asymptotes, if any.
(d) Give the equations of the horizontal asymptotes, if any.
(e) Give the numbers c, if any, at which the graph of g has a vertical tangent line.
(f) Give the numbers c, if any, at which the graph of g has a vertical cusp.

In Exercises 3–20, find the vertical and horizontal asymptotes.

3. $f(x) = \dfrac{x}{3x - 1}$.

4. $f(x) = \dfrac{x^3}{x + 2}$.

5. $f(x) = \dfrac{x^2}{x - 2}$.

6. $f(x) = \dfrac{4x}{x^2 + 1}$.

7. $f(x) = \dfrac{2x}{x^2 - 9}$.

8. $f(x) = \dfrac{\sqrt{x}}{4\sqrt{x} - x}$.

9. $f(x) = \left(\dfrac{2x - 1}{4 + 3x}\right)^2$.

10. $f(x) = \dfrac{4x^2}{(3x - 1)^2}$.

11. $f(x) = \dfrac{3x}{(2x - 5)^2}$.

12. $f(x) = \left(\dfrac{x}{1 - 2x}\right)^3$.

13. $f(x) = \dfrac{3x}{\sqrt{4x^2 + 1}}$.

14. $f(x) = \dfrac{x^{1/3}}{x^{2/3} - 4}$.

15. $f(x) = \dfrac{\sqrt{x}}{2\sqrt{x} - x - 1}$.

16. $f(x) = \dfrac{2x}{\sqrt{x^2 - 1}}$.

17. $f(x) = \sqrt{x + 4} - \sqrt{x}$.

18. $f(x) = \sqrt{x} - \sqrt{x - 2}$.

19. $f(x) = \dfrac{\sin x}{\sin x - 1}$.

20. $f(x) = \dfrac{1}{\sec x - 1}$.

In Exercises 21–34, determine whether or not the graph of f has a vertical tangent or a vertical cusp at c.

21. $f(x) = (x + 3)^{4/3}$; $c = -3$.

22. $f(x) = 3 + x^{2/5}$; $c = 0$.

23. $f(x) = (2 - x)^{4/5}$; $c = 2$.

24. $f(x) = (x + 1)^{-1/3}$; $c = -1$.

25. $f(x) = 2x^{3/5} - x^{6/5}$; $c = 0$.

26. $f(x) = (x - 5)^{7/5}$; $c = 5$.

27. $f(x) = (x + 2)^{-2/3}$; $c = -2$.

28. $f(x) = 4 - (2 - x)^{3/7}$; $c = 2$.

29. $f(x) = \sqrt{|x - 1|}$; $c = 1$.

30. $f(x) = x(x - 1)^{1/3}$; $c = 1$.

31. $f(x) = |(x + 8)^{1/3}|$; $c = -8$.

32. $f(x) = \sqrt{4 - x^2}$; $c = 2$.

33. $f(x) = \begin{cases} x^{1/3} + 2, & x \leq 0 \\ 1 - x^{1/5}, & x > 0; \end{cases}$ $c = 0$.

34. $f(x) = \begin{cases} 1 + \sqrt{-x}, & x \leq 0 \\ (4x - x^2)^{1/3}, & x > 0; \end{cases}$ $c = 0$.

In Exercises 35–38, sketch the graph of the given function, showing all asymptotes.

35. $f(x) = \dfrac{x + 1}{x - 2}$.

36. $f(x) = \dfrac{1}{(x + 1)^2}$.

37. $f(x) = \dfrac{x}{1 + x^2}$.

38. $f(x) = \dfrac{x - 2}{x^2 - 5x + 6}$.

In Exercises 39–44, find (a) the intervals on which f increases or decreases, and (b) the intervals of which the graph of f is concave up and the intervals on which it is concave down. Also, determine whether the graph of f has vertical tangents or vertical cusps. Confirm your results using a graphing utility. Then sketch the graph of f.

39. $f(x) = x - 3x^{1/3}$.

40. $f(x) = x^{2/3} - x^{1/3}$.

41. $f(x) = \frac{3}{5}x^{5/3} - 3x^{2/3}$.

42. $f(x) = \sqrt{|x|}$.

43. $f(x) = \dfrac{x^{2/3} - 1}{|x^{1/3} - 1|}$.

44. $f(x) = |(4 + x)^{3/5}|$.

In Exercises 45–50, sketch the graph of a continuous function f that satisfies the given conditions. Indicate whether the graph of f has any horizontal or vertical asymptotes, and whether the graph has any vertical tangent lines or vertical cusps. If no such function exists, explain why.

45. $f(3) = 0, f(0) = 4, f(-1) = 0, f(-2) = -3, f(x) \to \infty$ as $x \to 1^-$, $f(x) \to -\infty$ as $x \to 1^+$, $f(x) \to 2$ as $x \to \infty$, $f(x) \to 0$ as $x \to -\infty$; $f'(x) < 0$ if $x < -2$, $f'(x) > 0$ if $x > -2$, $x \neq 1$; $f''(x) < 0$ if $x > 1$ or if $x < -4$, $f''(x) > 0$ if $-4 < x < 1$.

46. $f(0) = 0$, $f(3) = f(-3) = 0$, $f(x) \to -\infty$ as $x \to 1$, $f(x) \to -\infty$ as $x \to -1$, $f(x) \to 1$ as $x \to \infty$, $f(x) \to 1$ as $x \to -\infty$; $f''(x) < 0$ for all x, $x \neq \pm 1$.

47. $f(x) \geq 1$ for all x, $f(0) = 1$; $f''(x) < 0$ for all $x \neq 0$; $f'(x) \to \infty$ as $x \to 0^+$, $f'(x) \to -\infty$ as $x \to 0^-$.

48. $f(0) = 1$, $f(x) \to 4$ as $x \to \infty$, $f(x) \to -\infty$ as $x \to -\infty$; $f'(x) \to \infty$ as $x \to 0$; $f''(x) > 0$ if $x < 0$, $f''(x) < 0$ if $x > 0$.

49. $f(0) = 0$, $f(x) \to -1$ as $x \to \infty$, $f(x) \to 1$ as $x \to -\infty$; $f'(x) \to -\infty$ as $x \to 0$; $f''(x) < 0$ if $x < 0$, $f''(x) > 0$ if $x > 0$; f is an odd function.

50. $f(0) = 1, f(2) = 0$; $f'(x) \to -\infty$ as $x \to 0^-$, $f'(x) \to 0$ as $x \to 0^+$; $f''(x) < 0$ for all $x \neq 0$.

51. Let p and q be positive integers with q odd and $p < q$. Let $f(x) = x^{p/q}$. Specify conditions on p and q so that:
(a) The graph of f will have a vertical tangent line at $x = 0$.
(b) The graph of f will have a vertical cusp at $x = 0$.

52. Let $r(x) = [p(x)/q(x)]$ be a rational function and suppose that (degree of p) = (degree of q) + 1.
(a) Show that r can be written in the form

$$r(x) = ax + b + \frac{Q(x)}{q(x)},$$

where (degree Q) < (degree q). HINT: divide q into p.
(b) Prove that $[r(x) - (ax + b)] \to 0$ as $x \to \infty$ and $[r(x) - (ax + b)] \to 0$ as $x \to -\infty$. Thus the graph of f "approaches" the line $y = ax + b$ as $x \to \pm\infty$. The line $y = ax + b$ is called an *oblique asymptote* of the graph of r.

In Exercises 53–56, sketch the graph of the given function showing all vertical and oblique asymptotes.

53. $f(x) = \dfrac{x^2 - 4}{x}$.

54. $f(x) = \dfrac{2x^2 + 3x - 2}{x + 1}$.

55. $f(x) = \dfrac{x^3}{(x - 1)^2}$.

56. $f(x) = \dfrac{1 + x - 3x^2}{x}$.

■ 4.8 SOME CURVE SKETCHING

During the course of the last few sections you have learned how to find the extreme values of a function, the intervals where it increases and the intervals where it decreases; you have seen how to determine the concavity of the graph and how to find the points of inflection; and in the preceding section you learned how to investigate the asymptotic properties of a function. This knowledge makes it possible to sketch an accurate graph of a somewhat complicated function without having to plot point, after point, after point.

Before attempting to sketch the graph of a function f, we try to gather the necessary information and record it in an organized form. Here is an outline of the procedure we will follow.

1. Domain of f: Determine the domain of f; identify endpoints; find the vertical asymptotes of f; determine the behavior of f as $x \to \pm\infty$, find the horizontal asymptotes.
2. Intercepts: Determine the x- and y-intercepts of the graph; the y-intercept is the value $f(0)$; the x-intercepts are the solutions of the equation $f(x) = 0$.
3. Symmetry and periodicity: If f is an *even function* $[f(-x) = f(x)]$, then the graph is symmetric with respect to the y-axis; if f is an *odd function* $[f(-x) = -f(x)]$, then the graph is symmetric with respect to the origin. A function f is *periodic* if there exists a positive number p such that $f(x + p) = f(x)$ whenever x and $x + p$ are in the domain of f.
4. Calculate f': Determine the critical numbers of f; examine the sign of f' to determine the intervals on which f increases and the intervals on which f decreases; determine vertical tangents and cusps.
5. Calculate f'': Examine the sign of f'' to determine the intervals on which the graph is concave up and the intervals on which the graph is concave down; determine the points of inflection.
6. Plot the points of interest in a preliminary sketch: intercepts, "extreme" points (i.e., local and absolute extrema), and points of inflection.
7. Sketch the graph by connecting the points in the preliminary sketch; make sure that the graph "rises," "falls," and "bends" in the proper way.

Figure 4.8.1 shows some of the elements that we might include in a preliminary sketch.

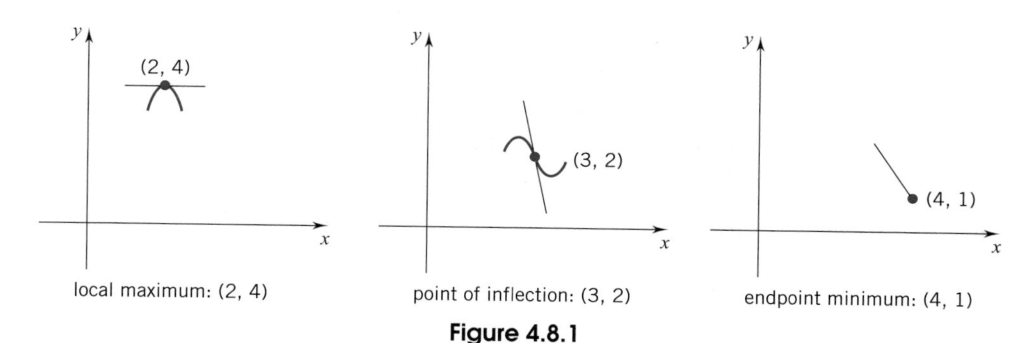

Figure 4.8.1

Example 1 Sketch the graph of the function

$$f(x) = \frac{1}{4}x^4 - 2x^2 + \frac{7}{4}.$$

SOLUTION

(1) Domain: This is a polynomial function, so its domain is the set of all real numbers. Since the leading term is $\frac{1}{4}x^4$, $f(x) \to \infty$ as $x \to \pm\infty$. There are no asymptotes.

(2) Intercepts: The y-intercept is $f(0) = \frac{7}{4}$. To find the x-intercepts, we solve the equation

$$f(x) = \tfrac{1}{4}x^4 - 2x^2 + \tfrac{7}{4} = 0$$
$$x^4 - 8x^2 + 7 = 0$$
$$(x^2 - 1)(x^2 - 7) = 0$$
$$(x + 1)(x - 1)(x + \sqrt{7})(x - \sqrt{7}) = 0.$$

Thus, the x-intercepts are $x = \pm 1$ and $x = \pm\sqrt{7}$.

(3) Symmetry/periodicity: Since $f(-x) = \frac{1}{4}(-x)^4 - 2(-x)^2 + \frac{7}{4} = \frac{1}{4}x^4 - 2x^2 + \frac{7}{4} = f(x)$, f is an even function and its graph is symmetric with respect to the y-axis; f is not a periodic function.

(4) First derivative:

$$f'(x) = x^3 - 4x = x(x^2 - 4) = x(x + 2)(x - 2).$$

The critical numbers of f are $x = 0, x = \pm 2$. The sign of f' and behavior of f are:

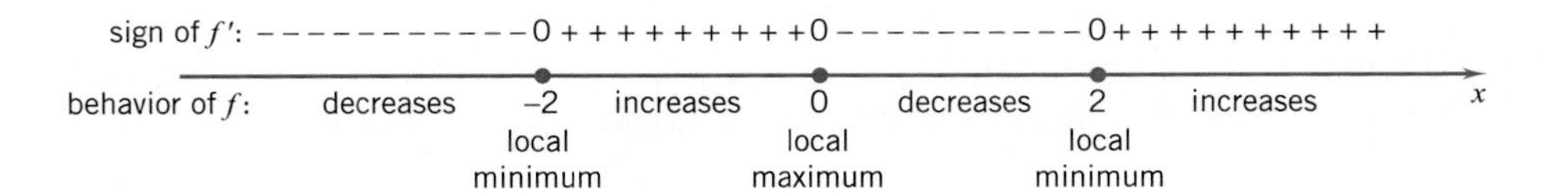

(5) Second derivative:

$$f''(x) = 3x^2 - 4 = 3\left(x - \frac{2}{\sqrt{3}}\right)\left(x + \frac{2}{\sqrt{3}}\right).$$

The sign of f'' and the behavior of the graph of f are:

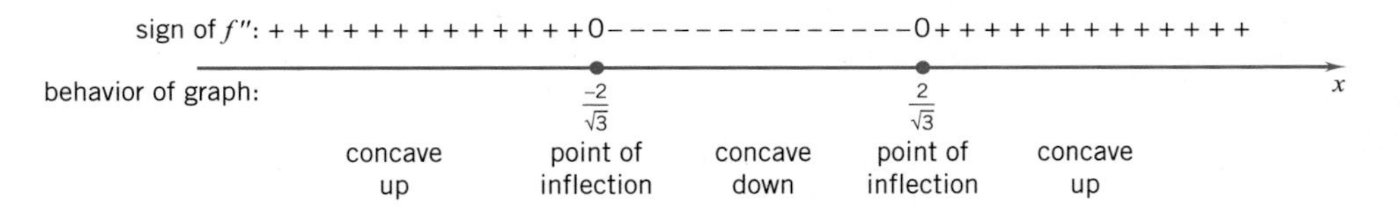

(6) Points of interest: (See Figure 4.8.2 for a preliminary sketch.)

$(0, \frac{7}{4})$:	y-intercept $\frac{7}{4}$.
$(-1, 0), (1, 0), (-\sqrt{7}, 0), (\sqrt{7}, 0)$:	x-intercepts.
$(0, \frac{7}{4})$:	$f(0) = \frac{7}{4}$ is a local maximum.
$(-2, -\frac{9}{4}), (2, -\frac{9}{4})$:	$f(-2) = f(2) = -\frac{9}{4}$ are local (and absolute) minimums.
$(-2/\sqrt{3}, -17/36), 2/\sqrt{3}, -17/36)$:	points of inflection; slopes $\pm 16/(3\sqrt{3}) \cong \pm 3.1$.

(7) Sketch the graph: Of course, since the graph is symmetric with respect to the y-axis, we can sketch the graph for $x \geq 0$, and then the graph for $x > 0$ is the mirror image of that in the y-axis. See Figure 4.8.3. ❑

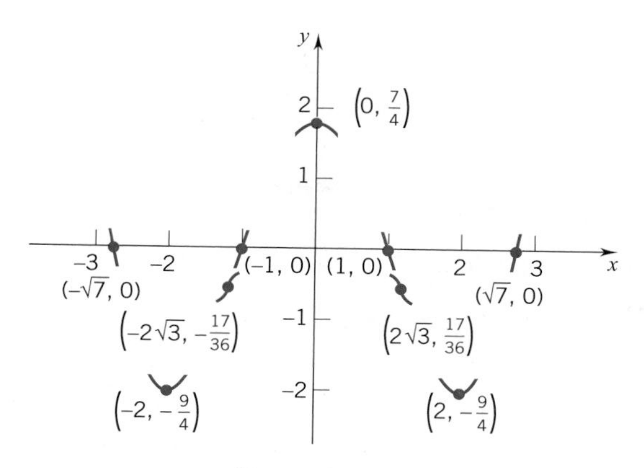

Figure 4.8.2

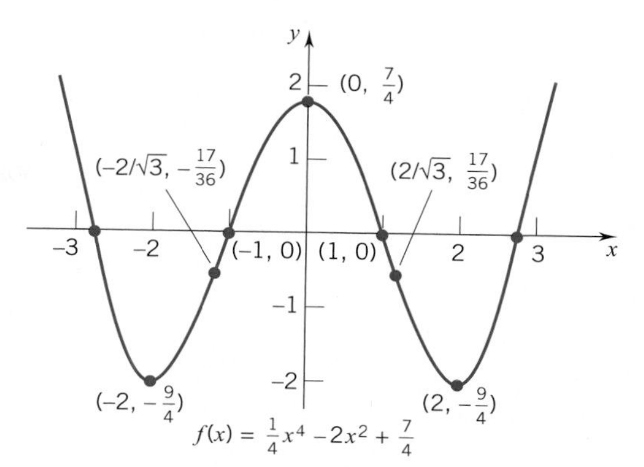

Figure 4.8.3

Example 2 Sketch the graph of the function

$$f(x) = x^4 - 4x^3 + 1 \qquad x \in [-1, 5).$$

SOLUTION

(1) Domain: The domain is given as $[-1, 5)$; -1 is the left endpoint; and 5 is a "missing right endpoint." There are no asymptotes and we do not consider the behavior of f as $x \to \pm\infty$ since f is defined only on $[-1, 5)$.

(2) Intercepts: The y-intercept is $f(0) = 1$. To find the x-intercepts, we must solve the equation

$$x^4 - 4x^3 + 1 = 0.$$

We cannot do this exactly, but you can verify that $f(0) > 0$ and $f(1) < 0$, and that $f(3) < 0$ and $f(4) > 0$. Thus there are x-intercepts on the intervals $(0, 1)$ and $(3, 4)$. We could find approximate values for these intercepts using the Newton–Raphson method but we do not need to do this since we are only trying to sketch the graph.

(3) Symmetry/periodicity: The graph is not symmetric with respect to the y-axis or the origin $[f(-x) \neq \pm f(x)]$; f is not periodic.

(4) First derivative: For $x \in (-1, 5)$

$$f'(x) = 4x^3 - 12x^2 = 4x^2(x - 3).$$

The critical numbers of f are $x = 0$ and $x = 3$.

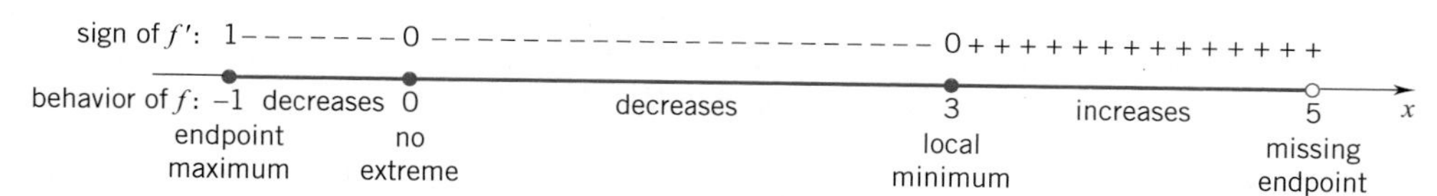

(5) Second derivative:

$$f''(x) = 12x^2 - 24x = 12x(x - 2).$$

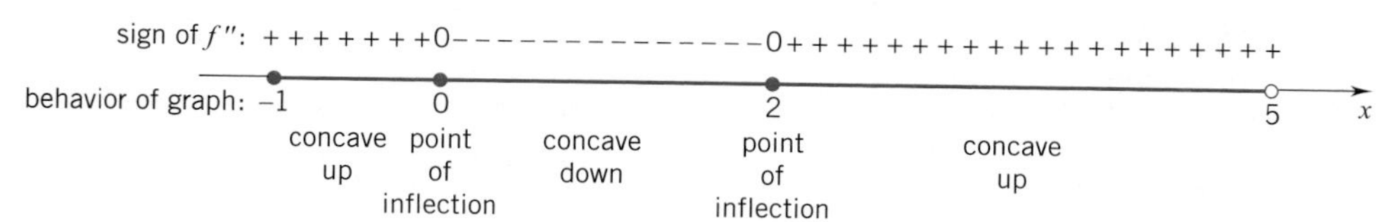

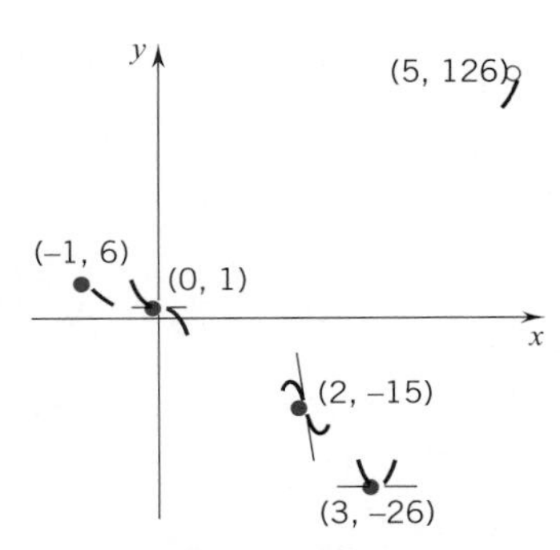

Figure 4.8.4

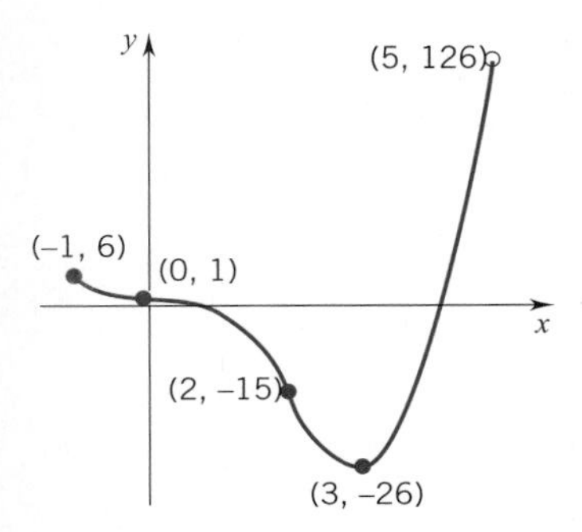

Figure 4.8.5

(6) Points of interest: (See Figure 4.8.4 for a preliminary sketch.)

$(0, 1)$: y-intercept; point of inflection with horizontal tangent.

$(-1, 6)$: $f(-1) = 6$ is an endpoint maximum.

$(2, -15)$: point of inflection; slope -16.

$(3, -26)$: $f(3) = -26$ is a local (and absolute) minimum.

As x approaches the missing endpoint 5 from the left, $f(x)$ increases toward a value of 126.

(7) Sketch the graph. Since the range of f makes a scale drawing impractical, we must be content with a rough sketch as in Figure 4.8.5. In cases like this, it is particularly important to give the coordinates of the points of interest. ❑

Example 3 Sketch the graph of the function

$$f(x) = \frac{x^2 - 3}{x^3}.$$

SOLUTION

(1) Domain: The domain of f is $\{x: x \neq 0\} = (-\infty, 0) \cup (0, \infty)$. The y-axis (the line $x = 0$) is a vertical asymptote since $f(x) \to \infty$ as $x \to 0^-$ and $f(x) \to -\infty$ as $x \to 0^+$. The x-axis (the line $y = 0$) is a horizontal asymptote: $f(x) \to 0$ as $x \to \pm\infty$.

(2) Intercepts: There is no y-intercept since $f(0)$ is not defined. The x-intercepts are $x = \pm\sqrt{3}$.

(3) Symmetry: Since

$$f(-x) = \frac{(-x)^2 - 3}{(-x)^3} = -\frac{x^2 - 3}{x^3} = -f(x),$$

the graph is symmetric with respect to the origin; f is not periodic.

(4) First derivative: It is easier to calculate f' if we first rewrite f as

$$f(x) = \frac{x^2 - 3}{x^3} = x^{-1} - 3x^{-3}.$$

Now,

$$f'(x) = -x^{-2} + 9x^{-4} = \frac{9 - x^2}{x^4}.$$

The critical numbers of f are $x = \pm 3$. (NOTE: $x = 0$ is not a critical number since 0 is not in the domain of f.)

The sign of f' and the behavior of f are:

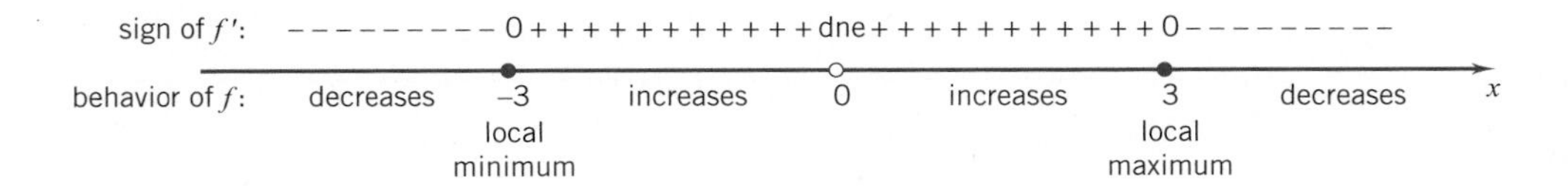

(5) Second derivative:

$$f''(x) = 2x^{-3} - 36x^{-5} = \frac{2(x^2 - 18)}{x^5} = \frac{2(x - 3\sqrt{2})(x + 3\sqrt{2})}{x^5}.$$

The sign of f'' and the behavior of the graph of f are:

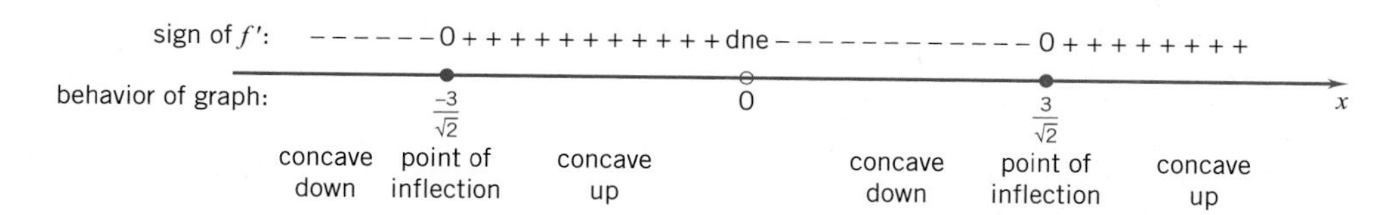

(6) Points of interest: (See Figure 4.8.6 for a preliminary sketch.)

$(-\sqrt{3}, 0), (\sqrt{3}, 0)$: x-intercepts.

$(-3, -2/9)$: local minimum.

$(3, 2/9)$: local maximum.

$(-3\sqrt{2}, -5\sqrt{2}/36), (3\sqrt{2}, 5\sqrt{2}/36)$: points of inflection.

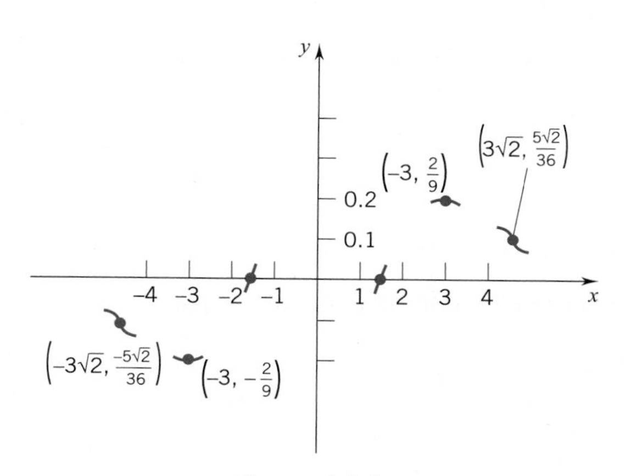

Figure 4.8.6

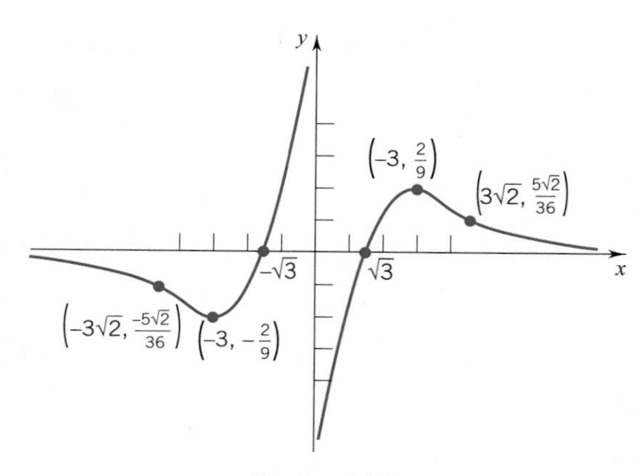

Figure 4.8.7

(7) Sketch the graph: see Figure 4.8.7. ❑

Example 4 Sketch the graph of the function

$$f(x) = \frac{3}{5}x^{5/3} - 3x^{2/3}.$$

SOLUTION

(1) Domain: The domain of f is the set of real numbers. Since we can write $f(x) = x^{2/3}(\frac{3}{5}x - 3)$, it is clear that, as $x \to \infty$, $f(x) \to \infty$, and as $x \to -\infty$, $f(x) \to -\infty$. There are no asymptotes.

(2) Intercepts: $f(0) = 0$ is both the y-intercept and an x-intercept; $x = 5$ is also an x-intercept.

(3) Symmetry/periodicity: There is no symmetry; f is not periodic.

(4) First derivative:

$$f'(x) = x^{2/3} - 2x^{-1/3} = \frac{x-2}{x^{1/3}}.$$

The critical numbers of f are $x = 0$ and $x = 2$. The sign of f' and the behavior of f are:

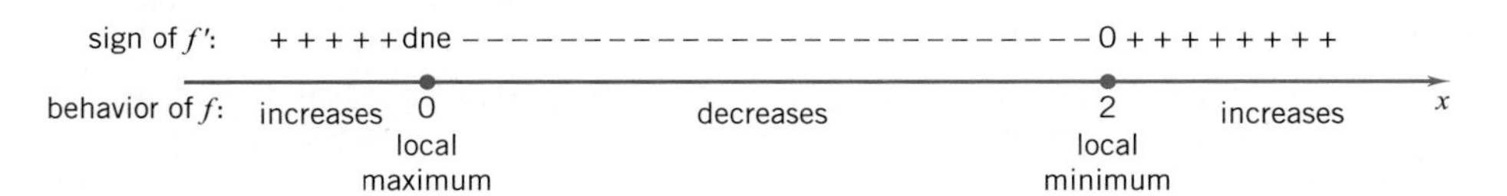

Note that, as $x \to 0^-, f'(x) \to \infty$, and as $x \to 0^+, f'(x) \to -\infty$. Thus the graph of f has a vertical cusp at $x = 0$.

(5) Second derivative:

$$f''(x) = \frac{2}{3}x^{-1/3} + \frac{2}{3}x^{-4/3} = \frac{2}{3}x^{-4/3}(x + 1).$$

The sign of f'' and the behavior of the graph of f are:

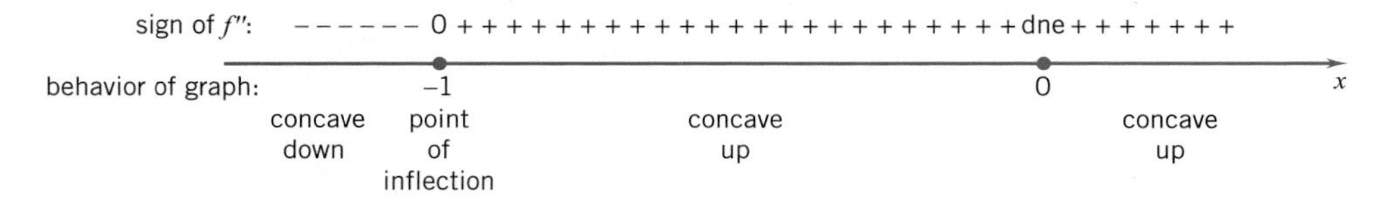

(6) Points of interest: (See Figure 4.8.8 for a preliminary sketch.)

$(0, 0)$:	y-intercept and a local maximum; vertical cusp at $(0, 0)$.
$(0, 0), (5, 0)$:	x-intercepts.
$(2, -9\sqrt[3]{4}/5)$:	$f(2) = -9\sqrt[3]{4}/5 \approx -2.9$ is a local minimum.
$(-1, -18/5)$:	point of inflection; slope 3.

(7) Sketch the graph: See Figure 4.8.9. ❑

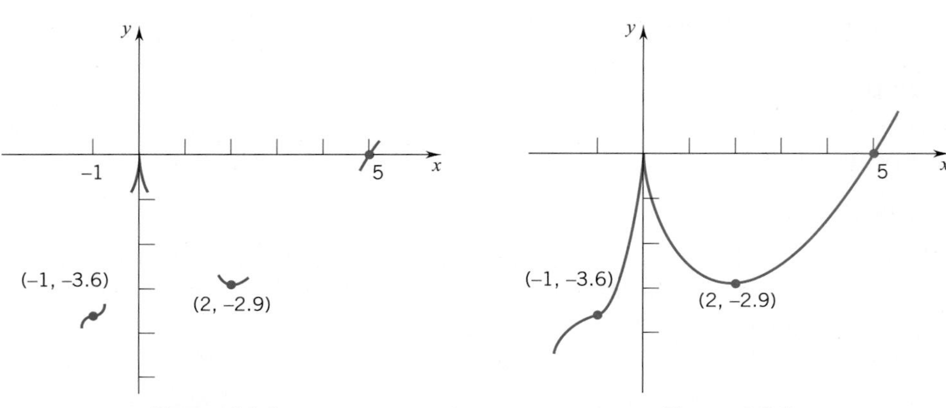

Figure 4.8.8

Figure 4.8.9

Example 5 Sketch the graph of

$$f(x) = \sin 2x - 2 \sin x.$$

SOLUTION

(1) Domain: The domain of f is the set of all real numbers. There are no asymptotes and you should verify that the graph of f oscillates between $\frac{3}{2}\sqrt{3}$ and $-\frac{3}{2}\sqrt{3}$ as $x \to \pm\infty$.

(2) Intercepts: The y-intercept is $f(0) = 0$. To find the x-intercepts, we set $f(x) = 0$:

$$\begin{aligned} \sin 2x - 2 \sin x &= 2 \sin x \cos x - 2 \sin x \\ &= 2 \sin x(\cos x - 1) = 0. \end{aligned}$$

Now, $\sin x = 0$ at $x = n\pi$, n an integer; and $\cos x = 1$ at $x = 2n\pi$, n an integer. Thus, the x-intercepts are $x = n\pi$, $n = 0, \pm 1, \pm 2, \ldots$.

(3) Symmetry/periodicity: Since $f(-x) = \sin 2(-x) - 2 \sin(-x) = -\sin 2x + 2 \sin x = -f(x)$, f is an odd function and the graph is symmetric with respect to the origin. Also, f is periodic with period 2π. Based on these two properties, it would be sufficient to sketch the graph of f on the interval $[0, \pi]$. The result could then be extended to the interval $[-\pi, 0]$ using the symmetry, and then to $(-\infty, \infty)$ using the periodicity. However, for purposes of illustration here, we will consider f and its derivatives on $[-\pi, \pi]$.

(4) First derivative:

$$\begin{aligned} f'(x) &= 2 \cos 2x - 2 \cos x \\ &= 2(2 \cos^2 x - 1) - 2 \cos x \\ &= 4 \cos^2 x - 2 \cos x - 2 \\ &= 2(2 \cos x + 1)(\cos x - 1). \end{aligned}$$

The critical points of f (on $[-\pi, \pi]$) are $x = -2\pi/3$, $2\pi/3$, and $x = 0$.

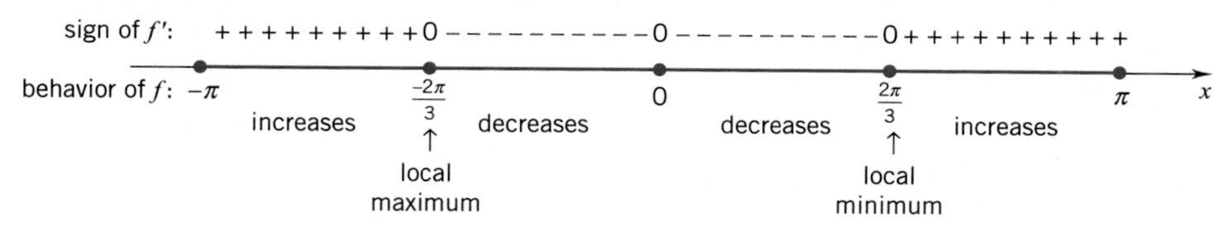

(5) Second derivative:

$$\begin{aligned} f''(x) &= -4 \sin 2x + 2 \sin x \\ &= -8 \sin x \cos x + 2 \sin x \\ &= 2 \sin x(-4 \cos x + 1). \end{aligned}$$

Now $f''(x) = 0$ at $x = -\pi, 0, \pi$, and at the values of x where $\cos x = \frac{1}{4}$, which yields $x \cong \pm 1.3$. The sign of f'' and the behavior of the graph (on $[-\pi, \pi]$) are:

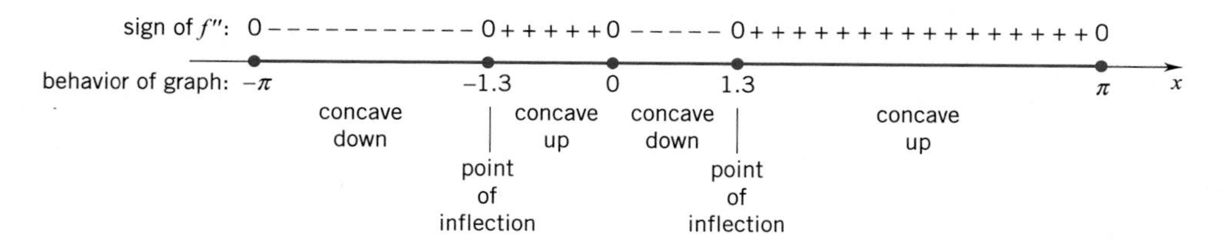

(6) Points of interest: (See Figure 4.8.10 for a preliminary sketch.)

$(0, 0)$:	y-intercept.
$(-\pi, 0), (0, 0), (\pi, 0)$:	x-intercepts; these are also points of inflection.
$(-\frac{2}{3}\pi, \frac{3}{2}\sqrt{3})$:	$f(-2\pi/3) = \frac{3}{2}\sqrt{3} \cong -2.6$ is a local (and absolute) maximum.
$(\frac{2}{3}\pi, -\frac{3}{2}\sqrt{3})$:	$f(2\pi/3) = -\frac{3}{2}\sqrt{3} \cong -2.6$ is a local (and absolute) minimum.
$(-1.3, 1.4), (1.3, -1.4)$:	points of inflection (approximately), in addition to those noted above.

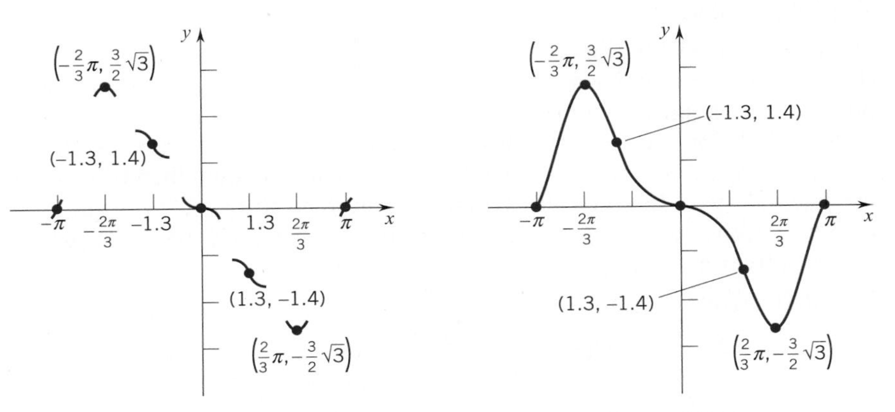

Figure 4.8.10

Figure 4.8.11

(7) Sketch the graph: The graph of f on the interval $[-\pi, \pi]$ is shown in Figure 4.8.11. An indication of the complete graph is shown in Figure 4.8.12

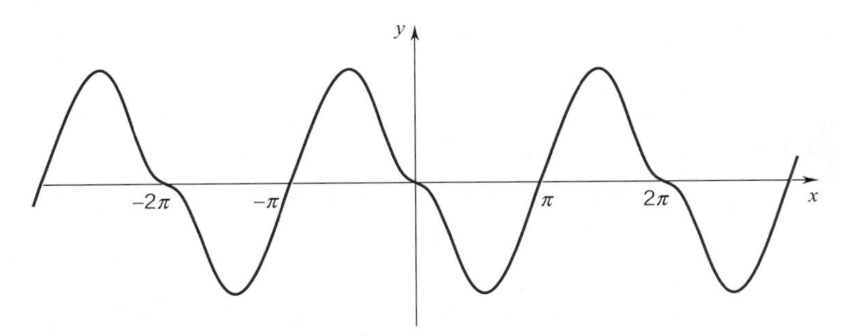

Figure 4.8.12

EXERCISES 4.8

In Exercises 1–54, sketch the graph of the function f using the approach presented in this section.

1. $f(x) = (x - 2)^2$.

2. $f(x) = 1 - (x - 2)^2$.

3. $f(x) = x^3 - 2x^2 + x + 1$.

4. $f(x) = x^3 - 9x^2 + 24x - 7$.

5. $f(x) = x^3 + 6x^2, \quad x \in [-4, 4]$.

6. $f(x) = x^4 - 8x^2, \quad x \in [0, \infty)$.

7. $f(x) = \frac{2}{3}x^3 - \frac{1}{2}x^2 - 10x - 1$.

8. $f(x) = x(x^2+4)^2$.

9. $f(x) = x^2 + \dfrac{2}{x}$.

10. $f(x) = x - \dfrac{1}{x}$.

11. $f(x) = \dfrac{x-4}{x^2}$.

12. $f(x) = \dfrac{x+2}{x^3}$.

13. $f(x) = 2\sqrt{x} - x, \quad x \in [0, 4]$.

14. $f(x) = \frac{1}{4}x - \sqrt{x}, \quad x \in [0, 9]$.

15. $f(x) = 2 + (x+1)^{6/5}$.

16. $f(x) = 2 + (x+1)^{7/5}$.

17. $f(x) = 3x^5 + 5x^3$.

18. $f(x) = 3x^4 + 4x^3$.

19. $f(x) = 1 + (x-2)^{5/3}$.

20. $f(x) = 1 + (x-2)^{4/3}$.

21. $f(x) = \dfrac{2x}{4x-3}$.

22. $f(x) = \dfrac{2x^2}{x+1}$.

23. $f(x) = \dfrac{x}{(x+3)^2}$.

24. $f(x) = \dfrac{x}{x^2+1}$.

25. $f(x) = \dfrac{x^2}{x^2-4}$.

26. $f(x) = \dfrac{2x}{x-4}$.

27. $f(x) = x\sqrt{1-x}$.

28. $f(x) = (x-1)^4 - 2(x-1)^2$.

29. $f(x) = x + \sin 2x, \quad x \in [0, \pi]$.

30. $f(x) = \cos^3 x + 6\cos x, \quad x \in [0, \pi]$.

31. $f(x) = \cos^4 x, \quad x \in [0, \pi]$.

32. $f(x) = \sqrt{3}x - \cos 2x, \quad x \in [0, \pi]$.

33. $f(x) = 2\sin^3 x + 3\sin x, \quad x \in [0, \pi]$.

34. $f(x) = \sin^4 x, \quad x \in [0, \pi]$.

35. $f(x) = (x+1)^3 - 3(x+1)^2 + 3(x+1)$.

36. $f(x) = x^3(x+5)^2$.

37. $f(x) = x^2(5-x)^3$.

38. $f(x) = 4 - |2x - x^2|$.

39. $f(x) = 3 - |x^2 - 1|$.

40. $f(x) = x - x^{1/3}$.

41. $f(x) = x(x-1)^{1/5}$.

42. $f(x) = x^2(x-7)^{1/3}$.

43. $f(x) = x^2 - 6x^{1/3}$.

44. $f(x) = \dfrac{2x}{\sqrt{x^2+1}}$.

45. $f(x) = \sqrt{\dfrac{x}{x-2}}$.

46. $f(x) = \sqrt{\dfrac{x}{x+4}}$.

47. $f(x) = \dfrac{x^2}{\sqrt{x^2-2}}$.

48. $f(x) = 3\cos 4x, \quad x \in [0, \pi]$.

49. $f(x) = 2\sin 3x, \quad x \in [0, \pi]$.

50. $f(x) = 3 + 2\cot x + \csc^2 x, \quad x \in (0, \frac{1}{2}\pi)$.

51. $f(x) = 2\tan x - \sec^2 x, \quad x \in (0, \frac{1}{2}\pi)$.

52. $f(x) = 2\cos x + \sin^2 x$.

53. $f(x) = \dfrac{\sin x}{1 - \sin x}, \quad x \in (-\pi, \pi)$.

54. $f(x) = \dfrac{1}{1 - \cos x}, \quad x \in (-\pi, \pi)$.

55. A function f is continuous on $(-\infty, \infty)$, differentiable for all $x \neq 0$, and $f(0) = 0$. The graph of the derivative of f is shown in the figure below.

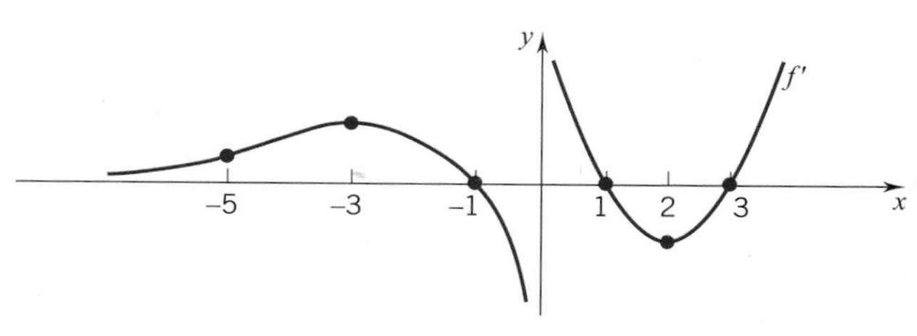

(a) Determine the intervals on which f increases and the intervals on which it decreases; find the critical numbers of f.

(b) Sketch the graph of $f''(x)$ and determine where the graph of f is concave up and where it is concave down.

(c) Does the graph of f have any horizontal asymptotes? Sketch the graph of f.

56. Set

$$F(x) = \begin{cases} \sin(1/x), & x \neq 0 \\ 0, & x = 0, \end{cases}$$

$$G(x) = \begin{cases} x\sin(1/x), & x \neq 0 \\ 0, & x = 0, \end{cases}$$

$$H(x) = \begin{cases} x^2\sin(1/x), & x \neq 0 \\ 0, & x = 0. \end{cases}$$

(a) Sketch a figure displaying the general nature of the graph of F.

(b) Sketch a figure displaying the general nature of the graph of G.

(c) Sketch a figure displaying the general nature of the graph of H.

(d) Which of these functions is continuous at 0?

(e) Which of these functions is differentiable at 0?

57. Show that the lines $y = (b/a)x$ and $y = -(b/a)x$ are oblique asymptotes of the hyperbola

$$\frac{x^2}{a^2} - \frac{y^2}{b^2} = 1.$$

■ CHAPTER HIGHLIGHTS

4.1 The Mean-Value Theorem

mean-value theorem (p. 211) Rolle's theorem (p. 213)

4.2 Increasing and Decreasing Functions

f increases on an interval I (p. 218) f decreases on an interval I (p. 218)

One can find the intervals on which a differentiable function increases or decreases or is constant by examining the sign of the derivative.

If two functions have the same derivative on an interval, then they differ by a constant on that interval (p. 222).

4.3 Local Extreme Values

local extreme: local maximum, local minimum (p. 226)
critical number (p. 227)   first-derivative test (p. 229)
second-derivative test (p. 231)

If f has a local extreme value at c, then either $f'(c) = 0$ or $f'(c)$ does not exist; the converse is false.

The first-derivative test requires that we examine the sign of the first derivative on both sides of the critical number; the test is conclusive provided the function is continuous at the critical number. The second-derivative test requires that we examine the sign of the second derivative at the critical number itself; the test is inconclusive if the second derivative is 0 at the critical number.

4.4 Endpoint and Absolute Extreme Values

endpoint maximum and endpoint minimum (p. 235)
absolute maximum and absolute minimum (p. 235)
$f(x) \to \pm\infty$ as $x \to \pm\infty$ (p. 237)

4.5 Optimization Problems

strategy for solving optimization problems (p. 244)

The *key steps* are to express the quantity to be maximized or minimized as a function of one variable and to specify the domain of the function.

4.6 Concavity and Points of Inflection

concave up and concave down (p. 252)
point of inflection (p. 253)

If $(c, f(c))$ is a point of inflection, then either $f''(c) = 0$ or $f''(c)$ does not exist; the converse is false.

4.7 Vertical and Horizontal Asymptotes; Vertical Tangents and Cusps

$f(x) \to \pm\infty$ as $x \to c$ (p. 257) $f(x) \to L$ as $x \to \pm\infty$ (p. 259)
vertical asymptote (p. 257)   horizontal asymptote (p. 259)
vertical tangent (p. 261)   vertical cusp (p. 262)

4.8 Curve Sketching

systematic procedure for sketching graphs (p. 265)

■ PROJECTS AND EXPLORATIONS USING TECHNOLOGY

To do these exercises you will need a graphics calculator or a computer with graphing capability. The majority of these problems are open-ended so different approaches may be used to solve them. You should be aware that different approaches can result in slight variations in the answers. Round your numerical answers to at least four decimal places. The rounding method that your calculator or computer uses also may cause variations in answers.

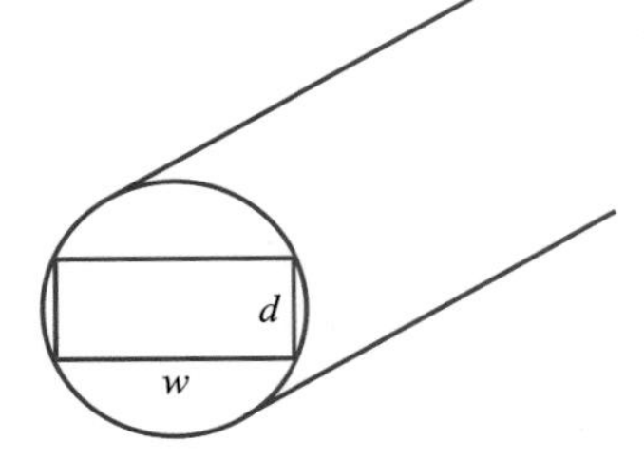

Figure A

4.1 A rectangular beam is to be cut from a cylindrical log of radius r (see Figure A).

(a) Show that the beam that has maximal cross-sectional area is a square beam.
(b) If rectangular planks are to be cut from the four pieces that remain after the square beam has been cut, find the ratio of the width to the depth that will give the plank of maximal cross-sectional area. See Figure B.
(c) Suppose that the strength of a beam is proportional to the product of the width and the square of the depth (see Figure A). Find the ratio of width to depth of the beam of maximum strength.
(d) If rectangular planks are to be cut from the four pieces that remain, find the ratio of width to depth that will give the plank of maximal strength.
(e) Suppose the log is elliptical instead of circular (Figure C). In particular, let the equation of the ellipse be

$$x^2 + 2y^2 = r^2.$$

Answer parts (a)–(d) for this log.

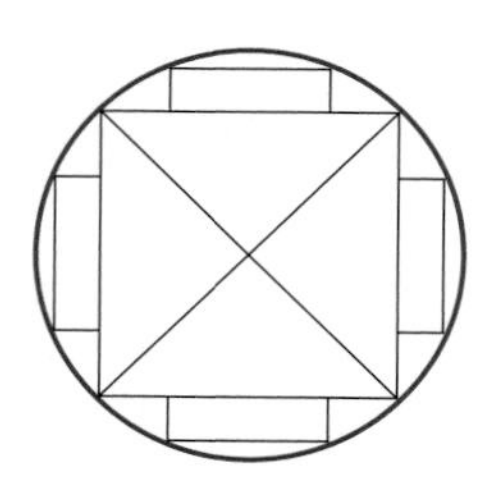

Figure B

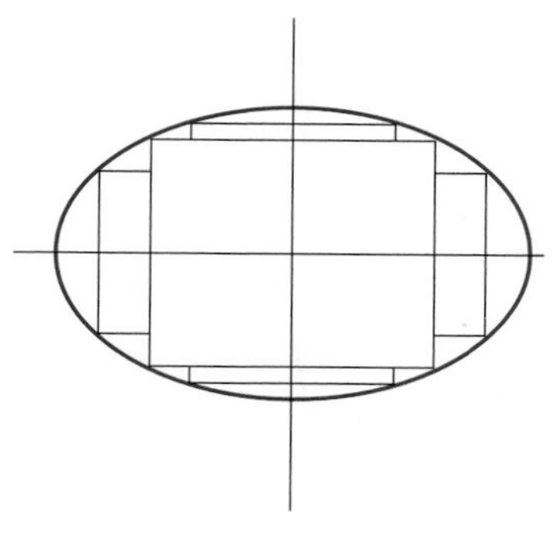

Figure C

4.2 In the Projects and Explorations section of Chapter 1 we introduced moving averages as a possible means of analyzing the stock market or seasonal sales figures. In this problem we consider a slight generalization of that moving average. Fix $a > 0$. For a given function f, define the moving average function g_a by

$$g_a(x) = \frac{f(x - a) + f(x)}{2}.$$

(a) Let f be a differentiable function. Show that the derivative of g_a is the moving average of f'.
(b) Set

$$f(x) = \frac{x^2 + 3x - 5}{x^3 - 6x^2 - 7}.$$

Let $x = 1$ and consider $g_a(1)$ as a function of a. Find the derivative of $g_a(1)$ with respect to a. Now fix any number x and use the chain rule to find the derivative with respect to a of $g_a(x)$.
(c) Suppose that f is an increasing function on an interval I. Prove that $g_a(x)$ is increasing on I.
(d) If $f'(b) = 0$, then we might conjecture that $g_a'[(b - a)/2] = 0$. Show that this conjecture is false. However, give an example of a function for which this conjecture is true.
(e) Suppose that f is twice differentiable and that its graph is concave up on an interval I. Prove that the graph of $g_a(x)$ is concave up on I.

4.3 Calculators and computers often approximate derivatives by using finite difference quotients. Fix $h > 0$. The *left quotient* is defined by

$$D_L(x) = \frac{f(x) - f(x - h)}{h};$$

the *right quotient* by

$$D_R(x) = \frac{f(x + h) - f(x)}{h};$$

and the *symmetric difference* by

$$D_C(x) = \frac{f(x + h) - f(x - h)}{2h}.$$

(a) Let $f(x) = x^3$ and set $x = 2$. Calculate $D_L(2)$, $D_R(2)$, and $D_C(2)$ for $h = 0.1, 0.01, 0.001$, and 0.0005. Which gives the best estimate for $f'(2)$?
(b) We would expect that the estimates will improve by taking smaller and smaller values of h. However, round-off error can cause the estimates to get worse for very small values of h. For your calculator, find some values of h that give large errors.

(c) Based on the concavity of $f(x) = x^3$, explain why the answers you obtained in part (a) were too small or too large.

(d) Consider the function

$$E(h) = D_L(2) - f'(2).$$

If you calculate the algebraic formula for $E(h)$, it follows that its graph is concave up. However, when you graph E (with h small) using your graphing utility this may not be as evident. Compare what the graph tells you with what the algebra tells you.

4.4 Let

$$f(x) = \frac{x^2 + x + 10}{x^4 + 7x^2 + 1}.$$

It is easy to show that f is defined and positive for all x, and that the x-axis is a horizontal asymptote of the graph of f. We want to find the rectangle of maximum area that has its upper vertices on the graph of f and its lower vertices on the x-axis. See the figure.

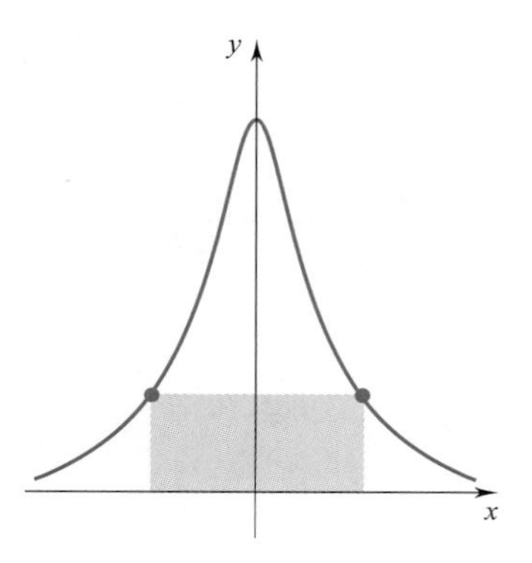

Solve this problem by using the following sequence of steps:

(a) Find f' and f''. Determine where f is increasing and where it is decreasing. Show that f has an absolute maximum value and find it. Determine where the graph of f is concave up and where it is concave down. Find the points of inflection.

(b) Verify that the x-axis is a horizontal asymptote of the graph of f and show that the graph is concave up when $|x|$ is large.

(c) Use the information from parts (a) and (b) to sketch the graph of f. Then verify that a horizontal line will either (1) not intersect the graph at all, (2) intersect the graph in exactly one point (namely, when the line passes through the absolute maximum point on the graph), or (3) intersect the graph in exactly two points.

(d) For several y values between 0 and the absolute maximum value of f, use a graphing utility to approximate the points of intersection of the horizontal lines and the graph.

(e) Using the points found in part (d), drop vertical lines to the x-axis to obtain inscribed rectangles. Calculate the areas of each. From these calculations, does it appear that the areas of the rectangles approach 0 as $y \to 0$ or as y approaches the absolute maximum value of f?

(f) Using the values found in part (e), make some new guesses of horizontal lines to use to find inscribed rectangles of larger area.

(g) After you have found your approximation to the rectangle of maximum area, evaluate f' at each of the upper vertices of the rectangle. What do you notice?

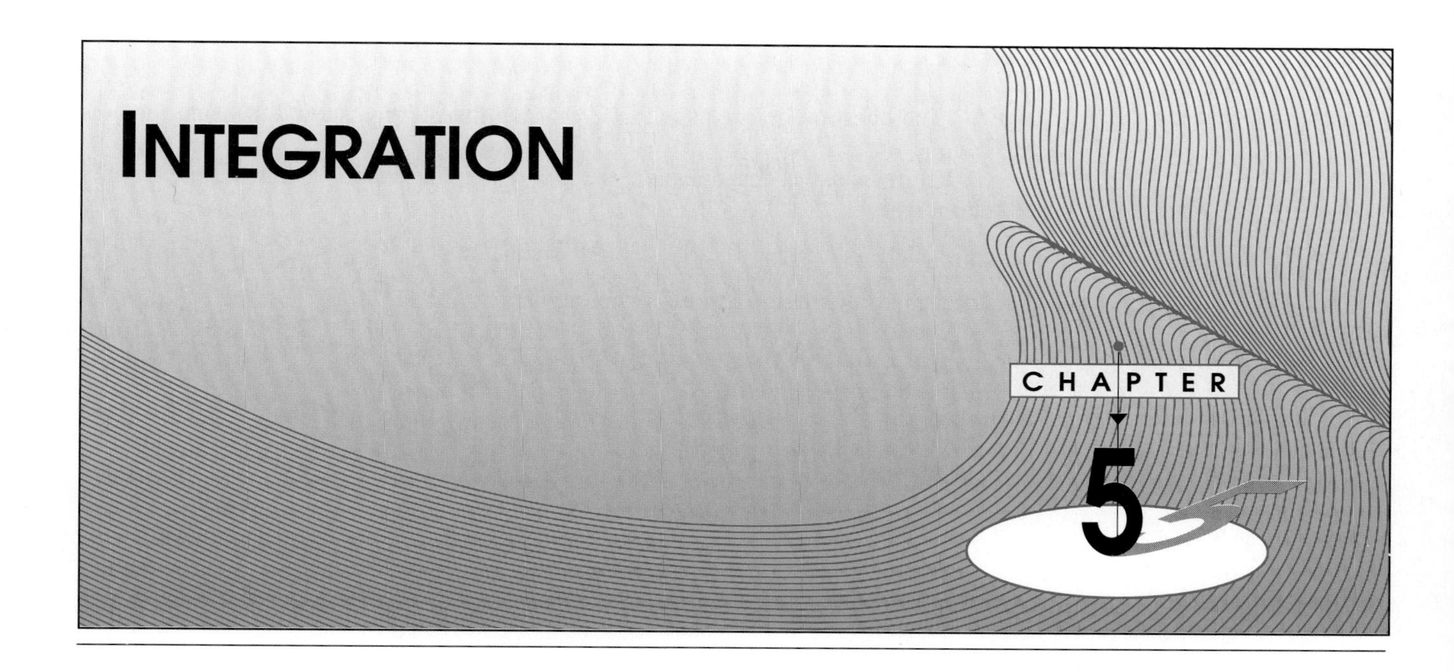

5.1 AN AREA PROBLEM; A SPEED–DISTANCE PROBLEM

An Area Problem

You are already familiar with the formulas for the area of such regular geometric figures as a triangle, rectangle, square and circle (Figure 5.1.1): In Figure 5.1.2 you can see a region Ω bounded above by the graph of a continuous nonnegative function f, bounded below by the x-axis, bounded to the left by $x = a$, and bounded to the right by

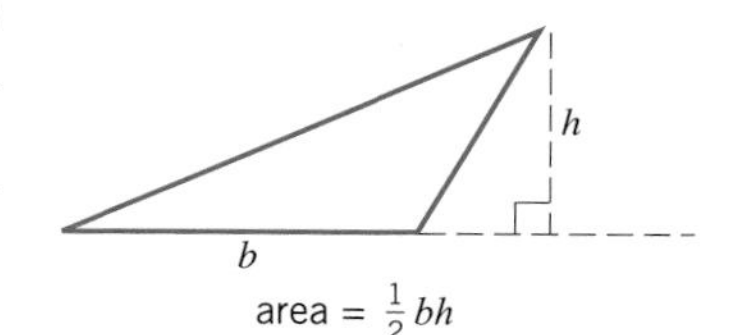

Figure 5.1.1A

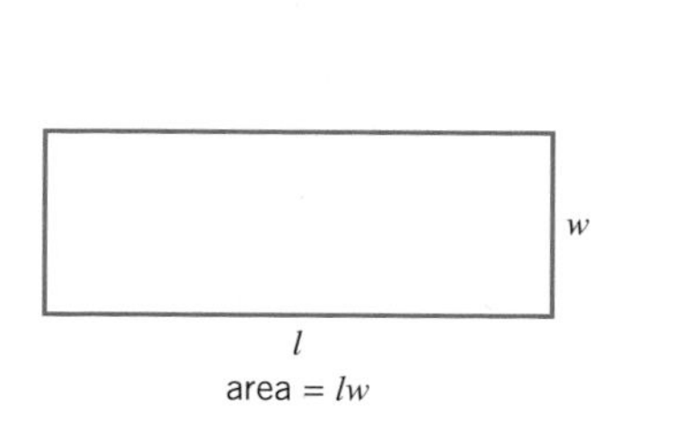

Figure 5.1.1B

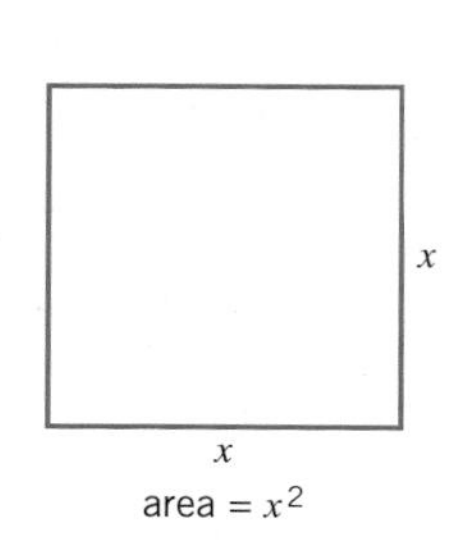

Figure 5.1.1C

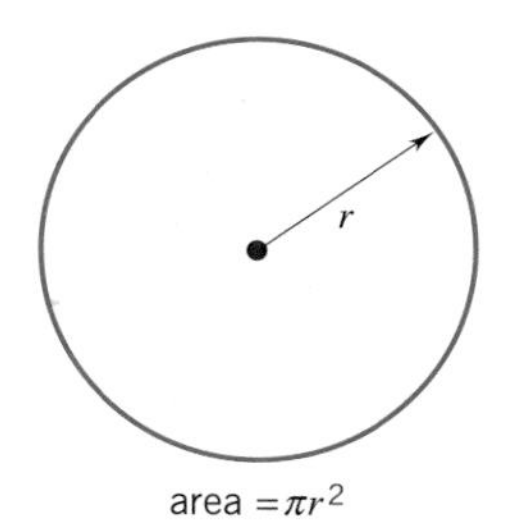

Figure 5.1.1D

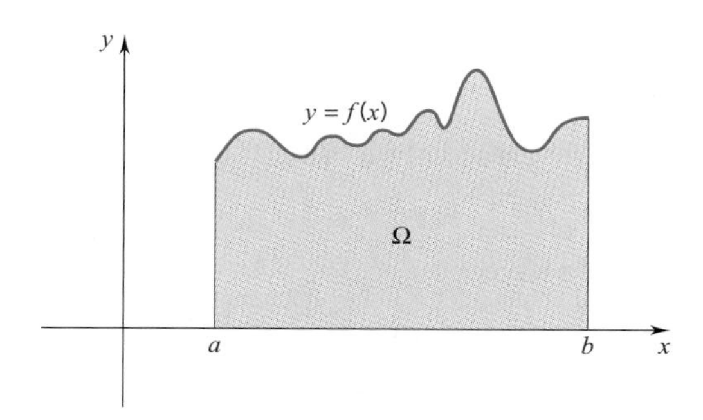

Figure 5.1.2

$x = b$. The question before us is this: What number, if any, should measure the area of Ω?

To begin to answer this question, we split up the interval $[a, b]$ into a finite number of subintervals

$$[x_0, x_1], [x_1, x_2], \ldots, [x_{n-1}, x_n] \quad \text{with} \quad a = x_0 < x_1 < \cdots < x_n = b.$$

This breaks up the region Ω into n subregions:

$$\Omega_1, \Omega_2, \ldots, \Omega_n. \qquad \text{(Figure 5.1.3)}$$

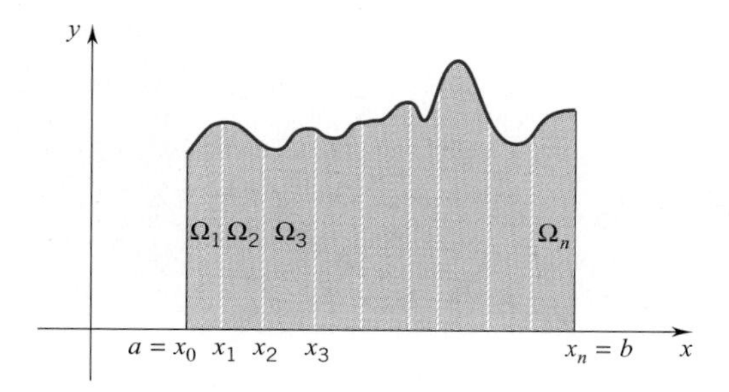

Figure 5.1.3

We can estimate the total area of Ω by estimating the area of each subregion Ω_i and adding up the results. Let's denote by M_i the maximum value of f on $[x_{i-1}, x_i]$ and by m_i the minimum value. (Recall Theorem 2.6.2.) Consider now the rectangles r_i and R_i of Figure 5.1.4. Since

$$r_i \subseteq \Omega_i \subseteq R_i,$$

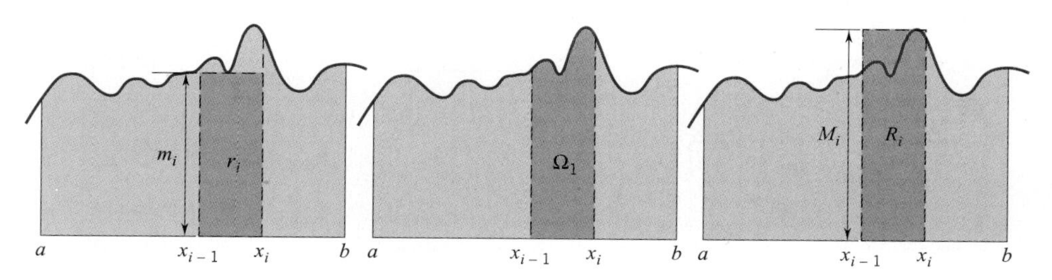

Figure 5.1.4

we must have

$$\text{area of } r_i \le \text{area of } \Omega_i \le \text{area of } R_i.$$

Since the area of a rectangle is the length times the width,

$$m_i(x_i - x_{i-1}) \le \text{area of } \Omega_i \le M_i(x_i - x_{i-1}).$$

Setting $x_i - x_{i-1} = \Delta x_i$ we have

$$m_i\,\Delta x_i \le \text{area of } \Omega_i \le M_i\,\Delta x_i.$$

This inequality holds for $i = 1, i = 2, \ldots, i = n$. Adding up these inequalities we get on the one hand

(5.1.1) $$m_1\Delta x_1 + m_2\Delta x_2 + \cdots + m_n\Delta x_n \le \text{area of } \Omega$$

and on the other hand

(5.1.2) $$\text{area of } \Omega \le M_1\Delta x_1 + M_2\Delta x_2 + \cdots + M_n\Delta x_n.$$

A sum of the form

$$m_1\Delta x_1 + m_2\Delta x_2 + \cdots + m_n\Delta x_n \qquad \text{(Figure 5.1.5)}$$

is called a *lower sum for f*. A sum of the form

$$M_1\Delta x_1 + M_2\Delta x_2 + \cdots + M_n\Delta x_n \qquad \text{(Figure 5.1.6)}$$

is called an *upper sum for f*.

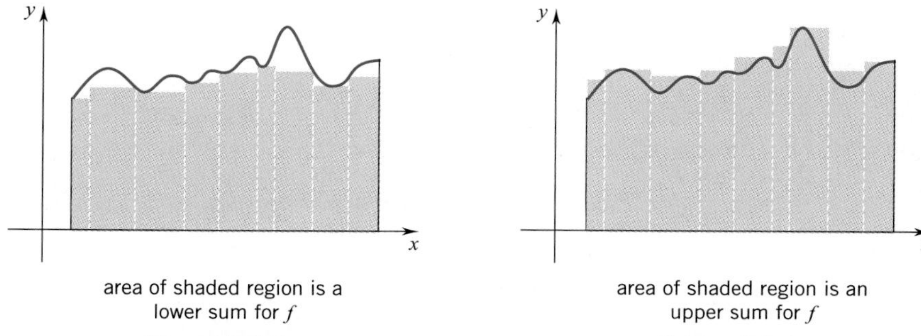

area of shaded region is a lower sum for f
Figure 5.1.5

area of shaded region is an upper sum for f
Figure 5.1.6

Inequalities (5.1.1) and (5.1.2) together tell us that for a number to be a candidate for the title "area of Ω" it must be greater than or equal to every lower sum for f, and it must be less than or equal to every upper sum. By an argument that we omit here, it can be proved that with f continuous on $[a, b]$ there is one and only one such number. This number we call *the area of* Ω.

Later we will return to the subject of area. At this point we turn to a speed–distance problem. As you will see, this new problem can be approached by the same technique that we just applied to the area problem.

A Speed–Distance Problem

If an object moves at a constant speed for a given period of time, then the total distance traveled is given by the familiar formula

$$\text{distance} = \text{speed} \times \text{time}.$$

Suppose now that during the course of the motion the speed does not remain constant but instead varies continuously. How can the total distance traveled be computed then?

To answer this question, we suppose that the motion begins at time a, ends at time b, and that during the time interval $[a, b]$ the speed varies continuously.

As in the case of the area problem we begin by breaking up the interval $[a, b]$ into a finite number of subintervals:

$$[t_0, t_1], [t_1, t_2], \ldots, [t_{n-1}, t_n], \quad \text{with} \quad a = t_0 < t_1 < \cdots < t_n = b.$$

On each subinterval $[t_{i-1}, t_i]$ the object attains a certain maximum speed M_i and a certain minimum speed m_i. (How do we know this?) If throughout the time interval $[t_{i-1}, t_i]$ the object were to move constantly at its minimum speed, m_i, then it would cover a distance of $m_i\Delta t_i$ units. If instead it were to move constantly at its maximum speed, M_i, then it would cover a distance of $M_i\Delta t_i$ units. As it is, the actual distance traveled, call it s_i, must lie somewhere in between; namely, we must have

$$m_i \Delta t_i \le s_i \le M_i \Delta t_i.$$

The total distance traveled during the time interval $[a, b]$, call it s, must be the sum of the distances traveled during the subintervals $[t_{i-1}, t_i]$. In other words we must have

$$s = s_1 + s_2 + \cdots + s_n.$$

Since

$$\begin{gathered} m_1 \Delta t_1 \le s_1 \le M_1 \Delta t_1 \\ m_2 \Delta t_2 \le s_2 \le M_2 \Delta t_2 \\ \vdots \\ m_n \Delta t_n \le s_n \le M_n \Delta t_n, \end{gathered}$$

it follows by the addition of these inequalities that

$$m_1 \Delta t_1 + m_2 \Delta t_2 + \cdots + m_n \Delta t_n \le s \le M_1 \Delta t_1 + M_2 \Delta t_2 + \cdots + M_n \Delta t_n.$$

A sum of the form

$$m_1 \Delta t_1 + m_2 \Delta t_2 + \cdots + m_n \Delta t_n$$

is called a *lower sum* for the speed function. A sum of the form

$$M_1 \Delta t_1 + M_2 \Delta t_2 + \cdots + M_n \Delta t_n$$

is called an *upper sum* for the speed function. The inequality we just obtained for s tells us that s must be greater than or equal to every lower sum for the speed function, and it must be less than or equal to every upper sum. As in the case of the area problem, it turns out that there is one and only one such number, and this is the total distance traveled.

■ 5.2 THE DEFINITE INTEGRAL OF A CONTINUOUS FUNCTION

The process we used to approach the two problems of Section 5.1 is called *integration,* and the end results of this process are called *definite integrals.* Our purpose here is to establish these notions more precisely.

(5.2.1) By a *partition* of the closed interval $[a, b]$ we mean a finite subset of $[a, b]$ which contains the points a and b.

We index the elements of a partition according to their natural order. Thus, if we write

$$P = \{x_0, x_1, \ldots, x_n\} \text{ is a partition of } [a, b],$$

you can conclude that

$$a = x_0 < x_1 < \cdots < x_n = b.$$

Example 1 The sets

$$\{0, 1\}, \quad \{0, \tfrac{1}{2}, 1\}, \quad \{0, \tfrac{1}{4}, \tfrac{1}{2}, 1\}, \quad \{0, \tfrac{1}{4}, \tfrac{1}{3}, \tfrac{1}{2}, \tfrac{5}{8}, 1\}$$

are all partitions of the interval $[0, 1]$. ❑

If $P = \{x_0, x_1, \ldots, x_n\}$ is a partition of $[a, b]$, then P breaks up $[a, b]$ into a finite number of subintervals

$$[x_0, x_1], [x_1, x_2], \ldots, [x_{n-1}, x_n]$$

of lengths $\Delta x_1, \Delta x_2, \ldots, \Delta x_n$, respectively.

Suppose now that f is continuous on $[a, b]$. Then on each interval $[x_{i-1}, x_i]$ the function f takes on a maximum value, M_i, and a minimum value, m_i.

(5.2.2)

> The number
>
> $$U_f(P) = M_1\Delta x_1 + M_2\Delta x_2 + \cdots + M_n\Delta x_n$$
>
> is called *the P upper sum for f*, and the number
>
> $$L_f(P) = m_1\Delta x_1 + m_2\Delta x_2 + \cdots + m_n\Delta x_n$$
>
> is called *the P lower sum for f.*

Example 2 The quadratic function

$$f(x) = x^2$$

is continuous on $[0, 1]$. The partition $P = \{0, \frac{1}{4}, \frac{1}{2}, 1\}$ breaks up $[0, 1]$ into three subintervals:

$$[x_0, x_1] = [0, \tfrac{1}{4}], \qquad [x_1, x_2] = [\tfrac{1}{4}, \tfrac{1}{2}], \qquad [x_2, x_3] = [\tfrac{1}{2}, 1]$$

of lengths

$$\Delta x_1 = \tfrac{1}{4} - 0 = \tfrac{1}{4}, \qquad \Delta x_2 = \tfrac{1}{2} - \tfrac{1}{4} = \tfrac{1}{4}, \qquad \Delta x_3 = 1 - \tfrac{1}{2} = \tfrac{1}{2},$$

respectively. Since f is increasing on $[0, 1]$, it takes on its maximum value at the right endpoint of each subinterval:

$$M_1 = f(\tfrac{1}{4}) = \tfrac{1}{16}, \qquad M_2 = f(\tfrac{1}{2}) = \tfrac{1}{4}, \qquad M_3 = f(1) = 1.$$

The minimum values of f are taken on at the left endpoints of each subinterval:

$$m_1 = f(0) = 0, \qquad m_2 = f(\tfrac{1}{4}) = \tfrac{1}{16}, \qquad m_3 = f(\tfrac{1}{2}) = \tfrac{1}{4}.$$

Thus,

$$U_f(P) = M_1\Delta x_1 + M_2\Delta x_2 + M_3\Delta x_3 = \tfrac{1}{16}(\tfrac{1}{4}) + \tfrac{1}{4}(\tfrac{1}{4}) + 1(\tfrac{1}{2}) = \tfrac{37}{64}$$

(see Figure 5.2.1*a*)

and

$$L_f(P) = m_1\Delta x_1 + m_2\Delta x_2 + m_3\Delta x_3 = 0(\tfrac{1}{4}) + \tfrac{1}{16}(\tfrac{1}{4}) + \tfrac{1}{4}(\tfrac{1}{2}) = \tfrac{9}{64}. \quad \square$$

(see Figure 5.2.1*b*)

Example 3 Now consider the function

$$f(x) = -2x - 1$$

on the closed interval $[-1, \frac{1}{2}]$. The partition $P = \{-1, -\frac{3}{4}, -\frac{1}{4}, \frac{1}{4}, \frac{1}{2}\}$ breaks up $[-1, \frac{1}{2}]$ into four subintervals: $[-1, -\frac{3}{4}]$, $[-\frac{3}{4}, -\frac{1}{4}]$, $[-\frac{1}{4}, \frac{1}{4}]$, and $[\frac{1}{4}, \frac{1}{2}]$. See Figure 5.2.2. Since f is a decreasing function, the maximum values of f will occur at the left

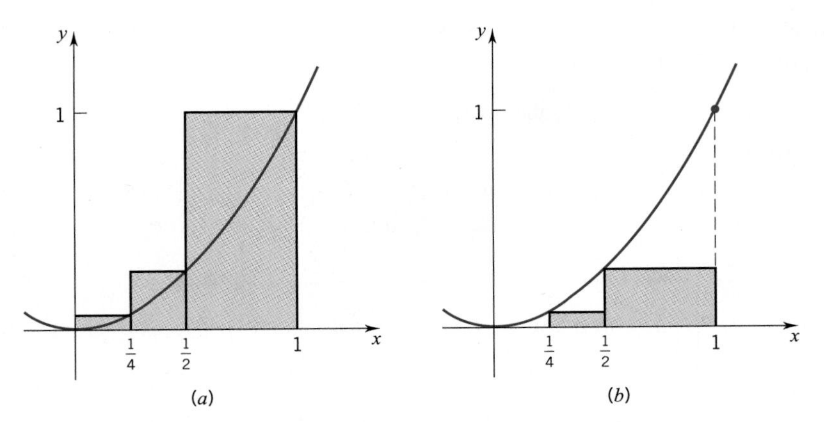

Figure 5.2.1

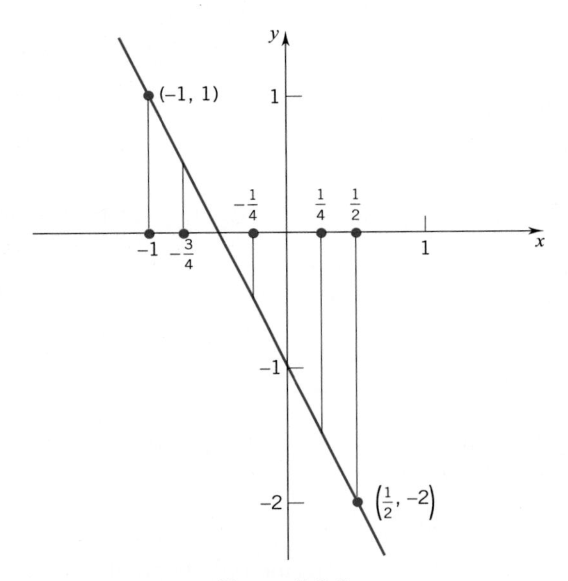

Figure 5.2.2

endpoints of the subintervals, and the minimum values will occur at the right endpoints. As you can check

$$U_f(P) = 1(\tfrac{1}{4}) + \tfrac{1}{2}(\tfrac{1}{2}) + (-\tfrac{1}{2})(\tfrac{1}{2}) + (-\tfrac{3}{2})(\tfrac{1}{4}) = -\tfrac{1}{8}$$

and

$$L_f(P) = \tfrac{1}{2}(\tfrac{1}{4}) + (-\tfrac{1}{2})(\tfrac{1}{2}) + (-\tfrac{3}{2})(\tfrac{1}{2}) + (-2)(\tfrac{1}{4}) = -\tfrac{11}{8}. \quad \square$$

By an argument that we omit here (it appears in Appendix B.4) it can be proved that, with f continuous on $[a, b]$, there is one and only one number I that satisfies the inequality

$$L_f(P) \leq I \leq U_f(P) \qquad \text{for } \textit{all} \text{ partitions } P \text{ of } [a, b].$$

This is the number we want.

DEFINITION 5.2.3 THE DEFINITE INTEGRAL

Let f be continuous on $[a, b]$. The unique number I that satisfies the inequality

$$L_f(P) \leq I \leq U_f(P) \qquad \text{for all partitions } P \text{ of } [a, b]$$

is called the *definite integral* (or more simply *the integral*) *of f from a to b* and is denoted by

$$\int_a^b f(x)\, dx.$$

The symbol $\int$ dates back to Leibniz and is called an *integral sign.* It is really an elongated S—as in *Sum.* The numbers a and b are called *the limits of integration, a* is the *lower limit* and b is the *upper limit,* and we'll speak of *integrating* a function f from a to b. The function f being integrated is called *the integrand.* This is not the only notation. Some mathematicians omit the dx and write simply $\int_a^b f$. We will keep the dx. As we go on, you will see that it does serve a useful purpose.

In the expression

$$\int_a^b f(x)\, dx$$

the letter x is a "dummy variable"; in other words, it can be replaced by any other letter not already engaged. Thus, for example,

$$\int_a^b f(x)\, dx, \qquad \int_a^b f(t)\, dt, \qquad \text{and} \qquad \int_a^b f(z)\, dz.$$

All denote exactly the same quantity, the definite integral of f from a to b.

Section 5.1 gives two immediate applications of the definite integral:

I. If f is nonnegative on $[a, b]$, then

$$A = \int_a^b f(x)\, dx$$

gives the area below the graph of f. See Figure 5.1.2.

II. If $|v(t)|$ is the speed of an object at time t, then

$$s = \int_a^b |v(t)|\, dt$$

gives the distance traveled from time a to time b.

We will come back to these ideas later. Right now we do some simple computations.

Example 4 If $f(x) = 3$ for all x in $[a, b]$ (see Figure 5.2.3),

$$\int_a^b f(x)\, dx = 3(b - a).$$

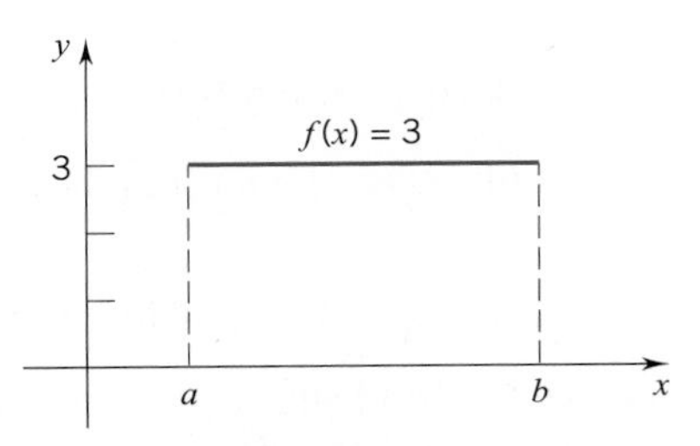

Figure 5.2.3

To see this, take $P = \{x_0, x_1, \ldots, x_n\}$ as an arbitrary partition of $[a, b]$. Since f is constantly 3 on $[a, b]$, it is constantly 3 on each subinterval $[x_{i-1}\ x_i]$. Thus, M_i and m_i are both 3. It follows that

$$\begin{aligned} U_f(P) &= 3\Delta x_1 + 3\Delta x_2 + \cdots + 3\Delta x_n \\ &= 3(\Delta x_1 + \Delta x_2 + \cdots + \Delta x_n) \\ &= 3[(x_1 - x_0) + (x_2 - x_1) + (x_3 - x_2) + \cdots + (x_{n-1} - x_{n-2}) + (x_n - x_{n-1})] \\ &= 3(b - a) \qquad \text{(recall } x_0 = a \text{ and } x_n = b) \end{aligned}$$

and

$$\begin{aligned} L_f(P) &= 3\Delta x_1 + 3\Delta x_2 + \cdots + 3\Delta x_n \\ &= 3(\Delta x_1 + \Delta x_2 + \cdots + \Delta x_n) = 3(b - a). \end{aligned}$$

Obviously then

$$L_f(P) \le 3(b - a) \le U_f(P)$$

Since this inequality holds for all partitions P of $[a, b]$, we can conclude that

$$\int_a^b f(x)\, dx = 3(b - a). \quad \square$$

Clearly, this same argument will hold with any constant k in place of 3. Thus, we have

(5.2.4)

$$\int_a^b k\, dx = k(b - a), \qquad \text{where } k \text{ is any constant.}$$

For example

$$\int_{-1}^{1} 3\, dx = 3[1 - (-1)] = 3(2) = 6 \qquad \text{and}$$

$$\int_4^{10} -2\, dx = -2[10 - 4] = -2(6) = -12.$$

If $k > 0$, the region between the graph and the x-axis is a rectangle of height k erected on the interval $[a, b]$ (Figure 5.2.4). The integral gives the area of the rectangle.

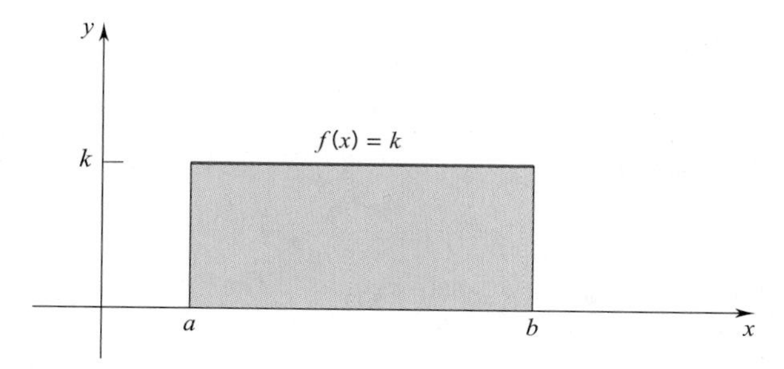

Figure 5.2.4

Example 5 Let $f(x) = x$ for all $x \in [a, b]$ (see Figure 5.2.5). Then

(5.2.5)
$$\int_a^b x\,dx = \tfrac{1}{2}(b^2 - a^2).$$

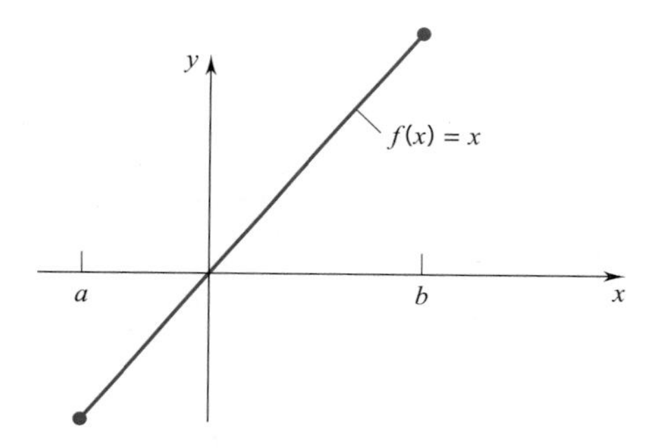

Figure 5.2.5

To see this, take $P = \{x_0, x_1, \ldots, x_n\}$ as an arbitrary partition of $[a, b]$. On each subinterval $[x_{i-1}, x_i]$ the function $f(x) = x$ has a maximum value M_i and a minimum value m_i. Since f is an increasing function, the maximum value $M_i = x_i$ occurs at the right endpoint of the subinterval and the minimum value $m_i = x_{i-1}$ occurs at the left endpoint. If follows that

$$U_f(P) = x_1\Delta x_1 + x_2\Delta x_2 + \cdots + x_n\Delta x_n$$

and

$$L_f(P) = x_0\Delta x_1 + x_1\Delta x_2 + \cdots + x_{n-1}\Delta x_n.$$

For each index i

(*) $$x_{i-i} \le \tfrac{1}{2}(x_i + x_{i-1}) \le x_i. \qquad \text{(explain)}$$

Multiplication by $\Delta x_i = x_i - x_{i-1}$ gives

$$x_{i-1}\Delta x_i \le \tfrac{1}{2}(x_i^2 - x_{i-1}^2) \le x_i\Delta x_i.$$

Summing from $i = 1$ to $i = n$, we find that

(**) $$L_f(P) \le \tfrac{1}{2}(x_1^2 - x_0^2) + \tfrac{1}{2}(x_2^2 - x_1^2) + \cdots + \tfrac{1}{2}(x_n^2 - x_{n-1}^2) \le U_f(P).$$

The sum in the middle collapses to

$$\tfrac{1}{2}(x_n^2 - x_0^2) = \tfrac{1}{2}(b^2 - a^2).$$

Consequently we have

$$L_f(P) \leq \tfrac{1}{2}(b^2 - a^2) \leq U_f(P).$$

Since P was chosen arbitrarily, we can conclude that this inequality holds for all partitions P of $[a, b]$. It follows then that

$$\int_a^b x\,dx = \tfrac{1}{2}(b^2 - a^2). \quad ❑$$

For example

$$\int_{-1}^{3} x\,dx = \tfrac{1}{2}[3^2 - (-1)^2] = \tfrac{1}{2}(8) = 4 \qquad \text{and} \qquad \int_{-2}^{2} x\,dx = \tfrac{1}{2}[2^2 - (-2)^2] = 0.$$

Remark You might be wondering why we chose the inequality $(*)$ in the preceding example instead of some other possible inequality, such as

$$x_{i-1} \leq \tfrac{2}{3}x_{i-1} + \tfrac{1}{3}x_i \leq x_i.$$

The reason is that the choice of $(*)$ resulted in the "collapsing" sum $(**)$. If we had not been clever in this way, the problem would have been very difficult. You will see this idea again in the next example. ❑

If the interval $[a, b]$ lies to the right of the origin, then the region below the graph of

$$f(x) = x, \qquad x \in [a, b]$$

is the trapezoid of Figure 5.2.6. The integral

$$\int_a^b x\,dx$$

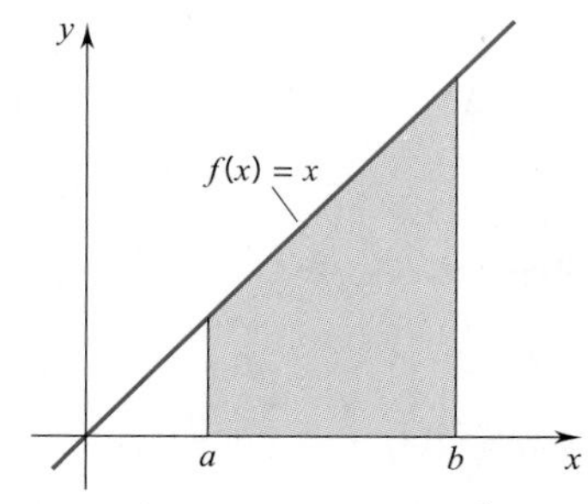

area of shaded region: $\int_a^b x\,dx = \frac{1}{2}(b^2 - a^2)$

Figure 5.2.6

gives the area of this trapezoid: $A = (b - a)[\tfrac{1}{2}(a + b)] = \tfrac{1}{2}(b^2 - a^2)$.

Example 6

$$\int_0^1 x^2\,dx = \tfrac{1}{3}. \qquad \text{(Figure 5.2.7)}$$

This time take $P = \{x_0, x_1, \ldots, x_n\}$ as a partition of $[0, 1]$. On each subinterval $[x_{i-1}, x_i]$ the function $f(x) = x^2$ has a maximum $M_i = x_i^2$ and a minimum $m_i = x_{i-1}^2$. It follows that

$$U_f(P) = x_1^2\Delta x_1 + \cdots + x_n^2\Delta x_n$$

and

$$L_f(P) = x_0^2\Delta x_1 + \cdots + x_{n-1}^2\Delta x_n.$$

For each index i,

$$x_{i-1}^2 \le \tfrac{1}{3}(x_{i-1}^2 + x_{i-1}x_i + x_i^2) \le x_i^2. \qquad \text{(verify this)}$$

(As we will see later in this chapter, $\frac{1}{3}(x_{i-1}^2 + x_{i-1}x_i + x_i^2)$ is the average value of x^2 on $[x_{i-1}, x_i]$.) We now multiply this inequality by $\Delta x_i = x_i - x_{i-1}$. The middle term is

$$\tfrac{1}{3}(x_{i-1}^2 + x_{i-1}x_i + x_i^2)(x_i - x_{i-1}) = \tfrac{1}{3}(x_i^3 - x_{i-1}^3),$$

showing us that

$$x_{i-1}^2\Delta x_i \le \tfrac{1}{3}(x_i^3 - x_{i-1}^3) \le x_i^2\Delta x_i.$$

The sum of the terms on the left is $L_f(P)$. The sum of all the middle terms collapses to $\frac{1}{3}$:

$$\tfrac{1}{3}(x_1^3 - x_0^3 + x_2^3 - x_1^3 + \cdots + x_n^3 - x_{n-1}^3) = \tfrac{1}{3}(x_n^3 - x_0^3) = \tfrac{1}{3}(1^3 - 0^3) = \tfrac{1}{3}.$$

The sum of the terms on the right is $U_f(P)$. Clearly then

$$L_f(P) \le \tfrac{1}{3} \le U_f(P).$$

Since P was chosen arbitrarily, we can conclude that this inequality holds for all partitions P of $[0, 1]$. It follows therefore that

$$\int_0^1 x^2\,dx = \tfrac{1}{3}. \quad \square$$

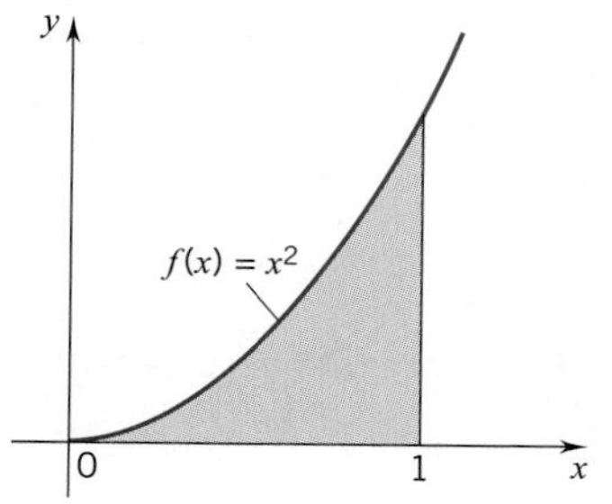

area of the shaded region: $\int_0^1 x^2\,dx = \frac{1}{3}$

Figure 5.2.7

EXERCISES 5.2

In Exercises 1–12, find $L_f(P)$ and $U_f(P)$.

1. $f(x) = 2x, \quad x \in [0, 1]; \quad P = \{0, \frac{1}{4}, \frac{1}{2}, 1\}.$

2. $f(x) = 1 - x, \quad x \in [0, 2]; \quad P = \{0, \frac{1}{3}, \frac{3}{4}, 1, 2\}.$

3. $f(x) = x^2, \quad x \in [-1, 0]; \quad P = \{-1, -\frac{1}{2}, -\frac{1}{4}, 0\}.$

4. $f(x) = 1 - x^2, \quad x \in [0, 1]; \quad P = \{0, \frac{1}{4}, \frac{1}{2}, 1\}.$

5. $f(x) = 1 + x^3, \quad x \in [0, 1]; \quad P = \{0, \frac{1}{2}, 1\}.$

6. $f(x) = \sqrt{x}, \quad x \in [0, 1]; \quad P = \{0, \frac{1}{25}, \frac{4}{25}, \frac{9}{25}, \frac{16}{25}, 1\}.$

7. $f(x) = |x|, \quad x \in [-1, 1]; \quad P = \{-1, -\frac{1}{2}, 0, \frac{1}{4}, 1\}.$

8. $f(x) = |x|, \quad x \in [-1, 1]; \quad P = \{-1, -\frac{1}{2}, -\frac{1}{4}, \frac{1}{2}, 1\}.$

9. $f(x) = x^2, \quad x \in [-1, 1]; \quad P = \{-1, -\frac{1}{4}, \frac{1}{4}, \frac{1}{2}, 1\}.$

10. $f(x) = x^2, \quad x \in [-1, 1]; \quad P = \{-1, -\frac{3}{4}, -\frac{1}{4}, \frac{1}{4}, \frac{1}{2}, 1\}.$

11. $f(x) = \sin x, \quad x \in [0, \pi]; \quad P = \{0, \frac{1}{6}\pi, \frac{1}{2}\pi, \pi\}.$

12. $f(x) = \cos x, \quad x \in [0, \pi]; \quad P = \{0, \frac{1}{3}\pi, \frac{1}{2}\pi, \pi\}.$

13. Let f be a continuous function on $[-1, 1]$. Explain why each of the following three statements is false. Take P as a partition of $[-1, 1]$.

(a) $L_f(P) = 3$ and $U_f(P) = 2$.

(b) $L_f(P) = 3, \quad U_f(P) = 6, \quad$ and $\quad \int_{-1}^1 f(x)\,dx = 2.$

(c) $L_f(P) = 3, \quad U_f(P) = 6, \quad \text{and} \quad \int_{-1}^{1} f(x)\,dx = 10.$

14. (a) Given that $P = \{x_0, x_1, \ldots, x_n\}$ is an arbitrary partition of $[a, b]$, find $L_f(P)$ and $U_f(P)$ for $f(x) = x + 3$.
(b) Use your answers to part (a) to evaluate

$$\int_a^b f(x)\,dx.$$

15. Repeat Exercise 14 with $f(x) = -3x$.

16. Repeat Exercise 14 with $f(x) = 1 + 2x$.

17. Evaluate

$$\int_0^1 x^3\,dx$$

by the methods of this section. HINT: $b^4 - a^4 = (b^3 + b^2a + ba^2 + a^3)(b - a)$.

18. Evaluate

$$\int_0^1 x^4\,dx$$

by the methods of this section.

19. A partition $P = \{x_0, x_1, x_2, \ldots, x_{n-1}, x_n\}$ of the interval $[a, b]$ is *regular* if the subintervals $[x_{i-1}, x_i]$, $i = 1, 2, \ldots, n$, all have the same length, namely $\Delta x = (b - a)/n$. Suppose that $P = \{x_0, x_1, \ldots, x_{n-1}, x_n\}$ is a regular partition of the interval $[a, b]$. Show that if f is continuous and increasing on $[a, b]$, then

$$U_f(P) - L_f(P) = [f(b) - f(a)]\,\Delta x.$$

20. Let $P = \{x_0, x_1, x_2, \ldots, x_{n-1}, x_n\}$ be a regular partition of the interval $[a, b]$. Show that if f is continuous and decreasing on $[a, b]$, then

$$U_f(P) - L_f(P) = [f(a) - f(b)]\,\Delta x.$$

21. Let f be continuous on the interval $[a, b]$ and let $I = \int_a^b f(x)\,dx$. Let $P = \{x_0, x_1, x_2, \ldots, x_{n-1}, x_n\}$ be an arbitrary partition of the interval $[a, b]$.
(a) Show that $I - L_f(P) \le U_f(P) - L_f(P)$. The difference $I - L_f(P)$ is sometimes called the *error* in using $L_f(P)$ to approximate I.
(b) Show that $U_f(P) - I \le U_f(P) - L_f(P)$ ($U_f(P) - I$ is the error in using $U_f(P)$ to approximate I).

22. Let $P = \{x_0, x_1, x_2, \ldots, x_{n-1}, x_n\}$ be a regular partition of the interval $[a, b]$ and let f be continuous. Show that if f is increasing on $[a, b]$, or if f is decreasing on $[a, b]$, then

$$I - L_f(P) \le |f(b) - f(a)|\,\Delta x \qquad \text{and}$$
$$U_f(P) - I \le |f(b) - f(a)|\,\Delta x.$$

C **23.** Let $f(x) = \sqrt{1 + x^2}$.
(a) Verify that f is increasing on $[0, 2]$.
(b) Use Exercise 22 to determine a value of n such that if $P = \{x_0, x_1, x_2, \ldots, x_n\}$ is a regular partition of $[0, 2]$, then

$$\int_0^2 f(x)\,dx - L_f(P) \le 0.1.$$

(c) Use a programmable calculator or computer to calculate $\int_0^2 f(x)\,dx$ with an error of less than 0.1.

C **24.** Let $f(x) = 1/(1 + x^2)$.
(a) Verify that f is decreasing on $[0, 1]$.
(b) Use Exercise 22 to determine a value of n such that if $P = \{x_0, x_1, x_2, \ldots, x_n\}$ is a regular partition of $[0, 1]$, then

$$U_f(P) - \int_0^1 f(x)\,dx \le 0.05.$$

(c) Use a programmable calculator or computer to calculate $\int_0^1 f(x)\,dx$ with an error of less than 0.05.
NOTE: You will see in Chapter 7 that the exact value of this integral is $\pi/4$.

25. Use mathematical induction to show that the sum of the first k positive integers is $\frac{1}{2}k(k + 1)$. That is,

$$1 + 2 + 3 + \cdots + k = \frac{k(k + 1)}{2},$$

where k is any positive integer.

26. Use mathematical induction to show that the sum of the squares of the first k positive integers is $\frac{1}{6}k(k + 1)(2k + 1)$. That is,

$$1^2 + 2^2 + 3^2 + \cdots + k^2 = \frac{k(k + 1)(2k + 1)}{6},$$

where k is any positive integer.

27. Let $P = \{x_0, x_1, x_2, \ldots, x_{n-1}, x_n\}$ be a regular partition of the interval $[0, b]$, and let $f(x) = x$.
(a) Show that

$$L_f(P) = \frac{b^2}{n^2}[0 + 1 + 2 + 3 + \cdots + (n - 1)].$$

(b) Show that

$$U_f(P) = \frac{b^2}{n^2}[1 + 2 + 3 + \cdots + n].$$

(c) Use Exercise 25 to show that $L_f(P), U_f(P) \to b^2/2$ as $n \to \infty$. Thus, $\int_0^b x\,dx = \frac{1}{2}b^2$. Compare with Example 5.

28. Let $P = \{x_0, x_1, x_2, \ldots, x_{n-1}, x_n\}$ be a regular partition of $[0, b]$, and let $f(x) = x^2$.

(a) Show that

$$L_f(P) = \frac{b^3}{n^3}[0^2 + 1^2 + 2^2 + \cdots + (n-1)^2].$$

(b) Show that

$$U_f(P) = \frac{b^3}{n^3}[1^2 + 2^2 + 3^2 + \cdots + n^2].$$

(c) Use Exercise 26 to show that $L_f(P)$, $U_f(P) \to (b^3/3)$ as $n \to \infty$. Thus, $\int_0^b x^2\,dx = \frac{1}{3}b^3$. Compare with Example 6.

29. (*Important*) The definition of integral that we have given (Definition 5.2.3) can be applied to functions with a finite number of discontinuities. Suppose that f is continuous on $[a, b]$. If g is defined on $[a, b]$ and differs from f only at a finite number of points, then g can be integrated on $[a, b]$ and

$$\int_a^b g(x)\,dx = \int_a^b f(x)\,dx.$$

For example, the function

$$g(x) = \begin{cases} 2, & x \in [0, 3) \cup (3, 4) \\ 7, & x = 3 \end{cases}$$

differs from the constant function

$$f(x) = 2, \qquad x \in [0, 4]$$

only at $x = 3$. Clearly

$$\int_0^4 f(x)\,dx = 8.$$

Show that

$$\int_0^4 g(x)\,dx = 8$$

by showing that 8 is the unique number I that satisfies the inequality

$$L_g(P) \le I \le U_g(P) \qquad \text{for all partitions } P \text{ of } [0, 4].$$

30. (*Important*) A function f is *piecewise continuous* on $[a, b]$ if it is continuous except, possibly, for a finite set of points at which it has a jump discontinuity. It can be shown that if f is piecewise continuous on $[a, b]$, then $\int_a^b f(x)\,dx$ exists. Let

$$f(x) = \begin{cases} x^2 & 0 \le x \le 2 \\ x & 2 < x \le 5. \end{cases}$$

(a) Sketch the graph of f.

(b) Find the area of the region bounded by the graph of f and the x-axis between $x = 0$ and $x = 5$. HINT: $\int_0^5 f(x)\,dx = \int_0^2 f(x)\,dx + \int_2^5 f(x)\,dx$, then use Exercise 28 and Example 5.

31. Consider the function f on $[2, 10]$:

$$f(x) = \begin{cases} 7 & \text{for } x \text{ rational} \\ 4 & \text{for } x \text{ irrational.} \end{cases}$$

(a) Show that for all partitions P of $[2, 10]$, $L_f(P) \le 40 \le U_f(P)$.

(b) Explain why we can *not* conclude that

$$\int_2^{10} f(x)\,dx = 40.$$

(c) Show that, for all partitions P of $[2, 10]$, $L_f(P) = 32$ and $U_f(P) = 56$. Hence we can conclude that $\int_2^{10} f(x)\,dx$ does not exist.

■ 5.3 THE FUNCTION $F(x) = \int_a^x f(t)\,dt$

The evaluation of a definite integral

$$\int_a^b f(x)\,dx$$

directly from its definition as the unique number I satisfying the inequality $L_f(P) \le I \le U_f(P)$ for all partitions P of $[a, b]$ is usually a laborious and difficult process. Try, for example, to evaluate

$$\int_2^5 \left(x^3 + x^{5/2} - \frac{2x}{1 - x^2}\right) dx \qquad \text{or} \qquad \int_{-1/2}^{1/4} \frac{x}{1 - x^2}\,dx$$

by this method. Theorem 5.4.2, called *the fundamental theorem of integral calculus,* gives us another way to evaluate such integrals. This other way depends on a connection between integration and differentiation described in Theorem 5.3.5. The main purpose of this section is to prove Theorem 5.3.5. Along the way we will pick up some information that is of interest in itself.

THEOREM 5.3.1

Let f be continuous on $[a, b]$ and let P and Q be partitions of the interval $[a, b]$. If $P \subseteq Q$, then

$$L_f(P) \leq L_f(Q) \quad \text{and} \quad U_f(Q) \leq U_f(P).$$

This result is easy to see. By adding points to a partition we make the subintervals $[x_{i-1}, x_i]$ smaller. This tends to make the minima, m_i, larger and the maxima, M_i, smaller. Thus the lower sums are made bigger, and the upper sums are made smaller. The idea is illustrated (for a positive function) in Figures 5.3.1 and 5.3.2.

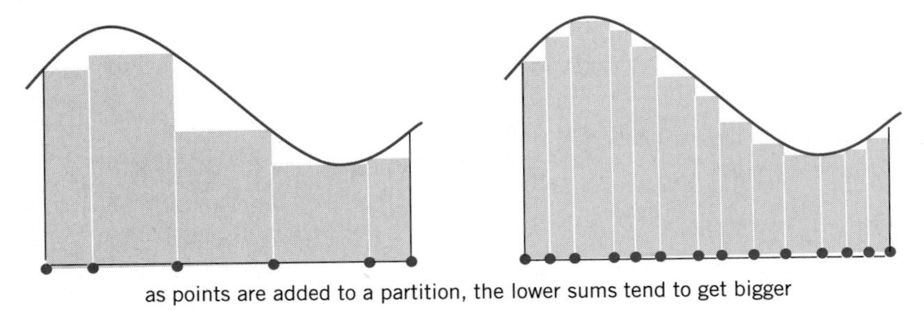

as points are added to a partition, the lower sums tend to get bigger

Figure 5.3.1

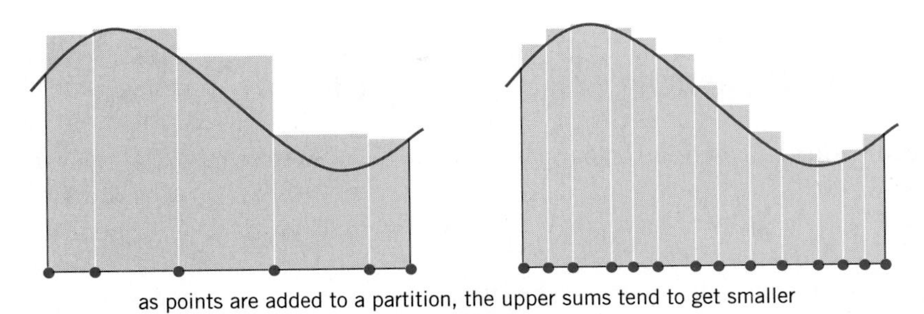

as points are added to a partition, the upper sums tend to get smaller

Figure 5.3.2

The next theorem says that the integral is *additive* on intervals.

THEOREM 5.3.2

If f is continuous on $[a, b]$ and $a < c < b$, then

$$\int_a^c f(t)\, dt + \int_c^b f(t)\, dt = \int_a^b f(t)\, dt.$$

For nonnegative functions f this theorem is easily understood in terms of area. The area of part I in Figure 5.3.3 is given by

$$\int_a^c f(t)\, dt;$$

the area of part II by

$$\int_c^b f(t)\, dt;$$

the area of the entire region by

$$\int_a^b f(t)\, dt.$$

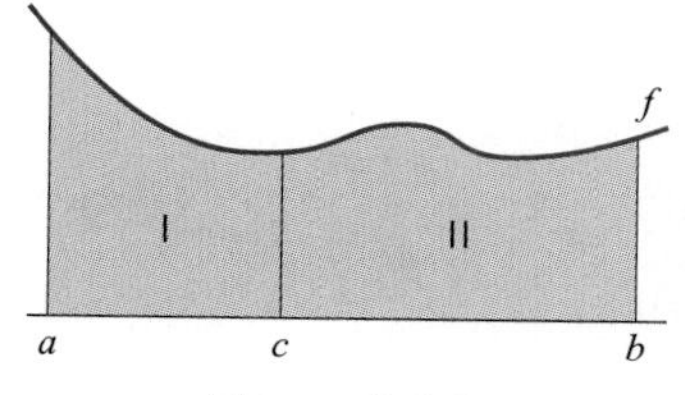

Figure 5.3.3

The theorem says that

the area of part I + the area of part II = the area of the entire region.

Theorem 5.3.2 can also be viewed in terms of speed and distance. With $f(t) \geq 0$, we can interpret $f(t)$ as the speed of an object at time t. With this interpretation,

$$\int_a^c f(t)\, dt = \text{distance traveled from time } a \text{ to time } c,$$

$$\int_c^b f(t)\, dt = \text{distance traveled from time } c \text{ to time } b,$$

$$\int_a^b f(t)\, dt = \text{distance traveled from time } a \text{ to time } b.$$

It is not surprising that the sum of the first two distances should equal the third.

The fact that the additivity theorem is so easy to understand does not relieve us of the necessity to prove it. Here is a proof.

PROOF OF THEOREM 5.3.2. To prove the theorem we need only show that for each partition P of $[a, b]$

$$L_f(P) \leq \int_a^c f(t)\, dt + \int_c^b f(t)\, dt \leq U_f(P) \qquad \text{(Why?)}$$

We begin with an arbitrary partition of $[a, b]$:

$$P = \{x_0, x_1, \ldots, x_n\}.$$

Since the partition $Q = P \cup \{c\}$ contains P, we know from Theorem 5.3.1 that

$$(1) \qquad L_f(P) \leq L_f(Q) \quad \text{and} \quad U_f(Q) \leq U_f(P).$$

The sets

$$Q_1 = Q \cap [a, c] \quad \text{and} \quad Q_2 = Q \cap [c, b]$$

are partitions of $[a, c]$ and $[c, b]$, respectively. Moreover

$$L_f(Q_1) + L_f(Q_2) = L_f(Q) \quad \text{and} \quad U_f(Q_1) + U_f(Q_2) = U_f(Q).$$

Since

$$L_f(Q_1) \leq \int_a^c f(t)\, dt \leq U_f(Q_1) \quad \text{and} \quad L_f(Q_2) \leq \int_c^b f(t)\, dt \leq U_f(Q_2),$$

we have

$$L_f(Q_1) + L_f(Q_2) \leq \int_a^c f(t)\, dt + \int_c^b f(t)\, dt \leq U_f(Q_1) + U_f(Q_2),$$

which is the same as

$$L_f(Q) \leq \int_a^c f(t)\, dt + \int_c^b f(t)\, dt \leq U_f(Q).$$

Therefore, by (1),

$$L_f(P) \leq \int_a^c f(t)\, dt + \int_c^b f(t)\, dt \leq U_f(P). \quad \square$$

Until now we have integrated only from left to right: from a number a to a number b greater than a. We integrate in the other direction by defining

(5.3.3)

$$\int_b^a f(t)\, dt = -\int_a^b f(t)\, dt.$$

The integral from any number to itself is defined to be zero:

(5.3.4)

$$\int_c^c f(t)\, dt = 0$$

for all real c. With these additional conventions, the additivity condition

$$\int_a^c f(t)\, dt + \int_c^b f(t)\, dt = \int_a^b f(t)\, dt$$

holds for all choices of a, b, c from an interval on which f is continuous, no matter what the order of a, b, c happens to be. We have left the proof of this to you as an exercise.

We are now ready to state the all-important connection that exists between integration and differentiation. Our first step is to point out that, if f is continuous on $[a, b]$, then for each x in $[a, b]$, the integral

$$\int_a^x f(t)\, dt$$

is a number, and consequently we can define a function F on $[a, b]$ by setting

$$F(x) = \int_a^x f(t)\, dt.$$

THEOREM 5.3.5

Let f be continuous on $[a, b]$. The function F defined on $[a, b]$ by

$$F(x) = \int_a^x f(t)\, dt$$

is continuous on $[a, b]$, differentiable on (a, b), and has derivative

$$F'(x) = f(x) \qquad \text{for all} \quad x \text{ in } (a, b).$$

PROOF We begin with x in the half-open interval $[a, b)$ and show that

$$\lim_{h \to 0^+} \frac{F(x + h) - F(x)}{h} = f(x).$$

(For a pictorial outline of the proof for f nonnegative, see Figure 5.3.4.)

If $x < x + h \leq b$, then

$$F(x + h) - F(x) = \int_a^{x+h} f(t)\, dt - \int_a^x f(t)\, dt.$$

It follows that

$$F(x + h) - F(x) = \int_x^{x+h} f(t)\, dt. \qquad \text{(justify this step)}$$

Now set

$$M_h = \text{maximum value of } f \text{ on } [x, x + h]$$

and

$$m_h = \text{minimum value of } f \text{ on } [x, x + h].$$

See Figure 5.3.5. Since

$$M_h[(x + h) - x] = M_h \cdot h$$

is an upper sum for f on $[x, x + h]$ and

$$m_h[(x + h) - x] = m_h \cdot h$$

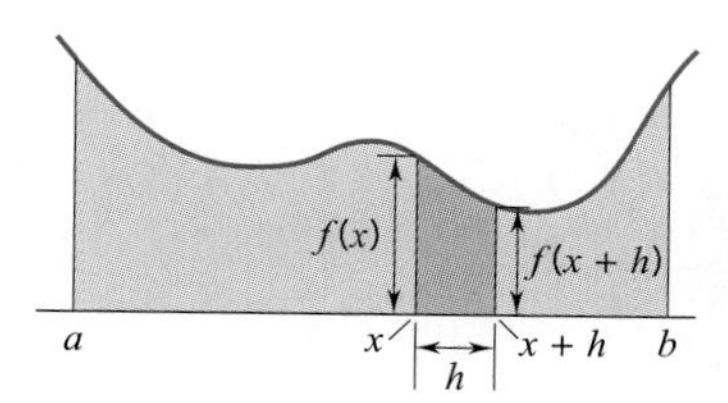

$F(x + h)$ = area from a to $x + h$
and
$F(x)$ = area from a to x.

Therefore

$F(x + h) - F(x)$
= area from x to $x + h \cong f(x)h$ if h is small

$\dfrac{F(x + h) - F(x)}{h} = \dfrac{\text{area from } x \text{ to } x + h}{h} \cong f(x)$

Figure 5.3.4

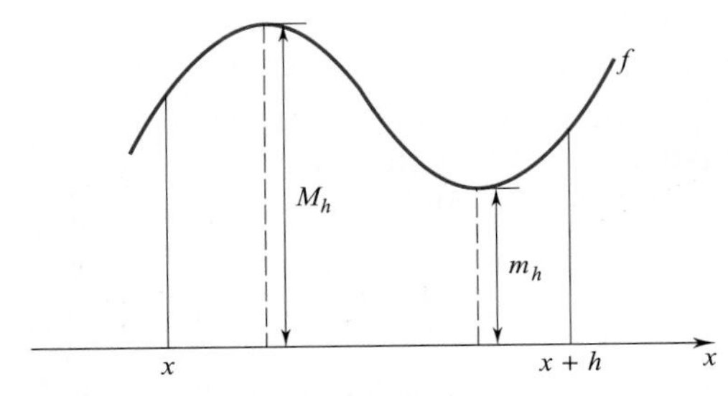

Figure 5.3.5

is a lower sum for f on $[x, x+h]$, we have

$$m_h \cdot h \le \int_x^{x+h} f(t)\, dt \le M_h \cdot h.$$

Now, using (1) and the fact that $h > 0$, it follows that

$$m_h \le \frac{F(x+h) - F(x)}{h} \le M_h.$$

Also, since f is continuous on $[x, x+h]$,

$$\lim_{h\to 0^+} m_h = f(x) = \lim_{h\to 0^+} M_h \tag{2}$$

and thus

$$\lim_{h\to 0^+} \frac{F(x+h) - F(x)}{h} = f(x) \tag{3}$$

by the "pinching theorem," Theorem 2.5.1.

In a similar manner we can prove that, for x in the half-open interval $(a, b]$,

$$\lim_{h\to 0^-} \frac{F(x+h) - F(x)}{h} = f(x). \tag{4}$$

For x in the open interval (a, b), both (3) and (4) hold, and we have

$$F'(x) = \lim_{h\to 0} \frac{F(x+h) - F(x)}{h} = f(x).$$

This proves that F is differentiable on (a, b) and has derivative $F'(x) = f(x)$.

All that remains to be shown is that F is continuous from the right at a and continuous from the left at b. Limit (3) at $x = a$ gives

$$\lim_{h\to 0^+} \frac{F(a+h) - F(a)}{h} = f(a).$$

Now, for $h > 0$,

$$F(a+h) - F(a) = \frac{F(a+h) - F(a)}{h} \cdot h$$

and so

$$\lim_{h\to 0^+} [F(a+h) - F(a)] = \lim_{h\to 0^+} \left[\frac{F(a+h) - F(a)}{h} \cdot h\right] = f(a) \cdot \lim_{h\to 0^+} h = 0.$$

Therefore

$$\lim_{h\to 0^+} F(a+h) = F(a).$$

This shows that F is continuous from the right at a. The continuity of F from the left at b can be shown by applying limit (4) at $x = b$. ❑

Example 1 If F is defined by

$$F(x) = \int_{-1}^{x} (2t + t^2)\,dt \qquad \text{for } -1 \leq x \leq 5,$$

then

$$F'(x) = 2x + x^2 \qquad \text{on } (-1, 5). \quad ❑$$

Example 2 If F is defined by

$$F(x) = \int_0^x \sin \pi t\,dt,$$

determine $F'(3/4)$.

SOLUTION By Theorem 5.3.5,

$$F'(x) = \sin \pi x.$$

Thus, $F'(3/4) = \sin(3\pi/4) = \sqrt{2}/2$. ❑

Example 3 Let F be defined by

$$F(x) = \int_0^x \frac{1}{1+t^2}\,dt, \qquad \text{where } x \text{ is any real number.}$$

(a) Find the critical numbers of F and determine the intervals on which F is increasing and the intervals on which F is decreasing.

(b) Determine the concavity of the graph of F and find the points of inflection (if any).

(c) Sketch the graph of F.

SOLUTION

(a) To find where F is increasing or decreasing, we need to examine the first derivative. By Theorem 5.3.5,

$$F'(x) = \frac{1}{1+x^2}.$$

Since $F'(x) > 0$ for all x, F is increasing on $(-\infty, \infty)$; there are no critical numbers.

(b) We use the second derivative to determine the concavity of the graph and to find the points of inflection:

$$F''(x) = \frac{-2x}{(1+x^2)^2}.$$

The sign of F'' and the behavior of the graph of F are:

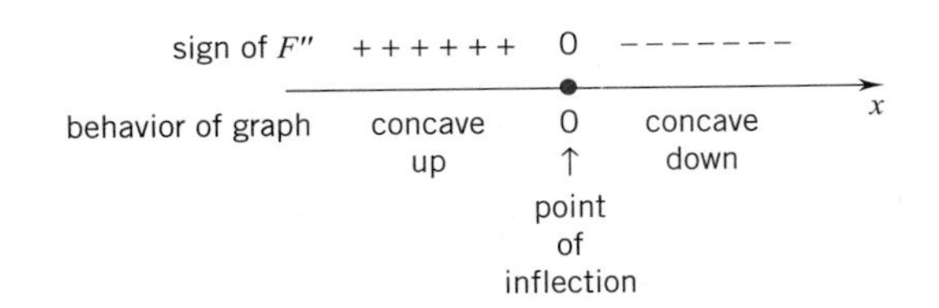

(c) Since $F(0) = 0$ and $F'(0) = 1$, the graph goes through the origin with slope 1. A sketch of the graph is shown in Figure 5.3.6. Although it is not at all clear now, you will see in Chapter 7 that $y = \pi/2$ and $y = -\pi/2$ are horizontal asymptotes of the graph. ❑

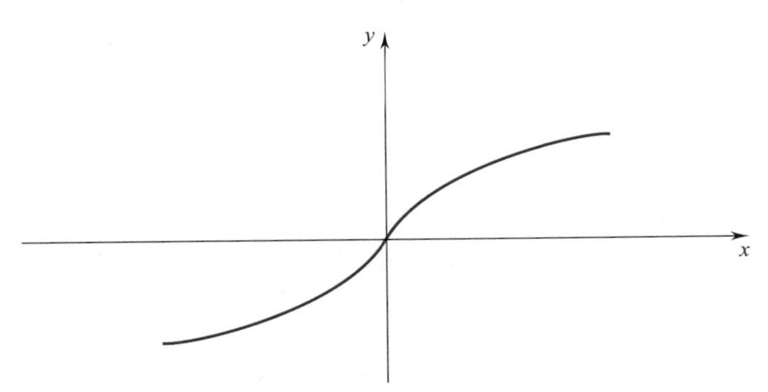

Figure 5.3.6

EXERCISES 5.3

1. Given that

$$\int_0^1 f(x)\,dx = 6, \qquad \int_0^2 f(x)\,dx = 4, \qquad \int_2^5 f(x)\,dx = 1,$$

find each of the following:

(a) $\int_0^5 f(x)\,dx.$ (b) $\int_1^2 f(x)\,dx.$ (c) $\int_1^5 f(x)\,dx.$

(d) $\int_0^0 f(x)\,dx.$ (e) $\int_2^0 f(x)\,dx.$ (f) $\int_5^1 f(x)\,dx.$

2. Given that

$$\int_1^4 f(x)\,dx = 5, \qquad \int_3^4 f(x)\,dx = 7, \qquad \int_1^8 f(x)\,dx = 11,$$

find each of the following:

(a) $\int_4^8 f(x)\,dx.$ (b) $\int_4^3 f(x)\,dx.$ (c) $\int_1^3 f(x)\,dx.$

(d) $\int_3^8 f(x)\,dx.$ (e) $\int_8^4 f(x)\,dx.$ (f) $\int_4^4 f(x)\,dx.$

3. Use upper and lower sums to show that

$$0.5 < \int_1^2 \frac{dx}{x} < 1.$$

4. Use upper and lower sums to show that

$$0.6 < \int_0^1 \frac{dx}{1 + x^2} < 1.$$

5. For $x > -1$ set $F(x) = \int_0^x t\sqrt{t+1}\,dt.$

(a) Find $F(0)$. (b) Find $F'(x)$. (c) Find $F'(2)$.
(d) Express $F(2)$ as an integral of $t\sqrt{t+1}$.
(e) Express $-F(x)$ as an integral of $t\sqrt{t+1}$.

6. Let $F(x) = \int_\pi^x t \sin t\,dt.$

(a) Find $F(\pi)$. (b) Find $F'(x)$. (c) Find $F'(\pi/2)$.
(d) Express $F(2\pi)$ as an integral of $t \sin t$.
(e) Express $-F(x)$ as an integral of $t \sin t$.

7. For $x > 0$, set $F(x) = \int_1^x (1/t)\,dt.$

(a) Find the critical numbers of F and determine the intervals on which F is increasing and the intervals on which F is decreasing.
(b) Determine the concavity of the graph of F and find the points of inflection, if any.
(c) Sketch the graph of F.

8. Let $F(x) = \int_0^x t(t-3)^2\,dt.$

(a) Find the critical numbers of F and determine the intervals on which F is increasing and the intervals on which F is decreasing.
(b) Determine the concavity of the graph of F and find the points of inflection, if any.
(c) Sketch the graph of F.

For the given function F in Exercises 9–14, compute the following:

(a) $F'(-1)$. (b) $F'(0)$. (c) $F'(\frac{1}{2})$. (d) $F''(x)$.

9. $F(x) = \int_0^x \frac{dt}{t^2+9}$.

10. $F(x) = \int_x^0 \sqrt{t^2+1}\,dt$.

11. $F(x) = \int_x^1 \sqrt{t^2+1}\,dt$.

12. $F(x) = \int_1^x \sin \pi t\,dt$.

13. $F(x) = \int_1^x \cos \pi t\,dt$.

14. $F(x) = \int_2^x (t+1)^3\,dt$.

15. Explain why each of the following statements must be false.

(a) $U_f(P_1) = 4$ for the partition $P_1 = \{0, 1, \frac{3}{2}, 2\}$ and $U_f(P_2) = 5$ for the partition $P_2 = \{0, \frac{1}{4}, 1, \frac{3}{2}, 2\}$.

(b) $L_f(P_1) = 5$ for the partition $P_1 = \{0, 1, \frac{3}{2}, 2\}$ and $L_f(P_2) = 4$ for the partition $P_2 = \{0, \frac{1}{4}, 1, \frac{3}{2}, 2\}$.

16. Show that, if f is continuous on an interval I, then

$$\int_a^c f(t)\,dt + \int_c^b f(t)\,dt = \int_a^b f(t)\,dt$$

for *every* choice of a, b, c from I. HINT: Assume $a < b$ and consider the four cases: $c = a$, $c = b$, $c < a$, $b < c$. Then consider what happens if $a > b$ or $a = b$.

In Exercises 17–22, find the derivative of the function F.

17. $F(x) = \int_0^{x^3} t \cos t\,dt$ HINT: Let $u = x^3$ and use the chain rule.

18. $F(x) = \int_1^{\cos x} \sqrt{1-t^2}\,dt$.

19. $F(x) = \int_{x^2}^1 (t - \sin^2 t)\,dt$.

20. $F(x) = \int_0^{\sqrt{x}} \frac{t^2}{1+t^4}\,dt$.

21. $F(x) = \int_{x^2}^{x^4} \sin \sqrt{t}\,dt$

HINT: $\int_{x^2}^{x^4} f(t)\,dt = \int_{x^2}^0 f(t)\,dt + \int_0^{x^4} f(t)\,dt$.

22. $F(x) = \int_{\sin x}^{3x+1} t(1+t^2)\,dt$.

23. Assume that f is a continuous function and that

$$\int_0^x f(t)\,dt = \frac{2x}{4+x^2}.$$

(a) Determine $f(0)$.
(b) Find the zeros of f, if any.

24. Assume that f is a continuous function and that

$$\int_0^x tf(t)\,dt = \sin x - x\cos x$$

(a) Determine $f(\pi/2)$.
(b) Find $f'(x)$.

25. (*A Mean-Value Theorem for Integrals*) Show that, if f is continuous on $[a, b]$, then there is at least one number c in (a, b) for which

$$\int_a^b f(x)\,dx = f(c)(b-a).$$

HINT: Apply the mean-value theorem to the function

$$F(x) = \int_a^x f(t)\,dt \quad \text{on } [a, b].$$

26. Show the validity of equation (2) in the proof of Theorem 5.3.5.

27. Extend Theorem 5.3.5 by showing that, if f is continuous on $[a, b]$ and c is *any* number in $[a, b]$, then the function

$$F(x) = \int_c^x f(t)\,dt$$

is continuous on $[a, b]$, differentiable on (a, b), and satisfies

$$F'(x) = f(x) \quad \text{for all } x \text{ in } (a, b).$$

HINT: $\int_c^x f(t)\,dt = \int_c^a f(t)\,dt + \int_a^x f(t)\,dt$.

28. Complete the proof of Theorem 5.3.5 by showing that

$$\lim_{h\to 0^-} \frac{F(x+h) - F(x)}{h} = f(x) \quad \text{for all } x \text{ in } (a, b].$$

■ 5.4 THE FUNDAMENTAL THEOREM OF INTEGRAL CALCULUS

DEFINITION 5.4.1 ANTIDERIVATIVE

Let f be continuous on $[a, b]$. A function G is called an *antiderivative* for f on $[a, b]$ iff

G is continuous on $[a, b]$ and $G'(x) = f(x)$ for all $x \in (a, b)$.

Theorem 5.3.5 says that if f is continuous on $[a, b]$, then

$$F(x) = \int_a^x f(t)\, dt$$

is an antiderivative for f on $[a, b]$. This gives us a prescription for constructing antiderivatives. It tells us that we can construct an antiderivative for f by integrating f.

The so-called "fundamental theorem" goes the other way. It gives us a prescription, not for finding antiderivatives, but for evaluating integrals. It tells us that we can evaluate

$$\int_a^b f(t)\, dt$$

by finding an antiderivative for f.

THEOREM 5.4.2 THE FUNDAMENTAL THEOREM OF INTEGRAL CALCULUS

Let f be continuous on $[a, b]$. If G is any antiderivative for f on $[a, b]$, then

$$\int_a^b f(t)\, dt = G(b) - G(a).$$

PROOF From Theorem 5.3.5 we know that the function

$$F(x) = \int_a^x f(t)\, dt$$

is an antiderivative for f on $[a, b]$. If G is also an antiderivative for f on $[a, b]$, then both F and G are continuous on $[a, b]$ and satisfy $F'(x) = G'(x)$ for all x in (a, b). From Theorem 4.2.4 we know that there exists a constant C such that

$$F(x) = G(x) + C \qquad \text{for all } x \text{ in } [a, b].$$

Since $F(a) = 0$,

$$G(a) + C = 0 \quad \text{and thus} \quad C = -G(a).$$

It follows that

$$F(x) = G(x) - G(a) \qquad \text{for all } x \text{ in } [a, b].$$

In particular,

$$\int_a^b f(t)\, dt = F(b) = G(b) - G(a). \quad \square$$

We now evaluate some integrals by applying the fundamental theorem. In each case we use the simplest antiderivative we can think of.

Example 1 Evaluate

$$\int_1^4 x^2\,dx.$$

SOLUTION As an antiderivative for $f(x) = x^2$, we can use the function

$$G(x) = \tfrac{1}{3}x^3. \qquad \text{(verify this)}$$

By the fundamental theorem

$$\int_1^4 x^2\,dx = G(4) - G(1) = \tfrac{1}{3}(4)^3 - \tfrac{1}{3}(1)^3 = \tfrac{64}{3} - \tfrac{1}{3} = 21.$$

NOTE: Any other antiderivative of $f(x) = x^2$ has the form $H(x) = \tfrac{1}{3}x^3 + C$ for some constant C. If we had chosen such an H instead of G, then

$$\int_1^4 x^2\,dx = H(4) - H(1) = [\tfrac{1}{3}(4)^3 + C] - [\tfrac{1}{3}(1)^3 + C] = \tfrac{64}{3} + C - \tfrac{1}{3} - C = 21;$$

the C "cancels out." ❑

Example 2 Evaluate

$$\int_0^{\pi/2} \sin x\,dx.$$

SOLUTION Here we use the antiderivative $G(x) = -\cos x$:

$$\int_0^{\pi/2} \sin x\,dx = G(\pi/2) - G(0) = -\cos(\pi/2) - [-\cos(0)] = 0 - (-1) = 1.$$

Expressions of the form $G(b) - G(a)$ are conveniently written

$$\Big[\,G(x)\,\Big]_a^b.$$

In this notation

$$\int_1^4 x^2\,dx = \Big[\,\tfrac{1}{3}x^3\,\Big]_1^4 = \tfrac{1}{3}(4)^3 - \tfrac{1}{3}(1)^3 = 21$$

and

$$\int_0^{\pi/2} \sin x\,dx = \Big[\,-\cos x\,\Big]_0^{\pi/2} = -\cos(\pi/2) - [-\cos(0)] = 1. \qquad ❑$$

Obviously the essential step in applying the fundamental theorem is the determination of an antiderivative G for the integrand f. Thus, we need to have a supply of antiderivatives of functions. From our study of differentiation, we know that

$$\frac{d}{dx}(x^r) = rx^{r-1}, \qquad r \text{ any rational number.}$$

It follows immediately from this formula that

$$G(x) = \frac{x^{r+1}}{r+1}, \qquad (r \neq -1)$$

is an antiderivative of $f(x) = x^r$. Similarly, the differentiation formula

$$\frac{d}{dx}(\sin x) = \cos x$$

implies that $G(x) = \sin x$ is an antiderivative of $f(x) = \cos x$. In general, each differentiation formula yields an immediate antidifferentiation formula. Table 5.4.3 lists some particular antiderivatives for our use in this section.

Table 5.4.3

Function	Antiderivative
x^r	$\dfrac{x^{r+1}}{r+1}$ (r a rational number $\neq -1$)
$\sin x$	$-\cos x$
$\cos x$	$\sin x$
$\sec^2 x$	$\tan x$
$\sec x \tan x$	$\sec x$
$\csc^2 x$	$-\cot x$
$\csc x \cot x$	$-\csc x$

You can verify that the derivative of each function in the right column is the corresponding function in the left column.

The following examples make use of the entries in Table 5.4.3.

$$\int_1^2 \frac{dx}{x^2} = \int_1^2 x^{-2}\,dx = \left[-x^{-1}\right]_1^2 = \left[-\frac{1}{x}\right]_1^2 = -\tfrac{1}{2} - (-1) = \tfrac{1}{2},$$

$$\int_0^1 t^{5/3}\,dt = \left[\tfrac{3}{8}t^{8/3}\right]_0^1 = \tfrac{3}{8}(1)^{8/3} - \tfrac{3}{8}(0)^{8/3} = \tfrac{3}{8}.$$

$$\int_{-\pi/4}^{\pi/3} \sec^2 x\,dx = \left[\tan x\right]_{-\pi/4}^{\pi/3} = \tan\frac{\pi}{3} - \tan\frac{-\pi}{4} = \sqrt{3} - (-1) = \sqrt{3} + 1.$$

$$\int_{\pi/6}^{\pi/2} \csc x \cot x\,dx = \left[-\csc x\right]_{\pi/6}^{\pi/2} = -\csc\frac{\pi}{2} - \left[-\csc\frac{\pi}{6}\right] = -1 - (-2)$$
$$= 1. \quad \square$$

Example 3 Evaluate

$$\int_0^1 (2x - 6x^4 + 5)\,dx.$$

SOLUTION As an antiderivative we use $G(x) = x^2 - \frac{6}{5}x^5 + 5x$:

$$\int_0^1 (2x - 6x^4 + 5)\,dx = \left[x^2 - \tfrac{6}{5}x^5 + 5x\right]_0^1 = 1 - \tfrac{6}{5} + 5 = 4\tfrac{4}{5}. \quad \square$$

Example 4 Evaluate

$$\int_{-1}^1 (x - 1)(x + 2)\,dx.$$

SOLUTION First we carry out the indicated multiplication:

$$(x - 1)(x + 2) = x^2 + x - 2.$$

As an antiderivative we use $G(x) = \frac{1}{3}x^3 + \frac{1}{2}x^2 - 2x$:

$$\int_{-1}^{1} (x-1)(x+2)\,dx = \left[\tfrac{1}{3}x^3 + \tfrac{1}{2}x^2 - 2x\right]_{-1}^{1} = -\tfrac{10}{3}. \quad \square$$

We now give some slightly more complicated examples. The essential step in each case is the determination of an antiderivative. Check each computation in detail.

$$\int_1^2 \frac{x^4+1}{x^2}\,dx = \int_1^2 (x^2 + x^{-2})\,dx = \left[\tfrac{1}{3}x^3 - x^{-1}\right]_1^2 = \tfrac{17}{6}.$$

$$\int_1^5 \sqrt{x-1}\,dx = \int_1^5 (x-1)^{1/2}\,dx = \left[\tfrac{2}{3}(x-1)^{3/2}\right]_1^5 = \tfrac{16}{3}.$$

$$\int_0^1 (4-\sqrt{x})^2\,dx = \int_0^1 (16 - 8\sqrt{x} + x)\,dx = \left[16x - \tfrac{16}{3}x^{3/2} + \tfrac{1}{2}x^2\right]_0^1 = \tfrac{67}{6}.$$

$$\int_1^2 -\frac{dt}{(t+2)^2} = \int_1^2 -(t+2)^{-2}\,dt = \left[(t+2)^{-1}\right]_1^2 = -\tfrac{1}{12}. \quad \square$$

The Linearity of the Integral

The preceding examples suggest some simple properties of the integral that are used regularly in computations. Throughout, take f and g as continuous functions and α and β as constants.

I. Constants may be factored through the integral sign:

(5.4.4)
$$\int_a^b \alpha f(x)\,dx = \alpha \int_a^b f(x)\,dx.$$

For example,

$$\int_1^4 \frac{3}{7}\sqrt{x}\,dx = \frac{3}{7}\int_1^4 x^{1/2}\,dx = \frac{3}{7}\left[\frac{x^{3/2}}{3/2}\right]_1^4 = \frac{2}{7}[(4)^{3/2} - (1)^{3/2}] = \frac{2}{7}[8-1]$$
$$= 2.$$

$$\int_0^{\pi/4} 2\cos x\,dx = 2\int_0^{\pi/4} \cos x\,dx = 2\left[\sin x\right]_0^{\pi/4} = 2\left[\sin\frac{\pi}{4} - \sin(0)\right]$$
$$= 2\frac{\sqrt{2}}{2} = \sqrt{2}. \quad \square$$

Remark We want to emphasize that only constants can be factored through the integral sign; an expression containing a variable cannot! For example,

$$\int_0^{\pi/2} x\cos x\,dx \neq x\int_0^{\pi/2} \cos x\,dx.$$

You can verify that an antiderivative for $h(x) = x\cos x$ is $H(x) = x\sin x + \cos x$, and so

$$\int_0^{\pi/2} x\cos x\,dx = \Big[x\sin x + \cos x\Big]_0^{\pi/2} = \tfrac{\pi}{2} - 1.$$

On the other hand,

$$x\int_0^{\pi/2} \cos x\,dx = x\Big[\sin x\Big]_0^{\pi/2} = x. \quad \square$$

II. The integral of a sum is the sum of the integrals:

(5.4.5)
$$\int_a^b [f(x) + g(x)]\,dx = \int_a^b f(x)\,dx + \int_a^b g(x)\,dx.$$

For example,

$$\begin{aligned}\int_1^2 \left[(x-1)^2 + \frac{1}{(x+2)^2}\right]dx &= \int_1^2 (x-1)^2\,dx + \int_1^2 \frac{dx}{(x+2)^2}\\ &= \left[\frac{1}{3}(x-1)^3\right]_1^2 + \left[-(x+2)^{-1}\right]_1^2\\ &= \frac{1}{3} - \frac{1}{4} + \frac{1}{3} = \frac{5}{12}. \quad \square\end{aligned}$$

III. The integral of a linear combination is the linear combination of the integrals:

(5.4.6)
$$\int_a^b [\alpha f(x) + \beta g(x)]\,dx = \alpha\int_a^b f(x)\,dx + \beta\int_a^b g(x)\,dx.$$

For example,

$$\begin{aligned}\int_0^1 (2x - 6x^4 + 5)\,dx &= 2\int_0^1 x\,dx - 6\int_0^1 x^4\,dx + \int_0^1 5\,dx\\ &= 2\left[\frac{x^2}{2}\right]_0^1 - 6\left[\frac{x^5}{5}\right]_0^1 + \Big[5x\Big]_0^1 = 1 - \tfrac{6}{5} + 5 = 4\tfrac{4}{5}\end{aligned}$$

as we saw in Example 3.

Properties I and II are particular instances of Property III. To prove III, let F be an antiderivative for f and let G be an antiderivative for g. Then, since

$$[\alpha F(x) + \beta G(x)]' = \alpha F'(x) + \beta G'(x) = \alpha f(x) + \beta g(x),$$

it follows that $\alpha F + \beta G$ is an antiderivative for $\alpha f + \beta g$. Therefore,

$$\begin{aligned}\int_a^b [\alpha f(x) + \beta g(x)]\, dx &= \Big[\alpha F(x) + \beta G(x)\Big]_a^b \\ &= [\alpha F(b) + \beta G(b)] - [\alpha F(a) + \beta G(a)] \\ &= \alpha[F(b) - F(a)] + \beta[G(b) - G(a)] \\ &= \alpha \int_a^b f(x)\, dx + \beta \int_a^b g(x)\, dx. \quad \square\end{aligned}$$

Example 5 Evaluate

$$\int_0^{\pi/4} \sec x\, [2 \tan x - 5 \sec x]\, dx.$$

SOLUTION

$$\begin{aligned}\int_0^{\pi/4} \sec x\, [2 \tan x - 5 \sec x]\, dx &= \int_0^{\pi/4} [2 \sec x \tan x - 5 \sec^2 x]\, dx \\ &= 2\int_0^{\pi/4} \sec x \tan x\, dx - 5\int_0^{\pi/4} \sec^2 x\, dx \\ &= 2\Big[\sec x\Big]_0^{\pi/4} - 5\Big[\tan x\Big]_0^{\pi/4} \\ &= 2\Big[\sec\frac{\pi}{4} - \sec(0)\Big] - 5\Big[\tan\frac{\pi}{4} - \tan(0)\Big] \\ &= 2[\sqrt{2} - 1] - 5[1 - 0] = 2\sqrt{2} - 7. \quad \square\end{aligned}$$

EXERCISES 5.4

In Exercises 1–40, evaluate the definite integral.

1. $\int_0^1 (2x - 3)\, dx.$
2. $\int_0^1 (3x + 2)\, dx.$
3. $\int_{-1}^0 5x^4\, dx.$
4. $\int_1^2 (2x + x^2)\, dx.$
5. $\int_1^4 2\sqrt{x}\, dx.$
6. $\int_0^4 \sqrt[3]{x}\, dx.$
7. $\int_1^5 2\sqrt{x - 1}\, dx.$
8. $\int_1^2 \left(\frac{3}{x^3} + 5x\right) dx.$
9. $\int_{-2}^0 (x + 1)(x - 2)\, dx.$
10. $\int_2^0 \frac{dx}{(x + 1)^2}.$
11. $\int_3^3 \sqrt{x}\, dx.$
12. $\int_{-1}^0 (t - 2)(t + 1)\, dt.$
13. $\int_0^1 \frac{dt}{(t + 2)^3}.$
14. $\int_1^0 (t^3 + t^2)\, dt.$
15. $\int_1^2 \left(3t + \frac{4}{t^2}\right) dt.$
16. $\int_{-1}^{-1} 7x^6\, dx.$
17. $\int_0^1 (x^{3/2} - x^{1/2})\, dx.$
18. $\int_0^1 (x^{3/4} - 2x^{1/2})\, dx.$
19. $\int_0^1 (x + 1)^{17}\, dx.$
20. $\int_0^a (a^2 x - x^3)\, dx.$
21. $\int_0^a (\sqrt{a} - \sqrt{x})^2\, dx.$
22. $\int_{-1}^1 (x - 2)^2\, dx.$
23. $\int_1^2 \frac{6 - t}{t^3}\, dt.$
24. $\int_1^2 \frac{2 - t}{t^3}\, dt.$
25. $\int_0^1 x^2(x - 1)\, dx.$
26. $\int_1^3 \left(x^2 - \frac{1}{x^2}\right) dx.$
27. $\int_1^2 2x(x^2 + 1)\, dx.$
28. $\int_0^1 3x^2(x^3 + 1)\, dx.$
29. $\int_0^{\pi/2} \cos x\, dx.$
30. $\int_0^{\pi} 3 \sin x\, dx.$
31. $\int_0^{\pi/4} 2 \sec^2 x\, dx.$
32. $\int_{\pi/6}^{\pi/3} \sec x \tan x\, dx.$
33. $\int_{\pi/6}^{\pi/4} \csc x \cot x\, dx.$
34. $\int_{\pi/4}^{\pi/3} -\csc^2 x\, dx.$
35. $\int_0^{2\pi} \sin x\, dx.$
36. $\int_0^{\pi} \frac{1}{2} \cos x\, dx.$

37. $\int_0^{\pi/3} \left(\frac{2}{\pi}x - 2\sec^2 x\right) dx.$

38. $\int_{\pi/4}^{\pi/2} \csc x\,(\cot x - 3\csc x)\,dx.$

39. $\int_0^3 \left[\frac{d}{dx}(\sqrt{4 + x^2})\right] dx.$

40. $\int_0^{\pi/2} \left[\frac{d}{dx}(\sin^3 x)\right] dx.$

41. Define a function F on $[1, 8]$ such that $F'(x) = 1/x$ and (a) $F(2) = 0$; (b) $F(2) = -3$.

42. Define a function F on $[0, 4]$ such that $F'(x) = \sqrt{1 + x^2}$ and (a) $F(3) = 0$; (b) $F(3) = 1$.

In Exercises 43–46, verify that the function f is nonnegative on the given interval, and then calculate the area between the graph of f and the x-axis on that interval.

43. $f(x) = 4x - x^2$; $[0, 4]$.

44. $f(x) = x\sqrt{x} + 1$; $[1, 9]$.

45. $f(x) = 2\cos x$; $[-\pi/2, \pi/4]$.

46. $f(x) = \sec x \tan x$; $[0, \pi/3]$.

In Exercises 47–50, evaluate the given integrals.

47. (a) $\int_2^5 (x - 3)\,dx.$ (b) $\int_2^5 |x - 3|\,dx.$

48. (a) $\int_{-4}^2 (2x + 3)\,dx.$ (b) $\int_{-4}^2 |2x + 3|\,dx.$

49. (a) $\int_{-2}^2 (x^2 - 1)\,dx.$ (b) $\int_{-2}^2 |x^2 - 1|\,dx.$

50. (a) $\int_{-\pi/2}^{\pi} \cos x\,dx.$ (b) $\int_{-\pi/2}^{\pi} |\cos x|\,dx.$

51. An object starts at the origin and moves along the x-axis with velocity

$$v(t) = 10t - t^2, \qquad 0 \le t \le 10.$$

(a) What is the position of the object at any time t, $0 \le t \le 10$?

(b) When is the object's velocity a maximum and what is its position at that time?

52. The velocity of a weight suspended on a spring is given by

$$v(t) = 3\sin t + 4\cos t, \qquad t \ge 0.$$

At time $t = 0$, the weight is one unit below the equilibrium position (see the figure).

(a) Determine the position of the weight at any time $t \ge 0$.

(b) What is the weight's maximum displacement from the equilibrium position?

(NOTE: The motion of the weight is called *simple harmonic motion.* Also see Exercises 61 and 62, Section 3.6.)

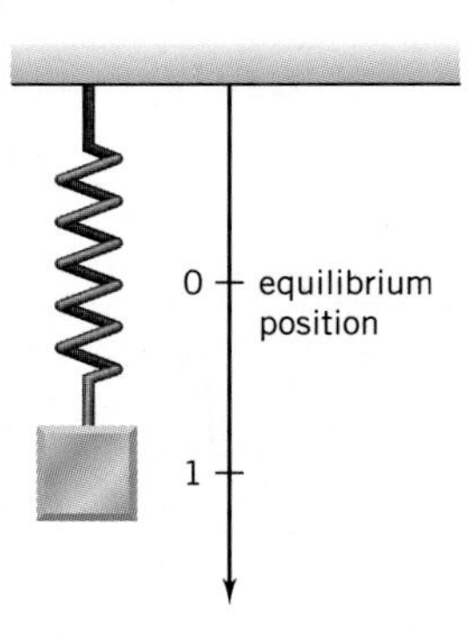

In Exercises 53–56, evaluate the given definite integral.

53. $\int_0^4 f(x)\,dx$, where $f(x) = \begin{cases} 2x + 1, & 0 \le x \le 1 \\ 4 - x, & 1 < x \le 4. \end{cases}$

54. $\int_{-2}^4 f(x)\,dx$, where $f(x) = \begin{cases} x^2, & -2 \le x < 0 \\ \frac{1}{2}x + 2, & 0 \le x \le 4. \end{cases}$

55. $\int_{-\pi/2}^{\pi} f(x)\,dx$, where $f(x) = \begin{cases} \cos x, & -\pi/2 \le x \le \pi/3 \\ (3/\pi)\,x + 1, & \pi/3 < x \le \pi. \end{cases}$

56. $\int_0^{3\pi/2} f(x)\,dx$, where $f(x) = \begin{cases} 2\sin x, & 0 \le x \le \pi/2 \\ \frac{1}{2}\cos x, & \pi/2 < x \le 3\pi/2. \end{cases}$

57. Let

$$f(x) = \begin{cases} x + 2, & -2 \le x \le 0 \\ 2, & 0 < x \le 1 \\ 4 - 2x, & 1 < x \le 2 \end{cases}$$

and let $g(x) = \int_{-2}^x f(t)\,dt.$

(a) Find an expression for $g(x)$.

(b) Sketch the graphs of f and g.

(c) Where is f continuous? Where is f differentiable? Where is g differentiable?

58. Let

$$f(x) = \begin{cases} 1 - x^2, & -1 \le x \le 1 \\ 1, & 1 < x < 3 \\ 2x - 5, & 3 \le x \le 5 \end{cases}$$

and let $g(x) = \int_{-1}^x f(t)\,dt.$

(a) Find an expression for $g(x)$.

(b) Sketch the graphs of f and g.

(c) Where is f continuous? Where is f differentiable? Where is g differentiable?

59. Compare $\dfrac{d}{dx}\left[\int_a^x f(t)\,dt\right]$ to $\int_a^x \dfrac{d}{dt}[f(t)]\,dt$.

60. Let $f_1, f_2, \ldots, f_n$ be continuous functions on $[a, b]$ and let $\alpha_1, \alpha_2, \ldots, \alpha_n$ be constants. Use mathematical induction to prove that

$$\int_a^b [\alpha_1 f_1(x) + \alpha_2 f_2(x) + \cdots + \alpha_n f_n(x)]\,dx = \alpha_1 \int_a^b f_1(x)\,dx + \alpha_2 \int_a^b f_2(x)\,dx + \cdots + \alpha_n \int_a^b f_n(x)\,dx.$$

■ 5.5 SOME AREA PROBLEMS

In Section 5.2 we noted that, if f is nonnegative and continuous on $[a, b]$, then the integral of f from a to b gives the area of the region below the graph of f, that is, the area of the region bounded by the graph of f and the x-axis between $x = a$ and $x = b$. With Ω as in Figure 5.5.1

(5.5.1)
$$\text{Area of } \Omega = \int_a^b f(x)\,dx.$$

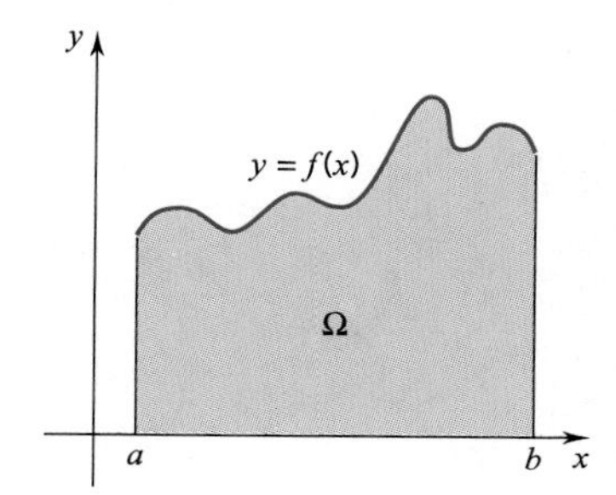

Figure 5.5.1

Example 1 Find the area below the graph of the square-root function from $x = 0$ to $x = 1$.

SOLUTION The graph is pictured in Figure 5.5.2. The area below the graph is $\frac{2}{3}$:

$$\int_0^1 \sqrt{x}\,dx = \int_0^1 x^{1/2}\,dx = \left[\tfrac{2}{3}x^{3/2}\right]_0^1 = \tfrac{2}{3}. \quad \square$$

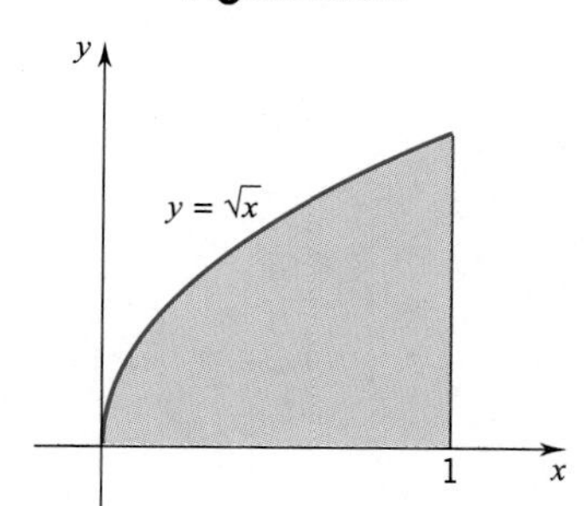

Figure 5.5.2

Example 2 Find the area of the region bounded above by the curve $y = 4 - x^2$ and below by the x-axis.

SOLUTION The curve intersects the x-axis at $x = -2$ and $x = 2$. See Figure 5.5.3. The area of the region is $\frac{32}{3}$:

$$\int_{-2}^2 (4 - x^2)\,dx = \left[4x - \tfrac{1}{3}x^3\right]_{-2}^2 = \tfrac{32}{3}.$$

NOTE: The region is symmetric with respect to the y-axis. Therefore, the area of the region is also $2\int_0^2 (4 - x^2)\,dx$, and this integral is a little easier to compute:

$$2\int_0^2 (4 - x^2)\,dx = 2\left[4x - \tfrac{1}{3}x^3\right]_0^2 = 2(8 - \tfrac{8}{3}) = 2(\tfrac{16}{3}) = \tfrac{32}{3}.$$

We will have more to say about symmetry in integration in Section 5.8. $\square$

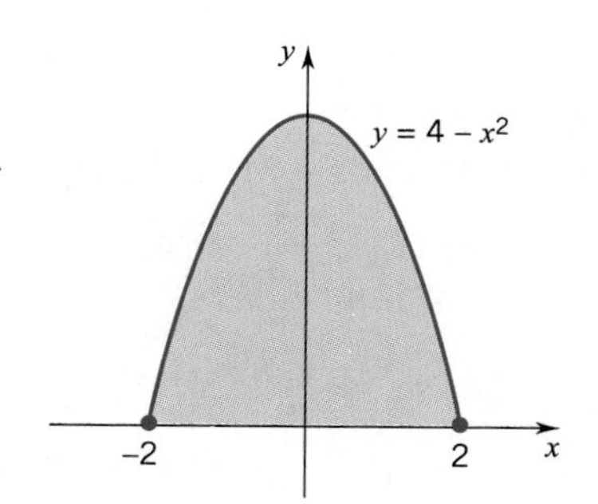

Figure 5.5.3

Now we calculate the areas of somewhat more complicated regions such as region Ω shown in Figure 5.5.4. To avoid excessive repetition, let's agree at the outset that throughout this section the symbols f, g, h represent continuous functions.

Look at the region Ω shown in Figure 5.5.4. The upper boundary of Ω is the graph of a nonnegative function f and the lower boundary is the graph of a nonnegative

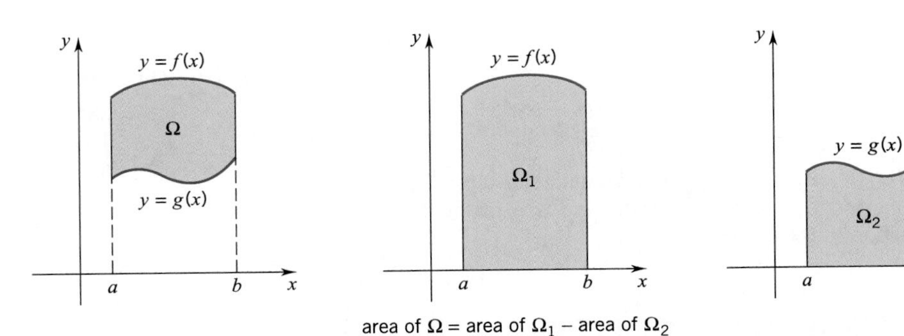

area of Ω = area of Ω_1 − area of Ω_2

Figure 5.5.4

function g. We can obtain the area of Ω by calculating the area of Ω_1 and subtracting off the area of Ω_2. Since

$$\text{Area of } \Omega_1 = \int_a^b f(x)\,dx \qquad \text{and} \qquad \text{Area of } \Omega_2 = \int_a^b g(x)\,dx,$$

we have

$$\text{Area of } \Omega = \int_a^b f(x)\,dx - \int_a^b g(x)\,dx.$$

We can combine the two integrals [by (5.4.6)] and write

(5.5.2)
$$\text{Area of } \Omega = \int_a^b [f(x) - g(x)]\,dx.$$

Example 3 Find the area of the region bounded above by $y = x + 2$ and below by $y = x^2$.

SOLUTION The region bounded by the graphs of the two equations is shown in Figure 5.5.5. Our first step is to find the points of intersection of the two curves:

$$x + 2 = x^2 \quad \text{iff} \quad x^2 - x - 2 = 0$$
$$\text{iff} \quad (x+1)(x-2) = 0,$$

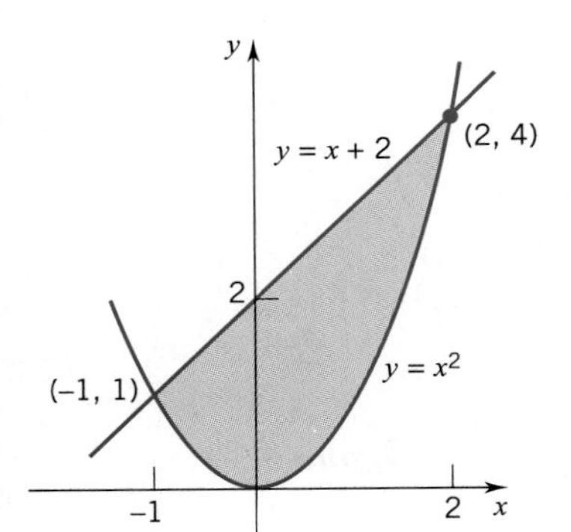

Figure 5.5.5

and so $x = -1, 2$. The curves intersect at $(-1, 1)$ and $(2, 4)$.

Now, the area of the region is:

$$\int_{-1}^{2} [(x+2) - x^2]\,dx = \left[\tfrac{1}{2}x^2 + 2x - \tfrac{1}{3}x^3\right]_{-1}^{2}$$
$$= (2 + 4 - \tfrac{8}{3}) - (\tfrac{1}{2} - 2 + \tfrac{1}{3})$$
$$= \tfrac{9}{2}. \quad ❑$$

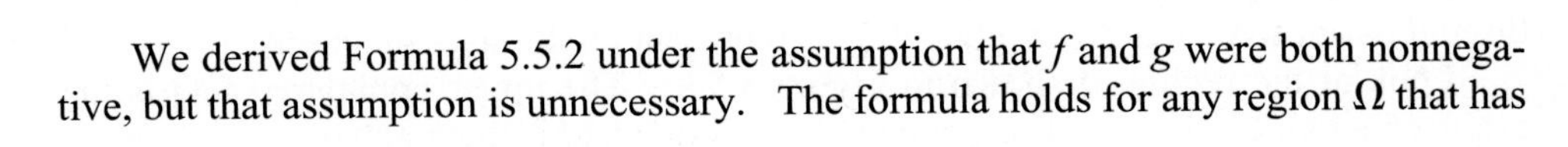

We derived Formula 5.5.2 under the assumption that f and g were both nonnegative, but that assumption is unnecessary. The formula holds for any region Ω that has

an upper boundary of the form $\quad y = f(x), \quad x \in [a, b]$

and

$$\text{a lower boundary of the form} \quad y = g(x), \quad x \in [a, b].$$

To see this, take Ω as in Figure 5.5.6. Obviously, Ω is congruent to the region marked Ω' in Figure 5.5.7; Ω' is Ω raised C units. Since Ω' lies entirely above the x-axis, the area of Ω' is given by the integral

$$\int_a^b \{[f(x) + C] - [g(x) + C]\}\, dx = \int_a^b [f(x) - g(x)]\, dx.$$

Since area of Ω = area of Ω',

$$\text{Area of } \Omega = \int_a^b [f(x) - g(x)]\, dx$$

as asserted.

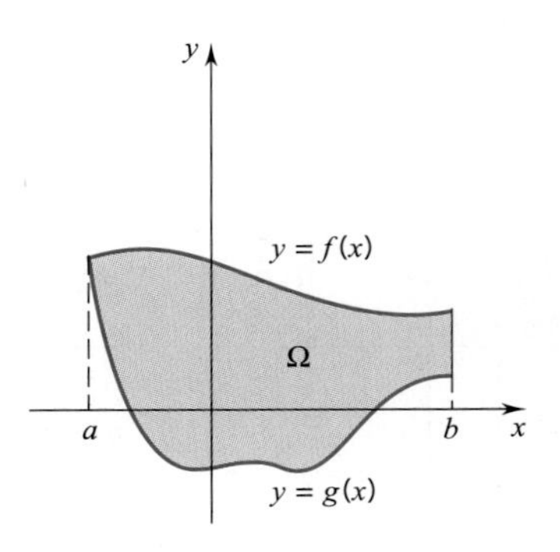

Figure 5.5.6

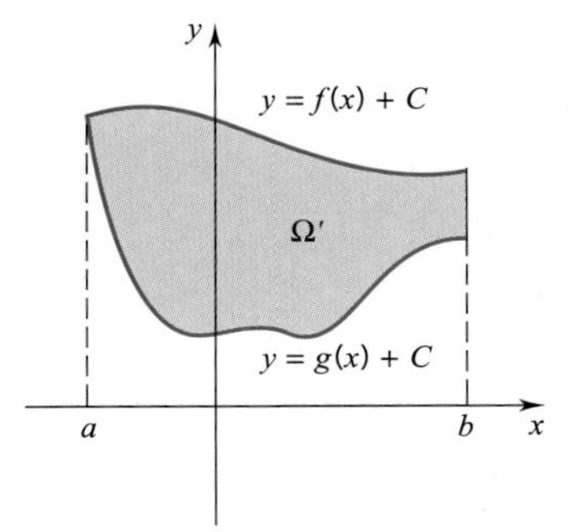

Figure 5.5.7

Example 4 Find the area of the region shown in Figure 5.5.8.

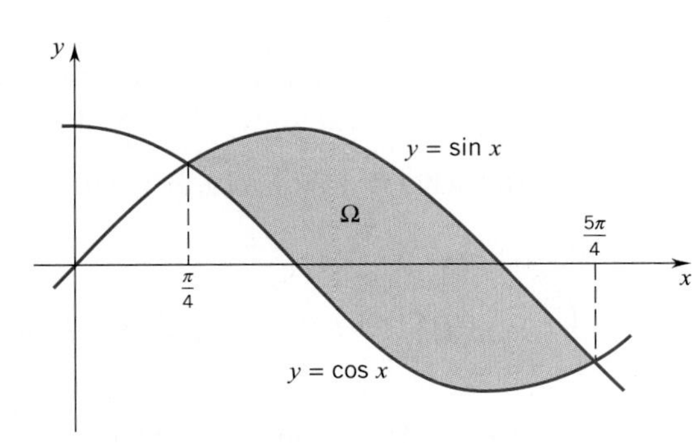

Figure 5.5.8

SOLUTION

$$\begin{aligned}\text{Area of } \Omega &= \int_{\pi/4}^{5\pi/4} [\sin x - \cos x]\, dx\\ &= \Big[-\cos x - \sin x\Big]_{\pi/4}^{5\pi/4}\\ &= 2\sqrt{2}. \quad \square\end{aligned}$$

Example 5 Find the area between the curves

$$y = 4x \qquad \text{and} \qquad y = x^3$$

from $x = -2$ to $x = 2$. See Figure 5.5.9.

SOLUTION Notice that $y = x^3$ is the upper boundary from $x = -2$ to $x = 0$, but it is the lower boundary from $x = 0$ to $x = 2$. Thus

$$\begin{aligned}\text{Area} &= \int_{-2}^{0} [x^3 - 4x]\, dx + \int_0^2 [4x - x^3]\, dx\\ &= \Big[\tfrac{1}{4}x^4 - 2x^2\Big]_{-2}^{0} + \Big[2x^2 - \tfrac{1}{4}x^4\Big]_0^2\\ &= [0 - (-4)] + [4 - 0] = 8. \quad \square\end{aligned}$$

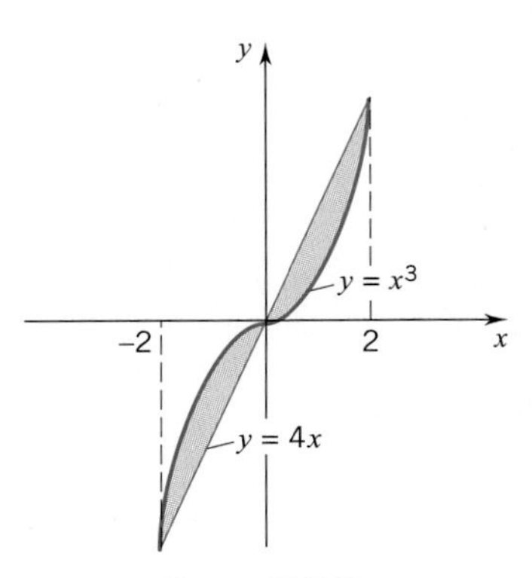

Figure 5.5.9

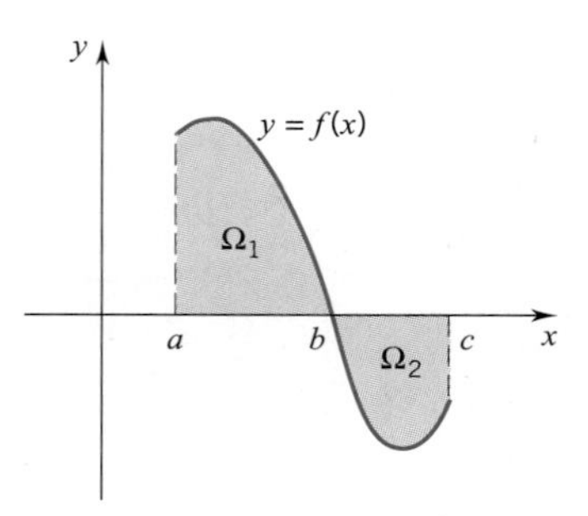

Figure 5.5.10

Example 6 Use integrals to represent the area of the region $\Omega = \Omega_1 \cup \Omega_2$ shaded in Figure 5.5.10.

SOLUTION From $x = a$ to $x = b$, the curve $y = f(x)$ is above the x-axis. Therefore

$$\text{Area of } \Omega_1 = \int_a^b f(x)\, dx.$$

From $x = b$ to $x = c$, the curve $y = f(x)$ is below the x-axis. The upper boundary for Ω_2 is the curve $y = 0$ (the x-axis) and the lower boundary is the curve $y = f(x)$. Thus

$$\text{Area of } \Omega_2 = \int_b^c [0 - f(x)]\, dx = -\int_b^c f(x)\, dx.$$

The area of Ω is the sum of these two areas:

$$\text{Area of } \Omega = \int_a^b f(x)\, dx - \int_b^c f(x)\, dx. \quad \square$$

Signed Area

The solution of Example 6 suggests a general geometric interpretation of the definite integral of a (continuous) function f. For example, if f is as denoted in Figure 5.5.10, then $\int_a^c f(x)\, dx$ is a number. But what does this number represent in terms of area? Since

$$\int_a^c f(x)\, dx = \int_a^b f(x)\, dx + \int_b^c f(x)\, dx = \text{area } \Omega_1 - \text{area } \Omega_2,$$

it follows that $\int_a^c f(x)\, dx$ is the area of the region above the x-axis *minus* the area of the region below the x-axis. This result is true in general: If f changes sign on the interval $[a, b]$, then $\int_a^b f(x)\, dx$ can be interpreted as the *difference* of two areas—the total area of the regions that lie above the x-axis *minus* the total area of the regions that lie below the x-axis. Consider the function f shown in Figure 5.5.11:

$$\begin{aligned}\int_a^b f(x)\, dx &= \int_a^c f(x)\, dx + \int_c^d f(x)\, dx + \int_d^e f(x)\, dx + \int_e^b f(x)\, dx \\ &= \text{area of } \Omega_1 - \text{area of } \Omega_2 + \text{area of } \Omega_3 - \text{area of } \Omega_4 \\ &= [\text{area of } \Omega_1 + \text{area of } \Omega_3] - [\text{area of } \Omega_2 + \text{area of } \Omega_4].\end{aligned}$$

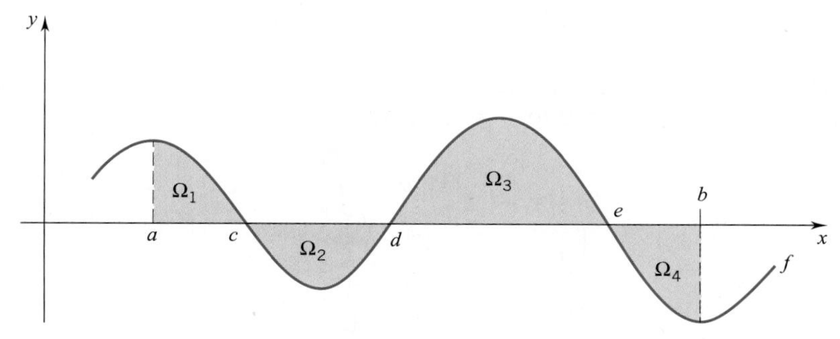

Figure 5.5.11

This value is sometimes referred to as the *signed area*; $\int_a^b f(x)\,dx$ can be interpreted geometrically as the signed area of the region bounded by the graph of f and the x-axis between $x = a$ and $x = b$.

Example 7 Evaluate

$$\int_{-1}^{3} (x^2 - 2x)\,dx$$

and interpret the result in terms of areas. Also, determine the area of the region bounded by the graph of $f(x) = x^2 - 2x$ and the x-axis between $x = -1$ and $x = 3$.

SOLUTION The graph of $f(x) = x^2 - 2x$ is shown in Figure 5.5.12. Now

$$\begin{aligned}\int_{-1}^{3} (x^2 - 2x)\,dx &= \left[\tfrac{1}{3}x^3 - x^2\right]_{-1}^{3} \\ &= \left[\tfrac{1}{3}(3)^3 - (3)^2\right] - \left[\tfrac{1}{3}(-1)^3 - (-1)^2\right] = \tfrac{4}{3}.\end{aligned}$$

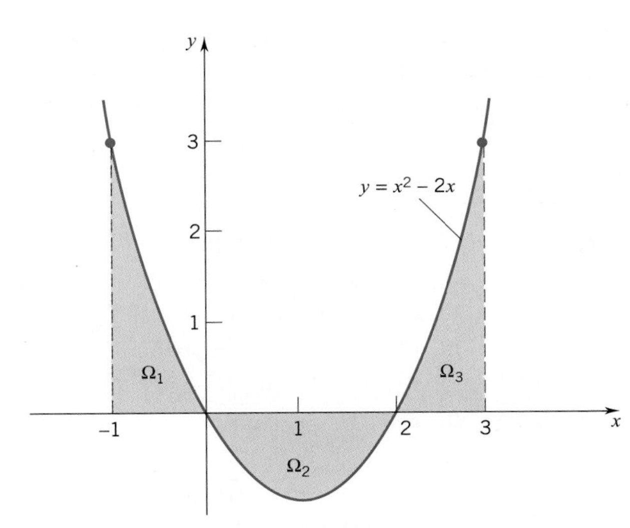

Figure 5.5.12

Thus, area of Ω_1 + area of Ω_3 − area of $\Omega_2 = \frac{4}{3}$. The area A of the region bounded by the graph of f and the x-axis from $x = -1$ to $x = 3$ is given by

$$\begin{aligned}A &= \int_{-1}^{0} (x^2 - 2x)\,dx + \int_{0}^{2} (2x - x^2)\,dx + \int_{2}^{3} (x^2 - 2x)\,dx \\ &= \left[\tfrac{1}{3}x^3 - x^2\right]_{-1}^{0} + \left[x^2 - \tfrac{1}{3}x^3\right]_{0}^{2} + \left[\tfrac{1}{3}x^3 - x^2\right]_{2}^{3} \\ &= \tfrac{4}{3} + \tfrac{4}{3} + \tfrac{4}{3} = 4. \quad ❑\end{aligned}$$

Example 8 Use integrals to represent the area of the region shaded in Figure 5.5.13.

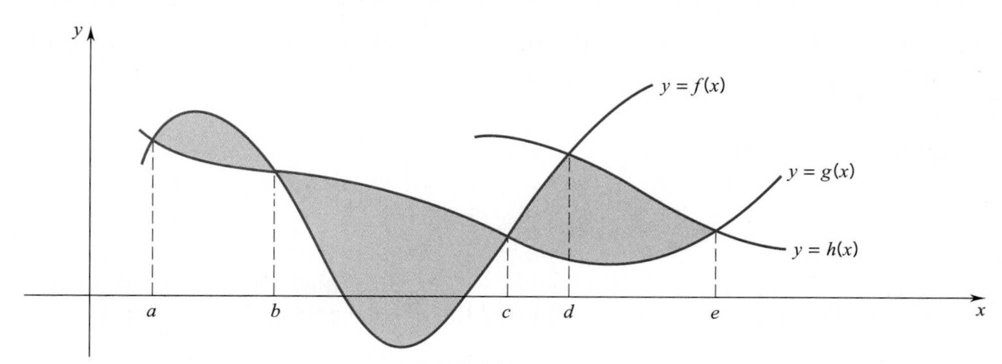

Figure 5.5.13

SOLUTION

$$\text{Area} = \int_a^b [f(x) - g(x)]\,dx + \int_b^c [g(x) - f(x)]\,dx + \int_c^d [f(x) - g(x)]\,dx + \int_d^e [h(x) - g(x)]\,dx. \quad \square$$

Remark The examples in this section also demonstrate the importance of having an accurate figure to illustrate the problem being considered. Drawing an accurate figure is usually the first and crucial step in effective problem solving. ❑

EXERCISES 5.5

In Exercises 1–10, find the area between the graph of f and the x-axis.

1. $f(x) = 2 + x^3, \quad x \in [0, 1]$.

2. $f(x) = (x + 2)^{-2}, \quad x \in [0, 2]$.

3. $f(x) = \sqrt{x + 1}, \quad x \in [3, 8]$.

4. $f(x) = x^2(3 + x), \quad x \in [0, 8]$.

5. $f(x) = (2x^2 + 1)^2, \quad x \in [0, 1]$.

6. $f(x) = \frac{1}{2}(x + 1)^{-1/2}, \quad x \in [0, 8]$.

7. $f(x) = x^2 - 4, \quad x \in [1, 2]$.

8. $f(x) = \cos x, \quad x \in [\frac{1}{6}\pi, \frac{1}{3}\pi]$.

9. $f(x) = \sin x, \quad x \in [\frac{1}{3}\pi, \frac{1}{2}\pi]$.

10. $f(x) = x^3 + 1, \quad x \in [-2, -1]$.

In Exercises 11–26, sketch the region bounded by the curves and find its area.

11. $y = \sqrt{x}, \quad y = x^2$.

12. $y = 6x - x^2, \quad y = 2x$.

13. $y = 5 - x^2, \quad y = 3 - x$.

14. $y = 8, \quad y = x^2 + 2x$.

15. $y = 8 - x^2, \quad y = x^2$.

16. $y = \sqrt{x}, \quad y = \frac{1}{4}x$.

17. $x^3 - 10y^2 = 0, \quad x - y = 0$.

18. $y^2 - 27x = 0, \quad x + y = 0$.

19. $x - y^2 + 3 = 0, \quad x - 2y = 0$.

20. $y^2 = 2x, \quad x - y = 4$.

21. $y = x, \quad y = 2x, \quad y = 4$.

22. $y = x^2, \quad y = -\sqrt{x}, \quad x = 4$.

23. $y = \cos x, \quad y = 4x^2 - \pi^2$.

24. $y = \sin x, \quad y = \pi x - x^2$.

25. $y = x, \quad y = \sin x, \quad x = \pi/2$.

26. $y = x + 1, \quad y = \cos x, \quad x = \pi$.

27. The graph of $f(x) = x^2 - x - 6$ is shown in the figure.

(a) Evaluate $\int_{-3}^{4} f(x)\,dx$ and interpret the result in terms of areas.

(b) Find the area of the region bounded by the graph of f and the x-axis for $x \in [-3, 4]$.

(c) Find the area of the region bounded by the graph of f and the x-axis for $x \in [-2, 3]$.

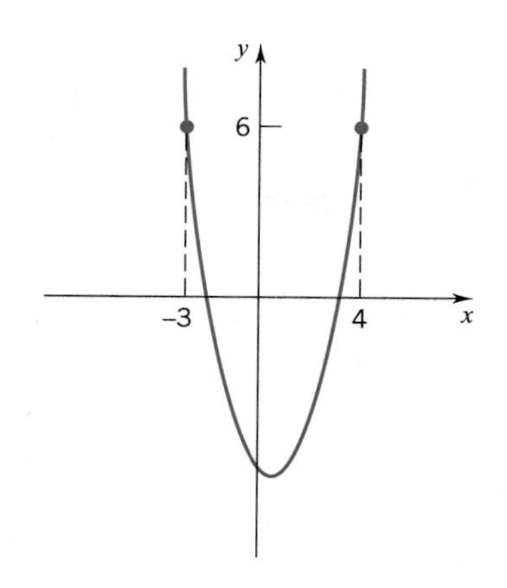

28. The graph of $f(x) = 2 \sin x$, $x \in [-\pi/2, 3\pi/4]$ is shown in the figure below.

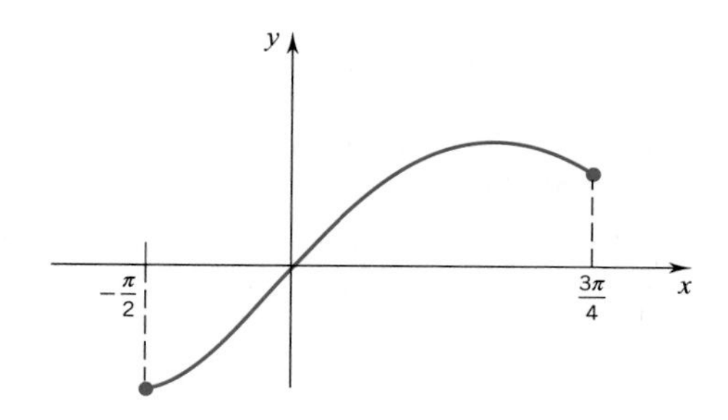

(a) Evaluate $\int_{-\pi/2}^{3\pi/4} f(x)\, dx$ and interpret the result in terms of areas.

(b) Find the area of the region bounded by the graph of f and the x-axis for $x \in [-\pi/2, 3\pi/4]$.

(c) Find the area of the region bounded by the graph of f and the x-axis for $x \in [-\pi/2, 0]$.

29. Let $f(x) = x^3 - x$.

(a) Evaluate $\int_{-2}^{2} f(x)\, dx$.

(b) Sketch the graph of f and find the area of the region bounded by the graph and the x-axis for $x \in [-2, 2]$.

30. Let $f(x) = \cos x + \sin x$, $x \in [-\pi, \pi]$.

(a) Evaluate $\int_{-\pi}^{\pi} f(x)\, dx$.

(b) Sketch the graph of f and find the area of the region bounded by the graph and the x-axis for $x \in [-\pi, \pi]$.

31. Let $f(x) = x^3 - 4x + 2$.

(a) Evaluate $\int_{-2}^{3} f(x)\, dx$.

(b) Use a graphing utility to graph f and estimate the area bounded by the graph and the x-axis for $x \in [-2, 3]$. Use two-decimal place accuracy in your approximations.

32. Let $f(x) = 3x^2 - 2 \cos x$.

(a) Evaluate $\int_{-\pi/2}^{\pi/2} f(x)\, dx$.

(b) Use a graphing utility to graph f and estimate the area bounded by the graph and the x-axis for $x \in [-\pi/2, \pi/2]$. Use two-decimal place accuracy in your approximations.

33. Let

$$f(x) = \begin{cases} x^2 + 1, & 0 \le x \le 1 \\ 3 - x, & 1 < x \le 3. \end{cases}$$

Sketch the graph of f and find the area of the region bounded by the graph and the x-axis.

34. Let

$$f(x) = \begin{cases} 3\sqrt{x}, & 0 \le x \le 1 \\ 4 - x^2, & 1 < x \le 2. \end{cases}$$

Sketch the graph of f and find the area of the region bounded by the graph and the x-axis.

35. Sketch the region bounded by the x-axis the curves $y = \sin x$ and $y = \cos x$, $x \in [0, \pi/2]$, and find its area.

36. Sketch the region bounded by the curves $y = 1$ and $y = 1 + \cos x$ for $x \in [0, \pi]$ and find its area.

37. Use a graphing utility to sketch the region bounded by the curves $y = x^3 + 2x$ and $y = 3x + 1$ for $x \in [0, 2]$, and estimate its area. Use two-decimal place accuracy in your approximations.

38. Use a graphing utility to sketch the region bounded by the curves $y = x^4 - 2x^2$ and $y = 4 - x^2$ for $x \in [-2, 2]$, and estimate its area. Use two-decimal place accuracy in your approximations.

■ 5.6 INDEFINITE INTEGRALS

We begin with a continuous function f. If F is an antiderivative for f on $[a, b]$, then

$$\int_a^b f(x)\, dx = \Big[F(x) \Big]_a^b. \tag{1}$$

If C is a constant, then

$$\left[F(x) + C\right]_a^b = [F(b) + C] - [F(a) + C] = F(b) - F(a) = \left[F(x)\right]_a^b.$$

Thus we can replace (1) by writing

$$\int_a^b f(x)\,dx = \left[F(x) + C\right]_a^b.$$

If we have no particular interest in the interval $[a, b]$, but wish instead to emphasize that F is an antiderivative for f on *some* interval, then we can omit the a and the b and simply write

$$\int f(x)\,dx = F(x) + C.$$

Antiderivatives expressed in this manner are called *indefinite integrals.* The constant C is called the *constant of integration;* it is an *arbitrary* constant since it can be assigned any real value.

The indefinite integral of a function f is actually a *family* of functions; a specific member of the family is determined by assigning a particular value to the constant of integration. This family has the property that each of its members is an antiderivative of f and, conversely, every antiderivative of f is a member of the family. The latter fact follows from Theorem 4.2.4.

Thus, for example,

$$\int x^2\,dx = \tfrac{1}{3}x^3 + C \qquad \text{and} \qquad \int \sqrt{s}\,ds = \tfrac{2}{3}s^{3/2} + C.$$

The graphs of some specific members of these families are sketched in Figures 5.6.1 and 5.6.2, respectively.

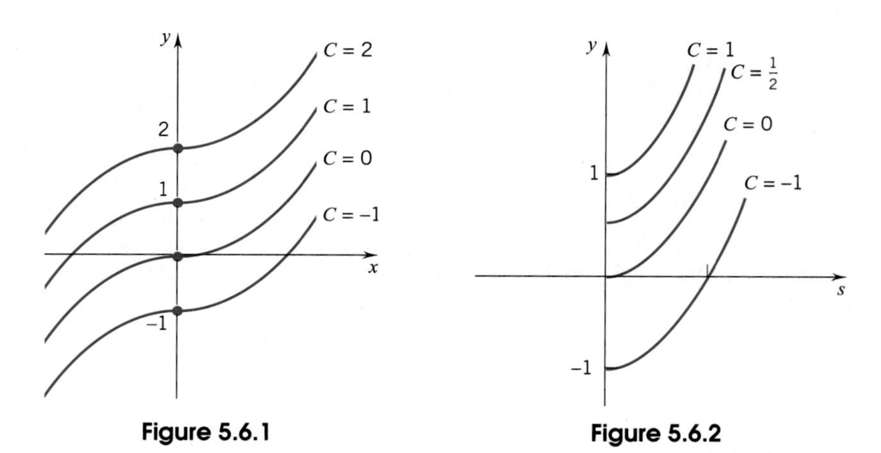

Figure 5.6.1

Figure 5.6.2

The Table of Antiderivatives (5.4.3) can be restated in terms of indefinite integrals:

5.6.1 TABLE OF INDEFINITE INTEGRALS

$$\int x^r\,dx = \frac{x^{r+1}}{r+1} + C \quad (r \text{ rational} \neq -1).$$

$$\int \sin x\,dx = -\cos x + C.$$

$$\int \cos x\,dx = \sin x + C.$$

$$\int \sec^2 x\,dx = \tan x + C.$$

$$\int \sec x \tan x\,dx = \sec x + C.$$

$$\int \csc^2 x\,dx = -\cot x + C.$$

$$\int \csc x \cot x\,dx = -\csc x + C.$$

The linearity properties of definite integrals (5.4.4), (5.4.5), and (5.4.6) also hold for indefinite integrals.

(5.6.2)
$$\int \alpha f(x)\,dx = \alpha \int f(x)\,dx, \quad \alpha \text{ a constant.}$$

(5.6.3)
$$\int [f(x) + g(x)]\,dx = \int f(x)\,dx + \int g(x)\,dx.$$

and, in general,

(5.6.4)
$$\int [\alpha f(x) + \beta g(x)]\,dx = \alpha \int f(x)\,dx + \beta \int g(x)\,dx,$$
where α and β are constants.

Example 1 Calculate

$$\int [5x^{3/2} - 2\csc^2 x]\,dx.$$

SOLUTION

$$\begin{aligned}\int [5x^{3/2} - 2\csc^2 x]\,dx &= 5\int x^{3/2}\,dx - 2\int \csc^2 x\,dx \\ &= 5(\tfrac{2}{5})\,x^{5/2} + C_1 - 2(-\cot x) + C_2 \\ &= 2x^{5/2} + 2\cot x + C,\end{aligned}$$

where $C = C_1 + C_2$. (We will routinely combine the constants of integration in this manner.) ❑

Example 2 Find f given that

$$f'(x) = x^3 + 2 \qquad \text{and} \qquad f(0) = 1.$$

SOLUTION Since f' is the derivative of f, f is an antiderivative for f'. Thus

$$f(x) = \int (x^3 + 2)\, dx = \tfrac{1}{4}x^4 + 2x + C,$$

for some value of the constant C. To evaluate C we use the fact that $f(0) = 1$. Since

$$f(0) = 1 \qquad \text{and} \qquad f(0) = \tfrac{1}{4}(0)^4 + 2(0) + C = C,$$

we see that $C = 1$. Therefore

$$f(x) = \tfrac{1}{4}x^4 + 2x + 1.$$

Some of the members of the family of curves $F(x) = \frac{1}{4}x^4 + 2x + C$ are shown in Figure 5.6.3. The graph of f is highlighted. ❑

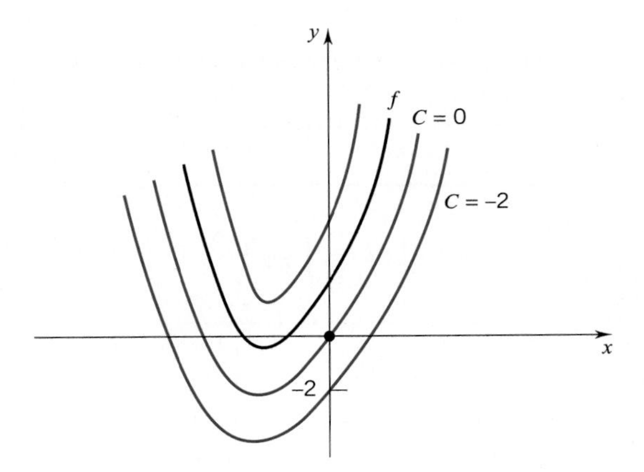

Figure 5.6.3

Example 3 Find f given that

$$f''(x) = 6x - 2, \qquad f'(1) = -5, \qquad \text{and} \qquad f(1) = 3.$$

SOLUTION First we get f' by integrating f'':

$$f'(x) = \int (6x - 2)\, dx = 3x^2 - 2x + C.$$

Since

$$f'(1) = -5 \qquad \text{and} \qquad f'(1) = 3(1)^2 - 2(1) + C = 1 + C,$$

we have

$$-5 = 1 + C \quad \text{and thus} \quad C = -6.$$

Therefore

$$f'(x) = 3x^2 - 2x - 6.$$

Now we get to f by integrating f':

$$f(x) = \int (3x^2 - 2x - 6)\, dx = x^3 - x^2 - 6x + K.$$

(We are writing the constant of integration as K because we used C before and it would be confusing to have C with two different values in the same problem.) Since

$$f(1) = 3 \qquad \text{and} \qquad f(1) = (1)^3 - (1)^2 - 6(1) + K = -6 + K,$$

we have

$$3 = -6 + K \quad \text{and thus} \quad K = 9.$$

Therefore

$$f(x) = x^3 - x^2 - 6x + 9. \quad ❑$$

Some Motion Examples

Example 4 An object moves along a coordinate line with velocity

$$v(t) = 2 - 3t + t^2 \quad \text{units per second.}$$

Its initial position (position at time $t = 0$) is 2 units to the right of the origin. Find the position of the object 4 seconds later.

SOLUTION Let $x(t)$ be the position (coordinate) of the object at time t. We are given that $x(0) = 2$. Since $x'(t) = v(t)$,

$$x(t) = \int v(t)\, dt = \int (2 - 3t + t^2)\, dt = 2t - \tfrac{3}{2}t^2 + \tfrac{1}{3}t^3 + C.$$

Since $x(0) = 2$ and $x(0) = 2(0) - \frac{3}{2}(0)^2 + \frac{1}{3}(0)^3 + C = C$, we have $C = 2$ and

$$x(t) = 2t - \tfrac{3}{2}t^2 + \tfrac{1}{3}t^3 + 2.$$

The position of the object at time $t = 4$ is the value of this function at $t = 4$:

$$x(4) = 2(4) - \tfrac{3}{2}(4)^2 + \tfrac{1}{3}(4)^3 + 2 = 7\tfrac{1}{3}.$$

At the end of 4 seconds the object is $7\frac{1}{3}$ units to the right of the origin.

The motion of the object is represented schematically in Figure 5.6.4. ❑

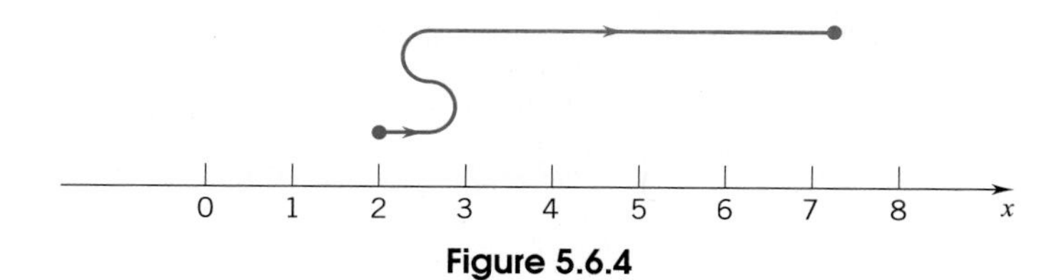

Figure 5.6.4

Recall that speed is the absolute value of velocity:

$$\text{Speed at time } t = |v(t)|$$

and the integral of the speed function gives the distance traveled:

$$\int_a^b |v(t)|\, dt = \text{distance traveled from time } t = a \text{ to time } t = b.$$

Example 5 An object moves along the x-axis with acceleration $a(t) = 2t - 2$ units per second per second. Its initial position (position at time $t = 0$) is 5 units to the right of the origin. One second later the object is moving left at the rate of 4 units per second.

(a) Find the position of the object at time $t = 4$ seconds.

(b) How far does the object travel during these 4 seconds?

SOLUTION (a) Let $x(t)$ and $v(t)$ denote the position and velocity of the object at time t. We are given that $x(0) = 5$ and $v(1) = -4$. Since $v'(t) = a(t)$,

$$v(t) = \int a(t)\,dt = \int (2t - 2)\,dt = t^2 - 2t + C.$$

Since

$$v(1) = -4 \qquad \text{and} \qquad v(1) = (1)^2 - 2(1) + C = -1 + C,$$

we have $C = -3$ and

$$v(t) = t^2 - 2t - 3.$$

Since $x'(t) = v(t)$,

$$x(t) = \int v(t)\,dt = \int (t^2 - 2t - 3)\,dt = \tfrac{1}{3}t^3 - t^2 - 3t + K.$$

Since

$$x(0) = 5 \qquad \text{and} \qquad x(0) = \tfrac{1}{3}(0)^3 - (0)^2 - 3(0) + K = K,$$

we have $K = 5$. Therefore

$$x(t) = \tfrac{1}{3}t^3 - t^2 - 3t + 5.$$

As you can check, $x(4) = -\frac{5}{3}$. At time $t = 4$ the object is $\frac{5}{3}$ units to the left of the origin.

(b) The distance traveled from time $t = 0$ to $t = 4$ is given by the integral

$$s = \int_0^4 |v(t)|\,dt = \int_0^4 |t^2 - 2t - 3|\,dt.$$

To evaluate this integral we first remove the absolute-value sign. As you can verify,

$$|t^2 - 2t - 3| = \begin{cases} -(t^2 - 2t - 3), & 0 \le t < 3 \\ t^2 - 2t - 3, & 3 \le t \le 4. \end{cases}$$

Thus

$$\begin{aligned} s &= \int_0^3 (3 + 2t - t^2)\,dt + \int_3^4 (t^2 - 2t - 3)\,dt \\ &= \left[3t + t^2 - \tfrac{1}{3}t^3\right]_0^3 + \left[\tfrac{1}{3}t^3 - t^2 - 3t\right]_3^4 = \tfrac{34}{3}. \end{aligned}$$

During the 4 seconds the object travels a distance of $\frac{34}{3}$ units.

The motion of the object is represented schematically in Figure 5.6.5. ❑

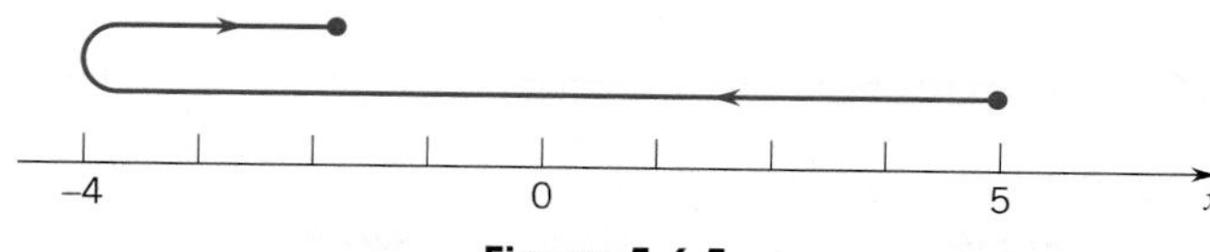

Figure 5.6.5

QUESTION: The object in Example 5 leaves $x = 5$ at time $t = 0$ and arrives at $x = -\frac{5}{3}$ at time $t = 4$. The separation between $x = 5$ and $x = -\frac{5}{3}$ is only $|5 - (-\frac{5}{3})| = \frac{20}{3}$. How is it possible that the object travels a distance of $\frac{34}{3}$ units?

ANSWER: The object does not maintain a fixed direction. It changes direction at time $t = 3$. You can see this by noting that the velocity function

$$v(t) = t^2 - 2t - 3 = (t - 3)(t + 1)$$

changes signs at $t = 3$. ❑

Example 6 Find the equation of motion for an object that moves along a straight line with constant acceleration a from an initial position x_0 with initial velocity v_0.

SOLUTION Call the line of the motion the x-axis. Here $a(t) = a$ at all times t. To find the velocity we integrate the acceleration:

$$v(t) = \int a\, dt = at + C.$$

The constant C is the initial velocity v_0:

$$v_0 = v(0) = a \cdot 0 + C = C.$$

We see therefore that

$$v(t) = at + v_0.$$

To find the position function we integrate the velocity:

$$x(t) = \int v(t)\, dt = \int (at + v_0)\, dt = \tfrac{1}{2}at^2 + v_0 t + K.$$

The constant K is the initial position x_0:

$$x_0 = x(0) = \tfrac{1}{2}a \cdot 0^2 + v_0 \cdot 0 + K = K.$$

The equation of motion is

(5.6.5) $$\boxed{x(t) = \tfrac{1}{2}at^2 + v_0 t + x_0.}$$ † ❑

† In the case of a free-falling body, $a = -g$ and we have Galileo's equation for free fall. See (3.4.5).

EXERCISES 5.6

In Exercises 1–18, calculate the given indefinite integral.

1. $\int \frac{dx}{x^4}$.

2. $\int (x-1)^2\,dx$.

3. $\int (ax+b)\,dx$.

4. $\int (ax^2+b)\,dx$.

5. $\int \frac{dx}{\sqrt{1+x}}$.

6. $\int \left(\frac{x^3+1}{x^5}\right)dx$.

7. $\int \left(\frac{x^3-1}{x^2}\right)dx$.

8. $\int \left(\sqrt{x}-\frac{1}{\sqrt{x}}\right)dx$.

9. $\int (t-a)(t-b)\,dt$.

10. $\int (t^2-a)(t^2-b)\,dt$.

11. $\int \frac{(t^2-a)(t^2-b)}{\sqrt{t}}\,dt$.

12. $\int (2-\sqrt{x})(2+\sqrt{x})\,dx$.

13. $\int g(x)g'(x)\,dx$.

14. $\int \sin x \cos x\,dx$.

15. $\int \tan x \sec^2 x\,dx$.

16. $\int \frac{g'(x)}{[g(x)]^2}\,dx$.

17. $\int \frac{4}{(4x+1)^2}\,dx$.

18. $\int \frac{3x^2}{(x^3+1)^2}\,dx$.

In Exercises 19–32, find f from the information given.

19. $f'(x) = 2x-1, \quad f(3) = 4$.

20. $f'(x) = 3-4x, \quad f(1) = 6$.

21. $f'(x) = ax+b, \quad f(2) = 0$.

22. $f'(x) = ax^2+bx+c, \quad f(0) = 0$.

23. $f'(x) = \sin x, \quad f(0) = 2$.

24. $f'(x) = \cos x, \quad f(\pi) = 3$.

25. $f''(x) = 6x-2, \quad f'(0) = 1, \quad f(0) = 2$.

26. $f''(x) = -12x^2, \quad f'(0) = 1, \quad f(0) = 2$.

27. $f''(x) = x^2-x, \quad f'(1) = 0, \quad f(1) = 2$.

28. $f''(x) = 1-x, \quad f'(2) = 1, \quad f(2) = 0$.

29. $f''(x) = \cos x, \quad f'(0) = 1, \quad f(0) = 2$.

30. $f''(x) = \sin x, \quad f'(0) = -2, \quad f(0) = 1$.

31. $f''(x) = 2x-3, \quad f(2) = -1, \quad f(0) = 3$.

32. $f''(x) = 5-4x, \quad f(1) = 1, \quad f(0) = -2$.

33. Compare $\frac{d}{dx}\left[\int f(x)\,dx\right]$ to $\int \frac{d}{dx}[f(x)]\,dx$.

34. Calculate

$$\int [f(x)g''(x) - g(x)f''(x)]\,dx.$$

35. An object moves along a coordinate line with velocity $v(t) = 6t^2 - 6$ units per second. Its initial position (position at time $t = 0$) is 2 units to the left of the origin. (a) Find the position of the object 3 seconds later. (b) Find the total distance traveled by the object during those 3 seconds.

36. An object moves along a coordinate line with acceleration $a(t) = (t+2)^3$ units per second per second. (a) Find the velocity function given that the initial velocity is 3 units per second. (b) Find the position function given that the initial velocity is 3 units per second and the initial position is the origin.

37. An object moves along a coordinate line with acceleration $a(t) = (t+1)^{-1/2}$ units per second per second. (a) Find the velocity function given that the initial velocity is 1 unit per second. (b) Find the position function given that the initial velocity is 1 unit per second and the initial position is the origin.

38. An object moves along a coordinate line with velocity $v(t) = t(1-t)$ units per second. Its initial position is 2 units to the left of the origin. (a) Find the position of the object 10 seconds later. (b) Find the total distance traveled by the object during those 10 seconds.

39. A car traveling at 60 mph decelerates at 20 feet per second per second. (a) How long does it take for the car to come to a complete stop? (b) What distance is required to bring the car to a complete stop?

40. An object moves along the x-axis with constant acceleration. Express the position $x(t)$ in terms of the initial position x_0, the initial velocity v_0, the velocity $v(t)$, and the elapsed time t.

41. An object moves along the x-axis with constant acceleration a. Verify that

$$[v(t)]^2 = v_0^2 + 2a[x(t) - x_0].$$

42. A bobsled moving at 60 mph decelerates at a constant rate to 40 mph over a distance of 264 feet and continues to decelerate at that same rate until it comes to a full stop. (a) What is the acceleration of the sled in feet per second per second? (b) How long does it take to reduce the speed to 40 mph? (c) How long does it take to bring the sled to a complete stop from 60 mph? (d) Over what distance does the sled come to a complete stop from 60 mph?

43. In the AB-run, minicars start from a standstill at point A, race along a straight track, and come to a full stop at point B one-half mile away. Given that the cars can accelerate uniformly to a maximum speed of 60 mph in 20 seconds and can brake at a maximum rate of 22 feet per second per second, what is the best possible time for the completion of the AB-run?

In Exercises 44–46, find the general law of motion of an object that moves in a straight line with acceleration $a(t)$. Write x_0 for initial position and v_0 for initial velocity.

44. $a(t) = \sin t$.

45. $a(t) = 2A + 6Bt$.

46. $a(t) = \cos t$.

47. As a particle moves about the plane, its x-coordinate changes at the rate of $t^2 - 5$ units per second and its y-coordinate changes at the rate of $3t$ units per second. If the particle is at the point (4, 2) when $t = 2$ seconds, where is the particle 4 seconds later?

48. As a particle moves about the plane, its x-coordinate changes at the rate of $t - 2$ units per second and its y-coordinate changes at the rate of $\sqrt{t}$ units per second. If the particle is at the point (3, 1) when $t = 4$ seconds, where is the particle 5 seconds later?

49. A particle moves along the x-axis with velocity $v(t) = At + B$. Determine A and B given that the initial velocity of the particle is 2 units per second and the position of the particle after 2 seconds of motion is 1 unit to the left of the initial position.

50. A particle moves along the x-axis with velocity $v(t) = At^2 + 1$. Determine A given that $x(1) = x(0)$. Compute the total distance traveled by the particle during the first second.

51. An object moves along a coordinate line with velocity $v(t) = \sin t$ units per second. The object passes through the origin at time $t = \pi/6$ seconds. When is the next time: (a) that the object passes through the origin? (b) that the object passes through the origin moving from left to right?

52. Repeat Exercise 51 with $v(t) = \cos t$.

53. An automobile with varying velocity $v(t)$ moves in a fixed direction for 5 minutes and covers a distance of 4 miles. What theorem would you invoke to argue that for at least one instant the speedometer must have read 48 miles per hour?

54. A speeding motorcyclist sees his way blocked by a haywagon some distance s ahead and slams on his brakes. Given that the brakes impart to the motorcycle a constant negative acceleration a and that the haywagon is moving with speed v_1 in the same direction as the motorcycle, show that the motorcyclist can avoid collision only if he is traveling at a speed less than $v_1 + \sqrt{2|a|s}$.

55. Find the velocity $v(t)$ given that $a(t) = 2[v(t)]^2$ and $v_0 \neq 0$.

■ 5.7 THE u-SUBSTITUTION; CHANGE OF VARIABLES

When we differentiate a composite function, we do so by the chain rule. In trying to calculate an indefinite integral, we are often called upon to apply the chain rule in reverse. This process can be facilitated by making what we call a "u-substitution."

An integral of the form

$$\int f(g(x))\, g'(x)\, dx$$

can be written

$$\int f(u)\, du$$

by setting

$$u = g(x), \qquad du = g'(x)\, dx.\dagger$$

If F is an antiderivative for f, then

$$[F(g(x))]' = F'(g(x))g'(x) = f(g(x))g'(x)$$

by the chain rule ↑ (first equality); ↑ $F' = f$ (second equality)

† Think of $du = g'(x)\, dx$ as a "formal differential," writing dx for h.

and so

$$\int f(g(x))g'(x)\,dx = \int [F(g(x))]'\,dx = F(g(x)) + C.$$

As illustrated below, we can obtain the same result by calculating

$$\int f(u)\,du$$

and then substituting $g(x)$ back in for u:

$$\int f(u)\,du = F(u) + C = F(g(x)) + C.$$

Example 1 Calculate

$$\int (x^2 - 1)^4\, 2x\,dx$$

and then check the result by differentiation.

SOLUTION Set

$$u = x^2 - 1 \qquad \text{so that} \qquad du = 2x\,dx.$$

Then

$$\int (x^2 - 1)^4\, 2x\,dx = \int u^4\,du = \tfrac{1}{5}u^5 + C = \tfrac{1}{5}(x^2 - 1)^5 + C.$$

CHECKING

$$\frac{d}{dx}\left[\frac{1}{5}(x^2 - 1)^5 + C\right] = \frac{5}{5}(x^2 - 1)^4\frac{d}{dx}(x^2 - 1) \overset{\checkmark}{=} (x^2 - 1)^4\, 2x. \quad \square$$

Example 2 Calculate

$$\int \sin x \cos x\,dx.$$

and then check the result by differentiation.

SOLUTION Set

$$u = \sin x, \qquad du = \cos x\,dx.$$

Then

$$\int \sin x \cos x\,dx = \int u\,du = \tfrac{1}{2}u^2 + C = \tfrac{1}{2}\sin^2 x + C.$$

CHECKING

$$\frac{d}{dx}[\tfrac{1}{2}\sin^2 x + C] = \tfrac{1}{2}(2)\sin x\frac{d}{dx}[\sin x] \overset{\checkmark}{=} \sin x \cos x.$$

ALTERNATE SOLUTION Since $\sin x \cos x = \frac{1}{2}\sin 2x$,

$$\int \sin x \cos x \, dx = \frac{1}{2}\int \sin 2x \, dx.$$

Set

$$u = 2x, \quad du = 2\,dx \quad \text{or} \quad dx = \tfrac{1}{2}\,du.$$

Then

$$\begin{aligned}\tfrac{1}{2}\int \sin 2x \, dx = \tfrac{1}{2}\int \sin u\left(\tfrac{1}{2}\,du\right) &= \tfrac{1}{4}\int \sin u \, du = -\tfrac{1}{4}\cos u + C\\ &= -\tfrac{1}{4}\cos 2x + C.\end{aligned}$$

We leave it to you to check this result and to reconcile it with the result given in the first solution. ❑

Example 3 Calculate

$$\int \frac{dx}{(3+5x)^2}.$$

and then check the result by differentiation.

SOLUTION Set

$$u = 3 + 5x \qquad \text{so that} \qquad du = 5\,dx.$$

Then

$$\frac{dx}{(3+5x)^2} = \frac{\frac{1}{5}\,du}{u^2} = \frac{1}{5}\frac{du}{u^2}$$

and

$$\int \frac{dx}{(3+5x)^2} = \frac{1}{5}\int \frac{du}{u^2} = -\frac{1}{5u} + C\dagger = -\frac{1}{5(3+5x)} + C.$$

CHECKING

$$\begin{aligned}\frac{d}{dx}\left[-\frac{1}{5(3+5x)} + C\right] &= \frac{d}{dx}\left[-\tfrac{1}{5}(3+5x)^{-1}\right]\\ &= \left(-\tfrac{1}{5}\right)(-1)(3+5x)^{-2}(5) \overset{\checkmark}{=} \frac{1}{(3+5x)^2}.\end{aligned}$$ ❑

In the remaining examples we leave the checking to you.

† We could write

$$\frac{1}{5}\int \frac{du}{u^2} = -\frac{1}{5}\left[\frac{1}{u} + C\right] = -\frac{1}{5u} - \frac{C}{5},$$

but, since C is arbitrary, $-C/5$ is arbitrary, and we can therefore write C instead.

Example 4 Calculate

$$\int x^2\sqrt{4+x^3}\,dx.$$

SOLUTION Set

$$u = 4 + x^3, \qquad du = 3x^2\,dx.$$

Then

$$x^2\sqrt{4+x^3}\,dx = \underbrace{\sqrt{4+x^3}}_{\sqrt{u}}\,\underbrace{x^2\,dx}_{\frac{1}{3}du} = \tfrac{1}{3}\sqrt{u}\,du$$

and

$$\int x^2\sqrt{4+x^3}\,dx = \tfrac{1}{3}\int \sqrt{u}\,du$$
$$= \tfrac{2}{9}u^{3/2} + C = \tfrac{2}{9}(4+x^3)^{3/2} + C. \quad \square$$

Example 5 Find

$$\int \sec^3 x \tan x\,dx.$$

SOLUTION We can write the integrand as $\sec^2 x \sec x \tan x\,dx$. Then, setting

$$u = \sec x, \qquad du = \sec x \tan x\,dx,$$

We have

$$\sec^3 x \tan x\,dx = \underbrace{\sec^2 x}_{u^2}\,\underbrace{(\sec x \tan x)\,dx}_{du} = u^2\,du$$

and

$$\int \sec^3 x \tan x\,dx = \int u^2 du = \tfrac{1}{3}u^3 + C = \tfrac{1}{3}\sec^3 x + C. \quad \square$$

The key step in making a u-substitution is to find a substitution $u = g(x)$ such that the expression $du = g'(x)\,dx$ appears in the original integral (at least up to a constant factor) and the new integral

$$\int f(u)\,du$$

is easier to calculate than the original integral. In most cases the form of the original integrand will suggest a good choice for u.

Example 6 Find

$$\int x \cos \pi x^2\,dx.$$

SOLUTION Set

$$u = \pi x^2, \qquad du = 2\pi x\,dx.$$

Then

$$x \cos \pi x^2\, dx = \underbrace{\cos \pi x^2}_{\cos u}\, \underbrace{x\, dx}_{\frac{1}{2\pi} du} = \frac{1}{2\pi} \cos u\, du$$

and

$$\int x \cos \pi x^2\, dx = \frac{1}{2\pi} \int \cos u\, du = \frac{1}{2\pi} \sin u + C = \frac{1}{2\pi} \sin \pi x^2 + C. \quad \square$$

Example 7 Evaluate the definite integral

$$\int_0^2 (x^2 - 1)(x^3 - 3x + 2)^3\, dx.$$

SOLUTION To evaluate this definite integral, we need to find an antiderivative for the integrand. The indefinite integral

$$\int (x^2 - 1)(x^3 - 3x + 2)^3\, dx$$

gives the set of all antiderivatives, and so we will calculate it first. Set

$$u = x^3 - 3x + 2, \qquad du = (3x^2 - 3)\, dx = 3(x^2 - 1)\, dx.$$

Then

$$(x^2 - 1)(x^3 - 3x + 2)^3\, dx = \underbrace{(x^3 - 3x + 2)^3}_{u^3}\, \underbrace{(x^2 - 1)\, dx}_{\frac{1}{3} du} = \tfrac{1}{3} u^3\, du$$

and

$$\int (x^2 - 1)(x^3 - 3x + 2)^3\, dx = \tfrac{1}{3} \int u^3\, du = \tfrac{1}{12} u^4 + C = \tfrac{1}{12} (x^3 - 3x + 2)^4 + C.$$

To evaluate the given definite integral, we need only one antiderivative, and so we will choose the one with $C = 0$. This gives

$$\int_0^2 (x^2 - 1)(x^3 - 3x + 2)^3\, dx = \left[\tfrac{1}{12} (x^3 - 3x + 2)^4 \right]_0^2 = 20. \quad \square$$

Later in this section we give another, more formal, approach to evaluating a definite integral when a substitution is involved in the integration step.

Example 8 Find

$$\int x^2 \csc^2 x^3 \cot^4 x^3\, dx.$$

SOLUTION We can effect some simplification by setting

$$u = x^3, \qquad du = 3x^2\, dx.$$

Then

$$x^2 \csc^2 x^3 \cot^4 x^3\, dx = \underbrace{\cot^4 x^3}_{\cot^4 u}\ \underbrace{\csc^2 x^3}_{\csc^2 u}\ \underbrace{x^2\, dx}_{\frac{1}{3}du} = \tfrac{1}{3} \cot^4 u \csc^2 u\, du$$

and

$$\int x^2 \csc^2 x^3 \cot^4 x^3\, dx = \tfrac{1}{3}\int \cot^4 u \csc^2 u\, du.$$

We can calculate the integral on the right by setting

$$t = \cot u, \qquad dt = -\csc^2 u\, du.$$

Then

$$\tfrac{1}{3}\int \cot^4 u \csc^2 u\, du = -\tfrac{1}{3}\int t^4\, dt = -\tfrac{1}{15}t^5 + C = -\tfrac{1}{15}\cot^5 u + C,$$

and thus

$$\int x^2 \csc^2 x^3 \cot^4 x^3\, dx = -\tfrac{1}{15}\cot^5 u + C = -\tfrac{1}{15}\cot^5 x^3 + C.$$

We arrived at this by making two consecutive substitutions. First we set $u = x^3$ and then we set $t = \cot u$. We could have saved ourselves some work by setting $u = \cot x^3$ at the very beginning. With

$$u = \cot x^3, \qquad du = -\csc^2 x^3 \cdot 3x^2\, dx$$

we have

$$x^2 \csc^2 x^3 \cot^4 x^3\, dx = \underbrace{\cot^4 x^3}_{u^4}\ \underbrace{\csc^2 x^3\, x^2\, dx}_{-\frac{1}{3}du} = -\tfrac{1}{3}u^4\, du$$

and

$$\int x^2 \csc^2 x^3 \cot^4 x^3\, dx = -\tfrac{1}{3}\int u^4\, du = -\tfrac{1}{15}u^5 + C = -\tfrac{1}{15}\cot^5 x^3 + C. \quad \square$$

Remark Thus far, all the integrals that we have calculated using a u-substitution can be calculated without substitution. All that is required is a good sense of the chain rule. For example:

$\int \sin x \cos x\, dx$ (Example 2). The cosine is the derivative of the sine. Thus

$$\int \sin x \cos x\, dx = \int \sin x \frac{d}{dx}(\sin x)\, dx = \tfrac{1}{2}\sin^2 x + C.$$

$\int x^2\sqrt{4 + x^3}\, dx$ (Example 4). The derivative of $4 + x^3$ is $3x^2$. Therefore

$$\frac{d}{dx}(4 + x^3) = 3x^2 \qquad \text{and} \qquad x^2 = \tfrac{1}{3}\frac{d}{dx}(4 + x^3).$$

Thus,

$$\int x^2\sqrt{4+x^3}\,dx = \tfrac{1}{3}\int (4+x^3)^{1/2}\frac{d}{dx}(4+x^3)\,dx$$
$$= \tfrac{1}{3}\cdot\tfrac{2}{3}(4+x^3)^{3/2} + C = \tfrac{2}{9}(4+x^3)^{3/2} + C.$$

$\int \sec^3 x \tan x\,dx$ (Example 5). Write the integrand as

$$\sec^2 x(\sec x \tan x) = \sec^2 x\frac{d}{dx}(\sec x).$$

Then

$$\int \sec^3 x \tan x\,dx = \int \sec^2 x\frac{d}{dx}(\sec x)\,dx = \tfrac{1}{3}\sec^3 x + C.$$

$\int x^2 \csc^2 x^3 \cot^4 x^3\,dx$ (Example 8). This integral may look complicated, but looked at in the right way, it becomes very simple. You know that the derivative of the cotangent is the negative of the cosecant squared. Therefore, by the chain rule,

$$\frac{d}{dx}(\cot x^3) = -\csc^2 x^3 \cdot 3x^2$$

and

$$x^2\csc^2 x^3 = -\frac{1}{3}\frac{d}{dx}(\cot x^3).$$

Thus

$$\int x^2\csc^2 x^3\cot^4 x^3\,dx = -\frac{1}{3}\int \cot^4 x^3\cdot\frac{d}{dx}(\cot x^3)\,dx = -\tfrac{1}{15}\cot^5 x^3 + C.$$

There is nothing wrong with calculating integrals by substitution. All we are saying is that, with some experience, you will be able to calculate many integrals without it. ❑

Substitution and Definite Integrals

In Example 7 we evaluated a definite integral by first calculating a corresponding indefinite integral. Then we chose one of the antiderivatives for the evaluation step. Here we present an alternative approach in which the limits of integration are changed along with the variable. This approach uses the *change of variables formula*:

(5.7.1)

$$\int_a^b f(g(x))g'(x)\,dx = \int_{g(a)}^{g(b)} f(u)\,du.$$

The formula holds provided that f and g' are both continuous. More precisely, g' must be continuous on $[a, b]$, and f must be continuous on the set of values taken on by g.

PROOF Let F be an antiderivative for f. Then $F' = f$ and

$$\int_a^b f(g(x))g'(x)\,dx = \int_a^b F'(g(x))g'(x)\,dx$$

$$= \Big[F(g(x))\Big]_a^b = F(g(b)) - F(g(a)) = \int_{g(a)}^{g(b)} f(u)\,du. \quad \square$$

We redo Example 7 using the change of variables formula.

Example 9 Evaluate

$$\int_0^2 (x^2 - 1)(x^3 - 3x + 2)^3\,dx.$$

SOLUTION As before, set

$$u = x^3 - 3x + 2, \qquad du = 3(x^2 - 1)\,dx.$$

Then

$$(x^2 - 1)(x^3 - 3x + 2)^3\,dx = \tfrac{1}{3}u^3\,du.$$

Now, at $x = 0$, $u = 2$. At $x = 2$, $u = 4$. Therefore,

$$\int_0^2 (x^2 - 1)(x^3 - 3x + 2)^3\,dx = \tfrac{1}{3}\int_2^4 u^3\,du = \Big[\tfrac{1}{12}u^4\Big]_2^4 = \tfrac{1}{12}(4)^4 - \tfrac{1}{12}(2)^4 = 20. \quad \square$$

Example 10 Evaluate

$$\int_0^{1/2} \cos^3 \pi x \sin \pi x\,dx.$$

SOLUTION Set

$$u = \cos \pi x, \qquad du = -\pi \sin \pi x\,dx.$$

Then

$$\cos^3 \pi x \sin \pi x\,dx = \underbrace{\cos^3 \pi x}_{u^3}\, \underbrace{\sin \pi x\,dx}_{-\frac{1}{\pi}du} = -\frac{1}{\pi}u^3\,du.$$

At $x = 0$, $u = 1$. At $x = 1/2$, $u = 0$. Therefore,

$$\int_0^{1/2} \cos^3 \pi x \sin \pi x\,dx = -\tfrac{1}{\pi}\int_1^0 u^3 du = \tfrac{1}{\pi}\int_0^1 u^3\,du = \tfrac{1}{\pi}\Big[\tfrac{1}{4}u^4\Big]_0^1 = \frac{1}{4\pi}. \quad \square$$

Up to this point our focus has been on substitutions connected with the chain rule. However, as we shall illustrate here and later in Chapter 8, "substitution" is a very general method of integration. Choosing the proper substitution is a matter of practice and experience. At this early stage, we will provide hints and guidance.

Example 11 Calculate

$$\int x(x-3)^5\,dx.$$

SOLUTION Set

$$u = x - 3, \qquad du = dx.$$

Then

$$x(x-3)^5\,dx = (u+3)u^5\,du = (u^6 + 3u^5)\,du$$

and

$$\int x(x-3)^5\,dx = \int (u^6 + 3u^5)\,du$$
$$= \tfrac{1}{7}u^7 + \tfrac{1}{2}u^6 + C = \tfrac{1}{7}(x-3)^7 + \tfrac{1}{2}(x-3)^6 + C. \quad \square$$

Example 12 Evalulate

$$\int_0^{\sqrt{3}} x^5\sqrt{x^2+1}\,dx.$$

SOLUTION Set

$$u = x^2 + 1, \qquad du = 2x\,dx.$$

Then

$$x^5\sqrt{x^2+1}\,dx = \underbrace{x^4}_{(u-1)^2}\,\underbrace{\sqrt{x^2+1}}_{\sqrt{u}}\,\underbrace{x\,dx}_{\frac{1}{2}du} = \tfrac{1}{2}(u-1)^2\sqrt{u}\,du.$$

At $x = 0$, $u = 1$. At $x = \sqrt{3}$, $u = 4$. Thus

$$\int_0^{\sqrt{3}} x^5\sqrt{x^2+1}\,dx = \tfrac{1}{2}\int_1^4 (u-1)^2\sqrt{u}\,du$$
$$= \tfrac{1}{2}\int_1^4 (u^{5/2} - 2u^{3/2} + u^{1/2})\,du$$
$$= \tfrac{1}{2}\left[\tfrac{2}{7}u^{7/2} - \tfrac{4}{5}u^{5/2} + \tfrac{2}{3}u^{3/2}\right]_1^4$$
$$= \left[u^{3/2}\left(\tfrac{1}{7}u^2 - \tfrac{2}{5}u + \tfrac{1}{3}\right)\right]_1^4 = \tfrac{848}{105}. \quad \square$$

EXERCISES 5.7

In Exercises 1–24, evaluate the integral by a u-substitution.

1. $\displaystyle\int \frac{dx}{(2-3x)^2}.$

2. $\displaystyle\int \frac{dx}{\sqrt{2x+1}}.$

3. $\displaystyle\int \sqrt{2x+1}\,dx$

4. $\displaystyle\int \sqrt{ax+b}\,dx.$

5. $\displaystyle\int (ax+b)^{3/4}\,dx.$

6. $\displaystyle\int 2ax(ax^2+b)^4\,dx.$

7. $\displaystyle\int \frac{t}{(4t^2+9)^2}\,dt.$

8. $\displaystyle\int \frac{3t}{(t^2+1)^2}\,dt.$

9. $\displaystyle\int t^2(5t^3+9)^4\,dt.$

10. $\displaystyle\int t(1+t^2)^3\,dt.$

11. $\int x^2(1+x^3)^{1/4}\,dx.$ **12.** $\int x^{n-1}\sqrt{a+bx^n}\,dx.$

13. $\int \frac{s}{(1+s^2)^3}\,ds.$ **14.** $\int \frac{2s}{\sqrt[3]{6-5s^2}}\,ds.$

15. $\int \frac{x}{\sqrt{x^2+1}}\,dx.$ **16.** $\int \frac{3ax^2-2bx}{\sqrt{ax^3-bx^2}}\,dx.$

17. $\int x^2(1-x^3)^{2/3}\,dx.$ **18.** $\int \frac{x^2}{(1-x^3)^{2/3}}\,dx.$

19. $\int 5x(x^2+1)^{-3}\,dx.$ **20.** $\int 2x^3(1-x^4)^{-1/4}\,dx.$

21. $\int x^{-3/4}(x^{1/4}+1)^{-2}\,dx.$ **22.** $\int \frac{4x+6}{\sqrt{x^2+3x+1}}\,dx.$

23. $\int \frac{b^3x^3}{\sqrt{1-a^4x^4}}\,dx.$ **24.** $\int \frac{x^{n-1}}{\sqrt{a+bx^n}}\,dx.$

In Exercises 25–32, evaluate the definite integral by a u-substitution.

25. $\int_0^1 x(x^2+1)^3\,dx.$ **26.** $\int_{-1}^0 3x^2(4+2x^3)^2\,dx.$

27. $\int_0^1 5x(1+x^2)^4\,dx.$ **28.** $\int_1^2 (6-x)^{-3}\,dx.$

29. $\int_{-1}^1 \frac{r}{(1+r^2)^4}\,dr.$ **30.** $\int_0^3 \frac{r}{\sqrt{r^2+16}}\,dr.$

31. $\int_0^a y\sqrt{a^2-y^2}\,dy.$ **32.** $\int_{-a}^0 y^2\left(1-\frac{y^3}{a^2}\right)^{-2}\,dy.$

In Exercises 33–44, evaluate the integral by a u-substitution.

33. $\int x\sqrt{x+1}\,dx.$ [set $u=x+1$]

34. $\int 2x\sqrt{x-1}\,dx.$ [set $u=x-1$]

35. $\int x\sqrt{2x-1}\,dx.$ **36.** $\int x^2\sqrt{x+1}\,dx.$

37. $\int y(y+1)^{12}\,dy.$ **38.** $\int t(2t+3)^8\,dt.$

39. $\int t^2(t-2)^{-5}\,dt.$ **40.** $\int x^2(2x-1)^{-2/3}\,dx.$

41. $\int_0^1 \frac{x+3}{\sqrt{x+1}}\,dx.$ **42.** $\int_0^1 \frac{x^2}{\sqrt{x+1}}\,dx.$

43. $\int_{-1}^0 x^3(x^2+1)^6\,dx.$ [set $u=x^2+1$]

44. $\int_1^{\sqrt{2}} x^3(x^2-1)^7\,dx$

In Exercises 45–64, calculate the indefinite integral.

45. $\int \cos(3x-1)\,dx.$ **46.** $\int \sin 2\pi x\,dx.$

47. $\int \csc^2 \pi x\,dx.$ **48.** $\int \sec 2x \tan 2x\,dx.$

49. $\int \sin(3-2x)\,dx.$ **50.** $\int \sin^2 x \cos x\,dx.$

51. $\int \cos^4 x \sin x\,dx.$ **52.** $\int x \sec^2 x^2\,dx.$

53. $\int x^{-1/2} \sin x^{1/2}\,dx.$

54. $\int \csc(1-2x)\cot(1-2x)\,dx.$

55. $\int \sqrt{1+\sin x}\cos x\,dx.$ **56.** $\int \frac{\sin x}{\sqrt{1+\cos x}}\,dx.$

57. $\int \frac{1}{\cos^2 x}\,dx.$ **58.** $\int (1+\tan^2 x)\sec^2 x\,dx.$

59. $\int x \sin^3 x^2 \cos x^2\,dx.$ **60.** $\int \sqrt{1+\cot x}\csc^2 x\,dx.$

61. $\int (1+\cot^2 x)\csc^2 x\,dx.$ **62.** $\int \frac{1}{\sin^2 x}\,dx.$

63. $\int \frac{\sec^2 x}{\sqrt{1+\tan x}}\,dx.$ **64.** $\int x^2 \sin(4x^3-7)\,dx.$

Evaluate the integral in Exercises 65–71.

65. $\int_{-\pi}^{\pi} \sin x\,dx.$ **66.** $\int_{-\pi/3}^{\pi/3} \sec x \tan x\,dx.$

67. $\int_{1/4}^{1/3} \sec^2 \pi x\,dx.$ **68.** $\int_0^1 \cos^2\frac{\pi}{2}x \sin\frac{\pi}{2}x\,dx.$

69. $\int_0^{\pi/2} \sin^3 x \cos x\,dx.$ **70.** $\int_0^{\pi} x \cos x^2\,dx$

71. $\int_{\pi/6}^{\pi/4} \csc x \cot x\,dx.$

72. Derive the formula

$$\int \cos^2 x\,dx = \tfrac{1}{2}x + \tfrac{1}{4}\sin 2x + C.$$

HINT: Recall the half-angle formula $\cos^2\theta = \frac{1}{2}(1+\cos 2\theta)$.

73. Derive the formula

$$\int \sin^2 x\,dx = \tfrac{1}{2}x - \tfrac{1}{4}\sin 2x + C.$$

HINT: Use the half-angle formula for the sine function.

Evaluate the integrals in Exercises 74–77.

74. $\int \sin^2 3x\,dx.$ **75.** $\int \cos^2 5x\,dx.$

76. $\int_0^{2\pi} \sin^2 x\,dx.$ **77.** $\int_0^{\pi/2} \cos^2 2x\,dx.$

In Exercises 78–81, find the area bounded by the following curves.

78. $y = \cos \pi x, \quad y = \sin \pi x, \quad x = 0, \quad x = \frac{1}{4}$.

79. $y = \cos^2 \pi x, \quad y = \sin^2 \pi x, \quad x = 0, \quad x = \frac{1}{4}$.
HINT: Use Exercises 72 and 73.

80. $y = \cos^2 \pi x, \quad y = -\sin^2 \pi x, \quad x = 0, \quad x = \frac{1}{4}$.

81. $y = \csc^2 \pi x, \quad y = \sec^2 \pi x, \quad x = \frac{1}{6}, \quad x = \frac{1}{4}$.

82. In Example 2 we found that

$$\int \sin x \cos x \, dx = \tfrac{1}{2} \sin^2 x + C$$

by setting $u = \sin x$. Calculate the integral by setting $u = \cos x$ and then reconcile the two answers.

83. Calculate

$$\int \sec^2 x \tan x \, dx$$

(a) Setting $u = \sec x$. (b) Setting $u = \tan x$.
(c) Reconcile your answers to (a) and (b).

84. (*The area of a circular region*) The circle $x^2 + y^2 = r^2$ encloses a circular disc of radius r. Justify the familiar formula $A = \pi r^2$ by integration. HINT: The quarter disc in the first quadrant is the region below the curve $y = \sqrt{r^2 - x^2}$, $x \in [0, r]$. Therefore

$$A = 4 \int_0^r \sqrt{r^2 - x^2} \, dx.$$

Evaluate the integral by setting $x = r \sin u$, $dx = r \cos u \, du$ and use Exercise 72.

85. Find the area enclosed by the ellipse $b^2x^2 + a^2y^2 = a^2b^2$.

■ 5.8 SOME FURTHER PROPERTIES OF THE DEFINITE INTEGRAL

In this section we feature some important general properties of the integral. The proofs are left mostly to you. You can assume throughout that the functions involved are continuous and that $a < b$.

I. The integral of a nonnegative function is nonnegative:

(5.8.1) if $f(x) \geq 0$ for all $x \in [a, b]$, then $\int_a^b f(x) \, dx \geq 0$.

The integral of a positive function is positive:

(5.8.2) if $f(x) > 0$ for all $x \in [a, b]$, then $\int_a^b f(x) \, dx > 0$.

The next property is an immediate consequence of Property I and linearity (5.4.6).

II. The integral is order-preserving:

(5.8.3) if $f(x) \leq g(x)$ for all $x \in [a, b]$, then $\int_a^b f(x) \, dx \leq \int_a^b g(x) \, dx$

and

(5.8.4)
$$\text{if}\quad f(x) < g(x) \text{ for all } x \in [a, b], \qquad \text{then} \quad \int_a^b f(x)\,dx < \int_a^b g(x)\,dx.$$

PROOF OF (5.8.3) Since $f(x) \leq g(x)$, it follows that $g(x) - f(x) \geq 0$. Therefore, using the linearity of the integral and (5.8.1), we have

$$\int_a^b g(x)\,dx - \int_a^b f(x)\,dx = \int_a^b [g(x) - f(x)]\,dx \geq 0.$$

and so

$$\int_a^b f(x)\,dx \leq \int_a^b g(x)\,dx.$$

The proof of (5.8.4) follows in exactly the same manner. ❑

III. Just as the absolute value of a sum is less than or equal to the sum of the absolute values,

$$|x_1 + x_2 + \cdots + x_n| \leq |x_1| + |x_2| + \cdots + |x_n|,$$

the absolute value of an integral is less than or equal to the integral of the absolute value:

(5.8.5)
$$\left| \int_a^b f(x)\,dx \right| \leq \int_a^b |f(x)|\,dx.$$

HINT FOR PROOF OF (5.8.5) Show that

$$\int_a^b f(x)\,dx \qquad \text{and} \qquad -\int_a^b f(x)\,dx$$

are both less than or equal to

$$\int_a^b |f(x)|\,dx. \quad ❑$$

You should already be familiar with the next property.

IV. If m is the minimum value of f on $[a, b]$ and M is the maximum value, then

(5.8.6)
$$m(b - a) \leq \int_a^b f(x)\,dx \leq M(b - a).$$

Recall Theorem (5.3.5), which states that

$$\frac{d}{dx}\left(\int_a^x f(t)\,dt\right) = f(x).$$

Our next result is a generalization of this theorem. Its importance will become apparent in Chapter 7.

V. If u is a differentiable function of x and f is continuous, then

(5.8.7)
$$\frac{d}{dx}\left(\int_a^u f(t)\,dt\right) = f(u)\,\frac{du}{dx}.$$

PROOF The function

$$F(u) = \int_a^u f(t)\,dt$$

is differentiable with respect to u and

$$\frac{d}{du}[F(u)] = f(u). \qquad \text{(Theorem 5.3.5)}$$

Thus

$$\frac{d}{dx}\left(\int_a^u f(t)\,dt\right) = \frac{d}{dx}[F(u)] = \frac{d}{du}[F(u)]\frac{du}{dx} = f(u)\frac{du}{dx}. \quad \square$$

chain rule (applied at the second equality)

Example 1 Find

$$\frac{d}{dx}\left(\int_0^{x^3} \frac{dt}{1+t}\right).$$

SOLUTION At this stage you would be hard put to carry out the integration; it requires the natural logarithm function, which is not introduced in this text until Chapter 7. But, for our purposes, that doesn't matter. We have $f(t) = 1/(1+t)$ and $u(x) = x^3$. By (5.8.7), you know that

$$\frac{d}{dx}\left(\int_0^{x^3} \frac{dt}{1+t}\right) = f(u)\frac{du}{dx} = \frac{1}{1+x^3}3x^2 = \frac{3x^2}{1+x^3}$$

without carrying out the integration. $\square$

Example 2 Find

$$\frac{d}{dx}\left(\int_x^{2x} \frac{dt}{1+t^2}\right).$$

SOLUTION The idea is to express the integral in terms of integrals that have constant lower limits of integration. Once we have done that, we can apply (5.8.7).

In this case, we choose 0 as a convenient lower limit. Then, by the additivity of the integral,

$$\int_0^x \frac{dt}{1+t^2} + \int_x^{2x} \frac{dt}{1+t^2} = \int_0^{2x} \frac{dt}{1+t^2}.$$

Thus

$$\int_x^{2x} \frac{dt}{1+t^2} = \int_0^{2x} \frac{dt}{1+t^2} - \int_0^x \frac{dt}{1+t^2}.$$

Differentiation gives

$$\frac{d}{dx}\left(\int_x^{2x} \frac{dt}{1+t^2}\right) = \frac{d}{dx}\left(\int_0^{2x} \frac{dt}{1+t^2}\right) - \frac{d}{dx}\left(\int_0^x \frac{dt}{1+t^2}\right)$$

$$= \frac{1}{1+(2x)^2}(2) - \frac{1}{1+x^2}(1) = \frac{2}{1+4x^2} - \frac{1}{1+x^2}. \quad \square$$

(5.8.7) ⮥

VI. Now a few words about the role of symmetry in integration. Suppose that f is a continuous function defined on some interval of the form $[-a, a]$, some closed interval that is symmetric about the origin.

(5.8.8)

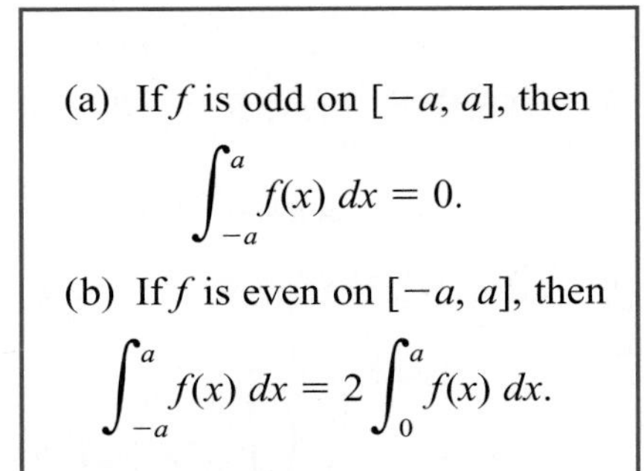
(a) If f is odd on $[-a, a]$, then

$$\int_{-a}^{a} f(x)\,dx = 0.$$

(b) If f is even on $[-a, a]$, then

$$\int_{-a}^{a} f(x)\,dx = 2\int_0^a f(x)\,dx.$$

f odd

Figure 5.8.1

These assertions can be verified by a simple change of variables (Exercise 40). Here we look at these assertions from the standpoint of area, referring to Figures 5.8.1 and 5.8.2.

For the odd function,

$$\int_{-a}^{a} f(x)\,dx = \int_{-a}^{0} f(x)\,dx + \int_0^a f(x)\,dx = \text{area of } \Omega_1 - \text{area of } \Omega_2 = 0.$$

For the even function,

$$\int_{-a}^{a} f(x)\,dx = \text{area of } \Omega_1 + \text{area of } \Omega_2 = 2(\text{area of } \Omega_2) = 2\int_0^a f(x)\,dx.$$

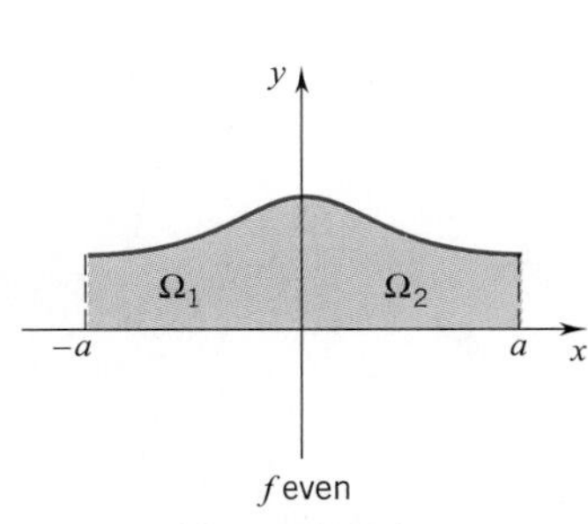

f even

Figure 5.8.2

Suppose we were asked to evaluate

$$\int_{-\pi}^{\pi} (\sin x - x\cos x)^3\,dx.$$

A laborious calculation would show that this integral is zero. We don't have to carry out that calculation. The integrand is an odd function, and thus we can tell immediately that the integral is zero.

EXERCISES 5.8

Assume: f and g continuous, $a < b$, and

$$\int_a^b f(x)\,dx > \int_a^b g(x)\,dx.$$

Answer questions 1–6, giving supporting reasons.

1. Does it necessarily follow that $\int_a^b [f(x) - g(x)]\,dx > 0$?
2. Does it necessarily follow that $f(x) > g(x)$ for all $x \in [a, b]$?
3. Does it necessarily follow that $f(x) > g(x)$ for at least some $x \in [a, b]$?
4. Does it necessarily follow that $\left|\int_a^b f(x)\,dx\right| > \left|\int_a^b g(x)\,dx\right|$?
5. Does it necessarily follow that $\int_a^b |f(x)|\,dx > \int_a^b |g(x)|\,dx$?
6. Does it necessarily follow that $\int_a^b |f(x)|\,dx > \int_a^b g(x)\,dx$?

Assume f continuous, $a < b$, and

$$\int_a^b f(x)\,dx = 0.$$

Answer questions 7–15, giving supporting reasons.

7. Does it necessarily follow that $f(x) = 0$ for all $x \in [a, b]$?
8. Does it necessarily follow that $f(x) = 0$ for at least some $x \in [a, b]$?
9. Does it necessarily follow that $\int_a^b |f(x)|\,dx = 0$?
10. Does it necessarily follow that $\left|\int_a^b f(x)\,dx\right| = 0$?
11. Must all upper sums $U_f(P)$ be nonnegative?
12. Must all upper sums $U_f(P)$ be positive?
13. Can a lower sum, $L_f(P)$, be positive?
14. Does it necessarily follow that $\int_a^b [f(x)]^2\,dx = 0$?
15. Does it necessarily follow that $\int_a^b [f(x) + 1]\,dx = b - a$?
16. Let u be a differentiable function of x and let f be continuous. Derive a formula for

$$\frac{d}{dx}\left(\int_u^b f(t)\,dt\right).$$

In Exercises 17–23, calculate the derivative.

17. $\dfrac{d}{dx}\left(\displaystyle\int_0^{1+x^2} \frac{dt}{\sqrt{2t+5}}\right)$.
18. $\dfrac{d}{dx}\left(\displaystyle\int_1^{x^2} \frac{dt}{t}\right)$.
19. $\dfrac{d}{dx}\left(\displaystyle\int_x^a f(t)\,dt\right)$.
20. $\dfrac{d}{dx}\left(\displaystyle\int_0^{x^3} \frac{dt}{\sqrt{1+t^2}}\right)$.
21. $\dfrac{d}{dx}\left(\displaystyle\int_{x^2}^3 \frac{\sin t}{t}\,dt\right)$.
22. $\dfrac{d}{dx}\left(\displaystyle\int_{\tan x}^4 \sin(t^2)\,dt\right)$.
23. $\dfrac{d}{dx}\left(\displaystyle\int_1^{\sqrt{x}} \frac{t^2}{1+t^2}\,dt\right)$.
24. Show that, if u and v are differentiable functions of x and f is continuous, then

$$\frac{d}{dx}\left(\int_u^v f(t)\,dt\right) = f(v)\frac{dv}{dx} - f(u)\frac{du}{dx}.$$

HINT: Take a number a from the domain of f. Express the integral as the difference of two integrals, each with lower limit a.

In Exercises 25–32, calculate the derivative using Exercise 24.

25. $\dfrac{d}{dx}\left(\displaystyle\int_x^{x^2} \frac{dt}{t}\right)$.
26. $\dfrac{d}{dx}\left(\displaystyle\int_{1-x}^{1+x} \frac{t-1}{t}\,dt\right)$.
27. $\dfrac{d}{dx}\left(\displaystyle\int_{x^{1/3}}^{2+3x} \frac{dt}{1+t^{3/2}}\right)$.
28. $\dfrac{d}{dx}\left(\displaystyle\int_{\sqrt{x}}^{x^2+x} \frac{dt}{2+\sqrt{t}}\right)$.
29. $\dfrac{d}{dx}\left(\displaystyle\int_{\sqrt{x}}^{x^2} t\cos t^2\,dt\right)$.
30. $\dfrac{d}{dx}\left(\displaystyle\int_{3x}^{1/x} \cos 2t\,dt\right)$.
31. $\dfrac{d}{dx}\left(\displaystyle\int_{\tan x}^{2x} t\sqrt{1+t^2}\,dt\right)$.
32. $\dfrac{d}{dx}\left(\displaystyle\int_{x^2}^{\sin x} \frac{t}{4+t^2}\,dt\right)$.
33. Verify (5.8.1): (a) by considering lower sums $L_f(P)$; (b) by using an antiderivative.

34. Verify (5.8.4). **35.** Verify (5.8.5).

36. (*Important*) Prove that, if f is continuous on $[a, b]$ and

$$\int_a^b |f(x)|\, dx = 0,$$

then $f(x) = 0$ for all x in $[a, b]$. HINT: Exercise 48, Section 2.4.

37. Find $H'(2)$ given that

$$H(x) = \int_{2x}^{x^3-4} \frac{t}{1+\sqrt{t}}\, dt.$$

38. Find $H'(3)$ given that

$$H(x) = \frac{1}{x}\int_3^x [2t - 3H'(t)]\, dt.$$

39. (a) Let f be continuous on $[-a, 0]$. Use a change of variable to prove that

$$\int_{-a}^0 f(x)\, dx = \int_0^a f(-x)\, dx.$$

(b) Let f be continuous on $[-a, a]$. Prove that

$$\int_{-a}^a f(x)\, dx = \int_0^a [f(x) + f(-x)]\, dx.$$

40. Let f be a continuous function on $[-a, a]$. Use Exercise 39 to prove that:

(a) $\displaystyle\int_{-a}^a f(x)\, dx = 0$ if f is odd.

(b) $\displaystyle\int_{-a}^a f(x)\, dx = 2\int_0^a f(x)\, dx$ if f is even.

In Exercises 41–44, use Exercise 40 to evaluate the given integral.

41. $\displaystyle\int_{-\pi/4}^{\pi/4} (x + \sin 2x)\, dx.$ **42.** $\displaystyle\int_{-3}^{3} \frac{t^3}{1+t^2}\, dt.$

43. $\displaystyle\int_{-\pi/3}^{\pi/3} (1 + x^2 - \cos x)\, dx.$

44. $\displaystyle\int_{-\pi/4}^{\pi/4} (x^2 - 2x + \sin x + \cos 2x)\, dx.$

■ 5.9 MEAN-VALUE THEOREMS FOR INTEGRALS; AVERAGE VALUES

We begin with a result that we asked you to prove in Exercise 25, Section 5.3.

THEOREM 5.9.1 THE FIRST MEAN-VALUE THEOREM FOR INTEGRALS

If f is continuous on $[a, b]$, then there is at least one number c in (a, b) for which

$$\int_a^b f(x)\, dx = f(c)(b - a).$$

This number $f(c)$ is called *the average* (or *mean*) *value of f on* $[a, b]$.

We then have the following identity:

(5.9.2) $$\int_a^b f(x)\, dx = (\text{the average value of } f \text{ on } [a, b]) \cdot (b - a).$$

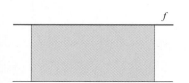

area = (the constant value of f)·$(b - a)$

Figure 5.9.1

This identity provides a powerful and intuitive way of viewing the definite integral. Think for a moment about area. If f is constant and positive on $[a, b]$, then Ω, the region below the graph, is a rectangle. Its area is given by the formula

$$\text{Area of } \Omega = (\text{the constant value of } f \text{ on } [a, b]) \cdot (b - a). \quad \text{(Figure 5.9.1)}$$

If f is now allowed to vary continuously on $[a, b]$, then we have

$$\text{Area of } \Omega = \int_a^b f(x)\,dx,$$

and the area formula reads

Area of Ω = *(the average value of* f *on* $[a, b]$*)* $\cdot$ $(b - a)$. (Figure 5.9.2)

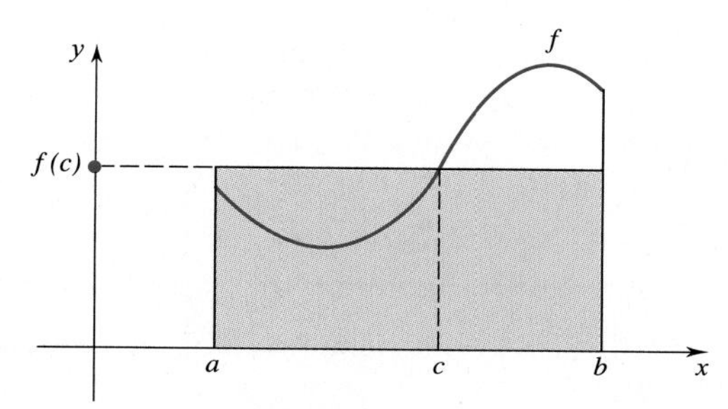

Figure 5.9.2

Think now about motion. If an object moves along a line with constant speed $|v|$ during the time interval $[a, b]$, then

$$\text{The distance traveled} = (\text{the constant value of } |v| \text{ on } [a, b]) \cdot (b - a).$$

If the speed $|v|$ varies, then we have

$$\text{Distance traveled} = \int_a^b |v(t)|\,dt$$

and the formula reads

Distance traveled = (the average value of $|v|$ *on* $[a, b]$*)* $\cdot$ $(b - a)$.

We take an interval $[a, b]$ and calculate the average value of the simplest possible functions on that interval. For convenience, let f_{avg} denote the average value of f on $[a, b]$. Solving identity (5.9.2) for f_{avg}, we have

$$f_{\text{avg}} = \frac{1}{b-a}\int_a^b f(x)\,dx.$$

The average value of a constant function $f(x) = k$ is, of course, k:

$$f_{\text{avg}} = \frac{1}{b-a}\int_a^b k\,dx = \frac{k}{b-a}\Big[\,x\,\Big]_a^b = \frac{k}{b-a}(b-a) = k.$$

The average value of $f(x) = x$ on $[a, b]$ is $\frac{1}{2}(b + a)$, the midpoint of the interval:

$$f_{\text{avg}} = \frac{1}{b-a}\int_a^b x\,dx = \frac{1}{b-a}\left[\frac{x^2}{2}\right]_a^b = \frac{1}{b-a}\left(\frac{b^2-a^2}{2}\right) = \frac{1}{2}(b+a).$$

What is the average value of $f(x) = x^2$?

$$\begin{aligned} f_{\text{avg}} &= \frac{1}{b-a}\int_a^b x^2\,dx = \frac{1}{b-a}\left[\frac{x^3}{3}\right]_a^b = \frac{1}{b-a}\left(\frac{b^3-a^3}{3}\right) \\ &= \frac{1}{b-a}\left[\frac{(b^2+ab+a^2)(b-a)}{3}\right] = \frac{1}{3}(b^2+ab+a^2). \end{aligned}$$

Thus, the average value of $f(x) = x^2$ on $[a, b]$ is $\frac{1}{3}(b^2 + ab + a^2)$ (see Example 6 in Section 5.2.). On $[1, 3]$ the values of x^2 range from 1 to 9; the average value is $\frac{13}{3}$.

There is an extension of Theorem 5.9.1 that, as you will see, is useful in applications.

> **THEOREM 5.9.3 THE SECOND MEAN-VALUE THEOREM FOR INTEGRALS**
>
> If f and g are continuous on $[a, b]$ and g is nonnegative, then there is a number c in (a, b) for which
>
> $$\int_a^b f(x)g(x)\,dx = f(c)\int_a^b g(x)\,dx.$$
>
> This number $f(c)$ is called *the g-weighted average of f on* $[a, b]$.

We will prove this theorem (and thereby have a proof of Theorem 5.9.1) at the end of the section. First a brief excursion into physics.

The Mass of a Rod Imagine a thin rod (a straight material wire of negligible thickness) lying on the x-axis from $x = a$ to $x = b$. If the *mass density* of the rod (the mass per unit length) is constant, then the total mass M of the rod is simply the density λ times the length of the rod: $M = \lambda(b - a)$.† If the density λ varies continuously from point to point, say $\lambda = \lambda(x)$, then the mass of the rod is the average density of the rod times the length of the rod:

$$M = (\text{average density}) \cdot (\text{length}).$$

This is an integral:

(5.9.4)
$$M = \int_a^b \lambda(x)\,dx.$$

† The symbol λ is the Greek letter "lambda."

The Center of Mass of a Rod Continue with that same rod. If the rod is homogeneous (constant density), then the center of mass x_M of the rod is simply the midpoint of the rod:

$$x_M = \tfrac{1}{2}(a + b). \qquad \text{(the average of } x \text{ from } a \text{ to } b\text{)}$$

If the rod is not homogeneous, the center of mass is still an average, but now a weighted average, *the density-weighted average of x from a to b*; namely, x_M is the point for which

$$x_M \int_a^b \lambda(x)\, dx = \int_a^b x\lambda(x)\, dx.$$

Since the integral on the left is simply M, we have

(5.9.5)

$$\boxed{x_M M = \int_a^b x\lambda(x)\, dx.}$$

Example 1 A rod of length L is placed on the x-axis from $x = 0$ to $x = L$. Find the mass of the rod and the center of mass if the density of the rod varies directly as the distance from the $x = 0$ endpoint of the rod.

SOLUTION Here $\lambda(x) = kx$ where k is some positive constant. Therefore

$$M = \int_0^L kx\, dx = \left[\tfrac{1}{2}kx^2\right]_0^L = \tfrac{1}{2}kL^2$$

and

$$x_M M = \int_0^L x(kx)\, dx = \int_0^L kx^2\, dx = \left[\tfrac{1}{3}kx^3\right]_0^L = \tfrac{1}{3}kL^3.$$

Division by M gives x_M gives $x_M = \tfrac{2}{3}L$.

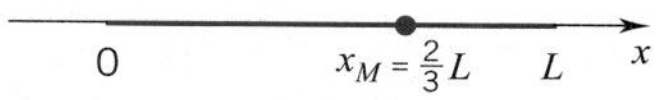

In this instance the center of mass is to the right of the midpoint. This makes sense. After all, the density increases from left to right. Thus there is more mass to the right of the midpoint than there is to the left of it. ❑

We go back now to Theorem 5.9.3 and prove it. [There is no reason to worry about Theorem 5.9.1. That is just Theorem 5.9.3 with $g(x)$ identically 1.]

PROOF OF THEOREM 5.9.3 Since f is continuous on $[a, b]$, f takes on a minimum value m on $[a, b]$ and a maximum value M. Since g is nonnegative on $[a, b]$,

$$mg(x) \le f(x)g(x) \le Mg(x) \qquad \text{for all } x \text{ in } [a, b].$$

Therefore

$$\int_a^b mg(x)\, dx \le \int_a^b f(x)g(x)\, dx \le \int_a^b Mg(x)\, dx$$

and

$$m\int_a^b g(x)\, dx \le \int_a^b f(x)g(x)\, dx \le M\int_a^b g(x)\, dx.$$

We know that $\int_a^b g(x)\,dx \geq 0$. If $\int_a^b g(x)\,dx = 0$, then $\int_a^b f(x)g(x)\,dx = 0$ and the theorem holds for all choices of c in (a, b). If $\int_a^b g(x)\,dx > 0$, then

$$m \leq \frac{\int_a^b f(x)g(x)\,dx}{\int_a^b g(x)\,dx} \leq M$$

and by the intermediate-value theorem (Theorem 2.6.1) there exists c in (a, b) for which

$$f(c) = \frac{\int_a^b f(x)g(x)\,dx}{\int_a^b g(x)\,dx}.$$

Obviously then

$$f(c)\int_a^b g(x)\,dx = \int_a^b f(x)g(x)\,dx. \quad \square$$

EXERCISES 5.9

In Exercises 1–12, determine the average value on the indicated interval and find a point in this interval at which the function takes on this average value.

1. $f(x) = mx + b, \quad x \in [0, c]$.

2. $f(x) = x^2, \quad x \in [-1, 1]$.

3. $f(x) = x^3, \quad x \in [-1, 1]$.

4. $f(x) = x^{-2}, \quad x \in [1, 4]$.

5. $f(x) = |x|, \quad x \in [-2, 2]$.

6. $f(x) = x^{1/3}, \quad x \in [-8, 8]$.

7. $f(x) = 2x - x^2, \quad x \in [0, 2]$.

8. $f(x) = 3 - 2x, \quad x \in [0, 3]$.

9. $f(x) = \sqrt{x}, \quad x \in [0, 9]$.

10. $f(x) = 4 - x^2, \quad x \in [-2, 2]$.

11. $f(x) = \sin x, \quad x \in [0, 2\pi]$.

12. $f(x) = \cos x, \quad x \in [0, \pi]$.

13. Let $f(x) = \sqrt{x}$ for $x \in [1, 9]$.
(a) Find the average value of f on this interval.
(b) Estimate with three decimal place accuracy a value c in the interval at which f takes on its average value.
(c) Use a graphing utility to illustrate your results with a figure similar to Figure 5.9.2.

14. Let $f(x) = x^3 - x + 1$ for $x \in [-1, 2]$.
(a) Find the average value of f on this interval.
(b) Estimate with three decimal place accuracy a value of c in the interval at which f takes on its average value.
(c) Use a graphing utility to illustrate your results with a figure similar to Figure 5.9.2.

15. Find $f(x) = \sin x$ for $x \in [0, \pi]$.
(a) Find the average value of f on this interval.
(b) Estimate with three decimal place accuracy the smallest value on the interval at which the average value occurs.
(c) Use a graphing utility to illustrate your results with a figure similar to Figure 5.9.2.

16. Let $f(x) = 2\cos 2x$ for $x \in [-\pi/4, \pi/6]$
(a) Find the average value of f on this interval.
(b) Estimate with three decimal place accuracy the smallest value on the interval at which the average value occurs.
(c) Use a graphing utility to illustrate your results with a figure similar to Figure 5.9.2.

17. Solve the following equation for A:

$$\int_a^b [f(x) - A]\,dx = 0.$$

18. Given that f is continuous on $[a, b]$, compare

$$f(b)(b-a) \quad \text{and} \quad \int_a^b f(x)\,dx$$

if f is: (a) constant on $[a, b]$; (b) increasing on $[a, b]$; (c) decreasing on $[a, b]$.

19. In Chapter 3, we viewed $[f(b) - f(a)]/(b - a)$ as the average rate of change of f on $[a, b]$ and $f'(t)$ as the instantaneous rate of change at time t. If our new sense of average is to be consistent with the old one, we must have

$$\frac{f(b) - f(a)}{b - a} = \text{average of } f' \text{ on } [a, b].$$

Prove that this is the case.

20. Determine whether the assertion is true or false.

(a) $$\begin{pmatrix}\text{the average of } f+g\\ \text{on } [a, b]\end{pmatrix} = \begin{pmatrix}\text{the average of } f\\ \text{on } [a, b]\end{pmatrix} + \begin{pmatrix}\text{the average of } g\\ \text{on } [a, b]\end{pmatrix}.$$

(b) $$\begin{pmatrix}\text{the average of } \alpha f\\ \text{on } [a, b]\end{pmatrix} = \alpha\begin{pmatrix}\text{the average of } f\\ \text{on } [a, b]\end{pmatrix}, \quad \alpha \text{ a constant.}$$

(c) $$\begin{pmatrix}\text{the average of } fg\\ \text{on } [a, b]\end{pmatrix} = \begin{pmatrix}\text{the average of } f\\ \text{on } [a, b]\end{pmatrix}\begin{pmatrix}\text{the average of } g\\ \text{on } [a, b]\end{pmatrix}.$$

21. Find the average distance of the parabolic arc

$$y = x^2, \quad x \in [0, \sqrt{3}]$$

from: (a) the x-axis; (b) the y-axis; (c) the origin.

22. Find the average distance of the line segment

$$y = mx, \qquad x \in [0, 1]$$

from: (a) the x-axis; (b) the y-axis; (c) the origin.

23. A stone falls from rest in a vacuum for t seconds (see Section 3.4). (a) Compare its terminal velocity to its average velocity; (b) compare its average velocity during the first $\frac{1}{2}t$ seconds to its average velocity during the next $\frac{1}{2}t$ seconds.

24. Let f be continuous. Show that, if f is an odd function, then its average value on every interval of the form $[-a, a]$ is zero.

25. An object starts from rest at the point x_0 and moves along the x-axis with constant acceleration a.
(a) Derive formulas for the velocity and position of the object at any time $t \geq 0$.
(b) Show that the average velocity over any time interval $[t_1, t_2]$ is the average of the initial and final velocities on that interval.

26. A rod 6 meters long is placed on the x-axis from $x = 0$ to $x = 6$. The mass density is $12/\sqrt{x+1}$ kilograms per meter.
(a) Find the mass of the rod and the center of mass.
(b) Find the average mass density of the rod.

27. A rod of length L is placed on the x-axis from $x = 0$ to $x = L$. Find the mass of the rod and the center of mass if the mass density of the rod varies directly: (a) as the square root of the distance from $x = 0$; (b) as the square of the distance from $x = L$.

28. A rod of varying density, mass M, and center of mass x_M extends from $x = a$ to $x = b$. A partition $P = \{x_0, x_1, \ldots, x_n\}$ of $[a, b]$ decomposes the rod into n pieces in the obvious way. Show that, if the n pieces have masses $M_1, M_2, \ldots, M_n$ and centers of mass x_{M_1}, $x_{M_2}, \ldots, x_{M_n}$, then

$$x_M M = x_{M_1}M_1 + x_{M_2}M_2 + \cdots + x_{M_n}M_n.$$

29. A rod that has mass M and extends from $x = 0$ to $x = L$ consists of two pieces with masses M_1, M_2. Given that the center of mass of the entire rod is at $x = \frac{1}{4}L$ and the center of mass of the first piece is at $x = \frac{1}{8}L$, determine the center of mass of the second piece.

30. A rod that has mass M and extends from $x = 0$ to $x = L$ consists of two pieces. Find the mass of each piece given that the center of mass of the entire rod is at $x = \frac{2}{3}L$, the center of mass of the first piece is at $x = \frac{1}{4}L$, and the center of mass of the second piece is at $x = \frac{7}{8}L$.

31. Prove Theorem 5.9.1 by applying the mean-value theorem of differential calculus to the function

$$F(x) = \int_a^x f(t)\,dt, \qquad x \in [a, b].$$

(This exercise was given before.)

32. Let f be continuous on $[a, b]$. Let $a < c < b$. Prove that

$$f(c) = \lim_{h \to 0^+} (\text{average value of } f \text{ on } [c-h, c+h]).$$

33. Prove that two distinct continuous functions cannot have the same average on every interval.

■ 5.10 THE INTEGRAL AS THE LIMIT OF RIEMANN SUMS

Let f be some function continuous on a closed interval $[a, b]$. With our approach to integration (at this point we ask you to review Section 5.2) the definite integral

$$\int_a^b f(x)\,dx$$

is the unique number that satisfies the inequality

$$L_f(P) \le \int_a^b f(x)\,dx \le U_f(P)$$

for all partitions P of $[a, b]$. This method of obtaining the definite integral by *squeezing* toward it with upper and lower sums is called the *Darboux method.*†

There is another way to obtain the integral that is frequently used. Take a partition $P = \{x_0, x_1, \ldots, x_n\}$ of $[a, b]$. Then P breaks up $[a, b]$ into n subintervals

$$[x_0, x_1], [x_1, x_2], \ldots, [x_{n-1}, x_n]$$

of lengths

$$\Delta x_1, \Delta x_2, \ldots, \Delta x_n.$$

Now pick a point x_1^* from $[x_0, x_1]$ and form the product $f(x_1^*)\,\Delta x_1$; pick a point x_2^* from $[x_1, x_2]$ and form the product $f(x_2^*)\,\Delta x_2$; go on in this manner until you have formed the products

$$f(x_1^*)\,\Delta x_1, f(x_2^*)\,\Delta x_2, \ldots, f(x_n^*)\,\Delta x_n.$$

The sum of these products

$$S^*(P) = f(x_1^*)\,\Delta x_1 + f(x_2^*)\,\Delta x_2 + \cdots + f(x_n^*)\,\Delta x_n$$

is called a *Riemann sum.*†† As an exercise, you are asked to show that if P is any partition of the interval $[a, b]$ and $S^*(P)$ is any corresponding Riemann sum, then

$$L_f(P) \le S^*(P) \le U_f(P).$$

Example 1 Let $f(x) = 1 - x^2$, $x \in [-1, 2]$. Take $P = \{-1, 0, \frac{1}{2}, \frac{3}{2}, 2\}$ and let $x_1^* = -\frac{1}{2}$, $x_2^* = \frac{1}{2}$, $x_3^* = \frac{5}{4}$, $x_4^* = \frac{7}{4}$. See Figure 5.10.1. Then $\Delta x_1 = 1$, $\Delta x_2 = \frac{1}{2}$, $\Delta x_3 = 1$, $\Delta x_4 = \frac{1}{2}$, and

$$\begin{aligned} S^*(P) &= f(-\tfrac{1}{2}) \cdot 1 + f(\tfrac{1}{2}) \cdot \tfrac{1}{2} + f(\tfrac{5}{4}) \cdot 1 + f(\tfrac{7}{4}) \cdot \tfrac{1}{2} \\ &= \tfrac{3}{4}(1) + \tfrac{3}{4}(\tfrac{1}{2}) + (-\tfrac{9}{16})(1) + (-\tfrac{33}{16})(\tfrac{1}{2}) \\ &= -\tfrac{15}{32}. \end{aligned}$$

You can verify that

$$L_f(P) = f(-1) \cdot 1 + f(\tfrac{1}{2}) \cdot \tfrac{1}{2} + f(\tfrac{3}{2}) \cdot 1 + f(2) \cdot \tfrac{1}{2} = \tfrac{-19}{8}$$

and

$$U_f(P) = f(0) \cdot 1 + f(0) \cdot \tfrac{1}{2} + f(\tfrac{1}{2}) \cdot 1 + f(\tfrac{3}{2}) \cdot \tfrac{1}{2} = \tfrac{13}{8}. \quad \square$$

Let f be continuous on the interval $[a, b]$. The definite integral of f can be viewed as the *limit* of Riemann sums in the following sense. For any partition P of $[a, b]$, define $\|P\|$, the *norm* of P, by setting

$$\|P\| = \max \Delta x_i, \qquad i = 1, 2, \ldots, n.$$

† After the French mathematician J.-G. Darboux (1842–1917).

†† After the German mathematician G. F. B. Riemann (1826–1866).

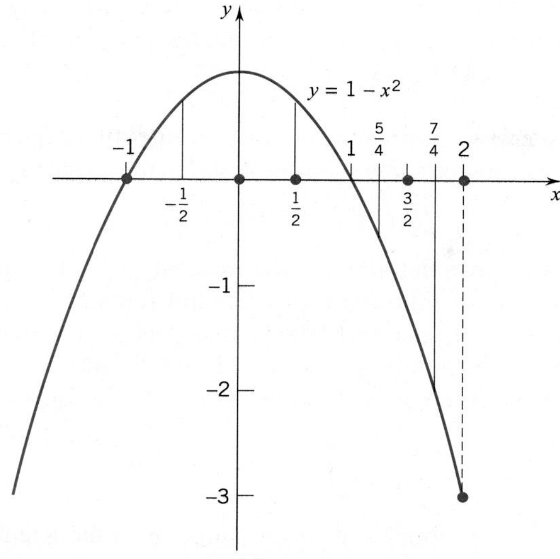

Figure 5.10.1

Then, given any $\epsilon > 0$, there exists a $\delta > 0$ such that

$$\text{if} \quad \|P\| < \delta \qquad \text{then} \quad \left| S^*(P) - \int_a^b f(x)\,dx \right| < \epsilon,$$

no matter how the x_i^* are chosen within $[x_{i-1}, x_i]$.

In symbols we write

(5.10.1)
$$\int_a^b f(x)\,dx = \lim_{\|P\|\to 0} [f(x_1^*)\,\Delta x_1 + f(x_2^*)\,\Delta x_2 + \cdots + f(x_n^*)\,\Delta x_n].$$

A proof of this assertion appears in Appendix B.5. Figure 5.10.2 illustrates the idea. Here the base interval is broken up into 8 subintervals. The point x_1^* is chosen from $[x_0, x_1]$, x_2^* from $[x_1, x_2]$, and so on. While the integral represents the area under the curve, the Riemann sum represents the sum of the areas of the shaded rectangles. The

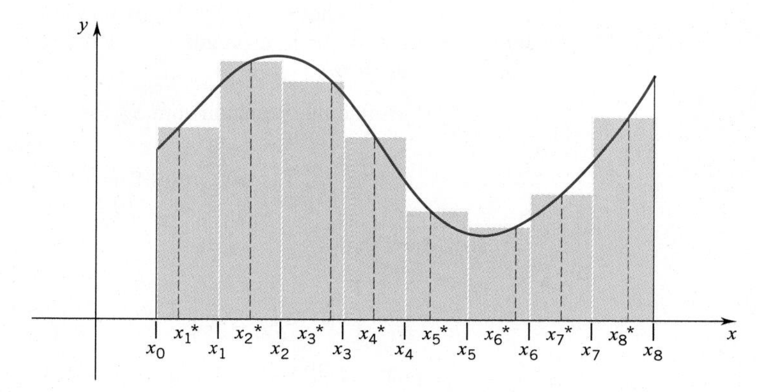

Figure 5.10.2

difference between the two can be made as small as we wish (less than ϵ) simply by making the maximum length of the base subintervals sufficiently small—that is, by making $\|P\|$ sufficiently small. ❑

You have now seen two approaches to the definite integral of a continuous function, through upper and lower sums, and through Riemann sums. Since

$$L_f(P) \leq S^*(P) \leq U_f(P)$$

for any partition P, it follows that the definite integral of f exists in the sense of upper and lower sums iff it exists in the sense of Riemann sums.

We have also pointed out that the definite integrals of certain discontinuous functions also exist. See, for example, Exercises 29 and 30 in Section 5.2. In general, a function f defined on the interval $[a, b]$ is *integrable on* $[a, b]$ if its definite integral

$$\int_a^b f(x)\,dx$$

exists either in the sense of upper and lower sums, or in the Riemann sense. Determining whether a given function is integrable is generally a very difficult problem. In addition to continuous functions, however, a function f is integrable on an interval $[a, b]$ if:

1. f is increasing on $[a, b]$, or f is decreasing on $[a, b]$.

2. f is bounded on $[a, b]$ and has at most a finite number of discontinuities.

EXERCISES 5.10

1. Let Ω be the region below the graph of $f(x) = x^2$, $x \in [0, 1]$. Draw a figure showing the Riemann sum $S^*(P)$ as an estimate for this area. Take

$$P = \{0, \tfrac{1}{4}, \tfrac{1}{2}, \tfrac{3}{4}, 1\} \quad \text{and}$$

$$x_1^* = \tfrac{1}{8}, \quad x_2^* = \tfrac{3}{8}, \quad x_3^* = \tfrac{5}{8}, \quad x_4^* = \tfrac{7}{8}.$$

2. Let Ω be the region below the graph of $f(x) = \frac{3}{2}x + 1$, $x \in [0, 2]$. Draw a figure showing the Riemann sum $S^*(P)$ as an estimate for this area. Take

$P = \{0, \tfrac{1}{4}, \tfrac{3}{4}, 1, \tfrac{3}{2}, 2\}$ and let the x_i^* be the midpoints of the subintervals.

3. Let $f(x) = 2x$, $x \in [0, 1]$. Take $P = \{0, \frac{1}{8}, \frac{1}{4}, \frac{1}{2}, \frac{3}{4}, 1\}$ and set

$$x_1^* = \tfrac{1}{16}, \quad x_2^* = \tfrac{3}{16}, \quad x_3^* = \tfrac{3}{8}, \quad x_4^* = \tfrac{5}{8}, \quad x_5^* = \tfrac{3}{4}.$$

Calculate the following:
(a) $\Delta x_1, \Delta x_2, \Delta x_3, \Delta x_4, \Delta x_5$.
(b) $\|P\|$.
(c) m_1, m_2, m_3, m_4, m_5.
(d) $f(x_1^*), f(x_2^*), f(x_3^*), f(x_4^*), f(x_5^*)$.
(e) M_1, M_2, M_3, M_4, M_5.
(f) $L_f(P)$.
(g) $S^*(P)$.
(h) $U_f(P)$.
(i) $\int_a^b f(x)\,dx$.

4. Let f be continuous on $[a, b]$, let $P = \{x_0, x_1, \ldots, x_n\}$ be a partition of $[a, b]$, and let $S^*(P)$ be any Riemann sum generated by P. Show that

$$L_f(P) \leq S^*(P) \leq U_f(P).$$

In numerical computations the base interval $[a, b]$ is usually broken up into n subintervals each of length $(b - a)/n$. The Riemann sums then take the form

$$S_n^* = \frac{b-a}{n}[f(x_1^*) + f(x_2^*) + \cdots + f(x_n^*)].$$

In Exercises 5–7, choose $\epsilon > 0$. Break up the interval $[0, 1]$ into n subintervals each of length $1/n$. Take $x_1^* = 1/n$, $x_2^* = 2/n, \ldots, x_n^* = n/n$.

5. (a) Determine the Riemann sum S_n^* for

$$\int_0^1 x\,dx. \quad \text{(See Exercise 27, Section 5.2.)}$$

(b) Show that

$$\left| S_n^* - \int_0^1 x\,dx \right| < \epsilon \qquad \text{if } n > 1/\epsilon.$$

HINT: $1 + 2 + \cdots + n = \frac{1}{2}n(n + 1)$.

6. (a) Determine the Riemann sum S_n^* for

$$\int_0^1 x^2\,dx. \quad \text{(See Exercise 28, Section 5.2.)}$$

(b) Show that

$$\left| S_n^* - \int_0^1 x^2\,dx \right| < \epsilon \qquad \text{if } n > 1/\epsilon.$$

HINT: $1^2 + 2^2 + \cdots + n^2 = \frac{1}{6}n(n+1)(2n+1)$.

7. (a) Determine the Riemann sum S_n^* for

$$\int_0^1 x^3\,dx.$$

(b) Show that

$$\left| S_n^* - \int_0^1 x^3\,dx \right| < \epsilon \qquad \text{if } n > 1/\epsilon.$$

HINT: $1^3 + 2^3 + \cdots + n^3 = (1 + 2 + \cdots + n)^2$.

8. Let f be a function continuous on $[a, b]$. Show that, if P is a partition of $[a, b]$, then $L_f(P)$, $U_f(P)$, and $\frac{1}{2}[L_f(P) + U_f(P)]$ are all Riemann sums.

9. Set $f(x) = x\cos x^2$, $x \in [0, 2]$. Take $P = \{0, \frac{1}{3}, \frac{2}{3}, 1, \frac{4}{3}, \frac{5}{3}, 2\}$ and set

$$x_1^* = \tfrac{1}{6},\quad x_2^* = \tfrac{3}{6},\quad x_3^* = \tfrac{5}{6},\quad x_4^* = \tfrac{7}{6},\quad x_5^* = \tfrac{9}{6},\quad x_6^* = \tfrac{11}{6}.$$

Calculate $S^*(P)$ and compare this to the value of the integral

$$\int_0^2 x\cos x^2\,dx.$$

10. Set $f(x) = \sec^2 x$, $x \in [0, 1]$. Take $P = \{0, \frac{2}{10}, \frac{4}{10}, \frac{6}{10}, \frac{8}{10}, 1\}$ and set

$$x_1^* = \tfrac{1}{10},\quad x_2^* = \tfrac{3}{10},\quad x_3^* = \tfrac{5}{10},\quad x_4^* = \tfrac{7}{10},\quad x_5^* = \tfrac{9}{10}.$$

Calculate $S^*(P)$ and compare this to the value of the integral

$$\int_0^1 \sec^2 x\,dx.$$

■ CHAPTER HIGHLIGHTS

5.1 An Area Problem; A Speed–Distance Problem

5.2 The Definite Integral of a Continuous Function

partition (p. 280) upper sum; lower sum (p. 281)
definite integral (p. 283) limits of integration; integrand (p. 283)

5.3 The Function $F(x) = \int_a^x f(t)\,dt$

additivity of the integral (p. 291)

$\dfrac{d}{dx}\left[\int_a^x f(t)\,dt\right] = f(x)$ provided f is continuous (p. 293)

5.4 The Fundamental Theorem of Integral Calculus

antiderivative (p. 297) fundamental theorem (p. 298)
Table of Antiderivatives (p. 300)
linearity of the integral (p. 301)

5.5 Some Area Problems

If f and g are continuous and $f(x) \geq g(x)$ for all x in $[a, b]$, then

$$\int_a^b [f(x) - g(x)]\,dx$$

gives the area between the graph of f and the graph of g over $[a, b]$.

signed area (p. 308)

5.6 Indefinite Integrals

indefinite integral; constant of integration (p. 312)
Table of Indefinite integrals (p. 313)
some motion problems (p. 315)

For an object that moves along a coordinate line with velocity $v(t)$

$$\int_a^b |v(t)|\, dt = \text{distance traveled from time } t = a \text{ to time } t = b$$

whereas

$$\int_a^b v(t)\, dt = \text{net displacement from time } t = a \text{ to time } t = b.$$

equation for linear motion with constant acceleration (p. 317)

5.7 The *u*-Substitution; Change of Variables

An integral of the form $\int f(g(x))g'(x)\, dx$ can be written $\int f(u)\, du$ by setting

$$u = g(x), \qquad du = g'(x)\, dx.$$

For definite integrals

$$\int_a^b f(g(x))g'(x)\, dx = \int_{g(a)}^{g(b)} f(u)\, du.$$

5.8 Some Further Properties of the Definite Integral

The integral of a nonnegative function is nonnegative; the integral of a positive function is positive; the integral is order-preserving (pp. 329–330)

$$\left| \int_a^b f(x)\, dx \right| \leq \int_a^b |f(x)|\, dx, \qquad \frac{d}{dx}\left[\int_a^u f(t)\, dt\right] = f(u)\frac{du}{dx}.$$

5.9 Mean-Value Theorems for Integrals; Average Values

first mean-value theorem for integrals (p. 334)
second mean-value theorem for integrals (p. 336)
mass of a rod (p. 336)
average value (p. 334)
weighted average (p. 336)
center of mass of a rod (p. 337)

5.10 The Integral as the Limit of Riemann Sums

Darboux method (p. 340)
Riemann sum (p. 340)

$$\int_a^b f(x)\, dx = \lim_{\|P\| \to 0} [f(x_1^*)\, \Delta x_1 + f(x_2^*)\, \Delta x_2 + \cdots + f(x_n^*)\, \Delta x_n].$$

■ PROJECTS AND EXPLORATIONS USING TECHNOLOGY

To do these exercises you will need a graphics calculator or a computer with graphing capability. The majority of these problems are open-ended so different approaches may be used to solve them. You should be aware that different approaches can result in slight variations in the answers. Round your numerical answers to at least four decimal places. The rounding method that your calculator or computer uses also may cause variations in answers.

5.1 There is no simple way to represent the function

$$F(x) = \int_0^x \sin^2(t^2)\, dt.$$

For example, you cannot represent it in terms of a sum, difference, product, quotient or composition of elementary functions. However, using the methods of this chapter and the preceding chapters, we can still obtain a lot of information about F.

(a) Give the domain of F.
(b) Find F' and F''.
(c) Explain why F is continuous.
(d) Show that F is a nondecreasing function. Is F one-to-one?
(e) Discuss the symmetry of the graph of F.
(f) Discuss the concavity of the graph of F, and find the points of inflection.
(g) Based on the information in parts (a)–(f), sketch the graph of F. Then use a calculator or graphing utility to check your graph.

5.2 Consider the family of functions given by

$$f_p(x) = 40x^6 - 24x^5 - 45x^4 + 50x^3 - 15x^2 + 3(2p - 1)x + 6,$$

where p is an arbitrary constant. The graph of $f_1(x)$ is shown in the figure.

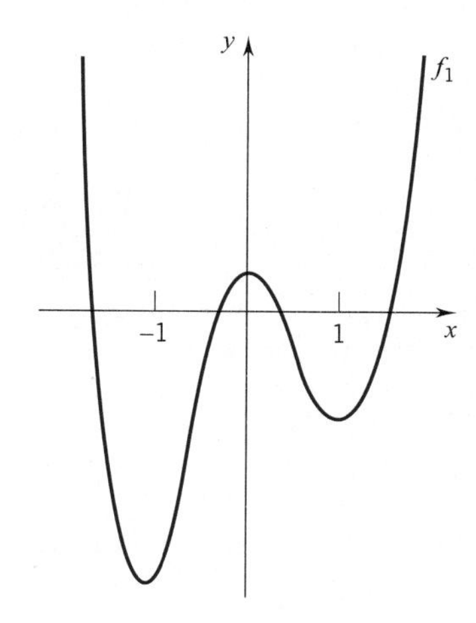

(a) Graph f_p for several values of p to get an idea about the graphs of the members of this family.

Let $g(p)$ be the x-coordinate of the largest (rightmost) local minimum of f_p. For example $g(1) \cong 1.02$. See the figure.

(b) Estimate $g(p)$ for the values of p used in part (a). HINT: Consider the zeros of f'_p.

(c) Based on the information obtained in parts (a) and (b), give the domain of g and determine where g is continuous.

Now consider the integral

$$G(z) = \int_0^z g(p)\, dp.$$

(d) Based on the theorems in this chapter, where does $G'(z) = g(z)$?
(e) Using the values of g calculated in part (b), approximate the values of G using Riemann sums. Estimate where G is increasing and where it is decreasing.

5.3 Assume that the velocity of a racing car is given by

$$v(t) = 40t - t^2,$$

where v is measured in feet per second and t is measured in seconds from the start of the race.

(a) Find T, the maximum value of t for which this formula is reasonable physically.
(b) Find the distance traveled from $t = 0$ to $t = T$.
(c) Find the average velocity from $t = 2$ to $t = 5.5$, that is, find

$$\frac{1}{3.5} \int_{2}^{5.5} (40t - t^2)\, dt.$$

(d) Find the instantaneous velocity of the car at $t = 2$ and at $t = 5.5$, and calculate the average of these two velocities.
(e) Find the instantaneous velocity at each half second starting at $t = 2$ and ending at $t = 5.5$. Then calculate the average of these velocities.
(f) Find the instantaneous velocity at $t = 2$, $t = 2.1$, $t = 2.2$, $t = 2.3$, . . . , $t = 5.4$, and $t = 5.5$. Then calculate the average of these velocities.
(g) Compare the results from parts (c)–(f) and explain their ordering.
(h) Approximate the average value of the velocity by generating 100 random values of t in the interval [2, 5.5] and averaging the instantaneous velocities at each of these times.

5.4 In previous chapters we considered moving averages of a function f as a means of smoothing data to find long-term trends. The examples given in the Projects and Explorations section of Chapter 1 averaged two values. More generally, we could average n values:

$$A(x) = \frac{1}{n} \sum_{i=1}^{n} f(x_i) = \frac{1}{a} \left[\frac{a}{n} \sum_{i=1}^{n} f(x_i) \right],$$

where $x_i = x - a + ia/n$, $i = 1, 2, \ldots, n$. Since the expression on the right is a Riemann sum for A, we can take the limit as $n \to \infty$ to obtain the moving average F of f given by

$$F(x) = \frac{1}{a} \int_{x-a}^{x} f(t)\, dt.$$

In the problems that follow, let $f(x) = \sin(3x + 2)$ and $a = \pi$.
(a) Graph $F(x)$. Also, graph

$$g(x) = \frac{f(x - \pi) + f(x)}{2}.$$

Analysts look for long-term "trends" rather than short-term "random fluctuations." Moving averages are one method for doing such analyses. Generally, a curve is considered to be "smooth" if it is differentiable and has few relative extrema or points of inflection. Which of these moving averages seems to do the best job of smoothing the data?

(b) Find $F'(x)$.
(c) Now find the moving average of f'. That is, find

$$F(x) = \frac{1}{\pi} \int_{x-\pi}^{x} f'(t)\, dt.$$

Is the moving average of f' equal to F'?
(d) Can you generalize this example to other functions?
(e) Repeat parts (a)–(c) for the greatest integer function $f(x) = [\![x]\!]$ with $a = 1$, and then with $a = 0.5$. How well does the moving average smooth a discontinuous function?

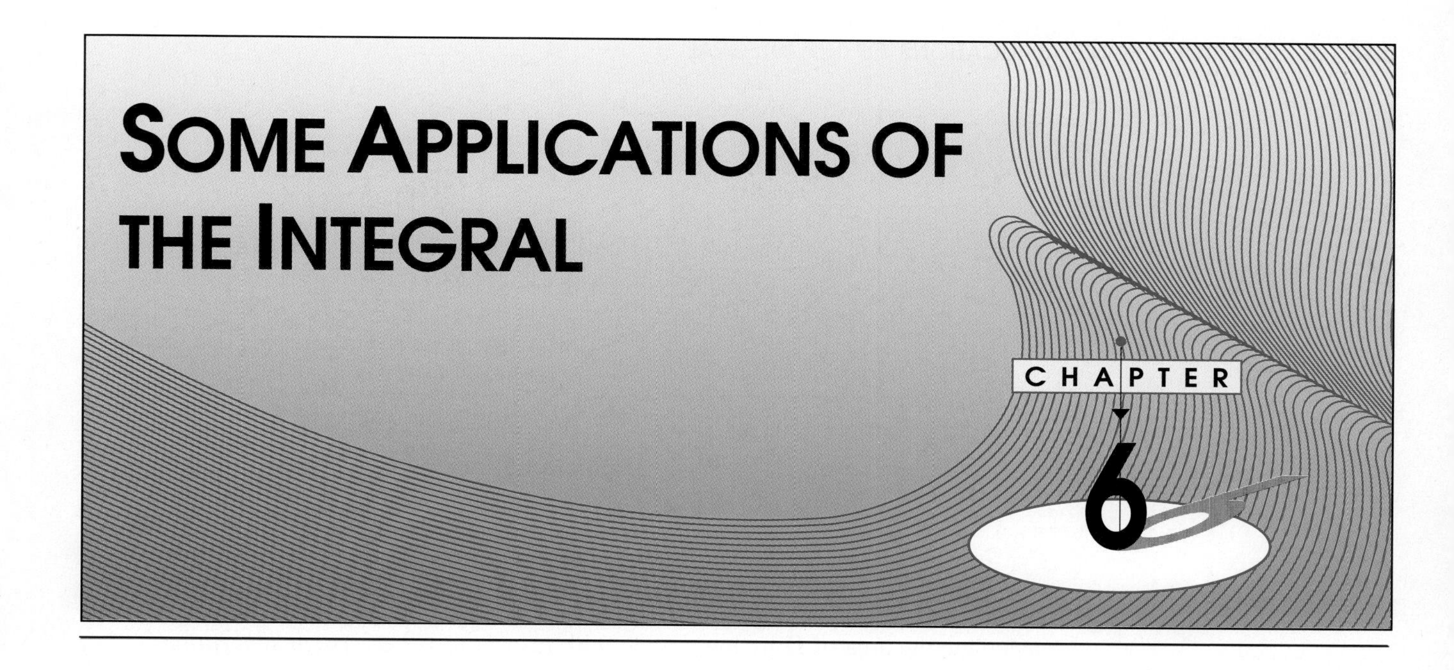

6.1 MORE ON AREA

Representative Rectangles

You have seen that the definite integral can be viewed as the limit of Riemann sums:

(1) $$\int_a^b f(x)\,dx = \lim_{\|P\|\to 0} [f(x_1^*)\,\Delta x_1 + f(x_2^*)\,\Delta x_2 + \cdots + f(x_n^*)\,\Delta x_n].$$

With x_i^* chosen arbitrarily from $[x_{i-1}, x_i]$, you can think of $f(x_i^*)$ as a *representative* value of f for that interval. If f is positive, then the product

$$f(x_i^*)\,\Delta x_i$$

gives the area of the *representative rectangle* shown in Figure 6.1.1. Formula (1) tells us that we can approximate the area under the curve as closely as we wish by adding up the areas of representative rectangles (Figure 6.1.2).

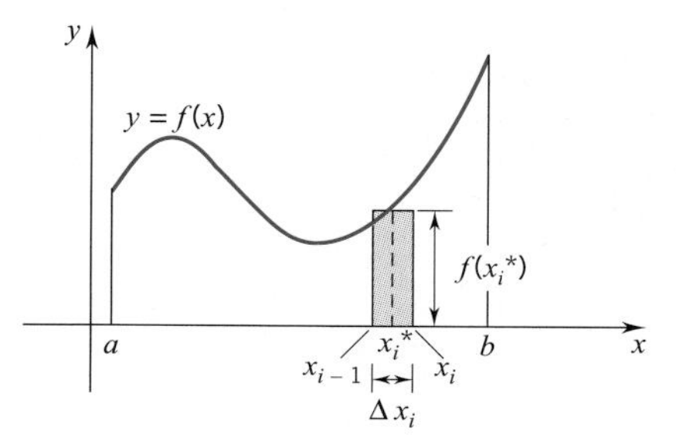

Figure 6.1.1

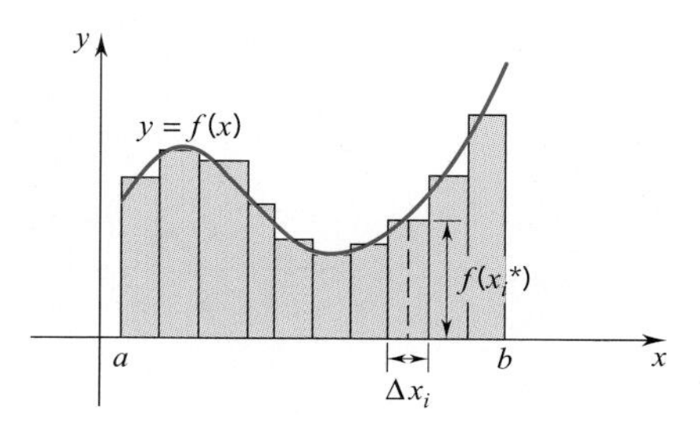

Figure 6.1.2

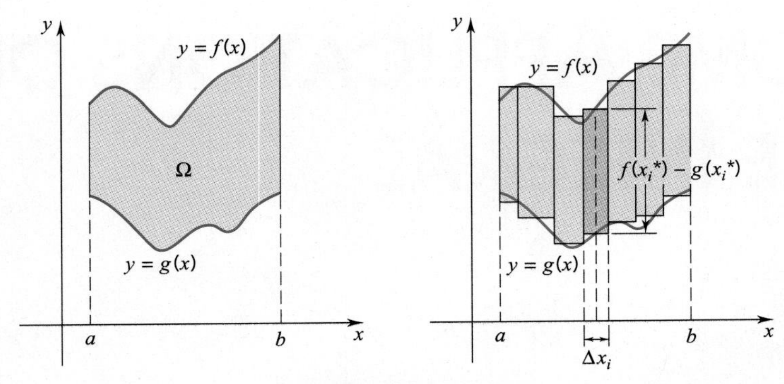

Figure 6.1.3

Figure 6.1.3 shows a region Ω bounded above by the graph of a function f and bounded below by the graph of a function g. As you have seen before, we can calculate the area of Ω by integrating with respect to x the *vertical separation*

$$f(x) - g(x)$$

from $x = a$ to $x = b$:

$$\text{area } (\Omega) = \int_a^b [f(x) - g(x)]\, dx.$$

In this case the approximating Riemann sums are of the form

$$[f(x_1^*) - g(x_1^*)]\,\Delta x_1 + [f(x_2^*) - g(x_2^*)]\,\Delta x_2 + \cdots + [f(x_n^*) - g(x_n^*)]\,\Delta x_n.$$

The dimensions of a representative rectangle are now

$$\text{“height”} = f(x_i^*) - g(x_i^*) \qquad \text{and} \qquad \text{“width”}\ \Delta x_i.$$

Example 1 Find the area of the shaded region shown in Figure 6.1.4.

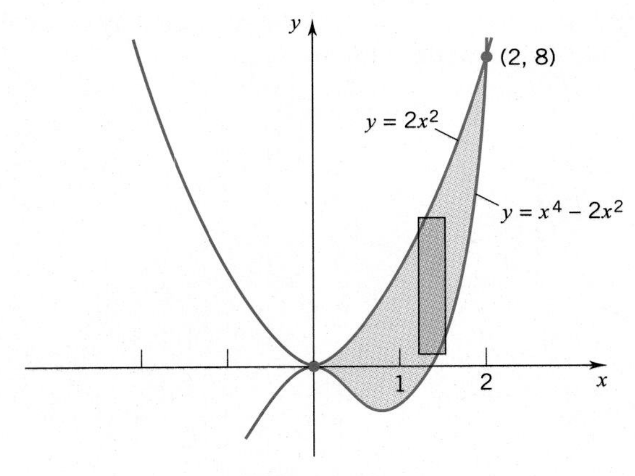

Figure 6.1.4

SOLUTION The area of the representative rectangle is $[2x^2 - (x^4 - 2x^2)]\,\Delta x$, $0 \le x \le 2$. Thus, the area of the shaded region is

$$A = \int_0^2 [2x^2 - (x^4 - 2x^2)]\,dx = \int_0^2 (4x^2 - x^4)\,dx$$
$$= \left[\tfrac{4}{3}x^3 - \tfrac{1}{5}x^5\right]_0^2 = \tfrac{64}{15}. \quad \square$$

Areas by Integration with Respect to y

In Figure 6.1.5 you can see a region, the boundaries of which are not functions of x but functions of y instead. In such a case we set the representative rectangles horizontally and we calculate the area of the region as the limit of Riemann sums of the form

$$[F(y_1^*) - G(y_1^*)]\,\Delta y_1 + [F(y_2^*) - G(y_2^*)]\,\Delta y_2 + \cdots + [F(y_n^*) - G(y_n^*)]\,\Delta y_n.$$

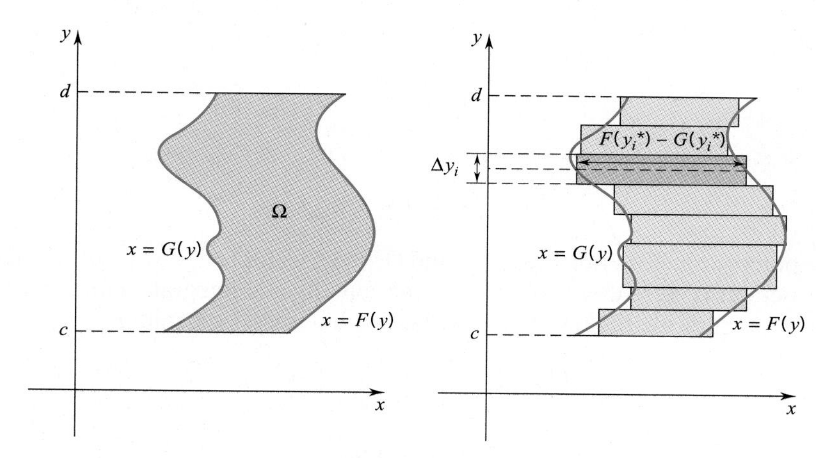

Figure 6.1.5

The area of the region is thus given by the integral

$$\int_c^d [F(y) - G(y)]\,dy.$$

Here we are integrating the *horizontal separation*

$$F(y) - G(y)$$

with respect to y.

Example 2 Find the area of the region bounded on the left by $x = y^2$ and on the right by $x = 3 - 2y^2$.

SOLUTION The region is sketched in Figure 6.1.6. The points of intersection are found by solving the two equations simultaneously:

$$x = y^2 \qquad \text{and} \qquad x = 3 - 2y^2$$

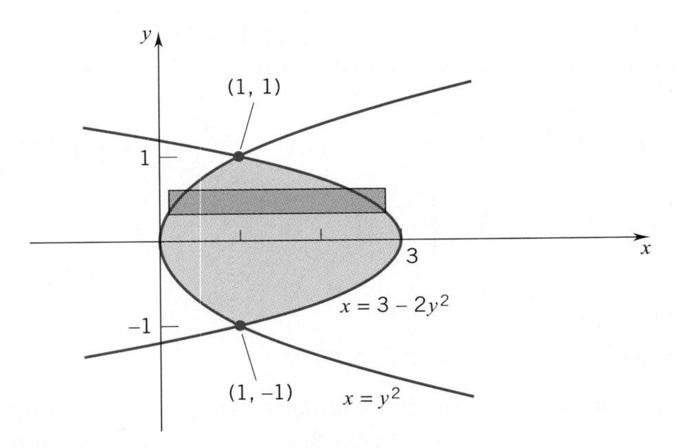

Figure 6.1.6

implies

$$y^2 = 3 - 2y^2$$
$$3y^2 = 3$$

and

$$y = \pm 1.$$

Thus the points of intersection are $(1, 1)$ and $(1, -1)$. The easiest way to calculate the area is to set our representative rectangles horizontally and integrate with respect to y. We find the area of the region by integrating the horizontal separation

$$(3 - 2y^2) - y^2 = 3 - 3y^2$$

from $y = -1$ to $y = 1$:

$$A = \int_{-1}^{1} (3 - 3y^2)\, dy = \left[3y - y^3\right]_{-1}^{1} = 4.$$

NOTE: In our solution we did not take advantage of the symmetry. The region is symmetric with respect to the x-axis (the integrand is an even function of y) and so the area is also given by

$$A = 2\int_{0}^{1} (3 - 3y^2)\, dy = 2\left[3y - y^3\right]_{0}^{1} = 4. \quad \square$$

Example 3 Calculate the area of the region bounded by the curves $x = y^2$ and $x - y = 2$ by integrating (a) with respect to x, (b) with respect to y.

SOLUTION You can verify that the two curves intersect at the points $(1, -1)$ and $(4, 2)$.

(a) To integrate with respect to x we set the representative rectangles vertically, and we solve the equations for y. Solving $x = y^2$ for y we get $y = \pm\sqrt{x}$; $y = \sqrt{x}$ is the upper half of the parabola and $y = -\sqrt{x}$ is the lower half. The equation of the line can be written $y = x - 2$. See Figure 6.1.7.

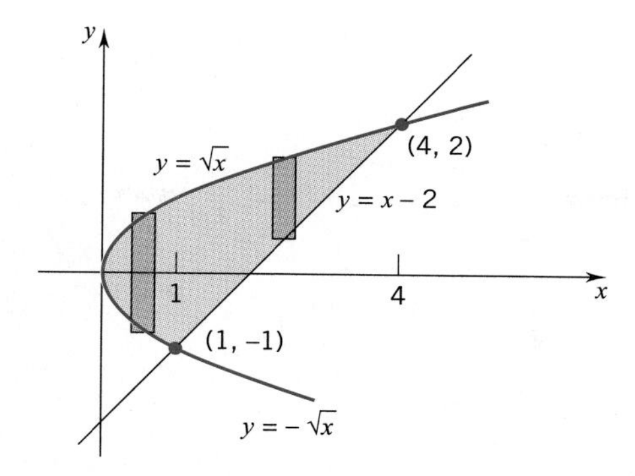

Figure 6.1.7

The upper boundary of the region is the curve $y = \sqrt{x}$. However, the lower boundary is described by two different equations: $y = -\sqrt{x}$ from $x = 0$ to $x = 1$, and $y = x - 2$ from $x = 1$ to $x = 4$. Thus, we need two integrals:

$$A = \int_0^1 [\sqrt{x} - (-\sqrt{x})]\, dx + \int_1^4 [\sqrt{x} - (x - 2)]\, dx$$

$$= 2\int_0^1 \sqrt{x}\, dx + \int_1^4 (\sqrt{x} - x + 2)\, dx = \left[\tfrac{4}{3}x^{3/2}\right]_0^1 + \left[\tfrac{2}{3}x^{3/2} - \tfrac{1}{2}x^2 + 2x\right]_1^4 = \tfrac{9}{2}.$$

(b) See Figure 6.1.8. To integrate with respect to y we set the representative rectangles horizontally. Now the right boundary is the line $x = y + 2$ and the left boundary is the curve $x = y^2$. Since y ranges from -1 to 2,

$$A = \int_{-1}^2 [(y + 2) - y^2]\, dy = \left[\tfrac{1}{2}y^2 + 2y - \tfrac{1}{3}y^3\right]_{-1}^2 = \tfrac{9}{2}. \quad \square$$

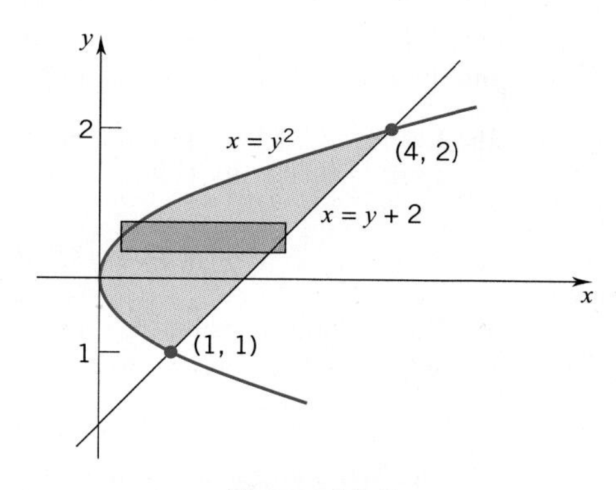

Figure 6.1.8

EXERCISES 6.1

In Exercises 1–14, sketch the region bounded by the curves. Represent the area of the region by one or more integrals (a) in terms of x; (b) in terms of y.

1. $y = x^2, \quad y = x + 2.$

2. $y = x^2, \quad y = -4x.$

3. $y = x^3, \quad y = 2x^2.$

4. $y = \sqrt{x}, \quad y = x^3.$

5. $y = -\sqrt{x}, \quad y = x - 6, \quad y = 0.$

6. $x = y^3, \quad x = 3y + 2.$

7. $y = |x|, \quad 3y - x = 8.$

8. $y = x, \quad y = 2x, \quad y = 3.$

9. $x + 4 = y^2, \quad x = 5.$

10. $x = |y|, \quad x = 2.$

11. $y = 2x, \quad x + y = 9, \quad y = x - 1.$

12. $y = x^3, \quad y = x^2 + x - 1.$

13. $y = x^{1/3}, \quad y = x^2 + x - 1.$

14. $y = x + 1, \quad y + 3x = 13, \quad 3y + x + 1 = 0.$

In Exercises 15–24, sketch the region bounded by the curves and find its area.

15. $4x = 4y - y^2, \quad 4x - y = 0.$

16. $x + y^2 - 4 = 0, \quad x + y = 2.$

17. $x = y^2, \quad x = 3 - 2y^2.$

18. $x + y = 2y^2, \quad y = x^3.$

19. $x + y - y^3 = 0, \quad x - y + y^2 = 0.$

20. $8x = y^3, \quad 8x = 2y^3 + y^2 - 2y.$

21. $y = \cos x, \quad y = \sec^2 x, \quad x \in [-\pi/4, \pi/4].$

22. $y = \sin^2 x, \quad y = \tan^2 x, \quad x \in [-\pi/4, \pi/4].$
HINT: See Exercise 73, Section 5.7.

23. $y = 2 \cos x, \quad y = \sin 2x, \quad [-\pi, \pi].$

24. $y = \sin x, \quad y = \sin 2x, \quad x \in [0, \pi/2].$

In Exercises 25 and 26, use integration to find the area of the triangle with the given vertices.

25. (0, 0), (1, 3), (3, 1).

26. (0, 1), (2, 0), (3, 4).

27. Use integration to find the area of the trapezoid with vertices (−2, −2), (1, 1), (5, 1), (7, −2).

28. Sketch the region bounded by the three curves $y = x^3$, $y = -x$, $y = 1$, and find its area.

29. Sketch the region bounded by the three curves $y = 6 - x^2$, $y = x$ $(x < 0)$, and $y = -x$ $(x > 0)$, and find its area.

30. Find the area of the region bounded by the parabolas $x^2 = 4py$ and $y^2 = 4px$, p a positive constant.

31. Sketch the region bounded by $y = x^2$ and $y = 4$. This region is divided into two subregions of equal area by the line $y = c$. Find c.

32. The region between $y = \cos x$ and the x-axis for $x \in [0, \pi/2]$ is divided into two subregions of equal area by the line $x = c$. Find c.

33. Use a graphing utility to sketch the region bounded by the curves $y = x^4 - 2x^2$ and $y = x + 2$. Then find (approximately) the area of the region.

34. Use a graphing utility to sketch the region bounded by the curves $y = \sin x$ and $y = |x - 1|$. Then find (approximately) the area of the region.

35. A section of rain gutter is 8 feet long. Vertical cross sections of the gutter are in the shape of the parabolic region bounded by $y = \frac{4}{9}x^2$ and $y = 4$, where these measurements are in inches. What is the volume of the rain gutter? HINT: $V =$ (cross-sectional area) × length.

Distribution of Income. Economists use a cumulative distribution called a *Lorenz curve* to measure the distribution of income between households in a given country. Typically, a Lorenz curve is defined on [0, 1], is continuous, increasing and concave up, and passes through (0, 0) and (1, 1). See the figure. The points on the curve are determined by ranking all households by income and then computing the percentage of households whose total income is less than or equal to a given percentage of the total income of the country. For example, the point (a, b) on the curve represents the fact that the bottom a% of the households receive less than or equal to b% of the total income. *Absolute equality* of income distribution would occur if the bottom a% of the households received a% of the total income. The Lorenz curve in this case would be the line $y = x$. The area of the region between the line $y = x$ and the Lorenz curve measures the extent to which the income distribution differs from absolute equality. The *coefficient of inequality* is the ratio of the area of the region between $y = x$ and the Lorenz curve to the area under $y = x$.

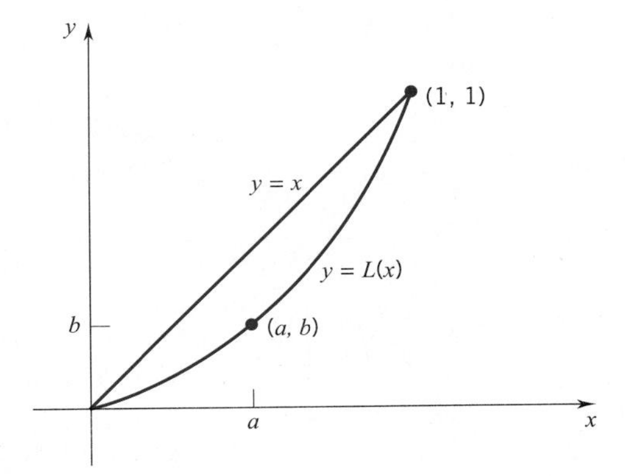

36. Let $y = L(x)$ be a Lorenz curve. Show that the coefficient of inequality C is given by

$$C = 2 \int_0^1 [x - L(x)]\, dx.$$

37. The income distribution of a certain country is represented by the Lorenz curve

$$L(x) = \tfrac{7}{12}x^2 + \tfrac{5}{12}x.$$

(a) What is the percentage of total income received by the bottom 50% of the households.

(b) Determine the coefficient of inequality.

38. Let

$$L(x) = \frac{5x^3}{4 + x^2}.$$

(a) Show that L has the characteristics of a Lorenz curve.

(b) Estimate the coefficient of inequality by calculating the Riemann sum for $f(x) = x - L(x)$ with the partition $P = \{0, 0.2, 0.4, 0.6, 0.8, 1\}$ and $x_1^* = 0.1, x_2^* = 0.3, x_3^* = 0.5, x_4^* = 0.7, x_5^* = 0.9$.

■ 6.2 VOLUME BY PARALLEL CROSS SECTIONS; DISCS AND WASHERS

In this section we use definite integrals to find the volumes of certain three-dimensional solids. To begin, let Ω be a plane region. A *right cylinder with cross section* Ω is a solid that is formed by translating Ω along a line, or *axis*, that is perpendicular to it. See Figure 6.2.1. Let A be the area of Ω. If a right cylinder is formed by translating the region Ω through a distance h, then the volume of the cylinder is given by

$$V = A \cdot h.$$

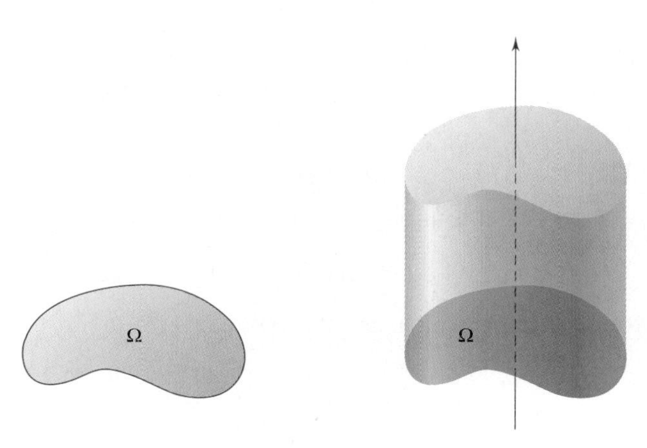

Figure 6.2.1

Some familiar examples are the right circular cylinder of radius r and height h, and a rectangular box of length l, width w, and height h (Figure 6.2.2).

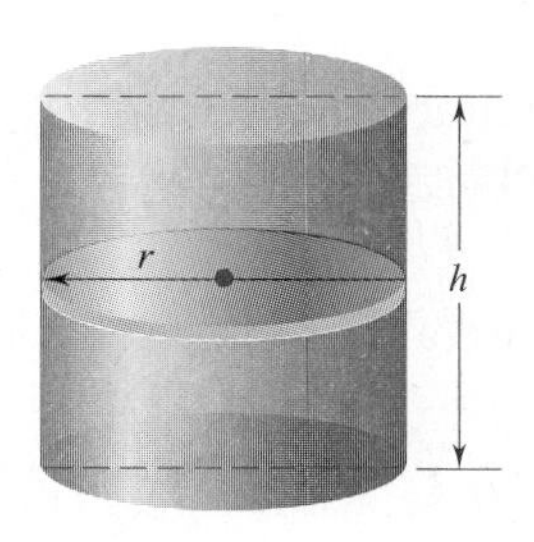

$V = \pi r^2 h =$ (cross-sectional area) · height

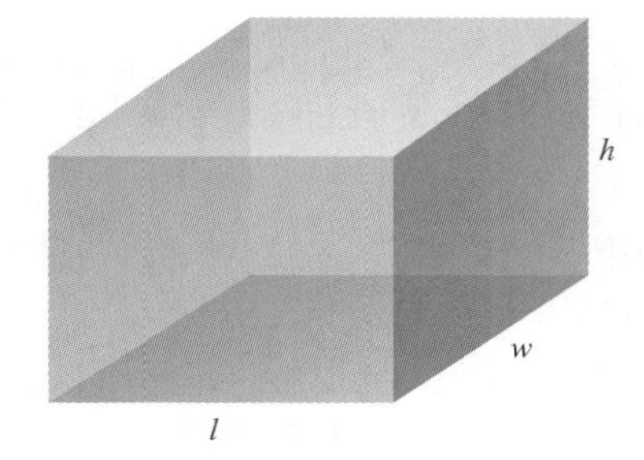

$V = l \cdot w \cdot h =$ (cross-sectional area) · height

Figure 6.2.2

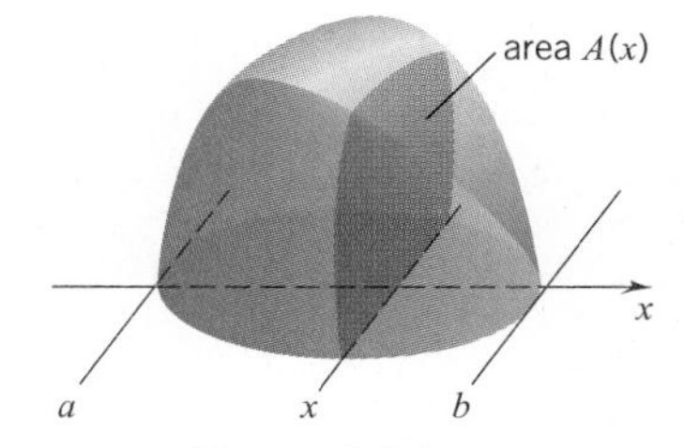

Figure 6.2.3

To calculate the volume of a more general solid, we introduce a coordinate axis and then examine the cross sections of the solid that are perpendicular to that axis. In Figure 6.2.3 we picture a solid and a coordinate axis that we label the x-axis. As in the figure, we suppose that the solid lies entirely between $x = a$ and $x = b$. The figure shows an arbitrary cross section perpendicular to the x-axis. We denote by $A(x)$ the area of the cross section that has coordinate x.

If the cross-sectional area, $A(x)$, varies continuously with x, then we can find the volume of the solid by integrating $A(x)$ from a to b:

(6.2.1)
$$V = \int_a^b A(x)\, dx.$$

To see this, let $P = \{x_0, x_1, \ldots, x_n\}$ be a partition of $[a, b]$. The solid from x_{i-1} to x_i is approximated by a cylinder whose cross section has area $A(x_i^*)$, where x_i^* is chosen arbitrarily from $[x_{i-1}, x_i]$, and whose thickness is Δx_i. The volume of this cylinder is

$$A(x_i^*)\, \Delta x_i. \qquad \text{(Figure 6.2.4)}$$

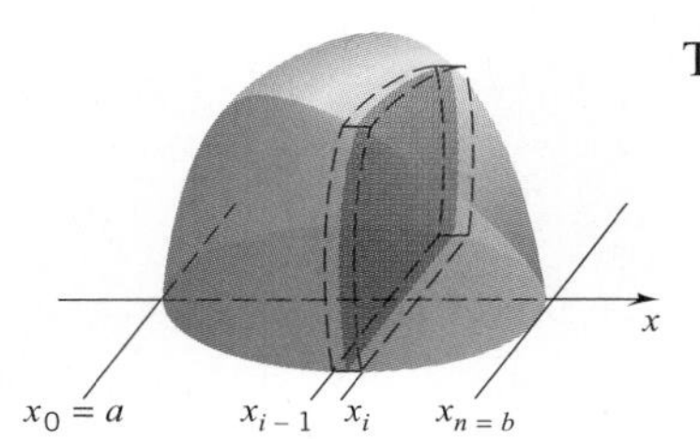

Figure 6.2.4

The volume of the entire solid can therefore be approximated by sums of the form

$$A(x_1^*)\, \Delta x_1 + A(x_2^*)\, \Delta x_2 + \cdots + A(x_n^*)\, \Delta x_n.$$

These are Riemann sums which, as $\|P\| \to 0$, converge to

$$\int_a^b A(x)\, dx. \quad ❑$$

Remark By the mean-value theorem for integrals (see 5.9.2), the formula for the volume of a solid with varying cross-sectional area $A = A(x)$ can also be written

(6.2.2)
$$V = (\text{the average cross-sectional area}) \cdot (b - a) = A_{\text{avg}}(b - a). \quad ❑$$

The calculation of volumes of solids of arbitrary shape is left for Chapter 16. Here we restrict our attention to solids with simple cross sections.

Example 1 Find the volume of a pyramid whose altitude is h and whose base is a square with sides of length r (Figure 6.2.5a).

SOLUTION Locate the x-axis perpendicular to the base, with origin at the center of the base. The cross section with coordinate x is a square. Let s denote the length of the side of that square. Then, by similar triangles (see Figure 6.2.5b), we have

$$\frac{\frac{1}{2}s}{h - x} = \frac{\frac{1}{2}r}{h},$$

and so

$$s = \frac{r}{h}(h - x).$$

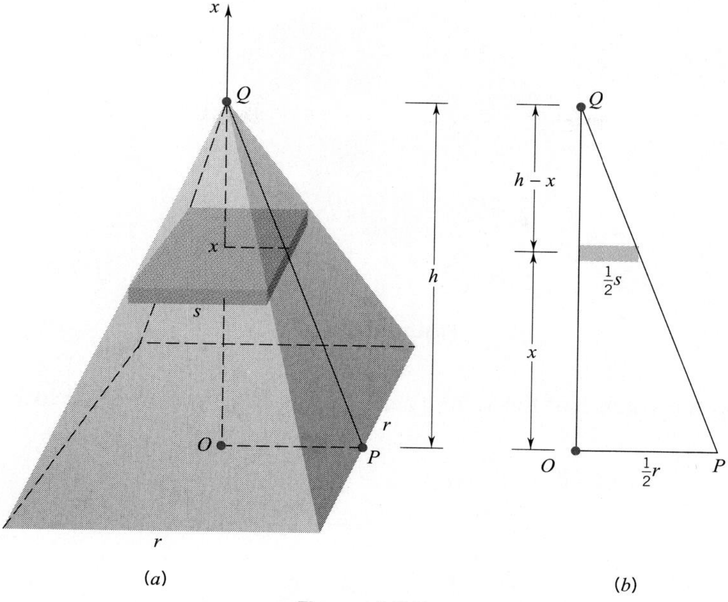

Figure 6.2.5

Now the area of the square at coordinate x is $A(x) = s^2 = (r^2/h^2)(h - x)^2$ and

$$V = \int_0^h A(x)\,dx = \frac{r^2}{h^2}\int_0^h (h-x)^2\,dx$$

$$= \frac{r^2}{h^2}\left[-\frac{(h-x)^3}{3}\right]_0^h = \tfrac{1}{3}r^2h. \quad \square$$

Example 2 The base of a solid is the region bounded by the ellipse

$$\frac{x^2}{a^2} + \frac{y^2}{b^2} = 1.$$

Find the volume of the solid given that each cross section perpendicular to the x-axis is an isosceles triangle with base in the region and altitude equal to one-half the base.

SOLUTION Take x as in Figure 6.2.6. The cross section with coordinate x is an isosceles triangle with base $\overline{PQ}$ and altitude $\frac{1}{2}\overline{PQ}$. The equation of the ellipse can be written

$$y^2 = \frac{b^2}{a^2}(a^2 - x^2).$$

Since

$$\text{length of } \overline{PQ} = 2y = \frac{2b}{a}\sqrt{a^2 - x^2},$$

the isosceles triangle has area

$$A(x) = \tfrac{1}{2}bh = \tfrac{1}{2}\left(\frac{2b}{a}\sqrt{a^2 - x^2}\right)\left(\frac{b}{a}\sqrt{a^2 - x^2}\right) = \frac{b^2}{a^2}(a^2 - x^2).$$

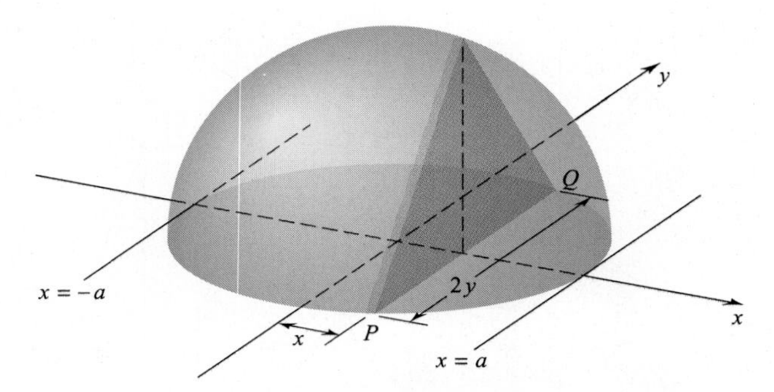

Figure 6.2.6

We can find the volume of the solid by integrating $A(x)$ from $x = -a$ to $x = a$:

$$V = \int_{-a}^{a} A(x)\,dx \underset{\text{by symmetry}}{=} 2\int_{0}^{a} A(x)\,dx$$

$$= \frac{2b^2}{a^2}\int_0^a (a^2 - x^2)\,dx$$

$$= \frac{2b^2}{a^2}\left[a^2x - \frac{x^3}{3}\right]_0^a = \tfrac{4}{3}ab^2. \quad \square$$

Example 3 The base of a solid is the region between the parabolas

$$x = y^2 \qquad \text{and} \qquad x = 3 - 2y^2. \quad \text{(See Example 2, Section 6.1)}$$

Find the volume of the solid given that the cross sections perpendicular to the x-axis are squares.

SOLUTION The two parabolas intersect at $(1, 1)$ and $(1, -1)$ (Figure 6.2.7). From $x = 0$ to $x = 1$, the cross section with coordinate x has area

$$A(x) = (2y)^2 = 4y^2 = 4x.$$

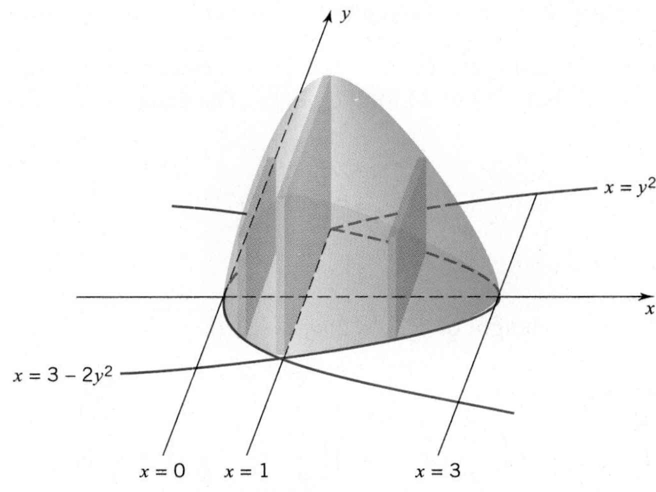

Figure 6.2.7

(Here we are measuring the span across the first parabola $x = y^2$.) The volume of the solid from $x = 0$ to $x = 1$ is

$$V_1 = \int_0^1 4x\,dx = \left[2x^2\right]_0^1 = 2.$$

From $x = 1$ to $x = 3$, the cross section with coordinate x has area

$$A(x) = (2y)^2 = 4y^2 = 2(3 - x) = 6 - 2x.$$

(Here we are measuring the span across the second parabola $x = 3 - 2y^2$.) The volume of the solid from $x = 1$ to $x = 3$ is

$$V_2 = \int_1^3 (6 - 2x)\,dx = \left[6x - x^2\right]_1^3 = 4.$$

The total volume is

$$V_1 + V_2 = 6. \quad ❑$$

Solids of Revolution: Disc Method

Suppose that f is nonnegative and continuous on $[a, b]$. (See Figure 6.2.8.) If we revolve the region bounded by the graph of f and the x-axis about the x-axis, we obtain a solid. The volume of this solid is given by the formula

(6.2.3)
$$V = \int_a^b \pi[f(x)]^2\,dx.$$

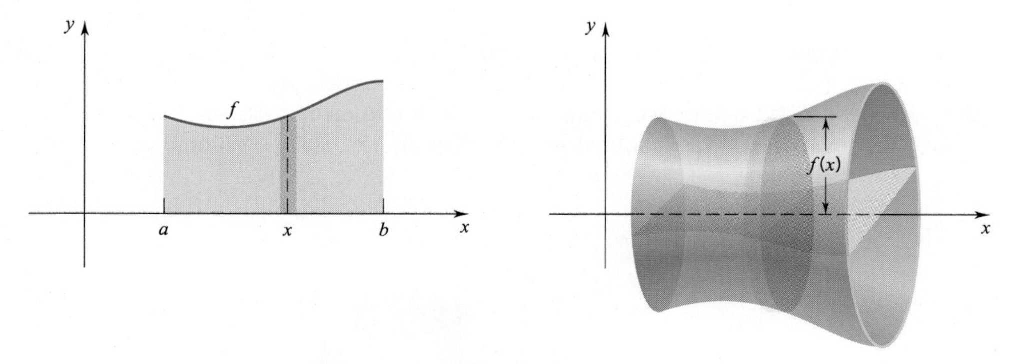

Figure 6.2.8

PROOF The cross section with coordinate x is a circular *disc* of radius $f(x)$. The cross-sectional area is thus $\pi[f(x)]^2$. ❑

Among the simplest solids of revolution are the cone and the sphere.

Example 4 We can generate a cone of base radius r and height h by revolving about the x-axis the region below the graph of

$$f(x) = \frac{r}{h}x, \qquad 0 \le x \le h. \qquad \text{(Figure 6.2.9)}$$

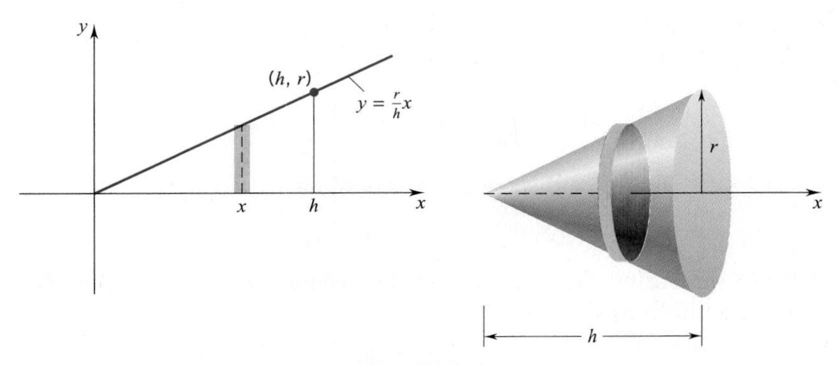

Figure 6.2.9

By Formula (6.2.3),

$$\text{volume of cone} = \int_0^h \pi\left[\frac{r}{h}x\right]^2 dx = \frac{\pi r^2}{h^2}\int_0^h x^2\,dx = \frac{\pi r^2}{h^2}\left[\frac{x^3}{3}\right]_0^h = \frac{1}{3}\pi r^2 h. \quad \square$$

Example 5 A sphere of radius r can be obtained by revolving about the x-axis the region below the graph of

$$f(x) = \sqrt{r^2 - x^2}, \qquad -r \le x \le r. \qquad \text{(draw a figure)}$$

Therefore

$$\text{volume of sphere} = \int_{-r}^{r} \pi(r^2 - x^2)\,dx = \pi\left[r^2x - \frac{1}{3}x^3\right]_{-r}^{r} = \frac{4}{3}\pi r^3.$$

This result was obtained by Archimedes in the third century B.C. ❑

Now suppose that $x = g(y)$ is continuous and nonnegative for $c \le y \le d$ (see Figure 6.2.10). If we revolve the region bounded by the graph of g and the y-axis about the y-axis, we obtain a solid such that each cross section perpendicular to the y-axis is a disc with area $A(y) = \pi[g(y)]^2$. Thus the volume of this solid

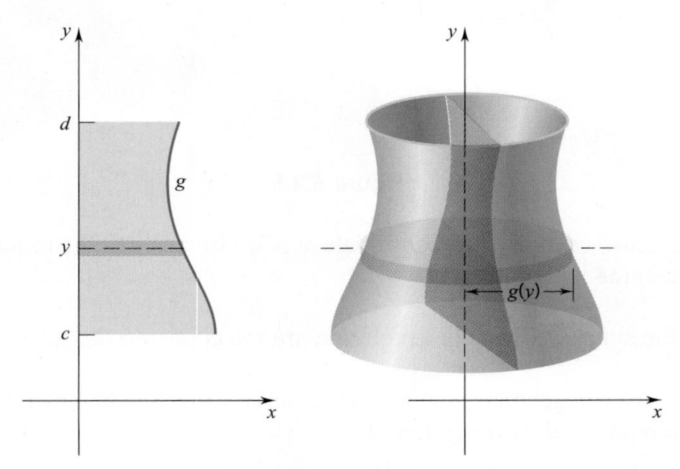

Figure 6.2.10

is given by

(6.2.4)
$$V = \int_c^d \pi[g(y)]^2\, dy.$$

Example 6 Find the volume of the solid generated by revolving the region bounded by $y = x^{2/3} + 1$, $0 \le x \le 8$, about the y-axis.

SOLUTION See Figure 6.2.11. We first solve $y = x^{2/3} + 1$, $0 \le x \le 8$, for x as a function of y:

$$x^{2/3} = y - 1$$
$$x = (y-1)^{3/2}, \qquad 1 \le y \le 5.$$

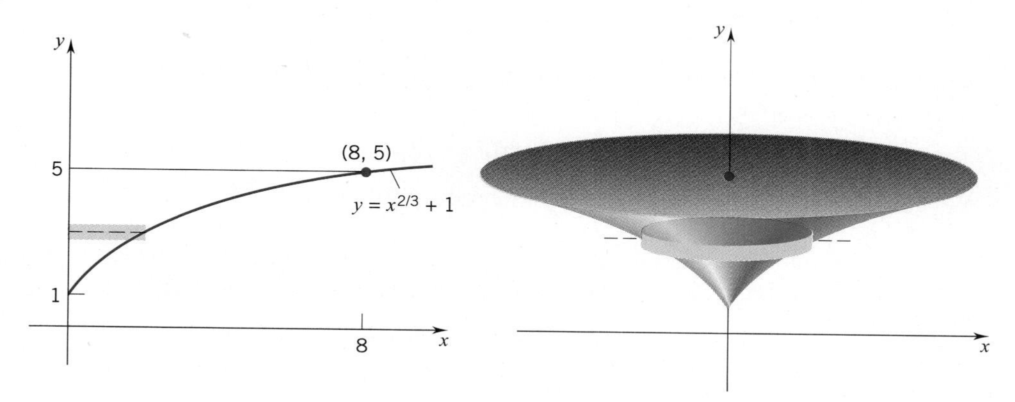

Figure 6.2.11

Now, applying (6.2.4), we have

$$\begin{aligned} V &= \int_1^5 \pi[g(y)]^2\, dy = \pi \int_1^5 [(y-1)^{3/2}]^2\, dy \\ &= \pi \int_1^5 (y-1)^3\, dy = \pi \left[\frac{(y-1)^4}{4}\right]_1^5 = 64\pi. \quad \square \end{aligned}$$

Solids of Revolution: Washer Method

The washer method is a slight generalization of the disc method. Suppose that f and g are nonnegative continuous functions with $g(x) \le f(x)$ for all x in $[a, b]$ (see Figure 6.2.12). If we revolve the region Ω about the x-axis, we obtain a solid. The volume of this solid is given by the formula

(6.2.5)
$$V = \int_a^b \pi([f(x)]^2 - [g(x)]^2)\, dx. \qquad \text{(washer method about } x\text{-axis)}$$

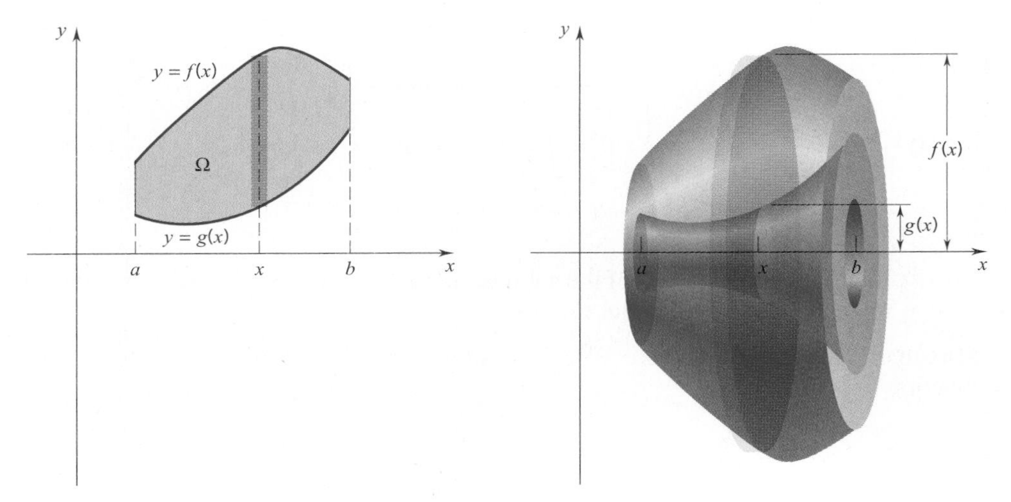

Figure 6.2.12

PROOF The cross section with coordinate x is a *washer* of outer radius $f(x)$, inner radius $g(x)$, and area

$$A(x) = \pi[f(x)]^2 - \pi[g(x)]^2 = \pi([f(x)]^2 - [g(x)]^2).$$

We can get the volume of the solid by integrating this function from a to b. ❑

Suppose now that the boundaries are functions of y rather than x (see Figure 6.2.13). By revolving Ω *about the y-axis,* we obtain a solid. It is clear from (6.2.4) that in this case

(6.2.6)
$$V = \int_c^d \pi([F(y)]^2 - [G(y)]^2)\, dy. \qquad \text{(washer method about } y\text{-axis)}$$

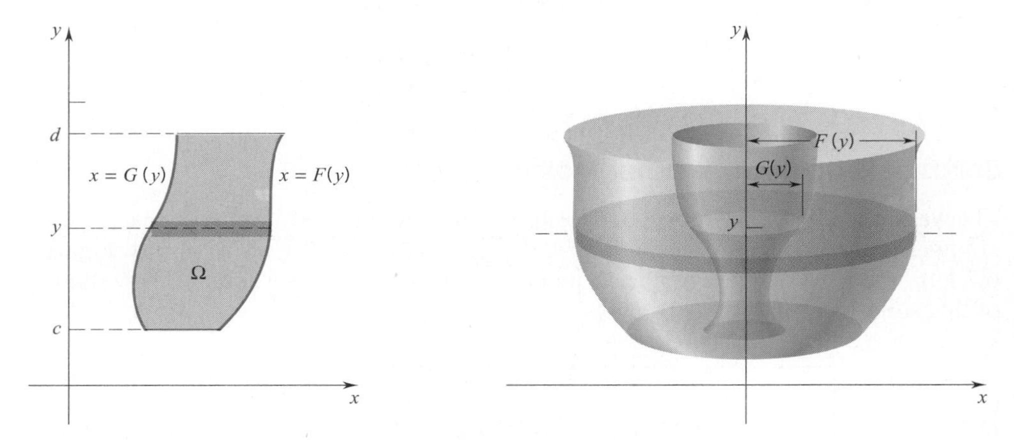

Figure 6.2.13

Example 7 Find the volume of the solid generated by revolving the region between $y = x^2$ and $y = 2x$ about the x-axis.

SOLUTION Setting $x^2 = 2x$, we get $x = 0, 2$. Thus, the curves intersect at the points $(0, 0)$ and $(2, 4)$. See Figure 6.2.14. For each x from 0 to 2, the x cross section is a washer of outer radius $2x$ and inner radius x^2. By (6.2.5)

$$V = \int_0^2 \pi[(2x)^2 - (x^2)^2]\,dx = \pi \int_0^2 (4x^2 - x^4)\,dx = \pi\left[\frac{4}{3}x^3 - \frac{1}{5}x^5\right]_0^2 = \frac{64}{15}\pi. \quad \square$$

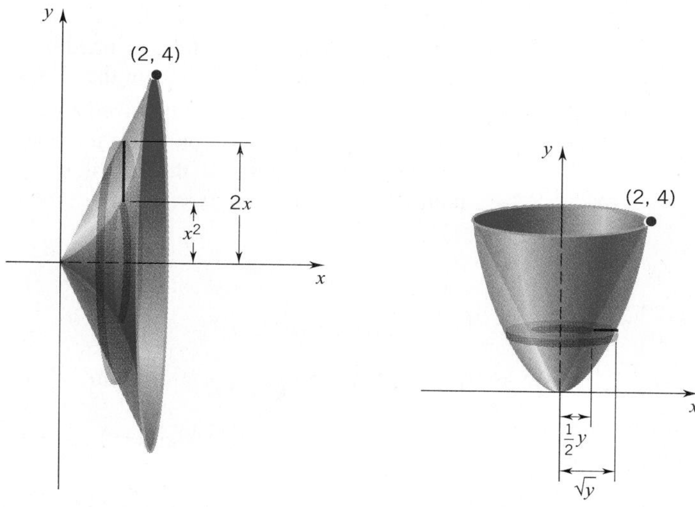

Figure 6.2.14

Figure 6.2.15

Example 8 Find the volume of the solid generated by revolving the region between $y = x^2$ and $y = 2x$ about the y-axis.

SOLUTION The solid is depicted in Figure 6.2.15. For each y from 0 to 4, the y cross section is a washer of outer radius $\sqrt{y}$ and inner radius $\frac{1}{2}y$. By (6.2.6)

$$\begin{aligned} V &= \int_0^4 \pi[(\sqrt{y})^2 - (\tfrac{1}{2}y)^2]\,dy = \pi \int_0^4 (y - \tfrac{1}{4}y^2)\,dy \\ &= \pi[\tfrac{1}{2}y^2 - \tfrac{1}{12}y^3]_0^4 = \tfrac{8}{3}\pi. \quad \square \end{aligned}$$

Remark These last two examples concerned solids generated by revolving the *same* region about *different* axes. Notice that the solids are different and have different volumes. ❑

EXERCISES 6.2

In Exercises 1–16, sketch the region Ω bounded by the curves and find the volume of the solid generated by revolving this region about the x-axis.

1. $y = x, \quad y = 0, \quad x = 1.$
2. $x + y = 3, \quad y = 0, \quad x = 0.$
3. $y = x^2, \quad y = 9.$
4. $y = x^3, \quad y = 8, \quad x = 0.$
5. $y = \sqrt{x}, \quad y = x^3.$
6. $y = x^2, \quad y = x^{1/3}.$
7. $y = x^3, \quad x + y = 10, \quad y = 1.$
8. $y = \sqrt{x}, \quad x + y = 6, \quad y = 1.$
9. $y = x^2, \quad y = x + 2.$
10. $y = x^2, \quad y = 2 - x.$
11. $y = \sqrt{4 - x^2}, \quad y = 0.$
12. $y = 1 - |x|, \quad y = 0.$
13. $y = \sec x, \quad x = 0, \quad x = \frac{1}{4}\pi, \quad y = 0.$
14. $y = \csc x, \quad x = \frac{1}{4}\pi, \quad x = \frac{3}{4}\pi, \quad y = 0.$
15. $y = \cos x, \quad y = x + 1, \quad x = \frac{1}{2}\pi.$
16. $y = \sin x, \quad x = \frac{1}{4}\pi, \quad x = \frac{1}{2}\pi, \quad y = 0.$

In Exercises 17–26, sketch the region Ω bounded by the curves and find the volume of the solid generated by revolving this region about the y-axis.

17. $y = 2x, \quad y = 4, \quad x = 0.$
18. $x + 3y = 6, \quad x = 0, \quad y = 0.$
19. $x = y^3, \quad x = 8, \quad y = 0.$
20. $x = y^2, \quad x = 4.$
21. $y = \sqrt{x}, \quad y = x^3.$
22. $y = x^2, \quad y = x^{1/3}.$
23. $y = x, \quad y = 2x, \quad x = 4.$
24. $x + y = 3, \quad 2x + y = 6, \quad x = 0.$
25. $x = y^2, \quad x = 2 - y^2.$
26. $x = \sqrt{9 - y^2}, \quad x = 0.$
27. The base of a solid is the circle $x^2 + y^2 = r^2$. Find the volume of the solid given that the cross sections perpendicular to the x-axis are: (a) squares; (b) equilateral triangles.
28. The base of a solid is the region bounded by the ellipse $b^2x^2 + a^2y^2 = a^2b^2$. Find the volume of the solid given that the cross sections perpendicular to the x-axis are: (a) isosceles right triangles each with hypotenuse on the xy-plane; (b) squares; (c) isosceles triangles of height 2.
29. The base of a solid is the region bounded by $y = x^2$ and $y = 4$. Find the volume of the solid given that the cross sections perpendicular to the x-axis are: (a) squares; (b) semicircles; (c) equilateral triangles.
30. The base of a solid is the region between the parabolas $x = y^2$ and $x = 3 - 2y^2$. Find the volume of the solid given that the cross sections perpendicular to the x-axis are: (a) rectangles of height h; (b) equilateral triangles; (c) isosceles right triangles each with hypotenuse on the xy-plane.
31. Exercise 29 with the cross sections perpendicular to the y-axis.
32. Exercise 30 with the cross sections perpendicular to the y-axis.
33. Find the volume of the solid generated by revolving the ellipse $b^2x^2 + a^2y^2 = a^2b^2$ about the x-axis.
34. Exercise 33 with the ellipse revolved about the y-axis.
35. Derive a formula for the volume of the frustum of a right circular cone in terms of the height h, the lower base radius R, and the upper base radius r. (See the figure.)

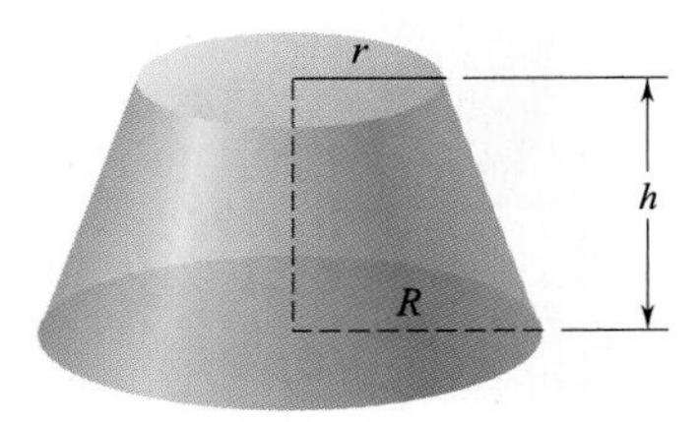

36. Find the volume enclosed by the surface that is generated by revolving the equilateral triangle with vertices $(0, 0)$, $(a, 0)$, $(\frac{1}{2}a, \frac{1}{2}\sqrt{3}a)$ about the x-axis.
37. A hemispheric basin of radius r feet is being used to store water. To what percent of capacity is it filled when the water is: (a) $\frac{1}{2}r$ feet deep? (b) $\frac{1}{3}r$ feet deep?
38. A sphere of radius r is cut by two parallel planes: one, a units above the equator; the other, b units above the equator. Find the volume of the portion of the sphere that lies between the two planes. Assume that $a < b$.
39. The region shown in the figure is revolved around the y-axis to form a parabolic-shaped container. The indi-

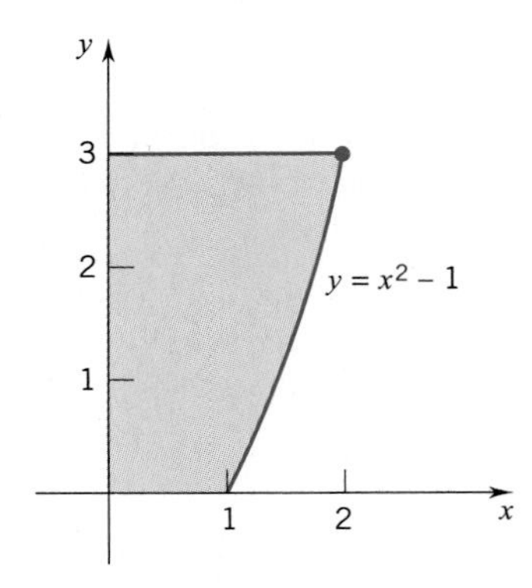

cated units are in feet. If a liquid is poured into the container at the rate of two cubic feet per minute, how fast is the level of the liquid rising at the instant its depth in the container is one foot? Two feet?

Rotation about Lines Parallel to a Coordinate Axis

In Exercises 40 and 41, let f be a continuous, nonnegative function on $[a, b]$ and let Ω be the region between the graph of f and the x-axis.

40. Let k be a positive constant. Prove that if the region Ω is revolved around the line $y = -k$ (Figure A), then the volume of the solid of revolution (Figure B) is given by

$$V = \int_a^b \pi([f(x) + k]^2 - k^2)\, dx.$$

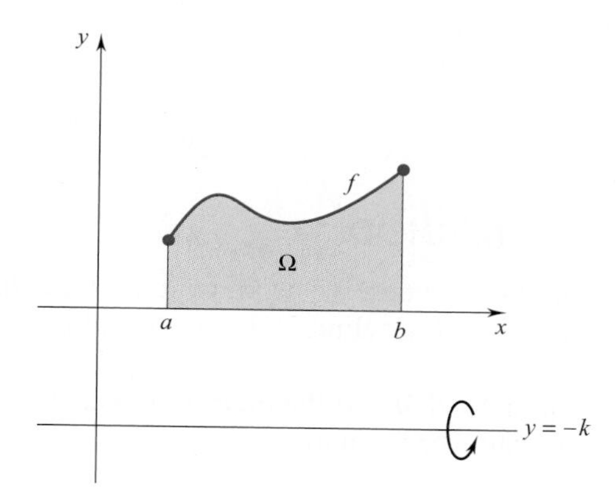

Figure A

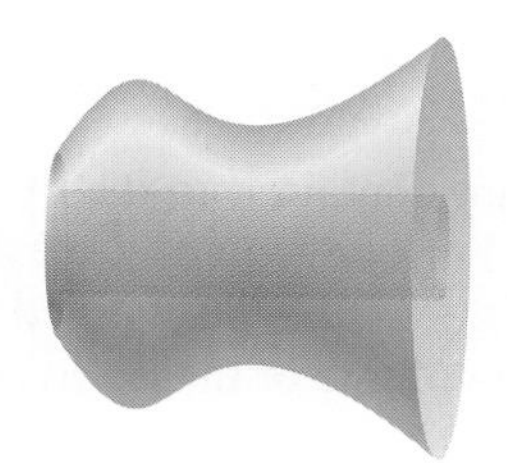

Figure B

41. Let k be a positive constant satisfying $k - f(x) \geq 0$ on $[a, b]$. Prove that if the region Ω is revolved around the line $y = k$ (Figure C), then the volume of the solid of revolution (Figure D) is given by

$$V = \int_a^b \pi(k^2 - [k - f(x)]^2)\, dx.$$

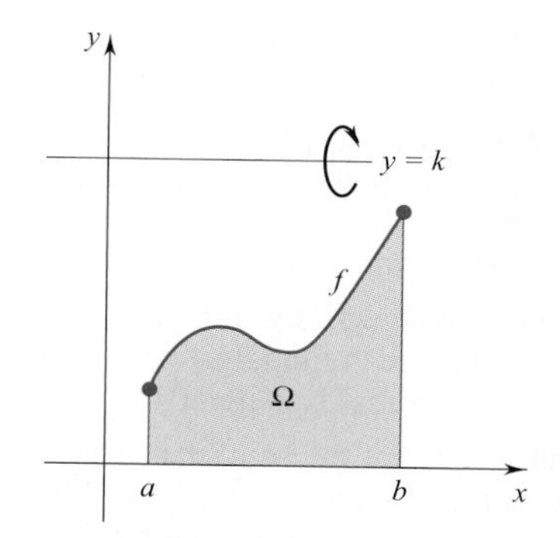

Figure C

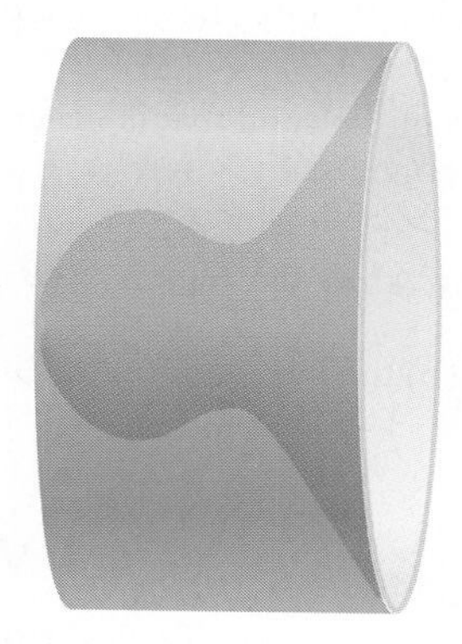

Figure D

42. Let f and g be continuous functions on $[a, b]$ and let k be a constant.

(a) Suppose that $f(x) \geq g(x) \geq k$ on $[a, b]$. Let Ω be the region between the graphs of f and g. Show that if Ω is revolved around the line $y = k$, then the volume of the resulting solid of revolution is given by

$$V = \int_a^b \pi([f(x) - k]^2 - [g(x) - k]^2)\, dx.$$

(b) Suppose that $k \geq f(x) \geq g(x)$ on $[a, b]$, and let Ω be the region between the graphs of f and g. Derive a formula for the volume of the solid which is generated by revolving Ω around the line $y = k$.

NOTE: Analogs of the formulas in Exercises 40–42 hold for functions $x = h(y)$ and regions revolved around vertical lines $x = l$.

43. The region between the graph of $f(x) = \sqrt{x}$ and the x-axis for $0 \leq x \leq 4$ is revolved around the line $y = 2$. Find the volume of the solid that is generated.

44. The region bounded by the curves $y = (x - 1)^2$ and $y = x + 1$ is revolved around the line $y = -1$. Find the volume of the solid that is generated.

45. The region between the graph of $y = \sin x$ and the x-axis, $0 \leq x \leq \pi$, is revolved around the line $y = 1$. Find the volume of the solid that is generated. HINT: See Exercise 73, Section 5.7.

46. The region bounded by $y = \sin x$ and $y = \cos x$, $\pi/4 \leq x \leq \pi$, is revolved around the line $y = 1$. Find the volume of the solid that is generated. HINT: Use a trigonometric identity.

47. The region bounded by the curves $y = x^2 - 2x$ and $y = 3x$ is revolved around the line $y = -1$. Find the volume of the solid that is generated.

48. Find the volume of the solid that is generated by revolving the region bounded by $y = x^2$ and $x = y^2$ around: (a) the line $x = -2$; (b) the line $x = 3$.

49. Find the volume of the solid generated by revolving the region bounded by $y^2 = 4x$ and $y = x$ around: (a) the x-axis; (b) the line $x = 4$.

50. Find the volume of the solid that is generated by revolving the region bounded by $y = x^2$ and $y = 4x$ around: (a) the line $x = 5$; (b) the line $x = -1$.

51. Find the volume when the region OAB in Figure E is revolved about:
(a) the x-axis; (b) the line AB; (c) the line CA; (d) the y-axis.

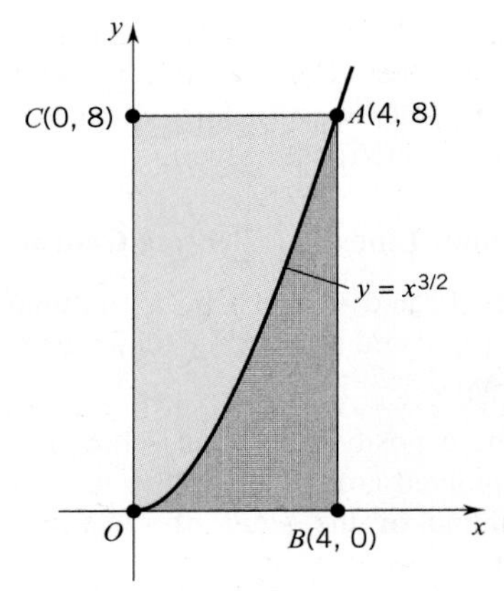

Figure E

52. Find the volume when the region OAC in Figure E is revolved about:
(a) the y-axis; (b) the line CA; (c) the line AB; (d) the x-axis.

■ 6.3 VOLUME BY THE SHELL METHOD

To describe the shell method of computing volumes, we begin with a solid cylinder of radius R and height h, and from it we cut out a cylindrical core of radius r (Figure 6.3.1).

Since the original cylinder had volume $\pi R^2 h$ and the piece removed had volume $\pi r^2 h$, the solid cylindrical shell that remains has volume

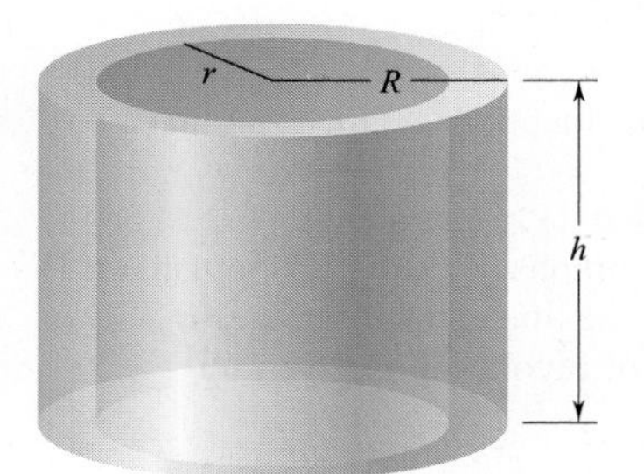

Figure 6.3.1

(6.3.1)
$$\pi R^2 h - \pi r^2 h = \pi h(R + r)(R - r).$$

We will use this shortly.

Now let $[a, b]$ be an interval that does not contain 0 in its interior, and let f be a nonnegative continuous function on $[a, b]$. For convenience, we assume that $a \geq 0$. If the region bounded by the graph of f and the x-axis is revolved around the y-axis, then a solid T is generated (Figure 6.3.2). The volume of this solid is given by the *shell method formula*:

(6.3.2)
$$V = \int_a^b 2\pi x f(x)\, dx.$$

To see this, we take a partition $P = \{x_0, x_1, \ldots, x_n\}$ of $[a, b]$ and concentrate on what's happening on the ith subinterval $[x_{i-1}, x_i]$. Recall that, when we form a Riemann sum, we are free to choose x_i^* as any point from $[x_{i-1}, x_i]$. For convenience

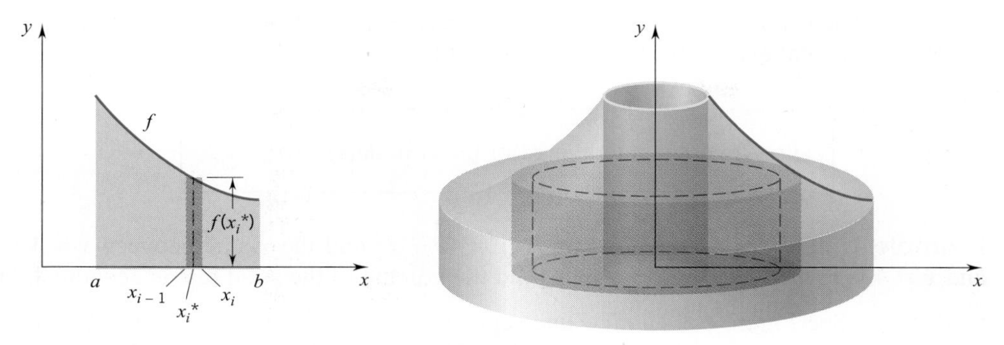

Figure 6.3.2

we take x_i^* as the midpoint $\frac{1}{2}(x_{i-1} + x_i)$. The representative rectangle of height $f(x_i^*)$ and base Δx_i (see Figure 6.3.2) generates a cylindrical shell of height $h = f(x_i^*)$, inner radius $r = x_{i-1}$, and outer radius $R = x_i$. We can calculate the volume of this shell by (6.3.1). Since

$$h = f(x_i^*) \quad \text{and} \quad R + r = x_i + x_{i-1} = 2x_i^* \quad \text{and} \quad R - r = \Delta x_i,$$

the volume of this shell is

$$\pi h(R + r)(R - r) = 2\pi x_i^* f(x_i^*)\Delta x_i.$$

The volume of the entire solid can be approximated by adding up the volumes of these shells:

$$V \cong 2\pi x_1^* f(x_1^*)\,\Delta x_1 + 2\pi x_2^* f(x_2^*)\,\Delta x_2 + \cdots + 2\pi x_n^* f(x_n^*)\,\Delta x_n.$$

The sum on the right is a Riemann sum which, as $\|P\| \to 0$, converge to

$$\int_a^b 2\pi x f(x)\,dx. \quad ❑$$

For a simple interpretation of Formula 6.3.2 we refer to Figure 6.3.3.

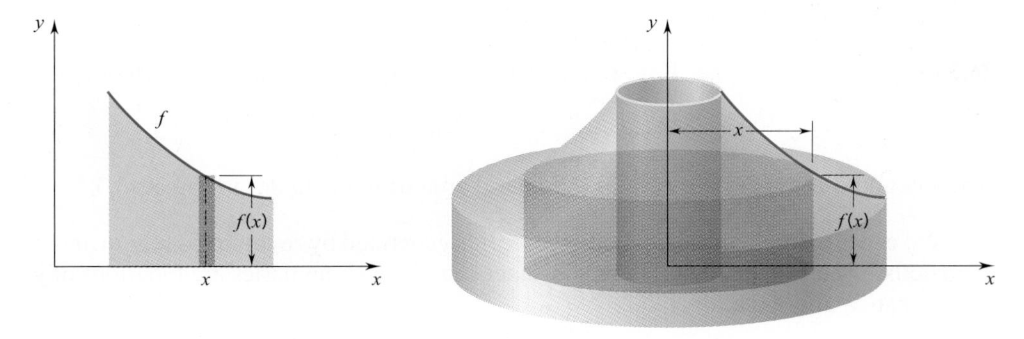

Figure 6.3.3

As the region below the graph of f is revolved about the y-axis, the line segment x units from the y-axis generates a cylinder of radius x, height $f(x)$, and lateral area

$2\pi x f(x)$. The shell method formula expresses the volume of a solid of revolution as the average lateral area of these cylinders times the length of the base interval:

(6.3.3) $$V = (\text{average lateral area of the cylinders}) \cdot (b - a).$$

Example 1 The region bounded by $f(x) = 4x - x^2$ and the x-axis between $x = 1$ and $x = 4$ is rotated about the y-axis. Find the volume of the solid that is generated.

SOLUTION See Figure 6.3.4. The line segment x units from the y-axis, $1 \le x \le 4$, generates a cylinder of radius x, height $f(x)$, and lateral area $2\pi x f(x)$. Thus, by (6.3.2)

$$V = \int_1^4 2\pi x(4x - x^2)\,dx = 2\pi \int_1^4 (4x^2 - x^3)\,dx = 2\pi \left[\frac{4x^3}{3} - \frac{x^4}{4}\right]_1^4$$

$$= 2\pi \left[\frac{256}{3} - 64\right] - 2\pi \left[\frac{4}{3} - \frac{1}{4}\right] = \frac{81\pi}{2}. \quad \square$$

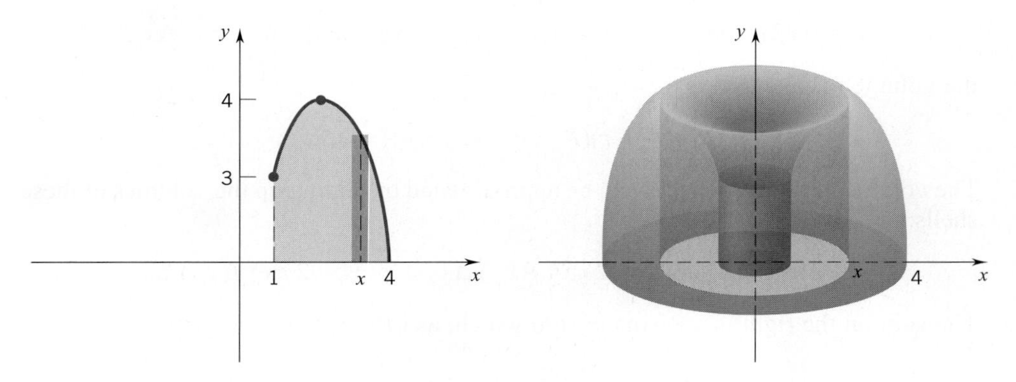

Figure 6.3.4

The shell method formula can be generalized. With Ω as in Figure 6.3.5, the volume generated by revolving Ω about the y-axis is given by the formula

(6.3.4) $$V = \int_a^b 2\pi x[f(x) - g(x)]\,dx.$$ (shell method about y-axis)

The integrand $2\pi x[f(x) - g(x)]$ is the lateral area of the cylinder in Figure 6.3.5.

We can also apply the shell method to solids generated by revolving a region about the x-axis. See Figure 6.3.6. In this instance the curved boundaries are functions of y rather than x, and we have

(6.3.5) $$V = \int_c^d 2\pi y[F(y) - G(y)]\,dy.$$ (shell method about x-axis)

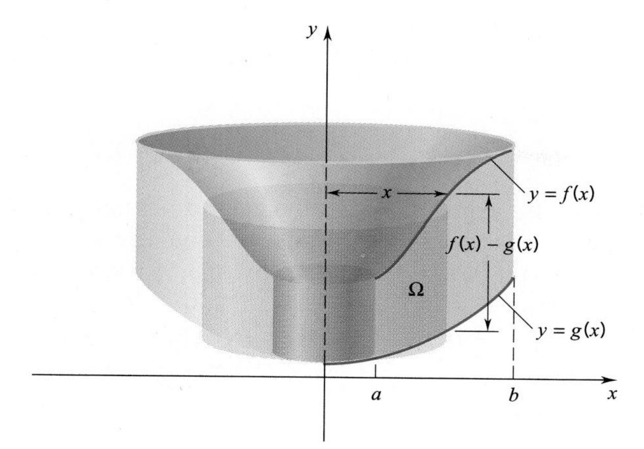

Figure 6.3.5

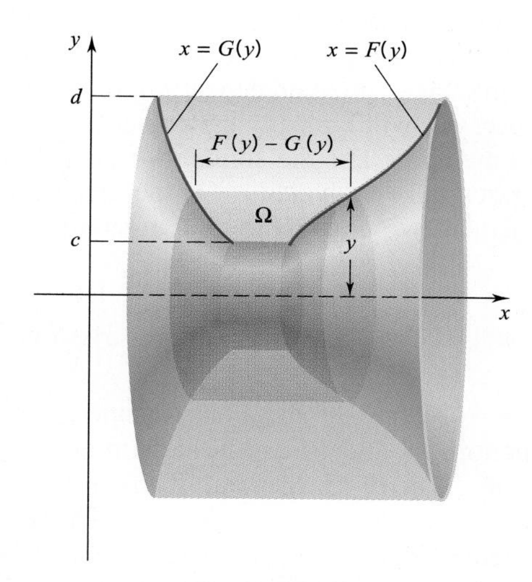

Figure 6.3.6

The integrand $2\pi y[F(y) - G(y)]$ is the lateral area of the cylinder in Figure 6.3.6.

Example 2 Find the volume of the solid generated by revolving the region between

$$y = x^2 \quad \text{and} \quad y = 2x$$

about the y-axis.

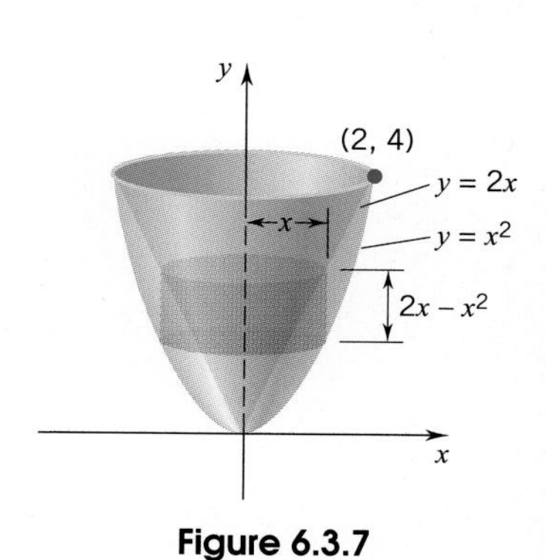

Figure 6.3.7

SOLUTION We refer to Figure 6.3.7. For each x from 0 to 2 the line segment x units from the y-axis generates a cylinder of radius x, height $(2x - x^2)$, and lateral area $2\pi x(2x - x^2)$. By (6.3.4)

$$V = \int_0^2 2\pi x(2x - x^2)\,dx = 2\pi \int_0^2 (2x^2 - x^3)\,dx = 2\pi \left[\tfrac{2}{3}x^3 - \tfrac{1}{4}x^4 \right]_0^2 = \tfrac{8}{3}\pi. \quad \square$$

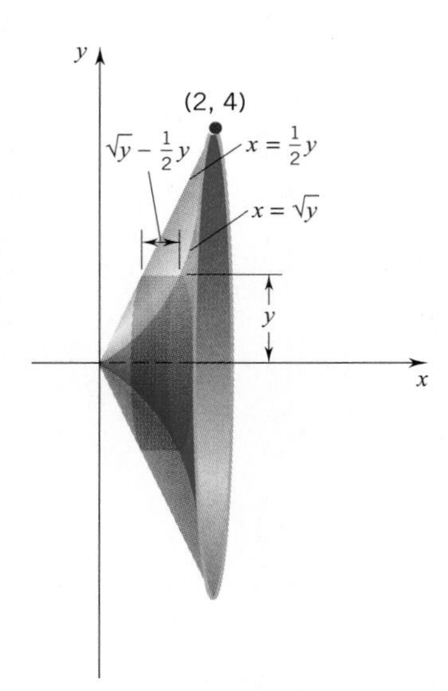

Figure 6.3.8

Example 3 Find the volume of the solid generated by revolving the region between

$$y = x^2 \quad \text{and} \quad y = 2x$$

about the x-axis.

SOLUTION We begin by expressing these boundaries as functions of y. We write $x = \sqrt{y}$ for the right boundary and $x = \frac{1}{2}y$ for the left boundary (see Figure 6.3.8). The shell of radius y has height $(\sqrt{y} - \frac{1}{2}y)$. Thus, by (6.3.5)

$$V = \int_0^4 2\pi y(\sqrt{y} - \tfrac{1}{2}y)\, dy = \pi \int_0^4 (2y^{3/2} - y^2)\, dy$$
$$= \pi \left[\tfrac{4}{5}y^{5/2} - \tfrac{1}{3}y^3 \right]_0^4 = \tfrac{64}{15}\pi. \quad ❑$$

Remark In Section 6.2 we calculated the volumes of the same solids (and got the same answers) by the washer method. ❑

Remark Examples 2 and 3 raise an obvious question: Which of the three methods —disc, washer, or shell—should you use to solve a particular problem? As you probably have guessed, the "best" method depends upon the problem. Since a disc is a special case of a washer, you will typically have to decide between the shell method and the washer or disc method. For the solids in Examples 2 and 3, the shell and washer methods were essentially equivalent. On the other hand, if you try to use the washer method to calculate the volume of the solid in Example 1, you will be led to the expression

$$V = \int_0^3 (\pi[g_1(y)]^2 - \pi[1]^2)\, dy + \int_3^4 (\pi[g_1(y)]^2 - \pi[g_2(y)]^2)\, dy, \quad \text{(see Figure 6.3.4)}$$

where $x = g_1(y)$ and $x = g_2(y)$ are the functions obtained by solving $y = 4x - x^2$ for x. While the two functions g_1 and g_2 can be determined and the integrals can be evaluated, it is not trivial. Try it! In general, it might not be possible to solve an equation $y = f(x)$ for x, or an equation $x = g(y)$ for y. The selection of the method to use on a particular problem will depend upon the region that is being revolved, the axis of revolution, and the flexibility you have in expressing the functions involved in terms of either x or y, whichever is required. ❑

Example 4 A round hole of radius r is drilled through the center of a hemisphere of radius a $(r < a)$. Find the volume of the portion of the hemisphere that remains.

SOLUTION A hemisphere of radius a can be generated by revolving the region in the first quadrant bounded by $x^2 + y^2 = a^2$ around the y-axis. See Figure 6.3.9a. The portion of the hemisphere that remains after the hole is drilled is the solid generated by revolving the shaded region around the y-axis (Figure 6.3.9b).

(a) Washer method. Refer to Figure 6.3.10.

$$V = \int_0^{\sqrt{a^2-r^2}} \left(\pi \left[\sqrt{a^2 - y^2} \right]^2 - \pi r^2 \right) dy = \pi \int_0^{\sqrt{a^2-r^2}} (a^2 - r^2 - y^2)\, dy$$
$$= \pi \left[(a^2 - r^2)y - \frac{y^3}{3} \right]_0^{\sqrt{a^2-r^2}} = \tfrac{2}{3}\pi(a^2 - r^2)^{3/2}.$$

(a)

(b)

Figure 6.3.9

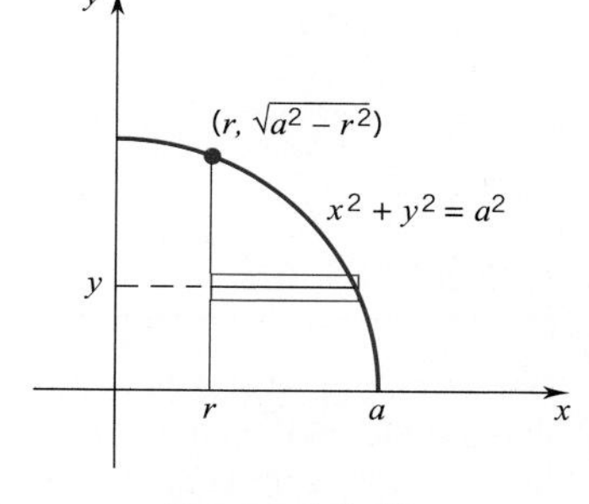

Figure 6.3.10

(b) Shell method. Refer to Figure 6.3.11.

$$V = \int_r^a 2\pi x\sqrt{a^2 - x^2}\, dx.$$

Let $u = a^2 - x^2$. Then $du = -2x\, dx$, and $u = a^2 - r^2$ at $x = r$, 0 at $x = a$. Thus

$$V = \int_r^a 2\pi x\sqrt{a^2 - x^2}\, dx = -\pi \int_{a^2 - r^2}^0 u^{1/2}\, du = \pi \int_0^{a^2 - r^2} u^{1/2}\, du$$

$$= \pi\left[\tfrac{2}{3}u^{3/2}\right]_0^{a^2 - r^2} = \tfrac{2}{3}\pi(a^2 - r^2)^{3/2}.$$

Note that if we let $r = 0$, then $V = \frac{2}{3}\pi a^3$, the volume of the hemisphere. ❑

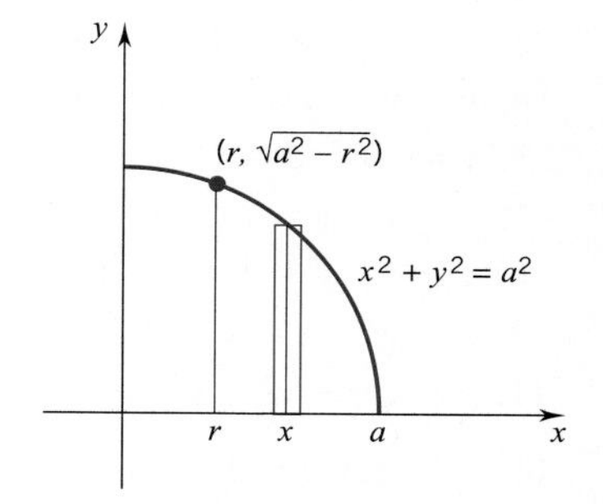

Figure 6.3.11

We conclude this section with an example in which we revolve a region about a line parallel to the y-axis.

Example 5 The region Ω between $y = \sqrt{x}$ and $y = x^2$, $0 \leq x \leq 1$, is revolved around the line $x = -2$ (see Figure 6.3.12). Find the volume of the solid which is generated.

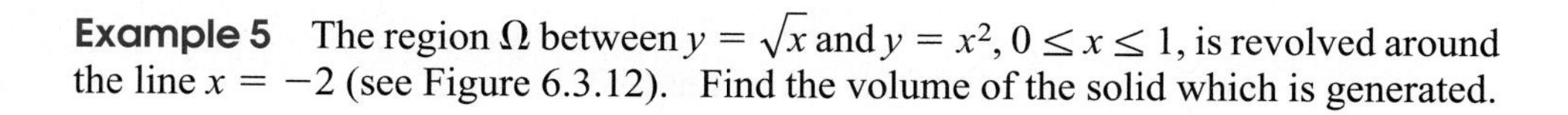

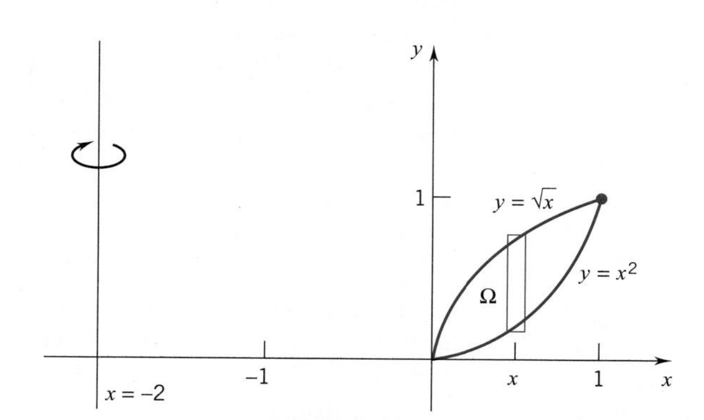

Figure 6.3.12

SOLUTION Using the shell method, for each x in $[0, 1]$, the line segment at x generates a cylinder of radius $x + 2$, height $\sqrt{x} - x^2$, and lateral area $2\pi(x + 2)(\sqrt{x} - x^2)$. Thus, the volume of the solid of revolution is

$$V = \int_0^1 2\pi(x + 2)(\sqrt{x} - x^2)\,dx$$
$$= 2\pi \int_0^1 (x^{3/2} + 2x^{1/2} - x^3 - 2x^2)\,dx$$
$$= 2\pi \left[\tfrac{2}{5}x^{5/2} + \tfrac{4}{3}x^{3/2} - \tfrac{1}{4}x^4 - \tfrac{2}{3}x^3\right]_0^1 = \tfrac{49}{30}\pi.$$

See Exercise 48, Section 6.2. ❑

EXERCISES 6.3

In Exercises 1–12, sketch the region Ω bounded by the curves and use the shell method to find the volume of the solid generated by revolving Ω about the y-axis.

1. $y = x, \quad y = 0, \quad x = 1.$
2. $x + y = 3, \quad y = 0, \quad x = 0.$
3. $y = \sqrt{x}, \quad x = 4, \quad y = 0.$
4. $y = x^3, \quad x = 2, \quad y = 0.$
5. $y = \sqrt{x}, \quad y = x^3.$
6. $y = x^2, \quad y = x^{1/3}.$
7. $y = x, \quad y = 2x, \quad y = 4.$
8. $y = x, \quad y = 1, \quad x + y = 6.$
9. $x = y^2, \quad x = y + 2.$
10. $x = y^2, \quad x = 2 - y.$
11. $x = \sqrt{9 - y^2}, \quad x = 0.$
12. $x = |y|, \quad x = 2 - y^2.$

In Exercises 13–24, sketch the region Ω bounded by the curves and use the shell method to find the volume of the solid generated by revolving Ω about the x-axis.

13. $x + 3y = 6, \quad y = 0, \quad x = 0.$
14. $y = x, \quad y = 5, \quad x = 0.$
15. $y = x^2, \quad y = 9.$
16. $y = x^3, \quad y = 8, \quad x = 0.$
17. $y = \sqrt{x}, \quad y = x^3.$
18. $y = x^2, \quad y = x^{1/3}.$
19. $y = x^2, \quad y = x + 2.$
20. $y = x^2, \quad y = 2 - x.$
21. $y = x, \quad y = 2x, \quad x = 4.$
22. $y = x, \quad x + y = 8, \quad x = 1.$
23. $y = \sqrt{1 - x^2}, \quad x + y = 1.$
24. $y = x^2, \quad y = 2 - |x|.$

The curves $y = x^2$, $y = \sqrt{x}$, $x = 0$, $x = 1$, $y = 0$, and $y = 1$ determine the three regions indicated in the figure. In Exercises 25–30, express the volume of the solid that is generated when the indicated region is revolved around the given line by: (a) a definite integral in which you will be integrating with respect to x; (b) a definite integral in which you will be integrating with respect to y. Then choose one of the two integrals and calculate the volume of the solid.

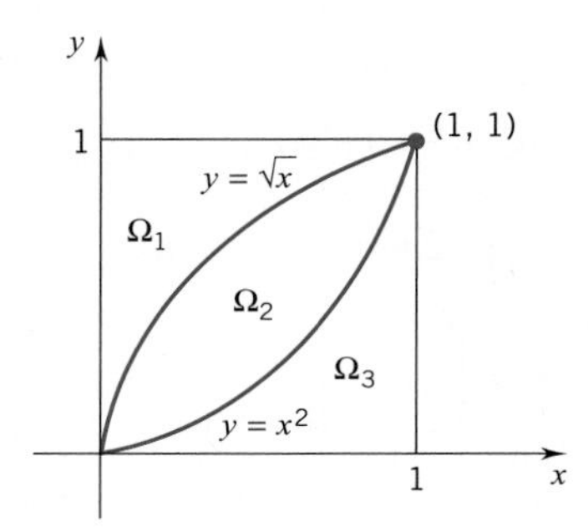

25. Ω_1 is revolved around the y-axis.
26. Ω_1 is revolved around the line $y = 2$.
27. Ω_2 is revolved around the x-axis.
28. Ω_2 is revolved around the line $x = -3$.
29. Ω_3 is revolved around the y-axis.
30. Ω_3 is revolved around the line $y = -1$.
31. Use the shell method to find the volume enclosed by the surface obtained by revolving the ellipse $b^2x^2 + a^2y^2 = a^2b^2$ about the y-axis.
32. Exercise 31 with the ellipse revolved about the x-axis.
33. Find the volume enclosed by the surface generated by revolving the equilateral triangle with vertices $(0, 0)$, $(a, 0)$, $(\frac{1}{2}a, \frac{1}{2}\sqrt{3}a)$ about the y-axis.
34. A ball of radius r is cut into two pieces by a horizontal plane a units above the center of the ball. Determine the volume of the upper piece by using the shell method.
35. Carry out Exercise 51 of Section 6.2, this time using the shell method.

36. Carry out Exercise 52 of Section 6.2, this time using the shell method.

37. (a) Verify that $F(x) = x \sin x + \cos x$ is an antiderivative of $f(x) = x \cos x$.
(b) Use the shell method to find the volume of the solid that is generated by revolving the region between $y = \cos x$ and the x-axis, $0 \le x \le \pi/2$, around the y-axis.

38. (a) Sketch the region in the right half-plane that is outside the parabola $y = x^2$ and is between the lines $y = x + 2$ and $y = 2x - 2$.
(b) The region in (a) is revolved around the y-axis. Use the method that you think is most practical to calculate the volume of the solid that is generated.

In Exercises 39–42, let

$$f(x) = \begin{cases} \sqrt{3}x & 0 \le x < 1 \\ \sqrt{4 - x^2} & 1 \le x \le 2, \end{cases}$$

and let Ω be the region between the graph of f and the x-axis (see the figure).

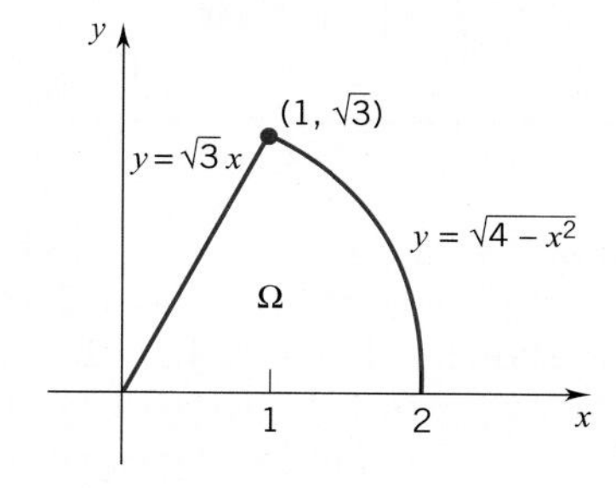

39. Revolve Ω around the y-axis. Express the volume of the solid that is generated as:
(a) A definite integral in which you will be integrating with respect to x.
(b) A definite integral in which you will be integrating with respect to y.
(c) Choose one of the two integrals and determine the volume of the solid.

40. Revolve Ω around the x-axis and repeat Exercise 39.

41. Revolve Ω around the line $x = 2$ and repeat parts (a) and (b) of Exercise 39.

42. Revolve Ω around the line $y = -1$ and repeat parts (a) and (b) of Exercise 39.

43. Let Ω be the circular disc $(x - b)^2 + y^2 \le a^2$, where $b > a > 0$. The doughnut-shaped region that is generated when Ω is revolved around the y-axis is called a *torus*. Express the volume V of the torus as:
(a) A definite integral in which you will be integrating with respect to x.
(b) A definite integral in which you will be integrating with respect to y.
(c) Show that $V = 2\pi^2 a^2 b$ by calculating one of the two definite integrals.

44. The circular disc $x^2 + y^2 \le a^2$, $a > 0$, is revolved around the line $x = a$. Find the volume of the solid that is generated.

■ 6.4 THE CENTROID OF A REGION; PAPPUS'S THEOREM ON VOLUMES

The Centroid of a Region

In Section 5.9 you saw how to calculate the center of mass of a thin rod. Suppose now that we have a thin distribution of matter, a *plate,* laid out in the xy-plane in the shape of some region Ω (Figure 6.4.1). If the mass density of the plate varies from point to point, then the determination of the center of mass of the plate requires the evaluation of a double integral (Chapter 16). If, however, the mass density of the plate is constant throughout Ω, then the center of mass of the plate depends only on the shape of Ω and falls on a point $(\bar{x}, \bar{y})$ that we call the *centroid.* Unless Ω has a very complicated shape, we can calculate its centroid by ordinary one-variable integration.

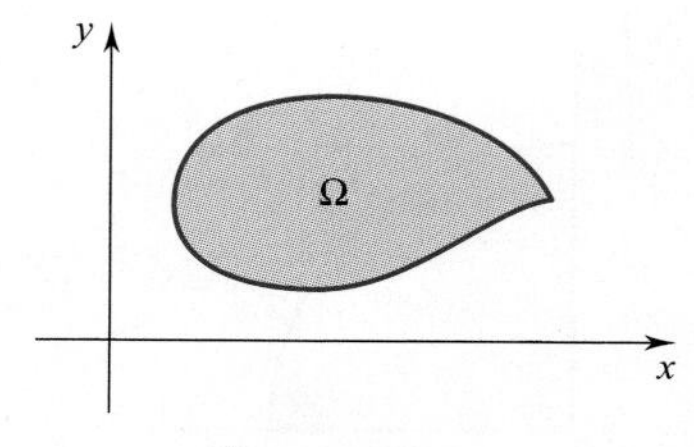

Figure 6.4.1

We will use two guiding principles to find the centroid of a plane region Ω. The first is obvious. The second we take from physics; this result is easily justified by double integration.

Principle 1: Symmetry If the region has an axis of symmetry, then the centroid $(\bar{x}, \bar{y})$ lies somewhere along that axis. In particular, if the region has a center, then that center is the centroid.

Principle 2: Additivity If the region, having area A, consists of a finite number of pieces with areas $A_1, \ldots, A_n$ and centroids $(\bar{x}_1, \bar{y}_1), \ldots, (\bar{x}_n, \bar{y}_n)$, then

(6.4.1) $$\bar{x}A = \bar{x}_1 A_1 + \cdots + \bar{x}_n A_n \quad \text{and} \quad \bar{y}A = \bar{y}_1 A_1 + \cdots + \bar{y}_n A_n.$$

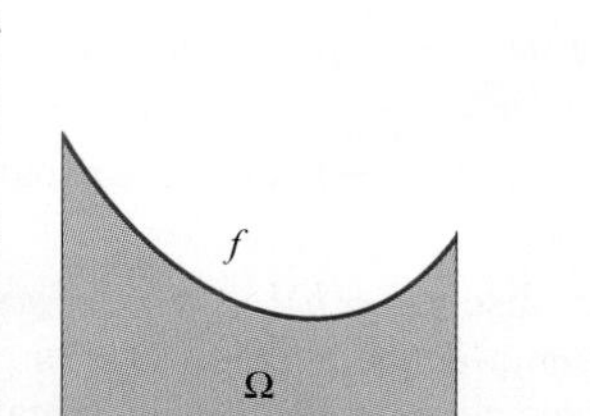

Figure 6.4.2

We are now ready to bring the techniques of calculus into play. Figure 6.4.2 shows the region Ω under the graph of a continuous function f. Let's denote the area of Ω by A. The centroid $(\bar{x}, \bar{y})$ of Ω can be obtained from the following formulas:

(6.4.2) $$\bar{x}A = \int_a^b x f(x)\, dx, \qquad \bar{y}A = \int_a^b \tfrac{1}{2}[f(x)]^2\, dx.$$

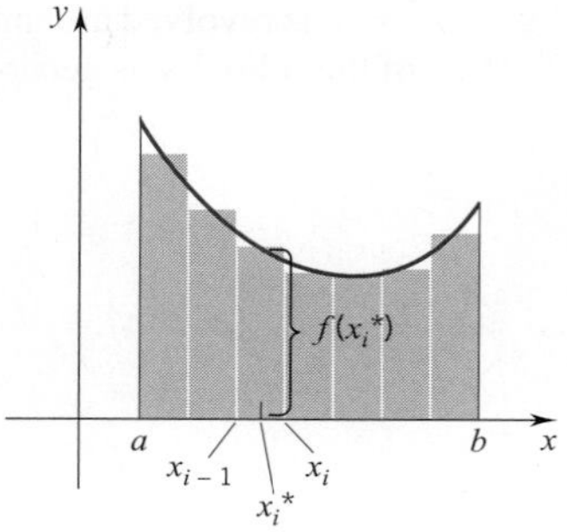

Figure 6.4.3

To derive these formulas we choose a partition $P = \{x_0, x_1, \ldots, x_n\}$ of $[a, b]$. This breaks up $[a, b]$ into n subintervals $[x_{i-1}, x_i]$. Choosing x_i^* as the midpoint of $[x_{i-1}, x_i]$, we form the midpoint rectangles R_i shown in Figure 6.4.3. The area of R_i is $f(x_i^*)\, \Delta x_i$ and the centroid of R_i is its center $(x_i^*, \frac{1}{2}f(x_i^*))$. By (6.4.1), the centroid $(\bar{x}_P, \bar{y}_P)$ of the union of all these rectangles satisfies the following equations:

$$\bar{x}_P A_P = x_1^* f(x_1^*)\, \Delta x_1 + \cdots + x_n^* f(x_n^*)\, \Delta x_n,$$
$$\bar{y}_P A_P = \tfrac{1}{2}[f(x_1^*)]^2\, \Delta x_1 + \cdots + \tfrac{1}{2}[f(x_n^*)]^2\, \Delta x_n.$$

(Here A_P represents the area of the union of the n rectangles.) As $\|P\| \to 0$, the union of rectangles tends to the shape of Ω and the equations we just derived tend to the formulas given in (6.4.2). ❑

Example 1 Find the centroid of the quarter-disc shown in Figure 6.4.4.

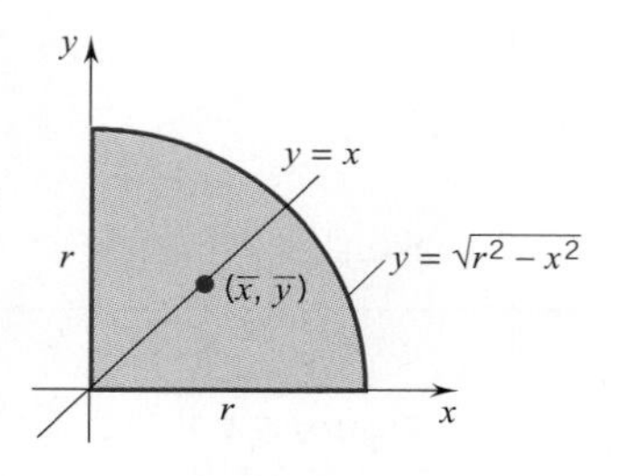

Figure 6.4.4

SOLUTION The quarter-disc is symmetric about the line $x = y$. We know therefore that $\bar{x} = \bar{y}$. Here

$$\bar{y}A = \int_0^r \tfrac{1}{2}[f(x)]^2\, dx = \int_0^r \tfrac{1}{2}(r^2 - x^2)\, dx = \tfrac{1}{2}\left[r^2 x - \frac{x^3}{3}\right]_0^r = \tfrac{1}{3}r^3.$$

$f(x) = \sqrt{r^2 - x^2}$

Since $A = \frac{1}{4}\pi r^2$,

$$\bar{y} = \frac{\frac{1}{3}r^3}{\frac{1}{4}\pi r^2} = \frac{4r}{3\pi}.$$

The centroid of the quarter-disc is the point

$$\left(\frac{4r}{3\pi}, \frac{4r}{3\pi}\right).$$

NOTE: It is almost as easy to calculate $\bar{x}A$:

$$\begin{aligned}
\bar{x}A &= \int_0^r x f(x)\,dx = \int_0^r x\sqrt{r^2 - x^2}\,dx \\
&= -\tfrac{1}{2}\int_{r^2}^{0} u^{1/2}\,du \qquad\qquad [u = (r^2 - x^2),\ du = -2x\,dx] \\
&= -\tfrac{1}{2}\left[\tfrac{2}{3}u^{3/2}\right]_{r^2}^{0} = \tfrac{1}{3}r^3. \quad \square
\end{aligned}$$

Example 2 Find the centroid of the right triangle shown in Figure 6.4.5

SOLUTION There is no symmetry that we can use here. The hypotenuse lies on the line

$$y = -\frac{h}{b}x + h.$$

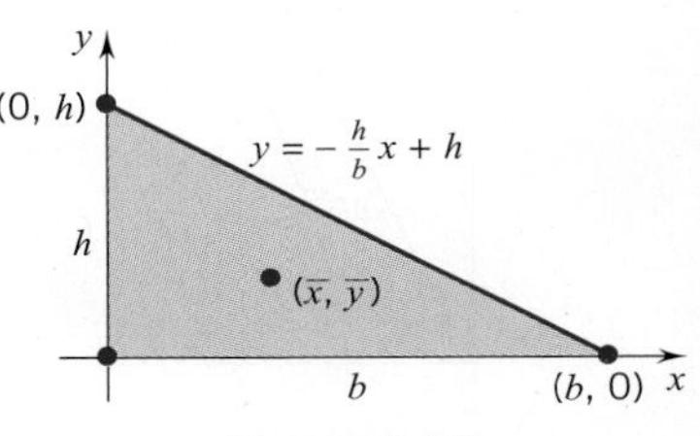

Figure 6.4.5

Hence

$$\bar{x}A = \int_0^b x f(x)\,dx = \int_0^b \left(-\frac{h}{b}x^2 + hx\right)dx = \tfrac{1}{6}b^2 h$$

and

$$\bar{y}A = \int_0^b \tfrac{1}{2}[f(x)]^2\,dx = \tfrac{1}{2}\int_0^b \left(\frac{h^2}{b^2}x^2 - \frac{2h^2}{b}x + h^2\right)dx = \tfrac{1}{6}bh^2.$$

Since $A = \frac{1}{2}bh$, we have

$$\bar{x} = \frac{\frac{1}{6}b^2h}{\frac{1}{2}bh} = \tfrac{1}{3}b \qquad \text{and} \qquad \bar{y} = \frac{\frac{1}{6}bh^2}{\frac{1}{2}bh} = \tfrac{1}{3}h.$$

The centroid is the point $(\frac{1}{3}b, \frac{1}{3}h)$. $\square$

Figure 6.4.6 shows the region Ω between the graphs of two continuous functions f and g. In this case, if Ω has area A and centroid $(\bar{x}, \bar{y})$, then

(6.4.3)

$$\boxed{\bar{x}A = \int_a^b x[f(x) - g(x)]\,dx, \qquad \bar{y}A = \int_a^b \tfrac{1}{2}([f(x)]^2 - [g(x)]^2)\,dx.}$$

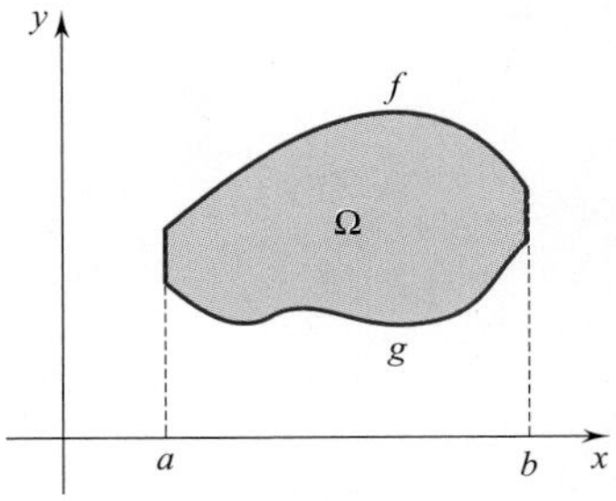

Figure 6.4.6

PROOF Let A_f be the area below the graph of f and let A_g be the area below the graph of g. Then in obvious notation

$$\bar{x}A + \bar{x}_g A_g = \bar{x}_f A_f \qquad \text{and} \qquad \bar{y}A + \bar{y}_g A_g = \bar{y}_f A_f. \qquad (6.4.1)$$

Therefore

$$\bar{x}A = \bar{x}_f A_f - \bar{x}_g A_g = \int_a^b xf(x)\,dx - \int_a^b xg(x)\,dx = \int_a^b x[f(x) - g(x)]\,dx$$

and

$$\bar{y}A = \bar{y}_f A_f - \bar{y}_g A_g = \int_a^b \tfrac{1}{2}[f(x)]^2\,dx - \int_a^b \tfrac{1}{2}[g(x)]^2\,dx$$

$$= \int_a^b \tfrac{1}{2}([f(x)]^2 - [g(x)]^2)\,dx. \quad \square$$

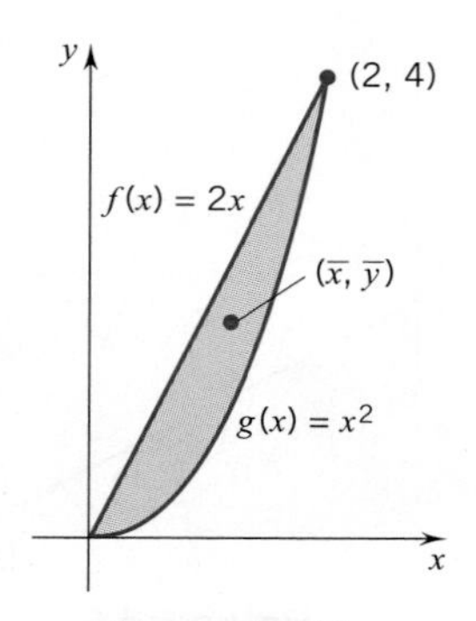

Figure 6.4.7

Example 3 Find the centroid of the region shown in Figure 6.4.7.

SOLUTION Here there is no symmetry we can appeal to. We must carry out the calculations.

$$A = \int_0^2 [f(x) - g(x)]\,dx = \int_0^2 (2x - x^2)\,dx = \left[x^2 - \tfrac{1}{3}x^3\right]_0^2 = \tfrac{4}{3},$$

$$\bar{x}A = \int_0^2 x[f(x) - g(x)]\,dx = \int_0^2 (2x^2 - x^3)\,dx = \left[\tfrac{2}{3}x^3 - \tfrac{1}{4}x^4\right]_0^2 = \tfrac{4}{3},$$

$$\bar{y}A = \int_0^2 \tfrac{1}{2}([f(x)]^2 - [g(x)]^2)\,dx = \tfrac{1}{2}\int_0^2 (4x^2 - x^4)\,dx = \tfrac{1}{2}\left[\tfrac{4}{3}x^3 - \tfrac{1}{5}x^5\right]_0^2 = \tfrac{32}{15}.$$

Therefore $\bar{x} = \frac{4}{3}/\frac{4}{3} = 1$ and $\bar{y} = \frac{32}{15}/\frac{4}{3} = \frac{8}{5}$. $\square$

Pappus's Theorem on Volumes

All the formulas that we have derived for volumes of solids of revolution are simple corollaries to an observation made by a brilliant ancient Greek, Pappus of Alexandria (circa 300 A.D.).

THEOREM 6.4.4 PAPPUS'S THEOREM ON VOLUMES†

A plane region is revolved about an axis that lies in its plane. If the region does not cross the axis, then the volume of the resulting solid of revolution is the area of the region multiplied by the circumference of the circle described by the centroid of the region:

$$V = 2\pi\bar{R}A$$

where A is the area of the region and $\bar{R}$ is the distance from the axis to the centroid of the region. See Figure 6.4.8.

† This theorem is found in Book VII of Pappus's "Mathematical Collection," largely a survey of ancient geometry to which Pappus made many original contributions (among them this theorem). Much of what we know today of Greek geometry we owe to Pappus.

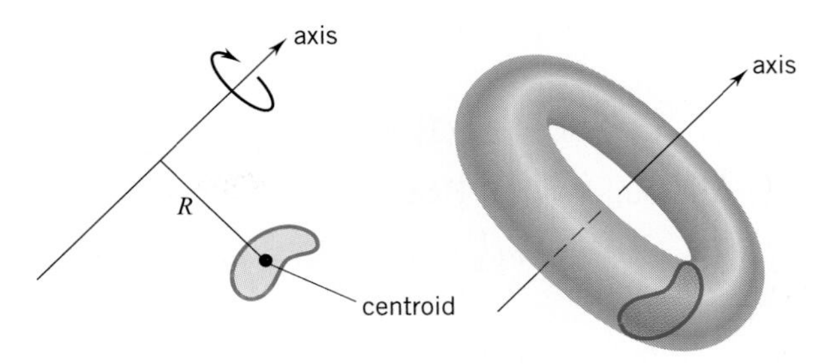

Figure 6.4.8

Basically we have derived only two formulas for the volumes of solids of revolution:

1. *The Washer Method Formula.* If the region Ω of Figure 6.4.6 is revolved about the x-axis, the resulting solid has volume

$$V_x = \int_a^b \pi([f(x)]^2 - [g(x)]^2)\,dx.$$

2. *The Shell Method Formula.* If the region Ω of Figure 6.4.6 is revolved about the y-axis, the resulting solid has volume

$$V_y = \int_a^b 2\pi x[f(x) - g(x)]\,dx.$$

Note that

$$V_x = \int_a^b \pi([f(x)]^2 - [g(x)]^2)\,dx = 2\pi \int_a^b \tfrac{1}{2}([f(x)]^2 - [g(x)]^2)\,dx = 2\pi\bar{y}A = 2\pi\bar{R}A$$

and

$$V_y = \int_a^b 2\pi x[f(x) - g(x)]\,dx = 2\pi\bar{x}A = 2\pi\bar{R}A,$$

exactly as predicted by Pappus.

Remark In stating Pappus's theorem we assumed a complete revolution. If Ω is only partially revolved about the given axis, then the volume of the resulting solid is simply the area of Ω multiplied by the length of the circular arc described by the centroid of Ω. ❑

Applications of Pappus's Theorem

Example 4 Earlier we saw that the region shown in Figure 6.4.7 has area $\frac{4}{3}$ and centroid $(1, \frac{8}{5})$. Find the volumes of the solids formed by revolving this region about: (a) the y-axis, and (b) the line $y = 5$.

SOLUTION (a) We have already calculated this volume two ways, by the washer method and by the shell method. (See Example 8, Section 6.2, and Example 2,

Section 6.3.) The result was $V = \frac{8}{3}\pi$. Now we will use Pappus's theorem. We have $\overline{R} = 1$ and $A = \frac{4}{3}$. Thus

$$V = 2\pi(1)(\tfrac{4}{3}) = \tfrac{8}{3}\pi.$$

(b) Here $\overline{R} = 5 - \frac{8}{5} = \frac{17}{5}$ and $A = \frac{4}{3}$. Hence

$$V = 2\pi(\tfrac{17}{5})(\tfrac{4}{3}) = \tfrac{136}{15}\pi \quad \square$$

Example 5 Find the volume of the torus generated by revolving the circular disc

$$(x-h)^2 + (y-k)^2 \leq c^2, \qquad h, k \geq c > 0 \qquad \text{(Figure 6.4.9)}$$

about (a) the x-axis, (b) the y-axis.

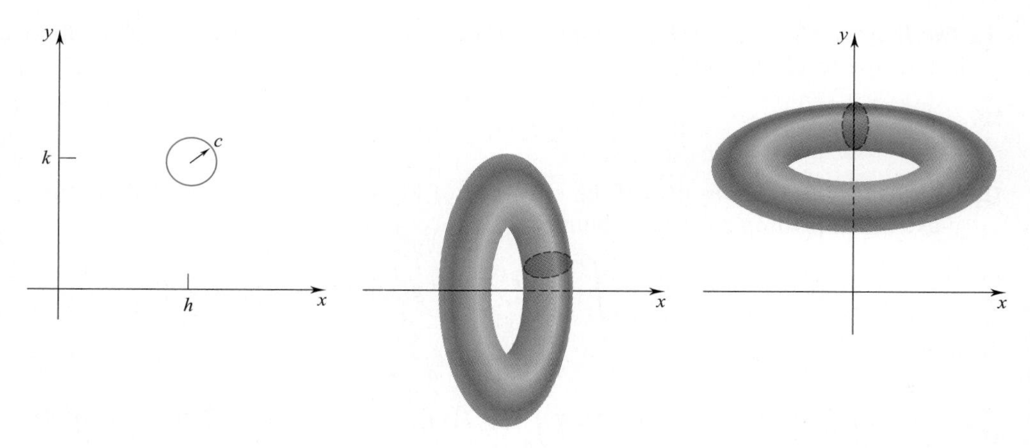

Figure 6.4.9

SOLUTION The centroid of the disc is the center (h, k). This lies k units from the x-axis and h units from the y-axis. The area of the disc is πc^2. Therefore
(a) $V_x = 2\pi(k)(\pi c^2) = 2\pi^2 kc^2$, (b) $V_y = 2\pi(h)(\pi c^2) = 2\pi^2 hc^2$. $\square$

Example 6 Find the centroid of the half-disc

$$x^2 + y^2 \leq r^2, \qquad y \geq 0$$

by appealing to Pappus's theorem.

SOLUTION Since the half-disc is symmetric about the y-axis, we know that $\bar{x} = 0$. All we need to find is $\bar{y}$.

If we revolve the half-disc about the x-axis, we obtain a solid ball of volume $\frac{4}{3}\pi r^3$. The area of the half-disc is $\frac{1}{2}\pi r^2$. By Pappus's theorem

$$\tfrac{4}{3}\pi r^3 = 2\pi\bar{y}(\tfrac{1}{2}\pi r^2).$$

Simple division gives $\bar{y} = 4r/3\pi$. $\square$

Remark Centroids of solids of revolution are discussed in the exercises. $\square$

EXERCISES 6.4

In Exercises 1–14, sketch the region bounded by the curves. Determine the centroid of the region and the volume generated by revolving the region about each of the coordinate axes.

1. $y = \sqrt{x}, \quad y = 0, \quad x = 4.$
2. $y = x^3, \quad y = 0, \quad x = 2.$
3. $y = x^2, \quad y = x^{1/3}.$
4. $y = x^3, \quad y = \sqrt{x}.$
5. $y = 2x, \quad y = 2, \quad x = 3.$
6. $y = 3x, \quad y = 6, \quad x = 1.$
7. $y = x^2 + 2, \quad y = 6, \quad x = 0.$
8. $y = x^2 + 1, \quad y = 1, \quad x = 3.$
9. $\sqrt{x} + \sqrt{y} = 1, \quad x + y = 1.$
10. $y = \sqrt{1 - x^2}, \quad x + y = 1.$
11. $y = x^2, \quad y = 0, \quad x = 1, \quad x = 2.$
12. $y = x^{1/3}, \quad y = 1, \quad x = 8.$
13. $y = x, \quad x + y = 6, \quad y = 1.$
14. $y = x, \quad y = 2x, \quad x = 3.$

In Exercises 15–24, find the centroid of the bounded region determined by the following curves.

15. $y = 6x - x^2, \quad y = x.$
16. $y = 4x - x^2, \quad y = 2x - 3.$
17. $x^2 = 4y, \quad x - 2y + 4 = 0.$
18. $y = x^2, \quad 2x - y + 3 = 0.$
19. $y^3 = x^2, \quad 2y = x.$
20. $y^2 = 2x, \quad y = x - x^2.$
21. $y = x^2 - 2x, \quad y = 6x - x^2.$
22. $y = 6x - x^2, \quad x + y = 6.$
23. $x + 1 = 0, \quad x + y^2 = 0.$
24. $\sqrt{x} + \sqrt{y} = \sqrt{a}, \quad x = 0, \quad y = 0.$
25. Let Ω be the annular region (ring) formed by the circles

$$x^2 + y^2 = \tfrac{1}{4} \quad \text{and} \quad x^2 + y^2 = 4.$$

(a) Locate the centroid of Ω. (b) Locate the centroid of the first-quadrant part of Ω. (c) Locate the centroid of the upper half of Ω.

26. The ellipse $b^2x^2 + a^2y^2 = a^2b^2$ encloses a region of area πab. Locate the centroid of the upper half of the region.
27. The rectangle in the figure is revolved about the line marked l. Find the volume of the resulting solid.

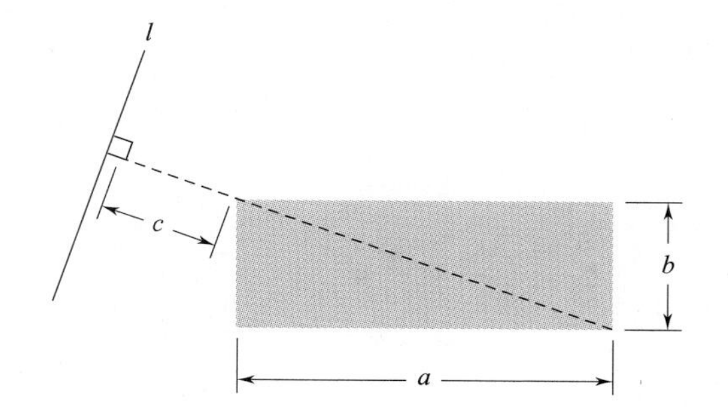

28. In Example 2 of this section you saw that the centroid of the triangle in Figure 6.4.5 is the point $(\frac{1}{3}b, \frac{1}{3}h)$.
    (a) Verify that the line segments that join the centroid to the vertices divide the triangle into three triangles of equal area.
    (b) Find the distance d from the centroid of the triangle to the hypotenuse.
    (c) Find the volume generated by revolving the triangle about the hypotenuse.
29. The triangular region in the figure is the union of two right triangles Ω_1, Ω_2. Find the centroid: (a) of Ω_1; (b) of Ω_2; (c) of the entire region.

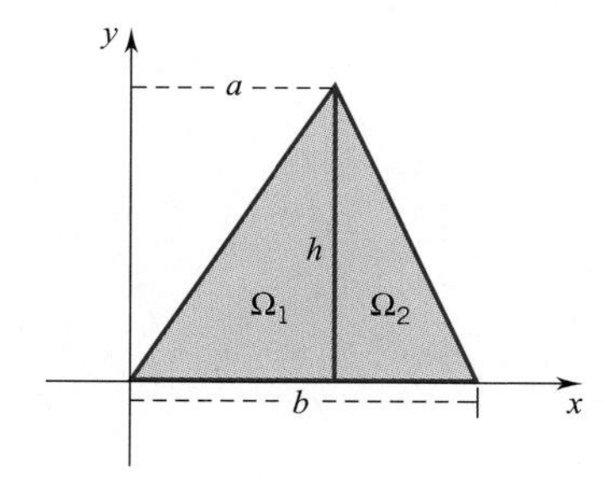

30. Find the volume of the solid generated by revolving the entire triangular region in the figure in Exercise 29 about: (a) the x-axis; (b) the y-axis.
31. (a) Find the volume of the ice-cream cone shown in the figure (a right circular cone topped by a solid hemisphere).

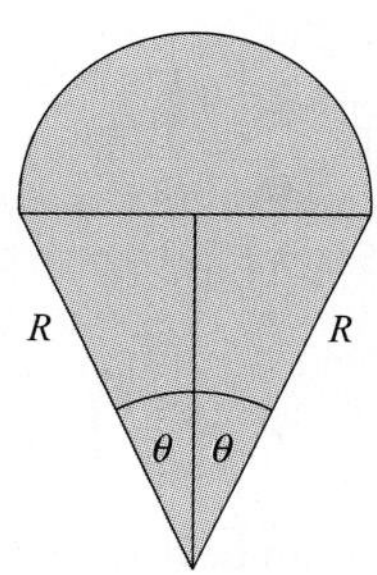

(b) Find $\bar{x}$ for the region Ω in the figure.

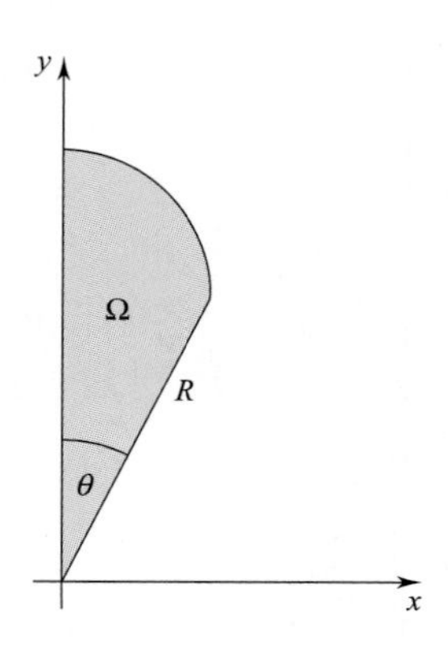

32. The region Ω in the figure consists of a square S of side $2r$ and four semidiscs of radius r. Find the centroid of each of the following.
(a) Ω. (b) Ω_1. (c) $S \cup \Omega_1$. (d) $S \cup \Omega_3$.
(e) $S \cup \Omega_1 \cup \Omega_3$. (f) $S \cup \Omega_1 \cup \Omega_2$.
(g) $S \cup \Omega_1 \cup \Omega_2 \cup \Omega_3$.

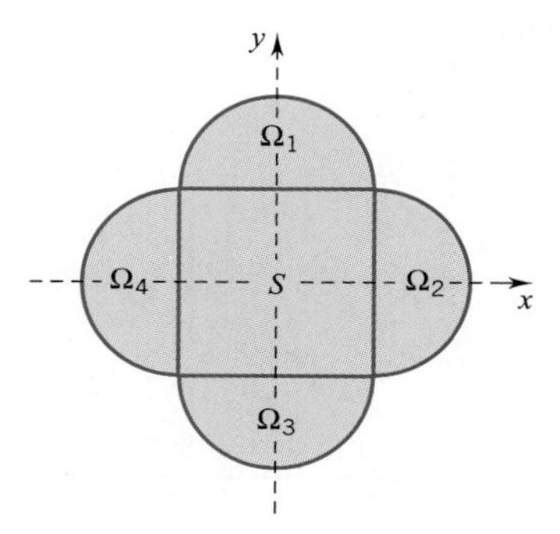

33. (a) In Section 5.9 we stated that the mass of a rod that extends from $x = a$ to $x = b$ is given by the formula

$$M = \int_a^b \lambda(x)\, dx$$

where λ is the mass density function. Derive this formula by taking an arbitrary partition $P = \{x_0, x_1, \ldots, x_n\}$ of $[a, b]$, estimating the mass contributed by $[x_{i-1}, x_i]$, adding up these contributions, and then taking the limit as $\|P\| \to 0$.

(b) In Section 5.9 we also gave the center of mass formula

$$x_M M = \int_a^b x\, \lambda(x)\, dx.$$

This formula can be derived from the physical premise that, if the rod is broken up into a finite number of pieces of masses $M_1, \ldots, M_n$ and centers of mass $x_{M_1}, \ldots, x_{M_n}$, then

$$x_M M = x_{M_1} M_1 + \cdots + x_{M_n} M_n.$$

Carry out the derivation.

34. (*The Centroid of a Solid of Revolution.*) If a solid is *homogeneous* (constant mass density), then the center of mass of the solid depends only on the shape of the solid and is called the *centroid.* The determination of the centroid of a solid of arbitrary shape requires triple integration (Chapter 16). However, if the solid is a solid of revolution, then the centroid can be found by ordinary one-variable integration.

(a) Let Ω be the region of Figure 6.4.2 and let T be the solid generated by revolving Ω about the x-axis. By symmetry the centroid of T lies on the x-axis. Thus the centroid of T is completely determined by its x-coordinate $\bar{x}$. Show that

(6.4.5)
$$\bar{x}V_x = \int_a^b \pi x[f(x)]^2\, dx$$

basing your argument on the following additivity principle: If a solid of volume V consists of a finite number of pieces with volumes $V_1, \ldots, V_n$ and the centroids of the pieces have x-coordinates $\bar{x}_1, \ldots, \bar{x}_n$, then

$$\bar{x}V = \bar{x}_1 V_1 + \cdots + \bar{x}_n V_n.$$

(b) Let Ω be the region of Figure 6.4.2 and let T be the solid generated by revolving Ω about the y-axis. By symmetry the centroid of T lies on the y-axis. Thus the centroid of T is completely determined by its y-coordinate $\bar{y}$. Show that

(6.4.6)
$$\bar{y}V_y = \int_a^b \pi x[f(x)]^2\, dx$$

basing your argument on the following additivity principle: If a solid of volume V consists of a finite number of pieces with volumes $V_1, \ldots, V_n$ and the centroids of the pieces have y-coordinates $\bar{y}_1, \ldots, \bar{y}_n$, then

$$\bar{y}V = \bar{y}_1 V_1 + \cdots + \bar{y}_n V_n.$$

In Exercises 35–40, locate the centroid of each of the given configurations.

35. A solid cone of base radius r and height h.

36. A solid hemisphere of radius r.

37. The solid generated by revolving the region below the graph of $f(x) = \sqrt{x}$, $x \in [0, 1]$ about: (a) the x-axis; (b) the y-axis.

38. The solid generated by revolving the region below the graph of $f(x) = 4 - x^2$, $x \in [0, 2]$ about: (a) the x-axis; (b) the y-axis.

39. The solid generated by revolving about the x-axis the first-quadrant part of the region enclosed by the ellipse $b^2x^2 + a^2y^2 = a^2b^2$.

40. A segment of height h cut from a ball of radius r. Take $h < r$.

■ 6.5 THE NOTION OF WORK

We begin with a constant force F that acts along some line that we call the x-axis. By convention F is positive if it acts in the direction of increasing x and negative if it acts in the direction of decreasing x (Figure 6.5.1).

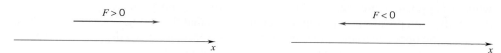

Figure 6.5.1

Suppose now that an object moves along the x-axis from $x = a$ to $x = b$ subject to this constant force F. The *work* done by F during the displacement is by definition the *force times the displacement*:

(6.5.1)
$$W = F \times (b - a).$$

It is not hard to see that, if F acts in the direction of the motion, then $W > 0$, but, if F acts against the motion, then $W < 0$. Thus, for example, if an object slides off a table and falls to the floor, then the work done by gravity is positive (after all, earth's gravity points down). But if an object is lifted from the floor and raised to tabletop level then the work done by gravity is by definition negative. However, the work done by the hand that lifts the object is positive.

To repeat, if an object moves from $x = a$ to $x = b$ subject to a constant force F, then the work done by F is the constant value of F times $b - a$. What is the work done by F if F does not remain constant but instead varies continuously as a function of x? As you would expect, we then define the work done by F as the *average value* of F times $b - a$:

(6.5.2)
$$W = \int_a^b F(x)\,dx.$$
(Figure 6.5.2)

a b
constant force
$W = F \times (b - a)$

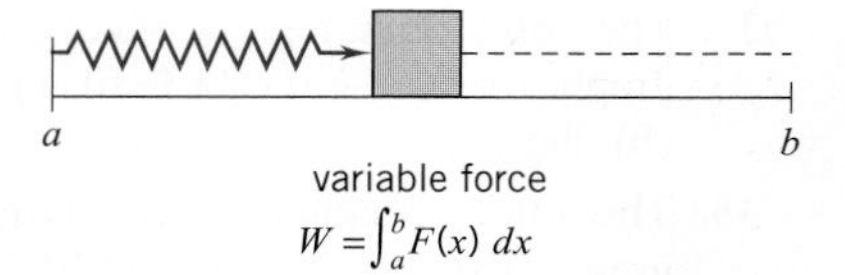

Figure 6.5.2

Hooke's Law

You can sense a variable force in the action of a steel spring. Stretch a spring within its elastic limit and you feel a pull in the opposite direction. The greater the stretching, the harder the pull of the spring. Compress a spring within its elastic limit and you feel a push against you. The greater the compression, the harder the push. According to Hooke's law (Robert Hooke, 1635–1703) the force exerted by the spring can be written

$$F(x) = -kx$$

where k is a positive number, called *the spring constant,* and x is the displacement from the equilibrium position. The minus sign indicates that the spring force always acts in the direction opposite to the direction in which the spring has been deformed (the force always acts so as to restore the spring to its equilibrium state).

Remark Hooke's law is only an approximation, but it is a good approximation for small displacements. In the problems that follow we assume that the restoring force of the spring is given by Hooke's law. ❑

Example 1 A spring of natural length L compressed to length $\frac{7}{8}L$ exerts a force F_0. Find the work done by the spring in restoring itself to natural length.

SOLUTION Place the spring on the x-axis so that the equilibrium point falls at the origin. View compression as a move to the left. See Figure 6.5.3.

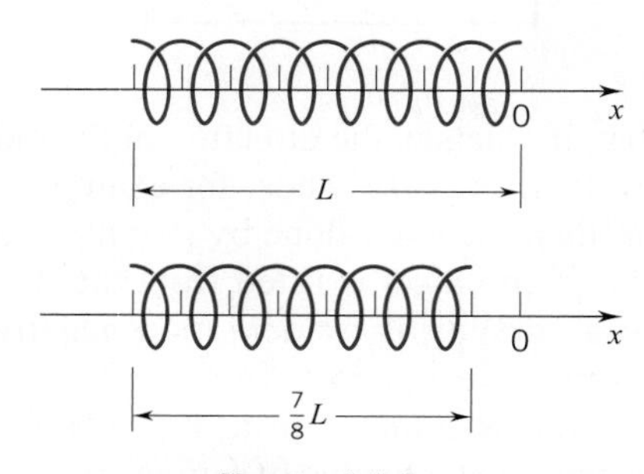

Figure 6.5.3

Compressed $\frac{1}{8}L$ units to the left, the spring exerts a force F_0. Thus by Hooke's law

$$F_0 = F(-\tfrac{1}{8}L) = -k(-\tfrac{1}{8}L) = \tfrac{1}{8}kL.$$

This tells us that $k = 8F_0/L$. Therefore the force law for this spring reads

$$F(x) = -\left(\frac{8F_0}{L}\right)x.$$

To find the work done by this spring in restoring itself to equilibrium, we integrate $F(x)$ from $x = -\frac{1}{8}L$ to $x = 0$:

$$W = \int_{-L/8}^{0} F(x)\,dx = \int_{-L/8}^{0} -\left(\frac{8F_0}{L}\right) x\,dx = -\frac{8F_0}{L}\left[\frac{x^2}{2}\right]_{-L/8}^{0} = \frac{LF_0}{16}. \quad \square$$

Example 2 For the spring in Example 1, what work must we do to stretch the spring to length $\frac{11}{10}L$?

SOLUTION Refer to Figure 6.5.4. To stretch the spring, we must counteract the force of the spring. The force exerted by the spring when stretched x units is

$$F(x) = -\left(\frac{8F_0}{L}\right) x. \qquad \text{(Example 1)}$$

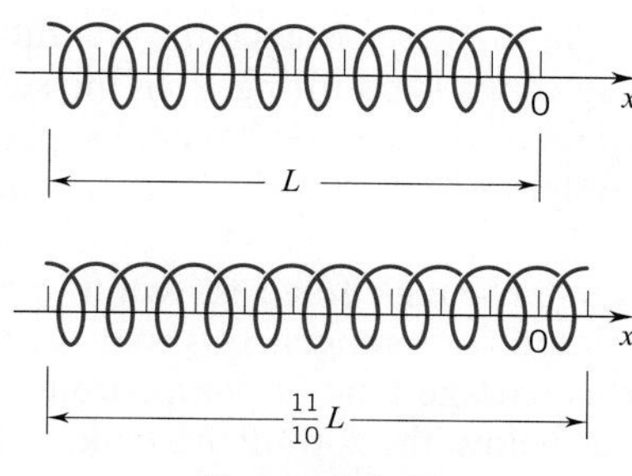

Figure 6.5.4

To counter this force we must apply the opposite force

$$-F(x) = \left(\frac{8F_0}{L}\right) x.$$

The work we must do to stretch the spring to length $\frac{11}{10}L$ can be found by integrating $-F(x)$ from $x = 0$ to $x = \frac{1}{10}L$:

$$W = \int_{0}^{L/10} -F(x)\,dx = \int_{0}^{L/10} \left(\frac{8F_0}{L}\right) x\,dx = \frac{8F_0}{L}\left[\frac{x^2}{2}\right]_{0}^{L/10} = \frac{LF_0}{25}. \quad \square$$

Remark about Units. One unit of work is the work done by a unit force in moving an object a unit distance. If force is measured in pounds and distance in feet, then the units of work are *foot-pounds.* In the metric system, the standard units of force are *newtons* (one newton is the force required to give a mass of 1 kilogram an acceleration of 1 meter per second per second) and *dynes* (one dyne is the force required to give a mass of 1 gram an acceleration of 1 centimeter per second per second). The units of work in the metric system are *newton-meters,* also called *joules,* and *dyne-centimeters,* also called *ergs.* $\square$

Example 3 Stretched $\frac{1}{3}$ meter beyond its natural length, a certain spring exerts a restoring force of 10 newtons. What work must we do to stretch the spring another $\frac{1}{3}$ meter?

SOLUTION Place the spring on the x-axis so that the equilibrium point falls at the origin. View stretching as moving to the right and assume Hooke's law: $F(x) = -kx$.

When the spring is stretched $\frac{1}{3}$ meter, it exerts a force of -10 newtons (10 newtons to the left). Therefore, $-10 = -k(\frac{1}{3})$ and $k = 30$ (that is, 30 newtons per meter).

To find the work we must do to stretch the spring another $\frac{1}{3}$ meter, we integrate the opposite force $-F(x) = 30x$ from $x = \frac{1}{3}$ to $x = \frac{2}{3}$:

$$W = \int_{1/3}^{2/3} 30x\,dx = 30\left[\tfrac{1}{2}x^2\right]_{1/3}^{2/3} = 5 \text{ joules.} \quad \square$$

Pumping Out a Tank

To lift an object we must counteract the force of gravity. Consequently, the work done in lifting an object is given by the formula

$$\text{Work} = (\text{weight of the object}) \times (\text{distance lifted}).$$

If we lift a leaking sand bag or pump out a water tank from above, the calculation of work is more complicated. In the first instance the weight varies during the motion. (There is less sand in the bag as we keep lifting.) In the second instance the distance varies. (Water at the top of the tank does not have to be pumped as far as water at the bottom.) The sand bag problem is left to the exercises. Here we take up the second problem.

Figure 6.5.5 depicts a storage tank filled to within a feet of the top with some liquid. Assume that the liquid is homogeneous and weighs σ† pounds per cubic foot. Suppose now that this storage tank is pumped out from above until the level of the liquid drops to b feet below the top of the tank. How much work has been done?

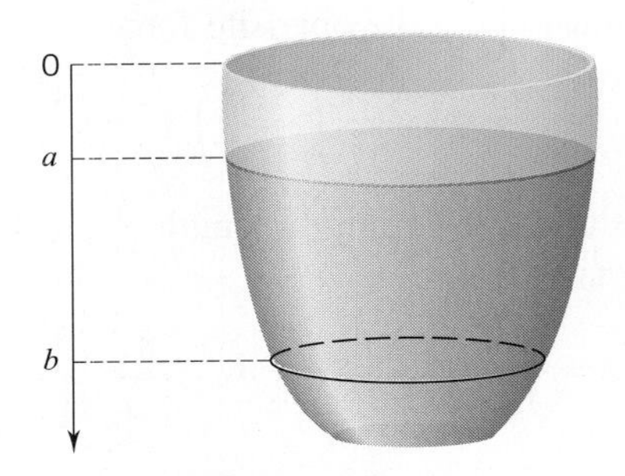

Figure 6.5.5

We can answer this question by the methods of integral calculus. For each $x \in [a, b]$, we let

$$A(x) = \text{cross-sectional area } x \text{ feet below the top of the tank,}$$

$$s(x) = \text{distance that the } x\text{-level must be lifted.}$$

We let $P = \{x_0, x_1, \ldots, x_n\}$ be an arbitrary partition of $[a, b]$ and focus our attention on the ith subinterval $[x_{i-1}, x_i]$ (Figure 6.5.6). Taking x_i^* as an arbitrary point in the ith subinterval, we have

† The symbol σ is the lower case Greek letter "sigma."

$$A(x_i^*)\,\Delta x_i = \text{approximate volume of the } i\text{th layer of liquid},$$
$$\sigma A(x_i^*)\,\Delta x_i = \text{approximate weight of this volume},$$
$$s(x_i^*) = \text{approximate distance this weight is to be moved.}$$

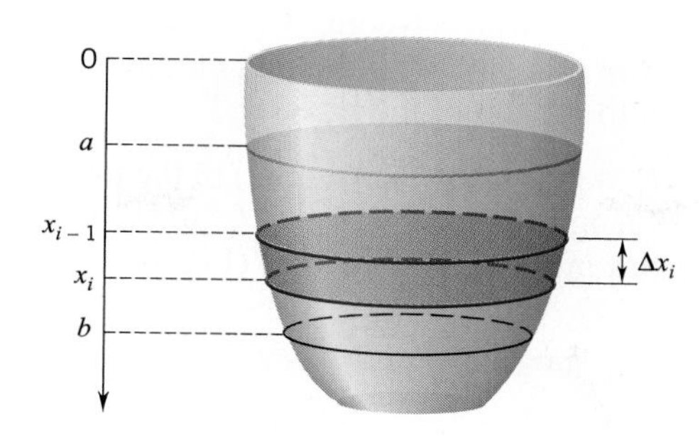

Figure 6.5.6

Therefore

$$\sigma s(x_i^*)A(x_i^*)\,\Delta x_i = \text{approximate work (weight} \times \text{distance) required to pump this layer of liquid to the top of the tank.}$$

The work required to pump out all the liquid can be approximated by adding up all these last terms:

$$W \cong \sigma s(x_1^*)A(x_1^*)\,\Delta x_1 + \sigma s(x_2^*)A(x_2^*)\,\Delta x_2 + \cdots + \sigma s(x_n^*)A(x_n^*)\,\Delta x_n.$$

The sum on the right is a Riemann sum which, as $\|P\| \to 0$, converge to give

(6.5.3)

$$W = \int_a^b \sigma s(x)A(x)\,dx.$$ ❑

We use this formula in the next problem.

Example 4 A hemispherical water tank of radius 10 feet is being pumped out (Figure 6.5.7). Find the work done in lowering the water level from 2 feet below the top of the tank to 4 feet below the top of the tank given that the pump is placed (a) at the top of the tank, (b) 3 feet above the top of the tank.

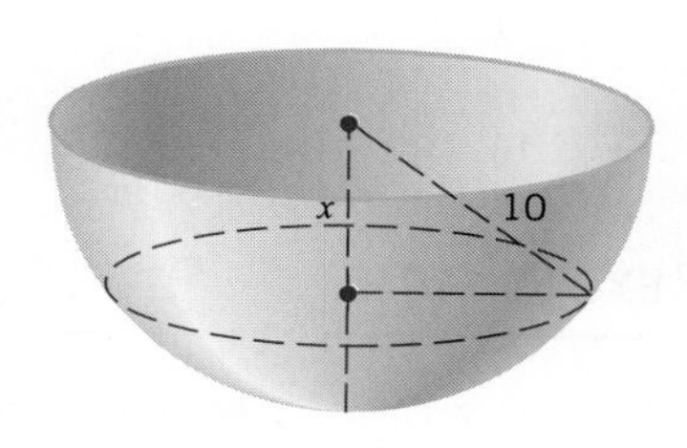

Figure 6.5.7

SOLUTION Take 62.5 pounds per cubic foot as the weight of water. It is not hard to see that the cross section x feet from the top of the tank is a disc of radius $\sqrt{100 - x^2}$. Its area is therefore

$$A(x) = \pi(100 - x^2).$$

For part (a) we have $s(x) = x$, so that

$$W = \int_2^4 62.5\pi x(100 - x^2)\,dx = 33{,}750\pi \cong 106{,}029 \text{ foot-pounds.}$$

For part (b) we have $s(x) = x + 3$, so that

$$W = \int_2^4 62.5\pi(x+3)(100 - x^2)\,dx = 67{,}750\pi \cong 212{,}843 \text{ foot-pounds.} \quad \square$$

Example 5 A 40-foot chain which weighs 4 pounds per foot hangs from a beam which projects from the top of a 50-foot building (Figure 6.5.8). How much work is done in pulling the chain to the top of the building?

SOLUTION A partition $P = \{x_0, x_1, x_2, \ldots, x_n\}$ of the interval [0, 40] will partition the chain into n pieces of length Δx_i, $i = 1, 2, \ldots, n$. For each integer i, $1 \le i \le n$, let x_i^* be an arbitrary point in the ith subinterval (Figure 6.5.9).

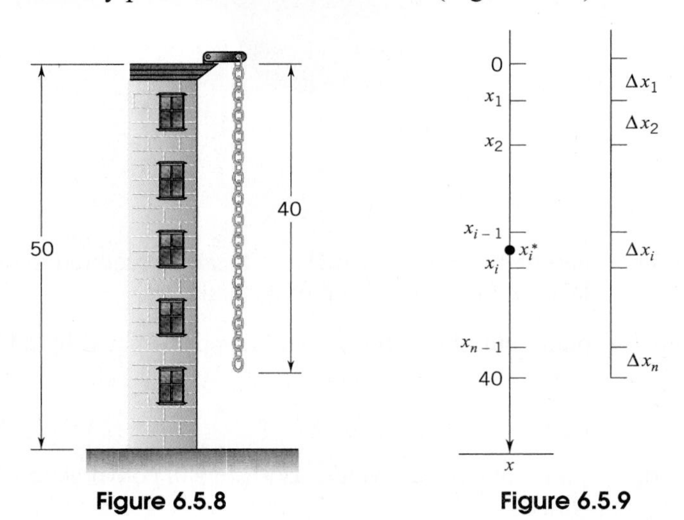

Figure 6.5.8　　Figure 6.5.9

Now, the ith piece of the chain weighs $4\,\Delta x_i$ and is approximately x_i^* feet from the top of the building. Thus, the work done in pulling this piece to the top is $x_i^*\,(4\,\Delta x_i) = 4\,x_i^*\,\Delta x_i$ foot-pounds. Adding the work done to pull each piece to the top, we have

$$4x_1^*\,\Delta x_1 + 4x_2^*\,\Delta x_2 + \cdots + 4x_n^*\,\Delta x_n \text{ foot-pounds.} \quad \square$$

This is a Riemann sum which, as $\|P\| \to 0$, converges to the definite integral $\int_0^{40} 4x\,dx$. Thus, the work done in pulling the chain to the top of the building is

$$\int_0^{40} 4x\,dx = \Big[2x^2\Big]_0^{40} = 3200 \text{ foot-pounds.}$$

EXERCISES 6.5

In Exercises 1 and 2, an object is moving along the x-axis under the action of a force of $F(x)$ pounds when it is x feet from the origin. Find the work done by F in moving the object from $x = a$ to $x = b$.

1. $F(x) = x(x^2 + 1)^2$;　$a = 1, b = 4$.

2. $F(x) = 2x\sqrt{x + 1}$;　$a = 3, b = 8$.

In Exercises 3 and 4, an object is moving along the x-axis under the action of a force of $F(x)$ newtons when it is x meters from the origin. Find the work done by F in moving the object from $x = a$ to $x = b$.

3. $F(x) = x + \sin 2x$;　$a = \dfrac{\pi}{6}, b = \pi$.

4. $F(x) = \dfrac{\cos 2x}{\sqrt{2 + \sin 2x}}$; $a = 0, b = \dfrac{\pi}{2}$.

5. A 600-pound force will compress a 10-inch automobile coil spring 1 inch. How much work must be done to compress the spring to 5 inches?

6. Five foot-pounds of work is needed to stretch a certain spring from 1 foot to 3 feet beyond its natural length. How far beyond its natural length will a 6-pound weight stretch the spring?

7. Stretched 4 feet beyond its natural length, a certain spring exerts a restoring force of 200 pounds. What work must we do to stretch the spring: (a) 1 foot beyond its natural length? (b) $1\frac{1}{2}$ feet beyond its natural length?

8. A certain spring has natural length L. Given that W is the work done in stretching the spring from L feet to $L + a$ feet, find the work done in stretching the spring: (a) from L feet to $L + 2a$ feet; (b) from L feet to $L + na$ feet; (c) from $L + a$ feet to $L + 2a$ feet; (d) from $L + a$ feet to $L + na$ feet.

9. Find the natural length of a heavy metal spring, given that the work done in stretching it from a length of 2 feet to a length of 2.1 feet is one-half of the work done in stretching it from a length of 2.1 feet to a length of 2.2 feet.

10. A vertical cylindrical tank of radius 2 feet and height 6 feet is full of water. Find the work done in pumping out the water: (a) to an outlet at the top of the tank; (b) to a level 5 feet above the top of the tank. (Assume that the water weighs 62.5 pounds per cubic foot.)

11. A horizontal cylindrical tank of radius 3 feet and length 8 feet is half full of oil weighing 60 pounds per cubic foot.

(a) Show that the work done in pumping out the oil to the top of the tank is given by the integral

$$960 \int_0^3 (x + 3)\sqrt{9 - x^2}\, dx.$$

Evaluate this integral by evaluating the integrals

$$\int_0^3 x\sqrt{9 - x^2}\, dx \qquad \text{and} \qquad \int_0^3 \sqrt{9 - x^2}\, dx$$

separately. HINT: Verify that the second integral represents one-quarter of the area of a circle of radius 3.

(b) What is the work done in pumping out the oil to a level 4 feet above the top of the tank?

12. What is the work done by gravity if the tank of Exercise 11 is completely drained through an opening at the bottom?

13. A conical container (vertex down) of radius r feet and height h feet is full of a liquid weighing σ pounds per cubic foot. Find the work done in pumping out the top $\frac{1}{2}h$ feet of liquid: (a) to the top of the tank; (b) to a level k feet above the top of the tank.

14. What is the work done by gravity if the tank of Exercise 13 is completely drained through an opening at the bottom?

15. The force of gravity exerted by the earth on a mass m at a distance r from the center of the earth is given by Newton's formula

$$F = -G\frac{mM}{r^2},$$

where M is the mass of the earth and G is the gravitational constant. Find the work done by gravity in pulling a mass m from $r = r_1$ to $r = r_2$.

16. A box that weights w pounds is dropped to the floor from a height of d feet. (a) What is the work done by gravity? (b) Show that the work is the same if the box slides to the floor along a smooth inclined plane.

17. A chain that weighs 15 pounds per foot is hanging from the top of an 80-foot building to the ground. How much work is done in pulling the chain to the top of the building?

18. An 8-pound monkey is attached to the end of a 20-foot rope that is hanging from a tree. The monkey climbs to the top of the rope. How much work has it done if the rope weighs 0.25 pound per foot?

19. A bucket of sand weighing 200 pounds is hoisted at a constant rate by a chain from the ground to the top of a building 100 feet high.

(a) How much work is required to lift the bucket if the weight of the chain is negligible?

(b) How much work is required to lift the bucket if the chain weighs 2 pounds per foot?

20. Suppose that the bucket in Exercise 19 has a hole in the bottom and the sand leaks out at a constant rate so that only 150 pounds of sand are left when the bucket reaches the top.

(a) How much work is required to lift the bucket if the weight of the chain is negligible?

(b) How much work is required to lift the bucket if the chain weighs 2 pounds per foot?

21. A 100-pound bag of sand is lifted for 2 seconds at the rate of 4 feet per second. Find the work done in lifting the bag if the sand leaks out at the rate of $1\frac{1}{2}$ pounds per second.

22. A water container initially weighing w pounds is hoisted by a crane at the rate of n feet per second. What is the work done if the tank is raised m feet and the water leaks out constantly at the rate of p gallons per second? (Assume that the water weighs 8.3 pounds per gallon.)

23. A rope of length l feet that weights σ pounds per foot is lying on the ground. What is the work done in lifting the rope so that it hangs from a beam: (a) l feet high; (b) $2l$ feet high?

24. A load of weight w is lifted from the bottom of a shaft h feet deep. Find the work done given that the rope used to hoist the load weighs σ pounds per foot.

25. An 800-pound steel beam hangs from a 50-foot cable which weighs 6 pounds per foot. Find the work done in winding 20 feet of the cable about a steel drum.

26. Two electrons repel each other with a force that is inversely proportional to the distance between them. Suppose one electron is located at a point A and the other is at a point B which is 8 centimeters from A. How much work is done against the repelling force if the electron at B is moved along the line connecting A and B to a point which is 4 centimeters from A? Express your answer in terms of the constant of proportionality.

27. An object of mass m is moving along the x-axis. At $x = a$, it has velocity v_a and at $x = b$ it has velocity v_b. Use Newton's second law of motion, $F = ma = m(dv/dt)$ to show that

$$W = \int_a^b F(x)\,dx = \tfrac{1}{2}mv_b^2 - \tfrac{1}{2}mv_a^2.$$

HINT: $\dfrac{dv}{dt} = \dfrac{dv}{dx}\dfrac{dx}{dt} = v\dfrac{dv}{dx}.$

Power. Power measures the rate at which work is done. Suppose an object starts at $x = a$ and moves along the x-axis under the action of a force F. Then the work done by F in moving the object to the position x is

$$W(x) = \int_a^x F(u)\,du.$$

If the position of the object is a function of time, that is, if $x = x(t)$, then

$$p = \frac{dW}{dt} = F[x(t)]\frac{dx}{dt} = F[x(t)]v(t)$$

is called the *power input* of the force F. The units of measurement of power are foot-pounds per second if force is measured in pounds, distance in feet, and time in seconds; joules per second, or *watts,* if force is measured in newtons, distance in meters, and time in seconds.

1 horsepower = 550 foot-pounds per second
= 746 watts.

28. (a) Assuming constant acceleration, what horsepower must an engine produce to accelerate a 3000-pound truck from 0 to 60 miles per hour (88 feet per second) in 15 seconds along a level road?
(b) What horsepower must the engine produce if the road rises 4 feet for every 100 feet of road? HINT: Integration is not required to answer these questions.

29. A vertical cylindrical tank with a height of 10 feet and a radius of 5 feet is half filled with water. If a 1 horsepower pump can do 550 foot-pounds of work per second, how long will it take a $\frac{1}{2}$-horsepower pump to:
(a) Pump the water out of the top of the tank?
(b) Pump the water to a point 5 feet above the top of the tank?

■ 6.6 FLUID PRESSURE AND FLUID FORCES

When an object is submerged in a liquid, it experiences a force caused by *pressure* from the liquid around it. By pressure we mean force per unit area. This force is perpendicular to the object at each of its points. The pressure P at a depth h below the surface is given by

$$P = \sigma h,$$

where σ is the *weight density* of the liquid (the weight per unit volume). Weight densities of liquids are determined experimentally.

When a horizontal surface is submerged in a liquid, the force against it is the total weight of the fluid above it (see Figure 6.6.1). If the horizontal surface has area A, if the depth is h, and if the weight density of the fluid is σ, then the force is given by the formula

(6.6.1) $$F = \sigma h A = PA.$$

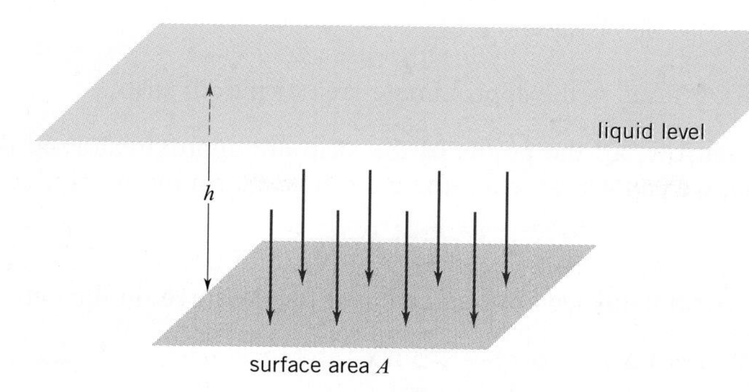

Figure 6.6.1

Remark about Units. If area is measured in square feet, depth in feet, and weight density in pounds per cubic foot, then the fluid force is measured in pounds. If, in the metric system, area is measured in square meters, depth in meters, and weight density in newtons per cubic meter, then the fluid force is measured in newtons. ❑

For example, the weight density of water is approximately 62.5 pounds per cubic foot, so the water pressure on a plate suspended horizontally in water at a depth of 6 feet is $(62.5)(6) = 375$ pounds per square foot. If the plate has an area of 4 square feet, then the force exerted on one side of the plate is $(375)(4) = 1500$ pounds.

Now suppose that a plate is suspended vertically in a liquid. The force on the plate is more difficult to determine in this case because the force exerted by the liquid at the bottom of the plate is greater than the force at the top. In Figure 6.6.2 we have depicted a vertical wall standing against a body of liquid. (Think of it as a dam or part of a container.) We want to calculate the force exerted by the liquid on this wall.

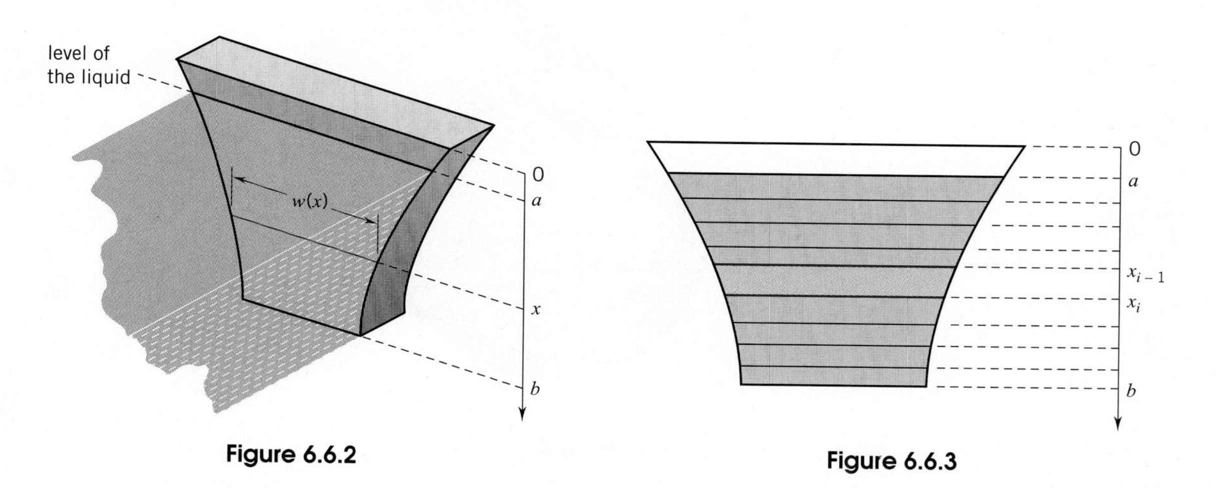

Figure 6.6.2

Figure 6.6.3

As in the figure we assume that the liquid sits from depth a to depth b, and we let $w(x)$ denote the width of the wall at depth x. A partition $P = \{x_0, x_1, \ldots, x_n\}$ of $[a, b]$ of small norm subdivides the wall into n narrow horizontal strips (Figure 6.6.3).

We can estimate the force on the ith strip by taking x_i^* as the midpoint of $[x_{i-1}, x_i]$. Then

$$w(x_i^*) = \text{the approximate width of the } i\text{th strip}$$

and

$$w(x_i^*)\,\Delta x_i = \text{the approximate area of the } i\text{th strip.}$$

Since this strip is narrow, all the points of the strip are approximately at depth x_i^*. Thus, using (6.6.1), we can estimate the force on the ith strip by the product

$$\sigma x_i^* w(x_i^*)\,\Delta x_i.$$

Adding up all these estimates we have an estimate for the force on the entire wall:

$$F \cong \sigma x_1^* w(x_1^*)\,\Delta x_1 + \sigma x_2^* w(x_2^*)\,\Delta x_2 + \cdots + \sigma x_n^* w(x_n^*)\,\Delta x_n.$$

The sum on the right is a Riemann sum for the integral

$$\int_a^b \sigma x w(x)\,dx$$

and as such converges to that integral as $\|P\| \to 0$. Thus we have

(6.6.2)

$$\text{force against wall} = \int_a^b \sigma x w(x)\,dx.$$

Example 1 A horizontal circular water main (Figure 6.6.4) 6 feet in diameter is capped when half full of water. What is the force of the water against the cap?

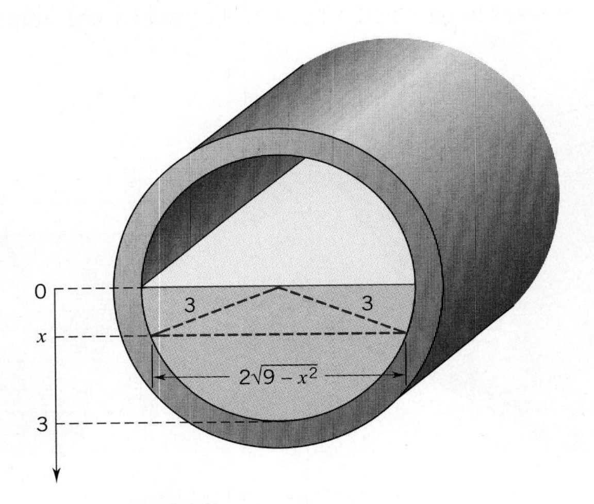

Figure 6.6.4

SOLUTION Here

$$w(x) = 2\sqrt{9 - x^2} \quad \text{and} \quad \sigma = 62.5 \text{ pounds per cubic foot.}$$

The force exerted against the cap is

$$F = \int_0^3 (62.5)x(2\sqrt{9 - x^2})\,dx = 62.5\int_0^3 2x\sqrt{9 - x^2}\,dx = 1125 \text{ pounds.} \quad \square$$

Example 2 A metal plate in the form of a trapezoid is affixed to a vertical dam as in Figure 6.6.5. The dimensions shown are given in meters; the weight density of water in the metric system is approximately 9800 newtons per cubic meter. Find the force on the plate.

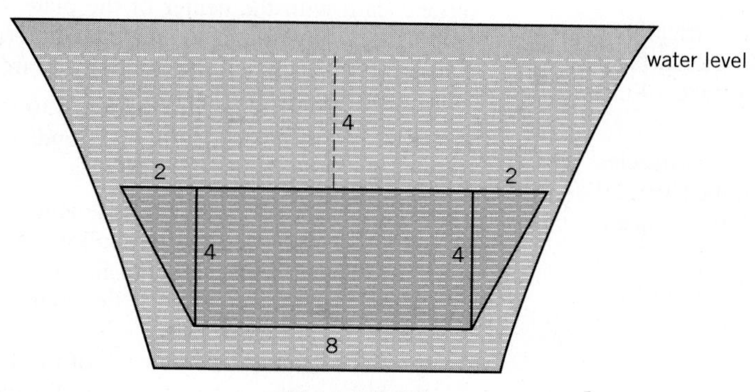

Figure 6.6.5

SOLUTION First we find the width of the plate x meters below the water level. By similar triangles (see Figure 6.6.6)

$$t = \tfrac{1}{2}(8 - x) \qquad \text{so that} \qquad w(x) = 8 + 2t = 16 - x.$$

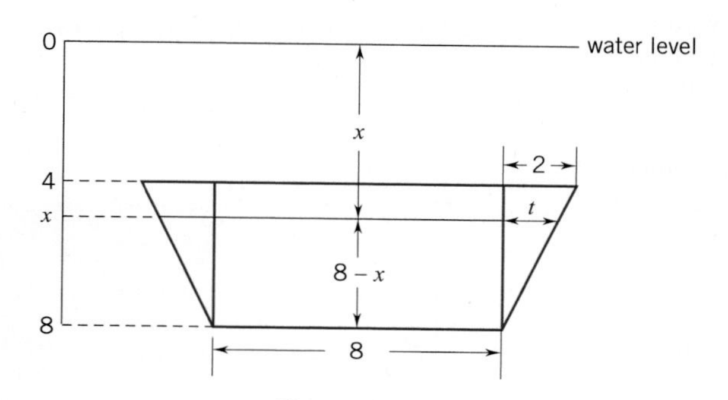

Figure 6.6.6

The force against the plate is

$$\int_4^8 9800x(16 - x)\,dx = 9800 \int_4^8 (16x - x^2)\,dx$$

$$= 9800 \left[8x^2 - \tfrac{1}{3}x^3 \right]_4^8 \cong 2{,}300{,}000 \text{ newtons.} \quad ❑$$

EXERCISES 6.6

1. A rectangular plate 8 feet wide by 6 feet deep is submerged vertically in a tank of water with the upper edge (that is, the 8-foot edge) at the surface of the water. Find the force of the water on one side of the plate.
2. A 6-foot by 6-foot square plate is submerged vertically in a tank of water with upper edge parallel to the surface. The center of the plate is 4 feet below the surface. Find the force of the water on one side of the plate.
3. A vertical dam in a river is in the shape of an isosceles trapezoid 100 meters across at the surface of the river, 60 meters across the bottom. If the river is 20 meters deep, find the force of the water on the dam.
4. A 5-meter by 5-meter square gate is at the bottom of the dam in Exercise 3. What is the force of the water on the gate?
5. A gate in the shape of an isosceles trapezoid 4 meters at the top, 6 meters at the bottom, and 3 meters high has its upper edge 10 meters below the top of the dam in Exercise 3. Find the force of the water on this gate.
6. A vertical dam in the shape of a rectangle is 1000 feet wide and 100 feet high. Calculate the force on the dam when:
   (a) the water level is 75 feet above the bottom.
   (b) the water level is 50 feet above the bottom.
7. Each end of a horizontal oil tank is an ellipse with horizontal axis 12 feet long and vertical axis 6 feet long. Calculate the force on one end when the tank is half full of oil weighing 60 pounds per cubic foot.
8. Each vertical end of a vat is a segment of a parabola (vertex down) 8 feet across the top and 16 feet deep. Calculate the force on an end when the vat is full of liquid weighing 70 pounds per cubic foot.
9. The vertical ends of a water trough are isosceles right triangles with the 90° angle at the bottom. Calculate the force on each triangle when the trough is full of water given that the legs of the triangle are 8 feet long.
10. The vertical ends of a water trough are isosceles triangles 5 feet across the top and 5 feet deep. Calculate the force on an end when the trough is full of water.
11. The ends of a water trough are semicircular discs with radius 2 feet. Find the force of the water on one end if the trough is full of water.
12. The ends of a water trough have the shape of the parabolic region bounded by $y = x^2 - 4$ and $y = 0$, where the measurements are in feet. Assuming that the trough is full of water, set up a definite integral whose value is the force of the water on one end.
13. A horizontal cylindrical tank of diameter 8 feet is half full of oil weighing 60 pounds per cubic foot. Calculate the force on one end.
14. Calculate the force on one end if the tank of Exercise 13 is full.
15. A rectangular metal plate 10 feet by 6 feet is affixed to a vertical dam with the center of the plate 11 feet below water level. Find the force on the plate if: (a) the 10-foot sides are horizontal; (b) the 6-foot sides are horizontal.
16. A vertical cylindrical tank of diameter 30 feet and height 50 feet is full of oil weighing 60 pounds per cubic foot. Find the force on the curved surface.
17. A swimming pool is 8 meters wide and 14 meters long. The pool is 1 meter deep at the shallow end and 3 meters deep at the other end; the depth increases linearly from the shallow to the deep end. If the pool is full of water, find:
    (a) the force of the water on each of the sides;
    (b) the force of the water on each of the ends.
18. Prove that if a plate is submerged in a liquid at an angle of θ with the vertical, $0 < \theta < \pi/2$, then the force on the plate is given by

$$F = \int_a^b \sigma x w(x) \sec\theta \, dx,$$

where σ is the weight density of the liquid, and x, $w(x)$, a, b are as shown in the figure below.

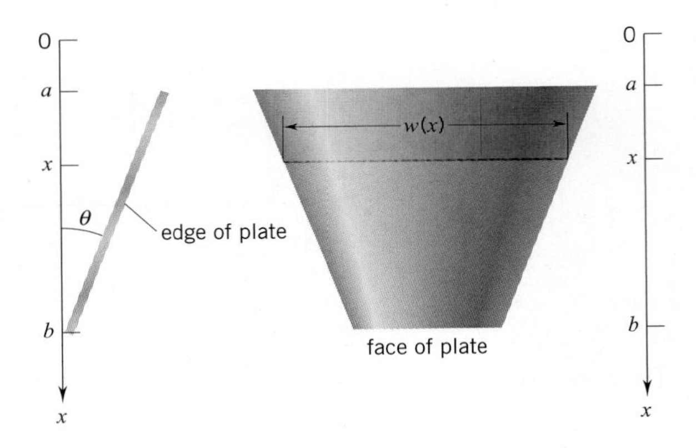

19. Find the force of the water on the bottom of the swimming pool in Exercise 17.
20. The face of a rectangular dam in a river is 1000 feet wide by 100 feet deep, and it makes an angle of 30° with the vertical. Find the force of the water on the face of the dam if:
    (a) the water level is at the top of the dam;
    (b) the water level is 75 feet from the bottom of the river.

21. Relate the force on a dam to the centroid of the submerged surface of the dam.

22. Two identical metal plates are affixed to a vertical dam. The centroid of the first plate is at depth h_1, the centroid of the second plate is at depth h_2. Compare the forces on the two plates. HINT: Exercise 21.

■ CHAPTER HIGHLIGHTS

6.1 More on Area

representative rectangle (p. 347)
area by integration with respect to y (p. 349)

6.2 Volume by Parallel Cross Sections; Discs and Washers

volume by parallel cross sections (p. 354)
disc method (p. 357)
washer method (x-axis) (p. 359)
washer method (y-axis) (p. 360)

6.3 Volume by the Shell Method

shell method (y-axis) (p. 364) shell method (x-axis) (p. 366)

6.4 The Centroid of a Region; Pappus's Theorem on Volumes

principles for finding the centroid of a region (p. 372)

$$\bar{x}A = \int_a^b x[f(x) - g(x)]\,dx, \qquad \bar{y}A = \int_a^b \tfrac{1}{2}([f(x)]^2 - [g(x)]^2)\,dx$$

Pappus's theorem on volumes (p. 374)

centroid of a solid of revolution: $\bar{x}V_x = \int_a^b \pi x[f(x)]^2\,dx, \quad \bar{y}V_y = \int_a^b \pi x[f(x)]^2\,dx$

6.5 The Notion of Work

Hooke's law (p. 380) pumping out a tank (p. 382)

6.6 The Force Against a Dam

force against a vertical wall $= \int_a^b \sigma x w(x)\,dx$

■ PROJECTS AND EXPLORATIONS USING TECHNOLOGY

To do these exercises you will need a graphics calculator or a computer with graphing capability. The majority of these problems are open-ended so different approaches may be used to solve them. You should be aware that different approaches can result in slight variations in the answers. Round your numerical answers to at least four decimal places. The rounding method that your calculator or computer uses also may cause variations in answers.

6.1 Let

$$f(x) = \frac{x^2 + 3x + 9}{x^6 - 2x^3 + x + 20},$$

and for each real number p, let

$$g_p(x) = px^2.$$

(a) Find the values of p such that f and g_p intersect at two points to form a bounded region.
(b) For each value of p found in part (a), let $A(p)$ be the area of the corresponding region. Evaluate $A(p)$ for several values of p.
(c) Find the values of p for which $A(p)$ is continuous.
(d) Calculate values of $A(p)$ to approximate $A'(p)$.

6.2 We are going to calculate areas and volumes in yet one more way.
(a) To find the area of a circle of radius r, subdivide it into a set of concentric rings as shown in the figure. Write a Riemann sum to approximate this area.
HINT: The area of a "ring" with inner radius r_1 and outer radius r_2 is given by $\pi(r_2^2 - r_1^2)$, which can be written $2\pi[\frac{1}{2}(r_1 + r_2)](r_2 - r_1)$.

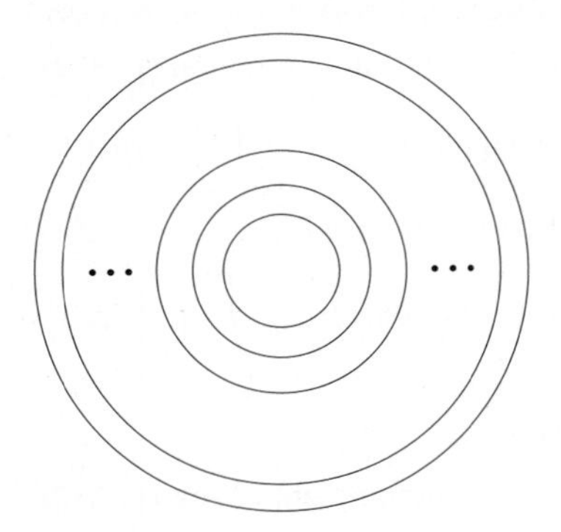

(b) Convert the Riemann sum from part (a) into a definite integral and evaluate it to find the area of the circle. When viewed in this manner, note that the derivative of the area with respect to the radius r is the circumference of the circle.
(c) Suppose that the density of a circular oil spill is $3/\sqrt{r^2 + 1}$ grams per square centimeter (we are ignoring the thickness) at a distance r from the center. Find the mass of the oil spill when the radius is 3 kilometers.
(d) Find a Riemann sum for approximating the volume of a sphere of radius r by subdividing it into spherical shells.
(e) Convert the Riemann sum from part (d) into a definite integral and evaluate it to find the volume of the sphere.
(f) If the density of the sphere is a function $\lambda(r)$, write the definite integral for finding the mass of the sphere. Evaluate this definite integral in the case where

$$\lambda(r) = \frac{1}{(r^3 + 3)^2} \quad \text{grams per square centimeter}$$

and the radius of the sphere is 3 centimeters.

6.3 The Great Pyramid in Giza, Egypt, has a square base 750 feet on a side, and a height of 400 feet. See Figure A.

(a) Find its volume.

(b) Suppose that the pyramid is solid stone, having a density of 175 pounds per cubic foot. Find the weight of the pyramid.

(c) Assume that the blocks were lifted into place from ground level. Find the work required to build the pyramid.

(d) Assume that the pyramid has a facade that is 10 feet thick and that the density of this stone is 225 pounds per cubic foot. Find the weight of the facade and the work required to construct it. Assume that once the side length near the top of the pyramid becomes less than 20 feet, the rest of the pyramid is all facade.

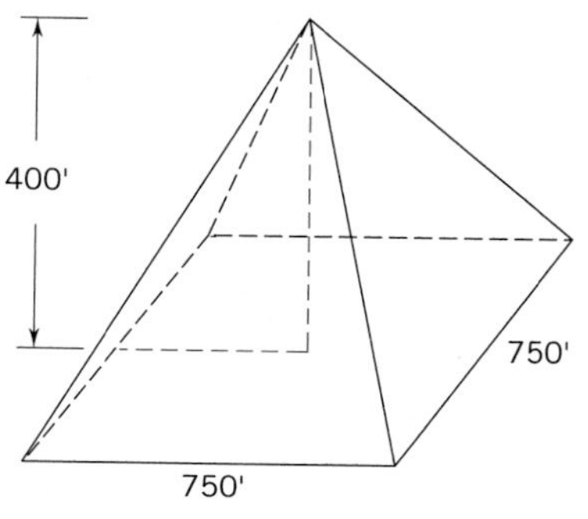

Figure A

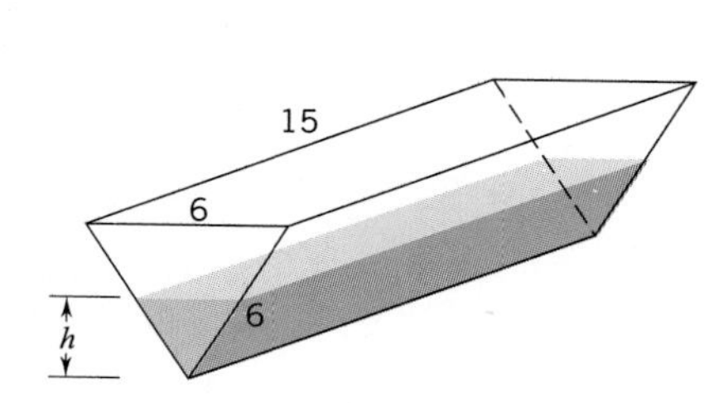

Figure B

6.4 Consider a water trough that is 15 feet long and has ends that are equilateral triangles of side length 6 feet.

(a) Find the function $V(h)$ that represents the volume of the trough from the bottom to a height of h feet. See Figure B.

(b) If water weighs 62.5 pounds per cubic foot, find the force of the water on an end of the trough if the height of the water in the trough is h feet.

(c) If water is running into the trough at the rate of 2 cubic feet per minute, find the rate at which the water level is rising.

(d) Find the rate at which the force on an end is increasing.

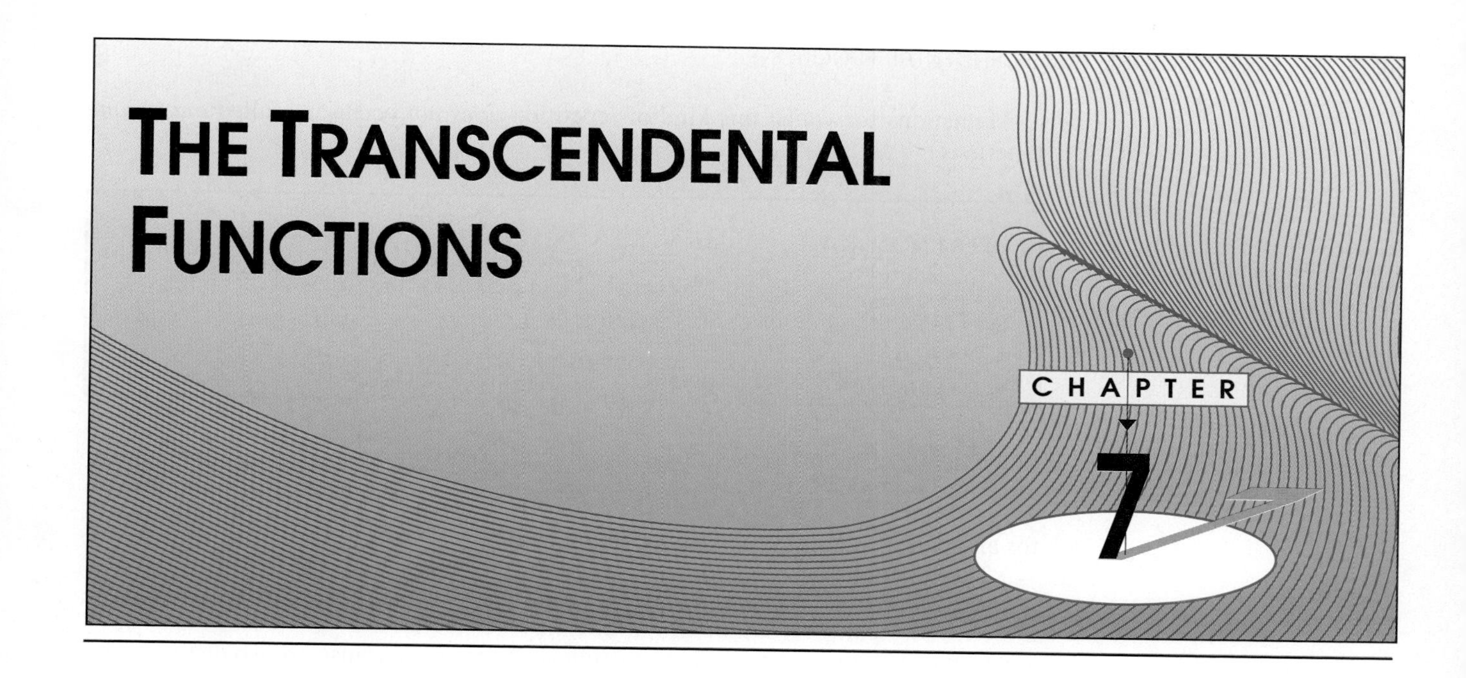

Some real numbers satisfy polynomial equations with integer coefficients:

$$\tfrac{3}{5} \text{ satisfies the equation } 5x - 3 = 0,$$
$$\sqrt{2} \text{ satisfies the equation } x^2 - 2 = 0.$$

Such numbers are called *algebraic*. There are, however, numbers that are not algebraic, among them π. Such numbers are called *transcendental*.

Some functions f satisfy polynomial equations with polynomial coefficients:

$$f(x) = \frac{x}{\pi x + \sqrt{2}} \quad \text{satisfies the equation} \quad (\pi x + \sqrt{2})f(x) - x = 0,$$

$$f(x) = 2\sqrt{x} - 3x^2 \quad \text{satisfies the equation} \quad [f(x)]^2 + 6x^2 f(x) + (9x^4 - 4x) = 0.$$

Such functions are called *algebraic*. There are, however, functions that are not algebraic; these are called *transcendental*. The functions that we study in this chapter (the logarithm, the exponential, and the trigonometric functions and their inverses) are all transcendental.

■ 7.1 ONE-TO-ONE FUNCTIONS; INVERSES

One-to-One Functions

A function can take on the same value at different points of its domain. Constant functions, for example, take on the same value at all points of their domains. The quadratic function $f(x) = x^2$ takes on the same value at $-c$ as it does at c; so does the absolute value function $g(x) = |x|$. The function

$$f(x) = 1 + (x - 3)(x - 5)$$

takes on the same value at $x = 5$ as it does at $x = 3$:

$$f(3) = 1, \qquad f(5) = 1.$$

Functions for which this kind of repetition *does not* occur are called *one-to-one* functions.

DEFINITION 7.1.1

A function f with domain D is said to be *one-to-one* iff no two distinct points in D have the same image under f, that is,

$$f(x_1) \neq f(x_2) \quad \text{whenever} \quad x_1 \neq x_2, \quad x_1, x_2 \in D.$$

Equivalently, f is one-to-one iff

$$f(x_1) = f(x_2) \quad \text{implies} \quad x_1 = x_2$$

for any pair of points x_1, x_2 in the domain of f.

The functions

$$f(x) = x^3 \quad \text{and} \quad g(x) = \sqrt{x}$$

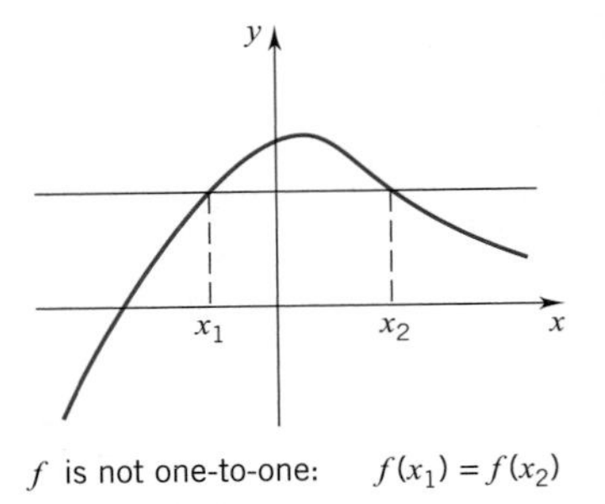

f is not one-to-one: $f(x_1) = f(x_2)$

Figure 7.1.1

are each one-to-one. The cubing function is one-to-one because no two distinct numbers have the same cube. The square-root function is one-to-one because no two distinct numbers have the same square root.

There is a simple geometric test to determine whether a function is one-to-one. Look at the graph of the function. If some horizontal line intersects the graph of the function more than once, then the function is not one-to-one (Figure 7.1.1). If, on the other hand, no horizontal line intersects the graph more than once, then the function is one-to-one (Figure 7.1.2).

Inverses

We begin with a theorem about one-to-one functions.

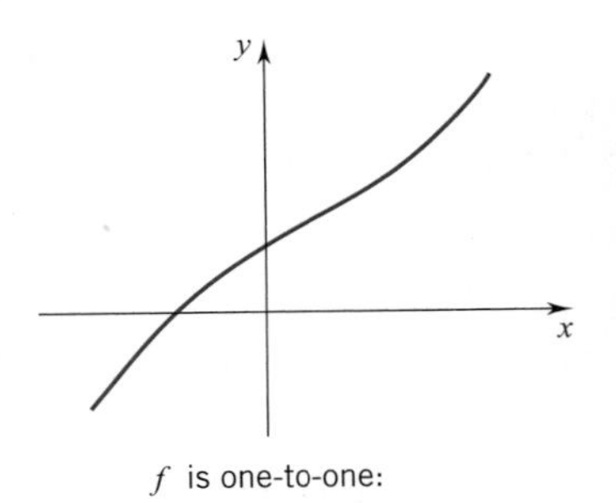

f is one-to-one:

Figure 7.1.2

THEOREM 7.1.2

If f is a one-to-one function, then there is one and only one function g that is defined on the range of f and satisfies the equation

$$f(g(x)) = x \qquad \text{for all } x \text{ in the range of } f.$$

PROOF The proof is easy. If x is in the range of f, then f must take on the value x at some number. Since f is one-to-one, there can be only one such number. We have called that number $g(x)$. ❑

The function that we have named g in the theorem is called the *inverse* of f and is usually denoted f^{-1}.

DEFINITION 7.1.3 INVERSE FUNCTION

Let f be a one-to-one function. The *inverse* of f, denoted by f^{-1}, is the unique function that is defined on the range of f and satisfies the equation

$$f(f^{-1}(x)) = x \qquad \text{for all } x \text{ in the range of } f.$$

Remark The notation f^{-1} for the inverse function of f is standard. Unfortunately, there is the danger of confusing f^{-1} with the reciprocal of f, that is, with $1/f(x)$. The "-1" in the notation for the inverse of f is *not an exponent*; $f^{-1}(x)$ *does not mean* $1/f(x)$. On those occasions when we want to express $1/f(x)$ using the exponent -1, we will write $[f(x)]^{-1}$.

Example 1 You have seen that the cubing function

$$f(x) = x^3$$

is one-to-one. Find the inverse.

SOLUTION We set $t = f^{-1}(x)$ and solve the equation $f(t) = x$ for t:

$$f(t) = x$$

$$t^3 = x \qquad (f \text{ is the cubing function})$$

$$t = x^{1/3}.$$

Substituting $f^{-1}(x)$ back in for t, we have

$$f^{-1}(x) = x^{1/3}.$$

The inverse of the cubing function is the cube-root function. The graphs of $f(x) = x^3$ and $f^{-1}(x) = x^{1/3}$ are shown in Figure 7.1.3. ❑

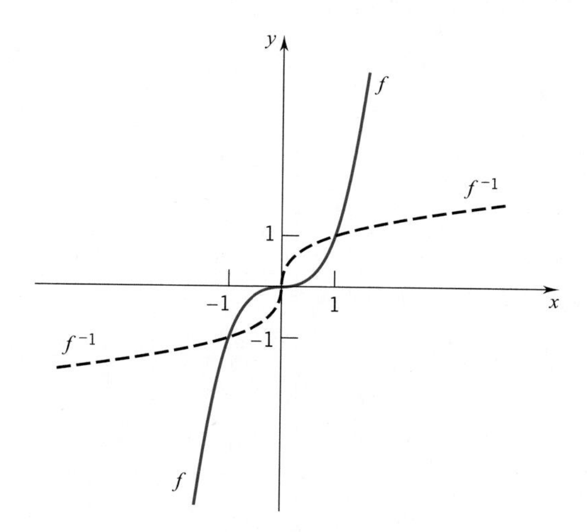

Figure 7.1.3

Remark We substitute t for $f^{-1}(x)$ merely to simplify the calculations. It is easier to work with the symbol t than with the string of symbols $f^{-1}(x)$. ❑

Example 2 Show that the linear function

$$f(x) = 3x - 5$$

is one-to-one. Then find the inverse.

SOLUTION To show that f is one-to-one, let's suppose that

$$f(x_1) = f(x_2).$$

Then

$$3x_1 - 5 = 3x_2 - 5$$
$$3x_1 = 3x_2$$
$$x_1 = x_2.$$

The function is one-to-one since

$$f(x_1) = f(x_2) \quad \text{implies} \quad x_1 = x_2.$$

(Another way to see that this function is one-to-one is to note that the graph is a line of slope 3 and as such cannot be intersected by a horizontal line more than once.)

Now let's find the inverse. To do this, we set $t = f^{-1}(x)$ and solve the equation $f(t) = x$ for t:

$$f(t) = x$$
$$3t - 5 = x$$
$$3t = x + 5$$
$$t = \tfrac{1}{3}x + \tfrac{5}{3}.$$

Substituting $f^{-1}(x)$ back in for t, we have

$$f^{-1}(x) = \tfrac{1}{3}x + \tfrac{5}{3}.$$

The graphs of f and f^{-1} are shown in Figure 7.1.4. ❑

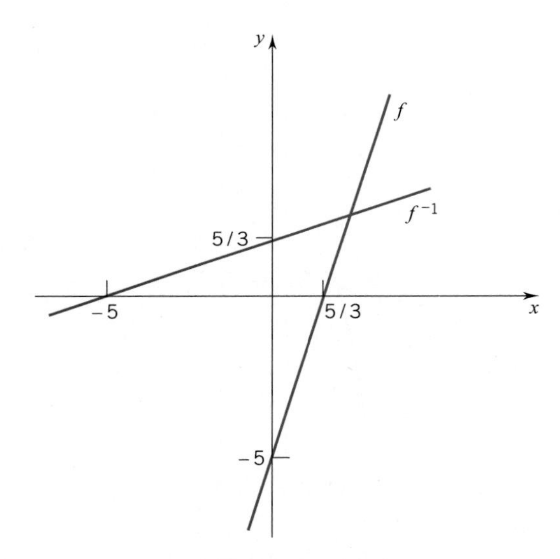

Figure 7.1.4

Example 3 Show that the function

$$f(x) = (1 - x^3)^{1/5} + 2$$

is one-to-one. What is the inverse?

SOLUTION First we show that f is one-to-one. Suppose that

$$f(x_1) = f(x_2).$$

Then

$$\begin{aligned}
(1 - x_1^3)^{1/5} + 2 &= (1 - x_2^3)^{1/5} + 2 \\
(1 - x_1^3)^{1/5} &= (1 - x_2^3)^{1/5} \\
1 - x_1^3 &= 1 - x_2^3 \\
x_1^3 &= x_2^3 \\
x_1 &= x_2.
\end{aligned}$$

The function is one-to-one since

$$f(x_1) = f(x_2) \quad \text{implies} \quad x_1 = x_2.$$

To find the inverse we set $t = f^{-1}(x)$ and solve the equation $f(t) = x$ for t:

$$\begin{aligned}
f(t) &= x \\
(1 - t^3)^{1/5} + 2 &= x \\
(1 - t^3)^{1/5} &= x - 2 \\
1 - t^3 &= (x - 2)^5 \\
t^3 &= 1 - (x - 2)^5 \\
t &= [1 - (x - 2)^5]^{1/3}.
\end{aligned}$$

Substituting $f^{-1}(x)$ back in for t, we have

$$f^{-1}(x) = [1 - (x - 2)^5]^{1/3}. \quad \square$$

In Chapter 4, Section 4.2, we studied increasing and decreasing functions: A function f is *increasing* if $x_1 < x_2$ implies $f(x_1) < f(x_2)$; f is *decreasing* if $x_1 < x_2$ implies $f(x_1) > f(x_2)$; x_1, x_2 in the domain of f.

THEOREM 7.1.4

If f is either an increasing function or a decreasing function, then f is one-to-one and hence has an inverse.

PROOF Suppose that f is increasing, and let x_1 and x_2 be points in the domain of f with $x_1 \neq x_2$. If $x_1 < x_2$, then $f(x_1) < f(x_2)$; if $x_1 > x_2$, then $f(x_1) > f(x_2)$. In either case, $f(x_1) \neq f(x_2)$ and so f is one-to-one. The proof for f as a decreasing function can be done in the same way. $\square$

Theorem 7.1.4 has an important application to differentiable functions whose domains are intervals. By Theorem 4.2.3, if f is continuous on an interval I, and if $f'(x) > 0$ $(f'(x) < 0)$ on the interior of I, then f is increasing (decreasing) on I. In Exercise 62, Section 4.2, you were asked to show that the result of Theorem 4.2.3 still holds if $f'(x) \geq 0$ $(f'(x) \leq 0)$ on the interior of I with $f'(x) = 0$ for at most finitely many values of x.

Example 4 Each of the following functions has an inverse.

(a) $f(x) = 3x - 5$ $\qquad f'(x) = 3 > 0.$

(b) $g(x) = 4 - 2x^3$ $\qquad \begin{cases} g'(x) = -6x^2 \leq 0; \\ g'(x) = 0 \text{ iff } x = 0. \end{cases}$

(c) $F(x) = x^5 + 2x^3 + 3x - 4$ $\qquad F'(x) = 5x^4 + 6x^2 + 3 > 0.$

(d) $G(x) = \cos x, \quad \text{for} \quad 0 \leq x \leq \pi$ $\qquad G'(x) = -\sin x < 0 \quad \text{for} \quad 0 < x < \pi.$

We found the inverse of f in Example 2. You can verify that $g^{-1}(x) = [(4 - x)/2]^{1/3}$. To find the inverse of F we would have to solve the equation

$$t^5 + 2t^3 + 3t - 4 = x$$

for t. It is not clear that this can be done. Thus, we have the situation that we know F^{-1} exists, but we may not be able to find a "formula" for it. The same is true for G. The inverse trigonometric functions are defined and studied in Section 7.8. ❑

Suppose that the function f has an inverse. Then, by definition, f^{-1} satisfies the equation

(7.1.5) $$f(f^{-1}(x)) = x \quad \text{for all } x \text{ in the range of } f.$$

It is also true that

(7.1.6) $$f^{-1}(f(x)) = x \quad \text{for all } x \text{ in the domain of } f.$$

PROOF Take x in the domain of f and set $y = f(x)$. Since y is in the range of f,

$$f(f^{-1}(y)) = y.$$

This means that

$$f(f^{-1}(f(x))) = f(x)$$

and tells us that f takes on the same value at $f^{-1}(f(x))$ as it does at x. With f one-to-one, this can only happen if

$$f^{-1}(f(x)) = x. \quad ❑$$

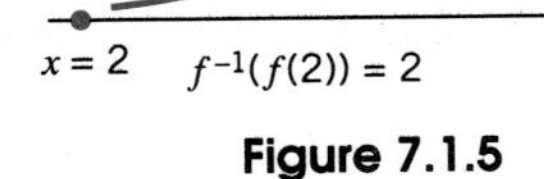

Figure 7.1.5

Equation (7.1.6) tells us that f^{-1} "undoes" the action of f:

if f takes x to $f(x)$, then f^{-1} takes $f(x)$ back to x.

Figure 7.1.5 illustrates this relationship in the case of $f(x) = x^3$, $f^{-1}(x) = x^{1/3}$ and $x = 2$.

Equation (7.1.5) tells us that f "undoes" the work of f^{-1}:

if f^{-1} takes x to $f^{-1}(x)$, then f takes $f^{-1}(x)$ back to x.

This is illustrated in Figure 7.1.6 with $f(x) = x^3$, $f^{-1}(x) = x^{1/3}$ and $x = 8$.

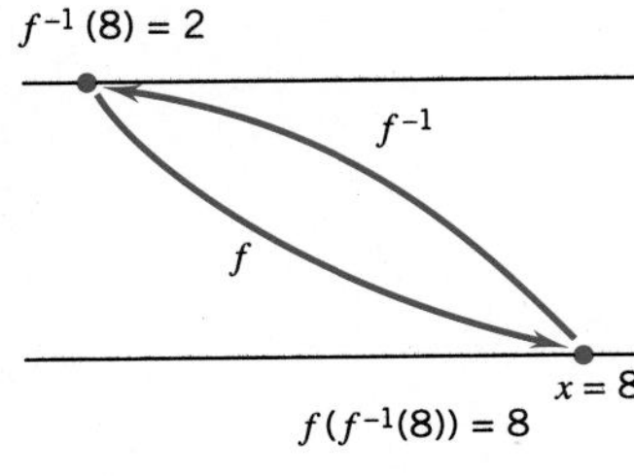

Figure 7.1.6

Finally, we emphasize that Equations (7.1.5) and (7.1.6) tell us that

$$\text{domain of } f^{-1} = \text{range of } f$$
$$\text{range of } f^{-1} = \text{domain of } f.$$

The Graphs of f and f^{-1}

There is an important relation between the graph of a one-to-one function f and the graph of f^{-1} (see Figure 7.1.7). The graph of f consists of points of the form $P(x, f(x))$. Since f^{-1} takes on the value x at $f(x)$, the graph of f^{-1} consists of points of the form $Q(f(x), x)$.

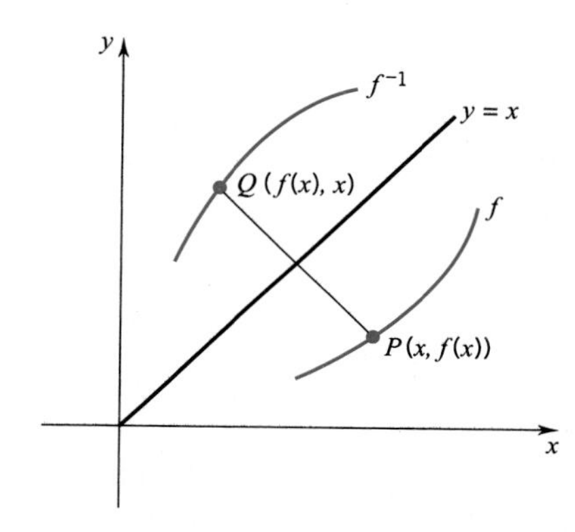

Figure 7.1.7

Now suppose that the scale on the y-axis is the same as that on the x-axis. Then, for each x, the points $P(x, f(x))$ and $Q(f(x), x)$ are symmetric with respect to the line $y = x$. Hence the graph of f^{-1} is the graph of f reflected in the line $y = x$.

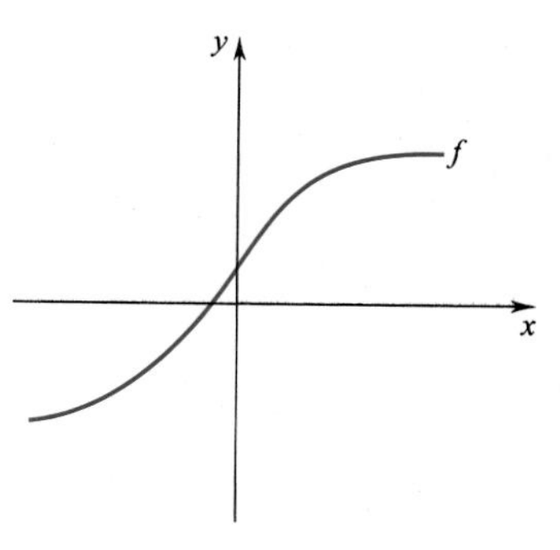

Figure 7.1.8

Example 5 Given the graph of f in Figure 7.1.8, sketch the graph of f^{-1}.

SOLUTION First draw in the line $y = x$. Then reflect the graph of f in that line. See Figure 7.1.9. ❑

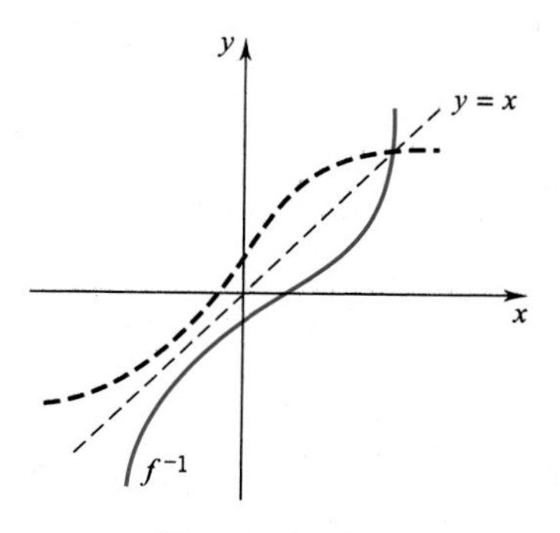

Figure 7.1.9

Continuity and Differentiability of Inverses

Let f be a one-to-one function. Then f has an inverse, f^{-1}. Suppose, in addition, that f is continuous. Since the graph of f has no "holes" or "gaps," and since the graph of f^{-1} is simply the reflection of the graph of f in the line $y = x$, we can conclude that the graph of f^{-1} has no holes or gaps. That is, f^{-1} must also be continuous. We state this result formally; a proof is given in Appendix B3.

THEOREM 7.1.7

Let f be a one-to-one function defined on an interval I. If f is continuous, then its inverse f^{-1} is also continuous.

Now suppose that f is differentiable. Is f^{-1} necessarily differentiable? The answer is yes, provided f' does not take on the value 0. We will assume this result for now and go on to describe how to calculate the derivative of f^{-1}.

To simplify the notation, we set $f^{-1} = g$. Then

$$f(g(x)) = x \qquad \text{for all} \quad x \text{ in the range of } f.$$

Differentiation gives

$$\frac{d}{dx}[f(g(x))] = 1,$$

and so

$$f'(g(x))g'(x) = 1 \qquad \text{(by the chain rule; we are assuming that } f \text{ and } g \text{ are differentiable).}$$

Therefore,

$$g'(x) = \frac{1}{f'(g(x))} \qquad \text{(we are assuming that } f' \text{ is never 0).}$$

Substituting f^{-1} back in for g, we have

(7.1.8)

$$(f^{-1})'(x) = \frac{1}{f'[f^{-1}(x)]}.$$

This is the standard formula for the derivative of an inverse.

Before illustrating this formula, we restate it in a more specific form:

(7.1.9)

Suppose that f has an inverse and is differentiable. Let a be a point in the domain of f and let $b = f(a)$. If $f'(a) \neq 0$, then $(f^{-1})'(b)$ exists and is given by

$$(f^{-1})'(b) = \frac{1}{f'(a)}.$$

Remark Be sure to notice that $(f^{-1})'(b)$ is obtained by evaluating f' at a, not at b. Notice, also, that $a = f^{-1}(b)$, so that

$$(f^{-1})'(b) = \frac{1}{f'(a)} = \frac{1}{f'[f^{-1}(b)]},$$

which is simply (7.1.8) with $x = b$. ❑

Example 6 Let $f(x) = x^3 + \frac{1}{2}x$. (a) Show that f has an inverse. (b) Show that 9 is in the range of f and calculate $(f^{-1})'(9)$.

SOLUTION (a) The domain of f is $(-\infty, \infty)$ and

$$f'(x) = 3x^2 + \tfrac{1}{2} > 0$$

for all x. Thus f has an inverse by Theorem 7.1.4.

(b) It is easy to verify that the range of f is also $(-\infty, \infty)$. Thus $9 \in \text{range}(f)$. To calculate $(f^{-1})'(9)$ we must first find the value a such that $f(a) = 9$. The solution of

$$a^3 + \tfrac{1}{2}a = 9$$

is $a = 2$. Thus

$$(f^{-1})'(9) = \frac{1}{f'(2)} = \frac{1}{3(2)^2 + \frac{1}{2}} = \frac{2}{25}.$$

The graphs of f and f^{-1} are shown in Figure 7.1.10. (These were obtained on a graphing calculator in parametric mode.) ❑

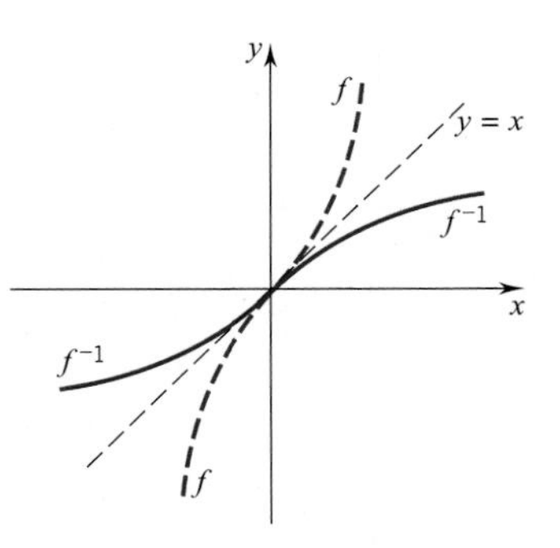

Figure 7.1.10

Formula (7.1.8) takes a simple form when it is expressed in Leibniz's notation. Suppose that the function f has an inverse and is differentiable. Let $y = f(x)$ so that $x = f^{-1}(y)$. Then

$$\frac{dx}{dy} = (f^{-1})'(y) \qquad \text{and} \qquad \frac{dy}{dx} = f'(x) = f'(f^{-1}(y)).$$

Substituting these expressions in (7.1.8), we get

(7.1.10)

$$\frac{dx}{dy} = \frac{1}{dy/dx}.$$

If $y = f(x)$, then the rate of change of x with respect to y is the reciprocal of the rate of change of y with respect to x. Of course, we have to be precise about where these derivatives are evaluated. The analog of (7.1.9) is

(7.1.11)

Suppose that f has an inverse and is differentiable, and set $y = f(x)$. Let a be a number in the domain of f and let $b = f(a)$. If

$$\frac{dy}{dx}(a) \neq 0, \qquad \text{then} \quad \frac{dx}{dy}(b) \text{ exists and is given by}$$

$$\frac{dx}{dy}(b) = \frac{1}{\dfrac{dy}{dx}(a)}$$

Example 7 The function defined by $y = x - \cos x$, for $-\pi \leq x \leq \pi$, is differentiable. (a) Show that this function has an inverse. (b) Calculate the derivative of the inverse function using Formula (7.1.10). (c) Evaluate the derivative of the inverse function at $y = -1$.

SOLUTION (a) Since $dy/dx = 1 + \sin x \geq 0$ for $-\pi < x < \pi$ and $1 + \sin x = 0$ only at $x = -\pi/2$, it follows that this function has an inverse.
(b) By Formula (7.1.10)

$$\frac{dx}{dy} = \frac{1}{dy/dx} = \frac{1}{1 + \sin x}.$$

(c) To calculate

$$\frac{dx}{dy}(-1)$$

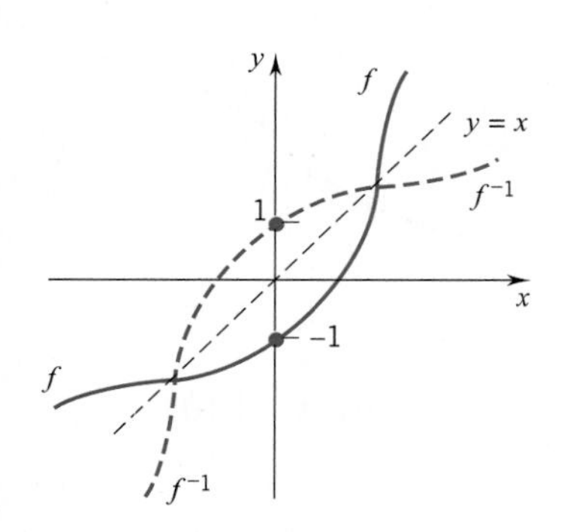

Figure 7.1.11

we must first find the value a such that $y = -1$ when $x = a$. Note that if we let $x = 0$, then $y = -1$. Thus

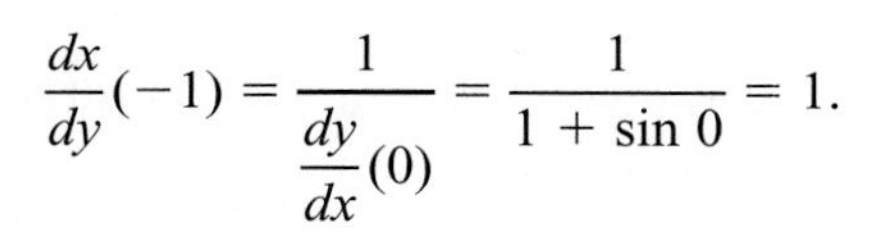

$$\frac{dx}{dy}(-1) = \frac{1}{\dfrac{dy}{dx}(0)} = \frac{1}{1 + \sin 0} = 1.$$

The graphs of $y = x - \cos x$ and its inverse are shown in Figure 7.1.11. (As in Example 6, these graphs were obtained on a graphing calculator in parametric mode.) ❑

For some geometric reassurance of the validity of Formula (7.1.8) we refer you to Figure 7.1.12. The graphs of f and f^{-1} are reflections of one another in the line $y = x$. The tangent lines l_1 and l_2 are also reflections of one another. From the figure

$$(f^{-1})'(x) = \text{slope of } l_1 = \frac{f^{-1}(x) - b}{x - b}, \qquad f'(f^{-1}(x)) = \text{slope of } l_2 = \frac{x - b}{f^{-1}(x) - b};$$

namely, $(f^{-1})'(x)$ and $f'(f^{-1}(x))$ are indeed reciprocals.

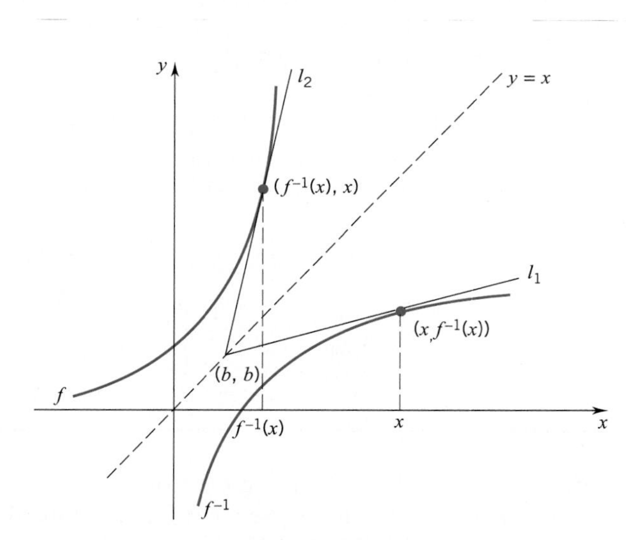

Figure 7.1.12

The figure also suggests the difficulty caused by having $f'(z) = 0$ at some point z. For if $f'(z) = 0$, then the graph of f has a horizontal tangent line at $(z, f(z))$. The reflection of a horizontal tangent in the line $y = x$ will be a vertical tangent line to the graph of f^{-1} at the point $(f(z), z)$ and this will be a point of nondifferentiability for f^{-1}.

EXERCISES 7.1

In Exercises 1–26, determine whether or not the given function is one-to-one and, if so, find its inverse.

1. $f(x) = 5x + 3.$

2. $f(x) = 3x + 5.$

3. $f(x) = 4x - 7.$

4. $f(x) = 7x - 4.$

5. $f(x) = 1 - x^2.$

6. $f(x) = x^5.$

7. $f(x) = x^5 + 1.$

8. $f(x) = x^2 - 3x + 2.$

9. $f(x) = 1 + 3x^3.$

10. $f(x) = x^3 - 1.$

11. $f(x) = (1 - x)^3.$

12. $f(x) = (1 - x)^4.$

13. $f(x) = (x + 1)^3 + 2.$

14. $f(x) = (4x - 1)^3.$

15. $f(x) = x^{3/5}.$

16. $f(x) = 1 - (x - 2)^{1/3}.$

17. $f(x) = (2 - 3x)^3.$

18. $f(x) = (2 - 3x^2)^3.$

19. $f(x) = \dfrac{1}{x}.$

20. $f(x) = \dfrac{1}{1 - x}.$

21. $f(x) = x + \dfrac{1}{x}.$

22. $f(x) = \dfrac{x}{|x|}.$

23. $f(x) = \dfrac{1}{x^3 + 1}.$

24. $f(x) = \dfrac{1}{1 - x} - 1.$

25. $f(x) = \dfrac{x + 2}{x + 1}.$

26. $\dfrac{1}{(x + 1)^{2/3}}.$

27. What is the relation between f and $(f^{-1})^{-1}$?

In Exercises 28–33, sketch the graph of f^{-1} given the graph of f.

28.

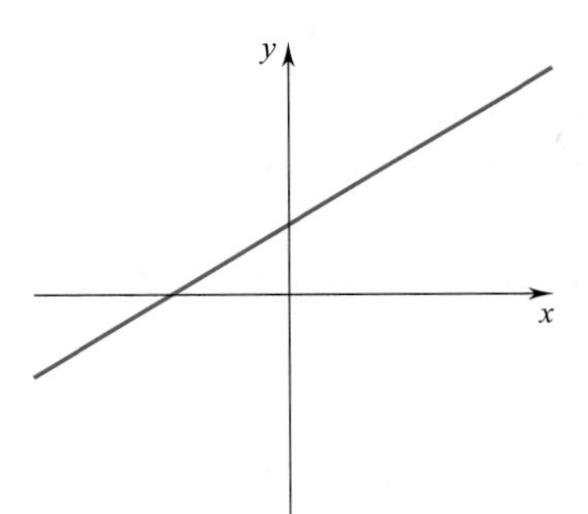

29.

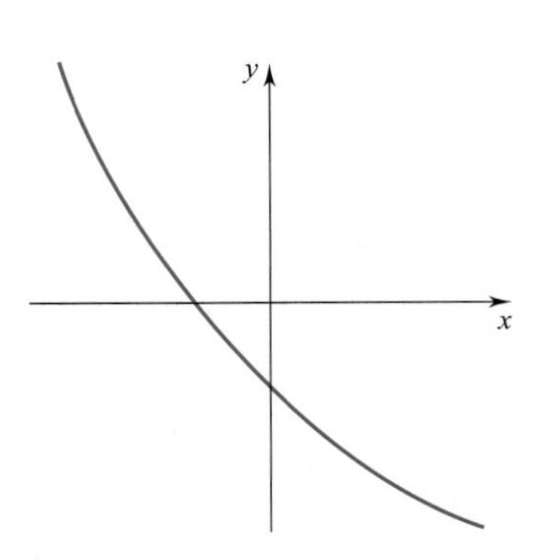

30.

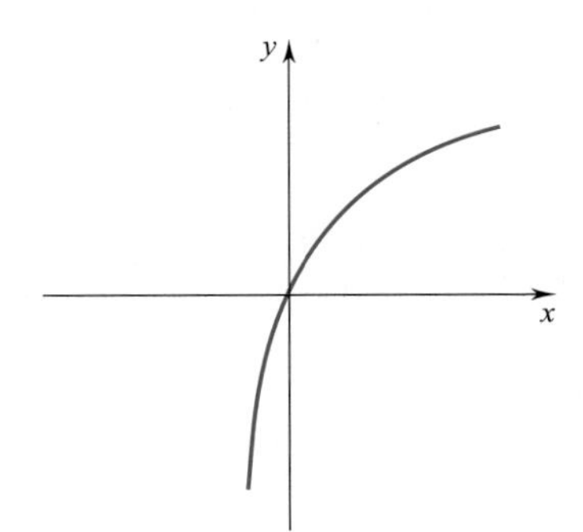

31.

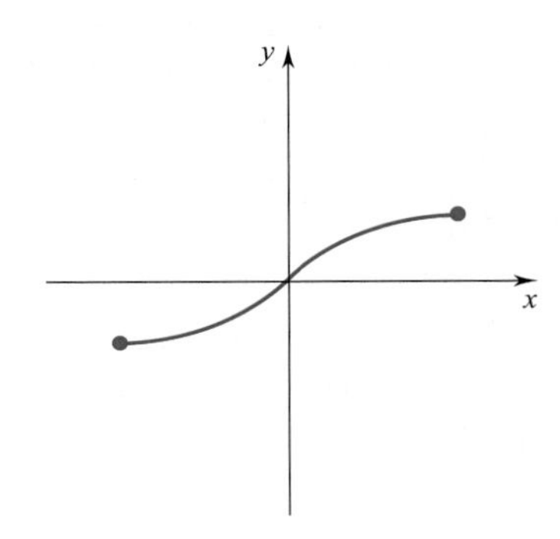

32.

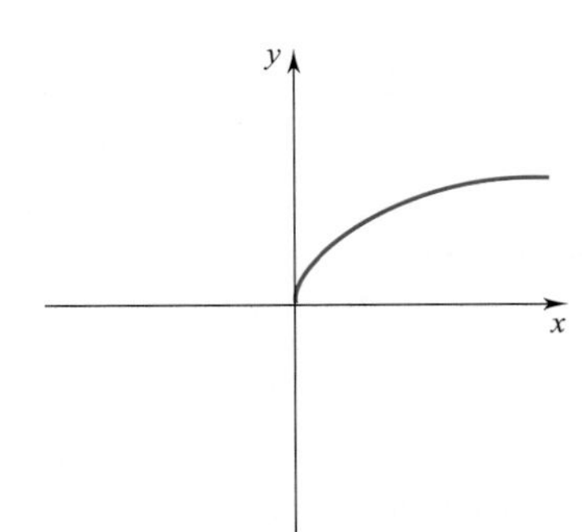

33.

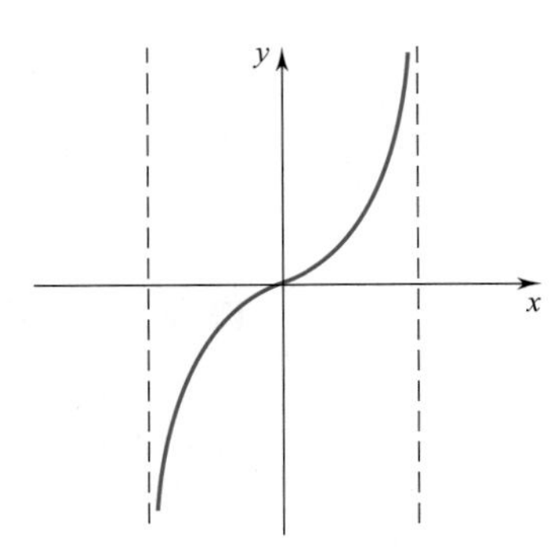

34. (a) Show that the composition of two one-to-one functions, f and g, is one-to-one.
(b) Express $(f \circ g)^{-1}$ in terms of f^{-1} and g^{-1}.

In Exercises 35–42, the given function f is differentiable. Verify that f has an inverse and find $(f^{-1})'(c)$.

35. $f(x) = x^3 + 1; \quad c = 9.$

36. $f(x) = 1 - 2x - x^3; \quad c = 4.$

37. $f(x) = x + 2\sqrt{x}, \quad x > 0; \quad c = 8.$

38. $f(x) = x + \sin x; \quad c = 0.$

39. $f(x) = 2x + \cos x; \quad c = \pi.$

40. $f(x) = \dfrac{x+3}{x-1}, \quad x > 1; \quad c = 3.$

41. $f(x) = \tan x, \quad -\frac{\pi}{2} < x < \frac{\pi}{2}; \quad c = \sqrt{3}.$

42. $f(x) = x^5 + 2x^3 + 2x; \quad c = -5.$

In Exercises 43–45, find a formula for $(f^{-1})'(x)$ given that f is one-to-one and its derivative satisfies the indicated equation.

43. $f'(x) = f(x).$

44. $f'(x) = 1 + [f(x)]^2.$

45. $f'(x) = \sqrt{1 - [f(x)]^2}.$

46. Let

$$f(x) = \begin{cases} x^3 - 1, & x < 0 \\ x^2, & x \geq 0. \end{cases}$$

(a) Sketch the graph of f and verify that f is one-to-one.
(b) Find f^{-1}.

47. Let

$$f(x) = \frac{ax+b}{cx+d}, \quad x \neq -\frac{d}{c}.$$

(a) Show that f is one-to-one iff $ad - bc \neq 0$.
(b) Suppose that $ad - bc \neq 0$. Find f^{-1}.

48. Let

$$f(x) = \int_2^x \sqrt{1+t^2}\, dt.$$

(a) Prove that f has an inverse.
(b) Find $(f^{-1})'(0)$.

49. Let

$$f(x) = \int_1^{2x} \sqrt{16+t^4}\, dt.$$

(a) Prove that f has an inverse.
(b) Find $(f^{-1})'(0)$.

50. Suppose that the function f has an inverse and fix a number $c \neq 0$.
(a) Let $g(x) = f(x+c)$ for all x such that $x + c \in \text{dom}(f)$. Prove that g has an inverse and find it.
(b) Let $h(x) = f(cx)$ for all x such that $cx \in \text{dom}(f)$. Prove that h has an inverse and find it.

51. Let f be a one-to-one, twice differentiable function and let $g = f^{-1}$.
(a) Show that

$$g''(x) = -\frac{f''[g(x)]}{(f'[g(x)])^3}.$$

(b) Suppose that the graph of f is concave upward (downward) on an interval I. What can you say about the graph of f^{-1} on I?

52. Let $P(x) = a_n x^n + a_{n-1}x^{n-1} + \cdots + a_1 x + a_0$ be a polynomial function.
(a) Can P have an inverse if its degree n is even? Explain why or why not.
(b) Can P have an inverse if n is odd? If so, give an example. Also, give an example of a polynomial function of odd degree that does not have an inverse.

53. Suppose that f is differentiable and has an inverse. If we let $y = f^{-1}(x)$, then $f(y) = x$ and we can find dy/dx by differentiating implicitly. This is how we derived (7.1.8). The function $f(x) = \sin x$, $-\frac{\pi}{2} \leq x \leq \frac{\pi}{2}$, is one-to-one and differentiable. Let $y = f^{-1}(x)$ and find dy/dx. Express your result as a function of x.

54. Repeat Exercise 53 for the function $f(x) = \tan x$, $-\frac{\pi}{2} < x < \frac{\pi}{2}$.

C In Exercises 55–58, use a graphing utility to draw the graph of f and verify that f is one-to-one. Then sketch the graph of f^{-1}.

55. $f(x) = x^3 + 3x + 2.$

56. $f(x) = x^{3/5} - 1.$

57. $f(x) = 4 \sin 2x, \quad -\frac{\pi}{4} \leq x \leq \frac{\pi}{4}.$

58. $f(x) = 2 - \cos 3x, \quad 0 \leq x \leq \frac{\pi}{3}.$

■ 7.2 THE LOGARITHM FUNCTION, PART I

If B is a positive number different from 1, the logarithm to the base B is defined in elementary mathematics by setting

$$C = \log_B A \quad \text{iff} \quad B^C = A.$$

Historically, the base 10 was chosen because our number system is based on the powers of 10. The defining relation then becomes:

$$C = \log_{10} A \quad \text{iff} \quad 10^C = A.$$

The basic properties of $\log_{10}$ can then be summarized as follows: with $X, Y > 0$,

$$\log_{10}(XY) = \log_{10}X + \log_{10}Y, \qquad \log_{10}1 = 0,$$
$$\log_{10}(1/Y) = -\log_{10}Y, \qquad \log_{10}(X/Y) = \log_{10}X - \log_{10}Y,$$
$$\log_{10}X^Y = Y\log_{10}X, \qquad \log_{10}10 = 1.$$

This elementary notion of logarithm is inadequate for calculus. It is unclear: What is meant by 10^C if C is irrational? It does not lend itself well to the methods of calculus: How would you differentiate $Y = \log_{10}X$ knowing only that $10^Y = X$?

Here we take an entirely different approach to logarithms. Instead of trying to tamper with the elementary definition, we discard it altogether. From our point of view the fundamental property of logarithms is that they transform multiplication into addition:

The log of a product = the sum of the logs.

Taking this as the central idea we are led to a general notion of logarithm that encompasses the elementary notion, lends itself well to the methods of calculus, and leads us naturally to a choice of base that simplifies many calculations.

DEFINITION 7.2.1

A *logarithm* function is a nonconstant differentiable function f defined on the set of positive numbers such that for all $x > 0$ and $y > 0$

$$f(xy) = f(x) + f(y).$$

Let's assume for the time being that such logarithm functions exist, and let's see what we can find out about them. In the first place, if f is such a function, then

$$f(1) = f(1 \cdot 1) = f(1) + f(1) = 2f(1) \quad \text{and so} \quad f(1) = 0.$$

Taking $y > 0$, we have

$$0 = f(1) = f(y \cdot 1/y) = f(y) + f(1/y)$$

and therefore

$$f(1/y) = -f(y).$$

Taking $x > 0$ and $y > 0$, we have

$$f(x/y) = f(x \cdot 1/y) = f(x) + f(1/y),$$

which, in view of the previous result, means that

$$f(x/y) = f(x) - f(y).$$

We are now ready to look for the derivative. (Remember, we are *assuming* that f is differentiable.) We begin by forming the difference quotient

$$\frac{f(x+h) - f(x)}{h},$$

where $x > 0$ is fixed. From what we have discovered about f,

$$f(x+h) - f(x) = f\left(\frac{x+h}{x}\right) = f(1 + h/x),$$

and therefore

$$\frac{f(x+h) - f(x)}{h} = \frac{f(1 + h/x)}{h}.$$

Multiplying the denominator by x/x and using the fact that $f(1) = 0$, we can write the difference quotient as

$$\frac{f(x+h) - f(x)}{h} = \frac{1}{x}\left[\frac{f(1 + h/x) - f(1)}{h/x}\right].$$

Now let $k = h/x$. Then we have

$$\frac{f(x+h) - f(x)}{h} = \frac{1}{x}\left[\frac{f(1 + k) - f(1)}{k}\right].$$

Since $k \to 0$ iff $h \to 0$ (remember, x is fixed), it follows that

$$\begin{aligned} f'(x) = \lim_{h\to 0} \frac{f(x+h) - f(x)}{h} &= \lim_{k\to 0} \frac{1}{x}\left[\frac{f(1 + k) - f(1)}{k}\right] \\ &= \frac{1}{x} \lim_{k\to 0} \left[\frac{f(1 + k) - f(1)}{k}\right] = \frac{1}{x} f'(1). \end{aligned}$$

In short,

(7.2.2)
$$\boxed{f'(x) = \frac{1}{x} f'(1).}$$

Thus we have shown that if f is a logarithm function and x is any positive number, then

$$f(1) = 0 \qquad \text{and} \qquad f'(x) = \frac{1}{x} f'(1).$$

We can't have $f'(1) = 0$, for that would make f constant. (Explain.) The most natural choice, the one that will keep calculations as simple as possible, is to set $f'(1) = 1$.† The derivative is then $1/x$.

This function, which takes on the value 0 at 1 and has derivative $1/x$ for $x > 0$, must, by the fundamental theorem of calculus, take the form

$$\int_1^x \frac{dt}{t}. \qquad \text{(verify this)}$$

† This, as you will see later, is tantamount to a choice of base.

DEFINITION 7.2.3

The function

$$L(x) = \int_1^x \frac{dt}{t}, \qquad x > 0$$

is called the (*natural*) *logarithm function*.

Here are some obvious properties of L:

(1) L is defined on $(0, \infty)$ with derivative

$$L'(x) = \frac{1}{x} \qquad \text{for all } x > 0.$$

L' is positive on $(0, \infty)$, and therefore L is an increasing function.

(2) L is continuous on $(0, \infty)$ since it is differentiable there.

(3) For $x > 1$, $L(x)$ gives the area of the region shaded in Figure 7.2.1.

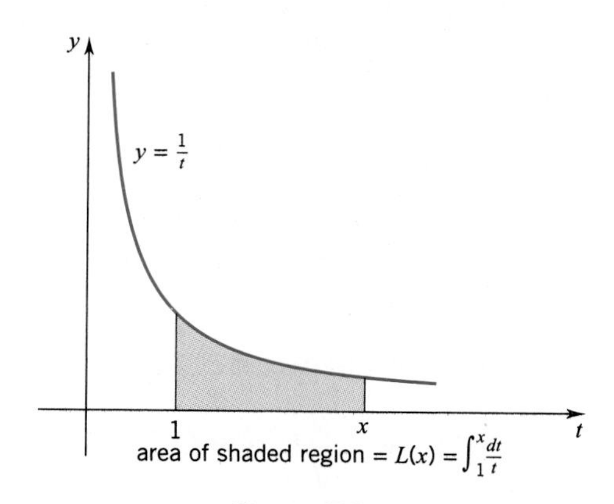

Figure 7.2.1

(4) $L(x)$ is negative if $0 < x < 1$, $L(1) = 0$, $L(x)$ is positive if $x > 1$.

The following result is fundamental; it establishes that L *is* a logarithm function by showing that it satisfies the equation in Definition 7.2.1.

THEOREM 7.2.4

If x and y are positive, then

$$L(xy) = L(x) + L(y).$$

PROOF Fix any positive number y. Then, since

$$\frac{d}{dx}[L(x)] = \frac{1}{x} \qquad \text{and} \qquad \frac{d}{dx}[L(xy)] \underset{\text{chain rule}}{=} \frac{y}{xy} = \frac{1}{x},$$

we can be sure that

$$L(xy) = L(x) + C,$$

for some constant C. We can evaluate the constant by taking $x = 1$:

$$L(y) = L(1 \cdot y) = L(1) + C \underset{L(1)=0}{=} C.$$

Therefore $L(xy) = L(x) + L(y)$ as asserted. ❑

We come now to another important result.

THEOREM 7.2.5

If x is positive and p/q is rational, then

$$L(x^{p/q}) = \frac{p}{q}L(x).$$

PROOF You have seen that $d/dx[L(x)] = 1/x$. By the chain rule

$$\frac{d}{dx}[L(x^{p/q})] = \frac{1}{x^{p/q}}\frac{d}{dx}(x^{p/q}) \underset{(3.7.1)}{=} \frac{1}{x^{p/q}}\left(\frac{p}{q}\right)x^{(p/q)-1} = \frac{p}{q}\left(\frac{1}{x}\right) = \frac{d}{dx}\left[\frac{p}{q}L(x)\right].$$

Since both functions have the same derivative, they differ by a constant:

$$L(x^{p/q}) = \frac{p}{q}L(x) + C.$$

Since both functions are zero at $x = 1$, $C = 0$ and the theorem is proved. ❑

The domain of L is $(0, \infty)$. What is the range of L?

THEOREM 7.2.6

The range of L is $(-\infty, \infty)$.

PROOF Since L is continuous on $(0, \infty)$, we know from the intermediate-value theorem that it "skips" no values. Thus, its range is an interval. To show that the interval is $(-\infty, \infty)$, we need only show that it is unbounded above and unbounded below. We can do this by taking M as an arbitrary positive number and showing that L takes on values greater than M and values less than $-M$.

Since

$$L(2) = \int_1^2 \frac{dt}{t}$$

is positive (explain), we know that some positive multiple of $L(2)$ must be greater than M; that is, since $L(2) > 0$, $nL(2) \to \infty$ as $n \to \infty$, and so we can choose a positive integer n big enough such that

$$nL(2) > M.$$

Multiplying this equation by -1 we have

$$-nL(2) < -M.$$

Since

$$nL(2) = L(2^n) \quad \text{and} \quad -nL(2) = L(2^{-n}),$$

we have

$$L(2^n) > M \quad \text{and} \quad L(2^{-n}) < -M.$$

This proves the unboundedness. ❑

The Number e

Since the range of L is $(-\infty, \infty)$ and L is an increasing function, we know that L takes on every value and does so only once. In particular, there is one and only one number at which the function L takes on the value 1. *This unique number is denoted by the letter e.*†

The number e can also be defined geometrically: e is the unique number with the property that the area under the graph of $f(t) = 1/t$ from $t = 1$ to $t = e$ is 1. See Figure 7.2.2.

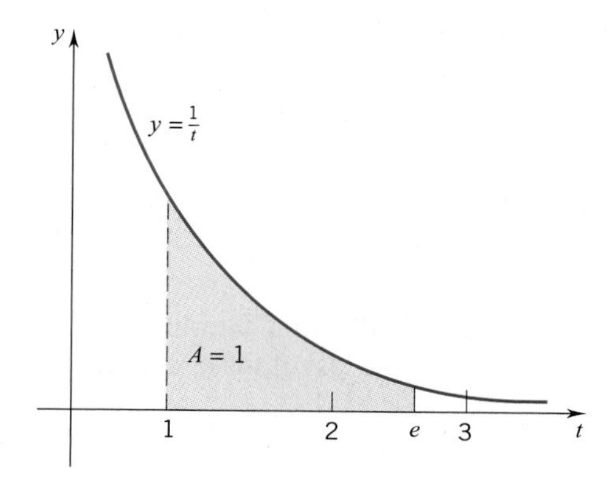

Figure 7.2.2

† After the Swiss mathematician Leonhard Euler (1707–1783), considered by many the greatest mathematician of the eighteenth century.

Since

(7.2.7)
$$L(e) = \int_1^e \frac{dt}{t} = 1,$$

it follows from Theorem 7.2.5 that

(7.2.8)
$$L(e^{p/q}) = \frac{p}{q} \qquad \text{for all rational numbers } \frac{p}{q}.$$

Because of this relation, we call L *the logarithm to the base e* and sometimes write

$$L(x) = \log_e x.$$

The number e arises naturally in many settings. Accordingly, we call $L(x)$ the natural logarithm and write

(7.2.9)
$$L(x) = \ln x. \quad \dagger$$

Here are the basic properties we have established for $\ln x$:

(7.2.10)

$\ln 1 = 0, \qquad \ln e = 1.$	
$\ln xy = \ln x + \ln y.$	$(x > 0, y > 0)$
$\ln 1/x = -\ln x.$	$(x > 0)$
$\ln x/y = \ln x - \ln y.$	$(x > 0, y > 0)$
$\ln x^r = r \ln x.$	$(x > 0, r \text{ rational})$

Notice how closely these rules parallel the familiar rules for common logarithms (base 10). Later we will show that the last of these rules also holds for irrational exponents.

† Logarithms to bases other than e will be taken up later [they arise by other choices of $f'(1)$], but by far the most important logarithm in calculus is the logarithm to the base e. So much so, that when we speak of the logarithm of a number x and don't specify the base, you can be sure that we are talking about the *natural logarithm* $\ln x$.

The Graph of the Logarithm Function

You know that the logarithm function

$$\ln x = \int_1^x \frac{dt}{t}$$

has domain $(0, \infty)$, range $(-\infty, \infty)$, and derivative

$$\frac{d}{dx}(\ln x) = \frac{1}{x} > 0.$$

For small x the derivative is large (near 0 the curve is steep); for large x the derivative is small (far out the curve flattens out). At $x = 1$ the logarithm is 0 and its derivative $1/x$ is 1. (The graph crosses the x-axis at the point $(1, 0)$, and the tangent line at that point is parallel to the line $y = x$.) The second derivative

$$\frac{d^2}{dx^2}(\ln x) = -\frac{1}{x^2}$$

is negative on $(0, \infty)$. (The graph is concave down throughout.) We have sketched the graph in Figure 7.2.3. The y-axis is a vertical asymptote:

$$\text{as } x \to 0^+, \quad \ln x \to -\infty.$$

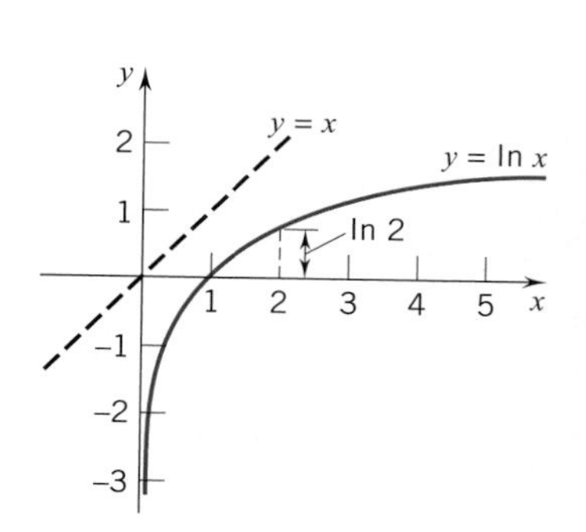

Figure 7.2.3

Example 1 Using upper and lower sums, estimate

$$\ln 2 = \int_1^2 \frac{dt}{t} \qquad \text{(Figure 7.2.4)}$$

from the partition

$$P = \{1 = \tfrac{10}{10}, \tfrac{11}{10}, \tfrac{12}{10}, \tfrac{13}{10}, \tfrac{14}{10}, \tfrac{15}{10}, \tfrac{16}{10}, \tfrac{17}{10}, \tfrac{18}{10}, \tfrac{19}{10}, \tfrac{20}{10} = 2\}.$$

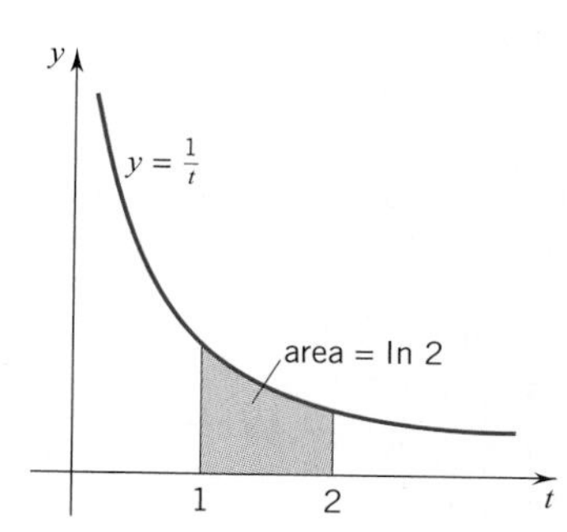

Figure 7.2.4

SOLUTION Using a calculator we find that

$$\begin{aligned} L_f(P) &= \tfrac{1}{10}(\tfrac{10}{11} + \tfrac{10}{12} + \tfrac{10}{13} + \tfrac{10}{14} + \tfrac{10}{15} + \tfrac{10}{16} + \tfrac{10}{17} + \tfrac{10}{18} + \tfrac{10}{19} + \tfrac{10}{20}) \\ &= \tfrac{1}{11} + \tfrac{1}{12} + \tfrac{1}{13} + \tfrac{1}{14} + \tfrac{1}{15} + \tfrac{1}{16} + \tfrac{1}{17} + \tfrac{1}{18} + \tfrac{1}{19} + \tfrac{1}{20} > 0.668 \end{aligned}$$

and

$$\begin{aligned} U_f(P) &= \tfrac{1}{10}(\tfrac{10}{10} + \tfrac{10}{11} + \tfrac{10}{12} + \tfrac{10}{13} + \tfrac{10}{14} + \tfrac{10}{15} + \tfrac{10}{16} + \tfrac{10}{17} + \tfrac{10}{18} + \tfrac{10}{19}) \\ &= \tfrac{1}{10} + \tfrac{1}{11} + \tfrac{1}{12} + \tfrac{1}{13} + \tfrac{1}{14} + \tfrac{1}{15} + \tfrac{1}{16} + \tfrac{1}{17} + \tfrac{1}{18} + \tfrac{1}{19} < 0.719. \end{aligned}$$

Thus we have

$$0.668 < L_f(P) \le \ln 2 \le U_f(P) < 0.719.$$

The average of these two estimates is

$$\tfrac{1}{2}(0.668 + 0.719) = 0.6935.$$

Rounded off to four decimal places, the value of ln 2 given on a calculator is 0.6931, so we are not far off. ❑

Table 7.2.1 gives the natural logarithms of the integers 1 through 10 rounded off to the nearest hundredth.

■ **Table 7.2.1**

n	$\ln n$	n	$\ln n$
1	0.00	6	1.79
2	0.69	7	1.95
3	1.10	8	2.08
4	1.39	9	2.20
5	1.61	10	2.30

Example 2 Use the properties of logarithms and Table 7.2.1 to estimate the following logarithms.

(a) $\ln 0.2$. **(b)** $\ln 0.25$. **(c)** $\ln 2.4$. **(d)** $\ln 90$.

SOLUTION

(a) $\ln 0.2 = \ln \frac{1}{5} = -\ln 5 \cong -1.61$. **(b)** $\ln 0.25 = \ln \frac{1}{4} = -\ln 4 \cong -1.39$.

(c) $\ln 2.4 = \ln \dfrac{12}{5} = \ln \dfrac{(3)(4)}{5} = \ln 3 + \ln 4 - \ln 5 \cong 0.88$.

(d) $\ln 90 = \ln[(9)(10)] = \ln 9 + \ln 10 \cong 4.50$. ❑

Example 3 Estimate e on the basis of Table 7.2.1.

SOLUTION So far all we know is that $\ln e = 1$. From the table you can see that

$$3 \ln 3 - \ln 10 \cong 1.$$

The expression on the left can be written

$$\ln 3^3 - \ln 10 = \ln 27 - \ln 10 = \ln \tfrac{27}{10} = \ln 2.7.$$

Thus $\ln 2.7 \cong 1$, and so $e \cong 2.7$. It can be shown that e is an irrational number; its decimal expansion to twelve decimal places is

$$e \cong 2.718281828459. \quad ❑$$

EXERCISES 7.2

In Exercises 1–12, estimate the given natural logarithm on the basis of Table 7.2.1; check your results on a calculator.

1. $\ln 20$.
2. $\ln 16$.
3. $\ln 1.6$.
4. $\ln 3^4$.
5. $\ln 0.1$.
6. $\ln 2.5$.
7. $\ln 7.2$.
8. $\ln \sqrt{630}$.
9. $\ln \sqrt{2}$.
10. $\ln 0.4$.
11. $\ln 2^5$.
12. $\ln 1.8$.

13. Interpret the equation $\ln n = \ln mn - \ln m$ in terms of area under the curve $y = 1/x$. Draw a figure.

14. Given that $0 < x < 1$, express as a logarithm the area under the curve $y = 1/t$ from $t = x$ to $t = 1$.

15. Estimate

$$\ln 1.5 = \int_1^{1.5} \frac{dt}{t}$$

using the approximation $\frac{1}{2}[L_f(P) + U_f(P)]$ with $P = \{1 = \frac{8}{8}, \frac{9}{8}, \frac{10}{8}, \frac{11}{8}, \frac{12}{8} = 1.5\}$.

16. Estimate

$$\ln 2.5 = \int_1^{2.5} \frac{dt}{t}$$

using the approximation $\frac{1}{2}[L_f(P) + U_f(P)]$ with $P = \{1 = \frac{4}{4}, \frac{5}{4}, \frac{6}{4}, \frac{7}{4}, \frac{8}{4}, \frac{9}{4}, \frac{10}{4} = \frac{5}{2}\}$.

17. Taking $\ln 5 \cong 1.61$, use differentials to estimate (a) $\ln 5.2$. (b) $\ln 4.8$. (c) $\ln 5.5$.

18. Taking $\ln 10 \cong 2.30$, use differentials to estimate (a) $\ln 10.3$. (b) $\ln 9.6$. (c) $\ln 11$.

In Exercises 19–24, solve the given equation for x.

19. $\ln x = 2$.
20. $\ln x = -1$.
21. $(2 - \ln x) \ln x = 0$.
22. $\frac{1}{2} \ln x = \ln(2x - 1)$.
23. $\ln[(2x + 1)(x + 2)] = 2 \ln(x + 2)$.
24. $2 \ln(x + 2) - \frac{1}{2} \ln x^4 = 1$.

25. Show that

$$\lim_{x \to 1} \frac{\ln x}{x - 1} = 1.$$

HINT: $\dfrac{\ln x}{x - 1} = \dfrac{\ln x - \ln 1}{x - 1}$; interpret the limit in terms of the derivative of $\ln x$.

26. (a) Use the mean-value theorem to show that

$$\frac{x - 1}{x} < \ln x < x - 1 \quad \text{for all } x > 0.$$

HINT: Consider the cases $x > 1$ and $0 < x < 1$ separately.

(b) Use the result in part (a) to give an alternative proof that

$$\lim_{x \to 1} \frac{\ln x}{x - 1} = 1.$$

27. (a) Show that for $n \geq 2$,

$$\frac{1}{2} + \frac{1}{3} + \cdots + \frac{1}{n} < \ln n < 1 + \frac{1}{2} + \frac{1}{3} + \cdots + \frac{1}{n-1}.$$

HINT: See the figure.

(b) Show that the area of the shaded region in the figure is given by

$$1 + \frac{1}{2} + \frac{1}{3} + \cdots + \frac{1}{n-1} - \ln n.$$

As $n \to \infty$, this area approaches the number γ known as *Euler's constant.*

(c) Use geometric reasoning to show that $\frac{1}{2} < \gamma < 1$. (To three decimal places, $\gamma \cong 0.577$).

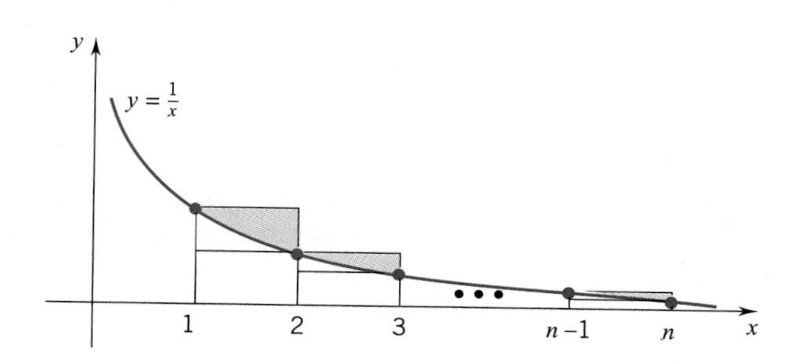

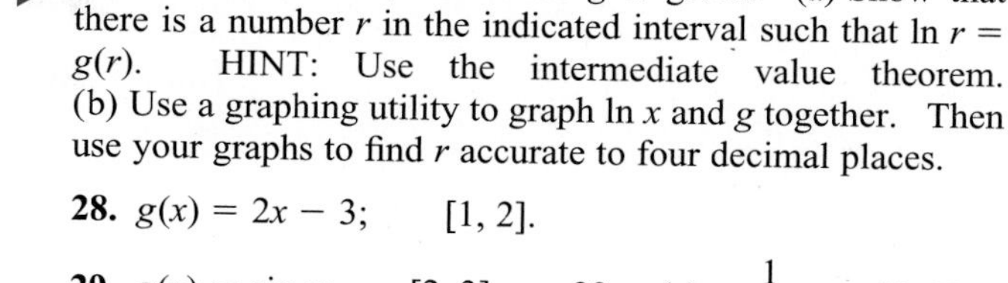

In Exercises 28–30, a function g is given. (a) Show that there is a number r in the indicated interval such that $\ln r = g(r)$. HINT: Use the intermediate value theorem. (b) Use a graphing utility to graph $\ln x$ and g together. Then use your graphs to find r accurate to four decimal places.

28. $g(x) = 2x - 3$; $[1, 2]$.

29. $g(x) = \sin x$; $[2, 3]$.

30. $g(x) = \dfrac{1}{x^2}$; $[1, 2]$.

In Exercises 31–33, estimate $\lim_{x \to a} f(x)$ numerically by evaluating f at the indicated values. Then use a graphing utility to zoom in on the graph near $x = a$ to justify your estimate.

31. $\lim_{x \to 1} \dfrac{\ln x}{x - 1}$; $x = 1 \pm 0.5, 1 \pm 0.1, 1 \pm 0.01, 1 \pm 0.001, 1 \pm 0.0001$.

32. $\lim_{x \to 0} x \ln |x|$; $x = \pm 0.5, \pm 0.1, \pm 0.01, \pm 0.001, \pm 0.0001$.

33. $\lim_{x \to 0^+} \sqrt{x} \ln x$; $x = 0.5, 0.1, 0.01, 0.001, 0.0001$.

■ 7.3 THE LOGARITHM FUNCTION, PART II

Differentiation and Graphing

We know that for $x > 0$

$$\frac{d}{dx}(\ln x) = \frac{1}{x}.$$

If u is a positive differentiable function of x, then by the chain rule

$$\frac{d}{dx}(\ln u) = \frac{d}{du}(\ln u)\frac{du}{dx} = \frac{1}{u}\frac{du}{dx}.$$

Thus, for example,

$$\frac{d}{dx}[\ln(1 + x^2)] = \frac{1}{1 + x^2} \cdot 2x = \frac{2x}{1 + x^2} \qquad \text{for all real } x$$

and

$$\frac{d}{dx}[\ln(1+3x)] = \frac{1}{1+3x}\cdot 3 = \frac{3}{1+3x} \qquad \text{for all } x > -\tfrac{1}{3}.$$

Example 1 Find the domain of f and find $f'(x)$ if

$$f(x) = \ln(x\sqrt{1+3x}).$$

SOLUTION For x to be in the domain of f, we must have $x\sqrt{1+3x} > 0$, and consequently, $x > 0$. The domain is the set of positive numbers.

Before differentiating f, we make use of the special properties of the logarithm:

$$f(x) = \ln(x\sqrt{1+3x}) = \ln x + \ln[(1+3x)^{1/2}] = \ln x + \tfrac{1}{2}\ln(1+3x).$$

From this we have

$$f'(x) = \frac{1}{x} + \frac{1}{2}\left(\frac{3}{1+3x}\right) = \frac{9x+2}{6x^2+2x}. \quad ❑$$

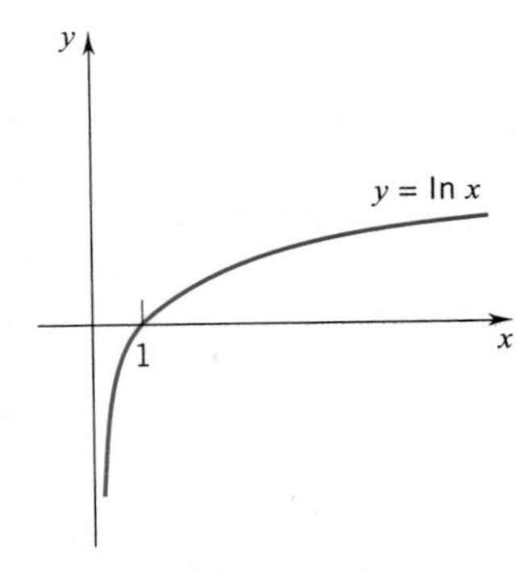

Figure 7.3.1

Example 2 Sketch the graph of

$$f(x) = \ln|x|.$$

SOLUTION The domain of f is all numbers $x \neq 0$, and f is an even function: $f(-x) = f(x)$ for all $x \neq 0$. Thus, the graph has two branches:

$$y = \ln(-x), \quad x < 0 \qquad \text{and} \qquad y = \ln x, \quad x > 0,$$

and is symmetric with respect to the y-axis. As we saw in the preceding section, the graph of $y = \ln x$, $x > 0$, is as shown in Figure 7.3.1. By symmetry, the graph of $y = \ln(-x)$, $x < 0$, is the mirror image of $y = \ln x$, $x > 0$, in the y-axis. Figure 7.3.2 shows the graph of $y = \ln|x|$. ❑

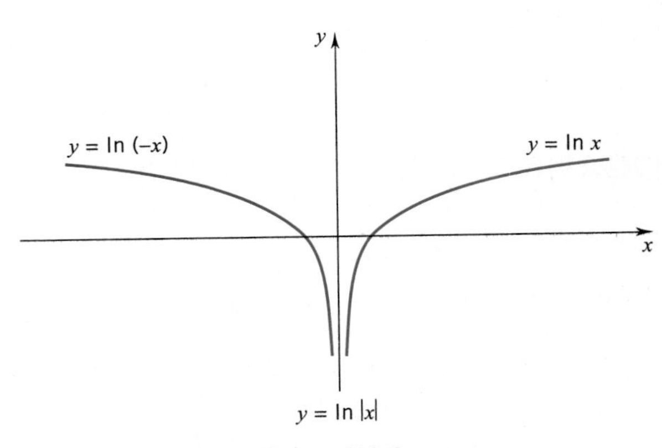

Figure 7.3.2

Example 3 Show that

(7.3.1)
$$\frac{d}{dx}(\ln|x|) = \frac{1}{x} \qquad \text{for all } x \neq 0.$$

SOLUTION For $x > 0$,

$$\frac{d}{dx}(\ln |x|) = \frac{d}{dx}(\ln x) = \frac{1}{x}.$$

For $x < 0$, we have $|x| = -x > 0$, so that

$$\frac{d}{dx}(\ln |x|) = \frac{d}{dx}[\ln(-x)] = \frac{1}{-x}\frac{d}{dx}(-x) = \left(\frac{1}{-x}\right)(-1) = \frac{1}{x}. \quad \square$$

Example 4 Find the domain of f and find $f'(x)$ given that

$$f(x) = (\ln x^2)^3.$$

SOLUTION Since the logarithm function is defined only for positive numbers, we must have $x^2 > 0$. Thus, the domain of f consists of all $x \neq 0$. Before differentiating, note that

$$f(x) = (\ln x^2)^3 = (2 \ln |x|)^3 = 8(\ln |x|)^3.$$

It follows that

$$f'(x) = 24(\ln |x|)^2 \frac{d}{dx}(\ln |x|) = 24(\ln |x|)^2 \frac{1}{x} = \frac{24}{x}(\ln |x|)^2. \quad \square$$

Example 5 Let $f(x) = x \ln x$.

(a) Specify the domain of f and find the intercepts, if any. **(b)** On what intervals does f increase? Decrease? **(c)** Find the extreme values of f. **(d)** Determine the concavity of the graph and find the points of inflection. **(e)** Sketch the graph of f.

SOLUTION Since the logarithm function is defined only for positive numbers, the domain of f is $(0, \infty)$ and there is no y-intercept. Since $f(1) = 1 \cdot \ln 1 = 0$, 1 is an x-intercept.

Differentiating f, we have

$$f'(x) = x \cdot \frac{1}{x} + \ln x = 1 + \ln x.$$

To find the critical numbers of f, we set $f'(x) = 0$:

$$1 + \ln x = 0, \qquad \ln x = -1, \qquad x = \frac{1}{e} \qquad \text{(verify this).}$$

Recalling that the logarithm function is increasing on $(0, \infty)$, and that $\ln x \to -\infty$ as $x \to 0^+$ and $\ln x \to \infty$ as $x \to \infty$, we have

f': - - - - - 0 +

f 0 $\frac{1}{e}$ 1 2 x

decreases increases

Therefore, f decreases on $(0, 1/e)$ and increases on $(1/e, \infty)$. By the first derivative test,

$$f(1/e) = \frac{1}{e}\ln\left(\frac{1}{e}\right) = -\frac{1}{e} \cong -0.368$$

is a local and absolute minimum of f.

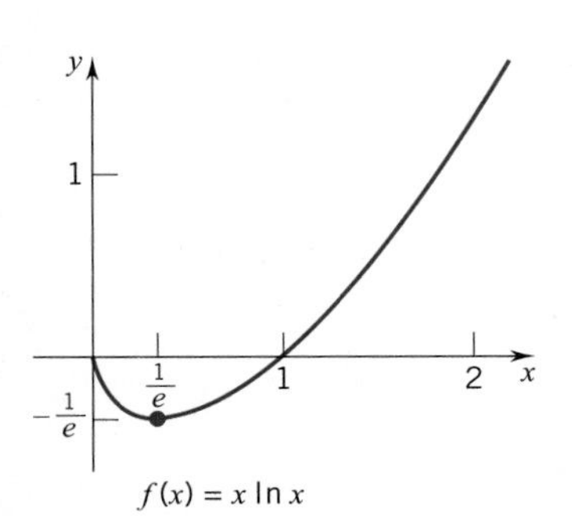

Figure 7.3.3

Since $f''(x) = \dfrac{1}{x} > 0$ for $x > 0$, the graph of f is concave up on $(0, \infty)$; there are no points of inflection.

You can verify numerically that $\lim_{x\to 0^+} x \ln x = 0$ (see Exercise 32, Section 7.2), and it is clear that $x \ln x \to \infty$ as $x \to \infty$. Also, $\lim_{x\to 0^+} f'(x) = \lim_{x\to 0^+} (1 + \ln x) = -\infty$.

A sketch of the graph of f is shown in Figure 7.3.3. ❑

Example 6 Show that, if u is a differentiable function of x, then for $u \neq 0$

(7.3.2)
$$\frac{d}{dx}(\ln |u|) = \frac{1}{u}\frac{du}{dx}.$$

SOLUTION

$$\frac{d}{dx}(\ln |u|) = \frac{d}{du}(\ln |u|)\frac{du}{dx} = \frac{1}{u}\frac{du}{dx}. \quad ❑$$

Here are two examples:

$$\frac{d}{dx}(\ln |1 - x^3|) = \frac{1}{1 - x^3}\frac{d}{dx}(1 - x^3) = \frac{-3x^2}{1 - x^3} = \frac{3x^2}{x^3 - 1}.$$

$$\frac{d}{dx}\left(\ln \left|\frac{x-1}{x-2}\right|\right) = \frac{d}{dx}(\ln |x - 1|) - \frac{d}{dx}(\ln |x - 2|) = \frac{1}{x-1} - \frac{1}{x-2}. \quad ❑$$

Example 7 Let

$$f(x) = \ln\left(\frac{x^4}{x-1}\right).$$

(a) Specify the domain of f. **(b)** On what intervals does f increase? Decrease? **(c)** Find the extreme values of f. **(d)** Determine the concavity of the graph and find the points of inflection. **(e)** Sketch the graph specifying the asymptotes if any.

SOLUTION Since the logarithm function is defined only for positive numbers, the domain of f is the open interval $(1, \infty)$.

Making use of the special properties of the logarithm we write

$$f(x) = \ln x^4 - \ln(x - 1) = 4 \ln x - \ln(x - 1).$$

Differentiation gives

$$f'(x) = \frac{4}{x} - \frac{1}{x-1} = \frac{3x - 4}{x(x-1)}$$

and

$$f''(x) = -\frac{4}{x^2} + \frac{1}{(x-1)^2} = -\frac{(x-2)(3x-2)}{x^2(x-1)^2}.$$

Since the domain of f is $(1, \infty)$, we consider only the values of x greater than 1.

It is not hard to see that

$$f'(x) \text{ is } \begin{cases} \text{negative,} & \text{for } 1 < x < \frac{4}{3} \\ 0, & \text{at } x = \frac{4}{3} \\ \text{positive,} & \text{for } x > \frac{4}{3}. \end{cases}$$

Thus f decreases on $(1, \frac{4}{3}]$ and increases on $[\frac{4}{3}, \infty)$. By the first derivative test

$$f(\tfrac{4}{3}) = 4 \ln 4 - 3 \ln 3 \cong 2.25$$

is a local and absolute minimum. There are no other extreme values.

Since

$$f''(x) \text{ is } \begin{cases} \text{positive,} & \text{for } 1 < x < 2 \\ 0, & \text{at } x = 2 \\ \text{negative,} & \text{for } x > 2, \end{cases}$$

the graph is concave up on $(1, 2)$ and concave down on $(2, \infty)$. The point

$$(2, f(2)) = (2, 4 \ln 2) \cong (2, 2.77)$$

is the only point of inflection.

Before sketching the graph, note that the derivative

$$f'(x) = \frac{4}{x} - \frac{1}{x-1}$$

is very large negative for x close to 1 and very close to 0 for x large. This means that the graph is very steep for x close to 1 and very flat for x large. See Figure 7.3.4. The line $x = 1$ is a vertical asymptote: as $x \rightarrow 1^+$, $f(x) \rightarrow \infty$. ❑

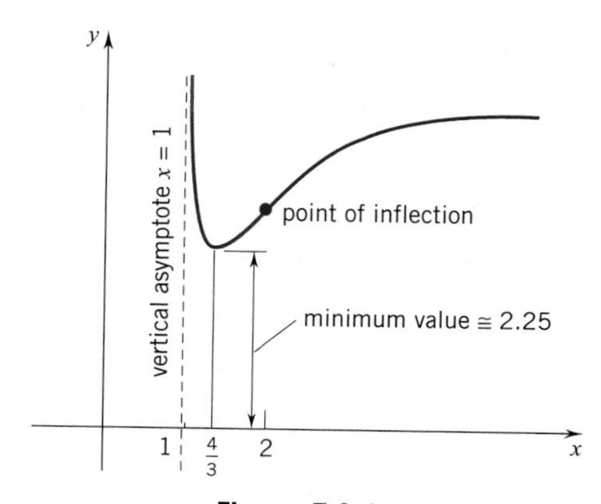

Figure 7.3.4

Integration

The integral counterpart of (7.3.1) takes the form

(7.3.3)

$$\int \frac{dx}{x} = \ln |x| + C, \qquad x \neq 0.$$

In practice

$$\int \frac{g'(x)}{g(x)}\,dx \quad \text{is reduced to} \quad \int \frac{du}{u}$$

by setting

$$u = g(x), \qquad du = g'(x)\,dx.$$

Thus, we have the important generalization of (7.3.3)

(7.3.4)

$$\int \frac{g'(x)}{g(x)}\,dx = \ln|g(x)| + C, \qquad g(x) \neq 0.$$

Example 8 Calculate

$$\int \frac{x^2}{1-4x^3}\,dx.$$

SOLUTION Set

$$u = 1 - 4x^3, \qquad du = -12x^2\,dx.$$

$$\int \frac{x^2}{1-4x^3}\,dx = -\frac{1}{12}\int \frac{du}{u} = -\frac{1}{12}\ln|u| + C = -\frac{1}{12}\ln|1-4x^3| + C. \quad \square$$

Example 9 Calculate

$$\int \frac{\ln x}{x}\,dx.$$

SOLUTION Set

$$u = \ln x, \qquad du = \frac{1}{x}\,dx.$$

$$\int \frac{\ln x}{x}\,dx = \int u\,du = \tfrac{1}{2}u^2 + C = \tfrac{1}{2}(\ln x)^2 + C. \quad \square$$

Example 10 Evaluate

$$\int_1^2 \frac{6x^2-2}{x^3-x+1}\,dx.$$

SOLUTION Set

$$u = x^3 - x + 1, \qquad du = (3x^2-1)\,dx.$$

At $x = 1$, $u = 1$; at $x = 2$, $u = 7$.

$$\int_1^2 \frac{6x^2-2}{x^3-x+1}\,dx = 2\int_1^7 \frac{du}{u} = 2\Big[\ln|u|\Big]_1^7 = 2(\ln 7 - \ln 1) = 2\ln 7. \quad \square$$

Logarithmic Differentiation

We can differentiate a lengthy product

$$g(x) = g_1(x)g_2(x) \cdots g_n(x)$$

by first writing

$$\begin{aligned} \ln |g(x)| &= \ln(|g_1(x)||g_2(x)| \cdots |g_n(x)|) \\ &= \ln|g_1(x)| + \ln |g_2(x)| + \cdots + \ln |g_n(x)| \end{aligned}$$

and then differentiating:

$$\frac{g'(x)}{g(x)} = \frac{g_1'(x)}{g_1(x)} + \frac{g_2'(x)}{g_2(x)} + \cdots + \frac{g_n'(x)}{g_n(x)}.$$

Multiplication by $g(x)$ gives

(7.3.5)
$$g'(x) = g(x)\left(\frac{g_1'(x)}{g_1(x)} + \frac{g_2'(x)}{g_2(x)} + \cdots + \frac{g_n'(x)}{g_n(x)}\right).$$

The process by which $g'(x)$ was obtained is called *logarithmic differentiation.* Logarithmic differentiation is valid at all points x where $g(x) \neq 0$. At points x where $g(x) = 0$, none of it makes sense.

A product of n factors

$$g(x) = g_1(x)g_2(x) \cdots g_n(x)$$

can, of course, also be differentiated by repeated applications of the product rule (3.2.5). The great advantage of logarithmic differentiation is that it gives us an explicit formula for the derivative, a formula that's easy to remember and easy to work with.

Example 11 Given that

$$g(x) = x(x-1)(x-2)(x-3)$$

find $g'(x)$ for $x \neq 0, 1, 2, 3$.

SOLUTION We can write down $g'(x)$ directly from Formula (7.3.5),

$$g'(x) = x(x-1)(x-2)(x-3)\left(\frac{1}{x} + \frac{1}{x-1} + \frac{1}{x-2} + \frac{1}{x-3}\right)$$

or we can go through the process by which we derived Formula (7.3.5):

$$\ln |g(x)| = \ln |x| + \ln |x-1| + \ln |x-2| + \ln |x-3|,$$

$$\frac{g'(x)}{g(x)} = \frac{1}{x} + \frac{1}{x-1} + \frac{1}{x-2} + \frac{1}{x-3},$$

$$g'(x) = x(x-1)(x-2)(x-3)\left(\frac{1}{x} + \frac{1}{x-1} + \frac{1}{x-2} + \frac{1}{x-3}\right). \quad \square$$

Logarithmic differentiation can also be used with quotients.

Example 12 Given that

$$g(x) = \frac{(x^2+1)^3(2x-5)^2}{(x^2+5)^2},$$

find $g'(x)$ for $x \neq \frac{5}{2}$.

SOLUTION Our first step is to write

$$g(x) = (x^2+1)^3(2x-5)^2(x^2+5)^{-2}.$$

Then according to Formula (7.3.5)

$$\begin{aligned} g'(x) &= \frac{(x^2+1)^3(2x-5)^2}{(x^2+5)^2}\left[\frac{3(x^2+1)^2(2x)}{(x^2+1)^3} + \frac{2(2x-5)(2)}{(2x-5)^2} + \frac{(-2)(x^2+5)^{-3}(2x)}{(x^2+5)^{-2}}\right] \\ &= \frac{(x^2+1)^3(2x-5)^2}{(x^2+5)^2}\left(\frac{6x}{x^2+1} + \frac{4}{2x-5} - \frac{4x}{x^2+5}\right). \end{aligned}$$

Equivalently, using the basic properties listed in (7.2.10), we have

$$\begin{aligned} \ln|g(x)| &= \ln(x^2+1)^3 + \ln(2x-5)^2 - \ln(x^2+5)^2 \\ &= 3\ln(x^2+1) + 2\ln|2x-5| - 2\ln(x^2+5). \end{aligned}$$

(We have omitted absolute value in the first and third terms since x^2+1 and x^2+5 are positive for all x.) Now

$$\frac{g'(x)}{g(x)} = \frac{3(2x)}{x^2+1} + \frac{2(2)}{2x-5} - \frac{2(2x)}{x^2+5}$$

and so

$$g'(x) = g(x)\left[\frac{6x}{x^2+1} + \frac{4}{2x-5} - \frac{4x}{x^2+5}\right]$$

as we saw above. ❑

EXERCISES 7.3

In Exercises 1–20, determine the domain and find the derivative.

1. $f(x) = \ln 4x$.
2. $f(x) = \ln(2x+1)$.
3. $f(x) = \ln(x^3+1)$.
4. $f(x) = \ln[(x+1)^3]$.
5. $f(x) = \ln\sqrt{1+x^2}$.
6. $f(x) = (\ln x)^3$.
7. $f(x) = \ln|x^4-1|$.
8. $f(x) = \ln(\ln x)$.
9. $f(x) = x^2 \ln x$.
10. $f(x) = \ln\left|\dfrac{x+2}{x^3-1}\right|$.
11. $f(x) = \dfrac{1}{\ln x}$.
12. $f(x) = \ln\sqrt[4]{x^2+1}$.
13. $f(x) = \dfrac{\ln(x+1)}{x+1}$.
14. $f(x) = \ln\sqrt{\dfrac{1-x}{2-x}}$.
15. $f(x) = \sin(\ln x)$.
16. $f(x) = \ln(\sin x)$.
17. $f(x) = \ln|\cos x|$.
18. $f(x) = \cos(\ln x)$.
19. $f(x) = \ln|\sec x + \tan x|$.
20. $f(x) = \ln|\csc x - \cot x|$.

In Exercises 21–38, calculate the given integral.

21. $\displaystyle\int \frac{dx}{x+1}$.
22. $\displaystyle\int \frac{dx}{3-x}$.
23. $\displaystyle\int \frac{x}{3-x^2}\,dx$.
24. $\displaystyle\int \frac{x+1}{x^2}\,dx$.
25. $\displaystyle\int \frac{x}{(3-x^2)^2}\,dx$.
26. $\displaystyle\int \frac{\ln(x+a)}{x+a}\,dx$.
27. $\displaystyle\int \frac{\sin x}{2+\cos x}\,dx$.
28. $\displaystyle\int \cot x\,dx$.
29. $\displaystyle\int \tan x\,dx$.
30. $\displaystyle\int \frac{\sec^2 2x}{4-\tan 2x}\,dx$.
31. $\displaystyle\int \left(\frac{1}{x+2} - \frac{1}{x-2}\right)dx$.

32. $\displaystyle\int \frac{x^2}{2x^3 - 1}\,dx.$

33. $\displaystyle\int \frac{dx}{x(\ln x)^2}.$

34. $\displaystyle\int \left(\frac{1}{x-a} - \frac{1}{x-b}\right) dx.$

35. $\displaystyle\int \frac{\sin x - \cos x}{\sin x + \cos x}\,dx.$

36. $\displaystyle\int \frac{1}{\sqrt{x}(1+\sqrt{x})}\,dx$ HINT: Let $u = \sqrt{x}$.

37. $\displaystyle\int \frac{\sqrt{x}}{1 + x\sqrt{x}}\,dx.$

38. $\displaystyle\int x\left(\frac{1}{x^2 - a^2} - \frac{1}{x^2 - b^2}\right) dx.$

In Exercises 39–46, evaluate the definite integral.

39. $\displaystyle\int_1^e \frac{dx}{x}.$

40. $\displaystyle\int_1^{e^2} \frac{dx}{x}.$

41. $\displaystyle\int_e^{e^2} \frac{dx}{x}.$

42. $\displaystyle\int_3^4 \frac{dx}{1-x}.$

43. $\displaystyle\int_4^5 \frac{x}{x^2 - 1}\,dx.$

44. $\displaystyle\int_1^e \frac{\ln x}{x}\,dx.$

45. $\displaystyle\int_0^1 \frac{\ln(x+1)}{x+1}\,dx.$

46. $\displaystyle\int_0^1 \left(\frac{1}{x+1} - \frac{1}{x+2}\right) dx.$

In Exercises 47–52, calculate the derivative by logarithmic differentiation.

47. $g(x) = (x^2+1)^2(x-1)^5x^3.$

48. $g(x) = x(x+a)(x+b)(x+c).$

49. $g(x) = \dfrac{x^4(x-1)}{(x+2)(x^2+1)}.$

50. $g(x) = \dfrac{(1+x)(2+x)x}{(4+x)(2-x)}.$

51. $g(x) = \sqrt{\dfrac{(x-1)(x-2)}{(x-3)(x-4)}}.$

52. $g(x) = \dfrac{x^2(x^2+1)(x^2+2)}{(x^2-1)(x^2-5)}.$

53. The region bounded by the graph of $f(x) = 1/\sqrt{1+x}$ and the x-axis for $0 \le x \le 8$ is revolved around the x-axis. Find the volume of the solid that is generated.

54. The region bounded by the graph of $f(x) = 3/(1+x^2)$ and the x-axis for $0 \le x \le 3$ is revolved around the y-axis. Find the volume of the solid that is generated.

In Exercises 55 and 56, find the area of the part of the first quadrant that lies between

55. $x + 4y - 5 = 0$ and $xy = 1$.

56. $x + y - 3 = 0$ and $xy = 2$.

57. A particle moves along a line with acceleration $a(t) = -(t+1)^{-2}$ feet per second per second. Find the distance traveled by the particle during the time interval $[0, 4]$, given that the initial velocity $v(0)$ is 1 foot per second.

58. Exercise 57 with $v(0)$ as 2 feet per second.

In Exercises 59–62, find a formula for the nth derivative.

59. $\dfrac{d^n}{dx^n}(\ln x).$

60. $\dfrac{d^n}{dx^n}[\ln(1-x)].$

61. $\dfrac{d^n}{dx^n}(\ln 2x).$

62. $\dfrac{d^n}{dx^n}\left(\ln\dfrac{1}{x}\right).$

63. (a) Verify that

$$\ln x = \int_1^x \frac{dt}{t} < \int_1^x \frac{dt}{\sqrt{t}} \qquad \text{for } x > 1.$$

(b) Use (a) to show that

$$0 < \ln x < 2(\sqrt{x} - 1) \qquad \text{for } x > 1.$$

(c) Use (b) to show that

$$2x\left(1 - \frac{1}{\sqrt{x}}\right) < x \ln x < 0 \qquad \text{for } 0 < x < 1.$$

(d) Use (c) to show that

$$\lim_{x \to 0^+} (x \ln x) = 0.$$

64. (a) Show that for $n = 2$ Formula (7.3.5) reduces to the product rule (3.2.5).
(b) Show that Formula (7.3.5) applied to

$$g(x) = \frac{g_1(x)}{g_2(x)}$$

reduces to the quotient rule (3.2.10).

In Exercises 65–71, (i) find the domain of f, (ii) find the intervals where the function increases and the intervals where it decreases, (iii) find the extreme values, (iv) determine the concavity of the graph and find the points of inflection, and finally, (v) sketch the graph, indicating asymptotes.

65. $f(x) = \ln 2x.$

66. $f(x) = x - \ln x.$

67. $f(x) = \ln(4 - x).$

68. $f(x) = \ln(4 - x^2).$

69. $f(x) = x^2 \ln x.$

70. $f(x) = \ln(8x - x^2).$

71. $f(x) = \ln\left(\dfrac{x}{1+x^2}\right).$

72. (a) Show that $f(x) = \ln 2x$ and $g(x) = \ln 3x$ have the same derivative.
(b) Calculate the derivative of $F(x) = \ln kx$, where k is any positive number.
(c) Explain these results in terms of the properties of logarithms.

In Exercises 73–76, use a graphing utility to graph f on the indicated interval. Estimate the x-intercepts of the graph of f and the values of x where f has either a local or absolute extremum (if any). Use four decimal place accuracy in your answers.

73. $f(x) = \sqrt{x} \ln x$; $[0, 10]$.

74. $f(x) = x^2 \ln x$; $[0, 2]$.

75. $f(x) = \sin(\ln x)$; $[1, 100]$.

76. $f(x) = x^2 \ln(\sin x)$; $[0, 2]$.

77. A particle moves along a line with acceleration $a(t) = 4 - 2(t + 1) + 3/(t + 1)$ feet per second per second for $0 \leq t \leq 3$.
(a) Find the velocity v of the particle at any time t during the time interval $[0, 3]$, given that the initial velocity $v(0) = 2$.
(b) Use a graphing utility to graph v and a together.
(c) Estimate the time $t \in [0, 3]$ at which the particle has maximum velocity and the time at which it has minimum velocity. Use four decimal place accuracy.

78. Repeat Exercise 77 for a particle whose acceleration is $a(t) = 2 \cos 2(t + 1) + 2/(t + 1)$ feet per second per second for $0 \leq t \leq 7$.

■ 7.4 THE EXPONENTIAL FUNCTION

Rational powers of e already have an established meaning: by $e^{p/q}$ we mean the qth root of e raised to the pth power. But what is meant by $e^{\sqrt{2}}$ or e^{π}?

Earlier we proved that each rational power $e^{p/q}$ had logarithm p/q:

(7.4.1)
$$\ln e^{p/q} = \frac{p}{q}.$$

The definition of e^z for z irrational is patterned after this relation.

DEFINITION 7.4.2

If z is irrational, then by e^z we mean the unique number which has logarithm z:

$$\ln e^z = z.$$

What is $e^{\sqrt{2}}$? It is the unique number that has logarithm $\sqrt{2}$. What is e^{π}? It is the unique number that has logarithm π. Note that e^x now has meaning for every real value of x: it is the unique number with logarithm x.

DEFINITION 7.4.3

The function

$$E(x) = e^x \qquad \text{for all real } x$$

is called the *exponential function*.

Some properties of the exponential function follow.

(1) In the first place

(7.4.4)
$$\ln e^x = x \qquad \text{for all real } x.$$

Writing $L(x) = \ln x$ and $E(x) = e^x$, we have

$$L(E(x)) = x \qquad \text{for all real } x.$$

This says that *the exponential function is the inverse of the logarithm function.*

(2) The graph of the exponential function appears in Figure 7.4.1. It can be obtained from the graph of the logarithm by reflection in the line $y = x$.

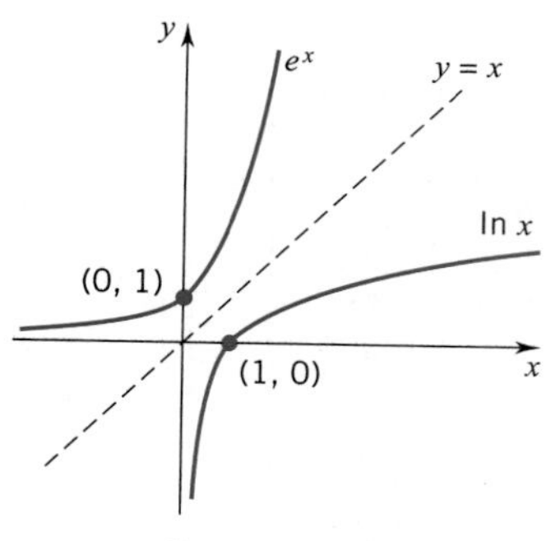

Figure 7.4.1

(3) Since the graph of the logarithm lies to the right of the y-axis, the graph of the exponential function lies above the x-axis:

(7.4.5)
$$e^x > 0 \qquad \text{for all real } x.$$

(4) Since the graph of the logarithm crosses the x-axis at $(1, 0)$, the graph of the exponential function crosses the y-axis at $(0, 1)$:

$$\ln 1 = 0 \quad \text{gives} \quad e^0 = 1.$$

(5) Since the y-axis is a vertical asymptote for the graph of the logarithm function, the x-axis is a horizontal asymptote for the graph of the exponential function:

$$\text{as } x \to -\infty, \quad e^x \to 0.$$

(6) Since the exponential function is the inverse of the logarithm function, the logarithm function is the inverse of the exponential function; namely, we must have

(7.4.6)
$$e^{\ln x} = x \qquad \text{for all } x > 0.$$

You can verify this equation directly by observing that both sides have the same logarithm:

$$\ln(e^{\ln x}) = \ln x$$

since, for all real t, $\ln e^t = t$. ❑

You know that for rational exponents

$$e^{(p/q + r/s)} = e^{p/q} \cdot e^{r/s}.$$

This property holds for all exponents, including irrational exponents.

THEOREM 7.4.7

$$e^{x+y} = e^x \cdot e^y \qquad \text{for all real } x \text{ and } y.$$

PROOF

$$\ln(e^x \cdot e^y) = \ln e^x + \ln e^y = x + y = \ln e^{x+y}.$$

Since the logarithm is one-to-one, we must have

$$e^{x+y} = e^x \cdot e^y. \quad ❑$$

We leave it to you to show that

(7.4.8)

$$e^{-y} = \frac{1}{e^y} \quad \text{and} \quad e^{x-y} = \frac{e^x}{e^y}.$$

■ Table 7.4.1

x	e^x	x	e^x
0.1	1.11	1.1	3.00
0.2	1.22	1.2	3.32
0.3	1.35	1.3	3.67
0.4	1.49	1.4	4.06
0.5	1.65	1.5	4.48
0.6	1.82	1.6	4.95
0.7	2.01	1.7	5.47
0.8	2.23	1.8	6.05
0.9	2.46	1.9	6.68
1.0	2.72	2.0	7.39

Table 7.4.1 gives some values of the exponential function rounded off to the nearest hundredth.

Example 1 Estimate the following powers of e using the approximations in Table 7.4.1.
(a) $e^{-0.2}$. **(b)** $e^{2.4}$. **(c)** $e^{3.1}$.

SOLUTION The idea, of course, is to use the laws of exponents.

(a) $e^{-0.2} = 1/e^{0.2} \cong 1/1.22 \cong 0.82.$
(b) $e^{2.4} = e^{2+0.4} = (e^2)(e^{0.4}) \cong (7.39)(1.49) \cong 11.01.$
(c) $e^{3.1} = e^{1.7+1.4} = (e^{1.7})(e^{1.4}) \cong (5.47)(4.06) \cong 22.21$ ❑

We come now to one of the most important results in calculus. It is marvelously simple.

THEOREM 7.4.9

The exponential function is its own derivative: for all real x

$$\frac{d}{dx}(e^x) = e^x.$$

PROOF The logarithm function is differentiable, and its derivative is never 0. It follows from our discussion in Section 7.1 that its inverse, the exponential function, is also differentiable. Knowing this, we can show that

$$\frac{d}{dx}(e^x) = e^x$$

by differentiating the identity

$$\ln e^x = x.$$

On the left-hand side, the chain rule gives

$$\frac{d}{dx}(\ln e^x) = \frac{1}{e^x}\frac{d}{dx}(e^x).$$

On the right-hand side, the derivative is 1:

$$\frac{d}{dx}(x) = 1.$$

Equating these derivatives, we have

$$\frac{1}{e^x}\frac{d}{dx}(e^x) = 1 \quad \text{and thus} \quad \frac{d}{dx}(e^x) = e^x. \quad \square$$

We frequently run across expressions of the form e^u where u is a function of x. If u is differentiable, then the chain rule gives

(7.4.10)

$$\boxed{\frac{d}{dx}(e^u) = e^u\frac{du}{dx}.}$$

PROOF

$$\frac{d}{dx}(e^u) = \frac{d}{du}(e^u)\frac{du}{dx} = e^u\frac{du}{dx}. \quad \square$$

Example 2

(a) $\dfrac{d}{dx}(e^{kx}) = e^{kx} \cdot k = ke^{kx}$.

(b) $\dfrac{d}{dx}(e^{\sqrt{x}}) = e^{\sqrt{x}}\dfrac{d}{dx}(\sqrt{x}) = e^{\sqrt{x}}\left(\dfrac{1}{2\sqrt{x}}\right) = \dfrac{1}{2\sqrt{x}}e^{\sqrt{x}}$.

(c) $\dfrac{d}{dx}(e^{-x^2}) = e^{-x^2}\dfrac{d}{dx}(-x^2) = e^{-x^2}(-2x) = -2xe^{-x^2}$. $\square$

The relation

$$\frac{d}{dx}(e^x) = e^x \quad \text{and its corollary} \quad \frac{d}{dx}(e^{kx}) = ke^{kx}$$

have important applications to engineering, physics, chemistry, biology, and economics. We discuss some of these applications in Section 7.6.

Example 3 Let

$$f(x) = xe^{-x} \qquad \text{for all real } x.$$

(a) On what intervals does f increase? Decrease? **(b)** Find the extreme values of f. **(c)** Determine the concavity of the graph and find the points of inflection. **(d)** Sketch the graph indicating the asymptotes if any.

SOLUTION We have

$$f(x) = xe^{-x},$$

$$f'(x) = xe^{-x}(-1) + e^{-x} = (1 - x)e^{-x},$$

$$f''(x) = (1 - x)e^{-x}(-1) - e^{-x} = (x - 2)e^{-x}.$$

Since $e^{-x} > 0$ for all real x,

$$f'(x) \text{ is } \begin{cases} \text{positive,} & \text{for} \quad x < 1 \\ 0, & \text{at} \quad x = 1 \\ \text{negative,} & \text{for} \quad x > 1. \end{cases}$$

The function f increases on $(-\infty, 1]$ and decreases on $[1, \infty)$. The number

$$f(1) = \frac{1}{e} \cong \frac{1}{2.72} \cong 0.368$$

is a local and absolute maximum. There are no other extreme values. Since

$$f''(x) \text{ is } \begin{cases} \text{negative,} & \text{for} \quad x < 2 \\ 0, & \text{at} \quad x = 2 \\ \text{positive,} & \text{for} \quad x > 2, \end{cases}$$

the graph is concave down on $(-\infty, 2)$ and concave up on $(2, \infty)$. The point

$$(2, f(2)) = (2, 2e^{-2}) \cong \left(2, \frac{2}{(2.72)^2}\right) \cong (2, 0.27)$$

is the only point of inflection. In Section 10.6, we prove that $f(x) = x/e^x \to 0$ as $x \to \infty$. Accepting this result for now, it follows that the x-axis is a horizontal asymptote. The graph is given in Figure 7.4.2.

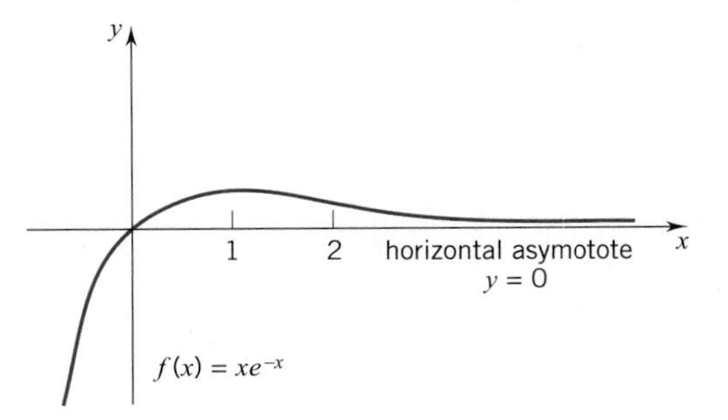

Figure 7.4.2

Example 4 Let

$$f(x) = e^{-x^2/2} \qquad \text{for all real } x.$$

(a) Determine the symmetry of the graph and the asymptotes, if any. **(b)** On what intervals does f increase? Decrease? **(c)** Find the extreme values. **(d)** Determine the concavity of the graph and find the points of inflection. **(e)** Sketch the graph.

NOTE. This function places a very important role in the mathematical fields of probability and statistics. As you will see after we complete (a)–(e), its graph is the familiar bell-shaped curve.

SOLUTION Since $f(-x) = e^{-(-x)^2/2} = e^{-x^2/2} = f(x)$, f is an even function so its graph is symmetric with respect to the y-axis. As $x \to \pm\infty$, $e^{-x^2/2} \to 0$. Therefore, the x-axis is a horizontal asymptote. There are no vertical asymptotes.

Differentiating f, we have

$$f'(x) = e^{-x^2/2}(-x) = -xe^{-x^2/2}$$

$$f''(x) = -x(-xe^{-x^2/2}) - e^{-x^2/2} = (x^2 - 1)e^{-x^2/2}.$$

Since $e^{-x^2/2} > 0$ for all x,

$$f'(x) \text{ is } \begin{cases} \text{positive,} & \text{for} \quad x < 0 \\ 0, & \text{at} \quad x = 0 \\ \text{negative,} & \text{for} \quad x > 0. \end{cases}$$

Thus, f is increasing on $(-\infty, 0]$ and decreasing on $[0, \infty)$. The value

$$f(0) = e^0 = 1$$

is a local and absolute maximum. There are no other extreme values.

Now consider $f''(x) = (x^2 - 1)e^{-x^2/2}$. The sign of f'' and the behavior of the graph of f are:

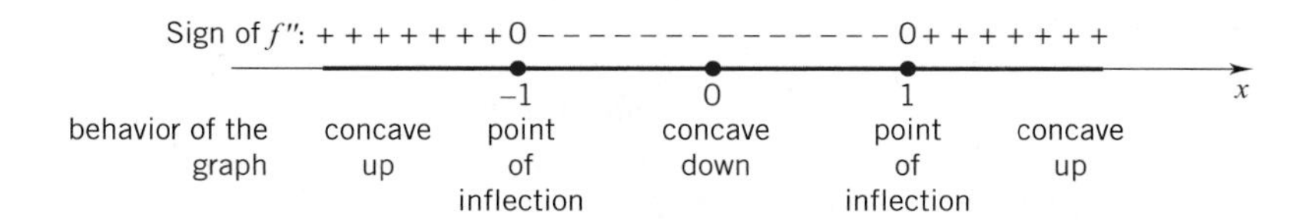

Thus, the graph of f is concave up on $(-\infty, -1)$ and $(1, \infty)$, the graph is concave down on $(-1, 1)$, and $(-1, e^{-1/2})$ and $(1, e^{-1/2})$ are points of inflection.

The graph of f is shown in Figure 7.4.3. ❑

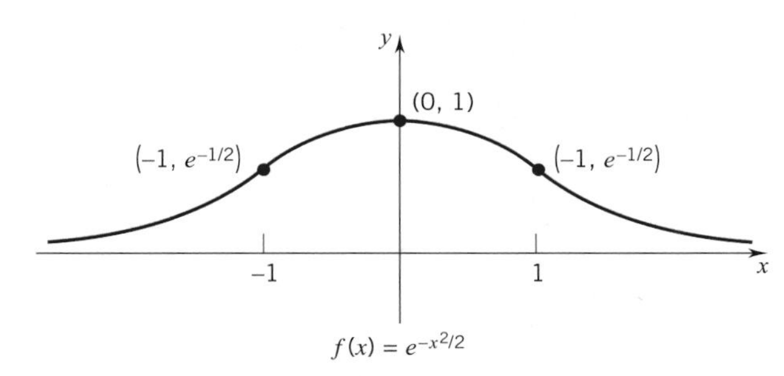

Figure 7.4.3

The integral counterpart of (7.4.9) takes the form

(7.4.11)
$$\boxed{\int e^x\,dx = e^x + C.}$$

In practice

$$\int e^{g(x)}g'(x)\,dx \quad \text{is reduced to} \quad \int e^u\,du$$

by setting

$$u = g(x), \quad du = g'(x)\,dx.$$

Thus, we have the important generalization of (7.4.11)

(7.4.12)
$$\int e^{g(x)} g'(x)\, dx = e^{g(x)} + C.$$

Example 5 Find

$$\int 9e^{3x}\, dx.$$

SOLUTION Set

$$u = 3x, \qquad du = 3\, dx.$$

$$\int 9e^{3x}\, dx = 3\int e^u\, du = 3e^u + C = 3e^{3x} + C.$$

If you recognize at the very beginning that

$$3e^{3x} = \frac{d}{dx}(e^{3x}),$$

then you can dispense with the u-substitution and simply write

$$\int 9e^{3x}\, dx = 3\int 3e^{3x}\, dx = 3e^{3x} + C. \quad \square$$

Example 6 Find

$$\int \frac{e^{\sqrt{x}}}{\sqrt{x}}\, dx.$$

SOLUTION Set

$$u = \sqrt{x}, \qquad du = \frac{1}{2\sqrt{x}}\, dx.$$

$$\int \frac{e^{\sqrt{x}}}{\sqrt{x}}\, dx = 2\int e^u\, du = 2e^u + C = 2e^{\sqrt{x}} + C.$$

If you recognize that

$$\tfrac{1}{2}\left(\frac{e^{\sqrt{x}}}{\sqrt{x}}\right) = \frac{d}{dx}(e^{\sqrt{x}}),$$

then you can dispense with the u-substitution and integrate directly:

$$\int \frac{e^{\sqrt{x}}}{\sqrt{x}}\, dx = 2\int \frac{1}{2}\left(\frac{e^{\sqrt{x}}}{\sqrt{x}}\right) dx = 2e^{\sqrt{x}} + C. \quad \square$$

Example 7 Find

$$\int \frac{e^{3x}}{e^{3x}+1}\,dx.$$

SOLUTION We can put this integral in the form

$$\int \frac{du}{u}$$

by setting

$$u = e^{3x} + 1, \qquad du = 3e^{3x}\,dx.$$

Then

$$\int \frac{e^{3x}}{e^{3x}+1}\,dx = \frac{1}{3}\int \frac{du}{u} = \frac{1}{3}\ln|u| + C = \frac{1}{3}\ln(e^{3x}+1) + C. \quad \square$$

Example 8 Integrals involving $xe^{-x^2/2}$ also play an important role in probability and statistics. Evaluate

$$\int_0^{\sqrt{2\ln 3}} xe^{-x^2/2}\,dx.$$

SOLUTION Set

$$u = -\tfrac{1}{2}x^2, \qquad du = -x\,dx.$$

At $x = 0$, $u = 0$; at $x = \sqrt{2\ln 3}$, $u = -\ln 3$. Thus

$$\int_0^{\sqrt{2\ln 3}} xe^{-x^2/2}\,dx = -\int_0^{-\ln 3} e^u\,du = -\left[e^u\right]_0^{-\ln 3} = 1 - e^{-\ln 3} = 1 - \tfrac{1}{3} = \tfrac{2}{3}. \quad \square$$

Example 9 Evaluate

$$\int_0^1 e^x(e^x+1)^{1/5}\,dx.$$

SOLUTION Set

$$u = e^x + 1, \qquad du = e^x\,dx.$$

At $x = 0$, $u = 2$; at $x = 1$, $u = e + 1$. Thus

$$\int_0^1 e^x(e^x+1)^{1/5}\,dx = \int_2^{e+1} u^{1/5}\,du = \left[\tfrac{5}{6}u^{6/5}\right]_2^{e+1} = \tfrac{5}{6}[(e+1)^{6/5} - 2^{6/5}]. \quad \square$$

Remark The u-substitution simplifies many calculations, but you will find that in many cases you can carry out the integration more quickly without it. $\square$

EXERCISES 7.4

In Exercises 1–30, differentiate the given function.

1. $y = e^{-2x}$.
2. $y = 3e^{2x+1}$.
3. $y = e^{x^2-1}$.
4. $y = 2e^{-4x}$.
5. $y = e^x \ln x$.
6. $y = x^2 e^x$.
7. $y = x^{-1}e^{-x}$.
8. $y = e^{\sqrt{x}+1}$.
9. $y = \frac{1}{2}(e^x + e^{-x})$.
10. $y = \frac{1}{2}(e^x - e^{-x})$.
11. $y = e^{\sqrt{x}} \ln \sqrt{x}$.
12. $y = (1 - e^{4x})^2$.
13. $y = (e^x + e^{-x})^2$.
14. $y = (3 - 2e^{-x})^3$.
15. $y = (e^{x^2} + 1)^2$.
16. $y = (e^{2x} - e^{-2x})^2$.
17. $y = (x^2 - 2x + 2)e^x$.
18. $y = x^2 e^x - xe^{x^2}$.
19. $y = \dfrac{e^x - 1}{e^x + 1}$.
20. $y = \dfrac{e^{2x} - 1}{e^{2x} + 1}$.
21. $y = \dfrac{e^{ax} - e^{bx}}{e^{ax} + e^{bx}}$.
22. $y = \ln e^{3x}$.
23. $y = e^{4 \ln x}$.
24. $y = e^{\sqrt{1-x^2}}$.
25. $f(x) = \sin(e^{2x})$.
26. $f(x) = e^{\sin 2x}$.
27. $f(x) = e^{-2x} \cos x$.
28. $f(x) = \ln(\cos e^{2x})$.
29. $f(x) = \tan\sqrt{e^{-3x}}$.
30. $f(x) = \sec(e^{\tan x})$.

In Exercises 31–52, calculate the indefinite integral.

31. $\int e^{2x}\, dx$.
32. $\int e^{-2x}\, dx$.
33. $\int e^{kx}\, dx$.
34. $\int e^{ax+b}\, dx$.
35. $\int xe^{x^2}\, dx$.
36. $\int xe^{-x^2}\, dx$.
37. $\int \dfrac{e^{1/x}}{x^2}\, dx$.
38. $\int \dfrac{e^{2\sqrt{x}}}{\sqrt{x}}\, dx$.
39. $\int (e^x + e^{-x})^2\, dx$.
40. $\int e^{\ln x}\, dx$.
41. $\int \ln e^x\, dx$.
42. $\int (e^{-x} - 1)^2\, dx$.
43. $\int \dfrac{4}{\sqrt{e^x}}\, dx$.
44. $\int \dfrac{e^x}{e^x + 1}\, dx$.
45. $\int \dfrac{e^x}{\sqrt{e^x + 1}}\, dx$.
46. $\int \dfrac{2e^x}{\sqrt[3]{e^x + 1}}\, dx$.
47. $\int \dfrac{e^{2x}}{2e^{2x} + 3}\, dx$.
48. $\int \dfrac{xe^{ax^2}}{e^{ax^2} + 1}\, dx$.
49. $\int \cos x\, e^{\sin x}\, dx$.
50. $\int \dfrac{\sin(e^{-2x})}{e^{2x}}\, dx$.
51. $\int e^{-x}[1 + \cos(e^{-x})]\, dx$.
52. $\int \sec^2 2x e^{\tan 2x}\, dx$.

In Exercises 53–64, evaluate the definite integral.

53. $\int_0^1 e^x\, dx$.
54. $\int_0^1 e^{-kx}\, dx$.
55. $\int_0^{\ln \pi} e^{-6x}\, dx$.
56. $\int_0^1 xe^{-x^2}\, dx$.
57. $\int_0^1 \dfrac{e^x + 1}{e^x}\, dx$.
58. $\int_0^1 \dfrac{4 - e^x}{e^x}\, dx$.
59. $\int_0^{\ln 2} \dfrac{e^x}{e^x + 1}\, dx$.
60. $\int_0^1 \dfrac{e^x + e^{-x}}{2}\, dx$.
61. $\int_{\ln 2}^{\ln 3} (e^x - e^{-x})^2\, dx$.
62. $\int_0^1 \dfrac{e^x}{4 - e^x}\, dx$.
63. $\int_0^1 x(e^{x^2} + 2)\, dx$.
64. $\int_1^2 (2 - e^{-x})^2\, dx$.

In Exercises 65–68, estimate on the basis of Table 7.4.1.

65. $e^{-0.4}$.
66. $e^{2.6}$.
67. $e^{2.8}$.
68. $e^{-2.1}$.

69. A particle moves on a coordinate line with its position at time t given by the function $x(t) = Ae^{ct} + Be^{-ct}$. Show that the acceleration of the particle is proportional to its position.

In Exercises 70–73, sketch the region bounded by the curves and find its area.

70. $x = e^{2y}, \quad x = e^{-y}, \quad x = 4$.
71. $y = e^x, \quad y = e^{2x}, \quad y = e^4$.
72. $y = e^x, \quad y = e, \quad y = x, \quad x = 0$.
73. $x = e^y, \quad y = 1, \quad y = 2, \quad x = 2$.

For each of the functions in Exercises 74–78, (i) find the domain, (ii) find the intervals where the function increases and the intervals where it decreases, (iii) find the extreme values, (iv) determine the concavity and find the points of inflection, and finally, (v) sketch the graph, indicating all asymptotes.

74. $f(x) = \frac{1}{2}(e^x - e^{-x})$.
75. $f(x) = \frac{1}{2}(e^x + e^{-x})$.
76. $f(x) = (1 - x)e^x$.
77. $f(x) = e^{(1/x)^2}$.
78. $f(x) = x^2 e^{-x}$.

79. Take $a > 0$ and refer to the figure.

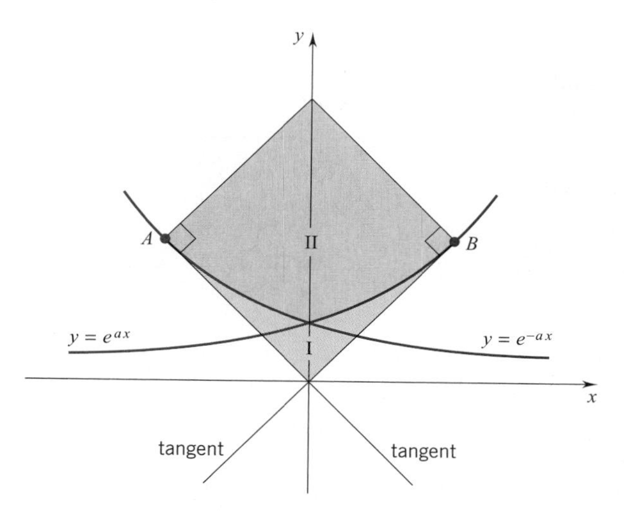

(a) Find the points of tangency, marked A and B.
(b) Find the area of region I.
(c) Find the area of region II.

80. Prove that for all $x > 0$ and all positive integers n

$$e^x > 1 + x + \frac{x^2}{2!} + \frac{x^3}{3!} + \cdots + \frac{x^n}{n!}.$$

Recall, $n! = n(n-1)(n-2)\cdots 3\cdot 2\cdot 1$.

HINT: $e^x = 1 + \int_0^x e^t\,dt > 1 + \int_0^x dt = 1 + x,$

$$e^x = 1 + \int_0^x e^t\,dt > 1 + \int_0^x (1+t)\,dt$$
$$= 1 + x + \frac{x^2}{2}, \quad \text{and so on.}$$

81. Prove that, if n is a positive integer, then

$$e^x > x^n \qquad \text{for all } x \text{ sufficiently large.}$$

HINT: Use Exercise 80.

In Exercises 82–84, evaluate the limit numerically and then justify your answers using a graphing utility.

82. $\lim_{x\to 0} \frac{e^{10x} - 1}{x}$.

83. $\lim_{x\to 1} \frac{e^{x^3} - e}{x - 1}$.

84. $\lim_{x\to 1} \frac{e^x - e}{\ln x}$.

HINT: Remember the alternative definition of derivative

$$f'(c) = \lim_{x\to c} \frac{f(x) - f(c)}{x - c}.$$

85. Let $f(x) = e^x$ and $g(x) = 4 - x^2$.
(a) Use a graphing utility to sketch the graphs of f and g.
(b) Estimate the x-coordinates of the points of intersection of the two curves. Use four decimal place accuracy.
(c) Estimate the area of the region bounded by the two curves.

86. Repeat Exercise 85 for the functions $f(x) = e^{-x^2}$ and $g(x) = x^2$.

In Exercises 87–90, use a graphing utility to graph the functions f and g. Your graphs should suggest that f and g are inverses of each other. Confirm this by using the methods in Section 7.1.

87. $f(x) = e^{2x}$; $\quad g(x) = \ln\sqrt{x}$.
88. $f(x) = e^{x/2}$; $\quad g(x) = \ln x^2$.
89. $f(x) = e^{x^2}$; $\quad g(x) = \sqrt{\ln x}$.
90. $f(x) = e^{x-2}$; $\quad g(x) = 2 + \ln x$.

■ 7.5 ARBITRARY POWERS; OTHER BASES; ESTIMATING e

Arbitrary Powers: The Function $f(x) = x^r$

The elementary notion of exponent applies only to rational numbers. Expressions such as

$$10^5, \quad 2^{1/3}, \quad 7^{-4/5}, \quad \pi^{-1/2}$$

make sense, but so far we have attached no meaning to expressions such as

$$10^{\sqrt{2}}, \quad 2^{\pi}, \quad 7^{-\sqrt{3}}, \quad \pi^e.$$

The extension of our sense of exponent to allow for irrational exponents is conveniently done by making use of the logarithm function and the exponential function. The heart of the matter is to observe that for $x > 0$ and p/q rational

$$x^{p/q} = e^{(p/q)\ln x}.$$

(To verify this, take the log of both sides.) We *define* x^z for irrational z by setting

$$x^z = e^{z \ln x}.$$

We then have the following result:

(7.5.1)

if $x > 0$, then $x^r = e^{r \ln x}$ for all real numbers r.

In particular

$$10^{\sqrt{2}} = e^{\sqrt{2} \ln 10}, \quad 2^{\pi} = e^{\pi \ln 2}, \quad 7^{-\sqrt{3}} = e^{-\sqrt{3} \ln 7}, \quad \pi^e = e^{e \ln \pi}. \quad \square$$

With this extended sense of exponent the usual laws of exponents

(7.5.2)

$$x^{r+s} = x^r x^s, \qquad x^{r-s} = \frac{x^r}{x^s}, \qquad (x^r)^s = x^{rs}$$

still hold:

$$x^{r+s} = e^{(r+s)\ln x} = e^{r \ln x} \cdot e^{s \ln x} = x^r x^s,$$

$$x^{r-s} = e^{(r-s)\ln x} = e^{r \ln x} \cdot e^{-s \ln x} = \frac{e^{r \ln x}}{e^{s \ln x}} = \frac{x^r}{x^s},$$

$$(x^r)^s = e^{s \ln x^r} = e^{rs \ln x} = x^{rs}. \quad \square$$

In Chapter 3, Section 3.7, we proved that $d(x^p)/dx = px^{p-1}$ for any rational number p. We can now extend this result to arbitrary powers. For *any real number r:*

(7.5.3)

$$\frac{d}{dx}(x^r) = rx^{r-1} \qquad \text{for all } x > 0.$$

PROOF

$$\frac{d}{dx}(x^r) = \frac{d}{dx}(e^{r \ln x}) = e^{r \ln x} \frac{d}{dx}(r \ln x) = x^r \frac{r}{x} = rx^{r-1}.$$

You can also write $f(x) = x^r$ and use logarithmic differentiation:

$$\ln f(x) = r \ln x$$

$$\frac{f'(x)}{f(x)} = \frac{r}{x}$$

$$f'(x) = \frac{rf(x)}{x} = \frac{rx^r}{x} = rx^{r-1}. \quad \square$$

Thus

$$\frac{d}{dx}(x^{\sqrt{2}}) = \sqrt{2}x^{\sqrt{2}-1}, \qquad \frac{d}{dx}(x^{\pi}) = \pi x^{\pi-1}.$$

If u is a positive differentiable function of x and r is any real number, then, by the chain rule,

(7.5.4)

$$\frac{d}{dx}(u^r) = ru^{r-1}\frac{du}{dx}.$$

PROOF

$$\frac{d}{dx}(u^r) = \frac{d}{du}(u^r)\frac{du}{dx} = ru^{r-1}\frac{du}{dx}. \quad \square$$

For example,

$$\frac{d}{dx}[(x^2+5)^{\sqrt{3}}] = \sqrt{3}(x^2+5)^{\sqrt{3}-1}(2x) = 2\sqrt{3}x(x^2+5)^{\sqrt{3}-1}.$$

Each derivative formula gives rise to a companion integral formula. The integral form of (7.5.3) is:

(7.5.5)

$$\int x^r\,dx = \frac{x^{r+1}}{r+1} + C, \qquad \text{for } r \neq -1.$$

Example 1 Find

$$\int \frac{x^3}{(2x^4+1)^{\pi}}\,dx.$$

SOLUTION Set

$$u = 2x^4 + 1, \qquad du = 8x^3\,dx.$$

$$\int \frac{x^3}{(2x^4+1)^{\pi}}\,dx = \frac{1}{8}\int \frac{du}{u^{\pi}} = \frac{1}{8}\left(\frac{u^{1-\pi}}{1-\pi}\right) + C = \frac{(2x^4+1)^{1-\pi}}{8(1-\pi)} + C. \quad \square$$

Example 2 Find

$$\frac{d}{dx}[(x^2+1)^{3x}].$$

SOLUTION One way to find the derivative is to observe that $(x^2+1)^{3x} = e^{3x\ln(x^2+1)}$ and then differentiate

$$\frac{d}{dx}[(x^2+1)^{3x}] = \frac{d}{dx}[e^{3x\ln(x^2+1)}] = e^{3x\ln(x^2+1)}\left[3x \cdot \frac{2x}{x^2+1} + \{\ln(x^2+1)\}(3)\right]$$

$$= (x^2+1)^{3x}\left[\frac{6x^2}{x^2+1} + 3\ln(x^2+1)\right].$$

Another way to find the derivative of $(x^2 + 1)^{3x}$ is to set $f(x) = (x^2 + 1)^{3x}$ and use logarithmic differentiation:

$$\ln f(x) = 3x \cdot \ln(x^2 + 1)$$

$$\frac{f'(x)}{f(x)} = 3x \cdot \frac{2x}{x^2+1} + [\ln(x^2 + 1)](3) = \frac{6x^2}{x^2+1} + 3\ln(x^2 + 1)$$

$$f'(x) = f(x)\left[\frac{6x^2}{x^2+1} + 3\ln(x^2 + 1)\right] = (x^2 + 1)^{3x}\left[\frac{6x^2}{x^2+1} + 3\ln(x^2 + 1)\right]. \quad \square$$

Base p: The Function $f(x) = p^x$

To form the function $f(x) = x^r$ we take a positive variable x and raise it to a constant power r. To form the function $f(x) = p^x$ we take a positive constant p and raise it to a variable power x. Since $1^x = 1$ for all x, the function is of interest only if $p \neq 1$.

Functions of the form $f(x) = p^x$ are called *exponential functions with base* p. The high status enjoyed by Euler's number e comes from the fact that

$$\frac{d}{dx}(e^x) = e^x.$$

For other bases the derivative has an extra factor:

(7.5.6)
$$\boxed{\frac{d}{dx}(p^x) = p^x \ln p.}$$

PROOF

$$\frac{d}{dx}(p^x) = \frac{d}{dx}(e^{x\ln p}) = e^{x\ln p}\ln p = p^x \ln p. \quad \square$$

Thus, for example,

$$\frac{d}{dx}(2^x) = 2^x \ln 2 \qquad \text{and} \qquad \frac{d}{dx}(10^x) = 10^x \ln 10.$$

If u is a differentiable function of x, then, by the chain rule,

(7.5.7)
$$\boxed{\frac{d}{dx}(p^u) = p^u \ln p \frac{du}{dx}.}$$

PROOF

$$\frac{d}{dx}(p^u) = \frac{d}{du}(p^u)\frac{du}{dx} = p^u \ln p \frac{du}{dx}. \quad \square$$

For example,

$$\frac{d}{dx}(2^{3x^2}) = 2^{3x^2}(\ln 2)(6x) = 6x\, 2^{3x^2} \ln 2.$$

The integral form of (7.5.6) is

(7.5.8)

$$\int p^x\,dx = \frac{1}{\ln p}p^x + C, \qquad p > 0, \quad p \neq 1.$$

For example,

$$\int 2^x\,dx = \frac{1}{\ln 2}2^x + C.$$

Example 3 Find

$$\int x5^{-x^2}\,dx.$$

SOLUTION Set

$$u = -x^2, \qquad du = -2x\,dx.$$

$$\int x5^{-x^2}\,dx = -\tfrac{1}{2}\int 5^u\,du = -\tfrac{1}{2}\left(\frac{1}{\ln 5}\right)5^u + C$$

$$= \frac{-1}{2\ln 5}5^{-x^2} + C. \quad \square$$

Example 4 Evaluate

$$\int_1^2 3^{2x-1}\,dx.$$

SOLUTION Set

$$u = 2x - 1, \qquad du = 2\,dx.$$

At $x = 1$, $u = 1$; at $x = 2$, $u = 3$. Thus

$$\int_1^2 3^{2x-1}\,dx = \tfrac{1}{2}\int_1^3 3^u\,du = \tfrac{1}{2}\left[\frac{1}{\ln 3}\cdot 3^u\right]_1^3 = \frac{12}{\ln 3} \cong 10.923. \quad \square$$

Base p: The Function $f(x) = \log_p x$

If $p > 0$, then

$$\ln p^t = t\ln p \qquad \text{for all } t.$$

If p is also different from 1, then $\ln p \neq 0$, and we have

$$\frac{\ln p^t}{\ln p} = t.$$

This indicates that the function

$$f(x) = \frac{\ln x}{\ln p}, \qquad x > 0,$$

satisfies the relation

$$f(p^t) = t \qquad \text{for all real } t.$$

In view of this we call

$$\frac{\ln x}{\ln p}$$

the logarithm of x to the base p and write:

(7.5.9)

$$\text{For } p > 0,\ p \neq 1, \quad \log_p x = \frac{\ln x}{\ln p}, \qquad x > 0.$$

For example,

$$\log_2 32 = \frac{\ln 32}{\ln 2} = \frac{\ln 2^5}{\ln 2} = \frac{5 \ln 2}{\ln 2} = 5$$

and

$$\log_{100}(\tfrac{1}{10}) = \frac{\ln(\frac{1}{10})}{\ln 100} = \frac{\ln 10^{-1}}{\ln 10^2} = \frac{-\ln 10}{2 \ln 10} = -\frac{1}{2}. \quad \square$$

We can obtain these same results more directly from the relation

(7.5.10)

$$\log_p p^t = t.$$

Accordingly

$$\log_2 32 = \log_2 2^5 = 5 \qquad \text{and} \qquad \log_{100}(\tfrac{1}{10}) = \log_{100}(100^{-1/2}) = -\tfrac{1}{2}. \quad \square$$

Differentiating (7.5.9) we have

(7.5.11)

$$\frac{d}{dx}(\log_p x) = \frac{1}{x \ln p}.\dagger$$

† The function $f(x) = \log_p x$ satisfies

$$f'(x) = \frac{1}{x \ln p}, \qquad f'(1) = \frac{1}{\ln p}.$$

This means that in general

$$f'(x) = \frac{1}{x} f'(1).$$

We predicted this from general considerations in Section 7.2. (See Formula (7.2.2).)

When p is e, the factor $\ln p$ is 1 and we have

$$\frac{d}{dx}(\log_e x) = \frac{1}{x}.$$

The logarithm to the base e, $\ln = \log_e$, is called the "*natural logarithm*" because it is the logarithm with the simplest derivative.

If u is a positive, differentiable function of x, then, by the chain rule

(7.5.12)

$$\frac{d}{dx}(\log_p u) = \frac{1}{u \ln p} \cdot \frac{du}{dx}.$$

Example 5 Find

(a) $\frac{d}{dx}[\log_2(3x^2 + 1)]$ and **(b)** $\frac{d}{dx}(\log_5|x|)$.

SOLUTION

(a) Using (7.5.12), we have

$$\frac{d}{dx}[\log_2(3x^2 + 1)] = \frac{1}{(3x^2 + 1)\ln 2} \cdot \frac{du}{dx} = \frac{6x}{(3x^2 + 1)\ln 2}.$$

Alternatively, we could have used the definition of $\log_p x$ given by (7.5.9):

$$\frac{d}{dx}[\log_2(3x^2 + 1)] = \frac{d}{dx}\left[\frac{\ln(3x^2 + 1)}{\ln 2}\right] = \frac{1}{\ln 2} \cdot \frac{1}{(3x^2 + 1)} \cdot 6x$$

$$= \frac{6x}{(3x^2 + 1)\ln 2}$$

(b) It is easier to use (7.5.9) here:

$$\frac{d}{dx}(\log_5|x|) = \frac{d}{dx}\left[\frac{\ln|x|}{\ln 5}\right] = \frac{1}{x \ln 5}.$$ ❑

There is no need to consider the companion integral form of (7.5.11) since the integral

$$\int \frac{1}{x \ln p}\,dx = \frac{1}{\ln p}\int \frac{1}{x}\,dx = \frac{\ln|x|}{\ln p} + C,$$

which can be expressed in terms of $\log_p$ by using (7.5.9) if we wished to do so. That is,

$$\frac{\ln|x|}{\ln p} + C = \log_p|x| + C.$$

Estimating the Number e

We defined the logarithm function by setting

$$\ln x = \int_1^x \frac{dt}{t}, \qquad x > 0.$$

We can derive a numerical estimate for e from this integral representation of the logarithm.

Since e is irrational, we cannot hope to express e as a repeating or terminating decimal. What we can do is describe a simple technique by which we can compute the value of e to any desired degree of accuracy.

THEOREM 7.5.13

For each positive integer n,

$$\left(1+\frac{1}{n}\right)^n \leq e \leq \left(1+\frac{1}{n}\right)^{n+1}.$$

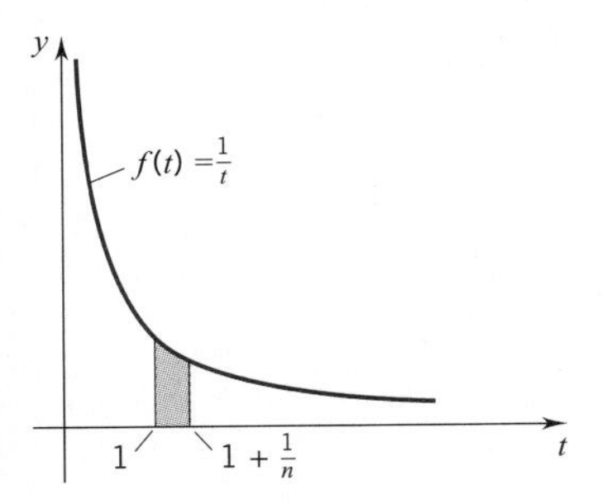

Figure 7.5.1

PROOF Refer throughout to Figure 7.5.1.

$$\ln\left(1+\frac{1}{n}\right) = \int_1^{1+1/n} \frac{dt}{t} \leq \int_1^{1+1/n} 1\, dt = \frac{1}{n}.$$

↑ since $\frac{1}{t} \leq 1$ throughout the interval of integration

$$\ln\left(1+\frac{1}{n}\right) = \int_1^{1+1/n} \frac{dt}{t} \geq \int_1^{1+1/n} \frac{dt}{1+1/n} = \frac{1}{1+1/n}\cdot\frac{1}{n} = \frac{1}{n+1}.$$

↑ since $\frac{1}{t} \geq \frac{1}{1+1/n}$ throughout the interval of integration

We have now shown that

$$\frac{1}{n+1} \leq \ln\left(1+\frac{1}{n}\right) \leq \frac{1}{n}.$$

From the inequality on the right

$$1+\frac{1}{n} \leq e^{1/n} \quad \text{and thus} \quad \left(1+\frac{1}{n}\right)^n \leq e.$$

From the inequality on the left

$$e^{1/(n+1)} \leq 1+\frac{1}{n} \quad \text{and thus} \quad e \leq \left(1+\frac{1}{n}\right)^{n+1}. \quad \square$$

The inequality

$$\left(1+\frac{1}{n}\right)^n \leq e \leq \left(1+\frac{1}{n}\right)^{n+1}$$

is an elegant characterization of e, but not a very efficient tool for calculating e. Reading from our calculator we find that

$$(1+\tfrac{1}{100})^{100} \cong 2.7048138 \quad \text{and} \quad (1+\tfrac{1}{100})^{101} \cong 2.7318619.$$

This gives e rounded off to one decimal place: $e \cong 2.7$. Going on to $n = 1000$, we find that

$$(1+\tfrac{1}{1000})^{1000} \cong 2.7169239 \quad \text{and} \quad (1+\tfrac{1}{1000})^{1001} \cong 2.7196409.$$

This gives a two-place estimate for e: $e \cong 2.72$. To get a five-place estimate for e, we have to go to about $n = 1{,}000{,}000$:

$$(1 + \tfrac{1}{1{,}000{,}000})^{1{,}000{,}000} \cong 2.7182805 \qquad \text{and} \qquad (1 + \tfrac{1}{1{,}000{,}000})^{1{,}000{,}001} \cong 2.7182832$$

and therefore, to five decimal places, $e \cong 2.71828$.

Based on these calculations, we are led to the conclusion that $(1 + 1/n)^n \to e$ as $n \to \infty$. This result is confirmed in Exercise 90. A more efficient way of estimating e is provided by Taylor series (Chapter 11). ❑

EXERCISES 7.5

In Exercises 1–12, find the indicated quantity.

1. $\log_2 64$.
2. $\log_2 \frac{1}{64}$.
3. $\log_{64} \frac{1}{2}$.
4. $\log_{10} 0.01$.
5. $\log_5 1$.
6. $\log_5 0.2$.
7. $\log_5 125$.
8. $\log_2 4^3$.
9. $\log_{32} 8$.
10. $\log_{100} 10^{-4/5}$.
11. $\log_{10} 100^{-4/5}$.
12. $\log_9 \sqrt{3}$.

In Exercises 13–16, show that the given identity holds.

13. $\log_p xy = \log_p x + \log_p y$.

14. $\log_p \dfrac{1}{x} = -\log_p x$.
15. $\log_p x^y = y \log_p x$.

16. $\log_p \dfrac{x}{y} = \log_p x - \log_p y$.

In Exercises 17–22, find those numbers x, if any, which satisfy the given equation.

17. $10^x = e^x$.
18. $\log_5 x = 0.04$.
19. $\log_x 10 = \log_4 100$.
20. $\log_x 2 = \log_3 x$.
21. $\log_2 x = \displaystyle\int_2^x \frac{dt}{t}$.
22. $\log_x 10 = \log_2(\frac{1}{10})$.

23. Estimate $\ln a$ given that $e^{t_1} < a < e^{t_2}$.

24. Estimate e^b given that $\ln x_1 < b < \ln x_2$.

In Exercises 25–36, find the derivative of the given function.

25. $f(x) = 3^{2x}$.
26. $g(x) = 4^{3x^2}$.
27. $f(x) = 2^{5x}\, 3^{\ln x}$.
28. $F(x) = 5^{-2x^2+x}$.
29. $g(x) = \sqrt{\log_3 x}$.
30. $h(x) = 7^{\sin x^2}$.
31. $f(x) = \tan(\log_5 x)$.
32. $g(x) = \dfrac{\log_{10} x}{x^2}$.
33. $F(x) = \cos(2^x + 2^{-x})$.
34. $f(x) = \log_5 \left(\dfrac{x}{x+1}\right)$.
35. $g(x) = \log_2[\log_4(2x + 1)]$.
36. $h(x) = a^{-x} \cos bx$.

In Exercises 37–45, calculate the following integrals.

37. $\displaystyle\int 3^x\, dx$.
38. $\displaystyle\int 2^{-x}\, dx$.
39. $\displaystyle\int (x^3 + 3^{-x})\, dx$.
40. $\displaystyle\int x 10^{-x^2}\, dx$.
41. $\displaystyle\int \frac{dx}{x \ln 5}$.
42. $\displaystyle\int \frac{\log_5 x}{x}\, dx$.
43. $\displaystyle\int \frac{\log_2 x^3}{x}\, dx$.
44. $\displaystyle\int \frac{\log_3 \sqrt{x} - 1}{x}\, dx$.
45. $\displaystyle\int x 10^{x^2}\, dx$.

46. Show that, if a, b, c are positive, then

$$\log_a c = \log_a b \log_b c$$

provided that a and b are both different from 1.

In Exercises 47–50, find $f'(e)$.

47. $f(x) = \log_3 x$.
48. $f(x) = x \log_3 x$.
49. $f(x) = \ln(\ln x)$.
50. $f(x) = \log_3(\log_2 x)$.

In Exercises 51 and 52, derive the formula for $f'(x)$ by logarithmic differentiation.

51. $f(x) = p^x$.
52. $f(x) = p^{g(x)}$.

In Exercises 53–64, find the derivative by logarithmic differentiation.

53. $\dfrac{d}{dx}[(x + 1)^x]$.
54. $\dfrac{d}{dx}[(\ln x)^x]$.
55. $\dfrac{d}{dx}[(\ln x)^{\ln x}]$.
56. $\dfrac{d}{dx}\left[\left(\dfrac{1}{x}\right)^x\right]$.
57. $\dfrac{d}{dx}[(x^2 + 2)^{\ln x}]$.
58. $\dfrac{d}{dx}[(\ln x)^{x^2+2}]$.
59. $\dfrac{d}{dx}[x^{\sin x}]$.
60. $\dfrac{d}{dx}[(\cos x)^{(x^2+1)}]$.
61. $\dfrac{d}{dx}[(\sin x)^{\cos x}]$.
62. $\dfrac{d}{dx}[x^{x^2}]$.

63. $\dfrac{d}{dx}[x^{2^x}]$.

64. $\dfrac{d}{dx}[(\tan x)^{\sec x}]$.

In Exercises 65–72, sketch figures in which you compare the following pairs of graphs.

65. $f(x) = e^x$ and $g(x) = 2^x$.

66. $f(x) = e^x$ and $g(x) = 3^x$.

67. $f(x) = e^x$ and $g(x) = e^{-x}$.

68. $f(x) = 2^x$ and $g(x) = 2^{-x}$.

69. $f(x) = \ln x$ and $g(x) = \log_3 x$.

70. $f(x) = \ln x$ and $g(x) = \log_2 x$.

71. $f(x) = 2^x$ and $g(x) = \log_2 x$.

72. $f(x) = 10^x$ and $g(x) = \log_{10} x$.

In Exercises 73–76, (i) specify the domain of f, (ii) find the intervals where f increases and those where it decreases, (iii) find the extreme values of f.

73. $f(x) = 10^{1-x^2}$.

74. $f(x) = 10^{1/(1-x^2)}$.

75. $f(x) = 10^{\sqrt{1-x^2}}$.

76. $f(x) = \log_{10}\sqrt{1-x^2}$.

In Exercises 77–85, evaluate the given integral.

77. $\displaystyle\int_1^2 2^{-x}\,dx$.

78. $\displaystyle\int_0^1 4^x\,dx$.

79. $\displaystyle\int_1^4 \frac{dx}{x\ln 2}$.

80. $\displaystyle\int_0^2 p^{x/2}\,dx$.

81. $\displaystyle\int_0^1 x10^{1+x^2}\,dx$.

82. $\displaystyle\int_1^3 \frac{\log_3 x}{x}\,dx$.

83. $\displaystyle\int_{10}^{100} \frac{dx}{x\log_{10} x}$.

84. $\displaystyle\int_0^1 \frac{5p^{\sqrt{x+1}}}{\sqrt{x+1}}\,dx$.

85. $\displaystyle\int_0^1 (2^x + x^2)\,dx$.

In Exercises 86–89, estimate the logarithm on the basis of Table 7.2.1.

86. $\log_{10} 4$.

87. $\log_{10} 7$.

88. $\log_{10} 12$.

89. $\log_{10} 45$.

90. Prove that

$$\lim_{h\to 0}(1+h)^{1/h} = e.$$

HINT: At $x = 1$ the logarithm function has derivative 1:

$$\lim_{h\to 0}\frac{\ln(1+h) - \ln 1}{h} = 1.$$

In Exercises 91–93, evaluate and then explain the result.

91. $5^{(\ln 17)/(\ln 5)}$.

92. $7^{1/\ln 7}$.

93. $16^{1/\ln 2}$.

■ 7.6 EXPONENTIAL GROWTH AND DECAY

We begin by comparing exponential change to linear change. Let y be a quantity that changes with time: $y = y(t)$.

If y is a linear function,

$$y(t) = kt + C, \qquad k, C \text{ constants},$$

then y changes by the *same amount during all periods of the same duration*:

$$y(t + \Delta t) = k[t + \Delta t] + C = (kt + C) + k\,\Delta t = y(t) + k\,\Delta t.$$

During every period of length Δt, y changes by the amount $k\,\Delta t$.

If, on the other hand, y is an exponential function,

$$y(t) = Ce^{kt},$$

then y changes by *the same factor during all periods of the same duration*:

$$y(t + \Delta t) = Ce^{k[t+\Delta t]} = Ce^{kt + k\Delta t} = Ce^{kt}\cdot e^{k\Delta t} = e^{k\Delta t}y(t).$$

During every period of length Δt, y changes by a factor of $e^{k\Delta t}$.

The exponential function

$$f(t) = Ce^{kt}$$

has the property that its derivative $f'(t)$ is proportional to $f(t)$:

$$f'(t) = Cke^{kt} = kCe^{kt} = kf(t).$$

Moreover it is the only such function:

THEOREM 7.6.1
If

$$f'(t) = kf(t) \qquad \text{for all } t \text{ in some interval } I,$$

then f is an exponential function of the form

$$f(t) = Ce^{kt} \qquad \text{for all } t \text{ in } I \text{ and some constant } C.$$

PROOF We are given that

$$f'(t) = kf(t)$$

or

$$f'(t) - kf(t) = 0.$$

Multiplying this equation by e^{-kt}, we have

$$(*) \qquad e^{-kt}f'(t) - ke^{-kt}f(t) = 0.$$

Now observe that the left-hand side of this equation is

$$\frac{d}{dt}[e^{-kt}f(t)] \qquad \text{(verify this).}$$

Therefore, equation (*) can be written

$$\frac{d}{dt}[e^{-kt}f(t)] = 0.$$

It now follows that

$$e^{-kt}f(t) = C \qquad \text{for some constant } C,$$

and

$$f(t) = Ce^{kt}. \quad \square$$

Remark According to Theorem 7.6.1, every solution of the equation $f'(t) = kf(t)$ has the form $f(t) = Ce^{kt}$ for some constant C. The value of C is often determined by evaluating f at 0 (assuming that $0 \in I$):

$$f(0) = Ce^0 = C \qquad \text{and} \qquad f(t) = f(0)e^{kt}.$$

The value $f(0)$ is called the *initial value of f*. ❑

Example 1 Find $f(t)$ given that $f'(t) = 2f(t)$ for all t and $f(0) = \sqrt{2}$.

SOLUTION The fact that $f'(t) = 2f(t)$ tells us that $f(t) = Ce^{2t}$ where C is some constant. The fact that $f(0) = \sqrt{2}$ tells us that $C = \sqrt{2}$. Thus $f(t) = \sqrt{2}\,e^{2t}$. ❑

Population Growth

Under ideal conditions (for example, unlimited space, adequate food supply, freedom from disease, and so on), the rate of increase of a population P at time t is proportional to the size of the population at time t. That is,

$$P'(t) = kP(t),$$

where $k > 0$ is a constant, called the *growth constant.*† Thus, applying Theorem 7.6.1, the size of the population at any time t is given by

$$P(t) = P(0)e^{kt},$$

and the population is said to grow *exponentially.* This is a model of uninhibited growth. In reality, the rate of increase of a population will not continue to be proportional to the size of the population. After some time has passed, factors such as limitations on space or food supply, diseases, and so forth, will affect the growth rate of the population.

Example 2 The number of bacteria in a certain culture is increasing at a rate proportional to the number present. Suppose that there are 1000 bacteria present initially, and that 1500 are present after 2 hours.

(a) Determine the number of bacteria in the culture at any time t. How many bacteria are in the culture after 5 hours?

(b) How long will it take for the number of bacteria in the culture to double?

SOLUTION Let $P(t)$ be the number of bacteria in the culture at time t. The basic equation $P'(t) = kP(t)$ gives

$$P(t) = P(0)e^{kt}.$$

Since $P(0) = 1000$, we have

$$P(t) = 1000e^{kt}. \tag{1}$$

We can use the fact that $P(2) = 1500$ to eliminate k from (1):

$$1500 = 1000e^{2k}, \qquad e^{2k} = \frac{1500}{1000} = \frac{3}{2}, \qquad 2k = \ln\left[\frac{3}{2}\right],$$

and so

$$k = \tfrac{1}{2}\ln\left[\frac{3}{2}\right] \cong 0.203. \tag{2}$$

(a) The number of bacteria in the culture at any time t is (approximately):

$$P(t) = 1000e^{0.203t}.$$

After 5 hours, the number of bacteria in the culture is $P(5) = 1000e^{0.203(5)} = 1000e^{1.015} \cong 2759$.

---

† The growth constant of a population is often expressed as a percentage. In such instances, the percentage must be converted to its decimal equivalent for use in the formula. For example, if the growth constant of a certain population is given as 1.7%, then $k = 0.017$.

(b) To find out how long it will take for the number of bacteria to double, we want to find the value of t for which $P(t) = 2000$.† Thus we set

$$1000e^{0.203t} = 2000,$$

$$e^{0.203t} = 2, \qquad 0.203t = \ln[2], \qquad \text{and} \qquad t = \frac{\ln[2]}{0.203} \cong 3.4 \text{ hours.} \quad \square$$

Remark There is an alternative way of expressing P which uses the exact value of k. From (2), we have $k = \frac{1}{2}\ln[\frac{3}{2}]$. Thus,

$$P(t) = 1000e^{(t/2)\ln[3/2]} = 1000e^{\ln[3/2]^{t/2}}$$

and

$$P(t) = 1000\left(\frac{3}{2}\right)^{t/2}.$$

Now, for example, to find the doubling time, we solve

$$1000\left(\frac{3}{2}\right)^{t/2} = 2000$$

for t:

$$\left(\frac{3}{2}\right)^{t/2} = 2, \qquad \left(\frac{t}{2}\right)\ln\left(\frac{3}{2}\right) = \ln(2), \qquad t = \frac{2\ln(2)}{\ln(3/2)} \cong 3.4. \quad \square$$

Example 3 The world population in 1960 was approximately 3 billion, and in 1980 it was approximately 4.5 billion. Assume that the world population is increasing at a rate proportional to the size of the population.

(a) Determine the growth constant and the (approximate) world population at any time t.

(b) How long will it take for the world population to reach 12 billion (quadruple the 1960 population)?

(c) The current (1994) world population is estimated to be 5.6. billion. What does the model determined in part (a) predict for the world population in 1994?

SOLUTION Let $P(t)$ be the world population (in billions) t years after 1960. The basic equation $P'(t) = kP(t)$, with $P(0) = 3$, gives

$$P(t) = 3e^{kt}.$$

(a) Since $P(20) = 4.5$, we have

$$3e^{20k} = 4.5, \qquad 20k = \ln\left(\frac{4.5}{3}\right), \qquad k = \frac{\ln(1.5)}{20} \cong 0.0203.$$

Thus the growth constant is approximately 0.0203 or 2.03%. The population at any time t is given (approximately) by

$$P(t) = 3e^{0.0203t}.$$

† The length of time that it takes for a population to double in size is called the *doubling time*.

(b) To find the value of t for which $P(t) = 12$, we set

$$3e^{0.0203t} = 12$$

$$e^{0.0203t} = 4, \qquad 0.0203t = \ln[4], \qquad t = \frac{\ln[4]}{0.0203} \cong 68.3.$$

Thus, based on the given data, the world population will reach 12 billion in just over 68 years (the year 2028).

(c) According to our model, $P(34) = 3e^{0.0203(34)} = 3e^{0.69} \cong 5.98$ billion. This is 380 million more than the current estimate.

Radioactive Decay

Experimental evidence shows that radioactive substances decay at a rate which is proportional to the amount of such substance present. Therefore, if $A(t)$ is the amount of a radioactive substance present at time t, then

$$A'(t) = kA(t),$$

where $k < 0$ is a constant, called the *decay constant.* Thus,

$$A(t) = A(0)e^{kt},$$

where $A(0)$ is the amount of the substance present initially.

The *half-life* of a radioactive substance is the length of time that it takes for half the initial amount to decay. If we let T denote the half-life and k the decay constant of a radioactive substance, then the relation between T and k is given by

(7.6.2)
$$\boxed{kT = -\ln 2.}$$

PROOF We have $A(T) = \frac{1}{2}A(0)$. Thus,

$$\tfrac{1}{2}A(0) = A(0)e^{kT}$$

$$e^{kT} = \tfrac{1}{2}$$

$$kT = \ln[\tfrac{1}{2}] = -\ln 2. \quad \square$$

Example 4 One-third of a radioactive substance decays every 5 years. Today we have $A(0) = A_0$ grams of the substance.

(a) Find the decay constant and determine how much will be left t years from now.

(b) What is the half-life of the substance?

SOLUTION Let $A(t)$ be the amount of the radioactive substance present at time t. Since $A'(t) = kA(t)$, we know that $A(t) = A_0e^{kt}$.

(a) At the end of 5 years, one-third of the substance will have decayed and therefore two-thirds of A_0 will remain:

$$A(5) = \tfrac{2}{3}A_0.$$

Now

$$\tfrac{2}{3}A_0 = A_0 e^{5k}$$
$$e^{5k} = \tfrac{2}{3}$$
$$5k = \ln[\tfrac{2}{3}]$$

and

$$k = \frac{\ln[\frac{2}{3}]}{5} \cong -0.081.$$

The amount of the substance that remains after t years is (approximately):

$$A(t) = A_0 e^{-0.081t}.$$

(b) We use Equation (7.6.2) to find the half-life of the substance:

$$T = \frac{-\ln 2}{k} = \frac{-\ln 2}{-0.081} \cong 8.56.$$

Thus, the half-life of the substance is approximately 8.56 years. ❑

Remark You can verify that the exact value of A in Example 4 is $A(t) = A_0(\frac{2}{3})^{t/5}$ and the exact value of the half-life is $T = (5 \ln 2)/(\ln 1.5)$. ❑

Example 5 Cobalt-60 is a radioactive substance that is used extensively in medical radiology. It has a half-life of 5.3 years. Suppose an initial sample of cobalt-60 has a mass of 100 grams.

(a) Find the decay constant and determine an expression for the amount of the sample that will remain t years from now.

(b) How long will it take for 90% of the sample to decay?

SOLUTION **(a)** The decay constant k is given by

$$k = \frac{-\ln 2}{T} = \frac{-\ln 2}{5.3} \cong -0.131.$$

With $A(0) = 100$, the amount of material that will remain after t years is

$$A(t) = 100e^{-0.131t}.$$

(b) If 90% of the sample decays, then 10%, or 10 grams, remains. Thus, we need to solve the equation

$$100e^{-0.131t} = 10$$

for t. We have

$$e^{-0.131t} = 0.1, \qquad -0.131t = \ln[0.1], \qquad t = \frac{\ln[0.1]}{-0.131} \cong 17.6.$$

It will take approximately 17.6 years for 90% of the sample to decay. ❑

Compound Interest

Consider money invested at interest rate r. If the accumulated interest is credited once a year, then the interest is said to be compounded annually; if twice a year, then semiannually; if four times a year, then quarterly. The idea can be pursued further. Interest can be credited every day, every hour, every second, every half-second, and so on. In the limiting case, interest is credited instantaneously. Economists call this *continuous compounding*.

The economists' formula for continuous compounding is a simple exponential:

(7.6.3) $$A(t) = A_0 e^{rt}.$$

Here t is measured in years,

$$A(t) = \text{the principal in dollars at time } t,$$

$$A_0 = A(0) = \text{the initial investment},$$

$$r = \text{the annual interest rate.}$$

The rate r is also called the *nominal* interest rate and it is conventional to express this rate as a percentage. However, as in the case of the growth and decay constants, the decimal equivalent of the percentage must be used in all calculations.

A DERIVATION OF THE COMPOUND INTEREST FORMULA Fix t and take h as a small time increment. Then

$$A(t+h) - A(t) = \text{interest earned from time } t \text{ to time } t+h.$$

Had the principal remained $A(t)$ from time t to time $t+h$, the interest earned during this time period would have been

$$rhA(t).$$

Had the principal been $A(t+h)$ throughout the time interval, the interest earned would have been

$$rhA(t+h).$$

The actual interest earned must be somewhere in between:

$$rhA(t) \leq A(t+h) - A(t) \leq rhA(t+h).$$

Dividing by h, we get

$$rA(t) \leq \frac{A(t+h) - A(t)}{h} \leq rA(t+h).$$

If A varies continuously, then, as h tends to zero, $rA(t+h)$ tends to $rA(t)$ and (by the pinching theorem) the difference quotient in the middle must also tend to $rA(t)$:

$$\lim_{h \to 0} \frac{A(t+h) - A(t)}{h} = rA(t).$$

This says that

$$A'(t) = rA(t),$$

that is, with continuous compounding, the principal increases at a rate proportional to the amount present and the growth constant is the nominal rate r. Now, it follows that

$$A(t) = Ce^{rt}.$$

With A_0 as the initial investment, we have $C = A_0$ and

$$A(t) = A_0 e^{rt}. \quad \square$$

Remark It is interesting to compare the accumulation of principal at a fixed nominal rate r using different compounding periods. For example, a principal of \$1000 invested at 6% compounded:

(a) annually (once per year) will have the value $A(1) = 1000(1 + 0.06) = \$1060$ at the end of one year.

(b) quarterly (4 times per year) will have the value $A(1) = 1000(1 + (0.06/4))^4 \cong$ \$1061.36 at the end of one year.

(c) monthly (12 times per year) will have the value $A(1) = 1000(1 + (0.06/12))^{12} \cong$ \$1061.67 at the end of one year.

(d) continuously will have the value $A(1) = 1000e^{0.06} \cong \$1061.84$ at the end of one year. ❑

Example 6 Suppose that \$1000 is deposited in a bank that pays 5% compounded continuously. How much money will be in the account after 5 years and what is the interest earned during this period?

SOLUTION Here we have $A_0 = 1000$ and $r = 0.05$. Thus, the amount of money in the account at any time t is given by

$$A(t) = 1000e^{0.05t}.$$

The amount of money in the account after 5 years is

$$A(5) = 1000e^{0.05(5)} = 1000e^{0.25} \cong \$1284.03,$$

and the interest earned during this period is \$284.03. ❑

Example 7 How long does it take for a sum of money to double at 6.5% compounded continuously?

SOLUTION In general

$$A(t) = A_0 e^{0.065t}.$$

We set

$$2A_0 = A_0 e^{0.065t}$$

and solve for t:

$$2 = e^{0.065t}, \qquad \ln 2 = 0.065t, \qquad t = \frac{\ln 2}{0.065} \cong 10.66.$$

It takes about 10 years and 8 months. ❑

Newton's Law of Cooling

Newton's law of cooling states that the rate of change of the temperature T of an object is proportional to the difference between T and the (constant) temperature τ of the surrounding medium, called the *ambient temperature*. Assuming that the temperature of the object is a differentiable function, the mathematical formulation of Newton's law is

$$\frac{dT}{dt} = m(T - \tau), \qquad \text{where } m \text{ is a constant.}$$

Remark The constant m in this model must be negative; for if the object is warmer than the ambient temperature $(T - \tau > 0)$, then its temperature will decrease $(dT/dt < 0)$, which implies $m < 0$; if the object is colder than the ambient temperature $(T - \tau < 0)$, its temperature will increase $(dT/dt > 0)$, which again implies $m < 0$. ❑

To emphasize the sign of the constant of proportionality, we write Newton's law of cooling as

(7.6.4)
$$\frac{dT}{dt} = -k(T - \tau), \qquad \text{where } k \text{ is a positive constant.}$$

Although this equation is not of the same form as the models for exponential growth or decay, it can be converted into that form by a change of variable. Let $y(t) = T(t) - \tau$. Then

$$\frac{dy}{dt} = \frac{d}{dt}(T - \tau) = \frac{dT}{dt} \qquad \text{since } \tau \text{ is a constant.}$$

With this change of variable, (7.6.4) becomes

$$\frac{dy}{dt} = -ky,$$

and, by Theorem 7.6.1,

$$y(t) = Ce^{-kt}.$$

Therefore

$$T(t) - \tau = Ce^{-kt},$$

and so

$$T(t) = \tau + Ce^{-kt}.$$

The constant C is determined by the initial temperature:

$$T(0) = \tau + C \qquad \text{and so} \qquad C = T(0) - \tau.$$

Thus, the solutions are given by

(7.6.5)
$$T(t) = \tau + [T(0) - \tau]e^{-kt}.$$

Example 8 A turkey at 325°F is removed from the oven and allowed to cool in a room of constant temperature 72°F. Three minutes later, the temperature of the turkey is 275°. How long will it take for the turkey to cool down to a normal eating temperature of 110°?

SOLUTION We know that $T(0) = 325$ and $\tau = 72$. Therefore, by Formula (7.6.5), $T(t) = 72 + 253e^{-kt}$. Since $T(3) = 275$, we have

$$275 = 72 + 253e^{-3k}$$
$$e^{-3k} = \frac{203}{253}$$

and

$$k = \frac{\ln[203/253]}{-3} \cong 0.073.$$

Thus

$$T(t) = 72 + 253e^{-0.073t}.$$

To find out how long it will take for the turkey to cool down to 110°, we solve the equation

$$72 + 253e^{-0.073t} = 110$$

for t:

$$253e^{-0.073t} = 38, \qquad -0.073t = \ln\left[\frac{38}{253}\right], \qquad t = \frac{\ln[38/253]}{-0.073} \cong 25.97.$$

It will take approximately 26 minutes for the temperature of the turkey to reach 110°. ❑

A Mixing Problem

In our approach to mixing problems of the sort presented below we assume that uniform concentration is maintained by some kind of "stirring."

Example 9 Water from a polluted reservoir (pollution level p_0 grams per liter) is being drawn off at the rate of n liters per hour and being replaced by less polluted water (pollution level p_1 grams per liter). Given that the reservoir contains M liters, how long will it take to reduce the pollution to p grams per liter, $p_1 < p < p_0$?

SOLUTION Let $A(t)$ be the total number of grams of pollution in the reservoir at time t. We want to find the time t at which

$$\frac{A(t)}{M} = p.$$

At time t, pollutants leave the reservoir at the rate of

$$n\frac{A(t)}{M} \quad \text{grams per hour}$$

and enter the reservoir at the rate of

$$np_1 \quad \text{grams per hour.}$$

It follows that

$$A'(t) = \text{rate in} - \text{rate out} = np_1 - n\frac{A(t)}{M}$$

and we have

$$A'(t) = -\frac{n}{M}[A(t) - Mp_1].$$

This equation has the same general form as the model for Newton's law of cooling. Thus, if we let $y = A(t) - Mp_1$, we have

$$\frac{dy}{dt} = -\frac{n}{M}y,$$

and so

$$y(t) = A(t) - Mp_1 = Ce^{-nt/M}.$$

Thus

$$\frac{A(t)}{M} = p_1 + \frac{C}{M}e^{-nt/M}.$$

The constant C is determined by the initial pollution level p_0:

$$p_0 = \frac{A(0)}{M} = p_1 + \frac{C}{M} \qquad \text{so that} \qquad C = M(p_0 - p_1).$$

Using this value for C, we have

$$\frac{A(t)}{M} = p_1 + (p_0 - p_1)e^{-nt/M}.$$

This equation gives the pollution level at each time t. To find the time at which the pollution level has been reduced to p, we set

$$p_1 + (p_0 - p_1)e^{-nt/M} = p$$

and solve for t:

$$e^{-nt/M} = \frac{p - p_1}{p_0 - p_1}, \qquad -\frac{n}{M}t = \ln\frac{p - p_1}{p_0 - p_1}, \qquad t = \frac{M}{n}\ln\frac{p_0 - p_1}{p - p_1}. \quad \square$$

Example 10 In the problem just considered, how long would it take to reduce the pollution by 50% if the reservoir contained 50,000,000 liters, the water was replaced at the rate of 25,000 liters per hour, and the replacement water was completely pollution free?

SOLUTION In general

$$t = \frac{M}{n}\ln\frac{p_0 - p_1}{p - p_1}.$$

Here

$$M = 50{,}000{,}000, \quad n = 25{,}000, \quad p_1 = 0, \quad p = \tfrac{1}{2}p_0,$$

so that

$$t = 2000 \ln 2 \cong 1386.$$

It would take about 1386 hours, or about $8\frac{1}{4}$ weeks. ❑

Differential Equations

An equation that involves an unknown function and one or more of its derivatives is called a *differential equation.* The *order* of a differential equation is the order of the highest derivative that appears in the equation. Thus, the equations that we have used in this section to model population growth, radioactive decay, and continuous compounding of interest,

$$\text{(1)} \qquad \frac{dy}{dt} = ky, \qquad k \text{ constant,}$$

are examples of first-order differential equations. The motion of a simple pendulum or a weight suspended by a spring (simple harmonic motion) is described by the second-order differential equation

$$\text{(2)} \qquad \frac{d^2y}{dx^2} + \omega^2 y = 0, \qquad \omega \text{ constant.}$$

See Exercises 61 and 62, Section 3.6.

A *solution* of a differential equation is simply a function f that satisfies the equation. As we have seen, the functions $f(t) = Ce^{kt}$, where C is a constant, are solutions of (1). The functions $y = A \sin \omega x + B \cos \omega x$, where A and B are constants, are solutions of (2).

Differential equations serve as mathematical models for a wide variety of phenomena in the physical and life sciences, engineering, the social sciences, and business. Chapter 18 is an introduction to the study of differential equations and their applications.

Equation (1) is a specific example of a class of first-order differential equations known as *separable equations.* In general, a separable equation is a first-order differential equation which can be written in the form

$$\frac{dy}{dx} = \frac{f(x)}{g(y)}.$$

An informal approach to solving this equation is as follows: Using differentials, the equation can be written

$$g(y)\, dy = f(x)\, dx. \qquad \text{(separate the variables)}$$

Integrating both sides, we have

$$\int g(y)\, dy = \int f(x)\, dx.$$

Now, if G is an antiderivative of g and F is an antiderivative of f, then

$$G(y) = F(x) + C, \qquad \text{where } C \text{ is an arbitrary constant,}$$

gives a relation between x and y which satisfies the differential equation.

Example 11 Solve the differential equation

$$\frac{dy}{dx} = r(y - a), \qquad \text{where } r \text{ and } a \text{ are constants,}$$

using the approach outlined above. Note that Newton's law of cooling and mixing problems led to this type of equation.

SOLUTION Separating the variables and using differentials, we can write the equation as

$$\frac{1}{y - a}\,dy = r\,dx \qquad (y \neq a).$$

Integrating both sides, we get

$$\int \frac{1}{y - a}\,dy = \int r\,dx$$

$$\ln|y - a| = rx + C, \qquad (C \text{ an arbitrary constant})$$

which gives y as an implicit function of x. In this case, we can solve for y explicitly as a function of x by taking the exponential of both sides. In Exercise 47 you are asked to show that the solutions of the differential equation can be written

$$y = a + Be^{rx} \qquad (B \text{ an arbitrary constant}).$$

You should verify that this result is consistent with the results given previously for Newton's law of cooling and the mixing problems. ❑

Again, our approach here has been informal. A more precise treatment is given in Section 18.3.

EXERCISES 7.6

Note: Most of these exercises require a calculator. Exercises 38 and 48 require a graphics calculator or a computer with graphing software.

1. Find the amount of interest earned by \$500 compounded continuously for 10 years (a) at 6%, (b) at 8%, (c) at 10%.
2. How long does it take for a sum of money to double when compounded continuously (a) at 6%? (b) at 8%? (c) at 10%?
3. At what rate r of continuous compounding does a sum of money triple in 20 years?
4. At what rate r of continuous compounding does a sum of money double in 10 years?
5. A certain species of viral bacteria is being grown in a culture. It is observed that the rate of growth of the bacteria population is proportional to the number present. If there were 1000 bacteria in the initial population and the number doubled after the first 30 minutes, how many bacteria will be present after 2 hours?
6. In a bacteria growing experiment, a biologist observes that the number of bacteria in a certain culture triples every 4 hours. After 12 hours, it is estimated that there are 1 million bacteria in the culture.
   (a) How many bacteria were present initially?
   (b) What is the doubling time for the bacteria population?
7. At a certain moment a 100-gallon mixing tank is full of brine containing 0.25 pound of salt per gallon. Find the amount of salt present t minutes later if the brine is being continuously drawn off at the rate of 3 gallons per minute and replaced by brine containing 0.2 pound of salt per gallon.

8. Water is pumped into a tank to dilute a saline solution. The volume of the solution, call it V, is kept constant by continuous outflow. The amount of salt in the tank, call it s, depends on the amount of water that has been pumped in, call this x. Given that

$$\frac{ds}{dx} = -\frac{s}{V},$$

find the amount of water that must be pumped into the tank to eliminate 50% of the salt. Take V as 10,000 gallons.

9. A 200-liter tank initially full of water develops a leak at the bottom. Given that 20% of the water leaks out in the first 5 minutes, find the amount of water left in the tank t minutes after the leak develops if the water drains off at a rate that is proportional to the product of the time elapsed and the amount of water present.

10. In Exercise 9 assume that the water drains off at a rate that is proportional to the amount of water present.

In Exercises 11–16 remember that the rate of decay of a radioactive substance is proportional to the amount of substance present.

11. What is the half-life of a radioactive substance if it takes 4 years for a quarter of the substance to decay?

12. A year ago there were 4 grams of a radioactive substance. Now there are 3 grams. How much was there 10 years ago?

13. Two years ago there were 5 grams of a radioactive substance. Now there are 4 grams. How much will remain 3 years from now?

14. What is the half-life of a radioactive substance if it takes 5 years for a third of the substance to decay?

15. Suppose the half-life of a radioactive substance is n years. What percentage of the substance present at the start of a year will decay during the ensuing year?

16. A radioactive substance weighed n grams at time $t = 0$. Today, 5 years later, the substance weighs m grams. How much will it weigh 5 years from now?

17. (*The power of exponential growth*) Imagine two racers competing on the x-axis (which has been calibrated in meters), a linear racer LIN [position function of the form $x_1(t) = kt + C$] and an exponential racer EXP [position function of the form $x_2(t) = e^{kt} + C$]. Suppose that both racers start out simultaneously from the origin, LIN at one million meters per second, EXP at only one meter per second. In the early stages of the race, fast-starting LIN will move far ahead of EXP, but in time EXP will catch up to LIN, pass her, and leave her hopelessly behind. Show that this is true as follows:

(a) Express the position of each racer as a function of time, measuring t in seconds.

(b) Show that LIN's lead over EXP starts to decline about 13.8 seconds into the race.

(c) Show that LIN is still ahead of EXP some 15 seconds into the race but far behind 3 seconds later. (Use $e^3 \cong 20$.)

(d) Show that, once EXP passes LIN, LIN can never catch up.

18. (*The weakness of logarithmic growth*) Having been soundly beaten in the race in Exercise 17, LIN finds an opponent she can beat, LOG, the logarithmic racer [position function $x_3(t) = k \ln(t + 1) + C$]. Once again the racetrack is the x-axis calibrated in meters. Both racers start out at the origin, LOG at one million meters per second, LIN at only one meter per second. (LIN is tired from the previous race.) In this race LOG will shoot ahead and remain ahead for a long time, but eventually LIN will catch up to LOG, pass her, and leave her permanently behind. Show that this is true as follows:

(a) Express the position of each racer as a function of time t, measuring t in seconds.

(b) Show that LOG's lead over LIN starts to decline $10^6 - 1$ seconds into the race.

(c) Show that LOG is still ahead of LIN $10^7 - 1$ seconds into the race but behind LIN $10^8 - 1$ seconds into the race.

(d) Show that, once LIN passes LOG, LOG can never catch up.

19. Atmospheric pressure p varies with altitude h according to the equation

$$\frac{dp}{dh} = kp, \qquad \text{where } k \text{ is a constant.}$$

Given that p is 15 pounds per square inch at sea level and 10 pounds per square inch at 10,000 feet, find p at: (a) 5000 feet; (b) 15,000 feet.

20. According to the compound interest formula (7.6.3), if P dollars are deposited in an account now at an interest rate r compounded continuously, then the amount of money in the account t years from now will be

$$Q = Pe^{rt}.$$

The quantity Q is sometimes called the *future value* of P (at the interest rate r). Solving this equation for P, we get

$$P = Qe^{-rt}.$$

In this formulation, the quantity P is called the *present value* of Q. Find the present value of \$20,000 at 6% compounded continuously for 4 years.

21. Find the interest rate r that is needed to have \$6000 be the present value of \$10,000 over an 8-year period.

22. You are 45 years old and are looking forward to an annual pension of \$50,000 per year at age 65. What is the present-day purchasing power (present value) of your pension if money can be invested over this period at a continuously compounded interest rate of (a) 4%? (b) 6%? (c) 8%?

23. The cost of the tuition, fees, room, and board at XYZ College is currently \$16,000 per year. What would you expect to pay 3 years from now if the costs at XYZ are rising at the continuously compounded rate of (a) 5%? (b) 8%? (c) 12%?

24. A boat moving in still water is subject to a retardation proportional to its velocity. Show that the velocity t seconds after the power is shut off is given by the formula $v = ce^{-kt}$ where c is the velocity at the instant the power is shut off.

25. A boat is drifting in still water at 4 miles per hour; 1 minute later, at 2 miles per hour. How far has the boat drifted in that 1 minute? (See Exercise 24.)

26. Population tends to grow with time at a rate roughly proportional to the population present. According to the Bureau of the Census, the population of the United States in 1970 was approximately 203 million and in 1980, 227 million. Use this information to estimate the population of 1960. (The actual figure was about 179 million.)

27. Use the data of Exercise 26 to predict the population for the year 2000.

28. Use the data of Exercise 26 to estimate how long it takes for the population to double.

29. The method of *carbon dating* makes use of the fact that all living organisms contain two isotopes of carbon, carbon-12, denoted ^{12}C, (a stable isotope) and carbon-14, denoted ^{14}C (a radioactive isotope). The ratio of the amount of ^{14}C to the amount of ^{12}C is essentially constant (approximately 1/10,000). When an organism dies, the amount of ^{12}C present remains unchanged, but the ^{14}C decays at a rate proportional to the amount present with a half-life of approximately 5700 years. This change in the amount of ^{14}C relative to the amount of ^{12}C makes it possible to estimate the time at which the organism lived. A fossil found in an archaeological dig was found to contain 25% of the original amount of ^{14}C. What is the approximate age of the fossil?

30. The Dead Sea Scrolls are approximately 2000 years old. How much of the original ^{14}C remains in them?

31. A lumber company finds from experience that the value of its standing timber increases with time according to the formula

$$V(t) = V_0(\tfrac{3}{2})^{\sqrt{t}}.$$

Here t is measured in years and V_0 is the value at planting time. How long should the company wait before cutting the timber? Neglect costs and assume continuous compounding at 5%.

32. Determine the time period during which $y = Ce^{kt}$ changes by a factor of q.

33. During the process of inversion, the amount A of raw sugar present decreases at a rate proportional to A. During the first 10 hours, 1000 pounds of raw sugar have been reduced to 800 pounds. How many pounds will remain after 10 more hours of inversion?

34. The number of bacteria present in a given culture increases at a rate proportional to the number present. When first observed, the culture contained n_0 bacteria, an hour later, n_1.

(a) Find the number present t hours after the observations began.

(b) How long did it take for the number of bacteria to double?

35. Let k_1, k_2 be constants, $k_1 \neq 0$. Show that

$$\text{if } f'(t) = k_1 f(t) + k_2 \qquad \text{then } f(t) = Ce^{k_1 t} - k_2/k_1$$

where C is an arbitrary constant.

36. Find $f(t)$ given that $f'(t) = k[2 - f(t)]$ and $f(0) = 0$.

37. An object falling from rest in air is subject to not only gravitational force but also air resistance. If we assume that the air resistance is proportional to the velocity and acts in a direction opposite to the motion, then the velocity of the object at time t satisfies

$$v'(t) + K\,v(t) = 32 \qquad \text{and} \qquad v(0) = 0.$$

(a) Find $v(t)$.

(b) Show that $v(t)$ cannot exceed $32/K$ and that $\lim_{t\to\infty} v(t) = 32/K$. The value $32/K$ is called the *terminal velocity* of the object.

(c) Sketch the graph of v.

38. The solutions of the differential equation that model Newton's law of cooling are given by (7.6.5):

$$T(t) = \tau + [T(0) - \tau]e^{-kt},$$

where τ is a constant representing the ambient temperature and $k > 0$ is the constant of proportionality.

(a) Show that $\lim_{t\to\infty} T(t) = \tau$.

(b) Set $\tau = 20$ and $k = 0.1$. Use a graphing utility to graph T in the cases where

$$\text{(i) } T(0) = 1 \qquad \text{and} \qquad \text{(ii) } T(0) = 40.$$

(c) For each of the cases in part (b), estimate from your graph how long it takes for T to be within 1 unit of 20. Use four decimal place accuracy.

39. A cup of coffee is served to you at 185°F in a room where the temperature is 65°F. Two minutes later the temperature of the coffee has dropped to 155°F. How many more minutes would you expect to wait for the coffee to cool to 105°F?

40. A thermometer that reads 72°F is taken outside where the air temperature is 32°F. Three minutes later, the thermometer reading is 40°F.
(a) What will the thermometer read 5 minutes after it is brought outside?
(b) How long will it take for the thermometer to be within $\frac{1}{2}$°F of the air temperature?

41. A small metal bar with an initial temperature of 20°C is dropped into a large container of boiling water (100°C). The water continues to boil and 20 seconds later, the temperature of the bar is 30°C.
(a) What will the temperature of the bar be after 1 minute?
(b) How long will it take for the temperature of the bar to exceed 98°C?

42. The current i in an electric circuit varies with time t according to the formula

$$L\frac{di}{dt} + Ri = E$$

where E (the voltage), L (the inductance), and R (the resistance) are positive constants. Measure time in seconds, current in amperes, and suppose that the initial current is 0.
(a) Find a formula for the current at each subsequent time t.
(b) What upper limit does the current approach as t increases?
(c) In how many seconds will the current reach 90% of its upper limit?

In Exercises 43–46, show that the given equation is a separable differential equation and then find the solutions.

43. $\dfrac{dy}{dx} = \dfrac{y}{x}$.

44. $\dfrac{dy}{dx} + y\cos x = 0$.

45. $\dfrac{dy}{dx} = e^y \cos x$.

46. $\dfrac{dy}{dx} = \dfrac{3x^2}{2y + \sin 2y}$.

47. Show that the solutions of the differential equation

$$\frac{dy}{dx} = r(y - a)$$

in Example 11 can be written in the form $y = a + Be^{rx}$, where B is an arbitrary constant.

48. The differential equation

$$\frac{dP}{dt} = (k\cos t)P, \quad k > 0 \text{ a constant,}$$

models a population that experiences periodic growth and decline.
(a) Find $P(t)$ if $P(0) = P_0$.
(b) Suppose that the population is measured in thousands, and $P(0) = 3$ and $k = 2$. Use a graphing utility to sketch the graph of P.
(c) Estimate the absolute maximum value and the absolute minimum value of P from your graph. Use four decimal place accuracy in your answers.

■ 7.7 MORE ON THE INTEGRATION OF THE TRIGONOMETRIC FUNCTIONS

In Section 5.6 you saw that

(7.7.1)

$$\int \sin x\,dx = -\cos x + C. \qquad \int \sec x \tan x\,dx = \sec x + C.$$

$$\int \cos x\,dx = \sin x + C. \qquad \int \csc^2 x\,dx = -\cot x + C.$$

$$\int \sec^2 x\,dx = \tan x + C. \qquad \int \csc x \cot x\,dx = -\cot x + C.$$

Now that you are familiar with the natural logarithm function, we can add four more basic formulas to the list

(7.7.2)

$$\textbf{(i)} \int \tan x\, dx = \ln|\sec x| + C.$$

$$\textbf{(ii)} \int \cot x\, dx = \ln|\sin x| + C.$$

$$\textbf{(iii)} \int \sec x\, dx = \ln|\sec x + \tan x| + C.$$

$$\textbf{(iv)} \int \csc x\, dx = \ln|\csc x - \cot x| + C.$$

Each of these formulas can be derived as follows:

$$\int \frac{g'(x)}{g(x)}dx = \int \frac{du}{u} = \ln|u| + C = \ln|g(x)| + C.$$

set $u = g(x)$, $du = g'(x)\, dx$

Derivation of Formulas (i)–(iii)

(i)
$$\int \tan x\, dx = \int \frac{\sin x}{\cos x}dx \qquad [\text{set } u = \cos x, \quad du = -\sin x\, dx]$$
$$= -\int \frac{du}{u} = -\ln|u| + C$$
$$= -\ln|\cos x| + C = \ln\left|\frac{1}{\cos x}\right| + C$$
$$= \ln|\sec x| + C. \quad \square$$

(ii)
$$\int \cot x\, dx = \int \frac{\cos x}{\sin x}dx \qquad [\text{set } u = \sin x, \quad du = \cos x\, dx]$$
$$= \int \frac{du}{u} = \ln|u| + C = \ln|\sin x| + C. \quad \square$$

(iii)
$$\int \sec x\, dx \overset{\dagger}{=} \int \sec x \frac{\sec x + \tan x}{\sec x + \tan x}dx$$
$$= \int \frac{\sec x \tan x + \sec^2 x}{\sec x + \tan x}dx$$
$$[\text{set } u = \sec x + \tan x, \quad du = (\sec x \tan x + \sec^2 x)\, dx]$$
$$= \int \frac{du}{u} = \ln|u| + C = \ln|\sec x + \tan x| + C. \quad \square$$

The derivation of formula (iv) is left to you.

† Only experience prompts us to multiply numerator and denominator by $\sec x + \tan x$.

Example 1 Calculate

$$\int \cot \pi x \, dx.$$

SOLUTION Set

$$u = \pi x, \qquad du = \pi \, dx$$

$$\int \cot \pi x \, dx = \frac{1}{\pi} \int \cot u \, du = \frac{1}{\pi} \ln|\sin u| + C = \frac{1}{\pi} \ln|\sin \pi x| + C. \quad \square$$

Remark The u-substitution simplifies many calculations, but you will find with experience that you can carry out many integrations without it. □

Example 2 Evaluate

$$\int_0^{\pi/8} \sec 2x \, dx.$$

SOLUTION It is easy to check that $\frac{1}{2}\ln|\sec 2x + \tan 2x|$ is an antiderivative of $\sec 2x$. Thus

$$\begin{aligned} \int_0^{\pi/8} \sec 2x \, dx &= \tfrac{1}{2}\Big[\ln|\sec 2x + \tan 2x|\Big]_0^{\pi/8} \\ &= \tfrac{1}{2}[\ln(\sqrt{2} + 1) - \ln 1] = \tfrac{1}{2}\ln(\sqrt{2} + 1) \cong 0.44. \quad \square \end{aligned}$$

When the integrand is a quotient, it is worth checking to see if the integral can be written in the form

$$\int \frac{du}{u}.$$

Example 3 Calculate

$$\int \frac{\sec^2 3x}{1 + \tan 3x} \, dx.$$

SOLUTION Set

$$u = 1 + \tan 3x, \qquad du = 3 \sec^2 3x \, dx.$$

$$\int \frac{\sec^2 3x}{1 + \tan 3x} \, dx = \frac{1}{3} \int \frac{du}{u} = \frac{1}{3} \ln|u| + C = \frac{1}{3} \ln|1 + \tan 3x| + C. \quad \square$$

Example 4 Calculate the area of the region bounded by the graph of

$$f(x) = \frac{\sin x}{2 + \cos x}$$

and the x-axis for $x \in [0, 2\pi/3]$.

SOLUTION The graph of f is shown in Figure 7.7.1. The area A is given by

$$A = \int_0^{2\pi/3} \frac{\sin x}{2 + \cos x} \, dx.$$

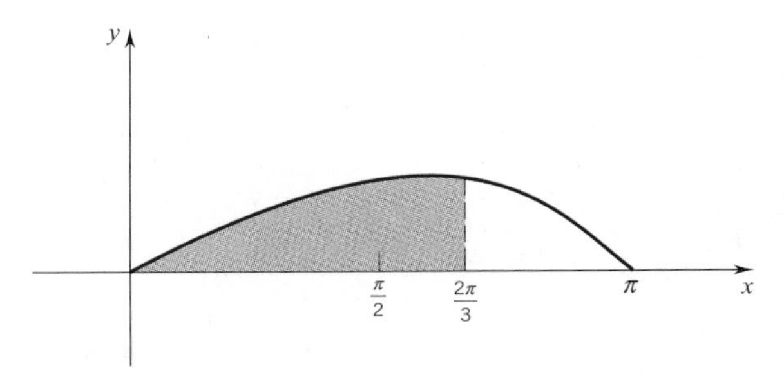

Figure 7.7.1

Set

$$u = 2 + \cos x, \qquad du = -\sin x\, dx.$$

At $x = 0$, $u = 3$, and at $x = 2\pi/3$, $u = 3/2$. Thus,

$$A = \int_0^{2\pi/3} \frac{\sin x}{2 + \cos x}\, dx = \int_3^{3/2} \frac{-du}{u} = \int_{3/2}^3 \frac{du}{u}$$

$$= \Big[\ln|u|\Big]_{3/2}^{3} = \ln(3) - \ln(3/2) = \ln(2) \cong 0.69. \quad \square$$

EXERCISES 7.7

In Exercises 1–20, calculate the indefinite integral.

1. $\int \tan 3x\, dx.$
2. $\int \sec \frac{1}{2}\pi x\, dx.$
3. $\int \csc \pi x\, dx.$
4. $\int \cot(\pi - x)\, dx.$
5. $\int e^x \cot e^x\, dx.$
6. $\int \frac{\csc^2 x}{2 + \cot x}\, dx.$
7. $\int \frac{\sin 2x}{3 - 2\cos 2x}\, dx.$
8. $\int e^{\csc x} \csc x \cot x\, dx.$
9. $\int e^{\tan 3x} \sec^2 3x\, dx.$
10. $\int e^x \cos e^x\, dx.$
11. $\int x \sec x^2\, dx.$
12. $\int \frac{\sec e^{-2x}}{e^{2x}}\, dx.$
13. $\int \cot x \ln(\sin x)\, dx.$
14. $\int \frac{\tan(\ln x)}{x}\, dx.$
15. $\int (1 + \sec x)^2\, dx.$
16. $\int \tan x \ln(\sec x)\, dx.$
17. $\int \left(\frac{\csc x}{1 + \cot x}\right)^2 dx.$
18. $\int (3 - \csc x)^2\, dx.$
19. $\int \frac{\sec x}{\sqrt{\ln|\sec x + \tan x|}}\, dx.$
20. $\int \frac{\sec 2x \tan 2x}{1 + \sec 2x}\, dx.$

In Exercises 21–26, evaluate the definite integral.

21. $\int_{\pi/6}^{\pi/2} \frac{\cos x}{1 + \sin x}\, dx.$
22. $\int_{\pi/4}^{\pi/2} (1 + \csc x)^2\, dx.$
23. $\int_{\pi/4}^{\pi/2} \cot x\, dx.$
24. $\int_{1/4}^{1/3} \tan \pi x\, dx.$
25. $\int_0^{\ln \pi/4} e^x \sec e^x\, dx.$
26. $\int_{\pi/4}^{\pi/2} \frac{\csc^2 x}{3 + \cot x}\, dx.$

In Exercises 27–32, sketch the region bounded by the curves and find its area.

27. $y = \sec x, \quad y = 2, \quad x = 0, \quad x = \pi/6.$
28. $y = \csc \frac{1}{2}\pi x, \quad y = x, \quad x = \frac{1}{2}.$
29. $y = \tan x, \quad y = 1, \quad x = 0.$
30. $y = \sec x, \quad y = \cos x, \quad x = 0, x = \frac{\pi}{4}.$
31. $y = \tan 2x, \quad y = 0, \quad x = -\pi/8, \quad x = \pi/6.$
32. $y = \sec^2 x, \quad y = \frac{1}{2}, \quad x = 0, \quad x = \pi/3.$
33. The region bounded by the graph of $f(x) = \sqrt{\sec x}$ and the x-axis for $-\pi/3 \le x \le \pi/3$ is revolved around the x-axis. Find the volume of the solid that is generated.

34. The region bounded by the graph of $f(x) = \tan x$ and the x-axis for $0 \le x \le \pi/4$ is revolved around the x-axis. Find the volume of the solid that is generated.

35. Find the average value of $f(x) = \dfrac{\sin x}{1 + 2\cos x}$ on $[0, \pi/2]$.

36. Find the average value of $f(x) = (1 + \tan x)^2$ on $[0, \pi/4]$.

37. Show that $\int \csc x\, dx = \ln|\csc x - \cot x| + C$ using the methods of this section.

38. (a) Explain why $\int_0^{\pi} \sqrt{1 - \sin^2 x}\, dx > 0$.
(b) Evaluate the integral in part (a).

■ 7.8 THE INVERSE TRIGONOMETRIC FUNCTIONS

Since none of the trigonometric functions are one-to-one, none of them have inverses. What then are the inverse trigonometric functions?

The Inverse Sine

The graph of $y = \sin x$ is shown in Figure 7.8.1. Since every horizontal line between -1 and 1 intersects the graph at infinitely many points, the sine function does not have an inverse. However, observe that if we restrict the domain to the interval $[-\frac{1}{2}\pi, \frac{1}{2}\pi]$ (the solid portion of the graph in Figure 7.8.1), then $y = \sin x$ is one-to-one, and on that interval it takes on as a value every number in $[-1, 1]$. Thus, if $x \in [-1, 1]$, there is one and only one number in the interval $[-\frac{1}{2}\pi, \frac{1}{2}\pi]$ at which the sine function has the value x. This number is called the *inverse sine of* x, or *the angle whose sine is* x, and is written $\sin^{-1}x$.

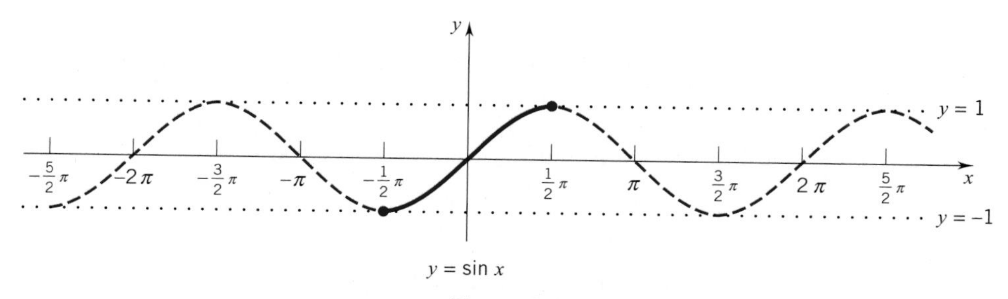

Figure 7.8.1

Remarks Another common notation for the inverse sine function is arcsin x, read "arc sine of x." For consistency in the treatment here, we use $\sin^{-1}x$ throughout, but we use both notations in the Exercises. Remember that the "-1" is not an exponent; do not confuse $\sin^{-1}x$ with the reciprocal $1/\sin x$. ❑

The *inverse sine function*

$$y = \sin^{-1}x, \qquad \text{domain: } [-1, 1], \qquad \text{range: } [-\tfrac{1}{2}\pi, \tfrac{1}{2}\pi]$$

is the inverse of the function

$$y = \sin x, \qquad \text{domain: } [-\tfrac{1}{2}\pi, \tfrac{1}{2}\pi], \qquad \text{range: } [-1, 1].$$

The graphs of these functions are shown in Figure 7.8.2. Each curve is the reflection of the other in the line $y = x$.

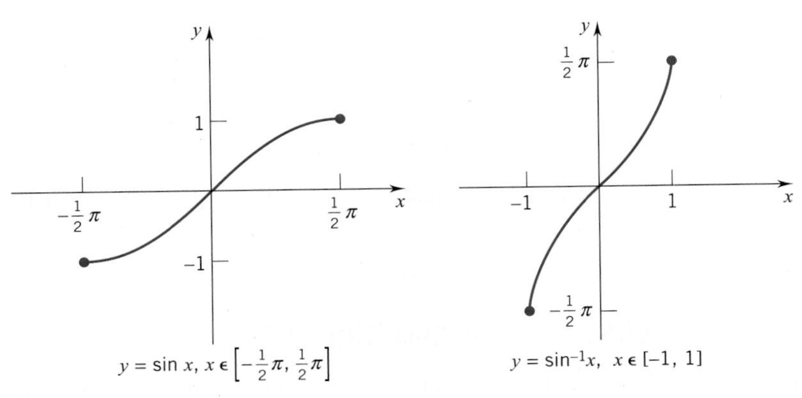

Figure 7.8.2

Because these functions are inverses,

(7.8.1) $$\text{for all } x \in [-1, 1], \qquad \sin(\sin^{-1}x) = x$$

and

(7.8.2) $$\text{for all } x \in [-\tfrac{1}{2}\pi, \tfrac{1}{2}\pi], \qquad \sin^{-1}(\sin x) = x.$$

■ **Table 7.8.1**

x	$\sin x$
$-\frac{1}{2}\pi$	-1
$-\frac{1}{3}\pi$	$-\frac{1}{2}\sqrt{3}$
$-\frac{1}{4}\pi$	$-\frac{1}{2}\sqrt{2}$
$-\frac{1}{6}\pi$	$-\frac{1}{2}$
0	0
$\frac{1}{6}\pi$	$\frac{1}{2}$
$\frac{1}{4}\pi$	$\frac{1}{2}\sqrt{2}$
$\frac{1}{3}\pi$	$\frac{1}{2}\sqrt{3}$
$\frac{1}{2}\pi$	1

Table 7.8.1 gives some representative values of the sine function from $x = -\frac{1}{2}\pi$ to $x = \frac{1}{2}\pi$. Reversing the order of the columns we have a table for the inverse sine (Table 7.8.2).

■ **Table 7.8.2**

x	$\sin^{-1} x$
-1	$-\frac{1}{2}\pi$
$-\frac{1}{2}\sqrt{3}$	$-\frac{1}{3}\pi$
$-\frac{1}{2}\sqrt{2}$	$-\frac{1}{4}\pi$
$-\frac{1}{2}$	$-\frac{1}{6}\pi$
0	0
$\frac{1}{2}$	$\frac{1}{6}\pi$
$\frac{1}{2}\sqrt{2}$	$\frac{1}{4}\pi$
$\frac{1}{2}\sqrt{3}$	$\frac{1}{3}\pi$
1	$\frac{1}{2}\pi$

On the basis of Table 7.8.2 one could guess that for all $x \in [-1, 1]$

$$\sin^{-1}(-x) = -\sin^{-1}x.$$

This is indeed the case. Being the inverse of an odd function ($\sin(-x) = -\sin x$ for all $x \in [-\frac{1}{2}\pi, \frac{1}{2}\pi]$), the inverse sine is itself an odd function. (Verify this.)

Example 1 Calculate if defined:

(a) $\sin^{-1}(\sin \frac{1}{16}\pi)$. **(b)** $\sin^{-1}(\sin \frac{8}{3}\pi)$. **(c)** $\sin(\sin^{-1}\frac{1}{3})$.
(d) $\sin^{-1}(\sin \frac{9}{5}\pi)$. **(e)** $\sin(\sin^{-1}2)$.

SOLUTION

(a) Since $\frac{1}{16}\pi$ is within the interval $[-\frac{1}{2}\pi, \frac{1}{2}\pi]$, we know by (7.8.2) that

$$\sin^{-1}(\sin \tfrac{1}{16}\pi) = \tfrac{1}{16}\pi.$$

(b) Since $\frac{8}{3}\pi$ is not in the interval $[-\pi/2, \pi/2]$, we cannot apply (7.8.2) directly. However, $\frac{8}{3}\pi = \frac{2}{3}\pi + 2\pi$ and $\sin(\frac{2}{3}\pi + 2\pi) = \sin(\frac{2}{3}\pi)$ (recall that the sine function is periodic with period 2π). Thus

$$\sin^{-1}(\sin \tfrac{8}{3}\pi) = \sin^{-1}(\sin[\tfrac{2}{3}\pi + 2\pi]) = \sin^{-1}(\sin \tfrac{2}{3}\pi) \underset{\text{by (7.8.2)}}{=} \tfrac{2}{3}\pi.$$

(c) By (7.8.1)

$$\sin(\sin^{-1}\tfrac{1}{3}) = \tfrac{1}{3}.$$

(d) Since $\frac{9}{5}\pi$ is not within the interval $[-\frac{1}{2}\pi, \frac{1}{2}\pi]$, we cannot apply (7.8.2) directly. However, $\frac{9}{5}\pi = 2\pi - \frac{1}{5}\pi$. Thus

$$\sin^{-1}(\sin \tfrac{9}{5}\pi) = \sin^{-1}(\sin(2\pi - \tfrac{1}{5}\pi)) = \sin^{-1}(\sin(-\tfrac{1}{5}\pi)) \underset{\text{by (7.8.2)}}{=} -\tfrac{1}{5}\pi.$$

(e) The expression $\sin(\sin^{-1}2)$ makes no sense since 2 is not in the domain of the inverse sine; there is *no* angle whose sine is 2. The inverse sine is defined only on $[-1, 1]$. ❑

If $0 < x < 1$, then $\sin^{-1}x$ is the radian measure of the acute angle whose sine is x. We can construct an angle of radian measure $\sin^{-1}x$ by drawing a right triangle with a leg of length x and a hypotenuse of length 1. See Figure 7.8.3.

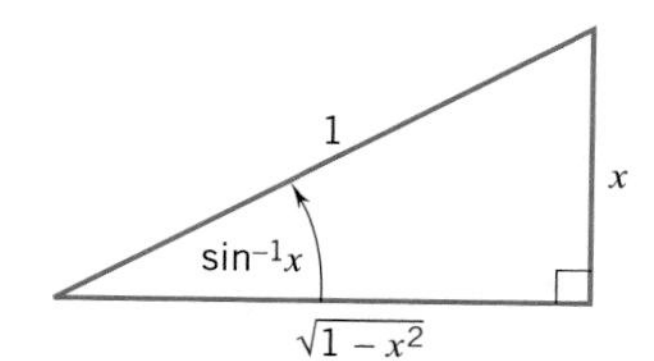

Figure 7.8.3

Reading from the figure we have

$$\sin(\sin^{-1}x) = x, \qquad \cos(\sin^{-1}x) = \sqrt{1-x^2}$$

$$\tan(\sin^{-1}x) = \frac{x}{\sqrt{1-x^2}}, \qquad \cot(\sin^{-1}x) = \frac{\sqrt{1-x^2}}{x}$$

$$\sec(\sin^{-1}x) = \frac{1}{\sqrt{1-x^2}}, \qquad \csc(\sin^{-1}x) = \frac{1}{x}.$$

Since the derivative of the sine function.,

$$\frac{d}{dx}(\sin x) = \cos x,$$

does not take on the value 0 on the *open* interval $(-\frac{1}{2}\pi, \frac{1}{2}\pi)$ (recall that $\cos x > 0$ for $-\frac{1}{2}\pi < x < \frac{1}{2}\pi$), the inverse sine function is differentiable on the open interval $(-1, 1)$.† We can find the derivative as follows:

$$y = \sin^{-1}x,$$

$$\sin y = x,$$

$$\cos y \frac{dy}{dx} = 1,$$

$$\frac{dy}{dx} = \frac{1}{\cos y} = \frac{1}{\sqrt{1-x^2}} \quad \text{(see the figure).}$$ ❑

In short

(7.8.3)

$$\boxed{\frac{d}{dx}(\sin^{-1}x) = \frac{1}{\sqrt{1-x^2}}.}$$

† See Section 7.1.

Example 2 Find

$$\frac{d}{dx}[\sin^{-1}(3x^2)].$$

NOTE: Although we have not made a point of it, we continue with the convention that if the domain of a function f is not specified explicitly, then it is understood to be the largest set of real numbers x for which $f(x)$ is a real number. In this case, the domain is the set of all real numbers x such that $-1 \le 3x^2 \le 1$, or $|x| \le 1/\sqrt{3}$.

SOLUTION In general, by the chain rule,

$$\frac{d}{dx}[\sin^{-1}u] = \frac{d}{du}[\sin^{-1}u]\frac{du}{dx} = \frac{1}{\sqrt{1-u^2}}\frac{du}{dx}.$$

Thus

$$\frac{d}{dx}[\sin^{-1}(3x^2)] = \frac{1}{\sqrt{1-(3x^2)^2}}\frac{d}{dx}(3x^2) = \frac{6x}{\sqrt{1-9x^4}}. \quad \square$$

Example 3 Show that for $a > 0$

(7.8.4)
$$\int \frac{dx}{\sqrt{a^2-x^2}} = \sin^{-1}\left(\frac{x}{a}\right) + C.$$

SOLUTION We change variables so that the a^2 becomes 1 and we can use (7.8.3). We set

$$au = x, \qquad a\,du = dx.$$

Then

$$\int \frac{dx}{\sqrt{a^2-x^2}} = \int \frac{a\,du}{\sqrt{a^2-a^2u^2}} = \int \frac{du}{\sqrt{1-u^2}} = \sin^{-1}u + C = \sin^{-1}\left(\frac{x}{a}\right) + C. \quad \square$$

since $a > 0$ ↑

Example 4 Evaluate

$$\int_0^{\sqrt{3}} \frac{dx}{\sqrt{4-x^2}}.$$

SOLUTION By (7.8.4)

$$\int \frac{dx}{\sqrt{4-x^2}} = \sin^{-1}\left(\frac{x}{2}\right) + C.$$

It follows that

$$\int_0^{\sqrt{3}} \frac{dx}{\sqrt{4-x^2}} = \left[\sin^{-1}\left(\frac{x}{2}\right)\right]_0^{\sqrt{3}} = \sin^{-1}\left(\frac{\sqrt{3}}{2}\right) - \sin^{-1}0 = \frac{\pi}{3} - 0 = \frac{\pi}{3}. \quad \square$$

The Inverse Tangent

Although not one-to-one on its full domain, the tangent function is one-to-one on the open interval $(-\frac{1}{2}\pi, \frac{1}{2}\pi)$ and on that interval it takes on as a value every real number (see Figure 7.8.4). Thus, for any real number x, there is one and only one number in the open interval $(-\frac{1}{2}\pi, \frac{1}{2}\pi)$ at which the tangent function has the value x. This number is called the *inverse tangent of* x, or *the angle whose tangent is* x, and is written $\tan^{-1}x$. Like the inverse sine function, the notation $\arctan x$ is also commonly used.

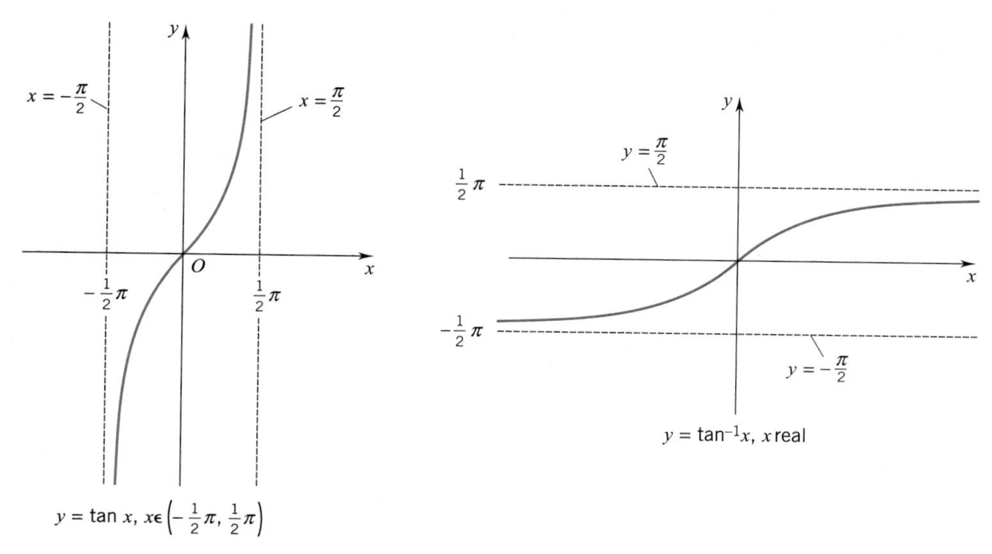

Figure 7.8.4

The *inverse tangent function*

$$y = \tan^{-1}x, \qquad \text{domain: } (-\infty, \infty), \qquad \text{range: } (-\tfrac{1}{2}\pi, \tfrac{1}{2}\pi)$$

is the inverse of the function

$$y = \tan x, \qquad \text{domain: } (-\tfrac{1}{2}\pi, \tfrac{1}{2}\pi), \qquad \text{range: } (-\infty, \infty).$$

The graphs of these two functions are given in Figure 7.8.4. Each curve is a reflection of the other in the line $y = x$. While the tangent has vertical asymptotes, the inverse tangent has horizontal asymptotes. Both functions are odd functions.

Because these functions are inverses,

(7.8.5) $$\text{for all real numbers } x, \qquad \tan(\tan^{-1}x) = x$$

and

(7.8.6) $$\text{for all } x \in (-\tfrac{1}{2}\pi, \tfrac{1}{2}\pi), \qquad \tan^{-1}(\tan x) = x.$$

It is hard to make a mistake with the first relation since it applies for all real numbers. The second relation requires the usual care:

$$\tan^{-1}(\tan \tfrac{1}{4}\pi) = \tfrac{1}{4}\pi \qquad \text{but} \qquad \tan^{-1}(\tan \tfrac{7}{5}\pi) \neq \tfrac{7}{5}\pi.$$

We can calculate $\tan^{-1}(\tan \frac{7}{5}\pi)$ as follows: $\frac{7}{5}\pi = \frac{2}{5}\pi + \pi$ and $\tan(\frac{2}{5}\pi + \pi) = \tan(\frac{2}{5}\pi)$ (recall that the tangent function is periodic with period π). The relation $\tan^{-1}(\tan \frac{2}{5}\pi) = \frac{2}{5}\pi$ is valid because $\frac{2}{5}\pi$ is within the interval $(-\frac{1}{2}\pi, \frac{1}{2}\pi)$.

If $x > 0$, then $\tan^{-1}x$ is the radian measure of the acute angle that has tangent x. We can construct an angle of radian measure $\tan^{-1}x$ by drawing a right triangle with legs of length x and 1 (Figure 7.8.5). Reading from the triangle, we have

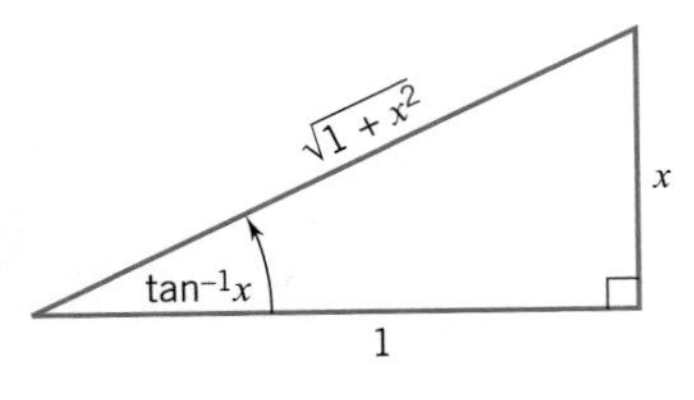

Figure 7.8.5

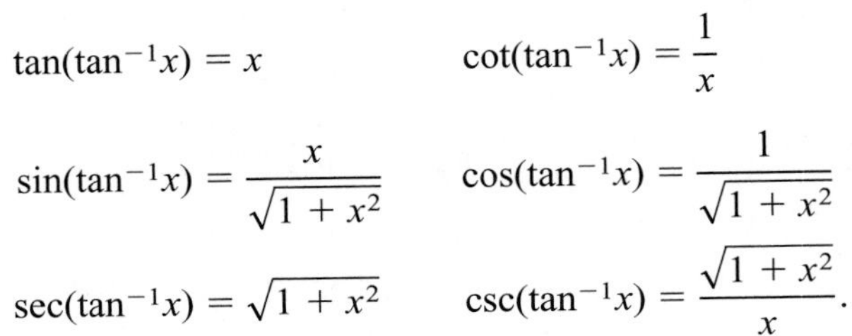

$$\tan(\tan^{-1}x) = x \qquad \cot(\tan^{-1}x) = \frac{1}{x}$$

$$\sin(\tan^{-1}x) = \frac{x}{\sqrt{1+x^2}} \qquad \cos(\tan^{-1}x) = \frac{1}{\sqrt{1+x^2}}$$

$$\sec(\tan^{-1}x) = \sqrt{1+x^2} \qquad \csc(\tan^{-1}x) = \frac{\sqrt{1+x^2}}{x}.$$

Since the derivative of the tangent function

$$\frac{d}{dx}(\tan x) = \sec^2 x = \frac{1}{\cos^2 x},$$

is never 0, the inverse tangent function is everywhere differentiable (Section 7.1). We can find the derivative as we did for the inverse sine:

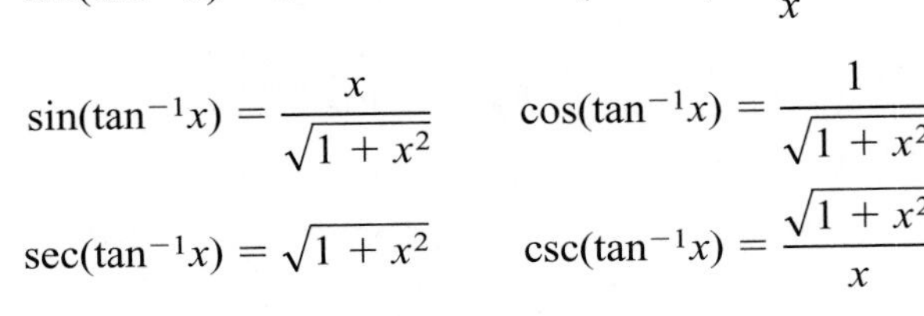

$$y = \tan^{-1}x,$$

$$\tan y = x,$$

$$\sec^2 y \frac{dy}{dx} = 1,$$

$$\frac{dy}{dx} = \frac{1}{\sec^2 y} = \cos^2 y = \frac{1}{1+x^2} \qquad \text{(see the figure).}$$

We have found that

(7.7.7)

$$\boxed{\frac{d}{dx}(\tan^{-1}x) = \frac{1}{1+x^2}.}$$

Example 5 Calculate

$$\frac{d}{dx}[\tan^{-1}(ax^2 + bx + c)].$$

SOLUTION In general, by the chain rule,

$$\frac{d}{dx}[\tan^{-1}u] = \frac{d}{du}[\tan^{-1}u]\frac{du}{dx} = \frac{1}{1+u^2}\frac{du}{dx}.$$

Thus

$$\frac{d}{dx}[\tan^{-1}(ax^2+bx+c)] = \frac{1}{1+(ax^2+bx+c)^2}\frac{d}{dx}(ax^2+bx+c)$$
$$= \frac{2ax+b}{1+(ax^2+bx+c)^2}. \quad \square$$

Example 6 Show that, for $a \neq 0$,

(7.8.8)
$$\int \frac{dx}{a^2+x^2} = \frac{1}{a}\tan^{-1}\left(\frac{x}{a}\right) + C.$$

SOLUTION We change variables so that a^2 is replaced by 1 and we can use (7.8.7). We set

$$au = x, \qquad a\,du = dx.$$

Then

$$\int \frac{dx}{a^2+x^2} = \int \frac{a\,du}{a^2+a^2u^2} = \frac{1}{a}\int \frac{du}{1+u^2}$$
$$= \frac{1}{a}\tan^{-1}u + C = \frac{1}{a}\tan^{-1}\left(\frac{x}{a}\right) + C. \quad \square$$

(7.8.7)⟶↑

Example 7 Evaluate

$$\int_0^2 \frac{dx}{4+x^2}.$$

SOLUTION By (7.8.8)

$$\int \frac{dx}{4+x^2} = \int \frac{dx}{2^2+x^2} = \frac{1}{2}\tan^{-1}\left(\frac{x}{2}\right) + C$$

so that

$$\int_0^2 \frac{dx}{4+x^2} = \left[\frac{1}{2}\tan^{-1}\left(\frac{x}{2}\right)\right]_0^2 = \frac{1}{2}\tan^{-1}1 - \frac{1}{2}\tan^{-1}0 = \frac{\pi}{8}. \quad \square$$

Inverse Secant

The procedure for defining the inverse secant function is the same as the procedure for the inverse sine and inverse tangent functions. The graph of $y = \sec x$ is indicated in Figure 7.8.6. Note that $|\sec x| \geq 1$ for all x in the domain. If we restrict the domain to the set $[0, \frac{1}{2}\pi) \cup (\frac{1}{2}\pi, \pi]$ (the solid portion of the graph in Figure 7.8.6), then the secant function is one-to-one, and for every real number x such that $|x| \geq 1$ there is one and only one number in $[0, \frac{1}{2}\pi) \cup (\frac{1}{2}\pi, \pi]$ at which the secant function has the value x. This number is called the *inverse secant of* x, or *the angle whose secant is* x, and is written $\sec^{-1}x$ or arcsec x.

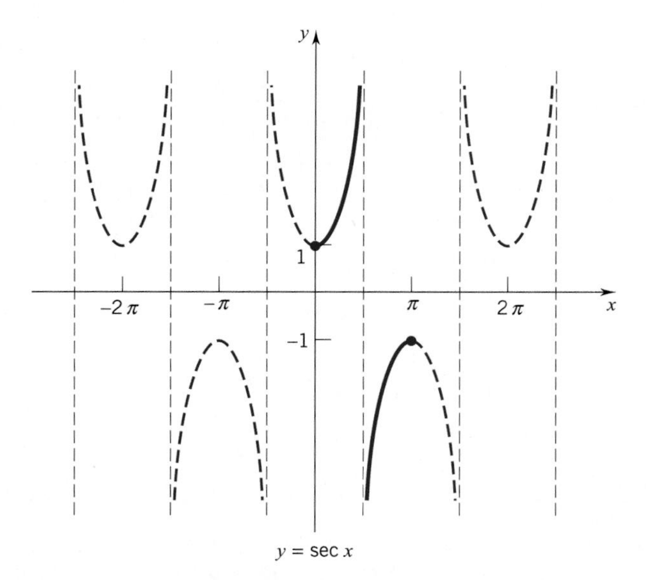

$y = \sec x$

Figure 7.8.6

The *inverse secant function*

$$y = \sec^{-1}x, \qquad \text{domain: } (-\infty, -1] \cup [1, \infty), \qquad \text{range: } [0, \tfrac{1}{2}\pi) \cup (\tfrac{1}{2}\pi, \pi]$$

is the inverse of the function

$$y = \sec x, \qquad \text{domain:} [0, \tfrac{1}{2}\pi) \cup (\tfrac{1}{2}\pi, \pi], \qquad \text{range: } (-\infty, -1] \cup [1, \infty).$$

The graphs of these two functions are given in Figure 7.8.7.

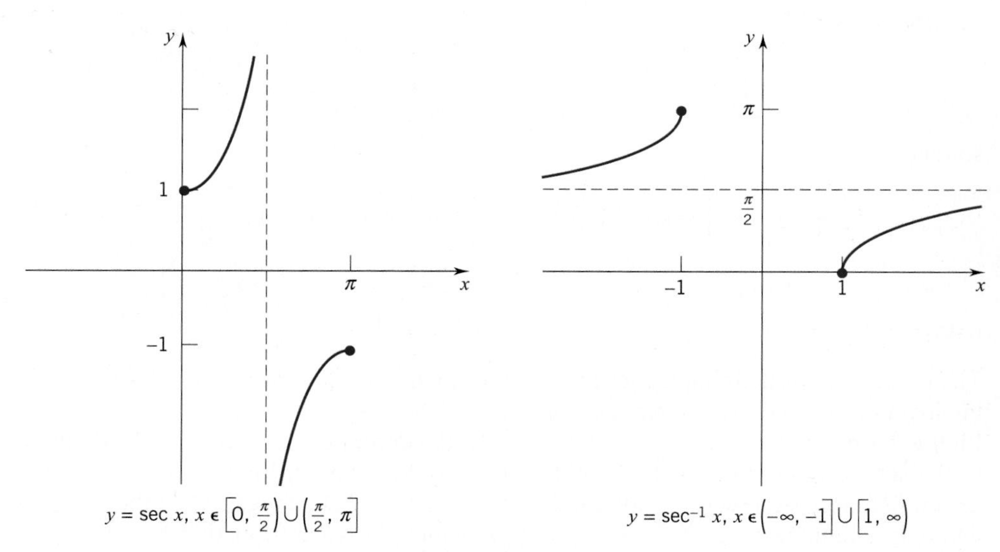

$y = \sec x,\ x \in \left[0, \frac{\pi}{2}\right) \cup \left(\frac{\pi}{2}, \pi\right]$

$y = \sec^{-1} x,\ x \in \left(-\infty, -1\right] \cup \left[1, \infty\right)$

Figure 7.8.7

The function–inverse function relationships for $\sec x$ and $\sec^{-1}x$ are:

(7.8.9) for all such that $|x| \geq 1$, $\quad \sec(\sec^{-1}x) = x$

(7.8.10) for all $x \in [0, \frac{1}{2}\pi) \cup (\frac{1}{2}\pi, \pi]$, $\quad \sec^{-1}(\sec x) = x.$

As we have seen previously, the second relation requires careful attention:

$$\sec^{-1}(\sec \tfrac{1}{3}\pi) = \tfrac{1}{3}\pi \qquad \text{but} \qquad \sec^{-1}(\sec \tfrac{7}{4}\pi) \neq \tfrac{7}{4}\pi.$$

We calculate $\sec^{-1}(\sec \frac{7}{4}\pi)$ as follows:

$$\sec^{-1}(\sec \tfrac{7}{4}\pi) = \sec^{-1}(\sec[2\pi - \tfrac{1}{4}\pi]) = \sec^{-1}(\sec \tfrac{1}{4}\pi) = \tfrac{1}{4}\pi.$$

If $x > 1$, then $\sec^{-1}x$ is the radian measure of the acute angle that has secant x. We can construct an angle of radian measure $\sec^{-1}x$ by drawing a right triangle with hypotenuse of length x and a side of length 1 (Figure 7.8.8). The values

$$\sec(\sec^{-1}x) = x \qquad \csc(\sec^{-1}x) = \frac{x}{\sqrt{x^2 - 1}}$$

$$\sin(\sec^{-1}x) = \frac{\sqrt{x^2 - 1}}{x} \qquad \cos(\sec^{-1}x) = \frac{1}{x}$$

$$\tan(\sec^{-1}x) = \sqrt{x^2 - 1} \qquad \cot(\sec^{-1}x) = \frac{1}{\sqrt{x^2 - 1}}$$

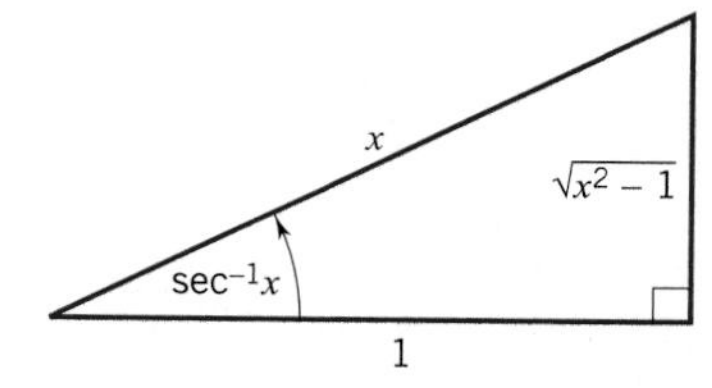

Figure 7.8.8

can all be read from this triangle.

The derivative of the secant function,

$$\frac{d}{dx}(\sec x) = \sec x \tan x,$$

is nonzero on $(0, \frac{1}{2}\pi) \cup (\frac{1}{2}\pi, \pi)$. Therefore, the inverse secant function is differentiable for $|x| > 1$. We can find the derivative as follows:

$$y = \sec^{-1}x,$$

$$\sec y = x,$$

$$\sec y \tan y \frac{dy}{dx} = 1,$$

$$\frac{dy}{dx} = \frac{1}{\sec y \tan y}.$$

To complete the calculation, we have to express the product $\sec y \tan y$ in terms of x. Clearly, $\sec y = x$, and from Figure 7.8.9 we see that $\tan y = \pm\sqrt{x^2 - 1}$. Now, if $x > 1$, then $0 < y < \frac{1}{2}\pi$ and $\tan y = \sqrt{x^2 - 1}$; if $x < -1$, then $\frac{1}{2}\pi < y < \pi$ and

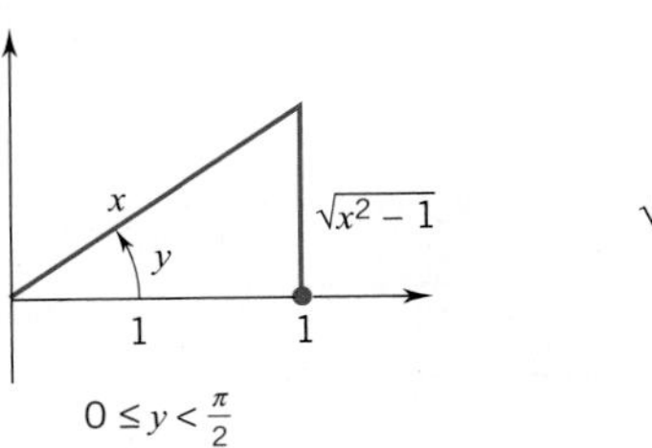

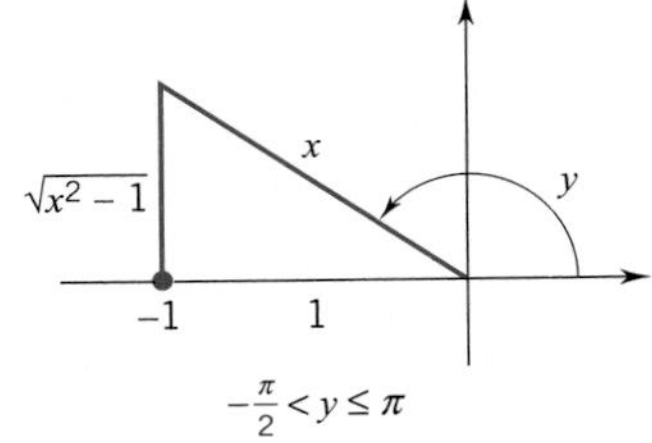

Figure 7.8.9

$\tan y = -\sqrt{x^2-1}$. In either case, the product $\sec y \tan y$ is *positive.* Therefore, we have

(7.8.11)
$$\frac{d}{dx}(\sec^{-1}x) = \frac{1}{|x|\sqrt{x^2-1}}.$$

Note that $dy/dx > 0$ on $(-\infty, -1)$ and on $(1, \infty)$, which implies that $y = \sec^{-1}x$ is an increasing function on each of these intervals. This is consistent with the graph shown in Figure 7.8.7.

Example 8 Find

$$\frac{d}{dx}[\sec^{-1}(2 \ln x)].$$

SOLUTION By the chain rule, we have

$$\frac{d}{dx}[\sec^{-1}u] = \frac{d}{du}[\sec^{-1}u]\frac{du}{dx} = \frac{1}{|u|\sqrt{u^2-1}}\frac{du}{dx}.$$

Thus

$$\frac{d}{dx}[\sec^{-1}(2 \ln x)] = \frac{1}{|2 \ln x|\sqrt{(2 \ln x)^2-1}}\frac{d}{dx}(2 \ln x) = \frac{1}{2|\ln x|\sqrt{4(\ln x)^2-1}} \cdot \frac{2}{x}$$
$$= \frac{1}{x|\ln x|\sqrt{4(\ln x)^2-1}}. \quad \square$$

Example 9 Show that

$$\int \frac{dx}{x\sqrt{x^2-1}} = \sec^{-1}|x| + C.$$

SOLUTION If $x > 1$, $\sec^{-1}|x| = \sec^{-1}x$ and

$$\frac{d}{dx}(\sec^{-1}x) = \frac{1}{|x|\sqrt{x^2-1}} = \frac{1}{x\sqrt{x^2-1}}.$$

If $x < -1$, $\sec^{-1}|x| = \sec^{-1}(-x)$ and

$$\frac{d}{dx}[\sec^{-1}(-x)] = \frac{1}{|-x|\sqrt{x^2-1}}(-1) = \frac{-1}{-x\sqrt{x^2-1}} = \frac{1}{x\sqrt{x^2-1}}.$$

Thus, it follows that $F(x) = \sec^{-1}|x|$ is an antiderivative for $f(x) = \dfrac{1}{x\sqrt{x^2-1}}$. ❑

We leave it as an exercise to show that if $a > 0$ is a constant, then

(7.8.12)
$$\int \frac{1}{x\sqrt{x^2-a^2}}\,dx = \frac{1}{a}\sec^{-1}\frac{|x|}{a} + C.$$

Example 10 Evaluate

$$\int_{2\sqrt{2}}^{4} \frac{dx}{x\sqrt{x^2-4}}.$$

SOLUTION Use (7.8.12) with $a = 2$:

$$\int_{2\sqrt{2}}^{4} \frac{dx}{x\sqrt{x^2-4}} = \left[\frac{1}{2}\sec^{-1}\frac{|x|}{2}\right]_{2\sqrt{2}}^{4} = \frac{1}{2}\sec^{-1}2 - \frac{1}{2}\sec^{-1}\sqrt{2}$$
$$= \frac{1}{2}\left(\frac{\pi}{3} - \frac{\pi}{4}\right) = \frac{\pi}{24}. \quad ❑$$

The Other Trigonometric Inverses

There are three other trigonometric inverses:

the *inverse cosine,* $y = \cos^{-1}x$, is the inverse of $y = \cos x$, $x \in [0, \pi]$;

the *inverse cotangent,* $y = \cot^{-1}x$, is the inverse of $y = \cot x$, $x \in (0, \pi)$;

the *inverse cosecant,* $y = \csc^{-1}x$, is the inverse of $y = \csc x$, $x \in [-\frac{1}{2}\pi, 0) \cup (0, \frac{1}{2}\pi]$.

Figure 7.8.10 illustrates each of these inverses between 0 and $\frac{1}{2}\pi$ in terms of right triangles.

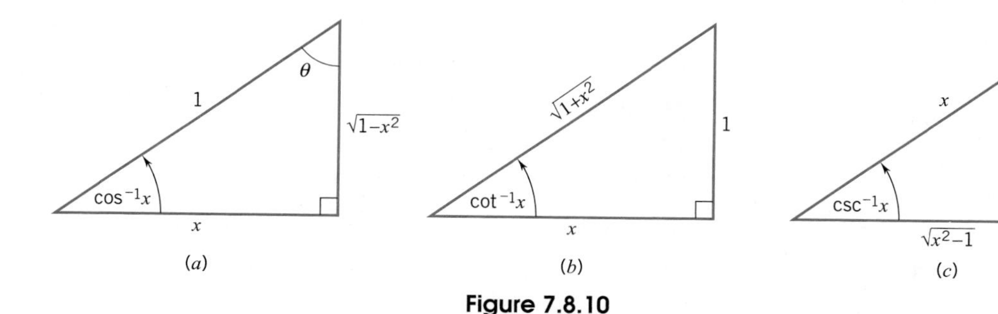

Figure 7.8.10

The differentiation formulas for these functions are:

$$\frac{d}{dx}(\cos^{-1}x) = \frac{-1}{\sqrt{1-x^2}} = -\frac{d}{dx}(\sin^{-1}x)$$

$$\frac{d}{dx}(\cot^{-1}x) = \frac{-1}{1+x^2} = -\frac{d}{dx}(\tan^{-1}x)$$

$$\frac{d}{dx}(\csc^{-1}x) = \frac{-1}{|x|\sqrt{x^2-1}} = -\frac{d}{dx}(\sec^{-1}x).$$

You are asked to verify these formulas in the Exercises. Since the derivatives of these functions differ from the derivatives of their corresponding cofunctions by a constant factor, they are not needed for finding antiderivatives.

Remark In Figure 7.8.10*a*, we let θ denote the other acute angle of the right triangle. Since the two acute angles in a right triangle are complementary, we have

$$\theta + \cos^{-1}x = \frac{\pi}{2}.$$

But note that $\sin\theta = x$, which means that $\theta = \sin^{-1}x$. Therefore,

$$\sin^{-1}x + \cos^{-1}x = \frac{\pi}{2} \qquad \text{or} \qquad \cos^{-1}x = \frac{\pi}{2} - \sin^{-1}x.$$

You can verify that the corresponding relations hold for the pairs $\{\tan x, \cot x\}$ and $\{\sec x, \csc x\}$:

$$\cot^{-1}x = \frac{\pi}{2} - \tan^{-1}x \qquad \text{and} \qquad \csc^{-1}x = \frac{\pi}{2} - \sec^{-1}x.$$

This is another reason why it is sufficient to restrict our attention to $\sin^{-1}x$, $\tan^{-1}x$, and $\sec^{-1}x$. ❑

EXERCISES 7.8

In Exercises 1–12, determine the exact value of the given expression.

1. $\tan^{-1}0$.
2. $\sin^{-1}(-\sqrt{3}/2)$.
3. arcsec 2.
4. $\tan^{-1}\sqrt{3}$.
5. $\cos^{-1}(-\frac{1}{2})$.
6. $\sec^{-1}(-\sqrt{2})$.
7. $\arcsin(\sin 7\pi/4)$.
8. $\arctan(\tan 11\pi/4)$.
9. $\sec[\sec^{-1}(-2/\sqrt{3})]$.
10. $\sin[\arccos(-\frac{1}{2})]$.
11. $\cos(\sec^{-1}2)$.
12. $\arctan(\sec 0)$.

C In Exercises 13–18, find the approximate value by using your calculator. Use four decimal place accuracy.

13. $\sin^{-1}0.918$.
14. $\arcsin(-0.795)$.
15. $\tan^{-1}(-0.493)$.
16. arctan 3.111.
17. arcsec (2.761).
18. $\cos^{-1}(-0.142)$.

In Exercises 19–24, take $x > 0$ and calculate the following from Figure 7.8.5.

19. $\cos(\tan^{-1}x)$.
20. $\sin(\tan^{-1}x)$.
21. $\tan(\tan^{-1}x)$.
22. $\cot(\tan^{-1}x)$.
23. $\sec(\tan^{-1}x)$.
24. $\csc(\tan^{-1}x)$.

In Exercises 25–30, take $0 < x < 1$ and calculate the following from an appropriate right triangle.

25. $\sin(\cos^{-1}x)$.
26. $\cos(\cot^{-1}x)$.
27. $\sec(\cot^{-1}x)$.
28. $\tan(\cos^{-1}x)$.
29. $\cot(\cos^{-1}x)$.
30. $\csc(\cot^{-1}x)$.

In Exercises 31–54, differentiate the given function.

31. $y = \tan^{-1}(x+1)$.
32. $y = \tan^{-1}\sqrt{x}$.
33. $f(x) = \sec^{-1}(2x^2)$.
34. $f(x) = e^x \sin^{-1}x$.

35. $f(x) = x \sin^{-1}2x$.

36. $f(x) = e^{\tan^{-1}x}$.

37. $u = (\sin^{-1}x)^2$.

38. $v = \tan^{-1}(e^x)$.

39. $y = \dfrac{\tan^{-1}x}{x}$.

40. $y = \sec^{-1}\sqrt{x^2+2}$.

41. $f(x) = \sqrt{\tan^{-1}2x}$.

42. $f(x) = \ln(\tan^{-1}x)$.

43. $y = \tan^{-1}(\ln x)$.

44. $g(x) = \sec^{-1}(\cos x + 2)$.

45. $\theta = \sin^{-1}(\sqrt{1-r^2})$.

46. $\theta = \sin^{-1}\left(\dfrac{r}{r+1}\right)$.

47. $g(x) = x^2\sec^{-1}\left(\dfrac{1}{x}\right)$.

48. $\theta = \tan^{-1}\left(\dfrac{1}{1+r^2}\right)$.

49. $y = \sin[\sec^{-1}(\ln x)]$.

50. $f(x) = e^{\sec^{-1}x}$.

51. $f(x) = \sqrt{c^2-x^2} + c\sin^{-1}\left(\dfrac{x}{c}\right),\ c > 0$.

52. $f(x) = \frac{1}{3}\sin^{-1}(3x-4x^2)$.

53. $y = \dfrac{x}{\sqrt{c^2-x^2}} - \sin^{-1}\left(\dfrac{x}{c}\right),\ c > 0$.

54. $y = x\sqrt{c^2-x^2} + c^2\sin^{-1}\left(\dfrac{x}{c}\right),\ c > 0$.

55. Show that for $a > 0$

$$\int \frac{dx}{\sqrt{a^2-(x+b)^2}} = \sin^{-1}\left(\frac{x+b}{a}\right) + C. \tag{7.8.13}$$

56. Show that for $a \neq 0$

$$\int \frac{dx}{a^2+(x+b)^2} = \frac{1}{a}\tan^{-1}\left(\frac{x+b}{a}\right) + C. \tag{7.8.14}$$

57. (a) Verify (7.8.12).
(b) Show that for $a > 0$

$$\int \frac{dx}{(x+b)\sqrt{(x+b)^2-a^2}} = \frac{1}{a}\sec^{-1}\frac{|x+b|}{a} + C$$

In Exercises 58–60, give the domain and range of each function, and verify the differentiation formula.

58. $f(x) = \cos^{-1}x$; $\quad \dfrac{d}{dx}(\cos^{-1}x) = \dfrac{-1}{\sqrt{1-x^2}}$.

59. $f(x) = \cot^{-1}x$; $\quad \dfrac{d}{dx}(\cot^{-1}x) = \dfrac{-1}{1+x^2}$.

60. $f(x) = \csc^{-1}x$; $\quad \dfrac{d}{dx}(\csc^{-1}x) = \dfrac{-1}{|x|\sqrt{x^2-1}}$.

In Exercises 61–74, evaluate the given integral.

61. $\displaystyle\int_0^1 \frac{dx}{1+x^2}$.

62. $\displaystyle\int_{-1}^1 \frac{dx}{1+x^2}$.

63. $\displaystyle\int_0^{1/\sqrt{2}} \frac{dx}{\sqrt{1-x^2}}$.

64. $\displaystyle\int_0^1 \frac{dx}{\sqrt{4-x^2}}$.

65. $\displaystyle\int_0^5 \frac{dx}{25+x^2}$.

66. $\displaystyle\int_1^2 \frac{dx}{x\sqrt{x^2-16}}$.

67. $\displaystyle\int_0^{3/2} \frac{dx}{9+4x^2}$.

68. $\displaystyle\int_2^5 \frac{dx}{9+(x-2)^2}$.

69. $\displaystyle\int_{3/4}^3 \frac{dx}{x\sqrt{16x^2-9}}$.

70. $\displaystyle\int_4^6 \frac{dx}{(x-3)\sqrt{x^2-6x+8}}$.

71. $\displaystyle\int_{-3}^{-2} \frac{dx}{\sqrt{4-(x+3)^2}}$.

72. $\displaystyle\int_{\ln 2}^{\ln 3} \frac{e^{-x}}{\sqrt{1-e^{-2x}}}\,dx$.

73. $\displaystyle\int_0^{\ln 2} \frac{e^x}{1+e^{2x}}\,dx$.

74. $\displaystyle\int_0^{1/2} \frac{1}{\sqrt{3-4x^2}}\,dx$.

In Exercises 75–84, calculate the indefinite integral.

75. $\displaystyle\int \frac{x}{\sqrt{1-x^4}}\,dx$.

76. $\displaystyle\int \frac{\sec^2x}{\sqrt{9-\tan^2x}}\,dx$.

77. $\displaystyle\int \frac{x}{1+x^4}\,dx$.

78. $\displaystyle\int \frac{dx}{\sqrt{4x-x^2}}$.

79. $\displaystyle\int \frac{\sec^2x}{9+\tan^2x}\,dx$.

80. $\displaystyle\int \frac{\cos x}{3+\sin^2x}\,dx$.

81. $\displaystyle\int \frac{\sin^{-1}x}{\sqrt{1-x^2}}\,dx$.

82. $\displaystyle\int \frac{\tan^{-1}x}{1+x^2}\,dx$.

83. $\displaystyle\int \frac{dx}{x\sqrt{1-(\ln x)^2}}$.

84. $\displaystyle\int \frac{dx}{x[1+(\ln x)^2]}$.

85. Find the area of the region bounded by graph of $y = 1/\sqrt{4-x^2}$ and the x-axis between $x = -1$ and $x = 1$.

86. Find the area of the region bounded by the graph of $y = 3/(9+x^2)$ and the x-axis between $x = -3$ and $x = 3$.

87. Sketch the region bounded by the graphs of $4y = x^2$ and $y = 8/(x^2+4)$ and find its area.

88. The region bounded by the graph of $y = 1/(x^2\sqrt{x^2-9})$ and the x-axis between $x = 2\sqrt{3}$ and $x = 6$ is revolved around the y-axis. Find the volume of the solid that is generated.

89. A billboard k feet wide is perpendicular to a straight road and is s feet from the road. At what point on the road would a motorist have the best view of the billboard; that is, at what point on the road is the angle θ subtended by the billboard a maximum (see the figure)?

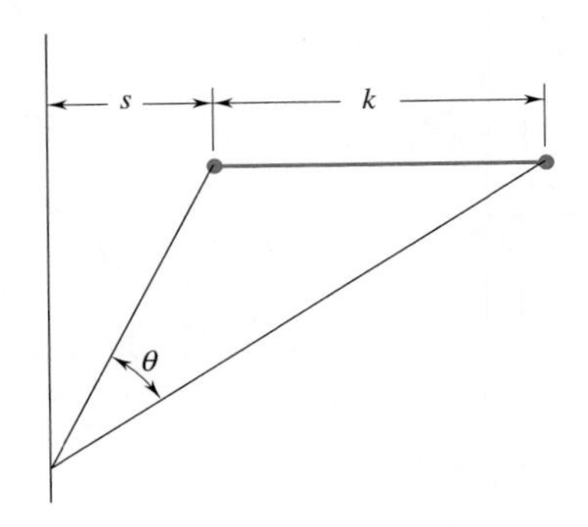

90. A person walking along a straight path at the rate of 6 feet per second is followed by a spotlight that is located 30 feet from the path. How fast is the spotlight turning at the instant the person is 50 feet past the point on the path that is closest to the spotlight?

91. Let

$$f(x) = \tan^{-1}\left(\frac{2+x}{1-2x}\right), \quad x \neq \tfrac{1}{2}.$$

(a) Use a graphing utility to graph f and $g(x) = \tan^{-1}x$. What do you notice about these two graphs?

(b) Estimate the limits

$$\lim_{x\to(1/2)^+} f(x) \quad \text{and} \quad \lim_{x\to(1/2)^-} f(x).$$

(c) Show that $f'(x) = \dfrac{1}{1+x^2}$, $x \neq \frac{1}{2}$.

(d) Show that there is no constant C such that $f(x) = \tan^{-1}x + C$ for all $x \neq \frac{1}{2}$.

(e) Find constants C_1 and C_2 such that

$$f(x) = \tan^{-1}x + C_1 \quad \text{for } x < 1/2$$

$$f(x) = \tan^{-1}x + C_2 \quad \text{for } x > 1/2.$$

92. Let

$$f(x) = \tan^{-1}\left(\frac{a+x}{1-ax}\right), \quad x \neq 1/a.$$

(a) Show that $f'(x) = \dfrac{1}{1+x^2}$, $x \neq 1/a$.

(b) Show that there is no constant C such that $f(x) = \tan^{-1}x + C$ for all $x \neq 1/a$.

(c) Find constants C_1 and C_2 such that

$$f(x) = \tan^{-1}x + C_1 \quad \text{for } x < 1/a$$

$$f(x) = \tan^{-1}x + C_2 \quad \text{for } x > 1/a.$$

93. Let $f(x) = \tan^{-1}(1/x)$, $x \neq 0$.

(a) Use a graphing utility to graph f and $g(x) = \tan^{-1}x$. What do you notice about these two graphs?

(b) Use your graph of f to estimate the limits

$$\lim_{x\to0^+} f(x) \quad \text{and} \quad \lim_{x\to0^-} f(x)$$

(c) Show that $f'(x) = \dfrac{-1}{1+x^2}$, $x \neq 0$.

(d) Show that there does not exist a constant C such that $f(x) + \tan^{-1}x = C$ for all x.

(e) Find constants C_1 and C_2 such that

$$f(x) + \tan^{-1}x = C_1 \quad \text{for } x > 0$$

$$f(x) + \tan^{-1}x = C_2 \quad \text{for } x < 0.$$

94. Evaluate

$$\lim_{x\to0} \frac{\sin^{-1}x}{x}$$

numerically. Justify your answer by other means.

95. Estimate the integral

$$\int_0^{0.5} \frac{1}{\sqrt{1-x^2}}\,dx$$

by using the partition $\{0, 0.1, 0.2, 0.3, 0.4, 0.5\}$ and the intermediate points

$$x_1^* = 0.05, \quad x_2^* = 0.15, \quad x_3^* = 0.25,$$

$$x_4^* = 0.35, \quad x_5^* = 0.45.$$

Note that the sine of your estimate is close to 0.5. Explain the reason for this.

■ 7.9 THE HYPERBOLIC SINE AND COSINE

Certain combinations of the exponential functions e^x and e^{-x} occur so frequently in mathematical applications that they are given special names. The *hyperbolic sine* (sinh) and *hyperbolic cosine* (cosh) are the functions defined by:

(7.9.1)
$$\sinh x = \tfrac{1}{2}(e^x - e^{-x}), \qquad \cosh x = \tfrac{1}{2}(e^x + e^{-x}).$$

The reasons for these names will become apparent as we go on.

Since

$$\frac{d}{dx}(\sinh x) = \frac{d}{dx}[\tfrac{1}{2}(e^x - e^{-x})] = \tfrac{1}{2}(e^x + e^{-x})$$

and

$$\frac{d}{dx}(\cosh x) = \frac{d}{dx}[\tfrac{1}{2}(e^x + e^{-x})] = \tfrac{1}{2}(e^x - e^{-x}),$$

we have

(7.9.2)

$$\frac{d}{dx}(\sinh x) = \cosh x, \qquad \frac{d}{dx}(\cosh x) = \sinh x.$$

In short, each of these functions is the derivative of the other.

The Graphs

We begin with the hyperbolic sine. Since

$$\sinh(-x) = \tfrac{1}{2}(e^{-x} - e^x) = -\tfrac{1}{2}(e^x - e^{-x}) = -\sinh x,$$

the hyperbolic sine is an odd function. The graph is therefore symmetric about the origin. Since

$$\frac{d}{dx}(\sinh x) = \cosh x = \tfrac{1}{2}(e^x + e^{-x}) > 0 \qquad \text{for all real } x,$$

the hyperbolic sine increases everywhere. Since

$$\frac{d^2}{dx^2}(\sinh x) = \frac{d}{dx}(\cosh x) = \sinh x = \tfrac{1}{2}(e^x - e^{-x}),$$

you can see that

$$\frac{d^2}{dx^2}(\sinh x) \quad \text{is} \quad \begin{cases} \text{negative}, & \text{for} \quad x < 0 \\ 0, & \text{at} \quad x = 0 \\ \text{positive}, & \text{for} \quad x > 0. \end{cases}$$

The graph is therefore concave down on $(-\infty, 0)$ and concave up on $(0, \infty)$. The point $(0, \sinh 0) = (0, 0)$ is the only point of inflection. The slope at the origin is $\cosh 0 = 1$. A sketch of the graph appears in Figure 7.9.1. ❑

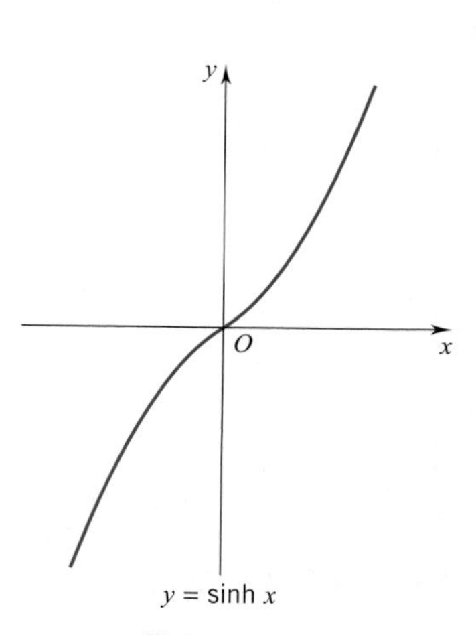

$y = \sinh x$

Figure 7.9.1

We turn now to the hyperbolic cosine. Since

$$\cosh(-x) = \tfrac{1}{2}(e^{-x} + e^x) = \tfrac{1}{2}(e^x + e^{-x}) = \cosh x,$$

the hyperbolic cosine is an even function. The graph is therefore symmetric about the y-axis. Since

$$\frac{d}{dx}(\cosh x) = \sinh x,$$

you can see that

$$\frac{d}{dx}(\cosh x) \quad \text{is} \quad \begin{cases} \text{negative,} & \text{for} \quad x < 0 \\ 0, & \text{at} \quad x = 0 \\ \text{positive,} & \text{for} \quad x > 0. \end{cases}$$

The function therefore decreases on $(-\infty, 0]$ and increases on $[0, \infty)$. The number

$$\cosh 0 = \tfrac{1}{2}(e^0 + e^{-0}) = \tfrac{1}{2}(1 + 1) = 1$$

is a local and absolute minimum. There are no other extreme values. Since

$$\frac{d^2}{dx^2}(\cosh x) = \frac{d}{dx}(\sinh x) = \cosh x > 0 \qquad \text{for all real } x,$$

the graph is everywhere concave up. (Figure 7.9.2) ❑

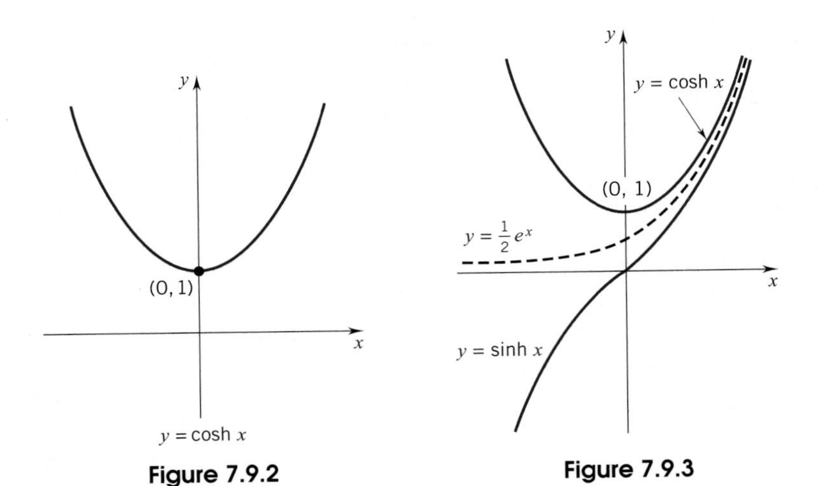

Figure 7.9.2

Figure 7.9.3

Figure 7.9.3 shows the graphs of three functions

$$y = \sinh x = \tfrac{1}{2}(e^x - e^{-x}), \qquad y = \tfrac{1}{2}e^x, \qquad y = \cosh x = \tfrac{1}{2}(e^x + e^{-x}).$$

For all real x

$$\sinh x < \tfrac{1}{2}e^x < \cosh x. \qquad (e^{-x} > 0)$$

Although markedly different for negative x, these functions are almost indistinguishable for large positive x. This follows from the fact that $e^{-x} \to 0$ as $x \to \infty$.

Applications

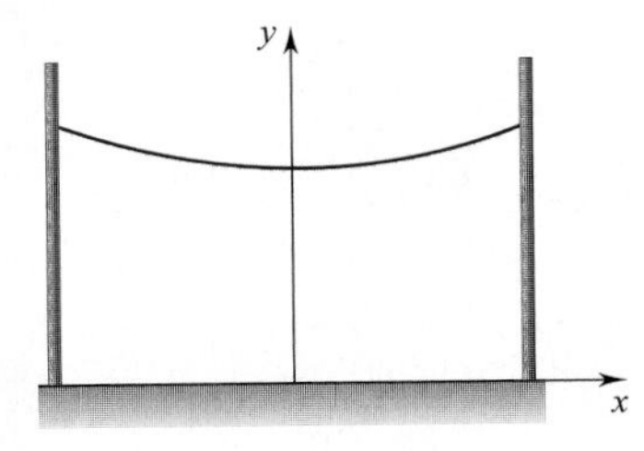

Figure 7.9.4

The hyperbolic functions have a variety of applications in science and engineering. Perhaps the best known application is the use of the hyperbolic cosine function to describe the shape of a flexible chain or cable that is suspended between two points. For example, think of a telephone wire or power line that sags under its own weight (Figure 7.9.4).

Suppose we have a flexible cable of uniform density that is suspended between two points of equal height. If we introduce an x-y coordinate system so that the lowest point of the cable is on the y-axis (see Figure 7.9.5), then it can be shown that the shape of the curve $y = f(x)$ must satisfy the differential equation

$$\frac{d^2y}{dx^2} = \frac{1}{a}\sqrt{1 + \left(\frac{dy}{dx}\right)^2},$$

where a is a constant that depends on the density of the cable and the tension or horizontal force at its two ends. In the Exercises you are asked to show that

$$y = a \cosh\left(\frac{x}{a}\right) + C$$

is a solution of this differential equation. The graph of this solution is called a *catenary*.

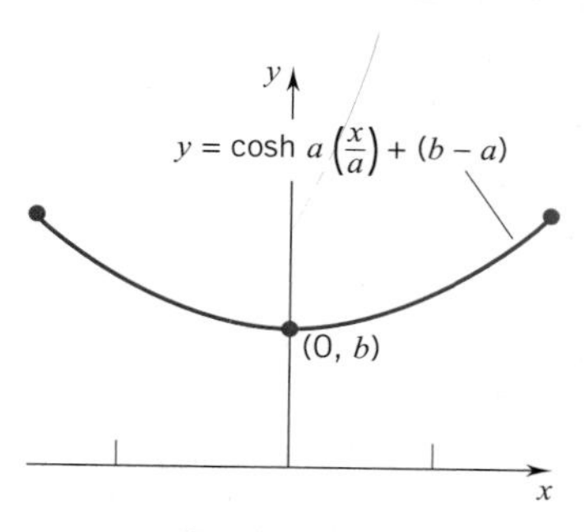

Figure 7.9.5

The Gateway to the West Arch in St. Louis, Missouri, is in the shape of an inverted catenary (see Figure 7.9.6). This arch is 630 feet high at its center, and it measures 630 across its base. The value of the constant a for this arch is approximately 127.7 and its equation is

$$y = -127.7 \cosh(x/127.7) + 757.7.$$

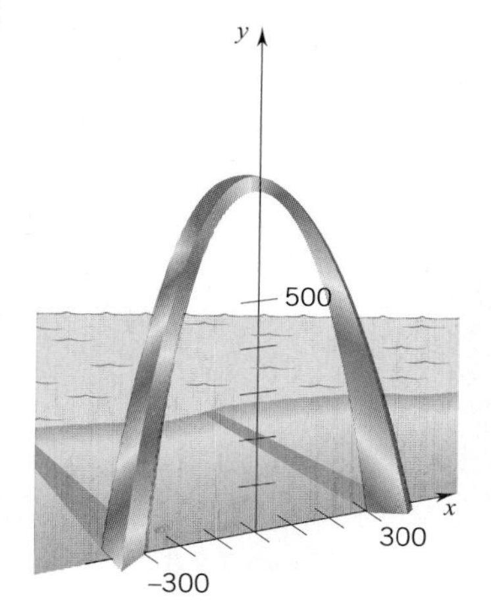

Figure 7.9.6

Identities

The hyperbolic sine and cosine functions satisfy identities similar to those satisfied by the "circular" sine and cosine.

(7.9.3)

$$\begin{aligned}
&\cosh^2 t - \sinh^2 t = 1,\\
&\sinh(t + s) = \sinh t \cosh s + \cosh t \sinh s,\\
&\cosh(t + s) = \cosh t \cosh s + \sinh t \sinh s,\\
&\sinh 2t = 2 \sinh t \cosh t,\\
&\cosh 2t = \cosh^2 t + \sinh^2 t.
\end{aligned}$$

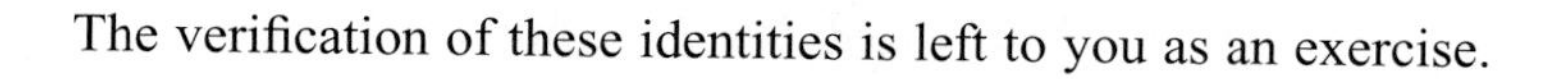

The verification of these identities is left to you as an exercise.

Relation to the Hyperbola $x^2 - y^2 = 1$

The hyperbolic sine and cosine are related to the hyperbola $x^2 - y^2 = 1$ much as the "circular" sine and cosine are related to the circle $x^2 + y^2 = 1$:

1. For each real t

$$\cos^2 t + \sin^2 t = 1,$$

and thus the point $(\cos t, \sin t)$ lies on the circle $x^2 + y^2 = 1$. For each real t

$$\cosh^2 t - \sinh^2 t = 1,$$

and thus the point $(\cosh t, \sinh t)$ lies on the hyperbola $x^2 - y^2 = 1$.

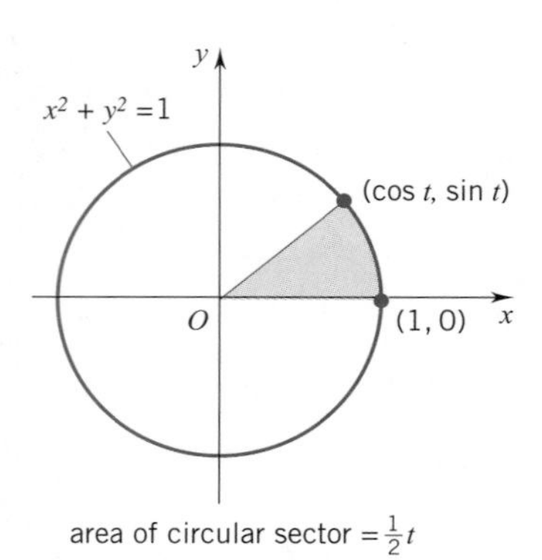

area of circular sector $=\frac{1}{2}t$

Figure 7.9.7

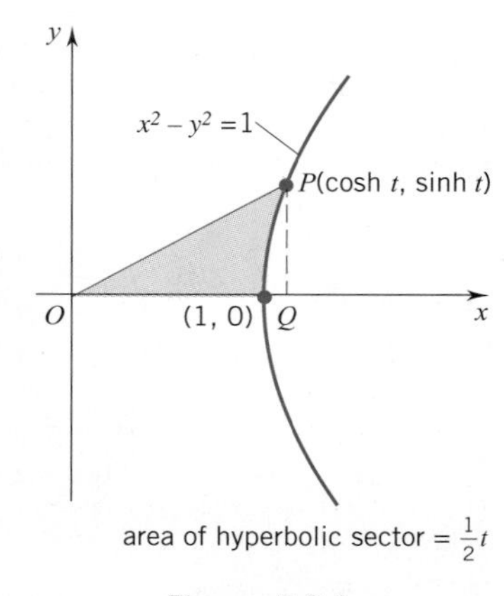

area of hyperbolic sector $=\frac{1}{2}t$

Figure 7.9.8

2. For each t in $[0, 2\pi]$ (see Figure 7.9.7), the number $\frac{1}{2}t$ gives the area of the circular sector generated by the circular arc that begins at $(1, 0)$ and ends at $(\cos t, \sin t)$. As we prove below, for each $t > 0$ (see Figure 7.9.8), the number $\frac{1}{2}t$ gives the area of the hyperbolic sector generated by the hyperbolic arc that begins at $(1, 0)$ and ends at $(\cosh t, \sinh t)$.

PROOF Let's call the area of the hyperbolic sector $A(t)$. It is not hard to see that

$$A(t) = \tfrac{1}{2}\cosh t \sinh t - \int_1^{\cosh t} \sqrt{x^2-1}\,dx.$$

The first term, $\frac{1}{2}\cosh t \sinh t$, gives the area of the triangle OPQ, and the integral

$$\int_1^{\cosh t} \sqrt{x^2-1}\,dx$$

gives the area of the unshaded portion of the triangle. We wish to show that

$$A(t) = \tfrac{1}{2}t \qquad \text{for all } t \geq 0.$$

We will do so by showing that

$$A'(t) = \tfrac{1}{2} \quad \text{for all } t > 0 \qquad \text{and} \qquad A(0) = 0.$$

Differentiating $A(t)$, we have

$$A'(t) = \tfrac{1}{2}\left[\cosh t\frac{d}{dt}(\sinh t) + \sinh t\frac{d}{dt}(\cosh t)\right] - \frac{d}{dt}\left(\int_1^{\cosh t}\sqrt{x^2-1}\,dx\right)$$

and therefore

$$(1) \qquad A'(t) = \tfrac{1}{2}(\cosh^2 t + \sinh^2 t) - \frac{d}{dt}\left(\int_1^{\cosh t}\sqrt{x^2-1}\,dx\right).$$

Now we differentiate the integral:

$$\frac{d}{dt}\left(\int_1^{\cosh t}\sqrt{x^2-1}\,dx\right) \underset{(5.8.7)}{=} \sqrt{\cosh^2 t - 1}\,\frac{d}{dt}(\cosh t) = \sinh t \cdot \sinh t = \sinh^2 t.$$

Substituting this last expression into Equation (1), we have

$$A'(t) = \tfrac{1}{2}(\cosh^2 t + \sinh^2 t) - \sinh^2 t = \tfrac{1}{2}(\cosh^2 t - \sinh^2 t) = \tfrac{1}{2}.$$

It is not hard to see that $A(0) = 0$:

$$A(0) = \tfrac{1}{2}\cosh 0 \sinh 0 - \int_1^{\cosh 0}\sqrt{x^2-1}\,dx = \tfrac{1}{2}(1)(0) - \int_1^1 \sqrt{x^2-1}\,dx = 0. \quad ❑$$

EXERCISES 7.9

In Exercises 1–18, differentiate the given function.

1. $y = \sinh x^2$.

2. $y = \cosh(x+a)$.

3. $y = \sqrt{\cosh ax}$.

4. $y = (\sinh ax)(\cosh ax)$.

5. $y = \dfrac{\sinh x}{\cosh x - 1}$.

6. $y = \dfrac{\sinh x}{x}$.

7. $y = a \sinh bx - b \cosh ax$.

8. $y = e^x(\cosh x + \sinh x)$.

9. $y = \ln|\sinh ax|$.

10. $y = \ln|1 - \cosh ax|$.

11. $y = \sinh(e^{2x})$.

12. $y = \cosh(\ln x^3)$.

13. $y = e^{-x}\cosh 2x$.

14. $y = \tan^{-1}(\sinh x)$.

15. $y = \ln(\cosh x)$.

16. $y = \ln(\sinh x)$.

17. $y = (\sinh x)^x$.

18. $y = x^{\cosh x}$.

In Exercises 19–25, verify the given identity

19. $\cosh^2 t - \sinh^2 t = 1$.

20. $\sinh(t + s) = \sinh t \cosh s + \cosh t \sinh s$.

21. $\cosh(t + s) = \cosh t \cosh s + \sinh t \sinh s$.

22. $\sinh 2t = 2 \sinh t \cosh t$.

23. $\cosh 2t = \cosh^2 t + \sinh^2 t = 2\cosh^2 t - 1$
$= 2\sinh^2 t + 1$.

24. $\cosh(-t) = \cosh t$; the hyperbolic cosine function is even.

25. $\sinh(-t) = -\sinh t$; the hyperbolic sine function is odd.

In Exercises 26–28, find the absolute extreme values.

26. $y = 5\cosh x + 4\sinh x$.

27. $y = -5\cosh x + 4\sinh x$.

28. $y = 4\cosh x + 5\sinh x$.

29. Show that for each positive integer n

$$(\cosh x + \sinh x)^n = \cosh nx + \sinh nx.$$

30. Verify that $y = A\cosh cx + B\sinh cx$ satisfies the differential equation $y'' - c^2 y = 0$.

31. Determine A, B, and c so that $y = A\cosh cx + B\sinh cx$ satisfies the conditions $y'' - 9y = 0$, $y(0) = 2$, $y'(0) = 1$. Take $c > 0$.

32. Determine A, B, and c so that $y = A\cosh cx + B\sinh cx$ satisfies the conditions $4y'' - y = 0$, $y(0) = 1$, $y'(0) = 2$. Take $c > 0$.

In Exercises 33–44, calculate the given indefinite integral.

33. $\int \cosh ax\, dx$.

34. $\int \sinh ax\, dx$.

35. $\int \sinh^2 ax \cosh ax\, dx$.

36. $\int \sinh ax \cosh^2 ax\, dx$.

37. $\int \frac{\sinh ax}{\cosh ax}\, dx$.

38. $\int \frac{\cosh ax}{\sinh ax}\, dx$.

39. $\int \frac{\sinh ax}{\cosh^2 ax}\, dx$.

40. $\int \sinh^2 x\, dx$.

41. $\int \cosh^2 x\, dx$.

42. $\int \sinh 2x\, e^{\cosh 2x}\, dx$.

43. $\int \frac{\sinh \sqrt{x}}{\sqrt{x}}\, dx$.

44. $\int \frac{\sinh x}{1 + \cosh x}\, dx$.

In Exercises 45 and 46, find the average value of the function on the indicated interval.

45. $f(x) = \cosh x$, $x \in [-1, 1]$.

46. $f(x) = \sinh 2x$, $x \in [0, 4]$.

47. Find the volume of the solid generated by revolving the region bounded by $y = \cosh x$ and $y = \sinh x$ between $x = 0$ and $x = 1$ about the x-axis.

48. (a) Evaluate the limit

$$\lim_{x\to\infty} \frac{\sinh x}{e^x}.$$

(b) Evaluate the limit

$$\lim_{x\to\infty} \frac{\cosh x}{e^{ax}}$$

when $0 < a < 1$ and when $a > 1$.

*■ 7.10 THE OTHER HYPERBOLIC FUNCTIONS

The hyperbolic tangent is defined by setting

$$\tanh x = \frac{\sinh x}{\cosh x} = \frac{e^x - e^{-x}}{e^x + e^{-x}}.$$

There is also a *hyperbolic cotangent*, a *hyperbolic secant*, and a *hyperbolic cosecant*:

$$\coth x = \frac{\cosh x}{\sinh x}, \qquad \operatorname{sech} x = \frac{1}{\cosh x}, \qquad \operatorname{csch} x = \frac{1}{\sinh x}.$$

The derivatives are as follows:

(7.10.1)

$$\frac{d}{dx}(\tanh x) = \operatorname{sech}^2 x, \qquad \frac{d}{dx}(\coth x) = -\operatorname{csch}^2 x,$$

$$\frac{d}{dx}(\operatorname{sech} x) = -\operatorname{sech} x \tanh x, \qquad \frac{d}{dx}(\operatorname{csch} x) = -\operatorname{csch} x \coth x.$$

These formulas are easy to verify. For instance,

$$\frac{d}{dx}(\tanh x) = \frac{d}{dx}\left(\frac{\sinh x}{\cosh x}\right) = \frac{\cosh x \frac{d}{dx}(\sinh x) - \sinh x \frac{d}{dx}(\cosh x)}{\cosh^2 x}$$

$$= \frac{\cosh^2 x - \sinh^2 x}{\cosh^2 x} = \frac{1}{\cosh^2 x} = \operatorname{sech}^2 x.$$

We leave it to you to verify the other formulas. ❑

Let's examine the hyperbolic tangent a little further. Since

$$\tanh(-x) = \frac{\sinh(-x)}{\cosh(-x)} = \frac{-\sinh x}{\cosh x} = -\tanh x,$$

the hyperbolic tangent is an odd function and thus the graph is symmetric about the origin. Since

$$\frac{d}{dx}(\tanh x) = \operatorname{sech}^2 x > 0 \qquad \text{for all real } x,$$

the function is everywhere increasing. From the relation

$$\tanh x = \frac{e^x - e^{-x}}{e^x + e^{-x}} = \frac{e^x - e^{-x}}{e^x + e^{-x}} \cdot \frac{e^x}{e^x} = \frac{e^{2x} - 1}{e^{2x} + 1} = \frac{e^{2x} + 1 - 2}{e^{2x} + 1} = 1 - \frac{2}{e^{2x} + 1},$$

you can see that $\tanh x$ always remains between -1 and 1. Moreover,

$$\text{as } x \to \infty, \quad \tanh x \to 1 \qquad \text{and} \qquad \text{as } x \to -\infty, \quad \tanh x \to -1.$$

The lines $y = 1$ and $y = -1$ are horizontal asymptotes. To check on the concavity of the graph, we take the second derivative:

$$\begin{aligned} \frac{d^2}{dx^2}(\tanh x) = \frac{d}{dx}(\operatorname{sech}^2 x) &= 2 \operatorname{sech} x \frac{d}{dx}(\operatorname{sech} x) \\ &= 2 \operatorname{sech} x\,(-\operatorname{sech} x \tanh x) \\ &= -2 \operatorname{sech}^2 x \tanh x. \end{aligned}$$

Since

$$\tanh x = \frac{e^x - e^{-x}}{e^x + e^{-x}} \quad \text{is} \quad \begin{cases} \text{negative,} & \text{for } x < 0 \\ 0, & \text{at } x = 0 \\ \text{positive,} & \text{for } x > 0, \end{cases}$$

you can see that

$$\frac{d^2}{dx^2}(\tanh x) \quad \text{is} \quad \begin{cases} \text{positive,} & \text{for } x < 0 \\ 0, & \text{at } x = 0 \\ \text{negative,} & \text{for } x > 0. \end{cases}$$

The graph is therefore concave up on $(-\infty, 0)$ and concave down on $(0, \infty)$. The point $(0, \tanh 0) = (0, 0)$ is a point of inflection. At the origin the slope is

$$\operatorname{sech}^2 0 = \frac{1}{\cosh^2 0} = 1.$$

The graph is shown in Figure 7.10.1.

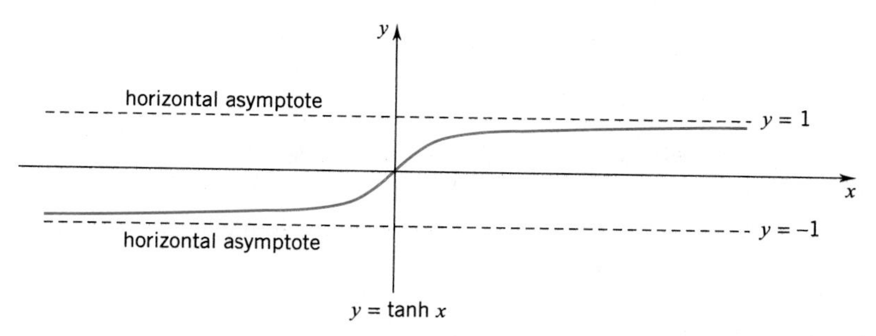

Figure 7.10.1

The Hyperbolic Inverses

Of the six hyperbolic functions, only the hyperbolic cosine and hyperbolic secant are not one-to-one (refer to the graphs of $y = \sinh x$, $y = \cosh x$, and $y = \tanh x$). Thus, the hyperbolic sine, hyperbolic tangent, hyperbolic cosecant, and hyperbolic cotangent functions all have inverses. If we restrict the domains of the hyperbolic cosine and hyperbolic secant functions to $x \geq 0$, then these functions will also have inverses. The hyperbolic inverses that are important to us are the *inverse hyperbolic sine*, the *inverse hyperbolic cosine*, and the *inverse hyperbolic tangent*. These functions

$$y = \sinh^{-1}x, \qquad y = \cosh^{-1}x, \qquad y = \tanh^{-1}x$$

are the inverses of

$$y = \sinh x, \quad y = \cosh x \quad (x \geq 0), \qquad y = \tanh x,$$

respectively.

THEOREM 7.10.2

(i) $\sinh^{-1}x = \ln\,(x + \sqrt{x^2 + 1}), \qquad x \text{ real}$

(ii) $\cosh^{-1}x = \ln\,(x + \sqrt{x^2 - 1}), \qquad x \geq 1$

(iii) $\tanh^{-1}x = \dfrac{1}{2}\ln\left(\dfrac{1 + x}{1 - x}\right), \qquad -1 < x < 1.$

PROOF To prove (i), we set $y = \sinh^{-1}x$ and note that

$$\sinh y = x.$$

This gives in sequence

$$\tfrac{1}{2}(e^y - e^{-y}) = x, \qquad e^y - e^{-y} = 2x, \qquad e^y - 2x - e^{-y} = 0, \qquad e^{2y} - 2xe^y - 1 = 0.$$

This last equation is a quadratic equation in e^y. From the quadratic formula we find that

$$e^y = \tfrac{1}{2}(2x \pm \sqrt{4x^2 + 4}) = x \pm \sqrt{x^2 + 1}.$$

Since $e^y > 0$, the minus sign on the right is impossible. Consequently, we have

$$e^y = x + \sqrt{x^2 + 1}$$

and, taking the natural log of both sides,

$$y = \ln\,(x + \sqrt{x^2 + 1}).$$

To prove (ii), we set

$$y = \cosh^{-1}x, \qquad x \geq 1$$

and note that

$$\cosh y = x \quad \text{and} \quad y \geq 0.$$

This gives in sequence

$$\tfrac{1}{2}(e^y + e^{-y}) = x, \qquad e^y + e^{-y} = 2x, \qquad e^{2y} - 2xe^y + 1 = 0.$$

Again we have a quadratic in e^y. Here the general quadratic formula gives

$$e^y = \tfrac{1}{2}(2x \pm \sqrt{4x^2 - 4}) = x \pm \sqrt{x^2 - 1}.$$

Since y is nonnegative,

$$e^y = x \pm \sqrt{x^2 - 1}$$

cannot be less than 1. This renders the negative sign impossible (check this out) and leaves

$$e^y = x + \sqrt{x^2 - 1}$$

as the only possibility. Taking the natural log of both sides, we get

$$y = \ln (x + \sqrt{x^2 - 1}).$$

The proof of (iii) is left as an exercise. ❑

EXERCISES *7.10

In Exercises 1–10, differentiate the given function.

1. $y = \tanh^2 x.$
2. $y = \tanh^2 3x.$
3. $y = \ln (\tanh x).$
4. $y = \tanh (\ln x).$
5. $y = \sinh (\tan^{-1} e^{2x}).$
6. $y = \operatorname{sech} (3x^2 + 1).$
7. $y = \coth (\sqrt{x^2 + 1}).$
8. $y = \ln (\operatorname{sech} x).$
9. $y = \dfrac{\operatorname{sech} x}{1 + \cosh x}.$
10. $y = \dfrac{\cosh x}{1 + \operatorname{sech} x}.$

In Exercises 11–13, verify the given differentiation formula.

11. $\dfrac{d}{dx}(\coth x) = -\operatorname{csch}^2 x.$
12. $\dfrac{d}{dx}(\operatorname{sech} x) = -\operatorname{sech} x \tanh x.$
13. $\dfrac{d}{dx}(\operatorname{csch} x) = -\operatorname{csch} x \coth x.$
14. Show that
$$\tanh (t + s) = \frac{\tanh t + \tanh s}{1 + \tanh t \tanh s}.$$
15. Given that $\tanh x_0 = \frac{4}{5}$, find (a) $\operatorname{sech} x_0$.
HINT: $1 - \tanh^2 x = \operatorname{sech}^2 x$. Then find (b) $\cosh x_0$, (c) $\sinh x_0$, (d) $\coth x_0$, (e) $\operatorname{csch} x_0$.
16. Given that $\tanh t_0 = -\frac{5}{12}$, evaluate the remaining hyperbolic functions at t_0.
17. Show that, if $x^2 \geq 1$, then $x - \sqrt{x^2 - 1} \leq 1$.
18. Show that
$$\tanh^{-1} x = \frac{1}{2} \ln \left(\frac{1 + x}{1 - x}\right), \qquad -1 < x < 1.$$
19. Show that
$$\textbf{(7.10.3)} \qquad \frac{d}{dx}(\sinh^{-1} x) = \frac{1}{\sqrt{x^2 + 1}}, \qquad x \text{ real.}$$
20. Show that
$$\textbf{(7.10.4)} \qquad \frac{d}{dx}(\cosh^{-1} x) = \frac{1}{\sqrt{x^2 - 1}}, \qquad x > 1.$$
21. Show that
$$\textbf{(7.10.5)} \qquad \frac{d}{dx}(\tanh^{-1} x) = \frac{1}{1 - x^2}, \qquad -1 < x < 1.$$
22. Show that
$$\frac{d}{dx}(\operatorname{sech}^{-1} x) = \frac{-1}{x\sqrt{1 - x^2}}, \qquad 0 < x < 1.$$

23. Show that

$$\frac{d}{dx}(\text{csch}^{-1}x) = \frac{-1}{|x|\sqrt{1+x^2}}, \quad x \neq 0.$$

24. Show that

$$\frac{d}{dx}(\coth^{-1}x) = \frac{1}{1-x^2}, \quad |x| > 1.$$

25. Sketch the graph of $y = \text{sech } x$, finding: (a) the extreme values; (b) the points of inflection; and (c) the concavity.

26. Sketch the graphs of (a) $y = \coth x$, (b) $y = \text{csch } x$.

27. Graph $y = \sinh x$ and $y = \sinh^{-1}x$ in the same coordinate system. Find all points of inflection.

28. Sketch the graphs of (a) $y = \cosh^{-1}x$. (b) $y = \tanh^{-1}x$.

29. Given that $\tan \phi = \sinh x$, show that

(a) $\dfrac{d\phi}{dx} = \text{sech } x$.

(b) $x = \ln(\sec \phi + \tan \phi)$.

(c) $\dfrac{dx}{d\phi} = \sec \phi$.

30. The region bounded by the graph of $y = \text{sech } x$ between $x = -1$ and $x = 1$ is revolved around the x-axis. Find the volume of the solid that is generated.

In Exercises 31–40, calculate the given integral.

31. $\displaystyle\int \tanh x \, dx.$

32. $\displaystyle\int \coth x \, dx.$

33. $\displaystyle\int \text{sech } x \, dx.$

34. $\displaystyle\int \text{csch } x \, dx.$

35. $\displaystyle\int \text{sech}^3 x \tanh x \, dx.$

36. $\displaystyle\int x \, \text{sech}^2 x^2 \, dx.$

37. $\displaystyle\int \tanh x \ln(\cosh x) \, dx.$

38. $\displaystyle\int \frac{1+\tanh x}{\cosh^2 x} dx.$

39. $\displaystyle\int \frac{\text{sech}^2 x}{1+\tanh x} dx.$

40. $\displaystyle\int \tanh^5 x \, \text{sech}^2 x \, dx.$

In Exercises 41–43, verify the given integration formula. In each case, assume that $a > 0$.

41. $\displaystyle\int \frac{1}{\sqrt{a^2+x^2}} dx = \sinh^{-1}\left(\frac{x}{a}\right) + C.$

42. $\displaystyle\int \frac{1}{\sqrt{x^2-a^2}} dx = \cosh^{-1}\left(\frac{x}{a}\right) + C.$

43. $$\int \frac{1}{a^2-x^2} dx = \begin{cases} \dfrac{1}{a}\tanh^{-1}\left(\dfrac{x}{a}\right) + C & \text{if } |x| < a. \\ \dfrac{1}{a}\coth^{-1}\left(\dfrac{x}{a}\right) + C & \text{if } |x| > a. \end{cases}$$

44. If a body of mass m falling from rest under the action of gravity encounters air resistance that is proportional to the square of its velocity, then the velocity $v(t)$ of the body at time t satisfies the differential equation

$$m\frac{dv}{dt} = mg - kv^2$$

where $k > 0$ is the constant of proportionality and g is the gravitational constant.

(a) Show that

$$v(t) = \sqrt{\frac{mg}{k}} \tanh\left(\sqrt{\frac{gk}{m}}\,t\right)$$

is a solution of the differential equation which satisfies $v(0) = 0$.

(b) Find

$$\lim_{t\to\infty} v(t).$$

This limit is called the *terminal velocity* of the body.

■ CHAPTER HIGHLIGHTS

7.1 Inverse Functions

one-to-one function; inverse function (p. 396)
one-to-one and increasing/decreasing functions; derivatives (p. 399)
relation between graph of f and the graph of f^{-1} (p. 401)
continuity and differentiability of inverse functions (p. 401)
derivative of an inverse (p. 402)

7.2 The Logarithm Function, Part I

definition of a logarithm function (p. 407)

natural logarithm: $\ln x = \displaystyle\int_1^x \frac{dt}{t}, \quad x > 0;$ domain $(0, \infty)$, range $(-\infty, \infty)$

$$\ln 1 = 0, \quad \ln e = 1$$

for $x, y > 0$ and r rational

$$\ln xy = \ln x + \ln y, \quad \ln \frac{1}{x} = -\ln x, \quad \ln \frac{x}{y} = \ln x - \ln y, \quad \ln x^r = r \ln x.$$

graph of $y = \ln x$ (p. 413)

7.3 The Logarithm Function, Part II

$$\frac{d}{dx}(\ln |u|) = \frac{1}{u}\frac{du}{dx} \qquad \int \frac{g'(x)}{g(x)}\,dx = \ln |g(x)| + C$$

logarithmic differentiation (p. 421)

7.4 The Exponential Function

The exponential function $y = e^x$ is the inverse of the logarithm function $y = \ln x$.

graph of $y = e^x$ (p. 425); domain $(-\infty, \infty)$, range $(0, \infty)$

$$\ln (e^x) = x, \qquad e^{\ln x} = x, \qquad e^0 = 1, \qquad e^{-x} = \frac{1}{e^x}, \qquad e^{x+y} = e^x e^y, \qquad e^{x-y} = \frac{e^x}{e^y}$$

$$\frac{d}{dx}(e^u) = e^u \frac{du}{dx} \qquad \int e^{g(x)} g'(x)\,dx = e^{g(x)} + C$$

7.5 Arbitrary Powers; Other Bases; Estimating *e*

$x^r = e^{r \ln x}$ for all $x > 0$, all real r

$$\frac{d}{dx}(p^u) = p^u \ln p \frac{du}{dx} \quad (p \text{ a positive constant}) \qquad \log_p x = \frac{\ln x}{\ln p}$$

$$\frac{d}{dx}(\log_p u) = \frac{1}{u \ln p}\frac{du}{dx}$$

$$\left(1 + \frac{1}{n}\right)^n \le e \le \left(1 + \frac{1}{n}\right)^{n+1} \qquad e \cong 2.71828$$

7.6 Exponential Growth and Decay

Linear functions of time change by the same *amount* during all periods of the same duration; exponential functions of time change by the same *factor* during all periods of the same duration.

All the functions that satisfy the equation $f'(t) = kf(t)$ are of the form $f(t) = Ce^{kt}$.

population growth (p. 444)
radioactive decay, half-life (p. 446)
compound interest, continuous compounding (p. 448)
Newton's Law of Cooling (p. 450)

7.7 More on the Integration of the Trigonometric Functions

$$\int \tan x\,dx = \ln |\sec x| + C, \qquad \int \sec x\,dx = \ln |\sec x + \tan x| + C,$$

$$\int \cot x\,dx = \ln |\sin x| + C, \qquad \int \csc x\,dx = \ln |\csc x - \cot x| + C.$$

7.8 The Inverse Trigonometric Functions

The inverse sine, $y = \sin^{-1} x$, is the inverse of $y = \sin x$, $x \in [-\frac{1}{2}\pi, \frac{1}{2}\pi]$.
The inverse tangent, $y = \tan^{-1} x$, is the inverse of $y = \tan x$, $x \in (-\frac{1}{2}\pi, \frac{1}{2}\pi)$.

The inverse secant, $y = \sec^{-1}x$, is the inverse of $y = \sec x$, $x \in [0, \frac{1}{2}\pi) \cup (\frac{1}{2}\pi, \pi]$.

graph of $y = \sin^{-1} x$ (p. 462)   graph of $y = \tan^{-1}x$ (p. 465)
graph of $y = \sec^{-1}x$ (p. 468)

$$\frac{d}{dx}(\sin^{-1}x) = \frac{1}{\sqrt{1-x^2}} \qquad \int \frac{dx}{\sqrt{a^2-x^2}} = \sin^{-1}\left(\frac{x}{a}\right) + C \quad (a>0)$$

$$\frac{d}{dx}(\tan^{-1}x) = \frac{1}{1+x^2} \qquad \int \frac{dx}{a^2+x^2} = \frac{1}{a}\tan^{-1}\left(\frac{x}{a}\right) + C \quad (a \neq 0)$$

$$\frac{d}{dx}(\sec^{-1}x) = \frac{1}{|x|\sqrt{x^2-1}} \qquad \int \frac{dx}{x\sqrt{x^2-a^2}} = \frac{1}{a}\sec^{-1}\left(\frac{|x|}{a}\right) + C \quad (a>0)$$

definition of the remaining inverse trigonometric functions (p. 471)

7.9 The Hyperbolic Sine and Cosine

$\sinh x = \frac{1}{2}(e^x - e^{-x})$, $\cosh x = \frac{1}{2}(e^x + e^{-x})$,

$$\frac{d}{dx}(\sinh x) = \cosh x, \qquad \frac{d}{dx}(\cosh x) = \sinh x.$$

graphs (pp. 475–476)   basic identities (p. 477)

*7.10 The Other Hyperbolic Functions

$$\tanh x = \frac{\sinh x}{\cosh x}, \qquad \coth x = \frac{\cosh x}{\sinh x},$$

$$\operatorname{sech} x = \frac{1}{\cosh x}, \qquad \operatorname{csch} x = \frac{1}{\sinh x}.$$

derivatives (p. 479)   hyperbolic inverses (p. 481)
derivatives of hyperbolic inverses (p. 482)

■ PROJECTS AND EXPLORATIONS USING TECHNOLOGY

To do these exercises you will need a graphics calculator or a computer with graphing capability. The majority of these problems are open-ended so different approaches may be used to solve them. You should be aware that different approaches can result in slight variations in the answers. Round your numerical answers to at least four decimal places. The rounding method that your calculator or computer uses also may cause variations in answers.

7.1 The functions $f(x) = a \ln x$, where a is a constant, have a number of applications, one of which will be considered in a later exercise.
(a) Find the values of a for which the graph of f is tangent to the line $y = x$.
(b) For each real number a, how many solutions will there be of $f(x) = x$? What is the value of f' at each solution of $f(x) = x$?
(c) How many solutions are there to $f[f(x)] = x$? What is the value of f' at each of these solutions?
(d) Represent f as a logarithm function in another base.

7.2 Let $A(t)$ denote the area of the rectangle of width $2t$ that has its lower vertices on the x-axis and its upper vertices on the graph of

$$f(x) = e^{(1-x^4)/(2+x^2)}.$$

See the figure.

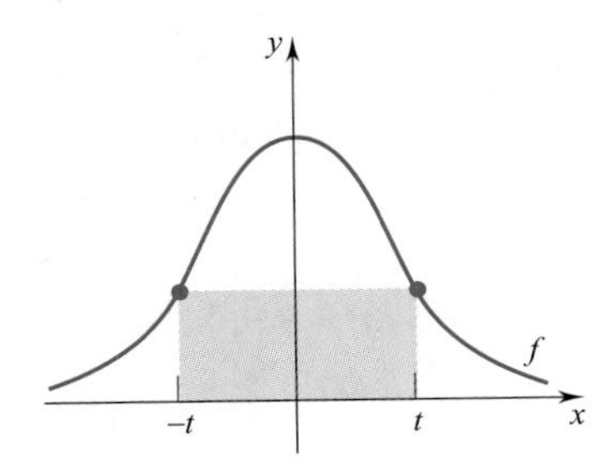

(a) Express A as a function of t and sketch its graph.
(b) Estimate the size of the rectangle of maximal area. Describe the properties of f at the value(s) of t where $A(t)$ is maximal.
(c) Discuss the concavity of the graph of A and find its points of inflection.
(d) Relate the values at which the graphs of A or f have points of inflection. For example, show that if f has a point of inflection at x_0, then the graph of A is concave down at x_0.

7.3 When we approximate the derivative of a function f on a calculator by the difference quotient

$$D(x, h) = \frac{f(x + h) - f(x)}{h},$$

we know that if h is "too small," then the calculator rounds off digits and the difference quotient fails to give accurate estimates. We will look at an unusual version of the difference quotient that will allow us to analyze different types of errors more effectively. Let

$$D_x(h) = \frac{f(x + e^{-h}) - f(x)}{e^{-h}}.$$

(a) Let f be a given function and fix a value of x. Explain why $y = f'(x)$ is a horizontal asymptote of the graph of D_x (as a function of h).
(b) Use your graphing technology to verify this for $f(x) = \log_5(x)$ at $x = 5$, $x = 7$, and $x = 100$.
(c) Fix x and let $E(h) = D_x(h) - f'(x)$; $E(h)$ is the error for this value of x at h. Is E strictly decreasing to 0 for all functions f?
(d) Find the smallest value of n such that

$$\lim_{h \to \infty} E(h)e^{nh}$$

is nonzero. Be careful about graphic and numeric solutions because there may be roundoff errors when using computer arithmetic. The importance of n is that it is a measure of how good $D(h)$ approximates $f'(x)$. That is, the bigger n is, the better the approximation.

7.4 The functions sinh x, cosh x, tanh x are used in the study of non-Euclidean geometry.
(a) Use the tangent line approximation:

$$f(x) \cong f(0) + f'(0)x$$

to show that

$$\sinh x \cong x, \qquad \cosh x \cong 1, \qquad \text{and} \qquad \tanh x \cong x,$$

for x small enough. From the concavity of these functions, determine whether these approximations are too small or too large.

Triangles are labeled by capital letters at the vertices and corresponding lower case letters for the sides opposite the angles (see the figure).

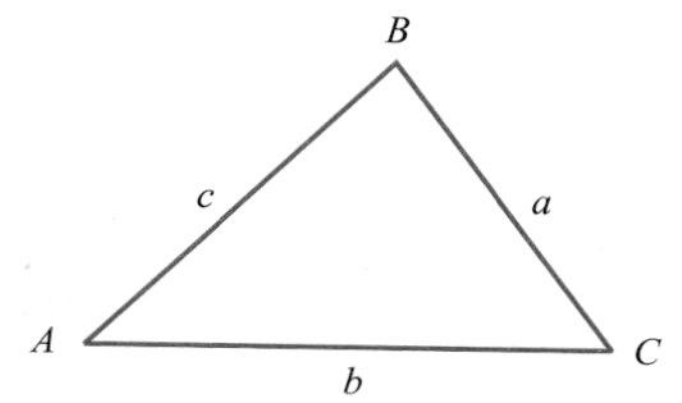

(b) Use the results in part (a) to show that the hyperbolic law of sines is approximately equal to the Euclidean law of sines for small side lengths. That is, show that

$$\frac{\sin A}{\sinh a} = \frac{\sin B}{\sinh b} = \frac{\sin C}{\sinh c} \quad \text{is approximately} \quad \frac{\sin A}{a} = \frac{\sin B}{b} = \frac{\sin C}{c}.$$

(c) A better approximation for $\cosh x$ is

$$\cosh x \cong 1 + \tfrac{1}{2}x^2,$$

for x small enough. Use this approximation to show that the hyperbolic law of cosines:

$$\cosh c = \cosh a \cosh b - \sinh a \sinh b \cos C$$

reduces to the Euclidean law of cosines for triangles with small sides.

(d) In Euclidean geometry if we fix the length of sides a and b, with $b > a$, and vary the angle C between them, then the angle A increases from 0 to a maximum value and then decreases back to 0. A way to find the maximum value is: Use the law of cosines to show that

(1)
$$\cos A = \frac{b - a \cos C}{\sqrt{a^2 + b^2 - 2ab \cos C}}$$

and then differentiate implicitly with respect to C to find dA/dC.

(e) In hyperbolic geometry, the equation corresponding to (1) is

$$\cos A = \frac{\sinh b \cosh a - \sinh a \cos C}{\sqrt{[\cosh a \cosh b - \sinh a \sinh b \cos C]^2 - 1}}.$$

Find the value of C for which A is maximal.

TECHNIQUES OF INTEGRATION

CHAPTER

8

8.1 REVIEW

We begin by listing the more important integrals with which you are already familiar. An extended table of integrals appears on the inside covers of the book.

1. $\int k\,dx = kx + C, \quad k$ constant.

2. $\int x^r\,dx = \dfrac{1}{r+1}x^{r+1} + C, \quad r \neq -1.$

3. $\int \dfrac{dx}{x} = \ln|x| + C.$

4. $\int e^x\,dx = e^x + C.$

5. $\int p^x\,dx = \dfrac{p^x}{\ln p} + C, \quad p > 0$ constant, $p \neq 1$.

6. $\int \sin x\,dx = -\cos x + C.$

7. $\int \cos x\,dx = \sin x + C.$

8. $\int \tan x\,dx = \ln|\sec x| + C.$

9. $\int \cot x\,dx = \ln|\sin x| + C.$

10. $\int \sec x\,dx = \ln|\sec x + \tan x| + C.$

11. $\int \csc x\,dx = \ln|\csc x - \cot x| + C.$

12. $\int \sec x \tan x\,dx = \sec x + C.$

13. $\int \csc x \cot x\,dx = -\csc x + C.$

14. $\int \sec^2 x\,dx = \tan x + C.$

15. $\int \csc^2 x\,dx = -\cot x + C.$

16. $\int \dfrac{dx}{\sqrt{a^2 - x^2}} = \sin^{-1}\left(\dfrac{x}{a}\right) + C, \quad a \neq 0$ constant.

17. $\displaystyle\int \frac{dx}{a^2 + x^2} = \frac{1}{a}\tan^{-1}\left(\frac{x}{a}\right) + C, \quad a \neq 0 \text{ constant.}$

18. $\displaystyle\int \frac{dx}{x\sqrt{x^2 - a^2}} = \frac{1}{a}\sec^{-1}\left|\frac{x}{a}\right| + C, \quad a \neq 0 \text{ constant.}$

19. $\displaystyle\int \sinh x\, dx = \cosh x + C.$ **20.** $\displaystyle\int \cosh x\, dx = \sinh x + C.$

For review we work out a few integrals involving u-substitutions.

Example 1 Find

$$\int x \tan x^2\, dx.$$

SOLUTION Set

$$u = x^2, \qquad du = 2x\, dx.$$

Then

$$\int x \tan x^2\, dx = \tfrac{1}{2}\int \tan u\, du = \tfrac{1}{2}\ln|\sec u| + C = \tfrac{1}{2}\ln|\sec x^2| + C. \quad \square$$

↑ Formula 8

Example 2 Compute

$$\int_0^1 \frac{e^x}{e^x + 2}\, dx.$$

SOLUTION Set

$$u = e^x + 2, \qquad du = e^x\, dx.$$

At $x = 0$, $u = 3$; at $x = 1$, $u = e + 2$. Thus

$$\int_0^1 \frac{e^x}{e^x + 2}\, dx = \int_3^{e+2} \frac{du}{u} = \Big[\ln|u|\Big]_3^{e+2}$$

$$= \ln(e + 2) - \ln 3 = \ln\left[\tfrac{1}{3}(e + 2)\right] \cong 0.45. \quad \square$$

Formula 3

Example 3 Find

$$\int \frac{\cos 2x}{(2 + \sin 2x)^{1/3}}\, dx.$$

SOLUTION Set

$$u = 2 + \sin 2x, \qquad du = 2\cos 2x\, dx.$$

Then

$$\int \frac{\cos 2x}{(2 + \sin 2x)^{1/3}}\, dx = \frac{1}{2}\int \frac{1}{u^{1/3}}\, du = \frac{1}{2}\int u^{-1/3}\, du = \frac{1}{2}\left(\frac{3}{2}\right)u^{2/3} + C$$

$$= \frac{3}{4}(2 + \sin 2x)^{2/3} + C. \quad \square$$

Formula 2

The final example involves a little algebra.

Example 4 Find

$$P = \int \frac{dx}{x^2 + 2x + 5}.$$

SOLUTION First we complete the square in the denominator:

$$P = \int \frac{dx}{x^2 + 2x + 5} = \int \frac{dx}{(x^2 + 2x + 1) + 4} = \int \frac{dx}{(x + 1)^2 + 2^2}.$$

We know that

$$\int \frac{du}{u^2 + a^2} = \frac{1}{a} \tan^{-1}\left(\frac{u}{a}\right) + C.$$

Setting

$$u = x + 1, \qquad du = dx, \qquad a = 2,$$

we have

$$P = \int \frac{du}{u^2 + 2^2} = \frac{1}{2}\tan^{-1}\left(\frac{u}{2}\right) + C = \frac{1}{2}\tan^{-1}\left(\frac{x + 1}{2}\right) + C. \quad \square$$

EXERCISES 8.1

In Exercises 1–48, evaluate the given integral.

1. $\int e^{2-x}\,dx.$
2. $\int \cos \frac{2}{3}x\,dx.$
3. $\int_0^1 \sin \pi x\,dx.$
4. $\int_0^t \sec \pi x \tan \pi x\,dx.$
5. $\int \sec^2(1 - x)\,dx.$
6. $\int \frac{dx}{5^x}.$
7. $\int_{\pi/6}^{\pi/3} \cot x\,dx.$
8. $\int_0^1 \frac{x^3}{1 + x^4}\,dx.$
9. $\int \frac{x}{\sqrt{1 - x^2}}\,dx.$
10. $\int_{-\pi/4}^{\pi/4} \frac{dx}{\cos^2 x}.$
11. $\int_{-\pi/4}^{\pi/4} \frac{\sin x}{\cos^2 x}\,dx.$
12. $\int \frac{e^{\sqrt{x}}}{\sqrt{x}}\,dx.$
13. $\int_1^2 \frac{e^{1/x}}{x^2}\,dx.$
14. $\int_{\pi/6}^{\pi/3} \csc x\,dx.$
15. $\int \frac{x}{x^2 + 1}\,dx.$
16. $\int \frac{x^3}{\sqrt{1 - x^4}}\,dx.$
17. $\int_0^c \frac{dx}{x^2 + c^2}.$
18. $\int a^x e^x\,dx.$
19. $\int \frac{\sec^2\theta}{\sqrt{3 \tan \theta + 1}}\,d\theta.$
20. $\int \frac{\sin \phi}{3 - 2 \cos \phi}\,d\phi.$
21. $\int \frac{e^x}{ae^x - b}\,dx.$
22. $\int_0^{\pi/4} \frac{\sec^2 x \tan x}{\sqrt{2 + \sec^2 x}}\,dx.$
23. $\int \frac{1 + \cos 2x}{\sin^2 2x}\,dx.$
24. $\int \frac{dx}{x^2 - 4x + 13}.$
25. $\int \frac{x}{(x + 1)^2 + 4}\,dx.$
26. $\int \frac{\ln x}{x}\,dx.$
27. $\int \frac{x}{\sqrt{1 - x^4}}\,dx.$
28. $\int \frac{e^x}{1 + e^{2x}}\,dx.$
29. $\int \frac{dx}{x^2 + 6x + 10}.$
30. $\int e^x \tan e^x\,dx.$
31. $\int x \sin x^2\,dx.$
32. $\int \frac{x}{9 + x^4}\,dx.$
33. $\int \tan^2 x\,dx.$
34. $\int \cosh 2x \sinh^3 2x\,dx.$
35. $\int_{-1}^2 \frac{x}{x^2 + 4}\,dx.$
36. $\int_0^\pi \sin\left(\frac{\pi}{2}x\right)\,dx.$
37. $\int_1^e \frac{\ln x^3}{x}\,dx.$
38. $\int_0^{\pi/4} \frac{\tan^{-1} x}{1 + x^2}\,dx.$

39. $\displaystyle\int \frac{\sin^{-1} x}{\sqrt{1-x^2}}\, dx.$

40. $\displaystyle\int (x+1)\cosh(x^2+2x+1)\, dx.$

41. $\displaystyle\int \frac{e^{2x}}{\sqrt{1-e^{4x}}}\, dx.$

42. $\displaystyle\int \tanh 2x\, dx.$

43. $\displaystyle\int \frac{1}{x\ln x}\, dx.$

44. $\displaystyle\int e^x \cosh(2-e^x)\, dx.$

45. $\displaystyle\int_1^4 \frac{\sqrt{\ln x}}{x}\, dx.$

46. $\displaystyle\int_{-1}^1 \frac{x^2}{x^2+1}\, dx.$ HINT: divide the denominator into the numerator.

47. $\displaystyle\int_0^{\pi/4} \frac{1+\sin x}{\cos^2 x}\, dx.$ HINT: write the integrand as two fractions.

48. $\displaystyle\int_0^{1/2} \frac{1+x}{\sqrt{1-x^2}}\, dx.$ (See the hint for Exercise 47.)

49. Evaluate $\displaystyle\int_0^{\pi} \sqrt{1+\cos x}\, dx.$
HINT: $\cos x = 2\cos^2(x/2) - 1$.

50. Evaluate $\displaystyle\int \sec^2 x \tan x\, dx$ in two ways.

(a) Let $u = \tan x$ and verify that the result is:

$$\int \sec^2 x \tan x\, dx = \tfrac{1}{2}\tan^2 x + C_1.$$

(b) Let $u = \sec x$ and verify that the result is:

$$\int \sec^2 x \tan x\, dx = \tfrac{1}{2}\sec^2 x + C_2.$$

(c) Reconcile the results in parts (a) and (b).

51. Verify that, for any positive integer n:

(a) $\displaystyle\int_0^{\pi} \sin^2 nx\, dx = \tfrac{1}{2}\pi.$

HINT: $\sin^2\theta = \dfrac{1-\cos 2\theta}{2}$.

(b) $\displaystyle\int_0^{\pi} \sin nx \cos nx\, dx = 0.$

(c) $\displaystyle\int_0^{\pi/n} \sin nx \cos nx\, dx = 0.$

52. (a) Evaluate $\displaystyle\int \sin^3 x\, dx.$ HINT: $\sin^2 x = 1 - \cos^2 x$.

(b) Evaluate $\displaystyle\int \sin^5 x\, dx.$

(c) Explain how to evaluate $\displaystyle\int \sin^{2k+1} x\, dx$ for any positive integer k.

53. (a) Evaluate $\displaystyle\int \tan^3 x\, dx.$ HINT: $\tan^2 x = \sec^2 x - 1$.

(b) Evaluate $\displaystyle\int \tan^5 x\, dx.$

(c) Evaluate $\displaystyle\int \tan^7 x\, dx.$

(d) Explain how to evaluate $\displaystyle\int \tan^{2k+1} x\, dx$ for any positive integer k.

54. (a) Sketch the region bounded on the left by the line $x = \pi/6$, and by the curves $y = \csc x$ and $y = \sin x$ for $\pi/6 \le x \le \pi/2$.

(b) Calculate the area of the region in part (a).

(c) The region in part (a) is rotated around the x-axis. Find the volume of the solid that is generated.

C **55.** (a) Use a graphing utility to sketch the graph of

$$f(x) = \frac{1}{\sin x + \cos x} \quad \text{for} \quad 0 \le x \le \frac{\pi}{2}.$$

(b) Find A and B such that $\sin x + \cos x = A\sin(x+B)$.

(c) Find the area of the region bounded by the graph of f and the x-axis. HINT: Use the result in part (b).

C **56.** (a) Use a graphing utility to sketch the graph of $f(x) = e^{-x^2}$.

(b) Fix a number $a > 0$ and rotate the region bounded by the graph of f and the y-axis on the interval $[0, a]$ around the y-axis. Find the volume of the solid that is generated.

(c) Find a such that the solid in part (b) has volume 2 cubic units.

C **57.** (a) Use a graphing utility to sketch the graphs of

$$f(x) = \frac{x^2+1}{x+1}, \quad x > -1, \quad \text{and} \quad x + 2y = 16$$

in the same coordinate system.

(b) The two graphs intersect at two points and determine a bounded region R. Estimate the x-coordinates of the two points of intersection accurate to two decimal places.

(c) Determine the approximate area of the region R.

C **58.** (a) Use a graphing utility to sketch the graph of

$$y^2 = x^2(1-x)$$

(b) Your sketch in part (a) should show that the curve forms a loop for $0 \le x \le 1$. Calculate the area of the loop. HINT: Use the symmetry of the curve.

■ 8.2 INTEGRATION BY PARTS

We begin with the formula for the derivative of a product

$$f(x)g'(x) + f'(x)g(x) = (f \cdot g)'(x).$$

Integrating both sides, we get

$$\int f(x)g'(x)\,dx + \int f'(x)g(x)\,dx = \int (f \cdot g)'(x)\,dx$$

Since

$$\int (f \cdot g)'(x)\,dx = f(x)g(x) + C,$$

we have

$$\int f(x)g'(x)\,dx + \int f'(x)g(x)\,dx = f(x)g(x) + C$$

and therefore

$$\int f(x)g'(x)\,dx = f(x)g(x) - \int f'(x)g(x)\,dx + C.$$

Since the computation of

$$\int f'(x)g(x)\,dx$$

will yield its own arbitrary constant, there is no reason to keep the constant C. We therefore drop it and write

(8.2.1)

$$\boxed{\int f(x)g'(x)\,dx = f(x)g(x) - \int f'(x)g(x)\,dx.}$$

This formula, called the formula for *integration by parts,* enables us to find

$$\int f(x)g'(x)\,dx$$

by computing

$$\int f'(x)g(x)\,dx$$

instead. Of course, it is of practical use only if the second integral is easier to compute than the first.

In practice we usually set

$$u = f(x), \qquad dv = g'(x)\,dx.$$

Then

$$du = f'(x)\,dx, \qquad v = g(x).$$

Now, with these substitutions, the formula for integration by parts can be written

(8.2.2)
$$\int u\,dv = uv - \int v\,du.$$

Success with this formula depends on choosing u and dv so that

$$\int v\,du \quad \text{is easier to calculate than} \quad \int u\,dv.$$

Example 1 Calculate

$$\int xe^x\,dx.$$

SOLUTION We want to separate x from e^x. Setting

$$u = x, \qquad dv = e^x\,dx,$$

we have

$$du = dx, \qquad v = e^x.$$

Accordingly,

$$\int xe^x\,dx = \int u\,dv = uv - \int v\,du = xe^x - \int e^x\,dx = xe^x - e^x + C.$$

Our choice of u and dv worked out very well. However, if we had set

$$u = e^x, \qquad dv = x\,dx,$$

then we would have had

$$du = e^x\,dx, \qquad v = \tfrac{1}{2}x^2.$$

Integration by parts would then have given

$$\int xe^x\,dx = \int u\,dv = uv - \int v\,du = \tfrac{1}{2}x^2e^x - \tfrac{1}{2}\int x^2e^x\,dx,$$

an integral more complicated than the one we started with. A good choice of u and dv is crucial. ❑

Example 2 Calculate

$$\int x \sin 2x\,dx.$$

SOLUTION Setting

$$u = x, \qquad dv = \sin 2x\,dx,$$

we have

$$du = dx, \qquad v = -\tfrac{1}{2}\cos 2x.$$

Therefore,

$$\int x \sin 2x \, dx = -\tfrac{1}{2}x \cos 2x - \int -\tfrac{1}{2} \cos 2x \, dx = -\tfrac{1}{2}x \cos 2x + \tfrac{1}{4} \sin 2x + C.$$

You can verify that if we had set

$$u = \sin 2x, \qquad dv = x \, dx,$$

then we would have run into the same kind of difficulty that we saw at the end of Example 1. ❑

In Examples 1 and 2 there was only one effective choice for u and dv. As the next example shows, however, there can be more than one way to choose u and dv.

Example 3 Calculate

$$\int x \ln x \, dx.$$

SOLUTION Setting

$$u = \ln x, \qquad dv = x \, dx,$$

we have

$$du = \frac{1}{x} dx, \qquad v = \frac{x^2}{2}.$$

Then

$$\begin{aligned} \int x \ln x \, dx &= \int u \, dv = uv - \int v \, du \\ &= \frac{x^2}{2} \ln x - \int \frac{1}{x} \frac{x^2}{2} dx = \tfrac{1}{2}x^2 \ln x - \tfrac{1}{4}x^2 + C. \end{aligned}$$

ALTERNATE SOLUTION This time we set

$$u = x \ln x, \qquad dv = dx,$$

so that

$$du = (1 + \ln x) \, dx, \qquad v = x.$$

Substituting these selections in

$$\int u \, dv = uv - \int v \, du,$$

we find that

$$(1) \qquad \int x \ln x \, dx = x^2 \ln x - \int x(1 + \ln x) \, dx.$$

It may seem that the integral on the right is more complicated than the one we started with. However, we can rewrite equation (1) as

$$\int x \ln x \, dx = x^2 \ln x - \int x \, dx - \int x \ln x \, dx.$$

If we let $P = \int x \ln x \, dx$, then this equation is

$$P = x^2 \ln x - \int x \, dx - P.$$

Solving for P, we have

$$2P = x^2 \ln x - \int x \, dx$$

and

$$P = \tfrac{1}{2}x^2 \ln x - \tfrac{1}{2}\int x \, dx = \tfrac{1}{2}x^2 \ln x - \tfrac{1}{4}x^2 + C.$$

Thus

$$\int x \ln x \, dx = \tfrac{1}{2}x^2 \ln x - \tfrac{1}{4}x^2 + C$$

as before. ❑

Remark Given the two successful approaches in Example 3, you might also be tempted to try setting

$$u = x, \qquad dv = \ln x \, dx.$$

This won't work, however, because you can't calculate v, that is, you don't know an antiderivative for $\ln x$. Actually, as you will see at the end of this section, $\int \ln x \, dx$ is, itself, calculated using integration by parts. ❑

Integration by parts is often used to calculate integrals where the integrand is a mixture of function types; for example, polynomials and exponentials, or polynomials and trigonometric functions, and so forth. Some integrands, however, are better left as mixtures; for example, it is easy to see that

$$\int 2xe^{x^2} \, dx = e^{x^2} + C \qquad \text{and} \qquad \int 3x^2 \cos x^3 \, dx = \sin x^3 + C.$$

Any attempt to separate these integrands for integration by parts is counterproductive. The mixtures in these integrands arise from the chain rule, and we need these mixtures to calculate the integrals.

Example 4 Calculate

$$\int x^5 e^{x^3} \, dx.$$

SOLUTION To integrate e^{x^3} we need an x^2 factor. So we will keep x^2 together with e^{x^3}. Setting

$$u = x^3, \qquad dv = x^2 e^{x^3} \, dx,$$

we have

$$du = 3x^2 \, dx, \qquad v = \tfrac{1}{3} e^{x^3}.$$

Thus

$$\int x^5 e^{x^3}\,dx = \int u\,dv = uv - \int v\,du$$
$$= \tfrac{1}{3}x^3 e^{x^3} - \int x^2 e^{x^3}\,dx$$
$$= \tfrac{1}{3}x^3 e^{x^3} - \tfrac{1}{3}e^{x^3} + C = \tfrac{1}{3}(x^3 - 1)e^{x^3} + C. \quad \square$$

Integration by parts can also be used in connection with definite integrals. Suppose that f' and g' are continuous on the interval $[a, b]$. Integrating the formula for the derivative of a product, we have

$$\int_a^b f(x)g'(x)\,dx + \int_a^b f'(x)g(x)\,dx = \int_a^b [f(x)\cdot g(x)]'\,dx.$$

However, by the fundamental theorem of calculus, Theorem 5.4.2,

$$\int_a^b [f(x)\cdot g(x)]'\,dx = \Big[f(x)\cdot g(x)\Big]_a^b.$$

Thus

$$\int_a^b f(x)g'(x)\,dx + \int_a^b f'(x)g(x)\,dx = \Big[f(x)\cdot g(x)\Big]_a^b$$

and

(8.2.3)
$$\int_a^b f(x)g'(x)\,dx = \Big[f(x)\cdot g(x)\Big]_a^b - \int_a^b f'(x)g(x)\,dx,$$

which is the definite integral version of Formula (8.2.1).

Example 5 Evaluate

$$\int_1^2 x^3 \ln x\,dx.$$

SOLUTION Setting

$$u = \ln x, \qquad dv = x^3\,dx,$$

we have

$$du = \frac{1}{x}\,dx, \qquad v = \tfrac{1}{4}x^4.$$

Substituting into (8.2.3) gives

$$\int_1^2 x^3 \ln x\,dx = \Big[\tfrac{1}{4}x^4 \ln x\Big]_1^2 - \tfrac{1}{4}\int_1^2 x^3\,dx$$
$$= 4\ln 2 - \tfrac{1}{16}\Big[x^4\Big]_1^2 = 4\ln 2 - \tfrac{15}{16}. \quad \square$$

To calculate some integrals you may have to integrate by parts more than once.

Example 6 Evaluate

$$\int_0^1 x^2 e^x \, dx.$$

SOLUTION First we calculate the indefinite integral

$$\int x^2 e^x \, dx$$

by separating x^2 from e^x. Setting

$$u = x^2, \qquad dv = e^x \, dx,$$

we have

$$du = 2x \, dx, \qquad v = e^x,$$

and thus

$$\int x^2 e^x \, dx = \int u \, dv = uv - \int v \, du = x^2 e^x - \int 2x \, e^x \, dx.$$

We now compute the integral on the right again by parts. This time we set

$$u = 2x, \qquad dv = e^x \, dx.$$

This gives

$$du = 2 \, dx, \qquad v = e^x$$

and thus

$$\int 2xe^x \, dx = \int u \, dv = uv - \int v \, du = 2xe^x - \int 2e^x \, dx = 2xe^x - 2e^x + C.$$

This together with our earlier calculation gives

$$\int x^2 e^x \, dx = x^2 e^x - 2xe^x + 2e^x + C.$$

It follows now that

$$\int_0^1 x^2 e^x \, dx = \Big[x^2 e^x - 2xe^x + 2e^x \Big]_0^1 = (e - 2e + 2e) - 2 = e - 2. \quad \square$$

Remark Note that we did not use Formula (8.2.3) in evaluating the definite integral in Example 6. Instead, we calculated the corresponding indefinite integral and then substituted the limits of integration at the end. This approach is sometimes easier than using (8.2.3), especially if several steps are involved. ❑

Example 7 Find

$$\int e^x \cos x \, dx.$$

SOLUTION Here we integrate by parts twice. First we write

$$u = e^x, \qquad dv = \cos x \, dx,$$

$$du = e^x \, dx, \qquad v = \sin x.$$

This gives

$$(1) \qquad \int e^x \cos x\, dx = \int u\, dv = uv - \int v\, du = e^x \sin x - \int e^x \sin x\, dx.$$

Now we work with the integral on the right. Setting

$$u = e^x, \qquad dv = \sin x\, dx,$$
$$du = e^x\, dx, \qquad v = -\cos x,$$

we have

$$(2) \qquad \int e^x \sin x\, dx = \int u\, dv = uv - \int v\, du = -e^x \cos x + \int e^x \cos x\, dx.$$

Substituting (2) into (1), we get

$$\int e^x \cos x\, dx = e^x \sin x + e^x \cos x - \int e^x \cos x\, dx,$$

and we have a situation similar to the one we had in the alternate solution in Example 3. "Solving" this equation for $\int e^x \cos x\, dx$ gives

$$2\int e^x \cos x\, dx = e^x(\sin x + \cos x),$$
$$\int e^x \cos x\, dx = \tfrac{1}{2}\, e^x(\sin x + \cos x).$$

Since this integral is an indefinite integral, we add an arbitrary constant C:

$$\int e^x \cos x\, dx = \tfrac{1}{2}\, e^x(\sin x + \cos x) + C. \quad ❑$$

Remark You should verify that if you switch the roles of u and dv in Example 7 by letting

$$u = \cos x, \qquad dv = e^x\, dx,$$
$$du = -\sin x\, dx, \qquad v = e^x,$$

then you will get precisely the same result in precisely the same way. ❑

Finally, the techniques of integration by parts enable us to integrate the logarithm function and the inverse trigonometric functions:

(8.2.4)
$$\boxed{\int \ln x\, dx = x \ln x - x + C.}$$

(8.2.5)
$$\boxed{\int \sin^{-1} x\, dx = x \sin^{-1} x + \sqrt{1 - x^2} + C.}$$

(8.2.6)
$$\int \tan^{-1}x\, dx = x \tan^{-1}x - \tfrac{1}{2}\ln(1 + x^2) + C.$$

(8.2.7)
$$\int \sec^{-1}x\, dx = x \sec^{-1}x - \ln|x + \sqrt{x^2 - 1}| + C.$$

We will derive (8.2.5) and leave the others to you as exercises. Set

$$u = \sin^{-1}x, \qquad dv = dx.$$

Then

$$du = \frac{dx}{\sqrt{1 - x^2}}, \qquad v = x,$$

and

$$\int \sin^{-1}x\, dx = x \sin^{-1}x - \int \frac{x}{\sqrt{1 - x^2}}\, dx.$$

The new integral is easy to evaluate by means of the substitution

$$w = 1 - x^2, \qquad dw = -2x\, dx,$$

which gives

$$\int \frac{x}{\sqrt{1 - x^2}}\, dx = -\tfrac{1}{2}\int \frac{dw}{\sqrt{w}} = -\sqrt{w} = -\sqrt{1 - x^2}.$$

Therefore,

$$\int \sin^{-1}x\, dx = x \sin^{-1}x + \sqrt{1 - x^2} + C. \quad \square$$

EXERCISES 8.2

In Exercises 1–46, evaluate the given integral.

1. $\int xe^{-x}\, dx.$

2. $\int \ln x\, dx.$ (8.2.4)

3. $\int_1^e x^2 \ln x\, dx.$

4. $\int_0^2 x\, 2^x\, dx.$

5. $\int x^2 e^{-x^3}\, dx.$

6. $\int x \ln x^2\, dx.$

7. $\int_0^1 x^2 e^{-x}\, dx.$

8. $\int x^3 e^{-x^2}\, dx.$

9. $\int \frac{x^2}{\sqrt{1 - x}}\, dx.$

10. $\int \frac{dx}{x(\ln x)^3}.$

11. $\int_1^{e^2} x \ln \sqrt{x}\, dx.$

12. $\int_0^3 x\sqrt{x + 1}\, dx.$

13. $\int \frac{\ln (x + 1)}{\sqrt{x + 1}}\, dx.$

14. $\int x^2(e^x - 1)\, dx.$

15. $\int (\ln x)^2\, dx.$

16. $\int x(x + 5)^{-14}\, dx.$

17. $\int x^3 3^x\, dx.$

18. $\int \sqrt{x} \ln x\, dx.$

19. $\int x(x+5)^{14}\,dx.$

20. $\int (2^x + x^2)^2\,dx.$

21. $\int_0^{1/2} x\cos \pi x\,dx.$

22. $\int_0^{\pi/2} x^2 \sin x\,dx.$

23. $\int x^2(x+1)^9\,dx.$

24. $\int x^2(2x-1)^{-7}\,dx.$

25. $\int e^x \sin x\,dx.$

26. $\int (e^x + 2x)^2\,dx.$

27. $\int_0^1 \ln(1+x^2)\,dx.$

28. $\int x\ln(x+1)\,dx.$

29. $\int x^n \ln x\,dx.\quad (n \neq -1)$

30. $\int e^{3x}\cos 2x\,dx.$

31. $\int x^3 \sin x^2\,dx.$

32. $\int x^3 \sin x\,dx.$

33. $\int x^4 e^x\,dx.$

34. $\int \dfrac{\sin^{-1}2x}{\sqrt{1-4x^2}}\,dx.$

35. $\int_0^{1/4} \sin^{-1}2x\,dx.$

36. $\int \tan^{-1}x\,dx.\quad (8.2.6)$

37. $\int \sec^{-1}x\,dx.\quad (8.2.7)$

38. $\int \cos\sqrt{x}\,dx.$ HINT: Let $u = \sqrt{x}$.

39. $\int_0^1 x\tan^{-1}(x^2)\,dx.$

40. $\int_{-1}^1 x\sinh(2x^2)\,dx.$

41. $\int x^2\cosh 2x\,dx.$

42. $\int \cos(\ln x)\,dx.$ HINT: Integrate by parts twice.

43. $\int \sin(\ln x)\,dx.$

44. $\int \cos x\tan^{-1}(\sin x)\,dx.$

45. $\int \dfrac{1}{x}\sin^{-1}(\ln x)\,dx.$

46. $\int_1^{2e} x^2(\ln x)^2\,dx.$

In Exercises 47 and 48, find the area of the region bounded by the graph of f and the x-axis.

47. $f(x) = \sin^{-1}x,\quad x \in [0, \frac{1}{2}].$

48. $f(x) = xe^{-2x},\quad x \in [0, 2].$

49. Let Ω be the region bounded by the graph of $f(x) = \ln x$ and the x-axis between $x = 1$ and $x = e$. (a) Find the area of Ω. (b) Find the centroid of Ω. (c) Find the volume of the solids generated by revolving Ω about each of the coordinate axes.

50. Let $f(x) = \dfrac{\ln x}{x}$ on $[1, 2e]$.

(a) Find the area of the region R bounded by the graph of f and the x-axis.
(b) Find the volume of the solid generated by revolving R around the x-axis.

In Exercises 51–54, find the centroid of the region under the graph.

51. $f(x) = e^x,\quad x \in [0, 1].$

52. $f(x) = e^{-x},\quad x \in [0, 1].$

53. $f(x) = \sin x,\quad x \in [0, \pi].$

54. $f(x) = \cos x,\quad x \in [0, \frac{1}{2}\pi].$

55. The mass density of a rod that extends from $x = 0$ to $x = 1$ is given by the function $\lambda(x) = e^{kx}$, where k is a constant. (a) Calculate the mass of the rod. (b) Find the center of mass of the rod.

56. The mass density of a rod that extends from $x = 2$ to $x = 3$ is given by the logarithm function $f(x) = \ln x$. (a) Calculate the mass of the rod. (b) Find the center of mass of the rod.

In Exercises 57–62, find the volume generated by revolving the region under the graph about the y-axis.

57. $f(x) = e^{ax},\quad x \in [0, 1].$

58. $f(x) = \sin \pi x,\quad x \in [0, 1].$

59. $f(x) = \cos \frac{1}{2}\pi x,\quad x \in [0, 1].$

60. $f(x) = x\sin x,\quad x \in [0, \pi].$

61. $f(x) = xe^x,\quad x \in [0, 1].$

62. $f(x) = x\cos x,\quad x \in [0, \frac{1}{2}\pi].$

63. Let Ω be the region under the curve $y = e^x$, $x \in [0, 1]$. Find the centroid of the solid generated by revolving Ω about the x-axis. (6.4.5)

64. Let Ω be the region under the curve $y = \sin x$, $x \in [0, \frac{1}{2}\pi]$. Find the centroid of the solid generated by revolving Ω about the x-axis. (6.4.5)

65. Let Ω be the region bounded by the graph of $y = \cosh x$ and the x-axis between $x = 0$ and $x = 1$. Find the area of Ω and determine the centroid.

66. Let Ω be the region given in Exercise 65. Find the centroid of the solid generated by revolving Ω about: (a) the x-axis; (b) the y-axis.

67. Let n be a positive integer. Use integration by parts to derive the formula

$$\int x^n e^{ax}\,dx = \frac{x^n e^{ax}}{a} - \frac{n}{a}\int x^{n-1}e^{ax}\,dx,\ a \neq 0.$$

Note: This formula is known as a *reduction formula* since the exponent n in the integrand has been reduced.

68. Let n be a positive integer. Use integration by parts to derive the reduction formula

$$\int (\ln x)^n\,dx = x\,(\ln x)^n - n\int (\ln x)^{n-1}\,dx.$$

HINT: Let $u = (\ln x)^n$.

In Exercises 69–72, use the reduction formulas in Exercises 67 and 68 to evaluate the given integral.

69. $\int x^3 e^{2x}\,dx.$ **70.** $\int x^2 e^{-x}\,dx.$

71. $\int (\ln x)^3\,dx.$ **72.** $\int (\ln x)^4\,dx.$

73. Let f and g have continuous second derivatives. Show that if $f(a) = g(a) = f(b) = g(b) = 0$, then

$$\int_a^b f(x)g''(x)\,dx = \int_a^b g(x)f''(x)\,dx.$$

74. Consider the identity

$$f(b) - f(a) = \int_a^b f'(x)\,dx.$$

(a) Assume that f has a continuous second derivative. Use integration by parts to derive the identity

$$f(b) - f(a) = f'(a)(b-a) - \int_a^b f''(x)(x-b)\,dx.$$

(b) Assume that f has a continuous third derivative. Use the result in part (a) and integration by parts to derive the identity

$$f(b) - f(a) = f'(a)(b-a) + \frac{f''(a)}{2}(b-a)^2 + \int_a^b \frac{f'''(x)}{2}(x-b)^2\,dx.$$

These results generalize the mean-value theorem. The identities begun here will be extended and used in Chapter 11.

Revenue Streams

In Section 7.6, we considered continuous compounding and saw that the present value of A dollars t years from now is given by the formula

$$P.V. = Ae^{-rt},$$

where r is the annual rate of continuous compounding.

75. Suppose that revenue flows continuously at a constant rate of R dollars per year for n years. Show that the present value of such a revenue stream is given by

$$P.V. = \int_0^n Re^{-rt}\,dt.$$

76. Given a constant revenue stream of $R = \$1000$ per year, what is the present value of the first four years of revenue if the annual rate of continuous compounding is: (a) 4%? (b) 8%?

Revenue streams usually do not flow at a constant rate R; rather they normally flow at a time-dependent rate $R(t)$. In general, $R(t)$ tends to increase when business is good and tends to decrease when business is poor. "Growth companies" are those that have a continually increasing flow rate. The so-called "cyclical companies" owe their name to a fluctuating flow rate. If we postulate a time dependent flow rate $R(t)$, then the present value of an n-year stream is given by the formula

$$P.V. = \int_0^n R(t)e^{-rt}\,dt.$$

In Exercises 77 and 78, assume a time-dependent revenue stream with $R(t) = 1000 + 60t$ dollars per year.

77. What is the present value of the first two years of revenue if the annual rate of continuous compounding is: (a) 5%? (b) 10%?

78. What is the present value of the third year of revenue if the annual rate of continuous compounding is: (a) 5%? (b) 10%?

■ 8.3 POWERS AND PRODUCTS OF SINES AND COSINES

I. We begin by explaining how to calculate integrals of the form

$$\int \sin^m x\,\cos^n x\,dx \qquad \text{with } m \text{ or } n \text{ an odd positive integer.}$$

Suppose that n is odd. If $n = 1$, we have

$$(1)\qquad \int \sin^m x\,\cos x\,dx = \frac{1}{m+1}\sin^{m+1} x + C, \qquad m \neq -1.$$

If $n > 1$, write

$$\cos^n x = \cos^{n-1} x \cos x.$$

Since $n - 1$ is even, $\cos^{n-1} x$ can be expressed in powers of $\sin^2 x$ by noting that $\cos^2 x = 1 - \sin^2 x$. The integral then takes the form

$$\int (\text{sum of powers of } \sin x) \cdot \cos x \, dx,$$

which can be broken up into integrals of the form (1).

Similarly if m is odd, write

$$\sin^m x = \sin^{m-1} x \sin x$$

and use the substitution $\sin^2 x = 1 - \cos^2 x$.

Example 1

$$\begin{aligned}
\int \sin^2 x \cos^5 x \, dx &= \int \sin^2 x \cos^4 x \cos x \, dx \\
&= \int \sin^2 x \, (1 - \sin^2 x)^2 \cos x \, dx \\
&= \int (\sin^2 x - 2 \sin^4 x + \sin^6 x) \cos x \, dx \\
&= \int \sin^2 x \cos x \, dx - 2 \int \sin^4 x \cos x \, dx + \int \sin^6 x \cos x \, dx \\
&= \tfrac{1}{3} \sin^3 x - \tfrac{2}{5} \sin^5 x + \tfrac{1}{7} \sin^7 x + C. \quad ❑
\end{aligned}$$

Example 2

$$\begin{aligned}
\int \sin^5 x \, dx &= \int \sin^4 x \sin x \, dx \\
&= \int (1 - \cos^2 x)^2 \sin x \, dx \\
&= \int (1 - 2 \cos^2 x + \cos^4 x) \sin x \, dx \\
&= \int \sin x \, dx - 2 \int \cos^2 x \sin x \, dx + \int \cos^4 x \sin x \, dx \\
&= -\cos x + \tfrac{2}{3} \cos^3 x - \tfrac{1}{5} \cos^5 x + C. \quad ❑
\end{aligned}$$

II. To calculate integrals of the form

$$\boxed{\int \sin^m x \cos^n x \, dx \qquad \text{with } m \text{ and } n \text{ both even positive integers}}$$

use the following trigonometric identities:

$$\sin x \cos x = \tfrac{1}{2} \sin 2x, \qquad \sin^2 x = \tfrac{1}{2} - \tfrac{1}{2} \cos 2x, \qquad \cos^2 x = \tfrac{1}{2} + \tfrac{1}{2} \cos 2x.$$

The first of these identities is derived from the double-angle formula for the sine function. The other two are the half-angle formulas for sine and cosine.

Example 3 (See Exercise 72, Section 5.7.)

$$\int \cos^2 x\, dx = \int (\tfrac{1}{2} + \tfrac{1}{2}\cos 2x)\, dx$$
$$= \tfrac{1}{2}\int dx + \tfrac{1}{2}\int \cos 2x\, dx = \tfrac{1}{2}x + \tfrac{1}{4}\sin 2x + C. \quad \square$$

Example 4

$$\int \sin^2 x \cos^2 x\, dx = \tfrac{1}{4}\int \sin^2 2x\, dx$$
$$= \tfrac{1}{4}\int (\tfrac{1}{2} - \tfrac{1}{2}\cos 4x)\, dx$$
$$= \tfrac{1}{8}\int dx - \tfrac{1}{8}\int \cos 4x\, dx = \tfrac{1}{8}x - \tfrac{1}{32}\sin 4x + C. \quad \square$$

Example 5

$$\int \sin^4 x \cos^2 x\, dx = \int (\sin x \cos x)^2 \sin^2 x\, dx$$
$$= \int \tfrac{1}{4}\sin^2 2x\, (\tfrac{1}{2} - \tfrac{1}{2}\cos 2x)\, dx$$
$$= \tfrac{1}{8}\int \sin^2 2x\, dx - \tfrac{1}{8}\int \sin^2 2x \cos 2x\, dx$$
$$= \tfrac{1}{8}\int (\tfrac{1}{2} - \tfrac{1}{2}\cos 4x)\, dx - \tfrac{1}{8}\int \sin^2 2x \cos 2x\, dx$$
$$= \tfrac{1}{16}x - \tfrac{1}{64}\sin 4x - \tfrac{1}{48}\sin^3 2x + C. \quad \square$$

III. Next we derive *reduction formulas* for integrals of the form

$$\int \sin^n x\, dx \quad \text{and} \quad \int \cos^n x\, dx, \qquad \text{where } n \text{ is a positive integer.}$$

We will use integration by parts to show that

(8.3.1)
$$\int \sin^n x\, dx = -\frac{1}{n}\sin^{n-1} x \cos x + \frac{n-1}{n}\int \sin^{n-2} x\, dx.$$

The corresponding formula for $\int \cos^n x\, dx$ is

(8.3.2)
$$\int \cos^n x\, dx = \frac{1}{n}\cos^{n-1} x \sin x + \frac{n-1}{n}\int \cos^{n-2} x\, dx.$$

The verification of this formula is left as an exercise.

Formulas such as (8.3.1) and (8.3.2) are called reduction formulas because the exponent is reduced (from n to $n - 2$ in each of these cases).

To establish (8.3.1), we write $\sin^n x$ as $\sin^{n-1}x \sin x$ and set

$$u = \sin^{n-1}x, \qquad dv = \sin x\, dx.$$

Then

$$du = (n-1)\sin^{n-2}x \cos x\, dx, \qquad v = -\cos x,$$

and

$$\begin{aligned}\int \sin^n x\, dx &= -\sin^{n-1}x \cos x + (n-1)\int \sin^{n-2}x \cos^2 x\, dx\\ &= -\sin^{n-1}x \cos x + (n-1)\int \sin^{n-2}x\,(1 - \sin^2 x)\, dx.\end{aligned}$$

Therefore,

$$\int \sin^n x\, dx = -\sin^{n-1}x \cos x + (n-1)\int \sin^{n-2}x\, dx - (n-1)\int \sin^n x\, dx.$$

Now, adding $(n-1)\int \sin^n x\, dx$ to both sides, we get

$$n\int \sin^n x\, dx = -\sin^{n-1}x \cos x + (n-1)\int \sin^{n-2}x\, dx$$

and (8.3.1) follows. ❑

Repeated applications of the formulas (8.3.1) or (8.3.2) will reduce a power of the sine or cosine to 0 or 1, depending upon whether n is even or odd.

Example 6

$$\begin{aligned}\int \sin^5 x\, dx &= -\tfrac{1}{5}\sin^4 x \cos x + \tfrac{4}{5}\int \sin^3 x\, dx\\ &= -\tfrac{1}{5}\sin^4 x \cos x + \tfrac{4}{5}\left[-\tfrac{1}{3}\sin^2 x \cos x + \tfrac{2}{3}\int \sin x\, dx\right]\\ &= -\tfrac{1}{5}\sin^4 x \cos x - \tfrac{4}{15}\sin^2 x \cos x - \tfrac{8}{15}\cos x + C.\end{aligned}$$ ❑

You should verify that this result is equivalent to the result in Example 2.

IV. Finally we come to integrals of the form

$$\int \sin mx \cos nx\, dx, \qquad \int \sin mx \sin nx\, dx, \qquad \int \cos mx \cos nx\, dx.$$

If $m = n$, there is no difficulty. For $m \neq n$ use the identities

$$\begin{aligned}\sin A \cos B &= \tfrac{1}{2}[\sin(A - B) + \sin(A + B)],\\ \sin A \sin B &= \tfrac{1}{2}[\cos(A - B) - \cos(A + B)],\\ \cos A \cos B &= \tfrac{1}{2}[\cos(A - B) + \cos(A + B)].\end{aligned}$$

These identities follow readily from the familiar addition formulas:

$$\sin(A+B) = \sin A \cos B + \cos A \sin B,$$
$$\sin(A-B) = \sin A \cos B - \cos A \sin B,$$
$$\cos(A+B) = \cos A \cos B - \sin A \sin B,$$
$$\cos(A-B) = \cos A \cos B + \sin A \sin B.$$

Example 7

$$\int \sin 5x \sin 3x \, dx = \int \tfrac{1}{2}[\cos(5x-3x) - \cos(5x+3x)]\, dx$$
$$= \tfrac{1}{2}\int [\cos 2x - \cos 8x]\, dx = \tfrac{1}{4}\sin 2x - \tfrac{1}{16}\sin 8x + C. \quad \square$$

EXERCISES 8.3

In Exercises 1–32, evaluate the given integral.

1. $\int \sin^3 x \, dx.$

2. $\int_0^{\pi/8} \cos^2 4x \, dx.$

3. $\int_0^{\pi/6} \sin^2 3x \, dx.$

4. $\int \cos^3 x \, dx.$

5. $\int \cos^4 x \sin^3 x \, dx.$

6. $\int \sin^3 x \cos^2 x \, dx.$

7. $\int \sin^3 x \cos^3 x \, dx.$

8. $\int \sin^2 x \cos^4 x \, dx.$

9. $\int \sin^2 x \cos^3 x \, dx.$

10. $\int \sin^4 x \cos^3 x \, dx.$

11. $\int_0^{\pi} \sin^4 x \, dx.$

12. $\int \cos^3 x \cos 2x \, dx.$

13. $\int \sin 2x \cos 3x \, dx.$

14. $\int_0^{\pi/2} \cos 2x \sin 3x \, dx.$

15. $\int \sin^2 x \sin 2x \, dx.$

16. $\int_0^{\pi/2} \cos^4 x \, dx.$

17. $\int \sin^4 x \cos^4 x \, dx.$

18. $\int \sin^7 x \, dx.$

19. $\int \sin^6 x \, dx.$

20. $\int \cos^5 x \sin^5 x \, dx.$

21. $\int \cos^7 x \, dx.$

22. $\int \cos^6 x \, dx.$

23. $\int_0^{\pi/2} \cos 3x \cos 2x \, dx.$

24. $\int \sin 3x \sin 2x \, dx.$

25. $\int \sin 5x \sin 2x \, dx.$

26. $\int_0^{\pi/4} \sin 5x \cos 2x \, dx.$

27. $\int_{-1/6}^{1/3} \sin^4 3\pi x \cos^3 3\pi x \, dx.$

28. $\int_0^{1/2} \cos \pi x \cos (\tfrac{\pi}{2} x) \, dx.$

29. $\int_0^{\pi/4} \cos 4x \sin 2x \, dx.$

30. $\int (\sin 3x - \sin x)^2 \, dx.$

31. $\int \sin(\tfrac{1}{2}x) \cos 2x \, dx.$

32. $\int_0^{2\pi} \sin^2 ax \, dx, \quad a \neq 0.$

In Exercises 33–36, calculate the integral by the indicated trigonometric substitution. Note: trigonometric substitutions will be studied in more detail in Section 8.5.

33. $\int \frac{dx}{(x^2+1)^3}.$ [set $x = \tan u$]

34. $\int \frac{dx}{(x^2+4)^3}.$ [set $x = 2 \tan u$]

35. $\int \frac{dx}{[(x+1)^2+1]^2}.$ [set $x + 1 = \tan u$]

36. $\int \frac{dx}{[(2x+1)^2+9]^2}.$ [set $2x + 1 = 3 \tan u$]

37. Find the area of the region R bounded by the graph of $f(x) = \sin^2 x$ and the x-axis, $x \in [0, \pi]$.

38. The region bounded by the graph of $f(x) = \cos x$ and the x-axis, $x \in [-\pi/2, \pi/2]$, is revolved around the x-axis. Find the volume of the solid that is generated.

39. The region R in Exercise 37 is revolved around the x-axis. Find the volume of the solid that is generated.

40. The region bounded by the y-axis and the graphs of $y = \sin x$ and $y = \cos x$, $0 \leq x \leq \pi/4$, is revolved around the x-axis. Find the volume of the solid that is generated.

41. Let m and n be positive integers. Prove that

$$\int \sin mx \sin nx \, dx = \begin{cases} \dfrac{\sin(m-n)x}{2(m-n)} - \dfrac{\sin(m+n)x}{2(m+n)} + C & \text{if } m \neq n \\ \dfrac{x}{2} - \dfrac{\sin 2mx}{4m} + C & \text{if } m = n. \end{cases}$$

42. Derive formulas corresponding to the formula in Exercise 41 for:

(a) $\displaystyle\int \sin mx \cos nx \, dx.$

(b) $\displaystyle\int \cos mx \cos nx \, dx.$

43. Use the result in Exercise 41 to prove that

$$\int_{-\pi}^{\pi} \sin mx \sin nx \, dx = \begin{cases} 0 & \text{if } m \neq n \\ \pi & \text{if } m = n. \end{cases}$$

44. (a) Evaluate

$$\int_{-\pi}^{\pi} \sin mx \cos nx \, dx.$$

(b) Evaluate

$$\int_{-\pi}^{\pi} \cos mx \cos nx \, dx.$$

Note: The formulas in Exercises 43 and 44 are important in applied mathematics. They arise in Fourier analysis where the approximation of functions by sums of sines and cosines is studied.

45. Verify the reduction formula (8.3.2).

46. (a) Use the reduction formula (8.3.1) to show that

$$\int_0^{\pi/2} \sin^n x \, dx = \frac{n-1}{n} \int_0^{\pi/2} \sin^{n-2} x \, dx.$$

(b) Show that

$$\int_0^{\pi/2} \sin^n x \, dx = \begin{cases} \left(\dfrac{(n-1)\cdots 5\cdot 3\cdot 1}{n\cdots 6\cdot 4\cdot 2}\right)\dfrac{\pi}{2}, & n \text{ even}, n \geq 2 \\ \dfrac{(n-1)\cdots 4\cdot 2}{n\cdots 5\cdot 3}, & n \text{ odd}, n \geq 3. \end{cases}$$

These formulas are known as the *Wallis sine formulas.*

(c) Show that

$$\int_0^{\pi/2} \cos^n x \, dx = \int_0^{\pi/2} \sin^n x \, dx.$$

Use the results in Exercise 46 to evaluate the integrals in Exercises 47 and 48.

47. $\displaystyle\int_0^{\pi/2} \sin^7 x \, dx.$

48. $\displaystyle\int_0^{\pi/2} \cos^6 x \, dx.$

■ 8.4 OTHER TRIGONOMETRIC POWERS

I. First we consider integrals of the form

$$\int \tan^n x \, dx, \qquad \int \cot^n x \, dx, \qquad n \geq 2 \text{ an integer.}$$

To integrate $\tan^n x$, set

$$\text{(1)} \quad \tan^n x = \tan^{n-2} x \tan^2 x = (\tan^{n-2} x)(\sec^2 x - 1) = \tan^{n-2} x \sec^2 x - \tan^{n-2} x.$$

If $n - 2 \geq 2$, then repeat the reduction to get

$$\begin{aligned} \tan^n x &= \tan^{n-2} x \sec^2 x - (\tan^{n-4} x \sec^2 x - \tan^{n-4} x) \\ &= \tan^{n-2} x \sec^2 x - \tan^{n-4} x \sec^2 x + \tan^{n-4} x. \end{aligned}$$

Continue until the final term is either $\pm \tan x$ (n odd) or ± 1 (n even). The idea is to obtain a sum of terms of the form $\tan^k x \sec^2 x$, pairing a power of $\tan x$ with $\sec^2 x$. These terms will be easy to integrate since they have the form $\int u^k \, du$, where $u = \tan x$, $du = \sec^2 x \, dx$.

Example 1

$$\begin{aligned}\int \tan^6 x\,dx &= \int (\tan^4 x \sec^2 x - \tan^4 x)\,dx \\ &= \int (\tan^4 x \sec^2 x - \tan^2 x \sec^2 x + \tan^2 x)\,dx \\ &= \int (\tan^4 x \sec^2 x - \tan^2 x \sec^2 x + \sec^2 x - 1)\,dx \\ &= \int \tan^4 x \sec^2 x\,dx - \int \tan^2 x \sec^2 x\,dx + \int \sec^2 x\,dx - \int dx \\ &= \tfrac{1}{5}\tan^5 x - \tfrac{1}{3}\tan^3 x + \tan x - x + C. \quad \square\end{aligned}$$

Remark The reduction formula

(8.4.1)
$$\int \tan^n x\,dx = \frac{1}{n-1}\tan^{n-1} x - \int \tan^{n-2} x\,dx, \qquad n \geq 2.$$

follows immediately from (1). ❑

Powers of the cotangent function are handled in the same manner:

$$\cot^n x = \cot^{n-2} x \cot^2 x = (\cot^{n-2} x)(\csc^2 x - 1) = \cot^{n-2} x \csc^2 x - \cot^{n-2} x.$$

The reduction formula for $\int \cot^n x\,dx$ is left as an exercise.

II. Next we consider the integrals

$$\int \sec^n x\,dx, \qquad \int \csc^n x\,dx, \qquad n \geq 2 \text{ an integer.}$$

A. n even.
For even powers write:

$$\sec^n x = \sec^{n-2} x \sec^2 x = (\tan^2 x + 1)^{(n-2)/2} \sec^2 x \qquad \text{and put } u = \tan x,$$
$$\csc^n x = \csc^{n-2} x \csc^2 x = (\cot^2 x + 1)^{(n-2)/2} \csc^2 x \qquad \text{and put } u = \cot x.$$

Example 2

$$\begin{aligned}\int \sec^6 x\,dx = \int \sec^4 x \sec^2 x\,dx &= \int (\tan^2 x + 1)^2 \sec^2 x\,dx \\ &= \int (u^2 + 1)^2\,du \qquad (u = \tan x,\ du = \sec^2 x\,dx) \\ &= \int (u^4 + 2u^2 + 1)\,du = \tfrac{1}{5}u^5 + \tfrac{2}{3}u^3 + u + C \\ &= \tfrac{1}{5}\tan^5 x + \tfrac{2}{3}\tan^3 x + \tan x + C. \quad \square\end{aligned}$$

B. n odd.
You can use integration by parts to integrate odd powers. For $\sec^n x$, set

$$u = \sec^{n-2} x, \qquad dv = \sec^2 x\, dx,$$

and when $\tan^2 x$ appears, use the identity $\tan^2 x = \sec^2 x - 1$. You can handle $\csc^n x$ in a similar manner.

Example 3 For

$$\int \sec^3 x\, dx,$$

set

$$u = \sec x, \qquad dv = \sec^2 x\, dx,$$
$$du = \sec x \tan x\, dx, \qquad v = \tan x.$$

Then

$$\begin{aligned}\int \sec^3 x\, dx &= \int \sec x \sec^2 x\, dx \\ &= \sec x \tan x - \int \tan^2 x \sec x\, dx \\ &= \sec x \tan x - \int (\sec^2 x - 1)\sec x\, dx \\ &= \sec x \tan x - \int \sec^3 x\, dx + \int \sec x\, dx.\end{aligned}$$

Therefore

$$\int \sec^3 x\, dx = \sec x \tan x - \int \sec^3 x\, dx + \ln|\sec x + \tan x|.$$

Adding $\int \sec^3 x\, dx$ to both sides, we get

$$2\int \sec^3 x\, dx = \sec x \tan x + \ln|\sec x + \tan x|$$

and

$$\int \sec^3 x\, dx = \tfrac{1}{2} \sec x \tan x + \tfrac{1}{2} \ln|\sec x + \tan x|.$$

Now add the arbitrary constant:

$$\int \sec^3 x\, dx = \tfrac{1}{2} \sec x \tan x + \tfrac{1}{2} \ln|\sec x + \tan x| + C.$$

This integral occurs so frequently in applications that you will find it listed on the inside covers of this text. ❑

Remark The integration by parts method used to integrate odd powers of the secant or cosecant can be used to derive reduction formulas that incorporate both the even and odd cases simultaneously.

Set

$$u = \sec^{n-2}x, \qquad dv = \sec^2 x\, dx.$$

Then

$$du = (n-2)\sec^{n-3}x \sec x \tan x\, dx = (n-2)\sec^{n-2}x \tan x\, dx, \qquad v = \tan x,$$

and

$$\begin{aligned}\int \sec^n x\, dx &= \sec^{n-2}x \tan x - (n-2)\int \sec^{n-2}x \tan^2 x\, dx \\ &= \sec^{n-2}x \tan x - (n-2)\int \sec^{n-2}x(\sec^2 x - 1)\, dx \\ &= \sec^{n-2}x \tan x - (n-2)\int \sec^n x\, dx + (n-2)\int \sec^{n-2}x\, dx.\end{aligned}$$

Now, solving for $\int \sec^n x\, dx$, we get

$$(n-1)\int \sec^n x\, dx = \sec^{n-2}x \tan x + (n-2)\int \sec^{n-2}x\, dx$$

from which it follows that

(8.4.2)
$$\int \sec^n x\, dx = \frac{1}{n-1}\sec^{n-2}x \tan x + \frac{n-2}{n-1}\int \sec^{n-2}x\, dx, \qquad n \geq 2.$$

For example,

$$\begin{aligned}\int \sec^6 x\, dx &= \tfrac{1}{5}\sec^4 x \tan x + \tfrac{4}{5}\int \sec^4 x\, dx \\ &= \tfrac{1}{5}\sec^4 x \tan x + \tfrac{4}{5}\left[\tfrac{1}{3}\sec^2 x \tan x + \tfrac{2}{3}\int \sec^2 x\, dx\right] \\ &= \tfrac{1}{5}\sec^4 x \tan x + \tfrac{4}{15}\sec^2 x \tan x + \tfrac{8}{15}\tan x + C.\end{aligned}$$

You should reconcile this result with that of Example 2. ❑

The reduction formula for $\int \csc^n x\, dx$ is left as an exercise.

III. Finally we come to integrals of the form

$$\int \tan^m x \sec^n x\, dx, \qquad \int \cot^m x \csc^n x\, dx, \qquad m, n \text{ positive integers.}$$

A. n even.
When n is even, write

$$\tan^m x \sec^n x = \tan^m x \sec^{n-2}x \sec^2 x$$

and express $\sec^{n-2}x$ entirely in terms of $\tan^2 x$ using $\sec^2 x = \tan^2 x + 1$.

Example 4

$$\begin{aligned}\int \tan^5 x \sec^4 x\, dx &= \int \tan^5 x \sec^2 x \sec^2 x\, dx\\ &= \int \tan^5 x(\tan^2 x + 1)\sec^2 x\, dx\\ &= \int \tan^7 x \sec^2 x\, dx + \int \tan^5 x \sec^2 x\, dx\\ &= \tfrac{1}{8}\tan^8 x + \tfrac{1}{6}\tan^6 x + C. \quad \square\end{aligned}$$

B. n, m odd.
When n and m are both odd, write

$$\tan^m x \sec^n x = \tan^{m-1} x \sec^{n-1} x \sec x \tan x$$

and express $\tan^{m-1} x$ entirely in terms of $\sec^2 x$ using $\tan^2 x = \sec^2 x - 1$.

Example 5

$$\begin{aligned}\int \tan^5 x \sec^3 x\, dx &= \int \tan^4 x \sec^2 x \sec x \tan x\, dx\\ &= \int (\sec^2 x - 1)^2 \sec^2 x \sec x \tan x\, dx\\ &= \int (\sec^6 x - 2\sec^4 x + \sec^2 x)\sec x \tan x\, dx\\ &= \tfrac{1}{7}\sec^7 x - \tfrac{2}{5}\sec^5 x + \tfrac{1}{3}\sec^3 x + C. \quad \square\end{aligned}$$

C. n odd, m even.
Finally, if n is odd and m is even, use $\tan^2 x = \sec^2 x - 1$ to write the product as a sum of odd powers of the secant. Then you can either use integration by parts as in II just given, or the reduction formula (8.4.2).

Example 6

$$\begin{aligned}\int \tan^2 x \sec x\, dx &= \int (\sec^2 x - 1)\sec x\, dx\\ &= \int (\sec^3 x - \sec x)\, dx = \int \sec^3 x\, dx - \int \sec x\, dx.\end{aligned}$$

We have calculated each of these integrals before:

$$\int \sec^3 x\, dx = \tfrac{1}{2}\sec x \tan x + \tfrac{1}{2}\ln|\sec x + \tan x| + C$$

and

$$\int \sec x\, dx = \ln|\sec x + \tan x| + C.$$

It follows that

$$\int \tan^2 x \sec x\, dx = \tfrac{1}{2}\sec x \tan x - \tfrac{1}{2}\ln|\sec x + \tan x| + C. \quad \square$$

You can handle integrals of $\cot^m x \csc^n x$ in a similar manner.

EXERCISES 8.4

In Exercises 1–26, evaluate the given integral.

1. $\int \tan^2 3x\,dx$.
2. $\int \cot^2 5x\,dx$.
3. $\int \sec^2 \pi x\,dx$.
4. $\int \csc^2 2x\,dx$.
5. $\int \tan^3 x\,dx$.
6. $\int \cot^3 x\,dx$.
7. $\int \tan^2 x \sec^2 x\,dx$.
8. $\int \cot^2 x \csc^2 x\,dx$.
9. $\int \csc^3 x\,dx$.
10. $\int \sec^3 \pi x\,dx$.
11. $\int \cot^4 x\,dx$.
12. $\int \tan^4 x\,dx$.
13. $\int \cot^3 x \csc^3 x\,dx$.
14. $\int \tan^3 x \sec^3 x\,dx$.
15. $\int \csc^4 2x\,dx$.
16. $\int \sec^4 3x\,dx$.
17. $\int \cot^2 x \csc x\,dx$.
18. $\int \csc^3(\frac{1}{2}x)\,dx$.
19. $\int \tan^5 3x\,dx$.
20. $\int \cot^5 2x\,dx$.
21. $\int \sec^5 x\,dx$.
22. $\int \csc^5 x\,dx$.
23. $\int \tan^4 x \sec^4 x\,dx$.
24. $\int \cot^4 x \csc^4 x\,dx$.
25. $\int e^{2x}\tan^2(e^{2x})\sec^2(e^{2x})\,dx$.
26. $\int \tan^2(\frac{1}{2}x)\sec^3(\frac{1}{2}x)\,dx$.

In Exercises 27–32, evaluate the definite integral.

27. $\int_0^{\pi/4} \tan^3 x \sec^2 x\,dx$.
28. $\int_{\pi/4}^{\pi/2} \csc^3 x \cot x\,dx$.
29. $\int_0^{\pi/6} \tan^2 2x\,dx$.
30. $\int_0^{\pi/3} \tan x \sec^{3/2} x\,dx$.
31. $\int_{\pi/6}^{\pi/3} \cot^3 x \csc^3 x\,dx$.
32. $\int_{\pi/6}^{\pi/2} \cot^2 x\,dx$.

33. The region bounded by the y-axis, the line $y = 1$ and the graph of $y = \tan x$, $x \in [0, \pi/4]$, is revolved around the x-axis. Find the volume of the solid that is generated.
34. The region bounded by the graph of $y = \tan^2 x$ and the x-axis, $x \in [0, \pi/4]$, is revolved around the x-axis. Find the volume of the solid that is generated.
35. The region bounded by the graph of $y = \tan x$ and the x-axis, $x \in [0, \pi/4]$, is revolved around the line $y = -1$. Find the volume of the solid that is generated.
36. The region bounded by the graph of $y = \sec^2 x$ and the x-axis, $x \in [0, \pi/4]$, is revolved around the x-axis. Find the volume of the solid that is generated.
37. Derive the reduction formula: for $n > 1$,
$$\int \cot^n x\,dx = \frac{-\cot^{n-1} x}{n-1} - \int \cot^{n-2} x\,dx.$$
38. Use Exercise 37 to evaluate the following integrals:
(a) $\int \cot^3 x\,dx$. (b) $\int \cot^4 x\,dx$. (c) $\int \cot^5 2x\,dx$.
39. Derive the reduction formula: for $n > 1$,
$$\int \csc^n x\,dx = \frac{-\csc^{n-2} x \cot x}{n-1} + \frac{n-2}{n-1}\int \csc^{n-2} x\,dx.$$
40. Use Exercise 39 to evaluate the integrals:
(a) $\int \csc^3 x\,dx$. (b) $\int \csc^4 x\,dx$. (c) $\int \csc^5 3x\,dx$.

■ 8.5 INTEGRALS INVOLVING $\sqrt{a^2 \pm x^2}$ and $\sqrt{x^2 - a^2}$; TRIGONOMETRIC SUBSTITUTIONS

Integrals involving one of the forms $\sqrt{a^2 - x^2}$, $\sqrt{a^2 + x^2}$, or $\sqrt{x^2 - a^2}$ can often be simplified by making a trigonometric substitution. These substitutions transform the integrand into a trigonometric form like those studied in the two preceding sections. The three cases with their suggested substitutions are:

1. For $\sqrt{a^2 - x^2}$ set $a \sin u = x$.
2. For $\sqrt{a^2 + x^2}$ set $a \tan u = x$.
3. For $\sqrt{x^2 - a^2}$ set $a \sec u = x$.

In each case, take $a > 0$.

To illustrate the idea behind these substitutions, consider the first case: $\sqrt{a^2 - x^2}$. If we set $a \sin u = x$, then

$$\sqrt{a^2 - x^2} = \sqrt{a^2 - a^2\sin^2 u} = a\sqrt{1 - \sin^2 u} = a\sqrt{\cos^2 u} = a\,|\cos u|.$$

Since $a \sin u = x$, we have $u = \sin^{-1}(x/a)$, which implies that $-\pi/2 \le u \le \pi/2$ (see Section 7.8). Hence, $\cos u \ge 0$, and

$$\sqrt{a^2 - x^2} = a \cos u.$$

Thus, the substitution $a \sin u = x$ has the effect of replacing the radical $\sqrt{a^2 - x^2}$ by $a \cos u$. The other two cases can be illustrated in a similar manner.

Right triangles provide an easy way to relate the forms with their corresponding substitutions:

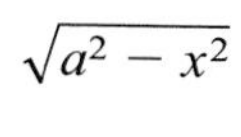

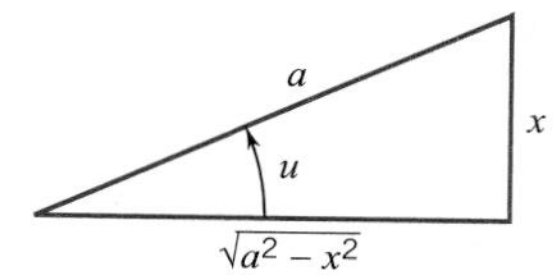

$\sin u = x/a$ or $x = a \sin u$

$\sqrt{a^2 - x^2} = a \cos u$

$\sqrt{a^2 + x^2}$

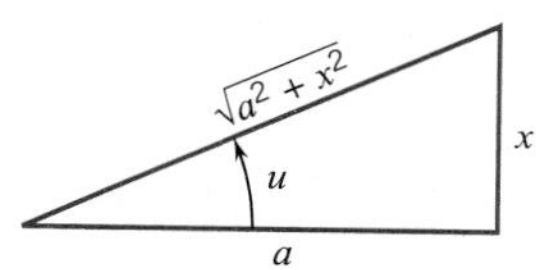

$\tan u = x/a$ or $x = a \tan u$

$\sqrt{a^2 + x^2} = a \sec u$

$\sqrt{x^2 - a^2}$

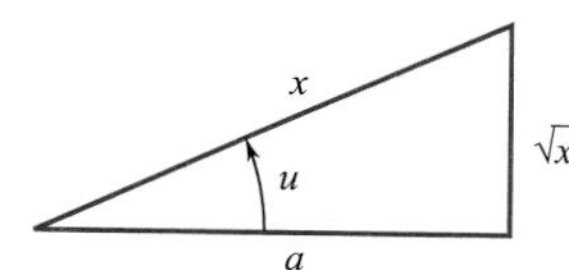

$\sec u = x/a$ or $x = a \sec u$

$\sqrt{x^2 - a^2} = a \tan u$

Example 1 Find

$$\int \frac{dx}{(9 - x^2)^{3/2}}.$$

SOLUTION First note that

$$\int \frac{dx}{(9 - x^2)^{3/2}} = \int \frac{dx}{[\sqrt{9 - x^2}]^3};$$

the integral involves $\sqrt{9 - x^2}$. Therefore, we set

$$3 \sin u = x, \qquad 3 \cos u\, du = dx.$$

Then

$$\sqrt{9 - x^2} = 3 \cos u. \qquad \text{(Figure 8.5.1)}$$

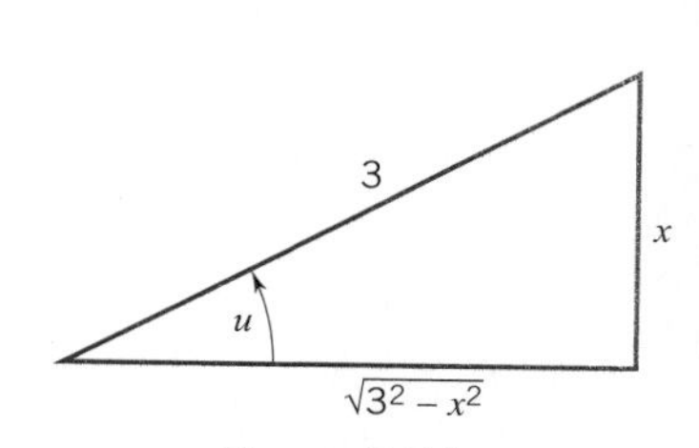

Figure 8.5.1

Now

$$\int \frac{dx}{(9-x^2)^{3/2}} = \int \frac{3\cos u\, du}{(3\cos u)^3}$$
$$= \tfrac{1}{9}\int \frac{\cos u}{\cos^3 u}\, du$$
$$= \tfrac{1}{9}\int \sec^2 u\, du$$
$$= \tfrac{1}{9}\tan u + C = \frac{x}{9\sqrt{9-x^2}} + C \quad \text{(see Figure 8.5.1).} \quad ❑$$

Example 2 Find

$$\int \sqrt{a^2+x^2}\, dx.$$

SOLUTION Set

$$a\tan u = x, \qquad a\sec^2 u\, du = dx.$$

Then

$$\sqrt{a^2+x^2} = a\sec u. \qquad \text{(Figure 8.5.2)}$$

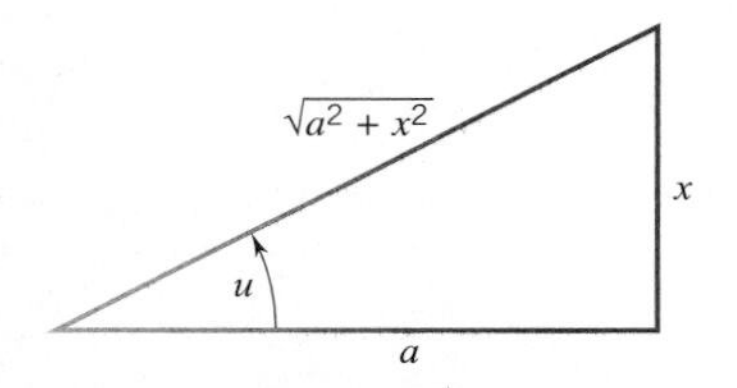

Figure 8.5.2

Now

$$\int \sqrt{a^2+x^2}\, dx = \int (a\sec u) a\sec^2 u\, du$$
$$= a^2 \int \sec^3 u\, du$$
$$= \frac{a^2}{2}(\sec u\tan u + \ln|\sec u + \tan u|) + C$$

Example 3, Section 8.4 ⟶

$$= \frac{a^2}{2}\left[\frac{\sqrt{a^2+x^2}}{a}\left(\frac{x}{a}\right) + \ln\left|\frac{\sqrt{a^2+x^2}}{a} + \frac{x}{a}\right|\right] + C$$
$$= \tfrac{1}{2}x\sqrt{a^2+x^2} + \tfrac{1}{2}a^2\ln(x+\sqrt{a^2+x^2}) - \tfrac{1}{2}a^2\ln a + C$$

(see Figure 8.5.2).

We can absorb the constant $-\frac{1}{2}a^2\ln a$ in C and write

(8.5.1) $$\int \sqrt{a^2+x^2}\, dx = \tfrac{1}{2}x\sqrt{a^2+x^2} + \tfrac{1}{2}\, a^2 \ln(x+\sqrt{a^2+x^2}) + C.$$

This is a standard integration formula. ❑

Example 3 Find

$$\int \frac{dx}{x^2\sqrt{x^2-4}}.$$

SOLUTION Set

$$2\sec u = x, \qquad 2\sec u \tan u\, du = dx.$$

Then

$$\sqrt{x^2-4} = 2\tan u. \qquad \text{(Figure 8.5.3)}$$

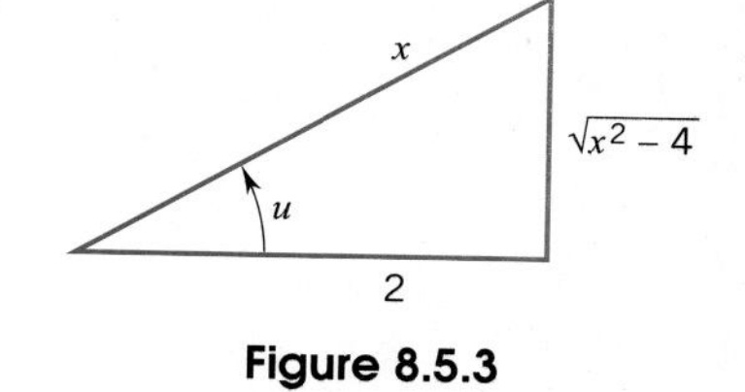

Figure 8.5.3

Now

$$\begin{aligned}\int \frac{dx}{x^2\sqrt{x^2-4}} &= \int \frac{2\sec u \tan u}{4\sec^2 u \cdot 2\tan u}\, du\\ &= \tfrac{1}{4}\int \frac{1}{\sec u}\, du\\ &= \tfrac{1}{4}\int \cos u\, du\\ &= \tfrac{1}{4}\sin u + C = \frac{\sqrt{x^2-4}}{4x} + C \qquad \text{(see Figure 8.5.3).} \quad ❑\end{aligned}$$

Remark Consider the integral

$$\int \frac{x}{\sqrt{a^2-x^2}}\, dx.$$

The appearance of $\sqrt{a^2-x^2}$ in the integrand suggests the substitution $a\sin u = x$, and if you carry through with this substitution, you will arrive at the correct result. However, there is a much simpler approach here. Set

$$u = a^2 - x^2, \qquad du = -2x\, dx.$$

Then

$$\int \frac{x}{\sqrt{a^2-x^2}}\, dx = -\tfrac{1}{2}\int \frac{1}{\sqrt{u}}\, du = -\sqrt{u} + C = -\sqrt{a^2-x^2} + C.$$

You should be aware that there will sometimes be more than one way to calculate an integral. In such cases, it will obviously be to your advantage to choose the easiest method. Only experience and practice can give you the insight to do this consistently. ❑

Now a slight variation.

Example 4 Find

$$\int \frac{dx}{x\sqrt{4x^2+9}}.$$

SOLUTION Set

$$3\tan u = 2x, \qquad 3\sec^2 u\, du = 2\, dx.$$

Then

$$\sqrt{4x^2 + 9} = 3 \sec u \qquad \text{(Figure 8.5.4)}$$

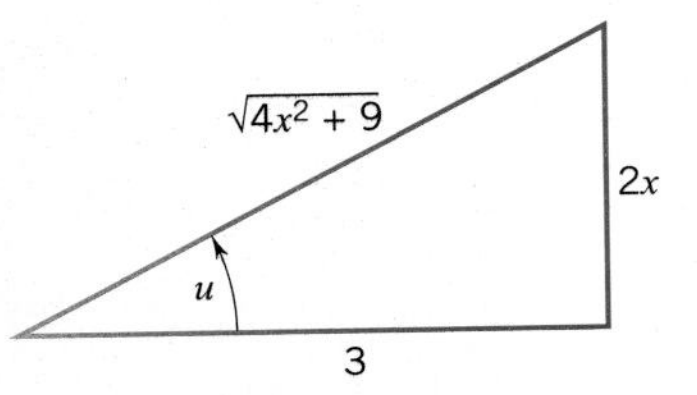

Figure 8.5.4

and

$$\begin{aligned}
\int \frac{dx}{x\sqrt{4x^2 + 9}} &= \int \frac{\frac{3}{2} \sec^2 u}{\frac{3}{2} \tan u \cdot 3 \sec u}\, du \\
&= \frac{1}{3} \int \frac{\sec u}{\tan u}\, du \\
&= \frac{1}{3} \int \csc u \, du \\
&= \frac{1}{3} \ln|\csc u - \cot u| + C \\
&= \frac{1}{3} \ln\left|\frac{\sqrt{4x^2 + 9} - 3}{2x}\right| + C \qquad \text{(see Figure 8.5.4).} \quad ❑
\end{aligned}$$

The next example requires that we first complete the square under the radical.

Example 5 Find

$$\int \frac{x}{\sqrt{x^2 + 2x - 3}}\, dx.$$

SOLUTION First note that

$$\int \frac{x}{\sqrt{x^2 + 2x - 3}}\, dx = \int \frac{x}{\sqrt{(x + 1)^2 - 4}}\, dx.$$

Now set

$$2 \sec u = x + 1, \qquad 2 \sec u \tan u \, du = dx.$$

Then

$$\sqrt{(x + 1)^2 - 4} = 2 \tan u \qquad \text{(Figure 8.5.5)}$$

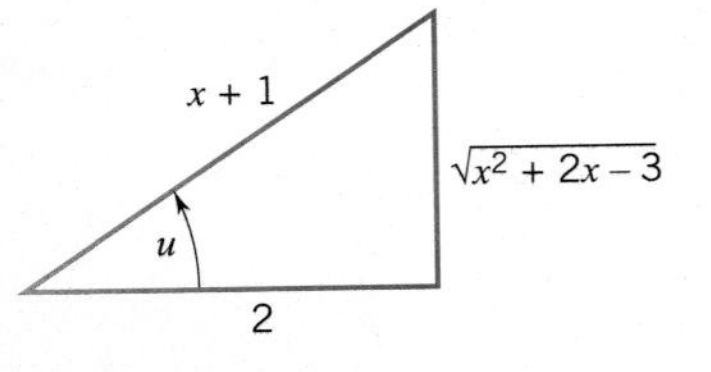

Figure 8.5.5

and

$$\begin{aligned}
\int \frac{x}{\sqrt{(x + 1)^2 - 4}}\, dx &= \int \frac{(2 \sec u - 1)2 \sec u \tan u}{2 \tan u}\, du \\
&= \int (2 \sec^2 u - \sec u)\, du \\
&= 2 \tan u - \ln|\sec u + \tan u| + C \\
&= \sqrt{x^2 + 2x - 3} - \ln\left|\frac{x + 1 + \sqrt{x^2 + 2x - 3}}{2}\right| + C. \quad ❑
\end{aligned}$$

Example 6 Find the area enclosed by the ellipse $(x^2/a^2) + (y^2/b^2) = 1$ (Figure 8.5.6).

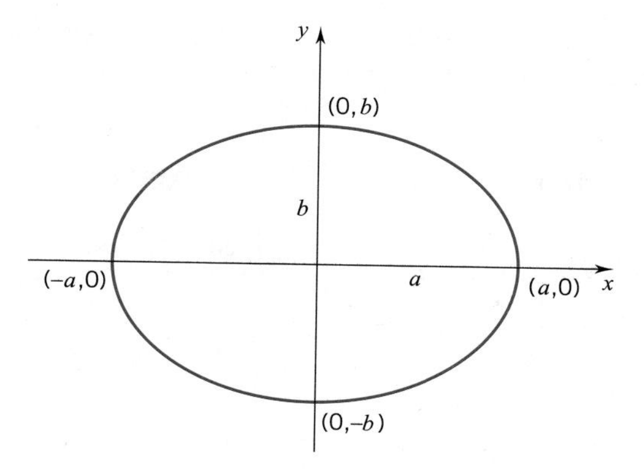

Figure 8.5.6

SOLUTION We first solve the equation of the ellipse for y:

$$y = \pm\frac{b}{a}\sqrt{a^2 - x^2}.$$

Since the graph is symmetric with respect to the x- and y-axes, the total area A enclosed by the ellipse is four times the area of the region in the first quadrant. This region is bounded by the graph of

$$y = \frac{b}{a}\sqrt{a^2 - x^2}, \qquad 0 \le x \le a.$$

Therefore

$$A = 4\int_0^a \frac{b}{a}\sqrt{a^2 - x^2}\,dx = \frac{4b}{a}\int_0^a \sqrt{a^2 - x^2}\,dx.$$

Set

$$a \sin u = x, \qquad a \cos u\,du = dx.$$

Then

$$\sqrt{a^2 - x^2} = a \cos u.$$

Also, $u = 0$ at $x = 0$ and $u = \pi/2$ at $x = a$. Thus

$$\begin{aligned} A &= \frac{4b}{a}\int_0^a \sqrt{a^2 - x^2}\,dx = \frac{4b}{a}\int_0^{\pi/2} (a \cos u) \cdot a \cos u\,du \\ &= 4ab\int_0^{\pi/2} \cos^2 u\,du = 4ab\int_0^{\pi/2} \left(\tfrac{1}{2} + \tfrac{1}{2}\cos 2u\right) du \\ &= 2ab\left[u + \tfrac{1}{2}\sin 2u\right]_0^{\pi/2} = \pi ab. \end{aligned}$$

Hence, the area enclosed by an ellipse with semiaxes a and b is πab. Note that if $a = b$, then the ellipse is a circle of radius $r = a = b$ and the area is the familiar formula $A = \pi r^2$. ❑

Finally, we note that a trigonometric substitution may also be effective in cases where the quadratic in the integrand is not under a radical. For example, the reduction formula

(8.5.2)
$$\int \frac{dx}{(x^2 + a^2)^n} = \frac{1}{a^{2n-1}} \int \cos^{2(n-1)} u \, du$$

follows from the substitution: $a \tan u = x$, $a \sec^2 u \, du = dx$. The proof is left to you as an exercise.

Hyperbolic Substitutions

Integrals involving $\sqrt{a^2 + x^2}$ or $\sqrt{x^2 - a^2}$, $a > 0$, can also be handled using a hyperbolic substitution. The ideas here are the same as the ideas supporting trigonometric substitutions, and are based on the fundamental identity

$$\cosh^2 u - \sinh^2 u = 1, \qquad \text{for all real } u. \qquad \text{[see (7.9.3)]}$$

For an integral involving $\sqrt{a^2 + x^2}$, set

$$a \sinh u = x.$$

Then

$$\sqrt{a^2 + x^2} = \sqrt{a^2 + a^2 \sinh^2 u} = a\sqrt{1 + \sinh^2 u} = a\sqrt{\cosh^2 u} = a \cosh u$$

since $\cosh u > 0$ for all u.

For an integral involving $\sqrt{x^2 - a^2}$, where $x \geq a$, set

$$a \cosh u = x, \qquad u \geq 0.$$

Then

$$\sqrt{x^2 - a^2} = \sqrt{a^2 \cosh^2 u - a^2} = a\sqrt{\cosh^2 u - 1} = a\sqrt{\sinh^2 u} = a \sinh u,$$

since $\sinh u \geq 0$ when $u \geq 0$.

Example 7 Use a hyperbolic substitution to find

$$\int \sqrt{a^2 + x^2} \, dx. \qquad \text{(see Example 2)}$$

SOLUTION Set $a \sinh u = x$. Then $dx = a \cosh u \, du$, $\sqrt{a^2 + x^2} = a \cosh u$, and

$$\begin{aligned}
\int \sqrt{a^2 + x^2} \, dx &= \int (a \cosh u)(a \cosh u) \, du \\
&= a^2 \int \cosh^2 u \, du \\
&= a^2 \int \left(\tfrac{1}{2} \cosh 2u + \tfrac{1}{2}\right) du \qquad \text{[see (7.9.3)]} \\
&= \frac{a^2}{2}\left[\tfrac{1}{2} \sinh 2u + u\right] + C
\end{aligned}$$

$$= \frac{a^2}{2}[\sinh u \cosh u + u] + C \qquad \text{[see (7.9.3)]}$$

$$= \frac{a^2}{2}\left[\frac{x}{a} \cdot \frac{\sqrt{a^2 + x^2}}{a} + \sinh^{-1}\left(\frac{x}{a}\right)\right] + C$$

$$= \frac{a^2}{2}\left[\frac{x\sqrt{a^2 + x^2}}{a^2} + \ln\left(\frac{x}{a} + \frac{\sqrt{a^2 + x^2}}{a}\right)\right] + C \qquad \text{(see Theorem 7.10.2)}$$

$$= \tfrac{1}{2} x\sqrt{a^2 + x^2} + \tfrac{1}{2} a^2 \ln(x + \sqrt{a^2 + x^2}) + C$$

as we saw in Example 2. ❑

EXERCISES 8.5

In Exercises 1–44, evaluate the given integral.

1. $\displaystyle\int \frac{dx}{\sqrt{a^2 - x^2}}$.
2. $\displaystyle\int \frac{dx}{(x^2 + 2)^{3/2}}$.
3. $\displaystyle\int_0^1 \frac{dx}{(5 - x^2)^{3/2}}$.
4. $\displaystyle\int_{5/2}^4 \frac{x}{\sqrt{x^2 - 4}}\, dx$.
5. $\displaystyle\int \sqrt{x^2 - 1}\, dx$.
6. $\displaystyle\int \frac{x}{\sqrt{4 - x^2}}\, dx$.
7. $\displaystyle\int \frac{x^2}{\sqrt{4 - x^2}}\, dx$.
8. $\displaystyle\int \frac{x^2}{\sqrt{x^2 - 4}}\, dx$.
9. $\displaystyle\int \frac{x}{(1 - x^2)^{3/2}}\, dx$.
10. $\displaystyle\int \frac{x^2}{\sqrt{4 + x^2}}\, dx$.
11. $\displaystyle\int_0^{1/2} \frac{x^2}{(1 - x^2)^{3/2}}\, dx$.
12. $\displaystyle\int \frac{x}{a^2 + x^2}\, dx$.
13. $\displaystyle\int x\sqrt{4 - x^2}\, dx$.
14. $\displaystyle\int_0^2 \frac{x^3}{\sqrt{16 - x^2}}\, dx$.
15. $\displaystyle\int_0^5 x^2\sqrt{25 - x^2}\, dx$.
16. $\displaystyle\int \frac{e^x}{\sqrt{9 - e^{2x}}}\, dx$.
17. $\displaystyle\int \frac{x^2}{(x^2 + 8)^{3/2}}\, dx$.
18. $\displaystyle\int \frac{\sqrt{1 - x^2}}{x^4}\, dx$.
19. $\displaystyle\int \frac{dx}{x\sqrt{a^2 - x^2}}$.
20. $\displaystyle\int \frac{dx}{\sqrt{x^2 + a^2}}$.
21. $\displaystyle\int \frac{dx}{\sqrt{x^2 - a^2}}$.
22. $\displaystyle\int_{-a}^a \sqrt{a^2 - x^2}\, dx$.
23. $\displaystyle\int e^x\sqrt{e^{2x} - 1}\, dx$.
24. $\displaystyle\int_0^2 \frac{1}{\sqrt{4 + x^2}}\, dx$.
25. $\displaystyle\int_0^3 \frac{x^3}{\sqrt{9 + x^2}}\, dx$.
26. $\displaystyle\int \frac{\sqrt{x^2 - 1}}{x}\, dx$.
27. $\displaystyle\int \frac{dx}{x^2\sqrt{a^2 + x^2}}$.
28. $\displaystyle\int \frac{dx}{x^2\sqrt{a^2 - x^2}}$.
29. $\displaystyle\int \frac{dx}{x^2\sqrt{x^2 - a^2}}$.
30. $\displaystyle\int_4^6 \frac{1}{x\sqrt{x^2 - 4}}\, dx$.
31. $\displaystyle\int_2^{2\sqrt{2}} \frac{\sqrt{x^2 - 4}}{x}\, dx$.
32. $\displaystyle\int \frac{dx}{e^x\sqrt{4 + e^{2x}}}$.
33. $\displaystyle\int \frac{dx}{e^x\sqrt{e^{2x} - 9}}$.
34. $\displaystyle\int \frac{dx}{\sqrt{x^2 - 2x - 3}}$.
35. $\displaystyle\int \frac{dx}{(x^2 - 4x + 4)^{3/2}}$.
36. $\displaystyle\int \frac{x}{\sqrt{6x - x^2}}\, dx$.
37. $\displaystyle\int x\sqrt{6x - x^2 - 8}\, dx$.
38. $\displaystyle\int \frac{x + 2}{\sqrt{x^2 + 4x + 13}}\, dx$.
39. $\displaystyle\int \frac{x}{(x^2 + 2x + 5)^2}\, dx$.
40. $\displaystyle\int \frac{x}{\sqrt{x^2 - 2x - 3}}\, dx$.
41. $\displaystyle\int \frac{x + 3}{\sqrt{x^2 + 4x + 13}}\, dx$.
42. $\displaystyle\int x\sqrt{x^2 + 6x}\, dx$.
43. $\displaystyle\int \sqrt{6x - x^2 - 8}\, dx$.
44. $\displaystyle\int \frac{dx}{(4 - x^2)^2}$.

45. Verify the reduction formula (8.5.2).

In Exercises 46 and 47, use the reduction formula (8.5.2) to evaluate the given integral.

46. $\displaystyle\int \frac{1}{(x^2 + 1)^2}\, dx$.
47. $\displaystyle\int \frac{1}{(x^2 + 1)^3}\, dx$.

48. Calculate

$$\int x \sin^{-1}x\, dx.$$

49. (a) Use the trigonometric substitution $x = a\tan u$ to show that

$$\int \frac{1}{\sqrt{x^2 + a^2}}\, dx = \ln(x + \sqrt{x^2 + a^2}) + C_1.$$

(b) Use the hyperbolic substitution $x = a \sinh u$ to show that

$$\int \frac{1}{\sqrt{x^2 + a^2}}\,dx = \sinh^{-1}\left(\frac{x}{a}\right) + C_2.$$

(c) Justify the results in parts (a) and (b). HINT: see Section 7.10.

50. (a) Use a trigonometric substitution to show that

$$\int \sqrt{x^2 - a^2}\,dx = \frac{1}{2}\left[x\sqrt{x^2 - a^2} - a^2 \ln(x + \sqrt{x^2 - a^2})\right] + C_1, \; x \geq a.$$

(b) Use a hyperbolic substitution to show that

$$\int \sqrt{x^2 - a^2}\,dx = \frac{1}{2}\left[x\sqrt{x^2 - a^2} - a^2 \cosh^{-1}\left(\frac{x}{a}\right)\right] + C_2.$$

(c) Reconcile the results in parts (a) and (b).

51. The region bounded by the graph of $f(x) = 1/(1 + x^2)$ and the x-axis between $x = 0$ and $x = 1$ is revolved around the x-axis. Find the volume of the solid that is generated.

52. In a disc of radius r a chord h units from the center generates a region of the disc called a *segment* (see the figure). Find a formula for the area of the segment.

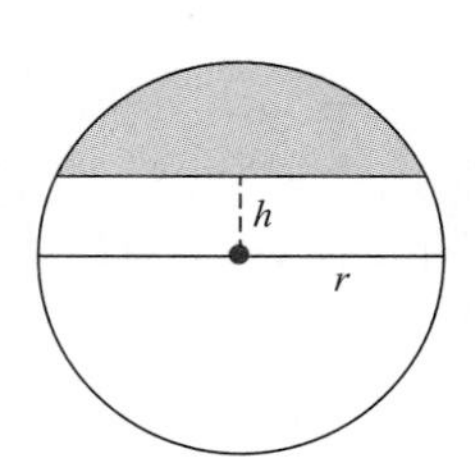

53. Derive the formula $A = \frac{1}{2}r^2\theta$ for the area of a sector of a circle of radius r and central angle θ (measured in radians). HINT: assume first that $0 < \theta < \frac{1}{2}\pi$ and subdivide the region as indicated in the figure. Then verify that the formula holds for any sector.

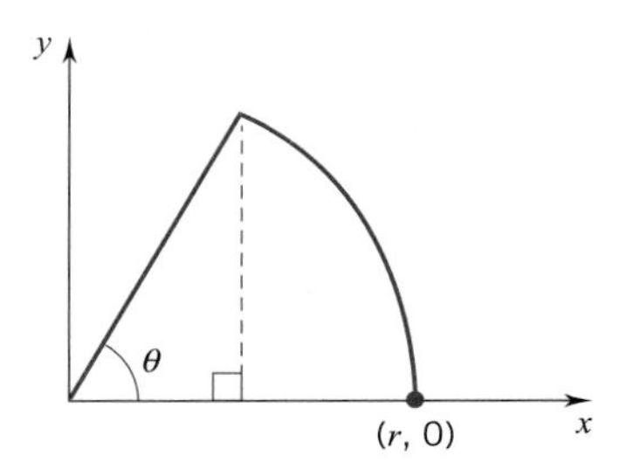

54. Find the area of the region bounded on the left and right by the two branches of the hyperbola $(x^2/a^2) - (y^2/b^2) = 1$, and above and below by the lines $y = \pm b$.

55. Calculate the mass and the center of mass of a rod that extends from $x = 0$ to $x = a > 0$ and has mass density $\lambda(x) = (x^2 + a^2)^{-1/2}$.

56. Calculate the mass and the center of mass of the rod in Exercise 55 if the mass density is given by $\lambda(x) = (x^2 + a^2)^{-3/2}$.

For Exercises 57–59, let Ω be the region under the curve $y = \sqrt{x^2 - a^2}$, $x \in [a, \sqrt{2}\,a]$.

57. Sketch Ω, find its area, and locate the centroid.

58. Find the volume of the solid generated by revolving Ω about the x-axis and determine the centroid of that solid.

59. Find the volume of the solid generated by revolving Ω about the y-axis and determine the centroid of that solid.

60. Let

$$f(x) = \frac{x^2}{\sqrt{1 - x^2}}.$$

(a) Use a graphing utility to sketch the graph of f.
(b) Find the area of the region bounded by the graph of f and the x-axis between $x = 0$ and $x = \frac{1}{2}$.
(c) Find the volume of the solid generated by revolving the region in part (b) around the y-axis.

61. Let

$$f(x) = \frac{\sqrt{x^2 - 9}}{x^2}, \quad x \geq 3.$$

(a) Use a graphing utility to sketch the graph of f.
(b) Find the area of the region bounded by the graph of f and the x-axis between $x = 3$ and $x = 6$.
(c) Find the centroid of the region.

■ 8.6 PARTIAL FRACTIONS

In this section we present a method for integrating rational functions. Recall that a rational function is, by definition, the quotient of two polynomials. For example,

$$\frac{1}{x^2-4},\qquad \frac{2x^2+3}{x(x-1)^2},\qquad \frac{-2x}{(x+1)(x^2+1)},\qquad \frac{1}{x(x^2+x+1)},$$

$$\frac{3x^4+x^3+20x^3+3x+31}{(x+1)(x^2+4)^2},\qquad \frac{x^5}{x^2-1}$$

are all rational functions, but

$$\frac{1}{\sqrt{x}},\qquad \ln x,\qquad \frac{|x-2|}{x^2}$$

are not rational functions.

A rational function $R(x) = P(x)/Q(x)$ is said to be *proper* if the degree of the numerator is less than the degree of the denominator. If the degree of the numerator is greater than or equal to the degree of the denominator, then the rational function is *improper.*† We will focus our attention on *proper rational functions* because any improper rational function can be written as the sum of a polynomial and a proper rational function

$$\frac{P(x)}{Q(x)} = p(x) + \frac{r(x)}{Q(x)}.\dagger\dagger$$

This is accomplished simply by dividing the denominator into the numerator (the polynomial p is called the quotient and the polynomial r is the remainder). Take, for example,

$$\frac{x^5+2}{x^2-1}.$$

Carrying out the suggested division, we have

$$\begin{array}{r@{}l} & x^3 + x \\ x^2-1\,\big) & x^5 \qquad\qquad + 2 \\ & \underline{x^5 - x^3} \\ & \quad x^3 \\ & \quad \underline{x^3 - x} \\ & \qquad\quad x + 2. \end{array}$$

This gives

$$\frac{x^5+2}{x^2-1} = x^3 + x + \frac{x+2}{x^2-1}.$$

It is shown in algebra that every proper rational function can be written in one and only one way as a sum of fractions of the form

(8.6.1)
$$\boxed{\frac{A}{(x-\alpha)^k} \quad\text{and}\quad \frac{Bx+C}{(x^2+\beta x+\gamma)^k}}$$

† These terms are taken from the familiar terms used to describe (numerical) fractions.

†† This is analogous to writing an improper fraction as a so-called *mixed number.*

with the quadratic $x^2 + \beta x + \gamma$ irreducible (that is, not factorable into linear factors with real coefficients; $\beta^2 - 4\gamma < 0$). Such fractions are called *partial fractions.* When a rational function $R(x)$ is expressed as a sum of partial fractions, the sum is called the *partial fraction decomposition of* R.

We begin by giving some examples of partial fraction decompositions. We then illustrate how these are used to calculate integrals.

Example 1 (*The denominator splits into distinct linear factors.*) The partial fraction decomposition of the rational function

$$\frac{1}{x^2 - 4} = \frac{1}{(x - 2)(x + 2)}$$

has the form

$$\frac{1}{x^2 - 4} = \frac{A}{x - 2} + \frac{B}{x + 2},$$

where A and B are constants whose values are to be determined. This is an identity that holds for all $x \neq 2, -2$. Clearing the fractions, we have

$$1 = A(x + 2) + B(x - 2), \tag{1}$$

which is still an identity, but now it holds for all x. We illustrate two methods for finding A and B.

METHOD 1 Substitute numbers for x in (1):

Setting $x = 2$, we get $1 = 4A$, which gives $A = \frac{1}{4}$;

Setting $x = -2$, we get $1 = -4B$, which gives $B = -\frac{1}{4}$.

Thus, the desired decomposition is

$$\frac{1}{x^2 - 4} = \frac{1}{4(x - 2)} - \frac{1}{4(x + 2)}.$$

You can verify this by carrying out the subtraction on the right.

METHOD 2 Clear the parentheses on the right-hand side of (1) and rewrite the equation as

$$1 = (A + B)x + 2A - 2B.$$

Now, equate the coefficients of the corresponding powers of x to produce the system of equations

$$A + B = 0$$

$$2A - 2B = 1.$$

We can then get A and B by solving these equations simultaneously. The solutions are, of course, $A = \frac{1}{4}$, $B = -\frac{1}{4}$. ❑

> In general, each distinct linear factor $x - \alpha$ in the denominator gives rise to a term of the form
>
> $$\frac{A}{x - \alpha}.$$

Example 2 (*The denominator has a repeated linear factor.*) For

$$\frac{2x^2 + 3}{x(x-1)^2},$$

we write

$$\frac{2x^2 + 3}{x(x-1)^2} = \frac{A}{x} + \frac{B}{x-1} + \frac{C}{(x-1)^2}.$$

This leads to

$$2x^2 + 3 = A(x-1)^2 + Bx(x-1) + Cx.$$

To determine the three coefficients A, B, C we need to substitute three values for x. We select 0 and 1 because for those values of x several terms on the right side will drop out. As a third value of x, any other number will do; we select 2 just to keep the arithmetic simple.

Setting $x = 0$, we get $3 = A$.

Setting $x = 1$, we get $5 = C$.

Setting $x = 2$, we get $11 = A + 2B + 2C$,

which, with $A = 3$ and $C = 5$, gives $B = -1$.

The decomposition is therefore

$$\frac{2x^2 + 3}{x(x-1)^2} = \frac{3}{x} - \frac{1}{x-1} + \frac{5}{(x-1)^2}. \quad \square$$

Remark You can verify that the alternative method for finding the constants A, B, and C leads to the system of equations

$$\begin{aligned} A + B \qquad &= 2 \\ -2A - B + C &= 0 \\ A \qquad\qquad &= 3. \end{aligned}$$

We can then get A, B, and C by solving these equations simultaneously. In general, this approach involves more algebra, and so we will emphasize the method of substituting *well-chosen* values of x in the examples which follow. It is also possible to combine the two methods and, with practice, you will see that this is often a convenient approach. We will illustrate this in the next example. ☐

> In general, each factor of the form $(x - \alpha)^k$ in the denominator gives rise to an expression of the form
>
> $$\frac{A_1}{x-\alpha} + \frac{A_2}{(x-\alpha)^2} + \cdots + \frac{A_k}{(x-\alpha)^k}.$$

Example 3 (*The denominator has an irreducible quadratic factor.*) For

$$\frac{x^2 + 5x + 2}{(x+1)(x^2+1)},$$

we write

$$\frac{x^2 + 5x + 2}{(x + 1)(x^2 + 1)} = \frac{A}{x + 1} + \frac{Bx + C}{x^2 + 1}$$

and obtain

$$x^2 + 5x + 2 = A(x^2 + 1) + (Bx + C)(x + 1).$$

This time we substitute $x = -1$, 0, and 1.

Setting $x = -1$, we get $-2 = 2A$, which gives $A = -1$.

Setting $x = 0$, we get $2 = A + C$, which gives $C = 3$.

Setting $x = 1$, we get $8 = 2A + 2B + 2C$,

which, with $A = -1$ and $C = 3$, gives $B = 2$.

Alternatively, after substituting -1 and 0, we could have noted that the coefficient of x^2 on the right-hand side is $A + B$ and on the left-hand side it's 1. Thus, $A + B = 1$ and $B = 1 - (-1) = 2$. This illustrates the "combined" approach. The decomposition reads

$$\frac{x^2 + 5x + 2}{(x + 1)(x^2 + 1)} = \frac{-1}{x + 1} + \frac{2x + 3}{x^2 + 1}. \quad \square$$

Example 4 (*The denominator has an irreducible quadratic factor.*) For

$$\frac{1}{x(x^2 + x + 1)}$$

we write

$$\frac{1}{x(x^2 + x + 1)} = \frac{A}{x} + \frac{Bx + C}{x^2 + x + 1}$$

and obtain

$$1 = A(x^2 + x + 1) + (Bx + C)x.$$

Again we select values of x that produce zeros or simple arithmetic on the right side.

$$1 = A \qquad (x = 0),$$

$$1 = 3A + B + C \qquad (x = 1),$$

$$1 = A + B - C \qquad (x = -1).$$

From this we find that

$$A = 1, \qquad B = -1, \qquad C = -1,$$

and therefore

$$\frac{1}{x(x^2 + x + 1)} = \frac{1}{x} - \frac{x + 1}{x^2 + x + 1}. \quad \square$$

In general, each irreducible quadratic factor $x^2 + \beta x + \gamma$ in the denominator gives rise to a term of the form

$$\frac{Ax + B}{x^2 + \beta x + \gamma}.$$

Example 5 (*The denominator has a repeated irreducible quadratic factor.*) For

$$\frac{3x^4 + x^3 + 20x^2 + 3x + 31}{(x + 1)(x^2 + 4)^2}$$

we write

$$\frac{3x^4 + x^3 + 20x^2 + 3x + 31}{(x + 1)(x^2 + 4)^2} = \frac{A}{x + 1} + \frac{Bx + C}{x^2 + 4} + \frac{Dx + E}{(x^2 + 4)^2}.$$

This gives

$$3x^4 + x^3 + 20x^2 + 3x + 31$$
$$= A(x^2 + 4)^2 + (Bx + C)(x + 1)(x^2 + 4) + (Dx + E)(x + 1).$$

This time we use -1, 0, 1, 2, and -2.

$$\begin{aligned}
50 &= 25A & &(x = -1),\\
31 &= 16A + 4C + E & &(x = 0),\\
58 &= 25A + 10B + 10C + 2D + 2E & &(x = 1),\\
173 &= 64A + 48B + 24C + 6D + 3E & &(x = 2),\\
145 &= 64A + 16B - 8C + 2D - E & &(x = -2).
\end{aligned}$$

With a little patience you can see that

$$A = 2, \quad B = 1, \quad C = 0, \quad D = 0, \quad \text{and} \quad E = -1.$$

This gives the decomposition

$$\frac{3x^4 + x^3 + 20x^2 + 3x + 31}{(x + 1)(x^2 + 4)^2} = \frac{2}{x + 1} + \frac{x}{x^2 + 4} - \frac{1}{(x^2 + 4)^2}. \quad \square$$

In general, each multiple irreducible quadratic factor $(x^2 + \beta x + \gamma)^k$ in the denominator gives rise to an expression of the form

$$\frac{A_1 x + B_1}{x^2 + \beta x + \gamma} + \frac{A_2 x + B_2}{(x^2 + \beta x + \gamma)^2} + \cdots + \frac{A_k x + B_k}{(x^2 + \beta x + \gamma)^k}.$$

As indicated at the beginning of this section, if the rational function is improper, then a polynomial will appear in the decomposition.

Example 6 (*The decomposition contains a polynomial.*) For the improper rational function introduced at the beginning of this section:

$$\frac{x^5+2}{x^2-1},$$

we saw that

$$\frac{x^5+2}{x^2-1} = x^3 + x + \frac{x+2}{x^2-1}.$$

Since the denominator of the fraction splits into linear factors, we write

$$\frac{x+2}{x^2-1} = \frac{A}{x+1} + \frac{B}{x-1}.$$

This gives

$$x + 2 = A(x-1) + B(x+1).$$

Substitution of $x = 1$ gives $B = \frac{3}{2}$; substitution of $x = -1$ gives $A = -\frac{1}{2}$. The decomposition takes the form

$$\frac{x^5+2}{x^2-1} = x^3 + x - \frac{1}{2(x+1)} + \frac{3}{2(x-1)}. \quad \square$$

We have been decomposing rational functions into partial fractions so as to be able to integrate them. Here we carry out the integrations, leaving some of the details to you.

Example 1′

$$\int \frac{dx}{x^2-4} = \frac{1}{4}\int \left(\frac{1}{x-2} - \frac{1}{x+2}\right) dx$$
$$= \frac{1}{4}(\ln|x-2| - \ln|x+2|) + C = \frac{1}{4}\ln\left|\frac{x-2}{x+2}\right| + C. \quad \square$$

Example 2′

$$\int \frac{2x^2+3}{x(x-1)^2}\, dx = \int \left[\frac{3}{x} - \frac{1}{x-1} + \frac{5}{(x-1)^2}\right] dx$$
$$= 3\ln|x| - \ln|x-1| - \frac{5}{x-1} + C$$
$$= \ln\left|\frac{x^3}{x-1}\right| - \frac{5}{x-1} + C \quad \square$$

Example 3′

$$\int \frac{x^2+5x+2}{(x+1)(x^2+1)}\, dx = \int \left(\frac{-1}{x+1} + \frac{2x+3}{x^2+1}\right) dx = -\int \frac{1}{x+1}\, dx + \int \frac{2x+3}{x^2+1}\, dx.$$

Since

$$-\int \frac{1}{x+1}\, dx = -\ln|x+1| + C_1$$

and

$$\int \frac{2x+3}{x^2+1}\,dx = \int \frac{2x}{x^2+1}\,dx + 3\int \frac{1}{x^2+1}\,dx = \ln(x^2+1) + 3\tan^{-1}x + C_2,$$

we have

$$\int \frac{x^2+5x+2}{(x+1)(x^2+1)}\,dx = -\ln|x+1| + \ln(x^2+1) + 3\tan^{-1}x + C$$

$$= \ln\left|\frac{x^2+1}{x+1}\right| + 3\tan^{-1}x + C. \quad \square$$

Example 4′

$$P = \int \frac{dx}{x(x^2+x+1)} = \int \left(\frac{1}{x} - \frac{x+1}{x^2+x+1}\right) dx = \ln|x| - \int \frac{x+1}{x^2+x+1}\,dx.$$

To compute the remaining integral, note that $(d/dx)(x^2+x+1) = 2x+1$, and so we manipulate the integrand to get a term of the form du/u, where $u = x^2+x+1$ and $du = (2x+1)\,dx$:

$$\frac{x+1}{x^2+x+1} = \frac{\frac{1}{2}[2x+1]+\frac{1}{2}}{x^2+x+1} = \frac{1}{2}\frac{2x+1}{x^2+x+1} + \frac{1}{2}\frac{1}{x^2+x+1}.$$

Therefore,

$$\int \frac{x+1}{x^2+x+1}\,dx = \frac{1}{2}\int \frac{2x+1}{x^2+x+1}\,dx + \frac{1}{2}\int \frac{1}{x^2+x+1}\,dx.$$

Now

$$\frac{1}{2}\int \frac{2x+1}{x^2+x+1}\,dx = \frac{1}{2}\ln(x^2+x+1) + C_1. \quad \text{(NOTE: } x^2+x+1>0 \text{ for all } x.)$$

You can verify that the second integral is an inverse tangent:

$$\frac{1}{2}\int \frac{dx}{x^2+x+1} = \frac{1}{2}\int \frac{dx}{(x+\frac{1}{2})^2+(\sqrt{3}/2)^2} = \frac{1}{\sqrt{3}}\tan^{-1}\left[\frac{2}{\sqrt{3}}\left(x+\frac{1}{2}\right)\right] + C_2.$$

Combining results, we have

$$P = \ln|x| - \frac{1}{2}\ln(x^2+x+1) - \frac{1}{\sqrt{3}}\tan^{-1}\left[\frac{2}{\sqrt{3}}\left(x+\frac{1}{2}\right)\right] + C. \quad \square$$

Example 5′

$$\int \frac{3x^4+x^3+20x^2+3x+31}{(x+1)(x^2+4)^2}\,dx = \int \left[\frac{2}{x+1} + \frac{x}{x^2+4} - \frac{1}{(x^2+4)^2}\right] dx.$$

The first two fractions are easy to integrate:

$$\int \frac{2}{x+1}\,dx = 2\ln|x+1| + C_1,$$

$$\int \frac{x}{x^2+4}\,dx = \frac{1}{2}\int \frac{2x}{x^2+4}\,dx = \frac{1}{2}\ln(x^2+4) + C_2.$$

The integral of the last fraction is of the form

$$\int \frac{dx}{(x^2 + a^2)^n}.$$

As we saw in the preceding section, such integrals can be calculated by using the trigonometric substitution $a \tan u = x$ [see (8.5.2)]. Here we have

$$\int \frac{dx}{(x^2+4)^2} = \frac{1}{8}\int \cos^2 u \, du \qquad (2 \tan u = x)$$

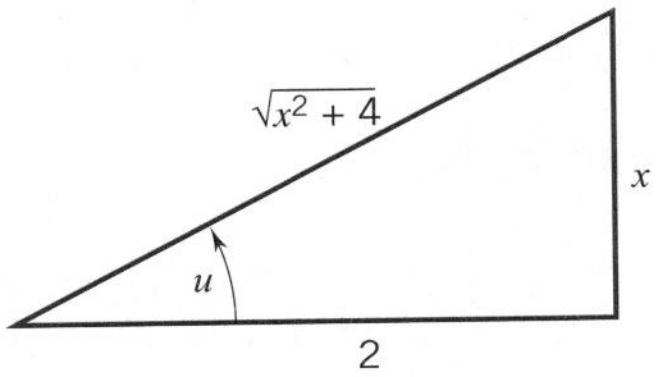

$$= \frac{1}{16}\int (1 + \cos 2u)\, du \qquad \text{(half-angle formula)}$$

$$= \frac{1}{16}u + \frac{1}{32}\sin 2u + C_3$$

$$= \frac{1}{16}u + \frac{1}{16}\sin u \cos u + C_3 \qquad (\sin 2u = 2 \sin u \cos u)$$

$$= \frac{1}{16}\tan^{-1}\frac{x}{2} + \frac{1}{16}\left(\frac{x}{\sqrt{x^2+4}}\right)\left(\frac{2}{\sqrt{x^2+4}}\right) + C_3$$

$$= \frac{1}{16}\tan^{-1}\frac{x}{2} + \frac{1}{8}\left(\frac{x}{x^2+4}\right) + C_3.$$

The integral we want is therefore equal to

$$2 \ln|x+1| + \frac{1}{2}\ln(x^2+4) - \frac{1}{8}\left(\frac{x}{x^2+4}\right) - \frac{1}{16}\tan^{-1}\frac{x}{2} + C. \quad ❑$$

Example 6′

$$\int \frac{x^5+2}{x^2-1}\,dx = \int \left[x^3 + x - \frac{1}{2(x+1)} + \frac{3}{2(x-1)}\right] dx$$

$$= \tfrac{1}{4}x^4 + \tfrac{1}{2}x^2 - \tfrac{1}{2}\ln|x+1| + \tfrac{3}{2}\ln|x-1| + C. \quad ❑$$

Partial fractions come up in a variety of applications. Here is an example from the life sciences.

Example 7 A theory used by scientists to describe the spread of a disease through a population states that the rate at which the disease spreads is proportional to the product of the number of individuals that are infected and the number that are not. If M denotes the total number of individuals in the population and $y(t)$ is the number of individuals that are infected at time t, then the mathematical formulation of the theory is

(8.6.2)
$$\frac{dy}{dt} = ky(M-y),$$

where $k > 0$ is the constant of proportionality. Note that the term "population" being used here does not necessarily mean "people." We might want to study the spread of a disease through a population of animals, or through a population of plants. Equation (8.6.2) is known as the *logistic equation* and its solutions are called *logistic functions*. The equation is a separable differential equation, and the solutions are found by "separating the variables" and integrating (see Section 7.6):

$$\frac{dy}{y(M-y)} = k\,dt$$

$$\int \frac{dy}{y(M-y)} = \int k\,dt = kt + C, \qquad C \text{ an arbitrary constant.}$$

The method of partial fractions is needed to calculate the integral on the left-hand side:

$$\frac{1}{y(M-y)} = \frac{A}{y} + \frac{B}{M-y}$$

$$1 = A(M-y) + By.$$

Setting $y = 0$ and $y = M$, we find that $A = B = 1/M$. Thus

$$\int \frac{dy}{y(M-y)} = \frac{1}{M}\int \frac{dy}{y} + \frac{1}{M}\int \frac{dy}{M-y} = \frac{1}{M}[\ln y - \ln(M-y)] = \frac{1}{M}\ln\left(\frac{y}{M-y}\right)$$

(note: absolute values are not needed since y and $M - y$ are both positive) and

$$\frac{1}{M}\ln\left(\frac{y}{M-y}\right) = kt + C.$$

Multiplying by M, we get

$$\ln\left(\frac{y}{M-y}\right) = Mkt + MC$$

or

$$\ln\left(\frac{y}{M-y}\right) = Mkt + C_1 \qquad (C_1 = MC \text{ is arbitrary}).$$

Taking the exponential of both sides of this equation, we find that

$$\frac{y}{M-y} = e^{Mkt + C_1} = e^{C_1}\,e^{Mkt} = C_2\,e^{Mkt}$$

If $y(0) = R$ individuals have the disease initially, then $C_2 = R/(M-R)$ and

$$\frac{y}{M-y} = \frac{R}{M-R}e^{Mkt}.$$

This equation can be solved for y. The result is

$$y(t) = \frac{MR}{R + (M-R)\,e^{-Mkt}}$$

and this gives the number of individuals that are infected at time t. The graph of y is shown in Figure 8.6.1. Note that the disease spreads rapidly in the beginning, and then "levels off." Note, also, that $y(t) \to M$ as $t \to \infty$, so that everyone will get the disease eventually unless something is introduced to stop it. ❑

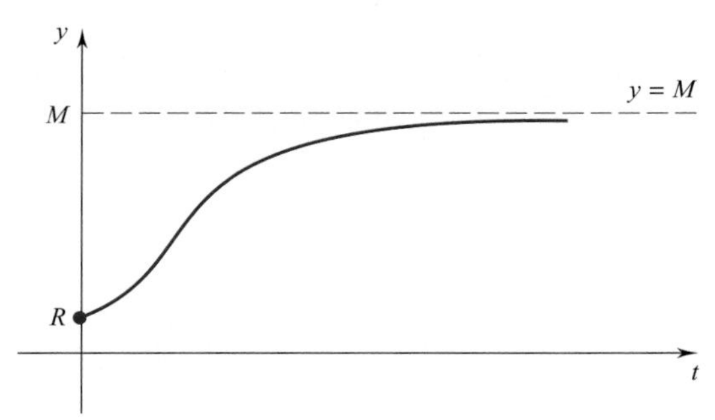

Figure 8.6.1

EXERCISES 8.6

In Exercises 1–8, determine the partial fraction decomposition of the given rational function.

1. $r(x) = \dfrac{1}{x^2 + 7x + 6}$.

2. $R(x) = \dfrac{x^2}{(x - 1)(x^2 + 4x + 5)}$.

3. $r(x) = \dfrac{x}{x^4 - 1}$.

4. $R(x) = \dfrac{x^4}{(x - 1)^3}$.

5. $r(x) = \dfrac{x^2 - 3x - 1}{x^3 + x^2 - 2x}$.

6. $R(x) = \dfrac{x^3 + x^2 + x + 2}{x^4 + 3x^2 + 2}$.

7. $r(x) = \dfrac{2x^2 + 1}{x^3 - 6x^2 + 11x - 6}$.

8. $R(x) = \dfrac{1}{x(x^2 + 1)^2}$.

In Exercises 9–30, evaluate the given integral.

9. $\displaystyle\int \frac{7}{(x - 2)(x + 5)}\, dx$.

10. $\displaystyle\int \frac{x}{(x + 1)(x + 2)(x + 3)}\, dx$.

11. $\displaystyle\int \frac{2x^2 + 3}{x^2(x - 1)}\, dx$.

12. $\displaystyle\int \frac{x^2 + 1}{x(x^2 - 1)}\, dx$.

13. $\displaystyle\int \frac{x^5}{(x - 2)^2}\, dx$.

14. $\displaystyle\int \frac{x^5}{x - 2}\, dx$.

15. $\displaystyle\int \frac{x + 3}{x^2 - 3x + 2}\, dx$.

16. $\displaystyle\int \frac{x^2 + 3}{x^2 - 3x + 2}\, dx$.

17. $\displaystyle\int \frac{dx}{(x - 1)^3}$.

18. $\displaystyle\int \frac{dx}{x^2 + 2x + 2}$.

19. $\displaystyle\int \frac{x^2}{(x - 1)^2(x + 1)}\, dx$.

20. $\displaystyle\int \frac{2x - 1}{(x + 1)^2(x - 2)^2}\, dx$.

21. $\displaystyle\int \frac{dx}{x^4 - 16}$.

22. $\displaystyle\int \frac{x}{x^3 - 1}\, dx$.

23. $\displaystyle\int \frac{x^3 + 4x^2 - 4x - 1}{(x^2 + 1)^2}\, dx$.

24. $\displaystyle\int \frac{dx}{(x^2 + 16)^2}$.

25. $\displaystyle\int \frac{dx}{x^4 + 4}$.†

26. $\displaystyle\int \frac{dx}{x^4 + 16}$.†

27. $\displaystyle\int \frac{x - 3}{x^3 + x^2}\, dx$.

28. $\displaystyle\int \frac{1}{(x - 1)(x^2 + 1)^2}\, dx$.

29. $\displaystyle\int \frac{x + 1}{x^3 + x^2 - 6x}\, dx$.

30. $\displaystyle\int \frac{x^3 + x^2 + x + 3}{(x^2 + 1)(x^2 + 3)}\, dx$.

In Exercises 31–36, evaluate the given definite integral.

31. $\displaystyle\int_0^2 \frac{x}{x^2 + 5x + 6}\, dx$.

32. $\displaystyle\int_1^3 \frac{1}{x^3 + x}\, dx$.

33. $\displaystyle\int_3^6 \frac{2x}{x^3 - 2x^2 - 4x + 8}\, dx$.

34. $\displaystyle\int_2^4 \frac{x^4 - x^3 - x - 1}{x^3 - x^2}\, dx$.

35. $\displaystyle\int_1^3 \frac{x^2 - 4x + 3}{x^3 + 2x^2 + x}\, dx$.

36. $\displaystyle\int_0^2 \frac{x^3}{(x^2 + 2)^2}\, dx$.

37. Show that if $y = \dfrac{1}{x^2 - 1}$, then

$$\frac{d^n y}{dx^n} = \frac{(-1)^n n!}{2}\left[\frac{1}{(x - 1)^{n+1}} - \frac{1}{(x + 1)^{n+1}}\right].$$

38. Calculate

$$\int x^3 \tan^{-1} x\, dx.$$

† HINT: With $a > 0$, $x^4 + a^2 = (x^2 + \sqrt{2a}x + a)(x^2 - \sqrt{2a}x + a)$.

39. Find the centroid of the region under the curve $y = (x^2 + 1)^{-1}$, $x \in [0, 1]$.

40. Find the centroid of the solid generated by revolving the region of Exercise 39 about: (a) the x-axis; (b) the y-axis.

**41.** Let

$$f(x) = \frac{x}{x^2 + 5x + 6}.$$

(a) Use a graphing utility to sketch the graph of f.
(b) Calculate the area of the region bounded by the graph of f and the x-axis between $x = 0$ and $x = 4$.

42. (a) The region in Exercise 41 is revolved around the y-axis. Find the volume of the solid that is generated.
(b) Find the centroid of the solid in part (a).

43. Let

$$f(x) = \frac{9 - x}{(x + 3)^2}.$$

(a) Use a graphing utility to sketch the graph of f.
(b) Find the area of the region bounded by the graph of f and the x-axis between $x = -2$ and $x = 9$.

44. (a) The region in Exercise 43 is revolved around the x-axis. Find the volume of the solid that is generated.
(b) Find the centroid of the solid in part (a).

45. Suppose that a chemical A combines with a chemical B to form a compound C. In addition, suppose that the rate at which C is produced at time t varies directly with the amounts of A and B present at time t. With this model, if A_0 grams of A are mixed with B_0 grams of B, then

$$\frac{dC}{dt} = k(A_0 - C)(B_0 - C).$$

(a) Find the amount of compound C present at time t if $A_0 = B_0$.
(b) Find the amount of compound C present at time t if $A_0 \neq B_0$.

46. A mathematical model for the growth of a certain strain of bacteria is

$$\frac{dP}{dt} = 0.0020P(800 - P),$$

where $P = P(t)$ denotes the number of bacteria present at time t. There are 100 bacteria present initially.
(a) Find the population $P = P(t)$ at any time t.
(b) Use a graphing utility to sketch the graph of P and dP/dt.
(c) Approximate the time at which the bacteria culture experiences its most rapid growth rate. That is, locate the maximum value of dP/dt. Use three decimal place accuracy. What point on the graph of P corresponds to the maximum growth rate?

47. When an object of mass m is moving through air or a viscous medium, it is acted upon by a frictional force that acts in a direction opposite to its motion. This frictional force depends on the velocity of the object and (within close approximation) is given by

$$F(v) = -\alpha v - \beta v^2,$$

where α and β are positive constants.
(a) From Newton's second law, $F = ma$, we have

$$m\frac{dv}{dt} = -\alpha v - \beta v^2.$$

Solve this differential equation to find $v = v(t)$.
(b) Find v if the object has initial velocity $v(0) = v_0$.
(c) What is $\lim_{t\to\infty} v(t)$?

48. A descending parachutist is acted on by two forces: a constant downward force mg and the upward force of air resistance, which (within close approximation) is of the form $-\beta v^2$ where β is a positive constant. (In this problem we are taking the downward direction as positive.)
(a) Express t in terms of the velocity v, the initial velocity v_0, and the constant $v_c = \sqrt{mg/\beta}$.
(b) Express v as a function of t.
(c) Express the acceleration a as a function of t. Verify that the acceleration never changes sign and in time tends to zero.
(d) Show that in time v tends to v_c. (This number v_c is called the *terminal velocity*.)

■ 8.7 SOME RATIONALIZING SUBSTITUTIONS

In this section we discuss two types of substitutions that lead to an integral of a rational function. Such substitutions are known as *rationalizing substitutions*.

First we consider integrals in which the integrand involves an expression of the form $\sqrt[n]{f(x)}$ for some function f. In such cases, the substitution $u = \sqrt[n]{f(x)}$, which is equivalent to $u^n = f(x)$, is sometimes effective. The idea behind this substitution is that it will replace fractional exponents with integer exponents; integer exponents are, in general, easier to handle.

Example 1 Find

$$\int \frac{dx}{1+\sqrt{x}}.$$

SOLUTION To rationalize the integrand, we set

$$u^2 = x, \qquad 2u\,du = dx.$$

Then $u = \sqrt{x}$ and

$$\begin{aligned}\int \frac{dx}{1+\sqrt{x}} = \int \frac{2u}{1+u}\,du &= \int \left(2 - \frac{2}{1+u}\right) du \\ &= 2u - 2\ln|1+u| + C \\ &= 2\sqrt{x} - 2\ln|1+\sqrt{x}| + C. \quad \square\end{aligned}$$

divide (annotation)

Example 2 Find

$$\int \frac{dx}{\sqrt[3]{x}+\sqrt{x}}.$$

SOLUTION Here the integrand contains two distinct roots, $x^{1/3}$ and $x^{1/2}$. If we set $u^6 = x$, then both terms will be rationalized simultaneously:

$$u^6 = x, \qquad 6u^5\,du = dx, \qquad \text{and} \qquad x^{1/3} = u^2, \qquad x^{1/2} = u^3.$$

Thus, we have

$$\begin{aligned}\int \frac{dx}{\sqrt[3]{x}+\sqrt{x}} = \int \frac{6u^5}{u^2+u^3}\,du &= 6\int \frac{u^3}{1+u}\,du \\ &= 6\int \left(u^2 - u + 1 - \frac{1}{1+u}\right) du \\ &= 6\left(\tfrac{1}{3}u^3 - \tfrac{1}{2}u^2 + u - \ln|1+u|\right) + C \\ &= 2\sqrt{x} - 3\sqrt[3]{x} + 6\sqrt[6]{x} - 6\ln|1+\sqrt[6]{x}| + C. \quad \square\end{aligned}$$

divide (annotation)

Example 3 Find

$$\int \sqrt{1-e^x}\,dx.$$

SOLUTION To rationalize the integrand we set

$$u^2 = 1 - e^x.$$

To find dx in terms of u and du we solve the equation for x:

$$1 - u^2 = e^x, \qquad \ln(1-u^2) = x, \qquad -\frac{2u}{1-u^2}\,du = dx.$$

The rest is straightforward:

$$\begin{aligned}
\int \sqrt{1 - e^x}\, dx &= \int u\left(-\frac{2u}{1 - u^2}\right) du \\
&= \int \frac{2u^2}{u^2 - 1} du = \int \left(2 + \frac{1}{u - 1} - \frac{1}{u + 1}\right) du \\
&= 2u + \ln|u - 1| - \ln|u + 1| + C \\
&= 2u + \ln\left|\frac{u - 1}{u + 1}\right| + C \\
&= 2\sqrt{1 - e^x} + \ln\left|\frac{\sqrt{1 - e^x} - 1}{\sqrt{1 - e^x} + 1}\right| + C. \quad \square
\end{aligned}$$

divide, then partial fractions

The second type of rationalizing substitution is applied to integrands which are rational expressions in sine and cosine. For example, suppose we want to calculate

$$\int \frac{\cos x}{\sin x + \sin^2 x} dx.$$

To convert a rational expression in sine and cosine to a rational function in u, we use the substitution

$$u = \tan\left(\frac{x}{2}\right), \qquad -\pi < x < \pi.$$

Then

$$\cos\left(\frac{x}{2}\right) = \frac{1}{\sec(x/2)} = \frac{1}{\sqrt{1 + \tan^2(x/2)}} = \frac{1}{\sqrt{1 + u^2}},$$

and

$$\sin\left(\frac{x}{2}\right) = \cos\left(\frac{x}{2}\right)\tan\left(\frac{x}{2}\right) = \frac{u}{\sqrt{1 + u^2}}.$$

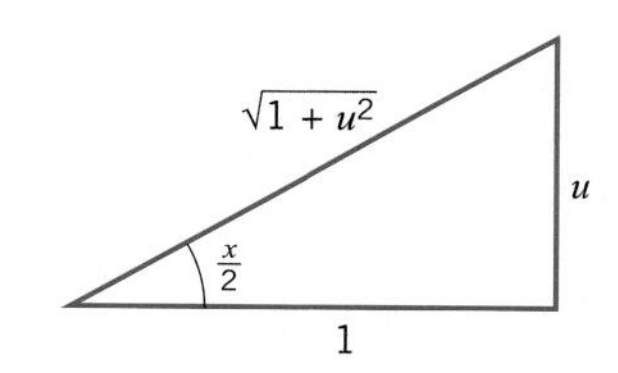

Figure 8.7.1

The right triangle in Figure 8.7.1 illustrates these relationships for $0 < x < \pi$.

Now,

$$\sin x = 2 \sin\left(\frac{x}{2}\right)\cos\left(\frac{x}{2}\right) = \frac{2u}{1 + u^2},$$

and

$$\cos x = \cos^2\left(\frac{x}{2}\right) - \sin^2\left(\frac{x}{2}\right) = \frac{1 - u^2}{1 + u^2}.$$

Also, since $u = \tan(x/2)$, $x = 2 \tan^{-1} u$ and so

$$dx = \frac{2}{1 + u^2} du.$$

To summarize, if an integrand is a rational expression in sine and cosine, then the substitution

$$\sin x = \frac{2u}{1 + u^2}, \qquad \cos x = \frac{1 - u^2}{1 + u^2}, \qquad dx = \frac{2}{1 + u^2} du,$$

where $u = \tan(x/2)$, $-\pi < x < \pi$, will convert the integrand into a rational function in u. The result can then be calculated using the methods of Section 8.6.

Example 4 Find

$$\int \frac{\cos x}{\sin x + \sin^2 x}\,dx.$$

SOLUTION If we let $u = \tan\left(\frac{x}{2}\right)$, then

$$\frac{\cos x}{\sin x + \sin^2 x} = \frac{(1-u^2)/(1+u^2)}{[2u/(1+u^2)] + [4u^2/(1+u^2)^2]} = \frac{(1-u^2)(1+u^2)}{2u^3 + 4u^2 + 2u}$$
$$= \frac{(1-u)(1+u^2)}{2u(1+u)}$$

and

$$\int \frac{\cos x}{\sin x + \sin^2 x}\,dx = \int \frac{(1-u)(1+u^2)}{2u(1+u)} \cdot \frac{2}{1+u^2}\,du = \int \frac{1-u}{u(1+u)}\,du.$$

Now

$$\frac{1-u}{u(1+u)} = \frac{1}{u} - \frac{2}{1+u}, \qquad \text{(partial fractions)}$$

and so

$$\int \frac{1-u}{u(1+u)}\,du = \int \frac{1}{u}\,du - 2\int \frac{1}{1+u}\,du$$
$$= \ln|u| - 2\ln|1+u| + C$$
$$= \ln\left|\frac{u}{(1+u)^2}\right| + C$$
$$= \ln\left|\frac{\tan(x/2)}{[1+\tan(x/2)]^2}\right| + C. \quad ❑$$

EXERCISES 8.7

In Exercises 1–30, evaluate the given integral.

1. $\int \frac{dx}{1-\sqrt{x}}$.
2. $\int \frac{\sqrt{x}}{1+x}\,dx$.
3. $\int \sqrt{1+e^x}\,dx$.
4. $\int \frac{dx}{x(x^{1/3}-1)}$.
5. $\int x\sqrt{1+x}\,dx$. [(a) set $u^2 = 1 + x$; (b) set $u = 1 + x$]
6. $\int x^2\sqrt{1+x}\,dx$. [(a) set $u^2 = 1 + x$; (b) set $u = 1 + x$]
7. $\int (x+2)\sqrt{x-1}\,dx$.
8. $\int (x-1)\sqrt{x+2}\,dx$.
9. $\int \frac{x^3}{(1+x^2)^3}\,dx$.
10. $\int x(1+x)^{1/3}\,dx$.
11. $\int \frac{\sqrt{x}}{\sqrt{x}-1}\,dx$.
12. $\int \frac{x}{\sqrt{x+1}}\,dx$.
13. $\int \frac{\sqrt{x-1}+1}{\sqrt{x-1}-1}\,dx$.
14. $\int \frac{1-e^x}{1+e^x}\,dx$.
15. $\int \frac{dx}{\sqrt{1+e^x}}$.
16. $\int \frac{dx}{1+e^{-x}}$.
17. $\int \frac{x}{\sqrt{x+4}}\,dx$.
18. $\int \frac{x+1}{x\sqrt{x-2}}\,dx$.

19. $\int 2x^2(4x+1)^{-5/2}\,dx.$ **20.** $\int x^2\sqrt{x-1}\,dx.$

21. $\int \frac{x}{(ax+b)^{3/2}}\,dx.$ **22.** $\int \frac{x}{\sqrt{ax+b}}\,dx.$

23. $\int \frac{1}{1+\cos x-\sin x}\,dx.$ **24.** $\int \frac{1}{2+\cos x}\,dx.$

25. $\int \frac{1}{2+\sin x}\,dx.$ **26.** $\int \frac{\sin x}{1+\sin^2 x}\,dx.$

27. $\int \frac{1}{\sin x+\tan x}\,dx.$ **28.** $\int \frac{1}{1+\sin x+\cos x}\,dx.$

29. $\int \frac{1-\cos x}{1+\sin x}\,dx.$ **30.** $\int \frac{1}{5+3\sin x}\,dx.$

In Exercises 31–36, evaluate the definite integral.

31. $\int_0^4 \frac{x^{3/2}}{x+1}\,dx.$ **32.** $\int_0^8 \frac{1}{1+\sqrt[3]{x}}\,dx.$

33. $\int_0^{\pi/2} \frac{\sin 2x}{2+\cos x}\,dx.$ **34.** $\int_0^{\pi/2} \frac{1}{1+\sin x}\,dx.$

35. $\int_0^{\pi/3} \frac{1}{\sin x-\cos x-1}\,dx.$ **36.** $\int_0^1 \frac{\sqrt{x}}{1+\sqrt{x}}\,dx.$

37. Use the method of this section to show that

$$\int \sec x\,dx = \int \frac{1}{\cos x}\,dx = \ln\left|\frac{1+\tan(x/2)}{1-\tan(x/2)}\right| + C.$$

38. (a) Another expression for $\int \sec x\,dx$ can be obtained as follows:

$$\int \sec x\,dx = \int \frac{\cos x}{\cos^2 x}\,dx = \int \frac{\cos x}{1-\sin^2 x}\,dx.$$

Use the method of this section to show that

$$\int \sec x\,dx = \ln\sqrt{\frac{1+\sin x}{1-\sin x}} + C.$$

(b) Show that the result in part (a) is equivalent to the familiar formula

$$\int \sec x\,dx = \ln|\sec x+\tan x| + C.$$

39. (a) Use the approach given in Exercise 38 (a) to show that

$$\int \csc x\,dx = \ln\sqrt{\frac{1-\cos x}{1+\cos x}} + C.$$

(b) Show that the result in part (a) is equivalent to the formula

$$\int \csc x\,dx = \ln|\csc x-\cot x| + C.$$

40. The integral of a rational function of $\sinh x$ and $\cosh x$ can be transformed into a rational function of u by means of the substitution $u=\tanh(x/2)$. With this substitution, show that

$$\sinh x = \frac{2u}{1-u^2}, \quad \cosh x = \frac{1+u^2}{1-u^2}, \quad dx = \frac{2}{1-u^2}\,du.$$

In Exercises 41–44, use the subsitution $u=\tanh(x/2)$ to evaluate the given integral.

41. $\int \operatorname{sech} x\,dx.$ **42.** $\int \frac{1}{1+\cosh x}\,dx.$

43. $\int \frac{1}{\sinh x+\cosh x}\,dx.$ **44.** $\int \frac{1-e^x}{1+e^x}\,dx.$

■ 8.8 NUMERICAL INTEGRATION

To evaluate a definite integral by the formula

$$\int_a^b f(x)\,dx = F(b) - F(a)$$

we must be able to find an antiderivative F and we must be able to evaluate this antiderivative both at a and at b. When this is not possible, the method fails.

The method fails even for such simple-looking integrals as

$$\int_0^1 \sqrt{x}\sin x\,dx \quad \text{and} \quad \int_0^1 e^{-x^2}\,dx.$$

There are no *elementary functions* with derivatives $\sqrt{x}\sin x$ and e^{-x^2}.

Here we take up some simple numerical methods for estimating definite integrals —methods that you can use whether or not you can find an antiderivative. All the methods we describe involve only simple arithmetic and are ideally suited to the computer.

We focus now on

$$\int_a^b f(x)\,dx.$$

We suppose that f is continuous on $[a, b]$ and, for pictorial convenience, assume that f is positive. Take a regular partition $P = \{x_0, x_1, x_2, \ldots, x_{n-1}, x_n\}$ of $[a, b]$, subdividing the interval into n subintervals each of length $(b - a)/n$:

$$[a, b] = [x_0, x_1] \cup \cdots \cup [x_{i-1}, x_i] \cup \cdots \cup [x_{n-1}, x_n],$$

with

$$\Delta x_i = \frac{b - a}{n}.$$

The region Ω_i pictured in Figure 8.8.1 can be approximated in many ways.

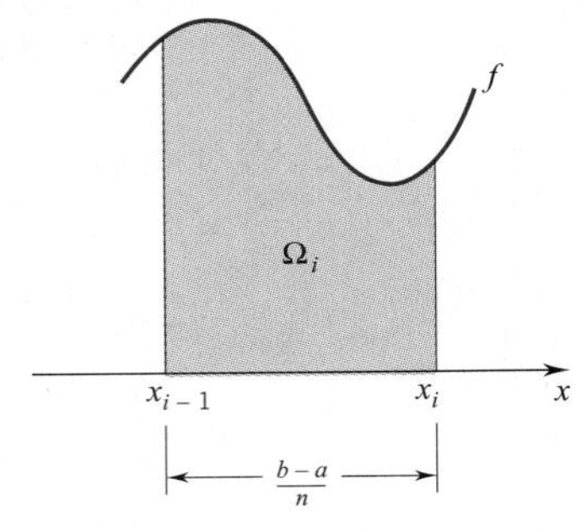

Figure 8.8.1

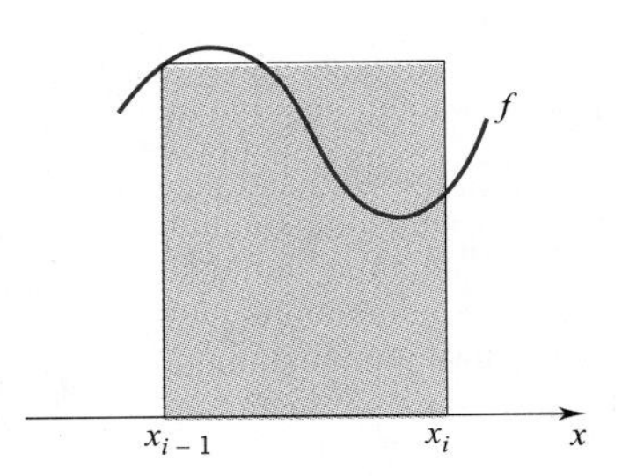

Figure 8.8.2

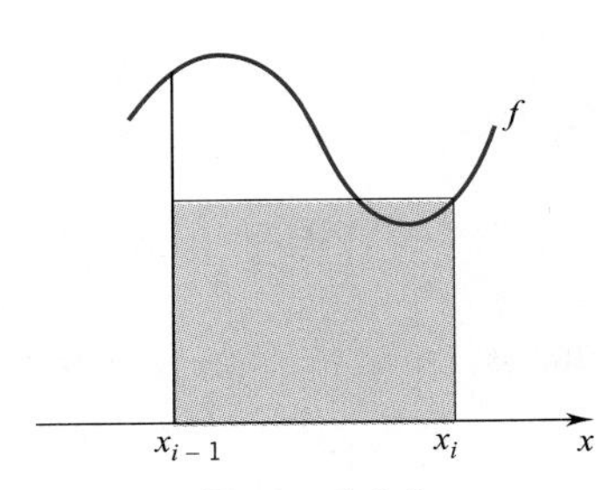

Figure 8.8.3

(1) By the left-endpoint rectangle (Figure 8.8.2):

$$\text{area} = f(x_{i-1})\,\Delta x_i = f(x_{i-1})\left(\frac{b - a}{n}\right).$$

(2) By the right-endpoint rectangle (Figure 8.8.3):

$$\text{area} = f(x_i)\,\Delta x_i = f(x_i)\left(\frac{b - a}{n}\right).$$

(3) By the midpoint rectangle (Figure 8.8.4):

$$\text{area} = f\left(\frac{x_{i-1} + x_i}{2}\right)\Delta x_i = f\left(\frac{x_{i-1} + x_i}{2}\right)\left(\frac{b - a}{n}\right)$$

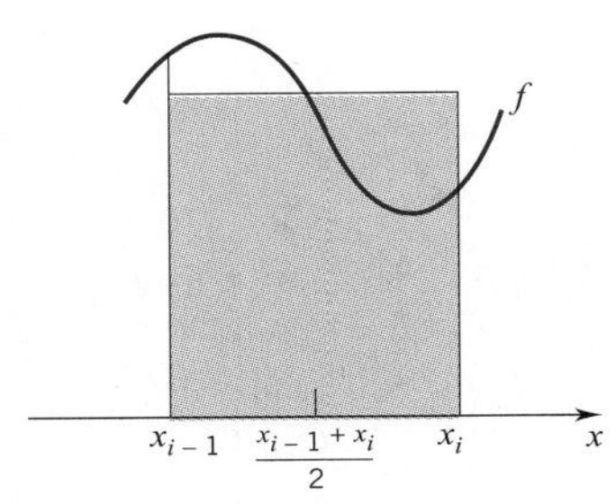

Figure 8.8.4

(4) By a trapezoid (Figure 8.8.5):

$$\text{area} = \frac{1}{2}[f(x_{i-1}) + f(x_i)]\,\Delta x_i$$
$$= \frac{1}{2}[f(x_{i-1}) + f(x_i)]\left(\frac{b-a}{n}\right).$$

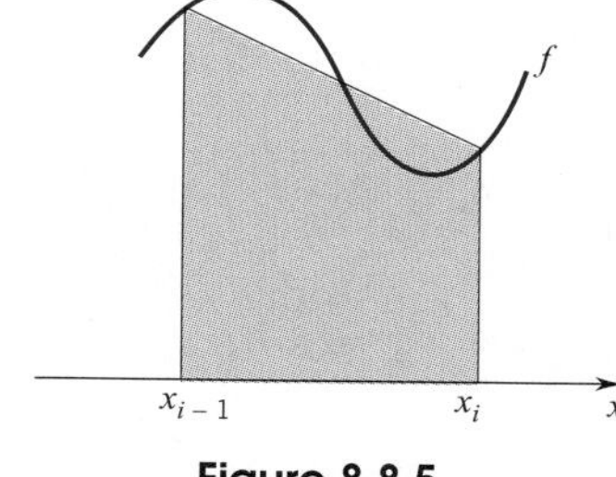

Figure 8.8.5

(5) By a parabolic region (Figure 8.8.6): take the parabola $y = Ax^2 + Bx + C$ that passes through the three points indicated.

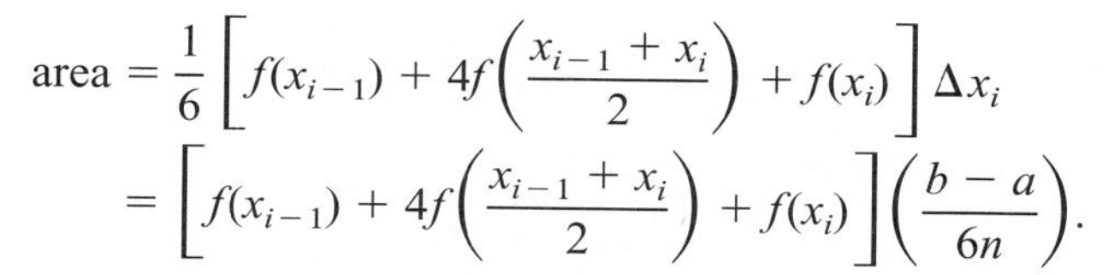

$$\text{area} = \frac{1}{6}\left[f(x_{i-1}) + 4f\left(\frac{x_{i-1}+x_i}{2}\right) + f(x_i)\right]\Delta x_i$$
$$= \left[f(x_{i-1}) + 4f\left(\frac{x_{i-1}+x_i}{2}\right) + f(x_i)\right]\left(\frac{b-a}{6n}\right).$$

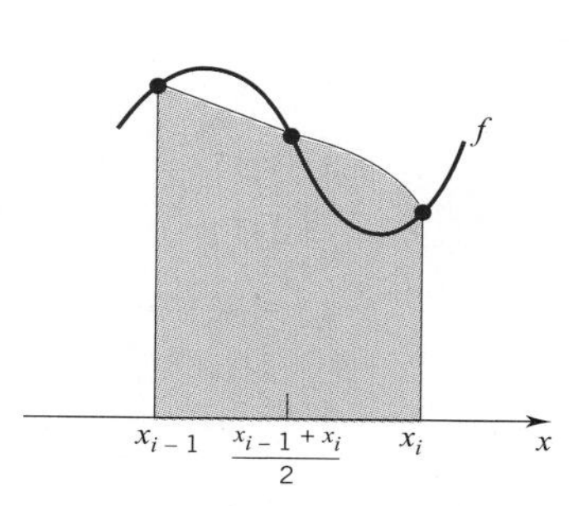

Figure 8.8.6

You can verify this formula for the area under the parabola by doing Exercises 11 and 12. (If the three points are collinear, the parabola degenerates to a straight line and the parabolic region becomes a trapezoid. The formula then gives the area of the trapezoid.)

The approximations to Ω_i just considered yield the following estimates for

$$\int_a^b f(x)\,dx.$$

(1) The left-endpoint estimate:

$$L_n = \frac{b-a}{n}[f(x_0) + f(x_1) + \cdots + f(x_{n-1})].$$

(2) The right-endpoint estimate:

$$R_n = \frac{b-a}{n}[f(x_1) + f(x_2) + \cdots + f(x_n)].$$

(3) The midpoint estimate:

$$M_n = \frac{b-a}{n}\left[f\left(\frac{x_0+x_1}{2}\right) + \cdots + f\left(\frac{x_{n-1}+x_n}{2}\right)\right].$$

(4) The trapezoidal estimate (*trapezoidal rule*):

$$T_n = \frac{b-a}{n}\left[\frac{f(x_0)+f(x_1)}{2} + \frac{f(x_1)+f(x_2)}{2} + \cdots + \frac{f(x_{n-1})+f(x_n)}{2}\right]$$
$$= \frac{b-a}{2n}[f(x_0) + 2f(x_1) + \cdots + 2f(x_{n-1}) + f(x_n)].$$

(5) The parabolic estimate (*Simpson's rule*):

$$S_n = \frac{b-a}{6n}\left\{f(x_0) + f(x_n) + 2[f(x_1) + \cdots + f(x_{n-1})]\right.$$
$$\left. + 4\left[f\left(\frac{x_0+x_1}{2}\right) + \cdots + f\left(\frac{x_{n-1}+x_n}{2}\right)\right]\right\}.$$

The first three estimates, L_n, R_n, M_n, are Riemann sums (Section 5.10); T_n and S_n, although not explicitly defined as Riemann sums, can be written as Riemann sums.

(See Exercise 26.) It follows from (5.10.1) that any one of these estimates can be used to approximate the integral as closely as we may wish. All we have to do is take n sufficiently large.

As an example, we will find the approximate value of

$$\ln 2 = \int_1^2 \frac{dx}{x}$$

by applying each of the five estimates. Here

$$f(x) = \frac{1}{x}, \qquad [a, b] = [1, 2].$$

Taking $n = 5$, we have

$$\frac{b-a}{n} = \frac{2-1}{5} = \frac{1}{5}.$$

The partition points are

$$x_0 = \tfrac{5}{5}, \quad x_1 = \tfrac{6}{5}, \quad x_2 = \tfrac{7}{5}, \quad x_3 = \tfrac{8}{5}, \quad x_4 = \tfrac{9}{5}, \quad x_5 = \tfrac{10}{5}. \qquad \text{(Figure 8.8.7)}$$

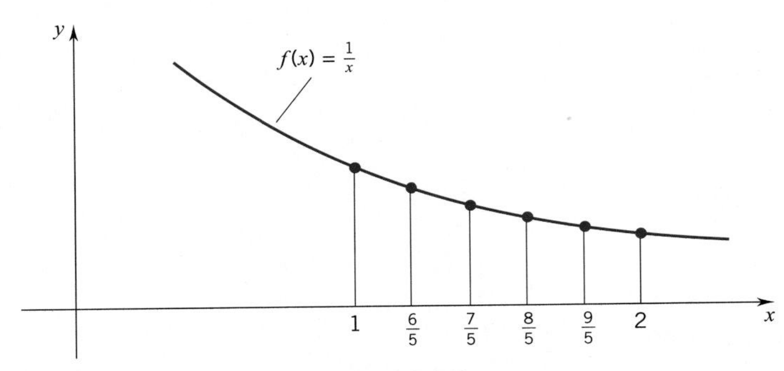

Figure 8.8.7

Using a calculator and rounding off to four decimal places, the five estimates are as follows:

$$L_5 = \tfrac{1}{5}(\tfrac{5}{5} + \tfrac{5}{6} + \tfrac{5}{7} + \tfrac{5}{8} + \tfrac{5}{9}) = (\tfrac{1}{5} + \tfrac{1}{6} + \tfrac{1}{7} + \tfrac{1}{8} + \tfrac{1}{9}) \cong 0.7457.$$

$$R_5 = \tfrac{1}{5}(\tfrac{5}{6} + \tfrac{5}{7} + \tfrac{5}{8} + \tfrac{5}{9} + \tfrac{5}{10}) = (\tfrac{1}{6} + \tfrac{1}{7} + \tfrac{1}{8} + \tfrac{1}{9} + \tfrac{1}{10}) \cong 0.6457.$$

$$M_5 = \tfrac{1}{5}(\tfrac{10}{11} + \tfrac{10}{13} + \tfrac{10}{15} + \tfrac{10}{17} + \tfrac{10}{19}) = 2(\tfrac{1}{11} + \tfrac{1}{13} + \tfrac{1}{15} + \tfrac{1}{17} + \tfrac{1}{19}) \cong 0.6919.$$

$$T_5 = \tfrac{1}{10}(\tfrac{5}{5} + \tfrac{10}{6} + \tfrac{10}{7} + \tfrac{10}{8} + \tfrac{10}{9} + \tfrac{5}{10}) = (\tfrac{1}{10} + \tfrac{1}{6} + \tfrac{1}{7} + \tfrac{1}{8} + \tfrac{1}{9} + \tfrac{1}{20}) \cong 0.6957.$$

$$S_5 = \tfrac{1}{30}[\tfrac{5}{5} + \tfrac{5}{10} + 2(\tfrac{5}{6} + \tfrac{5}{7} + \tfrac{5}{8} + \tfrac{5}{9}) + 4(\tfrac{10}{11} + \tfrac{10}{13} + \tfrac{10}{15} + \tfrac{10}{17} + \tfrac{10}{19})] \cong 0.6935.$$

Since the integrand $1/x$ decreases throughout the interval $[1, 2]$, you can expect the left-endpoint estimate, 0.7457, to be too large and you can expect the right-endpoint estimate, 0.6457, to be too small. The other estimates should be better.

The value of $\ln 2$ given on a calculator is $\ln 2 \cong 69314718$, or 0.6931 rounded to four decimal places. Thus S_5 is correct to the nearest thousandth.

Example 1 Find the approximate value of

$$\int_0^3 \sqrt{4 + x^3}\, dx.$$

by the trapezoid rule. Take $n = 6$.

SOLUTION Each subinterval has length

$$\frac{b - a}{n} = \frac{3 - 0}{6} = \frac{1}{2}.$$

The partition points are

$x_0 = 0, \quad x_1 = \frac{1}{2}, \quad x_2 = 1, \quad x_3 = \frac{3}{2}, \quad x_4 = 2, \quad x_5 = \frac{5}{2}, \quad x_6 = 3.$ (See Figure 8.8.8)

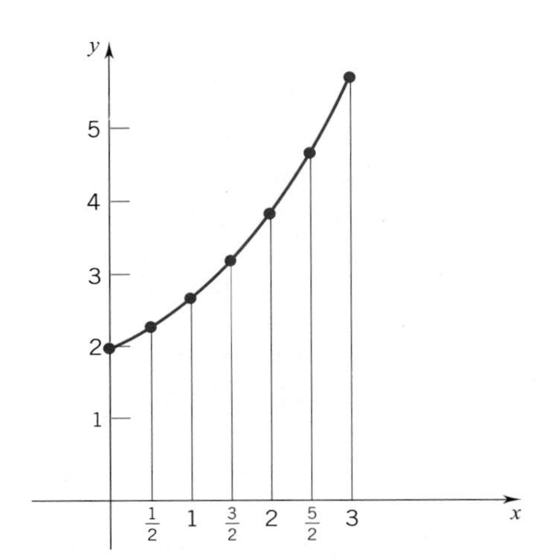

Figure 8.8.8

Now

$$T_6 = \tfrac{1}{4}[f(0) + 2f(\tfrac{1}{2}) + 2f(1) + 2f(\tfrac{3}{2}) + 2f(2) + 2f(\tfrac{5}{2}) + f(3)],$$

with $f(x) = \sqrt{4 + x^3}$. Using a calculator and rounding off to three decimal places, we have

$$f(0) = 2.000, \quad f(\tfrac{1}{2}) \cong 2.031, \quad f(1) \cong 2.236, \quad f(\tfrac{3}{2}) \cong 2.716,$$
$$f(2) \cong 3.464, \quad f(\tfrac{5}{2}) \cong 4.430, \quad f(3) \cong 5.568.$$

Thus

$$T_6 \cong \tfrac{1}{4}(2.000 + 4.062 + 4.472 + 5.432 + 6.928 + 8.860 + 5.568) \cong 9.331. \quad \square$$

Example 2 Find the approximate value of

$$\int_0^3 \sqrt{4 + x^3}\, dx$$

by Simpson's rule. Take $n = 3$.

SOLUTION There are three subintervals each of length

$$\frac{b - a}{n} = \frac{3 - 0}{3} = 1.$$

Here

$$x_0 = 0, \quad x_1 = 1, \quad x_2 = 2, \quad x_3 = 3,$$

$$\frac{x_0 + x_1}{2} = \frac{1}{2}, \quad \frac{x_1 + x_2}{2} = \frac{3}{2}, \quad \frac{x_2 + x_3}{2} = \frac{5}{2}.$$

Simpson's rule yields

$$S_3 = \tfrac{1}{6}[f(0) + f(3) + 2f(1) + 2f(2) + 4f(\tfrac{1}{2}) + 4f(\tfrac{3}{2}) + 4f(\tfrac{5}{2})],$$

with $f(x) = \sqrt{4 + x^3}$. Using the values of f from Example 1, we have

$$S_3 = \tfrac{1}{6}(2.000 + 5.568 + 4.472 + 6.928 + 8.124 + 10.864 + 17.72) \cong 9.279.$$

The value of this integral accurate to 5 decimal places is 9.27972. ❑

Error Estimates

A numerical estimate is useful only to the extent that we can gauge its accuracy. When we use any kind of approximation method, we face two forms of error: the error inherent in the method we use (we call this the *theoretical error*) and the error that accumulates from rounding off the decimals that arise during the course of computation (we call this the *round-off error*). The nature of round-off error is obvious. We will speak first about theoretical error.

We begin with a function f continuous and increasing on $[a, b]$. We subdivide $[a, b]$ into n nonoverlapping intervals, each of length $(b - a)/n$. We want to estimate

$$\int_a^b f(x)\, dx$$

by the left-endpoint method. What is the theoretical error? It should be clear from Figure 8.8.9 that the theoretical error does not exceed

$$[f(b) - f(a)]\left(\frac{b - a}{n}\right).$$

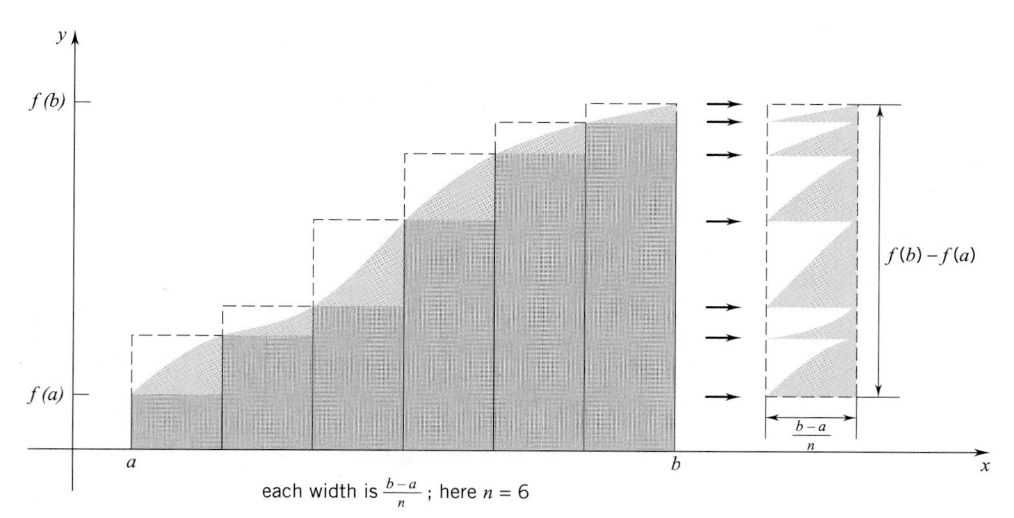

Figure 8.8.9

The error is represented by the sum of the areas of the shaded regions. These regions, when shifted to the right, all fit together within a rectangle of height $f(b) - f(a)$ and base $(b - a)/n$.

Similar reasoning shows that, under the same circumstances, the theoretical error associated with the trapezoidal method does not exceed

$$\frac{1}{2}[f(b) - f(a)]\left(\frac{b-a}{n}\right).$$

In this setting, at least, the trapezoidal estimate does a better job than the left-endpoint estimate.

This is a crude estimate of the error in the trapezoidal method. Even without the requirement that f be increasing, it can be improved considerably. It is shown in texts on numerical analysis that if f is continuous on $[a, b]$ and twice differentiable on (a, b), then the theoretical error of the trapezoidal rule

$$E_n^T = \int_a^b f(x)\,dx - T_n,$$

can be written

(8.8.1)
$$E_n^T = -\frac{(b-a)^3}{12n^2}f''(c),$$

where c is some number between a and b. Usually we cannot pinpoint c any further. However, if $f''(x)$ is bounded on $[a, b]$, that is, if there is a constant M such that $|f''(x)| \leq M$ for $a \leq x \leq b$, then it follows that

(8.8.2)
$$|E_n^T| \leq \frac{(b-a)^3}{12n^2}M.$$

This provides a useful method for analyzing the error in a trapezoidal estimate.

Recall the trapezoidal-rule estimate of ln 2 derived at the beginning of the section:

$$\ln 2 = \int_1^2 \frac{dx}{x} \cong 0.696.$$

We apply (8.8.2) to find the theoretical error. Here

$$f(x) = \frac{1}{x}, \qquad f'(x) = \frac{-1}{x^2}, \qquad f''(x) = \frac{2}{x^3}.$$

Now, it is easy to see that $|f''(x)| \leq 2$ for $1 \leq x \leq 2$. Therefore, with $a = 1$, $b = 2$, and $n = 5$, we have

$$|E_5^T| \leq \frac{(2-1)^3}{12 \cdot 5^2}2 = \frac{1}{150} < 0.007.$$

The estimate 0.696 is in theoretical error by less than 0.007.

Suppose, on the other hand, that we wanted an estimate for

$$\ln 2 = \int_1^2 \frac{dx}{x}$$

that is accurate to four decimal places. Then, using (8.8.2), we need

$$\frac{(b-a)^3}{12n^2} M < 0.00005. \tag{1}$$

Since

$$\frac{(b-a)^3}{12n^2} M \leq \frac{1}{12n^2} 2 = \frac{1}{6n^2},$$

we can guarantee that (1) holds by having

$$\frac{1}{6n^2} < 0.00005,$$

which is equivalent to

$$n^2 > 3333.$$

As you can check, $n = 58$ is the smallest integer that satisfies this inequality. Thus, the trapezoid rule requires a regular partition with at least 58 points to guarantee four-decimal-place accuracy in the approximation of $\int_1^2 (1/x)\,dx$.

Simpson's rule is more effective than the trapezoidal rule. If f is continuous on $[a, b]$ and if $f^{(4)}$ exists on (a, b), then the theoretical error for Simpson's rule

$$E_n^S = \int_a^b f(x)\,dx - S_n,$$

can be written

(8.8.3)

$$E_n^S = -\frac{(b-a)^5}{2880n^4} f^{(4)}(c),$$

where, as before, c is some number between a and b. Whereas (8.8.1) varies as $1/n^2$, this quantity varies as $1/n^4$. Thus, for comparable n, we can expect greater accuracy from Simpson's rule. In addition, if we assume that $f^{(4)}(x)$ is bounded on $[a, b]$, say $|f^{(4)}(x)| \leq M$ for $a \leq x \leq b$, then

(8.8.4)

$$|E_n^S| \leq \frac{(b-a)^5}{2880n^4} M,$$

and this can be used to analyze the theoretical error for Simpson's rule in the same way that we used (8.8.2) to analyze the error for the trapezoidal rule.

For example, to achieve four-decimal-place accuracy in estimating

$$\ln 2 = \int_1^2 \frac{dx}{x},$$

using Simpson's rule, we set $b = 2$ and $a = 1$ in (8.8.4), and then try to find n so that

$$|E_n^S| \leq \frac{(2-1)^5}{2880n^4} M = \frac{1}{2880n^4} M \leq 0.00005,$$

where M is a bound for

$$|f^{(4)}(x)| = \frac{24}{x^5} \quad \text{on } [1, 2].$$

Clearly, $|f^{(4)}(x)| \leq 24$ on $[1, 2]$, and so we want to find n such that

$$\frac{1}{2880n^4} 24 = \frac{1}{120n^4} < 0.00005.$$

This is equivalent to

$$n^4 > 167.$$

You can verify that $n = 4$ is the smallest positive integer which satisfies this inequality. Obviously this is a considerable improvement in efficiency over the trapezoidal rule.

Finally, a word about round-off error. Any numerical procedure requires careful consideration of round-off error. To illustrate this point, we rework our trapezoidal estimate for

$$\int_1^2 \frac{dx}{x},$$

again taking $n = 5$, but this time assuming that our computer or calculator can store only two significant digits. As before,

$$(2) \qquad \int_1^2 \frac{dx}{x} \cong \tfrac{1}{10}[\tfrac{1}{1} + 2(\tfrac{5}{6}) + 2(\tfrac{5}{7}) + 2(\tfrac{5}{8}) + 2(\tfrac{5}{9}) + (\tfrac{1}{2})].$$

Now our limited round-off machine goes to work:

$$\begin{aligned} \int_1^2 \frac{dx}{x} &\cong (0.10)[(1.0) + 2(0.83) + 2(0.71) + 2(0.62) + 2(0.44) + (0.50)] \\ &\text{“=”}\ (0.10)[(1.0) + (1.7) + (1.4) + (1.2) + (0.88) + (0.50)] \\ &= (0.10)[6.7] = 0.67. \end{aligned}$$

Earlier we used (8.8.1) to show that estimate (2) is in error by no more than 0.007 and found 0.70 as the approximation. Now with our limited round-off machine, we simplified (2) in a different way and obtained 0.67 as the approximation. The apparent error due to crude round off, $0.70 - 0.67 = 0.03$, exceeds the error of the approximation method itself. The lesson should be clear: round-off error is important.

EXERCISES 8.8

In Exercises 1–10, round your answers to four decimal places.

1. Estimate

$$\int_0^{12} x^2\,dx$$

using: (a) the left-endpoint estimate, $n = 12$; (b) the right-endpoint estimate, $n = 12$; (c) the midpoint estimate, $n = 6$; (d) the trapezoidal rule, $n = 12$; (e) Simpson's rule, $n = 6$. Check your results by performing the integration.

2. Estimate

$$\int_0^1 \sin^2 \pi x\,dx$$

using: (a) the midpoint estimate, $n = 3$; (b) the trapezoidal rule, $n = 6$; (c) Simpson's rule, $n = 3$. Check your results by performing the integration.

3. Estimate

$$\int_0^3 \frac{dx}{1 + x^3}$$

using: (a) the left-endpoint estimate, $n = 6$; (b) the right-endpoint estimate, $n = 6$; (c) the midpoint estimate, $n = 3$; (d) the trapezoidal rule, $n = 6$; (e) Simpson's rule, $n = 3$.

4. Estimate

$$\int_0^{\pi} \frac{\sin x}{\pi + x}\,dx$$

using: (a) the trapezoidal rule, $n = 6$; (b) Simpson's rule, $n = 3$.

5. Find the approximate value of π by estimating the integral

$$\frac{\pi}{4} = \tan^{-1} 1 = \int_0^1 \frac{dx}{1 + x^2}$$

using: (a) the trapezoidal rule, $n = 4$; (b) Simpson's rule, $n = 4$.

6. Estimate

$$\int_0^2 \frac{dx}{\sqrt{4 + x^3}}$$

using: (a) the trapezoidal rule, $n = 4$; (b) Simpson's rule, $n = 2$.

7. Estimate

$$\int_{-1}^{1} \cos(x^2)\,dx$$

using: (a) the midpoint estimate, $n = 4$; (b) the trapezoidal rule, $n = 8$; (c) Simpson's rule, $n = 4$.

8. Estimate

$$\int_1^2 \frac{e^x}{x}\,dx$$

using: (a) the midpoint estimate, $n = 4$; (b) the trapezoidal rule, $n = 8$; (c) Simpson's rule, $n = 4$.

9. Estimate

$$\int_0^2 e^{-x^2}\,dx$$

using: (a) the trapezoidal rule, $n = 10$; (b) Simpson's rule, $n = 5$.

10. Estimate

$$\int_2^4 \frac{1}{\ln x}\,dx$$

using: (a) the midpoint estimate, $n = 4$; (b) the trapezoidal rule, $n = 8$; (c) Simpson's rule, $n = 4$.

11. Show that there is one and only one curve of the form $y = Ax^2 + Bx + C$ through three distinct points with different x-coordinates.

12. Show that the function $g(x) = Ax^2 + Bx + C$ satisfies the condition

$$\int_a^b g(x)\,dx = \frac{b - a}{6}\left[g(a) + 4g\left(\frac{a + b}{2}\right) + g(b)\right]$$

for every interval $[a, b]$.

In Exercises 13–22, determine the values of n for which a theoretical error less than ϵ can be guaranteed if the integral is estimated using: (a) the trapezoidal rule; (b) Simpson's rule.

13. $\int_1^4 \sqrt{x}\,dx; \quad \epsilon = 0.01.$

14. $\int_1^3 x^5\,dx; \quad \epsilon = 0.01.$

15. $\int_1^4 \sqrt{x}\,dx; \quad \epsilon = 0.00001.$

16. $\int_1^3 x^5\,dx; \quad \epsilon = 0.00001.$

17. $\int_0^{\pi} \sin x\,dx; \quad \epsilon = 0.001$

18. $\int_0^{\pi} \cos x\,dx; \quad \epsilon = 0.001.$

19. $\int_1^3 e^x\,dx; \quad \epsilon = 0.01.$

20. $\int_1^e \ln x\,dx; \quad \epsilon = 0.01.$

21. $\int_0^2 e^{-x^2}\,dx; \quad \epsilon = 0.0001$

22. $\int_0^2 e^x\,dx; \quad \epsilon = 0.00001.$

23. Show that Simpson's rule is exact (theoretical error zero) for every polynomial of degree 3 or less.

24. Show that the trapezoidal rule is exact (theoretical error zero) if f is linear.

25. (a) Let $f(x) = x^2$ on [0, 1] and let $n = 2$. Show that

$$E_n^T = \frac{(b-a)^3}{12\,n^2}M$$

so that the inequality (8.8.2) can reduce to equality.

(b) Let $f(x) = x^4$ on [0, 1] and let $n = 1$. Show that

$$E_n^S = \frac{(b-a)^5}{2880\,n^4}M.$$

26. Show that, if f is continuous, then T_n and S_n can both be written as Riemann sums. HINT: Both

$$\frac{1}{2}[f(x_{i-1}) + f(x_i)] \qquad \text{and}$$

$$\frac{1}{6}\left[f(x_{i-1}) + 4f\left(\frac{x_{i-1}+x_i}{2}\right) + f(x_i)\right]$$

lie between m_i and M_i, the minimum and maximum values of f on $[x_{i-1}, x_i]$.

27. Let f be a twice differentiable function. Show that if $f(x) > 0$ and $f''(x) > 0$ on $[a, b]$, then

$$M_n \le \int_a^b f(x)\,dx \le T_n \qquad \text{for any } n,$$

HINT: See the figure. Show that the area of the rectangle $ABCD$ equals the area of the trapezoid $AEFD$.

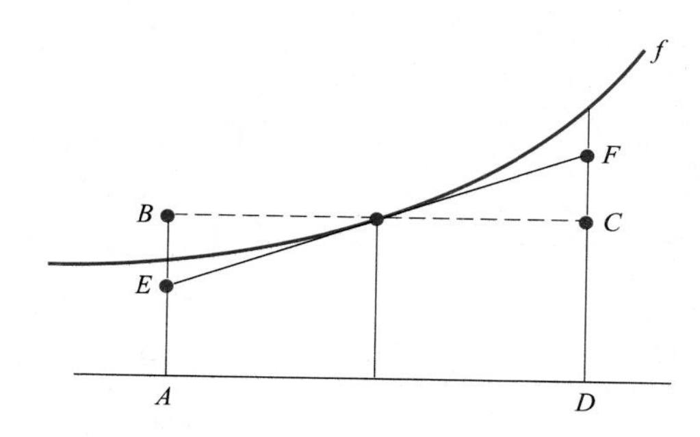

28. Show that $\frac{1}{3}T_n + \frac{2}{3}M_n = S_n$.

■ CHAPTER HIGHLIGHTS

8.1 Review

important integral formulas (p. 489)
a table of integrals appears on the inside covers

8.2 Integration by Parts

$$\int u\,dv = uv - \int v\,du$$

success with the technique depends on choosing u and dv so that $\int v\,du$ is easier to integrate than $\int u\,dv$

integration by parts is often used to calculate integrals when the integrand is a product of a mixture of functions; for example, polynomials and exponentials, or polynomials and trigonometric functions

to calculate some integrals, you may have to integrate by parts more than once

the integral of $\ln x$ and the integrals of the inverse trigonometric functions are calculated using integration by parts (p. 499)

8.3 Powers of Products of Sines and Cosines

Integrals of the form

$$\int \sin^m x \cos^n x\,dx$$

can be calculated by using the basic identity $\sin^2 x + \cos^2 x = 1$ and the double-angle formulas

$$\sin x \cos x = \tfrac{1}{2} \sin 2x, \qquad \sin^2 x = \tfrac{1}{2} - \tfrac{1}{2} \cos 2x, \qquad \cos^2 x = \tfrac{1}{2} + \tfrac{1}{2} \cos 2x.$$

Reduction formulas:

$$\int \sin^n x \, dx = -\frac{1}{n} \sin^{n-1} x \cos x + \frac{n-1}{n} \int \sin^{n-2} x \, dx$$

$$\int \cos^n x \, dx = \frac{1}{n} \cos^{n-1} x \sin x + \frac{n-1}{n} \int \cos^{n-2} x \, dx$$

8.4 Other Trigonometric Powers

the main tools for calculating such integrals are the identities

$$1 + \tan^2 x = \sec^2 x, \qquad 1 + \cot^2 x = \csc^2 x,$$

and integration by parts.

8.5 Integrals Involving $\sqrt{a^2 \pm x^2}$ and $\sqrt{x^2 - a^2}$; Trigonometric Substitutions

such integrals may be calculated by a trigonometric substitution:

for $\sqrt{a^2 - x^2}$ set $a \sin u = x$

for $\sqrt{a^2 + x^2}$ set $a \tan u = x$

for $\sqrt{x^2 - a^2}$ set $a \sec u = x$

integrals involving $\sqrt{a^2 + x^2}$ or $\sqrt{x^2 - a^2}$ may also be calculated by a hyperbolic substitution (p. 518)

it may be necessary to complete the square under the radical before making a trigonometric or hyperbolic substitution (p. 516)

8.6 Partial Fractions

a proper rational function may be integrated by writing it as a sum of fractions of the form

$$\frac{A}{(x - \alpha)^k} \quad \text{and} \quad \frac{Bx + C}{(x^2 + \beta x + \gamma)^k}$$

called the *partial fraction decomposition* (p. 522)

to integrate an improper rational function, express it as a polynomial plus a proper rational function by dividing the denominator into the numerator (p. 521)

8.7 Rationalizing Substitutions

for integrals involving $\sqrt[n]{f(x)}$ for some function f, let $u^n = f(x)$, $nu^{n-1} du = f'(x)\, dx$

for rational expressions in sine and cosine, let $u = \tan(x/2)$. Then

$$\sin x = \frac{2u}{1 + u^2}, \qquad \cos x = \frac{1 - u^2}{1 + u^2}, \qquad dx = \frac{2}{1 + u^2} du \qquad \text{(p. 533)}$$

8.8 Numerical Integration

left-endpoint, right-endpoint, and midpoint estimates; trapezoidal rule; Simpson's rule (p. 537)

the theoretical error in the trapezoidal rule varies as $1/n^2$

the theoretical error in Simpson's rule varies as $1/n^4$

■ PROJECTS AND EXPLORATIONS USING TECHNOLOGY

To do these exercises you will need a graphics calculator or a computer with graphing capability. The majority of these problems are open-ended so different approaches may be used to solve them. You should be aware that different approaches can result in slight variations in the answers. Round your numerical answers to at least four decimal places. The rounding method that your calculator or computer uses also may cause variations in answers.

8.1 The function

$$B(x) = \int_0^x e^{-t^2}\, dt, \qquad x \text{ real},$$

is of fundamental importance in science and engineering. (The function $(2/\sqrt{\pi})B(x)$ is called the *error function.*) While we know that B exists, we must approximate its values numerically. Nevertheless, we can determine many of the properties of this function.

(a) Explain why B is continuous on $(-\infty, \infty)$.
(b) Find B' and show that B is increasing on $(-\infty, \infty)$.
(c) Show that the graph of B has exactly one point of inflection and find it.
(d) Show that the graph of B has a horizontal asymptote and find it.
(e) Explain why B has an inverse and find $B^{-1}(0.5)$.
(f) Sketch the graph of B^{-1}.

8.2 Consider the definite integral

$$\int_1^3 e^{-x^2}\, dx.$$

(a) Obtain 20 approximations of this integral in the following manner: Evaluate the integrand at 100 random points between 1 and 3; calculate the average of these 100 values; then multiply by the length of the interval. Note that this procedure is equivalent to calculating 20 Riemann sums.
(b) Let $X_1, X_2, X_3, \ldots, X_{20}$ be the 20 values that you obtained in part (a). Find the mean m and standard deviation σ of these 20 values. The *mean* is the average of the values and the *standard deviation* is given by

$$\sigma = \sqrt{\sum_{n=1}^{20} \frac{(X_n - m)^2}{19}}.$$

(c) For each real number x, let $D(x)$ be the percentage of the values $X_1, X_2, X_3, \ldots, X_{20}$ which are less than or equal to x. Sketch the graph of D.
(d) Using the graph of D, sketch the graph of D' and discuss its properties.

8.3 Let F be the function defined by

$$F(x) = \int_0^x \cos[\sin t]\, dt,$$

and suppose that we want to approximate

$$\int_1^3 F(x)\, dx$$

correct to two decimal places using the trapezoidal rule. Since F changes its concavity on the interval $[1, 3]$, it is difficult to estimate the number of subintervals necessary to get this accuracy. The following procedure suggests one way.

(a) Find the points of inflection of F and use these to subdivide the interval $[1, 3]$ into subintervals on which the graph of F does not change its concavity.
(b) Approximate $\int F$ correct to two decimal places on each of these subintervals by both the trapezoid rule and the midpoint rule using the same n.

(c) Use these values to justify why your answer is within two decimal places. (Note that on each subinterval, the integral is between the trapezoidal approximation and the midpoint approximation. Upper and lower bounds for the whole interval can be obtained by adding the upper and lower bounds on each subinterval.

8.4 Let

$$f_n(x) = \cos\left(\frac{n}{x}\right)$$

for $n = 1, 2, 3, 4, 5$. Use six decimal place accuracy in the problems that follow.

(a) For each n, approximate

$$F(n) = \int_2^{10} f_n(x)\, dx$$

using Simpson's rule with $k = 2, 4, 8, 16$ subintervals. Denote these approximations by $S(n, k)$.

(b) Show that

$$F(n) - S(n, 16) \cong \frac{[F(n) - S(n, 8)]}{16}.$$

(c) Solve the approximation given in part (b) for $F(n)$ to get a final approximation for $F(n)$, $n = 1, 2, 3, 4, 5$.

(d) For each n compute a better approximation for $F(n)$ (for example, use Simpson's rule with $k > 16$ subintervals) and verify that the formula you obtained in part (c) did improve the approximation.

(e) Now estimate

$$\int_1^5 F(n)\, dn$$

by using Simpson's rule with 1 and 2 subintervals. Find a formula similar to that used in parts (b) and (c) for 1 and 2 subintervals instead of 8 and 16 subintervals. Then use this formula to get an improved approximation.

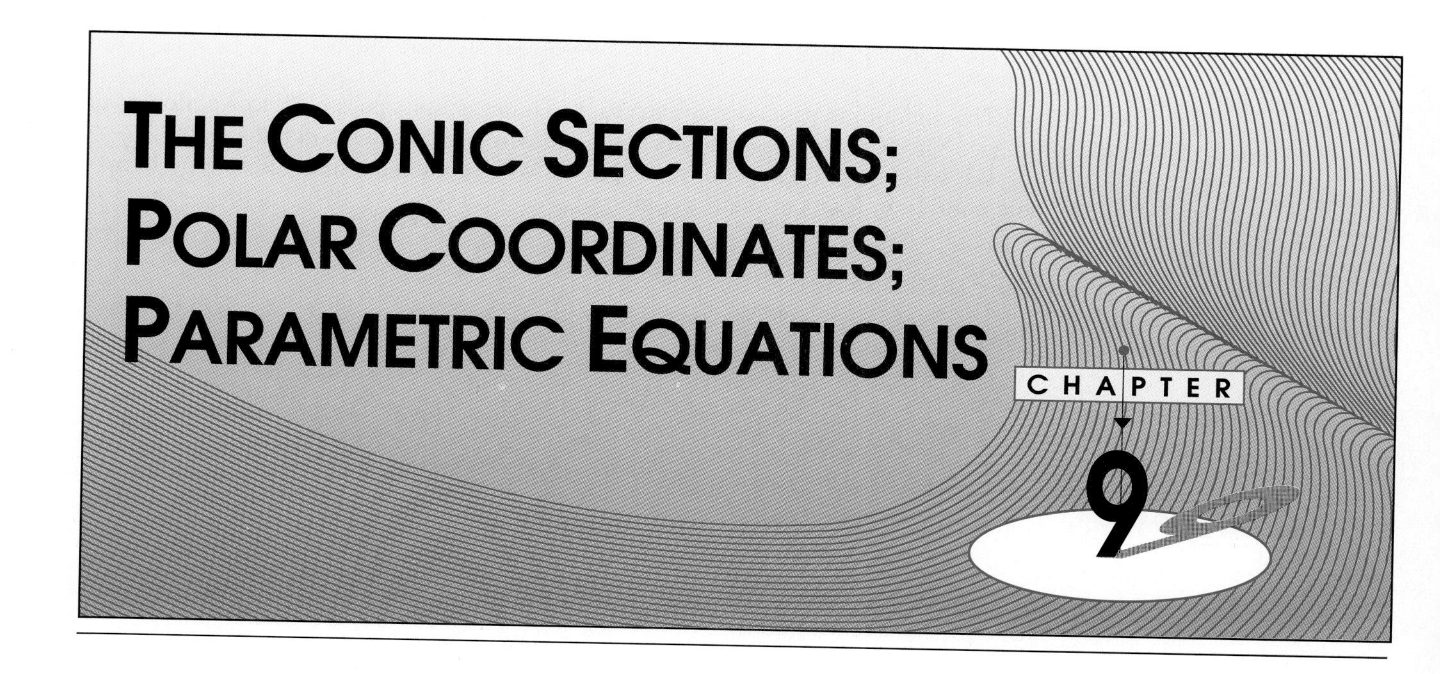

9.1 CONIC SECTIONS

Conic sections were introduced and discussed briefly in Section 1.4. We expand on that discussion here.

If a "double right circular cone" is cut by a plane, the resulting intersection is called a *conic section* or, more briefly, a *conic*. In Figure 9.1.1, we show a double cone and depict three important cases: the parabola, ellipse, and hyperbola.

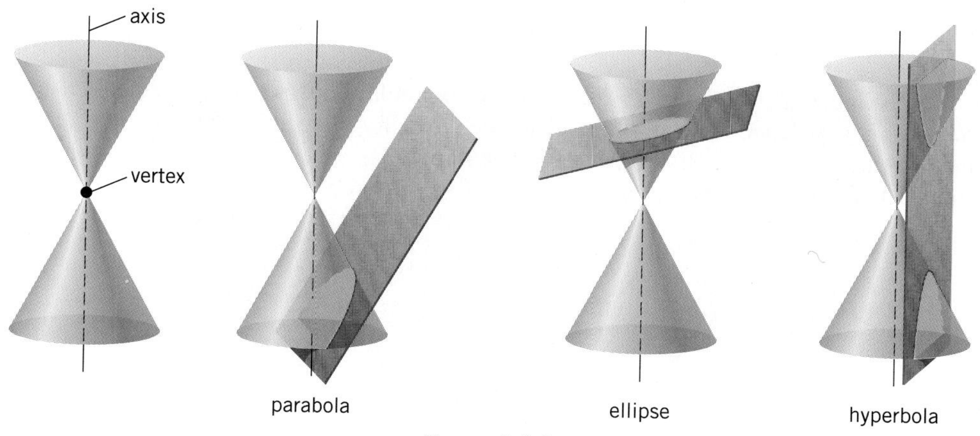

Figure 9.1.1

By choosing a plane perpendicular to the axis of the cone, we can obtain a circle. Other, less interesting, cases occur when the plane passes through the vertex of the cone. Depending on the orientation of the plane with respect to the axis of the cone, the intersection may either be a point, a line, or a pair of intersecting lines. Try to visualize these possibilities.

This three-dimensional approach to the conic sections goes back to Apollonius of Perga, a Greek of the third century B.C. He wrote eight books on the subject.

We will take a different approach. We will define parabola, ellipse, and hyperbola entirely in terms of plane geometry.

The Parabola

Figure 9.1.2 shows a line l and a point F not on l.

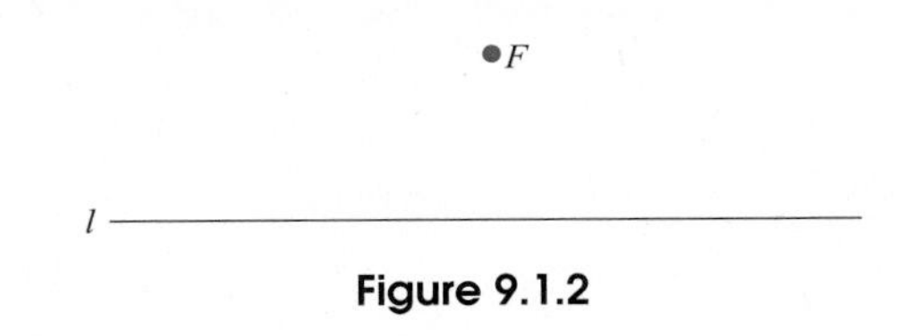

Figure 9.1.2

(9.1.1) The set of points P equidistant from F and l is called a *parabola.*

See Figure 9.1.3.

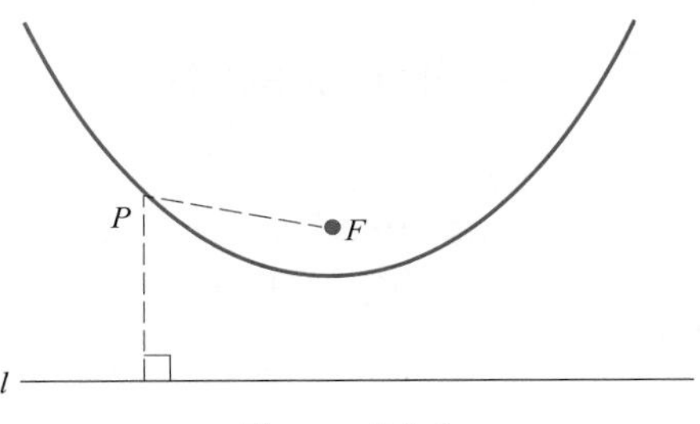

Figure 9.1.3

The line l is called the *directrix* of the parabola, and the point F is called the *focus.* (You will see why later on.) The line through F perpendicular to l is called the *axis* of the parabola. (It is the axis of symmetry.) The point at which the axis intersects the parabola is called the *vertex.* (See Figure 9.1.4.)

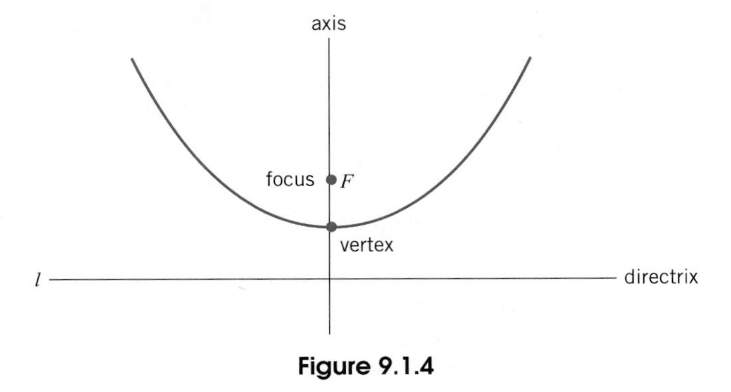

Figure 9.1.4

The equation of a parabola is particularly simple if we place the vertex at the origin and the focus on one of the coordinate axes. Suppose for the moment that the focus F is on the y-axis. Then F has coordinates of the form $(0, c)$. With the vertex at the origin, the directrix has equation $y = -c$. (See Figure 9.1.5.)

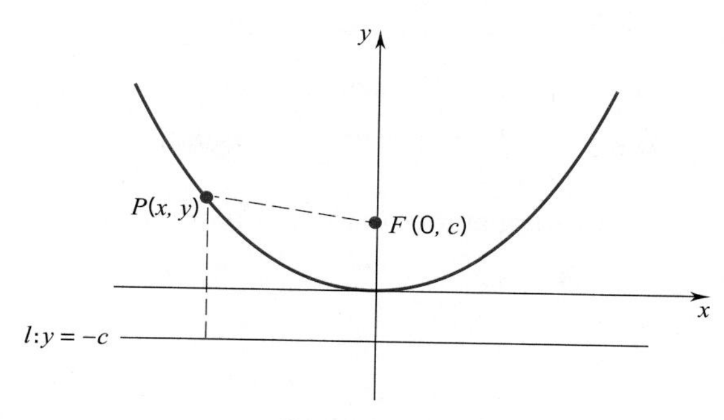

Figure 9.1.5

Every point $P(x, y)$ that lies on this parabola has the property that

$$d(P, F) = d(P, l).$$

Since

$$d(P, F) = \sqrt{x^2 + (y - c)^2} \qquad \text{and} \qquad d(P, l) = |y + c|,$$

you can see that

$$\begin{aligned} \sqrt{x^2 + (y - c)^2} &= |y + c|, \\ x^2 + (y - c)^2 &= |y + c|^2 = (y + c)^2, \\ x^2 + y^2 - 2cy + c^2 &= y^2 + 2cy + c^2, \\ x^2 &= 4cy. \quad \square \end{aligned}$$

You have just seen that the equation

(9.1.2)

$$\boxed{x^2 = 4cy}$$

(Figure 9.1.6)

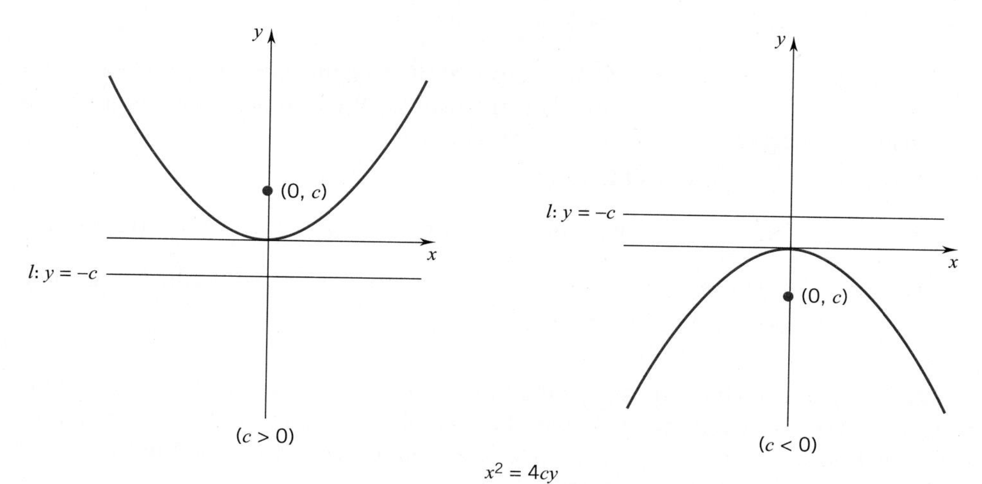

Figure 9.1.6

represents a parabola with vertex at the origin and focus at $(0, c)$. By interchanging the roles of x and y, you can see that the equation

(9.1.3)
$$y^2 = 4cx$$
(Figure 9.1.7)

represents a parabola with vertex at the origin and focus at $(c, 0)$.

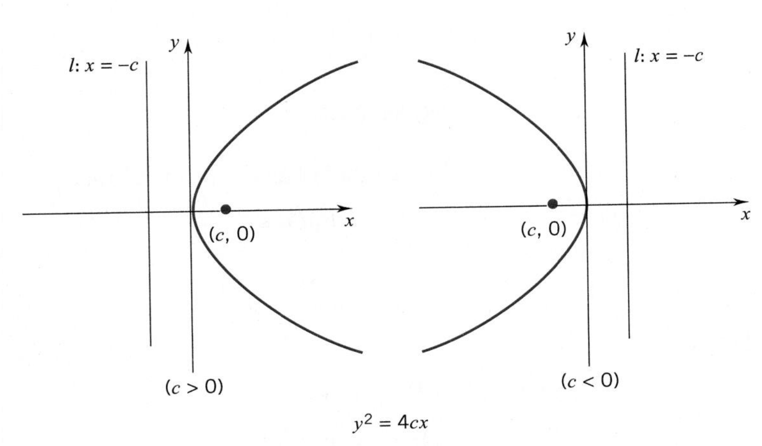

Figure 9.1.7

Example 1 Sketch the parabola specifying the vertex, focus, directrix, and axis:

(a) $x^2 = -4y$. (b) $y^2 = 3x$.

SOLUTION

(a) The equation $x^2 = -4y$ has the form

$$x^2 = 4cy \qquad \text{with} \quad c = -1.$$

The vertex is at the origin, and the focus is at $(0, -1)$; the directrix is the horizontal line $y = 1$; the axis of the parabola is the y-axis. We also plotted a few points to ensure the accuracy of our sketch. See Figure 9.1.8.

(b) The equation $y^2 = 3x$ has the form

$$y^2 = 4cx \qquad \text{with} \quad c = \tfrac{3}{4}.$$ (Figure 9.1.8)

The vertex is at the origin, and the focus is at $(\frac{3}{4}, 0)$; the directrix is the vertical line $x = -\frac{3}{4}$; the axis of the parabola is the x-axis. ❑

Every parabola with vertical axis is a translation of a parabola with equation of the form $x^2 = 4cy$, and every parabola with horizontal axis is a translation of a parabola with equation of the form $y^2 = 4cx$. We refer you to Section 1.4 for a discussion of translation of axes and shifting graphs.

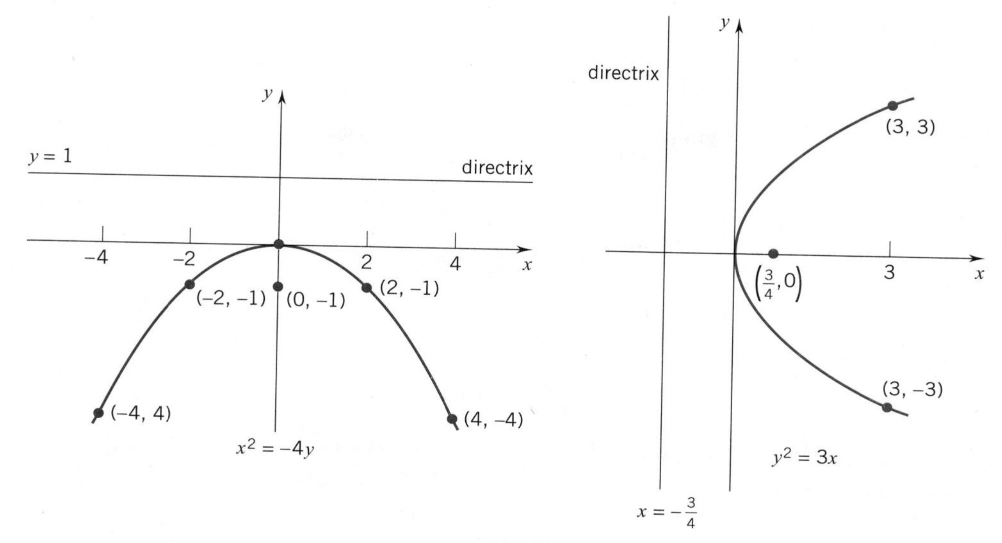

Figure 9.1.8

Example 2 Identify the curve

$$(x - 4)^2 = 8(y + 3).$$

SOLUTION The curve is the parabola

$$x^2 = 8y \qquad \text{(here } c = 2\text{)}$$

displaced 4 units right and 3 units down. The parabola $x^2 = 8y$ has vertex at the origin; the focus is at $(0, 2)$, and the directrix is the line $y = -2$. Thus, the parabola $(x - 4)^2 = 8(y + 3)$ has vertex at $(4, -3)$; the focus is at $(4, -1)$, and the directrix is the line $y = -5$. (See Figure 9.1.9.) ❑

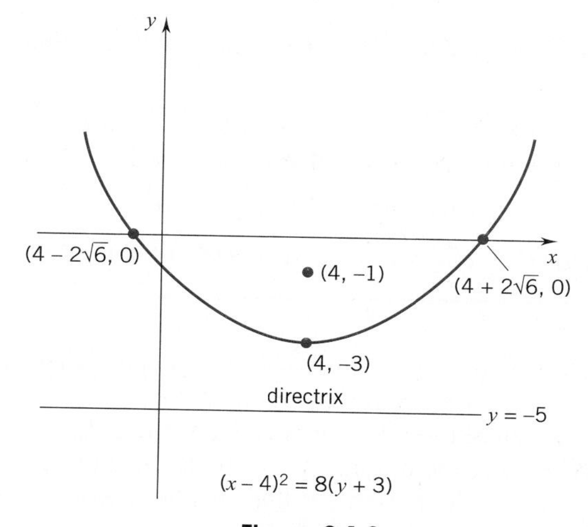

Figure 9.1.9

Example 3 Identify the curve

$$y^2 = 2y - 12x - 37.$$

SOLUTION First we gather the y terms on the left side of the equation and then complete the square:

$$y^2 - 2y = -12x - 37,$$
$$y^2 - 2y + 1 = -12x - 36,$$
$$(y - 1)^2 = -12(x + 3).$$

This is the parabola

$$y^2 = -12x \qquad \text{(here } c = -3\text{)}$$

displaced 3 units left, 1 unit up. The parabola $y^2 = -12x$ has vertex at the origin; the focus is at $(-3, 0)$ and the directrix is the vertical line $x = 3$. Thus, the parabola $(y - 1)^2 = -12(x + 3)$ has vertex at $(-3, 1)$; the focus is at $(-6, 1)$, and the directrix is the line $x = 0$ (the y-axis). See Figure 9.1.10. ❑

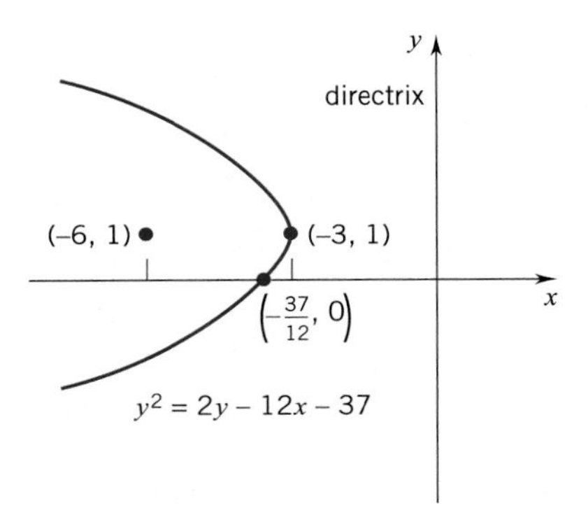

Figure 9.1.10

The Ellipse

Start with two points F_1, F_2 and a number k greater than the distance between them.

(9.1.4)

> The set of all points P for which
> $$d(P, F_1) + d(P, F_2) = k$$
> is called an *ellipse*. F_1 and F_2 are called the *foci*.

The idea is illustrated in Figure 9.1.11. A string is looped over tacks placed at the foci. The pencil placed in the loop traces out an ellipse.

Figure 9.1.12 shows an ellipse in what is called *standard position*: foci along the x-axis at equal distances from the origin. We now derive an equation for this ellipse setting $a = k/2$ or $k = 2a$. The reason for this change of constant will become apparent.

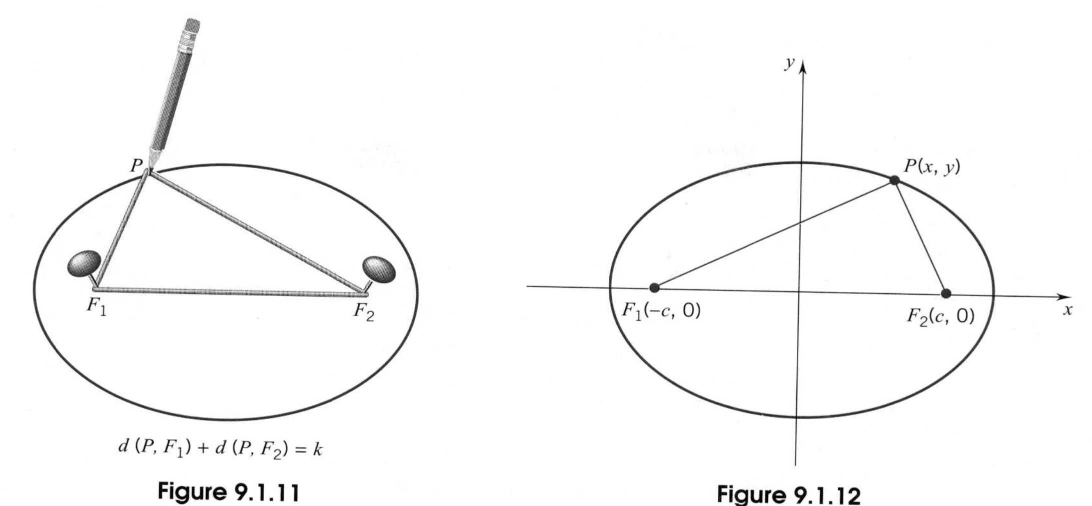

Figure 9.1.11

Figure 9.1.12

A point $P(x, y)$ lies on the ellipse iff

$$d(P, F_1) + d(P, F_2) = 2a.$$

With F_1 at $(-c, 0)$ and F_2 at $(c, 0)$, $c > 0$, we have

$$\sqrt{(x+c)^2 + y^2} + \sqrt{(x-c)^2 + y^2} = 2a.$$

Transferring the second term to the right-hand side and squaring both sides, we get

$$(x+c)^2 + y^2 = 4a^2 + (x-c)^2 + y^2 - 4a\sqrt{(x-c)^2 + y^2}.$$

This reduces to

$$4a\sqrt{(x-c)^2 + y^2} = 4(a^2 - cx).$$

Canceling the factor 4 and squaring again, we obtain

$$a^2(x^2 - 2cx + c^2 + y^2) = a^4 - 2a^2cx + c^2x^2.$$

This in turn reduces to

$$(a^2 - c^2)x^2 + a^2y^2 = a^2(a^2 - c^2),$$

which we write as

$$\frac{x^2}{a^2} + \frac{y^2}{a^2 - c^2} = 1.$$

Usually we set $b = \sqrt{a^2 - c^2}$ (you should verify that $a > c$). The equation for an ellipse in standard position then takes the form

(9.1.5)

$$\boxed{\frac{x^2}{a^2} + \frac{y^2}{b^2} = 1 \quad \text{with} \quad a > b.}$$

❑

The roles played by a, b, c can be read from Figure 9.1.13.

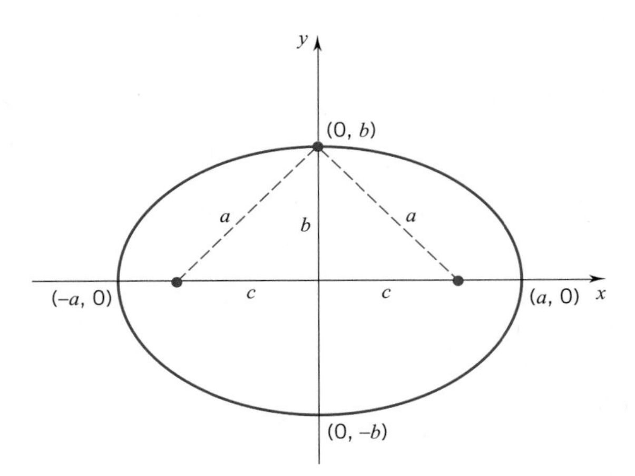

Figure 9.1.13

Every ellipse has four *vertices.* In Figure 9.1.13 these are marked $(a, 0)$, $(-a, 0)$, $(0, b)$, $(0, -b)$. The line segments that join opposite vertices are called the *axes* of the ellipse. The axis that contains the foci is called the *major axis,* the other the *minor axis.* In standard position the major axis is horizontal and has length $2a$; the minor axis is vertical and has length $2b$. The point at which the axes intersect is called the *center* of the ellipse. In standard position the center is at the origin.

Example 4 The equation $16x^2 + 25y^2 = 400$ can be written

$$\frac{x^2}{25} + \frac{y^2}{16} = 1. \qquad \text{(divide by 400)}$$

Here $a = 5$, $b = 4$, and $c = \sqrt{a^2 - b^2} = \sqrt{9} = 3$. The equation is in the form of (9.1.5). It is an ellipse in standard position with foci at $(-3, 0)$ and $(3, 0)$. The major axis has length $2a = 10$, and the minor axis has length $2b = 8$. The center is at the origin. The ellipse is sketched in Figure 9.1.14. ❑

Example 5 The equation

$$\frac{x^2}{16} + \frac{y^2}{25} = 1$$

does not represent an ellipse in standard position because $25 > 16$. This equation is the equation of Example 4 with x and y interchanged. It represents the ellipse of Example 4 reflected in the line $y = x$. (See Figure 9.1.15.) The foci are now on the y-axis, at $(0, -3)$ and $(0, 3)$. The major axis, now vertical, has length 10, and the minor axis has length 8. The center remains at the origin. ❑

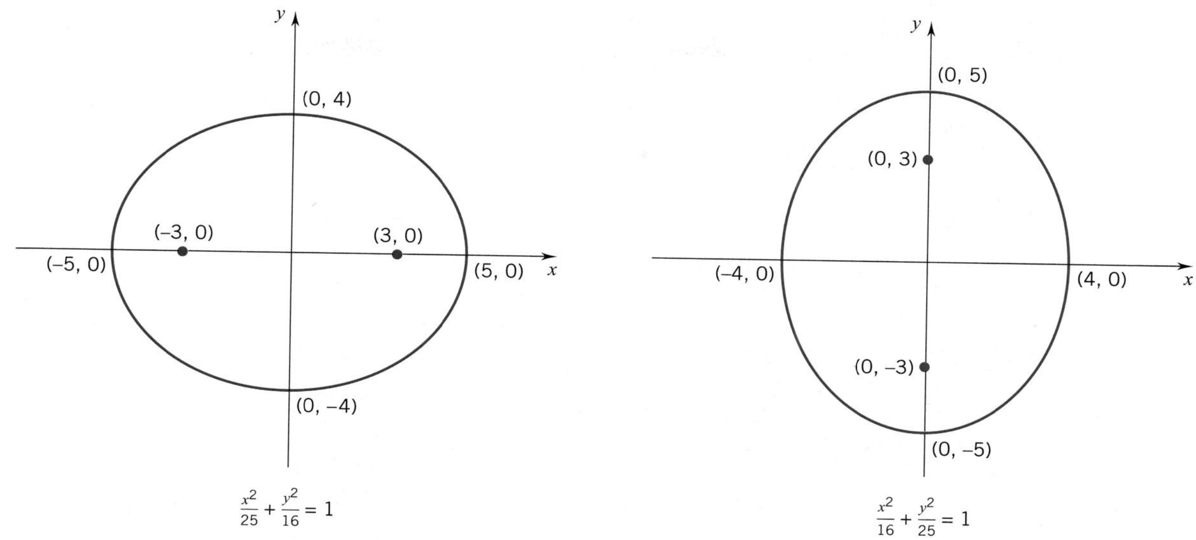

Figure 9.1.14

Figure 9.1.15

Example 6 Figure 9.1.16 shows two ellipses:

$$\frac{x^2}{25} + \frac{y^2}{9} = 1 \qquad \text{and} \qquad \frac{(x-1)^2}{25} + \frac{(y+4)^2}{9} = 1.$$

The first ellipse is in standard position. Here $a = 5$, $b = 3$, $c = \sqrt{a^2 - b^2} = 4$. The foci are at $(-4, 0)$ and $(4, 0)$. The major axis has length 10, and the minor axis has length 6.

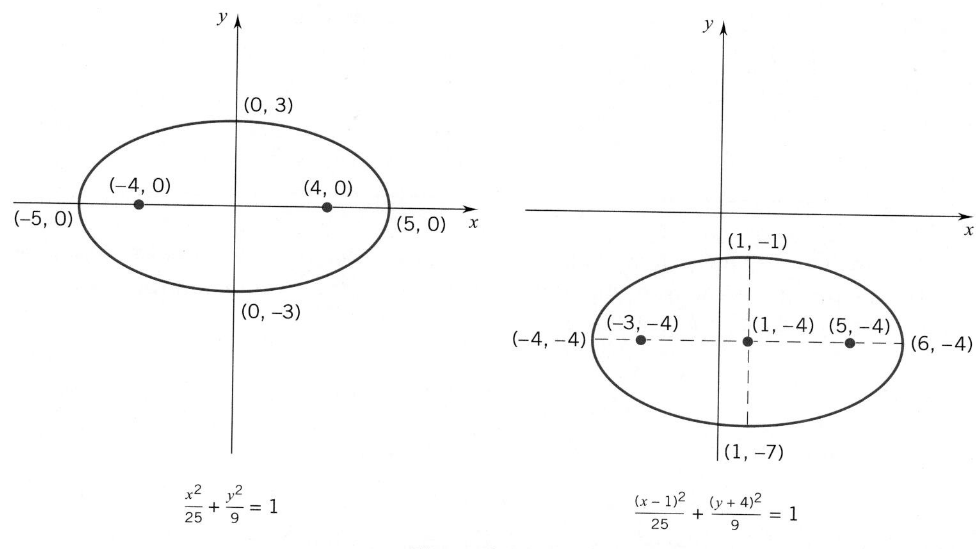

Figure 9.1.16

The second ellipse is the first ellipse displaced 1 unit right and 4 units down. The center is now at the point $(1, -4)$. The foci are at $(-3, -4)$ and $(5, -4)$. ❑

Example 7 To identify the curve

$$4x^2 - 8x + y^2 + 4y - 8 = 0.$$

we write

$$4(x^2 - 2x + \quad) + (y^2 + 4y + \quad) = 8,$$

and complete the squares within the parentheses. This gives

$$4(x^2 - 2x + 1) + (y^2 + 4y + 4) = 8 + 4 + 4,$$

$$4(x - 1)^2 + (y + 2)^2 = 16,$$

$$(*) \qquad \frac{(x - 1)^2}{4} + \frac{(y + 2)^2}{16} = 1.$$

This is the ellipse

$$(**) \qquad \frac{x^2}{16} + \frac{y^2}{4} = 1 \qquad (a = 4,\ b = 2,\ c = \sqrt{16 - 4} = 2\sqrt{3})$$

reflected in the line $y = x$ and then displaced 1 unit right and 2 units down. Since the foci of (**) are at $(-2\sqrt{3}, 0)$ and $(2\sqrt{3}, 0)$, the foci of (*) are at $(1, -2 - 2\sqrt{3})$ and $(1, -2 + 2\sqrt{3})$. The major axis, now vertical, has length 8; the minor axis has length 4. ❑

The Hyperbola

Start with two points F_1, F_2 and take a positive number k less than the distance between them.

(9.1.6)

> The set of all points P for which
>
> $$|d(P, F_1) - d(P, F_2)| = k$$
>
> is called a *hyperbola*. F_1 and F_2 are called the *foci*.

Figure 9.1.17 shows a hyperbola in what is called *standard position*: foci along the x-axis at equal distances from the origin. We will derive an equation for this hyperbola, again setting $a = k/2$ or $k = 2a$.

A point $P(x, y)$ lies on the hyperbola iff

$$|d(P, F_1) - d(P, F_2)| = 2a.$$

With F_1 at $(-c, 0)$ and F_2 at $(c, 0)$, $c > 0$, we have

$$\sqrt{(x + c)^2 + y^2} - \sqrt{(x - c)^2 + y^2} = \pm 2a. \qquad \text{(explain)}$$

Transferring the second term to the right and squaring both sides, we obtain

$$(x + c)^2 + y^2 = 4a^2 \pm 4a\sqrt{(x - c)^2 + y^2} + (x - c)^2 + y^2.$$

This equation reduces to

$$xc - a^2 = \pm a\sqrt{(x - c)^2 + y^2}.$$

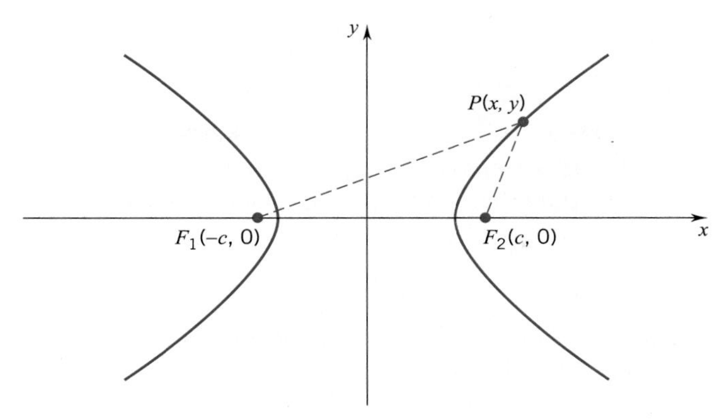

Figure 9.1.17

Squaring once more, we find that

$$x^2c^2 - 2a^2xc + a^4 = a^2(x^2 - 2xc + c^2 + y^2),$$

which reduces to

$$(c^2 - a^2)x^2 - a^2y^2 = a^2(c^2 - a^2),$$

and thus to

$$\frac{x^2}{a^2} - \frac{y^2}{c^2 - a^2} = 1.$$

Usually we set $b = \sqrt{c^2 - a^2}$ (verify that $c > a$). The equation for a hyperbola in standard position then takes the form

(9.1.7)
$$\boxed{\frac{x^2}{a^2} - \frac{y^2}{b^2} = 1.}$$ ❑

The roles played by a, b, c can be read from Figure 9.1.18. As the figure suggests, the

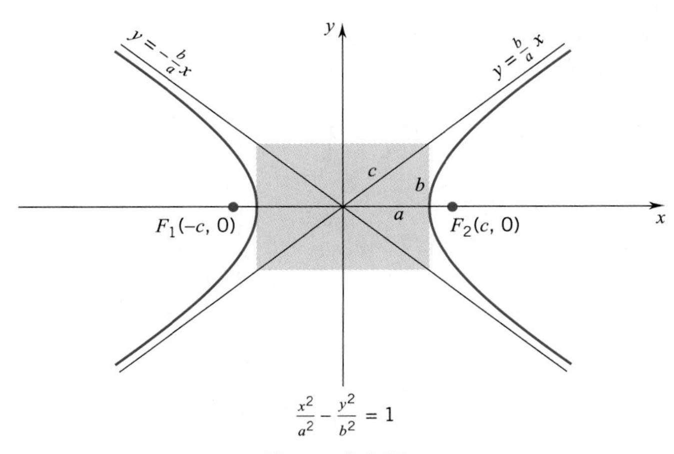

Figure 9.1.18

hyperbola remains between the lines

$$y = \frac{b}{a}x \qquad \text{and} \qquad y = -\frac{b}{a}x.$$

These lines are called the *asymptotes* of the hyperbola. They can be obtained from Equation (9.1.7) by replacing the 1 on the right-hand side by 0:

$$\frac{x^2}{a^2} - \frac{y^2}{b^2} = 0 \qquad \text{gives} \qquad y = \pm\frac{b}{a}x.$$

As $x \to \pm\infty$, the vertical separation between the hyperbola and the asymptotes tends to zero. To see this, solve the equation

$$\frac{x^2}{a^2} - \frac{y^2}{b^2} = 1$$

for y. This gives

$$y = \pm\sqrt{\frac{b^2}{a^2}x^2 - b^2} = \pm\frac{b}{a}\sqrt{x^2 - a^2}.$$

In all four quadrants the vertical separation between the hyperbola and the asymptotes can be written

$$\left|\frac{b}{a}|x| - \frac{b}{a}\sqrt{x^2 - a^2}\right| = \frac{b}{a}\left||x| - \sqrt{x^2 - a^2}\right|. \qquad \text{(check this)}$$

As $x \to \pm\infty$,

$$\begin{aligned}\frac{b}{a}\left||x| - \sqrt{x^2 - a^2}\right| &= \frac{b}{a}\left||x| - \sqrt{x^2 - a^2}\right|\frac{|x| + \sqrt{x^2 - a^2}}{|x| + \sqrt{x^2 - a^2}} \\ &= \frac{b}{a} \cdot \frac{x^2 - (x^2 - a^2)}{|x| + \sqrt{x^2 - a^2}} = \frac{b}{a} \cdot \frac{a^2}{|x| + \sqrt{x^2 - a^2}} \to 0.\end{aligned}$$

The line determined by the foci of a hyperbola intersects the hyperbola at two points, called the *vertices.* The line segment that joins the vertices is called the *transverse axis.* The midpoint of the transverse axis is called the *center* of the hyperbola.

In standard position (Figure 9.1.18), the vertices are $(\pm a, 0)$, the transverse axis has length $2a$, and the center is at the origin.

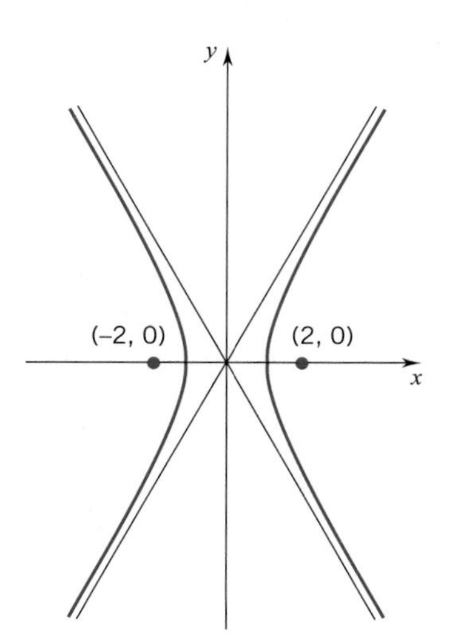

Figure 9.1.19

Example 8 The equation

$$\frac{x^2}{1} - \frac{y^2}{3} = 1 \qquad \text{(Figure 9.1.19)}$$

represents a hyperbola in standard position; here $a = 1$, $b = \sqrt{3}$, $c = \sqrt{1 + 3} = 2$. The center is at the origin. The foci are at $(-2, 0)$ and $(2, 0)$. The vertices are at $(-1, 0)$ and $(1, 0)$. The transverse axis has length 2. We can obtain the asymptotes by setting

$$\frac{x^2}{1} - \frac{y^2}{3} = 0.$$

The asymptotes are the lines $y = \pm\sqrt{3}x$. ❑

Example 9 The hyperbola

$$\frac{(x-4)^2}{1} - \frac{(y+5)^2}{3} = 1 \qquad \text{(Figure 9.1.20)}$$

is the hyperbola

$$\frac{x^2}{1} - \frac{y^2}{3} = 1$$

of Example 8 displaced 4 units right and 5 units down. The center of the hyperbola is now at the point $(4, -5)$. The foci are at $(2, -5)$ and $(6, -5)$. The vertices are at $(3, -5)$ and $(5, -5)$. The new asymptotes are the lines $y + 5 = \pm\sqrt{3}(x - 4)$. ❑

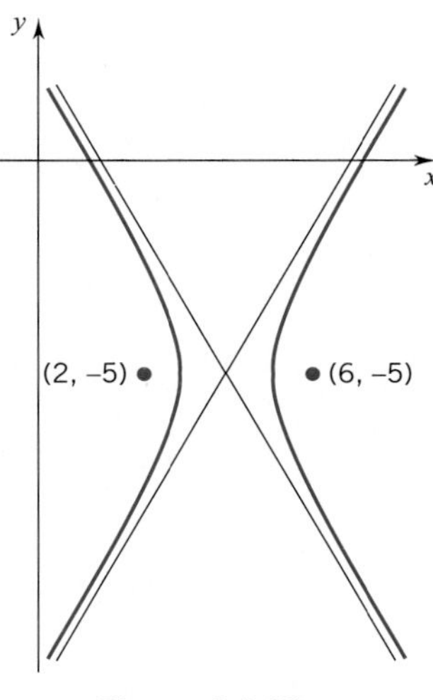

Figure 9.1.20

Example 10 The hyperbola

$$\frac{y^2}{1} - \frac{x^2}{3} = 1 \qquad \text{(Figure 9.1.21)}$$

is the hyperbola

$$\frac{x^2}{1} - \frac{y^2}{3} = 1$$

of Example 8 reflected in the line $y = x$. The center is still at the origin. The foci are now at $(0, -2)$ and $(0, 2)$. The vertices are at $(0, -1)$ and $(0, 1)$. The asymptotes are the lines $x = \pm\sqrt{3}y$. ❑

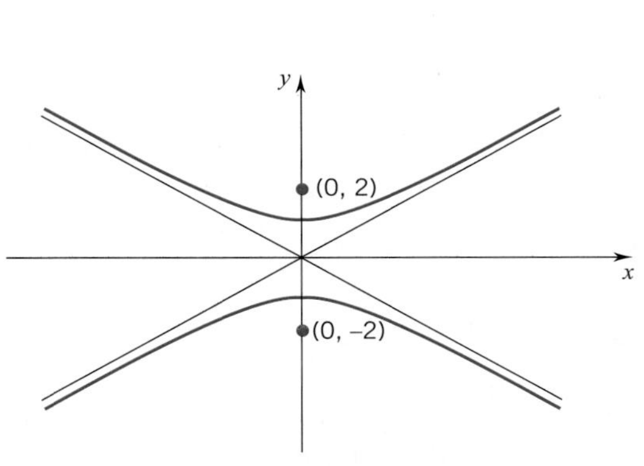

Figure 9.1.21

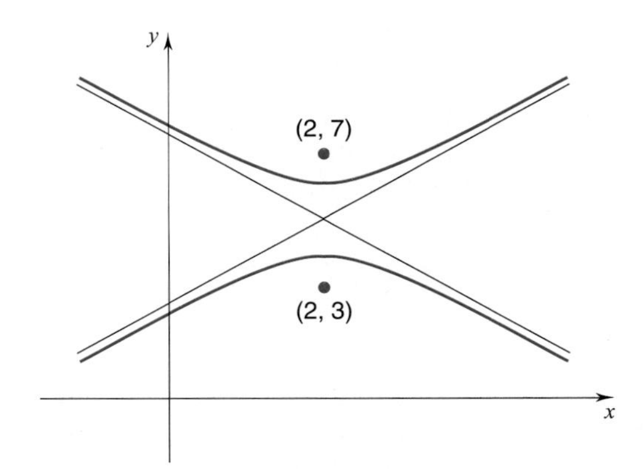

Figure 9.1.22

Example 11 The hyperbola

$$\frac{(y-5)^2}{1} - \frac{(x-2)^2}{3} = 1 \qquad \text{(Figure 9.1.22)}$$

is the hyperbola of Example 10 displaced 2 units right and 5 units up. ❑

Some Applications of Conic Sections

Parabolic Mirrors You are already familiar with the geometric principle of reflected light: that the angle of reflection equals the angle of incidence (Example 6, Section 4.5).

Now take a parabola and revolve it about its axis. This gives you a parabolic surface. A curved mirror of that form is called a *parabolic mirror.* Such mirrors are used in searchlights, automotive headlights, and reflecting telescopes. Our purpose here is to explain why.

We begin with a parabola and choose the coordinate system so that the equation takes the form $x^2 = 4cy$ with $c > 0$. We can express y in terms of x by writing

$$y = \frac{x^2}{4c}.$$

Since

$$\frac{dy}{dx} = \frac{2x}{4c} = \frac{x}{2c},$$

the tangent line at the point $P(x_0, y_0)$ has slope $m = x_0/2c$ and equation

$$(1) \qquad (y - y_0) = \frac{x_0}{2c}(x - x_0).$$

For the rest we refer to Figure 9.1.23.

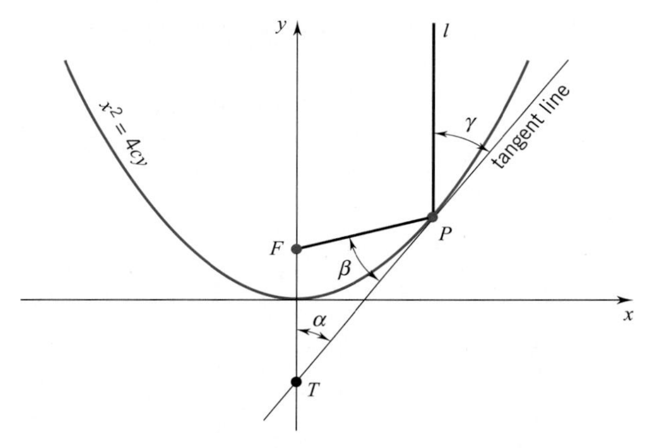

Figure 9.1.23

In the figure we have drawn a ray (a half-line) l parallel to the axis of the parabola, the y-axis. We want to show that the angles marked β and γ are equal.

Setting $x = 0$ in Equation (1), we find that

$$y = y_0 - \frac{x_0^2}{2c}.$$

Since the point (x_0, y_0) lies on the parabola, we have $x_0^2 = 4cy_0$, and thus

$$y_0 - \frac{x_0^2}{2c} = y_0 - \frac{4cy_0}{2c} = -y_0.$$

The y-coordinate of the point marked T is $-y_0$. Since the focus F is at $(0, c)$,

$$d(F, T) = y_0 + c.$$

The distance between F and P is also $y_0 + c$:

$$d(F, P) = \sqrt{x_0^2 + (y_0 - c)^2} = \sqrt{4cy_0 + (y_0 - c)^2} = \sqrt{(y_0 + c)^2} = y_0 + c.$$

$x_0^2 = 4cy_0$ ⟶ (second step); $y_0 + c > 0$ ⟶ (last step)

Since $d(F, T) = d(F, P)$, the triangle TFP is isosceles and the angles marked α and β are equal. Since l is parallel to the y-axis, $\alpha = \gamma$ and thus (and this is what we wanted to show)

$$\beta = \gamma.$$

The fact that $\beta = \gamma$ has important optical consequences. It means (Figure 9.1.24) that light emitted from a source at the focus of a parabolic mirror is reflected in a beam parallel to the axis of that mirror; this is the principle of the searchlight. It also means that light coming to a parabolic mirror in a beam parallel to the axis of the mirror is reflected entirely to the focus; this is the principle of the reflecting telescope.

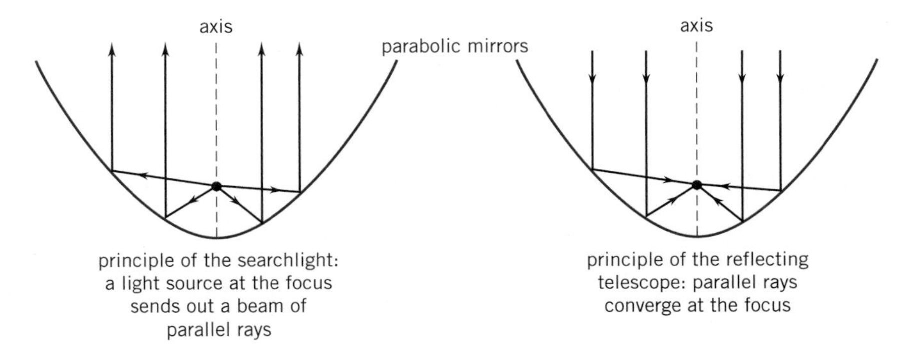
axis
parabolic mirrors
axis
principle of the searchlight: a light source at the focus sends out a beam of parallel rays
principle of the reflecting telescope: parallel rays converge at the focus

Figure 9.1.24

Elliptical Reflectors Like the parabola, the ellipse has an interesting reflecting property. To derive it, we consider the ellipse

$$\frac{x^2}{a^2} + \frac{y^2}{b^2} = 1.$$

Differentiating implicitly with respect to x, we get

$$\frac{2x}{a^2} + \frac{2y}{b^2}\frac{dy}{dx} = 0 \quad \text{and thus} \quad \frac{dy}{dx} = -\frac{b^2x}{a^2y}.$$

The slope at the point $P(x_0, y_0)$ is therefore

$$-\frac{b^2x_0}{a^2y_0}, \qquad y_0 \neq 0,$$

and the tangent line has equation

$$y - y_0 = -\frac{b^2x_0}{a^2y_0}(x - x_0).$$

We can rewrite this last equation as

$$(b^2x_0)x + (a^2y_0)y - a^2b^2 = 0.$$

We can now show the following:

(9.1.8) At each point P of the ellipse, the focal radii $\overline{F_1P}$ and $\overline{F_2P}$ make equal angles with the tangent.

PROOF If P lies on the x-axis, the focal radii are coincident and there is nothing to show. To visualize the argument for a point $P = P(x_0, y_0)$ not on the x-axis, see Figure 9.1.25.

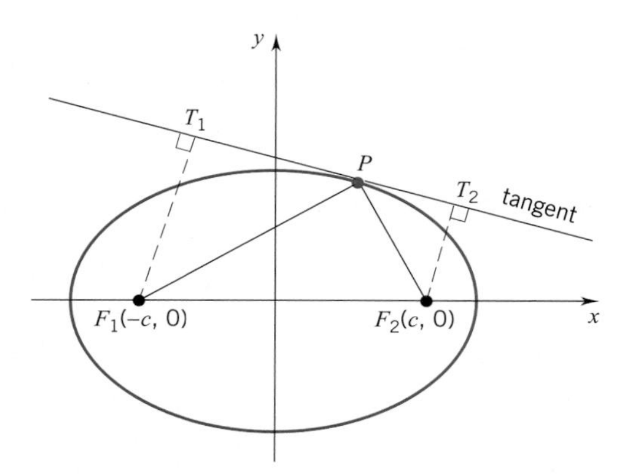

Figure 9.1.25

To show that $\overline{F_1P}$ and $\overline{F_2P}$ make equal angles with the tangent we need only show that the triangles PT_1F_1 and PT_2F_2 are similar. We can do this by showing that

$$\frac{d(T_1, F_1)}{d(F_1, P)} = \frac{d(T_2, F_2)}{d(F_2, P)}$$

or, equivalently, by showing that

$$\frac{|-b^2x_0c - a^2b^2|}{\sqrt{(x_0 + c)^2 + y_0^2}} = \frac{|b^2x_0c - a^2b^2|}{\sqrt{(x_0 - c)^2 + y_0^2}}. \qquad \text{(verify this)}$$

The validity of this last equation can be seen by canceling the factor b^2 and then squaring. This gives

$$\frac{(x_0c + a^2)^2}{(x_0 + c)^2 + y_0^2} = \frac{(x_0c - a^2)^2}{(x_0 - c)^2 + y_0^2},$$

which can be simplified to

$$(a^2 - c^2)x_0^2 + a^2y_0^2 = a^2(a^2 - c^2) \qquad \text{and thus to} \qquad \frac{x_0^2}{a^2} + \frac{y_0^2}{b^2} = 1.$$

This last equation holds since the point $P(x_0, y_0)$ is on the ellipse. ❑

The result we just proved has the following physical consequence:

(9.1.9) An elliptical reflector takes light or sound originating at one focus and converges it at the other focus.

In elliptical rooms called "whispering chambers," a whisper at one focus, inaudible nearby, is easily heard at the other focus. You will experience this phenomenon if you visit the Statuary Room in the Capitol in Washington, D.C.

Hyperbolic Reflectors A straightforward calculation that you are asked to carry out in the exercises shows that

(9.1.10) at each point P of a hyperbola, the tangent line bisects the angle between the focal radii $\overline{F_1P}$ and $\overline{F_2P}$.

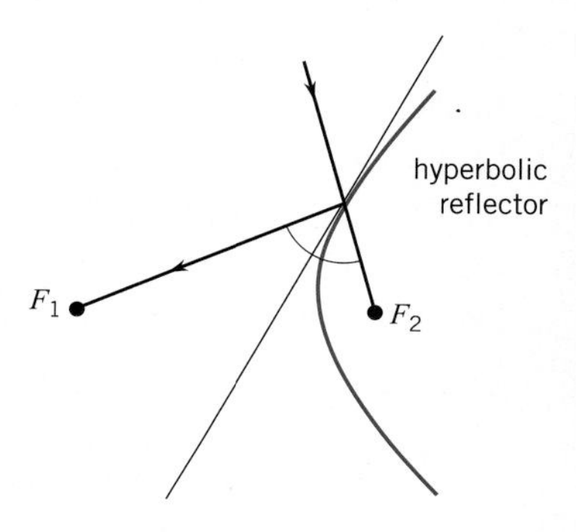

Figure 9.1.26

The optical consequences of this are illustrated in Figure 9.1.26. There you see the right branch of a hyperbola with foci F_1, F_2. Light or sound aimed at F_2 from any point to the left of the reflector is beamed to F_1.

An Application to Range Finding If observers, located at two listening posts at a known distance apart, time the firing of a cannon, the time difference multiplied by the velocity of sound gives the value of $2a$ and hence determines a hyperbola on which the cannon must be located. A third listening post gives two more hyperbolas. The cannon is found where the hyperbolas intersect.

EXERCISES 9.1

In Exercises 1–8, sketch the parabola and give an equation for it.

1. vertex $(0, 0)$, focus $(2, 0)$.

2. vertex $(0, 0)$, focus $(-2, 0)$.

3. vertex $(-1, 3)$, focus $(-1, 0)$.

4. vertex $(1, 2)$, focus $(1, 3)$.

5. focus $(1, 1)$, directrix $y = -1$.

6. focus $(2, -2)$, directrix $x = -5$.

7. focus $(1, 1)$, directrix $x = 2$.

8. focus $(2, 0)$, directrix $y = 3$.

In Exercises 9–16, find the vertex, focus, axis, and directrix; then sketch the parabola.

9. $y^2 = 2x$.

10. $x^2 = -5y$.

11. $2y = 4x^2 - 1$.

12. $y^2 = 2(x - 1)$.

13. $(x + 2)^2 = 12 - 8y$.

14. $y - 3 = 2(x - 1)^2$.

15. $x = y^2 + y + 1$.

16. $y = x^2 + x + 1$.

In Exercises 17–24, find (a) the center, (b) the foci, (c) the length of the major axis, and (d) the length of the minor axis for the given ellipse. Then sketch the graph.

17. $x^2/9 + y^2/4 = 1$.

18. $x^2/4 + y^2/9 = 1$.

19. $3x^2 + 2y^2 = 12$.

20. $3x^2 + 4y^2 - 12 = 0$.

21. $4x^2 + 9y^2 - 18y = 27$.

22. $4x^2 + y^2 - 6y + 5 = 0$.

23. $4(x - 1)^2 + y^2 = 64$.

24. $16(x - 2)^2 + 25(y - 3)^2 = 400$.

In Exercises 25–32, find an equation for the ellipse that satisfies the given conditions.

25. Foci at $(-1, 0)$, $(1, 0)$; major axis has length 6.

26. Foci at $(0, -1)$, $(0, 1)$; major axis has length 6.

27. Foci at $(1, 3)$, $(1, 9)$; minor axis has length 8.

28. Foci at (3, 1), (9, 1); minor axis has length 10.

29. Focus at (1, 1); center at (1, 3); major axis has length 10.

30. Center at (2, 1); vertices at (2, 6), (1, 1).

31. Major axis 10; vertices at (3, 2), (3, −4).

32. Focus at (6, 2); vertices at (1, 7), (1, −3).

In Exercises 33–40, find an equation for the indicated hyperbola.

33. Foci at (−5, 0), (5, 0); transverse axis has length 6.

34. Foci at (−13, 0), (13, 0); transverse axis has length 10.

35. Foci at (0, −13), (0, 13); transverse axis has length 10.

36. Foci at (0, −13), (0, 13); transverse axis has length 24.

37. Foci at (−5, 1), (5, 1); transverse axis has length 6.

38. Foci at (−3, 1), (7, 1); transverse axis has length 6.

39. Foci at (−1, −1), (−1, 1); transverse axis has length $\frac{1}{2}$.

40. Foci at (2, 1), (2, 5); transverse axis has length 3.

In Exercises 41–50, find the center, the vertices, the foci, the asymptotes, and the length of the transverse axis of the given hyperbola. Then sketch the graph.

41. $x^2 - y^2 = 1$.

42. $y^2 - x^2 = 1$.

43. $x^2/9 - y^2/16 = 1$.

44. $x^2/16 - y^2/9 = 1$.

45. $y^2/16 - x^2/9 = 1$.

46. $y^2/9 - x^2/16 = 1$.

47. $(x - 1)^2/9 - (y - 3)^2/16 = 1$.

48. $(x - 1)^2/16 - (y - 3)^2/9 = 1$.

49. $4x^2 - 8x - y^2 + 6y - 1 = 0$.

50. $-3x^2 + y^2 - 6x = 0$.

In Exercises 51–54, find an equation for the indicated parabola.

51. focus (1, 2), directrix $x + y + 1 = 0$.

52. vertex (2, 0), directrix $2x - y = 0$.

53. vertex (2, 0), focus (0, 2).

54. vertex (3, 0), focus (0, 1).

55. Show that every parabola has an equation of the form

$$(\alpha x + \beta y)^2 = \gamma x + \delta y + \epsilon \qquad \text{with} \quad \alpha^2 + \beta^2 \neq 0.$$

HINT: Take l: $Ax + By + C = 0$ as the directrix and $F(a, b)$ as the focus.

56. A line through the focus of a parabola (different from the axis) intersects the parabola at two points P and Q. Show that the tangent line through P meets the tangent line through Q at right angles.

57. A parabola intersects a rectangle of area A at two opposite vertices. Show that, if one side of the rectangle falls on the axis of the parabola, then the parabola subdivides the rectangle into two pieces, one of area $\frac{1}{3}A$, the other of area $\frac{2}{3}A$.

58. (a) Show that every parabola with axis parallel to the y-axis has an equation of the form $y = Ax^2 + Bx + C$ with $A \neq 0$. (b) Find the vertex, the focus, and the directrix of the parabola $y = Ax^2 + Bx + C$.

59. Find equations for all the parabolas that pass through the point (5, 6) and have directrix $y = 1$, axis $x = 2$.

60. Find an equation for the parabola that has horizontal axis, vertex (−1, 1), and passes through the point (−6, 13).

61. Find the distance between the foci of an ellipse of area A if the length of the major axis is $2a$.

62. Show that the set of all points $(a \cos t, b \sin t)$, with $a, b \neq 0$ and t real, lie on an ellipse.

63. Locate the foci of the ellipse given that the point (3, 4) lies on the ellipse and the ends of the major axis are at (0, 0) and (10, 0).

64. Show that in an ellipse the product of the distances between the foci and a tangent to the ellipse [$d(F_1, T_1)\, d(F_2, T_2)$ in Figure 9.1.25] is always the square of one-half the length of the minor axis.

The shape of an ellipse depends on its *eccentricity* e. This is half the distance between the foci divided by half the length of the major axis:

$$e = c/a.$$

For every ellipse $0 < e < 1$.

In Exercises 65–68, determine the eccentricity of the ellipse.

65. $x^2/25 + y^2/16 = 1$.

66. $x^2/16 + y^2/25 = 1$.

67. $(x - 1)^2/25 + (y + 2)^2/9 = 1$.

68. $(x + 1)^2/169 + (y - 1)^2/144 = 1$.

69. Suppose that E_1 and E_2 are both ellipses with the same major axis. Compare the shape of E_1 to the shape of E_2 if their eccentricities e_1, e_2 satisfy $e_1 < e_2$.

70. What happens to an ellipse with major axis $2a$ if its eccentricity e tends to zero?

71. What happens to an ellipse with major axis $2a$ if e tends to 1?

In Exercises 72 and 73, write an equation for the ellipse.

72. Major axis from (−3, 0) to (3, 0), eccentricity $\frac{1}{3}$.

73. Major axis from (−3, 0) to (3, 0), eccentricity $\frac{2}{3}\sqrt{2}$.

74. Let l be a line and let F be a point not on l. You have seen that the set of points P for which

$$d(F, P) = d(l, P)$$

is a parabola. Show that, if $0 < e < 1$, then the set of all points P for which

$$d(F, P) = e\, d(l, P)$$

is an ellipse of eccentricity e. HINT: Begin by choosing a coordinate system whereby F falls on the origin and l is a vertical line $x = d$.

75. Find the center, the vertices, the foci, the asymptotes, and the length of the transverse axis of the hyperbola with equation $xy = 1$. HINT: Define new XY-coordinates by setting $x = X + Y$ and $y = X - Y$.

For Exercises 76–78, we refer to the hyperbola in Figure 9.1.18.

76. Find functions $x = x(t)$, $y = y(t)$ such that, as t ranges over the set of real numbers, the points $(x(t), y(t))$ traverse: (a) the right branch of the hyperbola; (b) the left branch of the hyperbola.

77. Find the area of the region between the right branch of the hyperbola and the vertical line $x = 2a$.

78. Show that at each point P of the hyperbola the tangent at P bisects the angle between the focal radii $\overline{F_1P}$ and $\overline{F_2P}$.

The shape of a hyperbola is determined by its *eccentricity* e. This is half the distance between the foci divided by half the length of the transverse axis:

$$e = c/a.$$

For all hyperbolas $e > 1$.

In Exercises 79–82, determine the eccentricity of the hyperbola.

79. $x^2/9 - y^2/16 = 1$.

80. $x^2/16 - y^2/9 = 1$.

81. $x^2 - y^2 = 1$.

82. $x^2/25 - y^2/144 = 1$.

83. Suppose H_1 and H_2 are both hyperbolas with the same transverse axis. Compare the shape of H_1 to the shape of H_2 if their eccentricities e_1, e_2 satisfy $e_1 < e_2$.

84. What happens to a hyperbola if e tends to 1?

85. What happens to a hyperbola if its eccentricity e increases without bound?

86. (Compare to Exercise 74) Let l be a line and let F be a point not on l. Show that, if $e > 1$, then the set of all points P for which

$$d(F, P) = e\, d(l, P)$$

is a hyperbola of eccentricity e. HINT: Begin by choosing a coordinate system whereby F falls on the origin and l is a vertical line $x = d$.

The line that passes through the focus of a parabola and is parallel to the directrix intersects the parabola at two points A and B. The line segment $\overline{AB}$ is called the *latus rectum* of the parabola. In Exercises 87–90 we work with the parabola $x^2 = 4cy$, $c > 0$. By Ω we mean the region bounded below by the parabola and above by the latus rectum.

87. Find the length of the latus rectum.

88. What is the slope of the parabola at the endpoints of the latus rectum?

89. Determine the area of Ω and locate the centroid.

90. Find the volume of the solid generated by revolving Ω about the y-axis and determine the centroid of the solid (6.4.6).

91. Suppose that a flexible inelastic cable (see the figure) fixed at the ends supports a horizontal load. (Imagine a suspension bridge and think of the load on the cable as the roadway.) Show that, if the load has constant weight per unit length, then the cable hangs in the form of a parabola.

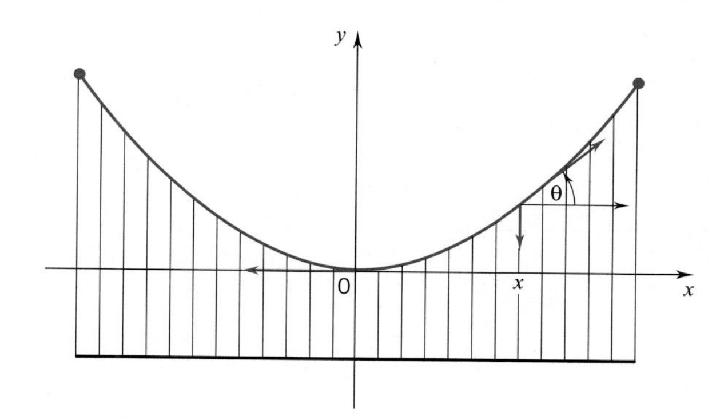

HINT: The part of the cable that supports the load from 0 to x is subject to the following forces:

(1) the weight of the load, which in this case is proportional to x

(2) the horizontal pull at 0: $p(0)$.

(3) the tangential pull at x: $p(x)$.

Balancing the vertical forces we have

$$kx = p(x) \sin\theta. \quad \text{[weight = vertical pull at } x]$$

Balancing the horizontal forces we have

$$p(0) = p(x) \cos\theta. \quad \text{[pull at 0 = horizontal pull at } x]$$

92. The parabolic mirror of a telescope gathers parallel light rays from a distant star and directs them all to the focus. Show that all the light paths to the focus are of the same length.

93. All equilateral triangles are similar; they differ only in scale. Show that the same is true of all parabolas.

94. A radio signal is received at the points marked P_1, P_2, P_3, P_4 in the figure on the next page. Suppose that the signal arrives at P_1 six hundred microseconds after it arrives at P_2 and arrives at P_4 eight hundred microseconds after it arrives at P_3. Locate the source of the signal given that radio waves travel at the speed of light, 186,000 miles per second. (A microsecond is a millionth of a second.)

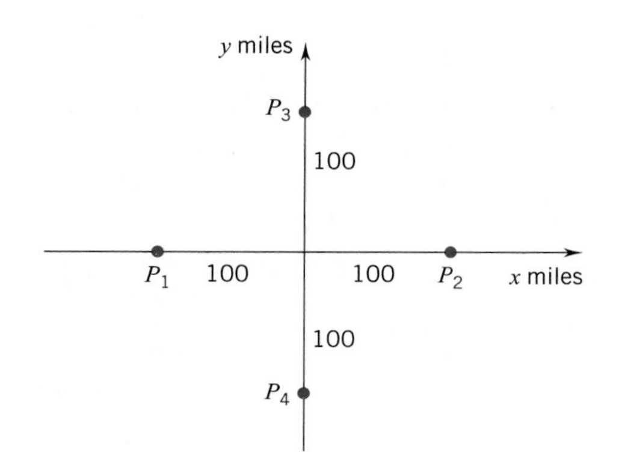

95. A meteor crashes somewhere in the hills that lie north of point A. The impact is heard at point A and four seconds later it is heard at point B. Two seconds still later it is heard at point C. Locate the point of impact given that A lies two miles due east of B and two miles due west of C. (Take 0.20 mile per second as the speed of sound.)

■ 9.2 POLAR COORDINATES

The purpose of coordinates is to fix position with respect to a frame of reference. When we use rectangular coordinates, our frame of reference is a pair of lines that intersect at right angles. For a *polar coordinate system,* the frame of reference is a point O that we call the *pole* and a ray that emanates from it that we call the *polar axis* (Figure 9.2.1).

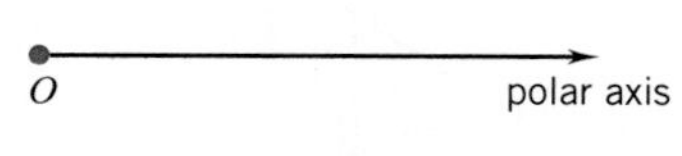

Figure 9.2.1

In Figure 9.2.2 we have drawn two more rays from the pole. One lies at an angle of θ radians from the polar axis; we call it *ray* θ. The opposite ray lies at an angle of $\theta + \pi$ radians; we call it *ray* $\theta + \pi$.

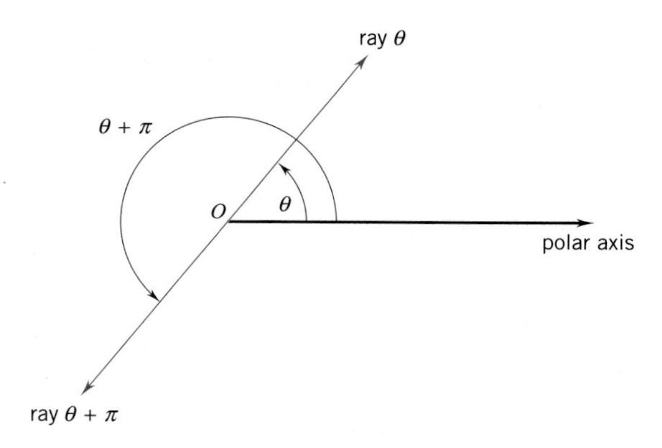

Figure 9.2.2

Figure 9.2.3 shows some points along these same rays, labeled with *polar coordinates.*

(9.2.1) In general, a point is given *polar coordinates* $[r, \theta]$ iff it lies at a distance $|r|$ from the pole

along the ray θ, if $r \geq 0$, and along the ray $\theta + \pi$, if $r < 0$.

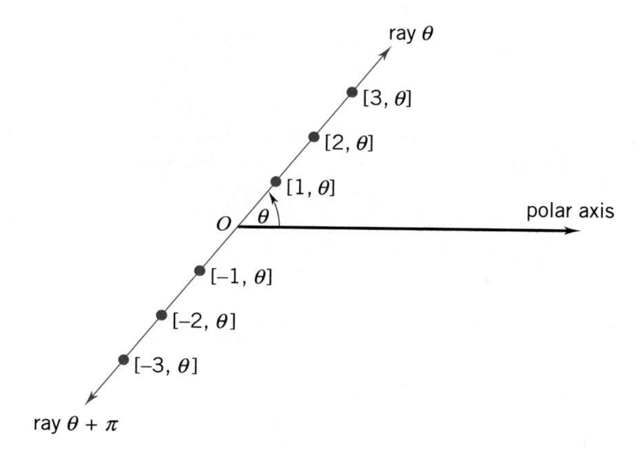

Figure 9.2.3

Figure 9.2.4 shows the point $[2, \frac{2}{3}\pi]$ at a distance of 2 units from the pole along the ray $\frac{2}{3}\pi$. The point $[-2, \frac{2}{3}\pi]$ also lies 2 units from the pole, not along the ray $\frac{2}{3}\pi$, but along the opposite ray.

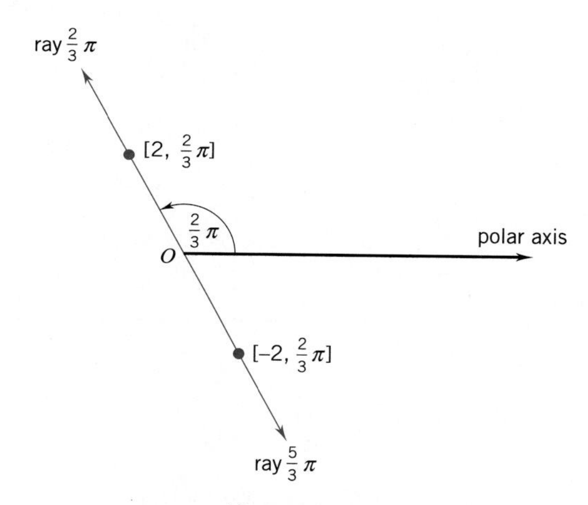

Figure 9.2.4

Polar coordinates are not unique. Many pairs $[r, \theta]$ can represent the same point.

1. If $r = 0$, it does not matter how we choose θ. The resulting point is still the pole:

(9.2.2)
$$O = [0, \theta] \qquad \text{for all } \theta.$$

2. Geometrically there is no distinction between angles that differ by an integer multiple of 2π. Consequently, as suggested in Figure 9.2.5,

(9.2.3)
$$[r, \theta] = [r, \theta + 2n\pi] \qquad \text{for all integers } n.$$

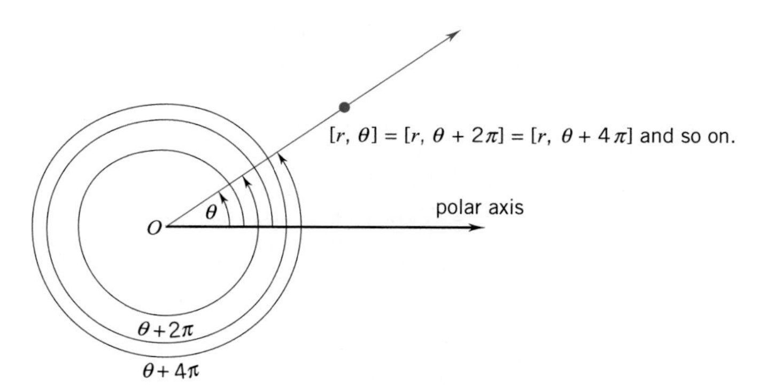

Figure 9.2.5

3. Adding π to the second coordinate is equivalent to changing the sign of the first coordinate:

(9.2.4) $$[r, \theta + \pi] = [-r, \theta].$$ (Figure 9.2.6)

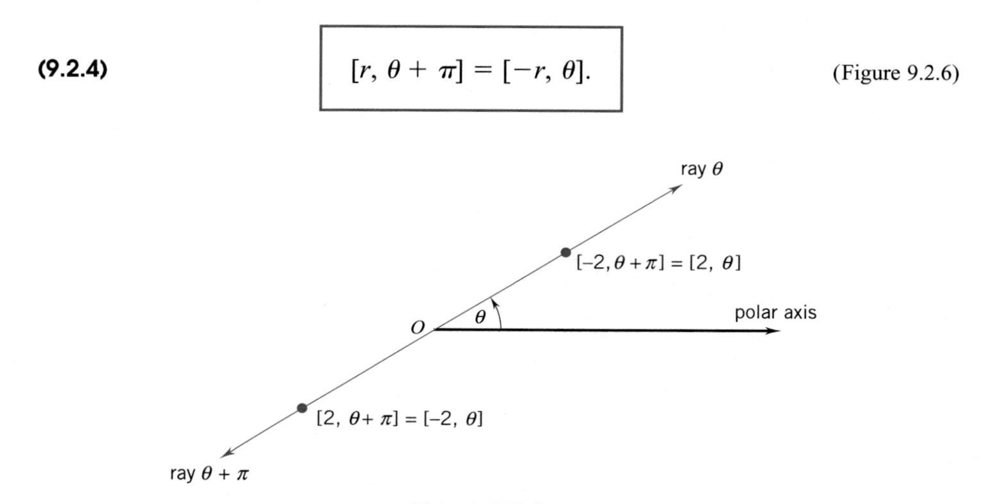

Figure 9.2.6

Remark Some authors do not allow r to take on negative values. There are some advantages to this approach. For example, the polar coordinates of a point are unique if θ is restricted to the interval $[0, 2\pi)$, or to $(-\pi, \pi]$. On the other hand, there are advantages in graphing and finding points of intersection that follow from letting r take on any real value. Since there is no convention on this issue, you should be aware that there are two approaches. ❑

Relation to Rectangular Coordinates

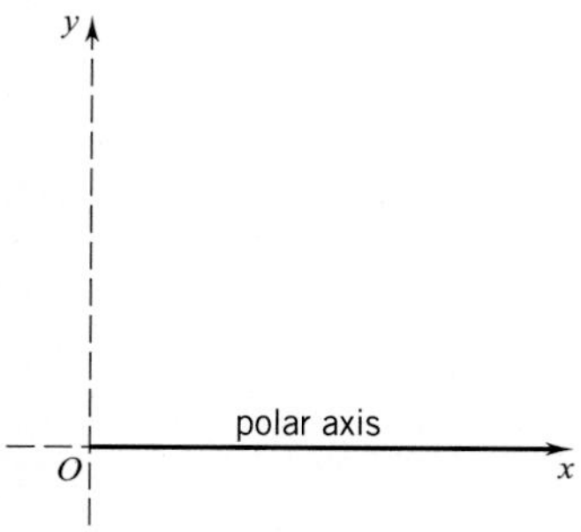

In Figure 9.2.7 we have superimposed a polar coordinate system on a rectangular coordinate system. We have placed the pole at the origin and the polar axis along the positive x-axis.

Figure 9.2.7

The relation between polar coordinates $[r, \theta]$ and rectangular coordinates (x, y) is given by the following equations:

(9.2.5)
$$x = r\cos\theta, \qquad y = r\sin\theta.$$

PROOF If $r = 0$, the formulas hold, since the point $[r, \theta]$ is then the origin and both x and y are 0:

$$0 = 0\cos\theta, \qquad 0 = 0\sin\theta.$$

For $r > 0$, we refer to Figure 9.2.8.† From the figure,

$$\cos\theta = \frac{x}{r}, \qquad \sin\theta = \frac{y}{r}$$

and therefore

$$x = r\cos\theta, \qquad y = r\sin\theta.$$

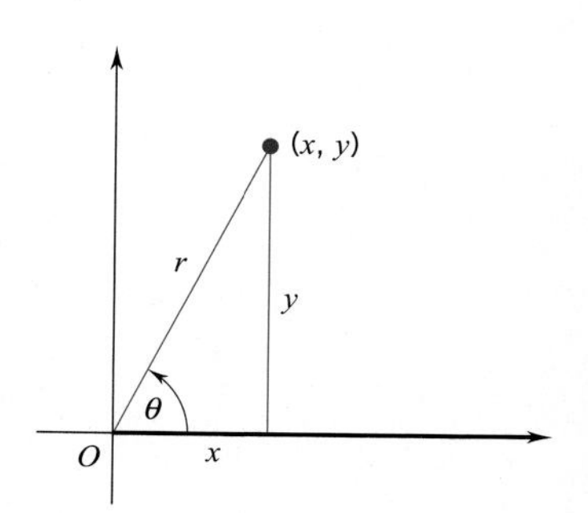

Figure 9.2.8

Suppose now that $r < 0$. Since $[r, \theta] = [-r, \theta + \pi]$ and $-r > 0$, we know from the previous case that

$$x = -r\cos(\theta + \pi), \qquad y = -r\sin(\theta + \pi).$$

Since

$$\cos(\theta + \pi) = -\cos\theta \qquad \text{and} \qquad \sin(\theta + \pi) = -\sin\theta,$$

once again we have

$$x = r\cos\theta, \qquad y = r\sin\theta. \quad \square$$

From the relations we just proved you can see that, unless $x = 0$,

(9.2.6)
$$\tan\theta = \frac{y}{x}$$

and, under all circumstances,

(9.2.7)
$$x^2 + y^2 = r^2. \qquad \text{(check this out)}$$

Example 1 Find the rectangular coordinates of the point P with polar coordinates $[-2, \frac{1}{3}\pi]$.

SOLUTION The relations

$$x = r\cos\theta, \qquad y = r\sin\theta$$

† For simplicity we have placed (x, y) in the first quadrant. A similar argument works in each of the other quadrants.

give

$$x = -2\cos\tfrac{1}{3}\pi = -2(\tfrac{1}{2}) = -1, \qquad y = -2\sin\tfrac{1}{3}\pi = -2(\tfrac{1}{2}\sqrt{3}) = -\sqrt{3}.$$

The point P has rectangular coordinates $(-1, -\sqrt{3})$. ❑

Example 2 Find all possible polar coordinates for the point P that has rectangular coordinates $(-2, 2\sqrt{3})$.

SOLUTION We know that

$$r\cos\theta = -2, \qquad r\sin\theta = 2\sqrt{3}.$$

We can get the possible values of r by squaring these expressions and then adding them:

$$r^2 = r^2\cos^2\theta + r^2\sin^2\theta = (-2)^2 + (2\sqrt{3})^2 = 16,$$

so that $r = \pm 4$.

Taking $r = 4$, we have

$$4\cos\theta = -2, \qquad 4\sin\theta = 2\sqrt{3}$$

$$\cos\theta = -\tfrac{1}{2}, \qquad \sin\theta = \tfrac{1}{2}\sqrt{3}.$$

These equations are satisfied by setting $\theta = \frac{2}{3}\pi$, or more generally, by setting

$$\theta = \tfrac{2}{3}\pi + 2n\pi.$$

The polar coordinates of P with first coordinate $r = 4$ are all pairs of the form

$$[4, \tfrac{2}{3}\pi + 2n\pi]$$

where n ranges over the set of all integers.

We could go through the same process again, this time taking $r = -4$, but there is no need to do so. Since $[r, \theta] = [-r, \theta + \pi]$, we know that

$$[4, \tfrac{2}{3}\pi + 2n\pi] = [-4, (\tfrac{2}{3}\pi + \pi) + 2n\pi].$$

The polar coordinates of P with first coordinate $r = -4$ are thus all pairs of the form

$$[-4, \tfrac{5}{3}\pi + 2n\pi]$$

where n again ranges over the set of all integers. ❑

Let's specify some simple sets in polar coordinates.

1. In rectangular coordinates the circle of radius a centered at the origin has equation

$$x^2 + y^2 = a^2.$$

The equation for this circle in polar coordinates is simply

$$r = a.$$

The interior of the circle is given by $0 \le r < a$ and the exterior by $r > a$.

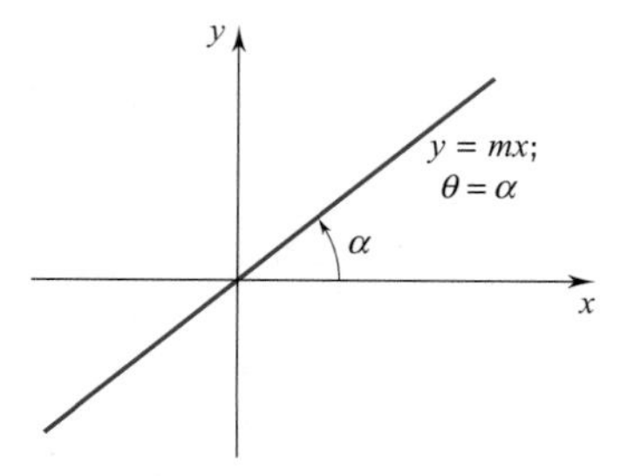

Figure 9.2.9

2. In rectangular coordinates the line through the origin with angle of inclination α has equation $y = mx$, where $m = \tan\alpha$ (Figure 9.2.9). The polar equation of this line is

$$\theta = \alpha. \qquad \text{(verify this)}$$

3. The vertical line $x = a$ becomes

$$r \cos \theta = a$$

and the horizontal line $y = b$ becomes

$$r \sin \theta = b.$$

4. The line $Ax + By + C = 0$ can be written

$$r(A \cos \theta + B \sin \theta) + C = 0.$$

Example 3 Find an equation in polar coordinates for the hyperbola $x^2 - y^2 = a^2$.

SOLUTION Setting $x = r \cos \theta$ and $y = r \sin \theta$, we have

$$r^2 \cos^2 \theta - r^2 \sin^2 \theta = a^2,$$

$$r^2(\cos^2 \theta - \sin^2 \theta) = a^2,$$

$$r^2 \cos 2\theta = a^2. \quad \square$$

Example 4 Show that the equation $r = 2a \cos \theta$, $a > 0$, represents a circle.

SOLUTION Multiplication by r gives

$$r^2 = 2ar \cos \theta,$$

$$x^2 + y^2 = 2ax,$$

$$x^2 - 2ax + y^2 = 0,$$

$$x^2 - 2ax + a^2 + y^2 = a^2,$$

$$(x - a)^2 + y^2 = a^2.$$

This is a circle of radius a centered at the point with rectangular coordinates $(a, 0)$. See Figure 9.2.10. $\square$

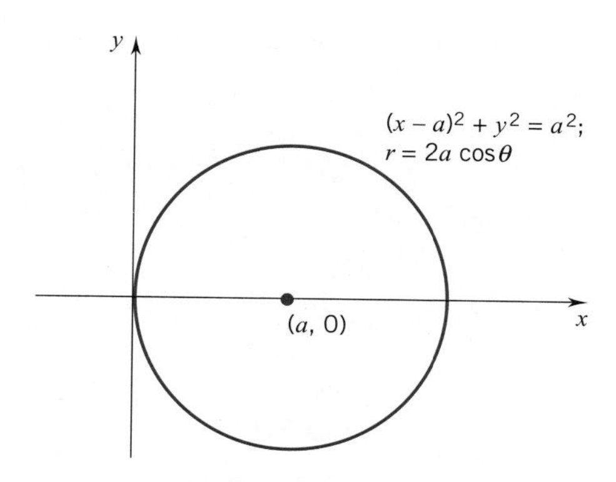

Figure 9.2.10

Symmetry

Symmetry with respect to each of the coordinate axes and with respect to the origin is illustrated in Figure 9.2.11. The coordinates marked are, of course, not the only ones possible. (The difficulties that this can cause are explained in Section 9.5.)

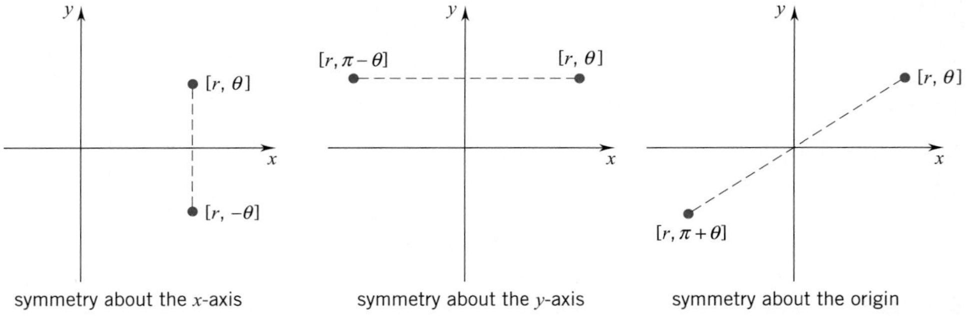

$[r, \theta]$ $[r, -\theta]$ symmetry about the x-axis

$[r, \pi-\theta]$ $[r, \theta]$ symmetry about the y-axis

$[r, \theta]$ $[r, \pi+\theta]$ symmetry about the origin

Figure 9.2.11

Example 5 Test the equation $r^2 = \cos 2\theta$ for symmetry.

SOLUTION Since

$$\cos [2(-\theta)] = \cos (-2\theta) = \cos 2\theta,$$

you can see that, if $[r, \theta]$ is on the curve, then so is $[r, -\theta]$. This says that the curve is symmetric about the x-axis. Since

$$\cos [2(\pi - \theta)] = \cos (2\pi - 2\theta) = \cos (-2\theta) = \cos 2\theta,$$

you can see that, if $[r, \theta]$ is on the curve, then so is $[r, \pi - \theta]$. The curve is therefore symmetric about the y-axis.

Being symmetric about both axes, the curve must also be symmetric about the origin. You can verify this directly by noting that

$$\cos [2(\pi + \theta)] = \cos (2\pi + 2\theta) = \cos 2\theta,$$

so that, if $[r, \theta]$ lies on the curve, then so does $[r, \pi + \theta]$. A sketch of the curve, which is called a *lemniscate,* appears in Figure 9.2.12. ❑

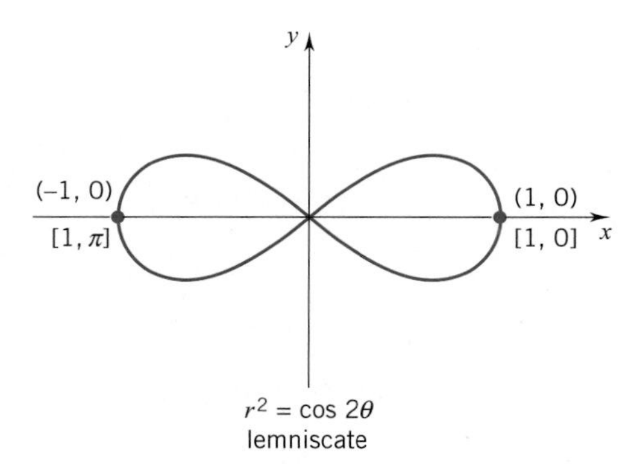

$r^2 = \cos 2\theta$
lemniscate

Figure 9.2.12

EXERCISES 9.2

In Exercises 1–8, plot the point given in polar coordinates.

1. $[1, \frac{1}{3}\pi]$. **2.** $[1, \frac{1}{2}\pi]$.

3. $[-1, \frac{1}{3}\pi]$. **4.** $[-1, -\frac{1}{3}\pi]$.

5. $[4, \frac{5}{4}\pi]$. **6.** $[-2, 0]$.

7. $[-\frac{1}{2}, \pi]$. **8.** $[\frac{1}{3}, \frac{2}{3}\pi]$.

In Exercises 9–16, find the rectangular coordinates of the given point.

9. $[3, \frac{1}{2}\pi]$. **10.** $[4, \frac{1}{6}\pi]$.

11. $[-1, -\pi]$.
12. $[-1, \frac{1}{4}\pi]$.
13. $[-3, -\frac{1}{3}\pi]$.
14. $[2, 0]$.
15. $[3, -\frac{1}{2}\pi]$.
16. $[2, 3\pi]$.

In Exercises 17–24, the points are given in rectangular coordinates. Find all possible polar coordinates for each point.

17. $(0, 1)$.
18. $(1, 0)$.
19. $(-3, 0)$.
20. $(4, 4)$.
21. $(2, -2)$.
22. $(3, -3\sqrt{3})$.
23. $(4\sqrt{3}, 4)$.
24. $(\sqrt{3}, -1)$.

25. Find a formula for the distance between $[r_1, \theta_1]$ and $[r_2, \theta_2]$.

26. Show that for $r_1 > 0$, $r_2 > 0$, $|\theta_1 - \theta_2| < \pi$ the distance formula you found in Exercise 25 is just the law of cosines.

In Exercises 27–30, find the point $[r, \theta]$ symmetric to the given point about: (a) the x-axis; (b) the y-axis; (c) the origin. Express your answer with $r > 0$ and $\theta \in [0, 2\pi)$.

27. $[\frac{1}{2}, \frac{1}{6}\pi]$.
28. $[3, -\frac{5}{4}\pi]$.
29. $[-2, \frac{1}{3}\pi]$.
30. $[-3, -\frac{7}{4}\pi]$.

In Exercises 31–36, test the given curve for symmetry about the coordinate axes and the origin.

31. $r = 2 + \cos\theta$.
32. $r = \cos 2\theta$.
33. $r(\sin\theta + \cos\theta) = 1$.
34. $r\sin\theta = 1$.
35. $r^2 \sin 2\theta = 1$.
36. $r^2 \cos 2\theta = 1$.

In Exercises 37–48, write the equation in polar coordinates.

37. $x = 2$.
28. $y = 3$.
39. $2xy = 1$.
40. $x^2 + y^2 = 9$.
41. $x^2 + (y - 2)^2 = 4$.
42. $(x - a)^2 + y^2 = a^2$.
43. $y = x$.
44. $x^2 - y^2 = 4$.
45. $x^2 + y^2 + x = \sqrt{x^2 + y^2}$.
46. $y = mx$.
47. $(x^2 + y^2)^2 = 2xy$.
48. $(x^2 + y^2)^2 = x^2 - y^2$.

In Exercises 49–58, identify the curve and write the equation in rectangular coordinates.

49. $r\sin\theta = 4$.
50. $r\cos\theta = 4$.
51. $\theta = \frac{1}{3}\pi$.
52. $\theta^2 = \frac{1}{9}\pi^2$.
53. $r = 2(1 - \cos\theta)^{-1}$.
54. $r = 4\sin(\theta + \pi)$.
55. $r = 3\cos\theta$.
56. $\theta = -\frac{1}{2}\pi$.
57. $\tan\theta = 2$.
58. $r = 2\sin\theta$.

In Exercises 59–62, write the equation in rectangular coordinates and identify it.

59. $r = \dfrac{4}{2 - \cos\theta}$
60. $r = \dfrac{6}{1 + 2\sin\theta}$
61. $r = \dfrac{4}{1 - \cos\theta}$
62. $r = \dfrac{2}{3 + 2\sin\theta}$

63. Show that if a and b are not both zero, then the curve

$$r = a\sin\theta + b\cos\theta$$

is a circle. Find the center and the radius.

64. Find a polar equation for the set of points $P[r, \theta]$ such that the distance from P to the pole equals the distance from P to the line $x = -d$. See the figure.

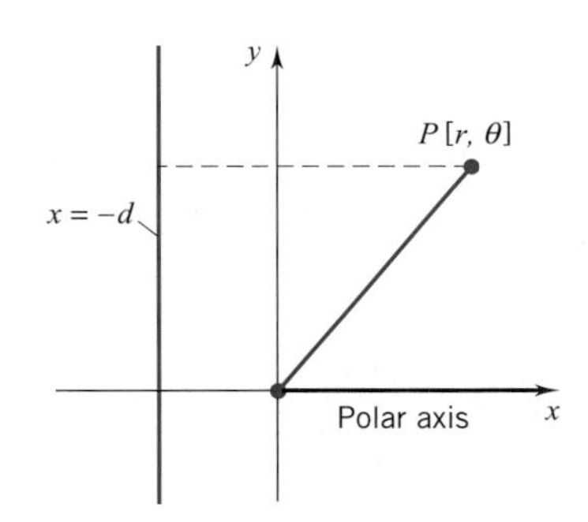

65. Find a polar equation for the set of points $P[r, \theta]$ such that the distance from P to the pole is half the distance from P to the line $x = -d$.

66. Find a polar equation for the set of points $P[r, \theta]$ such that the distance from P to the pole is twice the distance from P to the line $x = -d$.

■ 9.3 GRAPHING IN POLAR COORDINATES

We begin with the equation

$$r = \theta, \qquad \theta \geq 0.$$

The graph is a nonending spiral, part of the famous *spiral of Archimedes*. The curve is shown in detail from $\theta = 0$ to $\theta = 2\pi$ in Figure 9.3.1. At $\theta = 0$, $r = 0$; at $\theta = \frac{1}{4}\pi$, $r = \frac{1}{4}\pi$; at $\theta = \frac{1}{2}\pi$, $r = \frac{1}{2}\pi$; and so on.

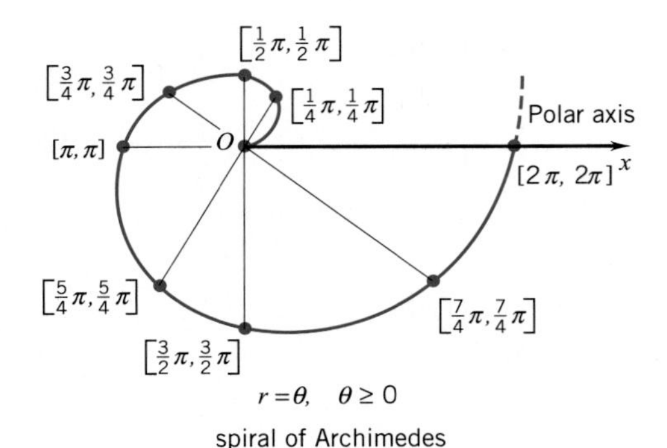

$r = \theta, \quad \theta \geq 0$

spiral of Archimedes

Figure 9.3.1

The next examples involve trigonometric functions.

Example 1 Sketch the curve

$$r = 1 - 2\cos\theta.$$

SOLUTION Since the cosine function is periodic, the curve $r = 1 - 2\cos\theta$ is a closed curve. We will draw it from $\theta = 0$ to $\theta = 2\pi$. The curve just repeats itself for values of θ outside the interval $[0, 2\pi]$.

We begin by calculating a table of values:

θ	0	$\pi/4$	$\pi/3$	$\pi/2$	$2\pi/3$	$3\pi/4$	π	$5\pi/4$	$4\pi/3$	$3\pi/2$	$5\pi/3$	$7\pi/4$	2π
r	-1	-0.41	0	1	2	2.41	3	2.41	2	1	0	-0.41	-1

These points are plotted in Figure 9.3.2. The values of θ for which $r = 0$ or $|r|$ is a local maximum are:

$r = 0$ at $\theta = \frac{1}{3}\pi, \frac{5}{3}\pi$ for then $\cos\theta = \frac{1}{2}$; $\quad |r|$ is a local maximum at $\theta = 0, \pi, 2\pi$.

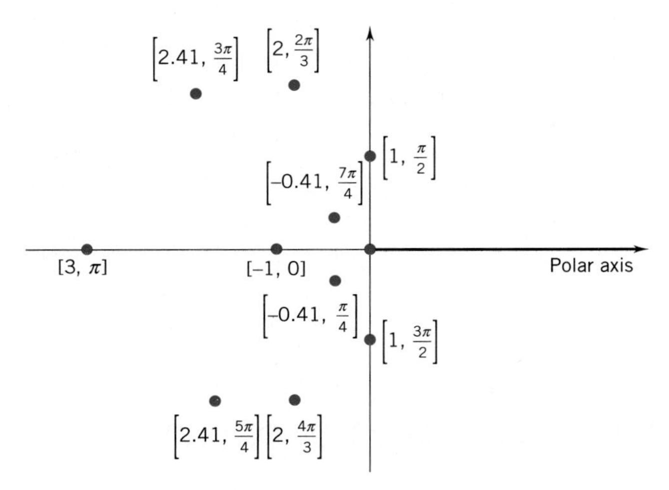

Figure 9.3.2

These five values of θ generate four intervals:

$$[0, \tfrac{1}{3}\pi], \quad [\tfrac{1}{3}\pi, \pi], \quad [\pi, \tfrac{5}{3}\pi], \quad [\tfrac{5}{3}\pi, 2\pi].$$

We sketch the curve in four stages. These stages are shown in Figure 9.3.3.

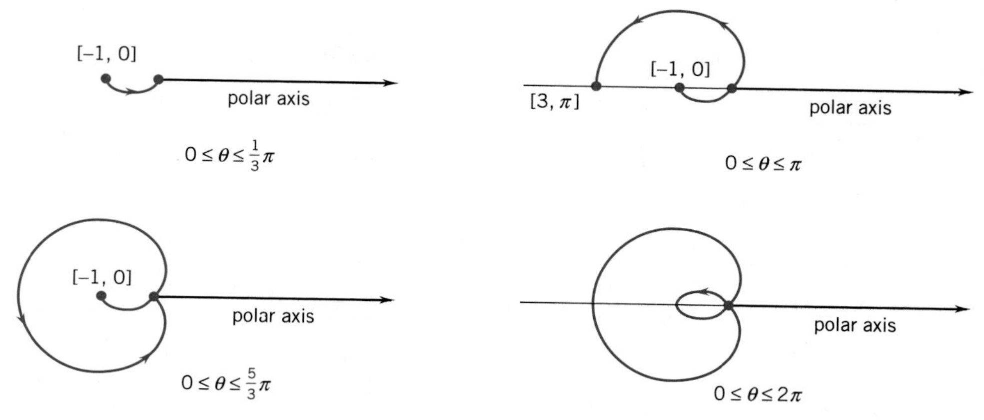

Figure 9.3.3

As θ increases from 0 to $\frac{1}{3}\pi$, $\cos\theta$ decreases from 1 to $\frac{1}{2}$ and $r = 1 - 2\cos\theta$ increases from -1 to 0.
As θ increases from $\frac{1}{3}\pi$ to π, $\cos\theta$ decreases from $\frac{1}{2}$ to -1 and r increases from 0 to 3.
As θ increases from π to $\frac{5}{3}\pi$, $\cos\theta$ increases from -1 to $\frac{1}{2}$ and r decreases from 3 to 0.
Finally, as θ increases from $\frac{5}{3}\pi$ to 2π, $\cos\theta$ increases from $\frac{1}{2}$ to 1 and r decreases from 0 to -1.

As we could have read from the equation, the curve is symmetric about the x-axis. ❑

EXAMPLE 2 Sketch the curve

$$r = \cos 2\theta, \qquad 0 \le \theta \le 2\pi.$$

SOLUTION As an alternative to compiling a table of values as we did in Example 1, we refer to the graph of $\cos 2\theta$ in rectangular coordinates for the values of r. See Figure 9.3.4. The values of θ for which r is zero or has an extreme value are as follows:

$$\theta = \tfrac{1}{4}n\pi, \qquad n = 0, 1, \ldots, 8.$$

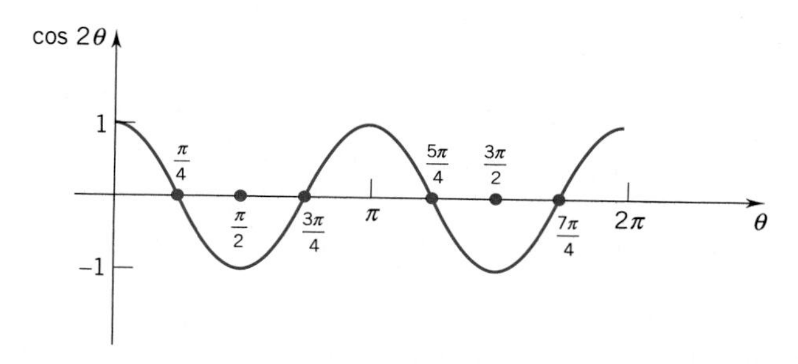

Figure 9.3.4

The curve drawn in the corresponding eight stages is shown in Figure 9.3.5. ❑

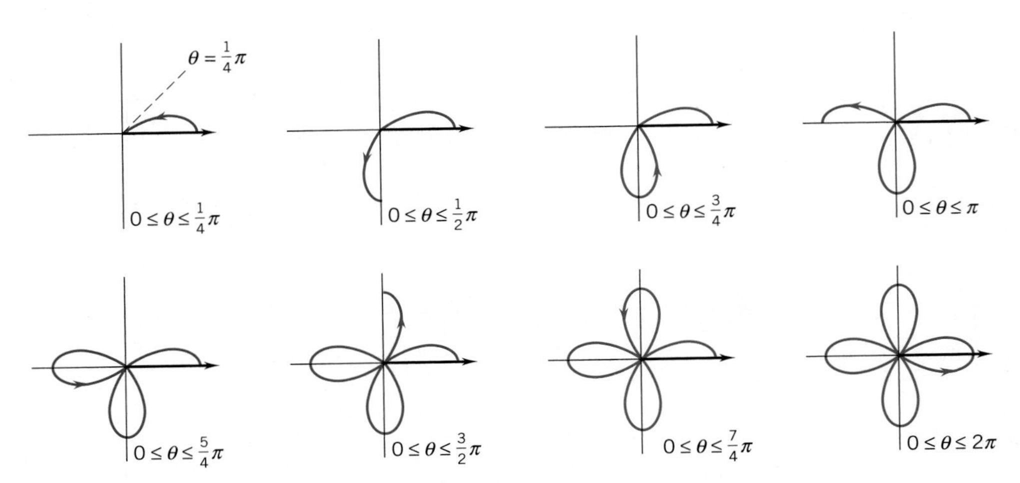

Figure 9.3.5

EXAMPLE 3 Figure 9.3.6 shows four *cardioids*, heart-shaped curves. Rotation of $r = 1 + \cos\theta$ by $\frac{1}{2}\pi$ radians, measured in the counterclockwise direction, gives

$$r = 1 + \cos(\theta - \tfrac{1}{2}\pi) = 1 + \sin\theta.$$

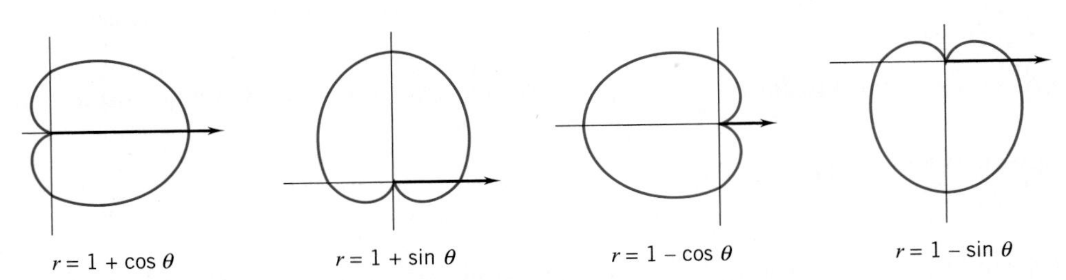

Figure 9.3.6

Rotation by another $\frac{1}{2}\pi$ radians gives

$$r = 1 + \cos(\theta - \pi) = 1 - \cos\theta.$$

Rotation by yet another $\frac{1}{2}\pi$ radians gives

$$r = 1 + \cos(\theta - \tfrac{3}{2}\pi) = 1 - \sin\theta.$$

Notice how easy it is to rotate axes in polar coordinates: each change

$$\cos\theta \to \sin\theta \to -\cos\theta \to -\sin\theta$$

represents a rotation by $\frac{1}{2}\pi$ radians. ❑

At this point we will try to give you a brief survey of some of the basic polar curves, leaving the parabola, the ellipse, and the hyperbola to Section *9.4. (The numbers a and b that appear below are to be interpreted as nonzero constants.)

Lines: $\theta = a, \qquad r = a \sec\theta, \qquad r = a \csc\theta.$ (Figure 9.3.7)

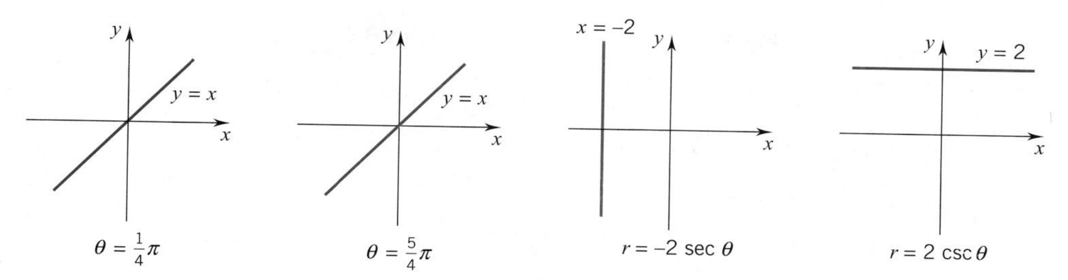

Figure 9.3.7

Circles: $r = a, \qquad r = a \sin\theta, \qquad r = a \cos\theta.$ (Figure 9.3.8)

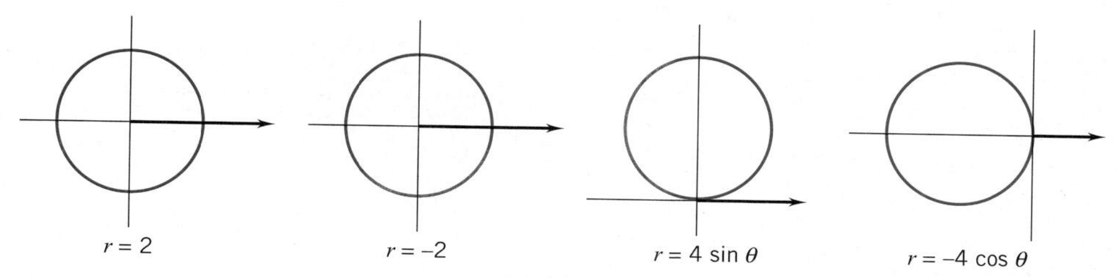

Figure 9.3.8

Limaçons:† $r = a + b \sin\theta, \qquad r = a + b\cos\theta.$ (Figure 9.3.9)

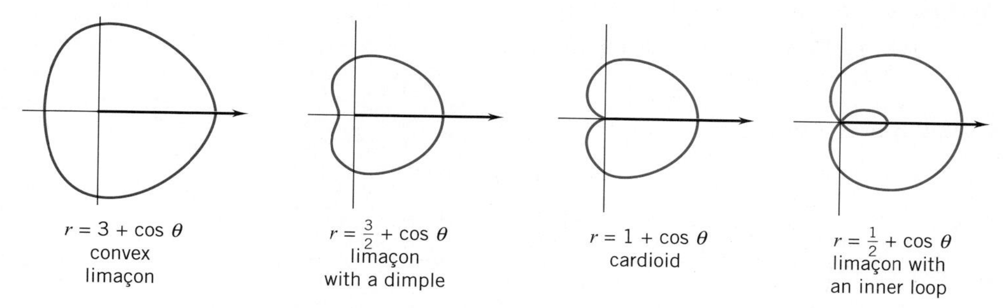

Figure 9.3.9

The general shape of the curve depends on the relative magnitudes of $|a|$ and $|b|$.

---

† From the French term for snail. The word is pronounced with a soft *c*.

Lemniscates:† $r^2 = a \sin 2\theta, \qquad r^2 = a \cos 2\theta.$ (Figure 9.3.10)

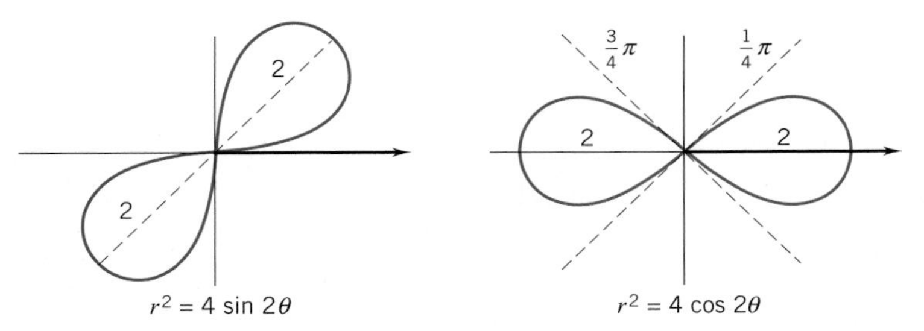

Figure 9.3.10

Petal Curves: $r = a \sin n\theta, \qquad r = a \cos n\theta \qquad$ integer n. (Figure 9.3.11)

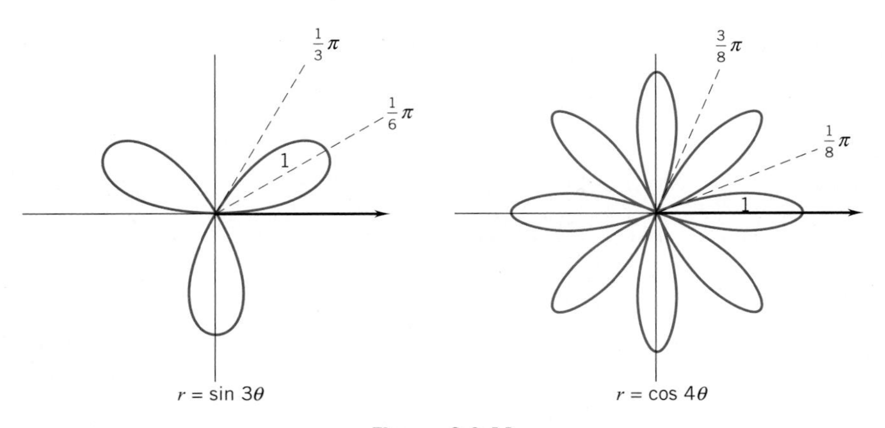

Figure 9.3.11

If n is *odd*, there are n petals. If n is *even*, there are $2n$ petals.

Spirals: $r = a\theta$, spiral of Archimedes. (Figure 9.3.12)

$r = e^{a\theta}$, $\ln r = a\theta$ logarithmic spiral. (Figure 9.3.13)

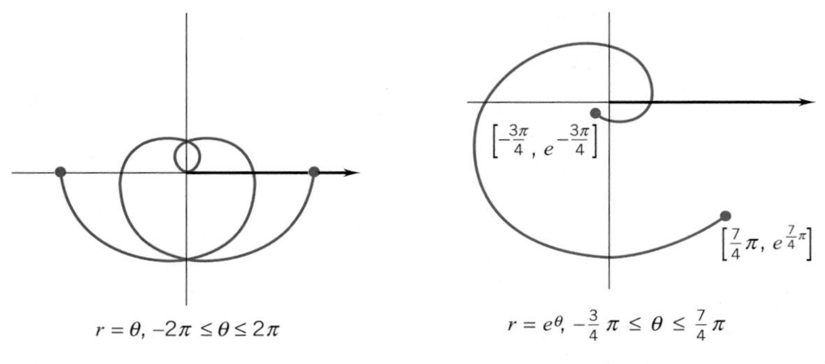

Figure 9.3.12 **Figure 9.3.13**

† From the Latin *lemniscatus*, meaning "adorned with pendent ribbons."

EXERCISES 9.3

In Exercises 1–34, sketch the given polar curve.

1. $\theta = -\frac{1}{4}\pi$.
2. $r = -3$.
3. $r = 4$.
4. $r = 3\cos\theta$.
5. $r = -2\sin\theta$.
6. $\theta = \frac{2}{3}\pi$.
7. $r\csc\theta = 3$.
8. $r = 1 - \cos\theta$.
9. $r = \theta, \quad -\frac{1}{2}\pi \le \theta \le \pi$.
10. $r\sec\theta = -2$.
11. $r = \sin 3\theta$.
12. $r^2 = \cos 2\theta$.
13. $r^2 = \sin 2\theta$.
14. $r = \cos 2\theta$.
15. $r^2 = 4, \quad 0 \le \theta \le \frac{3}{4}\pi$.
16. $r = \sin\theta$.
17. $r^3 = 9r$.
18. $\theta = -\frac{1}{4}\pi, \quad 1 \le r < 2$.
19. $r = -1 + \sin\theta$.
20. $r^2 = 4r$.
21. $r = \sin 2\theta$.
22. $r = \cos 3\theta, \quad 0 \le \theta \le \frac{1}{2}\pi$.
23. $r = \cos 5\theta, \quad 0 \le \theta \le \frac{1}{2}\pi$.
24. $r = e^{\theta}, \quad -\pi \le \theta \le \pi$.
25. $r = 2 + \sin\theta$.
26. $r = \cot\theta$.
27. $r = \tan\theta$.
28. $r = 2 - \cos\theta$.
29. $r = 2 + \sec\theta$.
30. $r = 3 - \csc\theta$.
31. $r = -1 + 2\cos\theta$.
32. $r = \dfrac{1}{1 + \sin\theta}$.
33. $r = \dfrac{2}{3 - 2\cos\theta}$.
34. $r = \dfrac{6}{1 - 2\cos\theta}$.

35. Suppose that you rotate the polar axis in a given polar coordinate system through an angle α. Show that if a point P has coordinates $[r, \theta]$ in the original system, then it has coordinates $[r', \theta']$ in the new system, where $r' = r$ and $\theta' = \theta - \alpha$. See the figure. Thus, the graph of a polar equation $r = f(\theta - \alpha)$ is simply the graph of the equation $r = f(\theta)$ rotated through the angle α.

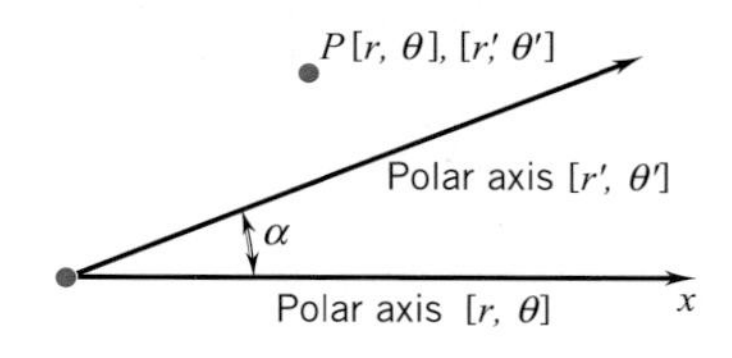

36. (a) By changing to rectangular coordinates, verify that the graph of

$$r = \frac{c}{1 - \cos\theta}$$

is a parabola with its focus at the pole and directrix $x = -c$. Find the vertex and the directrix.

(b) Describe the graphs of each of the following equations:

$$\text{(i) } r = \frac{c}{1 - \sin\theta}, \quad \text{(ii) } r = \frac{c}{1 + \cos\theta},$$

$$\text{(iii) } r = \frac{c}{1 + \sin\theta}.$$

HINT: Use Exercise 35.

37. (a) By changing to rectangular coordinates, verify that the graph of

$$r = \frac{ce}{1 - e\cos\theta}, \quad 0 < e < 1,$$

is an ellipse with one focus at the pole and eccentricity e. Find the center and the lengths of the major and minor axes.

(b) Describe the graphs of each of the following equations:

$$\text{(i) } r = \frac{ce}{1 - e\sin\theta}, \quad \text{(ii) } r = \frac{ce}{1 + e\cos\theta},$$

$$\text{(iii) } r = \frac{ce}{1 + e\sin\theta}, \quad 0 < e < 1.$$

38. (a) By changing to rectangular coordinates, verify that the graph of

$$r = \frac{ce}{1 - e\cos\theta}, \quad e > 1,$$

is a hyperbola with one focus at the pole and eccentricity e. Find the center, vertices, asymptotes, and length of the transverse axis.

(b) Describe the graphs of each of the following equations:

$$\text{(i) } r = \frac{ce}{1 - e\sin\theta}, \quad \text{(ii) } r = \frac{ce}{1 + e\cos\theta},$$

$$\text{(iii) } r = \frac{ce}{1 + e\sin\theta}, \quad e > 1.$$

39. Let

$$r = \frac{ed}{1 - e\cos\theta}.$$

(Again, the symbol "e" will be representing eccentricity here.)

(a) Let $d = 1$. Use a graphing utility to graph this equation for $e = 0.5, 0.1, 0.01, 0.001$. What is the "limiting curve" as $e \to 0^+$?

(b) Continuing to let $d = 1$, graph the equation for $e = 0.9, 0.99, 0.999$. What happens as $e \to 1^-$?

(c) Now let $e = 0.5$, and graph the equation for several values of d. Use both positive and negative numbers. What effect does d have on the graph?

40. Let

$$r = A \cos k\theta \quad \text{or} \quad r = A \sin k\theta$$

(a) Use a graphing utility to graph these equations for $A = 1$, 2, and 4, and $k = 2$, 3, and 4, respectively.
(b) What effect does the coefficient A have on the graphs? What is the difference between the "cosine" curves and the "sine" curves?
(c) Use your graphing utility to graph these equations for $A = 2$ and $k = 3/2$, $5/2$. Can you predict what the graph will look like for $k = m/2$ for any odd integer m?

41. Use a graphing utility to draw the graph

$$r = 1 + \sin k\theta + \cos^2(2k\theta)$$

for $k = 1, 2, 3, 4, 5$. If you were asked to give a name to this family of curves, what would you suggest?

*■ 9.4 THE CONIC SECTIONS IN POLAR COORDINATES

(This material is used later in this text only in Section *13.6 where we examine planetary motion.)

In Section 9.1, we defined a parabola in terms of a focus and a directrix, but our definitions of the ellipse and hyperbola were given in terms of two foci; there was no mention of a directrix for either of these conics. In this section, we give a unified approach to the conic sections that involves a focus and directrix for all three cases.

DEFINITION 9.4.1

Let F be a fixed point (the *focus*) and ℓ a fixed line (the *directrix*) in the plane. Let e be a fixed positive number (the *eccentricity*). The set of all points P in the plane such that

$$\frac{\text{distance from } P \text{ to } F}{\text{distance from } P \text{ to } \ell} = e$$

is an ellipse if $0 < e < 1$, a parabola if $e = 1$, and a hyperbola if $e > 1$.

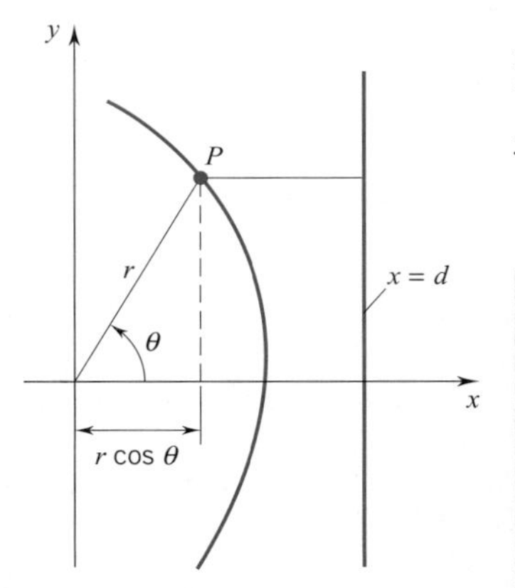

Figure 9.4.1

Figure 9.4.1 illustrates this definition. We have superimposed a polar and a rectangular coordinate system, and for convenience we refer to the pole as the origin. Without loss of generality, we have taken F at the origin and ℓ as the vertical line $x = d$.

We now have the following theorem.

THEOREM 9.4.2

Let e and d be positive numbers. The set of all points P such that

$$\frac{\text{distance from } P \text{ to the origin}}{\text{distance from } P \text{ to the line } x = d} = e$$

is described by the polar equation: $r = \dfrac{ed}{1 + e\cos\theta}$

THEOREM 9.4.2 (continued)

There are three cases (see Figure 9.4.2):

I. If $0 < e < 1$, the equation represents an ellipse of eccentricity e with right focus at the origin, major axis horizontal:

$$\frac{(x+c)^2}{a^2} + \frac{y^2}{a^2 - c^2} = 1 \qquad \text{with} \qquad a = \frac{ed}{1-e^2}, \quad c = ea.$$

II. If $e = 1$, the equation represents a parabola with focus at the origin and directrix $x = d$:

$$y^2 = -4\frac{d}{2}\left(x - \frac{d}{2}\right).$$

III. If $e > 1$, the equation represents a hyperbola of eccentricity e with left focus at the origin, transverse axis horizontal:

$$\frac{(x-c)^2}{a^2} - \frac{y^2}{c^2 - a^2} = 1 \qquad \text{with} \qquad a = \frac{ed}{e^2-1}, \quad c = ea.$$

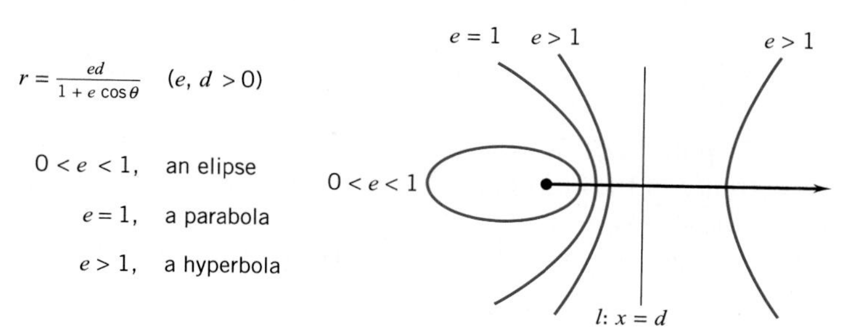

Figure 9.4.2

The derivation of the polar equation is left to you as an exercise.

To obtain an equation in rectangular coordinates, we first clear the denominator. Multiplication by $1 + e\cos\theta$ gives

$$r + er\cos\theta = ed.$$

Therefore

$$r = ed - er\cos\theta$$

$$r^2 = e^2d^2 - 2e^2dr\cos\theta + e^2r^2\cos^2\theta$$

$$(*) \qquad x^2 + y^2 = e^2d^2 - 2e^2dx + e^2x^2.$$

If $0 < e < 1$, equation $(*)$ can be written

$$(1 - e^2)x^2 + 2e^2dx + y^2 = e^2d^2.$$

By completing the square for x we can rearrange this equation to read

$$\left(x + \frac{e^2d}{1-e^2}\right)^2 + \frac{y^2}{1-e^2} = \frac{e^2d^2}{(1-e^2)^2}. \qquad \text{(carry out the details)}$$

Setting $a = ed/(1 - e^2)$ and $c = ea$, we have

$$(x + c)^2 + \frac{y^2}{1 - e^2} = a^2,$$

$$\frac{(x + c)^2}{a^2} + \frac{y^2}{(1 - e^2)a^2} = 1,$$

$$\frac{(x + c)^2}{a^2} + \frac{y^2}{a^2 - c^2} = 1.$$

This is an equation of an ellipse of eccentricity e with right focus at the origin, major axis horizontal.

If $e = 1$, equation (*) reads $y^2 = d^2 - 2dx$. This can be written

$$y^2 = -4\frac{d}{2}\left(x - \frac{d}{2}\right).$$

As you can check, this is an equation of a parabola with focus at the origin and directrix $x = d$.

The case $e > 1$ is left to you. ❑

Example 1 The ellipse

$$r = \frac{8}{4 + 3 \cos \theta}$$

has right focus at the pole, major axis horizontal. Without resorting to xy-coordinates, **(a)** find the eccentricity of the ellipse, **(b)** locate the ends of the major axis, **(c)** locate the center of the ellipse, **(d)** locate the second focus, **(e)** determine the length of the minor axis, **(f)** determine the width of the ellipse at the foci, and **(g)** sketch the ellipse.

SOLUTION

(a) Dividing numerator and denominator by 4, we have

$$r = \frac{2}{1 + \frac{3}{4} \cos \theta}.$$

The eccentricity of the ellipse is $\frac{3}{4}$.

(b) At the right end of the major axis, $\theta = 0$, $\cos \theta = 1$, and $r = \frac{8}{7}$. At the left end, $\theta = \pi$, $\cos \theta = -1$, and $r = 8$. One end of the major axis lies 8 units to the left of the pole, the other end lies $\frac{8}{7}$ units to the right of the pole.

(c) The center of the ellipse lies halfway between the endpoints of the major axis, in this case $\frac{24}{7}$ units to the left of the pole.

(d) In general the focal separation $2c$ divided by the length of the major axis $2a$ gives the eccentricity. Here $(2c) \cdot (\frac{7}{64}) = \frac{3}{4}$ and therefore $2c = \frac{48}{7}$. The second focus lies $\frac{48}{7}$ units to the left of the pole. (You can get the same result by symmetry: since the right focus lies $\frac{8}{7}$ units from the right end of the major axis, the other focus lies $\frac{8}{7}$ units from the left end of the major axis and thus $8 - \frac{8}{7} = \frac{48}{7}$ units to the left of the pole.)

(e) In general the length of the minor axis is $2b = 2\sqrt{a^2 - c^2}$. Here

$$2b = 2\sqrt{(\tfrac{32}{7})^2 - (\tfrac{24}{7})^2} = \tfrac{16}{7}\sqrt{7}.$$

The minor axis has length $\frac{16}{7}\sqrt{7} \cong 6.05$.

(f) The width of the ellipse at the foci is $2r$ where $\theta = \frac{1}{2}\pi$. The width at the foci is 4.

(g) A sketch of the ellipse appears in Figure 9.4.3. ❑

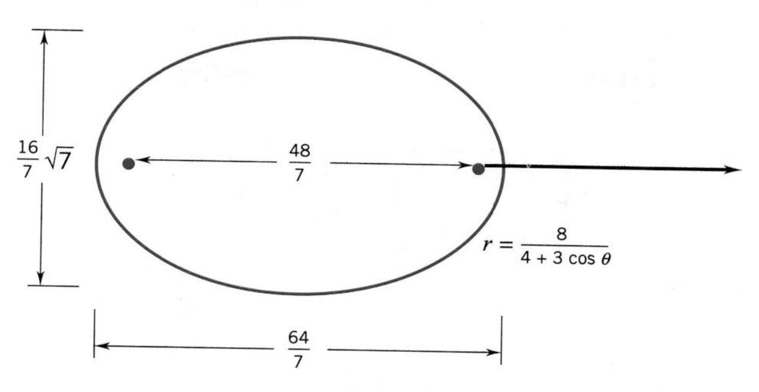

Figure 9.4.3

Example 2 Sketch the ellipse

$$r = \frac{8}{4 + 3\sin\theta}$$

SOLUTION No need to do much here. First,

$$r = \frac{8}{4 + 3\sin\theta} = \frac{2}{1 + \frac{3}{4}\sin\theta},$$

and since $\sin\theta = \cos(\theta - \pi/2)$, this ellipse is the ellipse in Example 1 rotated by $\frac{1}{2}\pi$ radians in the counterclockwise direction. See Figure 9.4.4. ❑

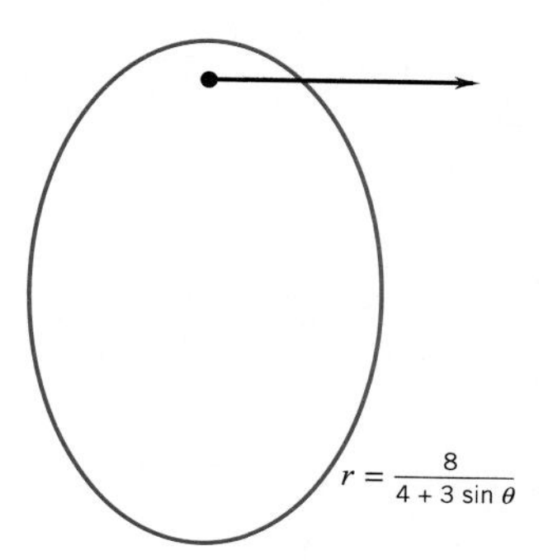

Figure 9.4.4

EXERCISES *9.4

1. The parabola

$$r = \frac{1}{1 + \cos\theta}$$

has focus at the pole and directrix $x = 1$. Without resorting to xy-coordinates, (a) locate the vertex of the parabola, (b) find the length of the latus rectum (the width of the parabola at the focus), and (c) sketch the parabola.

Sketch the parabolas in Exercises 2–4.

2. $r = \dfrac{1}{1 + \sin\theta}$.

3. $r = \dfrac{1}{1 - \cos\theta}$.

4. $r = \dfrac{1}{1 - \sin\theta}$.

5. The ellipse

$$r = \frac{2}{3 + 2\cos\theta}$$

has right focus at the pole, major axis horizontal. Without resorting to xy-coordinates, (a) find the eccentricity of the ellipse, (b) locate the ends of the major axis, (c) locate the center of the ellipse, (d) locate the second focus, (e) determine the length of the minor axis, (f) determine the width of the ellipse at the foci, and (g) sketch the ellipse.

Sketch the ellipses in Exercises 6–8.

6. $r = \dfrac{2}{3 - 2\sin\theta}$.

7. $r = \dfrac{2}{3 - 2\cos\theta}$.

8. $r = \dfrac{2}{3 + 2\sin\theta}$.

9. The hyperbola

$$r = \frac{6}{1 + 2\cos\theta}$$

has left focus at the pole, transverse axis horizontal. Without resorting to xy-coordinates, (a) find the eccentricity of the hyperbola, (b) locate the ends of the transverse axes (for one of these you need $r < 0$), (c) locate the center of the hyperbola, (d) locate the second focus, (e) determine the width of the hyperbola at the foci, and (f) sketch the hyperbola.

Sketch the hyperbolas in Exercises 10–12.

10. $r = \dfrac{6}{1 + 2\sin\theta}$.

11. $r = \dfrac{6}{1 - 2\sin\theta}$.

12. $r = \dfrac{6}{1 - 2\cos\theta}$.

In Exercises 13–15, identify the conic section and write an equation for it in rectangular coordinates.

13. $r = \dfrac{12}{2 + \sin\theta}$.

14. $r = \dfrac{4}{1 + 3\cos\theta}$.

15. $r = \dfrac{9}{5 - 4\sin\theta}$.

16. Let e and d be positive numbers and let ℓ be the vertical line $x = d$. Show that the equation

$$r = \frac{ed}{1 + e\cos\theta}$$

gives the set of all points P for which dist. $(P, O) = ed \cdot$ dist. (P, ℓ).

■ 9.5 THE INTERSECTION OF POLAR CURVES

The fact that a single point has many pairs of polar coordinates can cause complications. In particular, it means that a point $[r_1, \theta_1]$ can lie on a curve given by a polar equation although the coordinates r_1 and θ_1 do not satisfy the equation. For example, the coordinates of $[2, \pi]$ do not satisfy the equation $r^2 = 4\cos\theta$:

$$r^2 = 2^2 = 4 \quad \text{but} \quad 4\cos\theta = 4\cos\pi = -4.$$

Nevertheless the point $[2, \pi]$ does lie on the curve $r^2 = 4\cos\theta$. It lies on the curve because $[2, \pi] = [-2, 0]$ and the coordinates of $[-2, 0]$ satisfy the equation:

$$r^2 = (-2)^2 = 4, \qquad 4\cos\theta = 4\cos 0 = 4.$$

In general, a point $P[r_1, \theta_1]$ lies on a curve given by a polar equation if it has at least one polar coordinate representation $[r, \theta]$ which satisfies the equation. The difficulties are compounded when we deal with two or more curves. Here is an example.

Example 1 Find the points where the cardioids

$$r = a(1 - \cos\theta) \qquad \text{and} \qquad r = a(1 + \cos\theta) \qquad (a > 0)$$

intersect.

SOLUTION We begin by solving the two equations simultaneously. Adding these equations, we get $2r = 2a$ and thus $r = a$. This tells us that $\cos\theta = 0$ and therefore $\theta = \frac{1}{2}\pi + n\pi$. The points $[a, \frac{1}{2}\pi + n\pi]$ all lie on both curves. Not all of these points are distinct:

$$\text{for } n \text{ even, } [a, \tfrac{1}{2}\pi + n\pi] = [a, \tfrac{1}{2}\pi]; \qquad \text{for } n \text{ odd, } [a, \tfrac{1}{2}\pi + n\pi] = [a, \tfrac{3}{2}\pi].$$

In short, by solving the two equations simultaneously we have arrived at two common points:

$$[a, \tfrac{1}{2}\pi] = (0, a) \qquad \text{and} \qquad [a, \tfrac{3}{2}\pi] = (0, -a).$$

However, by sketching the graphs of the two curves (see Figure 9.5.1), we see that there is a third point at which the curves intersect; the two curves intersect at the origin, which clearly lies on both curves:

$$\text{for } r = a(1 - \cos\theta) \quad \text{take } \theta = 0, 2\pi, \ldots,$$

$$\text{for } r = a(1 + \cos\theta) \quad \text{take } \theta = \pi, 3\pi, \ldots.$$

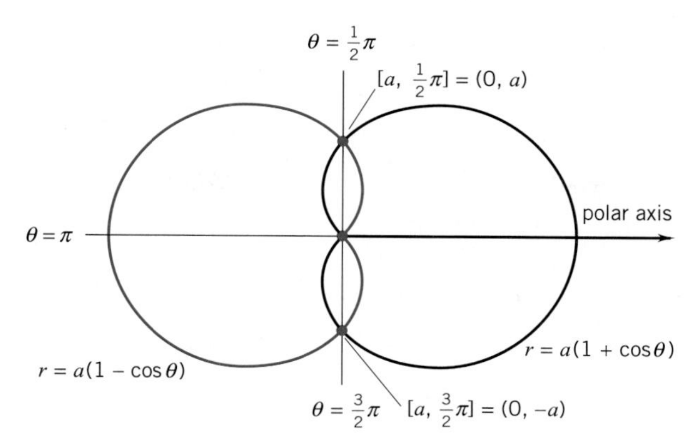

Figure 9.5.1

The reason that the origin does not appear when we solve the two equations simultaneously is that the curves do not pass through the origin "simultaneously"; that is, they do not pass through the origin for the same values of θ. Think of each of the equations

$$r = a(1 - \cos\theta) \qquad \text{and} \qquad r = a(1 + \cos\theta)$$

as giving the position of an object at time θ. At the points we found by solving the two equations simultaneously, the objects collide. (They both arrive there at the same time.) At the origin the situation is different. Both objects pass through the origin, but no collision takes place because the objects pass through the origin at *different* times. ❑

Example 2 Sketch the graphs of the polar curves

$$r = 2\sin\theta \qquad \text{and} \qquad r = 2\sin 2\theta,$$

and find their points of intersection, if any.

SOLUTION The graph of $r = 2\sin\theta$ is a circle with center on the ray $\theta = \pi/2$ and radius 1. The entire circle is traced out when θ varies from 0 to π. The graph of $r = 2\sin 2\theta$ is a petal curve with four petals. This curve is traced out when θ varies from 0 to 2π. The graphs are shown in Figure 9.5.2.

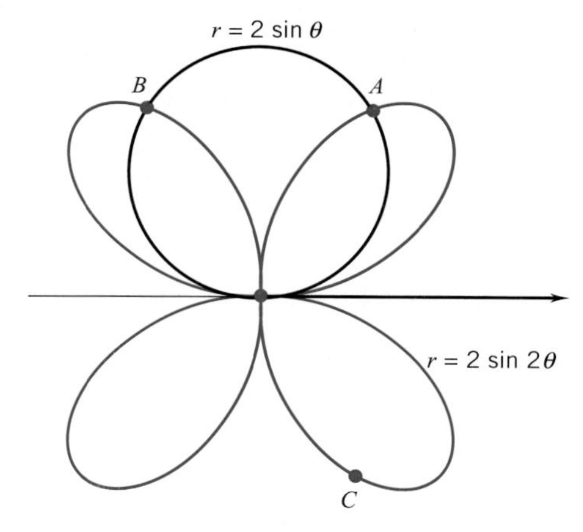

Figure 9.5.2

From the figure, we see that there are three points of intersection: the origin, the point labeled A, and the point labeled B. Solving the two equations simultaneously, we have

$$2 \sin 2\theta = 2 \sin \theta,$$

$$2 \sin \theta \cos \theta = \sin \theta,$$

and

$$\sin \theta(2 \cos \theta - 1) = 0.$$

Setting $\sin \theta = 0$, we get $\theta = n\pi$, n an integer. We can take $[0, 0]$ as the coordinates of this point of intersection.

Setting $2 \cos \theta - 1 = 0$, we get $\theta = (\pi/3) + 2n\pi$ and $\theta = (5\pi/3) + 2n\pi$. The point A clearly has coordinates $[\sqrt{3}, \pi/3]$, but the point B involves a complication. The coordinates $[\sqrt{3}, 2\pi/3]$ satisfy the equation for the circle, but they do not satisfy the equation for the petal curve: at $\theta = 2\pi/3$, $r = 2 \sin 2(2\pi/3) = -\sqrt{3}$ (the point labeled C in the figure).

You can verify that both the sets of coordinates $[\sqrt{3}, (2\pi/3) + 2n\pi]$ and $[-\sqrt{3}, (5\pi/3) + 2n\pi]$ satisfy $r = 2 \sin \theta$, while only $[-\sqrt{3}, (5\pi/3) + 2n\pi]$ satisfies $r = 2 \sin 2\theta$. Using the analogy in Example 1, if the equations describe the positions of two objects at time θ, one moving around the circle and the other around the petal curve, then at $\theta = 2\pi/3$ the object on the circle is at B and the object on the petal curve is at C. However, at time $\theta = 5\pi/3$ both objects are at the point B and they collide. The point $[-\sqrt{3}, (5\pi/3)]$ satisfies both equations simultaneously. ❑

Remark Problems of incidence (does such and such a point lie on the curve with the following polar equation?) and problems of intersection (where do such and such polar curves intersect?) can usually be analyzed by sketching the graphs. However, there are situations where such problems can be handled more readily by first changing to rectangular coordinates. For example, the rectangular equation of the curve $r^2 = 4 \cos \theta$ discussed at the beginning of this section is

$$(x^2 + y^2)^3 = 16x^2.$$

The point $P[2, \pi]$ has rectangular coordinates $(-2, 0)$, and it is easy to verify that the pair $x = -2$, $y = 0$ satisfies this equation.

In a similar manner, the symmetry properties of a curve can often be analyzed more easily in rectangular coordinates. To illustrate this, the curve C given by

$$r^2 = \sin \theta$$

is symmetric about the x-axis:

$$\text{(1)} \qquad \text{if } [r, \theta] \in C, \quad \text{then } [r, -\theta] \in C.$$

But this is not easy to see from the polar equation because, in general, if the coordinates of the first point satisfy the equation, the coordinates of the second point do not. One way to see that (1) is valid is to note that

$$[r, -\theta] = [-r, \pi - \theta]$$

and then verify that, if the coordinates of $[r, \theta]$ satisfy the equation, then so do the coordinates of $[-r, \pi - \theta]$. But all this is very cumbersome. The easiest way to see that the curve $r^2 = \sin\theta$ is symmetric about the x-axis is to write it as

$$(x^2 + y^2)^3 = y^2.$$

The other symmetries are then also clear. ❑

EXERCISES 9.5

In Exercises 1–4, determine whether the point lies on the curve.

1. $r^2 \cos\theta = 1$; $[1, \pi]$.

2. $r^2 = \cos 2\theta$; $[1, \frac{1}{4}\pi]$.

3. $r = \sin\frac{1}{3}\theta$; $[\frac{1}{2}, \frac{1}{2}\pi]$.

4. $r^2 = \sin 3\theta$; $[1, -\frac{5}{6}\pi]$.

5. Show that the point $[2, \pi]$ lies on both $r^2 = 4\cos\theta$ and $r = 3 + \cos\theta$.

6. Show that the point $[2, \frac{1}{2}\pi]$ lies on both $r^2 \sin\theta = 4$ and $r = 2\cos 2\theta$.

In Exercises 7–16, sketch the curves and find the points at which they intersect. Express your answers in rectangular coordinates.

7. $r = \sin\theta$, $r = -\cos\theta$.

8. $r^2 = \sin\theta$, $r = 2 - \sin\theta$.

9. $r = \cos^2\theta$, $r = -1$.

10. $r = 2\sin\theta$, $r = 2\cos\theta$.

11. $r = 1 - \cos\theta$, $r = \cos\theta$.

12. $r = 1 - \cos\theta$, $r = \sin\theta$.

13. $r = \dfrac{1}{1 - \cos\theta}$, $r\sin\theta = 2$.

14. $r = \dfrac{1}{2 - \cos\theta}$, $r\cos\theta = 1$.

15. $r = \cos 3\theta$, $r = \cos\theta$.

16. $r = \sin 2\theta$, $r = \sin\theta$.

17. (a) Use a graphing utility to draw the graphs of

$$r = 1 + \sin\theta \qquad \text{and} \qquad r^2 = 4\sin 2\theta$$

in the same coordinate system.

(b) Find all the points of intersection of the two graphs and estimate their coordinates accurate to three decimal places.

18. Repeat Exercise 17 for the pair of equations

$$r = 1 + \cos\theta \qquad \text{and} \qquad r = 1 + \cos(\tfrac{1}{2}\theta).$$

19. Repeat Exercise 17 for the pair of equations

$$r = 1 - 3\cos\theta \qquad \text{and} \qquad r = 2 - 5\sin\theta.$$

20. (a) The electrostatic charge distribution consisting of a charge q, $(q > 0)$ at the point $[r, 0]$, and a charge $-q$ at $[r, \pi]$ is called a *dipole*. The *lines of force* for the dipole are given by the equations

$$r = k\sin^2\theta.$$

Use a graphing utility to draw the lines of force for $k = 1, 2, 3$.

(b) The *equipotential lines* (the set of points with equal electric potential) for the dipole are given by the equations

$$r^2 = m\cos\theta.$$

Use a graphing utility to draw the equipotential lines for $m = 1, 2, 3$.

(c) Graph the curves $r = 2\sin^2\theta$ and $r^2 = 2\cos\theta$ in the same coordinate system and estimate the coordinates of their points of intersection. Use four decimal place accuracy.

■ 9.6 AREA IN POLAR COORDINATES

Here we develop a technique for calculating the area of a region the boundary of which is given in polar coordinates.

As a start, we suppose that α and β are two real numbers with $\alpha < \beta \leq \alpha + 2\pi$. We take ρ as a function that is continuous on $[\alpha, \beta]$ and keeps a constant sign on that interval. We want the area of the polar region Γ generated by the curve

$$r = \rho(\theta), \qquad \alpha \leq \theta \leq \beta.$$

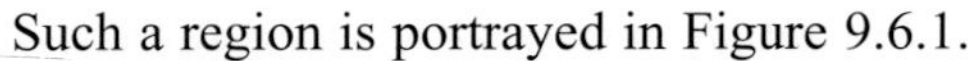

Such a region is portrayed in Figure 9.6.1.

In the figure $\rho(\theta)$ remains nonnegative. If $\rho(\theta)$ were negative, the region Γ would appear on the opposite side of the pole. In either case, the area of Γ is given by the formula

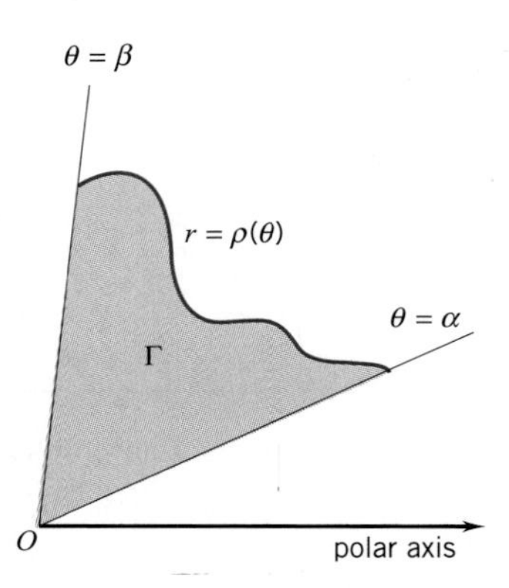

Figure 9.6.1

(9.6.1)
$$A = \int_{\alpha}^{\beta} \tfrac{1}{2}[\rho(\theta)]^2\, d\theta.$$

PROOF We consider the case where $\rho(\theta) \geq 0$. We take $P = \{\theta_0, \theta_1, \ldots, \theta_n\}$ as a partition of $[\alpha, \beta]$ and direct our attention to what happens between θ_{i-1} and θ_i. We set

$$r_i = \text{min value of } \rho \text{ on } [\theta_{i-1}, \theta_i] \qquad \text{and} \qquad R_i = \text{max value of } \rho \text{ on } [\theta_{i-1}, \theta_i].$$

The part of Γ that lies between θ_{i-1} and θ_i contains a circular sector of radius r_i and central angle $\Delta\theta_i = \theta_i - \theta_{i-1}$ and is contained in a circular sector of radius R_i and central angle $\Delta\theta_i$. (See Figure 9.6.2.) Its area A_i must therefore satisfy the inequality

$$\tfrac{1}{2}r_i^2\, \Delta\theta_i \leq A_i \leq \tfrac{1}{2}R_i^2\, \Delta\theta_i.\dagger$$

Figure 9.6.2

By summing these inequalities from $i = 1$ to $i = n$, you can see that the total area A must satisfy the inequality

(1) $$L_f(P) \leq A \leq U_f(P)$$

where

$$f(\theta) = \tfrac{1}{2}[\rho(\theta)]^2.$$

Since f is continuous and (1) holds for every partition P of $[a, b]$, we must have

$$A = \int_{\alpha}^{\beta} f(\theta)\, d\theta = \int_{\alpha}^{\beta} \tfrac{1}{2}[\rho(\theta)]^2\, d\theta. \quad \square$$

Example 1 Calculate the area enclosed by the cardioid

$$r = 1 - \cos\theta. \qquad \text{(Figure 9.6.3)}$$

SOLUTION The entire curve is traced out when θ increases from 0 to 2π and $1 - \cos\theta \geq 0$ on $[0, 2\pi]$. Thus, the area A is given by

$$A = \int_0^{2\pi} \tfrac{1}{2}(1 - \cos\theta)^2\, d\theta = \tfrac{1}{2}\int_0^{2\pi} (1 - 2\cos\theta + \cos^2\theta)\, d\theta$$

$$= \tfrac{1}{2}\int_0^{2\pi} \left(\tfrac{3}{2} - 2\cos\theta + \tfrac{1}{2}\cos 2\theta\right) d\theta.$$

↑ half-angle formula: $\cos^2\theta = \frac{1}{2} + \frac{1}{2}\cos 2\theta$

† The area of a circular sector of radius r and central angle α is $\frac{1}{2}r^2\alpha$.

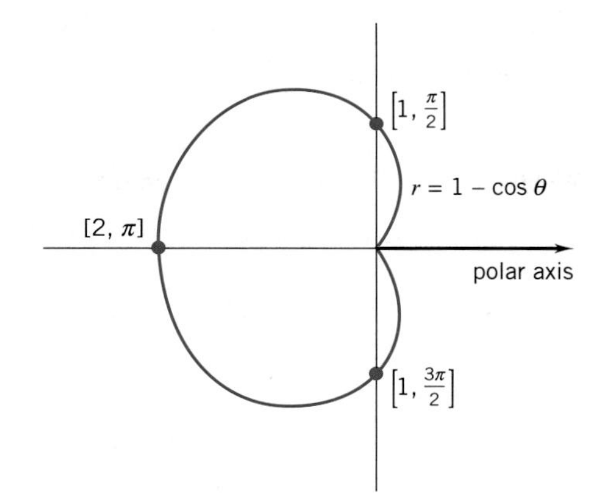

Figure 9.6.3

Since

$$\int_0^{2\pi} \cos\theta \, d\theta = 0 \qquad \text{and} \qquad \int_0^{2\pi} \cos 2\theta \, d\theta = 0,$$

we have

$$A = \tfrac{1}{2}\int_0^{2\pi} \tfrac{3}{2}\, d\theta = \tfrac{3}{4}\int_0^{2\pi} d\theta = \tfrac{3}{2}\pi. \quad \square$$

A slightly more complicated type of region is pictured in Figure 9.6.4. We approach the problem of calculating the area of the region Ω in the same way that we calculated the area between two curves in Section 5.5. That is, we calculate the area out to $r = \rho_2(\theta)$ and subtract from it the area out to $r = \rho_1(\theta)$. This gives the formula

(9.6.2)

$$\text{area of } \Omega = \int_\alpha^\beta \tfrac{1}{2}\left([\rho_2(\theta)]^2 - [\rho_1(\theta)]^2\right) d\theta.$$

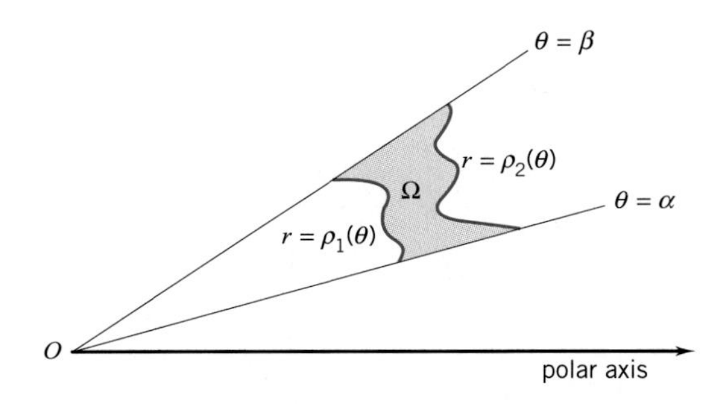

Figure 9.6.4

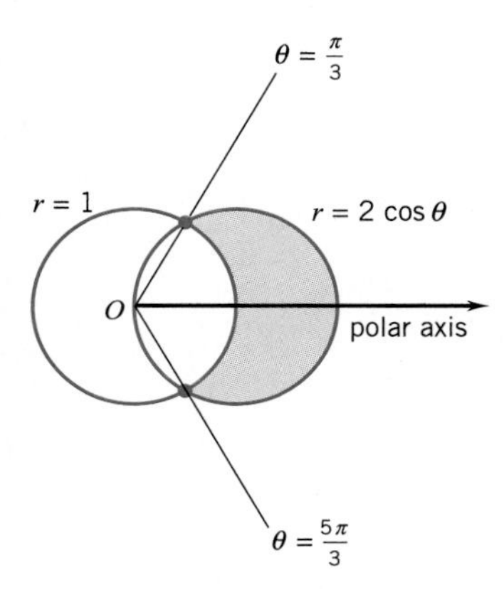

Figure 9.6.5

Example 2 Find the area of the region that consists of all points that lie within the circle $r = 2\cos\theta$ but outside the circle $r = 1$.

SOLUTION The region is shown in Figure 9.6.5. Our first step is to find values of θ for the two points where the circles intersect:

$$2\cos\theta = 1, \qquad \cos\theta = \tfrac{1}{2}, \qquad \theta = \tfrac{1}{3}\pi,\ \tfrac{5}{3}\pi.$$

Since the region is symmetric about the polar axis, the area below the polar axis equals the area above the polar axis. Thus

$$A = 2\int_0^{\pi/3} \tfrac{1}{2}([2\cos\theta]^2 - [1]^2)\,d\theta.$$

If you carry out the integration, you will see that $A = \frac{1}{3}\pi + \frac{1}{2}\sqrt{3} \cong 1.91.$ ❑

To find the area between two polar curves, we first determine the curves that serve as outer and inner boundaries of the region and the intervals of θ values over which these boundaries are traced out. Since the polar coordinates of a point are not unique, extra care is needed to determine these intervals of θ values.

Example 3 Find the area A of the region between the inner and outer loops of the limaçon

$$r = 1 - 2\cos\theta. \qquad \text{(Figure 9.6.6)}$$

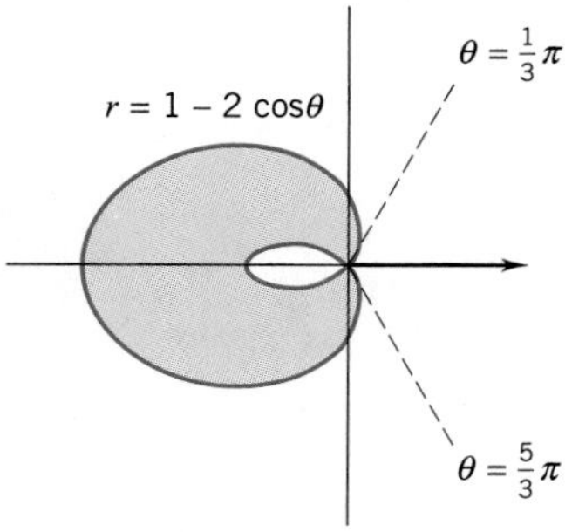

Figure 9.6.6

SOLUTION It is easy to verify that $r = 0$ when $\theta = \pi/3$ and when $\theta = 5\pi/3$. The outer loop is formed by having θ increase from $\pi/3$ to $5\pi/3$. Thus

$$\text{area within outer loop} = A_1 = \int_{\pi/3}^{5\pi/3} \tfrac{1}{2}[1 - 2\cos\theta]^2\,d\theta.$$

The lower half of the inner loop is formed when θ increases from 0 to $\pi/3$, and the upper half when θ increases from $5\pi/3$ to 2π (verify this). Therefore, we have

$$\text{area within inner loop} = A_2 = \int_0^{\pi/3} \tfrac{1}{2}[1 - 2\cos\theta]^2\,d\theta + \int_{5\pi/3}^{2\pi} \tfrac{1}{2}[1 - 2\cos\theta]^2\,d\theta.$$

Now

$$\begin{aligned}\int \tfrac{1}{2}[1 - 2\cos\theta]^2\,d\theta &= \tfrac{1}{2}\int [1 - 4\cos\theta + 4\cos^2\theta]\,d\theta \\ &= \tfrac{1}{2}\int [1 - 4\cos\theta + 2(1 + \cos 2\theta)]\,d\theta \\ &= \tfrac{1}{2}\int [3 - 4\cos\theta + 2\cos 2\theta]\,d\theta \\ &= \tfrac{1}{2}[3\theta - 4\sin\theta + \sin 2\theta] + C.\end{aligned}$$

Therefore,

$$A_1 = \tfrac{1}{2}\Big[3\theta - 4\sin\theta + \sin 2\theta\Big]_{\pi/3}^{5\pi/3} = 2\pi + \tfrac{3}{2}\sqrt{3}$$

and

$$A_2 = \tfrac{1}{2}\Big[3\theta - 4\sin\theta + \sin 2\theta\Big]_0^{\pi/3} + \tfrac{1}{2}\Big[3\theta - 4\sin\theta + \sin 2\theta\Big]_{5\pi/3}^{2\pi}$$
$$= \tfrac{1}{2}\pi - \tfrac{3}{4}\sqrt{3} + \tfrac{1}{2}\pi - \tfrac{3}{4}\sqrt{3} = \pi - \tfrac{3}{2}\sqrt{3}.$$

Thus,

$$A = A_1 - A_2 = 2\pi + \tfrac{3}{2}\sqrt{3} - (\pi - \tfrac{3}{2}\sqrt{3}) = \pi + 3\sqrt{3} \cong 8.34. \quad \square$$

Remark A more efficient approach to the solution in Example 3 would have been to use the symmetry of the region. The region is symmetric about the x-axis, and so the area A is also given by

$$A = 2\int_{\pi/3}^{\pi} \tfrac{1}{2}[1 - 2\cos\theta]^2\,d\theta - 2\int_0^{\pi/3} \tfrac{1}{2}[1 - 2\cos\theta]^2\,d\theta. \quad \square$$

Example 4 The region Ω common to the circle $r = 2\sin\theta$ and the limaçon $r = \tfrac{3}{2} - \sin\theta$ is indicated in Figure 9.6.7. The θ coordinates of the points of intersection can be found by solving the two equations simultaneously:

$$2\sin\theta = \tfrac{3}{2} - \sin\theta, \qquad \sin\theta = \tfrac{1}{2}, \qquad \text{and} \qquad \theta = \tfrac{\pi}{6}, \tfrac{5\pi}{6}.$$

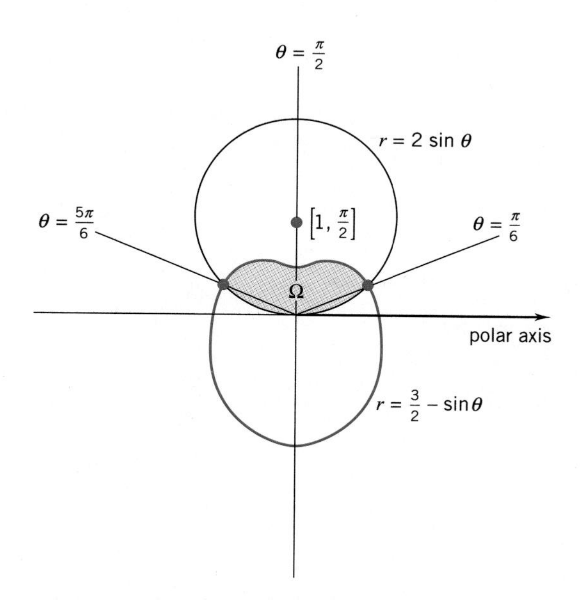

Figure 9.6.7

Thus, the area of Ω can be represented as follows:

$$\text{area of }\Omega = \int_0^{\pi/6} \tfrac{1}{2}[2\sin\theta]^2\,d\theta + \int_{\pi/6}^{5\pi/6} \tfrac{1}{2}[\tfrac{3}{2} - \sin\theta]^2\,d\theta + \int_{5\pi/6}^{\pi} \tfrac{1}{2}[2\sin\theta]^2\,d\theta$$

or, using the symmetry of the region,

$$\text{area of }\Omega = 2\int_0^{\pi/6} \tfrac{1}{2}[2\sin\theta]^2\,d\theta + 2\int_{\pi/6}^{\pi/2} \tfrac{1}{2}[\tfrac{3}{2} - \sin\theta]^2\,d\theta.$$

You can verify that the area of Ω is $\tfrac{5}{4}\pi - \tfrac{15}{8}\sqrt{3} \cong 0.68. \quad \square$

EXERCISES 9.6

In Exercises 1–6, calculate the area enclosed by the given curve. Take $a > 0$.

1. $r = a\cos\theta$ from $\theta = -\frac{1}{2}\pi$ to $\theta = \frac{1}{2}\pi$.
2. $r = a\cos 3\theta$ from $\theta = -\frac{1}{6}\pi$ to $\theta = \frac{1}{6}\pi$.
3. $r = a\sqrt{\cos 2\theta}$ from $\theta = -\frac{1}{4}\pi$ to $\theta = \frac{1}{4}\pi$.
4. $r = a(1 + \cos 3\theta)$ from $\theta = -\frac{1}{3}\pi$ to $\theta = \frac{1}{3}\pi$.
5. $r^2 = a^2\sin^2\theta$.
6. $r^2 = a^2\sin^2 2\theta$.

In Exercises 7–12, calculate the area of the given region.

7. $r = \tan 2\theta$ and the rays $\theta = 0$, $\theta = \frac{1}{8}\pi$.
8. $r = \cos\theta$, $r = \sin\theta$, and the rays $\theta = 0$, $\theta = \frac{1}{4}\pi$.
9. $r = 2\cos\theta$, $r = \cos\theta$, and the rays $\theta = 0$, $\theta = \frac{1}{4}\pi$.
10. $r = 1 + \cos\theta$, $r = \cos\theta$, and the rays $\theta = 0$, $\theta = \frac{1}{2}\pi$.
11. $r = a(4\cos\theta - \sec\theta)$ and the rays $\theta = 0$, $\theta = \frac{1}{4}\pi$.
12. $r = \frac{1}{2}\sec^2\frac{1}{2}\theta$ and the vertical line through the origin.

In Exercises 13–16, find the area of the given region.

13. $r = e^\theta$, $0 \le \theta \le \pi$; $r = \theta$, $0 \le \theta \le \pi$; the rays $\theta = 0$, $\theta = \pi$.
14. $r = e^\theta$, $2\pi \le \theta \le 3\pi$; $r = \theta$, $0 \le \theta \le \pi$; the rays $\theta = 0$, $\theta = \pi$.
15. $r = e^\theta$, $0 \le \theta \le \pi$; $r = e^{\theta/2}$, $0 \le \theta \le \pi$; the rays $\theta = 2\pi$, $\theta = 3\pi$.
16. $r = e^\theta$, $0 \le \theta \le \pi$; $r = e^\theta$, $2\pi \le \theta \le 3\pi$; the rays $\theta = 0$, $\theta = \pi$.

In Exercises 17–28, represent the indicated area by one or more integrals.

17. Outside $r = 2$, but inside $r = 4\sin\theta$.
18. Outside $r = 1 - \cos\theta$, but inside $r = 1 + \cos\theta$.
19. Inside $r = 4$, and to the right of $r = 2\sec\theta$.
20. Inside $r = 2$, but outside $r = 4\cos\theta$.
21. Inside $r = 4$, and between the lines $\theta = \frac{1}{2}\pi$ and $r = 2\sec\theta$.
22. Inside the inner loop of $r = 1 - 2\sin\theta$.
23. Inside one petal of $r = 2\sin 3\theta$.
24. Outside $r = 1 + \cos\theta$, but inside $r = 2 - \cos\theta$.
25. Interior to both $r = 1 - \sin\theta$ and $r = \sin\theta$.
26. Inside one petal of $r = 5\cos 6\theta$.
27. Outside $r = \cos 2\theta$, but inside $r = 1$.
28. Interior to both $r = 2a\cos\theta$ and $r = 2a\sin\theta$, $a > 0$.
29. Let n be a positive integer. Prove that the petal curves $r = a\cos 2n\theta$ and $r = a\sin 2n\theta$, $a > 0$, all enclose exactly the same area. Find the area.
30. Let n be a positive integer. Prove that the petal curves $r = a\cos([2n + 1]\theta)$ and $r = a\sin([2n + 1]\theta)$, $a > 0$, all enclose exactly the same area. Find the area.

Centroids in Polar Coordinates

Let Ω be the region bounded by the polar curve $r = f(\theta)$ between $\theta = \alpha$ and $\theta = \beta$. Since the centroid of a triangle lies on each median, two-thirds of the distance from the vertex to the opposite side, it follows that the x and y coordinates of the centroid of the "triangular" region shown in the figure are given approximately by

$$\bar{x} = \tfrac{2}{3} r\cos\theta, \qquad \bar{y} = \tfrac{2}{3} r\sin\theta.$$

These approximations improve as $\Delta\theta \to 0$.

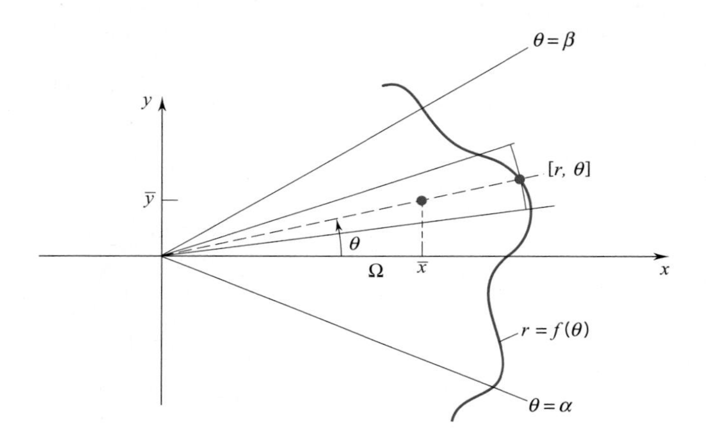

31. Following the approach used in Section 6.4, show that the rectangular coordinates of the centroid of Ω are

$$\bar{x} = \frac{\frac{2}{3}\int_\alpha^\beta r^3\cos\theta\, d\theta}{\int_\alpha^\beta r^2\, d\theta};$$

$$\bar{y} = \frac{\frac{2}{3}\int_\alpha^\beta r^3\sin\theta\, d\theta}{\int_\alpha^\beta r^2\, d\theta}.$$

In Exercises 32–34, use the result of Exercise 31 to find the rectangular coordinates of the centroid of the given region.

32. The region enclosed by $r = 4$ between $\theta = -\alpha$ and $\theta = \alpha$, $0 < \alpha < \pi/2$.
33. The region enclosed by the cardioid $r = 1 + \cos\theta$.
34. The region enclosed by $r = 2 + \sin\theta$.

35. The curve whose equation in rectangular coordinates is

$$y^2 = x^2\left(\frac{a-x}{a+x}\right), \quad a > 0,$$

is called a *strophoid*.

(a) Show that the polar equation of this curve has the form

$$r = a \cos 2\theta \sec \theta.$$

(b) Use a graphing utility to draw the graph of the curves for $a = 1$, 2, and 4.

(c) Let $a = 2$. Find the area inside the loop.

36. The curve whose equation in rectangular coordinates is

$$(x^2 + y^2)^2 = ax^2y, \quad a > 0,$$

is called a *bifolium*.

(a) Show that the polar equation of this curve has the form

$$r = a \sin \theta \cos^2 \theta.$$

(b) Use a graphing utility to draw the graph of the curves for $a = 1$, 2, and 4.

(c) Let $a = 2$. Find the area inside one of the loops.

■ 9.7 CURVES GIVEN PARAMETRICALLY

We begin with a pair of functions $x = x(t)$, $y = y(t)$ differentiable on the interior of an interval I. At the endpoints of I (if any) we require only continuity.

For each number t in I we can interpret $(x(t), y(t))$ as the point with x-coordinate $x(t)$ and y-coordinate $y(t)$. Then, as t ranges over I, the point $(x(t), y(t))$ traces out a path in the xy-plane. (Figure 9.7.1) We call such a path a *parametrized curve* and refer to t as the *parameter*.

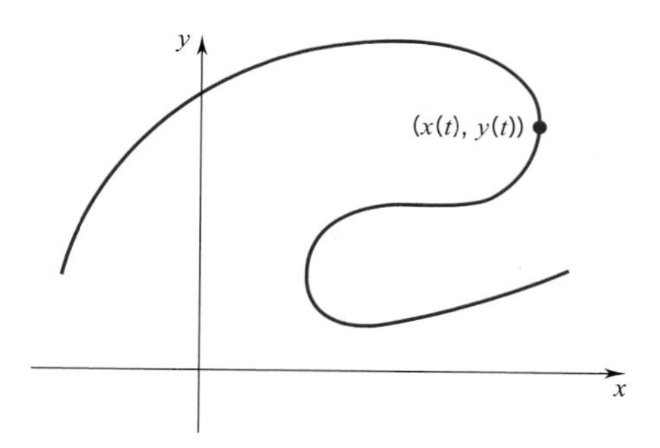

Figure 9.7.1

Example 1 Identify the curve parametrized by the functions

$$x(t) = t + 1, \quad y(t) = 2t - 5, \qquad t \text{ real}.$$

SOLUTION We can express $y(t)$ in terms of $x(t)$:

$$y(t) = 2[x(t) - 1] - 5 = 2x(t) - 7.$$

The functions parametrize the line $y = 2x - 7$: as t ranges over the set of real numbers, the point $(x(t), y(t))$ traces out the line $y = 2x - 7$. ❑

Example 2 Identify the curve parametrized by the functions

$$x(t) = 2t, \quad y(t) = t^2, \qquad t \in [0, \infty).$$

SOLUTION From the first equation $t = \frac{1}{2}x(t)$, and so

$$y(t) = \tfrac{1}{4}[x(t)]^2.$$

The functions parametrize that part of the parabola $y = \frac{1}{4}x^2$ which lies in the right half plane: as t ranges over the interval $[0, \infty)$, the point $(x(t), y(t))$ traces out the parabola $y = \frac{1}{4}x^2$ for $x \geq 0$ (Figure 9.7.2). ❑

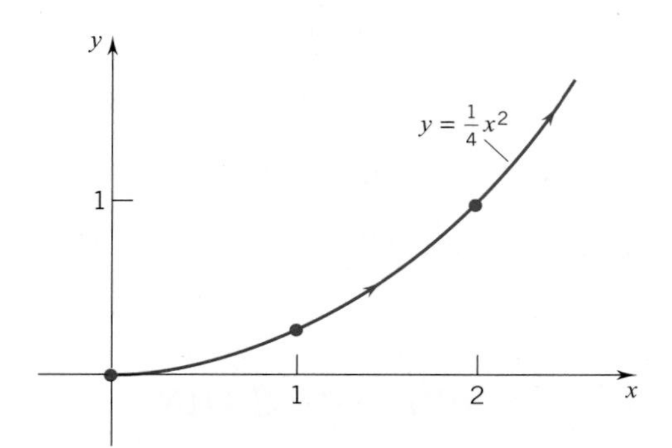

Figure 9.7.2

Example 3 Identify the curve parametrized by the functions

$$x(t) = \sin^2 t, \quad y(t) = \cos t, \qquad t \in [0, \pi].$$

SOLUTION Note first that

$$x(t) = \sin^2 t = 1 - \cos^2 t = 1 - [y(t)]^2.$$

The points $(x(t), y(t))$ all lie on the parabola

$$x = 1 - y^2. \qquad \text{(Figure 9.7.3)}$$

At $t = 0$, $x = 0$ and $y = 1$; at $t = \pi$, $x = 0$ and $y = -1$. As t ranges from 0 to π, the point $(x(t), y(t))$ traverses the parabolic arc

$$x = 1 - y^2, \qquad -1 \leq y \leq 1$$

from the point $(0, 1)$ to the point $(0, -1)$. ❑

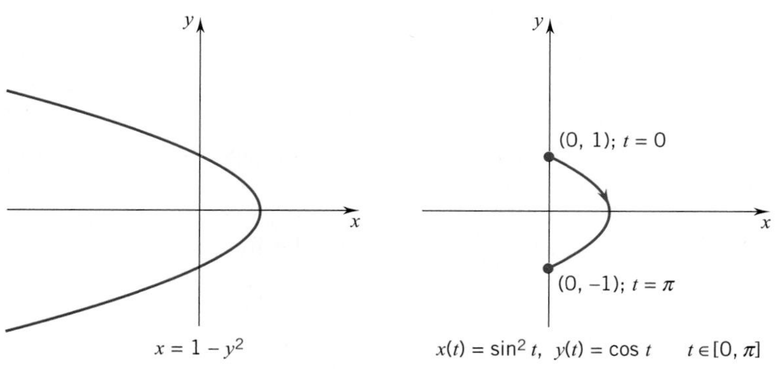

Figure 9.7.3

Remark Changing the domain in the previous problem to all real t does *not* give us any more of the parabola. For any given t we still have

$$0 \leq x(t) \leq 1 \qquad \text{and} \qquad -1 \leq y(t) \leq 1.$$

As t ranges over the set of real numbers, the point $(x(t), y(t))$ traces out that same parabolic arc back and forth an infinite number of times. ❑

The functions

(9.7.1) $$x(t) = x_0 + t(x_1 - x_0), \quad y(t) = y_0 + t(y_1 - y_0), \qquad t \text{ real}$$

parametrize the line that passes through (x_0, y_0) and (x_1, y_1), where $(x_0, y_0) \neq (x_1, y_1)$.

PROOF If $x_1 = x_0$, then we have

$$x(t) = x_0, \quad y(t) = y_0 + t(y_1 - y_0). \qquad (y_0 \neq y_1)$$

As t ranges over the set of real numbers, $x(t)$ remains constantly x_0 and $y(t)$ ranges over the set of real numbers. The functions parametrize the vertical line $x = x_0$. Since $x_1 = x_0$, both (x_0, y_0) and (x_1, y_1) lie on this vertical line.

If $x_1 \neq x_0$, then we can solve the first equation for t:

$$t = \frac{x(t) - x_0}{x_1 - x_0}.$$

Substituting this into the second equation we obtain the identity

$$y(t) - y_0 = \frac{y_1 - y_0}{x_1 - x_0}[x(t) - x_0].$$

The functions parametrize the line with equation

$$y - y_0 = \frac{y_1 - y_0}{x_1 - x_0}(x - x_0).$$

This is the line determined by (x_0, y_0) and (x_1, y_1). ❑

The functions $x(t) = a \cos t$, $y(t) = b \sin t$, where $a, b > 0$, satisfy the identity

$$\frac{[x(t)]^2}{a^2} + \frac{[y(t)]^2}{b^2} = 1.$$

As t ranges over any interval of length 2π, the point $(x(t), y(t))$ traces out the ellipse

$$\frac{x^2}{a^2} + \frac{y^2}{b^2} = 1.$$

Usually we let t range from 0 to 2π and parametrize the ellipse by setting

(9.7.2) $$x(t) = a \cos t, \quad y(t) = b \sin t, \qquad t \in [0, 2\pi].$$

If $b = a$, we have a circle. We can parametrize the circle

$$x^2 + y^2 = a^2$$

by setting

(9.7.3)
$$\boxed{x(t) = a \cos t, \quad y(t) = a \sin t, \quad t \in [0, 2\pi].}$$

The functions $x(t) = a \cosh t$, $y(t) = b \sinh t$, where $a, b > 0$, satisfy the identity

$$\frac{[x(t)]^2}{a^2} - \frac{[y(t)]^2}{b^2} = 1.$$

As t ranges over the real numbers, $x(t) = a \cosh t > 0$ and the point $(x(t), y(t))$ traces out the right branch of the hyperbola

$$\frac{x^2}{a^2} - \frac{y^2}{b^2} = 1. \qquad \text{(Figure 9.7.4)}$$

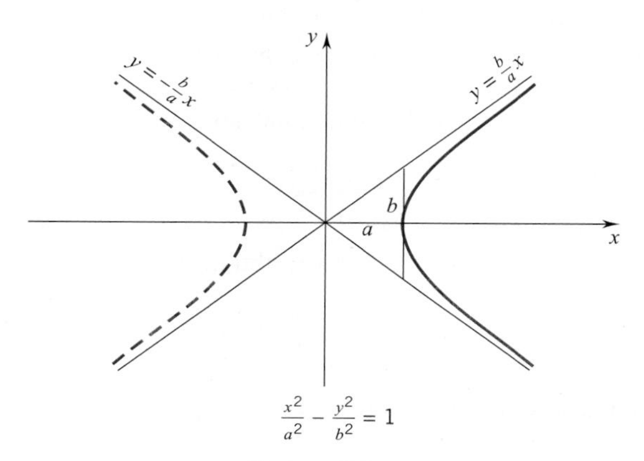

Figure 9.7.4

Thus, the right branch of the hyperbola is parametrized by

$$x(t) = a \cosh t, \quad y(t) = b \cosh t, \qquad t \text{ real.}$$

The left branch is parametrized by

$$x(t) = -a \cosh t, \quad y(t) = b \sinh t, \qquad t \text{ real.}$$

Interpreting the parameter t as time measured in seconds, we can think of a pair of parametric equations, $x = x(t)$ and $y = y(t)$, as describing the motion of a particle in the xy-plane. Different parametrizations of the same curve represent different ways of traversing that curve.

Example 4 The line that passes through the points (1, 2) and (3, 6) has equation $y = 2x$. The line segment that joins these same points is given by

$$y = 2x, \qquad 1 \le x \le 3.$$

We will parametrize this line segment in different ways and interpret each parametrization as the motion of a particle.

We begin by setting

$$x(t) = t, \quad y(t) = 2t, \qquad t \in [1, 3].$$

At time $t = 1$, the particle is at the point $(1, 2)$. It traverses the line segment and arrives at the point $(3, 6)$ at time $t = 3$.

Now we set

$$x(t) = t + 1, \quad y(t) = 2t + 2, \qquad t \in [0, 2].$$

At time $t = 0$, the particle is at the point $(1, 2)$. It traverses the line segment and arrives at the point $(3, 6)$ at time $t = 2$.

The equations

$$x(t) = 3 - t, \quad y(t) = 6 - 2t, \qquad t \in [0, 2]$$

represent a traversal of that same line segment in the opposite direction. At time $t = 0$, the particle is at $(3, 6)$. It arrives at $(1, 2)$ at time $t = 2$.

Set

$$x(t) = 3 - 4t, \quad y(t) = 6 - 8t, \qquad t \in [0, \tfrac{1}{2}].$$

Now the particle traverses the same line segment in only half a second. At time $t = 0$, the particle is at $(3, 6)$. It arrives at $(1, 2)$ at time $t = \frac{1}{2}$.

Finally we set

$$x(t) = 2 - \cos t, \quad y(t) = 4 - 2\cos t, \qquad t \in [0, 4\pi].$$

In this instance the particle begins and ends its motion at the point $(1, 2)$, having traced and retraced the line segment twice during a span of 4π seconds. See Figure 9.7.5. ❑

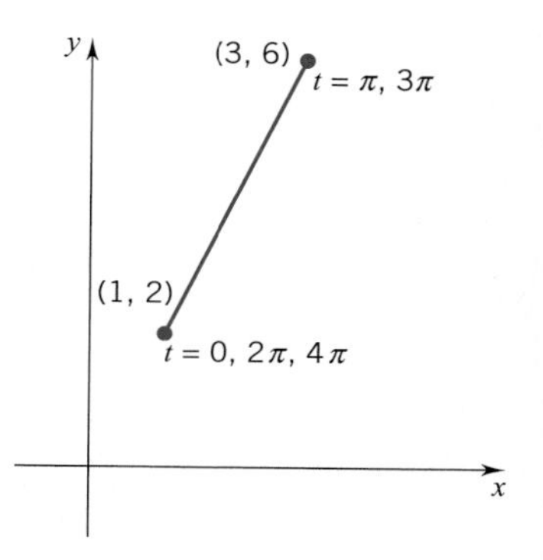

Figure 9.7.5

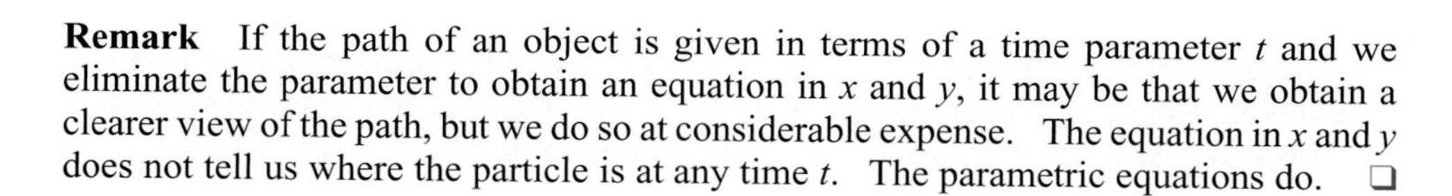

Remark If the path of an object is given in terms of a time parameter t and we eliminate the parameter to obtain an equation in x and y, it may be that we obtain a clearer view of the path, but we do so at considerable expense. The equation in x and y does not tell us where the particle is at any time t. The parametric equations do. ❑

Example 5 We return to the ellipse

$$\frac{x^2}{a^2} + \frac{y^2}{b^2} = 1.$$

A particle with position given by the equations

$$x(t) = a\cos t, \quad y(t) = b\sin t \qquad \text{with } t \in [0, 2\pi]$$

traverses the ellipse in a counterclockwise manner. It begins at the point $(a, 0)$ and makes a full circuit in 2π seconds. If the equations of motion are

$$x(t) = a\cos 2\pi t, \quad y(t) = -b\sin 2\pi t \qquad \text{with } t \in [0, 1],$$

the particle still travels the same ellipse, but in a different manner. Once again it starts at $(a, 0)$, but this time it moves clockwise and makes the full circuit in only one second. If the equations of motion are

$$x(t) = a\sin 4\pi t, \quad y(t) = b\cos 4\pi t \qquad \text{with } t \in [0, \infty),$$

the motion begins at $(0, b)$ and goes on in perpetuity. The motion is clockwise, a complete circuit taking place every half second. ❑

Intersections and Collisions

Example 6 Two particles start at the same instant, the first along the linear path

$$x_1(t) = \tfrac{16}{3} - \tfrac{8}{3}t, \quad y_1(t) = 4t - 5, \qquad t \geq 0$$

and the second along the elliptical path

$$x_2(t) = 2 \sin \tfrac{1}{2}\pi t, \quad y_2(t) = -3 \cos \tfrac{1}{2}\pi t, \qquad t \geq 0.$$

(a) At what points, if any, do the paths intersect?

(b) At what points, if any, do the particles collide?

SOLUTION To see where the paths intersect, we find equations for them in x and y. The linear path can be written

$$3x + 2y - 6 = 0, \qquad x \leq \tfrac{16}{3}$$

and the elliptical path

$$\frac{x^2}{4} + \frac{y^2}{9} = 1.$$

Solving the two equations simultaneously, we get

$$x = 2, \quad y = 0 \qquad \text{and} \qquad x = 0, \quad y = 3.$$

This means that the paths intersect at the points (2, 0) and (0, 3). This answers part (a).

Now for part (b). The first particle passes through (2, 0) only when

$$x_1(t) = \tfrac{16}{3} - \tfrac{8}{3}t = 2 \qquad \text{and} \qquad y_1(t) = 4t - 5 = 0.$$

As you can check, this happens only when $t = \frac{5}{4}$. When $t = \frac{5}{4}$, the second particle is elsewhere. Hence no collision takes place at (2, 0). There is, however, a collision at (0, 3) because both particles get there at exactly the same time, $t = 2$:

$$x_1(2) = 0 = x_2(2), \qquad y_1(2) = 3 = y_2(2). \quad ❑$$

EXERCISES 9.7

In Exercises 1–12, express the curve by an equation in x and y.

1. $x(t) = t^2, \quad y(t) = 2t + 1.$

2. $x(t) = 3t - 1, \quad y(t) = 5 - 2t.$

3. $x(t) = t^2, \quad y(t) = 4t^4 + 1.$

4. $x(t) = 2t - 1, \quad y(t) = 8t^3 - 5.$

5. $x(t) = 2 \cos t, \quad y(t) = 3 \sin t.$

6. $x(t) = \sec^2 t, \quad y(t) = 2 + \tan t.$

7. $x(t) = \tan t, \quad y(t) = \sec t.$

8. $x(t) = 2 - \sin t, \quad y(t) = \cos t.$

9. $x(t) = \sin t, \quad y(t) = 1 + \cos^2 t.$

10. $x(t) = e^t, \quad y(t) = 4 - e^{2t}.$

11. $x(t) = 4 \sin t, \quad y(t) = 3 + 2 \sin t.$

12. $x(t) = \csc t, \quad y(t) = \cot t.$

In Exercises 13–21, express the curve by an equation in x and y; then sketch the curve.

13. $x(t) = e^{2t}, \quad y(t) = e^{2t} - 1, \quad t \leq 0.$

14. $x(t) = 3 \cos t, \quad y(t) = 2 - \cos t, \quad 0 \leq t \leq \pi.$

15. $x(t) = \sin t, \quad y(t) = \csc t, \quad 0 < t \leq \frac{1}{4}\pi.$

16. $x(t) = 1/t, \quad y(t) = 1/t^2, \quad 0 < t < 3.$

17. $x(t) = 3 + 2t, \quad y(t) = 5 - 4t, \quad -1 \leq t \leq 2.$

18. $x(t) = \sec t, \quad y(t) = \tan t, \quad 0 \leq t \leq \frac{1}{4}\pi.$

19. $x(t) = \sin \pi t, \quad y(t) = 2t, \quad 0 \leq t \leq 4.$

20. $x(t) = 2 \sin t, \quad y(t) = \cos t, \quad 0 \leq t \leq \frac{1}{2}\pi.$

21. $x(t) = \cot t, \quad y(t) = \csc t, \quad \frac{1}{4}\pi \leq t < \frac{1}{2}\pi.$

22. (*Important*) Parametrize: (a) the curve $y = f(x)$, $x \in [a, b]$; (b) the polar curve $r = f(\theta)$, $\theta \in [\alpha, \beta]$.

23. A particle with position given by the equations

$$x(t) = \sin 2\pi t, \quad y(t) = \cos 2\pi t, \qquad t \in [0, 1],$$

starts at the point (0, 1) and traverses the unit circle $x^2 + y^2 = 1$ once in a clockwise manner. Write equations of the form

$$x(t) = f(t), \quad y(t) = g(t), \qquad t \in [0, 1],$$

so that the particle
(a) begins at (0, 1) and traverses the circle once in a counterclockwise manner;
(b) begins at (0, 1) and traverses the circle twice in a clockwise manner;
(c) traverses the quarter circle from (1, 0) to (0, 1);
(d) traverses the three-quarter circle from (1, 0) to (0, 1).

24. A particle with position given by the equations

$$x(t) = 3 \cos 2\pi t, \quad y(t) = 4 \sin 2\pi t, \qquad t \in [0, 1],$$

starts at the point (3, 0) and traverses the ellipse $16x^2 + 9y^2 = 144$ once in a counterclockwise manner. Write equations of the form

$$x(t) = f(t), \quad y(t) = g(t), \qquad t \in [0, 1],$$

so that the particle
(a) begins at (3, 0) and traverses the ellipse once in a clockwise manner;
(b) begins at (0, 4) and traverses the ellipse once in a clockwise manner;
(c) begins at (−3, 0) and traverses the ellipse twice in a counterclockwise manner;
(d) traverses the upper half of the ellipse from (3, 0) to (0, 3).

25. Find a parametrization

$$x = x(t), \quad y = y(t), \qquad t \in (-1, 1),$$

for the horizontal line $y = 2$.

26. Find a parametrization

$$x(t) = \sin f(t), \quad y(t) = \cos f(t), \ t \in (0, 1),$$

which traces out the unit circle infinitely often.

In Exercises 27–33, find a parametrization

$$x = x(t), \quad y = y(t), \qquad t \in [0, 1],$$

for the given curve.

27. The line segment from (3, 7) to (8, 5).

28. The line segment from (2, 6) to (6, 3).

29. The parabolic arc $x = 1 - y^2$ from (0, −1) to (0, 1).

30. The parabolic arc $x = y^2$ from (4, 2) to (0, 0).

31. The curve $y^2 = x^3$ from (4, 8) to (1, 1).

32. The curve $y^3 = x^2$ from (1, 1) to (8, 4).

33. The curve $y = f(x)$, $x \in [a, b]$.

34. (*Important*) Suppose that the curve

$$C: \quad x = x(t), \quad y = y(t), \qquad t \in [c, d],$$

is the graph of a nonnegative function $y = f(x)$ over an interval $[a, b]$. Suppose that $x'(t)$ and $y(t)$ are continuous, $x(c) = a$, and $x(d) = b$.
(a) (*The area under a parametrized curve*) Show that

(9.7.4) $$\text{the area below } C = \int_c^d y(t)x'(t)\,dt.$$

HINT: Since C is the graph of f, we know that $y(t) = f(x(t))$.
(b) (*The centroid of a region under a parametrized curve*) Show that, if the region under C has area A and centroid $(\bar{x}, \bar{y})$, then

(9.7.5) $$\bar{x}A = \int_c^d x(t)y(t)x'(t)\,dt,$$
$$\bar{y}A = \int_c^d \tfrac{1}{2}[y(t)]^2 x'(t)\,dt.$$

(c) (*The volume of the solid generated by revolving about a coordinate axis the region under a parametrized curve*) Show that

(9.7.6) $$V_x = \int_c^d \pi[y(t)]^2 x'(t)\,dt,$$
$$V_y = \int_c^d 2\pi x(t)y(t)x'(t)\,dt.$$
$\uparrow$ provided $x(c) \geq 0$

(d) (*The centroid of the solid generated by revolving about a coordinate axis the region under a parametrized curve*) Show that

(9.7.7) $$\bar{x}V_x = \int_c^d \pi x(t)[y(t)]^2 x'(t)\,dt,$$
$$\bar{y}V_y = \int_c^d \pi x(t)[y(t)]^2 x'(t)\,dt.$$
$\uparrow$ provided $x(c) \geq 0$

35. Sketch the curve

$$x(t) = at, \quad y(t) = a(1 - \cos t), \quad a > 0, \quad t \in [0, 2\pi]$$

and find the area below it.

36. Determine the centroid of the region under the curve in Exercise 35.

37. Find the volume generated by revolving the region in Exercise 36 about: (a) the x-axis; (b) the y-axis.

38. Find the centroid of the solid generated by revolving the region of Exercise 36 about: (a) the x-axis; (b) the y-axis.

39. Give a parametrization for the upper half of the ellipse $b^2x^2 + a^2b^2 = a^2b^2$ that satisfies the requirements of Exercise 34.

40. Use the parametrization you chose for Exercise 39 to find (a) the area of the region enclosed by the ellipse; (b) the centroid of the upper half of that region.

41. Two particles start at the same instant, the first along the ray

$$x(t) = 2t + 6, \quad y(t) = 5 - 4t, \qquad t \geq 0,$$

and the second along the circular path

$$x(t) = 3 - 5 \cos \pi t, \quad y(t) = 1 + 5 \sin \pi t, \qquad t \geq 0.$$

(a) At what points, if any, do these paths intersect?
(b) At what points, if any, will the particles collide?

42. Two particles start at the same instant, the first along the elliptical path

$$x_1(t) = 2 - 3 \cos \pi t, \quad y_1(t) = 3 + 7 \sin \pi t, \qquad t \geq 0,$$

and the second along the parabolic path

$$x_2(t) = 3t + 2, \quad y_2(t) = -\tfrac{7}{15}(3t + 1)^2 + \tfrac{157}{15}, \qquad t \geq 0.$$

(a) At what points, if any, do these paths intersect?
(b) At what points, if any, will the particles collide?

We can determine the points where a parametrized curve

$$C: \quad x = x(t), \quad y = y(t), \qquad t \in I$$

intersects itself by finding the numbers r and s in I $(r \neq s)$ for which

$$x(r) = x(s) \quad \text{and} \quad y(r) = y(s).$$

Use this method to find the point(s) of self-intersection for each of the curves in Exercises 43–46.

43. $x(t) = t^2 - 2t, \quad y(t) = t^3 - 3t^2 + 2t, \quad t$ real.

44. $x(t) = \cos t\,(1 - 2 \sin t), \quad y(t) = \sin t\,(1 - 2 \sin t),$ $t \in [0, \pi]$.

45. $x(t) = \sin 2\pi t, \quad y(t) = 2t - t^2, \quad t \in [0, 4]$.

46. $x(t) = t^3 - 4t, \quad y(t) = t^3 - 3t^2 - 3t^2 + 2t, \quad t$ real.

47. A *cycloid* is the curve traced out by a point P on a circle of radius a as it moves without slipping along the x-axis. See the figure. The parametric equations of the curve are

$$x(\theta) = a(\theta - \sin \theta),$$
$$y(\theta) = a(1 - \cos \theta).$$

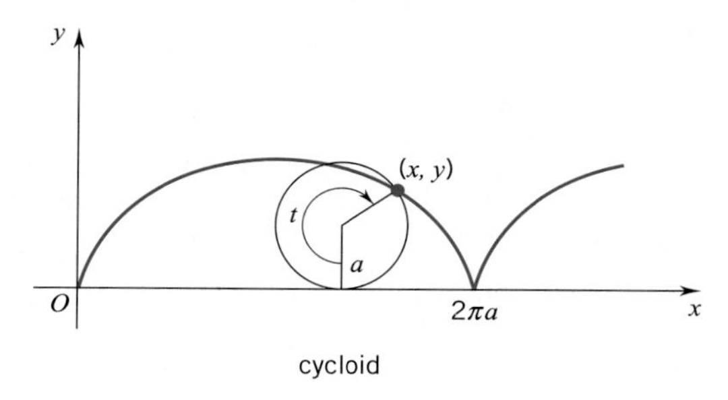

cycloid

(a) Use a graphing utility to draw the graph of the cycloid for $a = 1, 2$, and 4. What effect does the coefficient a have on the graph?
(b) The curve has a cusp at $\theta = 0, 2\pi$, and so on. Let $a = 2$. Use your graph to determine the slope of the curve for points near the cusp corresponding to $\theta = 2\pi$.
(c) Does the curve have a vertical tangent line at the cusp?

48. An *epicycloid* is the curve traced out by a point P on a circle of radius a as it rolls without slipping around the outside of a circle of radius R. See the figure. The parametric equations of the epicycloid are

$$x(\theta) = (R + a)\cos \theta - a \cos\left(\frac{R + a}{a}\right)\theta,$$

$$y(\theta) = (R + a)\sin \theta - a \sin\left(\frac{R + a}{a}\right)\theta, \quad 0 \leq \theta \leq 2\pi.$$

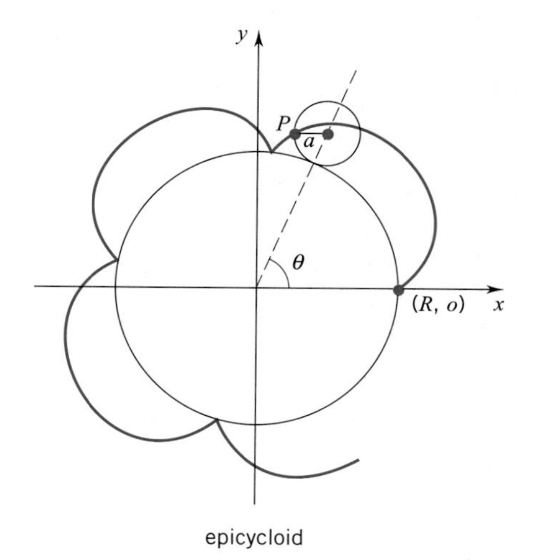

epicycloid

Use a graphing utility to draw the graph of the epicycloid for the following values of a and R:

(a) $a = 1, R = 4$.
(b) $a = 2, R = 6$.

49. A *hypocycloid* is the curve traced out by a point P on a circle of radius a as it rolls without slipping around the inside of a circle of radius R $(R > a)$. See the figure. The parametric equations of the hypocycloid are

$$x(\theta) = (R - a)\cos\theta + a\cos\left(\frac{R-a}{a}\right)\theta,$$

$$y(\theta) = (R - a)\sin\theta - a\sin\left(\frac{R-a}{a}\right)\theta, \quad 0 \le \theta \le 2\pi.$$

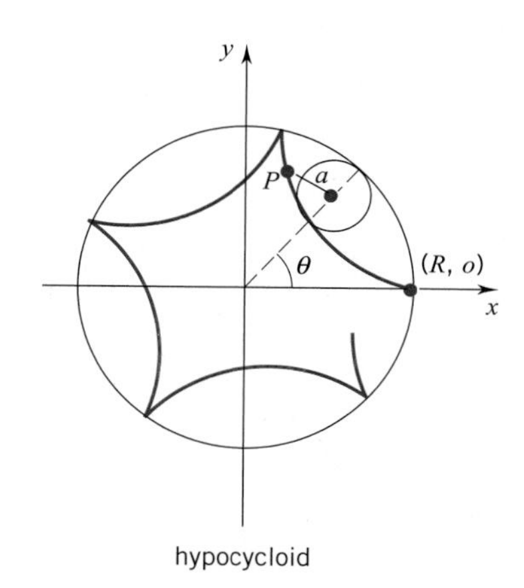

hypocycloid

Use a graphing utility to draw the graph of the hypocycloid for the following values of a and R.
(a) $a = 1, R = 4$.
(b) $a = 2, R = 6$.
(c) $a = 1, R = 8$.

50. Use a graphing utility to draw the graph of

$$x(\theta) = \cos\theta\,(a - b\sin\theta),$$

$$y(\theta) = \sin\theta\,(a - b\sin\theta), \qquad 0 \le \theta \le 2\pi,$$

for the following values of a and b:
(a) $a = 1, b = 2$
(b) $a = 2, b = 2$.
(c) $a = 2, b = 1$.
(d) In general, what can you say about the graphs when $a < b$, and $a > b$?

*SUPPLEMENT TO SECTION 9.7

PARABOLIC TRAJECTORIES

In the early part of the seventeenth century Galileo Galilei observed the motion of stones projected from the tower of Pisa and observed that their trajectory was parabolic. By simple calculus, together with some simplifying physical assumptions, we obtain results that agree with Galileo's observations.

Consider a projectile fired at angle θ, $0 < \theta < \frac{1}{2}\pi$, from a point (x_0, y_0) with initial velocity v_0 (Figure 9.7.6). The horizontal component of v_0 is $v_0 \cos\theta$, and the vertical component is $v_0 \sin\theta$ (Figure 9.7.7).

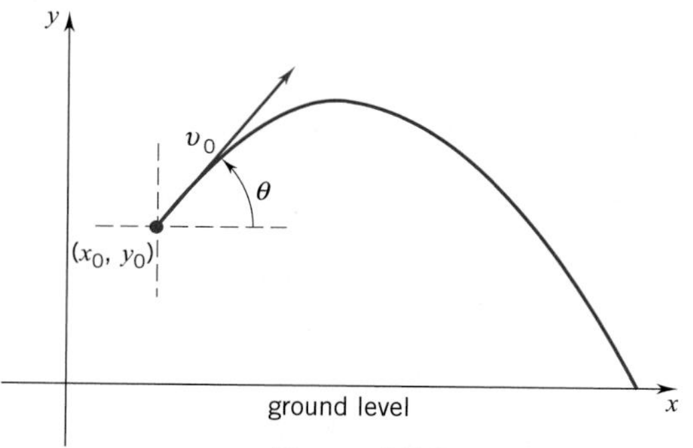

Figure 9.7.6

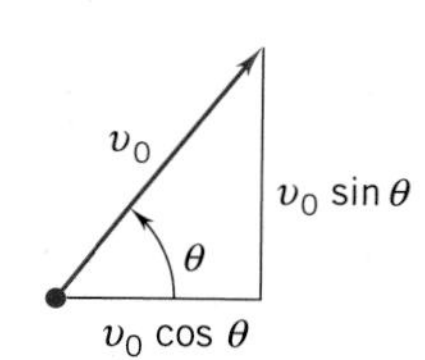

Figure 9.7.7

We neglect air resistance and the curvature of the earth. Under these circumstances there is no horizontal acceleration:

$$x''(t) = 0.$$

The only vertical acceleration is due to gravity:

$$y''(t) = -g.$$

From the first equation, we have

$$x'(t) = C$$

and, since $x'(0) = v_0 \cos \theta$,

$$x'(t) = v_0 \cos \theta.$$

Integrating again, we have

$$x(t) = (v_0 \cos \theta)t + C$$

and, since $x(0) = x_0$,

(1) $$x(t) = (v_0 \cos \theta)t + x_0.$$

The relation $y''(t) = -g$ gives

$$y'(t) = -gt + C,$$

and, since $y'(0) = v_0 \sin \theta$,

$$y'(t) = -gt + v_0 \sin \theta.$$

Integrating again, we find that

$$y(t) = -\tfrac{1}{2}gt^2 + (v_0 \sin \theta)t + C.$$

Since $y(0) = y_0$, we have

(2) $$y(t) = -\tfrac{1}{2}gt^2 + (v_0 \sin \theta)t + y_0.$$

Thus, the parametric equations of the path of the projectile are

$$x(t) = (v_0 \cos \theta)t + x_0, \quad y(t) = \tfrac{1}{2}gt^2 + (v_0 \sin \theta)t + y_0.$$

From the first equation

$$t = \frac{1}{v_0 \cos \theta}[x(t) - x_0].$$

Substitute this expression for t in the second equation and you will find that

$$y(t) = -\frac{g}{2v_0^2}\sec^2\theta\,[x(t) - x_0]^2 + \tan\theta\,[x(t) - x_0] + y_0.$$

The trajectory (the path followed by the projectile) is the curve

(9.7.8) $$y = -\frac{g}{2v_0^2}\sec^2\theta\,[x - x_0]^2 + \tan\theta\,[x - x_0] + y_0.$$

This is a quadratic in x and therefore a parabola. ❑

In Exercises 51–56, we neglect air resistance and the curvature of the earth. We measure distance in feet, time in seconds, and set $g = 32$ feet per second per second. We take O as the origin, the x-axis as ground level, and consider a projectile fired from O at an angle θ with initial velocity v_0.

51. Find an equation for the trajectory.

52. What is the maximum height attained by the projectile?

53. Find the range of the projectile.

54. How many seconds after firing does the impact take place?

55. How should θ be chosen so as to maximize the range?

56. How should θ be chosen so that the range becomes r?

■ 9.8 TANGENTS TO CURVES GIVEN PARAMETRICALLY

Let C be a curve which is parametrized by the functions

$$x = x(t), \quad y = y(t),$$

where x and y are defined on some interval I. Since C can intersect itself, it can have, at any given point:

(i) one tangent, (ii) two or more tangents, or (iii) no tangent at all.

We illustrate these possibilities in Figure 9.8.1.

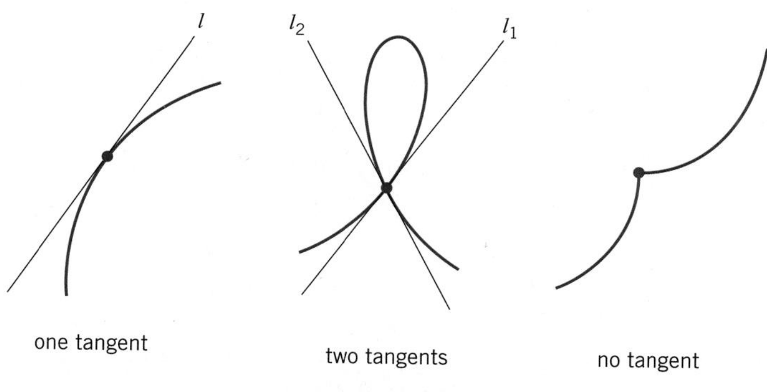

Figure 9.8.1

As before, we are assuming that $x'(t)$ and $y'(t)$ exist, at least on the interior of I. To make sure that at least one tangent line exists at each point of C, we will make the additional *assumption* that

(9.8.1)
$$[x'(t)]^2 + [y'(t)]^2 \neq 0.$$

(Without this assumption most anything can happen. See Exercises 31–35.)

Now choose a point (x_0, y_0) on the curve C and a time t_0 at which

$$x(t_0) = x_0 \quad \text{and} \quad y(t_0) = y_0.$$

We want the slope of the curve as it passes through the point (x_0, y_0) at time t_0.† To find this slope, we assume that $x'(t_0) \neq 0$. With $x'(t_0) \neq 0$, we can be sure that, for h sufficiently small, $h \neq 0$,

$$x(t_0 + h) - x(t_0) \neq 0. \qquad \text{(explain)}$$

For such h we can form the quotient

$$\frac{y(t_0 + h) - y(t_0)}{x(t_0 + h) - x(t_0)}.$$

This quotient is the slope of the secant line pictured in Figure 9.8.2. The limit of this quotient as h tends to zero is the slope of the tangent line and thus the slope of the curve. Since

$$\frac{y(t_0 + h) - y(t_0)}{x(t_0 + h) - x(t_0)} = \frac{(1/h)[y(t_0 + h) - y(t_0)]}{(1/h)[x(t_0 + h) - x(t_0)]} \rightarrow \frac{y'(t_0)}{x'(t_0)} \qquad \text{as } h \rightarrow 0,$$

you can see that

(9.8.2)
$$m = \frac{y'(t_0)}{x'(t_0)}.$$

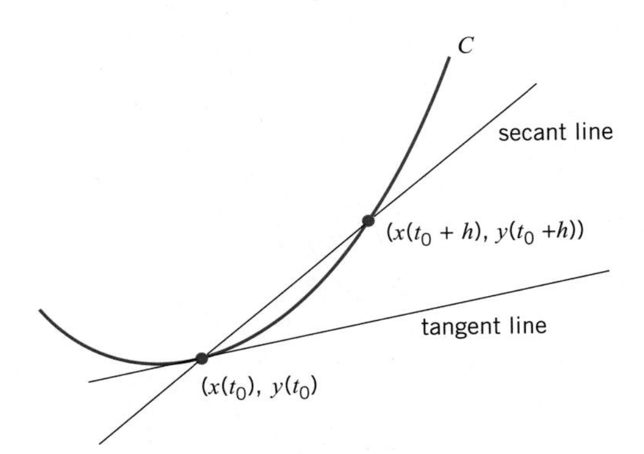

Figure 9.8.2

An equation for the tangent line can be written

$$y - y(t_0) = \frac{y'(t_0)}{x'(t_0)}[x - x(t_0)].$$

Multiplication by $x'(t_0)$ gives

$$y'(t_0)[x - x(t_0)] - x'(t_0)[y - y(t_0)] = 0,$$

† It could pass through the point (x_0, y_0) at other times also.

and thus

(9.8.3)
$$y'(t_0)[x - x_0] - x'(t_0)[y - y_0] = 0.$$

We derived this equation under the assumption that $x'(t_0) \neq 0$. If $x'(t_0) = 0$, Equation (9.8.3) still makes sense. It is simply $y'(t_0)[x - x_0] = 0$, which, since $y'(t_0) \neq 0$,† can be simplified to read

(9.8.4)
$$x = x_0.$$

In this instance the line is vertical, and we say that the curve has a *vertical tangent*.

Example 1 Find equation(s) for the tangent(s) to the curve

$$x(t) = t^3, \quad y(t) = 1 - t, \qquad t \text{ real},$$

at the point $(8, -1)$.

SOLUTION Since the curve passes through the point $(8, -1)$ only when $t = 2$, there can be only one tangent line at that point. With

$$x(t) = t^3, \quad y(t) = 1 - t,$$

we have

$$x'(t) = 3t^2, \quad y'(t) = -1,$$

and therefore

$$x'(2) = 12, \quad y'(2) = -1.$$

The tangent line has equation

$$(-1)[x - 8] - (12)[y - (-1)] = 0. \qquad (9.8.3)$$

This reduces to

$$x + 12y + 4 = 0. \quad \square$$

Example 2 Find the points of the curve

$$x(t) = 3 - 4 \sin t, \quad y(t) = 4 + 3 \cos t, \qquad t \text{ real},$$

at which there is (i) a horizontal tangent, (ii) a vertical tangent.

SOLUTION Observe first of all that the derivatives

$$x'(t) = -4 \cos t \quad \text{and} \quad y'(t) = -3 \sin t$$

are never 0 simultaneously.

† We are assuming that $[x'(t)]^2 + [y'(t)]^2$ is never 0. Since $x'(t_0) = 0$, $y'(t_0) \neq 0$.

To find the points at which there is a horizontal tangent, we set $y'(t) = 0$. This gives $t = n\pi$, $n = 0, \pm 1, \pm 2, \ldots$. Horizontal tangents occur at all points of the form $(x(n\pi), y(n\pi))$. Since

$$x(n\pi) = 3 - 4 \sin n\pi = 3 \qquad \text{and} \qquad y(n\pi) = 4 + 3 \cos n\pi = \begin{cases} 7, & n \text{ even} \\ 1, & n \text{ odd}, \end{cases}$$

there are horizontal tangents at (3, 7) and (3, 1).

To find the vertical tangents, we set $x'(t) = 0$. This gives $t = \frac{1}{2}\pi + n\pi$, $n = 0, \pm 1, \pm 2, \ldots$. Vertical tangents occur at all points of the form $(x(\frac{1}{2}\pi + n\pi), y(\frac{1}{2}\pi + n\pi))$. Since

$$x(\tfrac{1}{2}\pi + n\pi) = 3 - 4 \sin (\tfrac{1}{2}\pi + n\pi) = \begin{cases} -1, & n \text{ even} \\ 7, & n \text{ odd} \end{cases}$$

and

$$y(\tfrac{1}{2}\pi + n\pi) = 4 + 3 \cos (\tfrac{1}{2}\pi + n\pi) = 4,$$

there are vertical tangents at (−1, 4) and (7, 4). ❑

Remark The curve parametrized by the equations in Example 2 is the ellipse

$$\frac{(x-3)^2}{16} + \frac{(y-4)^2}{9} = 1$$

with center at (3, 4), major axis horizontal with length 8, and minor axis vertical with length 6. (See Figure 9.8.3). ❑

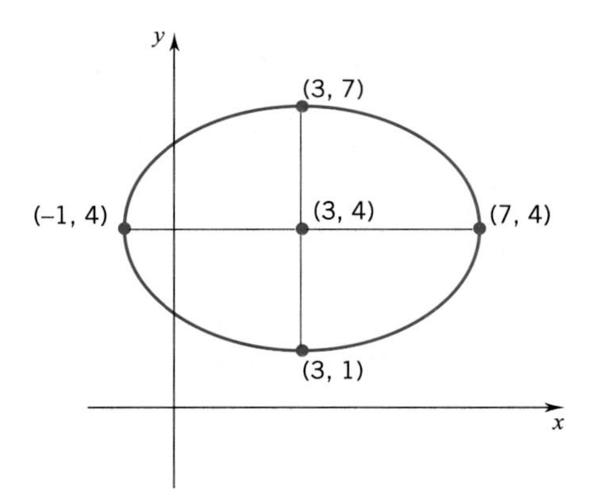

Figure 9.8.3

Example 3 The curve parametrized by

$$x(t) = \frac{1-t^2}{1+t^2}, \quad y(t) = \frac{t(1-t^2)}{1+t^2}, \qquad t \text{ real},$$

is called a *strophoid*. Its graph is shown in Figure 9.8.4. Find equations for the tangent lines to the curve at the origin. Also, find the points at which the tangent line is horizontal.

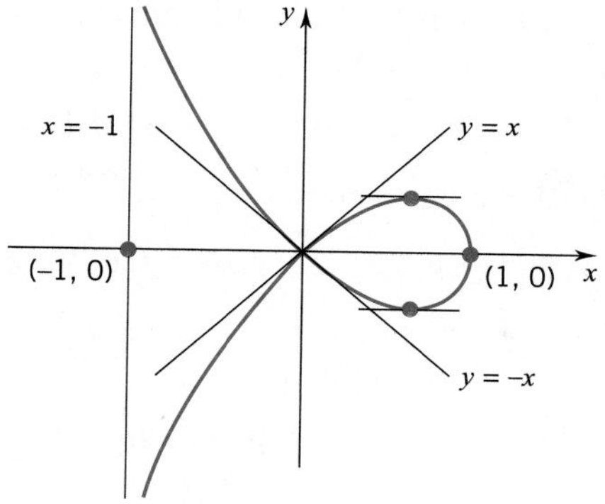

Figure 9.8.4

SOLUTION The curve passes through the origin when $t = -1$ and when $t = 1$ (verify this). Differentiating $x(t)$ and $y(t)$, we obtain

$$x'(t) = \frac{(1 + t^2)(-2t) - (1 - t^2)(2t)}{(1 + t^2)^2} = \frac{-4t}{(1 + t^2)^2},$$

and

$$y'(t) = \frac{(1 + t^2)(1 - 3t^2) - t(1 - t^2)(2t)}{(1 + t^2)^2} = \frac{1 - 4t^2 - t^4}{(1 + t^2)^2}.$$

When $t = -1$, $x'(-1) = 1$ and $y'(-1) = -1$. Thus, an equation for the tangent line is

$$-1(y - 0) - 1(x - 0) = 0 \qquad \text{or} \qquad x + y = 0. \qquad \text{(Figure 9.8.4)}$$

When $t = 1$, $x'(1) = -1$ and $y'(1) = -1$. Thus, an equation for the tangent line is

$$-1(y - 0) - (-1)(x - 0) = 0 \qquad \text{or} \qquad y - x = 0. \qquad \text{(Figure 9.8.4)}$$

To find the points at which there is a horizontal tangent, we set $y'(t) = 0$. This implies that

$$1 - 4t^2 - t^4 = 0.$$

Treating this equation as a quadratic in t^2, we have, by the quadratic formula,

$$t^2 = \frac{-4 \pm \sqrt{16 + 4}}{2} = -2 \pm \sqrt{5}.$$

Since $t^2 \geq 0$, it follows that $t^2 = \sqrt{5} - 2$, and therefore $t = \pm\sqrt{\sqrt{5} - 2}$. Now,

$$x\left(\pm\sqrt{\sqrt{5} - 2}\right) = \frac{1 - (\sqrt{5} - 2)}{1 + (\sqrt{5} - 2)} = \frac{\sqrt{5} - 1}{2} \cong 0.62$$

and

$$y\left(\pm\sqrt{\sqrt{5} - 2}\right) = \left(\pm\sqrt{\sqrt{5} - 2}\right)x = \left(\pm\sqrt{\sqrt{5} - 2}\right)\left(\frac{\sqrt{5} - 1}{2}\right) \cong \pm 0.30.$$

Thus, there is a horizontal tangent line at the points $(0.62, \pm 0.30)$ (approximately).

We can apply these ideas to a curve given in polar coordinates by an equation of the form $r = f(\theta)$. The coordinate transformations

$$x = r \cos \theta, \quad y = r \sin \theta$$

enable us to parametrize such a curve by setting

$$x(\theta) = f(\theta) \cos \theta, \quad y(\theta) = f(\theta) \sin \theta.$$

Example 4 Find the slope of the spiral

$$r = a\theta, \qquad \theta \in [0, \infty) \quad (a > 0)$$

at $\theta = \frac{1}{2}\pi$.

SOLUTION We write

$$x(\theta) = r \cos \theta = a\theta \cos \theta, \quad y(\theta) = r \sin \theta = a\theta \sin \theta.$$

Now we differentiate:

$$x'(\theta) = -a\theta \sin \theta + a \cos \theta, \quad y'(\theta) = a\theta \cos \theta + a \sin \theta.$$

Since

$$x'(\tfrac{1}{2}\pi) = -\tfrac{1}{2}\pi a \quad \text{and} \quad y'(\tfrac{1}{2}\pi) = a,$$

the slope of the curve at $\theta = \frac{1}{2}\pi$ is

$$\frac{y'(\frac{1}{2}\pi)}{x'(\frac{1}{2}\pi)} = -\frac{2}{\pi} \cong -0.64.$$

See Figure 9.8.5. ❑

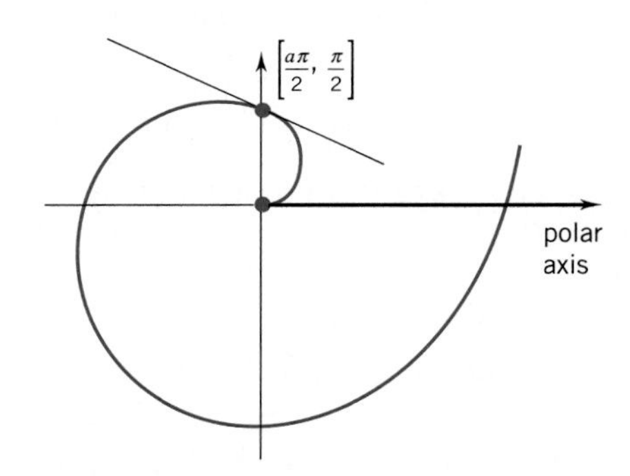

Figure 9.8.5

Example 5 Find the points of the cardioid

$$r = 1 - \cos \theta$$

at which the tangent line is vertical.

SOLUTION Since the cosine function has period 2π, we need only concern ourselves with θ in $[0, 2\pi)$. Parametrically we have

$$x(\theta) = (1 - \cos \theta) \cos \theta, \quad y(\theta) = (1 - \cos \theta) \sin \theta.$$

Differentiating and simplifying, we find that

$$x'(\theta) = (2\cos\theta - 1)\sin\theta, \quad y'(\theta) = (1 - \cos\theta)(1 + 2\cos\theta).$$

The only numbers in the interval $[0, 2\pi)$ at which x' is zero and y' is not zero are $\frac{1}{3}\pi$, π, and $\frac{5}{3}\pi$. The tangent line is vertical at

$$[\tfrac{1}{2}, \tfrac{1}{3}\pi], \quad [2, \pi], \quad [\tfrac{1}{2}, \tfrac{5}{3}\pi].$$

These points have rectangular coordinates

$(\frac{1}{4}, \frac{1}{4}\sqrt{3})$, $(-2, 0)$, $(\frac{1}{4}, -\frac{1}{4}\sqrt{3})$. ❑ (See Figure (9.8.6))

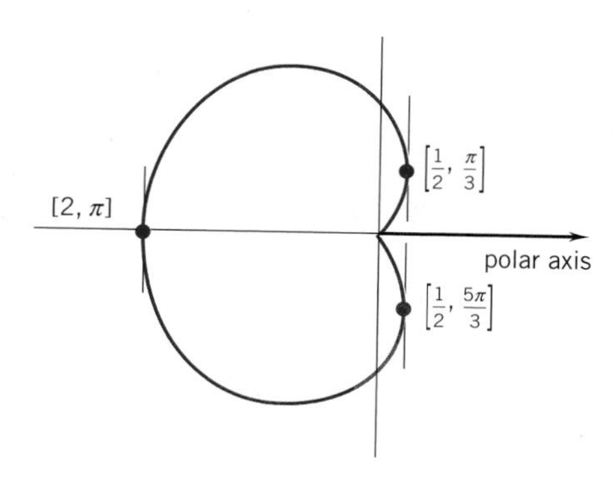

Figure 9.8 6

EXERCISES 9.8

In Exercises 1–8, find an equation in x and y for the line tangent to the curve.

1. $x(t) = t, \quad y(t) = t^3 - 1; \quad t = 1.$

2. $x(t) = t^2, \quad y(t) = t + 5; \quad t = 2.$

3. $x(t) = 2t, \quad y(t) = \cos \pi t; \quad t = 0.$

4. $x(t) = 2t - 1, \quad y(t) = t^4; \quad t = 1.$

5. $x(t) = t^2, \quad y(t) = (2 - t)^2; \quad t = \frac{1}{2}.$

6. $x(t) = 1/t, \quad y(t) = t^2 + 1; \quad t = 1.$

7. $x(t) = \cos^3 t, \quad y(t) = \sin^3 t; \quad t = \frac{1}{4}\pi.$

8. $x(t) = e^t, \quad y(t) = 3e^{-t}; \quad t = 0.$

In Exercises 9–14, find an equation in x and y for the line tangent to the polar curve.

9. $r = 4 - 2\sin\theta, \quad \theta = 0.$

10. $r = 4\cos 2\theta, \quad \theta = \frac{1}{2}\pi.$

11. $r = \dfrac{4}{5 - \cos\theta}, \quad \theta = \frac{1}{2}\pi.$

12. $r = \dfrac{5}{4 - \cos\theta}, \quad \theta = \frac{1}{6}\pi.$

13. $r = \dfrac{\sin\theta - \cos\theta}{\sin\theta + \cos\theta}, \quad \theta = 0.$

14. $r = \dfrac{\sin\theta + \cos\theta}{\sin\theta - \cos\theta}, \quad \theta = \frac{1}{2}\pi.$

In Exercises 15–18, parametrize the curve by a pair of differentiable functions

$$x = x(t), \quad y = y(t) \qquad \text{with} \quad [x'(t)]^2 + [y'(t)]^2 \neq 0.$$

Sketch the curve and determine the tangent line at the origin by the method of this section.

15. $y = x^3.$

16. $x = y^3.$

17. $y^5 = x^3.$

18. $y^3 = x^5.$

In Exercises 19–26, find the points (x, y) at which the curve has: (a) a horizontal tangent; (b) a vertical tangent. Then sketch the curve.

19. $x(t) = 3t - t^3, \quad y(t) = t + 1.$

20. $x(t) = t^2 - 2t, \quad y(t) = t^3 - 12t.$

21. $x(t) = 3 - 4\sin t, \quad y(t) = 4 + 3\cos t.$

22. $x(t) = \sin 2t, \quad y(t) = \sin t.$

23. $x(t) = t^2 - 2t, \quad y(t) = t^3 - 3t^2 + 2t.$

24. $x(t) = 2 - 5\cos t, \quad y(t) = 3 + \sin t.$

25. $x(t) = \cos t, \quad y(t) = \sin 2t.$

26. $x(t) = 3 + 2\sin t, \quad y(t) = 2 + 5\sin t.$

27. Find the tangent(s) to the curve

$$x(t) = -t + 2\cos\tfrac{1}{4}\pi t, \quad y(t) = t^4 - 4t^2$$

at the point (2, 0).

28. Find the tangent(s) to the curve.

$$x(t) = t^3 - t, \quad y(t) = t\sin\tfrac{1}{2}\pi t$$

at the point (0, 1).

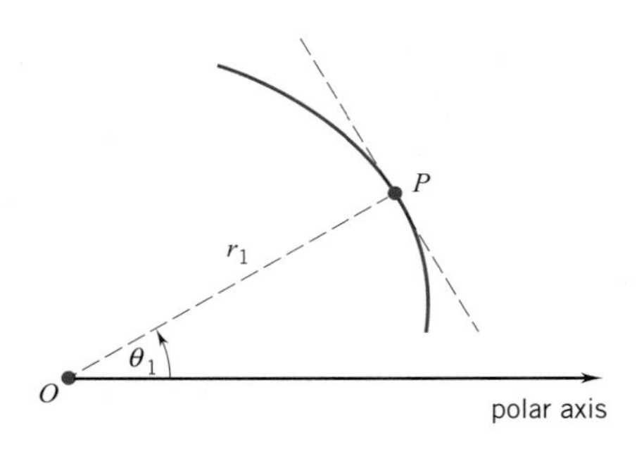

29. Let $P = [r_1, \theta_1]$ be a point on a polar curve $r = f(\theta)$ as in the figure. Show that, if $f'(\theta_1) = 0$ but $f(\theta_1) \neq 0$, then the tangent line at P is perpendicular to the line segment $\overline{OP}$.

30. If $0 < a < 1$, the polar curve $r = a - \cos\theta$ is a limaçon with an inner loop. Choose a so that the curve will intersect itself at the pole in a right angle.

In Exercises 31–35, verify that $x'(0) = y'(0) = 0$ and that the given description holds at the point where $t = 0$. Sketch the graph.

31. $x(t) = t^3, \quad y(t) = t^2;$ cusp.

32. $x(t) = t^3, \quad y(t) = t^5;$ horizontal tangent.

33. $x(t) = t^5, \quad y(t) = t^3;$ vertical tangent.

34. $x(t) = t^3 - 1, \quad y(t) = 2t^3;$ tangent with slope 2.

35. $x(t) = t^2, \quad y(t) = t^2 + 1;$ no tangent line.

36. Suppose that $x = x(t)$, $y = y(t)$ are twice differentiable functions that parametrize a curve. Take a point on the curve at which $x'(t) \neq 0$ and d^2y/dx^2 exists. Show that

(9.8.5)
$$\frac{d^2y}{dx^2} = \frac{x'(t)y''(t) - y'(t)x''(t)}{[x'(t)]^3}.$$

In Exercises 37–40, calculate d^2y/dx^2 at the indicated point without eliminating the parameter.

37. $x(t) = \cos t, \quad y(t) = \sin t; \quad t = \tfrac{1}{6}\pi.$

38. $x(t) = t^3, \quad y(t) = t - 2; \quad t = 1.$

39. $x(t) = e^t, \quad y(t) = e^{-t}; \quad t = 0.$

40. $x(t) = \sin^2 t, \quad y(t) = \cos t; \quad t = \tfrac{1}{4}\pi.$

■ 9.9 ARC LENGTH AND SPEED

Arc Length

We come now to the notion of arc length. In Figure 9.9.1 we have sketched a curve C which we assume is parametrized by a pair of *continuously differentiable* functions†

$$x = x(t), \quad y = y(t), \qquad t \in [a, b].$$

We want to determine the length of C.

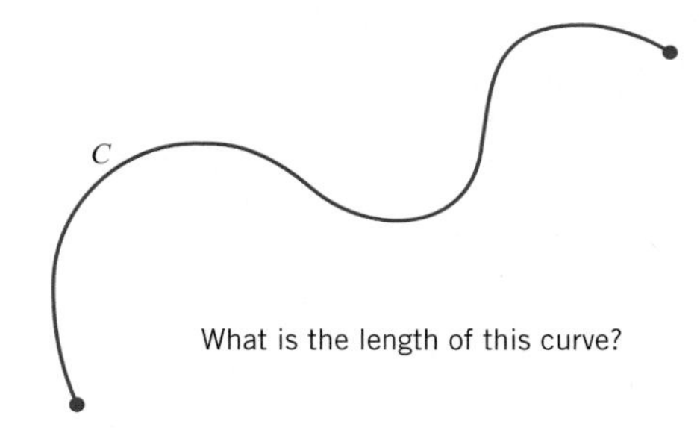

Figure 9.9.1

† By this we mean functions that have continuous first derivatives.

Here our experience in Chapter 5 can be used as a model. To decide what should be meant by the area of a region Ω, we approximated Ω by the union of a finite number of rectangles. To decide what should be meant by the length of C, we approximate C by the union of a finite number of line segments.

Each number t in $[a, b]$ gives rise to a point $P = P(x(t), y(t))$ that lies on C. By choosing a finite number of points in $[a, b]$,

$$a = t_0 < t_1 < \cdots < t_{i-1} < t_i < \cdots < t_{n-1} < t_n = b,$$

we obtain a finite number of points on C,

$$P_0, P_1, \ldots, P_{i-1}, P_i, \ldots, P_{n-1}, P_n.$$

We join these points consecutively by line segments and call the resulting path,

$$\gamma = \overline{P_0P_1} \cup \cdots \cup \overline{P_{i-1}P_i} \cup \cdots \cup \overline{P_{n-1}P_n},$$

a *polygonal path* inscribed in C. (See Figure 9.9.2.)

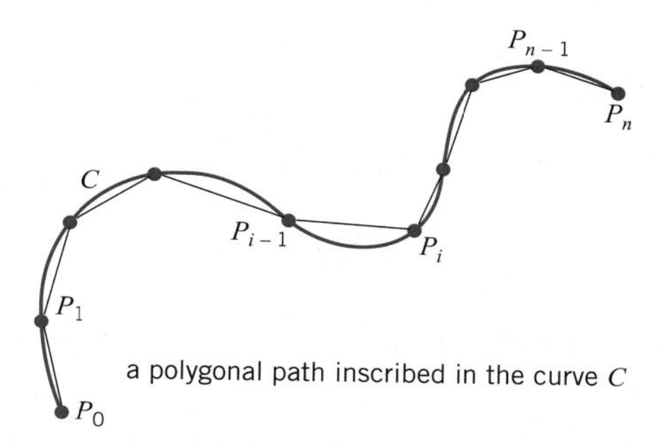

a polygonal path inscribed in the curve C

Figure 9.9.2

The length of such a polygonal path is the sum of the distances between consecutive vertices:

$$\text{length of } \gamma = L(\gamma) = d(P_0, P_1) + \cdots + d(P_{i-1}, P_i) + \cdots + d(P_{n-1}, P_n).$$

Now, the ith line segment $\overline{P_{i-1}P_i}$ has length

$$d(P_{i-1}, P_i) = \sqrt{[x(t_i) - x(t_{i-1})]^2 + [y(t_i) - y(t_{i-1})]^2}$$
$$= \sqrt{\left[\frac{x(t_i) - x(t_{i-1})}{t_i - t_{i-1}}\right]^2 + \left[\frac{y(t_i) - y(t_{i-1})}{t_i - t_{i-1}}\right]^2}\,(t_i - t_{i-1}).$$

By the mean-value theorem, there exist points t_i^* and t_i^{**}, both in the interval (t_{i-1}, t_i), such that

$$\frac{x(t_i) - x(t_{i-1})}{t_i - t_{i-1}} = x'(t_i^*) \quad \text{and} \quad \frac{y(t_i) - y(t_{i-1})}{t_i - t_{i-1}} = y'(t_i^{**}).$$

Letting $\Delta t_i = t_i - t_{i-1}$, we have

$$d(P_{i-1}, P_i) = \sqrt{[x'(t_i^*)]^2 + [y'(t_i^{**})]^2}\,\Delta t_i.$$

Thus,

$$L(\gamma) = \sqrt{[x'(t_1^*)]^2 + [y'(t_1^{**})]^2}\,\Delta t_1 + \sqrt{[x'(t_2^*)]^2 + y'(t_2^{**})]^2}\,\Delta t_2 + \cdots$$
$$+ \sqrt{[x'(t_n^*)]^2 + [y'(t_n^{**})]^2}\,\Delta t_n.$$

This is not a Riemann sum because it will not be true, in general, that $t_i^* = t_i^{**}$ for each i. However, this sum is "close" to a Riemann sum, close enough so that, as $\max \Delta t_i \to 0$, $i = 1, 2, \ldots, n$, it approaches the integral

$$\int_a^b \sqrt{[x'(t)]^2 + [y'(t)]^2}\, dt.$$

Arc Length Formulas

Based on the intuitive argument just given, we have a formula for calculating the length of a parametrized curve. If C is parametrized by a pair of continuously differentiable functions

$$x = x(t), \quad y = y(t), \qquad t \in [a, b],$$

then

(9.9.1)

$$\text{the length of } C = L(C) = \int_a^b \sqrt{[x'(t)]^2 + [y'(t)]^2}\, dt.$$

Obviously, this is not something to be taken on faith. It has to be proved. We will do so, but not until Chapter 13.

There is an implicit assumption underlying the arc length formula that we need to illustrate before going on. The functions

$$x(t) = a \cos t, \quad y = a \sin t, \qquad t \in [0, 2\pi],$$

parametrize the circle C of radius a centered at the origin. Since

$$x'(t) = -a \sin t, \quad y'(t) = a \cos t,$$

Formula (9.9.1) gives

$$L(C) = \int_0^{2\pi} \sqrt{a^2 \sin^2 t + a^2 \cos^2 t}\, dt = \int_0^{2\pi} a\, dt = 2\pi a.$$

This is, of course, correct; a circle of radius a has circumference $2\pi a$.

But notice that C can also be parametrized by the functions

$$x(t) = a \cos 2t, \quad y(t) = a \sin 2t, \qquad t \in [0, 2\pi].$$

Here we have

$$x'(t) = -2a \sin 2t, \quad y'(t) = 2a \cos 2t$$

and

$$L(C) = \int_0^{2\pi} \sqrt{4a^2 \sin^2 2t + 4a^2 \cos^2 2t}\, dt = \int_0^{2\pi} 2a\, dt = 4\pi a,$$

which is obviously not the length of C. The difference between the two cases is that with the first parametrization, the point $(x(t), y(t))$ traced out the circle exactly once as t increased from 0 to 2π; while in the second case the point $(x(t), y(t))$ traced out the curve twice as t increased from 0 to 2π.

In finding the arc length of a given curve C we will always use a parametrization that traverses the curve exactly once.

Suppose now that C is the graph of a continuously differentiable function

$$y = f(x), \qquad x \in [a, b].$$

Then we can parametrize C by setting

$$x(t) = t, \quad y(t) = f(t) \qquad t \in [a, b].$$

Since

$$x'(t) = 1 \qquad \text{and} \qquad y'(t) = f'(t),$$

Formula (9.9.1) gives

$$L(C) = \int_a^b \sqrt{1 + [f'(t)]^2}\, dt.$$

Replacing t by x we can write

(9.9.2)

$$\text{the length of the graph of } f = \int_a^b \sqrt{1 + [f'(x)]^2}\, dx.$$

A direct derivation of this formula is outlined in Exercise 42.

Example 1 If

$$f(x) = \tfrac{1}{6}x^3 + \tfrac{1}{2}x^{-1},$$

then

$$f'(x) = \tfrac{1}{2}x^2 - \tfrac{1}{2}x^{-2}.$$

Therefore

$$1 + [f'(x)]^2 = 1 + (\tfrac{1}{4}x^4 - \tfrac{1}{2} + \tfrac{1}{4}x^{-4}) = \tfrac{1}{4}x^4 + \tfrac{1}{2} + \tfrac{1}{4}x^{-4} = (\tfrac{1}{2}x^2 + \tfrac{1}{2}x^{-2})^2.$$

The length of the graph from $x = 1$ to $x = 3$ is

$$\int_1^3 \sqrt{1 + [f'(x)]^2}\, dx = \int_1^3 (\tfrac{1}{2}x^2 + \tfrac{1}{2}x^{-2})\, dx = \left[\tfrac{1}{6}x^3 - \tfrac{1}{2}x^{-1}\right]_1^3 = \tfrac{14}{3}. \quad \square$$

Example 2 The graph of the function

$$f(x) = x^2, \qquad x \in [0, 1]$$

is a parabolic arc. The length of this arc is given by

$$\int_0^1 \sqrt{1 + [f'(x)]^2}\, dx = \int_0^1 \sqrt{1 + 4x^2}\, dx = 2\int_0^1 \sqrt{(\tfrac{1}{2})^2 + x^2)}\, dx$$

$$\underset{\text{by (8.5.1)}}{=} \left[x\sqrt{(\tfrac{1}{2})^2 + x^2} + (\tfrac{1}{2})^2 \ln\left(x + \sqrt{(\tfrac{1}{2})^2 + x^2}\right) \right]_0^1$$

$$= \tfrac{1}{2}\sqrt{5} + \tfrac{1}{4}\ln(2 + \sqrt{5}) \cong 1.48. \quad \square$$

Suppose now that C is the graph of a polar function

$$r = \rho(\theta), \qquad \alpha \leq \theta \leq \beta,$$

where ρ is continuously differentiable.

We can parametrize C by setting

$$x(\theta) = \rho(\theta)\cos\theta, \quad y(\theta) = \rho(\theta)\sin\theta \qquad \theta \in [\alpha, \beta].$$

A straightforward calculation that we leave to you shows that

$$[x'(\theta)]^2 + [y'(\theta)]^2 = [\rho(\theta)]^2 + [\rho'(\theta)]^2.$$

The arc length formula now reads

(9.9.3)

$$L(C) = \int_\alpha^\beta \sqrt{[\rho(\theta)]^2 + [\rho'(\theta)]^2}\, d\theta.$$

Example 3 For fixed $a > 0$ the equation $r = a$ represents a circle of radius a. Here

$$\rho(\theta) = a \qquad \text{and} \qquad \rho'(\theta) = 0.$$

The circumference of this circle is

$$\int_0^{2\pi} \sqrt{[\rho(\theta)]^2 + [\rho'(\theta)]^2}\, d\theta = \int_0^{2\pi} \sqrt{a^2 + 0^2}\, d\theta = \int_0^{2\pi} a\, d\theta = 2\pi a. \quad \square$$

Example 4 In the case of the cardioid $r = a(1 - \cos\theta)$, $a > 0$, we have

$$\rho(\theta) = a(1 - \cos\theta), \qquad \rho'(\theta) = a\sin\theta.$$

Here

$$[\rho(\theta)]^2 + [\rho'(\theta)]^2 = a^2(1 - 2\cos\theta + \cos^2\theta) + a^2\sin^2\theta = 2a^2(1 - \cos\theta).$$

The identity

$$\tfrac{1}{2}(1 - \cos\theta) = \sin^2\tfrac{1}{2}\theta$$

gives

$$[\rho(\theta)]^2 + [\rho'(\theta)]^2 = 4a^2\sin^2\tfrac{1}{2}\theta.$$

The length of the cardioid is $8a$:

$$\int_0^{2\pi} \sqrt{[\rho(\theta)]^2 + [\rho'(\theta)]^2}\, d\theta = \int_0^{2\pi} 2a\sin\tfrac{1}{2}\theta\, d\theta = 4a\left[-\cos\tfrac{1}{2}\theta\right]_0^{2\pi} = 8a. \quad \square$$

The Geometric Significance of dx/ds and dy/ds

Figure 9.9.3 shows the graph of a function $y = f(x)$ which we assume to be continuously differentiable on (a, b). At the point (x, y) the tangent line has an inclination marked α_x.

The length of the graph from a to x can be written

$$s(x) = \int_a^x \sqrt{1 + [f'(t)]^2}\, dt.$$

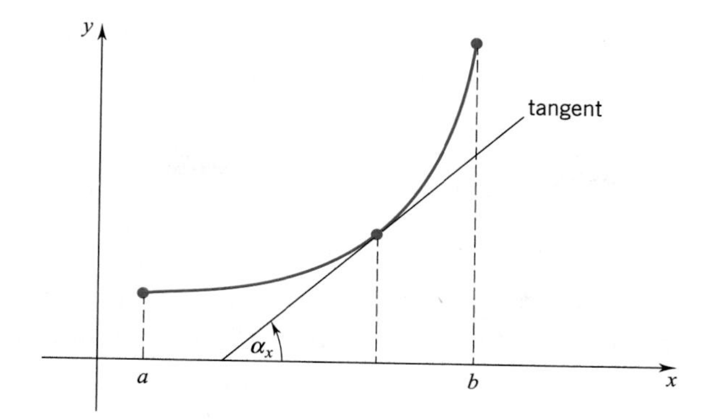

Figure 9.9.3

Differentiation with respect to x gives $s'(x) = \sqrt{1 + [f'(x)]^2}$ (Theorem 5.3.5). Using the Leibniz notation we have

$$\frac{ds}{dx} = \sqrt{1 + \left(\frac{dy}{dx}\right)^2} = \sqrt{1 + \tan^2 \alpha_x} = \sec \alpha_x.$$

($\sec \alpha_x > 0$ for $\alpha_x \in (-\frac{1}{2}\pi, \frac{1}{2}\pi)$)

By (7.1.10)

$$\frac{dx}{ds} = \frac{1}{\sec \alpha_x} = \cos \alpha_x.$$

To find dy/ds we note that

$$\tan \alpha_x = \frac{dy}{dx} = \frac{dy}{ds}\frac{ds}{dx} = \frac{dy}{ds}\sec \alpha_x.$$

(chain rule)

Multiplication by $\cos \alpha_x$ gives

$$\frac{dy}{ds} = \sin \alpha_x.$$

For the record

(9.9.4) $\frac{dx}{ds} = \cos \alpha_x$ and $\frac{dy}{ds} = \sin \alpha_x$ where α_x is the inclination of the tangent line at the point (x, y).

Speed Along a Plane Curve

So far we have talked about speed only in connection with straight-line motion. How can we calculate the speed of an object that moves along a curve? Imagine an object moving along some curved path. Suppose that $(x(t), y(t))$ gives the position of the object at time t. The distance traveled by the object from time zero to any later time t is simply the length of the path up to time t:

$$s(t) = \int_0^t \sqrt{[x'(u)]^2 + [y'(u)]^2}\, du.$$

The time rate of change of this distance is what we call the *speed* of the object. Denoting the speed of the object at time t by $\nu(t)$ we have

(9.9.5)
$$\nu(t) = s'(t) = \sqrt{[x'(t)]^2 + [y'(t)]^2}.$$

Example 5 The path of the projectile in Exercises 51–56 in the Supplement to Section 9.7 is given in terms of the time parameter t by the following equations:

$$x(t) = (v_0 \cos \theta)t, \qquad y(t) = -16t^2 + (v_0 \sin \theta)t.$$

From those exercises we know that the projectile impacts at time $t = \frac{1}{16}v_0 \sin \theta$. Calculate the speed at impact.

SOLUTION Since

$$x'(t) = v_0 \cos \theta \quad \text{and} \quad y'(t) = -32t + v_0 \sin \theta,$$

we have

$$\nu(t) = \sqrt{v_0^2 \cos^2\theta + (-32t + v_0 \sin \theta)^2}.$$

The speed at impact is therefore

$$\sqrt{v_0^2 \cos^2\theta + (-2v_0 \sin \theta + v_0 \sin \theta)^2} = \sqrt{v_0^2 \cos^2\theta + v_0^2 \sin^2\theta} = |v_0|,$$

which is exactly the speed with which the projectile was fired.† ❑

In the Leibniz notation the equation for speed reads

(9.9.6)
$$\nu = \frac{ds}{dt} = \sqrt{\left(\frac{dx}{dt}\right)^2 + \left(\frac{dy}{dt}\right)^2}.$$

If we know the speed of an object and we know its mass, then we can calculate its kinetic energy.

Example 6 A particle of mass m slides down a frictionless curve (see Figure 9.9.4) from a point (x_0, y_0) to a point (x_1, y_1) under the force of gravity. As discussed in Exercises 4.2, the particle has two forms of energy during the motion: gravitational potential energy mgy and kinetic energy $\frac{1}{2}m\nu^2$. Show that the sum of these two quantities remains constant:

$$\overset{\text{GPE}}{mgy} + \overset{\text{KE}}{\tfrac{1}{2}m\nu^2} = C.$$

SOLUTION The particle is subjected to a vertical force $-mg$ (a downward force of magnitude mg). Since the particle is constrained to remain on the curve, the effective force on the particle is tangential. The tangential component of the vertical force is

† This can be obtained from energy considerations. See Exercise 43.

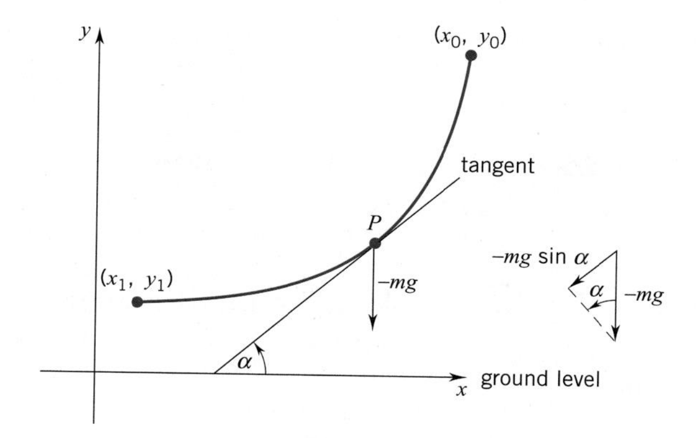

Figure 9.9.4

$-mg \sin \alpha$. (See Figure 9.9.4.) The speed of the particle is ds/dt and the tangential acceleration is d^2s/dt^2. (It is as if the particle were moving along the tangent line.) Therefore, by Newton's law $F = ma$, we have

$$m\frac{d^2s}{dt^2} = -mg \sin \alpha \underset{\text{by (9.9.4)}}{=} -mg\frac{dy}{ds}.$$

We can therefore write

$$mg\frac{dy}{ds} + m\frac{d^2s}{dt^2} = 0,$$

$$mg\frac{dy}{ds}\frac{ds}{dt} + m\frac{ds}{dt}\frac{d^2s}{dt^2} = 0, \qquad \left(\text{multiplied by } \frac{ds}{dt}\right)$$

$$mg\frac{dy}{dt} + mv\frac{dv}{dt} = 0. \qquad \text{(chain rule)}$$

Integrating with respect to t we have

$$mgy + \tfrac{1}{2}mv^2 = C$$

as asserted. ❑

EXERCISES 9.9

In Exercises 1–18, find the length of the graph and compare it to the straight-line distance between the endpoints of the graph.

1. $f(x) = 2x + 3, \quad x \in [0, 1]$.
2. $f(x) = 3x + 2, \quad x \in [0, 1]$.
3. $f(x) = (x - \frac{4}{9})^{3/2}, \quad x \in [1, 4]$.
4. $f(x) = x^{3/2}, \quad x \in [0, 44]$.
5. $f(x) = \frac{1}{3}\sqrt{x}\,(x - 3), \quad x \in [0, 3]$.
6. $f(x) = \frac{2}{3}(x - 1)^{3/2}, \quad x \in [1, 2]$.
7. $f(x) = \frac{1}{3}(x^2 + 2)^{3/2}, \quad x \in [0, 1]$.
8. $f(x) = \frac{1}{3}(x^2 - 2)^{3/2}, \quad x \in [2, 4]$.
9. $f(x) = \frac{1}{4}x^2 - \frac{1}{2}\ln x, \quad x \in [1, 5]$.
10. $f(x) = \frac{1}{8}x^2 - \ln x, \quad x \in [1, 4]$.
11. $f(x) = \frac{3}{8}x^{4/3} - \frac{3}{4}x^{2/3}, \quad x \in [1, 8]$.
12. $f(x) = \frac{1}{10}x^5 + \frac{1}{6}x^{-3}, \quad x \in [1, 2]$.
13. $f(x) = \ln(\sec x), \quad x \in [0, \frac{1}{4}\pi]$.
14. $f(x) = \frac{1}{2}x^2, \quad x \in [0, 1]$.
15. $f(x) = \frac{1}{2}x\sqrt{x^2 - 1} - \frac{1}{2}\ln(x + \sqrt{x^2 - 1}), \quad x \in [1, 2]$.
16. $f(x) = \cosh x, \quad x \in [0, \ln 2]$.
17. $f(x) = \frac{1}{2}x\sqrt{3 - x^2} + \frac{3}{2}\sin^{-1}(\frac{1}{3}\sqrt{3}x), \quad x \in [0, 1]$.
18. $f(x) = \ln(\sin x), \quad x \in [\frac{1}{6}\pi, \frac{1}{2}\pi]$.

In Exercises 19–24, the equations give the position of a particle at each time t during the time interval specified. Find the initial speed of the particle, the terminal speed, and the distance traveled.

19. $x(t) = t^2, \quad y(t) = 2t$ from $t = 0$ to $t = \sqrt{3}$.

20. $x(t) = t - 1, \quad y(t) = \frac{1}{2}t^2$ from $t = 0$ to $t = 1$.

21. $x(t) = t^2, \quad y(t) = t^3$ from $t = 0$ to $t = 1$.

22. $x(t) = a \cos^3 t, \quad y(t) = a \sin^3 t$ from $t = 0$ to $t = \frac{1}{2}\pi$.

23. $x(t) = e^t \sin t, \quad y(t) = e^t \cos t$ from $t = 0$ to $t = \pi$.

24. $x(t) = \cos t + t \sin t, \quad y(t) = \sin t - t \cos t$ from $t = 0$ to $t = \pi$.

25. The curve defined parametrically by

$$x(\theta) = a(\theta - \sin \theta), \qquad y(\theta) = a(1 - \cos \theta), \qquad a > 0,$$

for $0 \leq \theta \leq 2\pi$, is one arch of a cycloid (see Exercise 47, Section 9.7). Find the length of this curve.

26. Find the length of the epicycloid (see Exercise 48, Section 9.7)

$$x(\theta) = 2a \cos \theta - a \cos 2\theta,$$

$$y(\theta) = 2a \sin \theta - a \sin 2\theta,$$

where $a > 0$ and $0 \leq \theta \leq 2\pi$.

27. (a) Find the length of the hypocycloid (see Exercise 49, Section 9.7)

$$x(\theta) = 3a \cos \theta + a \cos 3\theta,$$

$$y(\theta) = 3a \sin \theta - a \sin 3\theta,$$

where $a > 0$ and $0 \leq \theta \leq 2\pi$.

(b) Show that this hypocycloid can also be represented by the parametric equations

$$x(\theta) = 4a \cos^3 \theta, \quad y(\theta) = 4a \sin^3 \theta, \quad 0 \leq \theta \leq 2\pi.$$

28. The curve defined parametrically by

$$x(\theta) = \theta \cos \theta, \qquad y(\theta) = \theta \sin \theta,$$

is called an *Archimedean spiral*. Find the length of this spiral for $0 \leq \theta \leq 2\pi$.

In Exercises 29–36, find the length of the polar curve.

29. $r = 1$ from $\theta = 0$ to $\theta = 2\pi$.

30. $r = 3$ from $\theta = 0$ to $\theta = \pi$.

31. $r = e^\theta$ from $\theta = 0$ to $\theta = 4\pi$. (logarithmic spiral)

32. $r = ae^\theta, a > 0$, from $\theta = -2\pi$ to $\theta = 2\pi$.

33. $r = e^{2\theta}$ from $\theta = 0$ to $\theta = 2\pi$.

34. $r = 1 + \cos \theta$ from $\theta = 0$ to $\theta = 2\pi$.

35. $r = 1 - \cos \theta$ from $\theta = 0$ to $\theta = \frac{1}{2}\pi$.

36. $r = 2a \sec \theta, a > 0$, from $\theta = 0$ to $\theta = \frac{1}{4}\pi$.

37. At time t a particle has position

$$x(t) = 1 + \tan^{-1} t, \quad y(t) = 1 - \ln \sqrt{1 + t^2}$$

Find the total distance traveled from time $t = 0$ to time $t = 1$. Give the initial speed and the terminal speed.

38. At time t a particle has position

$$x(t) = 1 - \cos t, \quad y(t) = t - \sin t.$$

Find the total distance traveled from time $t = 0$ to time $t = 2\pi$. Give the initial speed and the terminal speed.

39. Find c given that the length of the curve $y = \ln x$ from $x = 1$ to $x = e$ equals the length of the curve $y = e^x$ from $x = 0$ to $x = c$.

40. Find the length of the curve $y = x^{2/3}, x \in [1, 8]$.
HINT: Work with the mirror image $y = x^{3/2}, x \in [1, 4]$.

41. Show that the function $f(x) = \cosh x$ has the following property: for every interval $[a, b]$ the length of the graph equals the area under the graph.

42. The figure shows the graph of a continuously differentiable function f from $x = a$ to $x = b$ together with a polygonal approximation. Show that the length of this polygonal approximation can be written as the following Riemann sum:

$$\sqrt{1 + [f'(x_1^*)]^2}\, \Delta x_1 + \sqrt{1 + [f'(x_2^*)]^2}\, \Delta x_2 + \cdots + \sqrt{1 + [f'(x_n^*)]^2}\, \Delta x_n.$$

As $\|P\| = \max \Delta x_i$ tends to 0, such Riemann sums tend to

$$\int_a^b \sqrt{1 + [f'(x)]^2}\, dx.$$

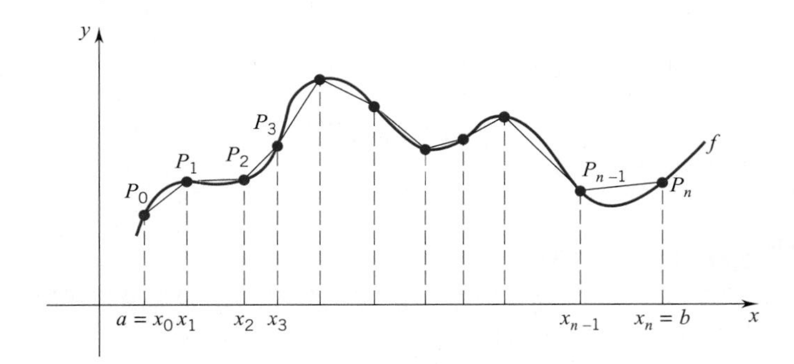

43. Rework Example 5, this time from energy considerations. (a) Verify that GPE + KE remains constant. (b) Then derive the speed at impact from an energy equation.

44. The path of the projectile discussed under the heading "Parabolic Trajectories" in Section 9.7 is given in terms of the time parameter t by the following equations:

$$x(t) = (v_0 \cos \theta)t + x_0, \; y(t) = -\tfrac{1}{2}gt^2 + (v_0 \sin \theta)t + y_0.$$

The projectile impacts at ground level, $y = 0$. Find the speed at impact.

45. Suppose that f is continuously differentiable from $x = a$ to $x = b$. Show that the

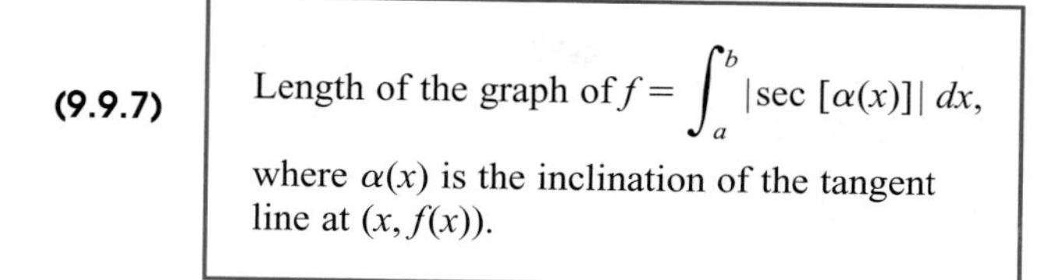

(9.9.7) Length of the graph of $f = \int_a^b |\sec[\alpha(x)]|\, dx$,

where $\alpha(x)$ is the inclination of the tangent line at $(x, f(x))$.

46. Show that a homogeneous, flexible, inelastic rope hanging from two fixed points assumes the shape of a *catenary*:

$$f(x) = a \cosh\left(\frac{x}{a}\right) = \frac{a}{2}(e^{x/a} + e^{-x/a}), \quad a > 0.$$

HINT: Refer to the figure below. The part of the rope that corresponds to the interval $[0, x]$ is subject to the following forces:

(1) its weight, which is proportional to its length;
(2) a horizontal pull at 0, $p(0)$;
(3) a tangential pull at x, $p(x)$.

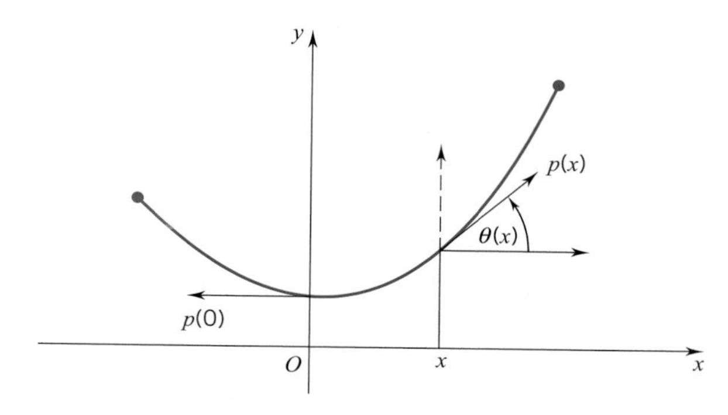

47. (a) Use a graphing utility to draw the graph of the curve defined parametrically by

$$x(t) = t^2, \qquad y(t) = t^3 - t, \qquad -\infty < t < \infty.$$

(b) Your graph in part (a) should show that the curve has a loop. Use Simpson's rule with $n = 3$ to find the approximate length of the loop. Round your answer to four decimal places.

48. Sketch the graph of the limaçon $r = 2 + \sin\theta$. Use Simpson's rule with $n = 4$ to estimate the length of this curve. Round your answer to four decimal places.

49. Elliptic integrals.
(a) Let $a > b > 0$. Show that the length of the ellipse

$$x(t) = a\cos t, \quad y(t) = b\sin t, \quad 0 \le t \le 2\pi,$$

is given by

$$L = 4a\int_0^{\pi/2} \sqrt{1 - e\cos^2 t}\, dt,$$

where $e = \sqrt{a^2 - b^2}/a$ is the eccentricity. The integrand does not have an elementary antiderivative.
(b) Let $a = 5$ and $b = 4$, and approximate the length of the ellipse using Simpson's rule with $n = 4$. Round your answer to four decimal places.

50. The curve defined parametrically by

$$x(t) = \frac{3t}{t^3 + 1}, \qquad y(t) = \frac{3t^2}{t^3 + 1}, \qquad t \neq -1,$$

is called the *folium of Descartes*.
(a) Use a graphing utility to draw the graph of this curve.
(b) Your graph in part (a) should show that the curve has a loop in the first quadrant. Use Simpson's rule with $n = 3$ to estimate the length of the loop. Round your answer to four decimal places. HINT: Use symmetry.

■ 9.10 THE AREA OF A SURFACE OF REVOLUTION; THE CENTROID OF A CURVE; PAPPUS'S THEOREM ON SURFACE AREA

The Area of a Surface of Revolution

In Figure 9.10.1 you can see the frustum of a cone; one radius is marked r, the other R, and the slant height is marked s. An interesting elementary calculation that we leave to you shows that the area of this slanted surface is given by the formula

(9.10.1) $$A = \pi(r + R)s. \qquad \text{(Exercise 21)}$$

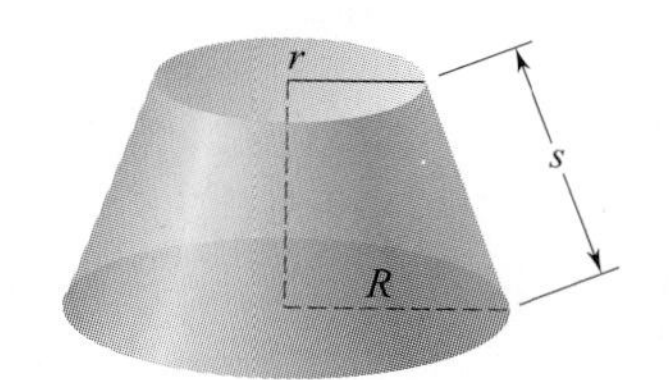

Figure 9.10.1

This little formula forms the basis for all that follows.

Let C be a curve in the upper half-plane (Figure 9.10.2). The curve can meet the x-axis, but only at a finite number of points. We will assume that C is parametrized by a pair of continuously differentiable functions

$$x = x(t), \quad y = y(t), \qquad t \in [c, d].$$

Furthermore, we will assume that C is *simple*: no two values of t between c and d give rise to the same point of C, that is, the curve does not intersect itself.

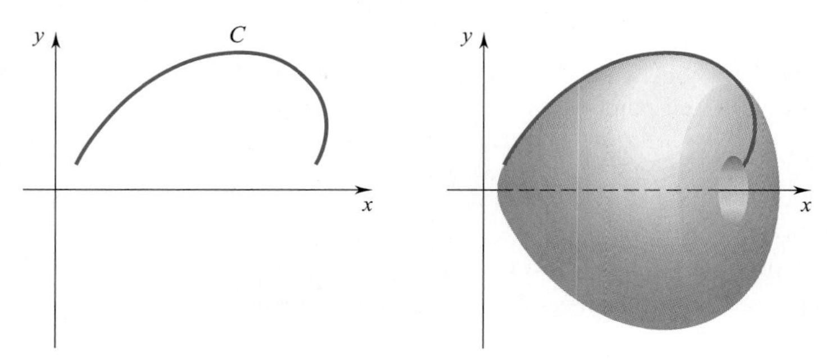

Figure 9.10.2

If we revolve C about the x-axis, we obtain a surface of revolution. The area of that surface is given by the formula

(9.10.2)
$$A = \int_c^d 2\pi y(t)\sqrt{[x'(t)]^2 + [y'(t)]^2}\, dt.$$

We will try to outline how this formula comes about. The argument is similar to the one given in Section 9.9 for the length of a curve.

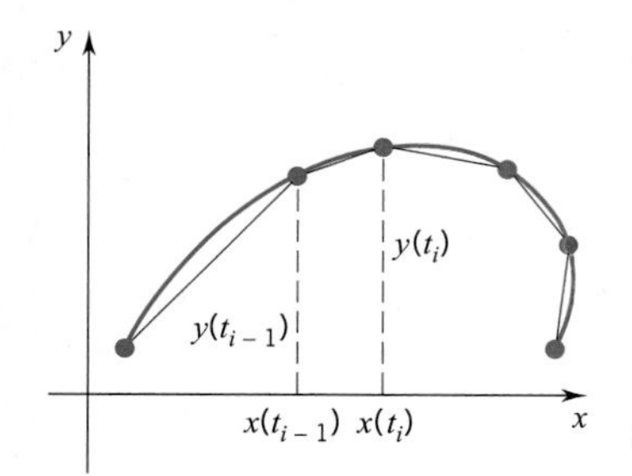

Figure 9.10.3

Each partition $P = \{c = t_0 < t_1 < \cdots < t_n = d\}$ of $[c, d]$ generates a polygonal approximation to C (Figure 9.10.3). Call this polygonal approximation C_p. By revolving C_p about the x-axis we get a surface made up of n conical frustums.

The ith frustum (Figure 9.10.4) has slant height

$$\begin{aligned} s_i &= \sqrt{[x(t_i) - x(t_{i-1})]^2 + [y(t_i) - y(t_{i-1})]^2} \\ &= \sqrt{\left[\frac{x(t_i) - x(t_{i-1})}{t_i - t_{i-1}}\right]^2 + \left[\frac{y(t_i) - y(t_{i-1})}{t_i - t_{i-1}}\right]^2}\,(t_i - t_{i-1}) \end{aligned}$$

and the lateral area $\pi[y(t_{i-1}) + y(t_i)]s_i$ (see Formula (9.10.1)) can be written

$$\pi[y(t_{i-1}) + y(t_i)]\sqrt{\left[\frac{x(t_i) - x(t_{i-1})}{t_i - t_{i-1}}\right]^2 + \left[\frac{y(t_i) - y(t_{i-1})}{t_i - t_{i-1}}\right]^2}\,(t_i - t_{i-1}).$$

There exist points t_i^*, t_i^{**}, t_i^{***} all in $[t_{i-1}, t_i]$ such that

$$y(t_i) + y(t_{i-1}) = 2y(t_i^*), \qquad \frac{x(t_i) - x(t_{i-1})}{t_i - t_{i-1}} = x'(t_i^{**}), \qquad \frac{y(t_i) - y(t_{i-1})}{t_i - t_{i-1}} = y'(t_i^{***}).$$

intermediate-value theorem (first equation); mean-value theorem (second and third equations)

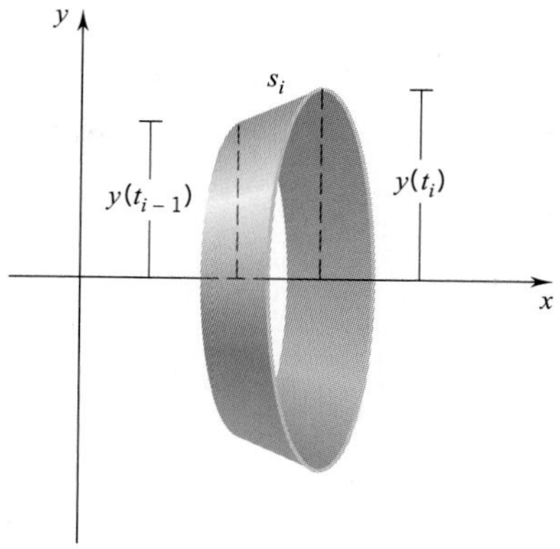

Figure 9.10.4

Let $\Delta t_i = t_i - t_{i-1}$. Then we can write the lateral area of the ith frustum as

$$2\pi y(t_i^*)\sqrt{[x'(t_i^{**})]^2 + [y'(t_i^{***})]^2}\,\Delta t_i.$$

The area generated by revolving all of C_p is the sum of these terms:

$$2\pi y(t_1^*)\sqrt{[x'(t_1^{**})]^2 + [y'(t_1^{***})]^2}\,\Delta t_1 + \cdots + 2\pi y(t_n^*)\sqrt{[x'(t_n^{**})]^2 + [y'(t_n^{***})]^2}\,\Delta t_n.$$

This is not a Riemann sum: we don't know that $t_i^* = t_i^{**} = t_i^{***}$. But it is "close" to a Riemann sum. Close enough that, as $\|P\| \to 0$, this "almost" Riemann sum tends to the integral

$$\int_c^d 2\pi y(t)\sqrt{[x'(t)]^2 + [y'(t)]^2}\,dt.$$

That this is so follows from a theorem of advanced calculus known as Duhamel's principle. We will not attempt to fill in the details. ❑

Example 1 Derive a formula for the surface area of a sphere from (9.10.2).

SOLUTION We can generate a sphere of radius r by revolving the arc

$$x(t) = r\cos t, \quad y(t) = r\sin t, \qquad t \in [0, \pi]$$

about the x-axis. Differentiation gives

$$x'(t) = -r\sin t, \quad y'(t) = r\cos t.$$

By Formula (9.10.2)

$$\begin{aligned} A &= 2\pi\int_0^\pi r\sin t\,\sqrt{r^2(\sin^2 t + \cos^2 t)}\,dt \\ &= 2\pi r^2\int_0^\pi \sin t\,dt = 2\pi r^2\Big[-\cos t\Big]_0^\pi = 4\pi r^2. \end{aligned}$$ ❑

Example 2 Find the area of the surface generated by revolving about the x-axis the curve

$$y^2 - 2\ln y = 4x \qquad \text{from } y = 1 \text{ to } y = 2.$$

SOLUTION We can represent the curve parametrically by setting

$$x(t) = \tfrac{1}{4}(t^2 - 2\ln t), \quad y(t) = t, \qquad t \in [1, 2].$$

Here

$$x'(t) = \tfrac{1}{2}(t - t^{-1}), \quad y'(t) = 1$$

and

$$[x'(t)]^2 + [y'(t)]^2 = [\tfrac{1}{2}(t + t^{-1})]^2. \qquad \text{(check this)}$$

It follows that

$$A = \int_1^2 2\pi t[\tfrac{1}{2}(t + t^{-1})]\, dt = \int_1^2 \pi(t^2 + 1)\, dt = \pi\left[\tfrac{1}{3}t^3 + t\right]_1^2 = \tfrac{10}{3}\pi. \quad ❑$$

Suppose now that f is some nonnegative function defined from $x = a$ to $x = b$. If f' is continuous, then the graph of f is a continuously differentiable curve in the upper half-plane. The area of the surface generated by revolving this graph about the x-axis is given by the formula

(9.10.3)
$$A = \int_a^b 2\pi f(x)\sqrt{1 + [f'(x)]^2}\, dx.$$

This follows readily from (9.10.2). Set

$$x = t, \quad y(t) = f(t), \qquad t \in [a, b].$$

Apply (9.10.2) and then replace the dummy variable t by x.

Example 3 Find the area of the surface generated by revolving about the x-axis the graph of the sine function from $x = 0$ to $x = \frac{1}{2}\pi$.

SOLUTION Setting $f(x) = \sin x$, we have $f'(x) = \cos x$ and therefore

$$A = \int_0^{\pi/2} 2\pi \sin x\sqrt{1 + \cos^2 x}\, dx.$$

To calculate this integral, we set

$$u = \cos x, \qquad du = -\sin x\, dx.$$

At $x = 0$, $u = 1$; at $x = \frac{1}{2}\pi$, $u = 0$. Therefore

$$A = -2\pi\int_1^0 \sqrt{1 + u^2}\, du = 2\pi\int_0^1 \sqrt{1 + u^2}\, du$$

$$= 2\pi\left[\tfrac{1}{2}u\sqrt{1 + u^2} + \tfrac{1}{2}\ln\left(u + \sqrt{1 + u^2}\right)\right]_0^1$$

by (8.5.1) ⟶

$$= \pi[\sqrt{2} + \ln(1 + \sqrt{2})] \cong 2.3\pi \cong 7.23. \quad ❑$$

The Centroid of a Curve

The centroid of a plane region Ω is the center of mass of a homogeneous plate in the shape of Ω. Likewise, the centroid of a solid of revolution T is the center of mass of a homogeneous solid in the shape of T. All this you know from Section 6.4.

What do we mean by the centroid of a plane curve C? Exactly what you would expect. By the *centroid* of a plane curve C, we mean the center of mass of a homogeneous wire in the shape of C. It should be noted, however, that the centroid of a plane curve does not necessarily lie on the curve. We can calculate the centroid of a curve from the following principles, which we take from physics.

Principle 1: Symmetry If a curve has an axis of symmetry, then the centroid $(\bar{x}, \bar{y})$ lies somewhere along that axis.

Principle 2: Additivity If a curve with length L is broken up into a finite number of pieces with arc lengths $\Delta s_1, \ldots, \Delta s_n$ and centroids $(\bar{x}_1, \bar{y}_1), \ldots, (\bar{x}_n, \bar{y}_n)$, then

$$\bar{x}L = \bar{x}_1\,\Delta s_1 + \cdots + \bar{x}_n\,\Delta s_n \quad\text{and}\quad \bar{y}L = \bar{y}_1\,\Delta s_1 + \cdots + \bar{y}_n\,\Delta s_n.$$

Figure 9.10.5 shows a curve C that begins at A and ends at B. Let's suppose that the curve is continuously differentiable and that the length of the curve is L. We want a formula for the centroid $(\bar{x}, \bar{y})$.

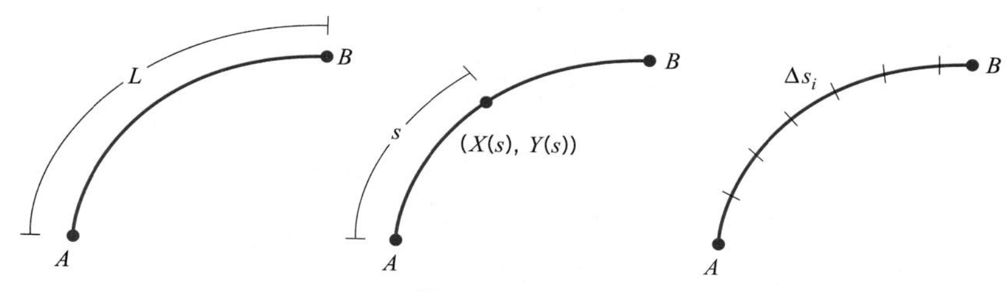

Figure 9.10.5

Let $(X(s), Y(s))$ be the point on C that is at an arc distance s from the initial point A. (What we are doing here is called *parametrizing C by arc length.*) A partition $P = \{0 = s_0 < s_1 < \cdots < s_n = L\}$ of $[0, L]$ breaks up C into n little pieces of lengths $\Delta s_1, \ldots, \Delta s_n$ and centroids $(\bar{x}_1, \bar{y}_1), \ldots, (\bar{x}_n, \bar{y}_n)$. From Principle 2 we know that

$$\bar{x}L = \bar{x}_1\,\Delta s_1 + \cdots + \bar{x}_n\Delta s_n \quad\text{and}\quad \bar{y}L = \bar{y}_1\,\Delta s_1 + \cdots + \bar{y}_n\,\Delta s_n.$$

Now for each i there exists s_i^* in $[s_{i-1}, s_i]$ for which $\bar{x}_i = X(s_i^*)$ and s_i^{**} in $[s_{i-1}, s_i]$ for which $\bar{y}_i = Y(s_i^{**})$. We can therefore write

$$\bar{x}L = X(s_1^*)\,\Delta s_1 + \cdots + X(s_n^*)\,\Delta s_n, \qquad \bar{y}L = Y(s_1^{**})\,\Delta s_1 + \cdots + Y(s_n^{**})\,\Delta s_n.$$

The sums on the right are Riemann sums tending to easily recognizable limits: letting $\|P\| \to 0$ we have

(9.10.4)
$$\bar{x}L = \int_0^L X(s)\,ds \quad\text{and}\quad \bar{y}L = \int_0^L Y(s)\,ds.$$

These formulas give the centroid of a curve in terms of the parameter arc length. It is but a short step from here to formulas of more ready applicability.

Suppose that the curve C is given parametrically by the functions

$$x = x(t), \quad y = y(t), \qquad t \in [c, d]$$

where t is now an arbitrary parameter. Then

$$s(t) = \int_c^t \sqrt{[x'(u)]^2 + [y'(u)]^2}\, du, \qquad ds = s'(t)\, dt = \sqrt{[x'(t)]^2 + [y'(t)]^2}\, dt.$$

At $s = 0$, $t = c$; at $s = L$, $t = d$. Changing variables in (9.10.4) from s to t, we have

$$\bar{x}L = \int_c^d X(s(t))s'(t)\, dt = \int_c^d X(s(t))\sqrt{[x'(t)]^2 + [y'(t)]^2}\, dt$$

and

$$\bar{y}L \int_c^d Y(s(t))s'(t)\, dt = \int_c^d Y(s(t))\sqrt{[x'(t)]^2 + [y'(t)]^2}\, dt.$$

A moment's reflection shows that

$$X(s(t)) = x(t) \quad \text{and} \quad Y(s(t)) = y(t).$$

We can then write

(9.10.5)

$$\bar{x}L = \int_c^d x(t)\sqrt{[x'(t)]^2 + [y'(t)]^2}\, dt,$$
$$\bar{y}L = \int_c^d y(t)\sqrt{[x'(t)]^2 + [y'(t)]^2}\, dt.$$

These are the centroid formulas in their most useful form.

Example 4 Find the centroid of the quarter-circle shown in Figure 9.10.6.

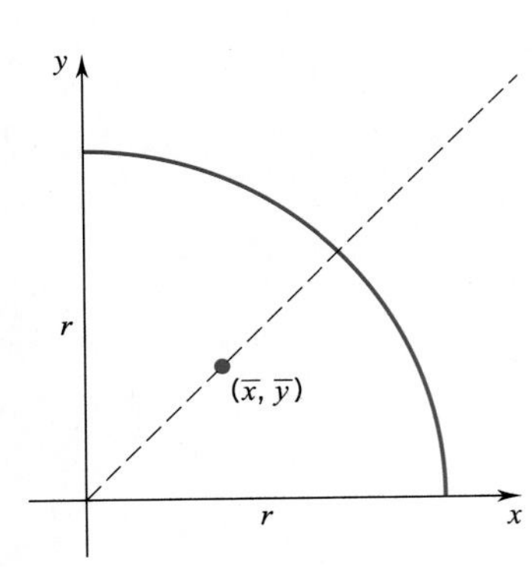

Figure 9.10.6

SOLUTION We can parametrize that quarter-circle by setting

$$x(t) = r\cos t, \quad y(t) = r\sin t, \quad t \in [0, \pi/2].$$

Since the curve is symmetric about the line $x = y$, we know that $\bar{x} = \bar{y}$. Here $x'(t) = -r\sin t$ and $y'(t) = r\cos t$. Therefore

$$\sqrt{[x'(t)]^2 + [y'(t)]^2} = \sqrt{r^2\sin^2 t + r^2\cos^2 t} = r.$$

By (9.10.5)

$$\bar{y}L = \int_0^{\pi/2} (r\sin t)r\, dt = r^2 \int_0^{\pi/2} \sin t\, dt = r^2\Big[-\cos t\Big]_0^{\pi/2} = r^2.$$

Note that $L = \pi r/2$. Therefore $\bar{y} = r^2/L = 2r/\pi$. The centroid of the quarter-circle is the point $(2r/\pi, 2r/\pi)$. [Note that this point is closer to the curve than the centroid of the quarter-disc. (Example 1, Section 6.4).] ❑

Example 5 Find the centroid of the cardioid $r = a(1 - \cos\theta)$, $a > 0$.

SOLUTION The curve (see Figure 9.5.1) is symmetric about the x-axis. Thus $\bar{y} = 0$. To find $\bar{x}$ we parametrize the curve as follows:

$$x(\theta) = r\cos\theta = a(1 - \cos\theta)\cos\theta,$$
$$y(\theta) = r\sin\theta = a(1 - \cos\theta)\sin\theta, \qquad \theta \in [0, 2\pi].$$

A straightforward calculation shows that

$$[x'(\theta)]^2 + [y'(\theta)]^2 = 4a^2\sin^2 \tfrac{1}{2}\theta.$$

Applying (9.10.5) we have

$$\bar{x}L = \int_0^{2\pi} [a(1 - \cos\theta)\cos\theta][2a \sin \tfrac{1}{2}\theta]\, d\theta = -\tfrac{32}{5}a^2.$$

check this out ↑

By Example 4 in Section 9.9, $L = 8a$. Thus $\bar{x} = (-\frac{32}{5}a^2)/8a = -\frac{4}{5}a$. The centroid of the curve is the point $(-\frac{4}{5}a, 0)$. ❑

If C is a curve of the form

$$y = f(x), \qquad x \in [a, b],$$

where f is continuously differentiable, then the formulas in (9.10.5) give

(9.10.6) $$\bar{x}L = \int_a^b x\sqrt{1 + [f'(x)]^2}\, dx, \qquad \bar{y}L = \int_a^b f(x)\sqrt{1 + [f'(x)]^2}\, dx.$$

The details are left to you.

Pappus's Theorem on Surface Area

That same Pappus who gave us that wonderful theorem on volumes of solids of revolution (Theorem 6.4.4) gave us the following equally marvelous result on surface area:

THEOREM 9.10.7 PAPPUS'S THEOREM ON SURFACE AREA

A plane curve is revolved about an axis that lies in its plane. The curve may meet the axis but, if so, only at a finite number of points. If the curve does not cross the axis, then the area of the resulting surface of revolution is the length of the curve multiplied by the circumference of the circle described by the centroid of the curve:

$$A = 2\pi\bar{R}L$$

where L is the length of the curve and $\bar{R}$ is the distance from the axis to the centroid of the curve.

Pappus did not have calculus to help him when he made his inspired guesses: he did his work 13 centuries before Newton or Leibniz were born. With the formulas that we have developed through calculus (through Newton and Leibniz, that is) Pappus's theorem is easily verified. Call the plane of the curve the xy-plane and call the axis of rotation the x-axis. Then $\bar{R} = \bar{y}$ and

$$A = \int_c^d 2\pi y(t)\sqrt{[x'(t)]^2 + [y'(t)]^2}\, dt$$

$$= 2\pi \int_c^d y(t)\sqrt{[x'(t)]^2 + [y'(t)]^2}\, dt = 2\pi\bar{y}L = 2\pi\bar{R}L. \quad ❑$$

EXERCISES 9.10

In Exercises 1–10, find the length of the curve, locate the centroid, and determine the area of the surface generated by revolving the curve about the x-axis.

1. $f(x) = 4, \quad x \in [0, 1]$.

2. $f(x) = 2x, \quad x \in [0, 1]$.

3. $y = \frac{4}{3}x, \quad x \in [0, 3]$.

4. $y = -\frac{12}{5}x + 12, \quad x \in [0, 5]$.

5. $x(t) = 3t, \quad y(t) = 4t; \quad t \in [0, 2]$.

6. $r = 5, \quad \theta \in [0, \frac{1}{4}\pi]$.

7. $x(t) = 2\cos t, \quad y(t) = 2\sin t; \quad t \in [0, \frac{1}{6}\pi]$.

8. $x(t) = \cos^3 t, \quad y(t) = \sin^3 t; \quad t \in [0, \frac{1}{2}\pi]$.

9. $x^2 + y^2 = a^2, \quad x \in [-\frac{1}{2}a, \frac{1}{2}a], \; a > 0$.

10. $r = 1 + \cos\theta, \quad \theta \in [0, \pi]$.

In Exercises 11–18, find the area of the surface generated by revolving the curve about the x-axis.

11. $f(x) = \frac{1}{3}x^3, \quad x \in [0, 2]$.

12. $f(x) = \sqrt{x}, \quad x \in [1, 2]$.

13. $4y = x^3, \quad x \in [0, 1]$.

14. $y^2 = 9x, \quad x \in [0, 4]$.

15. $y = \cos x, \quad x \in [0, \frac{1}{2}\pi]$.

16. $f(x) = 2\sqrt{1 - x}, \quad x \in [-1, 0]$.

17. $r = e^{\theta}, \quad \theta \in [0, \frac{1}{2}\pi]$.

18. $y = \cosh x, \quad x \in [0, \ln 2]$.

19. One arch of a cycloid is defined parametrically by

$$x(\theta) = a(\theta - \sin\theta), \qquad y(\theta) = a(1 - \cos\theta),$$
$$a > 0, \quad 0 \le \theta \le 2\pi.$$

(a) Find the area under the curve. HINT: See Exercise 47, Section 9.7.
(b) Find the area of the surface generated by revolving the arch around the x-axis.

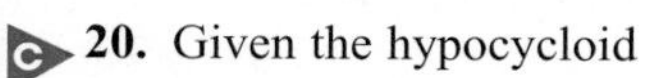

20. Given the hypocycloid

$$x(\theta) = 3a\cos\theta + a\cos 3\theta,$$
$$y(\theta) = 3a\sin\theta - a\sin 3\theta,$$
$$a > 0, \quad 0 \le \theta \le 2\pi.$$

See Exercise 49, Section 9.7.
(a) Use a graphing utility to draw the graph.
(b) Find the area enclosed by the hypocycloid.
(c) Set up a definite integral that gives the area of the surface generated by revolving the hypocycloid around the x-axis.

21. By cutting a cone of slant height s and base radius r along a lateral edge and laying the surface flat, we can form a sector of a circle of radius s. (See the figure.) Use this idea to verify Formula (9.10.1).

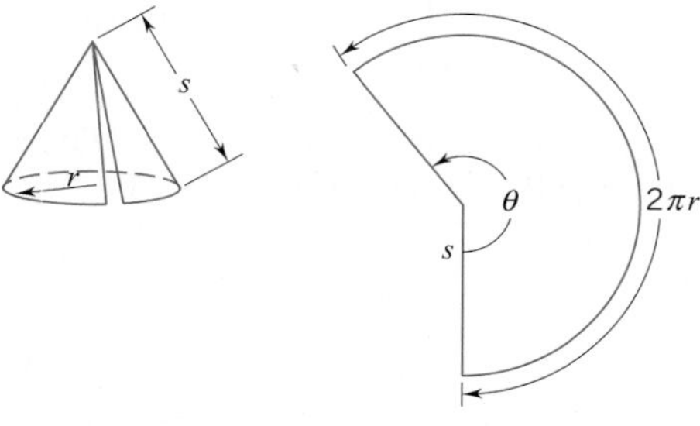

22. The figure shows a ring formed by two quarter-circles. Call the corresponding quarter-discs Ω_a and Ω_r. By Section 6.4, Ω_a has centroid $(4a/3\pi, 4a/3\pi)$ and Ω_r has centroid $(4r/3\pi, 4r/3\pi)$.

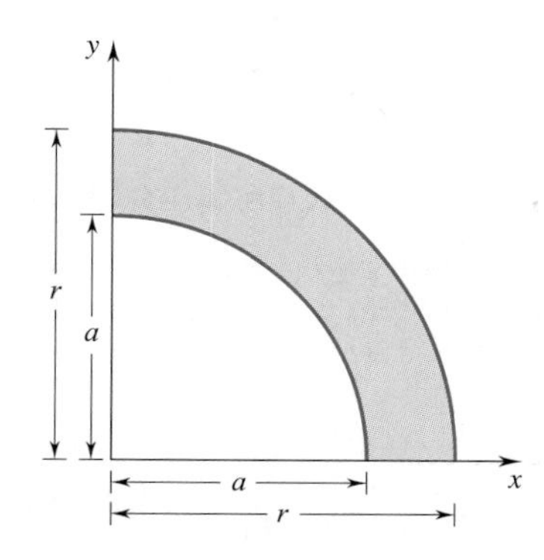

(a) Without integration calculate the centroid of the ring.
(b) Find the centroid of the outer arc from your answer to (a) by letting a tend to r.

23. (a) Find the centroid of each side of the triangle in the figure.

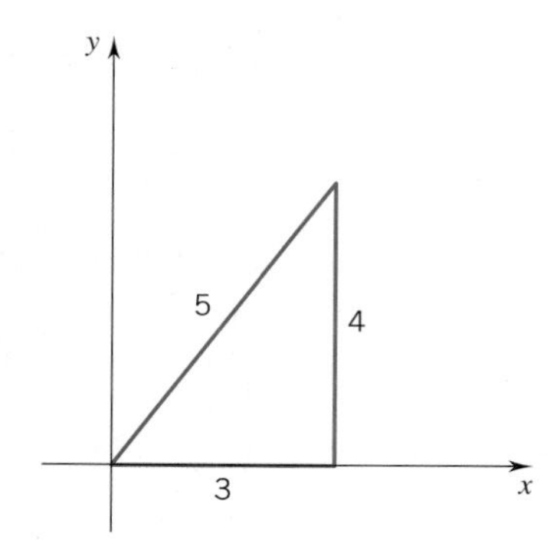

(b) Use your answers to (a) to calculate the centroid of the triangle.

(c) What is the centroid of the triangular region?
(d) What is the centroid of the curve consisting of sides 4 and 5?
(e) Use Pappus's theorem to find the slanted surface area of a cone of base radius 4 and height 3.
(f) Use Pappus's theorem to find the total surface area of the cone in (e). (This time include the base.)

24. Find the area of the surface generated by revolving about the x-axis the curve
(a) $2x = y\sqrt{y^2 - 1} + \ln|y - \sqrt{y^2 - 1}|, \quad y \in [2, 5]$.
(b) $6a^2xy = y^4 + 3a^4, \quad y \in [a, 3a]$.

25. Use Pappus's theorem to find the surface area of the *torus* generated by revolving about the x-axis the circle $x^2 + (y - b)^2 = a^2$. $(0 < a \leq b)$

26. (a) We calculated the total surface area of a sphere from (9.10.2) not (9.10.3). Could we just as well have used (9.10.3)? Explain.
(b) Verify that Formula (9.10.2) applied to

$$C: \quad x(t) = \cos t, \quad y(t) = r, \qquad \text{with } t \in [0, 2\pi],$$

gives $A = 8\pi r$. Note that the surface obtained by revolving C about the x-axis is a cylinder of base radius r and height 2, and therefore A should be $4\pi r$. What's wrong?

27. (*An interesting property of the sphere*) Slice a sphere along two parallel planes that are a fixed distance apart. Show that the surface area of the band so obtained is independent of where the cuts are made.

28. Locate the centroid of a first-quadrant circular arc

$$C: \quad x(t) = r \cos t, \quad y(t) = r \sin t, \qquad t \in [\theta_1, \theta_2].$$

29. Find the surface area of the ellipsoid obtained by revolving the ellipse

$$\frac{x^2}{a^2} + \frac{y^2}{b^2} = 1 \qquad (0 < b < a)$$

(a) about its major axis; (b) about its minor axis.

The Centroid of a Surface of Revolution

If a material surface of revolution is homogeneous (constant mass density), then the center of mass of that material surface is called the *centroid.* The determination of the centroid of a surface of arbitrary shape requires surface integration (Chapter 17). However, if the surface is a surface of revolution, then the centroid can be found by ordinary one-variable integration.

30. Let C be a simple curve in the upper half-plane parametrized by a pair of continuously differentiable functions.

$$x = x(t), \quad y = y(t), \qquad t \in [c, d].$$

By revolving C about the x-axis we obtain a surface of revolution, the area of which we denote by A. By symmetry the centroid of the surface lies on the x-axis. Thus the centroid is completely determined by its x-coordinate $\bar{x}$. Show that

(9.10.8)
$$\bar{x}A = \int_c^d 2\pi x(t)y(t)\sqrt{[x'(t)]^2 + [y'(t)]^2}\, dt$$

by assuming the following additivity principle: If the surface is broken up into n surfaces of revolution with areas $A_1, \ldots, A_n$ and the centroids of the surfaces have x-coordinates $\bar{x}_1, \ldots, \bar{x}_n$, then

$$\bar{x}A = \bar{x}_1 A_1 + \cdots + \bar{x}_n A_n.$$

31. Locate the centroid of a hemisphere of radius r.

32. Locate the centroid of a conical surface of base radius r and height h.

33. Where is the centroid of the lateral surface of the frustum of a cone of height h with base radii r and R?

34. (a) Show that the circle

$$(x - a)^2 + y^2 = b^2$$

can be parametrized by

$$x(t) = a + b \cos t, \qquad y(t) = b \sin t, \qquad 0 \leq t \leq 2\pi.$$

(b) Suppose that $0 < b < a$. The solid generated by revolving the circle around the y-axis is a torus. Find the volume of the torus.
(c) Find the surface area of the torus.

*■ 9.11 THE CYCLOID

(In this section the exercises are intertwined with the text.)
Take a wheel (a roll of tape will do) and fix your eyes on some point of the rim. Call that point P. Now roll the wheel slowly, keeping your eyes on P. The jumping-kangaroo path described by P is called a *cycloid.*

To obtain a mathematical characterization of the cycloid, call the radius of the wheel R and set the wheel on the x-axis so that the point P starts out at the origin. Figure 9.11.1 shows P after a turn of θ radians.

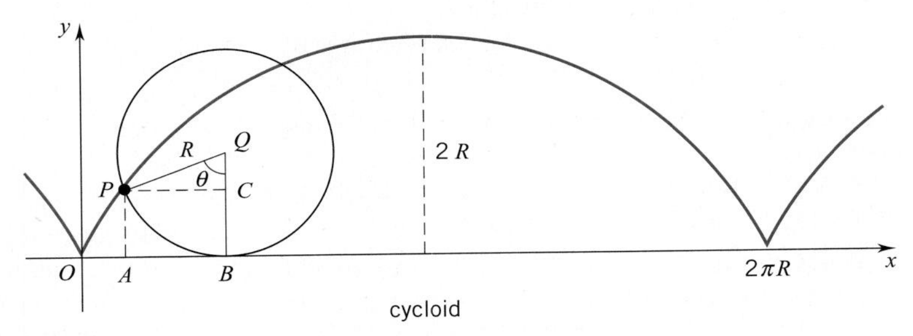

Figure 9.11.1

1. Show that the cycloid can be parametrized by the functions

(9.11.1) $$x(\theta) = R(\theta - \sin\theta), \quad y(\theta) = R(1 - \cos\theta).$$

HINT: Length of $\overline{OB}$ = length of $\widehat{PB} = R\theta$.

2. At the end of each arch the cycloid comes to a cusp. Verify that x' and y' are both 0 at the end of each arch.

3. Show that the line tangent to the cycloid at P (a) passes through the top of the circle and (b) intersects the x-axis at an angle of $\frac{1}{2}(\pi - \theta)$ radians.

4. Express the cycloid from $\theta = 0$ to $\theta = \pi$ by an equation in x and y. (The equation is quite complicated and not easy to work with. The properties of the cycloid are much easier to fathom from the parametric equations.)

5. Find the length of an arch of the cycloid. (9.9.1)

6. Show that the area under an arch of the cycloid is three times the area of the rolling circle. (9.7.4)

7. Locate the centroid of the region under the first arch of the cycloid. (9.7.5)

8. Find the volume of the solid generated by revolving about the x-axis the region under an arch of the cycloid. (9.7.6)

9. Find the volume of the solid generated by revolving about the y-axis the region under the first arch of the cycloid. (9.7.6)

10. Determine the centroid of the first arch of the cycloid. (9.10.5)

11. Find the area of the surface generated by revolving about the x-axis an arch of the cycloid. (9.10.2)

THE INVERTED CYCLOID

12. Show that the arch of the inverted cycloid shown in Figure 9.11.2 can be parametrized by setting

$$x = R(\phi + \sin\phi), \quad y = R(1 - \cos\phi), \qquad \phi \in [-\pi, \pi].$$

Proceed as follows:
(a) Reflect the arch of Figure 9.11.1 in the x-axis and find the θ parametrization for that.
(b) Raise the arch obtained in part (a) so that the low point rests on the x-axis and find the θ parametrization for that.

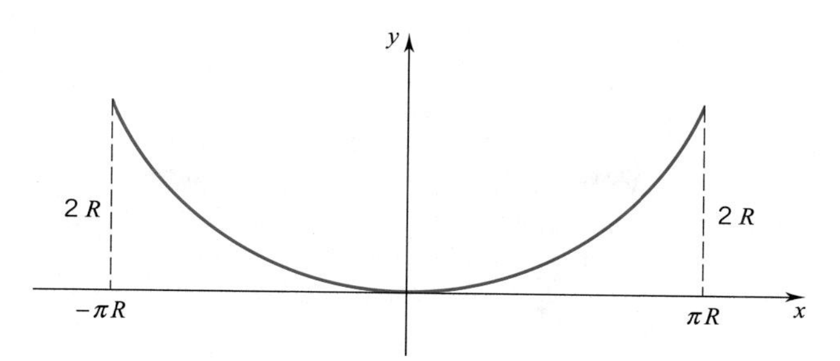

Figure 9.11.2

(c) Obtain the arch of Figure 9.11.2 by translating πR units to the left the arch obtained in part (b). Write down the θ parametrization for this arch and then set $\phi = \theta - \pi$.

13. Find the inclination α of the line tangent to the inverted arch at the point $(x(\phi), y(\phi))$.

14. Let s be the arc distance from the low point of the inverted arch to the point $(x(\phi), y(\phi))$ of that same arch. Show that $s = 4R \sin \frac{1}{2}\phi = 4R \sin \alpha$ where α is the inclination of the tangent line at $(x(\phi), y(\phi))$.

THE TAUTOCHRONE

Visualize two particles (beads if you prefer) sliding without friction down an arch of an inverted cycloid (Figure 9.11.3). If the two particles are released at the same time from different positions, which will reach the bottom first? Neither: They will both get there at exactly the same time.† Being the only curve that produces this effect, the inverted arch of a cycloid is known as *the tautochrone,* the *same-time* curve.

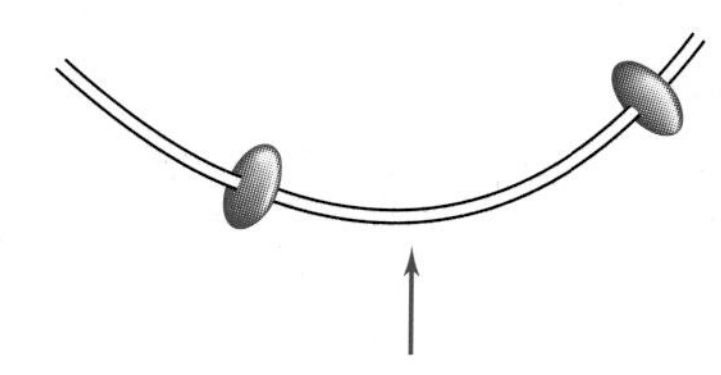

Figure 9.11.3

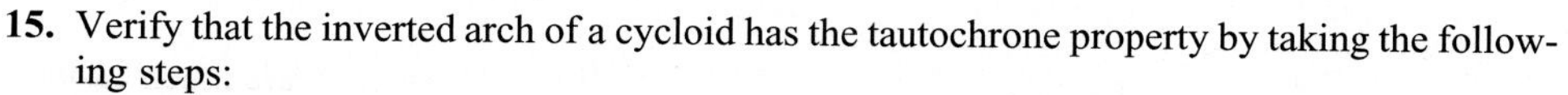

15. Verify that the inverted arch of a cycloid has the tautochrone property by taking the following steps:

(a) Show that the effective gravitational force on a particle of mass m is $-\,mg \sin \alpha$ where α is the inclination of the tangent line at the position of the particle. Conclude then that

$$(*) \qquad \frac{d^2s}{dt^2} = -g \sin \alpha.$$

(b) Combine $(*)$ with Exercise 14 to show that the particle is in simple harmonic motion with period

$$T = 4\pi\sqrt{R/g}.$$

(Thus, while the amplitude of the motion depends on the point of release, the frequency does not. Two particles released simultaneously from different points of the curve will reach the low point of the curve in exactly the same amount of time: $T/4 = \pi\sqrt{R/g}$.)

THE BRACHYSTOCHRONE

A particle is to descend without friction along a curve from some point A to a point B not directly below it. (See Figure 9.11.4.) What should be the shape of the curve so that the particle descends from A to B in the least possible time?

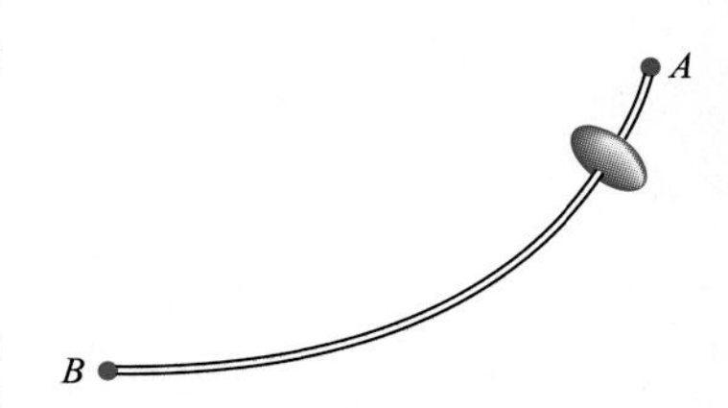

Figure 9.11.4

† This was understood in 1673 by Christian Huygens, the inventor of the pendulum clock. By tracking the bob of the pendulum along the arch of an inverted cycloid, Huygens was able to stabilize the frequency of the oscillations of the pendulum and thereby improve the accuracy of his clock.

This question was first formulated by Johann Bernoulli and posed by him as a challenge to the scientific community in 1696. The challenge was readily accepted and within months the answer was found—by Johann Bernoulli himself, by his brother Jacob, by Newton, by Leibniz, and by L'Hospital. The answer? Part of an inverted cycloid. Because of this, the inverted cycloid is heralded as the *brachystochrone,* the *least-time* curve.

A proof that the inverted cycloid is the least-time curve, the curve of quickest descent, is beyond our reach. The argument requires a sophisticated variant of calculus known as *the calculus of variations.* We can, however, compare the time of descent along a cycloid to the time of descent along a straight-line path.

16. You have seen that a particle descends along the inverted arch of a cycloid from $(\pi R, 2R)$ to $(0, 0)$ in time $t = T/4 = \pi\sqrt{R/g}$. What is the time of descent along a straight-line path?

■ CHAPTER HIGHLIGHTS

9.1 The Conic Sections

The Parabola

directrix, focus, axis, vertex (p. 550)
reflecting property (p. 562)

A parabola is the set of points equidistant from a fixed line and a fixed point not on that line.

The Ellipse

foci, standard position (p. 554)
vertices, axes, center (p. 556)
elliptical reflectors (p. 563)
eccentricity (p. 566)

An ellipse is the set of points the sum of whose distances from two fixed points is constant.

The Hyperbola

foci, standard position (p. 558)
asymptotes (p. 560)
vertices, transverse axis, center (p. 560)
hyperbolic reflectors (p. 565)
eccentricity (p. 567)

A hyperbola is the set of points the difference of whose distances from two fixed points is constant.

9.2 Polar Coordinates

Relation between rectangular coordinates (x, y) and polar coordinates $[r, \theta]$:

$$x = r\cos\theta, \qquad y = r\sin\theta \qquad \tan\theta = \frac{y}{x}, \qquad x^2 + y^2 = r^2$$

9.3 Graphing in Polar Coordinates

cardioids (p. 578)
lines (p. 579)
circles (p. 579)
limaçons (p. 579)
spirals (p. 580)
lemniscates (p. 580)
petal curves (p. 580)

*9.4 The Conic Sections in Polar Coordinates

$$r = \frac{ed}{1 + e\cos\theta}$$

ellipse $0 < e < 1$
parabola $e = 1$
hyperbola $e > 1$

9.5 The Intersection of Polar Curves

If we think of each of two polar curves as giving the position of an object at time θ, then the simultaneous solution of the two equations gives the points where the objects collide. A figure will often help to identify the points where two polar curves intersect, but do not collide; conversion to rectangular coordinates may also help to locate such points.

9.6 Area in Polar Coordinates

Let ρ_1 and ρ_2 be positive continuous functions defined on a closed interval $[\alpha, \beta]$ of length 2π, at most. If $\rho_1(\theta) \le \rho_2(\theta)$ for all θ in $[\alpha, \beta]$, then the area of the region between the polar curves $r = \rho_1(\theta)$ and $r = \rho_2(\theta)$ is given by the formula

$$A = \int_\alpha^\beta \tfrac{1}{2}([\rho_2(\theta)]^2 - [\rho_1(\theta)]^2)\, d\theta.$$

9.7 Curves Given Parametrically

Let $x = x(t), y = y(t)$ be a pair of functions differentiable on the interior of some interval I. At the endpoints of I (if any) we require only continuity. For each t in I we can interpret $(x(t), y(t))$ as the point with x-coordinate $x(t)$ and y-coordinate $y(t)$. Then, as t ranges over I, the point $(x(t), y(t))$ traces a path in the xy-plane. We call such a path a *parametrized curve* and refer to t as the *parameter*.

line: $x(t) = x_0 + t(x_1 - x_0), \quad y(t) = y_0 + t(y_1 - y_0), \quad t$ real.

circle: $x(t) = a\cos t, \quad y(t) = a\sin t, \quad t \in [0, 2\pi].$

ellipse: $x(t) = a\cos t, \quad y(t) = b\sin t, \quad t \in [0, 2\pi].$

9.8 Tangents to Curves Given Parametrically

Tangent line at (x_0, y_0): $y'(t_0)[x - x_0] - x'(t_0)[y - y_0] = 0$, (where $x(t_0) = x_0, y(t_0) = y_0$) provided that $[x'(t_0)]^2 + [y'(t_0)]^2 \ne 0$.

9.9 Arc Length and Speed

polygonal path (p. 613) definition of arc length (p. 614)
significance of dx/ds, dy/ds (p. 616)

length of a parametrized curve: $L = \int_a^b \sqrt{[x'(t)]^2 + [y'(t)]^2}\, dt$

length of a graph: $L = \int_a^b \sqrt{1 + [f'(x)]^2}\, dx$

speed along a curve parametrized by time t: $v(t) = \sqrt{[x'(t)]^2 + [y'(t)]^2}$

9.10 The Area of a Surface of Revolution; The Centroid of a Curve; Pappus's Theorem on Surface Area

revolution of a curve about x-axis: $A = \int_c^d 2\pi y(t) \sqrt{[x'(t)]^2 + [y'(t)]^2}\, dt$

revolution of a graph about x-axis: $A = \int_a^b 2\pi f(x) \sqrt{1 + [f'(x)]^2}\, dx$

parametrizing a curve by arc length (p. 625)

principles for finding the centroid of a plane curve (p. 625)

centroid of a plane curve: $\bar{x} L = \int_c^d x(t) \sqrt{[x'(t)]^2 + [y'(t)]^2}\, dt$

$$\bar{y} L = \int_c^d y(t) \sqrt{[x'(t)]^2 + [y'(t)]^2}\, dt$$

Pappus's theorem on surface area: $A = 2\pi \bar{R} L$

*9.11 The Cycloid

cycloid (p. 629) tautochrone (p. 631) brachystochrone (p. 631)

■ PROJECTS AND EXPLORATIONS USING TECHNOLOGY

To do these exercises you will need a graphics calculator or a computer with graphing capability. The majority of these problems are open-ended so different approaches may be used to solve them. You should be aware that different approaches can result in slight variations in the answers. Round your numerical answers to at least four decimal places. The rounding method that your calculator or computer uses also may cause variations in answers.

9.1 Consider the ellipse $x^2 + 4y^2 = 9$ and define the function $c = c(x)$ to be the cosine of the angle whose vertex is at the point $P(x, y)$ on the ellipse and whose sides pass through the foci of the ellipse. See the figure.

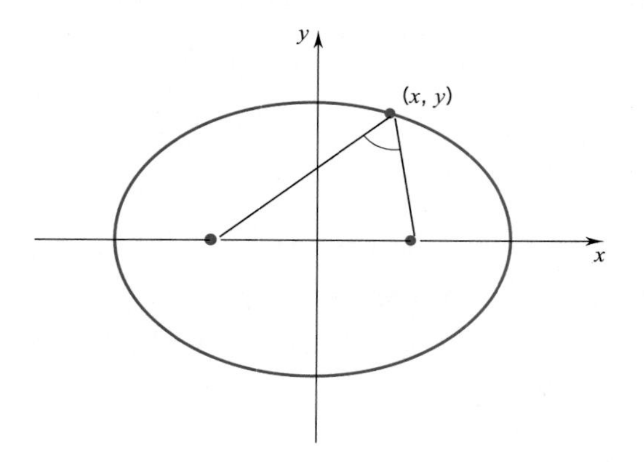

(a) Find a formula for $c(x)$.
(b) Use both the geometry and the formula that you found in part (b) to explain why c is an even function. As an even function, what can you say about the extrema of c?
(c) Suppose that the line segment extending between the two foci is a stage. Where on the ellipse should you sit to maximize your viewing angle?
(d) Repeat parts (a)–(c) for the ellipse

$$\frac{x^2}{a^2} + \frac{y^2}{b^2} = 1.$$

(e) Will the equation $c(x) = 0$ always have a solution? That is, is one of these angles always a right angle?

9.2 (*This problem requires software with an arc-length function.*) Let

$$f(x) = \sin x + x^{0.515}$$

and let $L(t)$ be the arc length of the graph of f from $x = 0$ to $x = t$.
(a) Graph L, L', and L''.
(b) Explain why L must have an inverse function and give an interpretation for it.
(c) Graph the inverse of L. What can you say about the points where L and its inverse intersect?
(d) Can you relate the local maxima, local minima, and points of inflection of f with those of L.

9.3 Consider using the summation

$$\sum_{i=1}^{n} \sqrt{(x_i - x_{i+1})^2 + (y_i - y_{i-1})^2}$$

as an approximation for arc length, where the points $P_i(x_i, y_i)$ are spaced along the curve. For convenience, either the x_i or the y_i are usually equally spaced. Not only is this method intuitively clear, but it can have advantages over approximations of appropriate integrals.

(a) Using arc length software or software to evaluate the arc length integral, find the arc length of the curve $f(x) = x^5$ on $[0, 1]$ to at least five decimal places. Then approximate the arc length using this summation with $n = 1, 2, 4, 8, 16, 32$.

(b) Using the values in part (a), plot n versus the difference between arc length calculated by software and the approximations. The differences should be positive. Why?

The summations

$$\sum_{i=1}^{n} \sqrt{1 + [f'(x_{i-1})]^2}\, \Delta x_i, \qquad \sum_{i=1}^{n} \sqrt{1 + [f'(m_i)]^2}\, \Delta x_i,$$

where $\Delta x_i = x_i - x_{i-1}$ and $m_i = (x_{i-1} + x_i)/2$ is the midpoint of $[x_{i-1}, x_i]$, can also be used to approximate arc length.

(c) Repeat part (a) using each of these summations and discuss the convergence of these approximations.

(d) Let $g(x) = \sin x$ on $[0, \pi/2]$. Use these summation formulas to approximate the arc length of the graph of g. Do you get the same type of results as you did for f?

SEQUENCES; INDETERMINATE FORMS; IMPROPER INTEGRALS

CHAPTER 10

10.1 THE LEAST UPPER BOUND AXIOM

In the next few sections and in the next chapter, we will be talking about limits of sequences. To be able to do that with precision we first need to look a little deeper into the real number system.

We begin with a nonempty set S of real numbers. As indicated in Chapter 1, a number M is called an *upper bound* for S iff

$$x \leq M \qquad \text{for all } x \in S.$$

Note that if M is an upper bound for S, then any number greater than M will also be an upper bound for S. Thus, if a set has *an* upper bound, then it has, in fact, infinitely many upper bounds. Of course, not all sets of real numbers have upper bounds. Those that do are said to be *bounded above.*

For example, the number 3 is an upper bound for the set $\{x\colon\ x^2 \leq 1\} = \{x\colon\ -1 \leq x \leq 1\}$; so are $\frac{3}{2}$ and 1. This set is bounded above; 1 and every number greater than 1 are upper bounds. The set $\{2, 4, 6, 8, \ldots, 2n, \ldots\}$ is not bounded above.

It is clear that every set that has a largest element has an upper bound: if b is the largest element of S, then

$$x \leq b \qquad \text{for all } x \in S;$$

this makes b an upper bound for S. The converse is false: the sets

$$(-\infty, 0) \quad \text{and} \quad \left\{\frac{1}{2}, \frac{2}{3}, \frac{3}{4}, \ldots, \frac{n}{n+1}, \ldots\right\}$$

both have upper bounds (2 for instance), but neither has a largest element.

Let's return to the first set, $(-\infty, 0)$. While $(-\infty, 0)$ does not have a largest element, the set of its upper bounds, $[0, \infty)$, does have a least element, 0. We call 0 the *least upper bound of* $(-\infty, 0)$.

Now let's reexamine the second set. While the set of quotients

$$\frac{n}{n+1} = 1 - \frac{1}{n+1}$$

does not have a greatest element, the set of its upper bounds, $[1, \infty)$, does have a least element, 1. We call 1 the *least upper bound* of that set of quotients.

In general, we have

DEFINITION 10.1.1 LEAST UPPER BOUND

Let S be a nonempty set of real numbers which is bounded above. A number M is the least upper bound of S iff

(i) M is an upper bound for S,

(ii) $M \leq K$, where K is any upper bound for S.

We are ready now to state explicitly one of the key *assumptions* that we make about the real number system. It is called the *least upper bound axiom.* Although we have not made an issue of it up to this point, this axiom has been an implicit assumption throughout the previous nine chapters; it underlies all of calculus.

AXIOM 10.1.2 THE LEAST UPPER BOUND AXIOM

Every nonempty set of real numbers that has an upper bound has a *least* upper bound.

Remark It is easy to see that if a set of real numbers S is bounded above, then its least upper bound is unique. For if L and M are least upper bounds of S, then $M \leq L$ and $L \leq M$ by property (ii) of Definition 10.1.1, and so $L = M$. Thus, it makes sense to use the term "*the* least upper bound" of S. ❑

To indicate the least upper bound of a set S, we will write lub S. Here are some examples:

1. $\text{lub}\,(-\infty, 0) = 0, \qquad \text{lub}\,(-\infty, 0] = 0.$

2. $\text{lub}\,(-4, -1) = -1, \qquad \text{lub}\,(-4, -1] = -1.$

3. $\text{lub}\left\{\frac{1}{2}, \frac{2}{3}, \frac{3}{4}, \dots, \frac{n}{n+1}, \dots\right\} = 1.$

4. $\text{lub}\left\{-\frac{1}{2}, -\frac{1}{8}, -\frac{1}{27}, \dots, -\frac{1}{n^3}, \dots\right\} = 0.$

5. $\text{lub}\,\{x\colon x^2 < 3\} = \text{lub}\,\{x\colon -\sqrt{3} < x < \sqrt{3}\} = \sqrt{3}.$ ❑

The least upper bound of a set has a special property that deserves particular attention. The idea is this: the fact that M is the least upper bound of the set S

does not tell us that M is in S (indeed, it need not be, as illustrated in the preceding examples), but it does tell us that we can approximate M as closely as we wish by elements of S.

THEOREM 10.1.3

If M is the least upper bound of the set S and ϵ is a positive number, then there is at least one number s in S such that

$$M - \epsilon < s \leq M.$$

PROOF Let $\epsilon > 0$. Since M is an upper bound for S, the condition $s \leq M$ is satisfied by all numbers s in S. All we have to show therefore is that there is some number s in S such that

$$M - \epsilon < s.$$

Suppose on the contrary that there is no such number in S. We then have

$$x \leq M - \epsilon \qquad \text{for all } x \in S.$$

This makes $M - \epsilon$ an upper bound for S. But this cannot be, for then $M - \epsilon$ is an upper bound for S that is *less* than M, and by assumption, M is the *least* upper bound. ❑

The theorem we just proved is illustrated in Figure 10.1.1. Take S as the set of points marked in the figure. If $M = \text{lub } S$, then S has at least one element in every half-open interval of the form $(M - \epsilon, M]$.

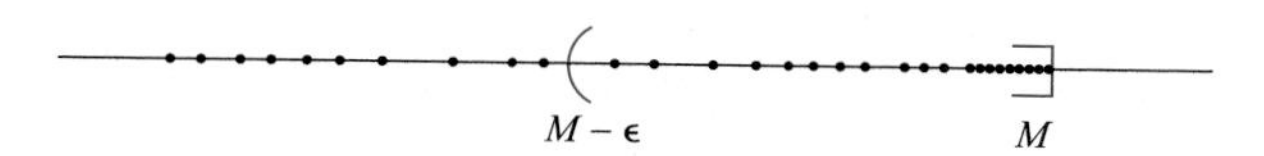

Figure 10.1.1

Example 1

(a) Let

$$S = \left\{\frac{1}{2}, \frac{2}{3}, \frac{3}{4}, \ldots, \frac{n}{n+1}, \ldots\right\}$$

and take $\epsilon = 0.0001$. Since 1 is the least upper bound of S, there must be a number s in S such that

$$1 - 0.0001 < s \leq 1.$$

There is: take, for example, $s = \frac{99999}{100000}$.

(b) Let

$$S = \{1, 2, 3\}$$

and take $\epsilon = 0.00001$. It is clear that 3 is the least upper bound of S. Therefore, there must be a number $s \in S$ such that

$$3 - 0.00001 < s \leq 3.$$

There is: $s = 3$. ❑

We come now to lower bounds. In the first place, a number m is called a *lower bound* for a nonempty set S iff

$$m \leq x \quad \text{for all } x \in S.$$

Sets that have lower bounds are said to be *bounded below.* Not all sets have lower bounds; those that do have *greatest lower bounds*. Paralleling the definition of least upper bound, if a nonempty set S is bounded below, then a number m is the greatest lower bound of S if (i) m is a lower bound, and (ii) $m \geq k$, where k is any lower bound for S.

The existence of a greatest lower bound of a set S which is bounded below does not need to be taken as an axiom. We can prove it as a theorem using the least upper bound axiom.

THEOREM 10.1.4

Every nonempty set of real numbers that has a lower bound has a *greatest* lower bound.

PROOF Suppose that S is nonempty and that it has a lower bound x. Then

$$x \leq s \qquad \text{for all } s \in S.$$

It follows that $-s \leq -x$ for all $s \in S$; that is,

$$\{-s\colon s \in S\} \quad \text{has an upper bound } -x.$$

From the least upper bound axiom we conclude that $\{-s\colon s \in S\}$ has a least upper bound; call it x_0. Since $-s \leq x_0$ for all $s \in S$, we can see that

$$-x_0 \leq s \qquad \text{for all } s \in S,$$

and thus $-x_0$ is a lower bound for S. We now assert that $-x_0$ is the greatest lower bound of the set S. To see this, note that, if there existed a number x_1 satisfying

$$-x_0 < x_1 \leq s \qquad \text{for all } s \in S,$$

then we would have

$$-s \leq -x_1 < x_0 \qquad \text{for all } s \in S,$$

and thus x_0 would not be the *least* upper bound of $\{-s\colon s \in S\}$.† ❑

† We proved Theorem 10.1.4 by assuming the least upper bound axiom. We could have proceeded the other way. We could have set Theorem 10.1.4 as an axiom, and then proved the least upper bound axiom as a theorem.

As in the case of the least upper bound, the greatest lower bound of a set is unique. Also, the greatest lower bound of a set need not be in the set, but can be approximated as closely as we wish by members of the set. In short, we have the following theorem, the proof of which is left as an exercise.

THEOREM 10.1.5

If m is the greatest lower bound of the set S and ϵ is a positive number, then there is at least one number s in S such that

$$m \leq s < m + \epsilon.$$

The theorem is illustrated in Figure 10.1.2. If $m = \text{glb } S$ (that is, if m is the greatest lower bound of the set S), then S has at least one element in every half-open interval of the form $[m, m + \epsilon)$.

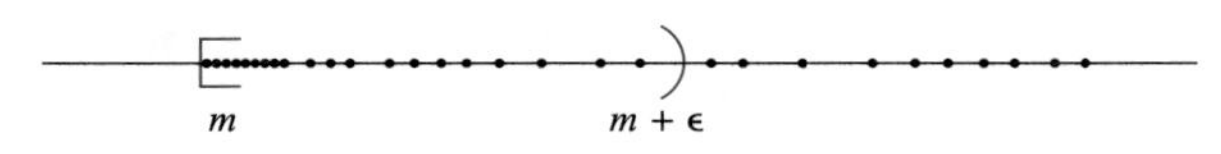

Figure 10.1.2

Remark Remember the intermediate-value theorem? It states that a continuous function skips no values. Remember the maximum-minimum theorem? It states that on a bounded closed interval a continuous function takes on both a maximum and a minimum value. We have been using these two results right along, but we have not proved them. Now that you understand least upper bounds and greatest lower bounds, you are in a position to follow proofs of both of these theorems. (See Appendix B.) Better still, try to prove the theorems yourself. ❑

EXERCISES 10.1

In Exercises 1–20, find the least upper bound (if it exists) and the greatest lower bound (if it exists) for the given set.

1. $(0, 2)$.

2. $[0, 2]$.

3. $(0, \infty)$.

4. $(-\infty, 1)$.

5. $\{x: x^2 < 4\}$.

6. $\{x: |x - 1| < 2\}$.

7. $\{x: x^3 \geq 8\}$.

8. $\{x: x^4 \leq 16\}$.

9. $\{2\frac{1}{2}, 2\frac{1}{3}, 2\frac{1}{4}, \ldots\}$.

10. $\{-1, -\frac{1}{2}, -\frac{1}{3}, -\frac{1}{4}, \ldots\}$.

11. $\{0.9, 0.99, 0.999, \ldots\}$.

12. $\{-2, 2, -2.1, 2.1, -2.11, 2.11, \ldots\}$.

13. $\{x: \ln x < 1\}$.

14. $\{x: \ln x > 0\}$.

15. $\{x: x^2 + x - 1 < 0\}$.

16. $\{x: x^2 + x + 2 \geq 0\}$.

17. $\{x: x^2 > 4\}$.

18. $\{x: |x - 1| > 2\}$.

19. $\{x: \sin x \geq -1\}$.

20. $\{x: e^x < 1\}$.

In Exercises 21–24, illustrate the validity of Theorem 10.1.5 taking S and ϵ as given.

21. $S = \{\frac{1}{11}, (\frac{1}{11})^2, (\frac{1}{11})^3, \ldots, (\frac{1}{11})^n, \ldots\}$, $\quad \epsilon = 0.001$.

22. $S = \{1, 2, 3, 4\}$, $\quad \epsilon = 0.0001$.

23. $S = \{\frac{1}{10}, \frac{1}{1000}, \frac{1}{100000}, \ldots, (\frac{1}{10})^{2n-1}, \ldots\}$,
$\epsilon = (\frac{1}{10})^k \quad (k \geq 1)$.

24. $S = \{\frac{1}{2}, \frac{1}{4}, \frac{1}{8}, \ldots, (\frac{1}{2})^n, \ldots\}$, $\quad \epsilon = (\frac{1}{4})^k \quad (k \geq 1)$.

25. Prove Theorem 10.1.5 by imitating the proof of Theorem 10.1.3.

26. Let $S = \{a_1, a_2, a_3, \ldots, a_n\}$ be a nonempty, finite set of real numbers.
(a) Prove that S is bounded.
(b) Prove that lub S and glb S are elements of S.

27. Suppose that b is an upper bound for a set S of real numbers. Prove that if $b \in S$, then $b =$ lub S.

28. Let S be a bounded set of real numbers and suppose that lub S = glb S. What can you conclude about S?

29. Suppose that S is a nonempty, bounded set of real numbers and that T is a nonempty subset of S.
(a) Prove that T is bounded.
(b) Prove that glb $S \leq$ glb $T \leq$ lub $T \leq$ lub S.

30. Let S and T be nonempty sets of real numbers such that $x \leq y$ for all $x \in S$ and all $y \in T$.
(a) Prove that lub $S \leq y$ for all $y \in T$.
(b) Prove that lub $S \leq$ glb T.

31. Let c be a positive number. Prove that the set $S = \{c, 2c, 3c, \ldots, nc, \ldots\}$ is not bounded above.

32. Prove that if a and b are any two positive numbers with $a < b$, then there exists a rational number r such that $a < r < b$. HINT: Show that there is a positive integer n such that $na > 1$ and $n(b - a) > 1$. Then show that there is a positive integer m such that $m > na$ and $m - 1 \leq na$. Let $r = m/n$.

33. Let S be the set of irrational numbers

$$S = \{\sqrt{2}, \sqrt{2\sqrt{2}}, \sqrt{2\sqrt{2\sqrt{2}}}, \ldots\}.$$

That is, $S = \{a_1, a_2, a_3, \ldots, a_n, \ldots\}$ where $a_1 = \sqrt{2}$, and for each positive integer n, $a_{n+1} = \sqrt{2a_n}$.
(a) Calculate the numbers $a_1, a_2, a_3, \ldots, a_{10}$.
(b) Use mathematical induction to prove that $a_n < 2$ for all n.
(c) Is 2 the least upper bound of S?
(d) Choose a positive number other than 2 and repeat this exercise. What can you conclude?

34. Let S be the set of irrational numbers

$$S = \{\sqrt{2}, \sqrt{2 + \sqrt{2}}, \sqrt{2 + \sqrt{2 + \sqrt{2}}}, \ldots\}$$

That is, $a_1 = \sqrt{2}$, and for each positive integer n, $a_{n+1} = \sqrt{2 + a_n}$.
(a) Calculate the numbers $a_1, a_2, a_3, \ldots, a_{10}$.
(b) Use mathematical induction to show that $a_n < 2$ for all n.
(c) Is 2 the least upper bound for S?

Choose a positive number other than 2 and repeat this exercise. What can you conclude?

■ 10.2 SEQUENCES OF REAL NUMBERS

So far in our study of calculus, our attention has been fixed on functions defined on an interval or on a union of intervals. Here we study functions defined on the set of positive integers.

DEFINITION 10.2.1 SEQUENCE OF REAL NUMBERS

A real-valued function defined on the set of positive integers is called a *sequence of real numbers*.

The functions defined on the set of positive integers by setting

$$a(n) = n^2, \quad b(n) = \frac{n}{n+1}, \quad c(n) = \sqrt{\ln n}, \quad d(n) = \frac{e^n}{n}, \quad \text{for } n = 1, 2, 3, \ldots$$

are examples of sequences of real numbers.

The notions developed for functions carry over to sequences. For example, if the functions a and b are sequences, then the linear combination $\alpha a + \beta b$ (α and β are real numbers) and their product ab are also sequences:

$$(\alpha a + \beta b)(n) = \alpha a(n) + \beta b(n) \quad \text{and} \quad (ab)(n) = a(n) \cdot b(n).$$

If the sequence b does not take on the value 0, then the reciprocal $1/b$ is a sequence and so is the quotient a/b:

$$\frac{1}{b}(n) = \frac{1}{b(n)} \qquad \text{and} \qquad \frac{a}{b}(n) = \frac{a(n)}{b(n)}.$$

Let a be a sequence. The numbers $a(1), a(2), a(3), \ldots$ are called the *terms* of a. In particular, $a(n)$ is called the *nth term of a*. In discussing sequences it is conventional to use subscript notation a_n rather than functional notation $a(n)$, $n = 1, 2, 3, \ldots$, to denote the terms, and the sequence itself is often written

$$\{a_1, a_2, a_3, \ldots\}$$

or, even more simply,

$$\{a_n\}.$$

For example, the sequence of reciprocals defined by setting

$$a_n = 1/n \qquad \text{for all } n$$

can be written

$$\{1, \tfrac{1}{2}, \tfrac{1}{3}, \ldots\} \qquad \text{or} \qquad \{1/n\}.$$

The sequence defined by setting

$$a_n = 10^{1/n} \qquad \text{for all } n$$

can be written

$$\{10, 10^{1/2}, 10^{1/3}, \ldots\} \qquad \text{or} \qquad \{10^{1/n}\}.$$

In this notation, we can write

$$\alpha\{a_n\} + \beta\{b_n\} = \{\alpha a_n + \beta b_n\}, \qquad \{a_n\}\{b_n\} = \{a_n b_n\}$$

and, provided that none of the b_n are zero,

$$\frac{1}{\{b_n\}} = \{1/b_n\} \qquad \text{and} \qquad \frac{\{a_n\}}{\{b_n\}} = \{a_n/b_n\}.$$

The following assertions will serve to illustrate the notation further.

(1) The sequence $\{1/2^n\}$ multiplied by 5 is the sequence $\{5/2^n\}$:

$$5\{1/2^n\} = \{5/2^n\}.$$

We can also write

$$5\{\tfrac{1}{2}, \tfrac{1}{4}, \tfrac{1}{8}, \tfrac{1}{16}, \ldots\} = \{\tfrac{5}{2}, \tfrac{5}{4}, \tfrac{5}{8}, \tfrac{5}{16}, \ldots\}.$$

(2) The sequence $\{n\}$ plus the sequence $\{1/n\}$ is the sequence $\{n + 1/n\}$:

$$\{n\} + \{1/n\} = \{n + 1/n\}.$$

In expanded form

$$\{1, 2, 3, \ldots\} + \{1, \tfrac{1}{2}, \tfrac{1}{3}, \ldots\} = \{2, 2\tfrac{1}{2}, 3\tfrac{1}{3}, \ldots\}.$$

(3) The sequence $\{n\}$ times the sequence $\{\sqrt{n}\}$ is the sequence $\{n\sqrt{n}\}$:

$$\{n\}\{\sqrt{n}\} = \{n\sqrt{n}\} = \{n^{3/2}\};$$

the sequence $\{2^n\}$ divided by the sequence $\{n\}$ is the sequence $\{2^n/n\}$:

$$\frac{\{2^n\}}{\{n\}} = \{2^n/n\}. \quad \square$$

The concept of boundedness for functions also carries over to sequences. In particular, a sequence a is said to be *bounded above* if $\{a_1, a_2, a_3, \ldots\}$ is bounded above; a is *bounded below* if $\{a_1, a_2, a_3, \ldots\}$ is bounded below. The sequence a is *bounded* if it is bounded above *and* below.

By the least upper bound axiom (10.1.2), we know that if the sequence a is bounded above, then $\{a_1, a_2, a_3, \ldots\}$ has a least upper bound which is called the *least upper bound of the sequence.* Similarly, if a is bounded below, then it has a *greatest lower bound.* The following assertions illustrate these ideas.

(1) The sequence $\{1/n\}$ is bounded above and below:

$$0 \le 1/n \le 1 \qquad \text{for } n = 1, 2, 3, \ldots.$$

The number 0 is the greatest lower bound of the set $\{1, \frac{1}{2}, \frac{1}{3}, \ldots\}$ and 1 is the least upper bound.

(2) The sequence $\{2^n\}$ is bounded below. For example, 1 is a lower bound:

$$1 \le 2^n \qquad \text{for } n = 1, 2, 3, \ldots.$$

The number 2 is the greatest lower bound for this sequence. It is not bounded above: there is no fixed number M that satisfies

$$2^n \le M \qquad \text{for all } n.$$

(3) The sequence $a_n = (-1)^n 2^n$ can be written

$$\{-2, 4, -8, 16, -32, 64, \ldots\}.$$

It is unbounded below and unbounded above. $\square$

DEFINITION 10.2.2

A sequence $\{a_n\}$ is said to be

(i) *increasing* iff $a_n < a_{n+1}$ for each positive integer n,
(ii) *nondecreasing* iff $a_n \le a_{n+1}$ for each positive integer n,
(iii) *decreasing* iff $a_n > a_{n+1}$ for each positive integer n,
(iv) *nonincreasing* iff $a_n \ge a_{n+1}$ for each positive integer n.

If any of these four properties holds, the sequence is said to be *monotonic*.

Remark An increasing sequence is also nondecreasing. However, a nondecreasing sequence will not, in general, be increasing. The sequence $\{1, 1, 1, \ldots\}$ is nondecreasing; it is not increasing. A corresponding relationship holds for decreasing and nonincreasing sequences. $\square$

The sequences

$$\{1, \tfrac{1}{2}, \tfrac{1}{3}, \ldots\}, \qquad \{2, 4, 8, 16, \ldots\}, \qquad \{2, 2, 4, 4, 8, 8, 16, 16, \ldots\}$$

are all monotonic, but the sequences

$$\{1, \tfrac{1}{2}, 1, \tfrac{1}{3}, 1, \tfrac{1}{4}, \ldots\} \qquad \text{and} \qquad \{-2, 4, -8, 16, -32, 64, \ldots\}$$

are not monotonic.

The following examples are less trivial.

Example 1 The sequence $\{a_n\}$ defined by

$$a_n = \frac{n}{n+1}$$

is increasing. It is bounded below by $\frac{1}{2}$ (the greatest lower bound) and above by 1 (the least upper bound).

PROOF Since

$$\frac{a_{n+1}}{a_n} = \frac{n+1}{n+2} \cdot \frac{n+1}{n} = \frac{n^2+2n+1}{n^2+2n} > 1$$

we have $a_n < a_{n+1}$. This confirms that the sequence is increasing. Since the sequence can be written

$$\{\tfrac{1}{2}, \tfrac{2}{3}, \tfrac{3}{4}, \tfrac{4}{5}, \tfrac{5}{6}, \ldots\}$$

it is easy to see that $\frac{1}{2}$ is the greatest lower bound. To show that 1 is the least upper bound, note first that

$$a_n = \frac{n}{n+1} < 1 \qquad \text{for all } n.$$

Thus, the sequence is bounded above and 1 is an upper bound. Now choose any number $K < 1$ and let p be a positive integer such that

$$\frac{1}{p+1} < 1 - K.$$

Then

$$K < 1 - \frac{1}{p+1} = \frac{p}{p+1} = a_p.$$

Thus, no number less than 1 can be an upper bound for the sequence. This means that 1 is the least upper bound. ❑

Example 2 The sequence $\{a_n\}$ defined by

$$a_n = \frac{2^n}{n!}\dagger$$

is nonincreasing. Moreover $a_n > a_{n+1}$ for $n \geq 2$.

PROOF The first two terms are equal:

$$a_1 = \frac{2^1}{1!} = 2 = \frac{2^2}{2!} = a_2.$$

† Recall that $n! = n(n-1)(n-2) \cdots 3 \cdot 2 \cdot 1$. See Section 1.2.

For $n \geq 2$ the sequence decreases:

$$\frac{a_{n+1}}{a_n} = \frac{2^{n+1}}{(n+1)!} \cdot \frac{n!}{2^n} = \frac{2}{n+1} < 1 \qquad \text{if } n \geq 2. \quad ❑$$

Remark The sequence in Example 2 is not decreasing since $a_1 = a_2 = 2$, but for $n \geq 2$, the terms *are* decreasing. This suggests an extension of Definition 10.2.2. We will say that a sequence $\{a_n\}$ is *increasing for* $n \geq k$ if $a_n < a_{n+1}$ for all $n \geq k$. The terms *nondecreasing for* $n \geq k$, *decreasing for* $n \geq k$, and *nonincreasing for* $n \geq k$ are defined similarly. ❑

Example 3 If $c > 1$, the sequence $\{a_n\}$ defined by

$$a_n = c^n$$

increases without bound.

PROOF Choose a number $c > 1$. Then

$$\frac{a_{n+1}}{a_n} = \frac{c^{n+1}}{c^n} = c > 1.$$

This shows that the sequence increases. To show the unboundedness, we take an arbitrary positive number M and show that there exists a positive integer k for which

$$c^k \geq M.$$

A suitable k is one such that

$$k \geq \frac{\ln M}{\ln c},$$

for then

$$k \ln c \geq \ln M, \qquad \ln c^k \geq \ln M, \qquad \text{and thus} \qquad c^k \geq M. \quad ❑$$

Since sequences are defined on the set of positive integers and not on an interval, they are not directly susceptible to the methods of calculus. Fortunately, we can sometimes circumvent this difficulty by dealing initially, not with the sequence itself, but with a function of a real variable x that agrees with the given sequence for all positive integers n.

Example 4 The sequence $\{a_n\}$ defined by

$$a_n = \frac{n}{e^n}$$

is decreasing. It is bounded above by $1/e$ and below by 0.

PROOF We will work with the function

$$f(x) = \frac{x}{e^x}. \qquad \text{(See Example 3, Section 7.4.)}$$

Note that $f(1) = 1/e = a_1, f(2) = 2/e^2 = a_2, f(3) = 3/e^3 = a_3$, and so on.

Differentiating f, we get

$$f'(x) = \frac{e^x - xe^x}{e^{2x}} = \frac{1 - x}{e^x}.$$

Since $f'(x) < 0$ for $x > 1$, f is decreasing on $[1, \infty)$. Thus $f(1) > f(2) > f(3) > \cdots$, that is $a_1 > a_2 > a_3 > \cdots$, and $\{a_n\}$ is decreasing.

It now follows that $a_1 > a_n$ for all $n \geq 2$, and so $a_1 = 1/e$ is an upper bound for the sequence. In fact, it is the least upper bound. Since all the terms in the sequence are positive, 0 is a lower bound.

Figure 10.2.1 illustrates the graphical relationship between $f(x) = x/e^x$ and $a_n = n/e^n$. The graph of f indicates that $a_n \to 0$ as $n \to \infty$, and so 0 is, in fact, the greatest lower bound. ❑

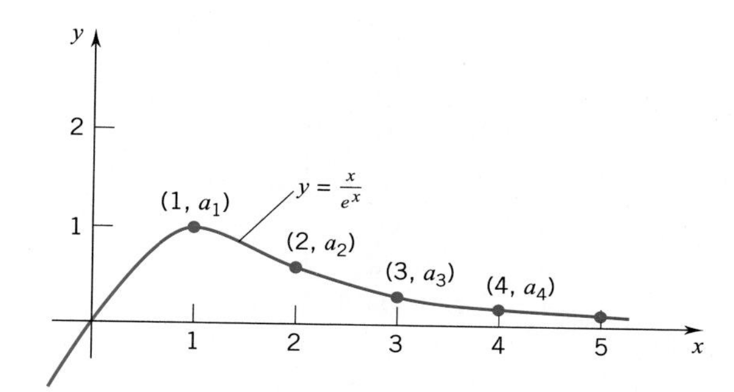

Figure 10.2.1

Example 5 The sequence $\{a_n\}$ defined by

$$a_n = n^{1/n}$$

decreases for $n \geq 3$.

PROOF We could compare a_n with a_{n+1} directly, but it is easier to consider the function

$$f(x) = x^{1/x}$$

instead. Since

$$f(x) = e^{(1/x)\ln x},$$

we have

$$f'(x) = e^{(1/x)\ln x}\frac{d}{dx}\left(\frac{1}{x}\ln x\right) = x^{1/x}\left(\frac{1 - \ln x}{x^2}\right).$$

For $x > e$, $f'(x) < 0$. This shows that f decreases on $[e, \infty)$. Since $3 > e$, f decreases on $[3, \infty)$. It follows that $\{a_n\}$ decreases for $n \geq 3$. ❑

Remark We must be careful when we examine a function of a real variable x in order to analyze the behavior of a sequence. The function $y = f(x)$ and the sequence $\{y_n\} = \{f(n)\}$ may behave differently. For example, the sequence $\{y_n\}$ defined by

$$y_n = f(n) = \frac{1}{n - 4.5}$$

is bounded (by 2 and -2) even though the function

$$f(x) = \frac{1}{x - 4.5}$$

is not bounded; the graph of f has a vertical asymptote at $x = 4.5$. See Figure 10.2.2. ❑

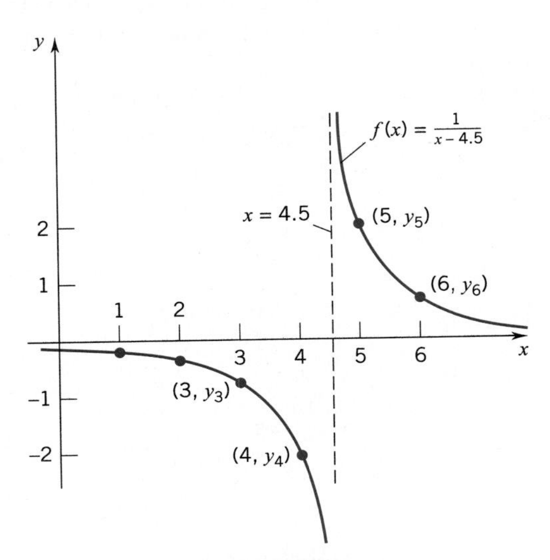

Figure 10.2.2

Clearly, the function $f(x) = \sin \pi x$ is not monotonic on $[0, \infty)$, but the sequence $\{u_n\}$ defined by $u_n = f(n) = \sin n\pi = 0$ for all n is monotonic. Indeed, it is constant. See Figure 10.2.3.

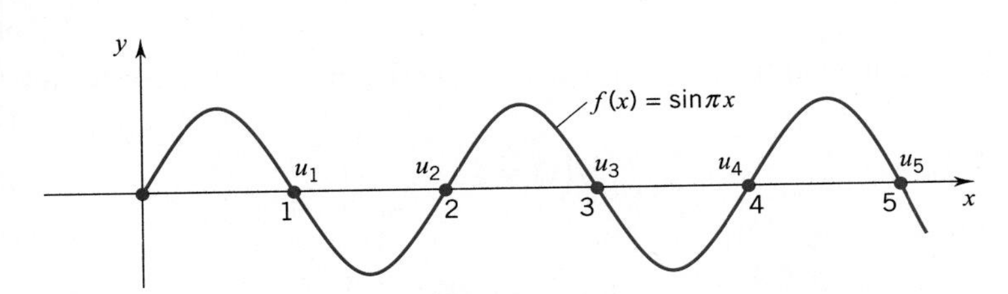

Figure 10.2.3

EXERCISES 10.2

In Exercises 1–8, the first several terms of a sequence $\{a_n\}$ are given. Assuming that the pattern continues as indicated, find an explicit formula for a_n.

1. $2, 5, 8, 11, 14, \ldots$

2. $2, 0, 2, 0, 2, \ldots$

3. $1, -\frac{1}{3}, \frac{1}{5}, -\frac{1}{7}, \frac{1}{9}, \ldots$

4. $\frac{1}{2}, \frac{3}{4}, \frac{7}{8}, \frac{15}{16}, \frac{31}{32}, \ldots$

5. $2, \frac{5}{2}, \frac{10}{3}, \frac{17}{4}, \frac{26}{5}, \ldots$

6. $-\frac{1}{4}, \frac{2}{9}, -\frac{3}{16}, \frac{4}{25}, -\frac{5}{36}, \ldots$

7. $1, \frac{1}{2}, 3, \frac{1}{4}, 5, \frac{1}{6}, \ldots$

8. $1, 2, \frac{1}{9}, 4, \frac{1}{25}, 6, \frac{1}{49}, \ldots$

In Exercises 9–44, determine the boundedness and monotonicity of the indicated sequence.

9. $\left\{\frac{2}{n}\right\}$.

10. $\left\{\frac{(-1)^n}{n}\right\}$.

11. $\{\sqrt{n}\}$.

12. $\{(1.001)^n\}$.

13. $\left\{\frac{n + (-1)^n}{n}\right\}$.

14. $\left\{\frac{n - 1}{n}\right\}$.

15. $\{(0.9)^n\}$.

16. $\{\sqrt{n^2 + 1}\}$.

17. $\left\{\frac{n^2}{n+1}\right\}$.

18. $\left\{\frac{2^n}{4^n+1}\right\}$.

19. $\left\{\frac{4n}{\sqrt{4n^2+1}}\right\}$.

20. $\left\{\frac{n+1}{n^2}\right\}$.

21. $\left\{\frac{4^n}{2^n+100}\right\}$.

22. $\left\{\frac{n^2}{\sqrt{n^3+1}}\right\}$.

23. $\left\{\frac{10^{10}\sqrt{n}}{n+1}\right\}$.

24. $\left\{\frac{n^2+1}{3n+2}\right\}$.

25. $\left\{\ln\left(\frac{2n}{n+1}\right)\right\}$.

26. $\left\{\frac{n+2}{3^{10}\sqrt{n}}\right\}$.

27. $\left\{\frac{(n+1)^2}{n^2}\right\}$.

28. $\{(-1)^n\sqrt{n}\}$.

29. $\left\{\sqrt{4-\frac{1}{n}}\right\}$.

30. $\left\{\ln\left(\frac{n+1}{n}\right)\right\}$.

31. $\{(-1)^{2n+1}\sqrt{n}\}$.

32. $\left\{\frac{\sqrt{n+1}}{\sqrt{n}}\right\}$.

33. $\left\{\frac{2^n-1}{2^n}\right\}$.

34. $\left\{\frac{1}{2n}-\frac{1}{2n+3}\right\}$.

35. $\left\{\sin\frac{\pi}{n+1}\right\}$.

36. $\{(-\frac{1}{2})^n\}$.

37. $\{(1.2)^{-n}\}$.

38. $\left\{\frac{n+3}{\ln(n+3)}\right\}$.

39. $\left\{\frac{1}{n}-\frac{1}{n+1}\right\}$.

40. $\{\cos n\pi\}$.

41. $\left\{\frac{\ln(n+2)}{n+2}\right\}$.

42. $\left\{\frac{(-2)^n}{n^{10}}\right\}$.

43. $\left\{\frac{3^n}{(n+1)^2}\right\}$.

44. $\left\{\frac{1-(\frac{1}{2})^n}{(\frac{1}{2})^n}\right\}$.

45. Show that the sequence $\{5^n/n!\}$ decreases for $n \geq 5$. Is the sequence nonincreasing?

46. Let M be a positive integer. Show that $\{M^n/n!\}$ decreases for $n \geq M$.

47. Show that, if $0 < c < d$, then the sequence

$$a_n = (c^n + d^n)^{1/n}$$

is bounded and monotonic.

48. Show that linear combinations and products of bounded sequences are bounded.

A sequence $\{a_n\}$ is said to be defined *recursively* if, for some $k \geq 1$, the terms $a_1, a_2, \ldots, a_k$ are given and a_n is specified in terms of $a_1, a_2, \ldots, a_{n-1}$ for each $n \geq k$. The formula specifying a_n for $n \geq k$ in terms of some (or all) of its predecessors is called a *recurrence relation.* In Exercises 49–64, write down the first six terms of the sequence and then give the general term.

49. $a_1 = 1;\quad a_{n+1} = \frac{1}{n+1}a_n$.

50. $a_1 = 1;\quad a_{n+1} = a_n + 3n(n+1) + 1$.

51. $a_1 = 1;\quad a_{n+1} = \frac{1}{2}(a_n + 1)$.

52. $a_1 = 1;\quad a_{n+1} = \frac{1}{2}a_n + 1$.

53. $a_1 = 1;\quad a_{n+1} = a_n + 2$.

54. $a_1 = 1;\quad a_{n+1} = \frac{n}{n+1}a_n$.

55. $a_1 = 1;\quad a_{n+1} = 3a_n + 1$.

56. $a_1 = 1;\quad a_{n+1} = 4a_n + 3$.

57. $a_1 = 1;\quad a_{n+1} = a_n + 2n + 1$.

58. $a_1 = 1;\quad a_{n+1} = 2a_n + 1$.

59. $a_1 = 1;\quad a_{n+1} = a_n + \cdots + a_1$.

60. $a_1 = 3;\quad a_{n+1} = 4 - a_n$.

61. $a_1 = 2, a_2 = 1, a_3 = 2;\quad a_{n+1} = 6 - (a_n + a_{n-1} + a_{n-2}),\ n \geq 3$.

62. $a_1 = 1, a_2 = 2;\quad a_{n+1} = 2a_n - a_{n-1},\ n \geq 2$.

63. $a_1 = 1, a_2 = 3;\quad a_{n+1} = 2a_n - a_{n-1},\ n \geq 2$.

64. $a_1 = 1, a_2 = 3;\quad a_{n+1} = 3a_n - 2n - 1,\ n \geq 2$.

In Exercises 65–68, use mathematical induction to prove the following assertions for all $n \geq 1$.

65. If $a_1 = 1$ and $a_{n+1} = 2a_n + 1$, then $a_n = 2^n - 1$.

66. If $a_1 = 3$ and $a_{n+1} = a_n + 5$, then $a_n = 5n - 2$.

67. If $a_1 = 1$ and $a_{n+1} = \frac{n+1}{2n}a_n$, then $a_n = \frac{n}{2^{n-1}}$.

68. If $a_1 = 1$ and $a_{n+1} = a_n - \frac{1}{n(n+1)}$, then $a_n = \frac{1}{n}$.

69. Let r be a real number, $r \neq 0$. Define a sequence $\{S_n\}$ by

$$S_1 = 1$$
$$S_2 = 1 + r$$
$$S_3 = 1 + r + r^2$$
$$\vdots$$
$$S_n = 1 + r + r^2 + \cdots + r^{n-1}$$
$$\vdots$$

(a) Suppose $r = 1$. What is S_n for $n = 1, 2, 3, \ldots$?
(b) Suppose $r \neq 1$. Find a formula for S_n that does not involve adding up the powers of r. HINT: Calculate $S_n - rS_n$.

70. Let $a_n = \dfrac{1}{n(n+1)}$, $n = 1, 2, 3, \ldots$, and let $\{S_n\}$ be the sequence defined by

$$\begin{aligned} S_1 &= a_1 \\ S_2 &= a_1 + a_2 \\ S_3 &= a_1 + a_2 + a_3 \\ &\ \vdots \\ S_n &= a_1 + a_2 + a_3 + \cdots + a_n \\ &\ \vdots \end{aligned}$$

Find a formula for S_n, $n = 1, 2, 3, \ldots$, that does not involve adding up the terms $a_1, a_2, a_3, \ldots$. HINT: Use partial fractions to write $1/k(k+1)$ as the sum of two fractions.

71. A ball is dropped from a height of 100 feet. Each time it hits the ground it rebounds to 75% of its previous height.
(a) Let S_n be the distance that the ball travels between the nth and $(n+1)$st bounce, $n = 1, 2, 3, \ldots$. Find a formula for S_n.
(b) Let T_n be the time that the ball is in the air between the nth and $(n+1)$st bounce, $n = 1, 2, 3, \ldots$. Find a formula for T_n.

72. Suppose that the number of bacteria in a culture is growing exponentially (see Section 7.6) and that the number doubles every 12 hours. Find a formula for the number P_n of bacteria in the culture after n hours, given that there are 500 bacteria initially.

C 73. Let $\{a_n\}$ be the sequence defined recursively by

$$a_1 = 1; \qquad a_n = 1 + \sqrt{a_{n-1}}, \quad n = 2, 3, 4, \ldots.$$

Use mathematical induction to show that:
(a) $\{a_n\}$ is an increasing sequence.
(b) $\{a_n\}$ is bounded above.
(c) Calculate $a_2, a_3, a_4, \ldots, a_{15}$. Can you estimate the least upper bound for $\{a_n\}$?

C 74. Let $\{a_n\}$ be the sequence defined recursively by

$$a_1 = 1; \qquad a_n = \sqrt{3a_{n-1}}, \quad n = 2, 3, 4, \ldots.$$

Use mathematical induction to show that:
(a) $\{a_n\}$ is an increasing sequence.
(b) $\{a_n\}$ is bounded above.
(c) Calculate $a_1, a_2, a_3, \ldots, a_{15}$. Can you estimate the least upper bound for $\{a_n\}$?

■ 10.3 LIMIT OF A SEQUENCE

The meaning of

$$\lim_{x \to c} f(x) = L$$

is that we can make $f(x)$ as close as we wish to the number L simply by requiring that x be sufficiently close to c. The meaning of

$$\lim_{n \to \infty} a_n = L$$

(read "the limit of a_n as n tends to infinity is L") is that we can make a_n as close as we wish to the number L simply by requiring that n be sufficiently large. For another analogy, recall Section 4.7 where we considered limits of the form

$$\lim_{x \to \infty} f(x)$$

in connection with the problem of finding the horizontal asymptotes for the graph of a function f; if $f(x) \to L$ as $x \to \infty$, then the line $y = L$ is a horizontal asymptote of the graph. The statement

$$\lim_{n \to \infty} a_n = L$$

is the same as saying $a_n \to L$ as $n \to \infty$.

DEFINITION 10.3.1 LIMIT

$$\lim_{n\to\infty} a_n = L$$

iff for each $\epsilon > 0$, there exists a positive integer K such that

$$\text{if} \quad n \geq K, \qquad \text{then} \qquad |a_n - L| < \epsilon.$$

Example 1

$$\lim_{n\to\infty} \frac{1}{n} = 0.$$

PROOF Let $\epsilon > 0$ and choose an integer $K > 1/\epsilon$. Then $1/K < \epsilon$.
Now, for any $n \geq K$, we have

$$0 < \frac{1}{n} \leq \frac{1}{K},$$

and so

$$\left|\frac{1}{n} - 0\right| = \left|\frac{1}{n}\right| = \frac{1}{n} \leq \frac{1}{K} < \epsilon. \quad \square$$

Example 2

$$\lim_{n\to\infty} \frac{2n-1}{n} = 2.$$

PROOF Let $\epsilon > 0$. We must show that there exists an integer K such that

$$\left|\frac{2n-1}{n} - 2\right| < \epsilon \qquad \text{for all } n \geq K.$$

Since

$$\left|\frac{2n-1}{n} - 2\right| = \left|\frac{2n-1-2n}{n}\right| = \left|-\frac{1}{n}\right| = \frac{1}{n},$$

again we need only choose $K > 1/\epsilon$. $\square$

The next example justifies the familiar statement

$$\tfrac{1}{3} = 0.333 \ldots.$$

Example 3 The decimal fractions

$$a_n = 0.\overbrace{33 \ldots 3}^{n}, \qquad n = 1, 2, 3, \ldots$$

tend to $\frac{1}{3}$ as a limit:

$$\lim_{n\to\infty} a_n = \tfrac{1}{3}.$$

PROOF Let $\epsilon > 0$. In the first place

$$(1)\qquad \left|a_n - \frac{1}{3}\right| = \left|0.\overbrace{33\ldots3}^{n} - \frac{1}{3}\right| = \left|\frac{0.\overbrace{99\cdots9}^{n} - 1}{3}\right| = \frac{1}{3}\cdot\frac{1}{10^n} < \frac{1}{10^n}.$$

Now choose K so that $1/10^K < \epsilon$. If $n \geq K$, then by (1)

$$\left|a_n - \frac{1}{3}\right| < \frac{1}{10^n} \leq \frac{1}{10^K} < \epsilon. \quad \square$$

Example 4 The limit

$$\lim_{n\to\infty} \frac{n}{\sqrt{n+1}}$$

does not exist.

PROOF Note that

$$\frac{n}{\sqrt{n+1}} = \frac{1}{\frac{1}{n}\sqrt{n+1}} = \frac{1}{\sqrt{\frac{1}{n} + \frac{1}{n^2}}}.$$

Since $1/n \to 0$ and $1/n^2 \to 0$ as $n \to \infty$, the limit does not exist. $\square$

Remark Following the ideas in Section 4.7, we will use the notation $a_n \to \infty$ as $n \to \infty$ to mean that corresponding to any positive number M there is a positive integer K such that $a_n > M$ for all $n \geq K$. Thus, in Example 4, we have

$$\frac{n}{\sqrt{n+1}} \to \infty \quad \text{as } n \to \infty.$$

The notation $a_n \to -\infty$ means that for any negative number Q there is a positive integer K such that $a_n < Q$ for all $n \geq K$. $\square$

Limit Theorems

The limit process for sequences is so similar to the limit process you have already studied that you may find you can prove many of the limit theorems yourself. In any case, try to come up with your own proofs and refer to these only if necessary.

THEOREM 10.3.2 UNIQUENESS OF LIMIT

If $\lim_{n\to\infty} a_n = L$ and $\lim_{n\to\infty} a_n = M$, then $L = M$.

A proof, similar to the proof of Theorem 2.3.1, is given in the supplement at the end of this section.

DEFINITION 10.3.3

A sequence that has a limit is said to be *convergent*. A sequence that has no limit is said to be *divergent*.

Instead of writing

$$\lim_{n\to\infty} a_n = L,$$

we will often write

$$a_n \to L \qquad \text{(read "}a_n\text{ converges to }L\text{")}$$

or more fully

$$a_n \to L \quad \text{as} \quad n \to \infty.$$

THEOREM 10.3.4

Every convergent sequence is bounded.

PROOF Assume that $a_n \to L$ and choose any positive number: 1, for instance. Using 1 as ϵ, you can see that there must exist a positive integer K such that

$$|a_n - L| < 1 \qquad \text{for all } n \geq K.$$

This means that

$$|a_n| < 1 + |L| \qquad \text{for all } n \geq K$$

and, consequently,

$$|a_n| \leq \max\{|a_1|, |a_2|, \ldots, |a_{K-1}|, 1 + |L|\} \qquad \text{for all } n.$$

This proves that $\{a_n\}$ is bounded. ❑

Since every convergent sequence is bounded, a sequence that is not bounded cannot be convergent; namely,

(10.3.5) every unbounded sequence is divergent.

The sequences

$$a_n = \tfrac{1}{2}n, \qquad b_n = \frac{n^2}{n+1}, \qquad c_n = n \ln n$$

are all unbounded. Each of these sequences is therefore divergent.

Boundedness does not imply convergence. As a counterexample, consider the oscillating sequence

$$\{1, 0, 1, 0, \ldots\} = \left\{\frac{1 + (-1)^{n+1}}{2}\right\}.$$

This sequence is certainly bounded (above by 1 and below by 0), but obviously it does not converge: the limit would have to be arbitrarily close to both 0 and 1 simultaneously.

Boundedness together with monotonicity does imply convergence.

THEOREM 10.3.6

A bounded nondecreasing sequence converges to its least upper bound; a bounded nonincreasing sequence converges to its greatest lower bound.

PROOF Suppose that $\{a_n\}$ is bounded and nondecreasing. If L is the least upper bound of this sequence, then

$$a_n \leq L \qquad \text{for all } n.$$

Now let ϵ be an arbitrary positive number. By Theorem 10.1.3 there exists a_k such that

$$L - \epsilon < a_k.$$

Since the sequence is nondecreasing,

$$a_k \leq a_n \qquad \text{for all } n \geq k.$$

It follows that

$$L - \epsilon < a_n \leq L \qquad \text{for all } n \geq k.$$

This shows that

$$|a_n - L| < \epsilon \qquad \text{for all } n \geq k$$

and proves that

$$a_n \to L.$$

The nonincreasing case can be handled in a similar manner. ❑

Example 5 Take the sequence

$$\{(3^n + 4^n)^{1/n}\}.$$

Since

$$3 = (3^n)^{1/n} < (3^n + 4^n)^{1/n} < (2 \cdot 4^n)^{1/n} = 2^{1/n} \cdot 4 \leq 8,$$

the sequence is bounded. Note that

$$\begin{aligned}(3^n + 4^n)^{(n+1)/n} &= (3^n + 4^n)^{1/n}(3^n + 4^n) \\ &= (3^n + 4^n)^{1/n}3^n + (3^n + 4^n)^{1/n}4^n.\end{aligned}$$

Since

$$(3^n + 4^n)^{1/n} > (3^n)^{1/n} = 3 \qquad \text{and} \qquad (3^n + 4^n)^{1/n} > (4^n)^{1/n} = 4,$$

it follows that

$$(3^n + 4^n)^{(n+1)/n} > 3 \cdot (3^n) + 4 \cdot (4^n) = 3^{n+1} + 4^{n+1}.$$

Taking the $(n + 1)$st root of the left and right sides of this inequality, we have

$$(3^n + 4^n)^{1/n} > (3^{n+1} + 4^{n+1})^{1/(n+1)}.$$

The sequence is decreasing. Being bounded, it must be convergent. (Later you will be asked to show that the limit is 4.) ❑

THEOREM 10.3.7

Let α be a real number. If $a_n \to L$ and $b_n \to M$, then

$$\text{(i) } a_n + b_n \to L + M, \qquad \text{(ii) } \alpha a_n \to \alpha L, \qquad \text{(iii) } a_n b_n \to LM.$$

If, in addition, $M \neq 0$ and $b_n \neq 0$ for all n, then

$$\text{(iv) } \frac{1}{b_n} \to \frac{1}{M} \quad \text{and} \quad \text{(v) } \frac{a_n}{b_n} \to \frac{L}{M}.$$

Proofs of parts (i) and (ii) are left as exercises. For proofs of parts (iii)–(v), see the supplement at the end of this section.

We are now in a position to handle any rational sequence $\{a_n\}$, where

$$\text{(2)} \qquad a_n = \frac{\alpha_k n^k + \alpha_{k-1} n^{k-1} + \cdots + \alpha_0}{\beta_j n^j + \beta_{j-1} n^{j-1} + \cdots + \beta_0}, \quad \alpha_k \neq 0,\ \beta_j \neq 0.$$

To determine the behavior of such a sequence we need only divide both numerator and denominator by the highest power of n that occurs.

Example 6

$$\frac{3n^4 - 2n^2 + 1}{n^5 - 3n^3} = \frac{3/n - 2/n^3 + 1/n^5}{1 - 3/n^2} \to \frac{0}{1} = 0. \quad ❑$$

Example 7

$$\frac{1 - 4n^7}{n^7 + 12n} = \frac{1/n^7 - 4}{1 + 12/n^6} \to \frac{-4}{1} = -4. \quad ❑$$

Example 8

$$\frac{n^4 - 3n^2 + n + 2}{n^3 + 7n} = \frac{1 - 3/n^2 + 1/n^3 + 2/n^4}{1/n + 7/n^3}.$$

Since the numerator tends to 1 and the denominator tends to 0, the sequence is unbounded. Therefore it cannot converge. ❑

In general, if a_n is given by (2), then,

$$\text{as} \quad n \to \infty, \qquad a_n \to \begin{cases} 0 & \text{if } k < j \\ \alpha_k/\beta_k & \text{if } k = j \\ \pm\infty & \text{if } k > j. \end{cases}$$

(See Exercise 40.) The corresponding result for the limit of a rational function $R(x) = P(x)/Q(x)$ as $x \to \infty$ was discussed in Section 4.7.

THEOREM 10.3.8

$$a_n \to L \quad \text{iff} \quad a_n - L \to 0 \quad \text{iff} \quad |a_n - L| \to 0.$$

We leave the proof to you.

THEOREM 10.3.9 THE PINCHING THEOREM FOR SEQUENCES

Suppose that there is a positive integer K such that for all $n \geq K$

$$a_n \leq b_n \leq c_n.$$

If $a_n \to L$ and $c_n \to L$, then $b_n \to L$.

Once again the proof is left to you.

As an immediate and obvious consequence of the pinching theorem we have the following corollary.

(10.3.10) Suppose that there is a positive integer K such that for all $n \geq K$

$$|b_n| \leq c_n.$$

If $c_n \to 0$, then $|b_n| \to 0$.

Example 9

$$\frac{\cos n}{n} \to 0 \quad \text{since} \quad \left|\frac{\cos n}{n}\right| \leq \frac{1}{n} \quad \text{and} \quad \frac{1}{n} \to 0. \quad ❑$$

Example 10

$$\sqrt{4 + \left(\frac{1}{n}\right)^2} \to 2$$

since

$$2 \leq \sqrt{4 + \left(\frac{1}{n}\right)^2} \leq \sqrt{4 + 4\left(\frac{1}{n}\right) + \left(\frac{1}{n}\right)^2} = 2 + \frac{1}{n} \quad \text{and} \quad 2 + \frac{1}{n} \to 2. \quad ❑$$

Example 11 Recall from Chapter 7 that there are several ways of defining the number e. Here we complete the argument begun in Theorem 7.5.13.

(10.3.11)
$$\lim_{n\to\infty}\left(1+\frac{1}{n}\right)^n = e.$$

PROOF You have already seen that, for all positive integers n,

$$\left(1+\frac{1}{n}\right)^n \leq e \leq \left(1+\frac{1}{n}\right)^{n+1}. \qquad \text{(Theorem 7.5.13)}$$

Dividing the right-hand inequality by $1 + 1/n$, we have

$$\frac{e}{1+1/n} \leq \left(1+\frac{1}{n}\right)^n.$$

Combining this with the left-hand inequality, we can write

$$\frac{e}{1+1/n} \leq \left(1+\frac{1}{n}\right)^n \leq e.$$

Since

$$\frac{e}{1+1/n} \to \frac{e}{1} = e,$$

we can conclude from the pinching theorem that

$$\left(1+\frac{1}{n}\right)^n \to e. \quad \square$$

The sequences

$$\left\{\cos\frac{\pi}{n}\right\}, \quad \left\{\ln\left(\frac{n}{n+1}\right)\right\}, \quad \{e^{1/n}\}, \quad \left\{\tan\left(\sqrt{\frac{\pi^2 n^2 - 8}{16n^2}}\right)\right\}$$

are all of the form $\{f(c_n)\}$ with f a continuous function. Such sequences are frequently easy to deal with. The basic idea is this: When a continuous function is applied to a convergent sequence, the result is itself a convergent sequence. More precisely, we have the following theorem.

THEOREM 10.3.12

Suppose that

$$c_n \to c$$

and that, for each n, c_n is in the domain of f. If f is continuous at c, then

$$f(c_n) \to f(c).$$

PROOF We assume that f is continuous at c and take $\epsilon > 0$. From the continuity of f at c we know that there exists $\delta > 0$ such that

$$\text{if} \quad |x - c| < \delta, \qquad \text{then} \quad |f(x) - f(c)| < \epsilon.$$

Since $c_n \to c$, we know that there exists a positive integer K such that

$$\text{if} \quad n \geq K, \qquad \text{then} \quad |c_n - c| < \delta.$$

It follows therefore that

$$\text{if} \quad n \geq K, \qquad \text{then} \quad |f(c_n) - f(c)| < \epsilon. \quad \square$$

The following examples illustrate the use of Theorem 10.3.12 in calculating limits.

Example 12 Since $\pi/n \to 0$ and the cosine function is continuous at 0,

$$\cos\left(\frac{\pi}{n}\right) \to \cos 0 = 1. \quad \square$$

Example 13 Since

$$\frac{n}{n+1} = \frac{1}{1 + 1/n} \to 1$$

and the logarithm function is continuous at 1,

$$\ln\left(\frac{n}{n+1}\right) \to \ln 1 = 0. \quad \square$$

Example 14 Since $1/n \to 0$ and the exponential function is continuous at 0,

$$e^{1/n} \to e^0 = 1. \quad \square$$

Example 15 Since

$$\frac{\pi^2 n^2 - 8}{16n^2} = \frac{\pi^2 - 8/n^2}{16} \to \frac{\pi^2}{16}$$

and the function $f(x) = \tan\sqrt{x}$ is continuous at $\pi^2/16$,

$$\tan\left(\sqrt{\frac{\pi^2 n^2 - 8}{16n^2}}\right) \to \tan\left(\sqrt{\frac{\pi^2}{16}}\right) = \tan\frac{\pi}{4} = 1. \quad \square$$

Example 16 Since

$$\frac{2n+1}{n} + \left(5 - \frac{1}{n^2}\right) \to 7$$

and the square-root function is continuous at 7,

$$\sqrt{\frac{2n+1}{n} + \left(5 - \frac{1}{n^2}\right)} \to \sqrt{7}. \quad \square$$

Example 17 Since the absolute-value function is everywhere continuous,

$$a_n \to L \quad \text{implies} \quad |a_n| \to |L|. \quad \square$$

Remark For some time now we have asked you to take on faith two fundamentals of integration: that continuous functions do have definite integrals and that these integrals can be expressed as limits of Riemann sums. We could not give you proofs of these assertions because we did not have the necessary tools. Now we do. Proofs are given in Appendix B. ❑

EXERCISES 10.3

In Exercises 1–36, state whether or not the sequence converges and, if it does, find the limit.

1. $\{2^n\}$.

2. $\left\{\frac{2}{n}\right\}$.

3. $\left\{\frac{(-1)^n}{n}\right\}$.

4. $\{\sqrt{n}\}$.

5. $\left\{\frac{n-1}{n}\right\}$.

6. $\left\{\frac{n+(-1)^n}{n}\right\}$.

7. $\left\{\frac{n+1}{n^2}\right\}$.

8. $\left\{\sin\frac{\pi}{2n}\right\}$.

9. $\left\{\frac{2^n}{4^n+1}\right\}$.

10. $\left\{\frac{n^2}{n+1}\right\}$.

11. $\{(-1)^n\sqrt{n}\}$.

12. $\left\{\frac{4n}{\sqrt{n^2+1}}\right\}$.

13. $\{(-\frac{1}{2})^n\}$.

14. $\left\{\frac{4^n}{2^n+10^6}\right\}$.

15. $\left\{\tan\frac{n\pi}{4n+1}\right\}$.

16. $\left\{\frac{10^{10}\sqrt{n}}{n+1}\right\}$.

17. $\left\{\frac{(2n+1)^2}{(3n-1)^2}\right\}$.

18. $\left\{\ln\left(\frac{2n}{n+1}\right)\right\}$.

19. $\left\{\frac{n^2}{\sqrt{2n^4+1}}\right\}$.

20. $\left\{\frac{n^4-1}{n^4+n-6}\right\}$.

21. $\{\cos n\pi\}$.

22. $\left\{\frac{n^5}{17n^4+12}\right\}$.

23. $\{e^{1/\sqrt{n}}\}$.

24. $\left\{\sqrt{4-\frac{1}{n}}\right\}$.

25. $\{(0.9)^{-n}\}$.

26. $\left\{\frac{2^n-1}{2^n}\right\}$.

27. $\{\ln n - \ln(n+1)\}$.

28. $\left\{\frac{1}{n}-\frac{1}{n+1}\right\}$.

29. $\left\{\frac{\sqrt{n+1}}{2\sqrt{n}}\right\}$.

30. $\{(0.9)^n\}$.

31. $\left\{\left(1+\frac{1}{n}\right)^{2n}\right\}$.

32. $\left\{\left(1+\frac{1}{n}\right)^{n/2}\right\}$.

33. $\left\{\frac{2^n}{n^2}\right\}$.

34. $\left\{\frac{(n+1)\cos\sqrt{n}}{n(1+\sqrt{n})}\right\}$.

35. $\left\{\frac{\sqrt{n}\sin(e^n\pi)}{n+1}\right\}$.

36. $\{2\ln 3n - \ln(n^2+1)\}$.

37. Prove that, if $a_n \to L$ and $b_n \to M$, then $a_n + b_n \to L + M$.

38. Let α be a real number. Prove that, if $a_n \to L$, then $\alpha a_n \to \alpha L$.

39. Prove that

$$\left(1+\frac{1}{n}\right)^{n+1} \to e \quad \text{given that} \quad \left(1+\frac{1}{n}\right)^n \to e.$$

40. Determine the convergence or divergence of a rational sequence

$$a_n = \frac{\alpha_k n^k + \alpha_{k-1}n^{k-1} + \cdots + \alpha_0}{\beta_j n^j + \beta_{j-1}n^{j-1} + \cdots + \beta_0}$$

$$\text{with } \alpha_k \neq 0,\ \beta_j \neq 0,$$

given that: (a) $k = j$; (b) $k < j$; (c) $k > j$. Justify your answers.

41. Prove that a bounded nonincreasing sequence converges to its greatest lower bound.

42. Let $\{a_n\}$ be a sequence of real numbers. Let $\{e_n\}$ be the sequence of even terms:

$$e_n = a_{2n}$$

and let $\{o_n\}$ be the sequence of odd terms:

$$o_n = a_{2n-1}.$$

Show that

$$a_n \to L \quad \text{iff} \quad e_n \to L \text{ and } o_n \to L.$$

43. Prove the pinching theorem for sequences.

44. Let $\{a_n\}$ and $\{b_n\}$ be sequences such that $a_n \to 0$ and $\{b_n\}$ is bounded. Prove that $a_n b_n \to 0$.

45. Let $\{a_n\}$ be a convergent sequence with limit L. Prove that if $a_n \le M$ for all n, then $L \le M$.

46. According to Example 17, if $a_n \to L$, then $|a_n| \to |L|$. Is the converse true? That is, if $|a_n| \to |L|$, does it follow that $a_n \to L$? Prove or give a counter-example.

47. Let f be a continuous function on $(-\infty, \infty)$ and let r be a real number. Define the sequence $\{a_n\}$ as follows:

$$a_1 = r,\ a_2 = f(r),\ a_3 = f[f(r)],\ a_4 = f\{f[f(r)]\},\ \ldots.$$

Prove that if $a_n \to L$, then $f(L) = L$; that is, L is a fixed point of f.

48. Show that

$$\frac{2^n}{n!} \to 0.$$

HINT: First show that

$$\frac{2^n}{n!} = \frac{2}{1}\cdot\frac{2}{2}\cdot\frac{2}{3}\cdot\cdots\cdot\frac{2}{n} \le \frac{4}{n}.$$

49. Prove that $(1/n)^{1/p} \to 0$ for all positive integers p.

50. Prove Theorem 10.3.8.

In Exercises 51–58, the sequences are defined recursively.† Determine in each case whether the sequence converges and, if so, find the limit. Start each sequence with $a_1 = 1$.

51. $a_{n+1} = \frac{1}{e}a_n$.

52. $a_{n+1} = 2^{n+1}a_n$.

53. $a_{n+1} = \frac{1}{n+1}a_n$.

54. $a_{n+1} = \frac{n}{n+1}a_n$.

55. $a_{n+1} = 1 - a_n$.

56. $a_{n+1} = -a_n$.

57. $a_{n+1} = \frac{1}{2}a_n + 1$.

58. $a_{n+1} = \frac{1}{3}a_n + 1$.

In Exercises 59–66, evaluate numerically the limit of each sequence as $n \to \infty$. Some of these sequences converge more rapidly than others. Determine for each sequence the least value of n for which the nth term differs from the limit by less than 0.001.

59. $\left\{\frac{1}{n^2}\right\}$.

60. $\left\{\frac{1}{\sqrt{n}}\right\}$.

61. $\left\{\frac{n}{10^n}\right\}$.

62. $\left\{\frac{n^{10}}{10^n}\right\}$.

63. $\left\{\frac{1}{n!}\right\}$.

64. $\left\{\frac{2^n}{n!}\right\}$.

65. $\left\{\frac{\ln n}{n^2}\right\}$.

66. $\left\{\frac{\ln n}{n}\right\}$.

67. (a) Find the exact value of the limit of the sequence $\{a_n\}$ given in Exercise 73, Section 10.2. HINT: Suppose $a_n \to L$, then $a_{n-1} \to L$.
(b) Find the exact value of the limit of the sequence $\{a_n\}$ given in Exercise 74, Section 10.2.

68. Let $\{a_n\}$ be the sequence defined recursively by

$$a_1 = 1, \quad a_n = \sqrt{6 + a_{n-1}}, \quad n = 2, 3, 4, \ldots.$$

(a) Approximate a_2, a_3, a_4, a_5, a_6. Round your answers to six decimal places.
(b) Use mathematical induction to show that $a_n \le 3$ for all n.
(c) Show that $\{a_n\}$ is an increasing sequence. HINT: $a_{n+1}^2 - a_n^2 = (3 - a_n)(2 + a_n)$.
(d) What is the limit of this sequence?

69. Let $\{a_n\}$ be the sequence defined recursively by

$$a_1 = 1, \quad a_n = \cos a_{n-1}, \quad n = 2, 3, 4, \ldots.$$

(a) Approximate $a_2, a_3, a_4, \ldots, a_{10}$. Round your answers to six decimal places.
(b) Assuming that $a_n \to L$, approximate L to six decimal places and interpret your result geometrically. HINT: Use Exercise 47.

70. Let $\{a_n\}$ be the sequence defined recursively by

$$a_1 = 1, \quad a_n = a_{n-1} + \cos a_{n-1}, \quad n = 2, 3, 4, \ldots.$$

(a) Approximate $a_2, a_3, a_4, \ldots, a_{10}$. Round your answers to six decimal places.
(b) Assuming that $a_n \to L$, approximate L to six decimal places and interpret your result geometrically.

71. Let R be a positive number. Approximations to $\sqrt{R}$ are generated by the sequence defined recursively by

$$a_1 = 1, \quad a_n = \frac{1}{2}\left(a_{n-1} + \frac{R}{a_{n-1}}\right), \quad n = 2, 3, 4, \ldots.$$

Let $R = 3$.
(a) Approximate $a_2, a_3, \cdots, a_8$. Round your answers to six decimal places.
(b) Assuming that $a_n \to L$, prove that $L = \sqrt{3}$.
(c) Show that the recursion relation given above is simply the sequence generated by the Newton-Raphson method applied to the function $f(x) = x^2 - R$ (see Section 3.9).

72. The Newton-Raphson method (Section 3.9) applied to a differentiable function f generates a sequence $\{x_n\}$ that,

† The notion was introduced in Exercises 10.2.

under certain conditions, converges to a zero of f. The recursion formula for the sequence is given by

$$x_{n+1} = x_n - \frac{f(x_n)}{f'(x_n)}, \quad n = 1, 2, 3, \ldots .$$

Determine whether the following sequences converge, and if so, give the limit. HINT: Identify the function that generates the sequence.

(a) $x_{n+1} = x_n - \dfrac{x_n^3 - 8}{3x_n^2}; \quad x_1 = 1.$

(b) $x_{n+1} = x_n - \dfrac{\sin x_n - 0.5}{\cos x_n}; \quad x_1 = 0.$

(c) $x_{n+1} = x_n - \dfrac{\ln x_n - 1}{1/x_n} = x_n[2 - \ln x_n]; \quad x_1 = 1.$

---

*SUPPLEMENT TO SECTION 10.3

PROOF OF THEOREM 10.3.2

If $L \neq M$, then

$$\tfrac{1}{2}|L - M| > 0.$$

The assumption that $\lim_{n\to\infty} a_n = L$ and $\lim_{n\to\infty} a_n = M$ gives the existence of K_1 such that

$$\text{if} \quad n \geq K_1, \qquad \text{then} \quad |a_n - L| < \tfrac{1}{2}|L - M|$$

and the existence of K_2 such that

$$\text{if} \quad n \geq K_2, \qquad \text{then} \quad |a_n - m| < \tfrac{1}{2}|L - M|.†$$

For $n \geq \max\{K_1, K_2\}$ we have

$$|a_n - L| + |a_n - M| < |L - M|.$$

By the triangle inequality we have

$$|L - M| = |(L - a_n) + (a_n - M)| \leq |L - a_n| + |a_n - M| = |a_n - L| + |a_n - M|.$$

Combining the last two statements, we have

$$|L - M| < |L - M|.$$

The hypothesis $L \neq M$ has led to an absurdity. We conclude that $L = M$. ❑

PROOF OF THEOREM 10.3.7 (iii)–(v)

To prove (iii), we set $\epsilon > 0$. For each n,

$$\begin{aligned} |a_n b_n - LM| &= |(a_n b_n - a_n M) + (a_n M - LM)| \\ &\leq |a_n||b_n - M| + |M||a_n - L|. \end{aligned}$$

Since $\{a_n\}$ is convergent, $\{a_n\}$ is bounded; that is, there exists $Q > 0$ such that

$$|a_n| \leq Q \qquad \text{for all } n.$$

Since $|M| < |M| + 1$, we have

$$(1) \qquad |a_n b_n - LM| \leq Q|b_n - M| + (|M| + 1)|a_n - L|.††$$

Since $b_n \to M$, we know that there exists K_1 such that

$$\text{if} \quad n \geq K_1, \qquad \text{then} \quad |b_n - M| < \frac{\epsilon}{2Q}.$$

---

† We can reach these conclusions from Definition 10.3.1 by taking $\tfrac{1}{2}|L - M|$ as ϵ.

†† Soon we will want to divide by the coefficient of $|a_n - L|$. We have replaced $|M|$ by $|M| + 1$ because $|M|$ can be zero.

Since $a_n \to L$, we know that there exists K_2 such that

$$\text{if} \quad n \geq K_2, \qquad \text{then} \quad |a_n - L| < \frac{\epsilon}{2(|M| + 1)}.$$

For $n \geq \max\{K_1, K_2\}$ both conditions hold, and consequently

$$Q|b_n - M| + (|M| + 1)|a_n - L| < \frac{\epsilon}{2} + \frac{\epsilon}{2} = \epsilon.$$

In view of (1), we can conclude that

$$\text{if} \quad n \geq \max\{K_1, K_2\}, \qquad \text{then} \quad |a_n b_n - LM| < \epsilon.$$

This proves that

$$a_n b_n \to LM. \quad \square$$

To prove (iv), once again we set $\epsilon > 0$. In the first place

$$\left|\frac{1}{b_n} - \frac{1}{M}\right| = \left|\frac{M - b_n}{b_n M}\right| = \frac{|b_n - M|}{|b_n||M|}.$$

Since $b_n \to M$ and $|M|/2 > 0$, there exists K_1 such that

$$\text{if} \quad n \geq K_1, \qquad \text{then} \quad |b_n - M| < \frac{|M|}{2}.$$

This tells us that for $n \geq K_1$ we have

$$|b_n| > \frac{|M|}{2} \qquad \text{and thus} \qquad \frac{1}{|b_n|} < \frac{2}{|M|}.$$

Thus for $n \geq K_1$ we have

$$\left|\frac{1}{b_n} - \frac{1}{M}\right| \leq \frac{2}{|M|^2}|b_n - M|. \tag{2}$$

Since $b_n \to M$ there exists K_2 such that

$$\text{if} \quad n \geq K_2, \qquad \text{then} \quad |b_n - M| < \frac{\epsilon|M|^2}{2}.$$

Thus for $n \geq K_2$ we have

$$\frac{2}{|M|^2}|b_n - M| < \epsilon.$$

In view of (2), we can be sure that

$$\text{if} \quad n \geq \max\{K_1, K_2\}, \qquad \text{then} \quad \left|\frac{1}{b_n} - \frac{1}{M}\right| < \epsilon.$$

This proves that

$$\frac{1}{b_n} \to \frac{1}{M}. \quad \square$$

The proof of (v) is now easy:

$$\frac{a_n}{b_n} = a_n \cdot \frac{1}{b_n} \to L \cdot \frac{1}{M} = \frac{L}{M}. \quad \square$$

■ 10.4 SOME IMPORTANT LIMITS

Our purpose here is to familiarize you with some limits that are particularly important in calculus and to give you more experience with limit arguments.

(10.4.1)

> If $x > 0$, then
> $$x^{1/n} \to 1 \qquad \text{as} \quad n \to \infty.$$

PROOF Fix any $x > 0$. Note that

$$\ln(x^{1/n}) = \frac{1}{n}\ln x \to 0 \qquad \text{as} \quad n \to \infty.$$

Since the exponential function is continuous at 0, it follows from Theorem 10.3.12 that

$$x^{1/n} = e^{(1/n)\ln x} \to e^0 = 1. \quad \square$$

(10.4.2)

> If $|x| < 1$, then
> $$x^n \to 0 \qquad \text{as} \quad n \to \infty.$$

PROOF The result clearly holds if $x = 0$. Fix any x with $|x| < 1$ and observe that $\{|x|^n\}$ is a decreasing sequence:

$$|x|^{n+1} = |x|\,|x|^n < |x|^n.$$

Now let $\epsilon > 0$. By (10.4.1)

$$\epsilon^{1/n} \to 1 \qquad \text{as} \quad n \to \infty.$$

Thus there exists an integer $k > 0$ such that

$$|x| < \epsilon^{1/k}. \qquad \text{(explain)}$$

Obviously, then, $|x|^k < \epsilon$. Since $\{|x|^n\}$ is a decreasing sequence,

$$|x^n| = |x|^n < \epsilon \qquad \text{for all } n \geq k. \quad \square$$

(10.4.3)

> For each $\alpha > 0$
> $$\frac{1}{n^\alpha} \to 0 \qquad \text{as} \quad n \to \infty.$$

PROOF Since $\alpha > 0$, there exists an odd positive integer p such that $1/p < \alpha$. Then

$$0 < \frac{1}{n^\alpha} = \left(\frac{1}{n}\right)^\alpha \leq \left(\frac{1}{n}\right)^{1/p}.$$

Since $1/n \to 0$ and $f(x) = x^{1/p}$ is continuous at 0, we have

$$\left(\frac{1}{n}\right)^{1/p} \to 0 \quad \text{and thus by the pinching theorem} \quad \frac{1}{n^\alpha} \to 0. \quad \square$$

(10.4.4)

For each real x

$$\frac{x^n}{n!} \to 0 \qquad \text{as} \quad n \to \infty.$$

PROOF Fix any real number x and choose an integer k such that $k > |x|$. For $n > k + 1$,

$$\frac{k^n}{n!} = \left(\frac{k^k}{k!}\right)\left[\frac{k}{k+1}\,\frac{k}{k+2}\cdots\frac{k}{n-1}\right]\left(\frac{k}{n}\right) < \left(\frac{k^{k+1}}{k!}\right)\left(\frac{1}{n}\right).$$

the middle term is less than 1

Since $k > |x|$, we have

$$0 < \frac{|x|^n}{n!} < \frac{k^n}{n!} < \left(\frac{k^{k+1}}{k!}\right)\left(\frac{1}{n}\right).$$

Since k is fixed and $1/n \to 0$, it follows from the pinching theorem that

$$\frac{|x|^n}{n!} \to 0 \qquad \text{and thus} \qquad \frac{x^n}{n!} \to 0. \quad \square$$

(10.4.5)

$$\frac{\ln n}{n} \to 0 \qquad \text{as} \quad n \to \infty.$$

PROOF A routine proof can be based on L'Hospital's rule (10.6.1), but that is not available to us yet. We will appeal to the pinching theorem and base our argument on the integral representation of the logarithm:

$$0 \le \frac{\ln n}{n} = \frac{1}{n}\int_1^n \frac{dt}{t} \le \frac{1}{n}\int_1^n \frac{dt}{\sqrt{t}} = \frac{2}{n}(\sqrt{n} - 1)$$
$$= 2\left(\frac{1}{\sqrt{n}} - \frac{1}{n}\right) \to 0. \quad \square$$

(10.4.6)

$$n^{1/n} \to 1 \qquad \text{as} \quad n \to \infty.$$

PROOF We know that

$$n^{1/n} = e^{(1/n)\ln n}.$$

Since

(11.1.1) $$(1/n)\ln n \to 0$$ (10.4.5)

and the exponential function is continuous at 0, it follows from Theorem 10.3.12 that

$$n^{1/n} \to e^0 = 1. \quad \square$$

(10.4.7) For each real x

$$\left(1 + \frac{x}{n}\right)^n \to e^x \quad \text{as} \quad n \to \infty.$$

PROOF For $x = 0$, the result is obvious. For $x \neq 0$,

$$\ln\left(1 + \frac{x}{n}\right)^n = n \ln\left(1 + \frac{x}{n}\right) = x\left[\frac{\ln(1 + x/n) - \ln 1}{x/n}\right].$$

The crux here is to recognize that the bracketed expression is a difference quotient for the logarithm function. Once we see this, we let $h = x/n$ and write

$$\lim_{n\to\infty}\left[\frac{\ln(1 + x/n) - \ln 1}{x/n}\right] = \lim_{h\to 0}\left[\frac{\ln(1 + h) - \ln 1}{h}\right] = 1.\dagger$$

It follows that

$$\ln\left(1 + \frac{x}{n}\right)^n \to x \quad \text{and therefore} \quad \left(1 + \frac{x}{n}\right)^n = e^{\ln(1 + x/n)^n} \to e^x. \quad \square$$

† For each $t > 0$

$$\lim_{h\to 0}\frac{\ln(t + h) - \ln t}{h} = \frac{d}{dt}(\ln t) = \frac{1}{t}.$$

EXERCISES 10.4

In Exercises 1–36, state whether or not the sequence converges as $n \to \infty$; if it does, find the limit.

1. $\{2^{2/n}\}$.
2. $\{e^{-\alpha/n}\}$.
3. $\left\{\left(\frac{2}{n}\right)^n\right\}$.
4. $\left\{\frac{\log_{10} n}{n}\right\}$.
5. $\left\{\frac{\ln(n + 1)}{n}\right\}$.
6. $\left\{\frac{3^n}{4^n}\right\}$.
7. $\left\{\frac{x^{100n}}{n!}\right\}$.
8. $\{n^{1/(n+2)}\}$.
9. $\{n^{\alpha/n}\}, \quad \alpha > 0$.
10. $\left\{\ln\left(\frac{n + 1}{n}\right)\right\}$.
11. $\left\{\frac{3^{n+1}}{4^{n-1}}\right\}$.
12. $\left\{\int_{-n}^{0} e^{2x}\,dx\right\}$.
13. $\{(n + 2)^{1/n}\}$.
14. $\left\{\left(1 - \frac{1}{n}\right)^n\right\}$.
15. $\left\{\int_0^n e^{-x}\,dx\right\}$.
16. $\left\{\frac{2^{3n-1}}{7^{n+2}}\right\}$.
17. $\left\{\int_{-n}^{n}\frac{dx}{1 + x^2}\right\}$.
18. $\left\{\int_0^n e^{-nx}\,dx\right\}$.
19. $\{(n + 2)^{1/(n+2)}\}$.
20. $\{n^2 \sin n\pi\}$.
21. $\left\{\frac{\ln n^2}{n}\right\}$.
22. $\left\{\int_{-1+1/n}^{1-1/n}\frac{dx}{\sqrt{1 - x^2}}\right\}$.

23. $\left\{n^2 \sin \dfrac{\pi}{n}\right\}$.

24. $\left\{\dfrac{n!}{2n}\right\}$.

25. $\left\{\dfrac{5^{n+1}}{4^{2n-1}}\right\}$.

26. $\left\{\left(1 + \dfrac{x}{n}\right)^{3n}\right\}$.

27. $\left\{\left(\dfrac{n+1}{n+2}\right)^n\right\}$.

28. $\left\{\displaystyle\int_{1/n}^{1} \frac{dx}{\sqrt{x}}\right\}$.

29. $\left\{\displaystyle\int_{n}^{n+1} e^{-x^2}\, dx\right\}$.

30. $\left\{\left(1 + \dfrac{1}{n^2}\right)^n\right\}$.

31. $\left\{\dfrac{n^n}{2^{n^2}}\right\}$.

32. $\left\{\displaystyle\int_{0}^{1/n} \cos e^x\, dx\right\}$.

33. $\left\{\left(1 + \dfrac{x}{2n}\right)^{2n}\right\}$.

34. $\left\{\left(1 + \dfrac{1}{n}\right)^{n^2}\right\}$.

35. $\left\{\displaystyle\int_{-1/n}^{1/n} \sin x^2\, dx\right\}$.

36. $\left\{\left(t + \dfrac{x}{n}\right)^n\right\}$, $t > 0$, $x > 0$.

37. Show that $\lim_{n\to\infty} (\sqrt{n+1} - \sqrt{n}) = 0$.

38. Show that $\lim_{n\to\infty} (\sqrt{n^2 + n} - n) = \frac{1}{2}$.

39. (a) Show that a regular polygon of n sides inscribed in a circle of radius r has perimeter $p_n = 2rn \sin(\pi/n)$.
(b) Find

$$\lim_{n\to\infty} p_n$$

and give a geometric interpretation of your result.

40. Show that

$$\text{if } 0 < c < d, \qquad \text{then } (c^n + d^n)^{1/n} \to d.$$

In Exercises 41–43, find the indicated limit.

41. $\displaystyle\lim_{n\to\infty} \frac{1 + 2 + \cdots + n}{n^2}$.

HINT: $1 + 2 + \cdots + n = \dfrac{n(n+1)}{2}$.

42. $\displaystyle\lim_{n\to\infty} \frac{1^2 + 2^2 + \cdots + n^2}{(1+n)(2+n)}$.

HINT: $1^2 + 2^2 + \cdots + n^2 = \dfrac{n(n+1)(2n+1)}{6}$.

43. $\displaystyle\lim_{n\to\infty} \frac{1^3 + 2^3 + \cdots + n^3}{2n^4 + n - 1}$.

HINT: $1^3 + 2^3 + \cdots + n^3 = \dfrac{n^2(n+1)^2}{4}$.

44. A sequence $\{a_n\}$ is said to be a *Cauchy sequence*† iff

(10.4.8)
> for each $\epsilon > 0$ there exists a positive integer K such that
> $|a_n - a_m| < \epsilon \qquad$ for all $m, n \geq K$.

Show that

(10.4.9)
> every convergent sequence is a Cauchy sequence.

It is also true that every Cauchy sequence is convergent, but this is more difficult to prove.

45. (*Arithmetic means*) For a given sequence $\{a_n\}$, let

$$m_n = \frac{1}{n}(a_1 + a_2 + \cdots + a_n).$$

(a) Prove that if $\{a_n\}$ is increasing, then $\{m_n\}$ is increasing.
(b) Prove that if $a_n \to 0$, then $m_n \to 0$.
HINT: Choose an integer $j > 0$ such that, if $n \geq j$, then $a_n < \epsilon/2$. Then for $n \geq j$,

$$|m_n| < \frac{|a_1 + a_2 + \cdots + a_j|}{n} + \frac{\epsilon}{2}\left(\frac{n-j}{n}\right).$$

46. (a) Let $\{a_n\}$ be a convergent sequence. Prove that

$$\lim_{n\to\infty} (a_n - a_{n-1}) = 0.$$

(b) What can you say about the converse? That is, suppose that $\{a_n\}$ is a sequence such that

$$\lim_{n\to\infty} (a_n - a_{n-1}) = 0.$$

Does $\{a_n\}$ necessarily converge? Prove or give a counter-example.

47. Let a and b be positive numbers with $b > a$. Define two sequences $\{a_n\}$ and $\{b_n\}$ as follows:

$$a_1 = \frac{a+b}{2} \text{ (the arithmetic mean of } a \text{ and } b\text{)},$$

$$b_1 = \sqrt{ab} \text{ (the geometric mean of } a \text{ and } b\text{)},$$

† After the French baron Augustin-Louis Cauchy (1789–1857), one of the most prolific mathematicians of all time.

and

$$a_n = \frac{a_{n-1} + b_{n-1}}{2},$$

$$b_n = \sqrt{a_{n-1}b_{n-1}}, \quad n = 2, 3, 4, \ldots .$$

(a) Use mathematical induction to show that

$$a_{n-1} > a_n > b_n > b_{n-1} \quad \text{for } n = 2, 3, 4, \ldots .$$

(b) Prove that $\{a_n\}$ and $\{b_n\}$ are convergent sequences and that $\lim_{n\to\infty} a_n = \lim_{n\to\infty} b_n$. The value of these limits is known as the *arithmetic-geometric mean* of a and b.

48. You have seen that for all real x

$$\lim_{n\to\infty}\left(1 + \frac{x}{n}\right)^n = e^x.$$

However, the rate of convergence is different for different x. Verify that at $n = 100$, $(1 + 1/n)^n$ is within 1% of its limit, while $(1 + 5/n)^n$ is still about 12% from its limit. Give comparable accuracy estimates for these two sequences at $n = 1000$.

49. Evaluate

$$\lim_{n\to\infty}\left(\sin\frac{1}{n}\right)^{1/n}$$

numerically and justify your answer by other means.

50. We have stated that

$$\lim_{n\to\infty}(\sqrt{n^2 + n} - n) = \tfrac{1}{2}. \quad \text{(Exercise 38)}$$

Evaluate numerically

$$\lim_{n\to\infty}[(n^3 + n^2)^{1/3} - n].$$

Formulate a conjecture about

$$\lim_{n\to\infty}[(n^k + n^{k-1})^{1/k} - n], \qquad k = 1, 2, 3, \ldots$$

and prove that your conjecture is valid.

51. The *Fibonacci sequence* is defined recursively by

$$a_{n+2} = a_{n+1} + a_n, \quad \text{with} \quad a_1 = a_2 = 1.$$

(a) Calculate $a_3, a_4, a_5, \ldots, a_{10}$.

(b) Now define the sequence $\{r_n\}$ by

$$r_n = \frac{a_{n+1}}{a_n}.$$

Calculate $r_1, r_2, \ldots, r_6$.

(c) Assuming that $r_n \to L$, find L. HINT: Show that $r_n = 1 + \dfrac{1}{r_{n-1}}$.

52. Let $\{a_n\}$ be the sequence defined by

$$a_n = \frac{1}{n^2} + \frac{2}{n^2} + \frac{3}{n^2} + \cdots + \frac{n}{n^2}.$$

Show that a_n is a Riemann sum for $\int_0^1 x\,dx$ for each $n \geq 1$. Does the sequence $\{a_n\}$ converge? If so, to what?

■ 10.5 THE INDETERMINATE FORM (0/0)

Recall that the quotient rule for evaluating the limit of a quotient $f(x)/g(x)$ fails when $f(x)$ and $g(x)$ both tend to zero (see Section 2.3). Some important examples of this behavior were

$$\lim_{x\to 0}\frac{\sin x}{x} \quad \text{and} \quad \lim_{x\to 0}\frac{1 - \cos x}{x}.$$

Also, the definition of the derivative:

$$\lim_{h\to 0}\frac{f(x + h) - f(x)}{h}$$

typically leads to a limit of this type. We called such limits *indeterminates of the form* 0/0. In this section we present a method for handling this kind of limit when elementary methods fail or are difficult to apply.

THEOREM 10.5.1 L'HOSPITAL'S RULE (0/0)†

Suppose that f and g are differentiable functions and that

$$f(x) \to 0 \quad \text{and} \quad g(x) \to 0$$

as $x \to c^+, x \to c^-, x \to c, x \to \infty$, or $x \to -\infty$.

$$\text{If} \quad \frac{f'(x)}{g'(x)} \to L, \quad \text{then} \quad \frac{f(x)}{g(x)} \to L.$$

$$\text{If} \quad \frac{f'(x)}{g'(x)} \to \infty \text{ or } -\infty, \quad \text{then} \quad \frac{f(x)}{g(x)} \to \infty \text{ or } -\infty, \text{ respectively.}$$

We will prove the validity of L'Hospital's rule later in the section. First we demonstrate its usefulness.

Example 1 Find

$$\lim_{x \to \pi/2} \frac{\cos x}{\pi - 2x}.$$

SOLUTION As $x \to \pi/2$, both the numerator $f(x) = \cos x$ and the denominator $g(x) = \pi - 2x$ tend to zero, but it is not at all obvious what happens to the quotient

$$\frac{f(x)}{g(x)} = \frac{\cos x}{\pi - 2x}.$$

Therefore we test the quotient of derivatives:

$$\frac{f'(x)}{g'(x)} = \frac{-\sin x}{-2} = \frac{\sin x}{2} \to \frac{1}{2} \quad \text{as} \quad x \to \frac{\pi}{2}.$$

It follows from L'Hospital's rule that

$$\frac{\cos x}{\pi - 2x} \to \frac{1}{2} \quad \text{as} \quad x \to \frac{\pi}{2}.$$

We can express all this on just one line using $*$ to indicate the differentiation of numerator and denominator:

$$\lim_{x \to \pi/2} \frac{\cos x}{\pi - 2x} \overset{*}{=} \lim_{x \to \pi/2} \frac{-\sin x}{-2} = \lim_{x \to \pi/2} \frac{\sin x}{2} = \frac{1}{2}. \quad \square$$

† Named after a Frenchman G. F. A. L'Hospital (1661–1704). The result was actually discovered by his teacher, Jakob Bernoulli (1654–1705).

Example 2 Find

$$\lim_{x\to 0^+} \frac{x}{\sin\sqrt{x}}.$$

SOLUTION As $x \to 0^+$, both numerator and denominator tend to 0. Since

$$\frac{f'(x)}{g'(x)} = \frac{1}{(\cos\sqrt{x})(1/2\sqrt{x})} = \frac{2\sqrt{x}}{\cos\sqrt{x}} \to 0 \quad \text{as} \quad x \to 0^+,$$

it follows from L'Hospital's rule that

$$\frac{x}{\sin\sqrt{x}} \to 0 \quad \text{as} \quad x \to 0^+.$$

For short we can write

$$\lim_{x\to 0^+} \frac{x}{\sin\sqrt{x}} \overset{*}{=} \lim_{x\to 0^+} \frac{2\sqrt{x}}{\cos\sqrt{x}} = 0. \quad \square$$

Remark It is important to understand that Theorem 10.5.1 does not apply to quotients in general; you should verify first that the numerator and denominator both tend to zero. For example,

$$\lim_{x\to 0} \frac{x}{x + \cos x} = \frac{0}{1} = 0,$$

but a blind application of L'Hospital's rule would lead to

$$\lim_{x\to 0} \frac{x}{x + \cos x} \overset{*}{=} \lim_{x\to 0} \frac{1}{1 - \sin x} = 1.$$

This is, of course, incorrect. $\square$

Sometimes it is necessary to differentiate numerator and denominator more than once. The next problem gives such an instance.

Example 3 Find

$$\lim_{x\to 0} \frac{e^x - x - 1}{x^2}.$$

SOLUTION As $x \to 0$, both numerator and denominator tend to 0. Here

$$\frac{f'(x)}{g'(x)} = \frac{e^x - 1}{2x}.$$

Since both numerator and denominator still tend to 0 as $x \to 0$, we differentiate again:

$$\frac{f''(x)}{g''(x)} = \frac{e^x}{2}.$$

Since this last quotient tends to $\frac{1}{2}$, we can conclude that

$$\frac{e^x - 1}{2x} \to \frac{1}{2} \quad \text{and therefore} \quad \frac{e^x - x - 1}{x^2} \to \frac{1}{2} \quad \text{as} \quad x \to 0.$$

For short we can write

$$\lim_{x\to 0} \frac{e^x - x - 1}{x^2} \stackrel{*}{=} \lim_{x\to 0} \frac{e^x - 1}{2x} \stackrel{*}{=} \lim_{x\to 0} \frac{e^x}{2} = \frac{1}{2}. \quad \square$$

In the next example we use L'Hospital's rule to find the limit of a sequence.

Example 4 Given the sequence

$$\left\{\frac{e^{2/n} - 1}{1/n}\right\}.$$

Find

$$\lim_{n\to\infty} \frac{e^{2/n} - 1}{1/n}.$$

SOLUTION This is an indeterminate of the form 0/0. To apply the methods of this section, we replace the integer variable n by the real variable x and examine the behavior of

$$\frac{e^{2/x} - 1}{1/x} \quad \text{as} \quad x \to \infty.$$

For an analogous example, see Example 4, Section 10.2. Applying L'Hospital's rule, we have

$$\lim_{x\to\infty} \frac{e^{2/x} - 1}{1/x} \stackrel{*}{=} \lim_{x\to\infty} \frac{e^{2/x}(-2/x^2)}{(-1/x^2)} = 2 \lim_{x\to\infty} e^{2/x} = 2.$$

This shows that

$$\frac{e^{2/x} - 1}{1/x} \to 2 \qquad \text{and so} \qquad \frac{e^{2/n} - 1}{1/n} \to 2 \qquad \text{as} \quad n \to \infty. \quad \square$$

To derive L'Hospital's rule, we need a generalization of the mean-value theorem.

THEOREM 10.5.2 THE CAUCHY MEAN-VALUE THEOREM†

Suppose that f and g are differentiable on (a, b) and continuous on $[a, b]$. If g' is never 0 in (a, b), then there is a number c in (a, b) for which

$$\frac{f'(c)}{g'(c)} = \frac{f(b) - f(a)}{g(b) - g(a)}.$$

PROOF We can prove this by applying Rolle's theorem (4.1.3) to the function

$$G(x) = [g(b) - g(a)][f(x) - f(a)] - [g(x) - g(a)][f(b) - f(a)].$$

† Another contribution of A. L. Cauchy, after whom Cauchy sequences were named.

Since

$$G(a) = 0 \quad \text{and} \quad G(b) = 0,$$

there exists (by Rolle's theorem) a number c in (a, b) for which

$$G'(c) = 0.$$

Since, in general,

$$G'(x) = [g(b) - g(a)]f'(x) - g'(x)[f(b) - f(a)],$$

we must have

$$[g(b) - g(a)]f'(c) - g'(c)[f(b) - f(a)] = 0,$$

and thus

$$[g(b) - g(a)]f'(c) = g'(c)[f(b) - f(a)].$$

Since g' is never 0 in (a, b),

$$g'(c) \neq 0 \quad \text{and} \quad g(b) - g(a) \neq 0.$$

↑ explain

We can therefore divide by these numbers and obtain

$$\frac{f'(c)}{g'(c)} = \frac{f(b) - f(a)}{g(b) - g(a)}. \quad \square$$

Now we prove L'Hospital's rule for the case $x \to c^+$. The case $x \to c^-$ is handled in the same way, and the two cases together can be used to prove the rule for the case $x \to c$.

Let f and g be defined on an interval (c, b) for some $b > c$. We assume that, as $x \to c^+$,

$$f(x) \to 0, \quad g(x) \to 0, \quad \text{and} \quad \frac{f'(x)}{g'(x)} \to L$$

and show that

$$\frac{f(x)}{g(x)} \to L.$$

PROOF The fact that

$$\frac{f'(x)}{g'(x)} \to L \qquad \text{as} \qquad x \to c^+$$

assures us that both f' and g' exist on a set of the form $(c, c + h]$ and that g' is not zero there. By setting $f(c) = 0$ and $g(c) = 0$, we ensure that f and g are both continuous on $[c, c + h]$. We can now apply the Cauchy mean-value theorem and conclude that there exists a number c_h between c and $c + h$ such that

$$\frac{f'(c_h)}{g'(c_h)} = \frac{f(c + h) - f(c)}{g(c + h) - g(c)} = \frac{f(c + h)}{g(c + h)}.$$

The result is now obtained by letting $h \to 0^+$. Since the left side tends to L, the right side tends to L. $\square$

Here is a proof of L'Hospital's rule for the case $x \to \infty$.

PROOF The key here is to set $x = 1/t$:

$$\lim_{x\to\infty}\frac{f'(x)}{g'(x)} = \lim_{t\to 0^+}\frac{f'(1/t)}{g'(1/t)} = \lim_{t\to 0^+}\frac{-t^{-2}f'(1/t)}{-t^{-2}g'(1/t)} = \lim_{t\to 0^+}\frac{f(1/t)}{g(1/t)} = \lim_{x\to\infty}\frac{f(x)}{g(x)}. \quad \square$$

(by L'Hospital's rule for the case $t \to 0^+$)

EXERCISES 10.5

In Exercises 1–32, find the indicated limit.

1. $\lim_{x\to 0^+}\frac{\sin x}{\sqrt{x}}$.

2. $\lim_{x\to 1}\frac{\ln x}{1-x}$.

3. $\lim_{x\to 0}\frac{e^x-1}{\ln(1+x)}$.

4. $\lim_{x\to 4}\frac{\sqrt{x}-2}{x-4}$.

5. $\lim_{x\to \pi/2}\frac{\cos x}{\sin 2x}$.

6. $\lim_{x\to a}\frac{x-a}{x^n-a^n}$.

7. $\lim_{x\to 0}\frac{2^x-1}{x}$.

8. $\lim_{x\to 0}\frac{\tan^{-1}x}{x}$.

9. $\lim_{x\to 1}\frac{x^{1/2}-x^{1/4}}{x-1}$.

10. $\lim_{x\to 0}\frac{e^x-1}{x(1+x)}$.

11. $\lim_{x\to 0}\frac{e^x-e^{-x}}{\sin x}$.

12. $\lim_{x\to 0}\frac{1-\cos x}{3x}$.

13. $\lim_{x\to 0}\frac{x+\sin \pi x}{x-\sin \pi x}$.

14. $\lim_{x\to 0}\frac{a^x-(a+1)^x}{x}$.

15. $\lim_{x\to 0}\frac{e^x+e^{-x}-2}{1-\cos 2x}$.

16. $\lim_{x\to 0}\frac{x-\ln(x+1)}{1-\cos 2x}$

17. $\lim_{x\to 0}\frac{\tan \pi x}{e^x-1}$.

18. $\lim_{x\to 0}\frac{\cos x-1+x^2/2}{x^4}$

19. $\lim_{x\to 0}\frac{1+x-e^x}{x(e^x-1)}$.

20. $\lim_{x\to 0}\frac{\ln(\sec x)}{x^2}$.

21. $\lim_{x\to 0}\frac{x-\tan x}{x-\sin x}$.

22. $\lim_{x\to 0}\frac{xe^{nx}-x}{1-\cos nx}$.

23. $\lim_{x\to 1^-}\frac{\sqrt{1-x^2}}{\sqrt{1-x^3}}$.

24. $\lim_{x\to 0}\frac{2x-\sin \pi x}{4x^2-1}$.

25. $\lim_{x\to \pi/2}\frac{\ln(\sin x)}{(\pi-2x)^2}$.

26. $\lim_{x\to 0^+}\frac{\sqrt{x}}{\sqrt{x}+\sin\sqrt{x}}$.

27. $\lim_{x\to 0}\frac{\cos x-\cos 3x}{\sin(x^2)}$

28. $\lim_{x\to 0}\frac{\sqrt{a+x}-\sqrt{a-x}}{x}$.

29. $\lim_{x\to \pi/4}\frac{\sec^2 x-2\tan x}{1+\cos 4x}$.

30. $\lim_{x\to 0}\frac{x-\sin^{-1}x}{\sin^3 x}$.

31. $\lim_{x\to 0}\frac{\tan^{-1}x}{\tan^{-1}2x}$.

32. $\lim_{x\to 0}\frac{\sin^{-1}x}{x}$.

In Exercises 33–36, find the limit of the sequence.

33. $\lim_{n\to\infty}\frac{(\pi/2-\tan^{-1}n)}{1/n}$.

34. $\lim_{n\to\infty}\frac{\ln(1-1/n)}{\sin(1/n)}$.

35. $\lim_{n\to\infty}\frac{1}{n[\ln(n+1)-\ln n]}$.

36. $\lim_{n\to\infty}\frac{\sinh \pi/n-\sin \pi/n}{\sin^3 \pi/n}$.

37. Find the fallacy:

$$\lim_{x\to 0}\frac{2+x+\sin x}{x^3+x-\cos x} \overset{*}{=} \lim_{x\to 0}\frac{1+\cos x}{3x^2+1+\sin x}$$
$$\overset{*}{=} \lim_{x\to 0}\frac{-\sin x}{6x+\cos x} = \frac{0}{1} = 0.$$

38. Show that, if $a > 0$, then

$$\lim_{n\to\infty} n(a^{1/n}-1) = \ln a.$$

39. Find values for a and b such that

$$\lim_{x\to 0}\frac{\cos ax-b}{2x^2} = -4.$$

40. Find values for a and b so that

$$\lim_{x\to 0}\frac{\sin 2x+ax+bx^3}{x^3} = 0$$

41. Given that f is continuous, use L'Hospital's rule to determine

$$\lim_{x\to 0}\left(\frac{1}{x}\int_0^x f(t)\,dt\right).$$

42. Let f be a twice differentiable function and fix a value of x.

(a) Prove that

$$\lim_{h\to 0}\frac{f(x+h)-f(x-h)}{2h}=f'(x).$$

(b) Prove that

$$\lim_{h\to 0}\frac{f(x+h)-2f(x)+f(x-h)}{h^2}=f''(x).$$

43. Let $A(b)$ be the area of the region bounded by the parabola $y=x^2$ and the horizontal line $y=b$ $(b>0)$, and let $T(b)$ be the area of the triangle AOB (see the figure). Find $\lim_{b\to 0^+} T(b)/A(b)$.

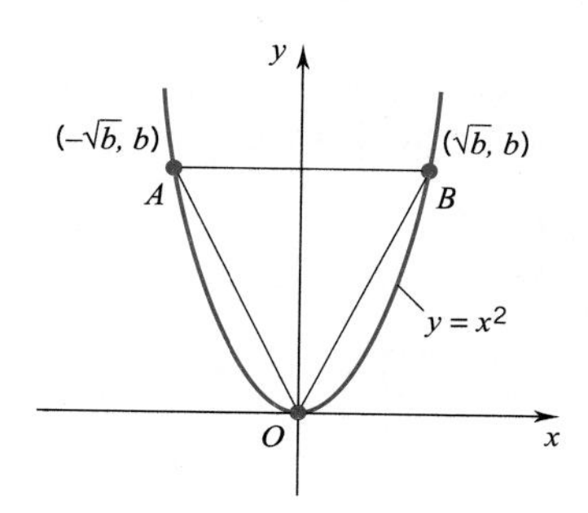

44. Choose an angle θ, $0<\theta<\pi/2$, in standard position as shown in the figure. Let $T(\theta)$ be the area of the triangle ABC, and let $S(\theta)$ be the area of the segment of the circle formed by the chord AB. Find $\lim_{\theta\to 0^+} T(\theta)/S(\theta)$.

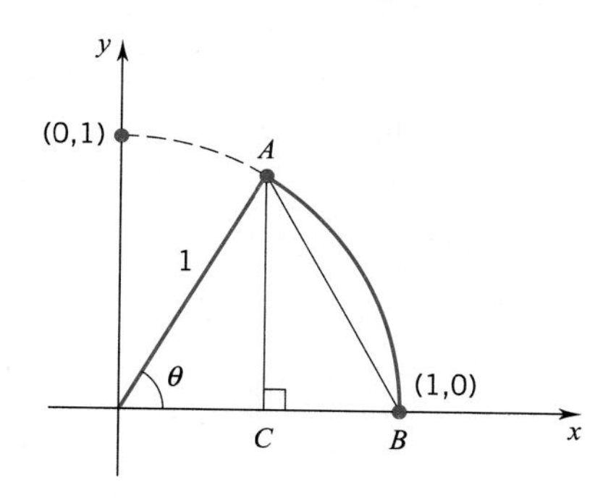

45. Let

$$f(x)=\frac{x^2-16}{\sqrt{x^2+9}-5}.$$

(a) Use a graphing utility to graph f. What is the behavior of the graph as $x\to\infty$ and as $x\to-\infty$?
(b) What is the behavior of f as $x\to 4$? Confirm your answer using L'Hospital's rule.

46. Let

$$f(x)=\frac{x-\sin x}{x^3}.$$

(a) Use a graphing utility to graph f. What is the behavior of the graph as $x\to\infty$ and as $x\to-\infty$?
(b) What is the behavior of f as $x\to 0$? Confirm your answer using *L*'Hospital's rule.

47. Let $f(x)=\dfrac{2^{\sin x}-1}{x}$.

(a) Use a graphing utility to graph f. Estimate

$$\lim_{x\to 0} f(x).$$

(b) Use L'Hospital's rule to confirm your estimate in part (a).

48. Let $g(x)=\dfrac{3^{\cos x}-3}{x^2}$.

(a) Use a graphing utility to graph g. Estimate

$$\lim_{x\to 0} g(x).$$

(b) Use L'Hospital's rule to confirm your estimate in part (a).

■ 10.6 THE INDETERMINATE FORM (∞/∞); OTHER INDETERMINATE FORMS

We come now to limits of quotients $f(x)/g(x)$ where numerator and denominator both tend to ∞. Such limits are called *indeterminates of the form* ∞/∞.

THEOREM 10.6.1 L'HOSPITAL'S RULE (∞/∞)

Suppose that f and g are differentiable, and that

$$f(x) \to \pm\infty \qquad \text{and} \qquad g(x) \to \pm\infty$$

as $x \to c^+$, $x \to c^-$, $x \to c$, $x \to \infty$, or $x \to -\infty$.

$$\text{If} \quad \frac{f'(x)}{g'(x)} \to L, \qquad \text{then} \quad \frac{f(x)}{g(x)} \to L.$$

$$\text{If} \quad \frac{f'(x)}{g'(x)} \to \infty \text{ or } -\infty, \qquad \text{then} \quad \frac{f(x)}{g(x)} \to \infty \text{ or } -\infty, \text{ respectively}$$

While the proof of L'Hospital's rule in this setting is a little more complicated than it was in the (0/0) case,† the application of the rule is much the same.

Example 1 Let α be any positive number. Show that

(10.6.2)
$$\lim_{x\to\infty} \frac{\ln x}{x^\alpha} = 0.$$

SOLUTION Both numerator and denominator tend to ∞ as $x \to \infty$. L'Hospital's rule gives

$$\lim_{x\to\infty} \frac{\ln x}{x^\alpha} \overset{*}{=} \lim_{x\to\infty} \frac{1/x}{\alpha x^{\alpha-1}} = \lim_{x\to\infty} \frac{1}{\alpha x^\alpha} = 0. \quad \square$$

For example,

$$\frac{\ln x}{x^{0.01}} \to 0 \qquad \text{and} \qquad \frac{\ln x}{x^{0.001}} \to 0$$

as $x \to \infty$.

Example 2 Let k be any positive integer. Show that

(10.6.3)
$$\lim_{x\to\infty} \frac{x^k}{e^x} = 0.$$

SOLUTION Here we differentiate numerator and denominator k times:

$$\lim_{x\to\infty} \frac{x^k}{e^x} \overset{*}{=} \lim_{x\to\infty} \frac{kx^{k-1}}{e^x} \overset{*}{=} \lim_{x\to\infty} \frac{k(k-1)x^{k-2}}{e^x} \overset{*}{=} \cdots \overset{*}{=} \lim_{x\to\infty} \frac{k!}{e^x} = 0. \quad \square$$

† We omit the proof.

For example,

$$\frac{x^{100}}{e^x} \to 0 \quad \text{and} \quad \frac{x^{1000}}{e^x} \to 0$$

as $x \to \infty$.

Remark The limits (10.6.2) and (10.6.3) tell us that $\ln x$ tends to infinity *more slowly than* any positive power of x and that e^x tends to infinity *faster than* any positive integral power of x. In the Exercises you are asked to show that e^x tends to infinity faster than *any* positive power of x and that *any* positive power of $\ln x$ tends to infinity more slowly than x. That is, for any positive number α,

$$\lim_{x\to\infty} \frac{x^\alpha}{e^x} = 0 \quad \text{and} \quad \lim_{x\to\infty} \frac{[\ln x]^\alpha}{x} = 0.$$

Comparisons of logarithmic and exponential growth were also given in Exercises 17 and 18 in Section 7.6. ❑

Example 3 Find the limit as $n \to \infty$ of the sequence $\{a_n\}$ given by

$$a_n = \frac{2^n}{n^2}.$$

SOLUTION To use the methods of calculus, we investigate

$$\lim_{x\to\infty} \frac{2^x}{x^2}.$$

Since both numerator and denominator tend to ∞ with x, we try L'Hospital's rule:

$$\lim_{x\to\infty} \frac{2^x}{x^2} \overset{*}{=} \lim_{x\to\infty} \frac{2^x \ln 2}{2x} \overset{*}{=} \lim_{x\to\infty} \frac{2^x (\ln 2)^2}{2} = \infty.$$

Therefore, the limit of the sequence must also be ∞. ❑

Other Indeterminate Forms: $0 \cdot \infty$, $\infty - \infty$, 0^0, 1^∞, ∞^0

If f tends to 0 and g tends to ∞ (or $-\infty$) as x approaches some number c (or $\pm\infty$), then it is not clear what the product $f \cdot g$ will do. For example, as $x \to 1$,

$$(x-1)^3 \to 0, \qquad \frac{1}{(x-1)^2} \to \infty, \qquad \text{and} \qquad (x-1)^3 \cdot \frac{1}{(x-1)^2} = (x-1) \to 0.$$

On the other hand,

$$\lim_{x\to 1} \left[(x-1)^3 \cdot \frac{1}{(x-1)^4}\right] = \lim_{x\to 1} \left[\frac{1}{x-1}\right] \qquad \text{does not exist.}$$

A limit of this type is called an *indeterminate of the form* $0 \cdot \infty$. We handle these indeterminates by writing the product $f \cdot g$ as a quotient

$$\frac{f}{1/g} \quad \text{or} \quad \frac{g}{1/f}.$$

In the first case, the result will be an indeterminate of the form 0/0, and in the second it will have the form ∞/∞.

Example 4 Find

$$\lim_{x\to 0^+} \sqrt{x}\ln x.$$

SOLUTION As $x \to 0^+$, $\sqrt{x} \to 0$ and $\ln x \to -\infty$. Thus the given limit is an indeterminate of the form $0 \cdot \infty$. Rewriting the product as a quotient, we have

$$\lim_{x\to 0^+} \sqrt{x}\ln x = \lim_{x\to 0^+} \frac{\ln x}{1/\sqrt{x}} \overset{*}{=} \lim_{x\to 0^+} \frac{1/x}{-\frac{1}{2}x^{-3/2}} = \lim_{x\to 0^+} -2\sqrt{x} = 0.$$

Therefore, $\lim_{x\to 0^+} \sqrt{x}\ln x = 0$.

Of course, we could have chosen to write $\sqrt{x}\ln x$ as the quotient

$$\frac{\sqrt{x}}{1/\ln x}.$$

Try to evaluate the limit using this quotient. ❑

If f and g both tend to ∞, or if both tend to $-\infty$, as x tends to c (or $\pm\infty$), then $\lim(f - g)$ is called an *indeterminate of the form* $\infty - \infty$. Like the preceding case, indeterminates of this type are handled by converting the difference to a quotient.

Example 5 Find

$$\lim_{x\to(\pi/2)^-} (\tan x - \sec x).$$

SOLUTION Both $\tan x$ and $\sec x$ tend to ∞ as x tends to $\pi/2$ from the left. We first rewrite the difference as a quotient:

$$\tan x - \sec x = \frac{\sin x}{\cos x} - \frac{1}{\cos x} = \frac{\sin x - 1}{\cos x}.$$

Now

$$\lim_{x\to(\pi/2)^-} \frac{\sin x - 1}{\cos x}$$

is an indeterminate of the form 0/0, and

$$\lim_{x\to(\pi/2)^-} \frac{\sin x - 1}{\cos x} \overset{*}{=} \lim_{x\to(\pi/2)^-} \frac{\cos x}{-\sin x} = \frac{0}{-1} = 0.$$

Thus, $\lim_{x\to(\pi/2)^-} (\tan x - \sec x) = 0$. ❑

Limits involving exponential expressions $[f(x)]^{g(x)}$ are indeterminate when: (1) f and g both tend to 0; (2) f tends to 1 and g tends to $\pm\infty$; and (3) f tends to $\pm\infty$ and g tends to 0. These cases are called *indeterminates of the forms* 0^0, 1^∞, and ∞^0, respectively. Exponential indeterminate forms are treated by taking natural logarithms:

$$\text{if} \quad y = [f(x)]^{g(x)}, \qquad \text{then} \quad \ln y = g(x)\ln[f(x)].$$

Now, $\lim \ln y = \lim g(x)\ln[f(x)]$ will be an indeterminate of the form $0 \cdot \infty$

Example 6 Show that

$$\lim_{x\to 0^+} x^x = 1. \tag{10.6.4}$$

SOLUTION Here we are dealing with an indeterminate of the form 0^0. Our first step is to take the logarithm of x^x. Then we apply L'Hospital's rule:

$$\lim_{x\to 0^+} \ln(x^x) = \lim_{x\to 0^+} (x \ln x) = \lim_{x\to 0^+} \frac{\ln x}{1/x} \overset{*}{=} \lim_{x\to 0^+} \frac{1/x}{-1/x^2} = \lim_{x\to 0^+} (-x) = 0.$$

Since $\ln(x^x) \to 0$ as $x \to 0^+$, $x^x = e^{\ln(x^x)} \to e^0 = 1$. ❑

Example 7 Find

$$\lim_{x\to 0^+} (1 + x)^{1/x}.$$

SOLUTION Here we are dealing with an indeterminate of the form 1^∞: as $x \to 0^+$, $1 + x \to 1$ and $1/x$ increases without bound. Taking the logarithm and then applying L'Hospital's rule, we have

$$\lim_{x\to 0^+} \ln(1 + x)^{1/x} = \lim_{x\to 0^+} \frac{\ln(1 + x)}{x} \overset{*}{=} \lim_{x\to 0^+} \frac{1}{1 + x} = 1.$$

Since $\ln(1 + x)^{1/x} \to 1$ as $x \to 0^+$, $(1 + x)^{1/x} = e^{\ln(1+x)^{1/x}} \to e^1 = e$. Note that if we set $x = 1/n$, we have the familiar result: $[1 + (1/n)]^n \to e$ as $n \to \infty$. ❑

Example 8 Show that

$$\lim_{x\to\infty} (x^2 + 1)^{1/\ln x} = e^2.$$

SOLUTION Here we have an indeterminate of the form ∞^0. Taking the logarithm and then applying L'Hospital's rule, we find that

$$\lim_{x\to\infty} \ln[(x^2 + 1)^{1/\ln x}] = \lim_{x\to\infty} \frac{\ln(x^2 + 1)}{\ln x} \overset{*}{=} \lim_{x\to\infty} \frac{2x/(x^2 + 1)}{1/x} = \lim_{x\to\infty} \frac{2x^2}{x^2 + 1} = 2.$$

It follows that

$$\lim_{x\to\infty} (x^2 + 1)^{1/\ln x} = e^2. \quad ❑$$

Concluding Remarks Suppose that $\lim(f/g)$ is an indeterminate form (either $0/0$ or ∞/∞). Both versions of L'Hospital's rule (Theorems 10.5.1 and 10.6.1) tell us that if

$$\lim \frac{f'}{g'} = L \text{ (or } \pm\infty), \qquad \text{then} \qquad \lim \frac{f}{g} = L \text{ (or } \pm\infty).$$

However, the rules do not provide any information when $\lim(f'/g')$ fails to exist; $\lim(f/g)$ may or may not exist. For example,

$$\lim_{x\to\infty} \frac{x + \cos x}{x}$$

is an indeterminate of the form ∞/∞. It is easy to show that this limit is 1 (divide numerator and denominator by x). On the other hand, if we try to apply Theorem 10.6.1, we consider the limit

$$\lim_{x\to\infty} \frac{f'(x)}{g'(x)} = \lim_{x\to\infty} \frac{1-\sin x}{1},$$

and this limit does not exist.

Finally, as noted in Section 10.5, you should always check first to make sure that a given limit actually involves an indeterminate form before trying to apply the methods of these sections. In the case of a quotient, L'Hospital's rule does not apply when either the numerator or denominator has a finite nonzero limit. For example,

$$\lim_{x\to 0^+} \frac{1+x}{\sin x} = \infty,$$

but a misapplication of L'Hospital's rule would lead to the limit

$$\lim_{x\to 0^+} \frac{1}{\cos x} = 1,$$

and the *incorrect conclusion* that

$$\lim_{x\to 0^+} \frac{1+x}{\sin x} = 1. \quad \square$$

EXERCISES 10.6

In Exercises 1–36, find the indicated limit.

1. $\lim_{x\to-\infty} \frac{x^2+1}{1-x}$.

2. $\lim_{x\to\infty} \frac{20x}{x^2+1}$.

3. $\lim_{x\to\infty} \frac{x^3}{1-x^3}$.

4. $\lim_{x\to\infty} \frac{x^3-1}{2-x}$.

5. $\lim_{x\to\infty} \left(x^2 \sin \frac{1}{x}\right)$.

6. $\lim_{x\to\infty} \frac{\ln x^k}{x}$.

7. $\lim_{x\to\pi/2^-} \frac{\tan 5x}{\tan x}$.

8. $\lim_{x\to 0} (x \ln|\sin x|)$.

9. $\lim_{x\to 0^+} x^{2x}$.

10. $\lim_{x\to\infty} \left(x \sin \frac{\pi}{x}\right)$.

11. $\lim_{x\to 0} [x(\ln|x|)^2]$.

12. $\lim_{x\to 0^+} \frac{\ln x}{\cot x}$.

13. $\lim_{x\to\infty} \left(\frac{1}{x}\int_0^x e^{t^2}\, dt\right)$.

14. $\lim_{x\to\infty} \frac{\sqrt{1+x^2}}{x}$.

15. $\lim_{x\to 0} \left[\frac{1}{\sin^2 x} - \frac{1}{x^2}\right]$.

16. $\lim_{x\to 0} |\sin x|^x$.

17. $\lim_{x\to 1} x^{1/(x-1)}$.

18. $\lim_{x\to 0^+} x^{\sin x}$.

19. $\lim_{x\to\infty} \left(\cos \frac{1}{x}\right)^x$.

20. $\lim_{x\to\pi/2} |\sec x|^{\cos x}$.

21. $\lim_{x\to 0} \left[\frac{1}{\ln(1+x)} - \frac{1}{x}\right]$.

22. $\lim_{x\to\infty} (x^2+a^2)^{(1/x)^2}$.

23. $\lim_{x\to 0} \left(\frac{1}{x} - \cot x\right)$.

24. $\lim_{x\to\infty} \ln\left(\frac{x^2-1}{x^2+1}\right)^3$.

25. $\lim_{x\to\infty} (\sqrt{x^2+2x} - x)$.

26. $\lim_{x\to\infty} \frac{1}{x}\int_0^x \sin\left(\frac{1}{t+1}\right) dt$.

27. $\lim_{x\to\infty} (x^3+1)^{1/\ln x}$.

28. $\lim_{x\to\infty} (e^x+1)^{1/x}$.

29. $\lim_{x\to\infty} (\cosh x)^{1/x}$.

30. $\lim_{x\to\infty} (x^4+1)^{1/\ln x}$.

31. $\lim_{x\to 0} (e^x+x)^{1/x}$.

32. $\lim_{x\to\infty} \left(1+\frac{1}{x}\right)^{3x}$.

33. $\lim_{x\to 0} \left(\frac{1}{\sin x} - \frac{1}{x}\right)$.

34. $\lim_{x\to 0} (e^x+3x)^{1/x}$.

35. $\lim_{x\to 1} \left(\frac{1}{\ln x} - \frac{x}{x-1}\right)$.

36. $\lim_{x\to 0} \left(\frac{1+2^x}{2}\right)^{1/x}$.

In Exercises 37–44, find the limit of the sequence.

37. $\lim_{n\to\infty} \left(\frac{1}{n}\ln\frac{1}{n}\right)$.

38. $\lim_{n\to\infty} \frac{n^k}{2^n}$.

39. $\lim_{n\to\infty} (\ln n)^{1/n}$.

40. $\lim_{n\to\infty} \dfrac{\ln n}{n^p}, \quad (p > 0)$.

41. $\lim_{n\to\infty} (n^2 + n)^{1/n}$.

42. $\lim_{n\to\infty} n^{\sin(\pi/n)}$.

43. $\lim_{n\to\infty} \dfrac{n^2 \ln n}{e^n}$

44. $\lim_{n\to\infty} (\sqrt{n} - 1)^{1/\sqrt{n}}$.

In Exercises 45–50, sketch the curve, specifying all vertical and horizontal asymptotes.

45. $y = x^2 - \dfrac{1}{x^3}$.

46. $y = \sqrt{\dfrac{x}{x-1}}$.

47. $y = xe^x$.

48. $y = xe^{-x}$.

49. $y = x^2e^{-x}$.

50. $y = \dfrac{\ln x}{x}$.

The graphs of two functions $y = f(x)$ and $y = g(x)$ are said to be *asymptotic as* $x \to \infty$ iff

$$\lim_{x\to\infty} [f(x) - g(x)] = 0;$$

they are said to be *asymptotic as* $x \to -\infty$ iff

$$\lim_{x\to-\infty} [f(x) - g(x)] = 0.$$

51. Show that the hyperbolic arc $y = (b/a)\sqrt{x^2 - a^2}$ is asymptotic to the line $y = (b/a)x$ as $x \to \infty$.

52. Show that the graphs of $y = \cosh x$ and $y = \sinh x$ are asymptotic.

53. Give an example of a function the graph of which is asymptotic to the parabola $y = x^2$ as $x \to \infty$ and crosses the graph of the parabola twice.

54. Give an example of a function the graph of which is asymptotic to the line $y = x$ as $x \to \infty$ and crosses the graph of the line infinitely often.

55. Find the fallacy:

$$\lim_{x\to0^+} \frac{x^2}{\sin x} \overset{*}{=} \lim_{x\to0^+} \frac{2x}{\cos x} \overset{*}{=} \lim_{x\to0^+} \frac{2}{-\sin x} = -\infty.$$

56. Let α be a positive number. Show that

$$\lim_{x\to\infty} \frac{x^\alpha}{e^x} = 0.$$

57. (a) Show by induction that, for each positive integer k,

$$\lim_{x\to\infty} \frac{(\ln x)^k}{x} = 0.$$

(b) Show that, for each positive number α,

$$\lim_{x\to\infty} \frac{(\ln x)^\alpha}{x} = 0.$$

58. (a) Try to evaluate

$$\lim_{x\to\infty} \frac{x}{\sqrt{x^2+1}}$$

using L'Hospital's rule and see what happens.
(b) Evaluate this limit by some other method.

59. (a) Try to evaluate

$$\lim_{x\to0} \frac{e^{-1/x^2}}{x}$$

using L'Hospital's rule and see what happens. Then rewrite the quotient in an equivalent form and show that the limit is 0.
(b) Define the function f by

$$f(x) = \begin{cases} e^{-1/x^2} & \text{if } x \neq 0 \\ 0 & \text{if } x = 0. \end{cases}$$

Show that f is differentiable at 0. What is $f'(0)$?

60. The differential equation governing the velocity of an object of mass m dropped from rest under the influence of gravity with air resistance directly proportional to the velocity is

$$(*) \qquad m\frac{dv}{dt} + kv = mg,$$

where $k > 0$ is the constant of proportionality, g is the gravitational constant and $v(0) = 0$. See Exercise 37, Section 7.6. The velocity of the object at time t is given by

$$v(t) = (mg/k)(1 - e^{-(k/m)t}).$$

(a) Fix t and find $\lim_{k\to0^+} v(t)$.
(b) Set $k = 0$ in equation (*) and solve

$$m\frac{dv}{dt} = mg, \quad v(0) = 0.$$

Does this result agree with the result found in part (a)?

61. Let $f(x) = (1 + x)^{1/x}$ and $g(x) = (1 + x^2)^{1/x}$ on $(0, \infty)$.
(a) Use a graphing utility to graph f and g in the same coordinate system. Estimate

$$\lim_{x\to0^+} g(x).$$

(b) Use L'Hospital's rule to confirm your estimate in part (a).

62. Let $f(x) = \sqrt{x^2 + 3x + 1} - x$.
(a) Use a graphing utility to graph f. Then use your graph to estimate

$$\lim_{x\to\infty} f(x)$$

(b) Use L'Hospital's rule to confirm your estimate. HINT: "Rationalize."

63. Let $g(x) = \sqrt[3]{x^3 - 5x^2 + 2x + 1} - x$.

(a) Use a graphing utility to graph g. Then use your graph to estimate

$$\lim_{x\to\infty} g(x).$$

(b) Use L'Hospital's rule to confirm your estimate.

64. Exercises 62 and 63 can be generalized as follows: Let n be a positive integer and let P be the polynomial

$$P(x) = x^n + b_1x^{n-1} + b_2x^{n-2} + \cdots + b_{n-1}x + b_n.$$

Prove that

$$\lim_{x\to\infty} ([P(x)]^{1/n} - x) = \frac{b_1}{n}.$$

■ 10.7 IMPROPER INTEGRALS

In all of our work involving the theory and applications of the definite integral

$$\int_a^b f(x)\,dx,$$

it has been assumed that the interval $[a, b]$ is finite and that the function f is bounded on $[a, b]$. In the context to be developed here, such integrals are said to be *proper.* In this section, we will use a limit process to calculate integrals in cases where the interval is infinite or where the function is unbounded. Such integrals are called *improper integrals.*

Integrals over Infinite Intervals

We begin with a function f continuous on an unbounded interval $[a, \infty)$. For each number $b > a$ we can form the definite integral

$$\int_a^b f(x)\,dx.$$

If, as b tends to ∞, this integral tends to a finite limit L,

$$\lim_{b\to\infty} \int_a^b f(x)\,dx = L,$$

then we write

$$\int_a^\infty f(x)\,dx = L$$

and say that

$$\textit{the improper integral} \quad \int_a^\infty f(x)\,dx \quad \textit{converges to } L.$$

Otherwise, we say that

$$\textit{the improper integral} \quad \int_a^\infty f(x)\,dx \quad \textit{diverges.}$$

In a similar manner, if f is continuous on the unbounded interval $(-\infty, b]$, then for each number $a < b$, we can form the definite integral

$$\int_a^b f(x)\,dx$$

and calculate

$$\lim_{a\to-\infty}\int_a^b f(x)\,dx.$$

If this limit exists and equals L, then

$$\textit{the improper integral}\quad \int_{-\infty}^b f(x)\,dx \quad \textit{converges to } L;$$

otherwise,

$$\textit{the improper integral}\quad \int_{-\infty}^b f(x)\,dx \quad \textit{diverges.}$$

Example 1

(a) $\displaystyle\int_0^\infty e^{-2x}\,dx = \tfrac{1}{2}.$

(b) $\displaystyle\int_1^\infty \frac{dx}{x}$ diverges.

(c) $\displaystyle\int_1^\infty \frac{dx}{x^2} = 1.$

(d) $\displaystyle\int_{-\infty}^1 \cos \pi x\,dx$ diverges.

VERIFICATION

(a) $\displaystyle\int_0^\infty e^{-2x}\,dx = \lim_{b\to\infty}\int_0^b e^{-2x}\,dx = \lim_{b\to\infty}\left[-\frac{e^{-2x}}{2}\right]_0^b = \lim_{b\to\infty}\left(\frac{1}{2}-\frac{1}{e^{2b}}\right) = \frac{1}{2}.$

(b) $\displaystyle\int_1^\infty \frac{dx}{x} = \lim_{b\to\infty}\int_1^b \frac{dx}{x} = \lim_{b\to\infty}\ln b = \infty.$

(c) $\displaystyle\int_1^\infty \frac{dx}{x^2} = \lim_{b\to\infty}\int_1^b \frac{dx}{x^2} = \lim_{b\to\infty}\left[-\frac{1}{x}\right]_1^b = \lim_{b\to\infty}\left(1-\frac{1}{b}\right) = 1.$

(d) Note first that

$$\int_a^1 \cos \pi x\,dx = \left[\frac{1}{\pi}\sin \pi x\right]_a^1 = -\frac{1}{\pi}\sin \pi a.$$

As a tends to $-\infty$, $\sin \pi a$ oscillates between -1 and 1. Therefore the integral oscillates between $1/\pi$ and $-1/\pi$ and does not converge. ❑

The usual formulas for area and volume are extended to the unbounded case by means of improper integrals.

Example 2 Let p be a positive number. If Ω is the region between the x-axis and the graph of

$$f(x) = \frac{1}{x^p}, \qquad x \geq 1 \qquad \text{(Figure 10.7.1)}$$

then

$$\text{area of } \Omega = \begin{cases} \dfrac{1}{p-1}, & \text{if } \ p > 1 \\ \infty, & \text{if } \ p \leq 1. \end{cases}$$

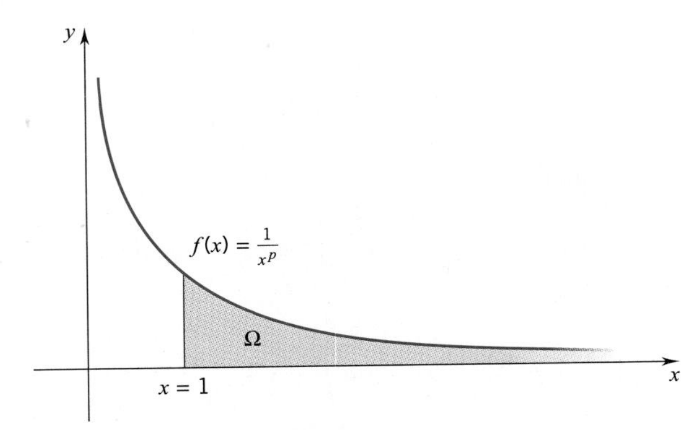

Figure 10.7.1

This comes about from setting

$$\text{area of } \Omega = \lim_{b\to\infty} \int_1^b \frac{dx}{x^p} = \int_1^\infty \frac{dx}{x^p}.$$

For $p \neq 1$,

$$\int_1^\infty \frac{dx}{x^p} = \lim_{b\to\infty} \int_1^b \frac{dx}{x^p} = \lim_{b\to\infty} \frac{1}{1-p}(b^{1-p} - 1) = \begin{cases} \dfrac{1}{p-1} & \text{if} \quad p > 1 \\ \infty & \text{if} \quad p < 1. \end{cases}$$

For $p = 1$,

$$\int_1^\infty \frac{dx}{x^p} = \int_1^\infty \frac{dx}{x} = \infty,$$

as you have seen already. ❑

Remark It is easy to verify that if $p \leq 0$ and Ω is the region between the x-axis and the graph of $f(x) = 1/x^p = x^{-p}$, $-p \geq 0$, then area of $\Omega = \infty$. Thus the conclusion in Example 2 actually holds for all real numbers p. ❑

Example 3 From the last example you know that the region below the graph of

$$f(x) = \frac{1}{x}, \qquad x \geq 1$$

has infinite area. Suppose that this region with infinite area is revolved about the x-axis (see Figure 10.7.2). What is the volume V of the resulting solid? It may surprise you somewhat, but the volume is not infinite. In fact, it is π. Using the disc method to calculate the volume (see Section 6.2), we have

$$V = \int_1^\infty \pi[f(x)]^2\, dx = \pi \int_1^\infty \frac{dx}{x^2} = \pi \lim_{b\to\infty} \int_1^b \frac{dx}{x^2}$$

$$= \pi \lim_{b\to\infty} \left[\frac{-1}{x}\right]_1^b = \pi \cdot 1 = \pi. \quad ❑$$

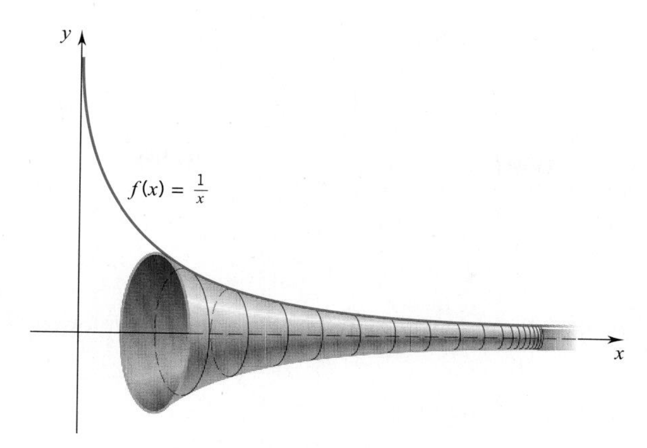

Figure 10.7.2

For future reference we record the following:

(10.7.1) $$\int_1^\infty \frac{dx}{x^p} \text{ converges for } p > 1 \text{ and diverges for } p \leq 1.$$

It is often difficult to determine the convergence or divergence of a given improper integral by direct methods, that is, by calculating the definite integral and evaluating the limit. In such cases we can sometimes gain information by comparison with integrals of known behavior.

(10.7.2) (*A comparison test*) Suppose that f and g are continuous and

$$0 \leq f(x) \leq g(x) \qquad \text{for all } x \in [a, \infty).$$

(i) If $\int_a^\infty g(x)\,dx$ converges, then $\int_a^\infty f(x)\,dx$ converges.

(ii) If $\int_a^\infty f(x)\,dx$ diverges, then $\int_a^\infty g(x)\,dx$ diverges.

A similar result holds for integrals from $-\infty$ to b. Figure 10.7.3 illustrates the comparison test (10.7.2). The proof of the result is left to you as an exercise.

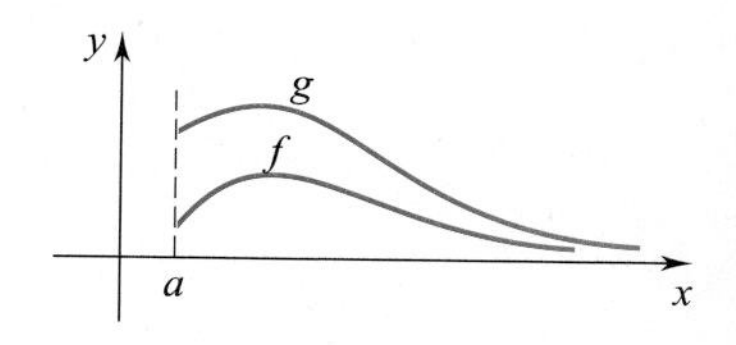

Figure 10.7.3

Example 4 The improper integral

$$\int_1^\infty \frac{dx}{\sqrt{1+x^3}}$$

converges since

$$\frac{1}{\sqrt{1+x^3}} < \frac{1}{x^{3/2}} \quad \text{for } x \in [1, \infty) \qquad \text{and} \qquad \int_1^\infty \frac{dx}{x^{3/2}} \text{ converges.}$$

In contrast, if we tried to evaluate

$$\lim_{b\to\infty}\int_1^b \frac{dx}{\sqrt{1+x^3}}$$

directly, we would have to calculate the integral

$$\int \frac{dx}{\sqrt{1+x^3}},$$

and this cannot be done by any of the methods we have developed so far. ❑

Example 5 The improper integral

$$\int_1^\infty \frac{dx}{\sqrt{1+x^2}}$$

diverges since

$$\frac{1}{1+x} \le \frac{1}{\sqrt{1+x^2}} \quad \text{for } x \in [1, \infty) \qquad \text{and} \qquad \int_1^\infty \frac{dx}{1+x} \text{ diverges.}$$

This result can also be obtained by evaluating

$$\int_1^b \frac{dx}{\sqrt{1+x^2}}$$

and then calculating the limit as $b \to \infty$. Try it. ❑

Suppose now that f is continuous on $(-\infty, \infty)$. The *improper integral*

$$\int_{-\infty}^\infty f(x)\,dx$$

is said to *converge* iff

$$\int_{-\infty}^0 f(x)\,dx \qquad \text{and} \qquad \int_0^\infty f(x)\,dx$$

both converge. We then set

$$\int_{-\infty}^\infty f(x)\,dx = L + M,$$

where

$$\int_{-\infty}^0 f(x)\,dx = L \qquad \text{and} \qquad \int_0^\infty f(x)\,dx = M.$$

Example 6 Determine whether the improper integral

$$\int_{-\infty}^\infty \frac{e^x}{1+e^{2x}}\,dx$$

converges or diverges. If it converges, give its value.

SOLUTION First consider the indefinite integral

$$\int \frac{e^x}{1 + e^{2x}}\,dx.$$

Let $u = e^x$ and $du = e^x\,dx$. Then

$$\int \frac{e^x}{1 + e^{2x}}\,dx = \int \frac{1}{1 + u^2}\,du = \tan^{-1}u + C = \tan^{-1}(e^x) + C.$$

Now,

$$\int_{-\infty}^{0} \frac{e^x}{1 + e^{2x}}\,dx = \lim_{a\to-\infty} \int_{a}^{0} \frac{e^x}{1 + e^{2x}}\,dx = \lim_{a\to-\infty} \Big[\tan^{-1}(e^x)\Big]_a^0$$

$$= \tan^{-1}(1) - \lim_{a\to-\infty} \tan^{-1}(e^a) = \frac{\pi}{4} - 0 = \frac{\pi}{4}$$

and

$$\int_{0}^{\infty} \frac{e^x}{1 + e^{2x}}\,dx = \lim_{b\to\infty} \int_{0}^{b} \frac{e^x}{1 + e^{2x}}\,dx = \lim_{b\to\infty} \Big[\tan^{-1}(e^x)\Big]_0^b$$

$$= \lim_{b\to\infty} \tan^{-1}(e^b) - \tan^{-1}(1) = \frac{\pi}{2} - \frac{\pi}{4} = \frac{\pi}{4}.$$

Therefore the improper integral

$$\int_{-\infty}^{\infty} \frac{e^x}{1 + e^{2x}}\,dx$$

converges. Its value is $\frac{1}{4}\pi + \frac{1}{4}\pi = \frac{1}{2}\pi$. ❑

Remark You might be wondering why we did not define

$$\int_{-\infty}^{\infty} f(x)\,dx$$

in terms of the limit

$$\lim_{b\to\infty} \int_{-b}^{b} f(x)\,dx.$$

In fact, it can be shown that if the integral from $-\infty$ to ∞ exists in the sense of the original definition, then the limit just given also exists and they are equal. On the other hand, if we let $f(x) = x$, then

$$\lim_{b\to\infty} \int_{-b}^{b} x\,dx = \lim_{b\to\infty} \left[\frac{x^2}{2}\right]_{-b}^{b} = \lim_{b\to\infty} \left[\frac{b^2}{2} - \frac{b^2}{2}\right] = 0,$$

while the calculation of

$$\int_{-\infty}^{\infty} x\,dx$$

using the definition will lead to an indeterminate of the form $\infty - \infty$. ❑

Integrals of Unbounded Functions

Improper integrals can also arise on bounded intervals. Suppose that f is continuous on the half-open interval $[a, b)$ but is unbounded there. See Figure 10.7.4. For each number $c < b$, we can form the definite integral

$$\int_a^c f(x)\, dx.$$

If

$$\lim_{c \to b^-} \int_a^c f(x)\, dx = L$$

exists, then we say that

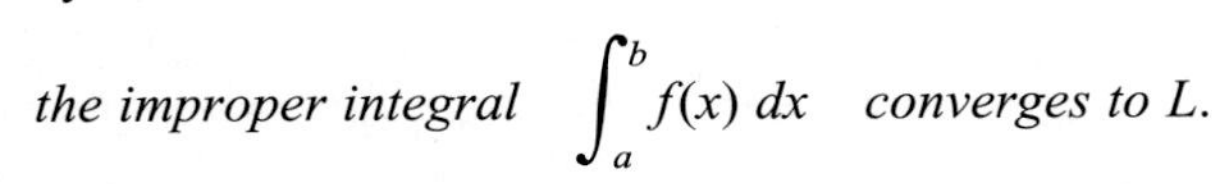

$$\textit{the improper integral} \quad \int_a^b f(x)\, dx \quad \textit{converges to } L.$$

Figure 10.7.4

Otherwise, *the improper integral diverges.*

In a similar manner, if f is continuous on $(a, b]$ and unbounded at a, then we consider the limit

$$\lim_{c \to a^+} \int_c^b f(x)\, dx.$$

If this limit exists and has the value L, then

$$\textit{the improper integral} \quad \int_a^b f(x)\, dx \quad \textit{converges to } L.$$

Otherwise, *the improper integral diverges.*

Example 7

(a) $\displaystyle\int_0^1 (1 - x)^{-2/3}\, dx = 3.$ **(b)** $\displaystyle\int_0^2 \frac{dx}{x}$ diverges.

VERIFICATION

(a)
$$\int_0^1 (1 - x)^{-2/3}\, dx = \lim_{c \to 1^-} \int_0^c (1 - x)^{-2/3}\, dx$$
$$= \lim_{c \to 1^-} \left[-3(1 - x)^{1/3} \right]_0^c = \lim_{c \to 1^-} [-3(1 - c)^{1/3} + 3] = 3.$$

(b)
$$\int_0^2 \frac{dx}{x} = \lim_{c \to 0^+} \int_c^2 \frac{dx}{x} = \lim_{c \to 0^+} [\ln 2 - \ln c] = \infty. \quad \square$$

Now suppose that f is continuous on an interval $[a, b]$ except at some point c in (a, b) where $f(x) \to \pm\infty$ as $x \to c^-$ or as $x \to c^+$. We say that the *improper integral*

$$\int_a^b f(x)\, dx$$

converges iff *both* of the integrals

$$\int_a^c f(x)\, dx \qquad \text{and} \qquad \int_c^b f(x)\, dx$$

converge. If

$$\int_a^c f(x)\,dx = L \qquad \text{and} \qquad \int_c^b f(x)\,dx = M,$$

then

$$\int_a^b f(x)\,dx = L + M.$$

Example 8 To evaluate

$$(*) \qquad \int_1^4 \frac{dx}{(x-2)^2}$$

we need to calculate

$$\lim_{c\to 2^-} \int_1^c \frac{dx}{(x-2)^2} \qquad \text{and} \qquad \lim_{c\to 2^+} \int_c^4 \frac{dx}{(x-2)^2}.$$

As you can verify, neither of these limits exists and thus improper integral (*) diverges.

Notice that, if we ignore the fact that integral (*) is improper, then we are led to the *incorrect conclusion* that

$$\int_1^4 \frac{dx}{(x-2)^2} = \left[\frac{-1}{x-2}\right]_1^4 = -\frac{3}{2}. \qquad \text{(see Figure 10.7.5)} \quad ❑$$

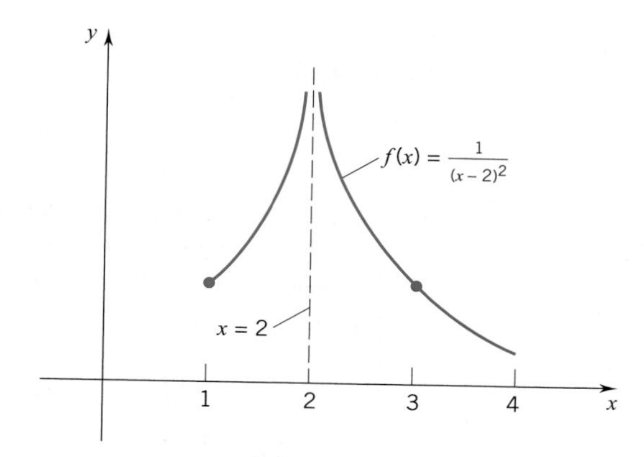

Figure 10.7.5

Example 9 Evaluate

$$\int_{-2}^1 \frac{dx}{x^{4/5}}.$$

SOLUTION Since $1/x^{4/5} \to \infty$ as $x \to 0^-$ and as $x \to 0^+$, the given integral is improper. Therefore, we need to calculate

$$\int_{-2}^0 \frac{dx}{x^{4/5}} \qquad \text{and} \qquad \int_0^1 \frac{dx}{x^{4/5}}.$$

Now

$$\int_{-2}^{0} \frac{dx}{x^{4/5}} = \lim_{c\to 0^-} \int_{-2}^{c} \frac{dx}{x^{4/5}} = \lim_{c\to 0^-} \left[5x^{1/5} \right]_{-2}^{c} = \lim_{c\to 0^-} [5c^{1/5} - 5(-2)^{1/5}] = 5(2^{1/5})$$

and

$$\int_{0}^{1} \frac{dx}{x^{4/5}} = \lim_{c\to 0^+} \int_{c}^{1} \frac{dx}{x^{4/5}} = \lim_{c\to 0^+} \left[5x^{1/5} \right]_{c}^{1} = \lim_{c\to 0^+} [5 - 5c^{1/5}] = 5.$$

Thus, the improper integral converges and

$$\int_{-2}^{1} \frac{dx}{x^{4/5}} = 5 + 5(2^{1/5}) \cong 10.74. \quad \square$$

EXERCISES 10.7

In Exercises 1–34, evaluate the improper integrals that converge.

1. $\displaystyle\int_{1}^{\infty} \frac{dx}{x^2}$.
2. $\displaystyle\int_{0}^{\infty} \frac{dx}{1+x^2}$.
3. $\displaystyle\int_{0}^{\infty} \frac{dx}{4+x^2}$.
4. $\displaystyle\int_{0}^{\infty} e^{-px}\,dx, \quad p > 0$.
5. $\displaystyle\int_{0}^{\infty} e^{px}\,dx, \quad p > 0$.
6. $\displaystyle\int_{0}^{1} \frac{dx}{\sqrt{x}}$.
7. $\displaystyle\int_{0}^{8} \frac{dx}{x^{2/3}}$.
8. $\displaystyle\int_{0}^{1} \frac{dx}{x^2}$.
9. $\displaystyle\int_{0}^{1} \frac{dx}{\sqrt{1-x^2}}$.
10. $\displaystyle\int_{0}^{1} \frac{dx}{\sqrt{1-x}}$.
11. $\displaystyle\int_{0}^{2} \frac{x}{\sqrt{4-x^2}}\,dx$.
12. $\displaystyle\int_{0}^{a} \frac{dx}{\sqrt{a^2-x^2}}$.
13. $\displaystyle\int_{e}^{\infty} \frac{\ln x}{x}\,dx$.
14. $\displaystyle\int_{e}^{\infty} \frac{dx}{x\ln x}$.
15. $\displaystyle\int_{0}^{1} x\ln x\,dx$.
16. $\displaystyle\int_{e}^{\infty} \frac{dx}{x(\ln x)^2}$.
17. $\displaystyle\int_{-\infty}^{\infty} \frac{dx}{1+x^2}$.
18. $\displaystyle\int_{2}^{\infty} \frac{dx}{x^2-1}$.
19. $\displaystyle\int_{-\infty}^{\infty} \frac{dx}{x^2}$.
20. $\displaystyle\int_{1/3}^{3} \frac{dx}{\sqrt[3]{3x-1}}$.
21. $\displaystyle\int_{1}^{\infty} \frac{dx}{x(x+1)}$.
22. $\displaystyle\int_{-\infty}^{0} xe^{x}\,dx$.
23. $\displaystyle\int_{3}^{5} \frac{x}{\sqrt{x^2-9}}\,dx$.
24. $\displaystyle\int_{1}^{4} \frac{dx}{x^2-4}$.
25. $\displaystyle\int_{-3}^{3} \frac{dx}{x(x+1)}$.
26. $\displaystyle\int_{1}^{\infty} \frac{x}{(1+x^2)^2}\,dx$.
27. $\displaystyle\int_{-3}^{1} \frac{dx}{x^2-4}$.
28. $\displaystyle\int_{0}^{\infty} \sinh x\,dx$.
29. $\displaystyle\int_{0}^{\infty} \cosh x\,dx$.
30. $\displaystyle\int_{1}^{4} \frac{dx}{x^2-5x+6}$.
31. $\displaystyle\int_{0}^{\infty} e^{-x}\sin x\,dx$.
32. $\displaystyle\int_{0}^{\infty} \cos^2 x\,dx$.
33. $\displaystyle\int_{0}^{1} \frac{e^{\sqrt{x}}}{\sqrt{x}}\,dx$.
34. $\displaystyle\int_{0}^{\pi/2} \frac{\cos x}{\sqrt{\sin x}}\,dx$.

35. The integral

$$\int_{0}^{1} \sin^{-1} x\,dx$$

is a "proper" definite integral. But the integration technique (integration by parts) will lead to an improper integral. Evaluate this integral.

36. (a) For what values of r is

$$\int_{0}^{\infty} x^r e^{-x}\,dx$$

convergent?

(b) Use mathematical induction to show that

$$\int_{0}^{\infty} x^n e^{-x}\,dx = n!, \quad n = 1, 2, 3, \ldots.$$

37. The integral

$$\int_{0}^{\infty} \frac{1}{\sqrt{x}(1+x)}\,dx$$

is improper for both of the reasons discussed in this section: the interval is infinite and the integrand is unbounded. If we rewrite the integral as

$$\int_0^\infty \frac{1}{\sqrt{x}(1+x)}\,dx = \int_0^1 \frac{1}{\sqrt{x}(1+x)}\,dx + \int_1^\infty \frac{1}{\sqrt{x}(1+x)}\,dx$$

then we have two improper integrals, the first having an unbounded integrand and the second an infinite interval. If each of these integrals converges with values L_1 and L_2, respectively, then the given integral converges and has the value $L_1 + L_2$. Evaluate the given integral.

38. Evaluate

$$\int_1^\infty \frac{1}{x\sqrt{x^2-1}}\,dx$$

using the method given in Exercise 37.

39. Let Ω be the region bounded by the coordinate axes, the graph of $y = 1/\sqrt{x}$, and the line $x = 1$. (a) Sketch Ω. (b) Show Ω has finite area and find it. (c) Show that if Ω is revolved about the x-axis, the solid obtained does not have finite volume.

40. Let Ω be the region between the graph of $y = 1/(1+x^2)$ and the x-axis, $x \geq 0$. (a) Sketch Ω. (b) Find the area of Ω. (c) Find the volume of the solid obtained by revolving Ω about the x-axis. (d) Find the volume of the solid obtained by revolving Ω about the y-axis.

41. Let Ω be the region bounded by the curve $y = e^{-x}$ and the x-axis, $x \geq 0$. (a) Sketch Ω. (b) Find the area of Ω. (c) Find the volume of the solid obtained by revolving Ω about the x-axis. (d) Find the volume obtained by revolving Ω about the y-axis. (e) Find the lateral surface area of the solid in part (c).

42. What point would you call the centroid of the region in Exercise 41? Does Pappus's theorem work in this instance?

43. Let Ω be the region bounded by the curve $y = e^{-x^2}$ and the x-axis, $x \geq 0$. (a) Show that Ω has finite area. (The area is actually $\frac{1}{2}\sqrt{\pi}$, as you will see in Chapter 16.) (b) Calculate the volume generated by revolving Ω about the y-axis.

44. Let Ω be the region bounded below by $y(x^2+1) = x$, above by $xy = 1$, and to the left by $x = 1$. (a) Find the area of Ω. (b) Show that the solid generated by revolving Ω about the x-axis has finite volume. (c) Calculate the volume generated by revolving Ω about the y-axis.

45. Let Ω be the region bounded by the curve $y = x^{-1/4}$ and the x-axis, $0 < x \leq 1$. (a) Sketch Ω. (b) Find the area of Ω. (c) Find the volume of the solid obtained by revolving Ω about the x-axis. (d) Find the volume of the solid obtained by revolving Ω about the y-axis.

46. Prove the validity of the comparison test (10.7.2).

In Exercises 47–52, use the comparison test (10.7.2) to determine which of the integrals converge.

47. $\displaystyle\int_1^\infty \frac{x}{\sqrt{1+x^5}}\,dx.$

48. $\displaystyle\int_1^\infty 2^{-x^2}\,dx.$

49. $\displaystyle\int_0^\infty (1+x^5)^{-1/6}\,dx.$

50. $\displaystyle\int_\pi^\infty \frac{\sin^2 2x}{x^2}\,dx.$

51. $\displaystyle\int_1^\infty \frac{\ln x}{x^2}\,dx.$

52. $\displaystyle\int_e^\infty \frac{dx}{\sqrt{x+1}\,\ln x}.$

53. Calculate the arc distance from the origin to the point $(x(\theta_1), y(\theta_1))$ along the exponential spiral $r = ae^{c\theta}$. (Take $a > 0$, $c > 0$.)

54. The function

$$f(x) = \frac{1}{\sqrt{2\pi}}\int_{-\infty}^{x} e^{-t^2/2}\,dt$$

is important in statistics. Prove that the integral on the right converges for all real x.

Exercises 55–58: **Laplace transforms.** Let f be continuous on $[0, \infty)$. The *Laplace transform* of f is the function F defined by

$$F(s) = \int_0^\infty e^{-sx} f(x)\,dx.$$

The domain of F is the set of all real numbers s such that the improper integral converges. Find the Laplace transform F of each of the following functions and give the domain of F.

55. $f(x) = 1.$

56. $f(x) = x.$

57. $f(x) = \cos 2x.$

58. $f(x) = e^{ax}.$

Exercises 59–62: **Probability density functions.** A nonnegative function f defined on $(-\infty, \infty)$ is a *probability density function* if

$$\int_{-\infty}^{\infty} f(x)\,dx = 1.$$

59. Show that the function f defined by

$$f(x) = \begin{cases} 6x/(1+3x^2)^2 & x \geq 0 \\ 0 & x < 0 \end{cases}$$

is a probability density function.

60. Show that the function f defined by

$$f(x) = \begin{cases} ke^{-kx} & x \geq 0 \\ 0 & x < 0, \end{cases} \quad k > 0,$$

is a probability density function. It is called the *exponential density function*.

61. If f is a probability density function, then its *mean* μ is given by

$$\mu = \int_{-\infty}^{\infty} x\,f(x)\,dx.$$

Calculate the mean for the exponential density function.

62. If f is a probability density function, then its *standard deviation* σ is given by

$$\sigma = \left[\int_{-\infty}^{\infty} (x-\mu)^2 f(x)\,dx\right]^{1/2}$$

where μ is the mean. Calculate the standard deviation for the exponential density function.

63. (*Useful later*) Let f be a continuous, positive, decreasing function on $[1, \infty)$. Show that

$$\int_1^{\infty} f(x)\,dx \quad \text{converges} \quad \text{iff} \left\{\int_1^n f(x)\,dx\right\} \text{converges.}$$

■ CHAPTER HIGHLIGHTS

10.1 The Least Upper Bound Axiom

Upper bound, bounded above, least upper bound (p. 637)
Least upper bound axiom (p. 638)
Lower bound, bounded below, greatest lower bound (p. 640)

10.2 Sequences of Real Numbers

Sequence (p. 642)
Bounded above, bounded below, bounded (p. 644)
Increasing, nondecreasing, decreasing, nonincreasing (p. 644)
Recurrence relation (p. 649)

It is sometimes possible to obtain useful information about a sequence $y_n = f(n)$ by applying the techniques of calculus to the function $y = f(x)$.

10.3 Limit of a Sequence

Limit of a sequence (p. 651) Uniqueness of the limit (p. 652)
Convergent, divergent (p. 653) Pinching theorem (p. 656)

Every convergent sequence is bounded (p. 653); thus, every unbounded sequence is divergent.

A bounded, monotonic sequence converges. (p. 654)

Suppose that $c_n \to c$ as $n \to \infty$, and all the c_n are in the domain of f. If f is continuous at c, then $f(c_n) \to f(c)$. (p. 657)

10.4 Some Important Limits

For $x > 0$, $\lim_{n\to\infty} x^{1/n} = 1$ For $|x| < 1$, $\lim_{n\to\infty} x^n = 0$

For each $\alpha > 0$, $\lim_{n\to\infty} \frac{1}{n^\alpha} = 0$ For each real x, $\lim_{n\to\infty} \frac{x^n}{n!} = 0$

$\lim_{n\to\infty} \frac{\ln n}{n} = 0$ $\lim_{n\to\infty} n^{1/n} = 1$

For each real x, $\lim_{n\to\infty} \left(1 + \frac{x}{n}\right)^n = e^x$ Cauchy sequence (p. 680)

10.5 The Indeterminate Form (0/0)

L'Hospital's rule (0/0) (p. 668) Cauchy mean-value theorem (p. 670)

10.6 The Indeterminate Form (∞/∞)

L'Hospital's rule (∞/∞) (p. 674)

$\lim_{x\to\infty} \frac{\ln x}{x^\alpha} = 0$ $\lim_{x\to\infty} \frac{x^k}{e^x} = 0$ $\lim_{x\to\infty} x^x = 1$

Other indeterminate forms: $0 \cdot \infty$, $\infty - \infty$, 0^0, 1^∞, ∞^0 (p. 675)

10.7 Improper Integrals

Integrals over infinite intervals (p. 680) convergent, divergent (p. 680)

$\int_1^\infty \frac{dx}{x^p}$ converges for $p > 1$ and diverges for $p \leq 1$.

A comparison test (p. 683)

Integrals of unbounded functions (p. 686) convergent, divergent (p. 686)

■ PROJECTS AND EXPLORATIONS USING TECHNOLOGY

To do these exercises you will need a graphics calculator or a computer with graphing capability. The majority of these problems are open-ended so different approaches may be used to solve them. You should be aware that different approaches can result in slight variations in the answers. Round your numerical answers to at least four decimal places. The rounding method that your calculator or computer uses also may cause variations in answers.

10.1 It is easy to show using the comparison test that the improper integral

$$\int_1^\infty \frac{1}{x^3 + 1}\, dx$$

converges. However, computer programs may have trouble evaluating this integral.

(a) Use the comparison test to show that this integral converges.
(b) Use the method of partial fractions to evaluate this integral.
(c) The function

$$F(x) = \int_1^x \frac{1}{t^3 + 1}\, dt$$

is an increasing function on $[1, \infty)$. Explain why.

(d) Evaluate F at a sequence of x values (for example, $x = 10$, $x = 100$, $x = 1000$, $x = 10{,}000$, and so on) using a computer program. Does F start to decrease as x increases? Can you explain why?
(e) Set up a procedure for checking the accuracy of the values that you compute.

10.2 This problem investigates the convergence of the improper integral

$$\int_1^\infty 1000xe^{-0.1\sqrt{x}}\, dx.$$

(a) Let $\{I_k\}$ be the sequence defined by

$$I_k = \int_1^k 1000xe^{-0.1\sqrt{x}}\, dx.$$

Calculate I_k for $k = 10, 100, 1000, 10{,}000$, and so on. Can you determine $L = \lim_{k\to\infty} I_k$?

(b) The difference $E(k) = |L - I_k|$ is the "error." The limit L can also be calculated using integration by parts or by software (be careful that the software gives reasonable answers). Sketch the graph of E to see how quickly the values I_k approach L.
(c) Let

$$a_k = \int_{10^{k-1}}^{10^k} 1000xe^{-0.1\sqrt{x}}\, dx.$$

Show that

$$I_{10^k} = a_0 + a_1 + a_2 + \cdots + a_k.$$

and compare the values for I_k obtained this way with the values calculated directly in part (a).

(d) Explain why a_k might have advantages over I_k in terms of approximating the integral.

10.3 When calculating compound interest, we work with the functions

$$f_r(n) = \left(1 + \frac{r}{n}\right)^n,$$

where r is the annual interest rate and n is the number of compounding periods per year.

(a) Let $r = 0.1$. Find and discuss the meaning of $f_r(4)$, $f_r(12)$, $f_r(365)$, $f_r(8760)$, $f_r(525{,}600)$, and $f_r(31{,}536{,}600)$.

(b) Repeat the calculations in part (a) using $r = 0.01, 0.05, 0.075$, and 0.12.

(c) Mathematically, f_r can be considered for all real numbers and not just for positive integers n. Based on your results in parts (a) and (b), does it appear that

$$\lim_{x \to \infty} f_r(x)$$

exists when r is a given number?

(d) Now let

$$g_n(r) = \left(1 + \frac{r}{n}\right)^n.$$

For $n = 4$, 12, 365, 8760, 525,600, and 31,536,000, respectively, approximate the following limits

$$\lim_{x \to 0^+} g_n(x).$$

What is the meaning of these limits?

(e) For the values of n and r given in parts (a) and (b), approximate $f_r'(n)$ and $g_n'(r)$. What conclusions can you draw? What are the relationships between these derivatives for a given n and r?

(f) Use a graphing utility to graph the functions $f_r(x)$ and $g_n(x)$ for the values of n and r given in parts (a) and (b). Are these functions increasing or decreasing? Do they have inverse functions?

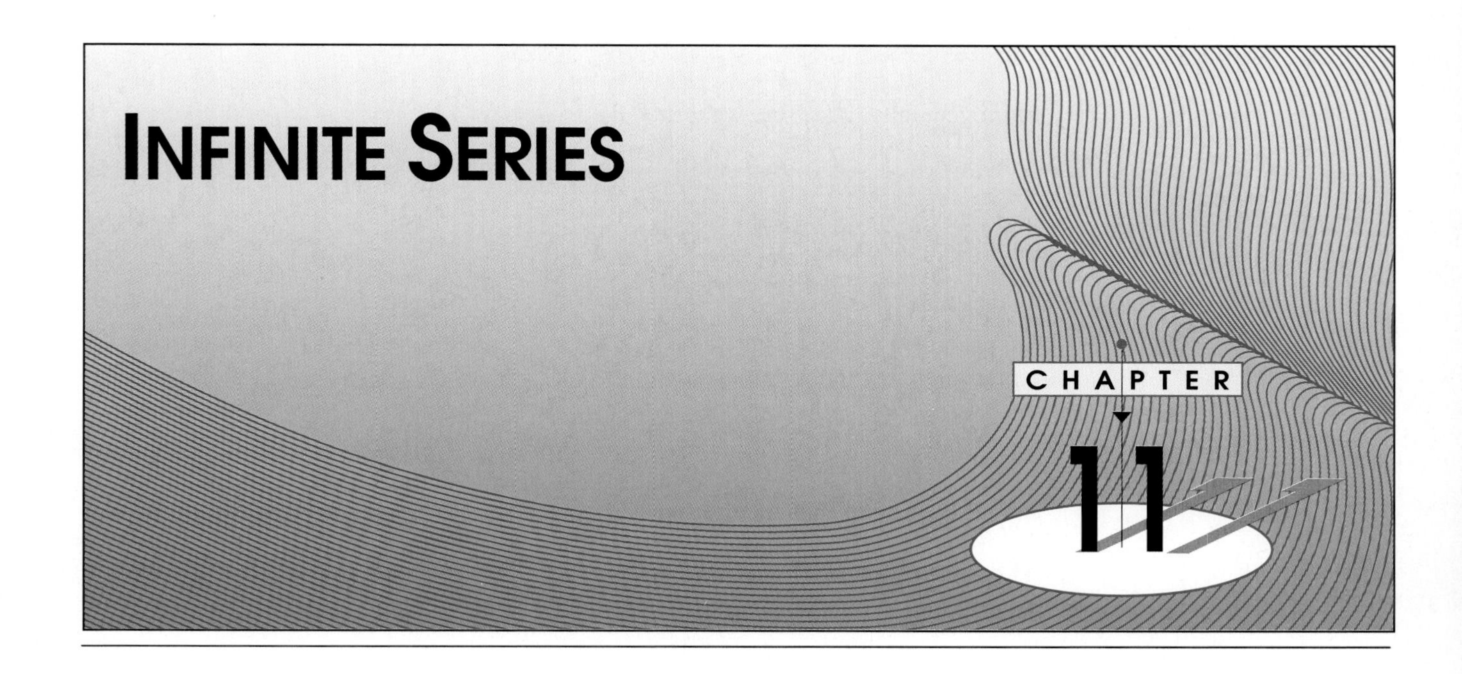

11.1 SIGMA NOTATION

In Section 10.2 we defined a sequence as a real-valued function whose domain is the set of positive integers. For example, to indicate the sequence

$$\{1, \tfrac{1}{2}, \tfrac{1}{4}, \tfrac{1}{8}, \ldots\}$$

we would set $a_n = (\frac{1}{2})^{n-1}$, $n = 1, 2, 3, \ldots$, and write

$$\{a_1, a_2, a_3, a_4, \ldots\}.$$

In this chapter, however, it will often be convenient to begin a sequence with an index other than 1. So, continuing with the example, we can also set $b_n = (\frac{1}{2})^n$, $n = 0, 1, 2, \ldots$, and write

$$\{b_0, b_1, b_2, b_3, \ldots\},$$

thereby beginning with the index 0. In general, we can set $c_n = (\frac{1}{2})^{n-p}$, where p is an integer and $n = p, p + 1, p + 2, \ldots$, and write

$$\{c_p, c_{p+1}, c_{p+2}, c_{p+3}, \ldots\},$$

so that we begin with the index p.

The symbol Σ is the capital Greek letter "sigma." We write

$$\sum_{k=0}^{n} a_k \qquad (1)$$

(read "the sum of the a sub k from k equals 0 to k equals n") to indicate the sum

$$a_0 + a_1 + \cdots + a_n.$$

More generally, if $n \geq m$, we write

(2) $$\sum_{k=m}^{n} a_k$$

to indicate the sum

$$a_m + a_{m+1} + \cdots + a_n.$$

In (1) and (2) the letter "k" is being used as a "dummy" variable. That is, it can be replaced by any letter not already engaged. For instance,

$$\sum_{i=3}^{7} a_i, \qquad \sum_{j=3}^{7} a_j, \qquad \sum_{k=3}^{7} a_k$$

can all be used to indicate the sum

$$a_3 + a_4 + a_5 + a_6 + a_7.$$

Translating

$$(a_0 + \cdots + a_n) + (b_0 + \cdots + b_n) = (a_0 + b_0) + \cdots + (a_n + b_n),$$
$$\alpha(a_0 + \cdots + a_n) = \alpha a_0 + \cdots + \alpha a_n,$$
$$(a_0 + \cdots + a_m) + (a_{m+1} + \cdots + a_n) = a_0 + \cdots + a_n$$

into the Σ-notation, we have

$$\sum_{k=0}^{n} a_k + \sum_{k=0}^{n} b_k = \sum_{k=0}^{n} (a_k + b_k), \qquad \alpha \sum_{k=0}^{n} a_k = \sum_{k=0}^{n} \alpha a_k,$$
$$\sum_{k=0}^{m} a_k + \sum_{k=m+1}^{n} a_k = \sum_{k=0}^{n} a_k.$$

At times it is convenient to change indices. In this connection note that

$$\sum_{k=j}^{n} a_k = \sum_{i=0}^{n-j} a_{i+j}. \qquad \text{(set } i = k - j\text{)}$$

Both expressions are abbreviations for $a_j + a_{j+1} + \cdots + a_n$.

You can familiarize yourself further with this notation by doing the exercises below, but first one more remark. If all the a_k are equal to some fixed number x, then

$$\sum_{k=0}^{n} a_k \quad \text{can be written} \quad \sum_{k=0}^{n} x.$$

Obviously then

$$\sum_{k=0}^{n} x = \overbrace{x + x + \cdots + x}^{n+1} = (n+1)x.$$

In particular

$$\sum_{k=0}^{n} 1 = n + 1.$$

EXERCISES 11.1

In Exercises 1–12, evaluate the given expression.

1. $\sum_{k=0}^{2}(3k+1)$.

2. $\sum_{k=1}^{4}(3k-1)$.

3. $\sum_{k=0}^{3}2^k$.

4. $\sum_{k=0}^{3}(-1)^k 2^k$.

5. $\sum_{k=0}^{3}(-1)^k 2^{k+1}$.

6. $\sum_{k=2}^{5}(-1)^{k+1}2^{k-1}$.

7. $\sum_{k=1}^{4}\frac{1}{2^k}$.

8. $\sum_{k=2}^{5}\frac{1}{k!}$.

9. $\sum_{k=3}^{5}\frac{(-1)^k}{k!}$.

10. $\sum_{k=2}^{4}\frac{1}{3^{k-1}}$.

11. $\sum_{k=0}^{3}(\frac{1}{2})^{2k}$.

12. $\sum_{k=1}^{3}(-1)^{k+1}(\frac{1}{2})^{2k-1}$.

In Exercises 13–28, express in sigma notation.

13. $1+3+5+7+\cdots+21$.

14. $1-3+5-7+\cdots-19$.

15. $2\cdot 1+2\cdot 2+2\cdot 3+\cdots+2\cdot 25$.

16. $1\cdot 2+2\cdot 3+3\cdot 4+\cdots+35\cdot 36$.

17. $1-\sqrt{2}+\sqrt{3}-2+\sqrt{5}-\cdots+9$.

18. $\frac{\tan 1}{2}+\frac{\tan 2}{5}+\frac{\tan 3}{10}+\frac{\tan 4}{17}+\cdots+\frac{\tan 10}{101}$.

19. The lower sum $m_1\Delta x_1+m_2\Delta x_2+\cdots+m_n\Delta x_n$.

20. The upper sum $M_1\Delta x_1+M_2\Delta x_2+\cdots+M_n\Delta x_n$.

21. The Riemann sum $f(x_1^*)\Delta x_1+f(x_2^*)\Delta x_2+\cdots+f(x_n^*)\Delta x_n$.

22. $a^5+a^4b+a^3b^2+a^2b^3+ab^4+b^5$.

23. $a^5-a^4b+a^3b^2-a^2b^3+ab^4-b^5$.

24. $a^n+a^{n-1}b+\cdots+ab^{n-1}+b^n$.

25. $a_0x^4+a_1x^3+a_2x^2+a_3x+a_4$.

26. $a_0x^n+a_1x^{n-1}+\cdots+a_{n-1}x+a_n$.

27. $1-2x+3x^2-4x^3+5x^4$.

28. $3x-4x^2+5x^3-6x^4$.

In Exercises 29–32, write the given sums as $\sum_{k=3}^{10}a_k$ and $\sum_{i=0}^{7}a_{i+3}$:

29. $\frac{1}{2^3}+\frac{1}{2^4}+\cdots+\frac{1}{2^{10}}$.

30. $\frac{3^3}{3!}+\frac{4^4}{4!}+\cdots+\frac{10^{10}}{10!}$.

31. $\frac{3}{4}-\frac{4}{5}+\cdots-\frac{10}{11}$.

32. $\frac{1}{3}+\frac{1}{5}+\frac{1}{7}+\cdots+\frac{1}{17}$.

In Exercises 33–36, verify by a change of indices that the two sums are identical.

33. $\sum_{k=2}^{10}\frac{k}{k^2+1}$; $\sum_{n=-1}^{7}\frac{n+3}{n^2+6n+10}$.

34. $\sum_{n=2}^{12}\frac{(-1)^n}{n-1}$; $\sum_{k=1}^{11}\frac{(-1)^{k+1}}{k}$.

35. $\sum_{k=4}^{25}\frac{1}{k^2-9}$; $\sum_{n=7}^{28}\frac{1}{n^2-6n}$.

36. $\sum_{k=0}^{15}\frac{3^{2k}}{k!}$; $81\sum_{n=-2}^{13}\frac{3^{2n}}{(n+2)!}$.

37. (a) (*Important*) Show that for $x \neq 1$

$$\sum_{k=0}^{n}x^k=\frac{1-x^{n+1}}{1-x}.$$

(b) Determine whether the sequence $a_n=\sum_{k=0}^{n}\frac{1}{3^k}$ converges and, if it does, find the limit.

38. Express $\sum_{k=1}^{n}\frac{a_k}{10^k}$ as a decimal fraction, given that each a_k is an integer from 0 to 9.

39. Let p be a positive integer. Show that, as $n\to\infty$,

$$a_n\to L \quad \text{iff} \quad a_{n-p}\to L.$$

40. Show that

$$\sum_{k=1}^{n}\frac{1}{\sqrt{k}}\geq\sqrt{n}.$$

In Exercises 41–44, verify by induction.

41. $\sum_{k=1}^{n}k=\frac{1}{2}(n)(n+1)$.

42. $\sum_{k=1}^{n}(2k-1)=n^2$.

43. $\sum_{k=1}^{n}k^2=\frac{1}{6}(n)(n+1)(2n+1)$.

44. $\sum_{k=1}^{n}k^3=\left(\sum_{k=1}^{n}k\right)^2$.

In Exercises 45–48, evaluate the sum.

45. $\sum_{k=1}^{10}(2k+3)$.

46. $\sum_{k=1}^{10}(2k^2+3k)$.

47. $\sum_{k=1}^{8}(2k-1)^2$.

48. $\sum_{k=1}^{n}k(k^2-5)$.

11.2 INFINITE SERIES

Introduction; Definitions

While it is possible to add two numbers, three numbers, a hundred numbers, or even a million numbers, how can we attach meaning to the sum of an infinite number of numbers? The theory of infinite series arose from attempts to answer this question. As you might expect, a limit process is involved.

To form an infinite series we begin with an infinite sequence of real numbers: $a_0, a_1, a_2, \ldots$. We can't form the sum of all the a_k (there is an infinite number of them), but we can form the *partial sums*

$$
\begin{aligned}
s_0 &= a_0 = \sum_{k=0}^{0} a_k, \\
s_1 &= a_0 + a_1 = \sum_{k=0}^{1} a_k, \\
s_2 &= a_0 + a_1 + a_2 = \sum_{k=0}^{2} a_k, \\
s_3 &= a_0 + a_1 + a_2 + a_3 = \sum_{k=0}^{3} a_k, \\
&\;\;\vdots \\
s_n &= a_0 + a_1 + a_2 + a_3 + \cdots + a_n = \sum_{k=0}^{n} a_k \\
&\;\;\vdots
\end{aligned}
$$

Continuing in this way, we are led to consider the "infinite sum" $\sum_{k=0}^{\infty} a_k$ which is called an *infinite series.* The corresponding sequence $\{s_n\}$ is called the *sequence of partial sums* of the series.

DEFINITION 11.2.1

Given the infinite series $\sum_{k=0}^{\infty} a_k$. If the sequence of partial sums $\{s_n\}$ converges to a finite limit L, then the series $\sum_{k=0}^{\infty} a_k$ is said to *converge* to L, written

$$\sum_{k=0}^{\infty} a_k = L.$$

The number L is called the *sum* of the series. If the sequence of partial sums diverges, then the series $\sum_{k=0}^{\infty} a_k$ *diverges.*

Remark It is important to note that the sum of a series is not a sum in the ordinary sense. It is a limit. ❑

Here are some examples.

Example 1 We begin with the series

$$\sum_{k=0}^{\infty} \frac{1}{(k+1)(k+2)}.$$

To determine whether or not this series converges we must examine the partial sums.

Since

$$\frac{1}{(k+1)(k+2)} = \frac{1}{k+1} - \frac{1}{k+2}, \qquad \text{(partial fraction decomposition, see Section 8.6)}$$

you can see that

$$\begin{aligned} s_n &= \frac{1}{1 \cdot 2} + \frac{1}{2 \cdot 3} + \cdots + \frac{1}{n(n+1)} + \frac{1}{(n+1)(n+2)} \\ &= \left(\frac{1}{1} - \frac{1}{2}\right) + \left(\frac{1}{2} - \frac{1}{3}\right) + \cdots + \left(\frac{1}{n} - \frac{1}{n+1}\right) + \left(\frac{1}{n+1} - \frac{1}{n+2}\right) \\ &= 1 - \frac{1}{2} + \frac{1}{2} - \frac{1}{3} + \cdots + \frac{1}{n} - \frac{1}{n+1} + \frac{1}{n+1} - \frac{1}{n+2}. \end{aligned}$$

Since all but the first and last terms occur in pairs with opposite signs, the sum collapses to give

$$s_n = 1 - \frac{1}{n+2}.$$

Obviously, as $n \to \infty$, $s_n \to 1$. This means that the series converges to 1:

$$\sum_{k=0}^{\infty} \frac{1}{(k+1)(k+2)} = 1. \quad ❑$$

Remark Infinite series with the special property illustrated in Example 1 (that is, except for the first and last term, the terms can be arranged in pairs with opposite signs) are called *telescoping series.* In general,

$$\sum_{k=p}^{n} \{f(k) - f(k+1)\} = f(p) - f(n+1) \qquad \text{and}$$

$$\sum_{k=p}^{n} \{f(k) - f(k-1)\} = f(n) - f(p-1). \qquad \text{(verify these)} \quad ❑$$

Example 2 Here we examine two divergent series

$$\sum_{k=0}^{\infty} 2^k \qquad \text{and} \qquad \sum_{k=1}^{\infty} (-1)^k.$$

The partial sums of the first series take the form

$$s_n = \sum_{k=0}^{n} 2^k = 1 + 2 + \cdots + 2^n.$$

The sequence $\{s_n\}$ is unbounded and therefore divergent (10.3.5). This means that the series diverges.

For the second series we have

$$s_n = -1 \quad \text{if } n \text{ is odd} \qquad \text{and} \qquad s_n = 0 \quad \text{if } n \text{ is even.}$$

The sequence of partial sums looks like this:

$$-1, 0, -1, 0, -1, 0, \ldots .$$

The series diverges since the sequence of partial sums diverges. ❑

Remark Example 2 illustrates two types of divergence. In the first case, $s_n \to \infty$ as $n \to \infty$. The notation $\Sigma_{k=0}^{\infty} a_k = \infty$ is sometimes used to denote this type of divergence. In the second case, s_n oscillates between -1 and 0. ❑

The Geometric Series

Fix a real number x. The sequence $1, x, x^2, x^3, \ldots = \{x^n\}, n = 0, 1, 2, 3, \ldots$, is called a *geometric progression*. Concerning convergence, we know that if $|x| < 1$, then $x^n \to 0$ ("special limit" (10.4.2)). If $x = 1$, then we have the constant sequence $1, 1, 1, \ldots$, which clearly converges to 1. Finally, it is easy to show that $\{x^n\}$ is divergent when $x = -1$ (the sequence oscillates between 1 and -1) and when $|x| > 1$ (the sequence is unbounded).

The sums

$$1, \quad 1 + x, \quad 1 + x + x^2, \quad 1 + x + x^2 + x^3, \ldots$$

generated by numbers in geometric progression are the partial sums of what is known as the *geometric series*:

$$\sum_{k=0}^{\infty} x^k.$$

The geometric series arises in so many contexts that it merits special attention.

The following result is fundamental:

(11.2.2)

$$\text{(i) if } |x| < 1, \quad \text{then } \sum_{k=0}^{\infty} x^k = \frac{1}{1-x};$$

$$\text{(ii) if } |x| \geq 1, \quad \text{then } \sum_{k=0}^{\infty} x^k \text{ diverges.}$$

PROOF The nth partial sum of the geometric series

$$\sum_{k=0}^{\infty} x^k$$

takes the form

(1) $$s_n = 1 + x + \cdots + x^n.$$

Multiplication by x gives

$$xs_n = x + x^2 + \cdots + x^{n+1}.$$

Subtracting the second equation from the first, we find that

$$(1 - x)s_n = 1 - x^{n+1}.$$

For $x \neq 1$, this gives

$$s_n = \frac{1 - x^{n+1}}{1 - x}. \tag{2}$$

If $|x| < 1$, then $x^{n+1} \to 0$ as $n \to \infty$ and thus, by Equation (2),

$$s_n \to \frac{1}{1 - x}.$$

This proves (i).

Now let's prove (ii). For $x = 1$, we use Equation (1) and deduce that $s_n = n + 1$. Obviously, $\{s_n\}$ diverges. For $x \neq 1$ with $|x| \geq 1$, we use Equation (2). Since in this instance $\{x^{n+1}\}$ diverges, $\{s_n\}$ diverges. ❑

You may have seen (11.2.2) before, written as

$$a + ar + ar^2 + \cdots + ar^n + \cdots = \begin{cases} \dfrac{a}{1 - r}, & |r| < 1 \\ \text{diverges}, & |r| \geq 1, \end{cases} \qquad (a \neq 0).$$

Taking $a = 1$ and $r = \frac{1}{2}$, we have

$$\sum_{k=0}^{\infty} \frac{1}{2^k} = \frac{1}{1 - \frac{1}{2}} = 2.$$

Begin the summation at $k = 1$ instead of at $k = 0$, and you see that

(11.2.3)

$$\boxed{\sum_{k=1}^{\infty} \frac{1}{2^k} = 1.}$$

The partial sums of this series

$$\begin{aligned} s_1 &= \tfrac{1}{2}, \\ s_2 &= \tfrac{1}{2} + \tfrac{1}{4} = \tfrac{3}{4}, \\ s_3 &= \tfrac{1}{2} + \tfrac{1}{4} + \tfrac{1}{8} = \tfrac{7}{8}, \\ s_4 &= \tfrac{1}{2} + \tfrac{1}{4} + \tfrac{1}{8} + \tfrac{1}{16} = \tfrac{15}{16}, \\ s_5 &= \tfrac{1}{2} + \tfrac{1}{4} + \tfrac{1}{8} + \tfrac{1}{16} + \tfrac{1}{32} = \tfrac{31}{32}, \\ &\;\vdots \end{aligned}$$

are illustrated in Figure 11.2.1. Each new partial sum lies halfway between the previous partial sum and the number 1.

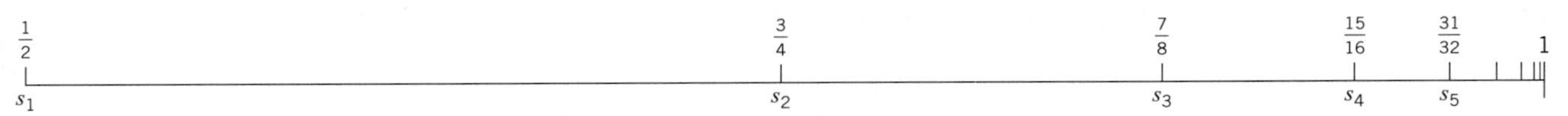

Figure 11.2.1

The convergence of the geometric series at $x = \frac{1}{10}$ enables us to assign a precise meaning to infinite decimals. Begin with the fact that

$$\sum_{k=0}^{\infty} \left(\frac{1}{10}\right)^k = \sum_{k=0}^{\infty} \frac{1}{10^k} = \frac{1}{1 - \frac{1}{10}} = \frac{10}{9}.$$

This gives

$$\sum_{k=1}^{\infty} \frac{1}{10^k} = \frac{1}{9}$$

and shows that the partial sums

$$s_n = \frac{1}{10} + \frac{1}{10^2} + \cdots + \frac{1}{10^n}$$

are all less than $\frac{1}{9}$. Now take a series of the form

$$\sum_{k=1}^{\infty} \frac{a_k}{10^k} \qquad \text{with} \qquad a_k = 0, 1, \ldots, \text{or } 9.$$

Its partial sums

$$t_n = \frac{a_1}{10} + \frac{a_2}{10^2} + \cdots + \frac{a_n}{10^n}$$

are all less than 1:

$$t_n = \frac{a_1}{10} + \frac{a_2}{10^2} + \cdots + \frac{a_n}{10^n} \le 9\left(\frac{1}{10} + \frac{1}{10^2} + \cdots + \frac{1}{10^n}\right) = 9s_n < 9\left(\frac{1}{9}\right) = 1.$$

Since $\{t_n\}$ is nondecreasing, as well as bounded above, $\{t_n\}$ is convergent; this means that the series

$$\sum_{k=1}^{\infty} \frac{a_k}{10^k}$$

is convergent. The sum of this series is what we mean by the infinite decimal

$$0.a_1a_2a_3 \cdots a_n \cdots.$$

Following are two simple examples that lead naturally to geometric series. You will find more in the exercises.

Example 3 An electric fan is turned off, and the blades begin to lose speed. Given that the blades turn N times during the first second of no power and lose at least $\sigma\%$ of their speed with the passing of each ensuing second, show that the blades cannot turn more than $100N\sigma^{-1}$ times after power shutdown.

SOLUTION The number of turns during the first second of no power is

$$N.$$

The number of turns during the first 2 seconds is at most

$$N + \left(1 - \frac{1}{100}\sigma\right)N;$$

during the first 3 seconds, at most

$$N + \left(1 - \frac{1}{100}\sigma\right)N + \left(1 - \frac{1}{100}\sigma\right)^2 N;$$

and, during the first $n + 1$ seconds, at most

$$N\sum_{k=0}^{n}\left(1 - \frac{1}{100}\sigma\right)^k.$$

The total number of turns after power shutdown cannot exceed the limiting value

$$N\sum_{k=0}^{\infty}\left(1 - \frac{1}{100}\sigma\right)^k.$$

This is a geometric series with $x = 1 - \frac{1}{100}\sigma$. Thus

$$N\sum_{k=0}^{\infty}\left(1 - \tfrac{1}{100}\sigma\right)^k = N\left[\frac{1}{1-(1-\frac{1}{100}\sigma)}\right] = \frac{100N}{\sigma} = 100N\sigma^{-1}. \quad \square$$

Example 4 According to Figure 11.2.2, it is 2 o'clock. At what time between 2 and 3 o'clock will the two hands coincide?

Figure 11.2.2

SOLUTION We will solve the problem by setting up a geometric series. We will then confirm our answer by approaching the problem from a different perspective.

The hour hand travels one-twelfth as fast as the minute hand. At 2 o'clock the minute hand points to 12 and the hour hand points to 2. By the time the minute hand reaches 2, the hour hand points to $2 + \frac{1}{6}$. By the time the minute hand reaches $2 + \frac{1}{6}$, the hour hand points to

$$2 + \frac{1}{6} + \frac{1}{6 \cdot 12}.$$

By the time the minute hand reaches

$$2 + \frac{1}{6} + \frac{1}{6 \cdot 12},$$

the hour hand points to

$$2 + \frac{1}{6} + \frac{1}{6 \cdot 12} + \frac{1}{6 \cdot 12^2}$$

and so on. In general, by the time the minute hand reaches

$$2 + \frac{1}{6} + \frac{1}{6 \cdot 12} + \cdots + \frac{1}{6 \cdot 12^{n-1}} = 2 + \frac{1}{6}\sum_{k=0}^{n-1}\frac{1}{12^k},$$

the hour hand points to

$$2 + \frac{1}{6} + \frac{1}{6 \cdot 12} + \cdots + \frac{1}{6 \cdot 12^{n-1}} + \frac{1}{6 \cdot 12^n} = 2 + \frac{1}{6} \sum_{k=0}^{n} \frac{1}{12^k}.$$

The two hands coincide when they both point to the limiting value

$$2 + \frac{1}{6} \sum_{k=0}^{\infty} \frac{1}{12^k} = 2 + \frac{1}{6}\left(\frac{1}{1 - \frac{1}{12}}\right) = 2 + \frac{2}{11}.$$

This happens at $2 + \frac{2}{11}$ o'clock, approximately 10 minutes and 55 seconds after 2.

We can confirm this as follows. Suppose that the two hands meet at hour $2 + x$. The hour hand will have moved a distance x and the minute hand a distance $2 + x$. Since the minute hand moves 12 times as fast as the hour hand, we have

$$12x = 2 + x, \qquad 11x = 2, \qquad \text{and thus} \qquad x = \frac{2}{11}. \quad \square$$

We will return to the geometric series later. Right now we turn our attention to series in general.

Some Basic Results

THEOREM 11.2.4

1. If $\sum_{k=0}^{\infty} a_k$ converges and $\sum_{k=0}^{\infty} b_k$ converges, then $\sum_{k=0}^{\infty} (a_k + b_k)$ converges.

Moreover, if $\sum_{k=0}^{\infty} a_k = L$ and $\sum_{k=0}^{\infty} b_k = M$, then $\sum_{k=0}^{\infty} (a_k + b_k) = L + M$.

2. If $\sum_{k=0}^{\infty} a_k$ converges, then $\sum_{k=0}^{\infty} \alpha a_k$ converges for each real number α.

Moreover, if $\sum_{k=0}^{\infty} a_k = L$, then $\sum_{k=0}^{\infty} \alpha a_k = \alpha L$.

PROOF Let

$$s_n = \sum_{k=0}^{n} a_k, \qquad t_n = \sum_{k=0}^{n} b_k, \qquad u_n = \sum_{k=0}^{n} (a_k + b_k), \qquad v_n = \sum_{k=0}^{n} \alpha a_k.$$

Note that

$$u_n = s_n + t_n \qquad \text{and} \qquad v_n = \alpha s_n.$$

If $s_n \to L$ and $t_n \to M$, then

$$u_n \to L + M \qquad \text{and} \qquad v_n \to \alpha L. \qquad \text{(Theorem 10.3.7)} \quad \square$$

THEOREM 11.2.5

Let j be a positive integer. The series $\sum_{k=0}^{\infty} a_k$ converges iff the series $\sum_{k=j}^{\infty} a_k$ converges. Moreover, if $\sum_{k=0}^{\infty} a_k = L$, then $\sum_{k=j}^{\infty} a_k = L - (a_0 + a_1 + a_2 + \cdots + a_{j-1})$; or if $\sum_{k=j}^{\infty} a_k = M$, then $\sum_{k=0}^{\infty} a_k = M + (a_0 + a_1 + a_2 + \cdots + a_{j-1})$.

It is important to understand what this theorem means, namely, the convergence (or divergence) of an infinite series is not affected by where you start the summation. In the case of a convergent series, however, the limit (sum) does depend upon where you begin the summation. The proof of the theorem is left to you as an exercise.

THEOREM 11.2.6

The *kth term* of a convergent series tends to 0; namely,

$$\text{if } \sum_{k=0}^{\infty} a_k \text{ converges, then } a_k \to 0 \text{ as } k \to \infty.$$

PROOF To say that the series converges is to say that the sequence of partial sums converges to some number L:

$$s_n = \sum_{k=0}^{n} a_k \to L.$$

Obviously, then, $s_{n-1} \to L$. Since $a_n = s_n - s_{n-1}$, we have $a_n \to L - L = 0$. A change in notation gives $a_k \to 0$. ❑

The next result is an obvious, but important, consequence of Theorem 11.2.6.

THEOREM 11.2.7 A DIVERGENCE TEST

$$\text{If } a_k \not\to 0 \text{ as } k \to \infty, \quad \text{then } \sum_{k=0}^{\infty} a_k \text{ diverges.}$$

Example 5

(a) Since $\dfrac{k}{k+1} \not\to 0$ as $k \to \infty$, the series

$$\sum_{k=0}^{\infty} \frac{k}{k+1} = 0 + \frac{1}{2} + \frac{2}{3} + \frac{3}{4} + \frac{4}{5} + \cdots \quad \text{diverges.}$$

(b) Since $\sin k \not\to 0$ as $k \to \infty$, the series

$$\sum_{k=0}^{\infty} \sin k = \sin 0 + \sin 1 + \sin 2 + \sin 3 + \cdots \quad \text{diverges.} \quad \square$$

CAUTION Theorem 11.2.6 does *not* say that, if $a_k \to 0$, then $\sum_{k=0}^{\infty} a_k$ converges. There are divergent series for which $a_k \to 0$. ☐

Example 6 In the case of

$$\sum_{k=1}^{\infty} \frac{1}{\sqrt{k}} = \frac{1}{\sqrt{1}} + \frac{1}{\sqrt{2}} + \frac{1}{\sqrt{3}} + \frac{1}{\sqrt{4}} + \cdots$$

we have

$$a_k = \frac{1}{\sqrt{k}} \to 0 \quad \text{as } k \to \infty,$$

but, since

$$s_n = \frac{1}{\sqrt{1}} + \frac{1}{\sqrt{2}} + \cdots + \frac{1}{\sqrt{n}} \geq \underbrace{\frac{1}{\sqrt{n}} + \frac{1}{\sqrt{n}} + \cdots + \frac{1}{\sqrt{n}}}_{n \text{ terms}} = \frac{n}{\sqrt{n}} = \sqrt{n},$$

the sequence of partial sums is unbounded, and therefore the series diverges. ☐

EXERCISES 11.2

In Exercises 1–18, find the sum of the series.

1. $\displaystyle\sum_{k=3}^{\infty} \frac{1}{(k+1)(k+2)}$.
2. $\displaystyle\sum_{k=0}^{\infty} \frac{1}{(k+3)(k+4)}$.
3. $\displaystyle\sum_{k=1}^{\infty} \frac{1}{2k(k+1)}$.
4. $\displaystyle\sum_{k=3}^{\infty} \frac{1}{k^2 - k}$.
5. $\displaystyle\sum_{k=1}^{\infty} \frac{1}{k(k+3)}$.
6. $\displaystyle\sum_{k=0}^{\infty} \frac{1}{(k+1)(k+3)}$.
7. $\displaystyle\sum_{k=0}^{\infty} \frac{3}{10^k}$.
8. $\displaystyle\sum_{k=0}^{\infty} \frac{12}{100^k}$.
9. $\displaystyle\sum_{k=0}^{\infty} \frac{67}{1000^k}$.
10. $\displaystyle\sum_{k=0}^{\infty} \frac{(-1)^k}{5^k}$.
11. $\displaystyle\sum_{k=0}^{\infty} \left(\frac{3}{4}\right)^k$.
12. $\displaystyle\sum_{k=0}^{\infty} \frac{3^k + 4^k}{5^k}$.
13. $\displaystyle\sum_{k=0}^{\infty} \frac{1 - 2^k}{3^k}$.
14. $\displaystyle\sum_{k=0}^{\infty} \left(\frac{25}{10^k} - \frac{6}{100^k}\right)$.
15. $\displaystyle\sum_{k=3}^{\infty} \frac{1}{2^{k-1}}$.
16. $\displaystyle\sum_{k=0}^{\infty} \frac{1}{2^{k+3}}$.
17. $\displaystyle\sum_{k=0}^{\infty} \frac{2^{k+3}}{3^k}$.
18. $\displaystyle\sum_{k=2}^{\infty} \frac{3^{k-1}}{4^{3k+1}}$.

In Exercises 19–26, write the decimal fraction as an infinite series and express the sum as the quotient of two integers.

19. 0.777 . . . .
20. 0.999 . . . .
21. 0.2424 . . . .
22. 0.8989 . . . .
23. 0.112112112 . . . .
24. 0.315315315 . . . .
25. 0.624545 . . . .
26. 0.112019019 . . . .
27. Using series, show that every repeating decimal represents a rational number (the quotient of two integers).
28. Prove Theorem 11.2.5.

In Exercises 29 and 30, derive the indicated result from the geometric series.

29. $\displaystyle\sum_{k=0}^{\infty} (-1)^k x^k = \frac{1}{1+x}, \quad |x| < 1.$
30. $\displaystyle\sum_{k=0}^{\infty} (-1)^k x^{2k} = \frac{1}{1+x^2}, \quad |x| < 1.$

In Exercises 31–36, find a series expansion for the given expression.

31. $\displaystyle\frac{x}{1-x}$ for $|x| < 1$.
32. $\displaystyle\frac{x}{1+x}$ for $|x| < 1$.

33. $\dfrac{x}{1+x^2}$ for $|x| < 1$.

34. $\dfrac{1}{4-x^2}$ for $|x| < 2$.

35. $\dfrac{1}{1+4x^2}$ for $|x| < \dfrac{1}{2}$.

36. $\dfrac{x^2}{1-x}$ for $|x| < 1$.

In Exercises 37–40, show that the given series diverges.

37. $1 + \dfrac{3}{2} + \dfrac{9}{4} + \dfrac{27}{8} + \dfrac{81}{16} + \cdots$.

38. $\sum_{k=0}^{\infty} (-1)^k$.

39. $\sum_{k=1}^{\infty} \left(\dfrac{k+1}{k}\right)^k$.

40. $\sum_{k=2}^{\infty} \dfrac{k^{k-2}}{3^k}$.

41. At some time between 4 and 5 o'clock the minute hand is directly above the hour hand. Express this time as a geometric series. What is the sum of this series?

42. Given that a ball dropped to the floor rebounds to a height proportional to the height from which it is dropped, find the total distance traveled by a ball dropped from a height of 6 feet if it rebounds initially to a height of 3 feet.

43. Exercise 42 under the supposition that the ball rebounds initially to a height of 2 feet.

44. In the setting of Exercise 42, to what height does the ball rebound initially if the total distance traveled by the ball is 21 feet?

45. How much money must you deposit at r% interest compounded annually to enable your descendants to withdraw n_1 dollars at the end of the first year, n_2 dollars at the end of the second year, n_3 dollars at the end of the third year, and so on in perpetuity? Assume that the sequence $\{n_k\}$ is bounded, $n_k \le N$ for all k, and express your answer as an infinite series.

46. Sum the series you obtained in Exercise 45 setting
(a) $r = 5$, $n_k = 5000(\frac{1}{2})^{k-1}$.
(b) $r = 6$, $n_k = 1000(0.8)^{k-1}$.
(c) $r = 5$, $n_k = N$.

47. Suppose that 90% of each dollar is recirculated into the economy. That is, suppose that when a dollar is put into circulation, 90% of it is spent; then 90% of that is spent, and so on. What is the total economic value of the dollar?

48. Consider the following sequence of steps. First, take the unit interval [0,1] and delete the open interval $(\frac{1}{3}, \frac{2}{3})$. Next delete the two open intervals $(\frac{1}{9}, \frac{2}{9})$ and $(\frac{7}{9}, \frac{8}{9})$ from the two intervals that remain after the first step. For the third step, delete the middle thirds from the four intervals that remain after the second step. Continue on in this manner. What is the sum of the lengths of the intervals that have been deleted? The set that remains after all of the "middle thirds" have been deleted is called the *Cantor middle third set*. Can you name some points that are in the Cantor set?

49. Start with a square whose sides are four units long. Join the midpoints of the sides of the square to form a second square inside the first. Then join the midpoints of the sides of the second square to form a third square, and so on. See the figure. Find the sum of the areas of the squares.

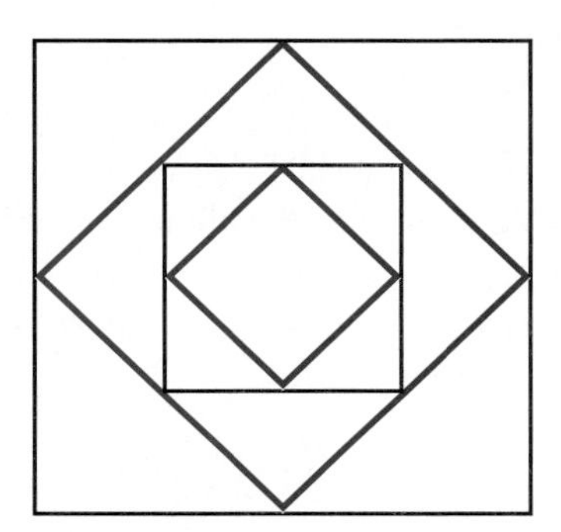

50. (a) Prove that if the series $\Sigma\, a_k$ converges and the series $\Sigma\, b_k$ diverges, then the series $\Sigma\, (a_k + b_k)$ diverges.
(b) Give examples to show that if $\Sigma\, a_k$ and $\Sigma\, b_k$ both diverge, then each of the series

$$\sum (a_k + b_k) \quad \text{and} \quad \sum (a_k - b_k)$$

may either converge or diverge.

51. Let $\Sigma_{k=0}^{\infty} a_k$ be a convergent series and let $R_n = \Sigma_{k=n+1}^{\infty} a_k$. Prove that $R_n \to 0$ as $n \to \infty$. Note that if s_n is the nth partial sum of the series, then $\Sigma_{k=0}^{\infty} a_k = s_n + R_n$; R_n is called the *remainder*.

52. (a) Prove that if $\Sigma_{k=0}^{\infty} a_k$ is a convergent series and $a_k \neq 0$ for all k, then $\Sigma_{k=0}^{\infty} (1/a_k)$ is divergent.
(b) Suppose that $a_k > 0$ for all k and $\Sigma_{k=0}^{\infty} a_k$ diverges. Show by means of examples that $\Sigma_{k=0}^{\infty} (1/a_k)$ may either converge or diverge.

53. Let $\{s_n\}$ be the sequence of partial sums of the series $\Sigma_{k=0}^{\infty} (-1)^k$. Find a formula for s_n. HINT: Find $s_0, s_1, s_2, \ldots$ and note the pattern.

54. Repeat Exercise 53 for the series $\sum_{k=1}^{\infty} \ln\left(\dfrac{k}{k+1}\right)$.

55. Show that

$$\sum_{k=1}^{\infty} \ln\left(\frac{k+1}{k}\right) \quad \text{diverges, even though} \quad \lim_{k\to\infty} \ln\left(\frac{k+1}{k}\right) = 0.$$

56. Show that

$$\sum_{k=1}^{\infty} \left(\frac{k}{k+1}\right)^k \quad \text{diverges.}$$

57. (a) Let $\{d_k\}$ be a sequence of real numbers that converges to 0. Show that

$$\sum_{k=1}^{\infty} (d_k - d_{k+1}) = d_1.$$

(b) Sum the following series:

(i) $\displaystyle\sum_{k=1}^{\infty} \frac{\sqrt{k+1}-\sqrt{k}}{\sqrt{k(k+1)}}$. (ii) $\displaystyle\sum_{k=1}^{\infty} \frac{2k+1}{2k^2(k+1)^2}$.

58. Show that

$$\sum_{k=1}^{\infty} kx^{k-1} = \frac{1}{(1-x)^2} \quad \text{for } |x| < 1.$$

HINT: Verify that s_n, the nth partial sum, satisfies the identity

$$(1-x)^2 s_n = 1 - (n+1)x^n + nx^{n+1}.$$

Speed of Convergence Suppose that $\Sigma_{k=0}^{\infty} a_k$ is a convergent series with sum L and let $\{s_n\}$ be its sequence of partial sums. It follows from Exercise 51 that $|L - s_n| = |R_n|$. In Exercises 59–62, find the smallest integer N such that $|L - s_N| < 0.0001$.

59. $\displaystyle\sum_{k=0}^{\infty} \frac{1}{4^k}$.

60. $\displaystyle\sum_{k=0}^{\infty} (0.9)^k$.

61. $\displaystyle\sum_{k=1}^{\infty} \frac{1}{k(k+2)}$.

62. $\displaystyle\sum_{k=0}^{\infty} \left(\frac{2}{3}\right)^k$.

63. Given the geometric series $\Sigma_{k=0}^{\infty} x^k$, with $|x| < 1$, and a positive number ϵ. Determine the smallest positive integer N such that $|L - s_N| < \epsilon$, where L is the sum of the series and s_n is the nth partial sum.

■ 11.3 THE INTEGRAL TEST; COMPARISON THEOREMS

Here and in the next section we direct our attention to *series with nonnegative terms*: $a_k \geq 0$ for all k. The significant feature of a series with nonnegative terms is the fact that its sequence of partial sums is nondecreasing:

$$s_{n+1} = \sum_{k=0}^{n+1} a_k = \sum_{k=0}^{n} a_k + a_{n+1} \geq \sum_{k=0}^{n} a_k = s_n, \qquad n = 0, 1, 2, \ldots.$$

The following theorem is fundamental.

THEOREM 11.3.1

A series with nonnegative terms converges iff the sequence of partial sums is bounded.

PROOF Assume that the series converges. Then the sequence of partial sums is convergent and therefore bounded (Theorem 10.3.4).

Suppose now that the sequence of partial sums is bounded. Since the terms are nonnegative, the sequence is nondecreasing. By being bounded and nondecreasing, the sequence of partial sums converges (Theorem 10.3.6). This means that the series converges. ❑

The convergence or divergence of a series can sometimes be deduced from the convergence or divergence of a closely related improper integral.

THEOREM 11.3.2 THE INTEGRAL TEST

If f is continuous, decreasing, and positive on $[1, \infty)$, then

$$\sum_{k=1}^{\infty} f(k) \quad \text{converges} \qquad \text{iff} \qquad \int_1^{\infty} f(x)\,dx \quad \text{converges.}$$

PROOF In Exercise 63, Section 10.7, you were asked to show that with f continuous, decreasing, and positive on $[1, \infty)$

$$\int_1^{\infty} f(x)\,dx \quad \text{converges} \qquad \text{iff} \qquad \text{the sequence} \quad \left\{\int_1^n f(x)\,dx\right\} \quad \text{converges.}$$

We assume this result and base our proof on the behavior of the sequence of integrals. To visualize our argument see Figure 11.3.1.

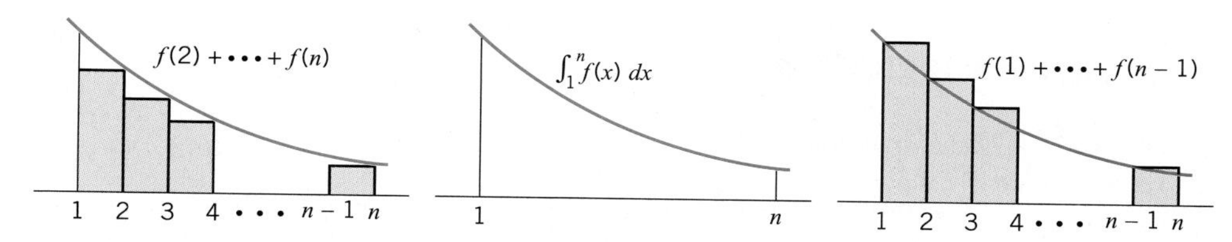

Figure 11.3.1

Since f decreases on the interval $[1, n]$,

$$f(2) + \cdots + f(n) \quad \text{is a lower sum for } f \text{ on } [1, n]$$

and

$$f(1) + \cdots + f(n-1) \quad \text{is an upper sum for } f \text{ on } [1, n].$$

Consequently

$$(1) \quad f(2) + \cdots + f(n) \leq \int_1^n f(x)\,dx \qquad \text{and} \qquad \int_1^n f(x)\,dx \leq f(1) + \cdots + f(n-1).$$

If the sequence of integrals converges, it is bounded. By the first inequality the sequence of partial sums is bounded and the series is therefore convergent.

Suppose now that the sequence of integrals diverges. Since f is positive, the sequence of integrals increases:

$$\int_1^n f(x)\,dx < \int_1^{n+1} f(x)\,dx.$$

Since this sequence diverges, it must be unbounded. By the second inequality, the sequence of partial sums must be unbounded and the series diverges. ❑

Remark The inequalities established in the proof of Theorem 11.3.2 lead to bounds on the sum of the infinite series

$$\sum_{k=1}^{\infty} f(k),$$

where f is continuous, decreasing, and positive on $[1, \infty)$. In particular, it follows from the second inequality in (1) that

$$\int_1^\infty f(x)\,dx \le \sum_{k=1}^\infty f(k),$$

and from the first inequality in (1),

$$\sum_{k=1}^\infty f(k) \le f(1) + \int_1^\infty f(x)\,dx.$$

Combining these two inequalities, we have

$$\int_1^\infty f(x)\,dx \le \sum_{k=1}^\infty f(k) \le f(1) + \int_1^\infty f(x)\,dx.$$

These inequalities also make clear the relation between the convergence of the infinite series and the convergence of the corresponding improper integral. ❑

Applying the Integral Test

Example 1 (*The Harmonic Series*)

(11.3.3)

$$\sum_{k=1}^\infty \frac{1}{k} = 1 + \frac{1}{2} + \frac{1}{3} + \frac{1}{4} + \cdots \quad \text{diverges.}$$

PROOF The function $f(x) = 1/x$ is continuous, decreasing, and positive on $[1, \infty)$. We know that

$$\int_1^\infty \frac{dx}{x} \quad \text{diverges.} \qquad (10.7.1)$$

By the integral test

$$\sum_{k=1}^\infty \frac{1}{k} \quad \text{diverges.} \quad ❑$$

The next example gives a more general result.

Example 2 (*The p-series*)

(11.3.4)

$$\sum_{k=1}^\infty \frac{1}{k^p} = 1 + \frac{1}{2^p} + \frac{1}{3^p} + \frac{1}{4^p} + \cdots \quad \text{converges} \quad \text{iff} \quad p > 1.$$

PROOF If $p \le 0$, then each term of the series is greater than or equal to 1. Therefore, the kth term does not have limit 0 and so, by the divergence test (11.2.7) the series cannot converge. (See, also, the remark following Example 2 in Section 10.7.) We assume, therefore, that $p > 0$. The function $f(x) = 1/x^p$ is then continuous, decreasing, and positive on $[1, \infty)$. Thus by the integral test

$$\sum_{k=1}^{\infty} \frac{1}{k^p} \quad \text{converges} \quad \text{iff} \quad \int_1^{\infty} \frac{dx}{x^p} \quad \text{converges.}$$

Earlier you saw that

$$\int_1^{\infty} \frac{dx}{x^p} \quad \text{converges} \quad \text{iff} \quad p > 1. \tag{10.7.1}$$

It follows that

$$\sum_{k=1}^{\infty} \frac{1}{k^p} \quad \text{converges} \quad \text{iff} \quad p > 1. \quad \square$$

Example 3 Here we show that the series

$$\sum_{k=1}^{\infty} \frac{1}{k \ln (k+1)} = \frac{1}{\ln 2} + \frac{1}{2 \ln 3} + \frac{1}{3 \ln 4} + \cdots$$

diverges. We begin by setting

$$f(x) = \frac{1}{x \ln (x+1)}.$$

Since f is continuous, decreasing, and positive on $[1, \infty)$, we can use the integral test. Note that

$$\begin{aligned} \int_1^b \frac{dx}{x \ln (x+1)} &> \int_1^b \frac{dx}{(x+1) \ln (x+1)} \\ &= \Big[\ln (\ln (x+1)) \Big]_1^b = \ln (\ln (b+1)) - \ln (\ln 2). \end{aligned}$$

As $b \to \infty$, $\ln (\ln (b+1)) \to \infty$. This shows that

$$\int_1^{\infty} \frac{dx}{x \ln (x+1)} \quad \text{diverges.}$$

It follows that the series diverges. $\square$

Remark on Notation You have seen that for each $j \geq 0$

$$\sum_{k=0}^{\infty} a_k \quad \text{converges} \quad \text{iff} \quad \sum_{k=j}^{\infty} a_k \quad \text{converges}$$

(Theorem 11.2.5). This tells you that, in determining whether or not a series converges, it does not matter where we begin the summation. Where detailed indexing would contribute nothing, we will omit it and write $\Sigma\, a_k$ without specifying where the summation begins. For instance, it makes sense to say that

$$\sum \frac{1}{k^2} \quad \text{converges} \quad \text{and} \quad \sum \frac{1}{k} \quad \text{diverges}$$

without specifying where we begin the summation. $\square$

The convergence or divergence of a series with nonnegative terms can sometimes be deduced by comparison with a series of known behavior.

THEOREM 11.3.5 THE BASIC COMPARISON TEST

Let $\Sigma\, a_k$ be a series with nonnegative terms.

(i) $\Sigma\, a_k$ converges if there exists a convergent series $\Sigma\, c_k$ with nonnegative terms such that $a_k \le c_k$ for all k sufficiently large;

(ii) $\Sigma\, a_k$ diverges if there exists a divergent series $\Sigma\, d_k$ with nonnegative terms such that $d_k \le a_k$ for all k sufficiently large.

PROOF The proof is just a matter of noting that, in the first instance, the partial sums of $\Sigma\, a_k$ are bounded and, in the second instance, unbounded. The details are left to you. ❑

Example 4

(a) $\sum \dfrac{1}{2k^3+1}$ converges by comparison with $\sum \dfrac{1}{k^3}$:

$$\frac{1}{2k^3+1} < \frac{1}{k^3} \qquad \text{and} \qquad \sum \frac{1}{k^3} \text{ converges.}$$

(b) $\sum \dfrac{1}{3k+1}$ diverges by comparison with $\sum \dfrac{1}{3(k+1)}$:

$$\frac{1}{3(k+1)} < \frac{1}{3k+1} \qquad \text{and} \qquad \sum \frac{1}{3(k+1)} = \frac{1}{3}\sum \frac{1}{k+1} \qquad \text{diverges}$$

(it's the series $\Sigma 1/k$ with a change of index).

(c) $\sum \dfrac{k^3}{k^5+5k^4+7}$ converges by comparison with $\sum \dfrac{1}{k^2}$:

$$\frac{k^3}{k^5+5k^4+7} < \frac{k^3}{k^5} = \frac{1}{k^2} \qquad \text{and} \qquad \sum \frac{1}{k^2} \quad \text{converges.} \quad ❑$$

Example 5 Show that

$$\sum \frac{1}{\ln(k+6)} \quad \text{diverges.}$$

SOLUTION We know that as $k \to \infty$

$$\frac{\ln k}{k} \to 0. \qquad \text{(see 10.4.5)}$$

It follows that

$$\frac{\ln(k+6)}{k+6} \to 0$$

and therefore that

$$\frac{\ln(k+6)}{k} = \frac{\ln(k+6)}{k+6}\left(\frac{k+6}{k}\right) \to 0.$$

Thus, for k sufficiently large,

$$\frac{\ln(k+6)}{k} < 1,$$

and so

$$\ln(k+6) < k \qquad \text{and} \qquad \frac{1}{k} < \frac{1}{\ln(k+6)}.$$

Since

$$\sum \frac{1}{k} \quad \text{diverges,}$$

we can conclude that

$$\sum \frac{1}{\ln(k+6)} \quad \text{diverges.} \quad \square$$

Remark Another way to show that $\ln(k+6) < k$ for large k is to examine the function $f(x) = x - \ln(x+6)$. At $x = 3$ the function is positive:

$$f(3) = 3 - \ln 9 \cong 3 - 2.197 > 0.$$

Since

$$f'(x) = 1 - \frac{1}{x+6} > 0 \qquad \text{for all } x > 0,$$

$f(x) > 0$ for all $x \geq 3$. It follows that

$$\ln(x+6) < x \qquad \text{for all } x \geq 3. \quad \square$$

The basic comparison test is algebraic in nature; it requires that certain inequalities hold. To apply the test to a series $\sum a_k$ you must show that the terms a_k are smaller than the corresponding terms c_k of a known convergent series to establish convergence, or larger than the terms d_k of a known divergent series to establish divergence. However, if the terms a_k are larger than the terms c_k, or smaller than the terms d_k, then the comparison test does not apply. For example, consider the series

$$\sum_{k=2}^{\infty} \frac{1}{k^3 - 1}.$$

It would be natural to compare this series with the convergent series

$$\sum_{k=2}^{\infty} \frac{1}{k^3},$$

but, unfortunately, the inequalities "go the wrong way:"

$$\frac{1}{k^3 - 1} > \frac{1}{k^3} \qquad \text{for all } k \geq 2.$$

Our next theorem is a more sophisticated comparison test. It is analytic in the sense that it involves the evaluation of a limit.

THEOREM 11.3.6 THE LIMIT COMPARISON TEST

Let $\Sigma\, a_k$ and $\Sigma\, b_k$ be series with positive terms. If $a_k/b_k \to L$, where L is some *positive* number, then either both series converge or both series diverge.

PROOF Choose ϵ between 0 and L. Since $a_k/b_k \to L$, we know that for all k sufficiently large (for all k greater than some k_0)

$$\left| \frac{a_k}{b_k} - L \right| < \epsilon.$$

For such k we have

$$L - \epsilon < \frac{a_k}{b_k} < L + \epsilon$$

and thus

$$(L - \epsilon) b_k < a_k < (L + \epsilon) b_k.$$

This last inequality is what we need:

if $\Sigma\, a_k$ converges, then $\Sigma\, (L - \epsilon) b_k$ converges, and thus $\Sigma\, b_k$ converges;
if $\Sigma\, b_k$ converges, then $\Sigma\, (L + \epsilon) b_k$ converges, and thus $\Sigma\, a_k$ converges. ❑

To apply the limit comparison theorem to a series $\Sigma\, a_k$, we must first find a series $\Sigma\, b_k$ of known behavior for which a_k/b_k converges to a positive number. To complete the example started earlier,

$$\sum_{k=2}^{\infty} \frac{1}{k^3 - 1}$$

converges since

$$\sum_{k=2}^{\infty} \frac{1}{k^3} \quad \text{converges} \qquad \text{and} \qquad \left(\frac{1}{k^3 - 1}\right) \div \left(\frac{1}{k^3}\right) = \frac{k^3}{k^3 - 1} \to 1 \quad \text{as } k \to \infty.$$

Example 6 Determine whether the series

$$\Sigma\, \frac{3k^2 + 2k + 1}{k^3 + 1}$$

converges or diverges.

SOLUTION For large k, the terms with the highest powers of k dominate. Here $3k^2$ dominates the numerator and k^3 dominates the denominator. Thus, for large k,

$$\frac{3k^2 + 2k + 1}{k^3 + 1} \quad \text{differs little from} \quad \frac{3k^2}{k^3} = \frac{3}{k}.$$

Since

$$\frac{3k^2 + 2k + 1}{k^3 + 1} \div \frac{3}{k} = \frac{3k^3 + 2k^2 + k}{3k^3 + 3} = \frac{1 + 2/(3k) + 1/(3k^2)}{1 + 1/k^3} \to 1$$

and

$$\sum \frac{3}{k} = 3 \sum \frac{1}{k} \quad \text{diverges,}$$

we know that the series diverges. ❑

Example 7 Determine whether the series

$$\sum \frac{5\sqrt{k} + 100}{2k^2\sqrt{k} + 9\sqrt{k}}$$

converges or diverges.

SOLUTION For large values of k, $5\sqrt{k}$ dominates the numerator and $2k^2\sqrt{k}$ dominates the denominator. Thus, for such k,

$$\frac{5\sqrt{k} + 100}{2k^2\sqrt{k} + 9\sqrt{k}} \quad \text{differs little from} \quad \frac{5\sqrt{k}}{2k^2\sqrt{k}} = \frac{5}{2k^2}.$$

Since

$$\frac{5\sqrt{k} + 100}{2k^2\sqrt{k} + 9\sqrt{k}} \div \frac{5}{2k^2} = \frac{10k^2\sqrt{k} + 200k^2}{10k^2\sqrt{k} + 45\sqrt{k}} = \frac{1 + 20/\sqrt{k}}{1 + 9/2k^2} \to 1 \quad \text{as } k \to \infty$$

and

$$\sum \frac{5}{2k^2} = \frac{5}{2} \sum \frac{1}{k^2} \quad \text{converges,}$$

the series converges. ❑

Example 8 Determine whether the series

$$\sum \sin \frac{\pi}{k}$$

converges or diverges.

SOLUTION Recall that

$$\text{as } x \to 0, \quad \frac{\sin x}{x} \to 1. \tag{2.5.5}$$

As $k \to \infty$, $\pi/k \to 0$ and thus

$$\frac{\sin(\pi/k)}{\pi/k} \to 1.$$

Since $\Sigma\ \pi/k$ diverges, $\Sigma \sin(\pi/k)$ diverges. ❑

Remark The question of what we can and cannot conclude by limit comparison if $a_k/b_k \to 0$ or if $a_k/b_k \to \infty$ is taken up in Exercises 45 and 46. ❑

EXERCISES 11.3

In Exercises 1–34, determine whether the series converges or diverges.

1. $\Sigma \dfrac{k}{k^3+1}$.

2. $\Sigma \dfrac{1}{3k+2}$.

3. $\Sigma \dfrac{1}{(2k+1)^2}$.

4. $\Sigma \dfrac{\ln k}{k}$.

5. $\Sigma \dfrac{1}{\sqrt{k+1}}$.

6. $\Sigma \dfrac{1}{k^2+1}$.

7. $\Sigma \dfrac{1}{\sqrt{2k^2-k}}$.

8. $\Sigma \left(\dfrac{5}{2}\right)^{-k}$.

9. $\Sigma \dfrac{\tan^{-1}k}{1+k^2}$.

10. $\Sigma \dfrac{\ln k}{k^3}$.

11. $\Sigma \dfrac{1}{k^{2/3}}$.

12. $\Sigma \dfrac{1}{(k+1)(k+2)(k+3)}$.

13. $\Sigma \left(\dfrac{3}{4}\right)^{-k}$.

14. $\Sigma \dfrac{1}{1+2\ln k}$.

15. $\Sigma \dfrac{\ln \sqrt{k}}{k}$.

16. $\Sigma \dfrac{2}{k\,(\ln k)^2}$.

17. $\Sigma \dfrac{1}{2+3^{-k}}$.

18. $\Sigma \dfrac{7k+2}{2k^5+7}$.

19. $\Sigma \dfrac{2k+5}{5k^3+3k^2}$.

20. $\Sigma \dfrac{k^4-1}{3k^2+5}$.

21. $\Sigma \dfrac{1}{k \ln k}$.

22. $\Sigma \dfrac{1}{2^{k+1}-1}$.

23. $\Sigma \dfrac{k^2}{k^4-k^3+1}$.

24. $\Sigma \dfrac{k^{3/2}}{k^{5/2}+2k-1}$.

25. $\Sigma \dfrac{2k+1}{\sqrt{k^4+1}}$.

26. $\Sigma \dfrac{2k+1}{\sqrt{k^3+1}}$.

27. $\Sigma \dfrac{2k+1}{\sqrt{k^5+1}}$.

28. $\Sigma \dfrac{1}{\sqrt{2k(k+1)}}$.

29. $\Sigma\, k e^{-k^2}$.

30. $\Sigma\, k^2 2^{-k^3}$.

31. $\Sigma \dfrac{2+\sin k}{k^2}$.

32. $\Sigma \dfrac{2+\cos k}{\sqrt{k+1}}$.

33. $\Sigma \dfrac{1}{1+2+3+\cdots+k}$.

34. $\Sigma \dfrac{k}{1+2^2+3^2+\cdots+k^2}$.

35. Find the values of p for which the series $\sum_{k=2}^{\infty} \dfrac{1}{k(\ln k)^p}$ converges.

36. Find the values of p for which the series $\sum_{k=2}^{\infty} \dfrac{\ln k}{k^p}$ converges.

37. (a) Prove that $\sum_{k=0}^{\infty} e^{-\alpha k}$ converges for any $\alpha > 0$.

(b) Prove that $\sum_{k=0}^{\infty} k e^{-\alpha k}$ converges for any $\alpha > 0$.

(c) In general, prove that $\sum_{k=0}^{\infty} k^n e^{-\alpha k}$ converges for any nonnegative integer n and any $\alpha > 0$.

38. Let $p > 1$. Use the integral test to show that

$$\frac{1}{(p-1)(n+1)^{p-1}} < \sum_{k=1}^{\infty} \frac{1}{k^p} - \sum_{k=1}^{n} \frac{1}{k^p} < \frac{1}{(p-1)n^{p-1}}.$$

This result gives bounds on the *error* (or remainder) R_n that results from using s_n to approximate L the sum of the convergent p-series.

In Exercises 39–40, (a) compute the sum of the first four terms of the given series; use four decimal place accuracy. (b) Use the result in Exercise 38 to give upper and lower bounds on R_4. (c) Use parts (a) and (b) to estimate the sum of the series.

39. $\sum_{k=1}^{\infty} \dfrac{1}{k^3}$.

40. $\sum_{k=1}^{\infty} \dfrac{1}{k^4}$.

In Exercises 41–44, use the error bounds given in Exercise 38.

41. (a) If you were to use s_{100} to approximate $\sum_{k=1}^{\infty} \dfrac{1}{k^2}$ what would be the bounds on your error?
(b) How large would you have to choose n to ensure that R_n is less than 0.0001?

42. (a) If you were to use s_{100} to approximate $\sum_{k=1}^{\infty} \dfrac{1}{k^3}$ what would be the bounds on your error?
(b) How large would you have to choose n to ensure that R_n is less than 0.0001?

43. (a) How many terms of the series $\sum_{k=1}^{\infty} \dfrac{1}{k^4}$ should you use to ensure that R_n is less than 0.0001?
(b) Estimate $\sum_{k=1}^{\infty} \dfrac{1}{k^4}$ to three decimal places.

44. Repeat Exercise 43 for the series $\sum_{k=1}^{\infty} \dfrac{1}{k^5}$.

45. Let $\Sigma\, a_k$ and $\Sigma\, b_k$ be series with positive terms and suppose that $a_k/b_k \to 0$.
(a) Show that, if $\Sigma\, b_k$ converges, then $\Sigma\, a_k$ converges.
(b) Show that, if $\Sigma\, a_k$ diverges, then $\Sigma\, b_k$ diverges.

(c) Show by example that, if $\Sigma\, a_k$ converges, then $\Sigma\, b_k$ may converge or diverge.
(d) Show by example that, if $\Sigma\, b_k$ diverges, then $\Sigma\, a_k$ may converge or diverge.
[Parts (c) and (d) explain why we stipulated $L > 0$ in Theorem 11.3.6.]

46. Let $\Sigma\, a_k$ and $\Sigma\, b_k$ be series with positive terms and suppose that $a_k/b_k \to \infty$.
(a) Show that if $\Sigma\, b_k$ diverges, then $\Sigma\, a_k$ diverges.
(b) Show that if $\Sigma\, a_k$ converges, then $\Sigma\, b_k$ converges.
(c) Show by example that if $\Sigma\, a_k$ diverges, then $\Sigma\, b_k$ may converge or diverge.
(d) Show by example that if $\Sigma\, b_k$ converges, then $\Sigma\, a_k$ may converge or diverge.

47. Let $\Sigma\, a_k$ be a series with positive terms.
(a) Prove that if $\Sigma\, a_k$ converges, then $\Sigma\, a_k^2$ converges.
(b) Suppose that $\Sigma\, a_k^2$ converges. Does $\Sigma\, a_k$ converge or diverge? Prove or give a counterexample.

48. Let $\Sigma\, a_k$ be a series with positive terms. Prove that if $\Sigma\, a_k^2$ converges, then $\Sigma(a_k/k)$ converges.

49. Let f be a continuous, positive, decreasing function on $[1, \infty)$ such that $\int_1^\infty f(x)\,dx$ converges. Then the series $\sum_{k=1}^{\infty} f(k)$ also converges. Prove that

$$0 < L - s_n < \int_n^\infty f(x)\,dx,$$

where L is the sum of the series and s_n is the nth partial sum.

In Exercises 50 and 51, use the result of Exercise 49 to determine the smallest integer N such that the difference between the sum of the given series and the Nth partial sum is less than 0.001.

50. $\displaystyle\sum_{k=1}^{\infty} \frac{1}{k^2+1}$.

51. $\displaystyle\sum_{k=1}^{\infty} ke^{-k^2}$.

52. All the results of this section were stated for series with nonnegative terms. Corresponding results hold for *series with nonpositive terms*: $a_k \le 0$ for all k.
(a) State a comparison theorem analogous to Theorem 11.3.5, this time for series with nonpositive terms.
(b) As stated, the integral test (Theorem 11.3.2) applies only to series with positive terms. State the equivalent result for series with negative terms.

53. This exercise demonstrates that we cannot always use the same testing series for both the basic comparison test and the limit comparison test.
(a) Show that

$$\Sigma \frac{\ln n}{n\sqrt{n}} \quad \text{converges by comparison with} \quad \Sigma \frac{1}{n^{5/4}}.$$

(b) Show that the limit comparison test does not apply.

11.4 THE ROOT TEST; THE RATIO TEST

We continue with our study of series with nonnegative terms. Comparison with geometric series

$$\Sigma\, x^k$$

and with the p-series

$$\Sigma \frac{1}{k^p}$$

leads to two important tests for convergence: the root test and the ratio test.

> **THEOREM 11.4.1 THE ROOT TEST**
>
> Let $\Sigma\, a_k$ be a series with nonnegative terms, and suppose that
>
> $$(a_k)^{1/k} \to \rho \quad \text{as} \quad k \to \infty.$$
>
> **(a)** If $\rho < 1$, then $\Sigma\, a_k$ converges.
>
> **(b)** If $\rho > 1$, then $\Sigma\, a_k$ diverges.
>
> **(c)** If $\rho = 1$, then the test is inconclusive; the series may either converge or diverge.

PROOF We suppose first that $\rho < 1$ and choose μ so that

$$\rho < \mu < 1.$$

Since $(a_k)^{1/k} \to \rho$, we have

$$(a_k)^{1/k} < \mu \qquad \text{for all } k \text{ sufficiently large.} \qquad \text{(explain)}$$

Thus

$$a_k < \mu^k \qquad \text{for all } k \text{ sufficiently large.}$$

Since $\Sigma\, \mu^k$ converges (a geometric series with $0 < \mu < 1$), we know by the basic comparison theorem that $\Sigma\, a_k$ converges.

We suppose now that $\rho > 1$. Since $(a_k)^{1/k} \to \rho$, we have

$$(a_k)^{1/k} > 1 \qquad \text{for all } k \text{ sufficiently large.} \qquad \text{(explain)}$$

Thus

$$a_k > 1 \qquad \text{for all } k \text{ sufficiently large.}$$

It now follows that $a_k \not\to 0$ as $k \to \infty$. Therefore $\Sigma\, a_k$ diverges by the divergence test (11.2.7).

To see the inconclusiveness of the root test when $\rho = 1$, consider the series $\Sigma(1/k^2)$ and $\Sigma(1/k)$. The first series converges and the second series diverges. However, in each case we have

$$(a_k)^{1/k} = \left(\frac{1}{k^2}\right)^{1/k} = \left(\frac{1}{k^{1/k}}\right)^2 \to 1^2 = 1 \quad \text{as } k \to \infty,$$

$$(a_k)^{1/k} = \left(\frac{1}{k}\right)^{1/k} = \frac{1}{k^{1/k}} \to 1 \quad \text{as } k \to \infty.$$

(Recall that $k^{1/k} \to 1$ as $k \to \infty$, see (10.4.6).) ❑

Applying the Root Test

Example 1 For the series

$$\Sigma \frac{1}{(\ln k)^k}$$

we have

$$(a_k)^{1/k} = \frac{1}{\ln k} \to 0.$$

The series converges. ❑

Example 2 For the series

$$\Sigma \frac{2^k}{k^3}$$

we have

$$(a_k)^{1/k} = 2\left(\frac{1}{k}\right)^{3/k} = 2\left[\left(\frac{1}{k}\right)^{1/k}\right]^3 = 2\left[\frac{1}{k^{1/k}}\right]^3 \to 2 \cdot 1^3 = 2 \quad \text{as } k \to \infty.$$

The series diverges. ❑

Example 3 In the case of

$$\Sigma \left(1 - \frac{1}{k}\right)^k,$$

we have

$$(a_k)^{1/k} = 1 - \frac{1}{k} \to 1.$$

Here the root test is inconclusive. It is also unnecessary: since $a_k = (1 - 1/k)^k$ converges to $1/e$ and not to 0 (10.4.7), the series diverges (11.2.7). ❑

> **THEOREM 11.4.2 THE RATIO TEST**
>
> Let Σa_k be a series with positive terms and suppose that
>
> $$\frac{a_{k+1}}{a_k} \to \lambda \quad \text{as } k \to \infty$$
>
> **(a)** If $\lambda < 1$, then $\Sigma\, a_k$ converges.
> **(b)** If $\lambda > 1$, then $\Sigma\, a_k$ diverges.
> **(c)** If $\lambda = 1$, then the test is inconclusive; the series may either converge or diverge.

PROOF We suppose first that $\lambda < 1$ and choose μ so that $\lambda < \mu < 1$. Since

$$\frac{a_{k+1}}{a_k} \to \lambda,$$

we know that there exists $k_0 > 0$ such that

$$\text{if } k \geq k_0, \quad \text{then} \quad \frac{a_{k+1}}{a_k} < \mu. \qquad \text{(explain)}$$

This gives

$$a_{k_0+1} < \mu a_{k_0}, \qquad a_{k_0+2} < \mu a_{k_0+1} < \mu^2 a_{k_0},$$

and more generally,

$$a_{k_0+j} < \mu^j a_{k_0}, \qquad j = 1, 2, \ldots .$$

For $k > k_0$ we have

(1)
$$a_k < \mu^{k-k_0} a_{k_0} = \frac{a_{k_0}}{\mu^{k_0}} \mu^k.$$

↑ set $j = k - k_0$

Since $\mu < 1$,

$$\Sigma \frac{a_{k_0}}{\mu^{k_0}} \mu^k = \frac{a_{k_0}}{\mu^{k_0}} \Sigma\, \mu^k \quad \text{converges.}$$

Recalling (1), you can see by the basic comparison theorem that Σa_k converges. The proof of the rest of the theorem is left to the exercises. ❑

Remark Contrary to some people's intuition the root and ratio tests are *not* equivalent. See Exercise 52. ❑

Applying the Ratio Test

Example 4 The ratio test shows that the series

$$\Sigma \frac{1}{k!}$$

converges:

$$\frac{a_{k+1}}{a_k} = \frac{1}{(k+1)!} \cdot \frac{k!}{1} = \frac{1}{k+1} \to 0 \qquad \text{as } k \to \infty. \quad \square$$

Example 5 For the series

$$\Sigma \frac{k}{10^k}$$

we have

$$\frac{a_{k+1}}{a_k} = \frac{k+1}{10^{k+1}} \cdot \frac{10^k}{k} = \frac{1}{10} \frac{k+1}{k} \to \frac{1}{10} \quad \text{as } k \to \infty.$$

The series converges.† ❑

Example 6 For the series

$$\Sigma \frac{k^k}{k!}$$

we have

$$\frac{a_{k+1}}{a_k} = \frac{(k+1)^{k+1}}{(k+1)!} \cdot \frac{k!}{k^k} = \left(\frac{k+1}{k}\right)^k = \left(1 + \frac{1}{k}\right)^k \to e \quad \text{as } k \to \infty.$$

Since $e > 1$, the series diverges. ❑

Example 7 For the series

$$\Sigma \frac{1}{2k+1}$$

the ratio test is inconclusive:

$$\frac{a_{k+1}}{a_k} = \frac{1}{2(k+1)+1} \cdot \frac{2k+1}{1} = \frac{2k+1}{2k+3} = \frac{2+1/k}{2+3/k} \to 1 \text{ as } k \to \infty.$$

Therefore, we have to look further. Comparison with the harmonic series shows that the series diverges:

$$\frac{1}{2k+1} \div \frac{1}{k} = \frac{k}{2k+1} \to \frac{1}{2} \qquad \text{and} \qquad \Sigma \frac{1}{k} \text{ diverges.} \quad \square$$

---

† This series can be summed explicitly. See Exercise 41.

Summary on Convergence Tests

In general, the root test is used only if powers are involved. The ratio test is particularly effective with factorials and with combinations of powers and factorials. If the terms are rational functions of k, the ratio test is inconclusive and the root test is difficult to apply. Rational terms are most easily handled by comparison or limit comparison with a p-series, $\Sigma\, 1/k^p$. If the terms have the configuration of a derivative, you may be able to apply the integral test. Finally, keep in mind that, if $a_k \not\to 0$, then there is no reason to apply any special convergence test; the series diverges by Theorem 11.2.7.

EXERCISES 11.4

In Exercises 1–40, determine whether the series converges or diverges.

1. $\Sigma \dfrac{10^k}{k!}$.
2. $\Sigma \dfrac{1}{k2^k}$.
3. $\Sigma \dfrac{1}{k^k}$.
4. $\Sigma \left(\dfrac{k}{2k+1}\right)^k$.
5. $\Sigma \dfrac{k!}{100^k}$.
6. $\Sigma \dfrac{(\ln k)^2}{k}$.
7. $\Sigma \dfrac{k^2+2}{k^3+6k}$.
8. $\Sigma \dfrac{1}{(\ln k)^k}$.
9. $\Sigma k\left(\dfrac{2}{3}\right)^k$.
10. $\Sigma \dfrac{1}{(\ln k)^{10}}$.
11. $\Sigma \dfrac{1}{1+\sqrt{k}}$.
12. $\Sigma \dfrac{2k+\sqrt{k}}{k^3+\sqrt{k}}$.
13. $\Sigma \dfrac{k!}{10^{4k}}$.
14. $\Sigma \dfrac{k^2}{e^k}$.
15. $\Sigma \dfrac{\sqrt{k}}{k^2+1}$.
16. $\Sigma \dfrac{2^k k!}{k^k}$.
17. $\Sigma \dfrac{k!}{(k+2)!}$.
18. $\Sigma \dfrac{1}{k}\left(\dfrac{1}{\ln k}\right)^{3/2}$.
19. $\Sigma \dfrac{1}{k}\left(\dfrac{1}{\ln k}\right)^{1/2}$.
20. $\Sigma \dfrac{1}{\sqrt{k^3-1}}$.
21. $\Sigma \left(\dfrac{k}{k+100}\right)^k$.
22. $\Sigma \dfrac{(k!)^2}{(2k)!}$.
23. $\Sigma k^{-(1+1/k)}$.
24. $\Sigma \dfrac{11}{1+100^{-k}}$.
25. $\Sigma \dfrac{\ln k}{e^k}$.
26. $\Sigma \dfrac{k!}{k^k}$.
27. $\Sigma \dfrac{\ln k}{k^2}$.
28. $\Sigma \dfrac{k!}{1\cdot 3\cdot \cdots \cdot (2k-1)}$.
29. $\Sigma \dfrac{2\cdot 4\cdot \cdots \cdot 2k}{(2k)!}$.
30. $\Sigma \dfrac{(2k+1)^{2k}}{(5k^2+1)^k}$.
31. $\Sigma \dfrac{k!(2k)!}{(3k)!}$.
32. $\Sigma \dfrac{\ln k}{k^{5/4}}$.
33. $\Sigma \dfrac{k^{k/2}}{k!}$.
34. $\Sigma \dfrac{k^k}{(3^k)^2}$.
35. $\Sigma \dfrac{k^k}{3^{(k^2)}}$.
36. $\Sigma(\sqrt{k}-\sqrt{k-1})^k$.
37. $\dfrac{1}{2}+\dfrac{2}{3^2}+\dfrac{4}{4^3}+\dfrac{8}{5^4}+\cdots$.
38. $1+\dfrac{1\cdot 2}{1\cdot 3}+\dfrac{1\cdot 2\cdot 3}{1\cdot 3\cdot 5}+\dfrac{1\cdot 2\cdot 3\cdot 4}{1\cdot 3\cdot 5\cdot 7}+\cdots$.
39. $\dfrac{1}{4}+\dfrac{1\cdot 3}{4\cdot 7}+\dfrac{1\cdot 3\cdot 5}{4\cdot 7\cdot 10}+\dfrac{1\cdot 3\cdot 5\cdot 7}{4\cdot 7\cdot 10\cdot 13}+\cdots$.
40. $\dfrac{2}{3}+\dfrac{2\cdot 4}{3\cdot 7}+\dfrac{2\cdot 4\cdot 6}{3\cdot 7\cdot 11}+\dfrac{2\cdot 4\cdot 6\cdot 8}{3\cdot 7\cdot 11\cdot 15}+\cdots$.
41. Find the sum of the series $\frac{1}{10}+\frac{2}{100}+\frac{3}{1000}+\frac{4}{10000}+\cdots$. HINT: Exercise 58 of Section 11.2.
42. Complete the proof of the ratio test.
    (a) Prove that, if $\lambda > 1$, then $\Sigma\, a_k$ diverges.
    (b) Prove that, if $\lambda = 1$, the ratio test is inconclusive. HINT: Consider $\Sigma\, 1/k$ and $\Sigma\, 1/k^2$.
43. Prove that the sequence $\left\{\dfrac{n!}{n^n}\right\}$ has limit 0. HINT: Consider the series $\Sigma \dfrac{k!}{k^k}$.
44. Let r be a positive number. Prove that the sequence $\left\{\dfrac{r^n}{n!}\right\}$ has limit 0.
45. Let $p \geq 2$ be an integer. Find the values of p (if any) such that $\Sigma \dfrac{(k!)^2}{(pk)!}$ converges.

46. Let r be a positive number. For what values of r (if any) does $\Sigma \dfrac{r^k}{k^r}$ converge?

In Exercises 47–50, find the values of x for which the given series converges.

47. $\displaystyle\sum_{k=1}^{\infty} \frac{|x|^k}{k}$.

48. $\displaystyle\sum_{k=1}^{\infty} \frac{|x|^k}{2^k}$.

49. $\displaystyle\sum_{k=1}^{\infty} \frac{2^k\,|x|^k}{k!}$.

50. $\displaystyle\sum_{k=2}^{\infty} \frac{|x-2|^k}{k\,3^k}$.

51. Let $\{a_k\}$ be a sequence of positive numbers and take $r > 0$. Use the root test to show that, if $(a_k)^{1/k} \rightarrow \rho$ and $\rho < 1/r$, then $\Sigma\, a_k r^k$ converges.

52. Consider the series $\frac{1}{2} + 1 + \frac{1}{8} + \frac{1}{4} + \frac{1}{32} + \frac{1}{16} + \cdots$ formed by rearranging a convergent geometric series. (a) Use the root test to show that the series converges. (b) Show that the ratio test does not apply.

■ 11.5 ABSOLUTE AND CONDITIONAL CONVERGENCE; ALTERNATING SERIES

In this section we consider series that have both positive and negative terms.

Absolute and Conditional Convergence

Let $\Sigma\, a_k$ be a series with both positive and negative terms. One way to show that $\Sigma\, a_k$ converges is to show that the series of absolute values, $\Sigma\,|a_k|$, converges.

> **THEOREM 11.5.1**
>
> If $\Sigma\,|a_k|$ converges, then $\Sigma\, a_k$ converges.

PROOF For each k,

$$-|a_k| \le a_k \le |a_k| \quad \text{and therefore} \quad 0 \le a_k + |a_k| \le 2|a_k|.$$

If $\Sigma\,|a_k|$ converges, then $\Sigma\, 2|a_k| = 2\,\Sigma\,|a_k|$ converges, and therefore, by the basic comparison theorem, $\Sigma\,(a_k + |a_k|)$ converges. Since

$$a_k = (a_k + |a_k|) - |a_k|,$$

we can conclude that $\Sigma\,|a_k|$ converges. ❑

> **DEFINITION 11.5.2 ABSOLUTE CONVERGENCE**
>
> A series $\Sigma\, a_k$ is *absolutely convergent* if the series of absolute values
>
> $$|a_1| + |a_2| + |a_3| + \cdots = \Sigma\,|a_k|$$
>
> is convergent

The theorem we have just proved says that *an absolutely convergent series is convergent.*

Example 1 Consider the series

$$1 - \frac{1}{2^2} + \frac{1}{3^2} - \frac{1}{4^2} + \frac{1}{5^2} - \frac{1}{6^2} + \cdots = \sum_{k=1}^{\infty} \frac{(-1)^{k+1}}{k^2}.$$

If we replace each term by its absolute value, we obtain the series

$$1 + \frac{1}{2^2} + \frac{1}{3^2} + \frac{1}{4^2} + \frac{1}{5^2} + \frac{1}{6^2} + \cdots = \sum_{k=1}^{\infty} \frac{1}{k^2}.$$

This is a p-series with $p = 2$. It is therefore convergent. This means that the initial series is absolutely convergent. ❑

Example 2 Consider the series

$$1 - \frac{1}{2} - \frac{1}{2^2} + \frac{1}{2^3} - \frac{1}{2^4} + \frac{1}{2^5} + \frac{1}{2^6} - \frac{1}{2^7} - \frac{1}{2^8} + \cdots.$$

If we replace each term by its absolute value, we obtain the series

$$1 + \frac{1}{2} + \frac{1}{2^2} + \frac{1}{2^3} + \frac{1}{2^4} + \frac{1}{2^5} + \frac{1}{2^6} + \frac{1}{2^7} + \frac{1}{2^8} + \cdots = \sum_{k=0}^{\infty} \frac{1}{2^k}.$$

This is a convergent geometric series. The initial series is therefore absolutely convergent. ❑

Example 3 As we will see after the next theorem, the series

$$1 - \frac{1}{2} + \frac{1}{3} - \frac{1}{4} + \frac{1}{5} - \frac{1}{6} + \cdots = \sum_{k=0}^{\infty} \frac{(-1)^k}{k+1}\dagger$$

is convergent, but it is not absolutely convergent: if we replace each term by its absolute value, we get the divergent harmonic series

$$1 + \frac{1}{2} + \frac{1}{3} + \frac{1}{4} + \frac{1}{5} + \frac{1}{6} + \cdots = \sum_{k=0}^{\infty} \frac{1}{k+1}.$$ ❑

DEFINITION 11.5.3 CONDITIONAL CONVERGENCE

A series $\Sigma\, a_k$ is *conditionally convergent* if it converges but $\Sigma\, |a_k|$ diverges.

Thus, the series $\Sigma\,(-1)^k/(k+1)$ is conditionally convergent.

Alternating Series

Series in which the consecutive terms have opposite signs are called *alternating series.* For example, the series

$$1 - \frac{1}{2} + \frac{1}{3} - \frac{1}{4} + \frac{1}{5} - \frac{1}{6} + \cdots = \sum_{k=0}^{\infty} \frac{(-1)^k}{k+1}$$

† In Section 11.6 we show that the series $1 - \frac{1}{2} + \frac{1}{3} - \frac{1}{4} + \frac{1}{5} - \frac{1}{6} + \cdots$ converges to ln 2.

and

$$-1 + \frac{1}{\sqrt{2}} - \frac{1}{\sqrt{3}} + \frac{1}{\sqrt{4}} - \frac{1}{\sqrt{5}} + \cdots = \sum_{k=1}^{\infty} \frac{(-1)^k}{\sqrt{k}}$$

are alternating series.

The series

$$1 - \frac{1}{2} - \frac{1}{3} + \frac{1}{4} - \frac{1}{5} - \frac{1}{6} + \cdots$$

is not an alternating series because there are consecutive terms with the same sign.

In general, an alternating series will either have the form

$$a_0 - a_1 + a_2 - a_3 + a_4 - \cdots = \sum_{k=0}^{\infty} (-1)^k a_k$$

or the form

$$-a_0 + a_1 - a_2 + a_3 - a_4 + \cdots = \sum_{k=0}^{\infty} (-1)^{k+1} a_k,$$

where $\{a_k\}$ is a sequence of positive numbers. Since the second form is simply the negative of the first, we will focus our attention on the first form.

THEOREM 11.5.4 ALTERNATING SERIES TEST†

Let $\{a_k\}$ be a sequence of positive numbers. If

(a) $a_{k+1} < a_k$ for all k; that is, if the sequence $\{a_k\}$ is decreasing, and

(b) $a_k \to 0$ as $k \to \infty$,

then $\sum_{k=0}^{\infty} (-1)^k a_k$ converges.

PROOF First we look at the even partial sums, s_{2m}. Since

$$s_{2m} = (a_0 - a_1) + (a_2 - a_3) + \cdots + (a_{2m-2} - a_{2m-1}) + a_{2m}$$

is the sum of positive numbers, the even partial sums are all positive. Since

$$s_{2m+2} = s_{2m} - (a_{2m+1} - a_{2m+2}) \qquad \text{and} \qquad a_{2m+1} - a_{2m+2} > 0,$$

we have

$$s_{2m+2} < s_{2m}.$$

This means that the sequence of even partial sums is decreasing. Being bounded below by 0, it is convergent; say,

$$s_{2m} \to L \quad \text{as } m \to \infty.$$

Now

$$s_{2m+1} = s_{2m} - a_{2m+1}.$$

† This theorem dates back to Leibniz. He proved the result in 1705.

Since $a_{2m+1} \to 0$ as $m \to \infty$, we also have

$$s_{2m+1} \to L.$$

Since both the even and the odd partial sums tend to L, the sequence of all partial sums tends to L (Exercise 42, Section 10.3). ❑

From this theorem you can see that the following series all converge:

$$1 - \frac{1}{2} + \frac{1}{3} - \frac{1}{4} + \frac{1}{5} - \frac{1}{6} + \cdots, \qquad 1 - \frac{1}{\sqrt{2}} + \frac{1}{\sqrt{3}} - \frac{1}{\sqrt{4}} + \frac{1}{\sqrt{5}} - \frac{1}{\sqrt{6}} + \cdots,$$

$$1 - \frac{1}{2!} + \frac{1}{3!} - \frac{1}{4!} + \frac{1}{5!} - \frac{1}{6!} + \cdots.$$

The first two series converge only conditionally; the third is absolutely convergent.

Remark In the proof of Theorem 11.5.4, we showed that the sequence of even partial sums, $\{s_{2m}\}$, is decreasing and bounded below (by 0). In Exercise 48 you are asked to show that the sequence of odd partial sums $\{s_{2m+1}\}$ is increasing and bounded above. This provides an alternative way to show that the sequence of odd partial sums converges, and since

$$s_{2m+1} - s_{2m} = -a_{2m+1} \to 0 \quad \text{as } m \to \infty,$$

each sequence has the same limit L. This is illustrated in Figure 11.5.1. ❑

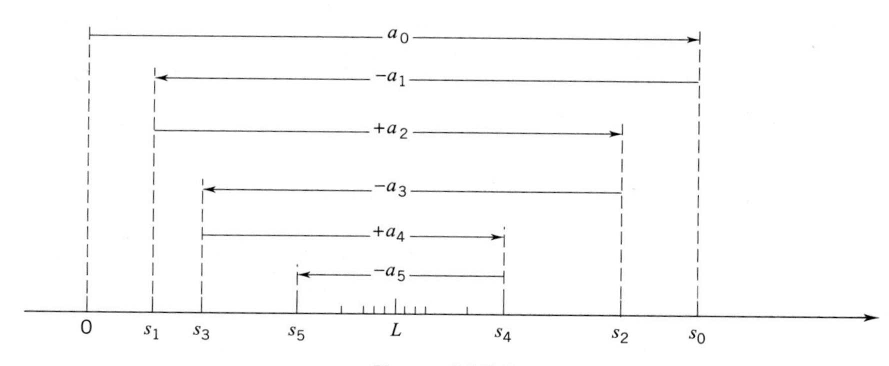

Figure 11.5.1

An Estimate for Alternating Series You have seen that if $\{a_k\}$ is a decreasing sequence of positive numbers that tends to 0, then

$$\sum_{k=0}^{\infty} (-1)^k a_k \quad \text{converges to a sum } L.$$

(11.5.5)

> The sum L of a convergent alternating series lies between consecutive partial sums s_n, s_{n+1}, and thus s_n approximates L to within a_{n+1}:
>
> $$|s_n - L| < a_{n+1}.$$

PROOF For all n

$$a_{n+1} > a_{n+2}.$$

If n is odd,

$$s_{n+2} = s_n + a_{n+1} - a_{n+2} > s_n;$$

if n is even,

$$s_{n+2} = s_n - a_{n+1} + a_{n+2} < s_n.$$

The odd partial sums increase toward L; the even partial sums decrease toward L.
For odd n

$$s_n < L < s_{n+1} = s_n + a_{n+1},$$

and for even n

$$s_n - a_{n+1} = s_{n+1} < L < s_n.$$

Thus, for all n, L lies between s_n and s_{n+1}, and s_n approximates L to within a_{n+1}. ❑

Example 5 Both

$$1 - \frac{1}{2} + \frac{1}{3} - \frac{1}{4} + \frac{1}{5} - \frac{1}{6} + \cdots \qquad \text{and} \qquad 1 - \frac{1}{2^2} + \frac{1}{3^2} - \frac{1}{4^2} + \frac{1}{5^2} - \frac{1}{6^2} + \cdots$$

are convergent alternating series. The nth partial sum of the first series approximates the sum of that series within $1/(n+1)$; the nth partial sum of the second series approximates the sum of the second series within $1/(n+1)^2$. The second series converges more rapidly than the first series. ❑

Example 6 Approximate the sum L of the alternating series

$$1 - \frac{1}{3!} + \frac{1}{5!} - \frac{1}{7!} + \cdots = \sum_{k=0}^{\infty} \frac{(-1)^k}{(2k+1)!}$$

within 0.001.

SOLUTION It is easy to verify that the series converges. In fact, it is absolutely convergent. The fourth term of the series, $1/7! \cong 0.0002$, is the first term which is less than 0.001. Thus, we have

$$|L - s_3| < a_4 < 0.001.$$

Now,

$$s_3 = 1 - \frac{1}{3!} + \frac{1}{5!} = 1 - \frac{1}{6} + \frac{1}{120} \cong 0.8417$$

and so $L = 0.842$ with an error of less than 0.001.† ❑

† You will see in Section 11.6 that the actual sum of this series is $\sin 1 \cong 0.8415$.

Rearrangements

A *rearrangement* of a series Σa_k is a series that has exactly the same terms but in a different order. Thus, for example,

$$1 + \frac{1}{3^3} - \frac{1}{2^2} + \frac{1}{5^5} - \frac{1}{4^4} + \frac{1}{7^7} - \frac{1}{6^6} + \cdots$$

and

$$1 + \frac{1}{3^3} + \frac{1}{5^5} - \frac{1}{2^2} - \frac{1}{4^4} + \frac{1}{7^7} + \frac{1}{9^9} - \cdots$$

are both rearrangements of

$$1 - \frac{1}{2^2} + \frac{1}{3^3} - \frac{1}{4^4} + \frac{1}{5^5} - \frac{1}{6^6} + \frac{1}{7^7} - \cdots.$$

In 1867 Riemann published a theorem on rearrangements of series that underscores the importance of distinguishing between absolute convergence and conditional convergence. According to this theorem all rearrangements of an absolutely convergent series converge absolutely to the same sum. In sharp contrast, a series that is only conditionally convergent can be rearranged to converge to any number we please. It can also be arranged to diverge to $+\infty$, or to diverge to $-\infty$, or even to oscillate between any two bounds we choose.†

Example 7 We have shown that the series

$$\sum_{k=0}^{\infty} \frac{(-1)^k}{k+1}$$

is conditionally convergent, and in the next section you will see that its sum is ln 2. Accepting this fact for now, we have

$$1 - \frac{1}{2} + \frac{1}{3} - \frac{1}{4} + \frac{1}{5} - \frac{1}{6} \cdots = \ln 2$$

and

$$\frac{1}{2} - \frac{1}{4} + \frac{1}{6} - \frac{1}{8} + \frac{1}{10} - \frac{1}{12} + \cdots = \tfrac{1}{2}\ln 2. \qquad \text{(multiply by } \tfrac{1}{2}\text{)}$$

Adding the two series, we get a rearrangement of the given series with a new sum:

$$1 + \frac{1}{3} - \frac{1}{2} + \frac{1}{5} + \frac{1}{7} - \frac{1}{4} + \cdots = \tfrac{3}{2}\ln 2. \quad \square$$

---

† For a complete proof see pp. 138–139, 318–320 in Konrad Knopp's *Theory and Applications of Infinite Series* (Second English Edition), Blackie & Son Limited, London, 1951.

EXERCISES 11.5

In Exercises 1–31, test these series for: (a) absolute convergence, (b) conditional convergence.

1. $1 + (-1) + 1 + \cdots + (-1)^k + \cdots.$

2. $\frac{1}{4} - \frac{1}{6} + \frac{1}{8} - \frac{1}{10} + \cdots + \frac{(-1)^k}{2k} + \cdots.$

3. $\frac{1}{2} - \frac{2}{3} + \frac{3}{4} - \frac{4}{5} + \cdots + (-1)^{k+1}\frac{k}{k+1} + \cdots.$

4. $\frac{1}{2\ln 2} - \frac{1}{3\ln 3} + \frac{1}{4\ln 4} - \frac{1}{5\ln 5} + \cdots + (-1)^k\frac{1}{k\ln k} + \cdots.$

5. $\Sigma(-1)^k\frac{\ln k}{k}.$

6. $\Sigma(-1)^k\frac{k}{\ln k}.$

7. $\Sigma\left(\frac{1}{k} - \frac{1}{k!}\right).$

8. $\Sigma\frac{k^3}{2^k}.$

9. $\Sigma(-1)^k\frac{1}{2k+1}.$

10. $\Sigma(-1)^k\frac{(k!)^2}{(2k)!}.$

11. $\Sigma\frac{k!}{(-2)^k}.$

12. $\Sigma\sin\left(\frac{k\pi}{4}\right).$

13. $\Sigma(-1)^k(\sqrt{k+1} - \sqrt{k}).$

14. $\Sigma(-1)^k\frac{k}{k^2+1}.$

15. $\Sigma\sin\left(\frac{\pi}{4k^2}\right).$

16. $\Sigma\frac{(-1)^k}{\sqrt{k(k+1)}}.$

17. $\Sigma(-1)^k\frac{k}{2^k}.$

18. $\Sigma\left(\frac{1}{\sqrt{k}} - \frac{1}{\sqrt{k+1}}\right).$

19. $\Sigma\frac{(-1)^k}{k - 2\sqrt{k}}.$

20. $\Sigma(-1)^k\frac{k+2}{k^2+k}.$

21. $\Sigma(-1)^k\frac{4^{k-2}}{e^k}.$

22. $\Sigma(-1)^k\frac{k^2}{2^k}.$

23. $\Sigma(-1)^k\, k\sin(1/k).$

24. $\Sigma(-1)^{k+1}\frac{k^k}{k!}.$

25. $\Sigma(-1)^k\, k\, e^{-k}.$

26. $\Sigma\frac{\cos\pi k}{k}.$

27. $\Sigma(-1)^k\frac{\cos\pi k}{k}.$

28. $\Sigma\frac{\sin(\pi k/2)}{k\sqrt{k}}.$

29. $\Sigma\frac{\sin(\pi k/4)}{k^2}.$

30. $\frac{1}{2} - \frac{1}{3} - \frac{1}{4} + \frac{1}{5} - \frac{1}{6} - \frac{1}{7} + \cdots + \frac{1}{3k+2} - \frac{1}{3k+3} - \frac{1}{3k+4} + \cdots.$

31. $\frac{2\cdot 3}{4\cdot 5} - \frac{5\cdot 6}{7\cdot 8} + \cdots + (-1)^k\frac{(3k+2)(3k+3)}{(3k+4)(3k+5)} + \cdots.$

In Exercises 32–35, estimate the error if the partial sum s_n is used to approximate the sum of the given alternating series.

32. $\sum_{k=1}^{\infty}(-1)^{k+1}\frac{1}{k}$; $\quad s_{20}$.

33. $\sum_{k=0}^{\infty}(-1)^k\frac{1}{\sqrt{k+1}}$; $\quad s_{80}$.

34. $\sum_{k=0}^{\infty}(-1)^k\frac{1}{(10)^k}$; $\quad s_4$.

35. $\sum_{k=1}^{\infty}(-1)^{k+1}\frac{1}{k^3}$; $\quad s_9$.

36. Let s_n be the nth partial sum of the series

$$\sum_{k=0}^{\infty}(-1)^k\frac{1}{10^k}.$$

Find the least value of n for which s_n approximates the sum of the series within:
(a) 0.001; (b) 0.0001.

37. Find the sum of the series in Exercise 36.

In Exercises 38 and 39, find the smallest integer N such that s_N will approximate the sum of the given alternating series to within the indicated accuracy.

38. $\sum_{k=1}^{\infty}(-1)^k\frac{(0.9)^k}{k}$; $\quad$ 0.001.

39. $\sum_{k=0}^{\infty}(-1)^k\frac{1}{\sqrt{k+1}}$; $\quad$ 0.005.

40. Verify that the series

$$1 - \frac{1}{2} + \frac{1}{2} - \frac{1}{3} + \frac{1}{2} - \frac{1}{3} - \frac{1}{4} + \frac{1}{3} - \frac{1}{4} + \frac{1}{3} - \frac{1}{4} + \cdots$$

diverges and explain how this does not violate the theorem on alternating series.

41. Let L be the sum of the series

$$\sum_{k=0}^{\infty}(-1)^k\frac{1}{k!}$$

and let s_n be the nth partial sum. Find the least value of n for which s_n approximates L within
(a) 0.01. (b) 0.001.

42. Let $\{a_k\}$ be a nonincreasing sequence of positive numbers that converges to 0. Does the alternating series $\Sigma(-1)^k a_k$ necessarily converge?

43. Can the hypothesis of Theorem 11.5.4 be relaxed to require only that $\{a_{2k}\}$ and $\{a_{2k+1}\}$ be decreasing sequences of positive numbers with limit zero?

44. Prove that if Σa_k is absolutely convergent and $|b_k| \leq |a_k|$ for all k, then Σb_k is absolutely convergent.

45. (a) Prove that if Σa_k is absolutely convergent, then Σa_k^2 is convergent.
(b) Show by means of an example that the converse of the result in part (a) is false.

46. Indicate how a conditionally convergent series can be rearranged (a) to converge to an arbitrary real number L; (b) to diverge to $+\infty$; (c) to diverge to $-\infty$. HINT: Collect the positive terms $p_1, p_2, p_3, \ldots$ and also the negative terms $n_1, n_2, n_3, \ldots$ in the order in which they appear in the original series.

47. In Section 11.8 we prove that, if $\Sigma a_k x_1^k$ converges, then $\Sigma a_k x^k$ converges absolutely for $|x| < |x_1|$. Try to prove this now.

48. Let $\sum_{k=0}^{\infty} (-1)^k a_k$ be an alternating series with $\{a_k\}$ a decreasing sequence. Prove that the sequence of odd partial sums $\{s_{2m+1}\}$ is increasing and bounded above.

49. Let a and b be positive numbers and consider the series

$$a - \frac{b}{2} + \frac{a}{3} - \frac{b}{4} + \frac{a}{5} - \frac{b}{6} + \cdots.$$

(a) Express this series in Σ notation.
(b) For what values of a and b is this series absolutely convergent? Conditionally convergent?

■ 11.6 TAYLOR POLYNOMIALS IN x; TAYLOR SERIES IN x

Taylor Polynomials in x

We begin with a function f continuous at 0 and set $P_0(x) = f(0)$. If f is differentiable at 0, the linear function that best approximates f at points close to 0 is the linear function

$$P_1(x) = f(0) + f'(0)x;$$

P_1 has the same value as f at 0 and also the same first derivative (the same rate of change):

$$P_1(0) = f(0), \quad P_1'(0) = f'(0).$$

(See Section 3.9, Exercise 48.) If f is twice differentiable at 0, then we can get a better approximation to f by using the quadratic polynomial

$$P_2(x) = f(0) + f'(0)x + \frac{f''(0)}{2!}x^2;$$

P_2 has the same value as f at 0 and the same first two derivatives:

$$P_2(0) = f(0), \quad P_2'(0) = f'(0), \quad P_2''(0) = f''(0).$$

If f has three derivatives at 0, we can form the cubic polynomial

$$P_3(x) = f(0) + f'(0)x + \frac{f''(0)}{2!}x^2 + \frac{f'''(0)}{3!}x^3;$$

P_3 has the same value as f at 0 and the same first three derivatives:

$$P_3(0) = f(0), \quad P_3'(0) = f'(0), \quad P_3''(0) = f''(0), \quad P_3'''(0) = f'''(0).$$

More generally, if f has n derivatives at 0, we can form the polynomial

$$P_n(x) = f(0) + f'(0)x + \frac{f''(0)}{2!}x^2 + \cdots + \frac{f^{(n)}(0)}{n!}x^n;$$

P_n is the polynomial of degree n that has the same value as f at 0 and the same first n derivatives:

$$P_n(0) = f(0), \quad P_n'(0) = f'(0), \quad P_n''(0) = f''(0), \quad \ldots, P_n^{(n)}(0) = f^{(n)}(0).$$

These approximating polynomials $P_0(x), P_1(x), P_2(x), \ldots, P_n(x)$ are called *Taylor polynomials* after the English mathematician Brook Taylor (1685–1731). In particular, for each nonnegative integer k, $P_k(x)$, is called the *Taylor polynomial of degree k for f* or the *kth Taylor polynomial of f.*

Example 1 The exponential function

$$f(x) = e^x$$

has derivatives

$$f'(x) = e^x, \quad f''(x) = e^x, \quad f'''(x) = e^x, \quad \text{and so on.}$$

Thus

$$f(0) = 1, \quad f'(0) = 1, \quad f''(0) = 1, \quad f'''(0) = 1, \quad \ldots, \quad f^{(n)}(0) = 1.$$

The nth Taylor polynomial takes the form

$$P_n(x) = 1 + x + \frac{x^2}{2!} + \frac{x^3}{3!} + \cdots + \frac{x^n}{n!}.$$

The graphs of $f(x) = e^x$, $P_0(x)$, $P_1(x)$, $P_2(x)$, and $P_3(x)$ are indicated in Figure 11.6.1. ❑

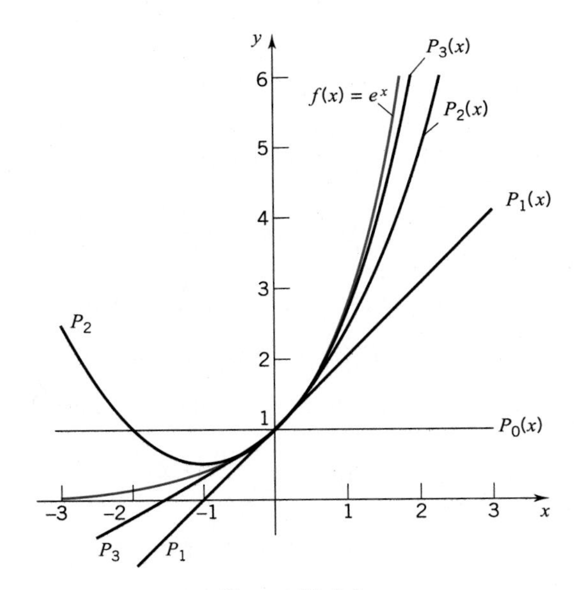

Figure 11.6.1

Example 2 To find the Taylor polynomials that approximate the sine function we write

$$f(x) = \sin x, \quad f'(x) = \cos x, \quad f''(x) = -\sin x, \quad f'''(x) = -\cos x.$$

The pattern now repeats itself:

$$f^{(4)}(x) = \sin x, \quad f^{(5)}(x) = \cos x, \quad f^{(6)}(x) = -\sin x, \quad f^{(7)}(x) = -\cos x.$$

At 0, the sine function and all its even derivatives are 0. The odd derivatives are alternately 1 and -1:

$$f'(0) = 1, \quad f'''(0) = -1, \quad f^{(5)}(0) = 1, \quad f^{(7)}(0) = -1, \quad \text{and so on.}$$

The Taylor polynomials are therefore as follows:

$$P_0(x) = 0$$

$$P_1(x) = P_2(x) = x$$

$$P_3(x) = P_4(x) = x - \frac{x^3}{3!}$$

$$P_5(x) = P_6(x) = x - \frac{x^3}{3!} + \frac{x^5}{5!}$$

$$P_7(x) = P_8(x) = x - \frac{x^3}{3!} + \frac{x^5}{5!} - \frac{x^7}{7!}, \quad \text{and so on.}$$

Only odd powers appear; you should relate this to the fact the $f(x) = \sin x$ is an *odd* function. The graphs of $f(x) = \sin x$, $P_1(x)$, $P_3(x)$, $P_5(x)$, $P_7(x)$ are indicated in Figure 11.6.2. ❑

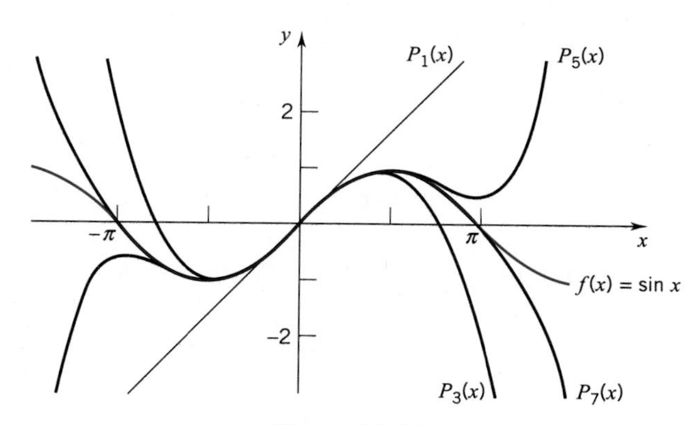

Figure 11.6.2

It is not enough to say that the Taylor polynomials

$$P_n(x) = f(0) + f'(0)x + \frac{f''(0)}{2!}x^2 + \cdots + \frac{f^{(n)}(0)}{n!}x^n$$

approximate $f(x)$. We must describe the accuracy of the approximation.

Our first step is to prove a result known as Taylor's theorem.

THEOREM 11.6.1 TAYLOR'S THEOREM

Suppose that f has $n + 1$ continuous derivatives on an open interval I containing 0. Then, for each $x \in I$,

$$f(x) = f(0) + f'(0)x + \frac{f''(0)}{2!}x^2 + \cdots + \frac{f^{(n)}(0)}{n!}x^n + R_{n+1}(x),$$

where the *remainder* $R_{n+1}(x)$ is given by the formula

$$R_{n+1}(x) = \frac{1}{n!}\int_0^x f^{(n+1)}(t)(x-t)^n\,dt.$$

PROOF Fix x in the interval I. Then

$$\int_0^x f'(t)\,dt = f(x) - f(0). \tag{1}$$

On the other hand, if we evaluate the integral using integration by parts with

$$u = f'(t) \qquad \text{and} \qquad dv = dt,$$

then

$$du = f''(t)\,dt \qquad \text{and} \qquad v = -(x-t) \qquad \text{(verify this expression for } v\text{)}$$

and

$$\int_0^x f'(t)\,dt = \Big[-f'(t)(x-t)\Big]_0^x + \int_0^x f''(t)(x-t)\,dt = f'(0)x + \int_0^x f''(t)(x-t)\,dt. \tag{2}$$

Thus, from Equations (1) and (2), we have

$$f(x) = f(0) + f'(0)x + \int_0^x f''(t)(x-t)\,dt.$$

Integrating by parts again [let $u = f''(t)$, $dv = (x-t)\,dt$, then $du = f'''(t)\,dt$, $v = -\frac{1}{2}(x-t)^2$], we get

$$f(x) = f(0) + f'(0)x + \frac{f''(0)}{2!}x^2 + \frac{1}{2!}\int_0^x f'''(t)(x-t)^2\,dt.$$

If we continue integrating by parts (see Exercise 53), we will get, after n steps,

$$f(x) = f(0) + f'(0)x + \frac{f''(0)}{2!}x^2 + \frac{f'''(0)}{3!}x^3 + \cdots + \frac{f^{(n)}(0)}{n!}x^n + \frac{1}{n!}\int_0^x f^{(n+1)}(t)(x-t)^n dt.$$

Thus,

$$f(x) = P_n(x) + \frac{1}{n!}\int_0^x f^{(n+1)}(t)(x-t)^n\,dt,$$

and

$$R_{n+1}(x) = \frac{1}{n!}\int_0^x f^{(n+1)}(t)(x-t)^n\,dt. \quad \square$$

To see how closely

$$P_n(x) = f(0) + f'(0)x + \frac{f''(0)}{2!}x^2 + \cdots + \frac{f^{(n)}(0)}{n!}x^n$$

approximates $f(x)$ we need an estimate for the remainder term $R_{n+1}(x)$. The following corollary to Taylor's theorem gives a more convenient form of the remainder. It was established by Joseph Lagrange in 1797, and it is known as the Lagrange formula for the remainder. The proof is left to you as an exercise.

COROLLARY 11.6.2 LAGRANGE FORMULA FOR THE REMAINDER

Suppose that f has $n + 1$ continuous derivatives on an open interval I containing 0. Let $x \in I$ and let $P_n(x)$ be the nth Taylor polynomial for f. Then

$$R_{n+1}(x) = \frac{f^{(n+1)}(c)}{(n+1)!}x^{n+1},$$

where c is some number between 0 and x.

Remark It is important to understand that the number c indicated in the corollary depends on x and, of course, on f and n. If we rewrite Taylor's theorem using the Lagrange form for the remainder, we have

$$f(x) = f(0) + f'(0)x + \frac{f''(0)}{2!}x^2 + \cdots + \frac{f^{(n)}}{n!}x^n + \frac{f^{(n+1)}(c)}{(n+1)!}x^{n+1}$$

where c is some number between 0 and x. This result is an extension of the mean-value theorem. Indeed, if $n = 0$, we get

$$f(x) = f(0) + f'(c)x \qquad \text{or} \qquad f(x) - f(0) = f'(c)(x - 0),$$

where c is between 0 and x. This is the mean-value theorem for f on the interval $[0, x]$. $\square$

The following estimate for $R_{n+1}(x)$ is an immediate consequence of Corollary 11.6.2. Let J be the interval joining 0 to x, $x \neq 0$. Then

(11.6.3)

$$|R_{n+1}(x)| \le \left(\max_{t\in J}|f^{(n+1)}(t)|\right)\frac{|x|^{n+1}}{(n+1)!}.$$

Example 3 The Taylor polynomials of the exponential function

$$f(x) = e^x$$

take the form

$$P_n(x) = 1 + x + \frac{x^2}{2!} + \cdots + \frac{x^n}{n!}. \qquad \text{(Example 1)}$$

We will show with our remainder estimate that for all real x

$$R_{n+1}(x) \to 0 \quad \text{as } n \to \infty,$$

and therefore we can approximate e^x as closely as we wish by Taylor polynomials.

We begin by fixing x and letting M be the maximum value of the exponential function on the interval J that joins 0 to x. (If $x > 0$, then $M = e^x$; if $x < 0$, $M = e^0 = 1$.) Since

$$f^{(n+1)}(t) = e^t \qquad \text{for all } n,$$

we have

$$\max_{t \in J} |f^{(n+1)}(t)| = M \qquad \text{for all } n.$$

Thus by (11.6.3)

$$|R_{n+1}(x)| \le M \frac{|x|^{n+1}}{(n+1)!}.$$

By (10.4.4) we know that

$$\frac{|x|^{n+1}}{(n+1)!} \to 0 \quad \text{as } n \to \infty.$$

It follows then that $R_{n+1}(x) \to 0$ as asserted. ❑

Example 4 We return to the sine function

$$f(x) = \sin x$$

and its Taylor polynomials

$$P_1(x) = P_2(x) = x$$

$$P_3(x) = P_4(x) = x - \frac{x^3}{3!}$$

$$P_5(x) = P_6(x) = x - \frac{x^3}{3!} + \frac{x^5}{5!}, \quad \text{and so on.}$$

The pattern of derivatives was established in Example 2; namely, for all k,

$$f^{(4k)}(x) = \sin x, \quad f^{(4k+1)}(x) = \cos x,$$
$$f^{(4k+2)}(x) = -\sin x, \quad f^{(4k+3)}(x) = -\cos x.$$

Thus, for all n and all real t,

$$|f^{(n+1)}(t)| \le 1.$$

It follows from our remainder estimate (11.6.3) that

$$|R_{n+1}(x)| \leq \frac{|x|^{n+1}}{(n+1)!}.$$

Since

$$\frac{|x|^{n+1}}{(n+1)!} \to 0 \qquad \text{for all real } x,$$

we see that $R_{n+1}(x) \to 0$ for all real x. Thus the sequence of Taylor polynomials converges to the sine function and therefore can be used to approximate $\sin x$ for any real number x as closely as we may wish. ❑

Taylor Series in x

By definition $0! = 1$. By adopting the convention that $f^{(0)} = f$, we can write Taylor polynomials

$$P_n(x) = f(0) + f'(0)x + \frac{f''(0)}{2!}x^2 + \cdots + \frac{f^{(n)}(0)}{n!}x^n$$

in Σ notation:

$$P_n(x) = \sum_{k=0}^{n} \frac{f^{(k)}(0)}{k!}x^k.$$

If f is infinitely differentiable on an open interval I containing 0, then we have

$$f(x) = \sum_{k=0}^{n} \frac{f^{(k)}(0)}{k!}x^k + R_{n+1}(x), \qquad x \in I,$$

for all positive integers n. If, as in the case of the exponential function and the sine function, $R_{n+1}(x) \to 0$ as $n \to \infty$ for each $x \in I$, then

$$\sum_{k=0}^{n} \frac{f^{(k)}(0)}{k!}x^k \to f(x).$$

In this case, we say that $f(x)$ can be expanded as a *Taylor series in* x and write

(11.6.4)

$$f(x) = \sum_{k=0}^{\infty} \frac{f^{(k)}(0)}{k!}x^k.$$

Taylor series in x are sometimes called Maclaurin series after Colin Maclaurin, a Scottish mathematician (1698–1746). In some circles the name Maclaurin remains attached to these series, although Taylor considered them some twenty years before Maclaurin.

From Example 3 it is clear that

(11.6.5)

$$e^x = \sum_{k=0}^{\infty} \frac{x^k}{k!} = 1 + x + \frac{x^2}{2!} + \frac{x^3}{3!} + \cdots \qquad \text{for all real } x.$$

From Example 4 we have

(11.6.6)
$$\sin x = \sum_{k=0}^{\infty} \frac{(-1)^k}{(2k+1)!} x^{2k+1} = x - \frac{x^3}{3!} + \frac{x^5}{5!} - \frac{x^7}{7!} + \cdots \quad \text{for all real } x.$$

Note that $\sin 1 = 1 - \frac{1}{3!} + \frac{1}{5!} - \frac{1}{7!} + \cdots$ as suggested in Section 11.5.

We leave it to you as an exercise to show that

(11.6.7)
$$\cos x = \sum_{k=0}^{\infty} \frac{(-1)^k}{(2k)!} x^{2k} = 1 - \frac{x^2}{2!} + \frac{x^4}{4!} - \frac{x^6}{6!} + \cdots \quad \text{for all real } x.$$

We come now to the logarithm function. Since $\ln x$ is not defined at $x = 0$, we cannot expand $\ln x$ in powers of x. We work instead with $\ln(1 + x)$.

(11.6.8)
$$\ln(1 + x) = \sum_{k=1}^{\infty} \frac{(-1)^{k+1}}{k} x^k = x - \frac{x^2}{2} + \frac{x^3}{3} - \cdots \quad \text{for } -1 < x \leq 1.$$

PROOF† The function

$$f(x) = \ln(1 + x)$$

is defined on $(-1, \infty)$ and has derivatives

$$f'(x) = \frac{1}{1+x}, \qquad f''(x) = -\frac{1}{(1+x)^2}, \qquad f'''(x) = \frac{2}{(1+x)^3},$$

$$f^{(4)}(x) = -\frac{3!}{(1+x)^4}, \qquad f^{(5)}(x) = \frac{4!}{(1+x)^5}, \qquad \text{and so on.}$$

For $k \geq 1$

$$f^{(k)}(x) = (-1)^{k+1}\frac{(k-1)!}{(1+x)^k}, \qquad f^{(k)}(0) = (-1)^{k+1}(k-1)!, \qquad \frac{f^{(k)}(0)}{k!} = \frac{(-1)^{k+1}}{k}.$$

Since $f(0) = 0$, the nth Taylor polynomial takes the form

$$P_n(x) = \sum_{k=1}^{n} (-1)^{k+1}\frac{x^k}{k} = x - \frac{x^2}{2} + \cdots + (-1)^{n+1}\frac{x^n}{n}.$$

All we have to show therefore is that

$$R_{n+1}(x) \to 0 \qquad \text{for } -1 < x \leq 1.$$

---

† The proof we give here illustrates the methods of this section. A much simpler way of obtaining this series expansion is given in Section 11.9.

Instead of trying to apply our usual remainder estimate (in this case, that estimate is not delicate enough to show that $R_{n+1}(x) \to 0$ for $-1 < x < -\frac{1}{2}$), we write the remainder in its integral form. From Taylor's theorem

$$R_{n+1}(x) = \frac{1}{n!}\int_0^x f^{(n+1)}(t)(x-t)^n\,dt,$$

so that in this case

$$R_{n+1}(x) = \frac{1}{n!}\int_0^x (-1)^{n+2}\frac{n!}{(1+t)^{n+1}}(x-t)^n\,dt = (-1)^n\int_0^x \frac{(x-t)^n}{(1+t)^{n+1}}\,dt.$$

For $0 \le x \le 1$ we have

$$|R_{n+1}(x)| = \int_0^x \frac{(x-t)^n}{(1+t)^{n+1}}\,dt \le \int_0^x (x-t)^n\,dt = \frac{x^{n+1}}{n+1} \to 0.$$

explain (arrow pointing to the $\le$)

For $-1 < x < 0$ we have

$$|R_{n+1}(x)| = \left|\int_0^x \frac{(x-t)^n}{(1+t)^{n+1}}\,dt\right| = \int_x^0 \left(\frac{t-x}{1+t}\right)^n \frac{1}{1+t}\,dt.$$

By the First Mean-Value Theorem for Integrals (5.9.1) there exists a number x_n between x and 0 such that

$$\int_x^0 \left(\frac{t-x}{1+t}\right)^n \frac{1}{1+t}\,dt = \left(\frac{x_n - x}{1+x_n}\right)^n\left(\frac{1}{1+x_n}\right)(-x).$$

Since $-x = |x|$ and $0 < 1 + x < 1 + x_n$, we can conclude that

$$|R_{n+1}(x)| < \left(\frac{x_n + |x|}{1+x_n}\right)^n\left(\frac{|x|}{1+x}\right).$$

Since $|x| < 1$ and $x_n < 0$, we have

$$x_n < |x|x_n, \qquad x_n + |x| < |x|x_n + |x| = |x|(1+x_n)$$

and thus

$$\frac{x_n + |x|}{1+x_n} < |x|.$$

It now follows that

$$|R_{n+1}(x)| < |x|^n\left(\frac{|x|}{1+x}\right)$$

and, since $|x| < 1$, that $R_{n+1}(x) \to 0$ as $n \to \infty$. ❑

Remark The series expansion for $\ln(1+x)$ that we have just verified for $-1 < x \le 1$ cannot be extended to other values of x. For $x \le -1$ neither side makes sense: $\ln(1+x)$ is not defined, and the series on the right diverges. For $x > 1$, $\ln(1+x)$ is defined, but the series on the right diverges and hence does not represent the function. At $x = 1$, the series gives the intriguing result that was mentioned in Section 11.5:

$$\ln 2 = 1 - \tfrac{1}{2} + \tfrac{1}{3} - \tfrac{1}{4} + \cdots.$$

The graphs of $f(x) = \ln(1 + x)$, $P_1(x)$, $P_2(x)$, and $P_3(x)$ are shown in Figure 11.6.3. Compare these approximations with the Taylor polynomial approximations of e^x and $\sin x$ shown in Figures 11.6.1 and 11.6.2. ❑

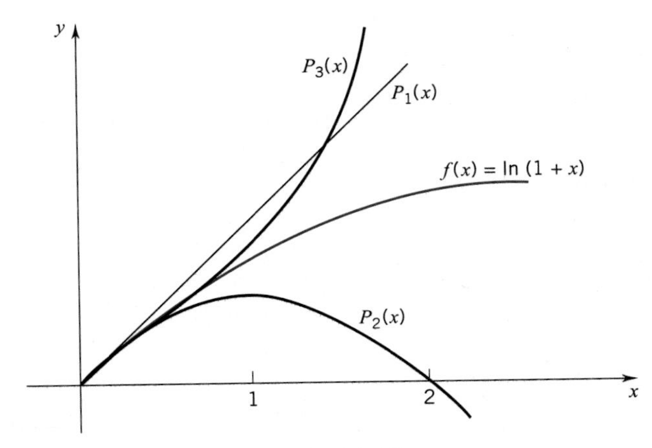

Figure 11.6.3

We want to emphasize again the role played by the remainder term $R_{n+1}(x)$. We can form a Taylor series

$$\sum_{k=0}^{\infty} \frac{f^{(k)}(0)}{k!} x^k$$

for any function f with derivatives of all orders at $x = 0$, but such a series need not converge at any number $x \neq 0$. Even if it does converge, the sum need not be $f(x)$. (See Exercise 55.) The Taylor series converges to $f(x)$ if and only if the remainder term $R_{n+1}(x)$ tends to 0.

Some Numerical Calculations

If the Taylor series converges to $f(x)$, we can use the partial sums (the Taylor polynomials) to calculate $f(x)$ as accurately as we wish. In what follows we show some sample calculations. For ready reference we list some values of $k!$ and $1/k!$ in Tables 11.6.1 and 11.6.2.

■ **Table 11.6.1**

$k!$
$2! = 2$
$3! = 6$
$4! = 24$
$5! = 120$
$6! = 720$
$7! = 5{,}040$
$8! = 40{,}320$

■ **Table 11.6.2**

$1/k!$	
$0.16666 < \frac{1}{3!} < 0.16667$	$0.00138 < \frac{1}{6!} < 0.00139$
$.04166 < \frac{1}{4!} < 0.04167$	$0.00019 < \frac{1}{7!} < 0.00020$
$0.00833 < \frac{1}{5!} < 0.00834$	$0.00002 < \frac{1}{8!} < 0.00003$

Example 5 Estimate e within 0.001.

SOLUTION For all x

$$e^x = 1 + x + \frac{x^2}{2!} + \cdots + \frac{x^n}{n!} + \cdots.$$

Taking $x = 1$ we have

$$e = 1 + 1 + \frac{1}{2!} + \cdots + \frac{1}{n!} + \cdots.$$

From Example 3 we know that the nth partial sum of this series, the nth Taylor polynomial of e^x evaluated at $x = 1$,

$$P_n(1) = 1 + 1 + \frac{1}{2!} + \cdots + \frac{1}{n!},$$

approximates e within

$$|R_{n+1}(1)| \leq e\frac{|1|^{n+1}}{(n+1)!} < \frac{3}{(n+1)!}.$$

(here $M = e^1 = e$; $e < 3$)

Since

$$\frac{3}{7!} = \frac{3}{5040} = \frac{1}{1680} < 0.001,$$

we can take $n = 6$ and be sure that

$$P_6(1) = 1 + 1 + \frac{1}{2!} + \frac{1}{3!} + \frac{1}{4!} + \frac{1}{5!} + \frac{1}{6!} = \frac{1957}{720}$$

differs from e by less than 0.001.

Our calculator gives

$$\frac{1957}{720} \cong 2.7180556 \qquad \text{and} \qquad e \cong 2.7182818. \quad \square$$

Example 6 Estimate $e^{0.2}$ within three decimal places (remainder less than 0.0005).

SOLUTION The exponential series at $x = 0.2$ gives

$$e^{0.2} = 1 + 0.2 + \frac{(0.2)^2}{2!} + \cdots + \frac{(0.2)^n}{n!} + \cdots.$$

From Example 3 we know that the nth partial sum of this series, the nth Taylor polynomial of e^x evaluated at $x = 0.2$

$$P_n(0.2) = 1 + 0.2 + \frac{(0.2)^2}{2!} + \cdots + \frac{(0.2)^n}{n!},$$

approximates $e^{0.2}$ within

$$|R_{n+1}(0.2)| \leq e^{0.2}\frac{|0.2|^{n+1}}{(n+1)!} < 3^{0.2}\frac{(0.2)^{n+1}}{(n+1)!} < 1.25\frac{(0.2)^{n+1}}{(n+1)!}.$$

(here $M = e^{0.2}$)

Since

$$1.25\frac{(0.2)^4}{4!} = \frac{1.25(0.0016)}{24} = 0.0000833\cdots < 0.00009,$$

we can take $n = 3$ and be sure that

$$P_3(0.2) = 1 + 0.2 + \frac{(0.2)^2}{2!} + \frac{(0.2)^3}{3!} = \frac{7.328}{6} \cong 1.22133\cdots$$

differs from $e^{0.2}$ by less than 0.00009, which is a much better result than the one we asked for. You can verify that $n = 2$ will not yield the desired accuracy.

It now follows that $1.22124 < e^{0.2} < 1.22143$. Our calculator gives $e^{0.2} \cong 1.2214028$. ❑

Example 7 Estimate sin 0.5 within 0.001.

SOLUTION At $x = 0.5$ the sine series gives

$$\sin 0.5 = 0.5 - \frac{(0.5)^3}{3!} + \frac{(0.5)^5}{5!} - \frac{(0.5)^7}{7!} + \cdots.$$

From Example 4 we know that $P_n(0.5)$, where $P_n(x)$ is the nth Taylor polynomial of $\sin x$, approximates sin 0.5 within

$$|R_{n+1}(0.5)| \leq \frac{(0.5)^{n+1}}{(n+1)!}.$$

Since

$$\frac{(0.5)^5}{5!} = \frac{1}{(2^5)(5!)} = \frac{1}{(32)(120)} = \frac{1}{3840} < 0.001,$$

we can be sure that

the coefficient of x^4 is 0

$$P_4(0.5) = P_3(0.5) = 0.5 - \frac{(0.5)^3}{3!} = \frac{23}{48}$$

approximates sin 0.5 within 0.001.

Our calculator gives

$$\frac{23}{48} \cong 0.4791666 \quad \text{and} \quad \sin 0.5 \cong 0.4794255. \quad ❑$$

Remark We could have solved the last problem without reference to the remainder estimate derived in Example 4. The series for sin 0.5 is a convergent alternating series with decreasing terms. By (11.5.5) we can conclude immediately that sin 0.5 lies between every two consecutive partial sums. In particular

$$0.5 - \frac{(0.5)^3}{3!} < \sin 0.5 < 0.5 - \frac{(0.5)^3}{3!} + \frac{(0.5)^5}{5!}. \quad ❑$$

Example 8 Estimate ln 1.4 within 0.01.

SOLUTION By (11.6.8)

$$\ln 1.4 = \ln(1 + 0.4) = 0.4 - \tfrac{1}{2}(0.4)^2 + \tfrac{1}{3}(0.4)^3 - \tfrac{1}{4}(0.4)^4 + \cdots.$$

This is a convergent alternating series with decreasing terms. Therefore ln 1.4 lies between every two consecutive partial sums.

The first term less than 0.01 is

$$\tfrac{1}{4}(0.4)^4 = \tfrac{1}{4}(0.0256) = 0.0064.$$

The relation

$$0.4 - \tfrac{1}{2}(0.4)^2 + \tfrac{1}{3}(0.4)^3 - \tfrac{1}{4}(0.4)^4 < \ln 1.4 < 0.4 - \tfrac{1}{2}(0.4)^2 + \tfrac{1}{3}(0.4)^3$$

gives

$$0.335 < \ln 1.4 < 0.341.$$

Within the prescribed limits of accuracy we can take $\ln 1.4 \cong 0.34$. ❑†

† A much more effective tool for computing logarithms is given in the exercises.

EXERCISES 11.6

In Exercises 1–4, find the Taylor polynomial $P_4(x)$ for the given function.

1. $f(x) = x - \cos x$.

2. $f(x) = \sqrt{1 + x}$.

3. $f(x) = \ln \cos x$.

4. $f(x) = \sec x$.

In Exercises 5–8, find the Taylor polynomial $P_5(x)$ for the given function.

5. $f(x) = (1 + x)^{-1}$.

6. $f(x) = e^x \sin x$.

7. $f(x) = \tan x$.

8. $f(x) = x \cos x^2$.

9. Determine $P_0(x)$, $P_1(x)$, $P_2(x)$, $P_3(x)$ for $f(x) = 1 - x + 3x^2 + 5x^3$.

10. Determine $P_0(x)$, $P_1(x)$, $P_2(x)$, $P_3(x)$ for $f(x) = (x + 1)^3$.

In Exercises 11–16, determine the nth Taylor polynomial $P_n(x)$ for the given function.

11. $f(x) = e^{-x}$.

12. $f(x) = \sinh x$.

13. $f(x) = \cosh x$.

14. $f(x) = \ln(1 - x)$.

15. $f(x) = e^{rx}$, r a real number.

16. $f(x) = \cos bx$, b a real number.

In Exercises 17–24, use Taylor polynomials to estimate the following within 0.01.

17. $\sqrt{e}$.

18. $\sin 0.3$.

19. $\sin 1$.

20. $\ln 1.2$.

21. $\cos 1$.

22. $e^{0.8}$.

23. $\sin 10°$.

24. $\cos 6°$.

In Exercises 25–32, find the Lagrange form of the remainder R_{n+1} for the given function and the indicated integer n.

25. $f(x) = e^{2x}$; $n = 4$.

26. $f(x) = \ln(1 + x)$; $n = 5$.

27. $f(x) = \cos 2x$; $n = 4$.

28. $f(x) = \sqrt{x + 1}$; $n = 3$.

29. $f(x) = \tan x$; $n = 2$.

30. $f(x) = \sin x$; $n = 5$.

31. $f(x) = \tan^{-1} x$; $n = 2$.

32. $f(x) = \dfrac{1}{1 + x}$; $n = 4$.

In Exercises 33–36, find the Lagrange form of the remainder R_{n+1} for the given function.

33. $f(x) = e^{-x}$.

34. $f(x) = \sin 2x$.

35. $f(x) = \dfrac{1}{1 - x}$.

36. $f(x) = \ln(1 + x)$.

37. Let $P_n(x)$ be the nth Taylor polynomial of

$$f(x) = \ln(1 + x).$$

Find the least integer n for which: (a) $P_n(0.5)$ approximates $\ln 1.5$ within 0.01; (b) $P_n(0.3)$ approximates $\ln 1.3$ within 0.01; (c) $P_n(1)$ approximates $\ln 2$ within 0.001.

38. Let $P_n(x)$ be the nth Taylor polynomial of

$$f(x) = \sin x.$$

Find the least integer n for which: (a) $P_n(1)$ approximates $\sin 1$ within 0.001; (b) $P_n(2)$ approximates $\sin 2$ within 0.001; (c) $P_n(3)$ approximates $\sin 3$ within 0.001.

39. Let $f(x) = e^x$.
(a) Find the Taylor polynomial P_n of f of least degree that will approximate $\sqrt{e}$ with four decimal place accuracy. Then evaluate $P_n(1/2)$ to obtain your approximation of $\sqrt{e}$.
(b) Find the Taylor polynomial P_n of f of least degree that will approximate $1/e$ with three decimal place accuracy. Then evaluate $P_n(-1)$ to obtain your approximation of $1/e$.

40. Let $g(x) = \cos x$.
(a) Find the Taylor polynomial P_n of g of least degree that will approximate $\cos(\pi/30)$ with three decimal place accuracy. Then evaluate $P_n(\pi/30)$ to obtain your approximation of $\cos(\pi/30)$.
(b) Find the Taylor polynomial P_n of g of least degree that will approximate $\cos 9°$ with four decimal place accuracy. Then evaluate P_n to obtain your approximation of $\cos 9°$. (Remember to convert to radian measure.)

41. Show that a polynomial $P(x) = a_0 + a_1x + \cdots + a_nx^n$ is its own Taylor series.

42. Show that

$$\cos x = \sum_{k=0}^{\infty} \frac{(-1)^k}{(2k)!}x^{2k} \qquad \text{for all real } x.$$

43. Show that

$$\sinh x = \sum_{k=0}^{\infty} \frac{1}{(2k+1)!}x^{2k+1} \qquad \text{for all real } x.$$

44. Show that

$$\cosh x = \sum_{k=0}^{\infty} \frac{1}{(2k)!}x^{2k} \qquad \text{for all real } x.$$

In Exercises 45–49, derive a series expansion in x for the given function and specify the numbers x for which the expansion is valid. Take $a > 0$.

45. $f(x) = e^{ax}$. HINT: Set $t = ax$ and expand e^t in powers of t.

46. $f(x) = \sin ax$.

47. $f(x) = \cos ax$.

48. $f(x) = \ln(1 - ax)$.

49. $f(x) = \ln(a + x)$.
HINT: $\ln(a + x) = \ln\{a(1 + x/a)\}$.

50. The series we derived for $\ln(1 + x)$ converges too slowly to be of much practical use. The following logarithm series converges much more quickly:

(11.6.9)
$$\ln\left(\frac{1+x}{1-x}\right) = 2\left(x + \frac{x^3}{3} + \frac{x^5}{5} + \cdots\right)$$
$$\text{for } -1 < x < 1.$$

Derive this series expansion.

51. Set $x = \frac{1}{3}$ and use the first three nonzero terms of (11.6.9) to estimate $\ln 2$.

52. Use the first two nonzero terms of (11.6.9) to estimate $\ln 1.4$.

53. Verify the identity

$$\frac{f^{(k)}(0)}{k!}x^k = \frac{1}{(k-1)!}\int_0^x f^{(k)}(t)(x-t)^{k-1}\,dt - \frac{1}{k!}\int_0^x f^{(k+1)}(t)(x-t)^k\,dt$$

by computing the second integral by parts.

54. Prove Corollary 11.6.2 and then derive the remainder estimate (11.6.3).

55. (a) Use a graphing utility to draw the graph of the function

$$f(x) = \begin{cases} e^{-1/x^2}, & x \neq 0 \\ 0, & x = 0 \end{cases}.$$

(b) Use L'Hospital's rule to show that for every positive integer n

$$\lim_{x \to 0} \frac{e^{-1/x^2}}{x^n} = 0.$$

(c) Use mathematical induction to prove that $f^{(n)}(0) = 0$ for all $n \geq 1$.
(d) What is the Taylor series of f?
(e) For what values of x does the Taylor series of f actually represent f?

56. Let $f(x) = \cos x$. Use a graphing utility to graph the Taylor polynomials $P_2(x)$, $P_4(x)$, $P_6(x)$, and $P_8(x)$ of f.

57. Let $g(x) = \ln(1 + x)$. Use a graphing utility to graph the Taylor polynomials $P_2(x)$, $P_3(x)$, $P_4(x)$, and $P_5(x)$ of g.

58. (*Important*) Show that e is irrational by following these steps.
(1) Take the expansion

$$e = \sum_{k=0}^{\infty} \frac{1}{k!}$$

and show that the qth partial sum

$$s_q = \sum_{k=0}^{q} \frac{1}{k!}$$

satisfies the inequality

$$0 < q!(e - s_q) < \frac{1}{q}.$$

(2) Show that $q!s_q$ is an integer and argue that, if e were of the form p/q, then $q!(e - s_q)$ would be a positive integer less than 1.

■ 11.7 TAYLOR POLYNOMIALS AND TAYLOR SERIES IN $x - a$

So far we have considered series expansions only in powers of x. Here we generalize to expansions in powers of $x - a$, where a is an arbitrary real number. We begin with a more general version of Taylor's theorem.

THEOREM 11.7.1 TAYLOR'S THEOREM

Suppose that g has $n + 1$ continuous derivatives on an open interval I containing the point a. Then, for each $x \in I$,

$$g(x) = g(a) + g'(a)(x - a) + \frac{g''(a)}{2!}(x - a)^2 + \cdots + \frac{g^{(n)}(a)}{n!}(x - a)^n + R_{n+1}(x),$$

where

$$R_{n+1}(x) = \frac{1}{n!}\int_a^x g^{(n+1)}(t)(x - t)^n\, dt.$$

The polynomial

$$P_n(x) = g(a) + g'(a)(x - a) + \frac{g''(a)}{2!}(x - a)^2 + \cdots + \frac{g^{(n)}(a)}{n!}(x - a)^n$$

is call the *nth Taylor polynomial for g in powers of* $(x - a)$. In this more general setting, the Lagrange formula for the remainder $R_{n+1}(x)$ is given by the following Corollary.

COROLLARY 11.7.2 LAGRANGE FORMULA FOR THE REMAINDER

Suppose that g has $n + 1$ continuous derivatives on an open interval I containing a. Let $x \in I$ and let P_n be the nth Taylor polynomial for g in powers of $(x - a)$. Then

$$R_{n+1}(x) = \frac{g^{(n+1)}(c)}{(n + 1)!}(x - a)^{n+1},$$

where c is some number between a and x.

Now let $x \in I$, $x \neq a$, and let J be the interval joining a and x. Then an estimate for the remainder can be written:

(11.7.3)
$$|R_{n+1}(x)| \leq \left(\max_{t \in J} |g^{(n+1)}(t)|\right) \frac{|x-a|^{n+1}}{(n+1)!}.$$

If $R_{n+1}(x) \to 0$, then we have

$$g(x) = g(a) + g'(a)(x-a) + \frac{g''(a)}{2!}(x-a)^2 + \cdots + \frac{g^{(n)}(a)}{n!}(x-a)^n + \cdots.$$

In sigma notation we have

(11.7.4)
$$g(x) = \sum_{k=0}^{\infty} \frac{g^{(k)}(a)}{k!}(x-a)^k.$$

This is known as the Taylor expansion of $g(x)$ in powers of $x - a$. The series on the right is called a *Taylor series in* $x - a$.

All this differs from what you saw before only by a translation. Define

$$f(x) = g(x + a).$$

Then obviously

$$f^{(k)}(x) = g^{(k)}(x + a) \qquad \text{and} \qquad f^{(k)}(0) = g^{(k)}(a).$$

The results of this section as stated for g can be derived by applying the results of Section 11.6 to the function f.

Example 1 Expand $g(x) = 4x^3 - 3x^2 + 5x - 1$ in powers of $x - 2$.

SOLUTION We need to evaluate g and its derivatives at $x = 2$.

$$\begin{aligned} g(x) &= 4x^3 - 3x^2 + 5x - 1 \\ g'(x) &= 12x^2 - 6x + 5 \\ g''(x) &= 24x - 6 \\ g'''(x) &= 24. \end{aligned}$$

All higher derivatives are identically 0.

Substitution gives $g(2) = 29$, $g'(2) = 41$, $g''(2) = 42$, $g'''(2) = 24$, and $g^{(k)}(2) = 0$ for all $k \geq 4$. Thus from (11.7.4)

$$\begin{aligned} g(x) &= 29 + 41(x-2) + \frac{42}{2!}(x-2)^2 + \frac{24}{3!}(x-2)^3 \\ &= 29 + 41(x-2) + 21(x-2)^2 + 4(x-2)^3. \quad \square \end{aligned}$$

Example 2 Expand $g(x) = x^2 \ln x$ in powers of $x - 1$.

SOLUTION We need to evaluate g and its derivatives at $x = 1$.

$$\begin{aligned}
g(x) &= x^2 \ln x \\
g'(x) &= x + 2x \ln x \\
g''(x) &= 3 + 2 \ln x \\
g'''(x) &= 2x^{-1} \\
g^{(4)}(x) &= -2x^{-2} \\
g^{(5)}(x) &= (2)(2)x^{-3} \\
g^{(6)}(x) &= -(2)(2)(3)x^{-4} = -2(3!)x^{-4} \\
g^{(7)}(x) &= (2)(2)(3)(4)x^{-5} = (2)(4!)x^{-5}, \quad \text{and so on.}
\end{aligned}$$

The pattern is now clear: for $k \geq 3$

$$g^{(k)}(x) = (-1)^{k+1}2(k-3)!x^{-k+2}.$$

Evaluation at $x = 1$ gives $g(1) = 0$, $g'(1) = 1$, $g''(1) = 3$ and, for $k \geq 3$,

$$g^{(k)}(1) = (-1)^{k+1}2(k-3)!.$$

The expansion in powers of $x - 1$ can be written

$$\begin{aligned}
g(x) &= (x-1) + \frac{3}{2!}(x-1)^2 + \sum_{k=3}^{\infty} \frac{(-1)^{k+1}(2)(k-3)!}{k!}(x-1)^k \\
&= (x-1) + \frac{3}{2}(x-1)^2 + 2\sum_{k=3}^{\infty} \frac{(-1)^{k+1}}{k(k-1)(k-2)}(x-1)^k. \quad ❑
\end{aligned}$$

Another way to expand $g(x)$ in powers of $x - a$ is to expand $g(t + a)$ in powers of t and then set $t = x - a$. This is the approach we take when the expansion in t is either known to us, or is easily available.

Example 3 We can expand $g(x) = e^{x/2}$ in powers of $x - 3$ by expanding

$$g(t+3) = e^{(t+3)/2} \qquad \text{in powers of } t$$

and then setting $t = x - 3$.

Note that

$$g(t+3) = e^{3/2}e^{t/2} = e^{3/2} \sum_{k=0}^{\infty} \frac{(t/2)^k}{k!} = e^{3/2} \sum_{k=0}^{\infty} \frac{1}{2^k k!} t^k.$$

exponential series ⤴

Setting $t = x - 3$, we have

$$g(x) = e^{3/2} \sum_{k=0}^{\infty} \frac{1}{2^k k!} (x-3)^k.$$

Since the expansion of $g(t + 3)$ is valid for all real t, the expansion of $g(x)$ is valid for all real x. ❑

Taking this same approach, we can prove:

(11.7.5)

For $a > 0$ and $0 < x \leq 2a$

$$\ln x = \ln a + \frac{1}{a}(x-a) - \frac{1}{2a^2}(x-a)^2 + \frac{1}{3a^3}(x-a)^3 - \cdots$$

$$= \ln a + \sum_{k=1}^{\infty} \frac{(-1)^{k+1}}{k\,a^k}(x-a)^k.$$

PROOF We will expand $\ln(a+t)$ in powers of t and then set $t = x - a$. In the first place

$$\ln(a+t) = \ln\left[a\left(1+\frac{t}{a}\right)\right] = \ln a + \ln\left(1+\frac{t}{a}\right).$$

From (11.6.8) it is clear that

$$\ln\left(1+\frac{t}{a}\right) = \frac{t}{a} - \frac{1}{2}\left(\frac{t}{a}\right)^2 + \frac{1}{3}\left(\frac{t}{a}\right)^3 - \cdots \quad \text{for } -1 < \frac{t}{a} \leq 1 \quad \text{or} \quad -a < t \leq a.$$

Adding $\ln a$ to both sides, we have

$$\ln(a+t) = \ln a + \frac{1}{a}t - \frac{1}{2a^2}t^2 + \frac{1}{3a^3}t^3 - \cdots \qquad \text{for } -a < t \leq a.$$

Setting $t = x - a$, we find that

$$\ln x = \ln a + \frac{1}{a}(x-a) - \frac{1}{2a^2}(x-a)^2 + \frac{1}{3a^3}(x-a)^3 - \cdots$$

for all x such that $-a < x - a \leq a$; that is, for all x such that $0 < x \leq 2a$. ❑

EXERCISES 11.7

In Exercises 1–6, find the Taylor polynomial of the function f for the given values of a and n, and give the Lagrange form of the remainder.

1. $f(x) = \sqrt{x}$; $\quad a = 4, \quad n = 3$.
2. $f(x) = \cos x$; $\quad a = \pi/3, \quad n = 4$.
3. $f(x) = \sin x$; $\quad a = \pi/4, \quad n = 4$.
4. $f(x) = \ln x$; $\quad a = 1, \quad n = 5$.
5. $f(x) = \tan^{-1} x$; $\quad a = 1, n = 3$.
6. $f(x) = \cos \pi x$; $\quad a = \frac{1}{2}, n = 4$.

In Exercises 7–22, expand $g(x)$ as indicated and specify the values of x for which the expansion is valid.

7. $g(x) = 3x^3 - 2x^2 + 4x + 1$ in powers of $x - 1$.
8. $g(x) = x^4 - x^3 + x^2 - x + 1$ in powers of $x - 2$.
9. $g(x) = 2x^5 + x^2 - 3x - 5$ in powers of $x + 1$.
10. $g(x) = x^{-1}$ in powers of $x - 1$.
11. $g(x) = (1 + x)^{-1}$ in powers of $x - 1$.
12. $g(x) = (b + x)^{-1}$ in powers of $x - a$, $\quad a \neq -b$.
13. $g(x) = (1 - 2x)^{-1}$ in powers of $x + 2$.
14. $g(x) = e^{-4x}$ in powers of $x + 1$.
15. $g(x) = \sin x$ in powers of $x - \pi$.
16. $g(x) = \sin x$ in powers of $x - \frac{1}{2}\pi$.
17. $g(x) = \cos x$ in powers of $x - \pi$.
18. $g(x) = \cos x$ in powers of $x - \frac{1}{2}\pi$.
19. $g(x) = \sin \frac{1}{2}\pi x$ in powers of $x - 1$.
20. $g(x) = \sin \pi x$ in powers of $x - 1$.
21. $g(x) = \ln(1 + 2x)$ in powers of $x - 1$.
22. $g(x) = \ln(2 + 3x)$ in powers of $x - 4$.

In Exercises 23–32, expand $g(x)$ as indicated.

23. $g(x) = x \ln x$ in powers of $x - 2$.

24. $g(x) = x^2 + e^{3x}$ in powers of $x - 2$.

25. $g(x) = x \sin x$ in powers of x.

26. $g(x) = \ln (x^2)$ in powers of $x - 1$.

27. $g(x) = (1 - 2x)^{-3}$ in powers of $x + 2$.

28. $g(x) = \sin^2 x$ in powers of $x - \frac{1}{2}\pi$.

29. $g(x) = \cos^2 x$ in powers of $x - \pi$.

30. $g(x) = (1 + 2x)^{-4}$ in powers of $x - 2$.

31. $g(x) = x^n$ in powers of $x - 1$.

32. $g(x) = (x - 1)^n$ in powers of x.

33. (a) Expand e^x in powers of $x - a$.
(b) Use the expansion to show that $e^{x_1 + x_2} = e^{x_1}e^{x_2}$.
(c) Expand e^{-x} in powers of $x - a$.

34. (a) Expand $\sin x$ and $\cos x$ in powers of $x - a$.
(b) Show that both series are absolutely convergent for all real x.
(c) As noted earlier (Section 11.5), Riemann proved that the order of the terms of an absolutely convergent series may be changed without altering the sum of the series. Use Riemann's discovery and the Taylor expansions of part (a) to derive the addition formulas

$$\sin (x_1 + x_2) = \sin x_1 \cos x_2 + \cos x_1 \sin x_2,$$
$$\cos (x_1 + x_2) = \cos x_1 \cos x_2 - \sin x_1 \sin x_2.$$

35. (a) Determine the Taylor polynomial P_n in powers of $(x - \pi/6)$ and of least degree that will approximate $\sin 35°$ with four decimal place accuracy.
(b) Evaluate the polynomial that you found in part (a) to obtain your approximation of $\sin 35°$.

36. (a) Determine the Taylor polynomial P_n in powers of $(x - \pi/3)$ and of least degree that will approximate $\cos 57°$ with four decimal place accuracy.
(b) Evaluate the polynomial that you found in part (a) to obtain your approximation of $\cos 57°$.

37. Choose an appropriate Taylor polynomial of $f(x) = \sqrt{x}$ to approximate $\sqrt{38}$ with three decimal place accuracy.

38. Choose an appropriate Taylor polynomial for $f(x) = \sqrt{x}$ to approximate $\sqrt{61}$ with three decimal place accuracy.

■ 11.8 POWER SERIES

You have become familiar with Taylor series

$$\sum_{k=0}^{\infty} \frac{f^{(k)}(0)}{k!} x^k \quad \text{and} \quad \sum_{k=0}^{\infty} \frac{f^{(k)}(a)}{k!} (x - a)^k.$$

Here we study series of the form

$$\sum_{k=0}^{\infty} a_k x^k \quad \text{and} \quad \sum_{k=0}^{\infty} a_k (x - a)^k$$

without regard to how the coefficients a_k have been generated. Such series are called *power series.* In particular, the first series is a *power series in powers of* x, and the second is a *power series in powers of* $x - a$.

Since a simple translation converts

$$\sum_{k=0}^{\infty} a_k (x - a)^k \quad \text{into} \quad \sum_{k=0}^{\infty} a_k x^k,$$

we can focus our attention on power series of the form

$$\sum_{k=0}^{\infty} a_k x^k.$$

When detailed indexing is unnecessary, we will omit it and write

$$\sum a_k x^k.$$

Note that if we replace the variable x by a real number c, then the result

$$\sum a_k c^k$$

is an infinite series, and so the ideas and methods of Sections 11.2–11.5 can be applied. In particular, our objective here is to determine the set of numbers c for which the resulting infinite series converges. We begin the discussion with a definition.

DEFINITION 11.8.1

A power series $\Sigma\, a_k x^k$ is said to converge

(i) at c iff $\Sigma\, a_k c^k$ converges;

(ii) on the set S iff $\Sigma\, a_k x^k$ converges for each $x \in S$.

The following result is fundamental.

THEOREM 11.8.2

If $\Sigma\, a_k x^k$ converges at $c \neq 0$, then it converges absolutely for $|x| < |c|$.
If $\Sigma\, a_k x^k$ diverges at c, then it diverges for $|x| > |c|$.

PROOF If $\Sigma\, a_k c^k$ converges, then $a_k c^k \to 0$ as $k \to \infty$. In particular, for k sufficiently large,

$$|a_k c^k| \leq 1$$

and thus

$$|a_k x^k| = |a_k c^k| \left|\frac{x}{c}\right|^k \leq \left|\frac{x}{c}\right|^k.$$

For $|x| < |c|$, we have

$$\left|\frac{x}{c}\right| < 1.$$

The convergence of $\Sigma |a_k x^k|$ follows by comparison with the geometric series. This proves the first statement.

Suppose now that $\Sigma\, a_k c^k$ diverges. By a similar argument, there cannot exist x with $|x| > |c|$ such that $\Sigma\, a_k x^k$ converges. The existence of such an x would imply the absolute convergence of $\Sigma\, a_k c^k$. This proves the second statement. ❑

From the theorem we just proved you can see that there are exactly three possibilities for a power series:

Case I. *The series converges only at $x = 0$.* This is what happens with

$$\sum k^k x^k.$$

For if $x \neq 0$, then $\lim_{k\to\infty} k^k x^k \neq 0$, and so the series cannot converge (Theorem 11.2.7).

Case II. The series is absolutely convergent for all real numbers x. This is what happens with the exponential series

$$\sum \frac{x^k}{k!}.$$

Case III. *There exists a positive integer r such that the series converges for $|x| < r$ and diverges for $|x| > r$.* This is what happens with the geometric series

$$\sum x^k.$$

In this instance, there is absolute convergence for $|x| < 1$ and divergence for $|x| > 1$.

Associated with each case is a *radius of convergence*:

In Case I, we say that the radius of convergence is 0.
In Case II, we say that the radius of convergence is ∞.
In Case III, we say that the radius of convergence is r.

The three cases are pictured in Figure 11.8.1.

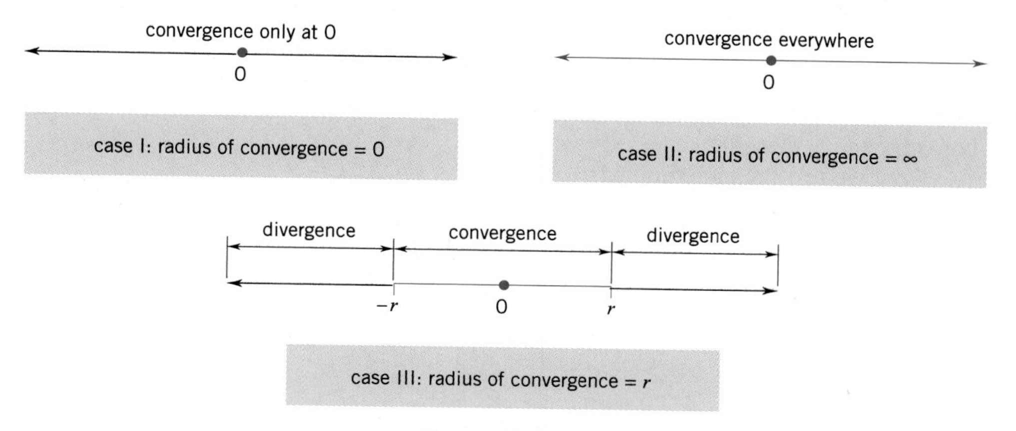

Figure 11.8.1

In general, the behavior of a power series at $-r$ and at r is not predictable. For example, the series

$$\sum x^k, \quad \sum \frac{(-1)^k}{k} x^k, \quad \sum \frac{1}{k} x^k, \quad \sum \frac{1}{k^2} x^k$$

all have radius of convergence 1, but the first series converges only on $(-1, 1)$ (verify this) and, as we show next, the second series converges on $(-1, 1]$, the third on $[-1, 1)$, and the fourth on $[-1, 1]$.

The maximal interval on which a power series converges is called the *interval of convergence*. For a series with infinite radius of convergence, the interval of convergence is $(-\infty, \infty)$. For a series with radius of convergence r, the interval of convergence can be $[-r, r]$, $(-r, r]$, $[-r, r)$, or $(-r, r)$. For a series with radius of convergence 0, the interval of convergence reduces to a point, $\{0\}$.

Example 1 Verify that the series

$$\sum \frac{(-1)^k}{k} x^k \tag{1}$$

has interval of convergence $(-1, 1]$.

SOLUTION First we show that the radius of convergence is 1 (that the series converges absolutely for $|x| < 1$ and diverges for $|x| > 1$). We do this by forming the series

$$\sum \left| \frac{(-1)^k}{k} x^k \right| = \sum \frac{1}{k} |x|^k \tag{2}$$

and applying the ratio test.

We set

$$b_k = \frac{1}{k} |x|^k$$

and note that

$$\frac{b_{k+1}}{b_k} = \frac{|x|^{k+1}/(k+1)}{|x|^k/k} = \frac{k}{k+1} \frac{|x|^{k+1}}{|x|^k} = \frac{k}{k+1}|x| \to |x| \quad \text{as } k \to \infty.$$

By the ratio test, series (2) converges for $|x| < 1$ and diverges for $|x| > 1$.† It follows that series (1) converges absolutely for $|x| < 1$ and diverges for $|x| > 1$. The radius of convergence is therefore 1.

Now we test the endpoints $x = -1$ and $x = 1$. At $x = -1$

$$\sum \frac{(-1)^k}{k} x^k \quad \text{becomes} \quad \sum \frac{(-1)^k}{k} (-1)^k = \sum \frac{1}{k}.$$

This is the harmonic series which, as you know, diverges. At $x = 1$

$$\sum \frac{(-1)^k}{k} x^k \quad \text{becomes} \quad \sum \frac{(-1)^k}{k}.$$

This is a convergent alternating series.

We have shown that series (1) converges absolutely for $|x| < 1$, diverges at -1, and converges at 1. The interval of convergence is $(-1, 1]$. ❑

---

† We could also have used the root test:

$$(b_k)^{1/k} = \left| \frac{1}{k} \right|^{1/k} |x| = \frac{1}{k^{1/k}} |x| \to |x|.$$

Remark The same arguments can be used to show that the series

$$\sum \frac{1}{k} x^k$$

converges on $[-1, 1)$. ❑

Example 2 Verify that the series

$$(1) \qquad \sum \frac{1}{k^2} x^k$$

has interval of convergence $[-1, 1]$.

SOLUTION First we examine the series

$$(2) \qquad \sum \left| \frac{1}{k^2} x^k \right| = \sum \frac{1}{k^2} |x|^k.$$

Here again we use the ratio test. We set

$$b_k = \frac{1}{k^2} |x|^k$$

and note that

$$\frac{b_{k+1}}{b_k} = \frac{k^2}{(k+1)^2} \frac{|x|^{k+1}}{|x|^k} = \left(\frac{k}{k+1}\right)^2 |x| \rightarrow |x| \quad \text{as } k \rightarrow \infty.$$

By the ratio test, (2) converges for $|x| < 1$ and diverges for $|x| > 1$.† This shows that (1) converges absolutely for $|x| < 1$ and diverges for $|x| > 1$. The radius of convergence is therefore 1.

Now for the endpoints. At $x = -1$,

$$\sum \frac{1}{k^2} x^k \quad \text{takes the form} \quad \sum \frac{(-1)^k}{k^2} = -1 + \tfrac{1}{4} - \tfrac{1}{9} + \tfrac{1}{16} - \cdots.$$

This is a convergent alternating series. At $x = 1$,

$$\sum \frac{1}{k^2} x^k \quad \text{becomes} \quad \sum \frac{1}{k^2}.$$

This is a convergent p-series. The interval of convergence is therefore the entire closed interval $[-1, 1]$. ❑

Example 3 Find the interval of convergence of

$$(1) \qquad \sum \frac{k}{6^k} x^k.$$

---

† Once again we could have used the root test:

$$(b_k) = \frac{1}{k^{2/k}} |x| \rightarrow |x| \quad \text{as } k \rightarrow \infty.$$

SOLUTION We begin by examining the series

$$\text{(2)} \qquad \sum \left| \frac{k}{6^k} x^k \right| = \sum \frac{k}{6^k} |x|^k.$$

We set

$$b_k = \frac{k}{6^k} |x|^k$$

and apply the root test. (The ratio test will also work.) Since

$$(b_k)^{1/k} = \tfrac{1}{6} k^{1/k} |x| \to \tfrac{1}{6} |x| \quad \text{as } k \to \infty \qquad \text{(recall } k^{1/k} \to 1\text{)}$$

you can see that (2) converges

$$\text{for } \tfrac{1}{6}|x| < 1, \qquad \text{that is, for } |x| < 6,$$

and diverges

$$\text{for } \tfrac{1}{6}|x| > 1, \qquad \text{that is, for } |x| > 6.$$

Testing the endpoints, we have:

at $x = 6$, $\quad \sum \frac{k}{6^k} 6^k = \sum k$, $\quad$ which is divergent;

at $x = -6$, $\quad \sum \frac{k}{6^k} (-6)^k = \sum (-1)^k k$, $\quad$ which is also divergent.

Thus, the interval of convergence is $(-6, 6)$. ❑

Example 4 Find the interval of convergence of

$$\text{(1)} \qquad \sum \frac{(2k)!}{(3k)!} x^k.$$

SOLUTION Again, we begin by examining the series

$$\text{(2)} \qquad \sum \left| \frac{(2k)!}{(3k)!} x^k \right| = \sum \frac{(2k)!}{(3k)!} |x|^k.$$

Set

$$b_k = \frac{(2k)!}{(3k)!} |x|^k.$$

Since factorials are involved, we will use the ratio test. Note that

$$\frac{b_{k+1}}{b_k} = \frac{\{2(k+1)\}!}{\{3(k+1)\}!} \frac{(3k)!}{(2k)!} \frac{|x|^{k+1}}{|x|^k} = \frac{(2k+2)(2k+1)}{(3k+3)(3k+2)(3k+1)} |x|.$$

Since

$$\frac{(2k+2)(2k+1)}{(3k+3)(3k+2)(3k+1)} \to 0 \quad \text{as } k \to \infty,$$

(the numerator is a quadratic in k, the denominator is a cubic), the ratio b_{k+1}/b_k tends to 0 no matter what x is. By the ratio test, (2) converges for all x and therefore (1) converges absolutely for all x. The radius of convergence is ∞ and the interval of convergence is $(-\infty, \infty)$. ❑

Example 5 Find the interval of convergence of

$$\sum \left(\tfrac{1}{2}k\right)^k x^k.$$

SOLUTION Since $(\frac{1}{2}k)^k x^k \to 0$ only if $x = 0$ (explain), there is no need to invoke the ratio test or the root test. By (11.2.7) the series can converge only at $x = 0$. There it converges trivially, that is, all the terms of the series are 0. ❑

Example 6 Find the interval of convergence of

$$\sum \frac{(-1)^k}{k^2 3^k}(x+2)^k.$$

SOLUTION We consider the series

$$\sum \left| \frac{(-1)^k}{k^2 3^k}(x+2)^k \right| = \sum \frac{1}{k^2 3^k}|x+2|^k.$$

Set

$$b_k = \frac{1}{k^2 3^k}|x+2|^k$$

and apply the ratio test (the root test will work equally as well):

$$\frac{b_{k+1}}{b_k} = \frac{|x+2|^{k+1}/(k+1)^2 3^{k+1}}{|x+2|^k/k^2 3^k} = \frac{k^2}{3(k+1)^2}|x+2| \to \tfrac{1}{3}|x+2| \quad \text{as } k \to \infty.$$

Thus, the series is absolutely convergent

$$\text{for} \quad \tfrac{1}{3}|x+2| < 1 \qquad \text{or} \qquad |x+2| < 3,$$

which is the same as $-5 < x < 1$.

We now check the endpoints. At $x = -5$:

$$\sum \frac{(-1)^k}{k^2 3^k}(-3)^k = \sum \frac{1}{k^2}.$$

This is a convergent p-series. At $x = 1$:

$$\sum \frac{(-1)^k}{k^2 3^k}(3)^k = \sum \frac{(-1)^k}{k^2},$$

and this is a convergent alternating series. Therefore, the interval of convergence is $[-5, 1]$ ❑

EXERCISES 11.8

In Exercises 1–38, find the interval of convergence.

1. $\sum kx^k$.
2. $\sum \frac{1}{k}x^k$.
3. $\sum \frac{1}{(2k)!}x^k$.
4. $\sum \frac{2^k}{k^2}x^k$.
5. $\sum (-k)^{2k}x^{2k}$.
6. $\sum \frac{(-1)^k}{\sqrt{k}}x^k$.
7. $\sum \frac{1}{k2^k}x^k$.
8. $\sum \frac{1}{k^2 2^k}x^k$.
9. $\sum \left(\frac{k}{100}\right)^k x^k$.
10. $\sum \frac{k^2}{1+k^2}x^k$.
11. $\sum \frac{2^k}{\sqrt{k}}x^k$.
12. $\sum \frac{1}{\ln k}x^k$.
13. $\sum \frac{k-1}{k}x^k$.
14. $\sum ka^k x^k$.

15. $\sum \frac{k}{10^k} x^k.$

16. $\sum \frac{3k^2}{e^k} x^k.$

17. $\sum \frac{x^k}{k^k}.$

18. $\sum \frac{7^k}{k!} x^k.$

19. $\sum \frac{(-1)^k}{k^k}(x-2)^k.$

20. $\sum k!\, x^k.$

21. $\sum (-1)^k \frac{2^k}{3^{k+1}} x^k.$

22. $\sum \frac{2^k}{(2k)!} x^k.$

23. $\sum (-1)^k \frac{k!}{k^3}(x-1)^k.$

24. $\sum \frac{(-e)^k}{k^2} x^k.$

25. $\sum \left(\frac{k}{k-1}\right) \frac{(x+2)^k}{2^k}.$

26. $\sum \frac{\ln k}{k}(x+1)^k.$

27. $\sum (-1)^k \frac{k^2}{(k+1)!}(x+3)^k.$

28. $\sum \frac{k^3}{e^k}(x-4)^k.$

29. $\sum \left(1+\frac{1}{k}\right)^k x^k.$

30. $\sum \frac{(-1)^k a^k}{k^2}(x-a)^k.$

31. $\sum \frac{\ln k}{2^k}(x-2)^k.$

32. $\sum \frac{1}{(\ln k)^k}(x-1)^k.$

33. $\sum (-1)^k (\tfrac{2}{3})^k (x+1)^k.$

34. $\sum \frac{2^{1/k}\pi^k}{k(k+1)(k+2)}(x-2)^k.$

35. $1 - \frac{x}{2} + \frac{2x^2}{4} - \frac{3x^3}{8} + \frac{4x^4}{16} - \cdots.$

36. $\frac{(x-1)}{5^2} + \frac{4}{5^4}(x-1)^2 + \frac{9}{5^6}(x-1)^3 + \frac{16}{5^8}(x-1)^4 + \cdots.$

37. $\frac{3x^2}{4} + \frac{9x^4}{9} + \frac{27x^6}{16} + \frac{81x^8}{25} + \cdots.$

38. $\frac{1}{16}(x+1) - \frac{2}{25}(x+1)^2 + \frac{3}{36}(x+1)^3 - \frac{4}{49}(x+1)^4 + \cdots.$

39. Let $\Sigma\, a_k x^k$ be a power series, and let r be its radius of convergence.
(a) Given that $|a_k|^{1/k} \to \rho$, show that, if $\rho \neq 0$, then $r = 1/\rho$ and, if $\rho = 0$, then $r = \infty$.
(b) Given that $|a_{k+1}/a_k| \to \lambda$, show that, if $\lambda \neq 0$, then $r = 1/\lambda$ and, if $\lambda = 0$, then $r = \infty$.

40. Find the interval of convergence of the series $\Sigma\, s_k x^k$ where s_k is the kth partial sum of the series

$$\sum_{n=1}^{\infty} \frac{1}{n}.$$

41. Let $\Sigma\, a_k x^k$ be a power series and let r, $0 < r < \infty$, be its radius of convergence. Prove that if the series is absolutely convergent at one endpoint of its interval of convergence, then it is absolutely convergent at the other also.

42. Let $\Sigma\, a_k x^k$ be a power series and let r, $0 < r < \infty$, be its radius of convergence. Prove that the power series $\Sigma\, a_k x^{2k}$ has radius of convergence $\sqrt{r}$.

■ 11.9 DIFFERENTIATION AND INTEGRATION OF POWER SERIES

Suppose that the power series

$$\sum_{k=0}^{\infty} a_k x^k$$

converges on the interval $(-c, c)$. Then, for each $x \in (-c, c)$, the infinite series

$$\sum_{k=0}^{\infty} a_k x^k$$

converges to a number L_x; the sum L_x depends upon x. Let f be the function defined on $(-c, c)$ by $f(x) = L_x$, that is,

$$f(x) = \sum_{k=0}^{\infty} a_k x^k \qquad \text{for } x \in (-c, c).$$

In this section we show that this function is both infinitely differentiable and integrable on $(-c, c)$. We begin with a simple but important result.

THEOREM 11.9.1

If

$$\sum_{k=0}^{\infty} a_k x^k = a_0 + a_1 x + a_2 x^2 + \cdots + a_n x^n + \cdots$$

converges on $(-c, c)$, then

$$\sum_{k=0}^{\infty} \frac{d}{dx}(a_k x^k) = \sum_{k=1}^{\infty} k a_k x^{k-1}$$
$$= a_1 + 2a_2 x + 3a_3 x^2 + \cdots + n a_n x^{n-1} + \cdots$$

also converges on $(-c, c)$.

PROOF Assume that

$$\sum_{k=0}^{\infty} a_k x^k \quad \text{converges on } (-c, c).$$

By Theorem 11.8.2 the series is absolutely convergent on this interval.

Now let x be some fixed number in $(-c, c)$ and choose $\epsilon > 0$ such that

$$|x| < |x| + \epsilon < c.$$

Since $|x| + \epsilon$ lies within the interval of convergence,

$$\sum_{k=0}^{\infty} |a_k(|x| + \epsilon)^k| \quad \text{converges.}$$

In Exercise 48 you are asked to show that, for all k sufficiently large,

$$|kx^{k-1}| \leq (|x| + \epsilon)^k.$$

It follows that for all such k

$$|k a_k x^{k-1}| \leq |a_k(|x| + \epsilon)^k|.$$

Since

$$\sum_{k=0}^{\infty} |a_k(|x| + \epsilon)^k| \quad \text{converges,}$$

we can conclude that

$$\sum_{k=0}^{\infty} \left| \frac{d}{dx}(a_k x^k) \right| = \sum_{k=1}^{\infty} |k a_k x^{k-1}| \quad \text{converges,}$$

and thus that

$$\sum_{k=0}^{\infty} \frac{d}{dx}(a_k x^k) = \sum_{k=1}^{\infty} k a_k x^{k-1} \quad \text{converges.} \quad \square$$

Repeated application of the theorem shows that

$$\sum_{k=0}^{\infty} \frac{d^2}{dx^2}(a_k x^k), \qquad \sum_{k=0}^{\infty} \frac{d^3}{dx^3}(a_k x^k), \qquad \sum_{k=0}^{\infty} \frac{d^4}{dx^4}(a_k x^k), \qquad \text{and so on,}$$

all converge on $(-c, c)$.

Example 1 Since the geometric series

$$\sum_{k=0}^{\infty} x^k = 1 + x + x^2 + x^3 + x^4 + x^5 + x^6 + \cdots$$

converges on $(-1, 1)$, the series

$$\sum_{k=0}^{\infty} \frac{d}{dx}(x^k) = \sum_{k=1}^{\infty} kx^{k-1} = 1 + 2x + 3x^2 + 4x^3 + 5x^4 + 6x^5 + \cdots,$$

$$\sum_{k=0}^{\infty} \frac{d^2}{dx^2}(x^k) = \sum_{k=2}^{\infty} k(k-1)x^{k-2} = 2 + 6x + 12x^2 + 20x^3 + 30x^4 + \cdots,$$

$$\sum_{k=0}^{\infty} \frac{d^3}{dx^3}(x^k) = \sum_{k=3}^{\infty} k(k-1)(k-2)x^{k-3} = 6 + 24x + 60x^2 + 120x^3 + \cdots,$$

$$\vdots$$

all converge on $(-1, 1)$. ❑

COROLLARY 11.9.2

If

$$\sum_{k=0}^{\infty} a_k x^k$$

has a radius of convergence r, then each of the series

$$\sum_{k=0}^{\infty} \frac{d}{dx}(a_k x^k), \qquad \sum_{k=0}^{\infty} \frac{d^2}{dx^2}(a_k x^k), \qquad \sum_{k=0}^{\infty} \frac{d^3}{dx^3}(a_k x^k), \qquad \text{and so on,}$$

has radius of convergence r.

Remark Even though $\Sigma\, a_k x^k$ and its "derivative" $\Sigma\, ka_k x^{k-1}$ have the same radius of convergence, their intervals of convergence may be different. For example, the interval of convergence of the series

$$\sum_{k=1}^{\infty} \frac{1}{k^2} x^k$$

is $[-1, 1]$, whereas the interval of convergence of its derivative

$$\sum_{k=1}^{\infty} \frac{1}{k} x^{k-1}$$

is $[-1, 1)$. Endpoints must always be checked separately. Obviously, this remark does not apply in the cases where $r = 0$ or $r = \infty$. ❑

Suppose now that

$$\sum_{k=0}^{\infty} a_k x^k \quad \text{converges on } (-c, c).$$

Then, as we have seen,

$$\sum_{k=0}^{\infty} \frac{d}{dx}(a_k x^k) \quad \text{also converges on } (-c, c).$$

As we noted at the beginning of this section, we can define a function f on $(-c, c)$ by setting

$$f(x) = \sum_{k=0}^{\infty} a_k x^k.$$

Using the second series, we can define a function g on $(-c, c)$ by setting

$$g(x) = \sum_{k=0}^{\infty} \frac{d}{dx}(a_k x^k).$$

The crucial point is that

$$f'(x) = g(x).$$

THEOREM 11.9.3 THE DIFFERENTIABILITY THEOREM

If

$$f(x) = \sum_{k=0}^{\infty} a_k x^k \qquad \text{for all } x \text{ in } (-c, c),$$

then f is differentiable on $(-c, c)$ and

$$f'(x) = \sum_{k=0}^{\infty} \frac{d}{dx}(a_k x^k) \qquad \text{for all } x \text{ in } (-c, c).$$

By applying this theorem to f', you can see that f' is itself differentiable. This in turn implies that f'' is differentiable, and so on. In short, f has derivatives of all orders.

The discussion up to this point can be summarized as follows:

In the interior of its interval of convergence a power series defines an infinitely differentiable function, the derivatives of which can be obtained by differentiating term by term:

$$\frac{d^n}{dx^n}\left(\sum_{k=0}^{\infty} a_k x^k\right) = \sum_{k=0}^{\infty} \frac{d^n}{dx^n}(a_k x^k) \qquad \text{for all } n.$$

For a detailed proof of the differentiability theorem see the supplement at the end of this section. We go on to examples.

Example 2 You know that

$$\frac{d}{dx}(e^x) = e^x.$$

You can see this directly by differentiating the exponential series:

$$\frac{d}{dx}(e^x) = \frac{d}{dx}\left(\sum_{k=0}^{\infty} \frac{x^k}{k!}\right) = \sum_{k=0}^{\infty} \frac{d}{dx}\left(\frac{x^k}{k!}\right) = \sum_{k=1}^{\infty} \frac{x^{k-1}}{(k-1)!} = \sum_{n=0}^{\infty} \frac{x^n}{n!} = e^x. \quad \square$$

set $n = k - 1$

Example 3 You have seen that

$$\sin x = x - \frac{x^3}{3!} + \frac{x^5}{5!} - \frac{x^7}{7!} + \frac{x^9}{9!} - \cdots$$

and

$$\cos x = 1 - \frac{x^2}{2!} + \frac{x^4}{4!} - \frac{x^6}{6!} + \frac{x^8}{8!} - \cdots.$$

The relations

$$\frac{d}{dx}(\sin x) = \cos x, \qquad \frac{d}{dx}(\cos x) = -\sin x$$

can be confirmed by differentiating the series term by term:

$$\frac{d}{dx}(\sin x) = 1 - \frac{3x^2}{3!} + \frac{5x^4}{5!} - \frac{7x^6}{7!} + \frac{9x^8}{9!} - \cdots$$

$$= 1 - \frac{x^2}{2!} + \frac{x^4}{4!} - \frac{x^6}{6!} + \frac{x^8}{8!} - \cdots = \cos x,$$

$$\frac{d}{dx}(\cos x) = -\frac{2x}{2!} + \frac{4x^3}{4!} - \frac{6x^5}{6!} + \frac{8x^7}{8!} - \cdots$$

$$= -x + \frac{x^3}{3!} - \frac{x^5}{5!} + \frac{x^7}{7!} - \cdots$$

$$= -\left(x - \frac{x^3}{3!} + \frac{x^5}{5!} - \frac{x^7}{7!} + \cdots\right) = -\sin x. \quad \square$$

Example 4 We can sum the series

$$\sum_{k=1}^{\infty} \frac{x^k}{k} \qquad \text{for all } x \text{ in } (-1, 1)$$

by setting

$$g(x) = \sum_{k=1}^{\infty} \frac{x^k}{k} \qquad \text{for all } x \text{ in } (-1, 1)$$

and noting that

$$g'(x) = \sum_{k=1}^{\infty} \frac{kx^{k-1}}{k} = \sum_{k=1}^{\infty} x^{k-1} = \sum_{n=0}^{\infty} x^n = \frac{1}{1-x}.$$

↑ the geometric series

With

$$g'(x) = \frac{1}{1-x} \quad \text{and} \quad g(0) = 0,$$

we can conclude that

$$g(x) = -\ln(1-x) = \ln\left(\frac{1}{1-x}\right).$$

It follows that

$$\sum_{k=1}^{\infty} \frac{x^k}{k} = \ln\left(\frac{1}{1-x}\right) \quad \text{for all } x \text{ in } (-1, 1). \quad \square$$

Power series can also be integrated term by term.

THEOREM 11.9.4 TERM-BY-TERM INTEGRATION

If $f(x) = \sum_{k=0}^{\infty} a_k x^k$ converges on $(-c, c)$, then

$$g(x) = \sum_{k=0}^{\infty} \frac{a_k}{k+1} x^{k+1} \quad \text{converges on } (-c, c) \text{ and} \quad \int f(x)\,dx = g(x) + C.$$

PROOF If $\sum_{k=0}^{\infty} a_k x^k$ converges on $(-c, c)$, then $\sum_{k=0}^{\infty} |a_k x^k|$ converges on $(-c, c)$ by Theorem 11.8.2. Since

$$\left|\frac{a_k}{k+1} x^k\right| \le |a_k x^k| \quad \text{for all } k,$$

we know by comparison that

$$\sum_{k=0}^{\infty} \left|\frac{a_k}{k+1} x^k\right| \quad \text{also converges on } (-c, c).$$

It follows that

$$x \sum_{k=0}^{\infty} \frac{a_k}{k+1} x^k = \sum_{k=0}^{\infty} \frac{a_k}{k+1} x^{k+1} \quad \text{converges on } (-c, c).$$

With

$$f(x) = \sum_{k=0}^{\infty} a_k x^k \quad \text{and} \quad g(x) = \sum_{k=0}^{\infty} \frac{a_k}{k+1} x^{k+1},$$

we know from the differentiability theorem that

$$g'(x) = f(x) \qquad \text{and therefore} \qquad \int f(x)\,dx = g(x) + C. \quad \square$$

Term-by-term integration can be expressed as follows:

(11.9.5)
$$\int \left(\sum_{k=0}^{\infty} a_k x^k \right) dx = \left(\sum_{k=0}^{\infty} \frac{a_k}{k+1} x^{k+1} \right) + C.$$

Remark It follows from Theorem 11.9.4 that if $\Sigma\, a_k x^k$ has radius of convergence r, then its "integral" $\Sigma\, [1/(k+1)]a_k x^{k+1}$ also has radius of convergence r. As in the case of differentiating power series, convergence at the endpoints (if any) has to be tested separately. $\square$

If a power series converges at c and converges at d, then it converges at all numbers in between and

(11.9.6)
$$\int_c^d \left(\sum_{k=0}^{\infty} a_k x^k \right) dx = \sum_{k=0}^{\infty} \left(\int_c^d a_k x^k \, dx \right) = \sum_{k=0}^{\infty} \frac{a_k}{k+1} (d^{k+1} - c^{k+1}).$$

Example 5 You are familiar with the series expansion

$$\frac{1}{1+x} = \frac{1}{1-(-x)} = \sum_{k=0}^{\infty} (-1)^k x^k.$$

It is valid for all x in $(-1, 1)$ and for no other x. Integrating term by term we have

$$\ln(1+x) = \int \left(\sum_{k=0}^{\infty} (-1)^k x^k \right) dx = \left(\sum_{k=0}^{\infty} \frac{(-1)^k}{k+1} x^{k+1} \right) + C$$

for all x in $(-1, 1)$. At $x = 0$ both $\ln(1+x)$ and the series on the right are 0. It follows that $C = 0$ and thus

$$\ln(1+x) = \sum_{k=0}^{\infty} \frac{(-1)^k}{k+1} x^{k+1} = x - \frac{x^2}{2} + \frac{x^3}{3} - \frac{x^4}{4} + \cdots$$

for all x in $(-1, 1)$. $\square$

In Section 11.6 we were able to prove that this expansion for $\ln(1+x)$ was valid on the half-closed interval $(-1, 1]$; this gave us an expansion for ln 2. Term-by-term integration gives us only the open interval $(-1, 1)$. Well, you may say, it's easy to see that the logarithm series also converges at $x = 1$.† True enough, but why to ln 2? This takes us back to consideration of the remainder term, the method of Section 11.6.

† An alternating series with $a_k \to 0$.

There is, however, another way to proceed. The great Norwegian mathematician Niels Henrik Abel (1802–1829) proved the following result: suppose that

$$\sum_{k=0}^{\infty} a_k x^k \quad \text{converges on } (-c, c) \text{ and there represents } f(x).$$

If f is continuous at one of the endpoints (c or $-c$) and the series converges there, then the series represents the function at that point. Using Abel's theorem it is evident that the series for $\ln(1+x)$ does represent the function at $x = 1$.

We come now to another important series expansion:

(11.9.7) $$\tan^{-1} x = x - \frac{x^3}{3} + \frac{x^5}{5} - \frac{x^7}{7} + \cdots \qquad \text{for } -1 \le x \le 1.$$

PROOF For x in $(-1, 1)$

$$\frac{1}{1+x^2} = \frac{1}{1-(-x^2)} = \sum_{k=0}^{\infty} (-1)^k x^{2k}$$

so that, by integration,

$$\tan^{-1} x = \int \left(\sum_{k=0}^{\infty} (-1)^k x^{2k} \right) dx = \left(\sum_{k=0}^{\infty} \frac{(-1)^k}{2k+1} x^{2k+1} \right) + C.$$

The constant C is 0 because the series on the right and the inverse tangent are both 0 at $x = 0$. Thus, for all x in $(-1, 1)$, we have

$$\tan^{-1} x = \sum_{k=0}^{\infty} \frac{(-1)^k}{2k+1} x^{2k+1} = x - \frac{x^3}{3} + \frac{x^5}{5} - \frac{x^7}{7} + \cdots.$$

That the series also represents the function at $x = -1$ and $x = 1$ follows directly from Abel's theorem: at both these points $\tan^{-1} x$ is continuous, and at both of these points the series converges. ❑

Since $\tan^{-1} 1 = \frac{1}{4}\pi$, we have

$$\tfrac{1}{4}\pi = 1 - \tfrac{1}{3} + \tfrac{1}{5} - \tfrac{1}{7} + \tfrac{1}{9} - \cdots.$$

This series was known to the Scottish mathematician James Gregory in 1671. It is an elegant formula for π, but it converges too slowly for computational purposes. A much more effective way of computing π is outlined in the supplement at the end of this section.

Term-by-term integration provides a method of calculating some (otherwise rather intractable) definite integrals. Suppose that you are trying to evaluate

$$\int_a^b f(x)\, dx,$$

but cannot find an antiderivative. If you can expand $f(x)$ in a convergent power series, then you can estimate the integral by forming the series and integrating term by term.

EXAMPLE 6 We will estimate

$$\int_0^1 e^{-x^2}\,dx$$

by expanding the integral in a power series and integrating term by term. Our starting point is the expansion

$$e^x = 1 + x + \frac{x^2}{2!} + \frac{x^3}{3!} + \frac{x^4}{4!} + \frac{x^5}{5!} + \frac{x^6}{6!} + \cdots \qquad \text{for all } x.$$

From this we see that

$$e^{-x^2} = 1 - x^2 + \frac{x^4}{2!} - \frac{x^6}{3!} + \frac{x^8}{4!} - \frac{x^{10}}{5!} + \frac{x^{12}}{6!} - \cdots \qquad \text{for all } x,$$

and therefore

$$\begin{aligned}\int_0^1 e^{-x^2}\,dx &= \left[x - \frac{x^3}{3} + \frac{x^5}{5(2!)} - \frac{x^7}{7(3!)} + \frac{x^9}{9(4!)} - \frac{x^{11}}{11(5!)} + \frac{x^{13}}{13(6!)} - \cdots\right]_0^1 \\ &= 1 - \frac{1}{3} + \frac{1}{5(2!)} - \frac{1}{7(3!)} + \frac{1}{9(4!)} - \frac{1}{11(5!)} + \frac{1}{13(6!)} - \cdots.\end{aligned}$$

This is an alternating series with declining terms. Therefore we know that the integral lies between consecutive partial sums. In particular it lies between

$$1 - \frac{1}{3} + \frac{1}{5(2!)} - \frac{1}{7(3!)} + \frac{1}{9(4!)} - \frac{1}{11(5!)}$$

and

$$\left[1 - \frac{1}{3} + \frac{1}{5(2!)} - \frac{1}{7(3!)} + \frac{1}{9(4!)} - \frac{1}{11(5!)}\right] + \frac{1}{13(6!)}.$$

As you can check, the first sum is greater than 0.74673 and the second one is less than 0.74684. It follows that

$$0.74673 < \int_0^1 e^{-x^2}\,dx < 0.74684.$$

The estimate 0.7468 approximates the integral within 0.0001. ❑

The integral of Example 6 was easy to estimate numerically because it could be expressed as an alternating series with decreasing terms. The next example requires more subtlety and illustrates a method more general than that used in Example 6.

EXAMPLE 7 We want to estimate

$$\int_0^1 e^{x^2}\,dx.$$

If we proceed exactly as in Example 6, we find that

$$\int_0^1 e^{x^2}\,dx = 1 + \frac{1}{3} + \frac{1}{5(2!)} + \frac{1}{7(3!)} + \frac{1}{9(4!)} + \frac{1}{11(5!)} + \frac{1}{13(6!)} + \cdots.$$

We now have a series expansion for the integral, but that expansion does not guide us directly to a numerical estimate for the integral. We know that s_n, the nth partial sum of the series, approximates the integral, but we don't know the accuracy of the approximation. We have no handle on the remainder left by s_n.

We start again, this time keeping track of the remainder. For $x \in [0, 1]$

$$0 \le e^x - \left(1 + x + \frac{x^2}{2!} + \cdots + \frac{x^n}{n!}\right) = R_{n+1}(x) \overset{(11.6.3)}{\le} e\left[\frac{x^{n+1}}{(n+1)!}\right] \le \frac{3}{(n+1)!}.$$

If $x \in [0, 1]$, then $x^2 \in [0, 1]$, and therefore

$$0 \le e^{x^2} - \left(1 + x^2 + \frac{x^4}{2!} + \cdots + \frac{x^{2n}}{n!}\right) \le \frac{3}{(n+1)!}$$

Integrating this inequality from $x = 0$ to $x = 1$, we have

$$0 \le \int_0^1 \left[e^{x^2} - \left(1 + x^2 + \frac{x^4}{2!} + \cdots + \frac{x^{2n}}{n!}\right)\right] dx \le \int_0^1 \frac{3}{(n+1)!}\, dx.$$

Carrying out the integration where possible, we see that

$$0 \le \int_0^1 e^{x^2}\, dx - \left[1 + \frac{1}{3} + \frac{1}{5(2!)} + \cdots + \frac{1}{(2n+1)(n!)}\right] \le \frac{3}{(n+1)!}.$$

We can use this inequality to estimate the integral as closely as we wish. Since

$$\frac{3}{7!} = \frac{1}{1680} < 0.0006,$$

we see that

$$\alpha = 1 + \frac{1}{3} + \frac{1}{5(2!)} + \frac{1}{7(3!)} + \frac{1}{9(4!)} + \frac{1}{11(5!)} + \frac{1}{13(6!)}$$

approximates the integral within 0.0006. Arithmetical computation shows that

$$1.4626 \le \alpha \le 1.4627.$$

It follows that

$$1.4626 \le \int_0^1 e^{x^2}\, dx \le 1.4627 + 0.0006 = 1.4633.$$

The estimate 1.463 approximates the integral within 0.0004. ❑

It is time to relate Taylor series

$$\sum_{k=0}^{\infty} \frac{f^{(k)}(0)}{k!} x^k$$

to power series in general. The relation is very simple:

On its interval of convergence a power series is the Taylor series of its sum.

That is, if you have a power series representation of a function f, then the series must be the Taylor series for f. To see this, all you have to do is differentiate

$$f(x) = a_0 + a_1 x + a_2 x^2 + \cdots + a_k x^k + \cdots$$

term by term. Do this and you will find that $f^{(k)}(0) = k!\, a_k$, and therefore

$$a_k = \frac{f^{(k)}(0)}{k!}.$$

The a_k are the Taylor coefficients of f.

We end this section by carrying out a few simple expansions.

Example 8 Expand $\cosh x$ and $\sinh x$ in powers of x.

SOLUTION There is no need to go through the labor of computing the Taylor coefficients

$$\frac{f^{(k)}(0)}{k!}$$

by differentiation. We know that

$$\cosh x = \tfrac{1}{2}(e^x + e^{-x}) \qquad \text{and} \qquad \sinh x = \tfrac{1}{2}(e^x - e^{-x}). \tag{7.9.1}$$

Since

$$e^x = 1 + x + \frac{x^2}{2!} + \frac{x^3}{3!} + \frac{x^4}{4!} + \frac{x^5}{5!} + \cdots,$$

we have

$$e^{-x} = 1 - x + \frac{x^2}{2!} - \frac{x^3}{3!} + \frac{x^4}{4!} - \frac{x^5}{5!} + \cdots.$$

Thus

$$\cosh x = \frac{1}{2}\left(2 + 2\frac{x^2}{2!} + 2\frac{x^4}{4!} + \cdots\right) = 1 + \frac{x^2}{2!} + \frac{x^4}{4!} + \cdots = \sum_{k=0}^{\infty} \frac{x^{2k}}{(2k)!}$$

and

$$\sinh x = \frac{1}{2}\left(2x + 2\frac{x^3}{3!} + 2\frac{x^5}{5!} + \cdots\right) = x + \frac{x^3}{3!} + \frac{x^5}{5!} + \cdots = \sum_{k=0}^{\infty} \frac{x^{2k+1}}{(2k+1)!}.$$

Both expansions are valid for all real x, since the exponential expansions are valid for all real x. ❑

Example 9 Expand $x^2 \cos x^3$ in powers of x.

SOLUTION

$$\cos x = 1 - \frac{x^2}{2!} + \frac{x^4}{4!} - \frac{x^6}{6!} + \cdots.$$

Thus

$$\cos x^3 = 1 - \frac{(x^3)^2}{2!} + \frac{(x^3)^4}{4!} - \frac{(x^3)^6}{6!} + \cdots = 1 - \frac{x^6}{2!} + \frac{x^{12}}{4!} - \frac{x^{18}}{6!} + \cdots,$$

and

$$x^2 \cos x^3 = x^2 - \frac{x^8}{2!} + \frac{x^{14}}{4!} - \frac{x^{20}}{6!} + \cdots.$$

This expansion is valid for all real x, since the expansion for $\cos x$ is valid for all real x. ❑

ALTERNATIVE SOLUTION Since

$$x^2 \cos x^3 = \frac{d}{dx}\left(\frac{1}{3}\sin x^3\right),$$

we can derive the expansion for $x^2 \cos x^3$ by expanding $\frac{1}{3}\sin x^3$ and then differentiating term by term. ❑

EXERCISES 11.9

In Exercises 1–6, expand f in powers of x, basing your calculations on the geometric series

$$\frac{1}{1-x} = 1 + x + x^2 + \cdots + x^n + \cdots.$$

1. $f(x) = \dfrac{1}{(1-x)^2}$.

2. $f(x) = \dfrac{1}{(1-x)^3}$.

3. $f(x) = \dfrac{1}{(1-x)^k}$.

4. $f(x) = \ln(1-x)$.

5. $f(x) = \ln(1-x^2)$.

6. $f(x) = \ln(2-3x)$.

In Exercises 7 and 8, expand f in powers of x, basing your calculations on the tangent series:

$$\tan x = x + \tfrac{1}{3}x^3 + \tfrac{2}{15}x^5 + \tfrac{17}{315}x^7 + \cdots.$$

7. $f(x) = \sec^2 x$.

8. $f(x) = \ln \cos x$.

In Exercises 9 and 10, find $f^{(9)}(0)$.

9. $f(x) = x^2 \sin x$.

10. $f(x) = x \cos x^2$.

In Exercises 11–22, expand f in powers of x.

11. $f(x) = \sin x^2$.

12. $f(x) = x^2 \tan^{-1} x$.

13. $f(x) = e^{3x^3}$.

14. $f(x) = \dfrac{1-x}{1+x}$.

15. $f(x) = \dfrac{2x}{1-x^2}$.

16. $f(x) = x \sinh x^2$.

17. $f(x) = \dfrac{1}{1-x} + e^x$.

18. $f(x) = \cosh x \sinh x$.

19. $f(x) = x \ln(1+x^3)$.

20. $f(x) = (x^2+x)\ln(1+x)$.

21. $f(x) = x^3 e^{-x^3}$.

22. $f(x) = x^5(\sin x + \cos 2x)$.

In Exercises 23–26, evaluate the given limit in two ways: (a) using L'Hospital's rule, and (b) using power series.

23. $\displaystyle\lim_{x\to 0} \frac{1-\cos x}{x^2}$.

24. $\displaystyle\lim_{x\to 0} \frac{\sin x - x}{x^2}$.

25. $\displaystyle\lim_{x\to 0} \frac{\cos x - 1}{x \sin x}$.

26. $\displaystyle\lim_{x\to 0} \frac{e^x - 1 - x}{x \tan^{-1} x}$.

In Exercises 27–30, find a powers series representation of the improper integral.

27. $\displaystyle\int_0^x \frac{\ln(1+t)}{t}\,dt$.

28. $\displaystyle\int_0^x \frac{1-\cos t}{t^2}\,dt$.

29. $\displaystyle\int_0^x \frac{\tan^{-1} t}{t}\,dt$.

30. $\displaystyle\int_0^x \frac{\sinh t}{t}\,dt$.

In Exercises 31–36, estimate within 0.01.

31. $\displaystyle\int_0^1 e^{-x^3}\,dx$.

32. $\displaystyle\int_0^1 \sin x^2\,dx$.

33. $\displaystyle\int_0^1 \sin\sqrt{x}\,dx$.

34. $\displaystyle\int_0^1 x^4 e^{-x^2}\,dx$.

35. $\displaystyle\int_0^1 \tan^{-1} x^2\,dx$.

36. $\displaystyle\int_1^2 \frac{1-\cos x}{x}\,dx$.

In Exercises 37–40, use a power series to estimate the integral within 0.0001.

37. $\displaystyle\int_0^1 \frac{\sin x}{x}\,dx$.

38. $\displaystyle\int_0^{0.5} \frac{1-\cos x}{x^2}\,dx$.

39. $\displaystyle\int_0^{0.5} \frac{\ln(1+x)}{x}\,dx$.

40. $\displaystyle\int_0^{0.2} x \sin x\,dx$.

In Exercises 41–43, sum the series.

41. $\displaystyle\sum_{k=0}^{\infty} \frac{1}{k!}x^{3k}$.

42. $\displaystyle\sum_{k=0}^{\infty} \frac{1}{k!}x^{3k+1}$.

43. $\sum_{k=1}^{\infty} \frac{3k}{k!} x^{3k-1}$.

44. Let $f(x) = \dfrac{e^x - 1}{x}$.
(a) Find a power series representation of f in powers of x.
(b) Differentiate the power series in part (a) and show that

$$\sum_{n=1}^{\infty} \frac{n}{(n+1)!} = 1.$$

45. Let $f(x) = xe^x$.
(a) Find a power series representation of f in powers of x.
(b) Integrate the power series in part (a) and show that

$$\sum_{n=1}^{\infty} \frac{1}{n!(n+2)} = \frac{1}{2}.$$

46. Deduce the differentiation formulas

$$\frac{d}{dx}(\sinh x) = \cosh x, \qquad \frac{d}{dx}(\cosh x) = \sinh x$$

from the expansions of $\sinh x$ and $\cosh x$ in powers of x.

47. Show that, if $\Sigma\, a_k x^k$ and $\Sigma\, b_k x^k$ both converge to the same sum on some interval, then $a_k = b_k$ for each k.

48. Show that, if $\epsilon > 0$, then

$$|kx^{k-1}| < (|x| + \epsilon)^k \qquad \text{for all } k \text{ sufficiently large.}$$

HINT: Take the kth root of the left side and let $k \to \infty$.

49. Suppose that the function f has the power series representation $f(x) = \Sigma_{k=0}^{\infty}\, a_k x^k$.
(a) Show that if f is an even function, then $a_{2k+1} = 0$ for all k.
(b) Show that if f is an odd function, then $a_{2k} = 0$ for all k.

50. Suppose that the function f is infinitely differentiable on an interval containing 0, and suppose that $f'(x) = -2f(x)$ and $f(0) = 1$. Use these properties to find the power series representation of f in powers of x. Do you recognize this function?

In Exercises 51–53, estimate within 0.001 by the method of this section and check your result by carrying out the integration directly.

51. $\displaystyle\int_0^{1/2} x \ln(1+x)\, dx$. **52.** $\displaystyle\int_0^1 x \sin x\, dx$.

53. $\displaystyle\int_0^1 xe^{-x}\, dx$.

54. Show that

$$0 \le \int_0^2 e^{x^2}\, dx - \left[2 + \frac{2^3}{3} + \frac{2^5}{5(2!)} + \cdots + \frac{2^{2n+1}}{(2n+1)n!}\right] < \frac{e^4 2^{2n+3}}{(n+1)!}.$$

*SUPPLEMENT TO SECTION 11.9

PROOF OF THEOREM 11.9.3

Set

$$f(x) = \sum_{k=0}^{\infty} a_k x^k \qquad \text{and} \qquad g(x) = \sum_{k=0}^{\infty} \frac{d}{dx}(a_k x^k) = \sum_{k=1}^{\infty} k a_k x^{k-1}.$$

Select x from $(-c, c)$. We want to show that

$$\lim_{h \to 0} \frac{f(x+h) - f(x)}{h} = g(x).$$

For $x + h$ in $(-c, c)$, $h \ne 0$, we have

$$\left| g(x) - \frac{f(x+h) - f(x)}{h} \right| = \left| \sum_{k=1}^{\infty} k a_k x^{k-1} - \sum_{k=0}^{\infty} \frac{a_k (x+h)^k - a_k x^k}{h} \right|$$
$$= \left| \sum_{k=1}^{\infty} k a_k x^{k-1} - \sum_{k=1}^{\infty} a_k \left[\frac{(x+h)^k - x^k}{h} \right] \right|.$$

By the mean-value theorem

$$\frac{(x+h)^k - x^k}{h} = k(t_k)^{k-1}$$

for some number t_k between x and $x + h$. Thus we can write

$$\left| g(x) - \frac{f(x+h) - f(x)}{h} \right| = \left| \sum_{k=1}^{\infty} ka_k x^{k-1} - \sum_{k=1}^{\infty} ka_k (t_k)^{k-1} \right|$$
$$= \left| \sum_{k=1}^{\infty} ka_k [x^{k-1} - (t_k)^{k-1}] \right|$$
$$= \left| \sum_{k=2}^{\infty} ka_k [x^{k-1} - (t_k)^{k-1}] \right|.$$

By the mean-value theorem

$$\frac{x^{k-1} - (t_k)^{k-1}}{x - t_k} = (k-1)(p_{k-1})^{k-2}$$

for some number p_{k-1} between x and t_k. Obviously, then,

$$|x^{k-1} - (t_k)^{k-1}| = |x - t_k||(k-1)(p_{k-1})^{k-2}|.$$

Since $|x - t_k| < |h|$ and $|p_{k-1}| \leq |\alpha|$ where $|\alpha| = \max\{|x|, |x+h|\}$,

$$|x^{k-1} - (t_k)^{k-1}| \leq |h||(k-1)\alpha^{k-2}|.$$

Thus

$$\left| g(x) - \frac{f(x+h) - f(x)}{h} \right| \leq |h| \sum_{k=2}^{\infty} |k(k-1)a_k \alpha^{k-2}|.$$

Since the series converges,

$$\lim_{h \to 0} \left(|h| \sum_{k=2}^{\infty} |k(k-1)a_k \alpha^{k-2}| \right) = 0.$$

This gives

$$\lim_{h \to 0} \left| g(x) - \frac{f(x+h) - f(x)}{h} \right| = 0 \quad \text{and thus} \quad \lim_{h \to 0} \frac{f(x+h) - f(x)}{h} = g(x). \quad \square$$

Calculating π

We base our computation of π on the inverse tangent series (11.9.7)

$$\tan^{-1} x = x - \frac{x^3}{3} + \frac{x^5}{5} - \frac{x^7}{7} + \cdots \qquad \text{for } -1 \leq x \leq 1$$

and the relation

(11.9.8)

$$\tfrac{1}{4}\pi = 4 \tan^{-1} \tfrac{1}{5} - \tan^{-1} \tfrac{1}{239}.\dagger$$

The inverse tangent series gives

$$\tan^{-1} \tfrac{1}{5} = \tfrac{1}{5} - \tfrac{1}{3}(\tfrac{1}{5})^3 + \tfrac{1}{5}(\tfrac{1}{5})^5 - \tfrac{1}{7}(\tfrac{1}{5})^7 + \cdots$$

† This relation was discovered in 1706 by John Machin, a Scotsman. It can be verified by repeated applications of the addition formula

$$\tan(A + B) = \frac{\tan A + \tan B}{1 - \tan A \tan B}.$$

First calculate $\tan(2 \tan^{-1} \frac{1}{5})$, then $\tan(4 \tan^{-1} \frac{1}{5})$, and finally $\tan(4 \tan^{-1} \frac{1}{5} - \tan^{-1} \frac{1}{239})$.

and

$$\tan^{-1}\tfrac{1}{239} = \tfrac{1}{239} - \tfrac{1}{3}(\tfrac{1}{239})^3 + \tfrac{1}{5}(\tfrac{1}{239})^5 - \tfrac{1}{7}(\tfrac{1}{239})^7 + \cdots.$$

These are alternating series $\Sigma\,(-1)^k a_k$ with a_k decreasing toward 0. Thus we know that

$$\tfrac{1}{5} - \tfrac{1}{3}(\tfrac{1}{5})^3 \le \tan^{-1}\tfrac{1}{5} \le \tfrac{1}{5} - \tfrac{1}{3}(\tfrac{1}{5})^3 + \tfrac{1}{5}(\tfrac{1}{5})^5$$

and

$$\tfrac{1}{239} - \tfrac{1}{3}(\tfrac{1}{239})^3 \le \tan^{-1}\tfrac{1}{239} \le \tfrac{1}{239}.$$

With these inequalities, together with relation (11.9.8), we can show that

$$3.14 < \pi < 3.147.$$

By using six terms of the series for $\tan^{-1}\frac{1}{5}$ and still only two of the series for $\tan^{-1}\frac{1}{239}$, we can show that

$$3.14159262 < \pi < 3.14159267.$$

Greater accuracy can be obtained by taking more terms into account. For instance, fifteen terms of the series for $\tan^{-1}\frac{1}{5}$ and just four terms of the series for $\tan^{-1}\frac{1}{239}$ determine π to twenty decimal places:

$$\pi \cong 3.14159\ 26535\ 89793\ 23846.$$

■ 11.10 THE BINOMIAL SERIES

Through a collection of problems we invite you to derive for yourself the basic properties of one of the most celebrated series of all—*the binomial series.*

Start with the binomial $1 + x$. Choose a real number $\alpha \neq 0$ and form the function

$$f(x) = (1 + x)^\alpha.$$

Note that if $\alpha = n$ is a positive integer, then

$$(1 + x)^n = 1 + nx + \frac{n(n-1)}{2!}x^2 + \cdots + nx^{n-1} + x^n$$

is the familiar binomial theorem. The binomial series is a generalization of the binomial theorem. You should also note that if $n = -1$, then $(1 + x)^{-1}$ is the sum of a geometric series

$$\frac{1}{1+x} = 1 - x + x^2 - x^3 + \cdots + (-1)^n x^n + \cdots$$

Problem 1 Show that

$$\frac{f^{(k)}(0)}{k!} = \frac{\alpha[\alpha - 1][\alpha - 2]\cdots[\alpha - (k-1)]}{k!}.$$

The number you just obtained is the coefficient of x^k in the expansion of $(1 + x)^\alpha$. It is called *the kth binomial coefficient* and is usually denoted by $\binom{\alpha}{k}$:

(11.10.1)
$$\binom{\alpha}{k} = \frac{\alpha[\alpha - 1][\alpha - 2]\cdots[\alpha - (k-1)]}{k!}.$$

For example, if $\alpha = 7$ and $k = 3$, then

$$\binom{7}{3} = \frac{7 \cdot 6 \cdot 5}{3!} = 35;$$

if $\alpha = 3/2$ and $k = 3$, then

$$\binom{3/2}{3} = \frac{(3/2)[(3/2) - 1][(3/2) - 2]}{3!} = \frac{(3/2)(1/2)(-1/2)}{6} = -\frac{1}{16}. \quad \square$$

Problem 2 Show that the binomial series

$$\sum \binom{\alpha}{k} x^k$$

has radius of convergence 1. HINT: Use the ratio test. ❑

From Problem 2 you know that the binomial series converges on the open interval $(-1, 1)$ and defines there an infinitely differentiable function. The next thing to show is that this function (the one defined by the series) is actually $(1 + x)^\alpha$. To do this, you first need some other results.

Problem 3 Verify the identity

$$(k + 1)\binom{\alpha}{k + 1} + k\binom{\alpha}{k} = \alpha\binom{\alpha}{k}. \quad \square$$

Problem 4 Use the identity of Problem 3 to show that the sum of the binomial series

$$\phi(x) = \sum_{k=0}^{\infty} \binom{\alpha}{k} x^k$$

satisfies the differential equation

$$(1 + x)\phi'(x) = \alpha\phi(x) \qquad \text{for all } x \text{ in } (-1, 1)$$

together with the side condition $\phi(0) = 1$. ❑

You are now in a position to prove the main result.

Problem 5 Show that

(11.10.2)
$$(1 + x)^\alpha = \sum_{k=0}^{\infty} \binom{\alpha}{k} x^k \qquad \text{for all } x \text{ in } (-1, 1).$$

You can probably get a better feeling for the series by writing out the first few terms:

(11.10.3)
$$(1 + x)^\alpha = 1 + \alpha x + \frac{\alpha(\alpha - 1)}{2!}x^2 + \frac{\alpha(\alpha - 1)(\alpha - 2)}{3!}x^3 + \cdots. \quad \square$$

EXERCISES 11.10

In Exercises 1–10, expand f in powers of x up to x^4.

1. $f(x) = \sqrt{1+x}$.
2. $f(x) = \sqrt{1-x}$.
3. $f(x) = \sqrt{1+x^2}$.
4. $f(x) = \sqrt{1-x^2}$.
5. $f(x) = \dfrac{1}{\sqrt{1+x}}$.
6. $f(x) = \dfrac{1}{\sqrt[3]{1+x}}$.
7. $f(x) = \sqrt[4]{1-x}$.
8. $f(x) = \dfrac{1}{\sqrt[4]{1+x}}$.
9. $f(x) = (4+x)^{3/2}$.
10. $f(x) = \sqrt{1+x^4}$.

11. (a) Use a binomial series to find the Taylor series of $f(x) = 1/\sqrt{1-x^2}$ in powers of x.
(b) Use the series for f in part (a) to find the Taylor series for $F(x) = \sin^{-1} x$ and give the radius of convergence.

12. (a) Use a binomial series to find the Taylor series of $f(x) = 1/\sqrt{1+x^2}$ in powers of x.
(b) Use the series for f in part (a) to find the Taylor series for $F(x) = \sinh^{-1} x$ and give the radius of convergence.

In Exercises 13–18, estimate by using the first three terms of a binomial expansion, rounding off your answer to four decimal places.

13. $\sqrt{98}$. HINT: $\sqrt{98} = (100-2)^{1/2} = 10(1-\frac{1}{50})^{1/2}$.
14. $\sqrt[5]{36}$.
15. $\sqrt[3]{9}$.
16. $\sqrt[4]{620}$.
17. $17^{-1/4}$.
18. $9^{-1/3}$.

In Exercises 19–22, approximate each integral to within 0.001.

19. $\displaystyle\int_0^{1/3} \sqrt{1+x^3}\,dx$.
20. $\displaystyle\int_0^{1/5} \sqrt{1+x^4}\,dx$.
21. $\displaystyle\int_0^{1/2} \frac{1}{\sqrt{1+x^2}}\,dx$.
22. $\displaystyle\int_0^{1/2} \frac{1}{\sqrt{1-x^3}}\,dx$.

■ CHAPTER HIGHLIGHTS

11.1 Sigma Notation

11.2 Infinite Series

partial sums (p. 696)
convergence, divergence (p. 696)
sum of a series (p. 696)
a divergence test (p. 703)

geometric series: $\displaystyle\sum_{k=0}^{\infty} x^k = \begin{cases} \dfrac{1}{1-x}, & |x| < 1 \\ \text{diverges}, & |x| \geq 1 \end{cases}$

If $\displaystyle\sum_{k=0}^{\infty} a_k$ converges, then $a_k \to 0$. The converse is false.

11.3 The Integral Test; Comparison Theorems

integral test (p. 707)
basic comparison (p. 710)
limit comparison (p. 712)

harmonic series: $\displaystyle\sum_{k=1}^{\infty} \frac{1}{k}$ diverges
p-series: $\displaystyle\sum_{k=1}^{\infty} \frac{1}{k^p}$ converges iff $p > 1$

11.4 The Root Test; The Ratio Test

root test (p. 715)
ratio test (p. 717)
summary on convergence tests (p. 719)

11.5 Absolute and Conditional Convergence; Alternating Series

absolutely convergent, conditionally convergent (pp. 720–721)
convergence theorem for alternating series (p. 722)
an estimate for alternating series (p. 723)
rearrangements (p. 725)

11.6 Taylor Polynomials in x; Taylor Series in x

Taylor polynomials in x (p. 728) remainder term $R_{n+1}(x)$ (p. 730)
remainder estimate (p. 731) Lagrange form of the remainder (p. 731)

Taylor series in x (Maclaurin series): $\sum_{k=0}^{\infty} \frac{f^{(k)}(0)}{k!} x^k$

$$e^x = \sum_{k=0}^{\infty} \frac{x^k}{k!}, \quad \text{all real } x \qquad \ln(1+x) = \sum_{k=1}^{\infty} \frac{(-1)^{k+1}}{k} x^k, \quad -1 < x \leq 1$$

$$\sin x = \sum_{k=0}^{\infty} \frac{(-1)^k}{(2k+1)!} x^{2k+1}, \quad \text{all real } x \qquad \cos x = \sum_{k=0}^{\infty} \frac{(-1)^k}{(2k)!} x^{2k}, \quad \text{all real } x$$

11.7 Taylor Polynomials and Taylor Series in $x - a$

Taylor series in $x - a$: $\sum_{k=0}^{\infty} \frac{g^{(k)}(a)}{k!}(x-a)^k$

11.8 Power Series

power series (p. 745) radius of convergence (p. 747)
interval of convergence (p. 748)

If a power series converges at $c \neq 0$, then it converges absolutely for $|x| < |c|$; if it diverges at c, then it diverges for $|x| > |c|$.

11.9 Differentiation and Integration of Power Series

$$\tan^{-1} x = \sum_{k=0}^{\infty} \frac{(-1)^k}{2k+1} x^{2k+1}, \quad -1 \leq x \leq 1$$

$$\cosh x = \sum_{k=0}^{\infty} \frac{x^{2k}}{(2k)!}, \quad \text{all real } x \qquad \sinh x = \sum_{k=0}^{\infty} \frac{x^{2k+1}}{(2k+1)!}, \quad \text{all real } x$$

On the interior of its interval of convergence, a power series can be differentiated and integrated term by term.

On its interval of convergence a power series is the Taylor series of its sum.

11.10 The Binomial Series

$$(1+x)^\alpha = \sum_{k=0}^{\infty} \binom{\alpha}{k} x^k = 1 + \alpha x + \frac{\alpha(\alpha-1)}{2!} x^2 + \cdots, \quad -1 < x < 1$$

■ PROJECTS AND EXPLORATIONS USING TECHNOLOGY

To do these exercises you will need a graphics calculator or a computer with graphing capability. The majority of these problems are open-ended so different approaches may be used to solve them. You should be aware that different approaches can result in slight variations in the answers. Round your numerical answers to at least four decimal places. The rounding method that your calculator or computer uses also may cause variations in answers.

11.1 The integral test is used to determine the convergence of a series of positive terms. In the case of convergent series, it can also be used to obtain reasonable approximations for the sum of the series. Let $\{a_n\}$ be the sequence defined by

$$a_n = \frac{[n^3 + \ln(n)]\ln(n)}{n^{4+\ln(n)} + 3n^3 + 7}.$$

(a) Find j such that the sequence is decreasing for $n \geq j$.
(b) Use the integral test to show that the series

$$\sum_{n=1}^{\infty} \frac{[n^3 + \ln(n)]\ln(n)}{n^{4+\ln(n)} + 3n^3 + 7}$$

converges. (You may need to start the series at a value $j > 1$.)
(c) Let

$$I_k = \int_j^k \frac{[x^3 + \ln(x)]\ln(x)}{x^{4+\ln(x)} + 3x^3 + 7}\,dx,$$

and let

$$S_k = \sum_{n=j}^{k} \frac{[n^3 + \ln(n)]\ln(n)}{n^{4+\ln(n)} + 3n^3 + 7}.$$

Show that $I_{k-1} < S_k < I_k$.
(d) How large must k be in order to get an estimate for the sum of the series that is accurate to three decimal places.
(e) Since you are approximating I_k, how can you be sure that you have enough accuracy for it so as to provide an accurate approximation of S_k?

11.2 The alternating series $\Sigma(-1)^n a_n$, ($a_n > 0$ for all n), converges if the sequence $\{a_n\}$ is decreasing and has limit zero. Moreover, if the series converges, then the sum S is always between two consecutive partial sums, S_k and S_{k+1}. The following procedure allows even better approximations for the sum. Consider the alternating series

$$\sum_{n=0}^{\infty} \frac{(-1)^n(n^2 + 1)}{n^3 + 1}.$$

(a) Estimate the sum of the first 10, 20, 30, . . . , 100 terms of this series.
(b) Determine k such that the partial sum S_k approximates the sum S with three decimal place accuracy.
(c) Using the values of the partial sums from part (a), estimate the differences between the partial sums and S as a function of n. Add the value given by this function to the partial sum S_{100} to get a new approximation for S. This process is called *extrapolation.*
(d) Now consider the following weighted average of consecutive terms

$$T_n = \frac{a_n S_n + a_{n-1} S_{n-1}}{a_n + a_{n-1}}$$

and calculate T_n for $n = 10, 20, 30, \ldots, 100$. Plot T_n versus n and estimate the error as a function of n.
(e) Generate the sequence $\{b_n\}$ where

$$b_n = \frac{(S - S_{n-1})}{(S_n - S_{n-1})}$$

and S is the sum of the series. For our example, what is $\lim_{n\to\infty} b_n$?
(f) Compare the convergence of $\{T_n\}$ with the convergence of the sequence $\{U_n\}$ given by

$$U_n = \frac{S_n + S_{n-1}}{2}.$$

11.3 Consider the function

$$F(z) = \int_0^1 [\cos zx^2 + \sin z^2x^3]\,dx.$$

(a) Using technology, graph the function F. Based on the graph, make conjectures about the properties of this function. For example, is it continuous? differentiable? Where is it increasing? decreasing? And so on.

(b) Determine the Taylor series in powers of x for each of the functions $\cos zx^2$ and $\sin z^2x^3$. Then approximate $F(z)$ using termwise integration. Can you verify the information that you found in part (a) from this expression for F?

(c) Now assume that the derivative of F with respect to z can be found by differentiating the integrand with respect to z, obtaining a series expansion in powers of x, and then integrating that series termwise. Do you get the same series?

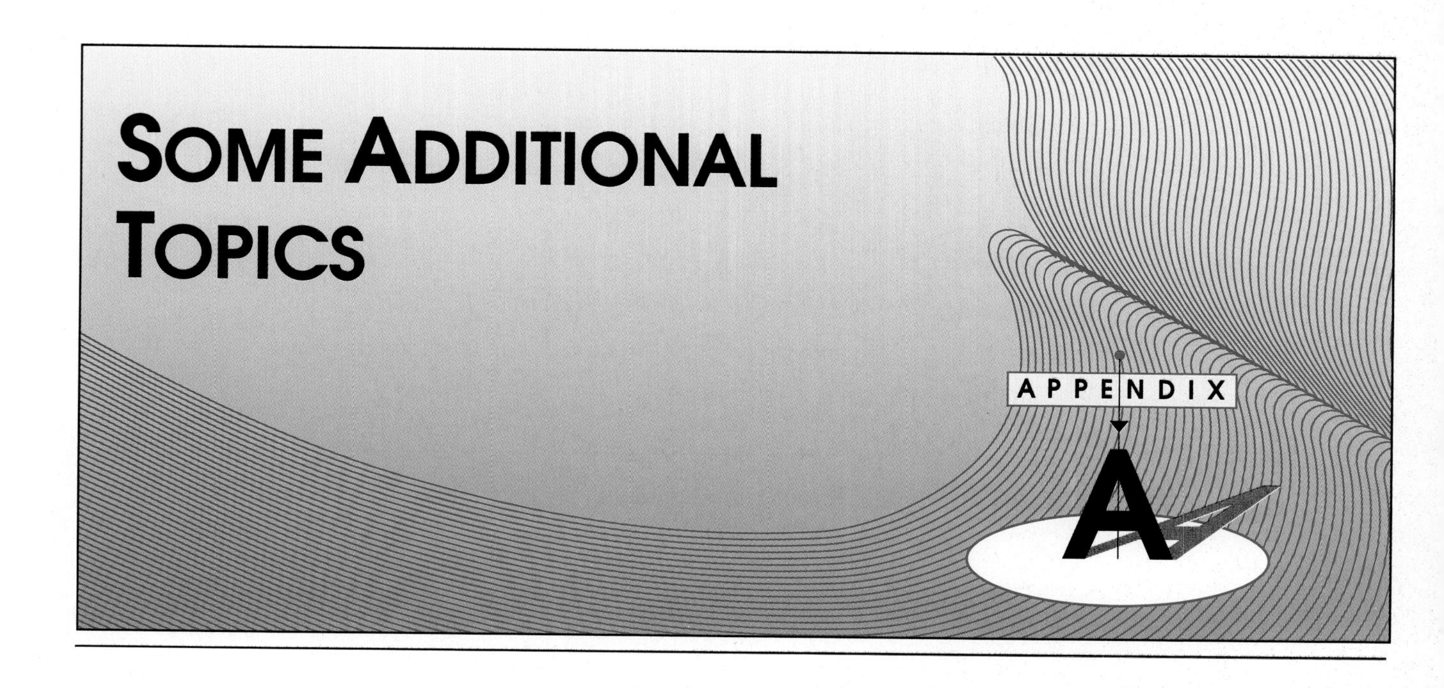

A.1 ROTATION OF AXES; EQUATIONS OF SECOND DEGREE

Rotation of Axes

We begin by referring to Figure A.1.1. From the figure,

$$\cos\theta = \frac{x}{r}, \qquad \sin\theta = \frac{y}{r}.$$

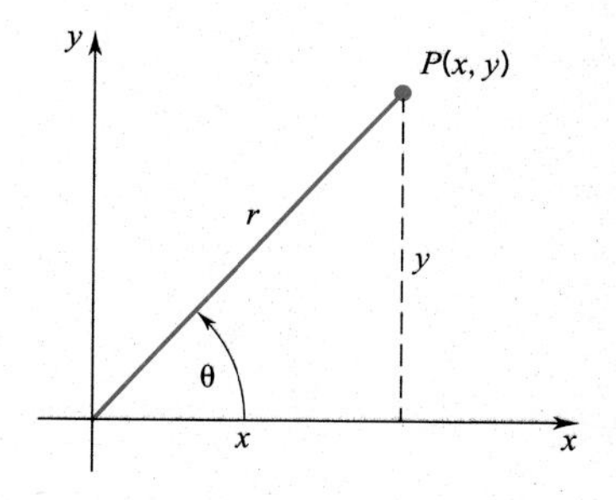

Figure A.1.1

Thus

(A.1.1) $$x = r\cos\theta, \qquad y = r\sin\theta.$$

Equations (A.1.1) come up repeatedly in calculus. In particular, these are the equations that you use to convert polar coordinates to rectangular coordinates (see Section 9.2).

Consider now a rectangular coordinate system Oxy. If we rotate this system counterclockwise α radians about the origin, we obtain a new coordinate system OXY. See Figure A.1.2.

A point P will now have two pairs of rectangular coordinates:

(x, y) in the Oxy system and (X, Y) in the OXY system.

Here we investigate the relation between (x, y) and (X, Y). With P as in Figure A.1.3,

$$x = r\cos(\alpha + \beta), \qquad y = r\sin(\alpha + \beta)$$

and

$$X = r\cos\beta, \qquad Y = r\sin\beta.$$

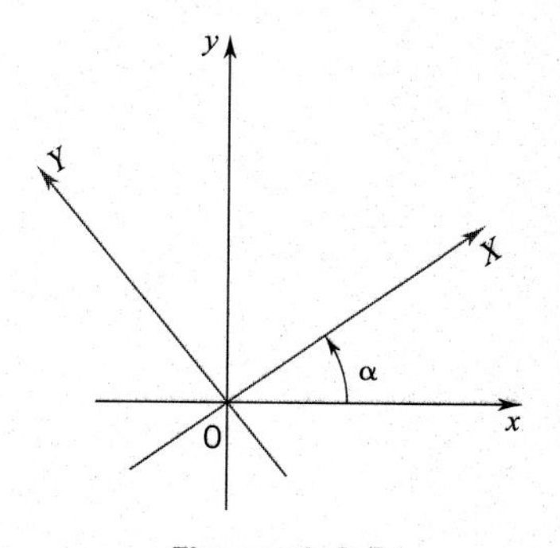

Figure A.1.2

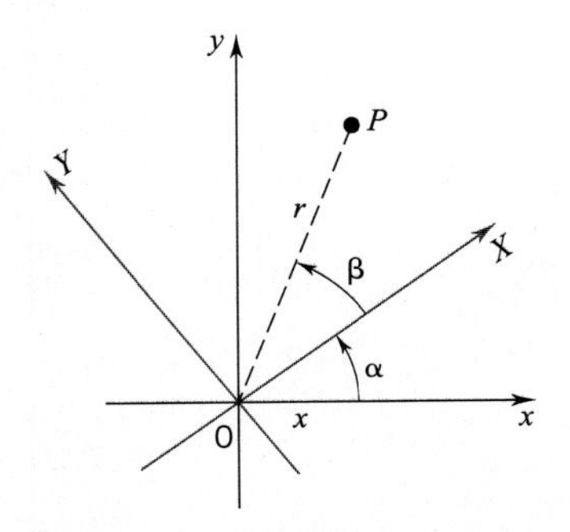

Figure A.1.3

Since

$$\cos(\alpha + \beta) = \cos\alpha\cos\beta - \sin\alpha\sin\beta,$$
$$\sin(\alpha + \beta) = \sin\alpha\cos\beta + \cos\alpha\sin\beta,$$

we have

$$x = r\cos(\alpha + \beta) = (\cos\alpha)\, r\cos\beta - (\sin\alpha)\, r\sin\beta,$$
$$y = r\sin(\alpha + \beta) = (\sin\alpha)\, r\cos\beta + (\cos\alpha)\, r\sin\beta,$$

and therefore

(A.1.2) $$x = (\cos\alpha)X - (\sin\alpha)Y, \qquad y = (\sin\alpha)X + (\cos\alpha)Y.$$

These formulas give the algebraic consequences of a counterclockwise rotation of α radians.

Equations of Second Degree

Equations of the form

(1) $$ax^2 + cy^2 + dx + ey + f = 0,$$

where a, c, d, e, f are constants and a and c are not both zero, are studied in Sections 1.4 and 9.1, and they occur throughout the text in the examples and exercises. Except for degenerate cases (for example, $x^2 + y^2 + 1 = 0$ or $x^2 - y^2 = 0$), the graph of (1) is a conic section: circle, ellipse, parabola, or hyperbola.

The *general equation of second degree in x and y* is an equation of the form

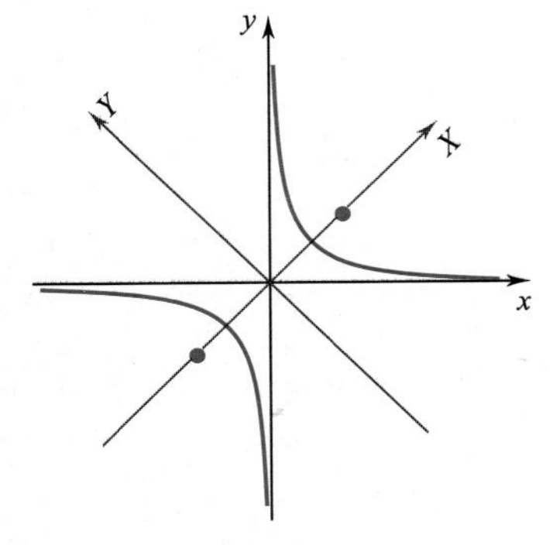

Figure A.1.4

(A.1.3) $$ax^2 + bxy + cy^2 + dx + ey + f = 0,$$

where a, b, c, d, e, f are constants and a, b, c are not all zero. The graph of such an equation is still a conic section (again, except for degenerate cases). For example, the graph of the equation

$$xy - 2 = 0$$

is the hyperbola shown in Figure A.1.4.

Eliminating the *xy*-Term

Rotations of the coordinate system enable us to simplify equations of the second degree by eliminating the xy-term. That is, if in the Oxy coordinate system, a curve S has an equation of the form

(2) $$ax^2 + bxy + cy^2 + dx + ey + f = 0 \qquad \text{with} \qquad b \neq 0,$$

then there exists a coordinate system OXY, differing from Oxy by a rotation α, where $0 < \alpha < \pi/2$, such that in the OXY system S has an equation of the form

(3) $$AX^2 + CY^2 + DX + EY + F = 0,$$

where A and C are not both zero. To see this, substitute

$$x = (\cos \alpha)X - (\sin \alpha)Y, \qquad y = (\sin \alpha)X + (\cos \alpha)Y$$

in equation (2). This will give you a second-degree equation in X and Y in which the coefficient of XY is

$$-2a \cos \alpha \sin \alpha + b(\cos^2 \alpha - \sin^2 \alpha) + 2c \cos \alpha \sin \alpha.$$

This can be simplified to

$$(c - a) \sin 2\alpha + b \cos 2\alpha.$$

To eliminate the XY term we must have this coefficient equal to zero, that is, we must have

$$b \cos 2\alpha = (a - c)\sin 2\alpha$$

or

$$\cot 2\alpha = \frac{a - c}{b} \qquad \text{(recall } b \neq 0\text{).}$$

Therefore,

$$2\alpha = \cot^{-1}\left(\frac{a - c}{b}\right)$$

and

$$\alpha = \frac{1}{2}\cot^{-1}\left(\frac{a - c}{b}\right).$$

Since the range of the inverse cotangent function is $(0, \pi)$, it follows that $0 < \alpha < \pi/2$.

We have shown that an equation of the form (2) can be transformed into an equation of the form (3) by rotating the axes through the angle α given by

(A.1.5)
$$\boxed{\alpha = \frac{1}{2}\cot^{-1}\left(\frac{a - c}{b}\right)}$$

We leave it as an exercise to show that the coefficients A and C in (3) are not both zero.

Example 1 In the case of

$$xy - 2 = 0,$$

we have $a = c = 0$, $b = 1$, and $\alpha = \frac{1}{2} \cot^{-1}(0) = \frac{1}{4}\pi$. Setting

$$x = (\cos \tfrac{1}{4}\pi)X - (\sin \tfrac{1}{4}\pi)Y = \tfrac{1}{2}\sqrt{2}\,(X - Y),$$
$$y = (\sin \tfrac{1}{4}\pi)X + (\cos \tfrac{1}{4}\pi)Y = \tfrac{1}{2}\sqrt{2}\,(X + Y),$$

we find that $xy - 2 = 0$ becomes

$$\tfrac{1}{2}(X^2 - Y^2) - 2 = 0,$$

which can be written

$$\frac{X^2}{4} - \frac{Y^2}{4} = 1.$$

This is the equation of a hyperbola in standard position in the OXY system. The hyperbola is shown in Figure A.1.4. ❑

Example 2 In the case of

$$11x^2 + 4\sqrt{3}xy + 7y^2 - 1 = 0,$$

we have $a = 11$, $b = 4\sqrt{3}$, and $c = 7$. Thus we can choose

$$\alpha = \tfrac{1}{2}\cot^{-1}\left(\frac{11-7}{4\sqrt{3}}\right) = \tfrac{1}{2}\cot^{-1}\left(\frac{1}{\sqrt{3}}\right) = \tfrac{1}{6}\pi$$

Setting

$$x = (\cos\tfrac{1}{6}\pi)X - (\sin\tfrac{1}{6}\pi)Y = \tfrac{1}{2}(\sqrt{3}X - Y),$$
$$y = (\sin\tfrac{1}{6}\pi)X + (\cos\tfrac{1}{6}\pi)Y = \tfrac{1}{2}(X + \sqrt{3}Y),$$

we find that our initial equation simplifies to $13X^2 + 5Y^2 - 1 = 0$, which we can write as

$$\frac{X^2}{(1/\sqrt{13})^2} + \frac{Y^2}{(1/\sqrt{5})^2} = 1.$$

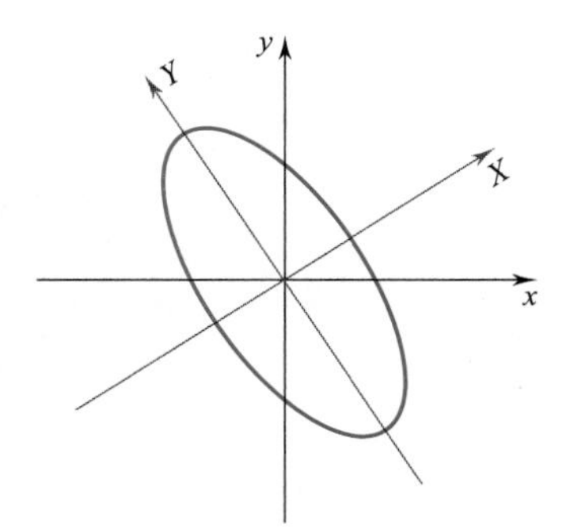

Figure A.1.5

This is the equation of an ellipse. The ellipse is pictured in Figure A.1.5. ❑

The Discriminant

It is possible to draw general conclusions about the graph of a second-degree equation

$$ax^2 + bxy + cy^2 + dx + ey + f = 0, \qquad a, b, c \text{ not all } 0,$$

just from the *discriminant* $\Delta = b^2 - 4ac$. There are three cases:

Case 1. If $\Delta < 0$, the graph is an ellipse, a circle, a point, or empty.

Case 2. If $\Delta > 0$, the graph is a hyperbola or a pair of intersecting lines.

Case 3. If $\Delta = 0$, the graph is a parabola, a line, a pair of lines, or empty.

Below we outline how these assertions can be verified. A useful first step is to rotate the coordinate system so that the equation takes the form

$$AX^2 + CY^2 + DX + EY + F = 0. \tag{4}$$

An elementary but time-consuming computation shows that the discriminant is unchanged by a rotation, so that in this instance we have

$$\Delta = b^2 - 4ac = -4AC.$$

Moreover, A and C cannot both be zero. If $\Delta < 0$, then $AC > 0$ and we can rewrite (4) as

$$\frac{X^2}{C} + \frac{D}{AC}X + \frac{Y^2}{A} + \frac{E}{AC}Y + \frac{F}{AC} = 0.$$

By completing the squares, we obtain an equation of the form

$$\frac{(X-\alpha)^2}{(\sqrt{|C|})^2}+\frac{(Y-\beta)^2}{(\sqrt{|A|})^2}=K.$$

If $K>0$, we have an ellipse or a circle. If $K=0$, we have the point (α, β). If $K<0$, the set is empty.

If $\Delta>0$, then $AC<0$. Proceeding as before, we obtain an equation of the form

$$\frac{(X-\alpha)^2}{(\sqrt{|C|})^2}-\frac{(Y-\beta)^2}{(\sqrt{|A|})^2}=K.$$

If $K\neq 0$, we have a hyperbola. If $K=0$, the equation becomes

$$\left(\frac{X-\alpha}{\sqrt{|C|}}-\frac{Y-\beta}{\sqrt{|A|}}\right)\left(\frac{X-\alpha}{\sqrt{|C|}}+\frac{Y-\beta}{\sqrt{|A|}}\right)=0,$$

so that we have a pair of lines intersecting at the point (α, β).

If $\Delta=0$, then $AC=0$, so that either $A=0$ or $C=0$. Since A and C are not both zero, there is no loss in generality in assuming that $A\neq 0$ and $C=0$. In this case equation (4) reduces to

$$AX^2+DX+EY+F=0.$$

Dividing by A and completing the square we have an equation of the form

$$(X-\alpha)^2=\beta Y+K.$$

If $\beta\neq 0$, we have a parabola. If $\beta=0$ and $K=0$, we have a line. If $\beta=0$ and $K>0$, we have a pair of parallel lines. If $\beta=0$ and $K<0$, the set is empty. ❑

EXERCISES A.1

In Exercises 1–8, (a) use the discriminant to give a possible identification of the graph of the equation; (b) find a rotation $\alpha\in(0, \pi/2)$ that eliminates the xy-term; (c) rewrite the equation in terms of the new coordinate system; (d) sketch the graph displaying both coordinate systems.

1. $xy=1$.

2. $xy-y+x=1$.

3. $11x^2+10\sqrt{3}xy+y^2-4=0$.

4. $52x^2-72xy+73y^2-100=0$.

5. $x^2-2xy+y^2+x+y=0$.

6. $3x^2+2\sqrt{3}xy+y^2-2x+2\sqrt{3}y=0$.

7. $x^2+2\sqrt{3}xy+3y^2+2\sqrt{3}x-2y=0$.

8. $2x^2+4\sqrt{3}xy+6y^2+(8-\sqrt{3})x+(8\sqrt{3}+1)y+8=0$.

In Exercises 9 and 10, find a rotation $\alpha\in(0, \pi/2)$ that eliminates the xy-term. Then find $\cos\alpha$ and $\sin\alpha$.

9. $x^2+xy+Kx+Ly+M=0$.

10. $5x^2+24xy+12y^2+Kx+Ly+M=0$.

11. Show that after a rotation of axes through an angle α, the coefficients in the equation

$$AX^2+BXY+CY^2+DX+EY+F=0$$

are related to the coefficients in the equation

$$ax^2+bxy+cy^2+dx+ey+f=0, \quad a, b, c \text{ not all } 0,$$

as follows:

$$\begin{aligned}
\mathrm{A}&=a\cos^2\alpha+b\cos\alpha\sin\alpha+c\sin^2\alpha,\\
\mathrm{B}&=2(c-a)\cos\alpha\sin\alpha+b(\cos^2\alpha-\sin^2\alpha),\\
\mathrm{C}&=a\sin^2\alpha-b\cos\alpha\sin\alpha+c\cos^2\alpha,\\
\mathrm{D}&=d\cos\alpha+e\sin\alpha,\\
\mathrm{E}&=e\cos\alpha-d\sin\alpha,\\
\mathrm{F}&=f.
\end{aligned}$$

12. Use the results of Exercise 11 to show:
(a) $B^2-4AC=b^2-4ac$.
(b) If $B=0$, then A and C cannot both be 0.

■ A.2 DETERMINANTS

By a *matrix* we mean a rectangular arrangement of numbers enclosed in parentheses. For example,

$$\begin{pmatrix} 2 & 4 \\ 3 & 1 \end{pmatrix} \quad \begin{pmatrix} 1 & 6 & 3 \\ 5 & 2 & 2 \end{pmatrix} \quad \begin{pmatrix} 2 & 4 & 0 \\ 4 & 7 & 1 \\ 0 & 1 & 1 \end{pmatrix}$$

are all matrices. The numbers occurring in a matrix are called the *entries.*

Each matrix has a certain number of rows and a certain number of columns. A matrix with m rows and n columns is called an *$m \times n$ matrix.* Thus the first matrix above is a 2×2 matrix, the second a 2×3 matrix, the third a 3×3 matrix. The first and third matrices are called *square*; they have the same number of rows as columns. Here we will be working with square matrices as these are the only ones that have determinants.

We could give a definition of determinant that is applicable to all square matrices, but the definition is complicated and would serve little purpose at this point. Our interest here is in the 2×2 case and the 3×3 case. We begin with the 2×2 case.

(A.2.1)

> The *determinant* of the matrix
>
> $$\begin{pmatrix} a_1 & a_2 \\ b_1 & b_2 \end{pmatrix}$$
>
> is the number $a_1b_2 - a_2b_1$.

We have a special notation for the determinant. We change the parentheses of the matrix to vertical bars:

$$\text{Determinant of } \begin{pmatrix} a_1 & a_2 \\ b_1 & b_2 \end{pmatrix} = \begin{vmatrix} a_1 & a_2 \\ b_1 & b_2 \end{vmatrix} = a_1b_2 - a_2b_1.$$

Thus, for example,

$$\begin{vmatrix} 5 & 8 \\ 4 & 2 \end{vmatrix} = (5 \cdot 2) - (8 \cdot 4) = 10 - 32 = -22$$

$$\text{and} \quad \begin{vmatrix} 4 & 0 \\ 0 & \frac{1}{4} \end{vmatrix} = (4 \cdot \tfrac{1}{4}) - (0 \cdot 0) = 1.$$

We remark on three properties of 2×2 determinants:

1. If the rows or columns of a 2×2 determinant are interchanged, the determinant changes sign:

$$\begin{vmatrix} b_1 & b_2 \\ a_1 & a_2 \end{vmatrix} = -\begin{vmatrix} a_1 & a_2 \\ b_1 & b_2 \end{vmatrix}, \qquad \begin{vmatrix} a_2 & a_1 \\ b_2 & b_1 \end{vmatrix} = -\begin{vmatrix} a_1 & a_2 \\ b_1 & b_2 \end{vmatrix}.$$

PROOF Just note that

$$b_1a_2 - b_2a_1 = -(a_1b_2 - a_2b_1) \quad \text{and} \quad a_2b_1 - a_1b_2 = -(a_1b_2 - a_2b_1). \quad \square$$

2. A common factor can be removed from any row or column and placed as a multiplier in front of the determinant:

$$\begin{vmatrix} \lambda a_1 & \lambda a_2 \\ b_1 & b_2 \end{vmatrix} = \lambda \begin{vmatrix} a_1 & a_2 \\ b_1 & b_2 \end{vmatrix}, \qquad \begin{vmatrix} \lambda a_1 & a_2 \\ \lambda b_1 & b_2 \end{vmatrix} = \lambda \begin{vmatrix} a_1 & a_2 \\ b_1 & b_2 \end{vmatrix}.$$

PROOF Just note that

$$(\lambda a_1)b_2 - (\lambda a_2)b_1 = \lambda(a_1 b_2 - a_2 b_1)$$

$$\text{and} \quad (\lambda a_1)b_2 - a_2(\lambda b_1) = \lambda(a_1 b_2 - a_2 b_1). \quad \square$$

3. If the rows or columns of a 2×2 determinant are the same, the determinant is 0.

PROOF

$$\begin{vmatrix} a_1 & a_2 \\ a_1 & a_2 \end{vmatrix} = a_1 a_2 - a_2 a_1 = 0, \qquad \begin{vmatrix} a_1 & a_1 \\ b_1 & b_1 \end{vmatrix} = a_1 b_1 - a_1 b_1 = 0. \quad \square$$

The determinant of a 3×3 matrix is harder to define. One definition is this:

$$\begin{vmatrix} a_1 & a_2 & a_3 \\ b_1 & b_2 & b_3 \\ c_1 & c_2 & c_3 \end{vmatrix} = a_1 b_2 c_3 - a_1 b_3 c_2 + a_2 b_3 c_1 - a_2 b_1 c_3 + a_3 b_1 c_2 - a_3 b_2 c_1.$$

The problem with this definition is that it is hard to remember. What saves us is that the expansion on the right can be conveniently written in terms of 2×2 determinants; namely, the expression on the right can be written

$$a_1(b_2 c_3 - b_3 c_2) - a_2(b_1 c_3 - b_3 c_1) + a_3(b_1 c_2 - b_2 c_1),$$

which turns into

$$a_1 \begin{vmatrix} b_2 & b_3 \\ c_2 & c_3 \end{vmatrix} - a_2 \begin{vmatrix} b_1 & b_3 \\ c_1 & c_3 \end{vmatrix} + a_3 \begin{vmatrix} b_1 & b_2 \\ c_1 & c_2 \end{vmatrix}.$$

We then have

$$\text{(A.2.2)} \qquad \begin{vmatrix} a_1 & a_2 & a_3 \\ b_1 & b_2 & b_3 \\ c_1 & c_2 & c_3 \end{vmatrix} = a_1 \begin{vmatrix} b_2 & b_3 \\ c_2 & c_3 \end{vmatrix} - a_2 \begin{vmatrix} b_1 & b_3 \\ c_1 & c_3 \end{vmatrix} + a_3 \begin{vmatrix} b_1 & b_2 \\ c_1 & c_2 \end{vmatrix}.$$

We will take this as our definition. It is called the *expansion of the determinant along the first row*. Note that the coefficients are the entries a_1, a_2, a_3 of the first row, that they occur alternately with + and − signs, and that each is multiplied by a determinant. You can remember which determinant goes with which entry a_i as follows: in the original matrix, mentally cross out the row and column in which the entry a_i is found, and take the determinant of the remaining 2×2 matrix. For example, the determinant that goes with a_3 is

$$\begin{vmatrix} \not{a_1} & \not{a_2} & \not{a_3} \\ b_1 & b_2 & \not{b_3} \\ c_1 & c_2 & \not{c_3} \end{vmatrix} = \begin{vmatrix} b_1 & b_2 \\ c_1 & c_2 \end{vmatrix}.$$

When first starting to work with specific 3×3 determinants, it is a good idea to set up the formula with blank 2×2 determinants:

$$\begin{vmatrix} a_1 & a_2 & a_3 \\ b_1 & b_2 & b_3 \\ c_1 & c_2 & c_3 \end{vmatrix} = a_1 \begin{vmatrix} & \\ & \end{vmatrix} - a_2 \begin{vmatrix} & \\ & \end{vmatrix} + a_3 \begin{vmatrix} & \\ & \end{vmatrix}$$

and then fill in the 2×2 determinants by the "crossing out" rule explained above.

Example 1

$$\begin{vmatrix} 1 & 2 & 1 \\ 0 & 3 & 4 \\ 6 & 2 & 5 \end{vmatrix} = 1 \begin{vmatrix} 3 & 4 \\ 2 & 5 \end{vmatrix} - 2 \begin{vmatrix} 0 & 4 \\ 6 & 5 \end{vmatrix} + 1 \begin{vmatrix} 0 & 3 \\ 6 & 2 \end{vmatrix}$$
$$= 1(15 - 8) - 2(0 - 24) + 1(0 - 18)$$
$$= 7 + 48 - 18 = 37. \quad \square$$

A straightforward (but somewhat laborious) calculation shows that 3×3 determinants have the same three properties we proved earlier for 2×2 determinants:

1. If two rows or columns are interchanged, the determinant changes sign.
2. A common factor can be removed from any row or column and placed as a multiplier in front of the determinant.
3. If two rows or columns are the same, the determinant is 0.

EXERCISES A.2

Evaluate the following determinants.

1. $\begin{vmatrix} 1 & 2 \\ 3 & 4 \end{vmatrix}$.

2. $\begin{vmatrix} 1 & -1 \\ -1 & 1 \end{vmatrix}$.

3. $\begin{vmatrix} 1 & 1 \\ a & a \end{vmatrix}$.

4. $\begin{vmatrix} a & b \\ b & d \end{vmatrix}$.

5. $\begin{vmatrix} 1 & 0 & 3 \\ 2 & 4 & 1 \\ 0 & 1 & 0 \end{vmatrix}$.

6. $\begin{vmatrix} 1 & 0 & 0 \\ 0 & 2 & 0 \\ 0 & 0 & 3 \end{vmatrix}$.

7. $\begin{vmatrix} 0 & 0 & 1 \\ 0 & 2 & 0 \\ 3 & 0 & 0 \end{vmatrix}$.

8. $\begin{vmatrix} a & 0 & 0 \\ b & c & 0 \\ d & e & f \end{vmatrix}$.

9. If A is a matrix, its *transpose* A^T is obtained by interchanging the rows and columns. Thus

$$\begin{pmatrix} a_1 & a_2 \\ b_1 & b_2 \end{pmatrix}^T = \begin{pmatrix} a_1 & b_1 \\ a_2 & b_2 \end{pmatrix}$$

and
$$\begin{pmatrix} a_1 & a_2 & a_3 \\ b_1 & b_2 & b_3 \\ c_1 & c_2 & c_3 \end{pmatrix}^T = \begin{pmatrix} a_1 & b_1 & c_1 \\ a_2 & b_2 & c_2 \\ a_3 & b_3 & c_3 \end{pmatrix}.$$

Show that the determinant of a matrix equals the determinant of its transpose: (a) the 2×2 case. (b) the 3×3 case.

Verify the assertions in Exercises 10–14.

10. $\begin{vmatrix} 1 & 2 & 3 \\ 4 & 5 & 6 \\ 7 & 8 & 9 \end{vmatrix} + \begin{vmatrix} 4 & 5 & 6 \\ 1 & 2 & 3 \\ 7 & 8 & 9 \end{vmatrix} = 0.$

11. $\begin{vmatrix} 1 & 2 & 3 \\ 4 & 5 & 6 \\ 7 & 8 & 9 \end{vmatrix} = \begin{vmatrix} 4 & 5 & 6 \\ 7 & 8 & 9 \\ 1 & 2 & 3 \end{vmatrix}.$

12. $\begin{vmatrix} 1 & 2 & 3 \\ 4 & 5 & 6 \\ 7 & 8 & 9 \end{vmatrix} + \begin{vmatrix} 1 & 2 & 3 \\ 1 & 2 & 3 \\ 7 & 8 & 9 \end{vmatrix} = \begin{vmatrix} 1 & 2 & 3 \\ 4 & 5 & 6 \\ 7 & 8 & 9 \end{vmatrix}.$

13. $\frac{1}{2}\begin{vmatrix} 1 & 0 & 7 \\ 3 & 4 & 5 \\ 2 & 4 & 6 \end{vmatrix} = \begin{vmatrix} 1 & 0 & 7 \\ 3 & 4 & 5 \\ 1 & 2 & 3 \end{vmatrix}.$

14. $\begin{vmatrix} 1 & 2 & 3 \\ x & 2x & 3x \\ 4 & 5 & 6 \end{vmatrix} = 0.$

15. (a) Verify that the equations

$$3x + 4y = 6$$

$$2x - 3y = 7$$

are solved by the prescription

$$x = \frac{\begin{vmatrix} 6 & 4 \\ 7 & -3 \end{vmatrix}}{\begin{vmatrix} 3 & 4 \\ 2 & -3 \end{vmatrix}}, \qquad y = \frac{\begin{vmatrix} 3 & 6 \\ 2 & 7 \end{vmatrix}}{\begin{vmatrix} 3 & 4 \\ 2 & -3 \end{vmatrix}}.$$

(b) More generally, verify that the equations

$$a_1 x + a_2 y = d$$

$$b_1 x + b_2 y = e$$

are solved by the prescription

$$x = \frac{\begin{vmatrix} d & a_2 \\ e & b_2 \end{vmatrix}}{\begin{vmatrix} a_1 & a_2 \\ b_1 & b_2 \end{vmatrix}}, \qquad y = \frac{\begin{vmatrix} a_1 & d \\ b_1 & e \end{vmatrix}}{\begin{vmatrix} a_1 & a_2 \\ b_1 & b_2 \end{vmatrix}}$$

provided that the determinant in the denominator is different from 0.

(c) Conjecture an analogous rule for solving three linear equations in three unknowns.

16. Show that a 3×3 determinant can be "expanded along the bottom row" as follows:

$$\begin{vmatrix} a_1 & a_2 & a_3 \\ b_1 & b_2 & b_3 \\ c_1 & c_2 & c_3 \end{vmatrix} = c_1 \begin{vmatrix} a_2 & a_3 \\ b_2 & b_3 \end{vmatrix} - c_2 \begin{vmatrix} a_1 & a_3 \\ b_1 & b_3 \end{vmatrix} + c_3 \begin{vmatrix} a_1 & a_2 \\ b_1 & b_2 \end{vmatrix}$$

HINT: You can check this directly by writing out the values of the determinants on the right, or you can interchange rows twice to bring the bottom row to the top and then use expansion along the top row.

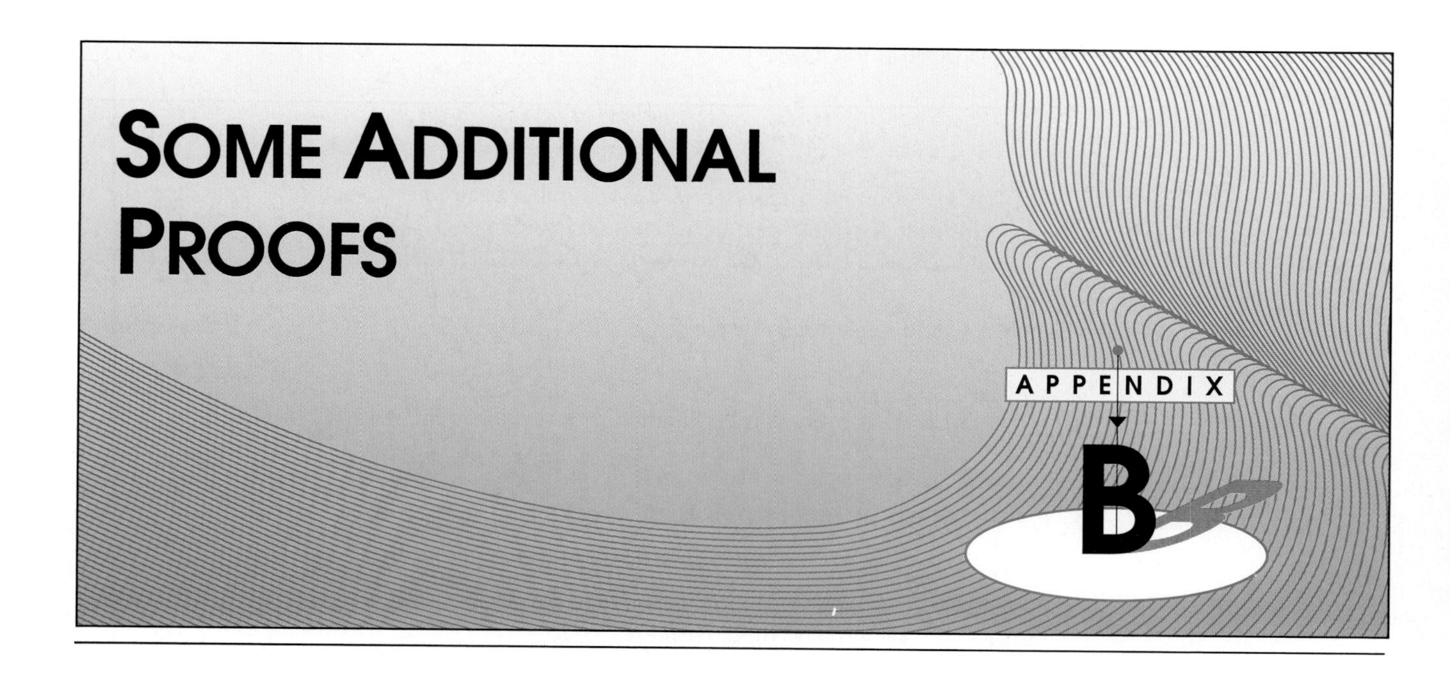

In this appendix we present some proofs that many would consider too advanced for the main body of the text. Some details are omitted. These are left to you.

The arguments presented in Sections B.1, B.2, and B.4 require some familiarity with the *least upper bound axiom.* This is discussed in Section 10.1. In addition, Section B.4 requires some understanding of *sequences,* for which we refer you to Sections 10.2 and 10.3.

B.1 THE INTERMEDIATE-VALUE THEOREM

LEMMA B.1.1

Let f be continuous on $[a, b]$. If $f(a) < 0 < f(b)$ or $f(b) < 0 < f(a)$, then there is a number c between a and b for which $f(c) = 0$.

PROOF Suppose that $f(a) < 0 < f(b)$. (The other case can be treated in a similar manner.) Since $f(a) < 0$, we know from the continuity of f that there exists a number ξ such that f is negative on $[a, \xi)$. Let

$$c = \text{lub}\ \{\xi: f \text{ is negative on } [a, \xi)\}.$$

Clearly, $c \le b$. We cannot have $f(c) > 0$, for then f would be positive on some interval extending to the left of c, and we know that, to the left of c, f is negative. Incidentally this argument excludes the possibility $c = b$ and means that $c < b$. We cannot have $f(c) < 0$, for then there would be an interval $[a, t)$, with $t > c$, on which f is negative, and this would contradict the definition of c. It follows that $f(c) = 0$. ❑

THEOREM B.1.2 THE INTERMEDIATE-VALUE THEOREM

If f is continuous on $[a, b]$ and C is a number between $f(a)$ and $f(b)$, then there is at least one number c between a and b for which $f(c) = C$.

PROOF Suppose for example that

$$f(a) < C < f(b).$$

(The other possibility can be handled in a similar manner.) The function

$$g(x) = f(x) - C$$

is continuous on $[a, b]$. Since

$$g(a) = f(a) - C < 0 \qquad \text{and} \qquad g(b) = f(b) - C > 0,$$

we know from the lemma that there is a number c between a and b for which $g(c) = 0$. Obviously, then, $f(c) = C$. ❑

■ B.2 THE MAXIMUM-MINIMUM THEOREM

LEMMA B.2.1

If f is continuous on $[a, b]$, then f is bounded on $[a, b]$.

PROOF Consider

$$\{x: x \in [a, b] \text{ and } f \text{ is bounded on } [a, x]\}.$$

It is easy to see that this set is nonempty and bounded above by b. Thus we can set

$$c = \text{lub } \{x: f \text{ is bounded on } [a, x]\}.$$

Now we argue that $c = b$. To do so, we suppose that $c < b$. From the continuity of f at c, it is easy to see that f is bounded on $[c - \epsilon, c + \epsilon]$ for some $\epsilon > 0$. Being bounded on $[a, c - \epsilon]$ and on $[c - \epsilon, c + \epsilon]$, it is obviously bounded on $[a, c + \epsilon]$. This contradicts our choice of c. We can therefore conclude that $c = b$. This tells us that f is bounded on $[a, x]$ for all $x < b$. We are now almost through. From the continuity of f, we know that f is bounded on some interval of the form $[b - \epsilon, b]$. Since $b - \epsilon < b$, we know from what we have just proved that f is bounded on $[a, b - \epsilon]$. Being bounded on $[a, b - \epsilon]$ and bounded on $[b - \epsilon, b]$, it is bounded on $[a, b]$. ❑

THEOREM B.2.2 THE MAXIMUM-MINIMUM THEOREM

If f is continuous on $[a, b]$, then f takes on both a maximum value M and a minimum value m on $[a, b]$.

PROOF By the lemma, f is bounded on $[a, b]$. Set

$$M = \text{lub}\,\{f(x) : x \in [a, b]\}.$$

We must show that there exists c in $[a, b]$ such that $f(c) = M$. To do this, we set

$$g(x) = \frac{1}{M - f(x)}.$$

If f does not take on the value M, then g is continuous on $[a, b]$ and thus, by the lemma, bounded on $[a, b]$. A look at the definition of g makes it clear that g cannot be bounded on $[a, b]$. The assumption that f does not take on the value M has led to a contradiction. (That f takes on a minimum value m can be proved in a similar manner.) ❑

■ B.3 INVERSES

> **THEOREM B.3.1 CONTINUITY OF THE INVERSE**
>
> Let f be a one-to-one function defined on an interval (a, b). If f is continuous, then its inverse f^{-1} is also continuous.

PROOF If f is continuous, then, being one-to-one, f either increases throughout (a, b) or it decreases throughout (a, b). The proof of this assertion we leave to you.

Let's suppose now that f increases throughout (a, b). Let's take c in the domain of f^{-1} and show that f^{-1} is continuous at c.

We first observe that $f^{-1}(c)$ lies in (a, b) and choose $\epsilon > 0$ sufficiently small so that $f^{-1}(c) - \epsilon$ and $f^{-1}(c) + \epsilon$ also lie in (a, b). We seek $\delta > 0$ such that

$$\text{if } c - \delta < x < c + \delta, \quad \text{then} \quad f^{-1}(c) - \epsilon < f^{-1}(x) < f^{-1}(c) + \epsilon.$$

This condition can be met by choosing δ to satisfy

$$f(f^{-1}(c) - \epsilon) < c - \delta \quad \text{and} \quad c + \delta < f(f^{-1}(c) + \epsilon)$$

for then, if $c - \delta < x < c + \delta$, then

$$f(f^{-1}(c) - \epsilon) < x < f(f^{-1}(c) + \epsilon),$$

and, since f^{-1} also increases,

$$f^{-1}(c) - \epsilon < f^{-1}(x) < f^{-1}(c) + \epsilon.$$

The case where f decreases throughout (a, b) can be handled in a similar manner. ❑

> **THEOREM B.3.2 DIFFERENTIABILITY OF THE INVERSE**
>
> Let f be a one-to-one function defined on an interval (a, b). If f is differentiable and its derivative does not take on the value 0, then f^{-1} is differentiable and
>
> $$(f^{-1})'(x) = \frac{1}{f'(f^{-1}(x))}.$$

PROOF (Here we use the characterization of derivative spelled out in Theorem 3.5.8.) Let x be in the domain of f^{-1}. We take $\epsilon > 0$ and show that there exists $\delta > 0$ such that

$$\text{if } 0 < |t - x| < \delta, \qquad \text{then} \qquad \left| \frac{f^{-1}(t) - f^{-1}(x)}{t - x} - \frac{1}{f'(f^{-1}(x))} \right| < \epsilon.$$

Since f is differentiable at $f^{-1}(x)$ and $f'(f^{-1}(x)) \neq 0$, there exists $\delta_1 > 0$ such that

$$\text{if } 0 < |y - f^{-1}(x)| < \delta_1, \qquad \text{then} \qquad \left| \frac{1}{\dfrac{f(y) - f(f^{-1}(x))}{y - f^{-1}(x)}} - \frac{1}{f'(f^{-1}(x))} \right| < \epsilon$$

and therefore

$$\left| \frac{y - f^{-1}(x)}{f(y) - f(f^{-1}(x))} - \frac{1}{f'(f^{-1}(x))} \right| < \epsilon.$$

By the previous theorem, f^{-1} is continuous at x and therefore there exists $\delta > 0$ such that

$$\text{if } 0 < |t - x| < \delta, \qquad \text{then} \qquad 0 < |f^{-1}(t) - f^{-1}(x)| < \delta_1.$$

It follows from the special property of δ_1 that

$$\left| \frac{f^{-1}(t) - f^{-1}(x)}{t - x} - \frac{1}{f'(f^{-1}(x))} \right| < \epsilon. \quad \square$$

■ B.4 THE INTEGRABILITY OF CONTINUOUS FUNCTIONS

The aim here is to prove that, if f is continuous on $[a, b]$, then there is one and only one number I that satisfies the inequality

$$L_f(P) \leq I \leq U_f(P) \qquad \text{for all partitions } P \text{ of } [a, b].$$

DEFINITION B.4.1

A function f is said to be *uniformly continuous* on $[a, b]$ iff for each $\epsilon > 0$ there exists $\delta > 0$ such that

$$\text{if } x, y \in [a, b] \text{ and } |x - y| < \delta, \qquad \text{then } |f(x) - f(y)| < \epsilon.$$

For convenience, let's agree to say that *the interval* $[a, b]$ *has the property* P_ϵ iff there exist sequences $\{x_n\}$, $\{y_n\}$ satisfying

$$x_n, y_n \in [a, b], \qquad |x_n - y_n| < 1/n, \qquad |f(x_n) - f(y_n)| \geq \epsilon.$$

LEMMA B.4.2

If f is not uniformly continuous on $[a, b]$, then $[a, b]$ has the property P_ϵ for some $\epsilon > 0$.

PROOF If f is not uniformly continuous on $[a, b]$, then there is no $\delta > 0$ such that

$$\text{if } x, y \in [a, b] \text{ and } |x - y| < \delta, \qquad \text{then } |f(x) - f(y)| < \epsilon.$$

The interval $[a, b]$ has the property P_ϵ for that choice of ϵ. The details of the argument are left to you. ❑

LEMMA B.4.3

Let f be continuous on $[a, b]$. If $[a, b]$ has the property P_ϵ, then at least one of the subintervals $[a, \frac{1}{2}(a + b)]$, $[\frac{1}{2}(a + b), b]$ has the property P_ϵ.

PROOF Let's suppose that the lemma is false. For convenience, we let $c = \frac{1}{2}(a + b)$, so that the halves become $[a, c]$ and $[c, b]$. Since $[a, c]$ fails to have the property P_ϵ, there exists an integer p such that

$$\text{if } x, y \in [a, c] \text{ and } |x - y| < 1/p, \qquad \text{then } |f(x) - f(y)| < \epsilon.$$

Since $[c, b]$ fails to have the property P_ϵ, there exists an integer q such that

$$\text{if } x, y \in [c, b] \text{ and } |x - y| < 1/q, \qquad \text{then} \qquad |f(x) - f(y)| < \epsilon.$$

Since f is continuous at c, there exists an integer r such that, if $|x - c| < 1/r$, then $|f(x) - f(c)| < \frac{1}{2}\epsilon$. Set $s = \max\{p, q, r\}$ and suppose that

$$x, y \in [a, b], \qquad |x - y| < 1/s.$$

If x, y are both in $[a, c]$ or both in $[c, b]$, then

$$|f(x) - f(y)| < \epsilon.$$

The only other possibility is that $x \in [a, c]$ and $y \in [c, b]$. In this case we have

$$|x - c| < 1/r, \qquad |y - c| < 1/r,$$

and thus

$$|f(x) - f(c)| < \tfrac{1}{2}\epsilon, \qquad |f(y) - f(c)| < \tfrac{1}{2}\epsilon.$$

By the triangle inequality, we again have

$$|f(x) - f(y)| < \epsilon.$$

In summary, we have obtained the existence of an integer s with the property that

$$x, y \in [a, b],\ |x - y| < 1/s \qquad \text{implies} \qquad |f(x) - f(y)| < \epsilon.$$

Hence $[a, b]$ does not have the property P_ϵ. This is a contradiction and proves the lemma. ❑

THEOREM B.4.4

If f is continuous on $[a, b]$, then f is uniformly continuous on $[a, b]$.

PROOF We suppose that f is not uniformly continuous on $[a, b]$ and base our argument on a mathematical version of "bisection."

By the first lemma of this section, we know that $[a, b]$ has the property P_ϵ for some $\epsilon > 0$. We bisect $[a, b]$ and note by the second lemma that one of the halves, say $[a_1, b_1]$, has the property P_ϵ. We then bisect $[a_1, b_1]$ and note that one of the halves, say $[a_2, b_2]$, has the property P_ϵ. Continuing in this manner, we obtain a sequence of intervals $[a_n, b_n]$, each with the property P_ϵ. Then for each n, we can choose x_n, $y_n \in [a_n, b_n]$ such that

$$|x_n - y_n| < 1/n \qquad \text{and} \qquad |f(x_n) - f(y_n)| \geq \epsilon.$$

Since

$$a \leq a_n \leq a_{n+1} < b_{n+1} \leq b_n \leq b,$$

we see that the sequences $\{a_n\}$ and $\{b_n\}$ are both bounded and monotonic. Thus they are convergent. Since $b_n - a_n \to 0$, we see that $\{a_n\}$ and $\{b_n\}$ both converge to the same limit, say L. From the inequality

$$a_n \leq x_n \leq y_n \leq b_n,$$

we conclude that

$$x_n \to L \quad \text{and} \quad y_n \to L.$$

This tells us that

$$|f(x_n) - f(y_n)| \to |f(L) - f(L)| = 0,$$

which contradicts the statement that $|f(x_n) - f(y_n)| \geq \epsilon$ for all n. ❑

LEMMA B.4.5

If P and Q are partitions of $[a, b]$, then $L_f(P) \leq U_f(Q)$.

PROOF $P \cup Q$ is a partition of $[a, b]$ that contains both P and Q. It is obvious then that

$$L_f(P) \leq L_f(P \cup Q) \leq U_f(P \cup Q) \leq U_f(Q). \quad ❑$$

From the last lemma it follows that the set of all lower sums is bounded above and has a least upper bound L. The number L satisfies the inequality

$$L_f(P) \leq L \leq U_f(P) \qquad \text{for all partitions } P$$

and is clearly the least of such numbers. Similarly, we find that the set of all upper sums is bounded below and has a greatest lower bound U. The number U satisfies the inequality

$$L_f(P) \leq U \leq U_f(P) \qquad \text{for all partitions } P$$

and is clearly the greatest of such numbers.

We are now ready to prove the basic theorem.

THEOREM B.4.6 THE INTEGRABILITY THEOREM

If f is continuous on $[a, b]$, then there exists one and only one number I that satisfies the inequality

$$L_f(P) \le I \le U_f(P) \qquad \text{for all partitions } P \text{ of } [a, b].$$

PROOF We know that

$$L_f(P) \le L \le U \le U_f(P) \qquad \text{for all } P,$$

so that existence is no problem. We will have uniqueness if we can prove that

$$L = U.$$

To do this, we take $\epsilon > 0$ and note that f, being continuous on $[a, b]$, is uniformly continuous on $[a, b]$. Thus there exists $\delta > 0$ such that, if

$$x, y \in [a, b] \text{ and } |x - y| < \delta, \qquad \text{then} \qquad |f(x) - f(y)| < \frac{\epsilon}{b - a}.$$

We now choose a partition $P = \{x_0, x_1, \ldots, x_n\}$ for which $\max \Delta x_i < \delta$. For this partition P, we have

$$\begin{aligned} U_f(P) - L_f(P) &= \sum_{i=1}^{n} M_i \,\Delta x_i - \sum_{i=1}^{n} m_i \,\Delta x_i \\ &= \sum_{i=1}^{n} (M_i - m_i) \,\Delta x_i \\ &< \sum_{i=1}^{n} \frac{\epsilon}{b - a} \Delta x_i = \frac{\epsilon}{b - a} \sum_{i=1}^{n} \Delta x_i = \frac{\epsilon}{b - a}(b - a) = \epsilon. \end{aligned}$$

Since

$$U_f(P) - L_f(P) < \epsilon \qquad \text{and} \qquad 0 \le U - L \le U_f(P) - L_f(P),$$

you can see that

$$0 \le U - L < \epsilon.$$

Since ϵ was chosen arbitrarily, we must have $U - L = 0$ and $L = U$. ❑

■ B.5 THE INTEGRAL AS THE LIMIT OF RIEMANN SUMS

For the notation we refer to Section 5.10.

THEOREM B.5.1

If f is continuous on $[a, b]$, then

$$\int_a^b f(x)\,dx = \lim_{\|P\| \to 0} S^*(P).$$

PROOF Let $\epsilon > 0$. We must show that there exists $\delta > 0$ such that

$$\text{if } \|P\| < \delta, \qquad \text{then} \qquad \left| S^*(P) - \int_a^b f(x)\, dx \right| < \epsilon.$$

From the proof of Theorem B.4.6 we know that there exists $\delta > 0$ such that

$$\text{if } \|P\| < \delta, \qquad \text{then} \qquad U_f(P) - L_f(P) < \epsilon.$$

For such P we have

$$U_f(P) - \epsilon < L_f(P) \leq S^*(P) \leq U_f(P) < L_f(P) + \epsilon.$$

This gives

$$\int_a^b f(x)\, dx - \epsilon < S^*(P) < \int_a^b f(x)\, dx + \epsilon,$$

and therefore

$$\left| S^*(P) - \int_a^b f(x)\, dx \right| < \epsilon. \quad \square$$

CHAPTER 1

SECTION 1.2

1. rational 3. rational 5. integer, rational 7. integer, rational 9. integer, rational 11. irrational 13. $\dfrac{13188}{999}$

15. $\dfrac{23}{99}$ 17. $\dfrac{5247}{999}$ 19. $=$ 21. $>$ 23. $<$ 25. 6 27. 4 29. 13 31. $5-\sqrt{5}$

33. $-\frac{3}{2}$ 0 x 35. -4 0 4 x 37. 0 x

39. 0 2 6 x 41. -3 0 x 43. 0 2 x

45. -2 0 5 x 47. 0 2 x 49. 0 x

51. bounded; lower bound 0, upper bound 4 53. not bounded 55. not bounded 57. bounded above, $\sqrt{2}$ is an upper bound.

59. $(x-5)^2$ 61. $8(x^2+2)(x^4-2x^2+4)$ 63. $(2x+3)^2$ 65. $2, -1$ 67. 3 69. none 71. $x=0$; $R(1)=0$

73. $R(x)$ is real for all x; $R(-1)=R(1)=0$ 75. $R(x)$ is real for all x; $R(-\frac{3}{2})=0$ 77. $\dfrac{1}{336}$ 79. 84

81. If r and $r+s$ are rational, then $s=(r+s)-r$ is rational.

83. The product could be either rational or irrational; $0\cdot\sqrt{2}=0$ is rational, $1\cdot\sqrt{2}=\sqrt{2}$ is irrational.

85. Suppose $\sqrt{2}=\dfrac{p}{q}$, where p, q are integers with no common divisor (other than ± 1). Then $p^2=2q^2$, which implies p^2 is even, which, in turn, implies $p=2r$ is even. Thus $4r^2=2q^2$, which implies q^2 and hence q are even. This is a contradiction.

87. If the length of a rectangle with perimeter P is x, $0<x<\frac{1}{2}P$, then the width is $\frac{1}{2}P-x$ and the area A is $A(x)=x(\frac{1}{2}P-x)=\left(\dfrac{P}{4}\right)^2-\left(x-\dfrac{P}{4}\right)^2$. Clearly A is a maximum when $x=P/4$. The width in this case is also $P/4$, and the rectangle is a square.

SECTION 1.3

1. $(-\infty, 1)$ 3. $(-\infty, -3]$ 5. $(-\infty, -\frac{1}{5})$ 7. $(-1, 1)$ 9. $(0,1)\cup(2,\infty)$ 11. $[0,\infty)$ 13. $(-\infty, 2-\sqrt{3})\cup(2+\sqrt{3},\infty)$

15. $(-\infty,0)\cup(0,\infty)$ 17. $(-1,0)\cup(1,\infty)$ 19. $(-\infty,0]\cup(5,\infty)$ 21. $(-\infty,-\frac{5}{3})\cup(5,\infty)$ 23. $(-3,-1)\cup(3,\infty)$

25. $(0,2)$ 27. $(-\infty,-6)\cup(2,\infty)$ 29. $(-2,0)\cup(2,\infty)$ 31. $(1,2)\cup(6,\infty)$ 33. $(-\infty,1)\cup(3,5)$

35. $(-\infty,0)\cup(0,1)\cup(3,\infty)$ 37. $(-2,2)$ 39. $(-\infty,-3)\cup(3,\infty)$ 41. $(\frac{3}{2},\frac{5}{2})$ 43. $(-1,0)\cup(0,1)$ 45. $(\frac{3}{2},2)\cup(2,\frac{5}{2})$

47. $(-5,3)\cup(3,11)$ 49. $(-\frac{5}{8},-\frac{3}{8})$ 51. $(-\infty,-4)\cup(-1,\infty)$ 53. $(-\infty,-\frac{8}{5})\cup(2,\infty)$ 55. $|x|<3$ 57. $|x-2|<5$

59. $|x+2|<5$ 61. $0<A\le\frac{3}{2}$ 63. $A\ge 6$ 65. $x<\sqrt{x}<1<\dfrac{1}{\sqrt{x}}<\dfrac{1}{x}$

67. $a<b$ and $ab>0$ implies

$$\frac{1}{b}-\frac{1}{a}=\frac{a-b}{ab}<0.$$

The result follows.

69. $b-a=(\sqrt{b}+\sqrt{a})(\sqrt{b}-\sqrt{a})$; since $\sqrt{b}+\sqrt{a}\ge 0$, $b-a$ and $\sqrt{b}-\sqrt{a}$ have the same sign.

71. $\big||a|-|b|\big|^2=(|a|-|b|)^2=|a|^2-2|a|\,|b|+|b|^2=a^2-2|a|\,|b|+b^2\le a^2-2ab+b^2=(a-b)^2$

↑—— $(ab\le|ab|)$

Thus, $\big||a|-|b|\big|\le\sqrt{(a-b)^2}=|a-b|$.

73. $0 \le a \le b$ implies

$$\frac{a}{1+a} - \frac{b}{1+b} = \frac{a-b}{(1+a)(1+b)} < 0.$$

The result follows.

75. $a < b$ implies

$$a - \frac{a+b}{2} = \frac{a-b}{2} < 0$$

and

$$\frac{a+b}{2} - b = \frac{a-b}{2} < 0.$$

Thus

$$a < \frac{a+b}{2} < b;$$

$(a+b)/2$ is the midpoint of the line segment $\overline{ab}$.

SECTION 1.4

1. 10 **3.** $4\sqrt{5}$ **5.** 6 **7.** (4, 6) **9.** $(\frac{9}{2}, -3)$ **11.** $\left(\frac{\sqrt{3}}{2}, \frac{\sqrt{3}}{2}\right)$ **13.** $-\frac{2}{3}$ **15.** 0 **17.** -1 **19.** $-y_0/x_0$

21. slope 2, y-intercept -4 **23.** slope $\frac{1}{3}$, y-intercept 2 **25.** slope undefined, no y-intercept **27.** slope $\frac{7}{3}$, y-intercept $\frac{4}{3}$

29. $y = 5x + 2$ **31.** $y = -5x + 2$ **33.** $y = 3$ **35.** $x = -3$ **37.** $y = 7$ **39.** $3y - 2x - 17 = 0$ **41.** $2y + 3x - 20 = 0$

43. 45° **45.** 90° **47.** approx. 143° **49.** $y = \frac{1}{3}\sqrt{3}\,x + 2$ **51.** $y = -\sqrt{3}\,x + 3$ **53.** $(\frac{1}{2}\sqrt{2}, \frac{1}{2}\sqrt{2}), (-\frac{1}{2}\sqrt{2}, -\frac{1}{2}\sqrt{2})$

55. (3, 4) **57.** (1, 1); approx. 39° **59.** $(-\frac{2}{23}, \frac{38}{23})$; approx. 17° **61.** (a) $\frac{2}{13}$ (b) $\frac{29}{13}$ **63.** (0, 1) is closest; $(-1, 1)$ is farthest away

65. $\frac{17}{2}$ **67.** parabola, vertex $(-1, -1)$ **69.** ellipse, center $(2, -1)$ **71.** hyperbola, center (1, 2) **73.** hyperbola, center $(3, -2)$

75. $-\frac{5}{12}$ **77.** $x - 2y - 3 = 0$ **79.** $3x + 13y - 40 = 0$ **81.** isosceles right triangle **83.** isosceles right triangle

85. The midpoint of the hypotenuse is $M(\frac{a}{2}, \frac{b}{2})$. The distance between M and the origin, M and $(0, b)$, and M and $(a, 0)$ is $\frac{1}{2}\sqrt{a^2 + b^2}$

87. $(1, \frac{10}{3})$ **89.** If $A\ (0, 0)$ and $B\ (a, 0)$ are two vertices of a parallelogram, and $C\ (b, c)$ is the vertex opposite B, then $D\ (a + b, c)$ is the vertex opposite A. The midpoint of the diagonal AD = midpoint of the diagonal BC = $(\frac{1}{2}(a + b, \frac{1}{2}c)$.

91. $F = \frac{9}{5}C + 32$; $-40°$

SECTION 1.5

1. $f(0) = 2$, $f(1) = 1$, $f(-2) = 16$, $f(\frac{3}{2}) = 2$ **3.** $f(0) = 0$, $f(1) = \sqrt{3}$, $f(-2) = 0, f(\frac{3}{2}) = \frac{\sqrt{21}}{2}$

5. $f(0) = 0$, $f(1) = \frac{1}{2}$, $f(-2) = -1$, $f(\frac{3}{2}) = \frac{12}{23}$ **7.** $f(-x) = x^2 + 2x$, $f\left(\frac{1}{x}\right) = \frac{1}{x^2} - \frac{2}{x}$, $f(a + b) = (a + b)^2 - 2\,(a + b)$

9. $f(-x) = \sqrt{1 + x^2}$, $f\left(\frac{1}{x}\right) = \frac{|x|}{\sqrt{1 + x^2}}$, $f(a + b) = \sqrt{a^2 + 2ab + b^2 + 1}$ **11.** $2a^2 + 4ah + 2h^2 - 3a - 3h$; $4a - 3 + 2h$

13. 1, 3 **15.** -2 **17.** 3, -3 **19.** dom $(f) = (-\infty, \infty)$; range $(f) = [0, \infty)$ **21.** dom $(f) = (-\infty, \infty)$; range $(f) = (-\infty, \infty)$

23. dom $(f) = (-\infty, 0) \cup (0, \infty)$; range $(f) = (0, \infty)$ **25.** dom $(f) = (-\infty, 1]$; range $(f) = [0, \infty)$

27. dom $(f) = (-\infty, 7]$; range $(f) = [-1, \infty)$ **29.** dom $(f) = (-\infty, 2)$; range $(f) = (0, \infty)$

31. horizontal line one unit above x-axis **33.** line through the origin with slope 2 **35.** line through the origin with slope $\frac{1}{2}$

37. line through (0, 2) with slope $\frac{1}{2}$ **39.** upper semicircle of radius 2 centered at the origin

41.

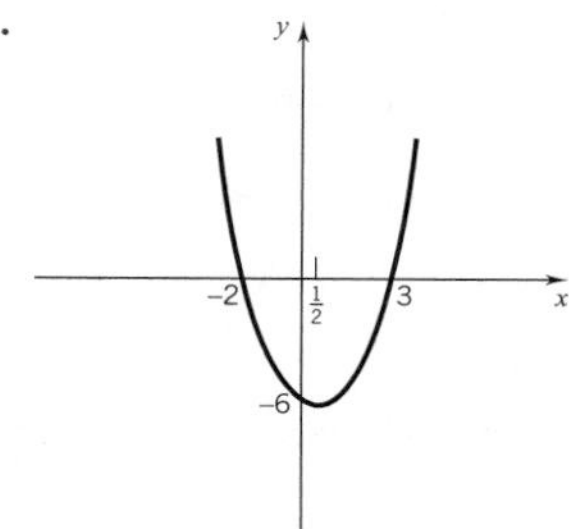

43.

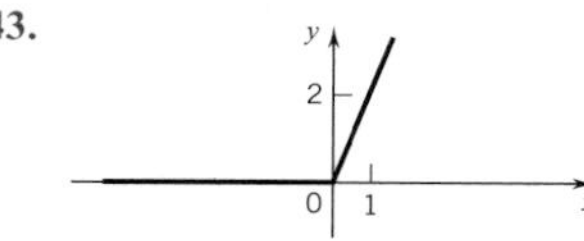

45. dom $(f) = (-\infty, 0) \cup (0, \infty)$; range $(f) = \{-1, 1\}$

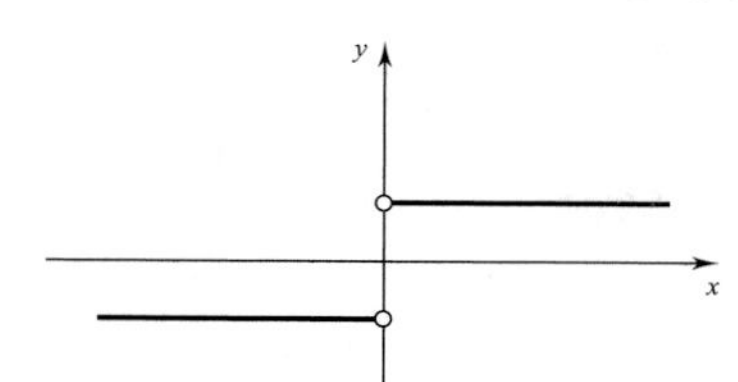

47. dom $(f) = [0, \infty)$; range $(f) = [1, \infty)$

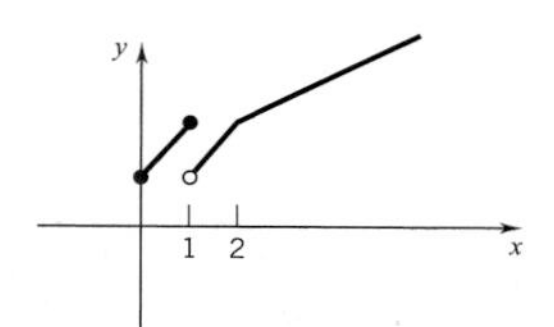

49. yes, dom$(f) = [-2, 2]$; range $(f) = [-2, 2]$ **51.** no **53.** odd **55.** neither **57.** even

59. (a)

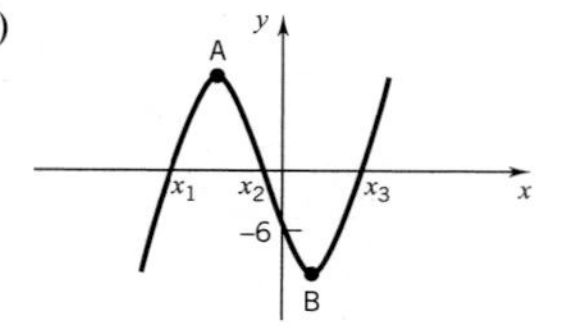

(b) $x_1 = -6.566, x_2 = -0.493, x_3 = 5.559$
(c) $A(-4, 28.667), B(3, 28.500)$

61.

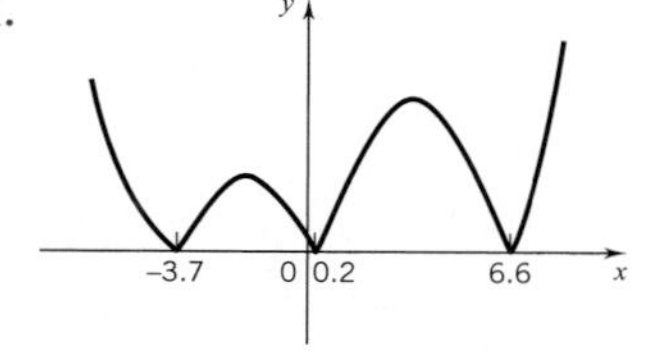

$-5 \le x \le 8,\ 0 \le y \le 100$

63. $A = \dfrac{C^2}{4\pi}$, where C is the circumference; dom $(A) = [0, \infty)$ **65.** $V = s^{3/2}$, where s is the area of a face; dom $V = [0, \infty)$

67. $S = 3d^2$, where d is the diagonal of a face; dom $(S) = [0, \infty)$ **69.** $A = \dfrac{\sqrt{3}}{4}x^2$, where x is the length of a side; dom $(A) = [0, \infty)$

71. $V = 108x^2 - 4x^3,\ 0 < x \le 27$ **73.** $A = \dfrac{15x}{2} - \dfrac{x^2}{2} + \dfrac{\pi x^2}{8},\ 0 < x < \dfrac{30}{\pi + 2}$ **75.** $A = x\sqrt{d^2 - x^2},\ 0 < x < d$

77. $A = bx - \dfrac{b}{a}x^2,\ 0 \le x \le a$ **79.** $A = \dfrac{P^2}{16} + \dfrac{(28 - P)^2}{4\pi},\ 0 \le P \le 28$

SECTION 1.6

1. polynomial, degree 0 **3.** rational function **5.** neither **7.** neither **9.** neither

11. dom $(f) = (-\infty, \infty)$

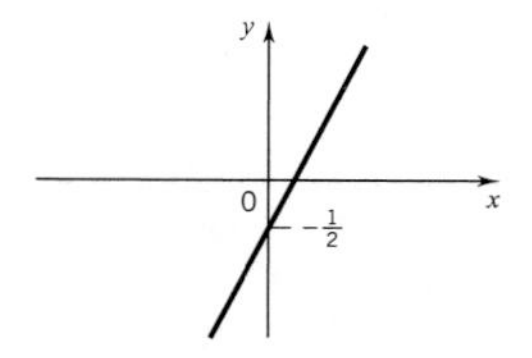

13. dom $(f) = (-\infty, \infty)$

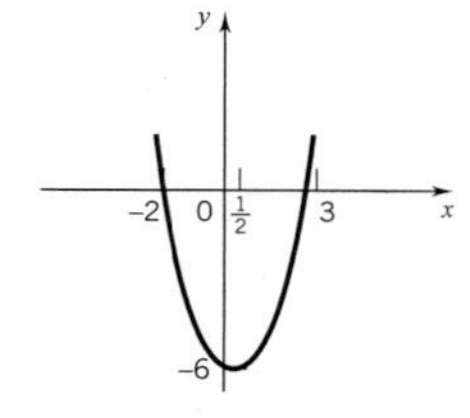

15. dom $(f) = (-\infty, -2) \cup (-2, 2) \cup (2, \infty)$

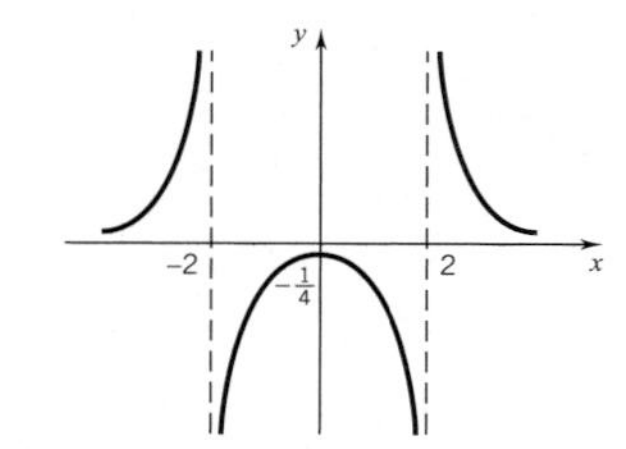

17. $\dfrac{5\pi}{4}$ **19.** $-\dfrac{5\pi}{3}$ **21.** $\dfrac{\pi}{12}$ **23.** $-270°$ **25.** $300°$ **27.** $114.59°$ **29.** $\dfrac{\pi}{6}, \dfrac{5\pi}{6}$ **31.** $\dfrac{\pi}{2}$ **33.** $\dfrac{\pi}{4}, \dfrac{7\pi}{4}$

35. $\dfrac{\pi}{4}, \dfrac{3\pi}{4}, \dfrac{5\pi}{4}, \dfrac{7\pi}{4}$ **37.** 0.7772 **39.** 0.7101 **41.** 3.1524 **43.** -3.8611 **45.** -2.8974 **47.** 0.5505 **49.** 1.4231

51. 1.7997 **53.** dom$(f) = (-\infty, \infty)$; range $(f) = [0, 1]$ **55.** dom$(f) = (-\infty, \infty)$; range $(f) = [-2, 2]$

57. dom $(f) = \left(k\pi - \dfrac{\pi}{2}, k\pi + \dfrac{\pi}{2}\right)$, $k = 0, \pm 1, \pm 2, \ldots$; range $(f) = [1, \infty)$

59.

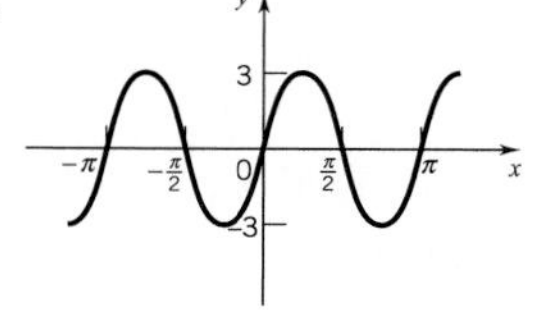

61.

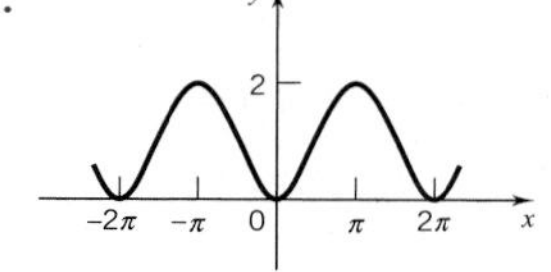

63.

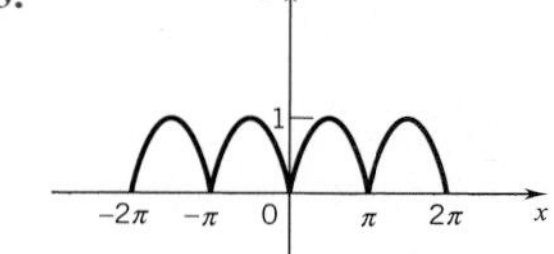

65. odd **67.** even **69.** odd **71.** Let $m_1 = \tan\theta_1$, $m_2 = \tan\theta_2$, $\alpha = |\theta_2 - \theta_1|$

$$\tan\alpha = |\tan(\theta_2 - \theta_1)| = \left|\frac{\tan\theta_2 - \tan\theta_1}{1 + \tan\theta_2\tan\theta_1}\right| = \left|\frac{m_2 - m_1}{1 + m_2 m_1}\right|$$

73. $b\sin A = a\sin B = h$, so $\dfrac{\sin A}{a} = \dfrac{\sin B}{b}$; similarly, $\dfrac{\sin A}{a} = \dfrac{\sin C}{c}$

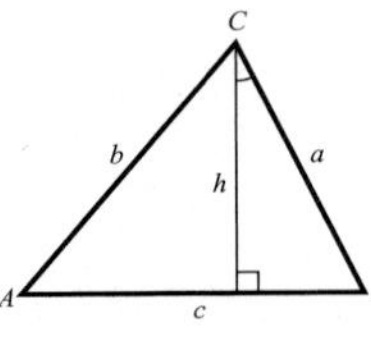

75. $A = \frac{1}{2}ah = \frac{1}{2}a^2 \sin\theta$

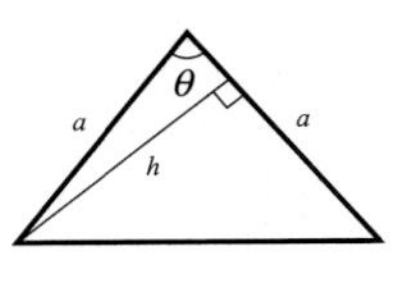

77.

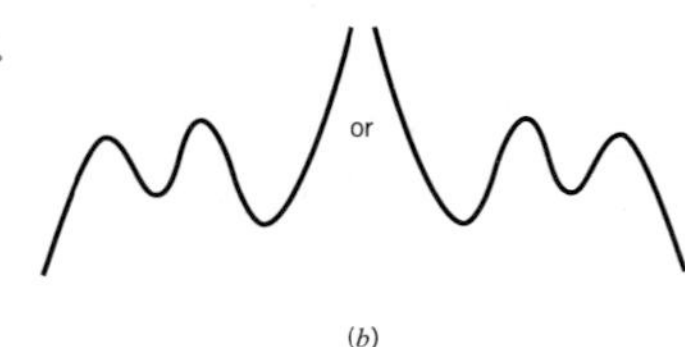

(b)

79.

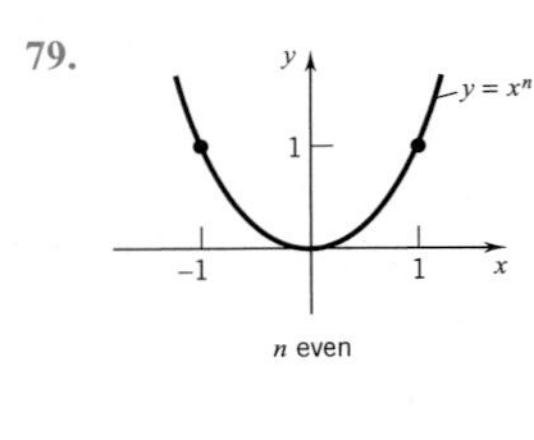

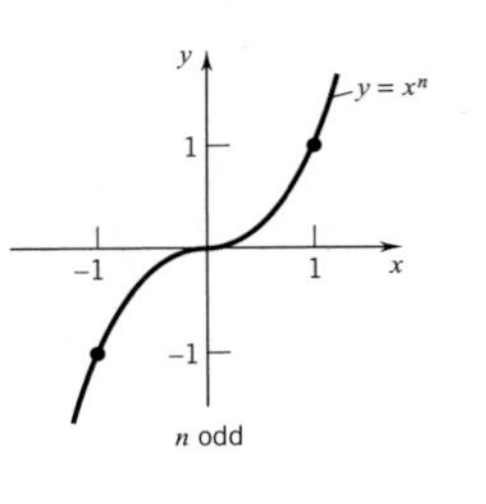

(b)

(c) $f_k(x) \geq f_{k+1}(x)$ on $[0, 1]$; $f_{k+1}(x) \geq f_k(x)$ on $[1, \infty)$

SECTION 1.7

1. $\frac{15}{2}$ **3.** $\frac{105}{2}$ **5.** $\frac{-27}{4}$ **7.** 3

9. $(f+g)(x) = x - 1$; domain $(-\infty, \infty)$
$(f-g)(x) = 3x - 5$; domain $(-\infty, \infty)$
$(f\cdot g)(x) = -2x^2 + 7x - 6$; domain $(-\infty, \infty)$
$\left(\dfrac{f}{g}\right)(x) = \dfrac{2x-3}{2-x}$; domain: all real numbers except $x = 2$

11. $(f+g)(x) = x + \sqrt{x-1} - \sqrt{x+1}$; domain $[1, \infty)$
$(f-g)(x) = \sqrt{x-1} + \sqrt{x+1} - x$; domain $[1, \infty)$
$(f\cdot g)(x) = \sqrt{x-1}\,(x - \sqrt{x+1}) = x\sqrt{x-1} - \sqrt{x^2-1}$; domain $[1, \infty)$
$\left(\dfrac{f}{g}\right)(x) = \dfrac{\sqrt{x-1}}{x - \sqrt{x+1}}$; domain $\left[1, \dfrac{1+\sqrt{5}}{2}\right) \cup \left(\dfrac{1+\sqrt{5}}{2}, \infty\right)$

13. (a) $(6f+3g)(x) = 6x - \dfrac{6}{\sqrt{x}}$, $x > 0$ (b) $(f-g)(x) = x + \dfrac{3}{\sqrt{x}} - \sqrt{x}$, $x > 0$ (c) $(f/g)(x) = \dfrac{x\sqrt{x}+1}{x-2}$, $x > 0$, $x \neq 2$.

15.

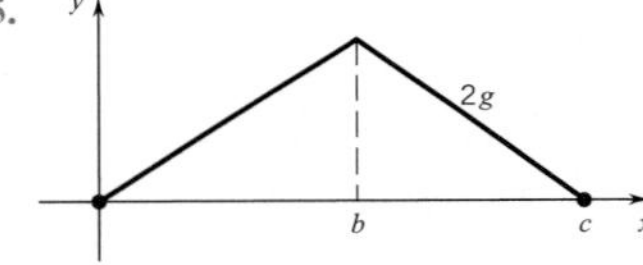

17.

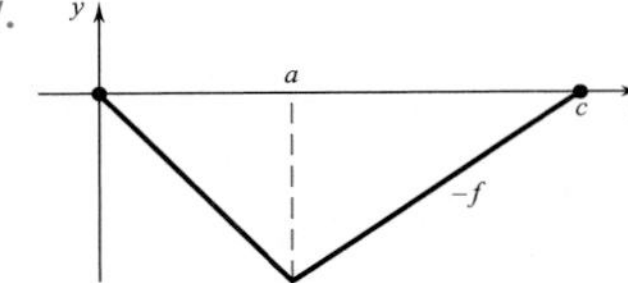

19.

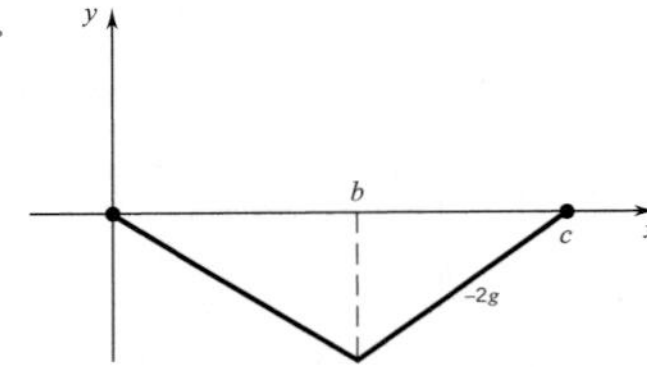

21.

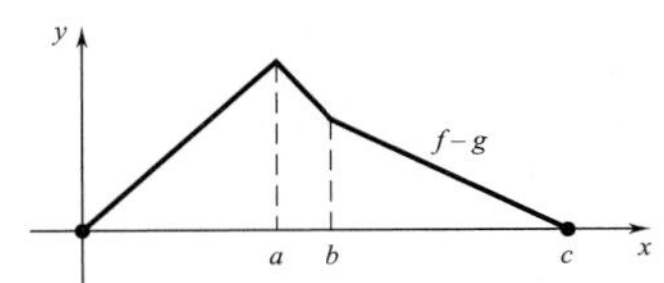

23. $(f\circ g)(x) = 2x^2 + 5$; domain $(-\infty, \infty)$ **25.** $(f\circ g)(x) = \sqrt{x^3+5}$; domain $(-\infty, \infty)$

27. $(f\circ g)(x) = \dfrac{x}{x-2}$; domain: all real numbers except $x = 0$, $x = 2$ **29.** $(f\circ g)(x) = \dfrac{1}{|x^2-1| - 3}$; domain: all real numbers except $x = \pm 2$

31. $(f\circ g)(x) = |\sin 2x|$; domain $(-\infty, \infty)$ **33.** $(f\circ g\circ h)(x) = 4(x^2 - 1)$; domain $(-\infty, \infty)$ **35.** $(f\circ g\circ h)(x) = 2x^2 + 1$; domain $(-\infty, \infty)$

37. $f(x) = \dfrac{1}{x}$ **39.** $f(x) = 2\sin x$ **41.** $g(x) = \left(1 - \dfrac{1}{x^4}\right)^{2/3}$ **43.** $g(x) = 2x^3 - 1$ **45.** $(f\circ g)(x) = |x|$; $(g\circ f)(x) = x$

47. $(f\circ g)(x) = \cos^2 x$; $(g\circ f)(x) = \sin(1 - x^2)$ **49.** $(f\circ g)(x) = x$; $(g\circ f)(x) = x$

51. fg is an even function since $(fg)(-x) = f(-x)g(-x) = f(x)g(x) = (fg)(x)$ **53.** (a) $f(x) = \begin{cases} -x, & -1 \leq x < 0 \\ 1, & x < -1 \end{cases}$ (b) $f(x) = \begin{cases} x, & -1 \leq x < 0 \\ -1 & x < -1 \end{cases}$

55. $g(-x) = f(-x) + f(x) = f(x) + f(-x) = g(x)$

57.

	f_1	f_2	f_3	f_4	f_5	f_6
f_1	f_1	f_2	f_3	f_4	f_5	f_6
f_2	f_2	f_1	f_4	f_3	f_6	f_5
f_3	f_3	f_5	f_1	f_6	f_2	f_4
f_4	f_4	f_6	f_2	f_5	f_1	f_3
f_5	f_5	f_3	f_6	f_1	f_4	f_2
f_6	f_6	f_4	f_5	f_2	f_3	f_1

59. $f(g(x)) = x$ and $g(f(x)) = x$

61. $f(g(x)) = x$ and $g(f(x)) = x$

63. (a) Varying a varies the x-coordinate of the vertex (a, b) of the parabola.
(b) Varying b varies the y-coordinate of the vertex (a, b) of the parabola.

65. (a) For $a > 0$ $(a < 0)$, the graph of $f(x - a)$ is the graph of $f(x)$ shifted horizontally $|a|$ units to the right (left).
(b) For $b > 1$, the graph of $f(bx)$ is compressed horizontally.
For $0 < b < 1$, the graph of $f(bx)$ is stretched horizontally.
For $-1 < b < 0$, the graph of $f(bx)$ is stretched horizontally and reflected in the y-axis.
For $b < -1$, the graph of $f(bx)$ is compressed horizontally and reflected in the y-axis.
(c) The graph of $f(x) + c$ is the graph of $f(x)$ shifted $|c|$ units up if $c > 0$ and shifted $|c|$ units down if $c < 0$.

67. (a) For $A > 0$, the graph of Af is the graph of f scaled vertically by the factor A.
For $A < 0$, the graph of Af is the graph of f scaled vertically by the factor $|A|$ and then reflected in the x-axis.
(b) See 65(b).

SECTION 1.8

1. Let S be the set of integers for which the statement is true. Since $2(1) \leq 2^1$, S contains 1. Assume now that $k \in S$. This tells us that $2k \leq 2^k$, and thus

$$2(k+1) = 2k + 2 \leq 2^k + 2 \leq 2^k + 2^k = 2(2^k) = 2^{k+1}.$$
$$\uparrow\!\!\!\text{———}\ (k \geq 1)$$

This places $k + 1$ in S.
We have shown that

$$1 \in S \quad \text{and that} \quad k \in S \ \text{ implies } \ k + 1 \in S.$$

It follows that S contains all the positive integers.

3. Let S be the set of integers for which the statement is true. Since $(1)(2) = 2$ is divisible by 2, $1 \in S$.
Assume now that $k \in S$. This tells us that $k(k + 1)$ is divisible by 2 and therefore

$$(k+1)(k+2) = k(k+1) + 2(k+1)$$

is also divisible by 2. This places $k + 1 \in S$.
We have shown that

$$1 \in S \quad \text{and that} \quad k \in S \ \text{ implies } \ k + 1 \in S.$$

It follows that S contains all the positive integers.

5. Use
$$\begin{aligned}1^2 + 2^2 + \cdots + k^2 + (k+1)^2 &= \tfrac{1}{6}k(k+1)(2k+1) + (k+1)^2\\ &= \tfrac{1}{6}(k+1)[k(2k+1) + 6(k+1)]\\ &= \tfrac{1}{6}(k+1)(2k^2 + 7k + 6)\\ &= \tfrac{1}{6}(k+1)(k+2)(2k+3) = \tfrac{1}{6}(k+1)[(k+1)+1][2(k+1)+1].\end{aligned}$$

7. By Exercise 6 and Example 1.

$$1^3 + 2^3 + \cdots + (n-1)^3 = [\tfrac{1}{2}(n-1)n]^2 = \tfrac{1}{4}(n-1)^2n^2 < \tfrac{1}{4}n^4$$

and

$$1^3 + 2^3 + \cdots + n^3 = [\tfrac{1}{2}n(n+1)]^2 = \tfrac{1}{4}n^2(n+1)^2 > \tfrac{1}{4}n^4.$$

9. Use $\dfrac{1}{\sqrt{1}} + \dfrac{1}{\sqrt{2}} + \dfrac{1}{\sqrt{3}} + \cdots + \dfrac{1}{\sqrt{n}} + \dfrac{1}{\sqrt{n+1}} > \sqrt{n} + \dfrac{1}{\sqrt{n+1}} > \sqrt{n} + \dfrac{1}{\sqrt{n+1}+\sqrt{n}}\left(\dfrac{\sqrt{n+1}-\sqrt{n}}{\sqrt{n+1}-\sqrt{n}}\right) = \sqrt{n+1}.$

11. Let S be the set of integers for which the statement is true. Since

$$3^{2(1)+1} + 2^{1+2} = 27 + 8 = 35$$

is divisible by 7, $1 \in S$.
Assume now that $k \in S$. This tells us that

$$3^{2k+1} + 2^{k+2} \text{ is divisible by 7.}$$

It follows that

$$3^{2(k+1)+1} + 2^{(k+1)+2} = 3^2 \cdot 3^{2k+1} + 2 \cdot 2^{k+2} = 9 \cdot 3^{2k+1} + 2 \cdot 2^{k+2} = 7 \cdot 3^{2k+1} + 2(3^{2k+1} + 2^{k+2})$$

is also divisible by 7. This places $k + 1 \in S$.

We have shown that

$$1 \in S \quad \text{and that} \quad k \in S \quad \text{implies} \quad k+1 \in S.$$

It follows that S contains all the positive integers.

13. For all positive integers $n \geq 2$,

$$\left(1 - \frac{1}{2}\right)\left(1 - \frac{1}{3}\right) \cdots \left(1 - \frac{1}{n}\right) = \frac{1}{n}.$$

To see this let S be the set of integers n for which the formula holds. Since $1 - \frac{1}{2} = \frac{1}{2}$, $2 \in S$. Suppose now that $k \in S$. This tells us that

$$\left(1 - \frac{1}{2}\right)\left(1 - \frac{1}{3}\right) \cdots \left(1 - \frac{1}{k}\right) = \frac{1}{k}$$

and therefore that

$$\left(1 - \frac{1}{2}\right)\left(1 - \frac{1}{3}\right) \cdots \left(1 - \frac{1}{k}\right)\left(1 - \frac{1}{k+1}\right) = \frac{1}{k}\left(1 - \frac{1}{k+1}\right) = \frac{1}{k}\left(\frac{k}{k+1}\right) = \frac{1}{k+1}.$$

This places $k + 1$ in S and verifies the formula for $n \geq 2$.

15. From the figure, observe that adding a vertex V_{N+1} to an N-sided polygon increases the number of diagonals by $(N - 2) + 1 = N - 1$.

Then use the identity

$$\tfrac{1}{2}N(N - 3) + (N - 1) = \tfrac{1}{2}(N + 1)(N + 1 - 3).$$

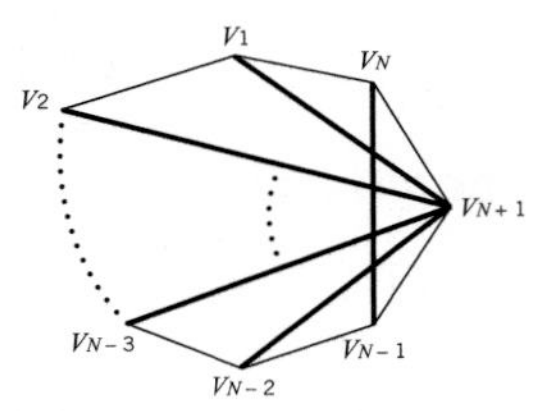

17. To go from k to $k + 1$, take $A = \{a_1, \ldots, a_{k+1}\}$ and $B = \{a_1, \ldots, a_k\}$. Assume that B has 2^k subsets: $B_1, B_2, \ldots, B_{2^k}$. The subsets of A are then $B_1, B_2, \ldots, B_{2^k}$ together with

$$B_1 \cup \{a_{k+1}\}, B_2 \cup \{a_{k+1}\}, \ldots, B_{2^k} \cup \{a_{k+1}\}.$$

This gives $2(2^k) = 2^{k+1}$ subsets for A.

CHAPTER 2

SECTION 2.1

1. (a) 2 (b) -1 (c) does not exist (d) -3 3. (a) does not exist (b) -3 (c) does not exist (d) -3
5. (a) does not exist (b) does not exist (c) does not exist (d) 1 7. (a) 2 (b) 2 (c) 2 (d) -1
9. (a) -1 (b) -1 (c) -1 (d) undefined 11. (a) 0 (b) 0 (c) 0 (d) 0 13. $c = 0, 6$ 15. -1 17. 4 19. 1
21. $\frac{3}{2}$ 23. does not exist 25. 2 27. does not exist 29. 1 31. does not exist 33. 2 35. 2 37. 0 39. 1
41. 16 43. does not exist 45. 4 47. 4 49. $1/\sqrt{2}$ 51. 4; $y = 4x - 4$ 53. -4; $y = -4x$
55. $\frac{1}{2}$; $y = \frac{1}{2}x + \frac{1}{2}$ 57. does not exist 59. (b) the limits do not exist (c)

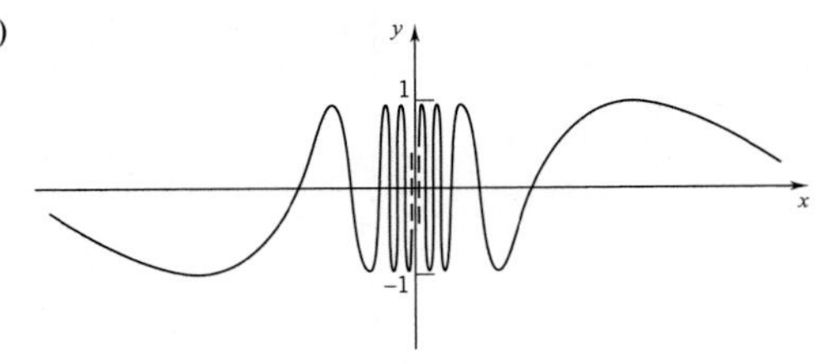

61. 2 63. $\frac{3}{2}$ 65. 2.71828 . . .

SECTION 2.2

1. $\frac{1}{2}$ 3. does not exist 5. $\frac{4}{5}\sqrt{5}$ 7. 4 9. does not exist 11. -1 13. does not exist 15. 0 17. 2 19. 1
21. 1 23. δ_1 and δ_2 25. $\frac{1}{2}\epsilon$ 27. 2ϵ 29. Take $\delta = \frac{1}{2}\epsilon$. If $0 < |x - 4| < \frac{1}{2}\epsilon$, then $|(2x - 5) - 3| = 2|x - 4| < \epsilon$.
31. Take $\delta = \frac{1}{6}\epsilon$. If $0 < |x - 3| < \frac{1}{6}\epsilon$, then $|(6x - 7) - 11| = 6|x - 3| < \epsilon$.
33. Take $\delta = \frac{1}{3}\epsilon$. If $0 < |x - 2| < \frac{1}{3}\epsilon$, then $\big||1 - 3x| - 5\big| \leq 3|x - 2| < \epsilon$. 35. Statements (b), (e), (g), and (i) are necessarily true.

37. (i) $\lim_{x\to 3} \frac{1}{x-1} = \frac{1}{2}$ (ii) $\lim_{h\to 0} \frac{1}{(3+h)-1} = \frac{1}{2}$ (iii) $\lim_{x\to 3} \left(\frac{1}{x-1} - \frac{1}{2}\right) = 0$ (iv) $\lim_{x\to 3} \left|\frac{1}{x-1} - \frac{1}{2}\right| = 0$

39. 5 **41.** $4 + 10x$ **43.** $-\frac{1}{(x+1)^2}$ **45.** (i) and (iv) of (2.2.5) with $L = 0$ **47.** $\delta = 0.001$ **49.** $\delta = 0.04$

51. Let $\epsilon > 0$. If $\lim_{x\to c} f(x) = L$, then there must exist $\delta > 0$ such that

$(*)$ if $0 < |x - c| < \delta$ then $|f(x) - L| < \epsilon$.

Suppose now that $0 < |h| < \delta$. Then $0 < |(c + h) - c| < \delta$, and thus by $(*)$, $|f(c + h) - L| < \epsilon$. This proves that, if $\lim_{x\to c} f(x) = L$, then $\lim_{h\to 0} f(c + h) = L$.

If, on the other hand, $\lim_{h\to 0} f(c + h) = L$, then there must exist $\delta > 0$ such that

$(**)$ if $0 < |h| < \delta$ then $|f(c + h) - L| < \epsilon$.

Suppose now that $0 < |x - c| < \delta$. Then by $(**)$, $|f(c + (x - c)) - L| < \epsilon$. More simply stated, $|f(x) - L| < \epsilon$. This proves that, if $\lim_{h\to 0} f(c + h) = L$, then $\lim_{x\to c} f(x) = L$.

53. (a) Set $\delta = \epsilon\sqrt{c}$. By the hint

$$\text{if } 0 < |x - c| < \epsilon\sqrt{c}, \quad \text{then } |\sqrt{x} - \sqrt{c}| < \frac{1}{\sqrt{c}}|x - c| < \epsilon.$$

(b) Set $\delta = \epsilon^2$. If $0 < x < \epsilon^2$, then $|\sqrt{x} - 0| = \sqrt{x} < \epsilon$.

55. Take δ = minimum of 1 and $\epsilon/7$. If $0 < |x - 1| < \delta$, then $0 < x < 2$ and $|x - 1| < \epsilon/7$. Therefore $|x^3 - 1| = |x^2 + x + 1||x - 1| < 7|x - 1| < 7(\epsilon/7) = \epsilon$.

57. Set $\delta = \epsilon^2$. If $3 - \epsilon^2 < x < 3$, then $-\epsilon^2 < x - 3$, $0 < 3 - x < \epsilon^2$ and therefore $|\sqrt{3 - x} - 0| < \epsilon$.

59. Suppose, on the contrary, that $\lim_{x\to c} f(x) = L$ for some particular c. Taking $\epsilon = \frac{1}{2}$, there must exist $\delta > 0$ such that,

if $0 < |x - c| < \delta$, then $|f(x) - L| < \frac{1}{2}$.

Let x_1 be a rational number satisfying $0 < |x_1 - c| < \delta$, and x_2 an irrational number satisfying $0 < |x_2 - c| < \delta$. (That such numbers exist follows from the fact that every interval contains both rational and irrational numbers.) Now $f(x_1) = 1$ and $f(x_2) = 0$. Thus we must have both $|1 - L| < \frac{1}{2}$ and $|0 - L| < \frac{1}{2}$. From the first inequality we conclude that $L > \frac{1}{2}$. From the second, we conclude that $L < \frac{1}{2}$. Clearly no such number L exists.

61. We begin by assuming that $\lim_{x\to c^+} f(x) = L$ and showing that $\lim_{h\to 0} f(c + |h|) = L$.

Let $\epsilon > 0$. Since $\lim_{x\to c^+} f(x) = L$, there exists $\delta > 0$ such that

$(*)$ if $c < x < c + \delta$ then $|f(x) - L| < \epsilon$.

Suppose now that $0 < |h| < \delta$. Then $c < c + |h| < c + \delta$ and, by $(*)$, $|f(c + |h|) - L| < \epsilon$ Thus $\lim_{h\to 0} f(c + |h|) = L$.

Conversely let's assume that $\lim_{h\to 0} f(c + |h|) = L$ and again take $\epsilon > 0$. Then there exists $\delta > 0$ such that

$(**)$ if $0 < |h| < \delta$ then $|f(c + |h|) - L| < \epsilon$.

Suppose now that $c < x < c + \delta$. Then $0 < x - c < \delta$ so that, by $(**)$,

$$|f(x) - L| = |f(c + (x - c)) - L| < \epsilon.$$

Thus $\lim_{x\to c^+} f(x) = L$.

63. (a) Let $\epsilon = L$. There exists $\gamma > 0$ such that if $0 < |x - c| < \gamma$, then $|f(x) - L| < L$, which is equivalent to $0 < f(x) < 2L$. Thus $f(x) > 0$ for all $x \in (c - \gamma, c + \gamma)$, $x \neq c$.
(b) Use the same argument.

65. (a) Suppose $\lim_{x\to c} f(x) = L$ and $\lim_{x\to c} g(x) = M$. Then $\lim_{x\to c} [g(x) - f(x)] = M - L$ exists. If $M - L < 0$, then, by Exercise 63(b), there exists $\gamma > 0$ such that $g(x) - f(x) < 0$ on $(c - \gamma, c + \gamma)$, $x \neq c$, which contradicts the fact that $f(x) \leq g(x)$ on $(c - p, c + p)$, $x \neq c$. Thus $M - L \geq 0$.
(b) No. Consider $f(x) = 1 - x^2$ and $g(x) = 1 + x^2$ on $(-1, 1)$, and let $c = 0$.

67. 5; $y = 5x - 8$ **69.** $\frac{1}{4}$; $y = \frac{1}{4}x + 1$ **71.** -2.7207 **73.** 1.0986

SECTION 2.3

1. (a) 3 (b) 4 (c) -2 (d) 0 (e) does not exist (f) $\frac{1}{3}$

3. $\lim_{x\to 4} \left[\left(\frac{1}{x} - \frac{1}{4}\right)\left(\frac{1}{x-4}\right)\right] = \lim_{x\to 4} \left[\left(\frac{4-x}{4x}\right)\left(\frac{1}{x-4}\right)\right] = \lim_{x\to 4} \frac{-1}{4x} = -\frac{1}{16}$; Theorem 2.3.2 does not apply since $\lim_{x\to 4} \frac{1}{x-4}$ does not exist.

5. 3 **7.** -3 **9.** 5 **11.** does not exist **13.** -1 **15.** does not exist **17.** 1 **19.** 4 **21.** $\frac{1}{4}$ **23.** $-\frac{2}{3}$

25. does not exist **27.** -1 **29.** 4 **31.** a/b **33.** $5/4$ **35.** does not exist **37.** 2

39. (a) 0 (b) $-\frac{1}{16}$ (c) 0 (d) does not exist **41.** (a) 4 (b) -2 (c) 2 (d) does not exist **43.** $f(x) = 1/x$, $g(x) = -1/x$, $c = 0$

45. True. Let $\lim_{x\to c} [f(x) + g(x)] = L$. If $\lim_{x\to c} g(x) = M$ exists, then $\lim_{x\to c} f(x) = \lim_{x\to c} [f(x) + g(x) - g(x)] = L - M$ also exists. Thus, $\lim_{x\to c} g(x)$ cannot exist.

47. True. If $\lim_{x\to c} \sqrt{f(x)} = L$ exists, then $\lim_{x\to c} f(x) = \lim_{x\to c} \sqrt{f(x)} \cdot \sqrt{f(x)} = L^2$ exists. **49.** False. Let $f(x) = x$, $c = 0$

51. False. Let $f(x) = 1 - x^2$, $g(x) = 1 + x^2$, and $c = 0$ **53.** If $\lim_{x\to c} f(x) = L$ and $\lim_{x\to c} g(x) = L$, then

$$\begin{aligned}\lim_{x\to c} h(x) &= \lim_{x\to c} \tfrac{1}{2}\{[f(x) + g(x)] - |f(x) - g(x)|\}\\ &= \lim_{x\to c} \tfrac{1}{2}[f(x) + g(x)] - \lim_{x\to c} |f(x) - g(x)| = \tfrac{1}{2}(L + L) - \tfrac{1}{2}|L - L| = L\end{aligned}$$

A similar argument works for H.

55. (a) Suppose $\lim_{x\to c} g(x) = k$ exists. Then $\lim_{x\to c} f(x) \cdot g(x) = 0 \cdot k = 0$. Thus $\lim_{x\to c} g(x)$ cannot exist. (b) $\lim_{x\to c} g(x) = \dfrac{1}{L}$

57. (a) 1 (b) $2x$ (c) $3x^2$ (d) $4x^3$ (e) nx^{n-1}

SECTION 2.4

1. (a) $x = -3$, $x = 0$, $x = 2$, $x = 6$ (b) At -3, neither; at 0, continuous from the right; at 2, neither; at 6, neither (c) removable discontinuity at $x = 2$; jump discontinuity at $x = 0$

3. continuous **5.** continuous **7.** continuous **9.** removable discontinuity **11.** jump discontinuity **13.** continuous

15. jump discontinuity **17.** removable discontinuity at 2 **19.** no discontinuities **21.** jump discontinuity at 1

23. no discontinuities **25.** no discontinuities **27.** jump discontinuities at 0 and 2

29. removable discontinuity at -2; jump discontinuity at 3 **31.** $f(1) = 2$ **33.** impossible **35.** 4 **37.** $A - B = 3$ with $B \neq 3$

39. $f(5) = \frac{1}{6}$ **41.** $f(5) = \frac{1}{3}$ **43.** nowhere **45.** $x = 0$, $x = 2$, and all nonintegral values of x

47. Refer to (2.2.5). Use the equivalence of (i) to (ii) setting $L = f(c)$.

49. Let $A = \{x_1, x_2, \ldots, x_n\}$ and let $\epsilon > 0$. In $A - \{c\}$ there is one point closest to c. Call it d and set $\delta_1 = |c - d|$. Note that if $0 < |x - c| < \delta_1$, then $f(x) = g(x)$.

Suppose now that g is continuous at c. Then there exists a positive number δ less than δ_1 such that

$$\text{if} \quad 0 < |x - c| < \delta, \quad \text{then} \quad |g(x) - g(c)| < \epsilon.$$

But for such x, $f(x) = g(x)$. Therefore we see that

$$\text{if} \quad 0 < |x - c| < \delta, \quad \text{then} \quad |f(x) - g(c)| < \epsilon.$$

This means that

$$\lim_{x\to c} f(x) = g(c)$$

and contradicts the assumption that f has a nonremovable discontinuity at c.

51. The hypothesis implies that f is defined at c and $B > 0$. Let $\epsilon > 0$ and choose $\delta = \min(p, B/\epsilon)$. Then $|x - c| < \delta$ implies

$$|f(x) - f(c)| < B\,|x - c| < B(\epsilon/B) = \epsilon$$

It now follows that f is continuous at c.

53. $\lim_{h\to 0} f(c + h) - f(c) = \lim_{h\to 0} \left[\dfrac{f(c + h) - f(c)}{h} \cdot h\right] = \lim_{h\to 0} \left[\dfrac{f(c + h) - f(c)}{h}\right] \cdot \lim_{h\to 0} h = L \cdot 0 = 0$. Therefore f is continuous at c by Exercise 47.

SECTION 2.5

1. 3 **3.** $\frac{3}{5}$ **5.** 2 **7.** 0 **9.** does not exist **11.** $\frac{9}{5}$ **13.** $\frac{2}{3}$ **15.** 1 **17.** $\frac{1}{2}$ **19.** -4 **21.** 1 **23.** $\frac{3}{5}$

25. 0 **27.** $\dfrac{2}{\pi}\sqrt{2}$ **29.** -1 **31.** 0

33. We will show that

$$\lim_{x\to c} \cos x = \cos c \quad \text{by showing that} \quad \lim_{h\to 0} \cos(c + h) = \cos c.$$

Note that $\cos(c + h) = \cos c \cos h - \sin c \sin h$. We know that

$$\lim_{h\to 0} \cos h = 1 \quad \text{and} \quad \lim_{h\to 0} \sin h = 0.$$

Therefore

$$\lim_{h\to 0} \cos(c+h) = (\cos c)(\lim_{h\to 0} \cos h) - (\sin c)(\lim_{h\to 0} \sin h) = (\cos c)(1) - (\sin c)(0) = \cos c.$$

Here is a different proof. The addition formula for the sine gives $\cos x = \sin(\frac{1}{2}\pi + x)$. Being the composition of functions that are everywhere continuous, the cosine function is itself everywhere continuous. Therefore

$$\lim_{x\to c} \cos x = \cos c \qquad \text{for all real } c.$$

35. $\dfrac{\sqrt{2}}{2}$; $y - \dfrac{\sqrt{2}}{2} = \dfrac{\sqrt{2}}{2}\left(x - \dfrac{\pi}{4}\right)$ 37. $-\sqrt{3}$; $y - \frac{1}{2} = -\sqrt{3}\left(x - \dfrac{\pi}{6}\right)$ 39. $0 \le |x \sin(1/x)| \le |x|$ for $x \ne 0$

41. $0 \le |x-1|\,|\sin x| \le |x-1|$ for all x 43. $0 \le |xf(x)| \le B|x|$ for $x \ne 0$ 45. $0 \le |f(x) - L| \le B|x - c|$ for $x \ne c$

SECTION 2.6

1.
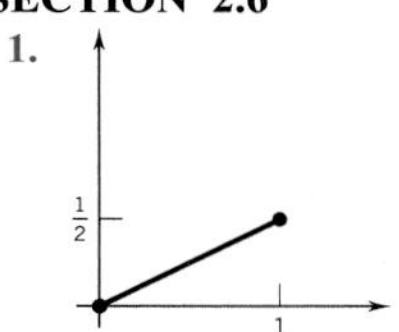

3.

5.
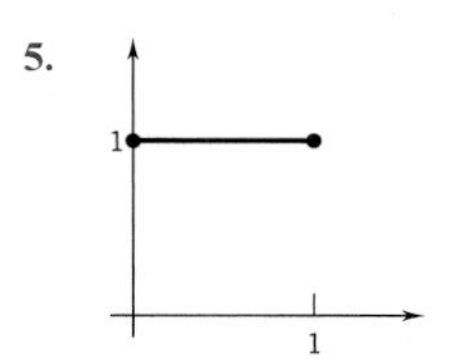

7.
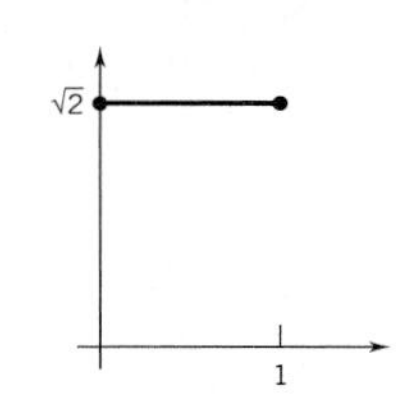

9. impossible by the intermediate-value theorem

11.
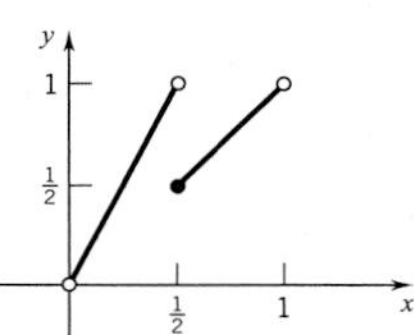

13.
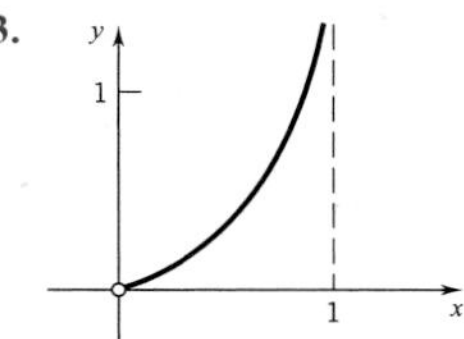

15.
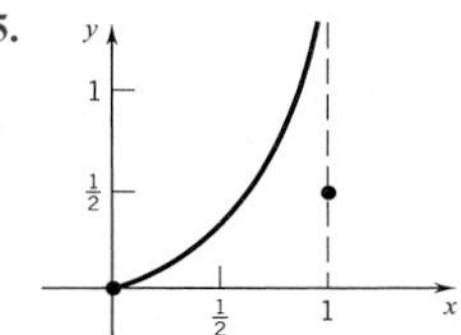

17. $f(1) = -1 < 0$ and $f(2) = 6 > 0$ 19. $\lim_{x\to 1^+} f(x) = \infty$ and $\lim_{x\to 4^-} f(x) = -\infty$

21. Set $g(x) = x - f(x)$. Since g is continuous on $[0, 1]$ and $g(0) \le 0 \le g(1)$, there exists c in $[0, 1]$ such that $g(c) = c - f(c) = 0$.

23. By Exercise 43, Section 2.5, $\lim_{x\to 0} x f(x) = 0$. Thus, $\lim_{x\to 0} |g(x) - g(0)| = \lim_{x\to 0} |x f(x)| = 0$ and g is continuous at 0.

25. The cubic polynomial $P(x) = x^3 + ax^2 + bx + c$ is continuous on $(-\infty, \infty)$. Writing P as

$$P(x) = x^3\left(1 + \frac{a}{x} + \frac{b}{x^2} + \frac{c}{x^3}\right),\ x \ne 0,$$

it follows that $P(x) < 0$ for large negative values of x and $P(x) > 0$ for large positive values of x. Thus, there exists a negative number N such that $P(x) < 0$ for $x < N$, and a positive number M such that $P(x) > 0$ for $x > M$. By the intermediate-value theorem, P has a zero on $[N, M]$.

27. A circle of radius r has area πr^2. Let $A(r) = \pi r^2$, $r \in [0, 10]$. Then $A(0) = 0$ and $A(10) = 100\pi \cong 314$. Since $0 < 250 < 314$, it follows from the intermediate-value theorem that there exists a number $c \in (0, 10)$ such that $A(c) = 250$.

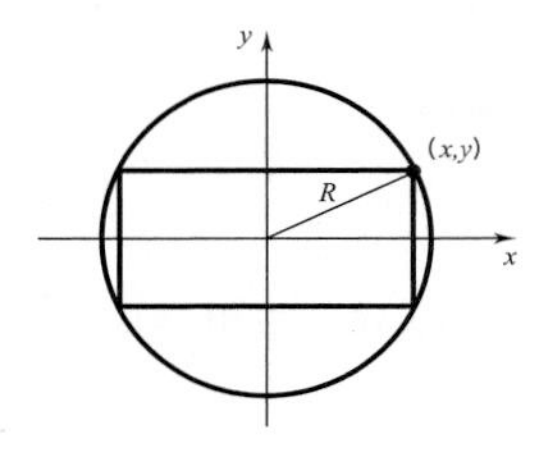

29. Inscribe a rectangle in a circle of radius R and then introduce a coordinate system as shown in the figure. Then the area of the rectangle is given by

$$A(x) = 4x\sqrt{R^2 - x^2},\ x \in [0, R]$$

Since A is continuous on $[0, R]$, A has a maximum value. (A also has a minimum value, namely 0.)

31. $m_{10} = 1.7314$ 33. $m_{10} = 0.7392$

35. f has a zero on $(-3, -2)$, $(0, 1)$ and $(1, 2)$. The zeros of f are: $r_1 = -2.4909$, $r_2 = 0.6566$, $r_3 = 1.8343$

37. f has a zero on $(-2, -1)$, $(0, 1)$, $(1, 2)$. The zeros of f are: $r_1 = -1.3482$, $r_2 = 0.2620$, $r_3 = 1.0816$

39. f is bounded; the maximum value of f is 1, the minimum value is -1

41. f is bounded; the maximum value of f is 0.5, the minimum value is approximately 0.3540

CHAPTER 3

SECTION 3.1

1. 0 **3.** -3 **5.** $5 - 2x$ **7.** $4x^3$ **9.** $\frac{1}{2}(x-1)^{-1/2}$ **11.** $-2x^{-3}$ **13.** -6 **15.** $-\frac{1}{4}$ **17.** $\frac{3}{2}$

19. tangent $y - 4x + 4 = 0$; normal $4y + x - 18 = 0$ **21.** tangent $y + 3x - 16 = 0$; normal $3y - x - 8 = 0$

23. tangent $y + x + 4 = 0$; normal $y - x - 2 = 0$

25. (a) Removable discontinuity at $c = -1$; jump discontinuity at $c = 1$ (b) f is continuous but not differentiable at $c = 0$ and $c = 3$

27. $x = -1$

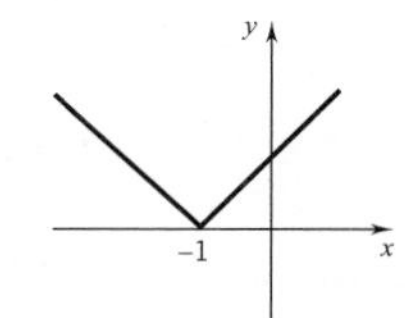

29. $x = 0$

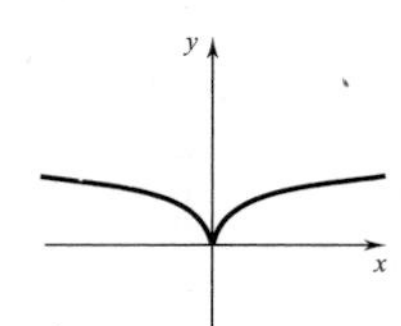

31. $x = 1$

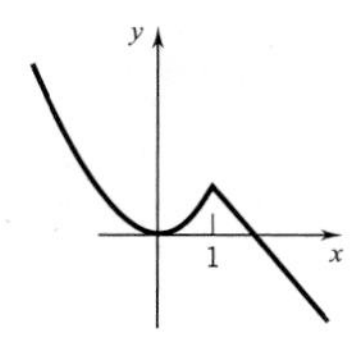

33. 4 **35.** does not exist

37.

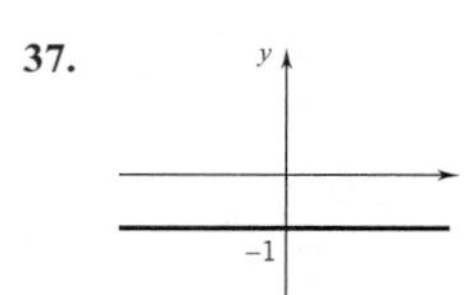

39.

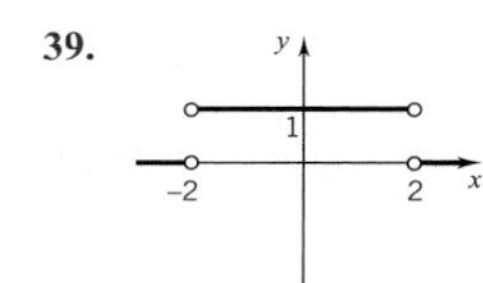

41.

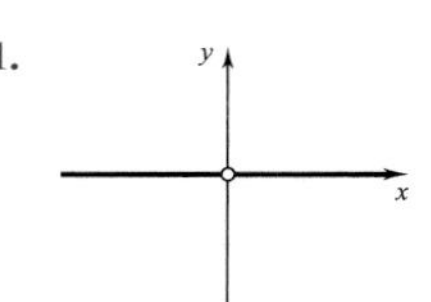

43. $f(x) = x^2$; $c = 1$ **45.** $f(x) = \sqrt{x}$; $c = 4$ **47.** $f(x) = \cos x$; $c = \pi$

49. Since $f(1) = 1$ and $\lim_{x \to 1^+} f(x) = 2$, f is not continuous at 1 and thus, by (3.1.4), is not differentiable at 1.

51. (a) $f'(x) = \begin{cases} 2(x+1), & x < 0 \\ 2(x-1), & x > 0 \end{cases}$ (b) $\lim_{h \to 0^-} \frac{f(0+h) - f(0)}{h} = \lim_{h \to 0^-} \frac{(h+1)^2 - 1}{h} = \lim_{h \to 0^-} (h+2) = 2$,

$\lim_{h \to 0^+} \frac{f(0+h) - f(0)}{h} = \lim_{h \to 0^+} \frac{(h-1)^2 - 1}{h} = \lim_{h \to 0^+} (h-2) = -2$

In 53–57, there are many possible answers. Here are some.

53. $f(x) = c$, c any constant **55.** $f(x) = |x+1|$; $f(x) = \begin{cases} 0, & x \neq -1 \\ 1, & x = -1 \end{cases}$ **57.** $f(x) = 2x + 5$

59. $f'(-x) = \lim_{h \to 0} \frac{f(-x+h) - f(-x)}{h} = \lim_{h \to 0} \frac{-f(x-h) + f(x)}{h} = \lim_{h \to 0} \frac{f[x + (-h)] - f(x)}{-h} = f'(x)$

61. (a) $\lim_{x \to 2^-} (x^2 - x) = \lim_{x \to 2^+} (2x - 2) = 2 = f(2)$ (b) $f'_-(2) = 3, f'_+(2) = 2$ (c) f is not differentiable at 2

63. (a) $f'(x) = \frac{-1}{2\sqrt{1-x}}$ (b) $f'_+(0) = -\frac{1}{2}$ (c) $f'_-(1)$ does not exist

65. Let $L = f'_-(c) = f'_+(c)$ and let $\epsilon > 0$. There exists $\delta_1 > 0$ such that

$$\left| \frac{f(c+h) - f(c)}{h} - L \right| < \epsilon$$

whenever $h \in (-\delta_1, 0)$. There exists $\delta_2 > 0$ such that

$$\left| \frac{f(c+h) - f(c)}{h} - L \right| < \epsilon$$

whenever $h \in (0, \delta_2)$. Let $\delta = \min(\delta_1, \delta_2)$. Then

$$\left| \frac{f(c+h) - f(c)}{h} - L \right| < \epsilon$$

whenever $|h| < \delta$, $h \neq 0$. Thus f is differentiable at c and $f'(c) = L$.

67. (a) $\lim_{x \to 0} x \sin \frac{1}{x} = 0 = f(0)$; $\lim_{x \to 0} x^2 \sin \frac{1}{x} = 0 = g(0)$

(b) $\lim_{h \to 0} \frac{h \sin(1/h) - 0}{h} = \lim_{h \to 0} \sin \frac{1}{h}$ does not exist

(c) $\lim_{h \to 0} \frac{h^2 \sin(1/h) - 0}{h} = \lim_{h \to 0} h \sin \frac{1}{h} = 0$; $g'(0) = 0$

69. $f'(1) = -1$ **71.** $f'(-1) = \frac{1}{3}$ **73.** (b) 7.071 (c) $D(0.001) \cong 7.074$; $D(-0.001) \cong 7.068$

75. (a) $T(x) = -\frac{11}{4}(x - \frac{3}{2}) + \frac{21}{8}$ (c) (1.453, 1.547)

SECTION 3.2

1. -1 **3.** $55x^4 - 18x^2$ **5.** $2ax + b$ **7.** $\dfrac{2}{x^3}$ **9.** $3x^2 - 6x - 1$ **11.** $\dfrac{3x^2 - 2x^3}{(1-x)^2}$ **13.** $\dfrac{2(x^2+3x+1)}{(2x+3)^2}$ **15.** $2x - 3$

17. $-\dfrac{2(3x^2 - x + 1)}{x^2(x-2)^2}$ **19.** $-80x^9 + 81x^8 - 64x^7 + 63x^6$ **21.** $f'(0) = -\frac{1}{4}; f'(1) = -1$ **23.** $f'(0) = 0; f'(1) = -1$

25. $f'(0) = \dfrac{ad - bc}{d^2}; f'(1) = \dfrac{ad - bc}{(c+d)^2}$ **27.** $f'(0) = 3$ **29.** $f'(0) = \frac{20}{9}$ **31.** $2y - x - 8 = 0$ **33.** $y - 4x + 12 = 0$

35. $9y + 4x - 18 = 0$ **37.** $(-1, 27), (3, -5)$ **39.** $(-1, -\frac{5}{2}), (1, \frac{5}{2})$ **41.** $(2, 3)$ **43.** $(-2, -10)$ **45.** $(-1, -2), (\frac{5}{3}, \frac{50}{27})$

47. $\frac{425}{8}$ **49.** $A = -1, B = 0, C = 4$ **51.** $x = -\dfrac{b}{2a}$ **53.** $c = -1, 1$

55. Let $a > 0$. An equation for the tangent line to the graph of $f(x) = 1/x$ at $x = a$ is $y = (-1/a^2)x + 2/a$. The y-intercept is $2/a$ and the x-intercept is $2a$. Thus, the area of the triangle formed by this line and the coordinate axes is $A = \frac{1}{2}(2/a)(2a) = 2$ square units.

57. Since f and $f + g$ are differentiable, $g = (f + g) - f$ is differentiable. The functions $f(x) = |x|$ and $g(x) = -|x|$ are not differentiable at 0; their sum $h(x) = 0$ is differentiable everywhere.

59. Since

$$\left(\frac{f}{g}\right)(x) = \frac{f(x)}{g(x)} = f(x) \cdot \frac{1}{g(x)},$$

it follows from the product and reciprocal rules that

$$\left(\frac{f}{g}\right)'(x) = \left(f \cdot \frac{1}{g}\right)'(x) = f(x)\left(-\frac{g'(x)}{[g(x)]^2}\right) + f'(x) \cdot \frac{1}{g(x)} = \frac{g(x)f'(x) - f(x)g'(x)}{[g(x)]^2}.$$

61. $F'(x) = 2x\left(1 + \dfrac{1}{x}\right)(2x^3 - x + 1) + (x^2 + 1)\left(-\dfrac{1}{x^2}\right)(2x^3 - x + 1) + (x^2 + 1)\left(1 + \dfrac{1}{x}\right)(6x^2 - 1)$

63. $g(x) = [f(x)]^2 = f(x) \cdot f(x); g'(x) = f(x)f'(x) + f(x)f'(x) = 2f(x)f'(x)$ **65.** $g'(x) = 3(x^3 - 2x^2 + x + 2)^2(3x^2 - 4x + 1)$

67. (a) $f'_+(-1) = -6; f'_-(3) = 2$ (b) $g'(x) = 2x - 4; g'(-1) = f'_+(-1); g'(3) = f'_-(3)$ **69.** $A = -2, B = -8$

71. (a) Let $D(h) = \dfrac{\sin(x+h) - \sin x}{h}$.

At $x = 0$, $D(0.001) \cong 0.99999$, $D(-0.001) \cong 0.99999$; at $x = \pi/6$, $D(0.001) \cong 0.86578$, $D(-0.001) \cong 0.86628$; at $x = \pi/4$, $D(0.001) \cong 0.70675$, $D(-0.001) \cong 0.70746$; at $x = \pi/3$, $D(0.001) \cong 0.49957$, $D(-0.001) \cong 0.50043$; at $x = \pi/2$, $D(0.001) \cong 0.0005$, $D(-0.001) \cong 0.0005$
(b) $\cos(0) = 1$, $\cos(\pi/6) \cong 0.866025$, $\cos(\pi/4) \cong 0.707107$, $\cos(\pi/3) = 0.5$, $\cos(\pi/2) = 1$ (c) $f'(x) = \cos x$.

73. (a) Let $D(h) = \dfrac{2^{x+h} - 2^x}{h}$.

At $x = 0$, $D(0.001) \cong 0.69339$, $D(-0.001) \cong 0.69291$; at $x = 1$, $D(0.001) \cong 1.38678$, $D(-0.001) \cong 1.38581$; at $x = 2$, $D(0.001) \cong 2.77355$, $D(-0.001) \cong 2.77163$; at $x = 3$, $D(0.001) \cong 5.54710$, $D(-0.001) \cong 5.54326$
(b) $\dfrac{f'(x)}{f(x)} \cong 0.693$ (c) $f'(x) \cong (0.693)\,2^x$

SECTION 3.3

1. $\dfrac{dy}{dx} = 12x^3 - 2x$ **3.** $\dfrac{dy}{dx} = 1 + \dfrac{1}{x^2}$ **5.** $\dfrac{dy}{dx} = \dfrac{1 - x^2}{(1 + x^2)^2}$ **7.** $\dfrac{dy}{dx} = \dfrac{2x - x^2}{(1-x)^2}$ **9.** $\dfrac{dy}{dx} = \dfrac{-6x^2}{(x^3 - 1)^2}$ **11.** 2

13. $18x^2 + 30x + 5x^{-2}$ **15.** $\dfrac{-4t}{(t^2 - 1)^2}$ **17.** $\dfrac{2t^3(t^3 - 2)}{(2t^3 - 1)^2}$ **19.** $\dfrac{2}{(1 - 2u)^2}$ **21.** $-\left[\dfrac{1}{(u-1)^2} + \dfrac{1}{(u+1)^2}\right] = -2\left[\dfrac{u^2 + 1}{(u^2 - 1)^2}\right]$

23. $2x\left[\dfrac{1}{(1 - x^2)^2} + \dfrac{1}{x^4}\right]$ **25.** $\dfrac{-2}{(x-1)^2}$ **27.** 47 **29.** $\frac{1}{4}$ **31.** $42x - 120x^3$ **33.** $-6x^{-3}$ **35.** $4 - 12x^{-4}$ **37.** 2

39. 0 **41.** $6 + 60x^{-6}$ **43.** $1 - 4x$ **45.** -24 **47.** -24 **49.** $p(x) = 2x^2 - 6x + 7$

51. (a) $n!$ (b) 0 (c) $f^{(k)}(x) = n(n-1)\cdots(n-k+1)\,x^{n-k}$

53. (a) $f'_+(0) = 0$ and $f'_-(0) = 0$. Thus $f'(0) = 0$.
(b) $f'(x) = \begin{cases} 2x, & x \geq 0 \\ 0, & x < 0 \end{cases}$
(c) $f''_+(0) = 2$ and $f''_-(0) = 0$. Thus $f''(0)$ does not exist.
(d)

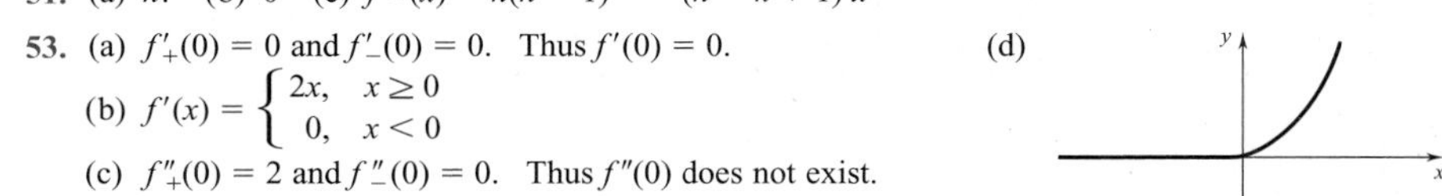

55. If $f(x) = g(x) = x$, then $(fg)(x) = x^2$, so that $(fg)''(x) = 2$, but $f(x)g''(x) + f''(x)g(x) = x \cdot 0 + 0 \cdot x = 0$.

57. (a) $x = 0$ (b) $x > 0$ (c) $x < 0$ 59. (a) $x = -2, x = 1$ (b) $x < -2, x > 1$ (c) $-2 < x < 1$

61. The result is true for $n = 1$:

$$\frac{d^1y}{dx^1} = \frac{dy}{dx} = -x^{-2} = (-1)^1 1!\, x^{-1-1}.$$

If the result is true for $n = k$:

$$\frac{d^ky}{dx^k} = (-1)^k k!\, x^{-k-1},$$

then the result is true for $n = k + 1$:

$$\frac{d^{k+1}y}{dx^{k+1}} = \frac{d}{dx}\left[\frac{d^ky}{dx^k}\right] = (-1)^k k!\,(-k-1)x^{-k-2} = (-1)^{k+1}(k+1)!\, x^{-(k+1)-1}.$$

63. (a) $(f \cdot g)'' = (f'g + fg')' = f''g + f'g' + f'g' + fg'' = f''g + 2f'g' + fg''$
(b) $(f \cdot g)''' = (f''g + 2f'g' + fg'')' = f'''g + f''g' + 2f''g' + 2f'g'' + f'g'' + fg''' = f'''g + 3f''g' + 3f'g'' + fg'''$

65. $\frac{d}{dx}(uvw) = vw\frac{du}{dx} + uw\frac{dv}{dx} + uv\frac{dw}{dx}$

67. (a) $f'(x) = 3x^2 + 2x - 4$ (c) The graph is "falling" when $f'(x) < 0$; the graph is rising when $f'(x) > 0$. 69. (a) $y = 4x + 1$ (c) $(6, 25)$

SECTION 3.4

1. $\frac{dA}{dr} = 2\pi r, 4\pi$ 3. $\frac{dA}{dz} = z, 4$ 5. $-\frac{5}{36}$ 7. $\frac{dV}{dr} = 4\pi r^2$ 9. $x_0 = \frac{3}{4}$ 11. (a) $\frac{3\sqrt{2}}{4}w^2$ (b) $\frac{\sqrt{3}}{3}z^2$

13. (a) $\frac{1}{2}r^2$ (b) $r\theta$ (c) $-4Ar^{-3} = -2\theta/r$ 15. $x = \frac{1}{2}$ 17. $x(5) = -6, v(5) = -7, a(5) = -2$, speed $= 7$

19. $x(2) = -4, v(2) = 6, a(2) = 12$, speed $= 6$ 21. $x(1) = 6, v(1) = -2, a(1) = \frac{4}{3}$, speed $= 2$

23. $x(1) = 0, v(1) = 18, a(1) = 54$, speed $= 18$ 25. never 27. at $-2 + \sqrt{5}$ 29. at $2\sqrt{2}$ 31. A 33. A 35. A and B

37. A 39. A and C 41. $(0, 2), (7, \infty)$ 43. $(0, 3), (4, \infty)$ 45. $(2, 5)$ 47. $(0, 2 - \frac{2}{3}\sqrt{3}), (4, \infty)$ 49. 576 ft 51. $v_0^2/2g$ meters

53. $y(t) = \frac{1}{2}gt^2 + v_0t + y_0$. If $y(t_1) = y(t_2)$, $t_1 \neq t_2$, then $v(t_1) = gt_1 + v_0 = -gt_2 - v_0 = -v(t_2)$. Thus $|v(t_1)| = |v(t_2)|$.

55. 9 ft/sec 57. (a) 2 sec (b) 16 ft (c) 48 ft/sec 59. (a) $\frac{1625}{16}$ ft (b) $\frac{6475}{64}$ ft (c) 100 ft

61. $C'(100) = 0.04$
$C(101) - C(100) = 0.0401$

63. $C'(100) = 0$
$C(101) - C(100) = 0$

65. $C'(10 = 23$
$C(11) - C(10) = 22.90$

67. (a) $\overline{C}'(x) = \frac{-200}{x^2} + 0.0001$ (b) $\overline{C}'(x) = \frac{-200}{x^2} - \frac{200}{x^3}$

69. (a) $v(t) = 3t^2 - 14t + 10$
(b) moving to the right when $0 < t < 0.88$ and when $3.79 < t < 5$; moving to the left when $0.88 < t < 3.79$
(c) the object stops at $t \cong 0.88$ and $t \cong 3.79$; the maximum speed is approximately 6.33
(d) $a(t) = 6t - 14$; slowing down when $0 < t < 0.88$ and when $2.33 < t < 3.79$; speeding up when $0.88 < t < 2.33$ and when $3.79 < t < 5$

SECTION 3.5

1. $f(x) = x^4 + 2x^2 + 1; f'(x) = 4x^3 + 4x = 4x(x^2 + 1)$
$f(x) = (x^2 + 1)^2; f'(x) = 2(x^2 + 1)(2x) = 4x(x^2 + 1)$

3. $f(x) = 8x^3 + 12x^2 + 6x + 1; f'(x) = 24x^2 + 24x + 6 = 6(2x + 1)^2$
$f(x) = (2x + 1)^3; f'(x) = 3(2x + 1)^2(2) = 6(2x + 1)^2$

5. $f(x) = x^2 + 2 + x^{-2}; f'(x) = 2x - 2x^{-3} = 2x(1 - x^{-4})$
$f(x) = (x + x^{-1})^2; f'(x) = 2(x + x^{-1})(1 - x^{-2}) = 2x(1 + x^{-2})(1 - x^{-2}) = 2x(1 - x^{-4})$

7. $2(1 - 2x)^{-2}$ 9. $20(x^5 - x^{10})^{19}(5x^4 - 10x^9)$ 11. $4\left(x - \frac{1}{x}\right)^3\left(1 + \frac{1}{x^2}\right)$ 13. $4(x - x^3 - x^5)^3(1 - 3x^2 - 5x^4)$

15. $200t(t^2 - 1)^{99}$ 17. $-4(t^{-1} + t^{-2})^3(t^{-2} + 2t^{-3})$ 19. $324x^3\left[\frac{1 - x^2}{(x^2 + 1)^5}\right]$ 21. $2(x^4 + x^2 + x)(4x^3 + 2x + 1)$

23. $-\left(\frac{x^3}{3} + \frac{x^2}{2} + \frac{x}{1}\right)^{-2}(x^2 + x + 1)$ 25. $3[(x + x^{-1})^2 - (x^2 + x^{-2})^{-1}]^2\,[2(x + x^{-1})(1 - x^{-2}) + (x^2 + x^{-2})^{-2}(2x - 2x^{-3})]$

27. -1 29. 0 31. $\frac{dy}{dt} = \frac{dy}{du}\frac{du}{dx}\frac{dx}{dt} = \frac{7(2t - 5)^4 + 12(2t + 5)^2 - 2}{[(2t - 5)^4 + 2(2t - 5)^2 + 2]^2}[4(2t - 5)]$ 33. 16 35. 1 37. 1 39. 1 41. 2

43. 0 45. $4(x^3 + x)^2[3(3x^2 + 1)^2 + 6x(x^3 + x)]$ 47. $\frac{6x(1 + x)}{(1 - x)^5}$ 49. $2xf'(x^2 + 1)$ 51. $2f(x)f'(x)$

53. (a) $x = 0$ (b) $x < 0$ (c) $x > 0$ 55. (a) $x = -1, x = 1$ (b) $-1 < x < 1$ (c) $x < -1, x > 1$ 57. at 3

59. at 2 and $2\sqrt{3}$ 61. $L'(x^2 + 1) = \frac{2x}{x^2 + 1}$ 63. $T'(x) = 0$

65. If $p(x) = (x - a)^2 q(x)$, where $q(a) \neq 0$, then $p(a) = p'(a) = 0$ and $p''(a) \neq 0$. Now suppose that $p(a) = p'(a) = 0$ and $p''(a) \neq 0$. Then $p(x) = (x - a)g(x)$ for some polynomial g. Since $p'(x) = (x - a)g'(x) + g(x)$ and $p'(a) = 0$, it follows that $g(a) = 0$. Therefore $g(x) = (x - a)q(x)$ for some polynomial q, and $p(x) = (x - a)^2 q(x)$. Finally $p''(a) \neq 0$ implies $q(a) \neq 0$.

67. Let $p(x)$ be a polynomial function of degree n. The number a is a zero of multiplicity k for p, $k < n$, if and only if $p(a) = p'(a) = \cdots = p^{(k-1)}(a) = 0$ and $p^{(k)}(a) \neq 0$.

69. 132.3 71. $800\,\pi$ cm^3/sec 73. $\dfrac{d(KE)}{dt} = mv\dfrac{dv}{dt}$

SECTION 3.6

1. $\dfrac{dy}{dx} = -3\sin x - 4\sec x \tan x$ 3. $\dfrac{dy}{dx} = 3x^2 \csc x - x^3 \csc x \cot x$ 5. $\dfrac{dy}{dt} = -2\cos t \sin t$ 7. $\dfrac{dy}{du} = 2u^{-1/2}\sin^3\sqrt{u}\cos\sqrt{u}$

9. $\dfrac{dy}{dx} = 2x\sec^2 x^2$ 11. $\dfrac{dy}{dx} = 4(1 - \pi\csc^2 \pi x)(x + \cot \pi x)^3$ 13. $\dfrac{d^2y}{dx^2} = -\sin x$ 15. $\dfrac{d^2y}{dx^2} = \cos x\,(1 + \sin x)^{-2}$

17. $\dfrac{d^2y}{du^2} = 12\cos 2u\,(2\sin^2 2u - \cos^2 2u)$ 19. $\dfrac{d^2y}{dt^2} = 8\sec^2 2t\tan 2t$ 21. $\dfrac{d^2y}{dx^2} = (2 - 9x^2)\sin 3x + 12x\cos 3x$ 23. $\dfrac{d^2y}{dx^2} = 0$

25. $\sin x$ 27. $(27t^3 - 12t)\sin 3t - 45t^2\cos 3t$ 29. $3\cos 3x\, f'(\sin 3x)$ 31. $y = x$ 33. $y - \sqrt{3} = -4(x - \frac{1}{6}\pi)$

35. $y - \sqrt{2} = \sqrt{2}(x - \frac{1}{4}\pi)$ 37. at π 39. at $\frac{1}{6}\pi, \frac{7}{6}\pi$ 41. at $\frac{1}{2}\pi, \pi, \frac{3}{2}\pi$ 43. at $\frac{1}{4}\pi, \frac{3}{4}\pi, \frac{5}{4}\pi, \frac{7}{4}\pi$ 45. at $\frac{7}{6}\pi, \frac{11}{6}\pi$

47. $(\frac{1}{2}\pi, \frac{2}{3}\pi), (\frac{7}{6}\pi, \frac{4}{3}\pi), (\frac{11}{6}\pi, 2\pi)$ 49. $(0, \frac{1}{4}\pi), (\frac{7}{4}\pi, 2\pi)$ 51. $(\frac{5}{6}\pi, \frac{3}{2}\pi)$

53. (a) $\dfrac{dy}{dt} = \dfrac{dy}{du}\dfrac{du}{dx}\dfrac{dx}{dt} = (2u)(\sec x\tan x)(\pi) = 2\pi\sec^2 \pi t\tan \pi t$ (b) $y = \sec^2 \pi t - 1$; $\dfrac{dy}{dt} = 2\sec \pi t\,(\sec \pi t\tan \pi t)\pi = 2\pi\sec^2 \pi t\tan \pi t$

55. (a) $\dfrac{dy}{dt} = \dfrac{dy}{du}\dfrac{du}{dx}\dfrac{dx}{dt} = 4[\frac{1}{2}(1 - u)]^3(-\frac{1}{2})\cdot(-\sin x)\cdot 2 = 4[\frac{1}{2}(1 - \cos 2t)]^3\sin 2t = (4\sin^6 t)(2\sin t\cos t) = 8\sin^7 t\cos t$

(b) $y = [\frac{1}{2}(1 - \cos 2t)]^4 = \sin^8 t$; $\dfrac{dy}{dt} = 8\sin^7 t\cos t$

57. $\dfrac{d^n}{dx^n}(\cos x) = \begin{cases} (-1)^{(n+1)/2}\sin x, & n \text{ odd} \\ (-1)^{n/2}\cos x, & n \text{ even} \end{cases}$

59. (a) $f'(x) = \sin\dfrac{1}{x} - \dfrac{1}{x}\cos\dfrac{1}{x}$; $g'(x) = 2x\sin\dfrac{1}{x} - \cos\dfrac{1}{x}$ (b) $\lim\limits_{x\to 0} g'(x) = \lim\limits_{x\to 0}\left(2x\sin\dfrac{1}{x} - \cos\dfrac{1}{x}\right) = \lim\limits_{x\to 0}\cos\dfrac{1}{x}$ does not exist

61. $y'' = -A\omega^2\sin\omega t - B\omega^2\cos\omega t = -\omega^2 y$. Thus $y'' + \omega^2 y = 0$. 63. $A(x) = \frac{1}{2}c^2\sin x$; $A'(x) = \frac{1}{2}c^2\cos x$ 65. $-\frac{1}{6}$ rad/sec

67. 0.74 rad/sec 69. by numerical work $f'(0) \cong 0$; by the chain rule $f'(x) = -2x\sin x^2$, $f'(0) = 0$

71. (b) $f(x) = 0$ at $x = 0$ and $x \cong 0.81$ (c) $f'(x) = 0$ at $x \cong -1.25$, $x \cong 0.68$, $x \cong 0.43$

73.

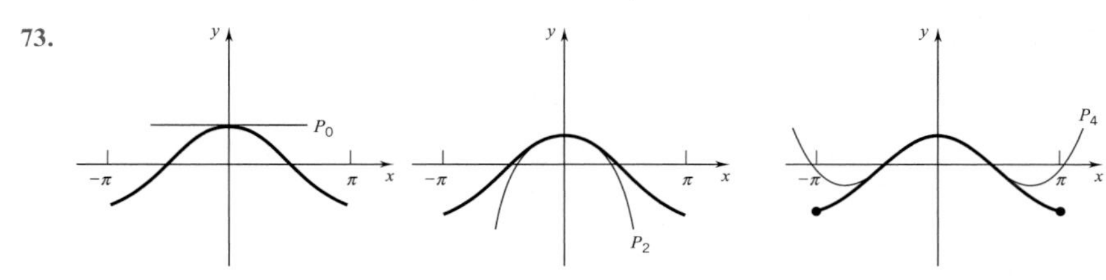

SECTION 3.7

1. $-\dfrac{x}{y}$ 3. $-\dfrac{4x}{9y}$ 5. $-\dfrac{x^2(x + 3y)}{x^3 + y^3}$ 7. $\dfrac{2(x - y)}{2(x - y) + 1}$ 9. $\dfrac{y - \cos(x + y)}{\cos(x + y) - x}$ 11. $\dfrac{16}{(x + y)^3}$ 13. $\dfrac{90}{(2y + x)^3}$

15. $\dfrac{d^2y}{dx^2} = \frac{3}{2}x\cos^2 y - \frac{9}{8}x^4\sin y\cos^3 y$ 17. $\dfrac{dy}{dx} = \dfrac{5}{8}$, $\dfrac{d^2y}{dx^2} = -\dfrac{9}{128}$ 19. $\dfrac{dy}{dx} = -\dfrac{1}{2}$, $\dfrac{d^2y}{dx^2} = 0$

21. tangent $2x + 3y - 5 = 0$; normal $3x - 2y + 12 = 0$ 23. tangent $x + 2y + 8 = 0$; normal $2x - y + 1 = 0$

25. tangent: $y = \dfrac{-2}{\sqrt{3}}x + \left(\dfrac{1}{\sqrt{3}} + \dfrac{\pi}{3}\right)$; normal: $y = \dfrac{\sqrt{3}}{2}x + \left(\dfrac{1}{\sqrt{3}} + \dfrac{\pi}{3}\right)$ 27. $\frac{3}{2}x^2(x^3 + 1)^{-1/2}$ 29. $\dfrac{2x^2 + 1}{\sqrt{x^2 + 1}}$ 31. $\dfrac{x}{(\sqrt[4]{2x^2 + 1})^3}$

33. $\dfrac{x(2x^2 - 5)}{\sqrt{2 - x^2}\sqrt{3 - x^2}}$ 35. $\dfrac{1}{2}\left(\dfrac{1}{\sqrt{x}} - \dfrac{1}{x\sqrt{x}}\right)$ 37. $\dfrac{1}{(\sqrt{x^2 + 1})^3}$ 39. $\dfrac{1}{3}\left[\dfrac{1}{(\sqrt[3]{x})^2} - \dfrac{1}{(\sqrt[3]{x})^4}\right]$

41. (a)

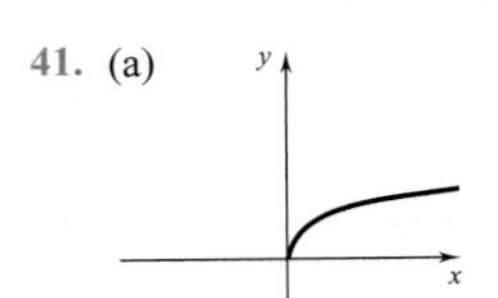

(b)

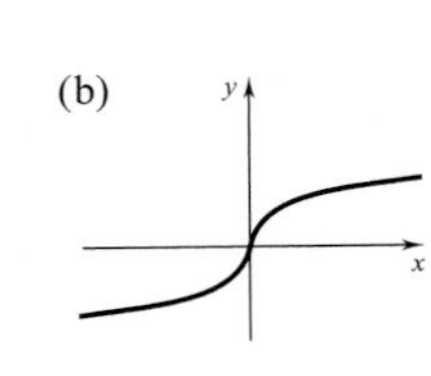

(c)

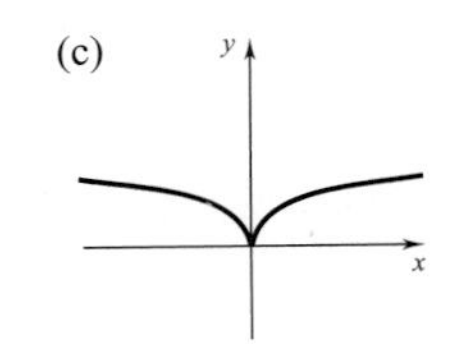

43. $-\dfrac{2b^2}{9(\sqrt[3]{a+bx})^5}$

45. $\dfrac{\sqrt{x}\sec^2\sqrt{x} - \tan\sqrt{x} + 2x\sec^2\sqrt{x}\tan\sqrt{x}}{4x\sqrt{x}}$

47. Let (x_0, y_0) be a point of the circle. If $x_0 = 0$, the normal line is the y-axis; if $y_0 = 0$, the normal line is the x-axis. For all other choices of (x_0, y_0) the normal line takes the form

$$y - y_0 = \frac{y_0}{x_0}(x - x_0) \qquad \text{which reduces to} \qquad y = \frac{y_0}{x_0}x.$$

In each case the normal line passes through the origin.

49. at right angles 51. at (1, 1), $\alpha = \pi/4$; at (0, 0), $\alpha = \pi/2$

53. The hyperbola and the ellipse intersect at the four points $(\pm 3, \pm 2)$. For the hyperbola, $\dfrac{dy}{dx} = \dfrac{x}{y}$. For the ellipse, $\dfrac{dy}{dx} = -\dfrac{4x}{9y}$. The product of these slopes is therefore $-\dfrac{4x^2}{9y^2}$. At each of the points of intersection this product is -1.

55. For the circles, $\dfrac{dy}{dx} = -\dfrac{x}{y}$, $y \neq 0$. For the straight lines $\dfrac{dy}{dx} = m = \dfrac{y}{x}$, $x \neq 0$. Thus, at a point of intersection of a circle $x^2 + y^2 = r^2$ and a line $y = mx$, we have $-\dfrac{x}{y} \cdot \dfrac{y}{x} = -1$ $(x \neq 0, y \neq 0)$.

57. $y - 2x + 12 = 0$, $y - 2x - 12 = 0$ 59. $\left(\dfrac{\sqrt{6}}{4}, \pm\dfrac{\sqrt{2}}{4}\right)$, $\left(-\dfrac{\sqrt{6}}{4}, \pm\dfrac{\sqrt{2}}{4}\right)$

61. Let (x_0, y_0) be a point on the graph. An equation for the tangent line at (x_0, y_0) is

$$y - y_0 = -\left(\frac{y_0}{x_0}\right)^{1/2}(x - x_0)$$

The y-intercept is: $(x_0y_0)^{1/2} + y_0$; the x-intercept is: $(x_0, y_0)^{1/2} + x_0$; and $(x_0y_0)^{1/2} + y_0 + (x_0y_0)^{1/2} + x_0 = (x_0^{1/2} + y_0^{1/2})^2 = (c^{1/2})^2 = c$

63. (a)

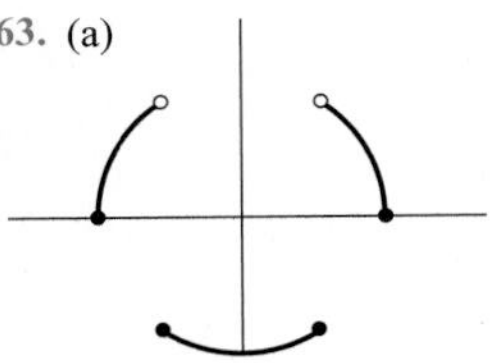

(b) $\sqrt{3}$, $-\sqrt{3}$, 0, respectively (c) $-\dfrac{1}{\sqrt{3}}$, $\dfrac{1}{\sqrt{3}}$, respectively

65. from numerical work, $f'(16) \cong 0.375$; from (3.7.1) $f'(x) = \frac{3}{4}x^{-1/4}$, $f'(16) = \frac{3}{8} = 0.375$

67.

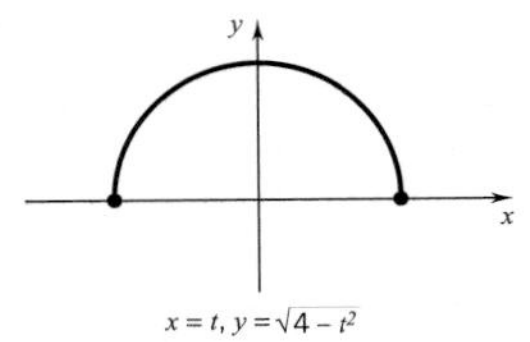

$x = t,\ y = \sqrt{4 - t^2}$

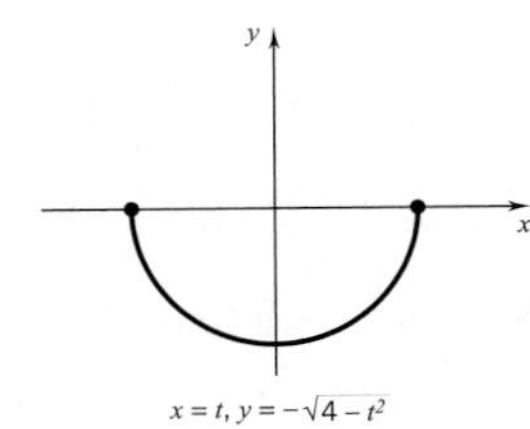

$x = t,\ y = -\sqrt{4 - t^2}$

69. (a)

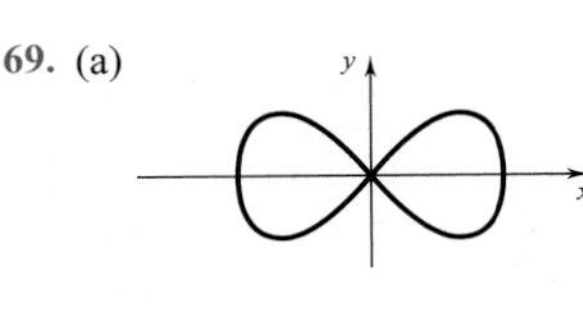

(b) $x = \pm\dfrac{\sqrt{2}}{2}$

SECTION 3.8

1. (a) -2 units/sec (b) 4 units/sec 3. 9 units/sec 5. $\dfrac{-12\sin t\cos t}{\sqrt{16\cos^2 t + 4\sin^2 t}}$; $-\frac{3}{5}\sqrt{10}$ 7. $-\frac{2}{27}$ m/min, $-\frac{8}{3}$ m²/min

9. $1/(50\pi)$ ft/min, $\frac{8}{5}$ ft²/min 11. (a) $A(\theta) = 50 \sin\theta$ (b) $\dfrac{25\pi}{8} \cong 4.36$ cm²/min (c) $\theta = \dfrac{\pi}{2}$

13. 6 cm 15. decreasing 7 in.²/min 17. boat A 19. 10 ft³/hr 21. $\frac{1600}{3}$ ft/min 23. 0.5634 lb/sec

25. dropping $1/2\pi$ in./min 27. decreasing 0.04 rad/min 29. 5π mi/min 31. decreasing 0.12 rad/sec 33. increasing $\frac{4}{101}$ rad/min

SECTION 3.9

1.
$$\begin{aligned} \Delta V &= (x+h)^3 - x^3 \\ &= (x^3 + 3x^2h + 3xh^2 + h^3) - x^3 \\ &= 3x^2h + 3xh^2 + h^3 \\ dV &= 3x^2h \\ \Delta V - dV &= 3xh^2 + h^3 \end{aligned}$$
(see figure)

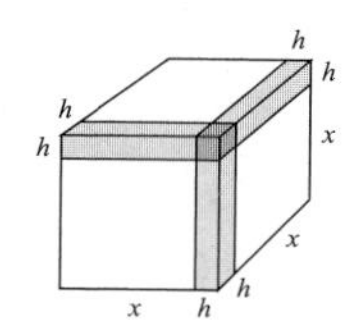

3. $10\frac{1}{30}$ taking $x = 1000$; 10.0332 5. $1\frac{31}{32}$ taking $x = 16$; 1.9680 7. 1.975 taking $x = 32$; 1.9744 9. 8.15 taking $x = 32$; 8.1491

11. 0.719; 0.7193 13. 0.531; 0.5317 15. 1.6 17. $2\pi rht$ 19. error ≤ 0.01 ft 21. 98 gallons

23. $P^2 = 4\pi^2 \dfrac{L}{g}$

$2P\dfrac{dP}{dL} = \dfrac{4\pi^2}{g} = \dfrac{P^2}{L}$

Thus $\dfrac{dP}{P} = \dfrac{1}{2}\dfrac{dL}{L}$

25. 0.00307 sec 27. within $\frac{1}{2}\%$ 29. (a) $x_{n+1} = \frac{1}{2}x_n + 12\left(\dfrac{1}{x_n}\right)$ (b) $x_4 \cong 4.89898$

31. (a) $x_{n+1} = \frac{2}{3}x_n + \frac{25}{3}\left(\dfrac{1}{x_n}\right)^2$ (b) $x_4 \cong 2.92402$ 33. (a) $x_{n+1} = \dfrac{x_n \sin x_n + \cos x_n}{\sin x_n + 1}$ (b) $x_4 \cong 0.73909$

35. (a) $x_{n+1} = \dfrac{2x_n \cos x_n - 2\sin x_n}{2\cos x_n - 1}$ (b) $x_4 \cong 1.89549$ 37. $x_2 = -2x_1, x_3 = 4x_1, x_4 = -8x_1, \ldots, x_n = (-2)^{n-1}x_1, \ldots$

39. (a) $x_1 = \frac{1}{2}, x_2 = -\frac{1}{2}, x_3 = \frac{1}{2}, \ldots, x_n = (-1)^{n-1}\frac{1}{2}, \ldots$ (b) $x_4 = 1.56165$ 41. (b) $x_4 \cong 2.84382$; $f(x_4) \cong -0.00114$

43. (a) Let $F(x) = \frac{1}{2}\cos x - x$. Then $F(0) = \frac{1}{2}$ and $F\left(\dfrac{\pi}{2}\right) = -\pi/2$. Thus, F has a zero in $(0, \pi/2)$. (b) $x_4 \cong 0.4502$; $f(x_4) \cong 0.45018$

45. (a) and (b) 47. $\displaystyle\lim_{h\to 0} \frac{g_1(h) + g_2(h)}{h} = \lim_{h\to 0}\frac{g_1(h)}{h} + \lim_{h\to 0}\frac{g_2(h)}{h} = 0 + 0 = 0$

$$\lim_{h\to 0}\frac{g_1(h)g_2(h)}{h} = \lim_{h\to 0} h\frac{g_1(h)g_2(h)}{h^2} = \left(\lim_{h\to 0} h\right)\left(\lim_{h\to 0}\frac{g_1(h)}{h}\right)\left(\lim_{h\to 0}\frac{g_2(h)}{h}\right) = (0)(0)(0) = 0$$

CHAPTER 4

SECTION 4.1

1. $c = \dfrac{\sqrt{3}}{3} \cong 0.577$ 3. $c = \dfrac{\pi}{4}, \dfrac{3\pi}{4}, \dfrac{5\pi}{4}, \dfrac{7\pi}{4}$ 5. $c = \frac{3}{2}$ 7. $c = \frac{1}{3}\sqrt{39}$ 9. $c = \frac{1}{2}\sqrt{2}$ 11. $c = 0$

13. No. By mean-value theorem there exists at least one number $c \in (0, 2)$ such that $f'(c) = \dfrac{f(2) - f(0)}{2 - 0} = \dfrac{3}{2}$.

15. $f'(x) = \begin{cases} 2, & x \le -1 \\ 3x^2 - 1, & x > -1; \end{cases}$ $-3 < c \le -1$ and $c = 1$

17. $\dfrac{f(b) - f(a)}{b - a} = A(b + a) + B$; $f'(c) = 2Ac + B$. Equating and solving for c gives $c = \dfrac{a + b}{2}$.

19. $\dfrac{f(1) - f(-1)}{(1) - (-1)} = 0$ and $f'(x)$ is never zero; f is not differentiable at 0

21. Set $P(x) = 6x^4 - 7x + 1$. If there existed three numbers $a < b < c$ at which $P(x) = 0$, then by Rolle's theorem $P'(x)$ would have to be zero for some x in (a, b) and also for some x in (b, c). This is not the case: $P'(x) = 24x^3 - 7$ is zero only at $x = (\frac{7}{24})^{1/3}$.

23. Set $P(x) = x^3 + 9x^2 + 33x - 8$. Note that $P(0) < 0$ and $P(1) > 0$. Thus by the intermediate-value theorem there exists some number c between 0 and 1 at which $P(x) = 0$. If the equation $P(x) = 0$ had an additional real root, then by Rolle's theorem there would have to be some real number at which $P'(x) = 0$. This is not the case: $P'(x) = 3x^2 + 18x + 33$ is never 0 since the discriminant $b^2 - 4ac = (18)^2 - 12(33) < 0$.

25. Let c and d be two consecutive roots of the equation $P'(x) = 0$. The equation $P(x) = 0$ cannot have two or more roots between c and d, for then, by Rolle's theorem, $P'(x)$ would have to be zero somewhere between these two roots and thus between c and d. In this case, c and d would no longer be consecutive roots of $P'(x) = 0$.

27. If $x_1, x_2 \in I$ are fixed points of f, then $g(x_1) = g(x_2) = 0$. By Rolle's theorem there exists a number $c \in (x_1, x_2)$ such that $g'(c) = 1 - f'(c) = 0$, which implies $f'(c) = 1$, contradicting the hypothesis.

29. If $x_1 = x_2$, then $|f(x_1) - f(x_2)|$ and $|x_1 - x_2|$ are both 0 and the inequality holds. If $x_1 \neq x_2$, then you know by the mean-value theorem that

$$\frac{f(x_1) - f(x_2)}{x_1 - x_2} = f'(c)$$

for some number c between x_1 and x_2. Since $|f'(c)| \leq 1$, you can conclude that

$$\left|\frac{f(x_1) - f(x_2)}{x_1 - x_2}\right| \leq 1 \qquad \text{and thus that} \qquad |f(x_1) - f(x_2)| \leq |x_1 - x_2|.$$

31. set, for instance, $f(x) = \begin{cases} 1, & a < x < b \\ 0, & x = a, b \end{cases}$

33. (a) By the mean-value theorem, there exists $c \in (a, b)$ such that $f(b) - f(a) = f'(c)(b - a)$. Since $f'(x) \leq M$ for all $x \in (a, b)$, it follows that $f(b) \leq f(a) + M(b - a)$. (b) Same argument as (a). (c) $|f'(x)| \leq L$ implies $-L \leq f'(x) \leq L$.

35. $f(x)g'(x) - g(x)f'(x) = \cos^2 x + \sin^2 x = 1$ for all $x \in I$. The result follows from Exercise 34.

37. $f'(x_0) = \lim_{y \to 0} \dfrac{f(x_0 + y) - f(x_0)}{y} = \lim_{y \to 0} \dfrac{f'(x_0 + \theta y)y}{y} = \lim_{y \to 0} f'(x_0 + \theta y) = \lim_{x \to x_0} f'(x) = L$

(first equality: by the hint; last equality: by (2.2.5))

39. (a) Between any two times that the object is at the origin there is at least one instant when the velocity is zero. (b) On any time interval there is at least one instant when the instantaneous velocity equals the average velocity over the interval.

41. Yes. Since $\dfrac{f(6) - f(0)}{6 - 0} = \dfrac{280}{6} \cong 46.67$ ft/sec, the driver's speed at some time $c \in (0, 6)$ was 46.67 ft/sec $\cong$ 32 mph, by the mean-value theorem. The driver's speed must have been greater than 32 mph at the instant the brakes were applied.

43. $\sqrt{65} \cong 8.0625$ **45.** (a) $f'(x) = 4 + 2 \sin x > 0$ for all x. (b) f has a zero in (0, 1). (c) $x_3 = 0.2361$. **47.** $c = 0.676$

49. $c = 2.205$

SECTION 4.2

1. increases on $(-\infty, -1]$ and $[1, \infty)$, decreases on $[-1, 1]$ **3.** increases on $(-\infty, -1]$ and $[1, \infty)$, decreases on $[-1, 0)$ and $(0, 1]$

5. increases on $[-\frac{3}{4}, \infty)$, decreases on $(-\infty, -\frac{3}{4}]$ **7.** increases on $[-1, \infty)$, decreases on $(-\infty, -1]$

9. increases on $(-\infty, 2)$, decreases on $(2, \infty)$ **11.** increases on $(-\infty, -1)$ and $(-1, 0]$, decreases on $[0, 1)$ and $(1, \infty)$

13. increases on $[-\sqrt{5}, 0]$ and $[\sqrt{5}, \infty)$, decreases on $(-\infty, -\sqrt{5}]$ and $[0, \sqrt{5}]$ **15.** increases on $(-\infty, -1)$ and $(-1, \infty)$

17. decreases on $[0, \infty)$ **19.** increases on $[0, \infty)$, decreases on $(-\infty, 0]$ **21.** decreases on $(0, 3]$ **23.** increases on $[0, 2\pi]$

25. increases on $[\frac{2}{3}\pi, \pi]$, decreases on $[0, \frac{2}{3}\pi]$ **27.** increases on $[0, \frac{2}{3}\pi]$ and $[\frac{5}{6}\pi, \pi]$, decreases on $[\frac{2}{3}\pi, \frac{5}{6}\pi]$

29. $f(x) = \frac{1}{3}x^3 - x + \frac{8}{3}$ **31.** $f(x) = x^5 + x^4 + x^3 + x^2 + x + 5$ **33.** $f(x) = \frac{3}{4}x^{4/3} - \frac{2}{3}x^{3/2} + 1, x \geq 0$ **35.** $f(x) = 2x - \cos x + 4$

37. increases on $(-\infty, -3)$ and $[-1, 1]$, decreases on $[-3, -1]$ and $[1, \infty)$ **39.** increases on $(-\infty, 0]$ and $[3, \infty)$, decreases on $[0, 1)$ and $[1, 3]$

41.

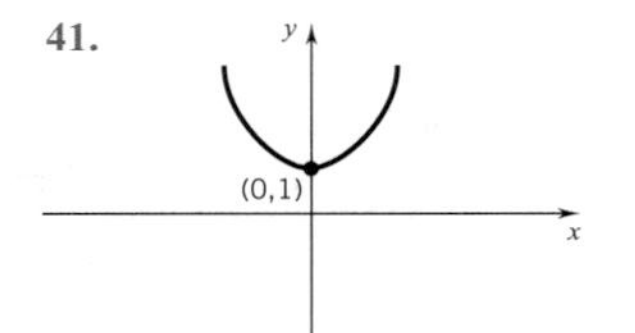

43.

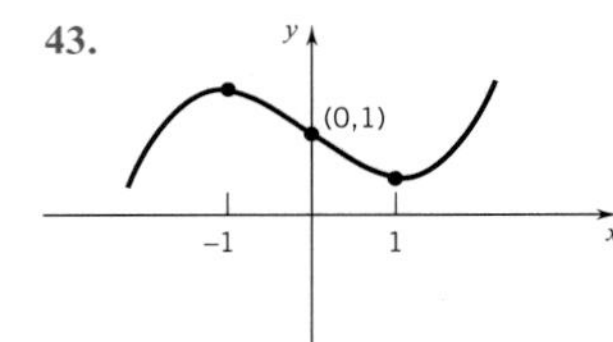

45. **47.**

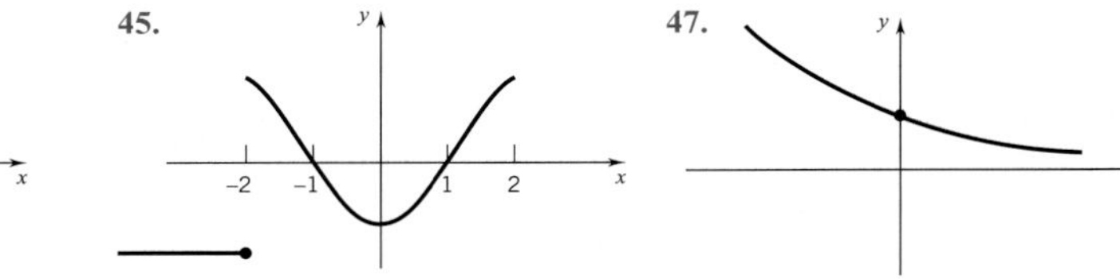

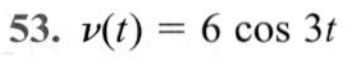

49. Not possible; f is increasing, so $f(2)$ must be greater than $f(-1)$.

51. $v(t) = 3t^2 - 12t + 9 = 3(t - 1)(t - 3)$

sign of v:

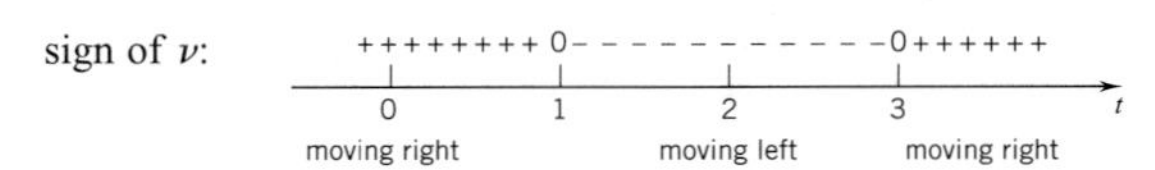

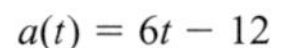

$a(t) = 6t - 12$

sign of a:

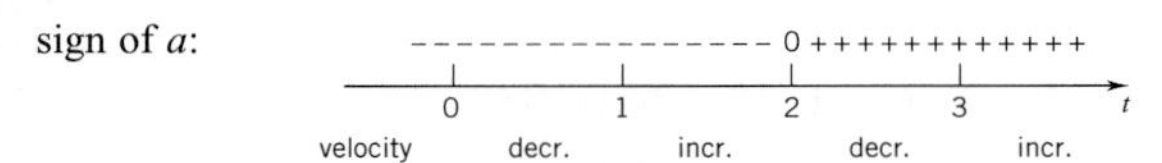

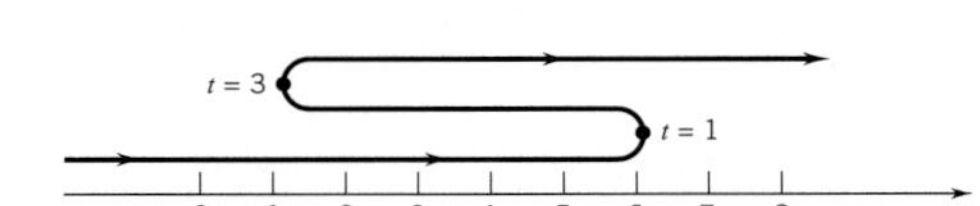

53. $v(t) = 6 \cos 3t$

sign of v:

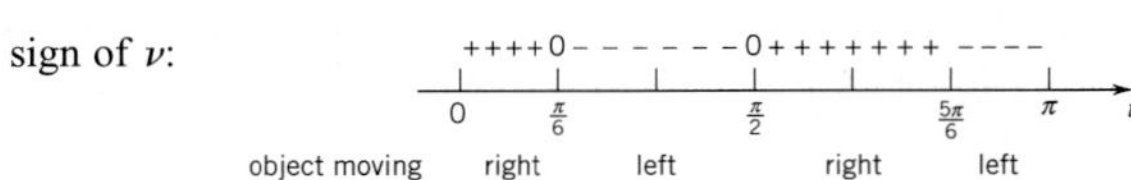

$a(t) = -18 \sin 3t$

sign of a:

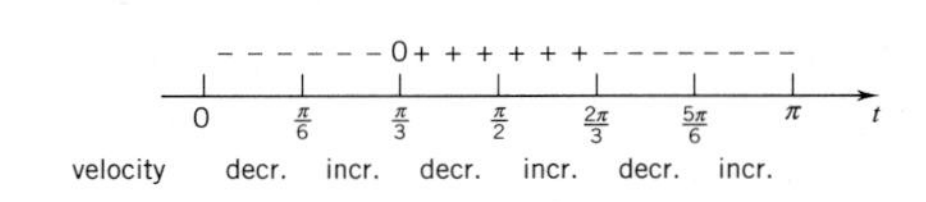

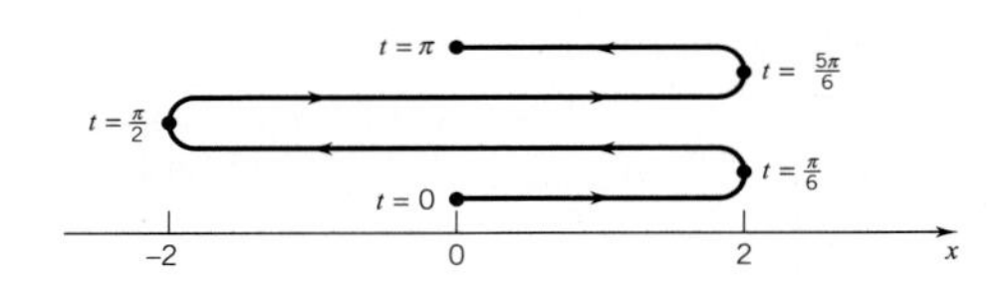

55. (a) $M \le L \le N$ (b) none (c) $M = L = N$ **57.** set, for instance, $f(x) = \begin{cases} 1, & x \text{ rational} \\ 0, & x \text{ irrational} \end{cases}$

59. (a) $f'(x) = 2 \sec x(\sec x \tan x) = 2 \sec^2 x \tan x \quad g'(x) = 2 \tan x (\sec^2 x) = 2 \sec^2 x \tan x$ (b) $C = 1$

61. (a) Let $h(x) = f^2(x) + g^2(x)$. Then $h'(x) = 0$; thus $h(x) = C$, constant.
(b) $f^2(a) + g^2(a) = C$ implies $C = 1$; $f(x) = \cos(x - a)$, $g(x) = \sin(x - a)$.

63. (a) $f'(x) = 1 - \cos x > 0$ except for $x = \dfrac{\pi}{2} + 2n\pi$, $n = 0, \pm 1, \pm 2, \ldots$. Thus f is increasing on $(-\infty, \infty)$. (b) Since f is increasing on $(-\infty, \infty)$ and $f(0) = 0$, $f(x) < 0$ on $(-\infty, 0)$ and $f(x) > 0$ on $(0, \infty)$. Thus $\sin x < x$ on $(-\infty, 0)$ and $\sin x > x$ on $(0, \infty)$.

65. Let $f(x) = \tan x$ and $g(x) = x$. Then $f(0) = g(0) = 0$ and $f'(x) = \sec^2 x > g'(x) = 1$ on $\left(0, \dfrac{\pi}{2}\right)$. The result follows from Exercise 64.

67. Choose an integer $n > 1$, and let $f(x) = (1 + x)^n$, $g(x) = 1 + nx$. Then $f(0) = g(0) = 1$ and $f'(x) = n(1 + x)^{n-1} > g'(x) = n$ since $(1 + x)^{n-1} > 1$. The result follows from Exercise 64.

69. $0.06975 < \sin 4° < 0.06981$

71. $f'(x) = 0$ at $x = -0.633, 0.5, 2.633$
f is decreasing on $[-2, -0.633]$ and $[0.5, 2.633]$
f is increasing on $[-0.633, 0.5]$ and $[2.633, 5]$

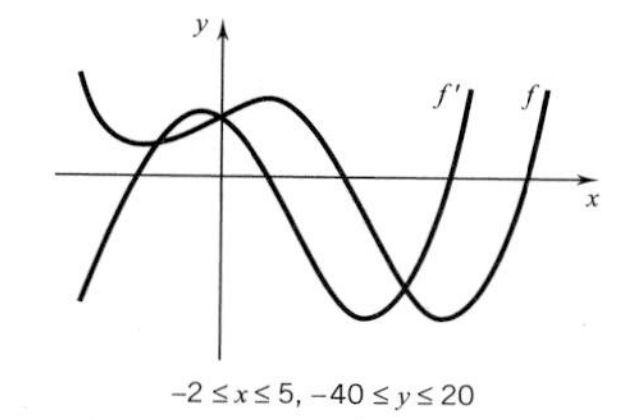

$-2 \le x \le 5,\ -40 \le y \le 20$

73. $f'(x) = 0$ at $x = 0.770, 2.155, 3.798, 5.812$
f is decreasing on $[0, 0.770]$, $[2.155, 3.798]$, and $[5.812, 6]$
f is increasing on $[0.770, 2.155]$ and $[3.798, 5.812]$

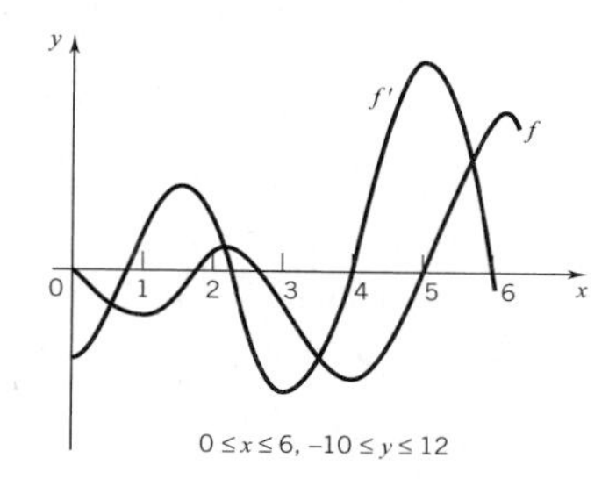

$0 \le x \le 6,\ -10 \le y \le 12$

75. $mgy_0 = mgy + \frac{1}{2}mv^2$; $|v| = \sqrt{2g(y_0 - y)}$. **77.** 54.22 m/sec

SECTION 4.3

1. no critical nos.; no local extreme values **3.** critical nos. ± 1; local max $f(-1) = -2$, local min $f(1) = 2$

5. critical nos. $0, \frac{2}{3}$; $f(0) = 0$ local min, $f(\frac{2}{3}) = \frac{4}{27}$ local max **7.** no critical nos.; no local extreme values

9. critical no. $-\frac{1}{2}$; local max $f(-\frac{1}{2}) = -8$ **11.** critical nos. $0, \frac{3}{5}, 1$; local max $f(\frac{3}{5}) = 2^2 3^3/5^5$, local min $f(1) = 0$

13. critical nos. $\frac{5}{8}, 1$; local max $f(\frac{5}{8}) = \frac{27}{2048}$ **15.** critical nos. $-2, 0$; local max $f(-2) = 4$, local min $f(0) = 0$

17. critical nos. $-2, -\frac{1}{2}, 1$; local max $f(-\frac{1}{2}) = \frac{9}{4}$, local min $f(-2) = f(1) = 0$

19. critical nos. $-2, -\frac{12}{7}, 0$; local max $f(-\frac{12}{7}) = \frac{144}{49}(\frac{2}{7})^{1/3}$, local min $f(0) = 0$ **21.** critical nos. $-\frac{1}{2}, 3$; local min $f(-\frac{1}{2}) = \frac{7}{2}$

23. critical no. 1; local min $f(1) = 3$ **25.** critical nos. $\frac{1}{4}\pi, \frac{5}{4}\pi$; local max $f(\frac{1}{4}\pi) = \sqrt{2}$, local min $f(\frac{5}{4}\pi) = -\sqrt{2}$

27. critical nos. $\frac{1}{3}\pi, \frac{1}{2}\pi, \frac{2}{3}\pi$; local max $f(\frac{1}{2}\pi) = 1 - \sqrt{3}$, local min $f(\frac{1}{3}\pi) = f(\frac{2}{3}\pi) = -\frac{3}{4}$

29. critical nos. $\frac{1}{3}\pi, \frac{5}{3}\pi$; local max $f(\frac{5}{3}\pi) = \frac{5}{4}\sqrt{3} + \frac{10}{3}\pi$, local min $f(\frac{1}{3}\pi) = -\frac{5}{4}\sqrt{3} + \frac{2}{3}\pi$

31. (i) f increases on $(c - \delta, c]$ and decreases on $[c, c + \delta)$. (ii) f decreases on $(c - \delta, c]$ and increases on $[c, c + \delta)$. (iii) If $f'(x) > 0$ on $(c - \delta, c) \cup (c, c + \delta)$, then, since f is continuous at c, f increases on $(c - \delta, c]$ and also on $[c, c + \delta)$. Therefore, in this case, f increases on $(c - \delta, c + \delta)$. A similar argument shows that, if $f'(x) < 0$ on $(c - \delta, c) \cup (c, c + \delta)$, then f decreases on $(c - \delta, c + \delta)$.

33. critical pts 1, 2, 3; local max $P(2) = -4$, local min $P(1) = P(3) = -5$

Since $P'(x) < 0$ for $x < 0$, P decreases on $(-\infty, 0]$. Since $P(0) > 0$, P does not take on the value 0 on $(-\infty, 0]$.

Since $P(0) > 0$ and $P(1) < 0$, P takes on the value 0 at least once on $(0, 1)$. Since $P'(x) < 0$ on $(0, 1)$, P decreases on $[0, 1]$. It follows that P takes on the value zero only once on $[0, 1]$.

Since $P'(x) > 0$ on $(1, 2)$ and $P'(x) < 0$ on $(2, 3)$, P increases on $[1, 2]$ and decreases on $[2, 3]$. Since $P(1)$, $P(2)$, $P(3)$ are all negative, P cannot take on the value 0 between 1 and 3.

Since $P(3) < 0$ and $P(100) > 0$, P takes on the value 0 at least once on $(3, 100)$. Since $P'(x) > 0$ on $(3, 100)$, P increases on $[3, 100]$. It follows that P takes on the value zero only once on $[3, 100]$.

Since $P'(x) > 0$ on $(100, \infty)$, P increases on $[100, \infty)$. Since $P(100) > 0$, P does not take on the value 0 on $[100, \infty)$.

35.

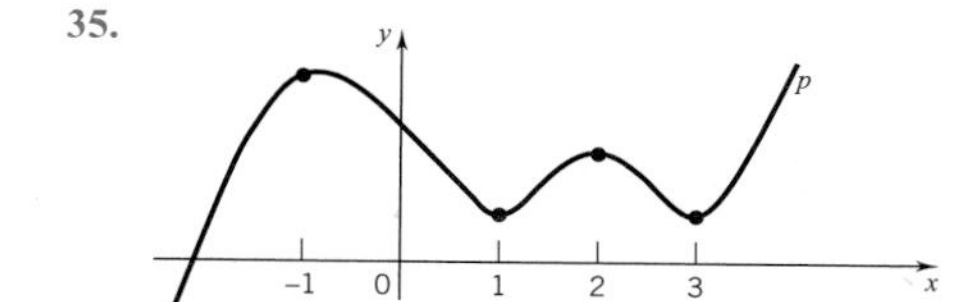

n odd

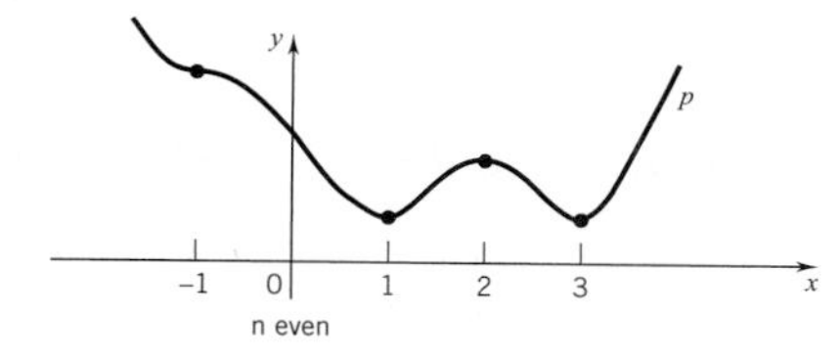

n even

37. $a = 4, b = \pm 2$ **39.** The function f takes on both positive and negative values in every open interval containing $x = 0$.

41. The line through (0, 0) and $(c, f(c))$ has equation $y = \dfrac{f(c)}{c}x$; the tangent line to the graph of f at c has equation $y - f(c) = f'(c)(x - c)$. If D has an extreme value at c, then $\dfrac{f(c)}{c} = -\dfrac{1}{f'(c)}$.

43. (a) $f'(x) = 4x^3 - 4x - 3$; $f'(1) = -3, f'(2) = 21$. Therefore f' has a zero in (1, 2). Since $f''(x) = 12x^2 - 4 > 0$ on [1, 2], f' has exactly one zero in (1, 2). (b) $c \cong 1.3125$; f has a local minimum at c.

45. (a) $f'(x) = 4x^3 - 14x - 8$; $f'(2) = -4, f'(3) = 58$. Therefore f' has a zero in (2, 3). Since $f''(x) = 12x^2 - 14 > 0$ on [2, 3], f' has exactly one zero in (2, 3). (b) $c \cong 2.1091$; f has a local minimum at c.

47. (a) $f'(x) = \cos x + x - 2$; $f'(2) = -0.4161, f'(3) = 0.01$. Therefore f' has a zero in (2, 3). Since $f''(x) = -\sin x + 1 \neq 0$ on [2, 3], f' has exactly one zero in (2, 3). (b) $c \cong 2.9883$; f has a local minimum at c.

49. (a) critical nos.: $-2.085, -1, 0.207, 1.096, 1.544$
local extreme values: $f(-2.085) \cong -6.255, f(-1) = 7, f(0.207) \cong 0.621, f(1.096) \cong 7.097, f(1.544) \cong 4.635$
(b) f is increasing on $[-2.085, -1]$, $[0.207, 1.096]$, $[1.544, 4]$; f is decreasing on $[-4, -2.085]$, $[-1, 0.207]$, $[1.096, 1.544]$

51. (a) critical nos.: $-2.204, -0.654, 0.654, 2.204$
local extreme values: $f(-2.204) \cong 2.226, f(-0.654) \cong -6.364, f(0.654) \cong 6.364, f(2.204) \cong -2.226$
(b) f is increasing on $[-3, -2.204]$, $[-0.654, 0.654]$, $[2.204, 3]$; f is decreasing on $[-2.204, -0.654]$, $[0.654, 2.204]$

53. critical nos.: $-1.326, 0, 1.816$ local maxima at -1.326 and 1.816, local minimum at 0

SECTION 4.4

1. critical no. -2; $f(-2) = 0$ endpt min and absolute min

3. critical nos. 0, 2, 3; $f(0) = 1$ endpt max and absolute max, $f(2) = -3$ local min and absolute min, $f(3) = -2$ endpt max

5. critical no. $2^{-1/3}$; $f(2^{-1/3}) = 3 \cdot 2^{-2/3}$ local min

7. critical nos. $\frac{1}{10}, 2^{-1/3}, 2$; $f(\frac{1}{10}) = 10\frac{1}{100}$ endpt max and absolute max; $f(2^{-1/3}) = 3 \cdot 2^{-2/3}$ local min and absolute min, $f(2) = 4\frac{1}{2}$ endpt max

9. critical nos. $0, \frac{3}{2}, 2$; $f(0) = 2$ endpt max and absolute max, $f(\frac{3}{2}) = -\frac{1}{4}$ local min and absolute min, $f(2) = 0$ endpt max

11. critical nos. $-3, -2, 1$; $f(-3) = -\frac{3}{13}$ endpt max, $f(-2) = -\frac{1}{4}$ local min and absolute min, $f(1) = \frac{1}{5}$ endpt max and absolute max

13. critical nos. $0, \frac{1}{4}, 1$; $f(0) = 0$ endpt min and absolute min, $f(\frac{1}{4}) = \frac{1}{16}$ local max, $f(1) = 0$ local min and absolute min

15. critical nos. 2, 3; $f(2) = 2$ local max and absolute max, $f(3) = 0$ endpt min

17. critical no. 1; no extreme values

19. critical nos. $0, \frac{5}{6}\pi, \pi$; $f(0) = -\sqrt{3}$ endpt min and absolute min, $f(\frac{5}{6}\pi) = \frac{7}{4}$ local max and absolute max, $f(\pi) = \sqrt{3}$ endpt min

21. critical nos. $0, \pi$; $f(0) = 5$ endpt max and absolute max, $f(\pi) = -5$ endpt min and absolute min

23. critical nos. $-\frac{1}{3}\pi, 0$; $f(-\frac{1}{3}\pi) = \frac{1}{3}\pi - \sqrt{3}$ endpt min and absolute min, no absolute max

25. critical nos. 0, 1, 4, 7; $f(0) = 0$ endpt min, $f(1) = -2$ local min and absolute min, $f(4) = 1$ local max and absolute max, $f(7) = -2$ endpt min and absolute min

27. critical nos. $-2, -1, 1, 3$; $f(-2) = 5$ endpt max, $f(-1) = 2$ local min and absolute min, $f(1) = 6$ local max and absolute max, $f(3) = 2$ local min and absolute min

29. critical nos. $-3, -1, 0, 2$; $f(-3) = 2$ endpt max and absolute max, $f(-1) = 0$ local min, $f(0) = 2$ local max and absolute max, $f(2) = -2$ local min and absolute min

31.

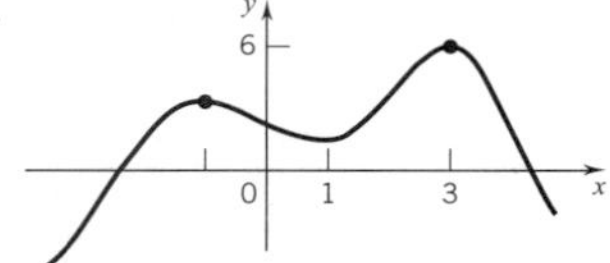

33. Not possible: $f(1) > 0$ and f is increasing on (1, 3). Therefore $f(3)$ cannot be 0.

35. The discriminant of the quadratic polynomial $p'(x) = 3x^2 + 2ax + b$ is $4a^2 - 12b = 4(a^2 - 3b)$. If $a^2 < 3b$, then p' has no real zeros.

37. By contradiction. If f is continuous at c, then $f(c)$ is not a local maximum by the first-derivative test (4.3.4).

39. If f is not differentiable on (a, b), then f has a critical number at each point c in (a, b) where $f'(c)$ does not exist. If f is differentiable on (a, b), then there exists c in (a, b) where $f'(c) = (f(b) - f(a))/(b - a)$ (mean-value theorem). With $f(b) = f(a)$, we have $f'(c) = 0$ and thus c is a critical point of f.

41.
$$P(x) - M \underset{\text{for } x > 0}{\geq} a_0x^n - (|a_1|x^{n-1} + \cdots + |a_{n-1}|x + |a_n| + M)$$
$$\underset{\text{for } x > 1}{\geq} a_0x^n - (|a_1| + \cdots + |a_{n-1}| + |a_n| + M) \geq 0 \quad \text{for } x \geq \left(\frac{|a_1| + \cdots + |a_{n-1}| + |a_n| + M}{a_0}\right)^{1/n}$$

43. Let x be the length and y the width of a rectangle with diagonal of length c. Then $y = \sqrt{c^2 - x^2}$ and the area $A = x\sqrt{c^2 - x^2}$. The maximum value of A occurs when $x = y = \dfrac{c}{\sqrt{2}}$.

45. (a) Use the wire to form a square. (b) Use $\dfrac{9L}{(4\sqrt{3} + 9)} \cong 0.57L$ for the triangle and the remainder for the square.

47. critical nos.: -1.452, local max; 0.727, local min.

49. critical nos.: -1.683, local max; -0.284, local min; 0.645, local max; 1.760, local min. absolute min at $-\pi$, absolute max at π.

SECTION 4.5

1. 400 3. 20 by 10 ft 5. 32 7. 100 by 150 ft with the divider 100 ft long

9. radius of semicircle $\dfrac{90}{12 + 5\pi} \cong 3.25$ ft; height of rectangle $\dfrac{90 + 30\pi}{12 + 5\pi} \cong 6.65$ ft 11. $x = 2, y = \frac{3}{2}$ 13. $-\frac{5}{2}$ 15. $\frac{10}{3}\sqrt{3}$ in. by $\frac{5}{3}\sqrt{3}$ in.

17. equilateral triangle with side 4 19. (1, 1) 21. height of rectangle $\frac{15}{11}(5 - \sqrt{3}) \cong 4.46$ in.; side of triangle $\frac{10}{11}(6 + \sqrt{3}) \cong 7.03$ in.

23. $\frac{5}{3}$ 25. $(0, \sqrt{3})$ 27. $5\sqrt{5}$ ft 29. 54 by 72 in. 31. (a) use it all for the circle (b) use $28\pi/(4 + \pi) \cong 12.32$ in. for the circle

33. base radius $\frac{10}{3}$ and height $\frac{8}{3}$ 35. 10 by 10 by 12.5 ft 37. equilateral triangle with side $2r\sqrt{3}$ 39. base radius $\frac{1}{3}R\sqrt{6}$ and height $\frac{2}{3}R\sqrt{3}$

41. base radius $\frac{2}{3}R\sqrt{2}$ and height $\frac{4}{3}R$ 43. \$160,000 45. $\tan\theta = m$ 47. $6\sqrt{6}$ ft 49. 125 customers

SECTION 4.6

1. (a) increasing on $[a, b]$, $[d, n]$, decreasing on $[b, d]$, $[n, p]$
   (b) concave up on (c, k), (l, m), concave down on (a, c), (k, l), (m, p); points of inflection at $x = c, k, l$, and m.

3. concave down on $(-\infty, 0)$, concave up on $(0, \infty)$ 5. concave down on $(-\infty, 0)$, concave up on $(0, \infty)$; pt of inflection (0, 2)

7. concave up on $(-\infty, -\frac{1}{3}\sqrt{3})$, concave down on $(-\frac{1}{3}\sqrt{3}, \frac{1}{3}\sqrt{3})$, concave up on $(\frac{1}{3}\sqrt{3}, \infty)$; pts of inflection $(-\frac{1}{3}\sqrt{3}\ -\frac{5}{36})$, $(\frac{1}{3}\sqrt{3}, -\frac{5}{36})$

9. concave down on $(-\infty, -1)$ and on $(0, 1)$, concave up on $(-1, 0)$ and on $(1, \infty)$; pt of inflection (0, 0)

11. concave up on $(-\infty, -\frac{1}{3}\sqrt{3})$, concave down on $(-\frac{1}{3}\sqrt{3}, \frac{1}{3}\sqrt{3})$, concave up on $(\frac{1}{3}\sqrt{3}, \infty)$; pts of inflection $(-\frac{1}{3}\sqrt{3}, \frac{4}{9})$, $(\frac{1}{3}\sqrt{3}, \frac{4}{9})$

13. concave up on $(0, \infty)$ 15. concave down on $(-\infty, -2)$, concave up on $(-2, \infty)$; pt of inflection $(-2, 0)$

17. concave up on $(0, \frac{1}{4}\pi)$, concave down on $(\frac{1}{4}\pi, \frac{3}{4}\pi)$, concave up on $(\frac{3}{4}\pi, \pi)$; pts of inflection $(\frac{1}{4}\pi, \frac{1}{2})$ and $(\frac{3}{4}\pi, \frac{1}{2})$

19. concave up on $(0, \frac{1}{12}\pi)$, concave down on $(\frac{1}{12}\pi, \frac{5}{12}\pi)$, concave up on $(\frac{5}{12}\pi, \pi)$; pts of inflection $(\frac{1}{12}\pi, \frac{1}{2} + \frac{1}{144}\pi^2)$ and $(\frac{5}{12}\pi, \frac{1}{2} + \frac{25}{144}\pi^2)$

21. (a) f increases on $(-\infty, -\sqrt{3}]$ and $[\sqrt{3}, \infty)$, decreases on $[-\sqrt{3}, \sqrt{3}]$
    (b) $f(-\sqrt{3}) \cong 10.39$ local max; $f(\sqrt{3}) \cong -10.39$ local min
    (c) concave down on $(-\infty, 0)$, concave up on $(0, \infty)$
    (d) (0, 0) is a point of inflection

23. (a) f decreases on $(-\infty, -1]$ and $[1, \infty)$, increases on $[-1, 1]$
    (b) $f(-1) = -1$ local min; $f(1) = 1$ local max
    (c) concave down on $(-\infty, -\sqrt{3})$ and $(0, \sqrt{3})$, concave up on $(-\sqrt{3}, 0)$ and $(\sqrt{3}, \infty)$
    (d) points of inflection $\left(-\sqrt{3}, -\dfrac{\sqrt{3}}{2}\right)$, (0, 0), $\left(\sqrt{3}, \dfrac{\sqrt{3}}{2}\right)$

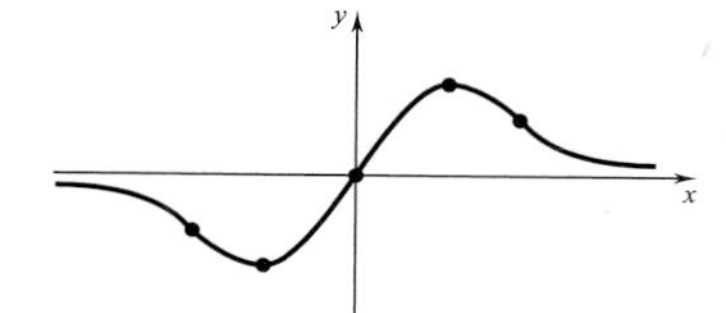

25. (a) increasing on $[-\pi, \pi]$
    (b) no local max or min
    (c) concave up on $(-\pi, 0)$, concave down on $(0, \pi)$
    (d) point of inflection (0, 0)

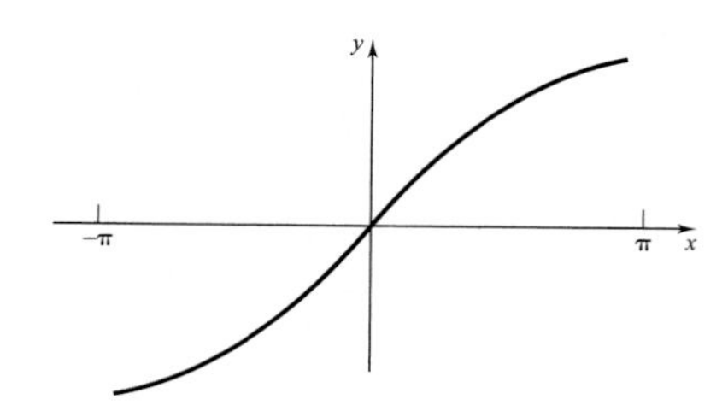

27. (a) increasing on $(-\infty, \infty)$
    (b) no local max or min
    (c) concave down on $(-\infty, 0)$, concave up on $(0, 1)$, no concavity on $(1, \infty)$
    (d) point of inflection (0, 0)

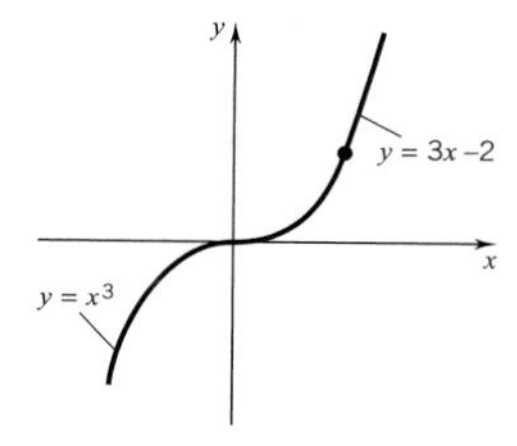

29.

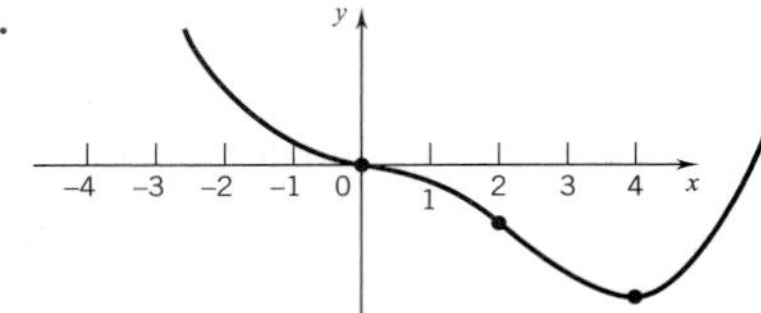

31.

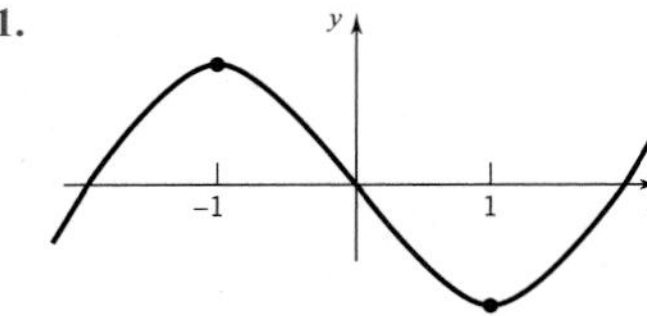

33. $d = \frac{1}{3}(a + b + c)$ 35. $a = -\frac{1}{2}, b = \frac{1}{2}$ 37. $A = 18, B = -4$ 39. $f(x) = x^3 - 3x^2 + 3x - 3$

41. (a) $p''(x) = 6x - 2a$ has exactly one zero
(b) $p'(x) = 3x^2 + 2ax + b$ has real zeros iff $a^2 \geq 3b$.

43. (a) concave up on $(-\infty, -0.913)$ and $(0.913, \infty)$; concave down on $(-0.913, 0.913)$
(b) points of inflection at $x \cong -0.913, 0.913$

45. (a) concave up on $(-\pi, -1.996)$ and $(-0.345, 2.550)$; concave down on $(-1.996, -0.345)$, $(2.550, \pi)$
(b) points of inflection at $x \cong -1.996, -0.345, 2.550$

47. (a) concave up on $(-2.726, 0.402)$ and $(1.823, 3)$; concave down on $(-3, -2.726)$, $(0.402, 1.823)$
(b) points of inflection at $x \cong -2.726, 0.402, 1.823$

SECTION 4.7

1. (a) ∞ (b) $-\infty$ (c) ∞ (d) 1 (e) 0 (f) $x = -1, x = -1$ (g) $y = 0, y = 1$

3. vertical: $x = \frac{1}{3}$; horizontal: $y = \frac{1}{3}$ 5. vertical: $x = 2$; horizontal: none 7. vertical: $x = \pm 3$; horizontal: $y = 0$

9. vertical: $x = -\frac{4}{3}$; horizontal: $y = \frac{4}{9}$ 11. vertical: $x = \frac{5}{2}$; horizontal: $y = 0$ 13. vertical: none; horizontal: $y = \pm\frac{3}{2}$

15. vertical: $x = 1$; horizontal: $y = 0$ 17. vertical: none; horizontal: $y = 0$ 19. vertical: $x = (2n + \frac{1}{2})\pi$; horizontal: none

21. neither 23. cusp 25. tangent 27. neither 29. cusp 31. cusp 33. neither

35.

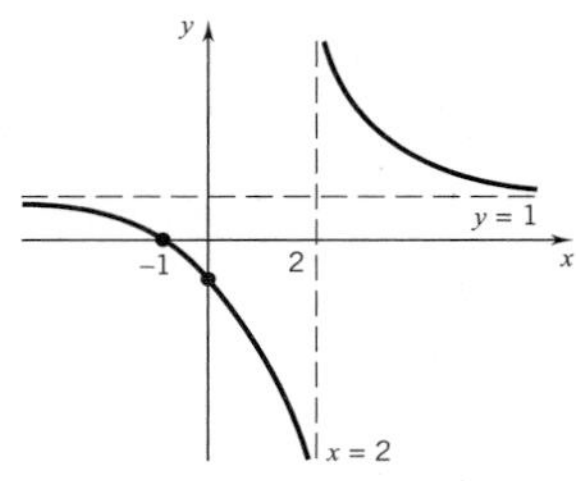

37.

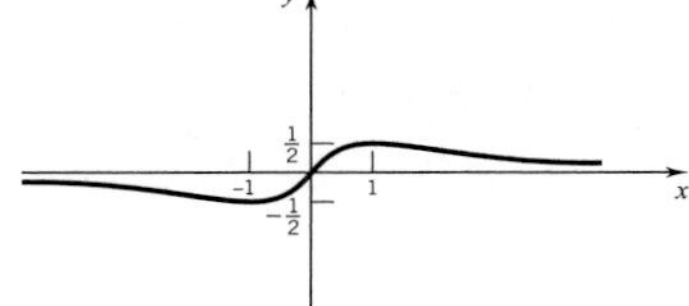

39. (a) increasing on $(-\infty, -1]$, $[1, \infty)$;
decreasing on $[-1, 1]$
(b) concave down on $(-\infty, 0)$;
concave up on $(0, \infty)$
vertical tangent at $x = 0$

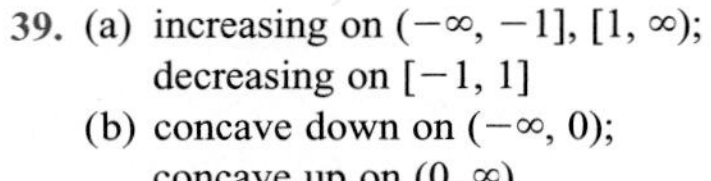

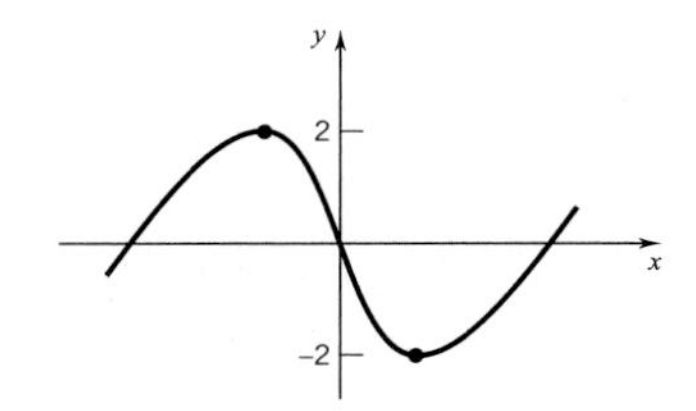

41. (a) decreasing on $[0, 2]$;
increasing on $(-\infty, 0]$, $[2, \infty)$
(b) concave up on $(-1, 0)$, $(0, \infty)$;
concave down on $(-\infty, -1)$
vertical cusp at $x = 0$

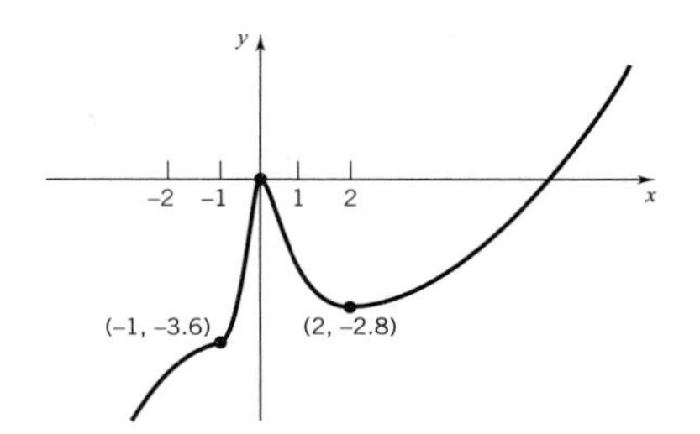

43. (a) decreasing on $[0, 1)$;
increasing on $(1, \infty)$
(b) concave up on $(0, 1)$;
concave down on $(1, \infty)$
vertical tangent line at $x = 0$

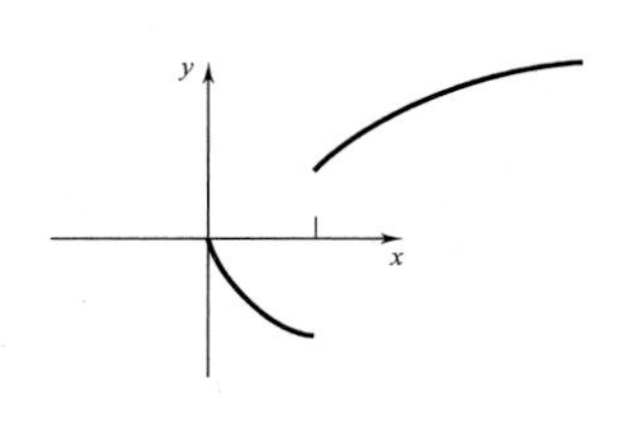

45. $x = 1$ vertical asymptote
$y = 0$ and $y = 2$ horizontal asymptotes

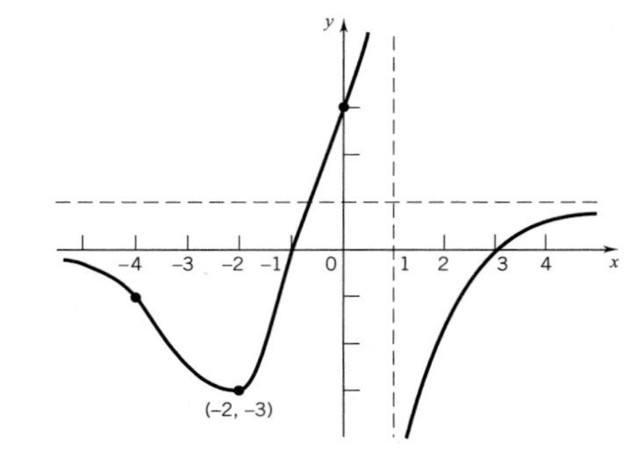

47. vertical cusp at $x = 0$

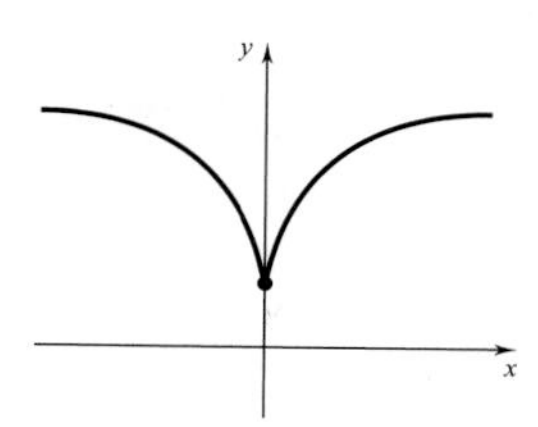

49. vertical tangent line at $x = 0$
$y = 1$ and $y = -1$ horizontal asymptotes

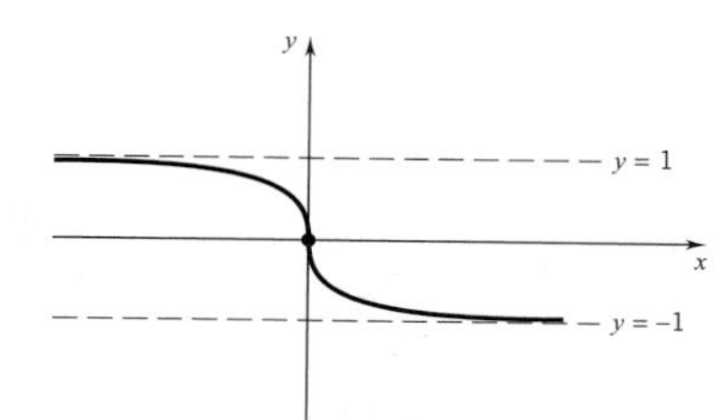

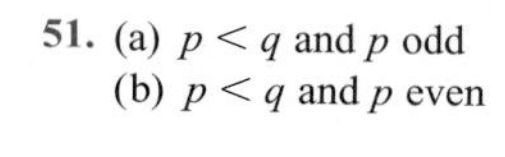

51. (a) $p < q$ and p odd
(b) $p < q$ and p even

53.

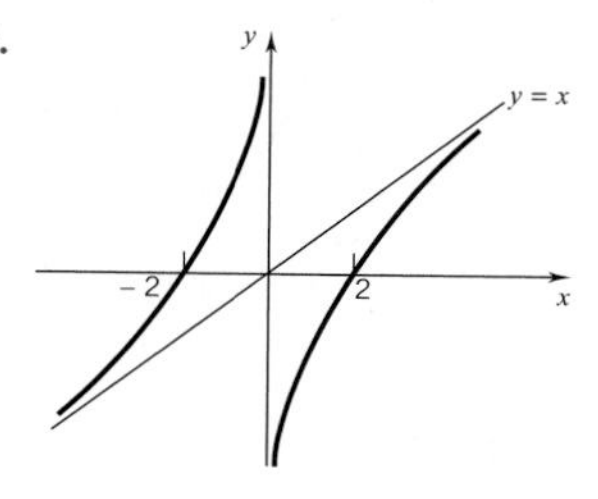

55.

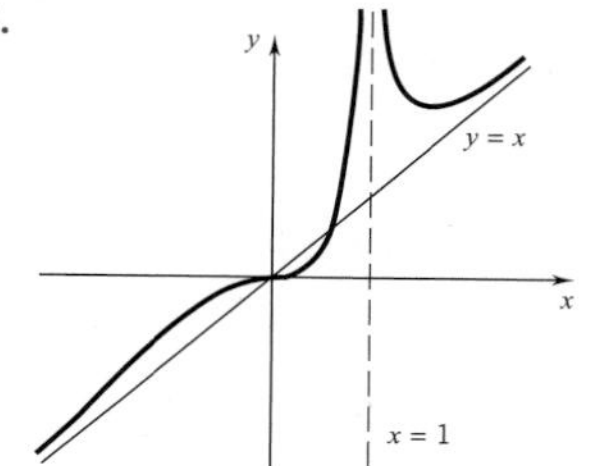

SECTION 4.8

1.

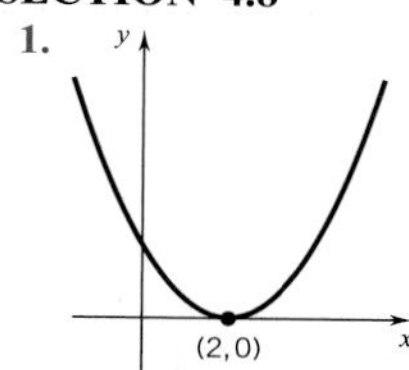

3.

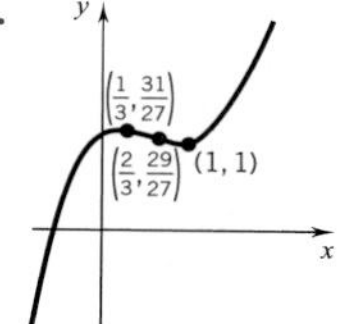

5.

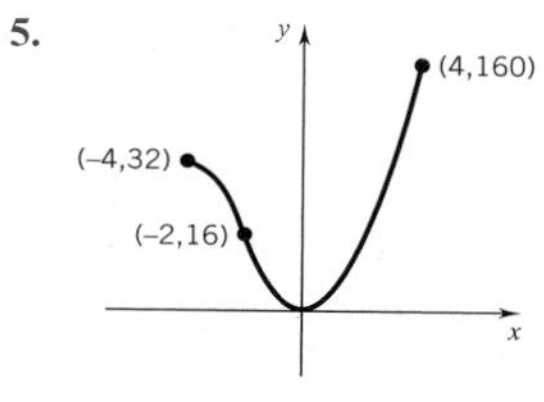

7.

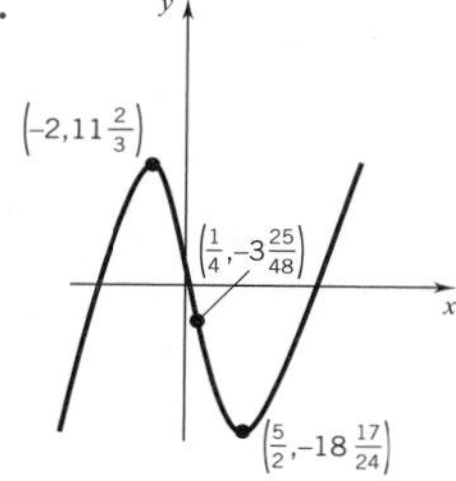

9.

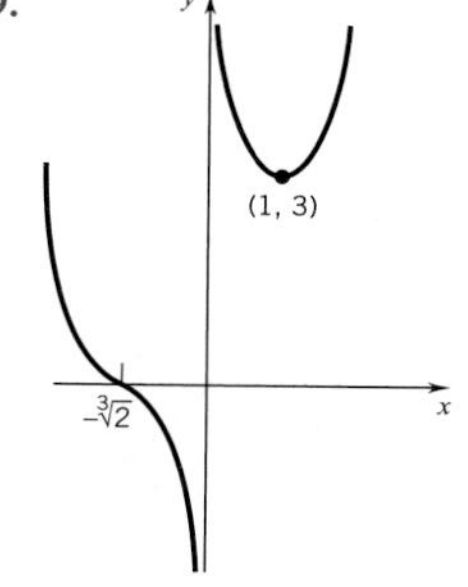

11.

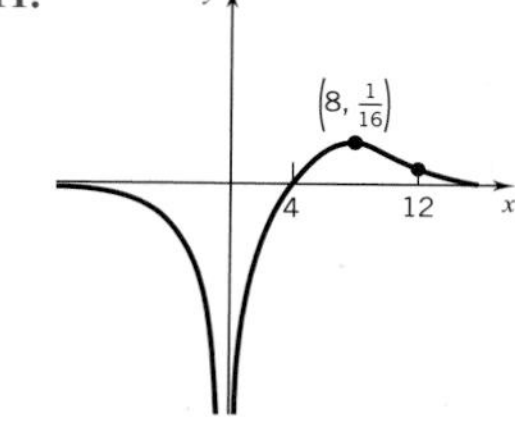

13.

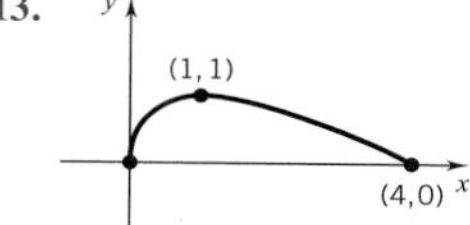

15.

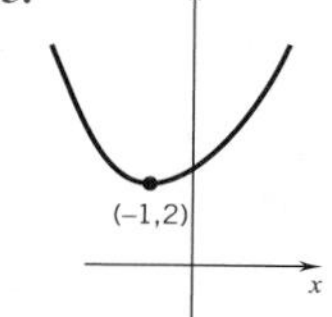

17.

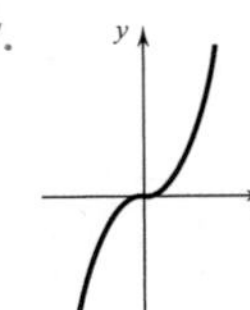

19.

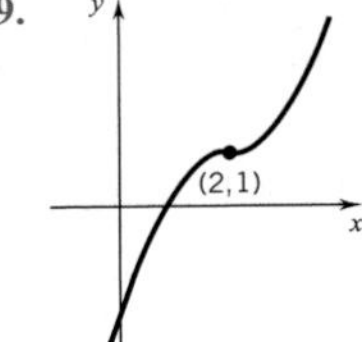

21.

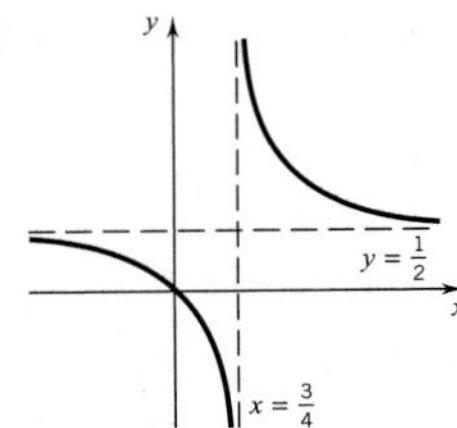

23.

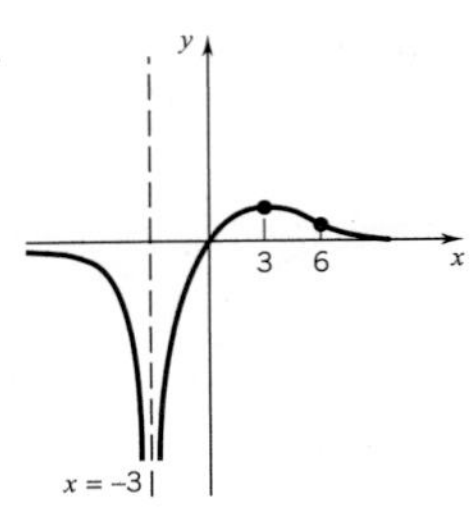

25.

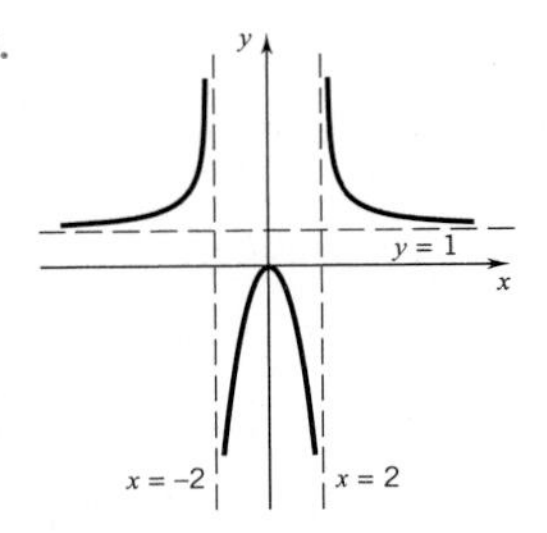

27.

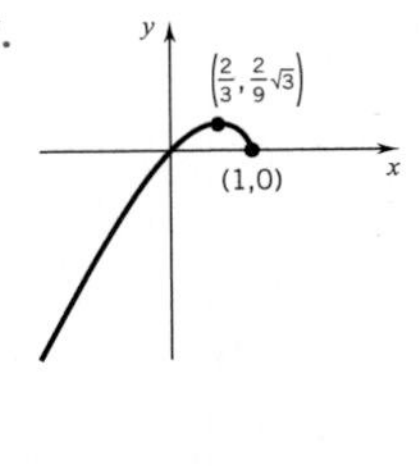

29.

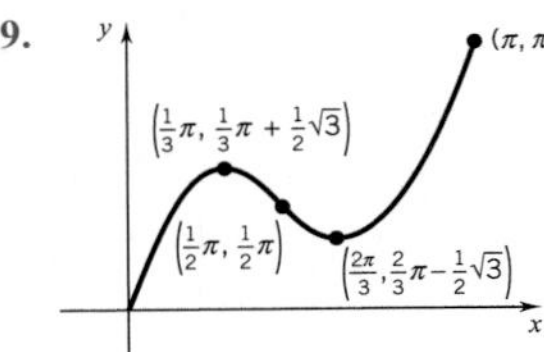

31.

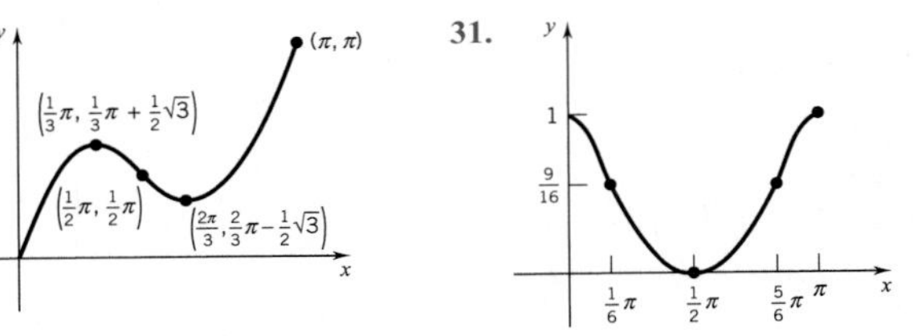

33.

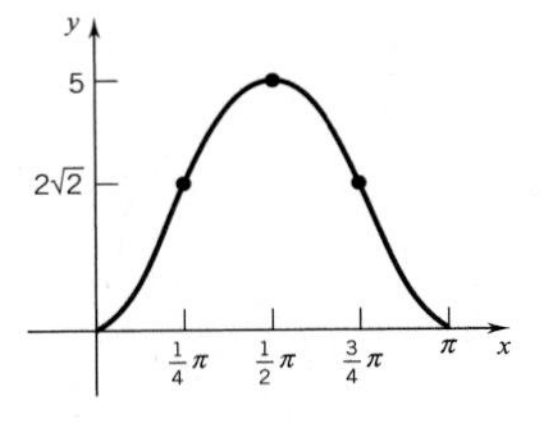

35.

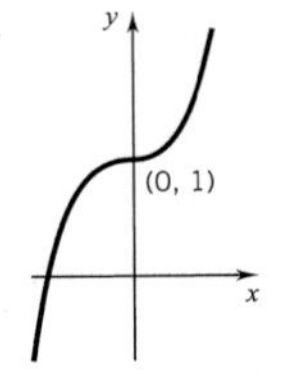

37.

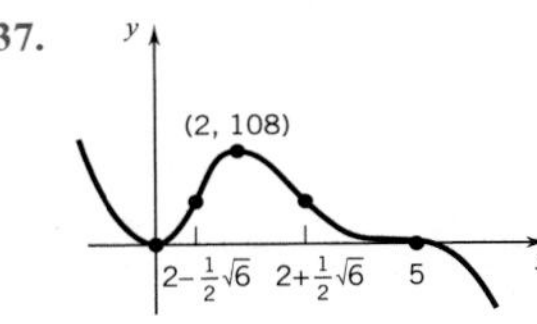

39.

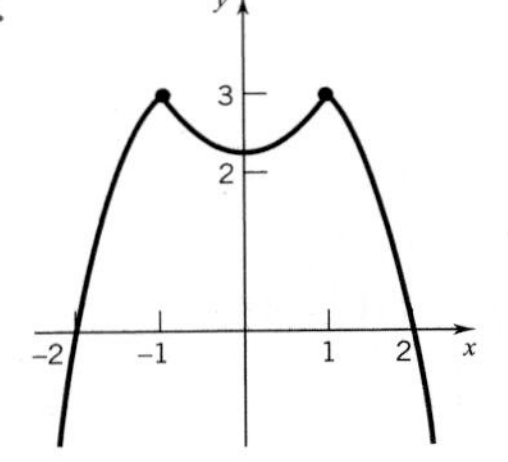

41.

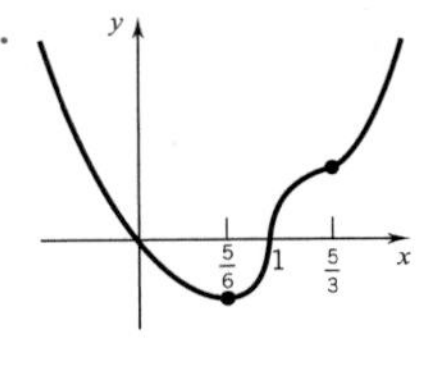

43.

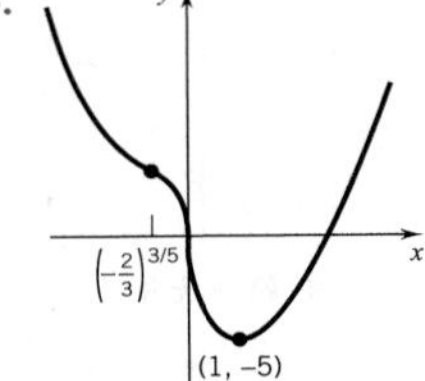

45.

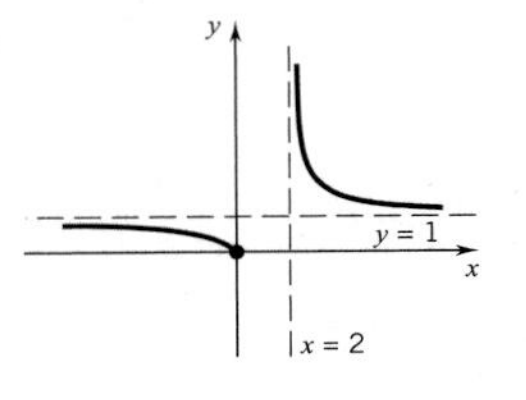

47.

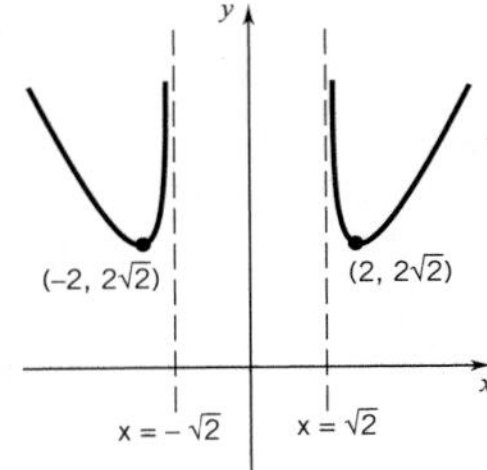

49.

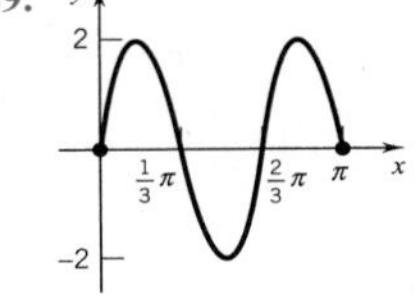

51.

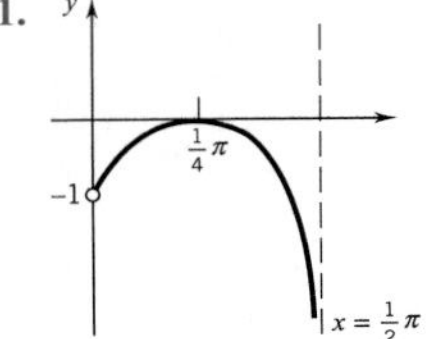

53.

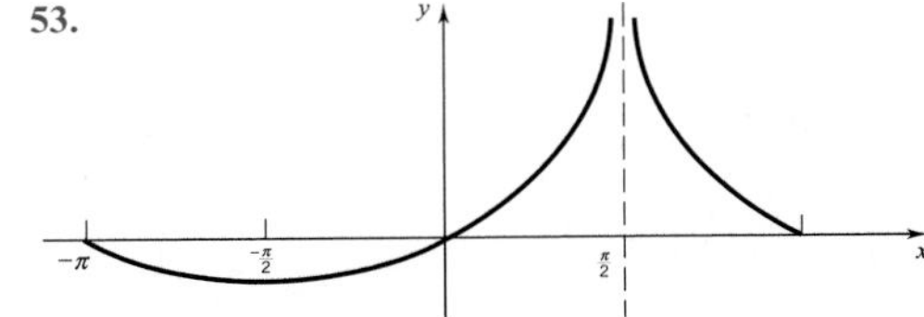

55. (a) increasing on $(-\infty, -1]$, $(0, 1]$, $[3, \infty)$; decreasing on $[-1, 0)$, $[1, 3]$ critical nos. $x = -1, 1, 3$
(b) concave up on $(-\infty, -3)$, $(2, \infty)$ concave down on $(-3, 0)$, $(0, 2)$ (c)

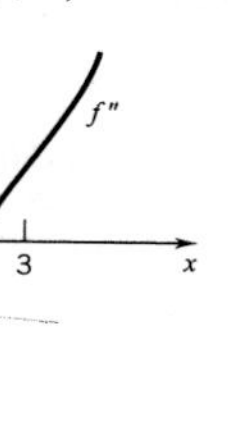

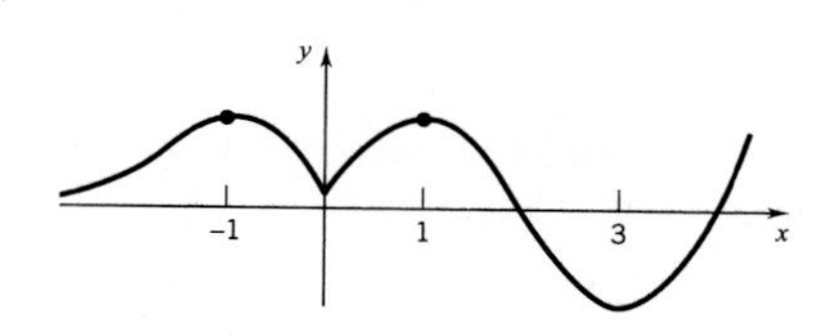

57. Solve for y: $y = \pm\dfrac{b}{a}x\sqrt{1 - \dfrac{a^2}{x^2}}$. For $|x|$ large, $y \cong \pm\dfrac{b}{a}x$.

CHAPTER 5

SECTION 5.2

1. $L_f(P) = \frac{5}{8}$, $U_f(P) = \frac{11}{8}$
3. $L_f(P) = \frac{9}{64}$, $U_f(P) = \frac{37}{64}$
5. $L_f(P) = \frac{17}{16}$, $U_f(P) = \frac{25}{16}$
7. $L_f(P) = \frac{7}{16}$, $U_f(P) = \frac{25}{16}$
9. $L_f(P) = \frac{3}{16}$, $U_f(P) = \frac{43}{32}$
11. $L_f(P) = \frac{1}{6}\pi$, $U_f(P) = \frac{11}{12}\pi$
13. (a) $L_f(P) \le U_f(P)$ but $3 \not\le 2$
(b) $L_f(P) \le \int_{-1}^{1} f(x)\,dx \le U_f(P)$ but $3 \not\le 2 \le 6$
(c) $L_f(P) \le \int_{-1}^{1} f(x)\,dx \le U_f(P)$ but $3 \le 10 \not\le 6$

15. (a) $L_f(P) = -3x_1(x_1 - x_0) - 3x_2(x_2 - x_1) - \cdots - 3x_n(x_n - x_{n-1})$
$U_f(P) = -3x_0(x_1 - x_0) - 3x_1(x_2 - x_1) - \cdots - 3x_{n-1}(x_n - x_{n-1})$
(b) $-\frac{3}{2}(b^2 - a^2)$

17. $L_f(P) = x_0^3(x_1 - x_0) + x_1^3(x_2 - x_1) + \cdots + x_{n-1}^3(x_n - x_{n-1})$
$U_f(P) = x_1^3(x_1 - x_0) + x_2^3(x_2 - x_1) + \cdots + x_n^3(x_n - x_{n-1})$
For each index i, $x_{i-1}^3 \le \frac{1}{4}(x_i^3 + x_i^2x_{i-1} + x_ix_{i-1}^2 + x_{i-1}^3) \le x_i^3$ and thus by the hint $x_{i-1}^3(x_i - x_{i-1}) \le \frac{1}{4}(x_i^4 - x_{i-1}^4) \le x_i^3(x_i - x_{i-1})$. Adding up these inequalities, we find that $L_f(P) \le \frac{1}{4}(x_n^4 - x_0^4) \le U_f(P)$. Since $x_n = 1$ and $x_0 = 0$, the middle term is $\frac{1}{4}$. Thus the integral is $\frac{1}{4}$.

19. Let f be continuous and increasing on $[a, b]$, and let P be a regular partition. Then

$$U_f(P) = [f(x_1) + f(x_2) + \cdots + f(x_n)]\Delta x$$
$$L_f(P) = [f(x_0) + f(x_1) + \cdots + f(x_{n-1})]\Delta x$$

and $U_f(P) - L_f(P) = [f(x_n) - f(x_0)]\Delta x = [f(b) - f(a)]\Delta x$.

21. (a) From Definition 5.2.3, $0 \le I - L_f(P) \le U_f(P) - L_f(P)$
(b) Use Definition 5.2.3

23. (a) Let $x_1, x_2 \in [0, 2]$, $x_2 \ge x_1$. Then

$$f(x_2) - f(x_1) = \frac{(x_2 + x_1)(x_2 - x_1)}{\sqrt{1 + x_2^2} + \sqrt{1 + x_1^2}} \ge 0.$$

(b) $n = 25$

25. Let S be the set of positive integers for which the statement is true. Since $1 = \dfrac{1(2)}{2}$, $1 \in S$. Assume $k \in S$. Then

$$\begin{aligned} 1 + 2 + \cdots + k + k + 1 &= (1 + 2 + \cdots + k) + k + 1 \\ &= \frac{k(k+1)}{2} + k + 1 = \frac{(k+1)(k+2)}{2}. \end{aligned}$$

Thus $k + 1 \in S$ and so S is the set of positive integers.

27. Let $f(x) = x$ on $[0, b]$ and let $P = \{x_0, x_1, \ldots, x_n\}$ be a regular partition.

(a) $L_f(P) = \left[0 \cdot \dfrac{b}{n} + 1 \cdot \dfrac{b}{n} + \cdots + (n-1) \cdot \dfrac{b}{n}\right]\dfrac{b}{n}$
$= \dfrac{b^2}{n^2}[1 + 2 + \cdots + (n-1)]$

(b) $U_f(P) = \left[1 \cdot \dfrac{b}{n} + 2 \cdot \dfrac{b}{n} + \cdots + n \cdot \dfrac{b}{n}\right]\dfrac{b}{n}$
$= \dfrac{b^2}{n^2}[1 + 2 + \cdots + n]$

(c) $L_f(P) = \dfrac{b^2}{n^2} \cdot \dfrac{(n-1)n}{2} \to \dfrac{b^2}{2}$ as $n \to \infty$
$U_f(P) = \dfrac{b^2}{n^2} \cdot \dfrac{n(n+1)}{2} \to \dfrac{b^2}{2}$ as $n \to \infty$
Thus $\int_0^1 x\,dx = \dfrac{b^2}{2}$.

29. Let P be an arbitrary partition of $[0, 4]$. Since each $m_i = 2$ and each $M_i \geq 2$.

$$L_g(P) = 2\Delta x_i + \cdots + 2\Delta x_n = 2(\Delta x_1 + \cdots + \Delta x_n) = 2 \cdot 4 = 8,$$
$$U_g(P) \geq 2\Delta x_1 + \cdots + 2\Delta x_n = 2(\Delta x_1 + \cdots + \Delta x_n) = 2 \cdot 4 = 8.$$

Thus $L_g(P) \leq 8 \leq U_g(P)$ for all partitions P of $[0, 4]$.
Uniqueness: Suppose that

$$(*) \qquad L_g(P) \leq I \leq U_g(P) \qquad \text{for all partitions } P \text{ of } [0, 4].$$

Since $L_g(P) = 8$ for all P, I is at least 8. Suppose now that $I > 8$ and choose a partition P of $[0, 4]$ with max $\Delta x_i < \frac{1}{5}(I - 8)$ and $0 = x_1 < \cdots < x_{i-1} < 3 < x_i < \cdots < x_n = 4$. Then

$$\begin{aligned} U_g(P) &= 2\Delta x_1 + \cdots + 2\Delta x_{i-1} + 7\Delta x_i + 2\Delta x_{i+1} + \cdots + 2\Delta x_n \\ &= 2(\Delta x_1 + \cdots + \Delta x_n) + 5\Delta x_i = 8 + 5\Delta x_i < 8 + \tfrac{5}{5}(I - 8) = I \end{aligned}$$

and I does not satisfy $(*)$. This contradiction proves that I is not greater than 8 and therefore $I = 8$.

31. Let $P = \{x_0, x_1, \ldots, x_n\}$ be any partition of $[2, 10]$.
(a) Since each subinterval of $[2, 10]$ contains both rational and irrational numbers,

$$L_f(P) = 4\Delta x_1 + 4\Delta x_2 + \cdots + 4\Delta x_n = 4(10 - 2) = 32$$
$$U_f(P) = 7\Delta x_1 + 7\Delta x_2 + \cdots + 7\Delta x_n = 7(10 - 2) = 56.$$

(b) There is more than one number I that satisfies $L_f(P) \leq I \leq U_f(P)$ for all partitions P.
(c) See (a).

SECTION 5.3

1. (a) 5 (b) -2 (c) -1 (d) 0 (e) -4 (f) 1

3. With $P = \{1, \frac{3}{2}, 2\}$ and $f(x) = \dfrac{1}{x}$, we have $0.5 < \frac{7}{12} = L_f(P) \leq \displaystyle\int_1^2 \frac{dx}{x} \leq U_f(P) = \frac{5}{6} < 1$.

5. (a) $F(0) = 0$ (b) $F'(x) = x\sqrt{x+1}$ (c) $F'(2) = 2\sqrt{3}$ (d) $F(2) = \displaystyle\int_0^2 t\sqrt{t+1}\,dt$ (e) $-F(x) = \displaystyle\int_x^0 t\sqrt{t+1}\,dt$

7. (a) $F'(x) = \dfrac{1}{x}$, $x > 0$; F is increasing on $(0, \infty)$. (c)

(b) $F''(x) = \dfrac{-1}{x^2}$, $x > 0$;
the graph of F is concave down on $(0, \infty)$.

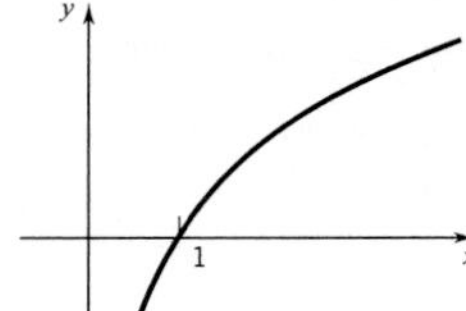

9. (a) $\frac{1}{10}$ (b) $\frac{1}{9}$ (c) $\frac{4}{37}$ (d) $\dfrac{-2x}{(x^2+9)^2}$

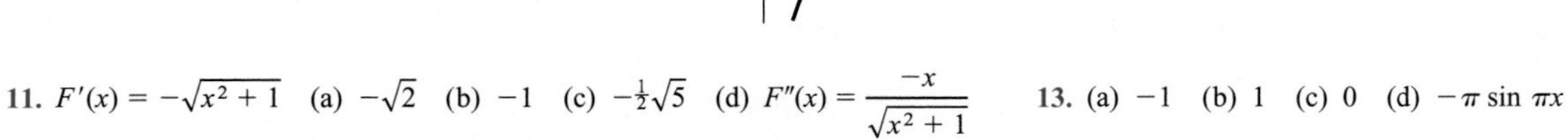

11. $F'(x) = -\sqrt{x^2+1}$ (a) $-\sqrt{2}$ (b) -1 (c) $-\frac{1}{2}\sqrt{5}$ (d) $F''(x) = \dfrac{-x}{\sqrt{x^2+1}}$

13. (a) -1 (b) 1 (c) 0 (d) $-\pi \sin \pi x$

15. (a) Since $P_1 \subseteq P_2$, $U_f(P_2) \leq U_f(P_1)$ but $5 \not\leq 4$. (b) Since $P_1 \subseteq P_2$, $L_f(P_1) \leq L_f(P_2)$ but $5 \not\leq 4$.

17. $F'(x) = 3x^5 \cos(x^3)$.

19. $F'(x) = 2x[\sin^2(x^2) - x^3]$.

21. $F'(x) = 4x^3 \sin(x^2) - 2x \sin x$ (assuming $x > 0$).

23. (a) $f(0) = \frac{1}{2}$ (b) $2, -2$

25. By the hint $F(b) - F(a) = F'(c)(b - a)$ for some c in (a, b). The desired result follows by observing that

$$F(b) = \int_a^b f(t)\,dt, \qquad F(a) = 0, \qquad \text{and} \qquad F'(c) = f(c).$$

27. Set $G(x) = \displaystyle\int_a^x f(t)\,dt$. Then $F(x) = \displaystyle\int_c^a f(t)\,dt + G(x)$. By (5.3.5) G, and thus F, is continuous on $[a, b]$, is differentiable on (a, b), and $F'(x) = G'(x) = f(x)$ for all x in (a, b).

SECTION 5.4

1. -2 **3.** 1 **5.** $\frac{28}{3}$ **7.** $\frac{32}{3}$ **9.** $\frac{2}{3}$ **11.** 0 **13.** $\frac{5}{72}$ **15.** $\frac{13}{2}$ **17.** $-\frac{4}{15}$ **19.** $\frac{1}{18}(2^{18} - 1)$ **21.** $\frac{1}{6}a^2$ **23.** $\frac{7}{4}$

25. $-\frac{1}{12}$ **27.** $\frac{21}{2}$ **29.** 1 **31.** 2 **33.** $2 - \sqrt{2}$ **35.** 0 **37.** $\dfrac{\pi}{9} - 2\sqrt{3}$ **39.** $\sqrt{13} - 2$

41. (a) $\displaystyle\int_2^x \frac{dt}{t}$ (b) $-3 + \displaystyle\int_2^x \frac{dt}{t}$ **43.** $\dfrac{32}{3}$ **45.** $2 + \sqrt{2}$ **47.** (a) $3/2$ (b) $5/2$ **49.** (a) $4/3$ (b) 4

51. (a) $x(t) = 5t^2 - \frac{1}{3}t^3, 0 \le t \le 10$ (b) At $t = 5$; $x(5) = \dfrac{250}{3}$ **53.** $\dfrac{13}{2}$ **55.** $\dfrac{2+\sqrt{3}}{2} + 2\pi$

57. (a) $g(x) = \begin{cases} x^2 + 2x + 2, & -2 \le x \le 0 \\ 2x + 2, & 0 < x \le 1 \\ -x^2 + 4x + 1, & 1 < x \le 2 \end{cases}$ (b)

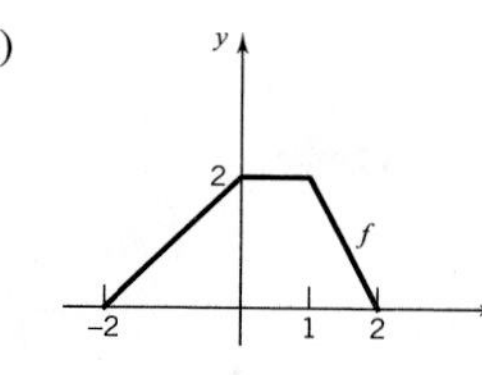

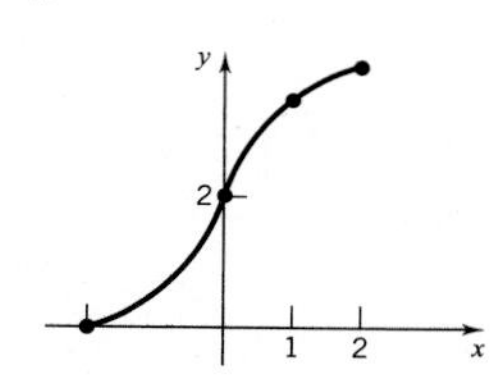

(c) f is differentiable on $(-2, 0)$, $(0, 1)$, $(1, 2)$; g is differentiable on $(-2, 2)$.

59. $f(x)$ and $f(x) - f(a)$, respectively

SECTION 5.5

1. $\frac{9}{4}$ **3.** $\frac{38}{3}$ **5.** $\frac{47}{15}$ **7.** $\frac{5}{3}$ **9.** $\frac{1}{2}$

11. area $= \frac{1}{3}$

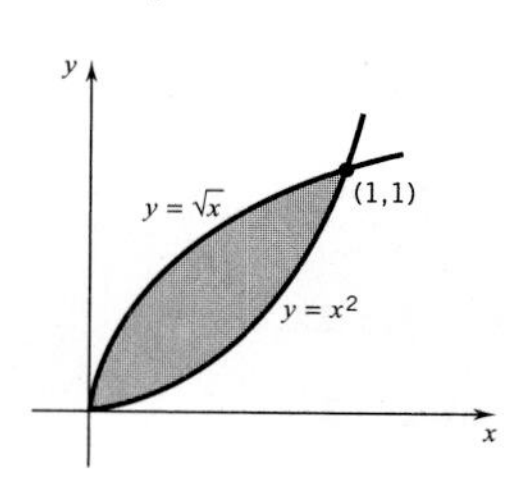

13. area $= \frac{9}{2}$

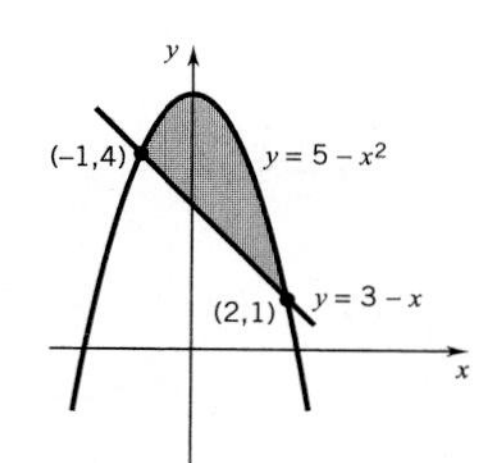

15. area $= \frac{64}{3}$

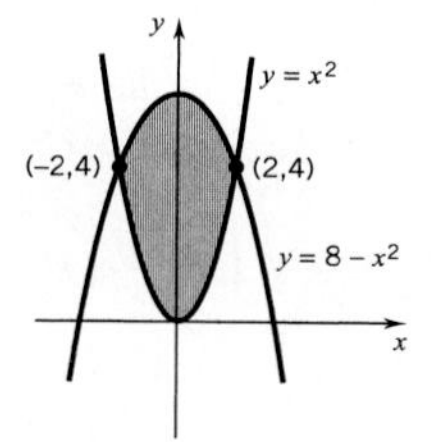

17. area $= 10$

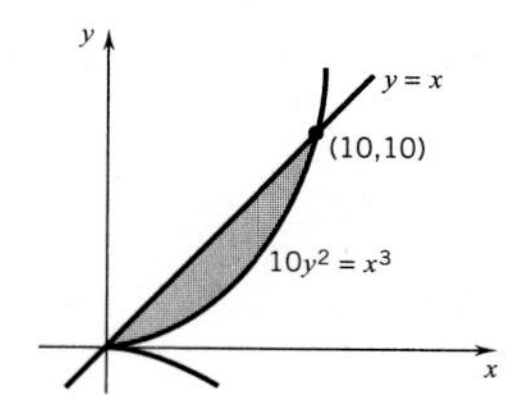

19. area $= \frac{32}{3}$

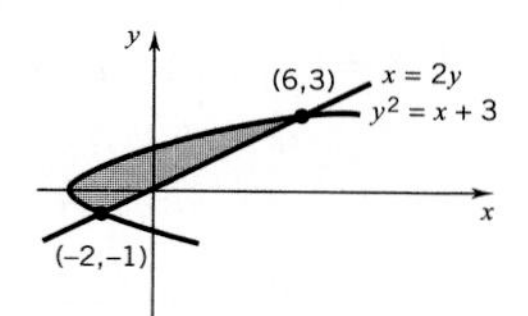

21. area $= 4$

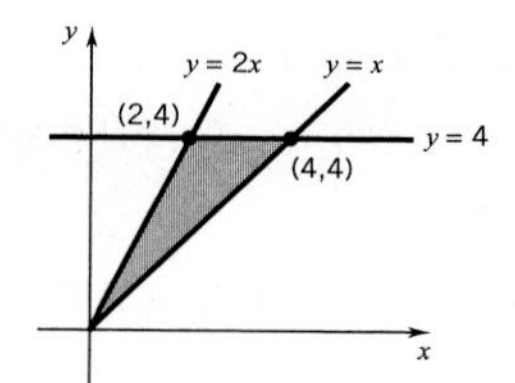

23. area $= 2 + \frac{2}{3}\pi^3$

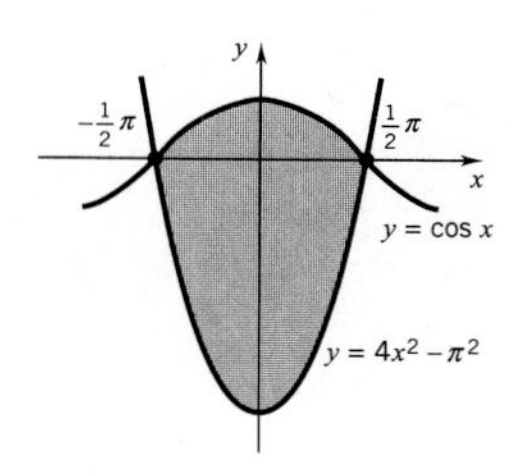

25. area $= \frac{1}{8}\pi^2 - 1$

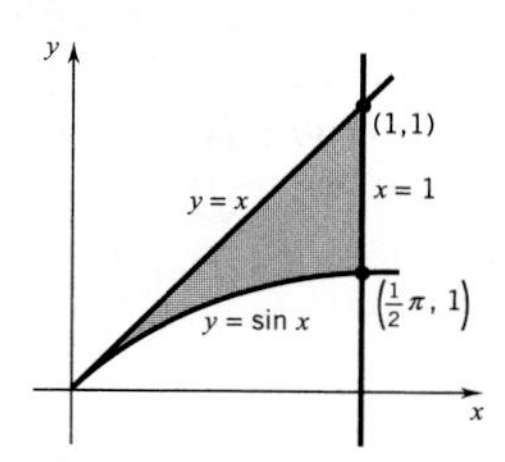

27. (a) $-\dfrac{91}{6}$, the area of the region bounded by f and the x-axis for $x \in [-3, -2] \cup [3, 4]$ minus the area of the region bounded by f and the x-axis for $x \in [-2, 3]$.
(b) $\dfrac{53}{2}$ (c) $\dfrac{125}{6}$

29. (a) 0 (b) 5

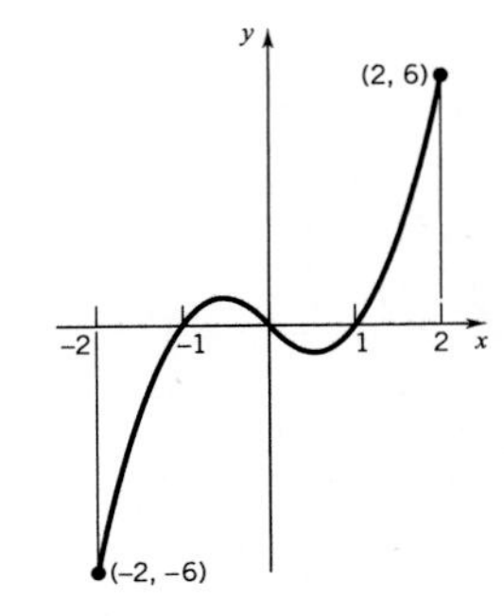

31. (a) $\frac{65}{4}$ (b) 17.87 33. $\frac{10}{3}$ 35. area $= 2 - \sqrt{2}$

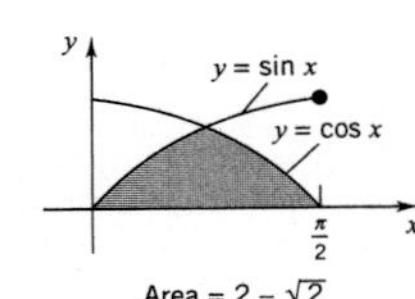

37. 2.86

SECTION 5.6

1. $-\frac{1}{3x^3} + C$ 3. $\frac{1}{2}ax^2 + bx + C$ 5. $2\sqrt{1+x} + C$ 7. $\frac{1}{2}x^2 + \frac{1}{x} + C$ 9. $\frac{1}{3}t^3 - \frac{1}{2}(a+b)t^2 + abt + C$

11. $\frac{2}{9}t^{9/2} - \frac{2}{5}(a+b)t^{5/2} + 2abt^{1/2} + C$ 13. $\frac{1}{2}[g(x)]^2 + C$ 15. $\frac{1}{2}\sec^2 x + C$ 17. $-\frac{1}{4x+1} + C$ 19. $x^2 - x - 2$

21. $\frac{1}{2}ax^2 + bx - 2a - 2b$ 23. $3 - \cos x$ 25. $x^3 - x^2 + x + 2$ 27. $\frac{1}{12}(x^4 - 2x^3 + 2x + 23)$ 29. $x - \cos x + 3$

31. $\frac{1}{3}x^3 - \frac{3}{2}x^2 - \frac{1}{3}x + 3$ 33. $\frac{d}{dx}\left(\int f(x)\,dx\right) = f(x),\quad \int \frac{d}{dx}[f(x)]\,dx = f(x) + C$

35. (a) 34 units to the right of the origin (b) 44 units 37. (a) $v(t) = 2(t+1)^{1/2} - 1$ (b) $x(t) = \frac{4}{3}(t+1)^{3/2} - t - \frac{4}{3}$

39. (a) 4.4 sec (b) 193.6 ft 41. $[v(t)]^2 = (at + v_0)^2 = a^2t^2 + 2av_0t + v_0^2 = v_0^2 + 2a(\frac{1}{2}at^2 + v_0t) = v_0^2 + 2a[x(t) - x_0]$ 43. 42 sec

$x(t) = \frac{1}{2}at^2 + v_0t + x_0$

45. $x(t) = x_0 + v_0t + At^2 + Bt^3$ 47. at $(\frac{160}{3}, 50)$ 49. $A = -\frac{5}{2}, B = 2$ 51. (a) at $t = \frac{11}{6}\pi$ sec (b) at $t = \frac{13}{6}\pi$ sec

53. mean-value theorem 55. $v(t) = v_0(1 - 2tv_0)^{-1}$

SECTION 5.7

1. $\frac{1}{3(2-3x)} + C$ 3. $\frac{1}{3}(2x+1)^{3/2} + C$ 5. $\frac{4}{7a}(ax+b)^{7/4} + C$ 7. $-\frac{1}{8(4t^2+9)} + C$ 9. $\frac{1}{75}(5t^3+9)^5 + C$

11. $\frac{4}{15}(1+x^3)^{5/4} + C$ 13. $-\frac{1}{4(1+s^2)^2} + C$ 15. $\sqrt{x^2+1} + C$ 17. $-\frac{1}{5}(1-x^3)^{5/3} + C$ 19. $-\frac{5}{4}(x^2+1)^{-2} + C$

21. $-4(x^{1/4}+1)^{-1} + C$ 23. $-\frac{b^3}{2a^4}\sqrt{1 - a^4x^4} + C$ 25. $\frac{15}{8}$ 27. $\frac{31}{2}$ 29. 0 31. $\frac{1}{3}|a|^3$ 33. $\frac{2}{5}(x+1)^{5/2} - \frac{2}{3}(x+1)^{3/2} + C$

35. $\frac{1}{10}(2x-1)^{5/2} + \frac{1}{6}(2x-1)^{3/2} + C$ 37. $\frac{1}{14}(y+1)^{14} - \frac{1}{13}(y+1)^{13} + C$ 39. $-\frac{1}{2}(t-2)^{-2} - \frac{4}{3}(t-2)^{-3} - (t-2)^{-4} + C$

41. $\frac{16}{3}\sqrt{2} - \frac{14}{3}$ 43. $-\frac{769}{112}$ 45. $\frac{1}{3}\sin(3x-1) + C$ 47. $-(\cot \pi x)/\pi + C$ 49. $\frac{1}{2}\cos(3-2x) + C$ 51. $-\frac{1}{5}\cos^5 x + C$

53. $-2\cos x^{1/2} + C$ 55. $\frac{2}{3}(1+\sin x)^{3/2} + C$ 57. $\tan x + C$ 59. $\frac{1}{8}\sin^4 x^2 + C$ 61. $-\cot x - \frac{1}{3}\cot^3 x + C$

63. $2(1+\tan x)^{1/2} + C$ 65. 0 67. $(\sqrt{3}-1)/\pi$ 69. $\frac{1}{4}$ 71. $2 - \sqrt{2}$

73. $\int \sin^2 x\,dx = \int \frac{1 - \cos 2x}{2}\,dx = \frac{1}{2}\int (1 - \cos 2x)\,dx = \frac{1}{2}x - \frac{1}{4}\sin 2x + C$ 75. $\frac{1}{2}x + \frac{1}{20}\sin 10x + C$ 77. $\frac{\pi}{4}$ 79. $1/2\pi$

81. $(4\sqrt{3} - 6)/3\pi$ 83. (a) $\frac{1}{2}\sec^2 x + C$ (b) $\frac{1}{2}\tan^2 x + C'$
(c) $\frac{1}{2}\sec^2 x + C = \frac{1}{2}(1 + \tan^2 x) + C = \frac{1}{2}\tan^2 x + (C + \frac{1}{2}) = \frac{1}{2}\tan^2 x + C'$; $C + \frac{1}{2}$ and C' each represent an arbitrary constant

85. πab

SECTION 5.8

1. yes; $\int_a^b [f(x) - g(x)]\,dx = \int_a^b f(x)\,dx - \int_a^b g(x)\,dx > 0$ 3. yes; otherwise we would have $f(x) \le g(x)$ for all $x \in [a, b]$, and it would follow that

$$\int_a^b f(x)\,dx \le \int_a^b g(x)\,dx$$

5. no; take $f(x) = 0$, $g(x) = -1$ on $[0, 1]$ 7. no; take, for example, any odd function on an interval of the form $[-c, c]$

9. no; $\int_{-1}^{1} x\,dx = 0$ but $\int_{-1}^{1} |x|\,dx \neq 0$ 11. yes; $U_f(P) \geq \int_a^b f(x)\,dx = 0$ 13. no; $L_f(P) \leq \int_a^b f(x)\,dx = 0$

15. yes; $\int_a^b [f(x) + 1]\,dx = \int_a^b f(x)\,dx + \int_a^b 1\,dx = 0 + b - a = b - a$ 17. $\dfrac{2x}{\sqrt{2x^2 + 7}}$ 19. $-f(x)$ 21. $-\dfrac{2\sin(x^2)}{x}$

23. $\dfrac{\sqrt{x}}{2(1+x)}$ 25. $\dfrac{1}{x}$ 27. $\dfrac{3}{1 + (2 + 3x)^{3/2}} - \dfrac{1}{3x^{2/3}(1 + x^{1/2})}$ 29. $2x^3 \cos(x^4) - \frac{1}{2}\cos x$ 31. $4x\sqrt{1 + 4x^2} - \tan x \sec^2 x\,|\sec x|$

33. (a) With P a partition of $[a, b]$

$$L_f(P) \leq \int_a^b f(x)\,dx.$$

If f is nonnegative on $[a, b]$, then $L_f(P$ is nonnegative and, consequently, so is the integral. If f is positive on $[a, b]$, then $L_f(P$ is positive and, consequently, so is the integral.

(b) Take F as an antiderivative of f on $[a, b]$. Observe that

$$F'(x) = f(x) \quad \text{on } (a, b) \qquad \text{and} \qquad \int_a^b f(x)\,dx = F(b) - F(a).$$

If $f(x) \geq 0$ on $[a, b]$, then F is nondecreasing on $[a, b]$ and $F(b) - F(a) \geq 0$.
If $f(x) > 0$ on $[a, b]$, then F is increasing on $[a, b]$ and $F(b) - F(a) > 0$.

35. For all $x \in [a, b]$

$$-f(x) \leq |f(x)| \quad \text{and} \quad f(x) \leq |f(x)|.$$

It follows from II that

$$\int_a^b -f(x)\,dx \leq \int_a^b |f(x)|\,dx \quad \text{and} \quad \int_a^b f(x)\,dx \leq \int_a^b |f(x)|\,dx,$$

and, consequently, that

$$\left| \int_a^b f(x)\,dx \right| \leq \int_a^b |f(x)|\,dx.$$

37. $H(x) = \int_{2x}^{x^3-4} \dfrac{x\,dt}{1 + \sqrt{t}} = x \int_{2x}^{x^3-4} \dfrac{dt}{1 + \sqrt{t}}$

$$H'(x) = x \cdot \left[\frac{3x^2}{1 + \sqrt{x^3 - 4}} - \frac{2}{1 + \sqrt{2x}} \right] + 1 \cdot \int_{2x}^{x^3-4} \frac{dt}{1 + \sqrt{t}}$$

$$H'(2) = 2\left[\frac{12}{3} - \frac{2}{3}\right] + \underbrace{\int_4^4 \frac{dt}{1 + \sqrt{t}}}_{=0} = \frac{20}{3}$$

39. (a) Let $u = -x$ (b) $\int_{-a}^{a} f(x)\,dx = \int_{-a}^{0} f(x)\,dx + \int_0^a f(x)\,dx = \int_0^a f(-x)\,dx + \int_0^a f(x)\,dx = \int_0^a [f(x) + f(-x)]\,dx$

41. 0 43. $\dfrac{2}{3}\pi + \dfrac{2}{81}\pi^3 - \sqrt{3}$

SECTION 5.9

1. $A.V. = \frac{1}{2}mc + b$, $x = \frac{1}{2}c$ 3. $A.V. = 0$, $x = 0$ 5. $A.V. = 1$, $x = \pm 1$ 7. $A.V. = \frac{2}{3}$, $x = 1 \pm \frac{1}{3}\sqrt{3}$

9. $A.V. = 2$, $x = 4$ 11. $A.V. = 0$, $x = 0, \pi, 2\pi$ 13. (a) $\dfrac{13}{6}$ (b) 4.694 15. (a) $\dfrac{2}{\pi} \cong 0.637$ (b) 0.691

17. A = average value of f on $[a, b] = \dfrac{1}{b - a}\int_a^b f'(x)\,dx$ 19. average of f' on $[a, b] = \dfrac{1}{b - a}\int_a^b f'(x)\,dx = \dfrac{1}{b - a}\Big[f(x)\Big]_a^b = \dfrac{f(b) - f(a)}{b - a}$

21. (a) 1 (b) $\frac{2}{3}\sqrt{3}$ (c) $\frac{7}{9}\sqrt{3}$ 23. (a) The terminal velocity is twice the average velocity.
(b) The average velocity during the first $\frac{1}{2}x$ seconds is one-third of the average velocity during the next $\frac{1}{2}x$ seconds.

25. (a) $v(t) = at$, $x(t) = \frac{1}{2}at^2 + x_0$ (b) $v_{avg} = \dfrac{1}{t_2 - t_1}\int_{t_1}^{t_2} at\,dt = \dfrac{at_1 + at_2}{2} = \dfrac{v(t_1) + v(t_2)}{2}$

27. (a) $M = \frac{2}{3}kL^{3/2}$, $x_M = \frac{3}{5}L$ (b) $M = \frac{1}{3}kL^3$, $x_M = \frac{3}{4}L$ 29. $x_{M_2} = (2M - M_1)L/8M_2$ 31. see answer to Exercise 25, Section 5.3

33. If f and g take on the same average value on every interval $[a, x]$, then

$$\frac{1}{x - a}\int_a^x f(t)\,dt = \frac{1}{x - a}\int_a^x g(t)\,dt.$$

Multiplication by $(x - a)$ gives

$$\int_a^x f(t)\,dt = \int_a^x g(t)\,dt.$$

Differentiation with respect to x gives $f(x) = g(x)$. This shows that, if the averages are the same on every interval, then the functions are everywhere the same.

SECTION 5.10

1.

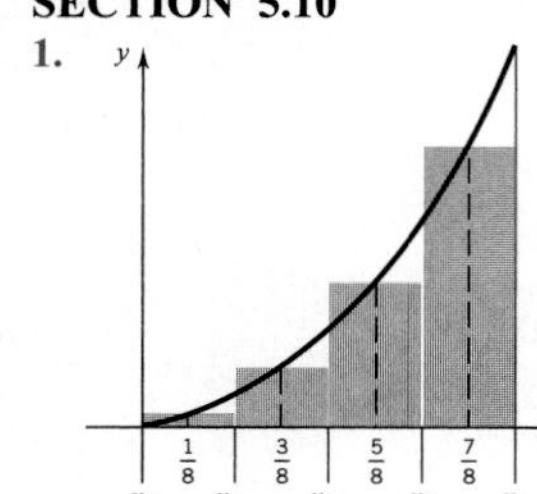

3. (a) $\Delta x_1 = \Delta x_2 = \frac{1}{8}$, $\Delta x_3 = \Delta x_4 = \Delta x_5 = \frac{1}{4}$
(b) $\|P\| = \frac{1}{4}$
(c) $m_1 = 0$, $m_2 = \frac{1}{4}$, $m_3 = \frac{1}{2}$, $m_4 = 1$, $m_5 = \frac{3}{2}$
(d) $f(x_1^*) = \frac{1}{8}, f(x_2^*) = \frac{3}{8}, f(x_3^*) = \frac{3}{4}, f(x_4^*) = \frac{5}{4}, f(x_5^*) = \frac{3}{2}$
(e) $M_1 = \frac{1}{4}$, $M_2 = \frac{1}{2}$, $M_3 = 1$, $M_4 = \frac{3}{2}$, $M_5 = 2$
(f) $L_f(P) = \frac{25}{32}$
(g) $S^*(P) = \frac{15}{16}$
(h) $U_f(P) = \frac{39}{32}$
(i) $\int_a^b f(x)\,dx = 1$

5. (a) $\frac{1}{n^2}(1 + 2 + \cdots + n) = \frac{1}{n^2}\left[\frac{n(n+1)}{2}\right] = \frac{1}{2} + \frac{1}{2n}$
(b) $S_n^* = \frac{1}{2} + \frac{1}{2n}, \qquad \int_0^1 x\,dx = \left[\frac{1}{2}x^2\right]_0^1 = \frac{1}{2}$
$\left|S_n^* - \int_0^1 x\,dx\right| = \frac{1}{2n} < \frac{1}{n} < \epsilon \quad \text{if} \quad n > \frac{1}{\epsilon}$

7. (a) $\frac{1}{n^4}(1^3 + 2^3 + \cdots + n^3) = \frac{1}{n^4}\left[\frac{n^2(n+1)^2}{4}\right] = \frac{1}{4} + \frac{1}{2n} + \frac{1}{4n^2}$
(b) $S_n^* = \frac{1}{4} + \frac{1}{2n} + \frac{1}{4n^2}, \qquad \int_0^1 x^3\,dx = \left[\frac{1}{4}x^4\right]_0^1 = \frac{1}{4}$
$\left|S_n^* - \int_0^1 x^3\,dx\right| = \frac{1}{2n} + \frac{1}{4n^2} < \frac{1}{n} < \epsilon \quad \text{if} \quad n > \frac{1}{\epsilon}$

9. $S^*(P) = \frac{1}{3}[\frac{1}{6}\cos(\frac{1}{6})^2 + \frac{3}{6}\cos(\frac{3}{6})^2 + \frac{5}{6}\cos(\frac{5}{6})^2 + \frac{7}{6}\cos(\frac{7}{6})^2 + \frac{9}{6}\cos(\frac{9}{6})^2 + \frac{11}{6}\cos(\frac{11}{6})^2] \cong -0.3991$
$\int_0^2 x\cos x^2\,dx = \frac{1}{2}\sin 4 \cong -0.3784$

CHAPTER 6

SECTION 6.1

1.

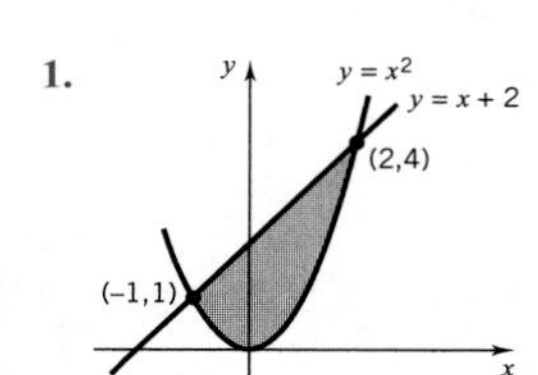

(a) $\int_{-1}^{2} [(x + 2) - x^2]\,dx$
(b) $\int_0^1 [\sqrt{y} - (-\sqrt{y})]\,dy + \int_1^4 [\sqrt{y} - (y - 2)]\,dy$

3.

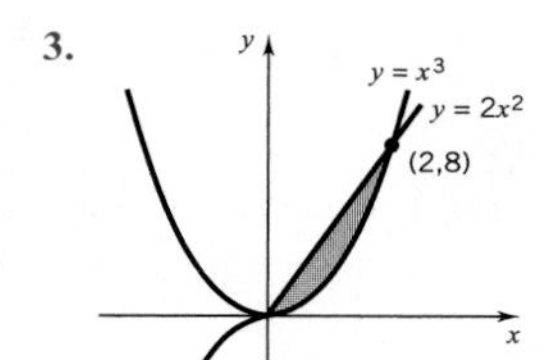

(a) $\int_0^2 [2x^2 - x^3]\,dx$
(b) $\int_0^8 \left[y^{1/3} - \left(\frac{1}{2}y\right)^{1/2}\right] dy$

5.

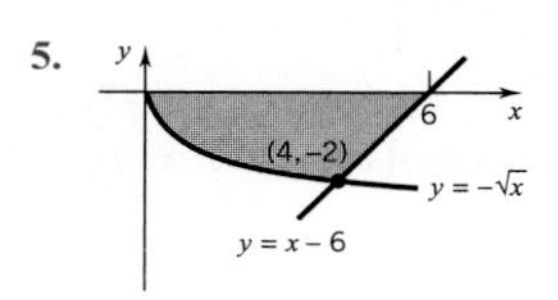

(a) $\int_0^4 [0 - (-\sqrt{x})]\,dx + \int_4^6 [0 - (x - 6)]\,dx$
(b) $\int_{-2}^{0} [(y + 6) - y^2]\,dy$

7.

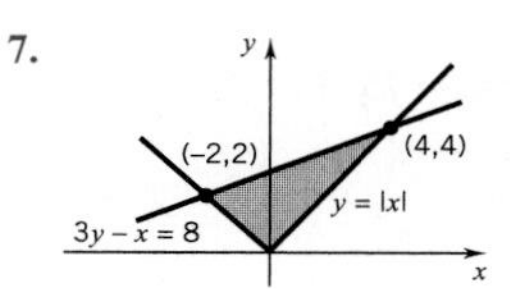

(a) $\int_{-2}^{0} \left[\frac{8 + x}{3} - (-x)\right] dx + \int_0^4 \left[\frac{8 + x}{3} - x\right] dx$
(b) $\int_0^2 [y - (-y)]\,dy + \int_2^4 [y - (3y - 8)]\,dy$

9.

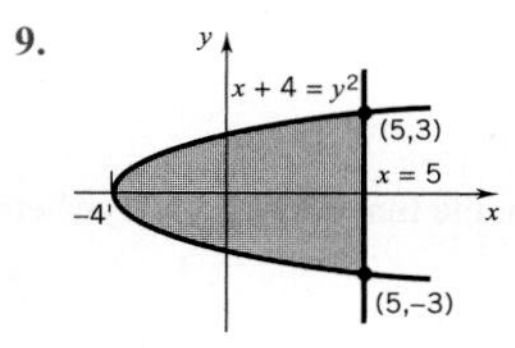

(a) $\int_{-4}^{5} [\sqrt{4 + x} - (-\sqrt{4 + x})]\,dx$
(b) $\int_{-3}^{3} [5 - (y^2 - 4)]\,dy$

11. (a) $\int_{-1}^{3} [2x - (x - 1)]\,dx + \int_{3}^{5} [(9 - x) - (x - 1)]\,dx$

(b) $\int_{-2}^{4} \left[(y + 1) - \frac{1}{2}y \right] dy + \int_{4}^{6} \left[(9 - y) - \frac{1}{2}y \right] dy$

13. 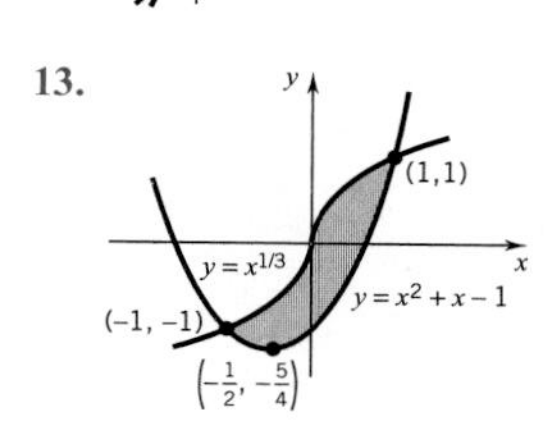

(a) $\int_{-1}^{1} [x^{1/3} - (x^2 + x - 1)]\,dx$

(b) $\int_{-5/4}^{-1} \left[\left(-\frac{1}{2} + \frac{1}{2}\sqrt{4y + 5} \right) - \left(-\frac{1}{2} - \frac{1}{2}\sqrt{4y + 5} \right) \right] dy + \int_{-1}^{1} \left[\left(-\frac{1}{2} + \frac{1}{2}\sqrt{4y + 5} \right) - y^3 \right] dy$

15. area $= \frac{9}{8}$

17. area $= 4$

19. area $= \frac{5}{12}$

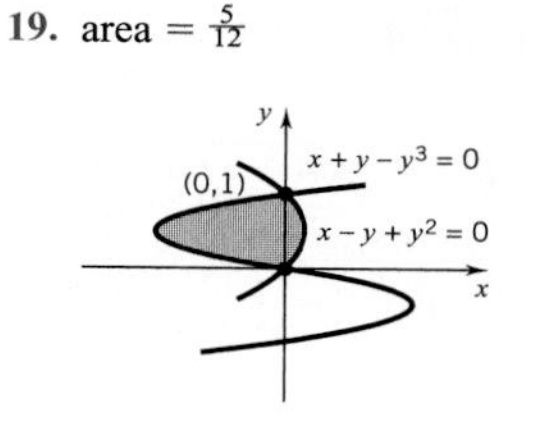

21. area $= 2 - \sqrt{2}$

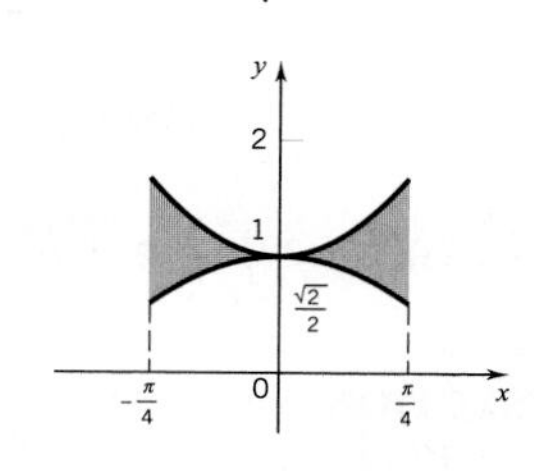

23. area $= 8$

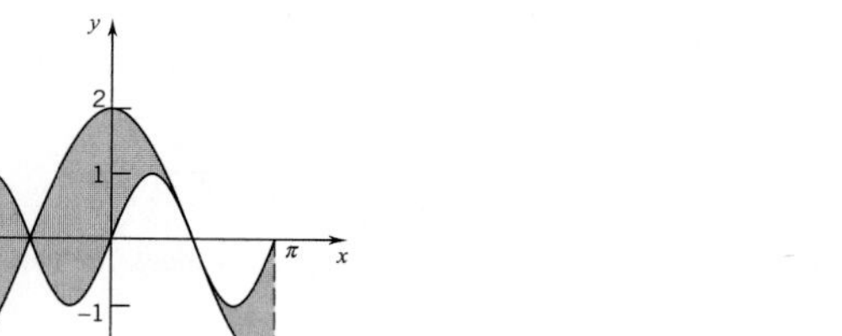

25. 4

27. $\frac{39}{2}$

29. area $= 27$

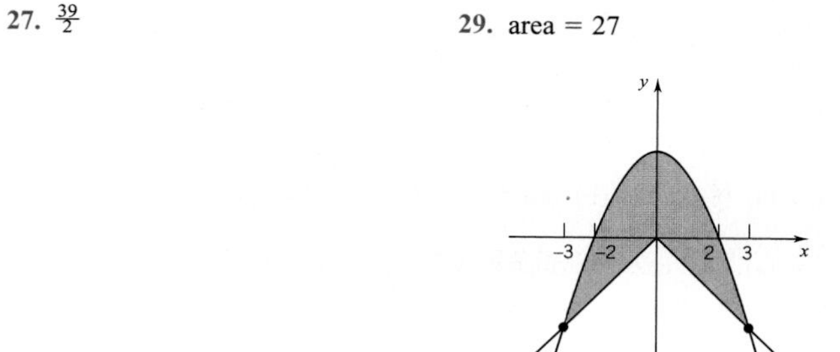

31. 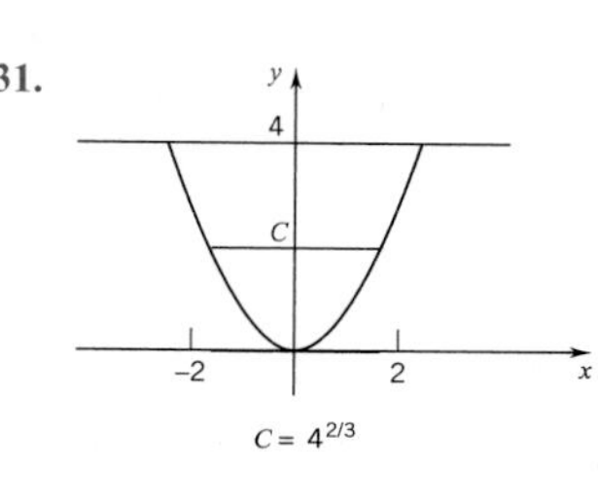

33. 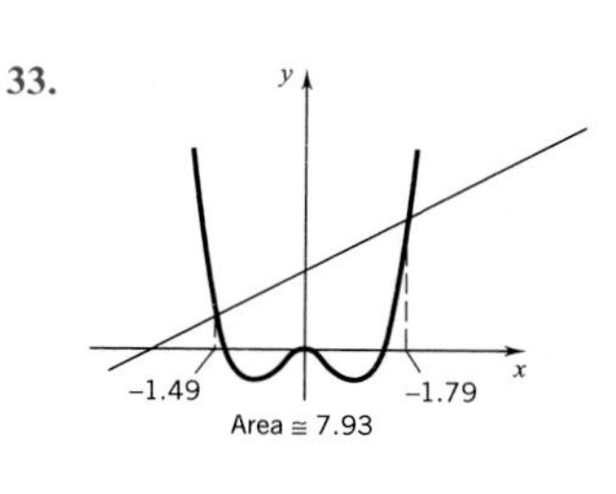

35. 1536 cu in. ≅ 0.89 cu ft

37. (a) 35.4% (b) 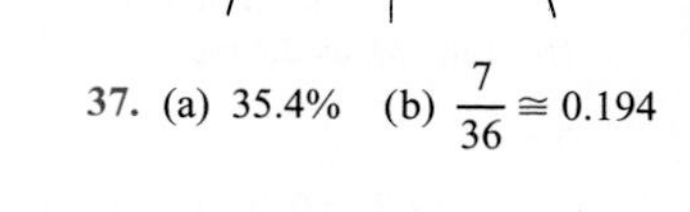 $\frac{7}{36} \cong 0.194$

SECTION 6.2

1. $\frac{1}{3}\pi$ 3. $\frac{1944}{5}\pi$ 5. $\frac{5}{14}\pi$ 7. $\frac{3790}{21}\pi$ 9. $\frac{72}{5}\pi$ 11. $\frac{32}{3}\pi$ 13. π 15. $\frac{\pi^2}{24}(\pi^2 + 6\pi + 6)$ 17. $\frac{16}{3}\pi$ 19. $\frac{768}{7}\pi$

21. $\frac{2}{5}\pi$ 23. $\frac{128}{3}\pi$ 25. $\frac{16}{3}\pi$ 27. (a) $\frac{16}{3}r^3$ (b) $\frac{4}{3}\sqrt{3}\,r^3$ 29. (a) $\frac{512}{15}$ (b) $\frac{64}{15}\pi$ (c) $\frac{128}{15}\sqrt{3}$

31. (a) 32 (b) 4π (c) $8\sqrt{3}$ 33. $\frac{4}{3}\pi ab^2$ 35. $\frac{1}{3}\pi h(R^2 + rR + r^2)$ 37. (a) $31\frac{1}{4}\%$ (b) $14\frac{22}{27}\%$

39. $\frac{1}{\pi}$ ft/min when the depth is 2 feet; $\frac{2}{3\pi}$ ft/min when the depth is 2 feet.

41. The cross section with coordinate x is a washer with outer radius k, inner radius $k - f(x)$, and area

$$A(x) = \pi k^2 - \pi[k - f(x)]^2 = 2\pi k f(x) - \pi[f(x)]^2$$

Thus,

$$V(x) = \int_a^b \pi(2kf(x) - [f(x)]^2)\,dx.$$

43. $\dfrac{40\pi}{3}$ 45. $4\pi - \dfrac{1}{2}\pi^2$ 47. 250π 49. (a) $\dfrac{32\pi}{3}$ (b) $\dfrac{64\pi}{5}$ 51. (a) 64π (b) $\frac{1024}{35}\pi$ (c) $\frac{704}{5}\pi$ (d) $\frac{512}{7}\pi$

SECTION 6.3

1. $\frac{2}{3}\pi$ 3. $\frac{128}{5}\pi$ 5. $\frac{2}{5}\pi$ 7. 16π 9. $\frac{72}{5}\pi$ 11. 36π 13. 8π 15. $\frac{1944}{5}\pi$ 17. $\frac{5}{14}\pi$ 19. $\frac{72}{5}\pi$ 21. 64π

23. $\frac{1}{3}\pi$ 25. (a) $V = \displaystyle\int_0^1 2\pi x(1-\sqrt{x})\,dx$ (b) $V = \displaystyle\int_0^1 \pi y^4\,dy$; $V = \dfrac{1}{5}\pi$

27. (a) $V = \displaystyle\int_0^1 \pi(x - x^4)\,dx$ (b) $V = \displaystyle\int_0^1 2\pi y(\sqrt{y} - y^2)\,dy$; $V = \dfrac{3}{10}\pi$

29. (a) $V = \displaystyle\int_0^1 2\pi x^3\,dx$ (b) $V = \displaystyle\int_0^1 \pi(1-y)\,dy$; $V = \dfrac{\pi}{2}$ 31. $\frac{4}{3}\pi ba^2$ 33. $\frac{1}{4}\pi a^3\sqrt{3}$

35. (a) 64π (b) $\frac{1024}{35}\pi$ (c) $\frac{704}{5}\pi$ (d) $\frac{512}{7}\pi$ 37. (a) $F'(x) = x\cos x$ (b) $V = \pi^2 - 2\pi$

39. (a) $V = \displaystyle\int_0^1 2\sqrt{3}\pi x^2\,dx + \int_1^2 2\pi x\sqrt{4-x^2}\,dx$ (b) $V = \displaystyle\int_0^{\sqrt{3}} \pi\left(4 - \tfrac{4}{3}y^2\right)dy$ (c) $V = \dfrac{8\pi\sqrt{3}}{3}$

41. (a) $V = \displaystyle\int_0^1 2\sqrt{3}\pi x(2-x)\,dx + \int_1^2 2\pi(2-x)\sqrt{4-x^2}\,dx$ (b) $V = \displaystyle\int_0^{\sqrt{3}} \pi\left[\left(2 - \frac{y}{\sqrt{3}}\right)^2 - (2 - \sqrt{4-y^2})^2\right]dy$

43. (a) $V = 2\displaystyle\int_{b-a}^{b+a} 2\pi x\sqrt{a^2 - (x-b)^2}\,dx$ (b) $V = \displaystyle\int_{-a}^{a} \pi[(b + \sqrt{a^2-y^2})^2 - (b - \sqrt{a^2-y^2})^2]\,dy$

SECTION 6.4

1. $(\frac{12}{5}, \frac{3}{4})$, $V_x = 8\pi$, $V_y = \frac{128}{5}\pi$ 3. $(\frac{3}{7}, \frac{12}{25})$, $V_x = \frac{2}{5}\pi$, $V_y = \frac{5}{14}\pi$ 5. $(\frac{7}{3}, \frac{10}{3})$, $V_x = \frac{80}{3}\pi$, $V_y = \frac{56}{3}\pi$ 7. $(\frac{3}{4}, \frac{22}{5})$, $V_x = \frac{704}{15}\pi$, $V_y = 8\pi$

9. $(\frac{2}{5}, \frac{2}{5})$, $V_x = \frac{4}{15}\pi$, $V_y = \frac{4}{15}\pi$ 11. $(\frac{45}{28}, \frac{93}{70})$, $V_x = \frac{31}{5}\pi$, $V_y = \frac{15}{2}\pi$ 13. $(3, \frac{5}{3})$, $V_x = \frac{40}{3}\pi$, $V_y = 24\pi$ 15. $(\frac{5}{2}, 5)$ 17. $(1, \frac{8}{5})$

19. $(\frac{10}{3}, \frac{40}{21})$ 21. $(2, 4)$ 23. $(-\frac{3}{5}, 0)$ 25. (a) $(0, 0)$ (b) $\left(\dfrac{14}{5\pi}, \dfrac{14}{5\pi}\right)$ (c) $\left(0, \dfrac{14}{5\pi}\right)$ 27. $V = \pi ab(2c + \sqrt{a^2 + b^2})$

29. (a) $(\frac{2}{3}a, \frac{1}{3}h)$ (b) $(\frac{2}{3}a + \frac{1}{3}b, \frac{1}{3}h)$ (c) $(\frac{1}{3}a + \frac{1}{3}b, \frac{1}{3}h)$ 31. (a) $\frac{1}{3}\pi R^3 \sin^2\theta\,(2\sin\theta + \cos\theta)$ (b) $\dfrac{2R\sin\theta\,(2\sin\theta + \cos\theta)}{3(\pi\sin\theta + 2\cos\theta)}$

33. (a) The mass contributed by $[x_{i-1}, x_i]$ is approximately $\lambda(x_i^*)\Delta x_i$ where x_i^* is the midpoint of $[x_{i-1}, x_i]$. The sum of these contributions,

$$\lambda(x_1^*)\Delta x_1 + \cdots + \lambda(x_n^*)\Delta x_n,$$

is a Riemann sum, which as $\|P\| \to 0$, tends to the given integral.
(b) Take M_i as the mass contributed by $[x_{i-1}, x_i]$. Then $x_{M_i}M_i \cong x_i^*\lambda(x_i^*)\Delta x_i$ where x_i^* is the midpoint of $[x_{i-1}, x_i]$. Therefore

$$x_M M = x_{M_1}M_1 + \cdots + x_{M_n}M_n \cong x_1^*\lambda(x_1^*)\Delta x_1 + \cdots + x_n^*\lambda(x_n^*)\Delta x_n.$$

As $\|P\| \to 0$, the sum on the right converges to the given integral.

35. on the axis of the cone at distance $\frac{3}{4}h$ from the vertex 37. (a) $(\frac{2}{3}, 0)$ (b) $(0, \frac{5}{12})$ 39. $(\frac{3}{8}a, 0)$

SECTION 6.5

1. 817.5 ft-lb 3. $\dfrac{35\pi^2}{72} - \dfrac{1}{4}$ newton-meters 5. 625 ft-lb 7. (a) 25 ft-lb (b) $\frac{225}{4}$ ft-lb 9. 1.95 ft

11. (a) $(6480\pi + 8640)$ ft-lb (b) $(15{,}120\pi + 8640)$ ft-lb 13. (a) $\frac{11}{192}\pi r^2h^2\sigma$ ft-lb (b) $(\frac{11}{192}\pi r^2h^2\sigma + \frac{7}{24}\pi r^2hk\sigma)$ ft-lb

15. $GmM\left(\dfrac{1}{r_2} - \dfrac{1}{r_1}\right)$ 17. 48,000 ft-lb 19. (a) 20,000 ft-lb (b) 30,000 ft-lb 21. 788 ft-lb 23. (a) $\frac{1}{2}\sigma l^2$ ft-lb (b) $\frac{3}{2}l^2\sigma$ ft-lb

25. 20,800 ft-lb 27. $W = \displaystyle\int_a^b ma\,dx = \int_a^b mv\,dv = \frac{1}{2}mv_b^2 - \frac{1}{2}mv_a^2$ 29. (a) 670 sec or 11 min, 10 sec (b) 1116 sec or 18 min, 36 sec

SECTION 6.6

1. 9000 lb 3. 1.437×10^8 newtons 5. 1.7502×10^6 newtons 7. 2160 lb 9. $\frac{8000}{3}\sqrt{2}$ lb 11. 333.33 lb 13. 2560 lb

15. (a) 41,250 lb (b) 41,250 lb 17. (a) 297,267 newtons (b) 39,200 newtons at the shallow end; 352,800 newtons at the deep end

19. 2.21749×10^6 newtons 21. $F = \sigma\bar{x}A$ where A is the area of the submerged surface and $\bar{x}$ is the depth of its centroid

CHAPTER 7

SECTION 7.1

1. $f^{-1}(x) = \frac{1}{5}(x - 3)$ 3. $f^{-1}(x) = \frac{1}{4}(x + 7)$ 5. not one-to-one 7. $f^{-1}(x) = (x - 1)^{1/5}$ 9. $f^{-1}(x) = [\frac{1}{3}(x - 1)]^{1/3}$

11. $f^{-1}(x) = 1 - x^{1/3}$ 13. $f^{-1}(x) = (x - 2)^{1/3} - 1$ 15. $f^{-1}(x) = x^{5/3}$ 17. $f^{-1}(x) = \frac{1}{3}(2 - x^{1/3})$ 19. $f^{-1}(x) = 1/x$

21. not one-to-one 23. $f^{-1}(x) = (1/x - 1)^{1/3}$ 25. $f^{-1}(x) = (2 - x)/(x - 1)$ 27. they are equal

29.

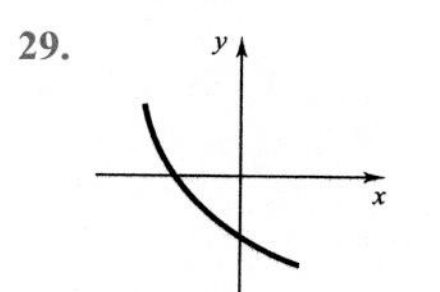

31.

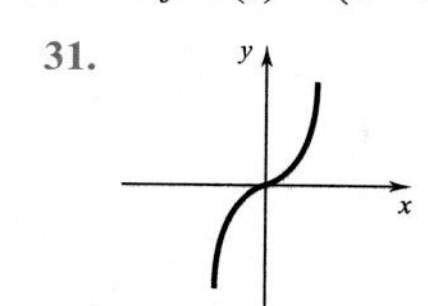

33.

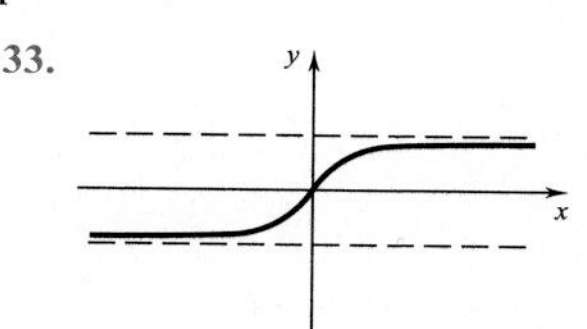

35. $f'(x) = 3x^2 \geq 0$ on $(-\infty, \infty)$, $f'(x) = 0$ only at $x = 0$; $(f^{-1})'(9) = \frac{1}{12}$ 37. $f'(x) = 1 + \dfrac{1}{\sqrt{x}} > 0$ on $(0, \infty)$; $(f^{-1})'(8) = \frac{2}{3}$

39. $f'(x) = 2 - \sin x > 0$ on $(-\infty, \infty)$; $(f^{-1})'(\pi) = 1$ 41. $f'(x) = \sec^2 x > 0$ on $(-\pi/2, \pi/2)$; $(f^{-1})'(\sqrt{3}) = \frac{1}{4}$

43. $(f^{-1})'(x) = \dfrac{1}{x}$ 45. $(f^{-1})'(x) = \dfrac{1}{\sqrt{1 - x^2}}$ 47. (a) $f'(x) = \dfrac{ad - bc}{(cx + d)^2} \neq 0$ iff $ad - bc \neq 0$ (b) $f^{-1}(x) = \dfrac{dx - b}{a - cx}$

49. (a) $f'(x) = 8\sqrt{1 + x^4} > 0$ (b) $(f^{-1})'(0) = \dfrac{\sqrt{17}}{34}$ 51. (a) $g'(x) = \dfrac{1}{f'(g(x))}$; $g''(x) = -\dfrac{f''(g(x))g'(x)}{[f'(g(x))]^2} = -\dfrac{f''(g(x))}{[f'(g(x))]^3}$

(b) If the graph of f is concave up (down), then the graph of $g = f^{-1}$ is concave down (up).

53. 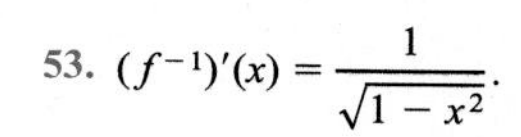 $(f^{-1})'(x) = \dfrac{1}{\sqrt{1 - x^2}}$.

55.

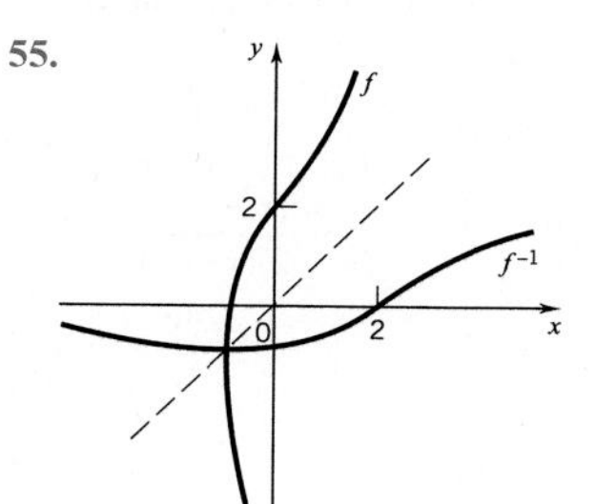

57.

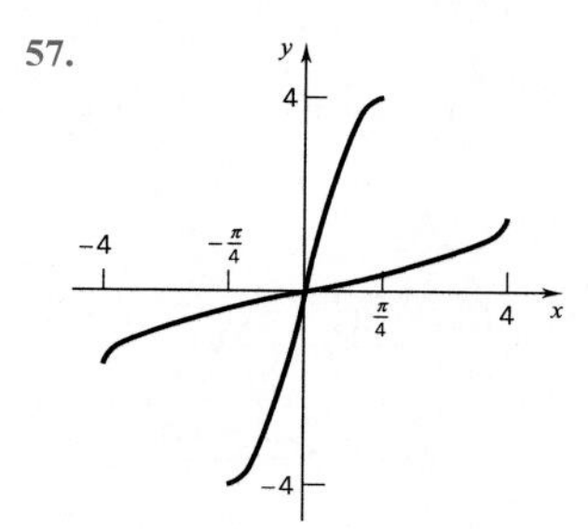

SECTION 7.2

1. $\ln 2 + \ln 10 \cong 2.99$

3. $2 \ln 4 - \ln 10 \cong 0.48$

5. $-\ln 10 \cong -2.30$

7. $\ln 8 + \ln 9 - \ln 10 \cong 1.98$

9. $\frac{1}{2} \ln 2 \cong 0.35$ 11. $5 \ln 2 \cong 3.45$

13.

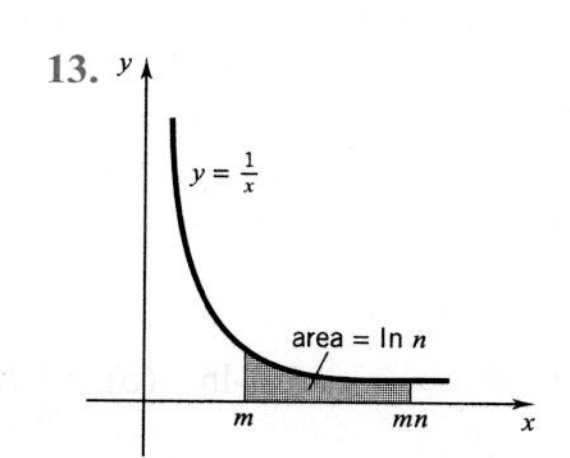

15. 0.406 17. (a) 1.65 (b) 1.57 (c) 1.71

19. $x = e^2$ 21. $x = 1, e^2$ 23. $x = 1$ 25. $\lim_{x \to 1} \dfrac{\ln x}{x - 1} = \dfrac{d(\ln x)}{dx}\Big|_{x=1} = 1$

27. (a) $P = \{1, 2, \ldots, n\}$ is a regular partition of $[1, n]$; $L_f(P) = \dfrac{1}{2} + \dfrac{1}{3} + \cdots + \dfrac{1}{n} < \displaystyle\int_1^n \dfrac{1}{t}\,dt = \ln n < 1 + \dfrac{1}{2} + \cdots + \dfrac{1}{n - 1} = U_f(P)$

(b) Sum of shaded areas $= U_f(P) - \displaystyle\int_1^n \dfrac{1}{t}\,dt = 1 + \dfrac{1}{2} + \cdots + \dfrac{1}{n - 1} - \ln n$.

(c) Connect the points $(1, 1), (2, \frac{1}{2}), \ldots, \left(n, \dfrac{1}{n}\right)$ by straight line segments. The sum of the areas of the triangles that are formed is

$$\frac{1}{2} \cdot 1\left[\left(1 - \frac{1}{2}\right) + \left(\frac{1}{2} - \frac{1}{3}\right) + \cdots + \left(\frac{1}{n - 1} - \frac{1}{n}\right)\right] = \frac{1}{2}\left(1 - \frac{1}{n}\right),$$

so

$$\frac{1}{2}\left(1-\frac{1}{n}\right) < \gamma.$$

The sum of the areas of the indicated rectangles is

$$1\left[\left(1-\frac{1}{2}\right)+\left(\frac{1}{2}-\frac{1}{3}\right)+\cdots+\left(\frac{1}{n-1}-\frac{1}{n}\right)\right] = 1-\frac{1}{n},$$

so

$$\gamma < 1-\frac{1}{n}.$$

Letting $n \to \infty$, we have $\frac{1}{2} < \gamma < 1$.

29. (a) $\ln 3 - \sin 3 \cong 0.96 > 0$; $\ln 2 - \sin 2 \cong -0.22 < 0$ (b) $r \cong 2.2191$ **31.** 1 **33.** 0

SECTION 7.3

1. domain $(0, \infty)$, $f'(x) = \dfrac{1}{x}$ **3.** domain $(-1, \infty)$, $f'(x) = \dfrac{3x^2}{x^3+1}$ **5.** domain $(-\infty, \infty)$, $f'(x) = \dfrac{x}{1+x^2}$

7. domain all $x \neq \pm 1$, $f'(x) = \dfrac{4x^3}{x^4-1}$ **9.** domain $(0, \infty)$, $f'(x) = x + 2x \ln x$ **11.** domain $(0, 1) \cup (1, \infty)$, $f'(x) = -\dfrac{1}{x(\ln x)^2}$

13. domain $(-1, \infty)$, $f'(x) = \dfrac{1-\ln(x+1)}{(x+1)^2}$ **15.** domain $(0, \infty)$; $f'(x) = \dfrac{1}{x}\cos(\ln x)$ **17.** domain all $x \neq (2k+1)\dfrac{\pi}{2}$; $f'(x) = -\tan x$

19. domain: the union of all intervals $\left(\dfrac{[2k-1]\pi}{2}, \dfrac{[2k+1]\pi}{2}\right)$, $k = 0, \pm 1, \pm 2, \ldots$; $f'(x) = \sec x$ **21.** $\ln|x+1| + C$

23. $-\frac{1}{2}\ln|3-x^2| + C$ **25.** $\dfrac{1}{2(3-x^2)} + C$ **27.** $-\ln(2+\cos x) + C$ **29.** $\ln|\sec x| + C$ **31.** $\ln\left|\dfrac{x+2}{x-2}\right| + C$

33. $\dfrac{-1}{\ln x} + C$ **35.** $-\ln|\sin x + \cos x| + C$ **37.** $\frac{2}{3}\ln|1 + x\sqrt{x}| + C$ **39.** 1 **41.** 1 **43.** $\frac{1}{2}\ln\frac{8}{5}$ **45.** $\frac{1}{2}(\ln 2)^2$

47. $g'(x) = (x^2+1)^2(x-1)^5x^3\left(\dfrac{4x}{x^2+1} + \dfrac{5}{x-1} + \dfrac{3}{x}\right)$ **49.** $g'(x) = \dfrac{x^4(x-1)}{(x+2)(x^2+1)}\left(\dfrac{4}{x} + \dfrac{1}{x-1} - \dfrac{1}{x+2} - \dfrac{2x}{x^2+1}\right)$

51. $g'(x) = \dfrac{1}{2}\sqrt{\dfrac{(x-1)(x-2)}{(x-3)(x-4)}}\left(\dfrac{1}{x-1} + \dfrac{1}{x-2} - \dfrac{1}{x-3} - \dfrac{1}{x-4}\right)$ **53.** $\pi \ln 9$ **55.** $\frac{15}{8} - \ln 4$

57. $\ln 5$ ft **59.** $(-1)^{n-1}\dfrac{(n-1)!}{x^n}$ **61.** $(-1)^{n-1}\dfrac{(n-1)!}{x^n}$

63. (a) for $t \in (1, x)$.

$$\frac{1}{t} < \frac{1}{\sqrt{t}}$$

(b) for $x > 1$

$$\ln x = \int_1^x \frac{dt}{t} < \int_1^x \frac{dt}{\sqrt{t}} = \left[2\sqrt{t}\right]_1^x = 2(\sqrt{x}-1)$$

(c) for $0 < x < 1$

$$0 < \ln\frac{1}{x} < 2\left(\sqrt{\frac{1}{x}} - 1\right) \quad \text{by (b)}$$

$$0 < -\ln x < 2\left(\frac{1}{\sqrt{x}} - 1\right)$$

$$2\left(1 - \frac{1}{\sqrt{x}}\right) < \ln x < 0$$

$$2x\left(1 - \frac{1}{\sqrt{x}}\right) < x \ln x < 0$$

(d) Use (c) and the pinching theorem for one-sided limits.

65. (i) domain $(0, \infty)$
(ii) increases on $(0, \infty)$
(iii) no extreme values
(iv) concave down on $(0, \infty)$; no pts of inflection
(v)

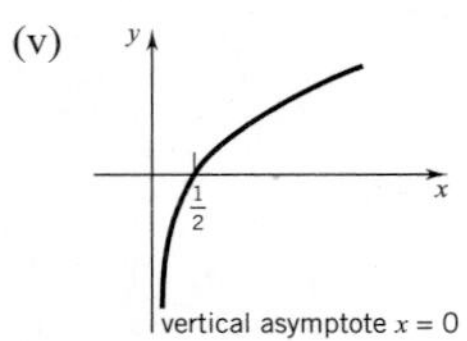

67. (i) domain $(-\infty, 4)$
(ii) decreases throughout
(iii) no extreme values
(iv) concave down throughout; no pts of inflection
(v)

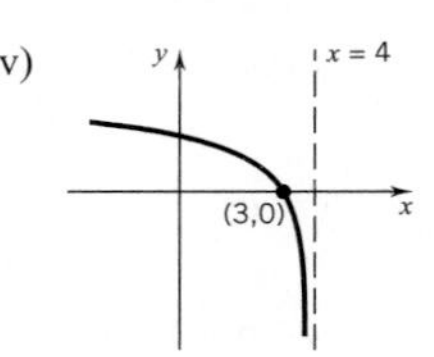

69. (i) domain $(0, \infty)$
(ii) decreases on $(0, e^{-1/2}]$, increases on $[e^{-1/2}, \infty)$
(iii) $f(e^{-1/2}) = -\frac{1}{2}e$ local and absolute min.
(iv) concave up throughout; no pts of inflection
(v)

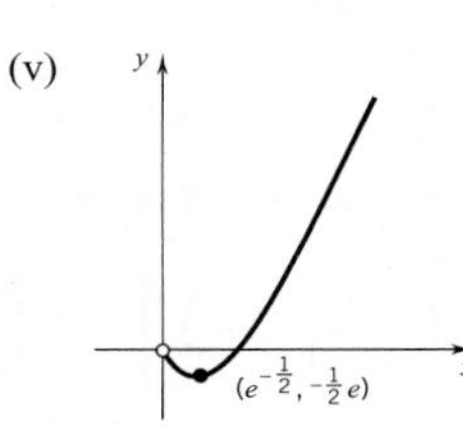

71. (i) domain $(0, \infty)$
(ii) increases on $(0, 1]$, decreases on $[1, \infty)$
(iii) $f(1) = -\ln 2$ local and absolute max
(iv) concave down on $(0, \sqrt{2+\sqrt{5}})$, concave up on $(\sqrt{2+\sqrt{5}}, \infty)$; pt of inflection at $x = \sqrt{2+\sqrt{5}}$
(v)

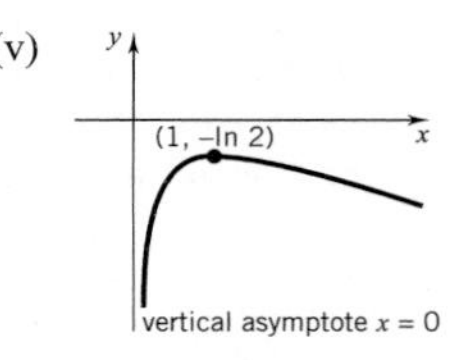

73. x-intercept: 1; absolute min at $x = \frac{1}{4}$, absolute max at $x = 10$

75. x-intercepts: 1, 23.1407; absolute max at $x \cong 4.8105$, absolute min at $x = 100$

77. (a) $v(t) = 2 + 2t - t^2 + 3\ln(t+1)$ (c) max velocity at $t \cong 1.5811$; min velocity at $t = 0$

SECTION 7.4

1. $\frac{dy}{dx} = -2e^{-2x}$ 3. $\frac{dy}{dx} = 2xe^{x^2-1}$ 5. $\frac{dy}{dx} = e^x\left(\frac{1}{x} + \ln x\right)$ 7. $\frac{dy}{dx} = -(x^{-1} + x^{-2})e^{-x}$ 9. $\frac{dy}{dx} = \frac{1}{2}(e^x - e^{-x})$

11. $\frac{dy}{dx} = \frac{1}{2}e^{\sqrt{x}}\left(\frac{1}{x} + \frac{\ln\sqrt{x}}{\sqrt{x}}\right)$ 13. $\frac{dy}{dx} = 2(e^{2x} - e^{-2x})$ 15. $\frac{dy}{dx} = 4xe^{x^2}(e^{x^2}+1)$ 17. $\frac{dy}{dx} = x^2e^x$ 19. $\frac{dy}{dx} = \frac{2e^x}{(e^x+1)^2}$

21. $\frac{dy}{dx} = 2(a-b)\frac{e^{(a+b)x}}{(e^{ax}+e^{bx})^2}$ 23. $\frac{dy}{dx} = 4x^3$ 25. $f'(x) = 2\cos(e^{2x})e^{2x}$ 27. $f'(x) = -e^{-2x}(2\cos x + \sin x)$

29. $f'(x) = -\frac{3}{2}\sqrt{e^{-3x}}\sec^2\sqrt{e^{-3x}}$ 31. $\frac{1}{2}e^{2x} + C$ 33. $\frac{1}{k}e^{kx} + C$ 35. $\frac{1}{2}e^{x^2} + C$ 37. $-e^{1/x} + C$ 39. $\frac{1}{2}e^{2x} - \frac{1}{2}e^{-2x} + 2x + C$

41. $\frac{1}{2}x^2 + C$ 43. $-8e^{-x/2} + C$ 45. $2\sqrt{e^x+1} + C$ 47. $\frac{1}{4}\ln(2e^{2x}+3) + C$ 49. $e^{\sin x} + C$ 51. $-\sin(e^{-x}) - e^{-x} + C$

53. $e - 1$ 55. $\frac{1}{6}(1 - \pi^{-6})$ 57. $2 - \frac{1}{e}$ 59. $\ln\frac{3}{2}$ 61. $\frac{185}{72} + \ln\frac{4}{9}$ 63. $\frac{1}{2}e + \frac{1}{2}$ 65. $e^{-0.4} = \frac{1}{e^{0.4}} \cong \frac{1}{1.49} \cong 0.67$

67. $e^{2.8} = (e^2)(e^{0.8}) \cong (7.39)(2.23) \cong 16.48$ 69. $x''(t) = Ac^2e^{ct} + Bc^2e^{-ct} = c^2(Ae^{ct} + Be^{-ct}) = c^2x(t)$ 71. $\frac{1}{2}(3e^4 + 1)$ 73. $e^2 - e - 2$

75. (i) domain $(-\infty, \infty)$
(ii) decreases on $(-\infty, 0]$, increases on $[0, \infty)$
(iii) $f(0) = 1$ local and absolute min
(iv) concave up everywhere
(v)

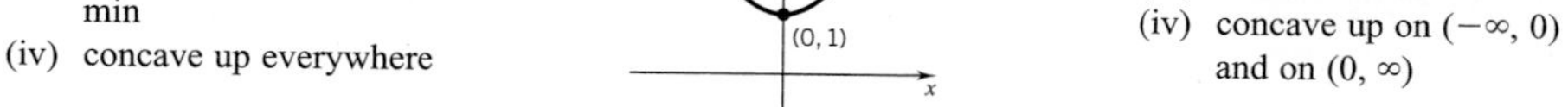

77. (i) domain $(-\infty, 0) \cup (0, \infty)$
(ii) increases on $(-\infty, 0)$, decreases on $(0, \infty)$
(iii) no extreme values
(iv) concave up on $(-\infty, 0)$ and on $(0, \infty)$
(v)

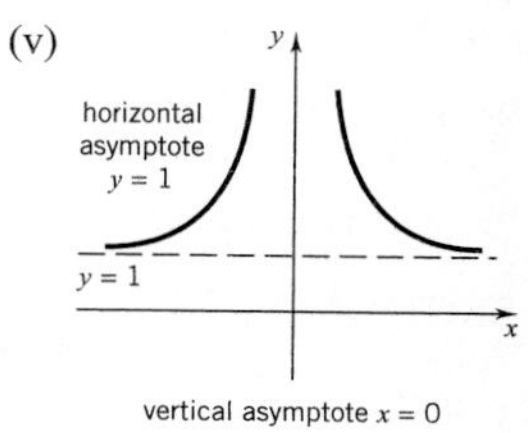

79. (a) $\left(\pm\frac{1}{a}, e\right)$ (b) $\frac{1}{a}(e-2)$ (c) $\frac{1+2a^2e}{a^3e}$ 81. for $x > (n+1)!$

$$e^x > 1 + x + \cdots + \frac{x^{n+1}}{(n+1)!} > \frac{x^{n+1}}{(n+1)!} = x^n\left[\frac{x}{(n+1)!}\right] > x^n$$

83. Numerically, $8.15 \le L \le 8.16$. The limit is the derivative of $f(x) = e^{x^3}$ at $x = 1$; note that $f'(x) = 3x^2e^{x^3}$ and $f'(1) = 3e \cong 8.15485$.

85. (a)

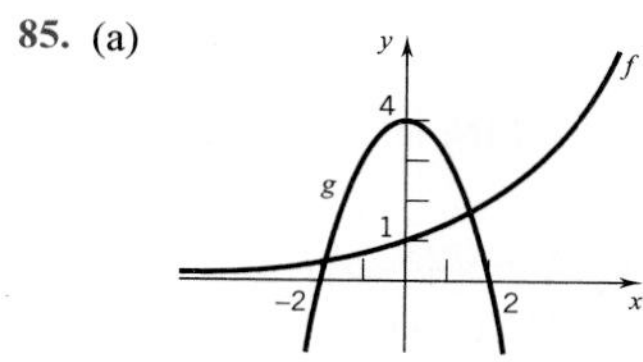

(b) $x = -1.9646$; $x = 1.0580$
(c) 6.4240

87.

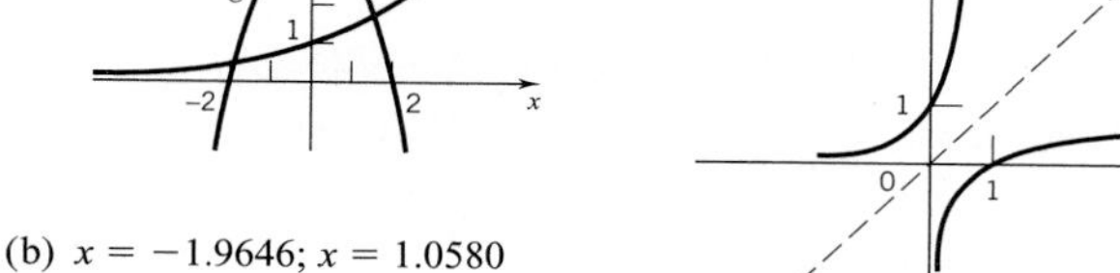

89.

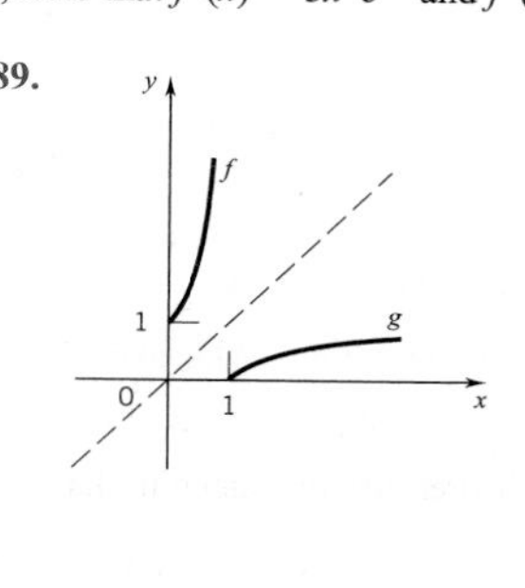

SECTION 7.5

1. 6 3. $-\frac{1}{6}$ 5. 0 7. 3 9. $\frac{3}{5}$ 11. $-\frac{8}{5}$ 13. $\log_p xy = \frac{\ln xy}{\ln p} = \frac{\ln x + \ln y}{\ln p} = \frac{\ln x}{\ln p} + \frac{\ln y}{\ln p} = \log_p x + \log_p y$

15. $\log_p x^y = \frac{\ln x^y}{\ln p} = y\frac{\ln x}{\ln p} = y\log_p x$ 17. 0 19. 2 21. e^c, where $c = \frac{(\ln 2)^2}{\ln 2 - 1}$ 23. $t_1 < \ln a < t_2$

25. $f'(x) = 2(\ln 3)3^{2x}$ 27. $f'(x) = \left(5\ln 2 + \frac{\ln 3}{x}\right)2^{5x}3^{\ln x}$ 29. $g'(x) = \frac{1}{2\ln 3}\cdot\frac{1}{x\sqrt{\log_3 x}}$ 31. $f'(x) = \frac{\sec^2(\log_5 x)}{x\ln 5}$

33. $F'(x) = \ln 2\,(2^{-x} - 2^x)\sin(2^x + 2^{-x})$ 35. $g'(x) = \frac{2}{\ln 2\,\log_4(2x+1)\ln 4\,(2x+1)}$ 37. $\frac{3^x}{\ln 3} + C$

39. $\frac{1}{4}x^4 - \frac{3^{-x}}{\ln 3} + C$ 41. $\log_5|x| + C$ 43. $\frac{3}{\ln 4}(\ln x)^2 + C$ 45. $\frac{10^{x^2}}{2\ln 10} + C$ 47. $\frac{1}{e\ln 3}$ 49. $\frac{1}{e}$

51. $f(x) = p^x$
$\ln f(x) = x \ln p$
$\dfrac{f'(x)}{f(x)} = \ln p$
$f'(x) = p^x \ln p$

53. $(x+1)^x \left[\dfrac{x}{x+1} + \ln(x+1)\right]$

55. $(\ln x)^{\ln x}\left[\dfrac{1 + \ln(\ln x)}{x}\right]$

57. $(x^2+2)^{\ln x}\left[\dfrac{2x \ln x}{x^2+2} + \dfrac{\ln(x^2+2)}{x}\right]$

59. $x^{\sin x}\left(\cos x \ln x + \dfrac{\sin x}{x}\right)$

61. $(\sin x)^{\cos x}\left[\dfrac{\cos^2 x}{\sin x} - \sin x \ln(\sin x)\right]$

63. $x^{2^x}\left[\dfrac{2^x}{x} + 2^x(\ln x)(\ln 2)\right]$

65.
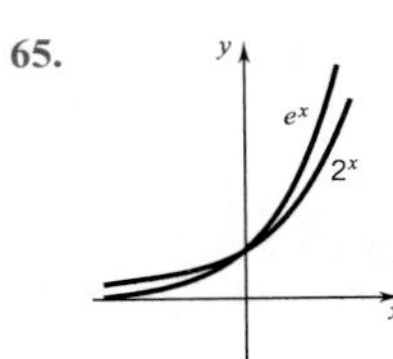

67.
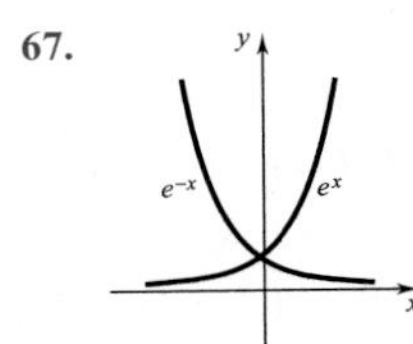

69.
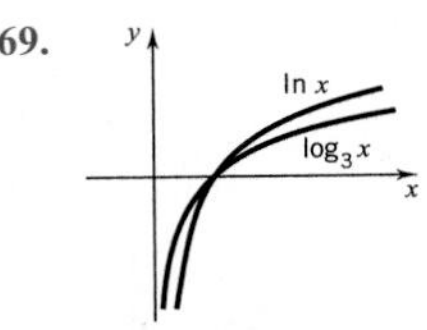

71.
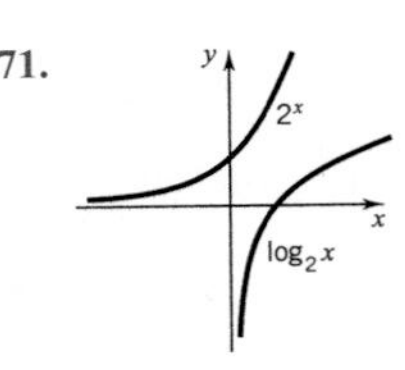

73. domain $(-\infty, \infty)$; increasing on $(-\infty, 0]$, decreasing on $[0, \infty)$; $f(0) = 10$ local and absolute max

75. domain $[-1, 1]$; increasing on $[-1, 0]$, decreasing on $[0, 1]$; $f(0) = 10$ local and absolute max, $f(-1) = f(1) = 1$ endpt and absolute min

77. $\dfrac{1}{4 \ln 2}$

79. 2

81. $\dfrac{45}{\ln 10}$

83. $(\ln 2)(\ln 10)$

85. $\dfrac{1}{3} + \dfrac{1}{\ln 2}$

87. $\log_{10} 7 = \dfrac{\ln 7}{\ln 10} \cong \dfrac{1.95}{2.30} \cong 0.85$

89. $\log_{10} 45 = \dfrac{\ln 45}{\ln 10} = \dfrac{\ln 9 + \ln 5}{\ln 10} \cong \dfrac{2.20 + 1.61}{2.30} \cong 1.66$

91. approx. 16.999999; $5^{(\ln 17)/(\ln 5)} = (e^{\ln 5})^{(\ln 17)/(\ln 5)} = e^{\ln 17} = 17$

93. approx. 54.59815; $16^{1/\ln 2} = (e^{\ln 16})^{1/\ln 2} = e^{(\ln 16)/(\ln 2)} = e^{4(\ln 2)/(\ln 2)} = e^4 \cong 54.59815$

SECTION 7.6

1. (a) \$411.06 (b) \$612.77 (c) \$859.14

3. about $5\frac{1}{2}\%$: $(\ln 3)/20 \cong 0.0549$

5. 16,000

7. $20 + 5e^{-0.03t}$ lb

9. $200(\frac{4}{5})^{t^2/25}$ liters

11. a little more than $9\frac{1}{2}$ years: $\dfrac{4 \ln \frac{1}{2}}{\ln \frac{3}{4}} \cong 9.64$

13. $5(\frac{4}{5})^{5/2} \cong 2.86$ gms

15. $100[1 - (\frac{1}{2})^{1/n}]\%$

17. (a) $x_1(t) = 10^6 t$, $x_2(t) = e^t - 1$
(b) $\dfrac{d}{dt}[x_1(t) - x_2(t)] = \dfrac{d}{dt}[10^6 t - (e^t - 1)] = 10^6 - e^t.$
This derivative is zero at $t = 6 \ln 10 \cong 13.8$. After that the derivative is negative.
(c) $x_2(15) < e^{15} = (e^3)^5 \cong 20^5 = 2^5(10^5) = 3.2(10^6) < 15(10^6) = x_1(15)$
$x_2(18) = e^{18} - 1 = (e^3)^6 - 1 \cong 20^6 - 1 = 64(10^6) - 1 > 18(10^6) = x_1(18)$
$x_2(18) - x_1(18) \cong 64(10^6) - 1 - 18(10^6) \cong 46(10^6)$
(d) If by time t_1 EXP has passed LIN, then $t_1 > 6 \ln 10$. For all $t \geq t_1$ the speed of EXP is greater than the speed of LIN: for $t \geq t_1 > 6 \ln 10$, $v_2(t) = e^t > 10^6 = v_1(t)$.

19. (a) $15(\frac{2}{3})^{1/2} \cong 12.25$ lb/in.^{2} (b) $15(\frac{2}{3})^{3/2} \cong 8.16$ lb/in.^{2}

21. 6.4%

23. (a) \$18,589.35 (b) \$20,339.99 (c) \$22,933.27

25. $176/\ln 2 \cong 254$ ft

27. about 284 million: $203(\frac{227}{203})^3 \cong 283.85$

29. 11,400 years

31. about $16\frac{1}{2}$ years: $100(\ln \frac{9}{8})^2 \cong 16.44$

33. 640 lb

35. Proceeding from the hint, we know from Theorem 7.6.1 that

$$f(t) + k_2/k_1 = Ae^{k_1(t + k_2/k_1)} = Ae^{k_1 t + k_2} = (Ae^{k_2})e^{k_1 t} = Ce^{k_1 t}.$$

(where A is an arbitrary constant; set $Ae^{k_2} = C$)

Therefore $f(t) = Ce^{k_1 t} - k_2/k_1$, C an arbitrary constant.

37. (a) From Exercise 35 you can determine that

$$v(t) = \frac{32}{K}(1 - e^{-Kt}).$$

(b) At each time t, $1 - e^{-Kt} < 1$. With $K > 0$,

$$v(t) = \frac{32}{K}(1 - e^{-Kt}) < \frac{32}{K}.$$

39. $\dfrac{2 \ln 3}{\ln \frac{4}{3}} - 2 \cong 5.64$ min

41. (a) 46.48°C (b) 9.2 minutes

43. $y = cx$ (c constant)

45. $y = -\ln|c - \sin x|$ (c constant)

47. $\ln|y - a| = rx + C$
$|y - a| = e^C e^{rx}$
$y = a + Be^{rx}$ $(B = \pm e^C)$

SECTION 7.7

1. $\frac{1}{3}\ln|\sec 3x| + C$ 3. $(1/\pi)\ln|\csc \pi x - \cot \pi x| + C$ 5. $\ln|\sin e^x| + C$ 7. $\frac{1}{4}\ln|3 - 2\cos 2x| + C$ 9. $\frac{1}{3}e^{\tan 3x} + C$
11. $\frac{1}{2}\ln|\sec x^2 + \tan x^2| + C$ 13. $\frac{1}{2}(\ln|\sin x|)^2 + C$ 15. $x + 2\ln|\sec x + \tan x| + \tan x + C$ 17. $(1 + \cot x)^{-1} + C$
19. $2\sqrt{\ln|\sec x + \tan x|} + C$ 21. $\ln\frac{4}{3}$ 23. $\frac{1}{2}\ln 2$ 25. $\ln[(1 + \sqrt{2})/(\sec 1 + \tan 1)] \cong -0.345$ 27. $\frac{1}{3}\pi - \frac{1}{2}\ln 3$ 29. $\frac{1}{4}\pi - \frac{1}{2}\ln 2$
31. $\frac{3}{4}\ln 2$ 33. $2\pi\ln(2 + \sqrt{3})$ 35. $\frac{\ln 3}{\pi}$ 37.

$$\int \csc x\,dx = \int \frac{\csc x(\csc x - \cot x)}{\csc x - \cot x}dx$$

Let $u = \csc x - \cot x$, $du = \csc x(\csc x - \cot x)\,dx$. Then

$$\int \csc x\,dx = \int \frac{1}{u}\,du = \ln|u| + C = \ln|\csc x - \cot x| + C.$$

SECTION 7.8

1. 0 3. $\frac{\pi}{3}$ 5. $\frac{2\pi}{3}$ 7. $-\frac{\pi}{4}$ 9. $-\frac{2}{\sqrt{3}}$ 11. $\frac{1}{2}$ 13. 1.1630 15. -0.4580 17. 1.2002 19. $1/\sqrt{1 + x^2}$
21. x 23. $\sqrt{1 + x^2}$ 25. $\sqrt{1 - x^2}$ 27. $\frac{1}{x}\sqrt{1 + x^2}$ 29. $\frac{x}{\sqrt{1 - x^2}}$ 31. $\frac{1}{x^2 + 2x + 2}$ 33. $\frac{2}{x\sqrt{4x^4 - 1}}$
35. $\frac{2x}{\sqrt{1 - 4x^2}} + \sin^{-1} 2x$ 37. $\frac{2\sin^{-1} x}{\sqrt{1 - x^2}}$ 39. $\frac{x - (1 + x^2)\tan^{-1} x}{x^2(1 + x^2)}$ 41. $\frac{1}{(1 + 4x^2)\sqrt{\tan^{-1} 2x}}$ 43. $\frac{1}{x[1 + (\ln x)^2]}$
45. $-\frac{r}{|r|\sqrt{1 - r^2}}$ 47. $2x\sec^{-1}\left(\frac{1}{x}\right) - \frac{x^2}{\sqrt{1 - x^2}}$ 49. $\cos[\sec^{-1}(\ln x)] \cdot \frac{1}{x|\ln x|\sqrt{(\ln x)^2 - 1}}$ 51. $\sqrt{\frac{c - x}{c + x}}$ 53. $\frac{x^2}{(c^2 - x^2)^{3/2}}$
55. Set $au = x + b$, $a\,du = dx$.

$$\int \frac{dx}{\sqrt{a^2 - (x + b)^2}} = \int \frac{a\,du}{\sqrt{a^2 - a^2u^2}} = \int \frac{du}{\sqrt{1 - u^2}} = \sin^{-1} u + C = \sin^{-1}\left(\frac{x + b}{a}\right) + C$$

57. Set $au = x + b$, $a\,du = dx$.

$$\int \frac{dx}{(x + b)\sqrt{(x + b)^2 - a^2}} = \int \frac{a\,du}{au\sqrt{a^2u^2 - a^2}} = \frac{1}{a}\int \frac{du}{u\sqrt{u^2 - 1}} = \frac{1}{a}\sec^{-1}\frac{|x + b|}{a} + C.$$

59. domain $(-\infty, \infty)$, range $(0, \pi)$ 61. $\frac{1}{4}\pi$ 63. $\frac{1}{4}\pi$ 65. $\frac{1}{20}\pi$ 67. $\frac{1}{24}\pi$ 69. $\frac{1}{3}\sec^{-1} 4 \cong 0.439$ 71. $\frac{1}{6}\pi$
73. $\tan^{-1} 2 - \frac{1}{4}\pi \cong 0.322$ 75. $\frac{1}{2}\sin^{-1} x^2 + C$ 77. $\frac{1}{2}\tan^{-1} x^2 + C$ 79. $\tan^{-1}(\frac{1}{3}\tan x) + C$ 81. $\frac{1}{2}(\sin^{-1} x)^2 + C$
83. $\sin^{-1}(\ln x) + C$ 85. $\frac{\pi}{3}$ 87. $2\pi - \frac{4}{3}$ 89. $\sqrt{s^2 + sk}$ feet from the point where the line of the sign intersects the road
91. (a) There exists constants C_1, C_2 such that

$$f(x) = g(x) + C_1 \quad \text{for } x < 1/2; \qquad f(x) = g(x) + C_2 \quad \text{for } x > 1/2.$$

(b) $\lim_{x \to 1/2^+} f(x) = -\frac{\pi}{2}$; $\lim_{x \to 1/2^-} f(x) = \frac{\pi}{2}$ (e) $C_1 = \tan^{-1} 2$; $C_2 = \tan^{-1} 3 - \frac{\pi}{4}$
93. (a) There exist constants C_1, C_2 such that

$$f(x) + g(x) = C_1 \quad \text{for } x < 0; \qquad f(x) + g(x) = C_2 \quad \text{for } x > 0.$$

(b) $\lim_{x \to 0^+} f(x) = \frac{\pi}{2}$; $\lim_{x \to 0^-} f(x) = -\frac{\pi}{2}$ (e) $C_1 = \frac{\pi}{2}$; $C_2 = -\frac{\pi}{2}$
95. estimate $\cong 0.523$, $\sin 0.523 \cong 0.499$ explanation: the integral $= \sin^{-1} 0.5$; therefore sin (integral) $= 0.5$

SECTION 7.9

1. $2x\cosh x^2$ 3. $\frac{a\sinh ax}{2\sqrt{\cosh ax}}$ 5. $\frac{1}{1 - \cosh x}$ 7. $ab(\cosh bx - \sinh ax)$ 9. $\frac{a\cosh ax}{\sinh ax}$ 11. $2e^{2x}\cosh(e^{2x})$
13. $-e^{-x}\cosh 2x + 2e^{-x}\sinh 2x$ 15. $\tanh x$ 17. $(\sinh x)^x[\ln(\sinh x) + x\coth x]$
19. $\cosh^2 t - \sinh^2 t = \left(\frac{e^t + e^{-t}}{2}\right)^2 - \left(\frac{e^t - e^{-t}}{2}\right)^2 = \frac{e^{2t} + 2 + e^{-2t}}{4} - \frac{e^{2t} - 2 + e^{-2t}}{4} = 1$
21. $\cosh t\cosh s + \sinh t\sinh s = \left(\frac{e^t + e^{-t}}{2}\right)\left(\frac{e^s + e^{-s}}{2}\right) + \left(\frac{e^t - e^{-t}}{2}\right)\left(\frac{e^s - e^{-s}}{2}\right)$

$$= \tfrac{1}{4}(e^{t+s} + e^{s-t} + e^{t-s} + e^{-t-s} + e^{t+s} - e^{s-t} - e^{t-s} + e^{-t-s})$$
$$= \tfrac{1}{2}(e^{t+s} + e^{-(t+s)}) = \cosh(t + s)$$

23. $\cosh^2 t + \sinh^2 t = \left(\dfrac{e^t + e^{-t}}{2}\right)^2 + \left(\dfrac{e^t - e^{-t}}{2}\right)^2 = \tfrac{1}{4}(e^{2t} + 2 + e^{-2t} + e^{2t} - 2 - e^{-2t}) = \dfrac{e^{2t} + e^{-2t}}{2} = \cosh 2t$

25. $\sinh(-t) = \dfrac{e^{-t} - e^{-(-t)}}{2} = \dfrac{e^{-t} - e^{t}}{2} = -\sinh t$ **27.** absolute max -3

29. $[\cosh x + \sinh x]^n = \left[\dfrac{e^x + e^{-x}}{2} + \dfrac{e^x - e^{-x}}{2}\right]^n = [e^x]^n = e^{nx} = \dfrac{e^{nx} + e^{-nx}}{2} + \dfrac{e^{nx} - e^{-nx}}{2} = \cosh nx + \sinh nx$

31. $A = 2, B = \tfrac{1}{3}, c = 3$ **33.** $\dfrac{1}{a}\sinh ax + C$ **35.** $\dfrac{1}{3a}\sinh^3 ax + C$ **37.** $\dfrac{1}{a}\ln(\cosh ax) + C$ **39.** $-\dfrac{1}{a\cosh ax} + C$

41. $\tfrac{1}{2}(\sinh x\cosh x + x) + C$ **43.** $2\cosh\sqrt{x} + C$ **45.** $\dfrac{e^2 - 1}{2e} \cong 1.175$ **47.** π

SECTION 7.10

1. $2\tanh x\,\mathrm{sech}^2 x$ **3.** $\mathrm{sech}\,x\,\mathrm{csch}\,x$ **5.** $\dfrac{2e^{2x}\cosh(\tan^{-1} e^{2x})}{1 + e^{4x}}$ **7.** $\dfrac{-x\,\mathrm{csch}^2(\sqrt{x^2+1})}{\sqrt{x^2+1}}$ **9.** $\dfrac{-\mathrm{sech}\,x(\tanh x + 2\sinh x)}{(1 + \cosh x)^2}$

11. $\dfrac{d}{dx}(\coth x) = \dfrac{d}{dx}\left(\dfrac{\cosh x}{\sinh x}\right) = \dfrac{\sinh^2 x - \cosh^2 x}{\sinh^2 x} = \dfrac{-1}{\sinh^2 x} = -\mathrm{csch}^2 x$

13. $\dfrac{d}{dx}(\mathrm{csch}\,x) = \dfrac{d}{dx}\left(\dfrac{1}{\sinh x}\right) = -\dfrac{\cosh x}{\sinh^2 x} = -\mathrm{csch}\,x\coth x$ **15.** (a) $\tfrac{3}{5}$ (b) $\tfrac{5}{3}$ (c) $\tfrac{4}{3}$ (d) $\tfrac{5}{4}$ (e) $\tfrac{3}{4}$

17. If $x \le 0$, the result is obvious. Suppose then that $x > 0$. Since $x^2 \ge 1$, we have $x \ge 1$. Consequently,

$$x - 1 = \sqrt{x-1}\,\sqrt{x-1} \le \sqrt{x-1}\,\sqrt{x+1} = \sqrt{x^2 - 1} \quad \text{and therefore} \quad x - \sqrt{x^2-1} \le 1.$$

19. $\dfrac{d}{dx}(\sinh^{-1} x) = \dfrac{d}{dx}[\ln(x + \sqrt{x^2+1})] = \dfrac{1 + \dfrac{x}{\sqrt{x^2+1}}}{x + \sqrt{x^2+1}} = \dfrac{1}{\sqrt{x^2+1}}$

21. $\dfrac{d}{dx}(\tanh^{-1} x) = \dfrac{1}{2}\dfrac{d}{dx}\left[\ln\left(\dfrac{1+x}{1-x}\right)\right] = \dfrac{1}{2}\cdot\dfrac{1}{\left(\dfrac{1+x}{1-x}\right)}\cdot\dfrac{2}{(1-x)^2} = \dfrac{1}{1-x^2}$

23. Let $y = \mathrm{csch}^{-1} x$. Then $\mathrm{csch}\,y = x$ and $\sinh y = \dfrac{1}{x}$. Thus $\cosh y \cdot y' = -\dfrac{1}{x^2}$ and

$$y' = -\frac{1}{x^2\cosh y} = -\frac{1}{x^2\sqrt{1 + \left(\dfrac{1}{x}\right)^2}} = -\frac{1}{|x|\sqrt{1+x^2}}.$$

25. (a) absolute max $(0, 1)$
(b) points of inflection at $x = \ln(1 + \sqrt{2}) \cong 0.881$, $x = -\ln(1 + \sqrt{2}) \cong -0.881$
(c) concave up on $(-\infty, -\ln(1+\sqrt{2})) \cup (\ln(1+\sqrt{2}), \infty)$; concave down on $(-\ln(1+\sqrt{2}), \ln(1+\sqrt{2}))$
(d)

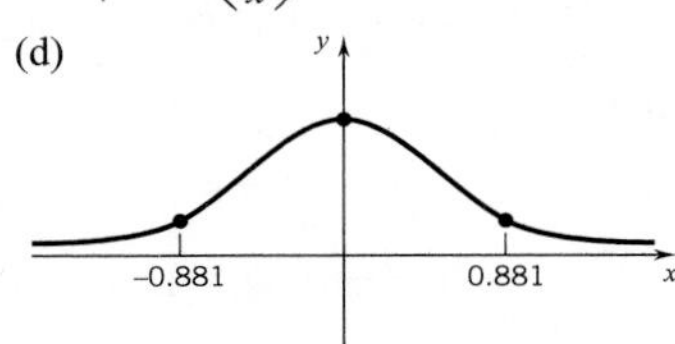

27.

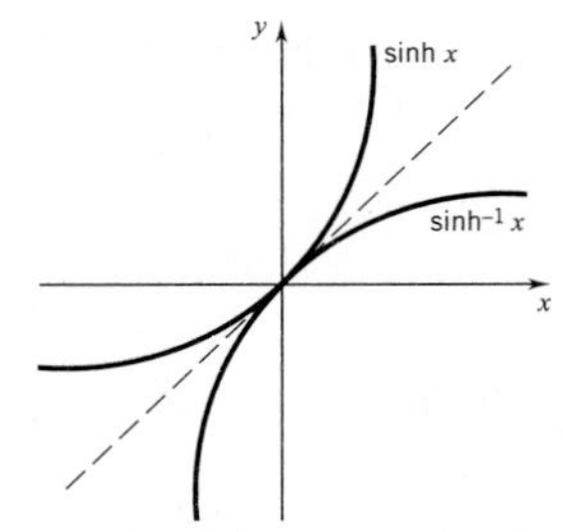

(0, 0) is a point of inflection for both graphs

29. (a)
$$\tan\phi = \sinh x$$
$$\phi = \tan^{-1}(\sinh x)$$
$$\frac{d\phi}{dx} = \frac{\cosh x}{1 + \sinh^2 x} = \frac{\cosh x}{\cosh^2 x} = \frac{1}{\cosh x} = \mathrm{sech}\,x$$

(b)
$$\sinh x = \tan\phi$$
$$x = \sinh^{-1}(\tan\phi)$$
$$= \ln(\tan\phi + \sqrt{\tan^2\phi + 1})$$
$$= \ln(\tan\phi + \sec\phi) = \ln(\sec\phi + \tan\phi)$$

(c)
$$x = \ln(\sec\phi + \tan\phi)$$
$$\frac{dx}{d\phi} = \frac{\sec\phi\tan\phi + \sec^2\phi}{\tan\phi + \sec\phi} = \sec\phi$$

31. $\ln|\cosh x| + C$ **33.** $2\tan^{-1}(e^x) + C$ **35.** $-\tfrac{1}{3}\mathrm{sech}^3 x + C$ **37.** $\tfrac{1}{2}[\ln(\cosh x)]^2 + C$ **39.** $\ln|1 + \tanh x| + C$

41. Let $x = a\sinh u$, $dx = a\cosh u\,du$. Then

$$\int \frac{dx}{\sqrt{a^2 + x^2}} = \int \frac{a\cosh u}{\sqrt{a^2 + a^2\sinh^2 u}}\,du = \int du = \sinh^{-1}\left(\frac{x}{a}\right) + C.$$

43. Suppose $|x| < a$. Let $x = a \tanh u$, $dx = a\,\mathrm{sech}^2 u\, du$. Then

$$\int \frac{dx}{a^2 - x^2} = \int \frac{a\,\mathrm{sech}^2 u}{a^2 - a^2 \tanh^2 u}\, du = \frac{1}{a}\int du = \frac{1}{a}\tanh^{-1}\left(\frac{x}{a}\right) + C.$$

The other case is done in the same way.

CHAPTER 8

SECTION 8.1

1. $-e^{2-x} + C$ **3.** $2/\pi$ **5.** $-\tan(1-x) + C$ **7.** $\frac{1}{2}\ln 3$ **9.** $-\sqrt{1-x^2} + C$ **11.** 0 **13.** $e - \sqrt{e}$ **15.** $\frac{1}{2}\ln(x^2+1) + C$
17. $\pi/4c$ **19.** $\frac{2}{3}\sqrt{3\tan\theta + 1} + C$ **21.** $(1/a)\ln|ae^x - b| + C$ **23.** $-\frac{1}{2}(\cot 2x + \csc 2x) + C$
25. $\frac{1}{2}\ln[(x+1)^2 + 4] - \frac{1}{2}\tan^{-1}(\frac{1}{2}[x+1]) + C$ **27.** $\frac{1}{2}\sin^{-1}x^2 + C$ **29.** $\tan^{-1}(x+3) + C$ **31.** $-\frac{1}{2}\cos x^2 + C$ **33.** $\tan x - x + C$
35. $\frac{1}{2}\ln(\frac{8}{5})$ **37.** $3/2$ **39.** $\frac{1}{2}(\sin^{-1}x)^2 + C$ **41.** $\frac{1}{2}\sin^{-1}(e^{2x}) + C$ **43.** $\ln|\ln x| + C$ **45.** $\frac{2}{3}(\ln 4)^{3/2}$ **47.** $\sqrt{2}$ **49.** $2\sqrt{2}$

51. (a) $\displaystyle\int_0^\pi \sin^2 nx\, dx = \int_0^\pi \left(\frac{1}{2} - \frac{\cos 2nx}{2}\right) dx = \left[\frac{1}{2}x - \frac{\sin 2nx}{4n}\right]_0^\pi = \frac{\pi}{2}$

(b) $\displaystyle\int_0^\pi \sin nx \cos nx\, dx = \frac{1}{n}\int_0^0 u\, du = 0 \quad \left(u = \sin nx,\ du = \frac{1}{n}\cos nx\, dx\right)$ (c) $\displaystyle\int_0^{\pi/n} \sin nx \cos nx\, dx = \frac{1}{n}\int_0^0 u\, du = 0.$

53. (a) $\frac{1}{2}\tan^2 x - \ln|\sec x| + C$ (b) $\frac{1}{4}\tan^4 x - \frac{1}{2}\tan^2 x + \ln|\sec x| + C$ (c) $\frac{1}{6}\tan^6 x - \frac{1}{4}\tan^4 x + \frac{1}{2}\tan^2 x - \ln|\sec x| + C$

(d) $\displaystyle\int \tan^{2k+1}x\, dx = \frac{1}{2k}\tan^{2k}x - \int \tan^{2k-1}x\, dx$

55. (b) $A = \sqrt{2}$, $B = \frac{\pi}{4}$ (c) $\displaystyle\frac{\sqrt{2}}{2}\ln\left(\frac{\sqrt{2}+1}{\sqrt{2}-1}\right)$ **57.** (b) -0.80, 5.80 (c) 27.60

SECTION 8.2

1. $-xe^{-x} - e^{-x} + C$ **3.** $\frac{2}{9}e^3 - \frac{1}{9}$ **5.** $-\frac{1}{3}e^{-x^3} + C$ **7.** $2 - 5e^{-1}$ **9.** $-2x^2(1-x)^{1/2} - \frac{8}{3}x(1-x)^{3/2} - \frac{16}{15}(1-x)^{5/2} + C$
11. $\frac{3}{8}e^4 - \frac{1}{8}$ **13.** $2\sqrt{x+1}\ln(x+1) - 4\sqrt{x+1} + C$ **15.** $x(\ln x)^2 - 2x\ln x + 2x + C$

17. $3x\left(\dfrac{x^3}{\ln 3} - \dfrac{3x^2}{(\ln 3)^2} + \dfrac{6x}{(\ln 3)^3} - \dfrac{6}{(\ln 3)^4}\right) + C$ **19.** $\frac{1}{15}x(x+5)^{15} - \frac{1}{240}(x+5)^{16} + C$ **21.** $\dfrac{1}{2\pi} - \dfrac{1}{\pi^2}$

23. $\frac{1}{10}x^2(x+1)^{10} - \frac{1}{55}x(x+1)^{11} + \frac{1}{660}(x+1)^{12} + C$ **25.** $\frac{1}{2}e^x(\sin x - \cos x) + C$ **27.** $\ln 2 + \dfrac{\pi}{2} - 2$ **29.** $\dfrac{x^{n+1}}{n+1}\ln x - \dfrac{x^{n+1}}{(n+1)^2} + C$

31. $-\frac{1}{2}x^2\cos x^2 + \frac{1}{2}\sin x^2 + C$ **33.** $e^x(x^4 - 4x^3 + 12x^2 - 24x + 24) + C$ **35.** $\dfrac{\pi}{24} + \dfrac{\sqrt{3}-2}{4}$ **37.** $x\sec^{-1}x - \ln|x + \sqrt{x^2-1}| + C$

39. $\dfrac{\pi}{8} - \dfrac{1}{4}\ln 2$ **41.** $\frac{1}{2}x^2\sinh 2x - \frac{1}{2}x\cosh 2x + \frac{1}{4}\sinh 2x + C$ **43.** $\dfrac{x}{2}[\sin(\ln x) - \cos(\ln x)] + C$ **45.** $\ln x\sin^{-1}(\ln x) + \sqrt{1-(\ln x)^2} + C$

47. $\dfrac{\pi}{12} + \dfrac{\sqrt{3}-2}{2}$ **49.** (a) 1 (b) $\bar{x} = \dfrac{e^2}{4} + \dfrac{1}{4}$, $\bar{y} = e - 2$ (c) x-axis: $\pi(e-2)$, y-axis: $\dfrac{\pi}{2}(e^2+1)$ **51.** $\bar{x} = 1/(e-1)$, $\bar{y} = (e+1)/4$

53. $\bar{x} = \frac{1}{2}\pi$, $\bar{y} = \frac{1}{8}\pi$ **55.** (a) $M = (e^k - 1)/k$ (b) $x_M = [(k-1)e^k + 1]/[k(e^k - 1)]$ **57.** $V = (2\pi/\alpha^2)(\alpha e^\alpha - e^\alpha + 1)$

59. $V = 4 - 8/\pi$ **61.** $V = 2\pi(e-2)$ **63.** $\bar{x} = (e^2+1)/[2(e^2-1)]$ **65.** area $= \sinh 1 = \dfrac{e^2-1}{2e}$; $\bar{x} = \dfrac{2}{e+1}$, $\bar{y} = \dfrac{e^4 + 4e^2 - 1}{8e(e^2-1)}$

67. Let $u = x^n$, $dv = e^{ax}dx$. Then $du = nx^{n-1}\,dx$, $v = \dfrac{1}{a}e^{ax}$. **69.** $(\frac{1}{2}x^3 - \frac{3}{4}x^2 + \frac{3}{4}x - \frac{3}{8})e^{2x} + C$ **71.** $x[(\ln x)^3 - 3(\ln x)^2 + 6\ln x - 6] + C$

73. Let $u = f(x)$, $dv = g''(x)\,dx$. Then $du = f'(x)\,dx$, $v = g'(x)$, and

$$\int_a^b f(x)g''(x)\,dx = [f(x)g'(x)]_a^b - \int_a^b f'(x)g'(x)\,dx.$$

Now let $u = f'(x)$, $dv = g'(x)\,dx$ and integrate by parts again. The result follows.

75. The present value of R dollars at the interest rate r on the interval $[t, t + \Delta t]$ is approximately $Re^{-rt}\Delta t$. **77.** (a) about \$2016 (b) about \$1918

SECTION 8.3

1. $\frac{1}{3}\cos^3 x - \cos x + C$ **3.** $\dfrac{\pi}{12}$ **5.** $-\frac{1}{5}\cos^5 x + \frac{1}{7}\cos^7 x + C$ **7.** $\frac{1}{4}\sin^4 x - \frac{1}{6}\sin^6 x + C$ **9.** $\frac{1}{3}\sin^3 x - \frac{1}{5}\sin^5 x + C$

11. $\frac{3}{8}\pi$ **13.** $\frac{1}{2}\cos x - \frac{1}{10}\cos 5x + C$ **15.** $\frac{1}{2}\sin^4 x + C$ **17.** $\frac{3}{128}x - \frac{1}{128}\sin 4x + \frac{1}{1024}\sin 8x + C$
19. $\frac{5}{16}x - \frac{1}{4}\sin 2x + \frac{3}{64}\sin 4x + \frac{1}{48}\sin^3 2x + C$ **21.** $\sin x - \sin^3 x + \frac{3}{5}\sin^5 x - \frac{1}{7}\sin^7 x + C$

23. $\frac{3}{5}$ **25.** $\frac{1}{6}\sin 3x - \frac{1}{14}\sin 7x + C$ **27.** $\dfrac{2}{105\pi}$ **29.** $-1/6$ **31.** $\frac{1}{3}\cos(\frac{3}{2}x) - \frac{1}{5}\cos(\frac{5}{2}x) + C$

33. $\frac{3}{8}\tan^{-1}x + \frac{x}{2(x^2+1)} + \frac{x(1-x^2)}{8(x^2+1)^2} + C$ 35. $\frac{1}{2}\left[\tan^{-1}(x+1) + \frac{x+1}{x^2+2x+2}\right] + C$ 37. $\pi/2$ 39. $\frac{3\pi^2}{8}$

41. $\sin mx \sin nx = \frac{1}{2}[\cos(m-n)x - \cos(m+n)x], \quad m \neq n$

$$\sin mx \sin nx = \sin^2 mx = \frac{1 - \cos 2mx}{2}, \quad m = n$$

45. Let $u = \cos^{n-1}x$, $dv = \cos x\,dx$. Then $du = (n-1)\cos^{n-2}x(-\sin x)\,dx$, $\quad v = \sin x$.   47. $\frac{16}{35}$

$$\int \cos^n x\,dx = \int \cos^{n-1}x \cos x\,dx = \cos^{n-1}x \sin x + (n-1)\int \cos^{n-2}x \sin^2 x\,dx$$

$$= \cos^{n-1}x \sin x + (n-1)\int (\cos^{n-2}x - \cos^n x)\,dx$$

Now solve for $\int \cos^n x\,dx$.

SECTION 8.4

1. $\frac{1}{3}\tan 3x - x + C$ 3. $(1/\pi)\tan \pi x + C$ 5. $\frac{1}{2}\tan^2 x + \ln|\cos x| + C$ 7. $\frac{1}{3}\tan^3 x + C$

9. $-\frac{1}{2}\csc x \cot x + \frac{1}{2}\ln|\csc x - \cot x| + C$ 11. $-\frac{1}{3}\cot^3 x + \cot x + x + C$ 13. $-\frac{1}{5}\csc^5 x + \frac{1}{3}\csc^3 x + C$

15. $-\frac{1}{6}\cot^3 2x - \frac{1}{2}\cot 2x + C$ 17. $-\frac{1}{2}\cot x \csc x - \frac{1}{2}\ln|\csc x - \cot x| + C$ 19. $\frac{1}{12}\tan^4 3x - \frac{1}{6}\tan^2 3x + \frac{1}{3}\ln|\sec 3x| + C$

21. $\frac{1}{4}\sec^3 x \tan x + \frac{3}{8}\sec x \tan x + \frac{3}{8}\ln|\sec x + \tan x| + C$ 23. $\frac{1}{7}\tan^7 x + \frac{1}{5}\tan^5 x + C$ 25. $\frac{1}{6}\tan^3(e^{2x}) + C$ 27. $\frac{1}{4}$ 29. $\frac{\sqrt{3}}{2} - \frac{\pi}{6}$

31. $\frac{8\sqrt{3} + 504}{135}$ 33. $\frac{\pi^2}{2} - \pi$ 35. $\pi\left[1 - \frac{\pi}{4} + \ln 2\right]$ 37. $\int \cot^n x\,dx = \int \cot^{n-2}x(\csc^2 x - 1)\,dx = -\frac{\cot^{n-1}x}{n-1} - \int \cot^{n-2}x\,dx$

39. $\int \csc^n x\,dx = \int \csc^{n-2}x \csc^2 x\,dx$. Now let $u = \csc^{n-2}x$, $dv = \csc^2 x\,dx$ and use integration by parts.

SECTION 8.5

1. $\sin^{-1}\left(\frac{x}{a}\right) + C$ 3. $\frac{1}{10}$ 5. $\frac{1}{2}x\sqrt{x^2-1} - \frac{1}{2}\ln|x + \sqrt{x^2-1}| + C$ 7. $2\sin^{-1}\left(\frac{x}{2}\right) - \frac{1}{2}x\sqrt{4-x^2} + C$

9. $\frac{1}{\sqrt{1-x^2}} + C$ 11. $\frac{2\sqrt{3} - \pi}{6}$ 13. $-\frac{1}{3}(4-x^2)^{3/2} + C$ 15. $\frac{625\pi}{16}$ 17. $\ln(\sqrt{8+x^2} + x) - \frac{x}{\sqrt{8+x^2}} + C$

19. $\frac{1}{a}\ln\left|\frac{a - \sqrt{a^2-x^2}}{x}\right| + C$ 21. $\ln|x + \sqrt{x^2-a^2}| + C$ 23. $\frac{1}{2}e^x\sqrt{e^{2x}-1} - \frac{1}{2}\ln(e^x + \sqrt{e^{2x}-1}) + C$ 25. $18 - 9\sqrt{2}$

27. $-\frac{1}{a^2x}\sqrt{a^2+x^2} + C$ 29. $\frac{1}{a^2x}\sqrt{x^2-a^2} + C$ 31. $2 - \frac{\pi}{2}$ 33. $\frac{1}{9}e^{-x}\sqrt{e^{2x}-9} + C$ 35. $-\frac{1}{2(x-2)^2} + C$

37. $-\frac{1}{3}(6x - x^2 - 8)^{3/2} + \frac{3}{2}\sin^{-1}(x-3) + \frac{3}{2}(x-3)\sqrt{6x-x^2-8} + C$ 39. $\frac{x^2+x}{8(x^2+2x+5)} - \frac{1}{16}\tan^{-1}\left(\frac{x+1}{2}\right) + C$

41. $\sqrt{x^2+4x+13} + \ln(x + 2 + \sqrt{x^2+4x+13}) + C$ 43. $\frac{1}{2}(x-3)\sqrt{6x-x^2-8} + \frac{1}{2}\sin^{-1}(x-3) + C$

45. Let $x = a\tan u$, $dx = a\sec^2 u\,du$; $\sqrt{x^2+a^2} = a\sec u$. Then $(x^2+a^2)^n = a^{2n}\sec^{2n}u$ and the result follows by substitution.

47. $\frac{3}{8}\tan^{-1}x + \frac{x}{2(x^2+1)} + \frac{x(1-x^2)}{8(x^2+1)^2} + C$

49. (a) Let $x = a\tan u$. Then $dx = a\sec^2 u\,du$ and $\sqrt{x^2+a^2} = a\sec u$. Thus $\int \frac{1}{\sqrt{x^2+a^2}}\,dx = \int \sec u\,du$ and the result follows.

(b) Let $x = a\sinh u$. Then $dx = a\cosh u\,du$ and $\sqrt{x^2+a^2} = a\cosh u$. Thus $\int \frac{1}{\sqrt{x^2+a^2}}\,dx = \int du$ and the result follows.

(c) See Theorem 7.10.2

51. $\frac{\pi^2}{8} + \frac{\pi}{4}$ 53. $A = \frac{1}{2}r^2\sin\theta\cos\theta + \int_{r\cos\theta}^{r}\sqrt{r^2-x^2}\,dx = \frac{1}{2}r^2\theta$ 55. $M = \ln(1+\sqrt{2})$, $\quad x_M = \frac{(\sqrt{2}-1)a}{\ln(1+\sqrt{2})}$

57. $A = \frac{1}{2}a^2[\sqrt{2} - \ln(\sqrt{2}+1)]$; $\quad \bar{x} = \frac{2a}{3[\sqrt{2} - \ln(\sqrt{2}+1)]}$, $\quad \bar{y} = \frac{(2-\sqrt{2})a}{3[\sqrt{2} - \ln(\sqrt{2}+1)]}$ 59. $V_y = \frac{2}{3}\pi a^3$, $\quad \bar{y} = \frac{3}{8}a$

61. (b) $\ln(2+\sqrt{3}) - \frac{\sqrt{3}}{2}$ (c) $\bar{x} = \frac{2(3\sqrt{3} - \pi)}{2\ln(2+\sqrt{3}) - \sqrt{3}}$, $\quad \bar{y} = \frac{5}{72[\ln(2+\sqrt{3}) - \sqrt{3}]}$

SECTION 8.6

1. $\dfrac{1/5}{x+1} - \dfrac{1/5}{x+6}$ **3.** $\dfrac{1/4}{x-1} + \dfrac{1/4}{x+1} - \dfrac{x/2}{x^2+1}$ **5.** $\dfrac{1/2}{x} + \dfrac{3/2}{x+2} - \dfrac{1}{x-1}$ **7.** $\dfrac{3/2}{x-1} - \dfrac{9}{x-2} + \dfrac{19/2}{x-3}$ **9.** $\ln\left|\dfrac{x-2}{x+5}\right| + C$

11. $5\ln|x-1| - 3\ln|x| + \dfrac{3}{x} + C$ **13.** $\dfrac{1}{4}x^4 + \dfrac{4}{3}x^3 + 6x^2 + 32x - \dfrac{32}{x-2} + 80\ln|x-2| + C$ **15.** $5\ln|x-2| - 4\ln|x-1| + C$

17. $\dfrac{-1}{2(x-1)^2} + C$ **19.** $\dfrac{3}{4}\ln|x-1| - \dfrac{1}{2(x-1)} + \dfrac{1}{4}\ln|x+1| + C$ **21.** $\dfrac{1}{32}\ln\left|\dfrac{x-2}{x+2}\right| - \dfrac{1}{16}\tan^{-1}\dfrac{x}{2} + C$

23. $\dfrac{1}{2}\ln(x^2+1) + \dfrac{3}{2}\tan^{-1}x + \dfrac{5(1-x)}{2(x^2+1)} + C$ **25.** $\dfrac{1}{16}\ln\left[\dfrac{x^2+2x+2}{x^2-2x+2}\right] + \dfrac{1}{8}\tan^{-1}(x+1) + \dfrac{1}{8}\tan^{-1}(x-1) + C$

27. $\dfrac{3}{x} + 4\ln\left|\dfrac{x}{x+1}\right| + C$ **29.** $-\frac{1}{6}\ln|x| + \frac{3}{10}\ln|x-2| - \frac{2}{15}\ln|x+3| + C$ **31.** $\ln\left(\frac{125}{108}\right)$ **33.** $\frac{1}{4}\ln\left(\frac{5}{2}\right) + \frac{3}{4}$ **35.** $\ln\left(\frac{27}{4}\right) - 2$

37. Note that

$$y = \frac{1}{x^2-1} = \frac{1}{2}\left[\frac{1}{x-1} - \frac{1}{x+1}\right] \quad \text{and thus} \quad \frac{d^0y}{dx^0} = \left(\frac{1}{2}\right)(-1)^0\,0!\left[\frac{1}{(x-1)^{0+1}} - \frac{1}{(x+1)^{0+1}}\right].$$

The rest is a routine induction.

39. $\bar{x} = (2\ln 2)/\pi, \quad \bar{y} = (\pi+2)/\pi$ **41.** (b) $3\ln 7 - 5\ln 3$ **43.** (b) $11 - \ln 12$

45. (a) $C(t) = \dfrac{kA_0^2 t}{1 + kA_0 t}$ (b) $C(t) = \dfrac{A_0B_0(e^{kA_0t} - e^{kB_0t})}{B_0e^{kA_0t} - A_0e^{kB_0t}}$ **47.** (a) $v(t) = \dfrac{\alpha}{Ce^{(\alpha/m)t} - \beta}$, where C is an arbitrary constant.

(b) $v(t) = \dfrac{\alpha v_0}{(\alpha + \beta v_0)e^{(\alpha/m)t} - \beta v_0} = \dfrac{\alpha v_0 e^{-(\alpha/m)t}}{\alpha + \beta v_0 - \beta v_0 e^{-(\alpha/m)t}}$

(c) $\lim_{t\to\infty} v(t) = 0$

49. $F = \dfrac{2GmM}{L^2}\left[\dfrac{L}{h} + \ln\left(\dfrac{h}{L+h}\right)\right]$

SECTION 8.7

1. $-2(\sqrt{x} + \ln|1-\sqrt{x}|) + C$ **3.** $2\ln(\sqrt{1+e^x} - 1) - x + 2\sqrt{1+e^x} + C$ **5.** $\frac{2}{5}(1+x)^{5/2} - \frac{2}{3}(1+x)^{3/2} + C$

7. $\frac{2}{5}(x-1)^{5/2} + 2(x-1)^{3/2} + C$ **9.** $-\dfrac{1+2x^2}{4(1+x^2)^2} + C$ **11.** $x + 2\sqrt{x} + 2\ln|\sqrt{x}-1| + C$ **13.** $x + 4\sqrt{x-1} + 4\ln|\sqrt{x-1}-1| + C$

15. $2\ln(\sqrt{1+e^x} - 1) - x + C$ **17.** $\frac{2}{3}(x-8)\sqrt{x+4} + C$ **19.** $\frac{1}{16}(4x+1)^{1/2} + \frac{1}{8}(4x+1)^{-1/2} - \frac{1}{48}(4x+1)^{-3/2} + C$ **21.** $\dfrac{4b+2ax}{a^2\sqrt{ax+b}} + C$

23. $-\ln\left|1 - \tan\dfrac{x}{2}\right| + C$ **25.** $\dfrac{2}{\sqrt{3}}\tan^{-1}\left[\dfrac{1}{\sqrt{3}}\left(2\tan\dfrac{x}{2} + 1\right)\right] + C$ **27.** $\dfrac{1}{2}\ln\left|\tan\dfrac{x}{2}\right| - \dfrac{1}{4}\tan^2\dfrac{x}{2} + C$

29. $\ln\left|\dfrac{1}{1+\sin x}\right| - \dfrac{2}{1+\tan(x/2)} + C$ **31.** $\frac{4}{3} + 2\tan^{-1}2$ **33.** $2 + 4\ln\frac{2}{3}$ **35.** $\ln\left(\dfrac{\sqrt{3}-1}{\sqrt{3}}\right)$

37. Let $u = \tan\dfrac{x}{2}$. Then $\displaystyle\int \frac{1}{\cos x}\,dx = 2\int \frac{du}{1-u^2}$ and the result follows.

39. $\displaystyle\int \csc x\,dx = \int \frac{\sin x}{\sin^2 x}\,dx = \int \frac{\sin x}{1-\cos^2 x}\,dx = -\int \frac{du}{1-u^2}$ where $u = \cos x$. The result follows. **41.** $2\tan^{-1}\left(\tanh\dfrac{x}{2}\right) + C$

43. $\dfrac{-2}{\tanh\dfrac{x}{2} + 1} + C$

SECTION 8.8

1. (a) 506 (b) 650 (c) 575 (d) 578.8 (e) 576 **3.** (a) 1.394 (b) 0.9122 (c) 1.1852 (d) 1.1533 (e) 1.1614

5. (a) $\frac{1}{4}\pi \cong 0.7828$ (b) $\frac{1}{4}\pi \cong 0.7854$ **7.** (a) 1.8440 (b) 1.7915 (c) 1.8090 **9.** (a) 0.8818 (b) 0.8821

11. Such a curve passes through the three points $(a_1, b_1), (a_2, b_2), (a_3, b_3)$ iff

$$b_1 = a_1^2A + a_1B + C, \qquad b_2 = a_2^2A + a_2B + C, \qquad b_3 = a_3^2A + a_3B + C,$$

which happens iff

$$A = \frac{b_1(a_2-a_3) - b_2(a_1-a_3) + b_3(a_1-a_2)}{(a_1-a_3)(a_1-a_2)(a_2-a_3)}, \qquad B = -\frac{b_1(a_2^2-a_3^2) - b_2(a_1^2-a_3^2) + b_3(a_1^2-a_2^2)}{(a_1-a_3)(a_1-a_2)(a_2-a_3)},$$

$$C = \frac{a_1^2(a_2b_3 - a_3b_2) - a_2^2(a_1b_3 - a_3b_1) + a_3^2(a_1b_2 - a_2b_1)}{(a_1-a_3)(a_1-a_2)(a_2-a_3)}.$$

13. (a) $n \geq 8$ (b) $n \geq 4$ **15.** (a) $n \geq 238$ (b) $n \geq 19$ **17.** (a) $n \geq 51$ (b) $n \geq 7$ **19.** (a) $n \geq 37$ (b) $n \geq 5$

21. (a) 78 (b) 7 **23.** $f^{(4)}(x) = 0$ for all x; therefore by (8.7.2) the theoretical error is zero

25. (a) $\left| T_2 - \int_0^1 x^2\,dx \right| = \frac{3}{8} - \frac{1}{3} = \frac{1}{24} = E_2^T$ (b) $\left| S_1 - \int_0^1 x^4\,dx \right| = \frac{5}{24} - \frac{1}{5} = \frac{1}{120} = E_1^S$

27. Using the hint, $M_n = \text{area } ABCD = \text{area } AEFD \leq \int_a^b f(x)\,dx \leq T_n$.

CHAPTER 9

SECTION 9.1

1. $y^2 = 8x$

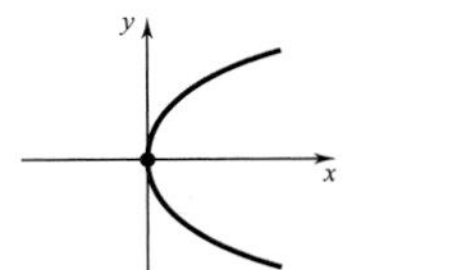

3. $(x + 1)^2 = -12(y - 3)$

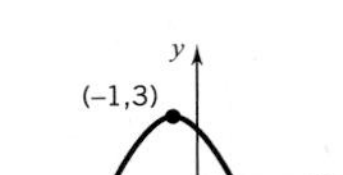
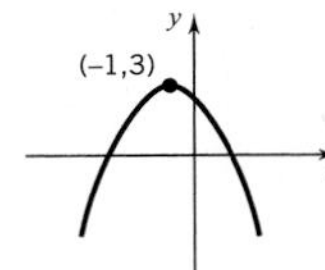

5. $4y = (x - 1)^2$

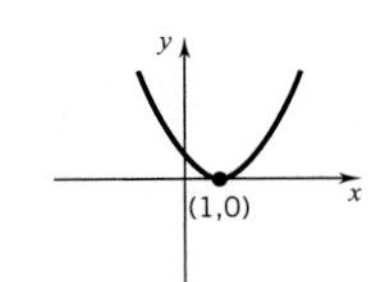

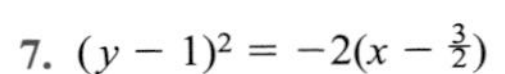
7. $(y - 1)^2 = -2(x - \frac{3}{2})$

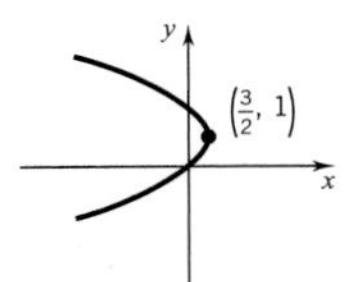

9. vertex $(0, 0)$
focus $(\frac{1}{2}, 0)$
axis $y = 0$
directrix $x = -\frac{1}{2}$

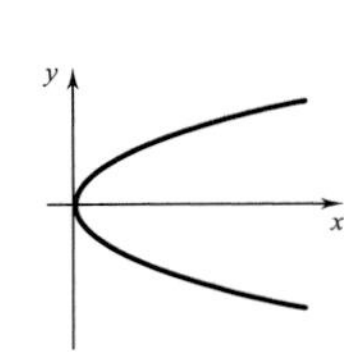

11. vertex $(0, -\frac{1}{2})$
focus $(0, -\frac{3}{8})$
axis $x = 0$
directrix $y = -\frac{5}{8}$

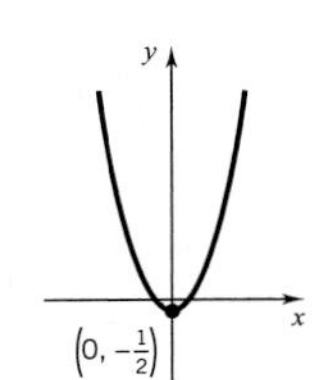

13. vertex $(-2, \frac{3}{2})$
focus $(-2, -\frac{1}{2})$
axis $x = -2$
directrix $y = \frac{7}{2}$

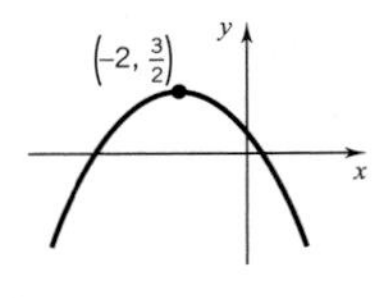

15. vertex $(\frac{3}{4}, -\frac{1}{2})$
focus $(1, -\frac{1}{2})$
axis $y = -\frac{1}{2}$
directrix $x = \frac{1}{2}$

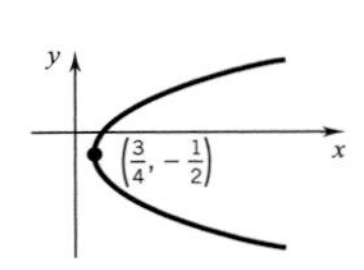

17. center $(0, 0)$
foci $(\pm\sqrt{5}, 0)$
length of major axis 6
length of minor axis 4

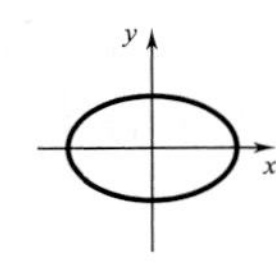

19. center $(0, 0)$
foci $(0, \pm\sqrt{2})$
length of major axis $2\sqrt{6}$
length of minor axis 4

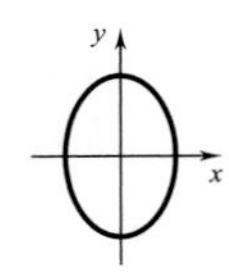

21. center $(0, 1)$
foci $(\pm\sqrt{5}, 1)$
length of major axis 6
length of minor axis 4

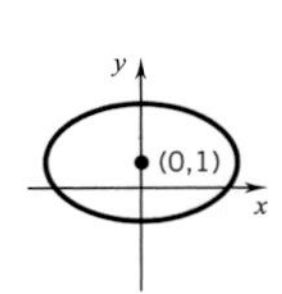

23. center $(1, 0)$
foci $(1, \pm 4\sqrt{3})$
length of major axis 16
length of minor axis 8

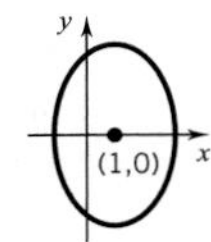

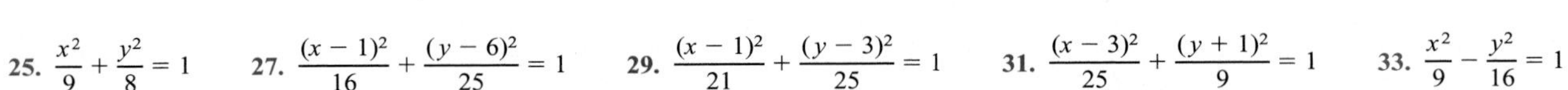
25. $\frac{x^2}{9} + \frac{y^2}{8} = 1$ **27.** $\frac{(x-1)^2}{16} + \frac{(y-6)^2}{25} = 1$ **29.** $\frac{(x-1)^2}{21} + \frac{(y-3)^2}{25} = 1$ **31.** $\frac{(x-3)^2}{25} + \frac{(y+1)^2}{9} = 1$ **33.** $\frac{x^2}{9} - \frac{y^2}{16} = 1$

35. $\frac{y^2}{25} - \frac{x^2}{144} = 1$ **37.** $\frac{x^2}{9} - \frac{(y-1)^2}{16} = 1$ **39.** $16y^2 - \frac{16}{15}(x+1)^2 = 1$

41. center $(0, 0)$
transverse axis 2
vertices $(\pm 1, 0)$
foci $(\pm\sqrt{2}, 0)$
asymptotes $y = \pm x$

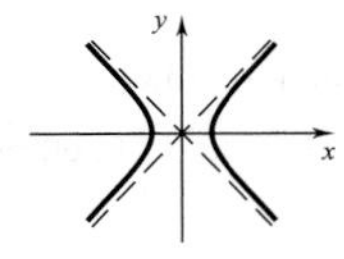

43. center $(0, 0)$
transverse axis 6
vertices $(\pm 3, 0)$
foci $(\pm 5, 0)$
asymptotes $y = \pm\frac{4}{3}x$

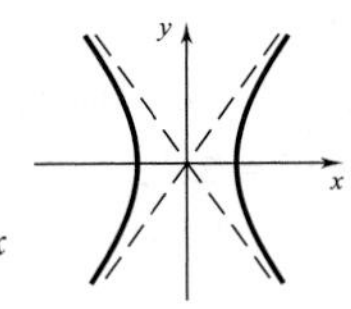

45. center $(0, 0)$
transverse axis 8
vertices $(0, \pm 4)$
foci $(0, \pm 5)$
asymptotes $y = \pm\frac{4}{3}x$

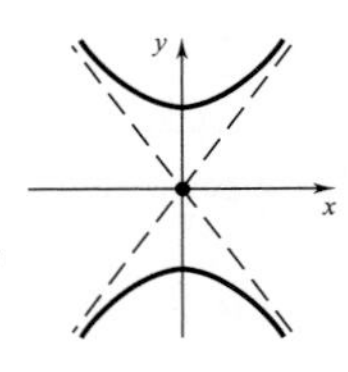

47. center $(1, 3)$
transverse axis 6
vertices $(4, 3)$ and $(-2, 3)$
foci $(6, 3)$ and $(-4, 3)$
asymptotes $y = \pm\frac{4}{3}(x - 1) + 3$

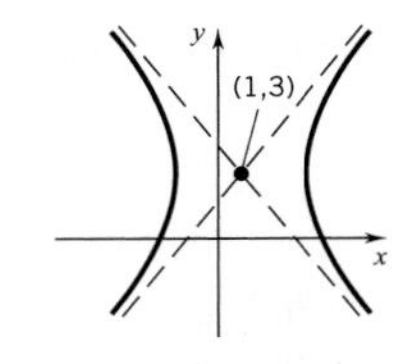

49. center $(1, 3)$
transverse axis 4
vertices $(1, 5)$ and $(1, 1)$
foci $(1, 3 \pm \sqrt{5})$
asymptotes $y = 2x + 1$, $y = -2x + 5$

51. $(x - y)^2 = 6x + 10y - 9$ **53.** $(x + y)^2 = -12x + 20y + 28$

55. $P(x, y)$ is on the parabola with directrix l: $Ax + By + C = 0$ and focus $F(a, b)$ iff

$$d(P, l) = d(P, F) \quad \text{which happens iff} \quad \frac{|Ax + By + C|}{\sqrt{A^2 + B^2}} = \sqrt{(x-a)^2 + (y-b)^2}.$$

Square this last equation and simplify.

57. We can choose the coordinate system so that the parabola has an equation of the form $y = \alpha x^2$, $\alpha > 0$. One of the points of intersection is then the origin and the other is of the form $(c, \alpha c^2)$. We will assume that $c > 0$.

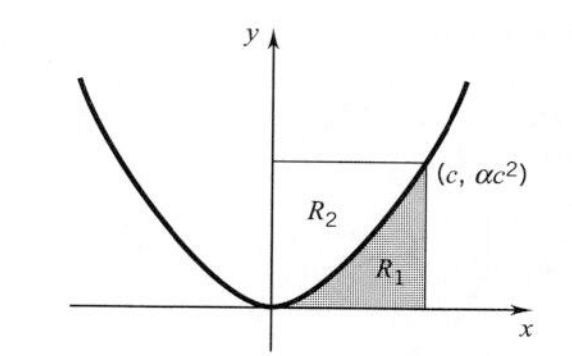

$$\text{area of } R_1 = \int_0^c \alpha x^2\, dx = \tfrac{1}{3}\alpha c^3 = \tfrac{1}{3}A$$
$$\text{area of } R_2 = A - \tfrac{1}{3}A = \tfrac{2}{3}A.$$

59. $2y = x^2 - 4x + 7$, $18y = x^2 - 4x + 103$ **61.** $2\sqrt{\pi^2 a^4 - A^2}/\pi a$ **63.** $(5 \pm \frac{5}{21}\sqrt{5}, 0)$ **65.** $\frac{3}{5}$ **67.** $\frac{4}{5}$

69. E_1 is fatter than E_2, more like a circle **71.** the ellipse tends to a line segment of length $2a$ **73.** $x^2/9 + y^2 = 1$

75. center $(0, 0)$, vertices $(1, 1)$ and $(-1, 1)$, foci $(\sqrt{2}, \sqrt{2})$ and $(-\sqrt{2}, \sqrt{2})$, asymptotes $x = 0$ and $y = 0$, transverse axis $2\sqrt{2}$

77. $[2\sqrt{3} - \ln(2 + \sqrt{3})]ab$ **79.** $\frac{5}{3}$ **81.** $\sqrt{2}$ **83.** the branches of H_1 open up less quickly than the branches of H_2

85. the hyperbola tends to a pair of parallel lines separated by the transverse axis **87.** $4c$ **89.** $A = \frac{8}{3}c^3$; $\bar{x} = 0$, $\bar{y} = \frac{3}{5}c$

91. $\dfrac{kx}{p(0)} = \tan\theta = \dfrac{dy}{dx}$, $y = \dfrac{k}{2p(0)}x^2 + C$

In our figure, $C = y(0) = 0$. Thus the equation of the cable is $y = kx^2/2p(0)$, the equation of a parabola.

93. Start with any two parabolas γ_1, γ_2. By moving them we can see to it that they have equations of the following form:

$$\gamma_1\colon\ x^2 = 4c_1 y,\quad c_1 > 0; \qquad \gamma_2\colon\ x^2 = 4c_2 y,\quad c_2 > 0.$$

Now we change the scale for γ_2 so that the equation for γ_2 will look exactly like the equation for γ_1. Set $X = (c_1/c_2)x$, $Y = (c_1/c_2)y$. Then

$$x^2 = 4c_2 y \Rightarrow (c_2/c_1)^2 X^2 = 4c_2(c_2/c_1)Y \Rightarrow X^2 = 4c_1 Y.$$

Now γ_2 has exactly the same equation as γ_1; only the scale, the units by which we measure distance, has changed.

95. about 0.25 mile west and 1.5 miles north of point A.

SECTION 9.2

1–7. See figure to the right. **9.** $(0, 3)$ **11.** $(1, 0)$ **13.** $(-\frac{3}{2}, \frac{3}{2}\sqrt{3})$ **15.** $(0, -3)$

17. $[1, \frac{1}{2}\pi + 2n\pi]$, $[-1, \frac{3}{2}\pi + 2n\pi]$ **19.** $[3, \pi + 2n\pi]$, $[-3, 2n\pi]$

21. $[2\sqrt{2}, \frac{7}{4}\pi + 2n\pi]$, $[-2\sqrt{2}, \frac{3}{4}\pi + 2n\pi]$ **23.** $[8, \frac{1}{6}\pi + 2n\pi]$, $[-8, \frac{7}{6}\pi + 2n\pi]$

25. $\sqrt{r_1^2 + r_2^2 - 2r_1 r_2 \cos(\theta_1 - \theta_2)}$ **27.** (a) $[\frac{1}{2}, \frac{11}{6}\pi]$ (b) $[\frac{1}{2}, \frac{5}{6}\pi]$ (c) $[\frac{1}{2}, \frac{7}{6}\pi]$

29. (a) $[2, \frac{2}{3}\pi]$ (b) $[2, \frac{5}{3}\pi]$ (c) $[2, \frac{1}{3}\pi]$ **31.** symmetry about the x-axis

33. no symmetry about the coordinate axes; no symmetry about the origin

35. symmetry about the origin **37.** $r\cos\theta = 2$ **39.** $r^2 \sin 2\theta = 1$ **41.** $r = 4\sin\theta$

43. $\theta = \pi/4$ **45.** $r = 1 - \cos\theta$ **47.** $r^2 = \sin 2\theta$ **49.** the horizontal line $y = 4$ **51.** the line $y = \sqrt{3}x$

53. the parabola $y^2 = 4(x + 1)$ **55.** the circle $x^2 + y^2 = 3x$ **57.** the line $y = 2x$ **59.** $3x^2 + 4y^2 - 8x = 16$, ellipse

61. $y^2 = 8x + 16$, parabola **63.** $\left(x - \dfrac{b}{2}\right)^2 + \left(y - \dfrac{a}{2}\right)^2 = \dfrac{a^2 + b^2}{4}$; center: $\left(\dfrac{b}{2}, \dfrac{a}{2}\right)$, radius: $\dfrac{\sqrt{a^2 + b^2}}{2}$ **65.** $r = \dfrac{d}{2 - \cos\theta}$

SECTION 9.3

1.

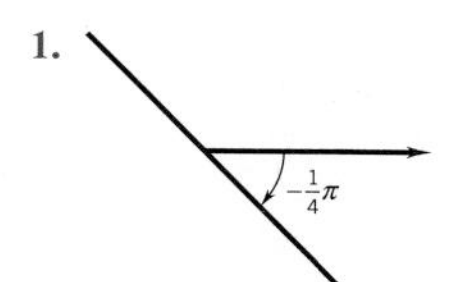

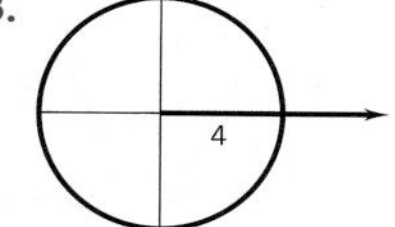

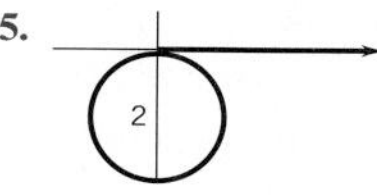

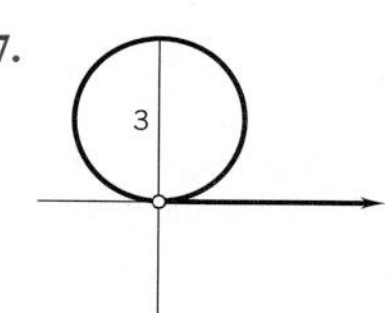

9.

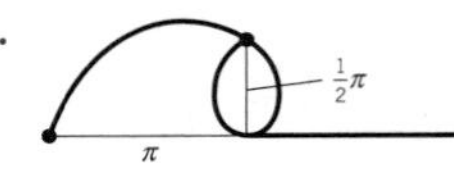

11.

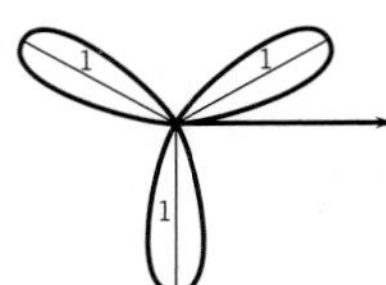

13.

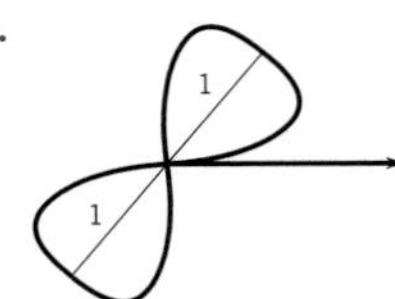

15.

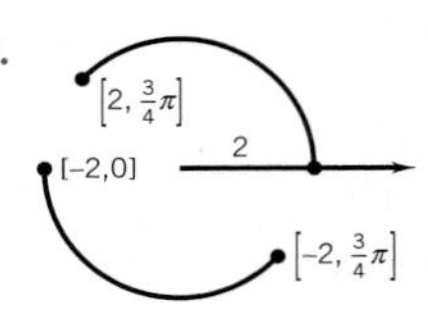

17.

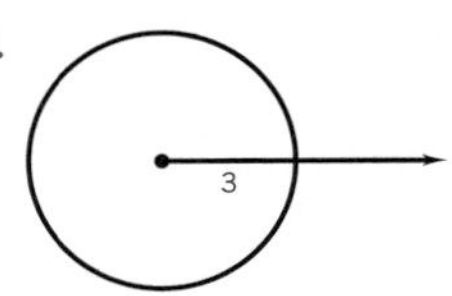

19.

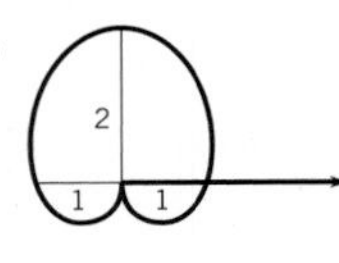

21.

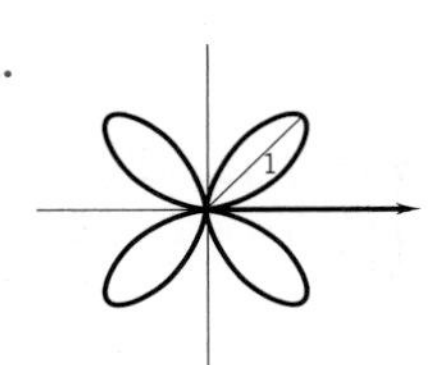

23.

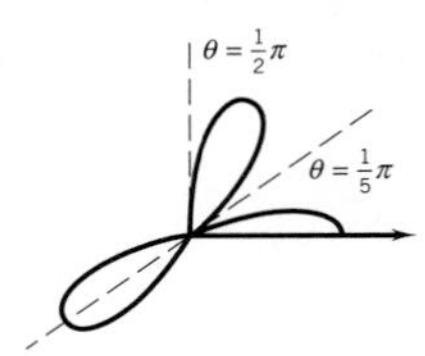

25.

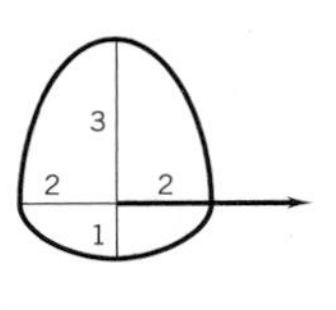

27.

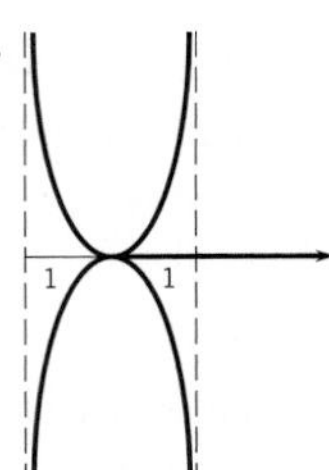

29.

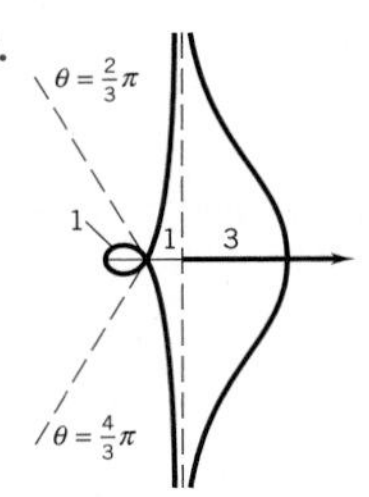

31.

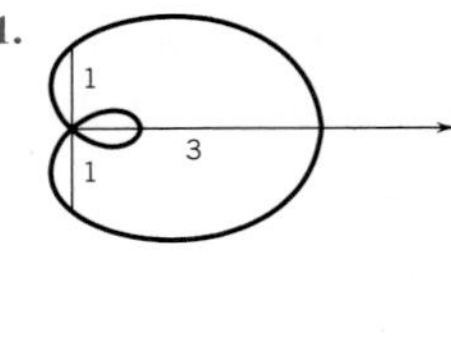

33.

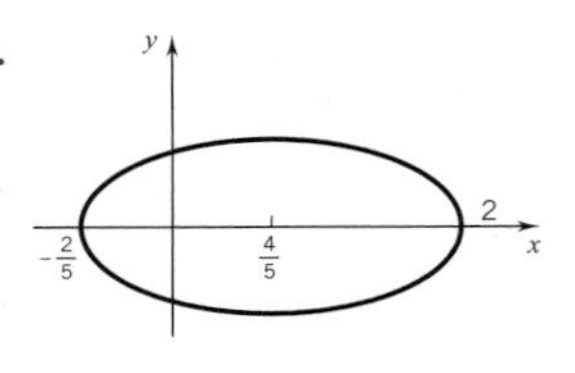

37. (a) center: $\left(\dfrac{e^2c}{1-e^2}, 0\right)$ length of major axis: $\dfrac{2ec}{1-e^2}$; length of minor axis: $\dfrac{2ec}{\sqrt{1-e^2}}$
(b) (i) the ellipse in (a) rotated $\pi/2$ radians in the counterclockwise direction
(ii) the ellipse in (a) rotated π radians in the counterclockwise direction
(iii) the ellipse in (a) rotated $\pi/2$ radians in the clockwise direction

39. (a) a point, the origin (b) parabola (c) ellipse, center at $\left(\dfrac{2d}{3}, 0\right)$, length of major axis $\dfrac{8}{3}d$ **41.** Butterfly

SECTION *9.4

1. (a) $\frac{1}{2}$ unit to the right of the pole (b) 2 units (c) Figure A. **3.** the parabola of Figure A rotated by π radians

5. (a) $e = \frac{2}{3}$ (b) 2 units to the left of the pole and $\frac{2}{5}$ units to the right of the pole (c) $\frac{4}{5}$ units to the left of the pole
(d) $\frac{8}{5}$ units to the left of the pole (e) $\frac{4}{5}\sqrt{5}$ units (about 1.79 units) (f) $\frac{4}{3}$ units (g) Figure B

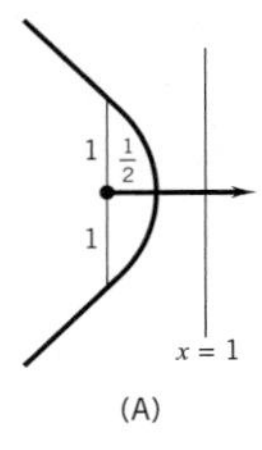

(A)

$\frac{4}{5}\sqrt{5}$ $\frac{2}{5}$ $\frac{2}{5}$ $\frac{12}{5}$

(B)

7. the ellipse of Figure B rotated by $\frac{1}{2}\pi$ radians

9. (a) $e = 2$
(b) 2 units to the right of the pole and 6 units to the right of the pole
(c) 4 units to the right of the pole
(d) 8 units to the right of the pole
(e) 12 units

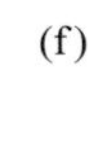

(f)

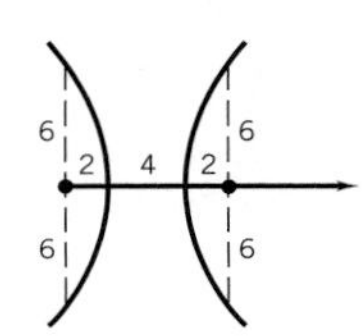

11. the hyperbola of Exercise 9(f) rotated by $\frac{3}{2}\pi$ radians **13.** ellipse: $x^2/48 + (y+4)^2/64 = 1$ **15.** ellipse: $x^2/9 + (y-4)^2/25 = 1$

SECTION 9.5

1. yes; $[1, \pi] = [-1, 0]$ and the pair $r = -1$, $\theta = 0$ satisfies the equation **3.** yes; the pair $r = \frac{1}{2}$, $\theta = \frac{1}{2}\pi$ satisfies the equation

5. $[2, \pi] = [-2, 0]$. The coordinates of $[-2, 0]$ satisfy the equation $r^2 = 4\cos\theta$, and the coordinates of $[2, \pi]$ satisfy the equation $r = 3 + \cos\theta$.

7. $(0, 0), (-\frac{1}{2}, \frac{1}{2})$ **9.** $(-1, 0), (1, 0)$ **11.** $(0, 0), (\frac{1}{4}, \pm\frac{1}{4}\sqrt{3})$ **13.** $(\frac{3}{2}, 2)$ **15.** $(0, 0), (1, 0)$

17. (b) The curves intersect at the pole and at (1.172, 0.173), (1.86, 1.036), (0.90, 3.245).

19. (b) The curves intersect at the pole and at:

$r = 1 - 3\sin\theta$	$r = 2 - 5\sin\theta$
(−2, 0)	(2, π)
(3.800, 3.510)	(3.800, 3.510)
(2.412, 4.223)	(−2.412, 1.081)
(−1.267, 0.713)	(−1.267, 0.713)

SECTION 9.6

1. $\frac{1}{4}\pi a^2$ **3.** $\frac{1}{2}a^2$ **5.** $\frac{1}{2}\pi a^2$ **7.** $\frac{1}{4} - \frac{1}{16}\pi$ **9.** $\frac{3}{16}\pi + \frac{3}{8}$ **11.** $\frac{5}{2}a^2$ **13.** $\frac{1}{12}(3e^{2\pi} - 3 - 2\pi^3)$ **15.** $\frac{1}{4}(e^{2\pi} + 1 - 2e^{\pi})$

17. $\displaystyle\int_{\pi/6}^{5\pi/6} \tfrac{1}{2}([4\sin\theta]^2 - [2]^2)\,d\theta$ **19.** $\displaystyle\int_{-\pi/3}^{\pi/3} \tfrac{1}{2}([4]^2 - [2\sec\theta]^2)\,d\theta$ **21.** $\displaystyle 2\left[\int_0^{\pi/3} \tfrac{1}{2}(2\sec\theta)^2\,d\theta + \int_{\pi/3}^{\pi/2} \tfrac{1}{2}(4)^2\,d\theta\right]$

23. $\displaystyle\int_0^{\pi/3} \tfrac{1}{2}(2\sin 3\theta)^2\,d\theta$ **25.** $\displaystyle 2\left[\int_0^{\pi/6} \tfrac{1}{2}(\sin\theta)^2\,d\theta + \int_{\pi/6}^{\pi/2} \tfrac{1}{2}(1 - \sin\theta)^2\,d\theta\right]$ **27.** $\displaystyle \pi - 8\int_0^{\pi/4} \tfrac{1}{2}(\cos 2\theta)^2\,d\theta$

29. For $r = a\cos 2n\theta$, area of one petal is $\displaystyle a^2\int_0^{\pi/4n} \cos^2 2n\theta\,d\theta = \frac{\pi a^2}{8n}$. For $r = a\sin 2n\theta$, area of one petal is $\displaystyle a^2\int_0^{\pi/4n} \sin^2 2n\theta\,d\theta = \frac{\pi a^2}{8n}$.

Total area $= \dfrac{\pi a^2}{2}$.

33. $(5/6, 0)$ **35.** (a) Substitute $x = r\cos\theta$, $y = r\sin\theta$ into the equation and solve for r. (c) $8 - 2\pi$

SECTION 9.7

1. $4x = (y-1)^2$ **3.** $y = 4x^2 + 1$, $x \geq 0$ **5.** $9x^2 + 4y^2 = 36$ **7.** $1 + x^2 = y^2$ **9.** $y = 2 - x^2$, $-1 \leq x \leq 1$

11. $2y - 6 = x$, $-4 \leq x \leq 4$

13. $y = x - 1$

15. $xy = 1$

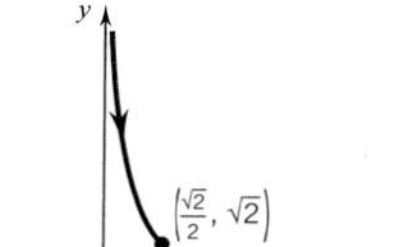

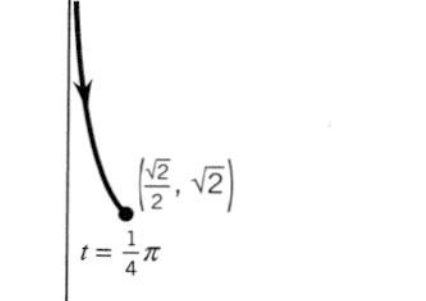

17. $y + 2x = 11$

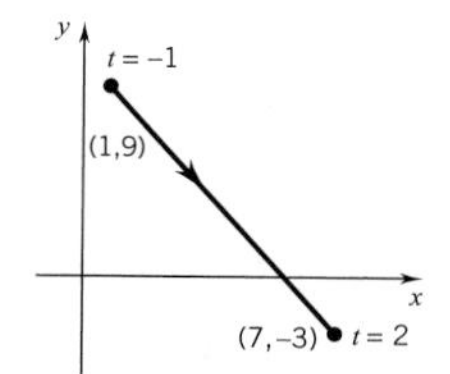

19. $x = \sin\frac{1}{2}\pi y$

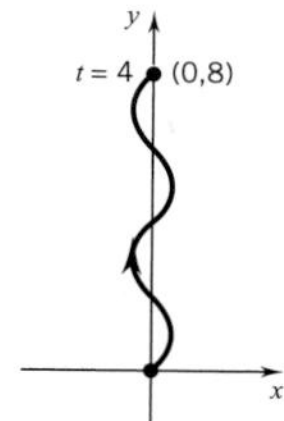

21. $y^2 = x^2 + 1$

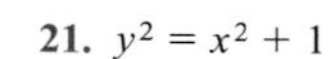

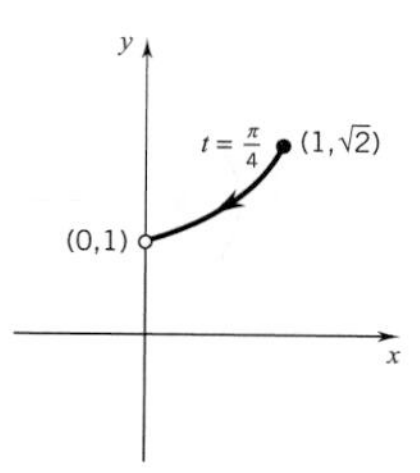

23. (a) $x(t) = -\sin 2\pi t$, $y(t) = \cos 2\pi t$ (b) $x(t) = \sin 4\pi t$, $y(t) = \cos 4\pi t$ (c) $x(t) = \cos\frac{1}{2}\pi t$, $y(t) = \sin\frac{1}{2}\pi t$
(d) $x(t) = \cos\frac{3}{2}\pi t$, $y(t) = -\sin\frac{3}{2}\pi t$

25. $x(t) = \tan\frac{1}{2}\pi t$, $y(t) = 2$ **27.** $x(t) = 3 + 5t$, $y(t) = 7 - 2t$ **29.** $x(t) = \sin^2\pi t$, $y(t) = -\cos\pi t$

31. $x(t) = (2-t)^2$, $y(t) = (2-t)^3$ **33.** $x(t) = t(b-a) + a$, $y(t) = f(t(b-a) + a)$ **35.** $A = 2\pi a^2$

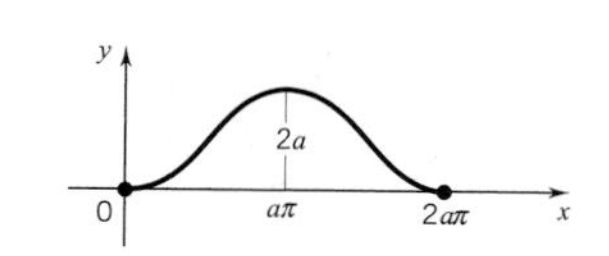

37. (a) $V_x = 3\pi^2 r^3$ (b) $V_y = 4\pi^3 r^3$ **39.** $x(t) = -a \cos t,\ \ y(t) = b \sin t;\ \ t \in [0, \pi]$

41. (a) paths intersect at (6, 5) and (8, 1) (b) particles collide at (8, 1) **43.** curve intersects itself at (0, 0)

45. curve intersects itself at (0, 0) and $(0, \frac{3}{4})$

47. (a) The coefficient a determines the amplitude and the period. (b) $\dfrac{dy}{dx} \to -\infty$ as $\theta \to 2\pi^-$; $\dfrac{dy}{dx} \to \infty$ as $\theta \to 2\pi^+$

49.

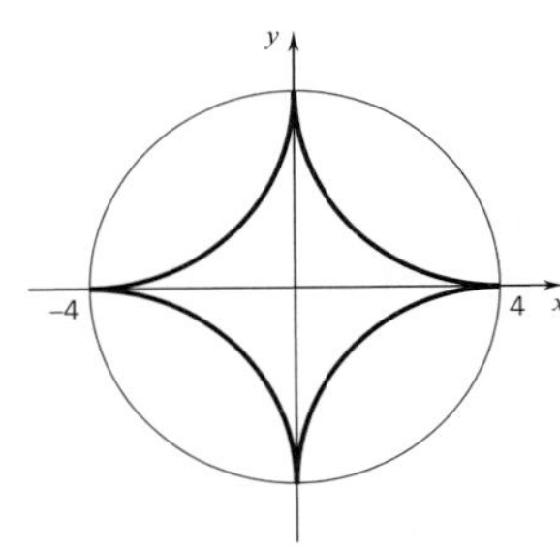

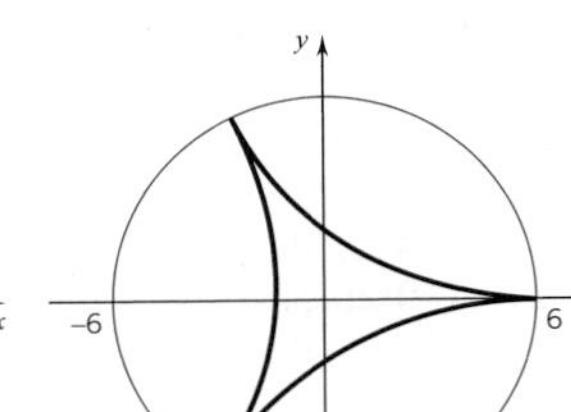

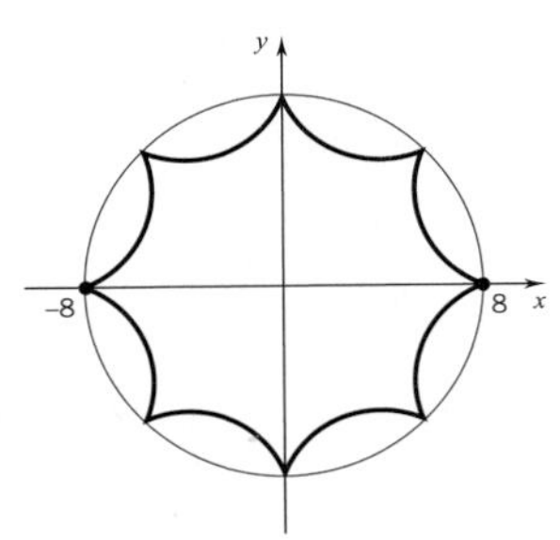

51. $y = -\dfrac{16}{v_0^2}(\sec^2\theta)x^2 + (\tan\theta)x$ **53.** $\dfrac{1}{16} v_0^2 \cos\theta \sin\theta$ feet **55.** $\pi/4$

SECTION 9.8

1. $3x - y - 3 = 0$ **3.** $y = 1$ **5.** $3x + y - 3 = 0$ **7.** $2x + 2y - \sqrt{2} = 0$ **9.** $2x + y - 8 = 0$ **11.** $x - 5y + 4 = 0$

13. $x + 2y + 1 = 0$

15. $x(t) = t,\ \ y(t) = t^3$; tangent line $y = 0$

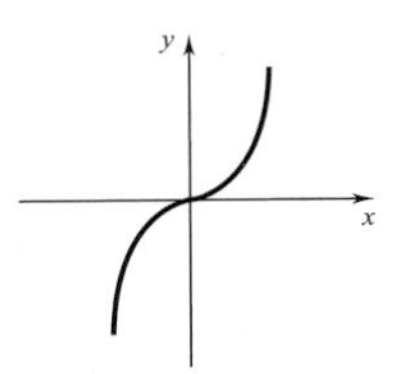

17. $x(t) = t^{5/3},\ \ y(t) = t$; tangent line $x = 0$

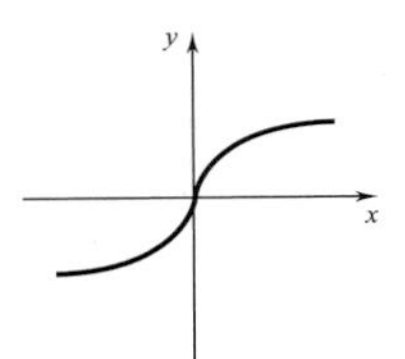

19. (a) none; (b) at (2, 2) and (−2, 0)

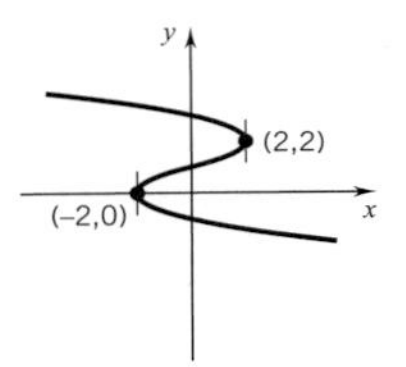

21. (a) at (3, 7) and (3, 1); (b) at (−1, 4) and (7, 4)

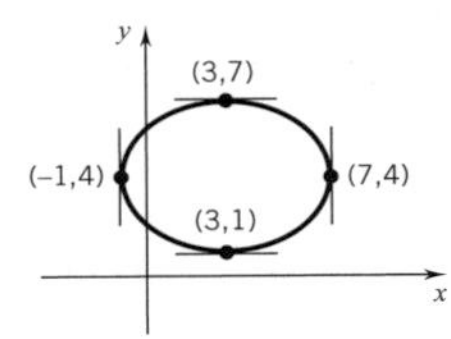

23. (a) at $(-\frac{2}{3}, \pm\frac{2}{9}\sqrt{3})$; (b) at (−1, 0)

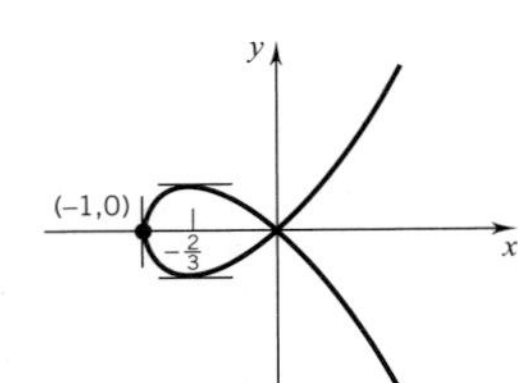

25. (a) at $(\pm\frac{1}{2}\sqrt{2}, \pm 1)$; (b) at $(\pm 1, 0)$

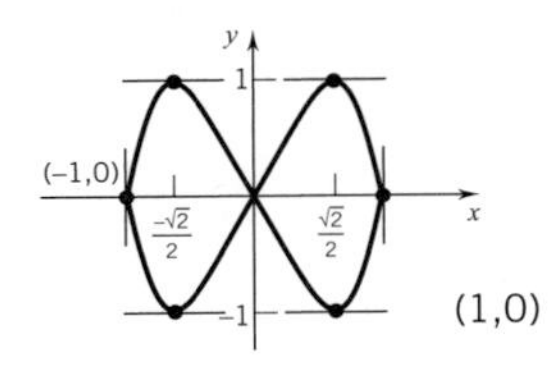

27. $y = 0,\ \ (\pi - 2)y + 32x - 64 = 0$

29. The slope of $\overline{OP}$ is $\tan\theta_1$. The curve $r = f(\theta)$ can be parametrized by setting

$$x(\theta) = f(\theta)\cos\theta, \qquad y(\theta) = f(\theta)\sin\theta.$$

Differentiation gives

$$x'(\theta) = -f(\theta)\sin\theta + f'(\theta)\cos\theta, \qquad y'(\theta) = f(\theta)\cos\theta + f'(\theta)\sin\theta.$$

If $f'(\theta_1) = 0$, then

$$x'(\theta_1) = -f(\theta_1)\sin\theta_1, \qquad y'(\theta_1) = f(\theta_1)\cos\theta_1.$$

Since $f(\theta_1) \neq 0$, we have

$$m = \frac{y'(\theta_1)}{x'(\theta_1)} = -\cot\theta_1 = -\frac{1}{\text{slope of } \overline{OP}}.$$

31. 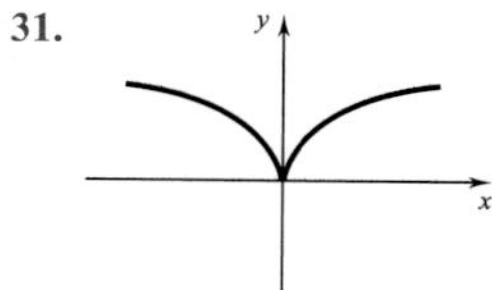

33. 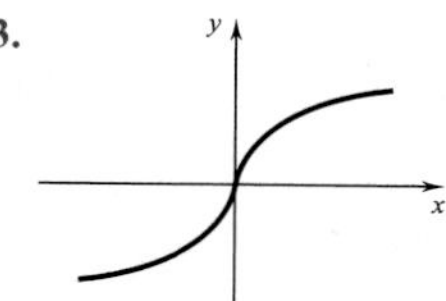

35. 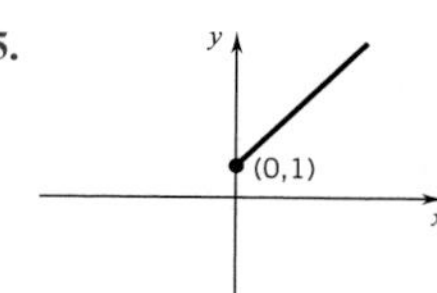

37. -8 39. 2

SECTION 9.9

1. $\sqrt{5}$ 3. 7 5. $2\sqrt{3}$ 7. $\frac{4}{3}$ 9. $6 + \frac{1}{2}\ln 5$ 11. $\frac{63}{8}$ 13. $\ln(1+\sqrt{2})$ 15. $\frac{3}{2}$ 17. $\frac{1}{3}\pi + \frac{1}{2}\sqrt{3}$

19. initial speed 2, terminal speed 4; $s = 2\sqrt{3} + \ln(2+\sqrt{3})$ 21. initial speed 0, terminal speed $\sqrt{13}$; $x = \frac{1}{27}(13\sqrt{13} - 8)$

23. initial speed $\sqrt{2}$, terminal speed $\sqrt{2}e^{\pi}$; $s = \sqrt{2}(e^{\pi} - 1)$ 25. $8a$

27. (a) $24a$ (b) use the identities $\cos 3\theta = 4\cos^3\theta - 3\cos\theta$, $\sin 3\theta = 3\sin\theta - 4\sin^3\theta$ 29. 2π 31. $\sqrt{2}(e^{4\pi} - 1)$

33. $\frac{1}{2}\sqrt{5}(e^{4\pi} - 1)$ 35. $4 - 2\sqrt{2}$ 37. $\ln(1+\sqrt{2})$ 39. $c = 1$ 41. $L = \int_a^b \sqrt{1+\sinh^2 x}\,dx = \int_a^b \sqrt{\cosh^2 x}\,dx = \int_a^b \cosh x\,dx = A$

43. (a) Express GPE + KE as a function of t and verify that the derivative with respect to t is zero.
(b) From (a) we know that throughout the motion

$$32my + \tfrac{1}{2}mv^2 = C.$$

At the time of firing $y = 0$ and $v = |\mathbf{v}_0| = v_0$. Therefore

$$32my + \tfrac{1}{2}mv^2 = \tfrac{1}{2}mv_0^2.$$

At impact $y = 0$, $\frac{1}{2}mv^2 = \frac{1}{2}mv_0^2$, and $v = v_0$.

45. $\sqrt{1+[f'(x)]^2} = \sqrt{1+\tan^2[\alpha(x)]} = |\sec[\alpha(x)]|$ 47. (a) 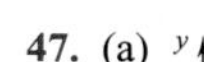 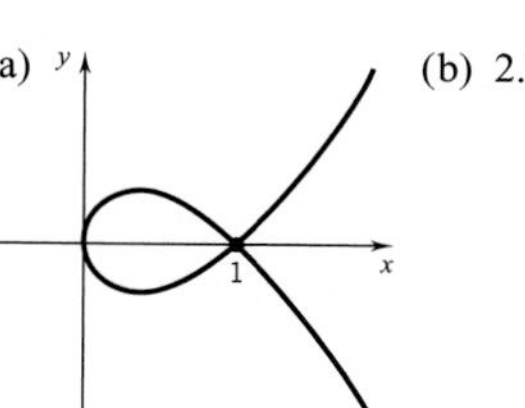

(b) 2.7156 49. (b) 28.3617

SECTION 9.10

1. $L = 1$, $(\bar{x}, \bar{y}) = (\frac{1}{2}, 4)$, $A_x = 16\pi$ 3. $L = 5$, $(\bar{x}, \bar{y}) = (\frac{3}{2}, 2)$, $A_x = 20\pi$ 5. $L = 10$, $(\bar{x}, \bar{y}) = (3, 4)$, $A_x = 80\pi$

7. $L = \frac{1}{3}\pi$; $\bar{x} = 6/\pi$, $\bar{y} = 6(2-\sqrt{3})/\pi$; $A_x = 4\pi(2-\sqrt{3})$ 9. $L = \frac{1}{3}\pi a$; $\bar{x} = 0$, $\bar{y} = 3a/\pi$; $A_x = 2\pi a^2$ 11. $\frac{1}{9}\pi(17\sqrt{7} - 1)$

13. $\frac{61}{432}\pi$ 15. $\pi[\sqrt{2} + \ln(1+\sqrt{2})]$ 17. $\frac{2}{5}\sqrt{2}\pi(2e^{\pi} + 1)$ 19. (a) $3\pi a^2$ (b) $\dfrac{64\pi a^2}{3}$

21. See the figure.

$$\begin{aligned} A &= \tfrac{1}{2}\theta s_2^2 - \tfrac{1}{2}\theta s_1^2 \\ &= \tfrac{1}{2}(\theta s_2 + \theta s_1)(s_2 - s_1) \\ &= \tfrac{1}{2}(2\pi R + 2\pi r)s = \pi(R+r)s \end{aligned}$$

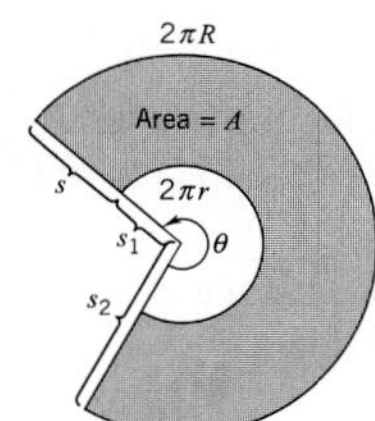

23. (a) the 3, 4, 5 sides have centroids $(\frac{3}{2}, 0), (4, 2), (\frac{3}{2}, 2)$
(b) $\bar{x} = 2$, $\bar{y} = \frac{3}{2}$
(c) $\bar{x} = 2$, $\bar{y} = \frac{4}{3}$
(d) $\bar{x} = \frac{13}{6}$, $\bar{y} = 2$
(e) $A = 20\pi$
(f) $A = 36\pi$

25. $4\pi^2 ab$

27. The band can be obtained by revolving about the x-axis the graph of a function

$$f(x) = \sqrt{r^2 - x^2}, \quad x \in [a, b].$$

A straightforward calculation shows that the surface area of the band is $2\pi r(b-a)$.

29. (a) $2\pi b^2 + \dfrac{2\pi ab}{e}\sin^{-1}e$ (b) $2\pi a^2 + \dfrac{\pi b^2}{e}\ln\left|\dfrac{1+e}{1-e}\right|$, where e is the eccentricity $c/a = \sqrt{a^2 - b^2}/a$

31. at the midpoint of the axis of the hemisphere 33. on the axis of the cone $\left(\dfrac{2R+r}{R+r}\right)\dfrac{h}{3}$ units from the base of radius r

SECTION 9.11

1. $x(\theta) = \overline{OB} - \overline{AB} = R\theta - R\sin\theta = R(\theta - \sin\theta), \quad y(\theta) = \overline{BQ} - \overline{QC} = R - R\cos\theta = R(1 - \cos\theta)$

3. (a) The slope at P is

$$m = \frac{y'(\theta)}{x'(\theta)} = \frac{\sin\theta}{1 - \cos\theta}.$$

The line tangent to the cycloid at P has equation

$$y - R(1 - \cos\theta) = \frac{\sin\theta}{1 - \cos\theta}[x - R(\theta - \sin\theta)].$$

The top of the circle is the point $(R\theta, 2R)$. Its coordinates satisfy the equation for the tangent:

$$2R - R(1 - \cos\theta) \stackrel{?}{=} \frac{\sin\theta}{1 - \cos\theta}[R\theta - R(\theta - \sin\theta)]$$

$$R(1 + \cos\theta) \stackrel{?}{=} \frac{R\sin^2\theta}{1 - \cos\theta}$$

$$1 - \cos^2\theta \stackrel{\checkmark}{=} \sin^2\theta.$$

(b) In view of the symmetry and repetitiveness of the curve we can assume that $\theta \in (0, \pi)$. Then

$$\tan\alpha = \frac{\sin\theta}{1 - \cos\theta} = \frac{\sin\theta}{2\sin^2\frac{1}{2}\theta} = \frac{\sin\frac{1}{2}\theta\cos\frac{1}{2}\theta}{\sin^2\frac{1}{2}\theta} = \cot\tfrac{1}{2}\theta$$

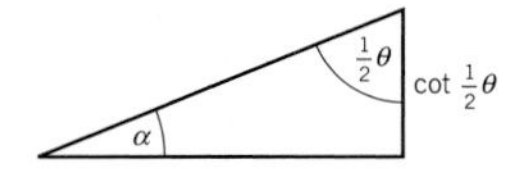

and $\alpha = \frac{1}{2}\pi - \frac{1}{2}\theta = \frac{1}{2}(\pi - \theta)$.

5. $8R$ **7.** $\bar{x} = \pi R, \quad \bar{y} = \frac{5}{6}R$ **9.** $V_y = 6\pi^3R^3$ **11.** $A = \frac{64}{3}\pi R^2$ **13.** $\alpha = \frac{1}{2}\phi$ (radian measure)

15. (a) already shown more generally in Example 6, Section 9.9
(b) Combining $d^2s/dt^2 = -g\sin\alpha$ with $s = 4R\sin\alpha$, we have

$$\frac{d^2s}{dt^2} = -\frac{g}{4R}s.$$

This is simple harmonic motion with angular frequency $\omega = \frac{1}{2}\sqrt{g/R}$ and period $T = 2\pi/\omega = 4\pi\sqrt{R/g}$.

CHAPTER 10

SECTION 10.1

1. lub = 2; glb = 0 **3.** no lub; glb = 0 **5.** lub = 2; glb = −2 **7.** no lub; glb = 2 **9.** lub = $2\frac{1}{2}$; glb = 2

11. lub = 1; glb = 0.9 **13.** lub = e; glb = 0 **15.** lub = $\frac{1}{2}(-1 + \sqrt{5})$; glb = $\frac{1}{2}(-1 - \sqrt{5})$ **17.** no lub; no glb

19. no lub; no glb **21.** glb $S = 0, \quad 0 \le (\frac{1}{11})^3 < 0 + 0.001$ **23.** glb $S = 0, \quad 0 \le (\frac{1}{10})^{2n-1} < 0 + (\frac{1}{10})^k, \quad n > \frac{1}{2}(k + 1)$

25. Let $\epsilon > 0$. The condition $m \le s$ is satisfied by all numbers s in S. All we have to show therefore is that there is some number s in S such that $s < m + \epsilon$. Suppose on the contrary that there is no such number in S. We then have $m + \epsilon \le x$ for all $x \in S$, so that $m + \epsilon$ becomes a lower bound for S. But this cannot happen, for it makes $m + \epsilon$ a lower bound that is *greater* than m, and by assumption, m is the *greatest* lower bound.

27. Let c = lub S. Since $b \in S$, $b \le c$. Since b is an upper bound for S, $c \le b$. Thus $b = c$.

29. (a) Any upper bound for S is an upper bound for T; any lower bound for S is a lower bound for T.
(b) Let a = glb S. Then $a \le t$ for all $t \in T$. Therefore $a \le$ glb T. Similarly, if b = lub S, then $t \le b$ for all $t \in T$, so lub $T \le b$. It now follows that glb $S \le$ glb $T \le$ lub $T \le$ lub S.

31. Let M be any positive number and consider M/c. Since the set of positive integers is not bounded above, there exists a positive integer k such that $k \ge M/c$. This implies $kc \ge M$. Since $kc \in S$, it follows that S is not bounded above.

33. (a)

a_1	a_2	a_3	a_4	a_5	a_6	a_7	a_8	a_9	a_{10}
1.4142	1.6818	1.8340	1.9152	1.9571	1.9785	1.9892	1.9946	1.9973	1.9986

(b) Let S be the set of positive integers for which $a_n < 2$. Then $1 \in S$ since $a_1 = \sqrt{2} \cong 1.4142 < 2$. Assume that $k \in S$. Now $a_{k+1}^2 = 2a_k < 4$, which implies $a_{k+1} < 2$. Thus $k + 1 \in S$ and S is the set of positive integers.
(c) yes (d) The number you chose is the least upper bound of the corresponding set S.

SECTION 10.2

1. $a_n = 2 + 3(n - 1), \quad n = 1, 2, 3, \ldots$ **3.** $a_n = \dfrac{(-1)^{n-1}}{2n - 1}, \quad n = 1, 2, 3, \ldots$ **5.** $a_n = \dfrac{n^2 + 1}{n}, \quad n = 1, 2, 3, \ldots$

7. $a_n = \begin{cases} n & \text{if } n = 2k-1, \\ \frac{1}{n} & \text{if } n = 2k, \end{cases}$ where $k = 1, 2, 3, \ldots$ **9.** decreasing; bounded below by 0 and above by 2

11. increasing; bounded below by 1 but not bounded above **13.** not monotonic; bounded below by 0 and above by $\frac{3}{2}$

15. decreasing; bounded below by 0 and above by 0.9 **17.** increasing; bounded below by $\frac{1}{2}$ but not bounded above

19. increasing; bounded below by $\frac{4}{5}\sqrt{5}$ and above by 2 **21.** increasing; bounded below by $\frac{2}{51}$ but not bounded above

23. decreasing; bounded below by 0 and above by $\frac{1}{2}(10^{10})$ **25.** increasing; bounded below by 0 and above by ln 2

27. decreasing; bounded below by 1 and above by 4 **29.** increasing; bounded below by $\sqrt{3}$ and above by 2

31. decreasing; bounded above by -1 but not bounded below **33.** increasing; bounded below by $\frac{1}{2}$ and above by 1

35. decreasing; bounded below by 0 and above by 1 **37.** decreasing; bounded below by 0 and above by $\frac{5}{6}$

39. decreasing; bounded below by 0 and above by $\frac{1}{2}$ **41.** decreasing; bounded below by 0 and above by $\frac{1}{3}\ln 3$

43. increasing; bounded below by $\frac{3}{4}$ but not bounded above **45.** For $n \geq 5$

$$\frac{a_{n+1}}{a_n} = \frac{5^{n+1}}{(n+1)!} \cdot \frac{n!}{5^n} = \frac{5}{n+1} < 1 \quad \text{and thus} \quad a_{n+1} < a_n.$$

Sequence is not nonincreasing: $a_1 = 5 < \frac{25}{2} = a_2$.

47. boundedness: $0 < (c^n + d^n)^{1/n} < (2d^n)1/n = 2^{1/n}\, d \leq 2d.$
monotonicity: $a_{n+1}^{n+1} = c^{n+1} + d^{n+1} = cc^n + dd^n < (c^n + d^n)^{1/n} c^n + (c^n + d^n)^{1/n} d^n$
$= (c^n + d^n)^{1+(1/n)} = (c^n + d^n)^{(n+1)/n} = a_n^{n+1}.$
Taking the $n+1$-th root of each side we have $a_{n+1} < a_n$. The sequence is monotonic decreasing.

49. $a_1 = 1, a_2 = \frac{1}{2}, a_3 = \frac{1}{6}, a_4 = \frac{1}{24}, a_5 = \frac{1}{120}, a_6 = \frac{1}{720};\quad a_n = 1/n!$ **51.** $a_1 = a_2 = a_3 = a_4 = a_5 = a_6 = 1;\quad a_n = 1$

53. $a_1 = 1, a_2 = 3, a_3 = 5, a_4 = 7, a_5 = 9, a_6 = 11;\quad a_n = 2n - 1$ **55.** $a_1 = 1, a_2 = 4, a_3 = 13, a_4 = 40, a_5 = 121, a_6 = 364;\quad a_n = \frac{1}{2}(3^n - 1)$

57. $a_1 = 1, a_2 = 4, a_3 = 9, a_4 = 16, a_5 = 25, a_6 = 36;\quad a_n = n^2$ **59.** $a_1 = 1, a_2 = 3, a_3 = 4, a_4 = 8, a_5 = 16, a_6 = 32;\quad a_n = 2^{n-1}\quad (n \geq 3)$

61. $a_1 = 2, a_2 = 1, a_3 = 2, a_4 = 1, a_5 = 2, a_6 = 1;\quad a_n = \frac{1}{2}[3 - (-1)^n]$ **63.** $a_1 = 1, a_2 = 3, a_3 = 5, a_4 = 7, a_5 = 9, a_6 = 11;\quad a_n = 2n - 1$

65. First $a_1 = 2^1 - 1 = 1$. Next suppose $a_k = 2^k - 1$ for some $k \geq 1$. Then

$$a_{k+1} = 2a_k + 1 = 2(2^k - 1) + 1 = 2^{k+1} - 1.$$

67. First $a_1 = \dfrac{1}{2^0} = 1$. Next suppose $a_k = \dfrac{k}{2^{k-1}}$ for some $k \geq 1$. Then $a_{k+1} = \dfrac{k+1}{2k} a_k = \dfrac{k+1}{2k}\dfrac{k}{2^{k-1}} = \dfrac{k+1}{2^k}$.

69. (a) n (b) $\dfrac{1 - r^n}{1 - r}$ **71.** (a) $150(\frac{3}{4})^{n-1}$ (b) $\dfrac{5\sqrt{3}}{2}\left(\dfrac{3}{4}\right)^{\frac{n-1}{2}}$

73. (a) $a_2 = 1 + \sqrt{a_1} = 2 > 1 = a_1$. Assume $a_k = 1 + \sqrt{a_{k-1}} > a_{k-1}$. Then $a_{k+1} = 1 + \sqrt{a_k} > 1 + \sqrt{a_{k-1}} = a_k$. Thus $\{a_n\}$ is an increasing sequence.
(b) $a_n = 1 + \sqrt{a_{n-1}} < 1 + \sqrt{a_n}$, since $a_{n-1} < a_n$.
$a_n - \sqrt{a_n} - 1 < 0$, or $(\sqrt{a_n})^2 - \sqrt{a_n} - 1 < 0$, which implies (solve the inequality) that $\sqrt{a_n} < \dfrac{1+\sqrt{5}}{2}$, hence $a_n < \dfrac{3+\sqrt{5}}{2}$ for all n.
(c) lub $\{a_n\} \cong 2.6180$.

SECTION 10.3

1. diverges **3.** converges to 0 **5.** converges to 1 **7.** converges to 0 **9.** converges to 0 **11.** diverges **13.** converges to 0

15. converges to 1 **17.** converges to $\frac{4}{9}$ **19.** converges to $\frac{1}{2}\sqrt{2}$ **21.** diverges **23.** converges to 1 **25.** diverges

27. converges to 0 **29.** converges to $\frac{1}{2}$ **31.** converges to e^2 **33.** diverges **35.** converges to 0

37. Use $|(a_n + b_n) - (L + M)| \leq |a_n - L| + |b_n - M|$. **39.** $\left(1 + \dfrac{1}{n}\right)^{n+1} = \left(1 + \dfrac{1}{n}\right)^n\left(1 + \dfrac{1}{n}\right)$. Note that $\left(1 + \dfrac{1}{n}\right)^n \to e$ and $\left(1 + \dfrac{1}{n}\right) \to 1$.

41. Imitate the proof given for the nondecreasing case in Theorem 10.3.6. **43.** Let $\epsilon > 0$. Choose k so that, for $n \geq k$,

$$L - \epsilon < a_n < L + \epsilon, \quad L - \epsilon < c_n < L + \epsilon \quad \text{and} \quad a_n \leq b_n \leq c_n.$$

For such n,

$$L - \epsilon < b_n < L + \epsilon.$$

45. Let $\epsilon > 0$. Since $a_n \to L$, there exists a positive integer N such that $L - \epsilon < a_n < L + \epsilon$ for all $n \geq N$. Now $a_n \leq M$ for all n, so $L - \epsilon < M$, or $L < M + \epsilon$. Since ϵ is arbitrary, $L \leq M$.

47. By the continuity of f, $f(L) = f(\lim_{n\to\infty} a_n) = \lim_{n\to\infty} f(a_n) = \lim_{n\to\infty} a_{n+1} = L.$ **49.** Use Theorem 10.3.12 with $f(x) = x^{1/p}$. **51.** converges to 0

53. converges to 0 **55.** diverges **57.** converges to 2 **59.** $L = 0, n = 32$ **61.** $L = 0, n = 4$ **63.** $L = 0, n = 7$

65. $L = 0, n = 65$ **67.** (a) $\dfrac{3+\sqrt{5}}{2}$ (b) 3

69. (a)

a_2	a_3	a_4	a_5	a_6	a_7	a_8	a_9	a_{10}
0.540302	0.857553	0.654290	0.793480	0.701369	0.763960	0.722102	0.750418	0.731404

(b) 0.739085; it is the fixed point of $f(x) = \cos x$.

71. (a)

a_2	a_3	a_4	a_5	a_6	a_7	a_8
2.000000	1.750000	1.732143	1.732051	1.732051	1.732051	1.732051

(b) $L = \dfrac{1}{2}\left(L + \dfrac{3}{L}\right)$, which implies $L^2 = 3$, or $L = \sqrt{3}$.

SECTION 10.4

1. converges to 1 **3.** converges to 0 **5.** converges to 0 **7.** converges to 0 **9.** converges to 1 **11.** converges to 0

13. converges to 1 **15.** converges to 1 **17.** converges to π **19.** converges to 1 **21.** converges to 0 **23.** diverges

25. converges to 0 **27.** converges to e^{-1} **29.** converges to 0 **31.** converges to 0 **33.** converges to e^x **35.** converges to 0

37. $\sqrt{n+1} - \sqrt{n} = \dfrac{\sqrt{n+1}-\sqrt{n}}{\sqrt{n+1}+\sqrt{n}}(\sqrt{n+1}+\sqrt{n}) = \dfrac{1}{\sqrt{n+1}+\sqrt{n}} \to 0$

39. (b) $2\pi r$. As $n \to \infty$, the perimeter of the polygon tends to the circumference of the circle. **41.** $\frac{1}{2}$ **43.** $\frac{1}{8}$

45. (a) $m_{n+1} - m_n = \dfrac{1}{n+1}(a_1 + \cdots + a_n + a_{n+1}) - \dfrac{1}{n}(a_1 + \cdots + a_n)$

$= \dfrac{1}{n(n+1)}\left[na_{n+1} - (a_1 + \cdots + a_n)\right] > 0$ since $\{a_n\}$ is increasing.

(b) We begin with the hint $m_n < \dfrac{|a_1 + \cdots + a_j|}{n} + \dfrac{\epsilon}{2}\left(\dfrac{n-j}{n}\right)$. Since j is fixed, $\dfrac{|a_1 + \cdots + a_j|}{n} \to 0$, and therefore for n sufficiently large $\dfrac{|a_1 + \cdots + a_j|}{n} < \dfrac{\epsilon}{2}$. Since $\dfrac{\epsilon}{2}\left(\dfrac{n-j}{n}\right) < \dfrac{\epsilon}{2}$, we see that, for n sufficiently large, $|m_n| < \epsilon$. This shows that $m_n \to 0$.

47. (a) Let S be the set of positive integers n $(n \ge 2)$ for which the inequalities hold. Since $(\sqrt{b})^2 - 2\sqrt{ab} + (\sqrt{a})^2 = (\sqrt{b} - \sqrt{a})^2 > 0$, it follows that $\dfrac{a+b}{2} > \sqrt{ab}$ and $a_1 > b_1$. Now $a_2 = \dfrac{a_1+b_1}{2} < a_1$ and $b_2 = \sqrt{a_1b_1} < b_1$. Also, by the argument above, $a_2 = \dfrac{a_1+b_1}{2} > \sqrt{a_1b_1} = b_2$, and so $a_1 > a_2 > b_2 > b_1$. Thus $2 \in S$. Assume that $k \in S$. Then $a_{k+1} = \dfrac{a_k+b_k}{2} < \dfrac{a_k+a_k}{2} = a_k$, $b_{k+1} = \sqrt{a_kb_k} > \sqrt{b_k^2} = b_k$, and $a_{k+1} = \dfrac{a_k+b_k}{2} > \sqrt{a_kb_k} = b_{k+1}$. Thus $k+1 \in S$. Therefore the inequalities hold for all $n \ge 2$.

(b) $\{a_n\}$ is a decreasing sequence which is bounded below.
$\{b_n\}$ is an increasing sequence which is bounded above.
Let $L_a = \lim\limits_{n\to\infty} a_n$, $L_b = \lim\limits_{n\to\infty} b_n$. Then $a_n = \dfrac{a_{n-1}+b_{n-1}}{2}$ implies $L_a = \dfrac{L_a + L_b}{2}$ and $L_a = L_b$.

49. The numerical work suggests $L \cong 1$. Justification: Set $f(x) = \sin x - x^2$. Note that $f(0) = 0$ and for x close to 0, $f'(x) = \cos x - 2x > 0$. Therefore $\sin x - x^2 > 0$ for x close to 0 and $\sin(1/n) - 1/n^2 > 0$ for n large. Thus, for n large,

$$\frac{1}{n^2} < \sin\frac{1}{n} < \frac{1}{n} \qquad \text{(}|\sin x| \le |x| \text{ for all } x\text{)}$$

$$\left(\frac{1}{n^2}\right)^{1/n} < \left(\sin\frac{1}{n}\right)^{1/n} < \left(\frac{1}{n}\right)^{1/n}$$

$$\left(\frac{1}{n^{1/n}}\right)^2 < \left(\sin\frac{1}{n}\right)^{1/n} < \frac{1}{n^{1/n}}.$$

As $n \to \infty$ both bounds tend to 1 and therefore the middle term also tends to 1.

51. (a)

a_3	a_4	a_5	a_6	a_7	a_8	a_9	a_{10}
2	3	5	8	13	21	34	55

(b)

r_1	r_2	r_3	r_4	r_5	r_6
1	2	1.5	1.6667	1.6000	1.625

(c) $L = \dfrac{1+\sqrt{5}}{2} \cong 1.618033989$

SECTION 10.5

1. 0 **3.** 1 **5.** $\frac{1}{2}$ **7.** ln 2 **9.** $\frac{1}{4}$ **11.** 2 **13.** $\dfrac{1+\pi}{1-\pi}$ **15.** $\frac{1}{2}$ **17.** π **19.** $-\frac{1}{2}$ **21.** -2 **23.** $\frac{1}{3}\sqrt{6}$

25. $-\frac{1}{8}$ **27.** 4 **29.** $\frac{1}{2}$ **31.** $\frac{1}{2}$ **33.** 1 **35.** 1 **37.** $\lim_{x\to 0}(2+x+\sin x) \neq 0,\ \lim_{x\to 0}(x^3+x-\cos x)\neq 0$

39. $a = \pm 4, \quad b = 1$ **41.** $f(0)$ **43.** $\frac{3}{4}$ **45.** (a) $f(x) \to \infty$ as $x \to \pm\infty$ (b) 10 **47.** (b) $\ln 2 \cong 0.6931$

SECTION 10.6

1. ∞ **3.** -1 **5.** ∞ **7.** $\frac{1}{5}$ **9.** 1 **11.** 0 **13.** ∞ **15.** $\frac{1}{3}$ **17.** e **19.** 1 **21.** $\frac{1}{2}$ **23.** 0 **25.** 1

27. e^3 **29.** e **31.** e^2 **33.** 0 **35.** $-\frac{1}{2}$ **37.** 0 **39.** 1 **41.** 1 **43.** 0

45. y-axis vertical asymptote

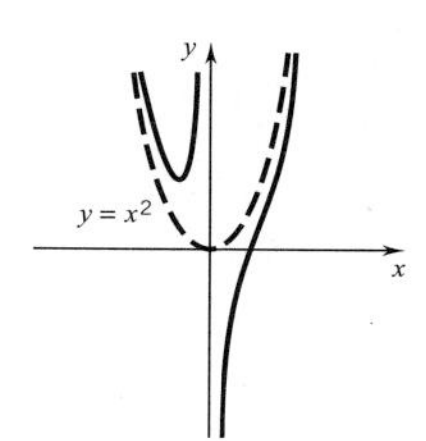

47. x-axis horizontal asymptote

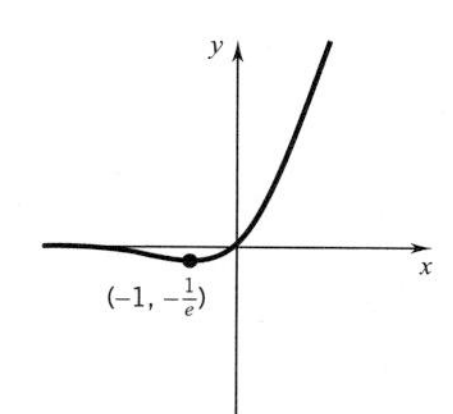

49. x-axis horizontal asymptote

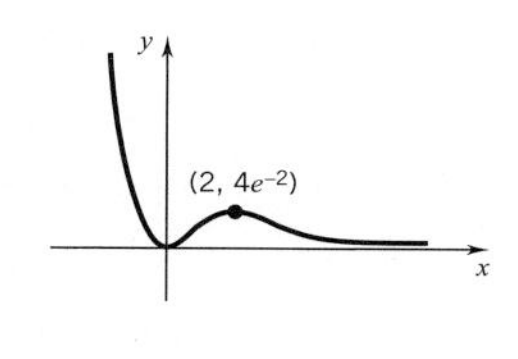

51. $\dfrac{b}{a}\sqrt{x^2-a^2} - \dfrac{b}{a}x = \dfrac{\sqrt{x^2-a^2}+x}{\sqrt{x^2-a^2}+x}\left(\dfrac{b}{a}\right)(\sqrt{x^2-a^2}-x) = \dfrac{-ab}{\sqrt{x^2-a^2}+x} \to 0$ as $x \to \infty$ **53.** Example: $f(x) = x^2 + \dfrac{(x-1)(x-2)}{x^3}$

55. $\lim_{x\to 0^+} \cos x \neq 0$

57. (a) Let S be the set of positive integers for which the statement is true. Since $\lim_{x\to\infty} \dfrac{\ln x}{x} = 0$, $1 \in S$. Assume that $k \in S$. By L'Hospital's rule,

$$\lim_{x\to\infty} \frac{(\ln x)^{k+1}}{x} \overset{*}{=} \lim_{x\to\infty} \frac{(k+1)(\ln x)^k}{x} = 0 \qquad (\text{since } k \in S).$$

Thus $k+1 \in S$, and S is the set of positive integers.
(b) Choose any positive number α. Let $k-1$ and k be positive integers such that $k - 1 \le \alpha \le k$. Then, for $x > e$,

$$\frac{(\ln x)^{k-1}}{x} \le \frac{(\ln x)^{\alpha}}{x} \le \frac{(\ln x)^k}{x}$$

and the result follows by the pinching theorem.

59. (a) L'Hospital's rule applied to the given limit results in $\lim_{x\to 0} \dfrac{2e^{-1/x^2}}{x^3}$. Rewrite the quotient as $\dfrac{1/x}{e^{1/x^2}} \to 0$ as $x \to 0$. (b) $f'(0) = 0$.

61. $\lim_{x\to 0^+}(1+x^2)^{1/x} = 1$. **63.** $\lim_{x\to\infty} g(x) = -5/3$.

SECTION 10.7

1. 1 **3.** $\frac{1}{4}\pi$ **5.** diverges **7.** 6 **9.** $\frac{1}{2}\pi$ **11.** 2 **13.** diverges **15.** $-\frac{1}{4}$ **17.** π **19.** diverges **21.** ln 2

23. 4 **25.** diverges **27.** diverges **29.** diverges **31.** $\frac{1}{2}$ **33.** $2e-2$ **35.** $\dfrac{\pi}{2} - 1$ **37.** π

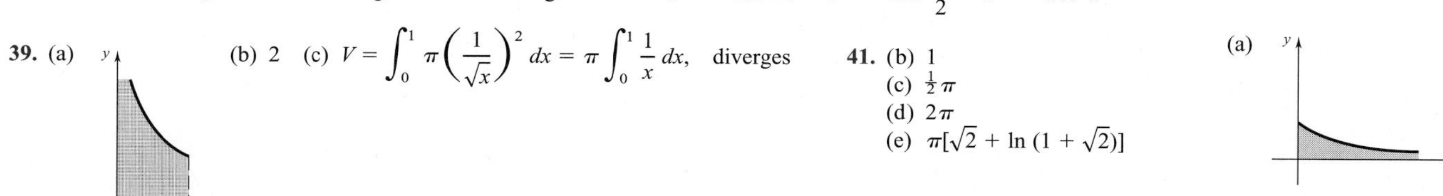

39. (a) (b) 2 (c) $V = \displaystyle\int_0^1 \pi\left(\frac{1}{\sqrt{x}}\right)^2 dx = \pi\int_0^1 \frac{1}{x}\,dx$, diverges

41. (a) (b) 1
(c) $\frac{1}{2}\pi$
(d) 2π
(e) $\pi[\sqrt{2} + \ln(1+\sqrt{2})]$

43. (a) The interval [0, 1] causes no problem. For $x \geq 1$, $e^{-x^2} \leq e^{-x}$ and $\int_1^\infty e^{-x}\,dx$ is finite.

(b) $V_y = \int_0^\infty 2\pi x e^{-x^2}\,dx = \pi$

45. (b) $\frac{4}{3}$ (c) 2π (d) $\frac{8}{7}\pi$ (a) [graph]

47. converges by comparison with $\int_0^\infty \frac{dx}{x^{3/2}}$

49. diverges since for x large the integrand is greater than $\frac{1}{x}$ and $\int_1^\infty \frac{1}{x}\,dx$ diverges **51.** converges by comparison with $\int_1^\infty \frac{dx}{x^{3/2}}$

53. $L = (a\sqrt{1+c^2}/c)e^{c\theta_1}$ **55.** $\frac{1}{s}$; $\text{dom}(F) = (0, \infty)$ **57.** $\frac{s}{s^2+4}$; $\text{dom}(F) = (0, \infty)$

59. $f(x) \geq 0$ for all x and $\int_{-\infty}^\infty f(x)\,dx = \int_0^\infty \frac{6x}{(1+3x^2)^2}\,dx = 1$ **61.** $\frac{1}{k}$

CHAPTER 11

SECTION 11.1

1. 12 **3.** 15 **5.** -10 **7.** $\frac{15}{16}$ **9.** $-\frac{2}{15}$ **11.** $\frac{85}{64}$ **13.** $\sum_{n=1}^{11} 2n-1$ **15.** $\sum_{n=1}^{25} 2n$ **17.** $\sum_{n=1}^{81} (-1)^{n-1}\sqrt{n}$ **19.** $\sum_{k=1}^{n} m_k \Delta x_k$

21. $\sum_{k=1}^{n} f(x_k^*)\Delta x_k$ **23.** $\sum_{k=0}^{5} (-1)^k a^{5-k} b_k$ **25.** $\sum_{k=0}^{4} a_k x^{4-k}$ **27.** $\sum_{k=0}^{4} (-1)^k (k+1)x^k$ **29.** $\sum_{k=3}^{10} \frac{1}{2^k}$, $\sum_{i=0}^{7} \frac{1}{2^{i+3}}$

31. $\sum_{k=3}^{10} (-1)^{k+1} \frac{k}{k+1}$, $\sum_{i=0}^{7} (-1)^i \frac{i+3}{i+4}$ **33.** let $k = n+3$ **35.** let $k = n-3$

37. (a) $(1-x)\sum_{k=0}^{n} x^k = \sum_{k=0}^{n} (x^k - x^{k+1})$

$$= (1-x) + (x-x^2) + (x^2-x^3) + \cdots + (x^n - x^{n+1}) = 1 - x^{n+1}.$$

(b) converges to $\frac{3}{2}$

39. $|a_n - L| < \epsilon$ for $n \geq k$ iff $|a_{n-p} - L| < \epsilon$ for $n \geq k+p$.

41. True for $n = 1$: $\sum_{k=1}^{1} k = 1 = \frac{1}{2}(1)(2)$. Suppose true for $n = p$. Then

$$\sum_{k=1}^{p+1} k = \sum_{k=1}^{p} k + (p+1) = \tfrac{1}{2}(p)(p+1) + (p+1) = \tfrac{1}{2}(p+1)(p+2) = \tfrac{1}{2}(p+1)[(p+1)+1]$$

and thus true for $n = p+1$.

43. True for $n = 1$: $\sum_{k=1}^{1} k^2 = 1 = \frac{1}{6}(1)(2)(3)$. Suppose true for $n = p$. Then

$$\sum_{k=1}^{p+1} k^2 = \sum_{k=1}^{p} k^2 + (p+1)^2 = \tfrac{1}{6}(p)(p+1)(2p+1) + (p+1)^2$$

$$= \tfrac{1}{6}(p+1)(2p^2+7p+6) = \tfrac{1}{6}(p+1)(p+2)(2p+3) = \tfrac{1}{6}(p+1)[(p+1)+1][2(p+1)+1]$$

and thus true for $n = p+1$.

45. 140 **47.** 680

SECTION 11.2

1. $\frac{1}{4}$ **3.** $\frac{1}{2}$ **5.** $\frac{11}{18}$ **7.** $\frac{10}{3}$ **9.** $\frac{67000}{999}$ **11.** 4 **13.** $-\frac{3}{2}$ **15.** $\frac{1}{2}$ **17.** 24 **19.** $\sum_{k=1}^{\infty} \frac{7}{10^k} = \frac{7}{9}$ **21.** $\sum_{k=1}^{\infty} \frac{24}{100^k} = \frac{8}{33}$

23. $\sum_{k=1}^{\infty} \frac{112}{1000^k} = \frac{112}{999}$ **25.** $\frac{62}{100} + \frac{1}{100}\sum_{k=1}^{\infty} \frac{45}{100^k} = \frac{687}{1100}$

27. Let $x = .\overparen{a_1 a_2 \cdots a_n}\overparen{a_1 a_2 \cdots a_n} \cdots$. Then

$$x = \sum_{k=1}^{\infty} \frac{a_1 a_2 \cdots a_n}{(10^n)^k} = a_1 a_2 \cdots a_n \sum_{k=1}^{\infty} \left(\frac{1}{10^n}\right)^k = a_1 a_2 \cdots a_n \left[\frac{1}{1 - \frac{1}{10^n}} - 1\right] = \frac{a_1 a_2 \cdots a_n}{10^n - 1}.$$

29. $\dfrac{1}{1+x} = \dfrac{1}{1-(-x)} = \sum_{k=0}^{\infty}(-x)^k = \sum_{k=0}^{\infty}(-1)^k x^k$ 31. $\sum_{k=0}^{\infty} x^{k+1}$ 33. $\sum_{k=0}^{\infty}(-1)^k x^{2k+1}$ 35. $\sum_{k=0}^{\infty}(-1)^k(2x)^{2k}$

37. $\sum_{n=0}^{\infty}\left(\dfrac{3}{2}\right)^n$; geometric series with $r = \dfrac{3}{2} > 1$ 39. $\lim_{k\to\infty}\left(\dfrac{k}{k+1}\right)^k = \dfrac{1}{e} \neq 0$ 41. $4 + \dfrac{1}{3}\sum_{k=0}^{\infty}\left(\dfrac{1}{12}\right)^k = 4 + \dfrac{4}{11}$ o'clock 43. 12 ft

45. $\sum_{k=1}^{\infty} n_k\left(1+\dfrac{r}{100}\right)^{-k}$ 47. \$9 49. 32 51. $\lim_{n\to\infty} s_n = L = \sum_{k=0}^{\infty} a_k$. Thus $\lim_{n\to\infty} R_n = \lim_{n\to\infty}\left(s_n - \sum_{k=0}^{\infty} a_k\right) = \lim_{n\to\infty} s_n - L = 0$.

53. $\dfrac{1+(-1)^n}{2}$, $n = 0, 1, 2, \ldots$ 55. $s_n = \sum_{k=1}^{n}\ln\left(\dfrac{k+1}{k}\right) = \sum_{k=1}^{n}[\ln(k+1) - \ln k] = \ln(n+1) \to \infty$

57. (a) $s_n = \sum_{k=1}^{n}(d_k - d_{k+1}) = d_1 - d_{n+1} \to d_1$

(b) (i) $\sum_{k=1}^{\infty}\dfrac{\sqrt{k+1}-\sqrt{k}}{\sqrt{k(k+1)}} = \sum_{k=1}^{\infty}\left(\dfrac{1}{\sqrt{k}} - \dfrac{1}{\sqrt{k+1}}\right) = 1$ (ii) $\sum_{k=1}^{\infty}\dfrac{2k+1}{2k^2(k+1)^2} = \sum_{k=1}^{\infty}\dfrac{1}{2}\left(\dfrac{1}{k^2} - \dfrac{1}{(k+1)^2}\right) = \dfrac{1}{2}$

59. $N = 6$ 61. $N = 9999$ 63. $N = \left[\!\left[\dfrac{\ln(\epsilon[1-x])}{\ln|x|}\right]\!\right]$, where $[\![\ ]\!]$ denotes the greatest integer function.

SECTION 11.3

1. converges; comparison $\Sigma\, 1/k^2$ 3. converges; comparison $\Sigma\, 1/k^2$ 5. diverges; comparison $\Sigma\, 1/(k+1)$ 7. diverges; limit comparison $\Sigma\, 1/k$
9. converges; integral test 11. diverges; p-series with $p = \frac{2}{3} \le 1$ 13. diverges; $a_k \not\to 0$ 15. diverges; comparison $\Sigma\, 1/k$
17. diverges; $a_k \not\to 0$ 19. converges; limit comparison $\Sigma\, 1/k^2$ 21. diverges; integral test 23. converges; limit comparison $\Sigma\, 1/k^2$
25. diverges; limit comparison $\Sigma\, 1/k$ 27. converges; limit comparison $\Sigma\, 1/k^{3/2}$ 29. converges; integral test
31. converges; comparison $\Sigma\, 3/k^2$ 33. converges; comparison $\Sigma\, 2/k^2$ 35. $p > 1$

37. (a) The improper integral $\int_0^{\infty} e^{-\alpha x}\,dx = \dfrac{1}{\alpha}$ converges. (b) The improper integral $\int_0^{\infty} xe^{-\alpha x}\,dx = \dfrac{1}{\alpha^2}$ converges.
(c) The improper integral $\int_0^{\infty} x^n e^{-\alpha x}\,dx = \dfrac{1}{\alpha^{n+1}}$ converges.

39. (a) 1.1777 (b) $0.02 < R_4 < 0.0313$ (c) $1.1977 < \sum_{k=1}^{\infty}\dfrac{1}{k^3} < 1.209$ 41. (a) $1/101 < R_{100} < 1/100$ (b) 10,001 43. (a) 15 (b) 1.082

45. (a) If $a_k/b_k \to 0$, then $a_k/b_k < 1$ for all $k \ge K$ for some K. But then $a_k < b_k$ for all $k \ge K$ and, since Σb_k converges, Σa_k converges. [The basic comparison test, Theorem 11.3.5.]
(b) Similar to (a) except that this time we appeal to part (ii) of Theorem 11.3.5.
(c) $\Sigma a_k = \Sigma\dfrac{1}{k^2}$ converges, $\Sigma b_k = \Sigma\dfrac{1}{k^{3/2}}$ converges, $\dfrac{1/k^2}{1/k^{3/2}} = \dfrac{1}{\sqrt{k}} \to 0$
$\Sigma a_k = \Sigma\dfrac{1}{k^2}$ converges, $\Sigma b_k = \Sigma\dfrac{1}{\sqrt{k}}$ diverges, $\dfrac{1/k^2}{1/\sqrt{k}} = \dfrac{1}{k^{3/2}} \to 0$
(d) $\Sigma b_k = \Sigma\dfrac{1}{\sqrt{k}}$ diverges, $\Sigma a_k = \Sigma\dfrac{1}{k^2}$ converges, $\dfrac{1/k^2}{1/\sqrt{k}} = \dfrac{1}{k^{3/2}} \to 0$
$\Sigma b_k = \Sigma\dfrac{1}{\sqrt{k}}$ diverges, $\Sigma a_k = \Sigma\dfrac{1}{k}$ diverges, $\dfrac{1/k}{1/\sqrt{k}} = \dfrac{1}{\sqrt{k}} \to 0$

47. (a) Since Σa_k converges, $\lim_{k\to\infty} a_k = 0$. Therefore there exists a positive integer N such that $0 < a_k < 1$ for $k \ge N$.
Thus, for $k \ge N$, $a_k^2 < a_k$ and so Σa_k^2 converges by the comparison test.
(b) Σa_k may either converge or diverge.
$\Sigma\, 1/k^4$ converges and $\Sigma\, 1/k^2$ converges; $\Sigma\, 1/k^2$ converges and $\Sigma\, 1/k$ diverges.

49. $0 < L - \sum_{k=1}^{n} f(k) = L - s_n = \sum_{k=n+1}^{\infty} f(k) < \int_n^{\infty} f(x)\,dx$ 51. $N = 3$.

53. (a) Set $f(x) = x^{1/4} - \ln x$. Then $f'(x) = \dfrac{1}{4}x^{-3/4} - \dfrac{1}{x} = \dfrac{1}{4x}(x^{1/4} - 4)$. Since $f(e^{12}) = e^3 - 12 > 0$ and $f'(x) > 0$ for $x > e^{12}$, we know that $n^{1/4} > \ln n$ and therefore $\dfrac{1}{n^{5/4}} > \dfrac{\ln n}{n^{3/2}}$ for sufficiently large n. Since $\Sigma\dfrac{1}{n^{5/4}}$ is a convergent p-series, $\Sigma\dfrac{\ln n}{n^{3/2}}$ converges by the basic comparison test.
(b) By L'Hospital's rule $\lim_{x\to\infty}\left[\left(\dfrac{\ln x}{x^{3/2}}\right)\Big/\left(\dfrac{1}{x^{5/4}}\right)\right] = 0$

SECTION 11.4

1. converges; ratio test **3.** converges; root test **5.** diverges; $a_k \not\to 0$ **7.** diverges; limit comparison $\Sigma\, 1/k$ **9.** converges; root test
11. diverges; limit comparison $\Sigma\, 1/\sqrt{k}$ **13.** diverges; ratio test **15.** converges; comparison $\Sigma\, 1/k^{3/2}$ **17.** converges; comparison $\Sigma\, 1/k^2$
19. diverges; integral test **21.** diverges; $a_k \to e^{-100} \neq 0$ **23.** diverges; limit comparison $\Sigma\, 1/k$ **25.** converges; ratio test
27. converges; comparison $\Sigma\, 1/k^{3/2}$ **29.** converges; ratio test **31.** converges; ratio test: $a_{k+1}/a_k \to \frac{4}{27}$ **33.** converges; ratio test
35. converges; root test **37.** converges; root test **39.** converges; ratio test **41.** $\frac{10}{81}$

43. The series $\Sigma \dfrac{k!}{k^k}$ converges. Therefore $\lim\limits_{k\to\infty} \dfrac{k!}{k^k} = 0$ by Theorem 11.2.6. **45.** $p \geq 2$ **47.** $|x| < 1$ **49.** converges for all x

51. Set $b_k = a_k r^k$. If $(a_k)^{1/k} \to \rho$ and $\rho < 1/r$, then
$$(b_k)^{1/k} = (a_k r^k)^{1/k} = (a_k)^{1/k} r \to \rho r < 1$$
and thus, by the root test, $\Sigma b_k = \Sigma a_k r^k$ converges.

SECTION 11.5

1. diverges; $a_k \not\to 0$ **3.** diverges; $a_k \not\to 0$ **5.** (a) does not converge absolutely; integral test (b) converges conditionally; Theorem 11.5.4
7. diverges; limit comparison $\Sigma\, 1/k$ **9.** (a) does not converge absolutely; limit comparison $\Sigma\, 1/k$ (b) converges conditionally; Theorem 11.5.4
11. diverges; $a_k \not\to 0$ **13.** (a) does not converge absolutely; comparison $2\,\Sigma\, 1/\sqrt{k+1}$ (b) converges conditionally; Theorem 11.5.4

15. converges absolutely (terms already positive); $\Sigma \sin\left(\dfrac{\pi}{4k^2}\right) \leq \Sigma \dfrac{\pi}{4k^2} = \dfrac{\pi}{4}\Sigma \dfrac{1}{k^2}$ $(|\sin x| \leq |x|)$ **17.** converges absolutely; ratio test

19. (a) does not converge absolutely; limit comparison $\Sigma\, 1/k$ (b) converges conditionally; Theorem 11.5.4 **21.** diverges; $a_k \not\to 0$

23. diverges; $a_k \not\to 0$ **25.** converges absolutely; ratio test **27.** diverges; $a_k = \dfrac{1}{k}$ for all k **29.** converges absolutely; comparison $\Sigma\, 1/k^2$

31. diverges; $a_k \not\to 0$ **33.** 0.1104 **35.** 0.001 **37.** $\frac{10}{11}$ **39.** $N = 39{,}998$ **41.** (a) 4 (b) 6
43. No. For instance, set $a_{2k} = 2/k$ and $a_{2k+1} = 1/k$.

45. (a) Since $\Sigma|a_k|$ converges, $\Sigma|a_k|^2 = \Sigma a_k^2$ converges (Exercise 47, Section 11.3). (b) $\Sigma\, 1/k^2$ is convergent, $\Sigma \dfrac{(-1)^k}{k}$ is not absolutely convergent.

47. See the proof of Theorem 11.8.2. **49.** (a) $\sum\limits_{k=1}^{\infty} \dfrac{(-1)^{k-1}(a+b)+(a-b)}{2k}$ (b) if $a = b = 0$, absolutely convergent (vacuously)
if $a = b \neq 0$, conditionally convergent
if $a \neq b$, divergent

SECTION 11.6

1. $-1 + x + \frac{1}{2}x^2 - \frac{1}{24}x^4$ **3.** $-\frac{1}{2}x^2 - \frac{1}{12}x^4$ **5.** $1 - x + x^2 - x^3 + x^4 - x^5$ **7.** $x + \frac{1}{3}x^3 + \frac{2}{15}x^5$
9. $P_0(x) = 1,\quad P_1(x) = 1 - x,\quad P_2(x) = 1 - x + 3x^2,\quad P_3(x) = 1 - x + 3x^2 + 5x^3$

11. $\sum\limits_{k=0}^{n} (-1)^k \dfrac{x^k}{k!}$ **13.** $\sum\limits_{k=0}^{m} \dfrac{x^{2k}}{(2k!)}$ where $m = \dfrac{n}{2}$ and n is even **15.** $\sum\limits_{k=0}^{n} \dfrac{r^k}{k!}x^k$ **17.** 79/48 (79/48 ≅ 1.646)

19. 5/6 (5/6 ≅ 0.833) **21.** 13/24 (13/24 ≅ 0.542) **23.** 0.17 **25.** $\dfrac{4e^{2c}}{15}x^5,\ |c| < |x|$ **27.** $\dfrac{-4\sin 2c}{15}x^5,\ |c| < |x|$

29. $\dfrac{3\sec^4 c - 2\sec^2 c}{3}x^3,\ |c| < |x|$ **31.** $\dfrac{3c^2 - 1}{3(1+c^2)^3}x^3,\ |c| < |x|$ **33.** $\dfrac{(-1)^{n+1}e^{-c}}{(n+1)!}x^{n+1},\ |c| < |x|$ **35.** $\dfrac{1}{(1-c)^{n+2}}x^{n+1},\ |c| < |x|$

37. (a) 4 (b) 2 (c) 999 **39.** (a) 1.649 (b) 0.368 **41.** For $0 \leq k \leq n$, $P^{(k)}(0) = k!a_k$; for $k > n$, $P^{(k)}(0) = 0$. Thus $P(x) = \sum\limits_{k=0}^{\infty} P^{(k)}(0)\dfrac{x^k}{k!}$.

43. $\left.\dfrac{d^{2k}(\sinh x)}{dx^{2k}}\right|_{x=0} = \sinh(0) = 0;\quad \left.\dfrac{d^{2k+1}(\sinh x)}{dx^{2k+1}}\right|_{x=0} = \cosh(0) = 1$

Therefore $\sinh x = x + \dfrac{x^3}{3!} + \dfrac{x^5}{5!} + \cdots = \sum\limits_{k=0}^{\infty} \dfrac{1}{(2k+1)!}x^{2k+1}$

45. $\sum\limits_{k=0}^{\infty} \dfrac{a^k}{k!}x^k,\ (-\infty, \infty)$ **47.** $\sum\limits_{k=0}^{\infty} \dfrac{(-1)^k a^{2k}}{(2k)!}x^{2k},\ (-\infty, \infty)$ **49.** $\ln a + \sum\limits_{k=1}^{\infty} \dfrac{(-1)^{k-1}}{ka^k}x^k,\ (-a, a]$

51. $\ln 2 = \ln\left(\dfrac{1+\frac{1}{3}}{1-\frac{1}{3}}\right) \cong 2\left[\dfrac{1}{3} + \dfrac{1}{3}\left(\dfrac{1}{3}\right)^3 + \dfrac{1}{5}\left(\dfrac{1}{3}\right)^5\right] = \dfrac{842}{1215}$ $\left(\dfrac{842}{1215} \cong 0.693\right)$ **53.** routine; use $u = (x-t)^k$ and $dv = f^{(k+1)}(t)\,dt$

55. (b) $f(x) = \dfrac{x^{-n}}{e^{1/x^2}}$ and $\lim\limits_{x\to 0} f(x)$ has the form ∞/∞. Successive applications of L'Hospital's rule will finally produce a quotient of the form

$\dfrac{cx^k}{e^{1/x^2}}$, where k is a nonnegative integer and c is a constant. It follows that $\lim\limits_{x\to 0} f(x) = 0$.

(c) $f'(0) = \lim_{x\to 0} \dfrac{e^{-1/x^2} - 0}{x} = 0$ by part (b). Assume that $f^{(k)}(0) = 0$. Then

$$f^{(k+1)}(0) = \lim_{x\to 0} \frac{f^{(k)}(x) - 0}{x} = \lim_{x\to 0} \frac{f^{(k)}(x)}{x}.$$

Now, $\dfrac{f^{(k)}(x)}{x}$ is a sum of terms of the form $\dfrac{ce^{-1/x^2}}{x^n}$, where n is a positive integer and c is a constant. By part (b), $f^{(k+1)}(0) = 0$. Therefore $f^{(n)}(0) = 0$ for all n.

(d) 0 (e) $x = 0$

57. (a) $p_1(x) = 0,\quad R_2(x) = \dfrac{35c^{4/3}}{9}x^2$ (b) no; $f^{(4)}(0)$ does not exist

$p_2(x) = 0,\quad R_3(x) = \dfrac{140c^{1/3}}{81}x^3$

$p_3(x) = 0,\quad R_4(x) = \dfrac{35}{243c^{2/3}}x^4$

59.

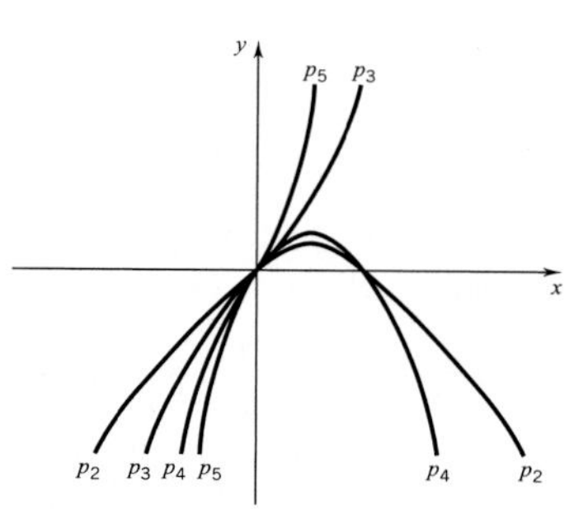

SECTION 11.7

1. $P_3(x) = 2 + \frac{1}{4}(x-4) - \frac{1}{64}(x-4)^2 + \frac{1}{512}(x-4)^3$

$R_4(x) = \dfrac{-5}{128c^{7/2}}x^4,\quad |c| < |x|$

3. $P_4(x) = \dfrac{\sqrt{2}}{2} + \dfrac{\sqrt{2}}{2}\left(x - \dfrac{\pi}{4}\right) - \dfrac{\sqrt{2}}{4}\left(x - \dfrac{\pi}{4}\right)^2 - \dfrac{\sqrt{2}}{12}\left(x - \dfrac{\pi}{4}\right)^3 + \dfrac{\sqrt{2}}{48}\left(x - \dfrac{\pi}{4}\right)^4$

$R_5(x) = \dfrac{\cos c}{120}\left(x - \dfrac{\pi}{4}\right)^5,\quad \left|c - \dfrac{\pi}{4}\right| < \left|x - \dfrac{\pi}{4}\right|$

5. $P_3(x) = \dfrac{\pi}{4} + \dfrac{1}{2}(x-1) - \dfrac{1}{4}(x-1)^2 + \dfrac{1}{12}(x-1)^3\qquad R_4(x) = \dfrac{c(1-c^2)}{(1+c^2)^3}(x-1)^4,\quad |c-1| < |x-1|$

7. $6 + 9(x-1) + 7(x-1)^2 + 3(x-1)^3,\quad (-\infty, \infty)$ **9.** $-3 + 5(x+1) - 19(x+1)^2 + 20(x+1)^3 - 10(x+1)^4 + 2(x+1)^5,\quad (-\infty, \infty)$

11. $\sum_{k=0}^{\infty} (-1)^k \left(\dfrac{1}{2}\right)^{k+1}(x-1)^k,\quad (-1, 3)$ **13.** $\dfrac{1}{5}\sum_{k=0}^{\infty}\left(\dfrac{2}{5}\right)^k (x+2)^k,\quad \left(-\dfrac{9}{2}, \dfrac{1}{2}\right)$ **15.** $\sum_{k=0}^{\infty} \dfrac{(-1)^{k+1}}{(2k+1)!}(x-\pi)^{2k+1},\quad (-\infty, \infty)$

17. $\sum_{k=0}^{\infty} \dfrac{(-1)^{k+1}}{(2k)!}(x-\pi)^{2k},\quad (-\infty, \infty)$ **19.** $\sum_{k=0}^{\infty} \dfrac{(-1)^k}{(2k)!}\left(\dfrac{\pi}{2}\right)^{2k}(x-1)^{2k},\quad (-\infty, \infty)$ **21.** $\ln 3 + \sum_{k=1}^{\infty} \dfrac{(-1)^{k+1}}{k}\left(\dfrac{2}{3}\right)^k (x-1)^k,\quad \left(-\dfrac{1}{2}, \dfrac{5}{2}\right]$

23. $2\ln 2 + (1 + \ln 2)(x-2) + \sum_{k=2}^{\infty} \dfrac{(-1)^k}{k(k-1)2^{k-1}}(x-2)^k$ **25.** $\sum_{k=0}^{\infty} \dfrac{(-1)^k}{(2k+1)!}x^{2k+2}$ **27.** $\sum_{k=0}^{\infty} (k+2)(k+1)\dfrac{2^{k-1}}{5^{k+3}}(x+2)^k$

29. $1 + \sum_{k=1}^{\infty} \dfrac{(-1)^k 2^{2k-1}}{(2k)!}(x-\pi)^{2k}$ **31.** $\sum_{k=0}^{\infty} \dfrac{n!}{(n-k)!\,k!}(x-1)^k$

33. (a) $\dfrac{e^x}{e^a} = e^{x-a} = \sum_{k=0}^{\infty} \dfrac{(x-a)^k}{k!},\quad e^x = e^a \sum_{k=0}^{\infty} \dfrac{(x-a)^k}{k!}$ (b) $e^{a+(x-a)} = e^x = e^a \sum_{k=0}^{\infty} \dfrac{(x-a)^k}{k!},\quad e^{x_1+x_2} = e^{x_1}\sum_{k=0}^{\infty} \dfrac{x_2^k}{k!} = e^{x_1}e^{x_2}$

(c) $e^{-a}\sum_{k=0}^{\infty} (-1)^k \dfrac{(x-a)^k}{k!}$

35. (a) $P_2(x) = \dfrac{1}{2} + \dfrac{\sqrt{3}}{2}\left(x - \dfrac{\pi}{6}\right) - \dfrac{1}{4}\left(x - \dfrac{\pi}{6}\right)^2 - \dfrac{\sqrt{3}}{12}\left(x - \dfrac{\pi}{6}\right)^3$ (b) 0.5736

37. $P_2(x) = 6 + \frac{1}{12}(x-36) - \frac{1}{1728}(x-36)^2;\quad \sqrt{38} \cong 6.164$

SECTION 11.8

1. $(-1, 1)$ **3.** $(-\infty, \infty)$ **5.** $\{0\}$ **7.** $[-2, 2)$ **9.** $\{0\}$ **11.** $[-\frac{1}{2}, \frac{1}{2})$ **13.** $(-1, 1)$ **15.** $(-10, 10)$ **17.** $(-\infty, \infty)$

19. $(-\infty, \infty)$ **21.** $(-3/2, 3/2)$ **23.** converges only at $x = 1$ **25.** $(-4, 0)$ **27.** $(-\infty, \infty)$ **29.** $(-1, 1)$ **31.** $(0, 4)$

33. $(-\frac{5}{2}, \frac{1}{2})$ **35.** $(-2, 2)$ **37.** $\left[-\dfrac{1}{\sqrt{3}}, \dfrac{1}{\sqrt{3}}\right]$

39. Examine the convergence of $\Sigma\, |a_k x^k|$; for (a) use the root test and for (b) use the ratio test. **41.** $\Sigma\, |a_k(-r)^k| = \Sigma\, |a_k r^k|$

SECTION 11.9

1. $1 + 2x + 3x^2 + \cdots + nx^{n-1} + \cdots$ **3.** $1 + kx + \dfrac{(k+1)k}{2!}x^2 + \cdots + \dfrac{(n+k-1)!}{n!(k-1)!}x^n + \cdots$

5. $\ln(1-x^2) = -x^2 - \dfrac{1}{2}x^4 - \dfrac{1}{3}x^6 - \cdots - \dfrac{1}{n+1}x^{2n+2} - \cdots$ **7.** $1 + x^2 + \frac{2}{3}x^4 + \frac{17}{45}x^6 + \cdots$ **9.** -72 **11.** $\sum_{k=0}^{\infty} \dfrac{(-1)^k}{(2k+1)!}x^{4k+2}$

13. $\sum_{k=0}^{\infty} \frac{3^k}{k!} x^{3k}$ **15.** $2 \sum_{k=0}^{\infty} x^{2k+1}$ **17.** $\sum_{k=0}^{\infty} \frac{(k!+1)}{k!} x^k$ **19.** $\sum_{k=1}^{\infty} \frac{(-1)^{k+1}}{k} x^{3k+1}$ **21.** $\sum_{k=0}^{\infty} \frac{(-1)^k}{k!} x^{3k+3}$ **23.** $\frac{1}{2}$ **25.** $-\frac{1}{2}$

27. $\sum_{k=1}^{\infty} \frac{(-1)^{k-1}}{k^2} x^k, \quad -1 \le x \le 1$ **29.** $\sum_{k=1}^{\infty} \frac{(-1)^{k-1}}{(2k-1)^2} x^{2k-1}$ **31.** $0.804 \le I \le 0.808$ **33.** $0.600 \le I \le 0.603$ **35.** $0.294 \le I \le 0.304$

37. 0.9461 **39.** 0.4485 **41.** e^{x^3} **43.** $3x^2 e^{x^3}$ **45.** (a) $\sum_{n=0}^{\infty} \frac{1}{n!} x^{n+1}$ (b) $\int_0^1 xe^x\,dx = 1 = \int_0^1 \left(\sum_{n=0}^{\infty} \frac{1}{n!} x^{n+1}\right) dx = \sum_{n=0}^{\infty} \frac{1}{n!(n+2)}$

47. Let $f(x)$ be the sum of these series; a_k and b_k are both $f^{(k)}(0)/k!$.

49. (a) If f is even, then $f^{(2k-1)}$ is odd for $k = 1, 2, \ldots$. This implies that $f^{(2k-1)}(0) = (0)$, and so $a_{2k-1} = \frac{f^{(2k-1)}(0)}{(2k-1)!} = 0$ for all k.

(b) If f is odd, then $f^{(2k)}$ is odd for $k = 1, 2, \ldots$, which implies $a_{2k} = 0$ for all k.

51. $0.0352 \le I \le 0.0359$; $I = \frac{3}{6} - \frac{3}{8} \ln 1.5 \cong 0.0354505$ **53.** $0.2640 \le I \le 0.2643$; $I = 1 - 2/e \cong 0.2642411$

SECTION 11.10

1. $1 + \frac{1}{2}x - \frac{1}{8}x^2 + \frac{1}{16}x^3 - \frac{5}{128}x^4$ **3.** $1 + \frac{1}{2}x^2 - \frac{1}{8}x^4$ **5.** $1 - \frac{1}{2}x + \frac{3}{8}x^2 - \frac{5}{16}x^3 + \frac{35}{128}x^4$ **7.** $1 - \frac{1}{4}x - \frac{3}{32}x^2 - \frac{7}{128}x^3 - \frac{77}{2048}x^4$

9. $8 + 3x + \frac{3}{16}x^2 - \frac{1}{128}x^3 + \frac{3}{4096}x^4$ **11.** (a) $\sum_{k=0}^{\infty} (-1)^k \binom{-1/2}{k} x^{2k}$ (b) $\sum_{k=0}^{\infty} (-1)^k \binom{-1/2}{k} \frac{1}{2k+1} x^{2k+1}, \quad R = 1$ **13.** 9.8995

15. 2.0799 **17.** 0.4925 **19.** 0.3349 **21.** 0.4815

INDEX

(continued from the front)

INVERSE TRIGONOMETRIC FUNCTIONS

64. $\int \sin^{-1}u\,du = u\sin^{-1}u + \sqrt{1-u^2} + C$

65. $\int \cos^{-1}u\,du = u\cos^{-1}u - \sqrt{1-u^2} + C$

66. $\int \tan^{-1}u\,du = u\tan^{-1}u - \frac{1}{2}\ln(1+u^2) + C$

67. $\int \cot^{-1}u\,du = u\cot^{-1}u + \frac{1}{2}\ln(1+u^2) + C$

68. $\int \sec^{-1}u\,du = u\sec^{-1}u - \ln|u + \sqrt{u^2-1}| + C$

69. $\int \csc^{-1}u\,du = u\csc^{-1}u + \ln|u + \sqrt{u^2-1}| + C$

70. $\int u\sin^{-1}u\,du = \frac{1}{4}(2u^2-1)\sin^{-1}u + u\sqrt{1-u^2} + C$

71. $\int u\tan^{-1}u\,du = \frac{1}{2}(u^2+1)\tan^{-1}u - \frac{1}{2}u + C$

72. $\int u\cos^{-1}u\,du = \frac{1}{4}(2u^2-1)\cos^{-1}u - u\sqrt{1-u^2} + C$

73. $\int u^n\sin^{-1}u\,du = \frac{1}{n+1}\left[u^{n+1}\sin^{-1}u - \int\frac{u^{n+1}\,du}{\sqrt{1-u^2}}\right],\ n \neq -1$

74. $\int u^n\cos^{-1}u\,du = \frac{1}{n+1}\left[u^{n+1}\cos^{-1}u + \int\frac{u^{n+1}\,du}{\sqrt{1-u^2}}\right],\ n \neq -1$

75. $\int u^n\tan^{-1}u\,du = \frac{1}{n+1}\left[u^{n+1}\tan^{-1}u - \int\frac{u^{n+1}\,du}{1+u^2}\right],\ n \neq -1$

$\sqrt{a^2 + u^2},\ a > 0$

76. $\int \frac{du}{a^2+u^2} = \frac{1}{a}\tan^{-1}\frac{u}{a} + C$

77. $\int \frac{du}{\sqrt{a^2+u^2}} = \ln|u + \sqrt{a^2+u^2}| + C$

78. $\int \sqrt{a^2+u^2}\,du = \frac{u}{2}\sqrt{a^2+u^2} + \frac{a^2}{2}\ln|u + \sqrt{a^2+u^2}| + C$

79. $\int u^2\sqrt{a^2+u^2}\,du = \frac{u}{8}(a^2+2u^2)\sqrt{a^2+u^2} - \frac{a^4}{8}\ln|u + \sqrt{a^2+u^2}| + C$

80. $\int \frac{\sqrt{a^2+u^2}}{u}\,du = \sqrt{a^2+u^2} - a\ln\left|\frac{a+\sqrt{a^2+u^2}}{u}\right| + C$

81. $\int \frac{\sqrt{a^2+u^2}}{u^2}\,du = -\frac{\sqrt{a^2+u^2}}{u} + \ln|u + \sqrt{a^2+u^2}| + C$

82. $\int \frac{u^2\,du}{\sqrt{a^2+u^2}} = \frac{u}{2}\sqrt{a^2+u^2} - \frac{a^2}{2}\ln|u + \sqrt{a^2+u^2}| + C$

83. $\int \frac{du}{u\sqrt{a^2+u^2}} = -\frac{1}{a}\ln\left|\frac{\sqrt{a^2+u^2}+a}{u}\right| + C$

84. $\int \frac{du}{u^2\sqrt{a^2+u^2}} = -\frac{\sqrt{a^2+u^2}}{a^2u} + C$

85. $\int \frac{du}{(a^2+u^2)^{3/2}} = \frac{u}{a^2\sqrt{a^2+u^2}} + C$

$\sqrt{a^2 - u^2},\ a > 0$

86. $\int \frac{du}{\sqrt{a^2-u^2}} = \sin^{-1}\frac{u}{a} + C$

87. $\int \sqrt{a^2-u^2}\,du = \frac{u}{2}\sqrt{a^2-u^2} + \frac{a^2}{2}\sin^{-1}\frac{u}{a} + C$

88. $\int u^2\sqrt{a^2-u^2}\,du = \frac{u}{8}(2u^2-a^2)\sqrt{a^2-u^2} + \frac{a^4}{8}\sin^{-1}\frac{u}{a} + C$